# Thermal Engineering

## Second Edition

# Thermal Engineering

## Second Edition

**Ajoy Kumar**
**G N Sah**

# Alpha Science International Ltd.
Oxford, U.K.

**Thermal Engineering**
*Second Edition*
964 pgs.| 568 fgs.| 41 tbls.

**Ajoy Kumar**
Professor and Head
Department of Automobile Engineering
M.H. Saboo Siddik College of Engineering
Mumbai, India

**G N Sah**
Department of Mechanical Engineering
Birsa Institute of Technology
Sindri, India

Copyright © 2004, 2010

Second Edition 2010

ALPHA SCIENCE INTERNATIONAL LTD.

7200 The Quorum, Oxford Business Park North
Garington Road, Oxford OX4 2JZ, U.K.

**www.alphasci.com**



ISBN 978-1-84265-440-8

Printed in India.

**Dedicated to**

**Our Parents**
**Late Shri Devi Chand Sah**
**&**
**Smt. Sona Devi**

**and**

**Technocub**

# Preface to the Second Edition

Many colleges, universities and institutions at home and abroad have used this book of Thermal Engineering since publication of its first edition. Most universities have recently revised the structure of their under graduate courses .This is more particularly so in the area of thermoscience. Moreever, the colleagues and students of the authors have made valuable suggestion for the improvement of the book. The goal of the second edition of this book is the same as that of the first edition with updated material in several areas that reflects the ever advancing technology of thermal engineering.

The present edition of this book is divided into same number of chapters, but the chapters Boiler, Boiler Draught and Boiler Performance have been clubbed together to a single chapter 9 as **Boiler.** Moreover, two new chapters 10 and 11 entitled **Air Compressor** and **Refrigeration & Air Conditioning** respectively, have been introduced with same moderated quality as rest are, to meet the requirement of a complete thermal group collection.

The authors would like to thank the people who received the original book manuscript of the first edition, and those who received the book in preparation of this second edition. The suggestion for additions and improvements that these people made have resulted in a better text. We wish to express our sincere thanks to these people at home and abroad for recommending the book to their students and friends. We hope that they will continue to patronize this book in the future also. We would like to express our gratitude and appreciation to our Publisher and Managing Director, Mr. N.K. Mehra and Anupama Jauhry(AGM-Production ) for taking extra pains in printing the book very systematically. The authors' grateful thanks are due to Smt. Malti Sah, Smt. Rekha Kumari, Pankaj Keshri, Smita Keshri, Kesarinandan and Abhinandan for extending full co-operation and encouragement during preparation of the manuscript of this revised edition.

We would greatly appreciate constructive criticism and suggestion for improvement for our readers, which will be greatly acknowledged.

AUTHORS

# Preface to the First Edition

Thermal Engineering, a core subject in many universities, is offered to engineering students under various names, such as Mechanical Engineering, Heat Engineering, Applied Thermodynamics and Heat Power Engineering. Having observed that students face difficulty in understanding clearly the basic principles, fundamental concepts and theory without a proper explanation of the topics, a conscious effort has been made to stress the points where students generally make mistakes. At the end of every chapter, solved problems, objective and descriptive questions and class-tested numerical problems have been included. All the topics in this book have been developed lucidly from the fundamentals. The treatment of topics like basic concepts and definitions, absolute temperature scale, work, heat, internal energy, enthalpy, laws of Thermodynamics, perfect gases, gas power cycle, vapors cycles have been carried out in a logical sequence maintaining clarity of concepts. The Boiler section has been divided into three chapters and numerical problems asked. The chapter on steam turbine gives a clear-cut method to draw velocity triangle without which numericals cannot be solved. Chapters 14-16 deal with thermal energy release (fuel and combustion), nuclear energy and direct energy conversion based on the fundamentals given in previous chapters. The last chapter on heat and mass transfer gives a general view of the subject.

Written in simple language the whole book in S.I. units will be useful for degree, diploma as well as AMIE students of mechanical engineering.

The authors express their sense of gratitude to professors Dr. B.N. Prasad (Prof. & Head, NIT Jamshedpur) and Dr. R.L. Singh (Prof. B.I.T. Sindri) for their guidance. Our brothers namely Er. B.N. Sah, Er. Rajendra Kumar and Er. Vijoy Kumar deserve a special word of appreciation for motivating and encouraging us during the book writing process. We cannot forget other family members namely, late Kumud Sah, Malati, Gayatri, Anjana, Asha and Rekha for providing an ideal book-writing atmosphere at home. The authors are extremely thankful to the students, teachers, co-workers and many learned professors from different parts of the country who helped directly or indirectly in the preparation of this text with their constructive criticism.

Last but not the least the authors are highly indebted to Chairman Shri P.P. Chhabria, Hope Foundations', Finolex Academy of Management and Technology, Ratnagiri (Finolex Group of Industries) and Director of B.I.T. Sindri for giving permission and motivating us to write the book.

# Nomenclature

| | |
|---|---|
| $a$ | Acceleration, m/s$^2$; sound velocity, m/s |
| $A$ | Area, m$^2$; Ampere (electric current) |
| $C$ | Celsius scale; constant; Liminus intensity; conductance, kw/k |
| $C_p$, $C_v$ | Constant pressure and constant volume specific heats, kJ/kg k |
| COP | Coefficient of Performance |
| $C_v$ | Calorific value, kJ/kg |
| $C_n$ | Polytropic specific heat, kJ/kg k |
| $dl$ | Small displacement, m |
| $ds$ | Change in entropy, kJ/kg k |
| $dv$ | Displacement in volume, m$^3$ |
| $d\theta$ | Angle of rotation, degree |
| $D$ | Diameter, m; Diffusivity, m$^2$/hr |
| $e$ | Energy per kg, kJ/kg |
| $E$ | Total energy, kJ; Emmissive power, W/m$^2$, Macroscopic energy, kJ |
| $f$ | Fraction of thermal neutron; fuel |
| $F$ | Force, $N$; Fahrenheit scale; shape factor |
| $g$ | Local gravitational acceleration, m/s$^2$ |
| $g_c$ | Constant for defining unit of force numerically equal to the value of g at sea level, N $-$ sec$^2$ (in S.I. Unit) |
| $h$ | Heat transfer co-efficient, kw/m$^2$ k; Specific enthalpy, kJ/kg |
| $H$ | Toal enthalpy, kJ |
| $Hg$ | Mercury |
| $I$ | Irreversibility; Irreversible engine; Current (ampere); Intensity of radiation, W/m$^2$ |
| $J$ | Joule; Mechanical equivalent of heat, 1J = 1 N-m |
| $J_o$ | Current density, ampere/m$^2$ |
| $K$ | Kelvin scale; Thermal conductivity, kw/mk; Boltzman constant = $1.55*10^{-4}$ ev/density, w kg/m$^3$ |
| $KE$ | Kinetic energy, kJ |
| $L$ | Length, m; stroke length, m |
| $m$ | Mass, kg |
| mep | Mean Effective Pressure, bar |
| $M$ | Molecular mass; Mach number |
| $n$ | number of moles; Index of polytropic expansion; number |
| $N$ | Newton (force); Revolution per minute |
| $Nu$ | Nusselt number |

| | |
|---|---|
| $P$ | Pressure, bar; Probability of non-leakage of thermal neutrons |
| $PE$ | Potential energy, kJ |
| $q$ | Heat flux, $kJ/m^2$ |
| $Q$ | Heat energy, kJ; Discharge rate, $m^3/s$ |
| $Q_H$ | Heat exchange with high temperature body |
| $Q_L$ | Heat exchange with low temperature body |
| $Q_S$ | Energy released by Sun, kJ/s |
| $r$ | Compression ratio; radius, m |
| $r_c$ | Cut-off ratio |
| $r_p$ | Pressure ratio |
| $R$ | Gas constant, kJ/kg k; Rankine scale; Reversible engine; Resistance-Ohm gas |
| $R_o$ | Universal gas constant, kJ/kg k mole; Original resistance, Ohm |
| $R_e$ | Reynold number |
| $s$ | Second; entropy per kg, kJ/kg k |
| $S$ | Total entropy, kJ/kg; Solar constant, $w/m^2$ |
| $S_r$ | Solid angle, Steredian |
| SSC | Specific Steam Consumption, kg/kw hr |
| $S_t$ | Stanton number |
| $t$ | Celsius temperature, °C |
| $T$ | Absolute temperature, k; Torque, N-m |
| $T_c$ | Critical temperature |
| $T_r$ | Reduced parameter of temperature |
| $T_{tp}$ | Triple point temperature, °C |
| $u$ | Velocity; Blade velocity, m/s, Specific internal energy, kJ/kg |
| $U$ | Total internal energy, kJ; overall heat transfer co-efficient, $kw/m^2k$ |
| $v$ | Specific volume, $m^3/kg$ |
| $V$ | Total volume, $m^3$; velocity, m/s |
| $w$ | Specific work, kJ/kg; Hour angle; weight, N |
| $W$ | Total work, kJ |
| $x$ | Quality/dryness fraction of system |
| $Z$ | Height of potential, m; compressibility factor = pv/RT; Actual air fuel ratio; figure merit |
| $Z_s$ | Theoretical air fuel ratio |

## GREEK LETTERS

| | |
|---|---|
| $\alpha$ | Expansion ratio or compression ratio; co-efficient of linear thermal expansion; Thermal diffusivity; Angle (degree) |
| $\beta$ | Co-efficient of volumetric thermal expansion; Angle (degree) |
| $\gamma$ | Index of adiabatic expansion |
| $\varepsilon$ | Emmissivity; fast Fission factor |
| $\delta$ | Degree; declination angle; hydraulic boundary layer |
| $\delta_t$ | Thermal boundary layer |
| $\eta$ | Efficiency |
| $\mu$ | Dynamic viscocity = $vg$, $N\text{-}S/m^2$ |
| $\upsilon$ | Kinematic viscocity, $m^2/s$ |
| $\rho$ | Density, $kg/m^3$; Electrical resistivity |

| | |
|---|---|
| $\phi$ | Latitude angle; work function; Neutron flux |
| $\sigma$ | Stefan-Boltzman constant = $5.67 \times 10^{-8}$ w/m$^2$k$^4$ |
| $\tau$ | Shear stress, N/m$^2$, Thomson co-efficient |
| $\lambda$ | Wavelength, $\mu$m; Decay constant |
| $\theta_A$ | Azimuth angle |
| $\theta_Z$ | Zenith angle |
| $C_r$ | Skin friction co-efficient |
| $\xi$ | c.o.p (co-efficient of performace) |

# SUBSCRIPTS

| | |
|---|---|
| $a$ | Actual; air |
| $b$ | Back; Black; Boiler |
| $c$ | Critical; clearance; condensate |
| $d$ | Candela |
| $e$ | Earth |
| $f$ | Friction; Flow; property of saturated liquid |
| $fg$ | Vaporization |
| $g$ | Gas; Generated |
| $i$ | Inlet; ice |
| I.T. | Indicated thermal |
| I.P. | Intermediate pressure |
| $Irr$ | Irreversible |
| L.P. | Low pressure |
| $m$ | Mean; Mechanical |
| $ma$ | Mean actual |
| $mi$ | Mean ideal |
| $n$ | Polytropic |
| $N$ | Net |
| $O$ | Overall; Original; out (exit) |
| $P$ | Constant pressure; potential; pump |
| $Pw$ | Constant pressure for water |
| $r$ | Reduced; rejected; relative |
| $S$ | Steam; entropy; supplied; saturation |
| sup | Superheated |
| $t$ | Total; Thermal |
| $T$ | Turbine |
| $W$ | Whirl |
| $H.P$ | Heat Pump |
| $R$ | Refrigerator |

# Contents

Flow Rotary Compressors—Centrifugal Compressor—Axial Flow Compressor—Surging and Choking of Compressors—Difference between Centrifugal and axial Flow Compressor—Static and Total Head (or Satagnation) Quantities in Rotary Compressors—Isentropic Efficiency in Rotary Compressor.

1

# Sources of Energy

## 1.1 INTRODUCTION

The degree of industrial advancement and prosperity of a nation is directly related to the per capita energy consumption of its people. The industrially more advanced countries consume more energy than the developing and under-developed countries of the world. Simitarly, in India, the energy demand is growing at 7 percent per annum due to increasing rate of industrialisation, urbanisation and agricultural activity. At present there is a energy and peak demand shortage of about 8 percent and 19 percent respectively. Large-scale expansion of energy production will place a serious strain on the earth's limited energy resources.

Therefore, it is the responsibility of scientists, power engineers and technicians to locate, develop and exploit the new sources. In order to accomplish this, these people should have an indepth knowledge of the various energy forms sources, conversion techniques, and conservation methods, along with their limitations and inherent problems.

In the first half of the twentieth century, energy sources were exploited with the primary consideration given to economics—the low cost. But today the power engineer must them selves with the three "E's"—energy, economy and Ecology; the modern engineer must try to develop systems that produce large quantities of energy at low cost with minimal impact on the environment. The proper balancing of these "E's" is a major technological challenge before engineers.

## 1.2 ENERGY AND MAN

Man needs energy to sustain himself. This energy is required in many forms—he needs light to see in the dark, heat energy to prepare food, mechanical and electrical energy to run millions of machines which the civilization demands, energy for transportation and energy for comfort. Man needs energy for economic growth, make drugs, medicines and cloths. In fact transportation and economic growth are two great hallmarks of modern civilization. The per capita energy consumed in transportation and industry is the highest in the most advanced countries and lowest in the less developed nations. Energy is required in all respects of human life.

## 1.3  ENERGY SOURCES

The Sun is the source of all energy. Solar energy includes direct solar energy as well as indirect solar energy, for example, energy from fossil fuels, hydraulic energy, wind energy, tidal energy, geo thermal and thermo electric energy. Solar energy is also an outcome of nuclear fusion The discussion in detail will be follow in the subsequent articles.

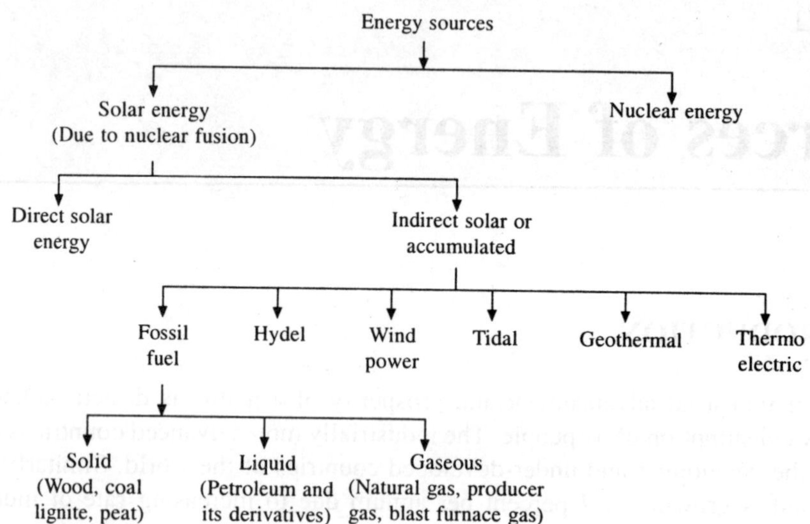


## 1.4  MEASUREMENT OF ENERGY

Different units of energy are:

   (i) Joule, 1 Joule = 1 Newton–meter, SI unit of energy
  (ii) kg meter, 1 kg meter = 9.81 Newton-meter = 9.81 Joule
 (iii) Kilocalorie (K-Cal), 1 Kcal = 427 kg meter
  (iv) Kilowatt (hour kwh), 1 kWh = 1000 watt $\times$ 3,600 second = $36 \times 10^5$ Joule.
   (v) Megawatt hour (Mwh), 1 Mwh = $36 \times 10^8$ Joule
  (vi) Electron-Volt (eV), 1 eV = $1.6 \times 10^{-19}$ Joule
 (vii) Mega electron Volt (MeV), 1 MeV = $1.6 \times 10^{-13}$ Joule
(viii) Ton of TNT, 1 ton of T N T = $4.2 \times 10^9$ Joule

$$1 \text{ kWh} = 860 \text{ kcal}$$

$$1 \text{ Kcal} = 4.187 \text{ KJ}$$

Energy equivalent of 1 ton of TNT = $4.2 \times 10^9$ Joule.

## 1.5   ENERGY DEMAND

In 1960, the energy consumed in the world from all sources was estimated at $34.5 \times 10^{12}$ kWh and by 1970 it had gone up to $57.5 \times 10^{12}$ kWh and is expected to further shoot up to $194 \times 10^{12}$ kWh in the year 2010 (237 percent increase over 1970 value). The total energy demand forcasts for India and the USA are:

| Year | Energy demand for India | Energy demand for the USA |
|------|------|------|
| 1982 | $4.2 \times 10^{12}$ kWh | $40.55 \times 10^{12}$ kWh |
| 1992 | $6.27 \times 10^{12}$ kWh | $52.7 \times 10^{12}$ kWh |
| 2000 | $8.82 \times 10^{12}$ kWh | $73.6 \times 10^{12}$ kWh |
| 2008 | $10.65 \times 10^{12}$ kWh | $95.5 \times 10^{12}$ kWh |

## 1.6   ESTIMATED ENERGY RESERVES OF THE WORLD

| | Source | Amount | Type |
|------|------|------|------|
| 1. | Total available tidal power | $6.7 \times 10^{10}$ W | Mechanical |
| 2. | Total available water power | $300 \times 10^{10}$ W | Mechanical |
| 3. | Wind power | $2000.0 \times 10^{10}$ W | Mechanical |
| 4. | Recoverable geothermal | $0.4 \times 10^{21}$ J | Thermal |
| 5. | Shale oil reserves | $1.2 \times 10^{21}$ J | Chemical |
| 6. | Tarsand oil reserves | $1.8 \times 10^{21}$ J | Chemical |
| 7. | Natural gas reserves | $9.5 \times 10^{21}$ J | Chemical |
| 8. | Petroleum reserves | $1.7 \times 10^{21}$ J | Chemical |
| 9. | Uranium $U^{235}$ reserves | $13.7 \times 10^{21}$ J | Nuclear |
| 10. | Solar power | $297.36 \times 10^{16}$ W | Thermal |
| 11. | Coal and Lignite | $200.00 \times 10^{21}$ J | Chemical |
| 12. | Uranium 238 | $1800.00 \times 10^{21}$ J | Nuclear |

## 1.   DIRECT SOLAR ENERGY

### 1.7.1   Introduction

The most abundant and continous energy source available is solar energy specially, the electro magnetic energy emitted by the sun. The US on an average receives 150 times more energy than is consumed there. India has been blessed with plenty of solar energy amount varying from 5 to 7 kWhr per square meter on daily basis, received energy is $5 \times 10^5$ kWhr p.a. While solar energy is not being used as a primary source of fuel energy at the present, a large research and development effort is under way to develop economical systems to harness solar energy as a major source of fuel energy, particularly for heating and cooling of the buildings. Solar energy as an option is lucrative because it is:

(a) non-polluting, (b) non depletable, (c) clean and safe, (d) reliable and free. However, the two principal disadvantages of solar energy are: (i) it is available in dilute form and a large system is required to run on solar energy, (ii) it is available only intermittently, and a facility to store it usually required.

### 1.7.2   Solar Energy Calculations

The sun is nearly spherical, (diameter $= 1.39 \times 10^6$ km) its area and volume may be calculated as $6.09 \times 10^{18}$ m$^2$ and $1.41 \times 10^{27}$ m$^3$ respectively. The surface of the sunemits radiation like a black body at an approximate temperature of 6,000 K. The temperature at the core of the sun is several millions of degree. The sun subtends an angle of 32′ at the centre of the earth that has 109.44 times larger diameter. The sun releases tremendous amounts of energy through a process of slow fusion in which hydrogen (four protons) combine to form helium, (one helium nucleus), the mass of the helium nucleus is less than that of four protons, mass being lost in the reaction converts into energy. The sun is 3,32,000 times heavier than the earth.

In order to compute the energy released by the sun and incident on the earth's surface, experiments have been conducted with balloons, artificial satellites and other space vehicles fitted with gadgets and instruments. These experiments have shown that the intensity of solar radiation at the outer limit of earth's atmosphere varies between 1.36 to 1.43 kW/m$^2$.

An useful term, solar constant, is introduced in the analysis of solar energy which is defined as *"when the earth is at its mean distance from the sun, the solar radiation incident upon a surface per unit area per unit time normal to the sun's ray placed at the outer limit of earth's atmosphere is called solar constant."* The estimated vlaue is 1353 w/m$^2$ with an error of $\pm$ 1.5%. The world radiation centre (WRC) has adopted a value of 1367 w/m$^2$ with an uncertainty of the order of 1%. 1367 W/m$^2$ = 1.960 cal/cm$^2$ min = 433 Btu/ft$^2$hr = 4.921 MJ/m$^2$hr. To accept the value of solar constant, the radiation in the range of 0.3 to 3 $\mu$m (90% of total radiation) has been considered.

In order to estimate the energy incident on the outer limits of the earth atmosphere, it may be assumed that the radiation intensity is equal to the solar constant "$S$", over sphere whose radius is equal to the mean earth — sun distance, $R_{s-e} = 149.7 \times 10^6$ km. The total energy released by the sun is $Q_s = (4\pi R_{s-e}^2).\, S = 4\,\pi\,(149.7 \times 10^9 \times 10^3)^2 \times 1367 \times 10^{-3} = 384.96 \times 10^{27}$ kJ/second. This quantity is inconceivably huge. We can never utilize all this energy anyway, being confined to the surface of the earth. To see how much solar energy is incident on the earth surface, we compute:

$$Q_{s-e} = (\pi R_e^2).\, S = \pi\,(6356.9 \times 10^3)^2 \times 1367 \times 10^{-3} = 173.61 \times 10^{12}\ \text{kJ/second}$$

where $R_e$ = earth radius = 6356.9 km.

About 0.1% of energy produced by the sun is sufficient to meet all the energy requirement for the whole world in one year.

The above figures do not mean that the entire solar energy incident outside the earth's atmosphere is available for utilization by man. In reality, about 30 percent of the incident energy never reaches the earth's land surface but is absorbed by the outer layers of the earth's atmosphere. There are 250–300 days of useful sunshine per year in most part of India. The solar intensity at ground level compared with that outside the atmosphere is reduced due to following four factors:

   (i)  absorption in the outer layers of the earth's atmosphere,
  (ii)  atmospheric pollution, the presence of dust and soot.
 (iii)  the degree of cloudiness or haze, and
 (iv)  the variation of sun's position in the sky with the time of the day.

When the losses due to all these causes are subtracted, it is found that the intensity of normal solar radiation incident in the northern hemisphere on a clear cloudless day amounts to roughly 1 kw/m$^2$. In spite of this depletion in the available intensity, the total solar radiation incident on the earth's surfaces is more than sufficient to meet the requirement for atleast a few hundred years.

### 1.7.3  Solar Radiation at Earth's Surface

The earth's surface receives solar radiation after being subjected to the mechanisms of attenuation, reflection and scattering in the earth's atmosphere.

*Terminology*

1. Solar radiation — Energy radiated by the sun
2. Solar irradiation — Radiated energy received by the earth—surface.
3. Solar insolation — Solar radiation received on a flat horizontal surface on the earth.
4. Beam (direct) radiation—The portion of the incident solar radiation which comes directly from the apparent solar disc, without being reflected, absorbed or scattered, is called a beam or direct radiation.
5. Diffused (scattered) radiation—The radiation received from the sun after its direction has been changed by reflection and scattered by the atmospheric clouds, dust, gases and vapours is called diffused or scattered radiation.
6. Global (total) radiation—The total solar radiation received at any point on the earth's surface is the sum of the direct and diffused radiation. Radiation received by a collector surface is always global or total radiation. Total does not include radiation that has been absorbed by matter and then re-emitted, because most of this radiation is at longer wavelengths 3 $\mu$m.
7. Extraterrestrial solar radiation—The solar radiation incident on the outer atmosphere of the earth. This is not affected by earth's atmosphere, clouds, gases etc. It is a beam radiation. Solar constant refers to the extraterrestrial.
8. Solar spectrum—Graph of radiated energy versus wavelength.
9. Earth's polar axis (P.A)—The axis through the north pole to south pole around which the earth spins resulting in day and night.
10. Elliptical plane—Plane in which the earth rotates (angle = 23.5°; measured between polar axis and vertical line passing through centre) around the sun. One rotation is completed in one year (365 days).

*Terms related with earth's geography*

1. Lines of longitude—Imaginary lines on surface of the earth between the north pole and south pole.
2. Lines of latitude—Imaginary lines on surface of the earth, parallel to the equatorial lines.
3. Latitude angle ($\phi$)—The latitude angle for a particular place is the angle made by the radial line joining the location to the centre of the earth with the projection of the line on an equatorial plane. For northern hemisphere, it is measured as positive and for southern hemisphere, negative by convention. It ranges from + 90° to –90°. $\phi = \angle$AOP (Fig. 1.1).

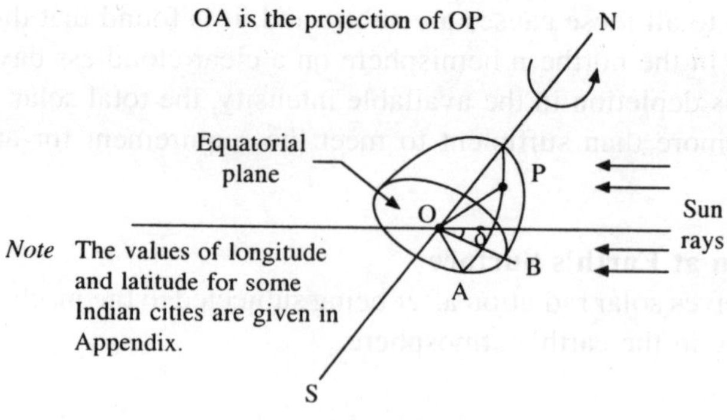


**Fig. 1.1**

4. Declination angle ($\delta$)—If is the angle made by the line joining the centres of the sun and the earth with its projection on the equatorial plane.
   $-23.45°$ (on December 21) $\angle \delta \angle +23.45°$ (on June 21) $\delta = 0$ on March 21 and September 22

5. Hour angle ($\omega$)—It is an angular measure of time and is equivalent to $15° = \dfrac{360}{24}$ per hour.
   It is measured from noon based on local apparent time (LAT), being positive in the morning and negative in the afternoon.

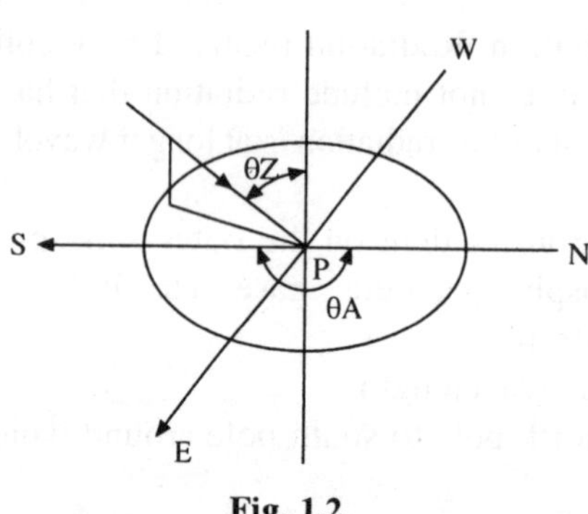


**Fig. 1.2**

6. Zenith angle ($\theta_z$)—Angle between the sun's says and the vertical line at any point $P$ (Fig. 1.2).
7. Azimuth angle ($\theta_A$)—Angle measured from the north direction to the projection of the sun's rays in the horizontal plane.
8. Meridian—It is necessary to select some reference location on the earth for helping in locating a particular position. The location of Royal observatory Greenwich, outside London, has been universally accepted as a reference point. An imaginary great circle passing through this point and the two poles, intersecting the equator at right angles, is called the Prime (or Greenwich) Meridian. Similar great circles have been drawn at intervals of 15° through the two poles.
9. Longitude—It is the angular distance of the location, measured east or west from the prime meridian.

## 1.7.4 Geographical Regions on the Earth and its Positions with Respect to The Sun

The earth is a globe with imaginary lines of latitude drawn parallel to the equator at the interval of 15° and meridians of longitude drawn at same interval between north and south pole indicating zero degree longitude at Greenwich (Fig. 1.3).

Solar insolation is maximum for 0° latitude (equator) and minimum for 90° latitude (north and south pole). India lies in the northern hemisphere within latitudes of 7° and 37.5° N and longitudes of 68.5° and 97.1° E. The average solar radiation values for India are between 12.5 and 22.7 MJ/m$^2$ day. The peak solar radiation in India occurs in some parts of Rajasthan and Gujarat and is equal to 25 MJ/m$^2$ day.

Figure 1.4 shows the geographical regions of the earth namely, torrid zone, temperate zone and frigid zone in each hemisphere between north and south pole.

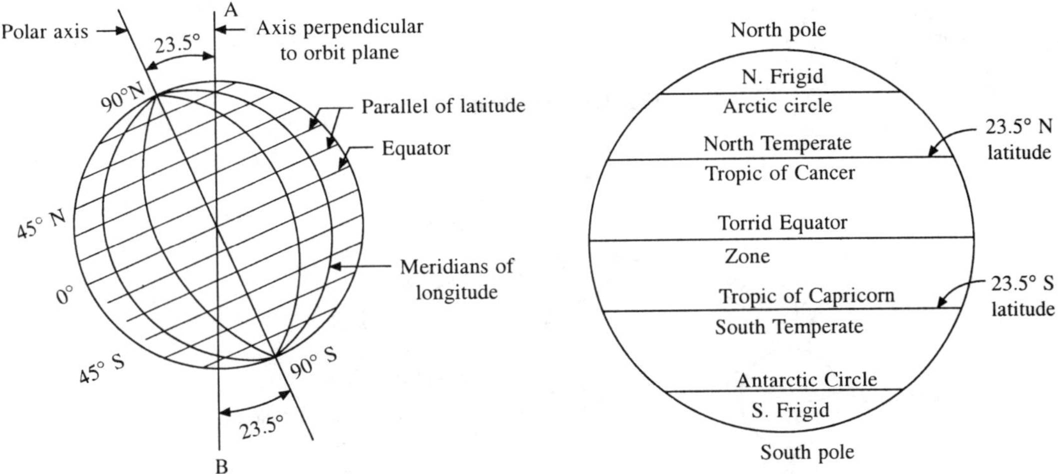


**Fig. 1.3    Parallels of latitude and meridians of longitude on the earth's surface**

**Fig. 1.4    Geographical regions of the earth caused by uneven angle of solar rays**

The earth revolves round the sun in an elliptical orbit by rotating about its polar axis with a complete rotation in 23 hours and 56 minutes. Its axis of rotation is, however, tilted at 23.5° with respect to its orbit around the sun as shown in Fig. 1.5. This angle of rotation is essentially responsible for the distribution of solar radiation over the earth's surface and, consequently, the change of seasons. Figure 1.6 illustrates the cause for four seasons for the northern hemisphere. December 21 and June 21 are the extreme points for the northern hemisphere for seasonal variation of the sunlight. They are called solstices*. At the winter solstice (December 21), the north pole is tilted at 23.5° away from the sun. All points on the earth's surface north of 66.5° north latitude are in total darkness on this date for 24 hours. So even during the daily spinning the Arctic regions do not receive the sunlight. The intensity of the sunlight in the northern hemisphere is lesser. At the same time, on December 21, all the regions within 23.5° of the south pole (Antarctic regions) receive continuous sunlight for 24 hours of the day. December 21 is called winter solstice for the northern hemisphere. At the time of summer solstice (June 21), the situation is reversed. Northern hemisphere receives maximum sunlight with a very long day, while the southern hemisphere has winter with a very short day. This happens due to turning of north poles towards the sun. At the

---

* Solstice (Latin word) means "sun standing still". For maximum or minimum declination the sun appears to stand still

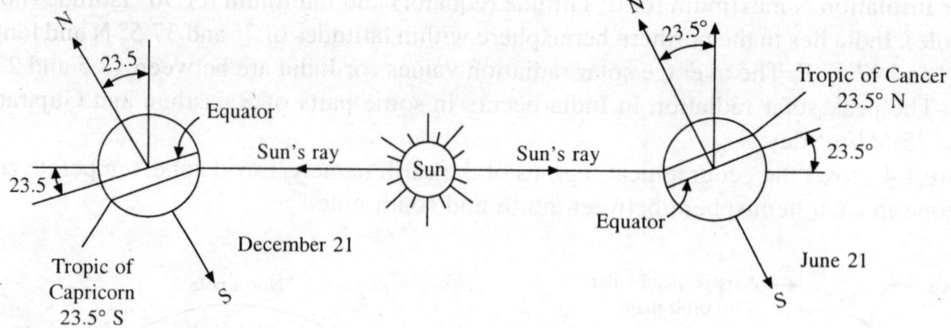

**Fig. 1.5   The tropics**

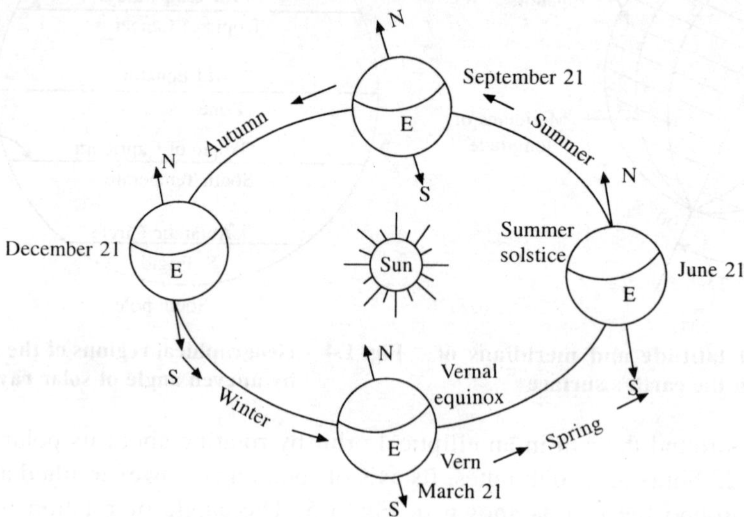

**Fig. 1.6   Earth's position with respect to the sun during seansons**

time of the two equinoxes* (March 21 and September 21), both the poles are equidistant from the sun and all points on the earth's surface have 12 hours of daylight and 12 hours of darkness. Thus we get the four prominent seasons in the northern hemisphere, with demarcation by the two equinoxes and two solstices.

## *Facts about the Earth*

| | | |
|---|---|---|
| Estimated age | — | 4,600 million years |
| Mean distance to the sun | — | $149.7 \times 10^6$ km |
| Mean distance to the moon | — | 3,81,597.5 km |
| Equatorial diameter | — | 12,756.8 km |
| Polar diameter | — | 12,713.8 km |

---

* Equinox (Latin word) means "Equal nights". The nights are equal when the declination of sun is zero.

| Equatorial circumference | — | 40,075 km |
|---|---|---|
| Mass (in metric tons) | — | $597.4 \times 10^{19}$ |
| Period of revolution | — | 365 days 5 hours 48 minutes and 46 seconds |
| Speed of revolution | — | 1,07,160 km/hour |
| Period of rotation | — | 23 hours, 56 minutes |
| Total area | — | 51,00,66,000 sq.km |
| Land area | — | 14,84,29,000 sq. km (29.1%) |
| Water area | — | 36,16,37,000 sq. km (70.9%) |
| Volume | — | $10,83,230 \times 10^6$ cubic m |

### 1.7.5  Solar Time

Solar radiation calculations are made in terms of the solar time. It is always desirable to convert clock time into solar time. Solar time is measured with respect to solar noon, which is the time when the sun is crossing the observer's meridian (longitude). The difference between two such consecutive solar noons defines a solar day. Solar time does not coincide with local clock time. A civil day is exactly 24 hours whereas, a solar day is not exactly so. So one can not use a clock according to solar time. Therefore, it is necessary to convert standard time to solar time for a particular locality by applying two corrections. First, there is a constant correction for the difference in longitude between the observer's longitude and the longitude on which the local standard time is based. The sun takes 4 minutes to transverse 1° of longitude. The positive sign is used if the observer is in the western hemisphere, while the negative sign is used for eastern hemisphere. India lies in the eastern hemisphere with reference to Greenwich. The second correction is from the equation of time $E$, which takes into account due to variation in the length of the solar day through the year. Solar time is converted to standard time by the equation:

$$\text{Solar time} = \text{Standard time} \pm 4\,(L_{st} - L_{loc}) + E \tag{1.1}$$

where $L_{st}$ is the standard meridian for the local time zone. To arrive at Indian Standard Time (IST) a longitude (meridian) of 82.5° passing through the city of Allahabad has been selected as standard meridian. $L_{loc}$ is the longitude of the location in question. The equation of time $E$ (in minutes) is determined by following equation:

$$E = 9.87 \sin 2B - 7.53 \cos B - 1.5 \sin B \tag{1.2}$$

where
$$B = \frac{360\,(n - 81)}{365}$$

and $n$ = day of the year, $1 \leq n \leq 365$
care must be exercised while using Equation (1.1) which contains two corrections where values are in minutes. Local solar noon = 12 – equation of time in hours on that day

$$\pm \ \frac{\text{longitude of reference meridian} - \text{observer's longitude}}{15}$$

**Example 1.1**　Determine the local solar time corresponding to 11.00 AM IST on February 20 at Mumbai (19°07′ N, 72°51′ E).

*Solution*　Local longitude, $L_{loc} = 72° + \dfrac{51}{60} = 72.85°$ E

Local solar time = Standard time – 4 (82.5 – 72.85) + $E$

Equation of time $E = 9.87 \sin 2B – 7.53 \cos B – 1.5 \sin B$

on February 20, $n = 31 + 20 = 51$

and
$$B = \frac{360\,(51 – 81)}{365} = – 29.58$$

$\therefore\ E = 9.87 \sin (2 – 29.58) – 7.53 \cos (–29.58) – 1.5 \sin (–29.58)$

$$= – 8.474 – 6.548 + 0.740$$

$$= –14.28$$

$\therefore$ Local solar time = 11.00 hr – 4 (82.5 – 72.85) – 14.28

$$= 11.00 \text{ hour} – 52.88 \text{ minute}$$

$$= 10 \text{ hour } 7.12 \text{ minute}$$

$$= 10 \text{ hour } 7 \text{ minute } 7 \text{ second } Ans$$

### 1.7.6　The Solar Energy: A SWOT Analysis

Solar energy itself is not renewable energy as its hydrogen mass fuses into a mass of helium and the remaining converts into heat energy, which is lost. This energy is not renewed again.

SWOT analysis means analysis of strength, weakness, opportunity and threat for any concept of process we propose. This analysis is necessary for getting right decision before planning and design of the system to be utilized.

SWOT analysis of solar energy is given below:

*STRENGTHS*

1. Source of energy is abundant
2. Totally pollution free
3. Economically self sufficient
4. Can be utilized in any form of energy and for all purposes
5. Easy to operate
6. Less hazardous
7. Saves fossil fuel deposit

WEAKNESS

1. Not available on cloudy days and at nights
2. Intensity varies according to season or weather
3. Collection and storage problem

4.  Initial investment is high
5.  Needs subsidy

*OPPORTUNITY*
1.  By bringing down the prices, it can be a boon for low income group people
2.  Can be used for producing electrical power, cooking and industrial applications etc.
3.  Chance of advertant exploitation of energy consumers.
4.  Scope for utilizing magnetic energy from solar wind

*THREAT*
1.  Threat from oil and coal lobby
2.  Opposition from different forces due to subsidy
3.  Common consumer lacks knowledge
4.  Fluctuation in intensity due to season or weather may discourage consumers.

### 1.7.7  Solar Energy Applications
1.  Solar water heating
2.  Space heating or heating of building
3.  Cooling of building
4.  Power generation
5.  Solar distillation on a small community scale
6.  Solar drying of agricultural and industrial purpose
7.  Solar cooking
8.  Solar refrigeration and airconditioning
9.  Photo-voltaic conversion
10.  Salt production by evaporation of sea water or inland brines
11.  Solar energy for water pumping
12.  Solar furnaces
13.  Solar thermal power generation
14.  Indirect source of solar energy conversion, that is, in the form of wind, through bio gas and ocean thermal electric conversion.
15.  Solar timber seasoning
16.  Solar car
17.  Solar green house

### 1.7.8  Technologies for Converting Solar Energy to Electricity
Availability of cheap power is an index of technological advancement and standard of living of a country. Conversion of solar energy into electrical power has been a subject of research for the last nearly three centuries. Solar energy is either used directly as source of heat to achieve the end objective (as in crop drying or production of salt) or is used to heat water, and the energy thus captured in water is used to run certain machines (as in solar space heating or air conditioning). Solar cell (photovoltaic cell) is the only device to convert solar energy into electrical power directly. Detail of solar cell has been given in chapter 18.

But the systems that have received maximum attention and research and development investments in recent years are the ones in which solar energy is converted to electrical power. Solar thermal power cycles can be classified as low, medium and high temperature cycles which are given below with temperature range, cycle efficiency and type of collector to be used:

(i) Low temperature cycles work at maximum temperatures of about 100°C by using flat plate collectors or solar ponds for collecting solar energy, with actual cycle efficiency about 8 to 10 percent.

(ii) Medium temperature cycles work at maximum temperatures of about 400°C by using line focussing parabolic collector, with actual cycle efficiency about 20 to 25 percent.

(iii) High temperature cycles work at temperatures above 400°C by using either paraboloidal dish collectors or central receivers located at the top of towers, with actual cycle efficiency more than 25 percent depending on the value of maximum temperature achieved.

Figure 1.7 provides a general view of the plan of a solar thermal power plant. The solar thermal collectors receive solar energy and the heat transport fluid (primary fluid) is heated. The heat transport fluid is circulated through the receiver. The thermal storage tank stores heat in the form of hot water and solids. The heat exchanger transfers heat to the working fluid (secondary fluid) of the Rankine cycle for electrical power generation. Some heat may be transferred from heat exchanger in getting hot water or steam to the processplant. Water, synthetic oil and liquid metal (or molten salt) are genarally used as primary fluid in low temperature cycle, medium temperature cycle and high temperature cycle respectively. However, there is no any secondary fluid and heat

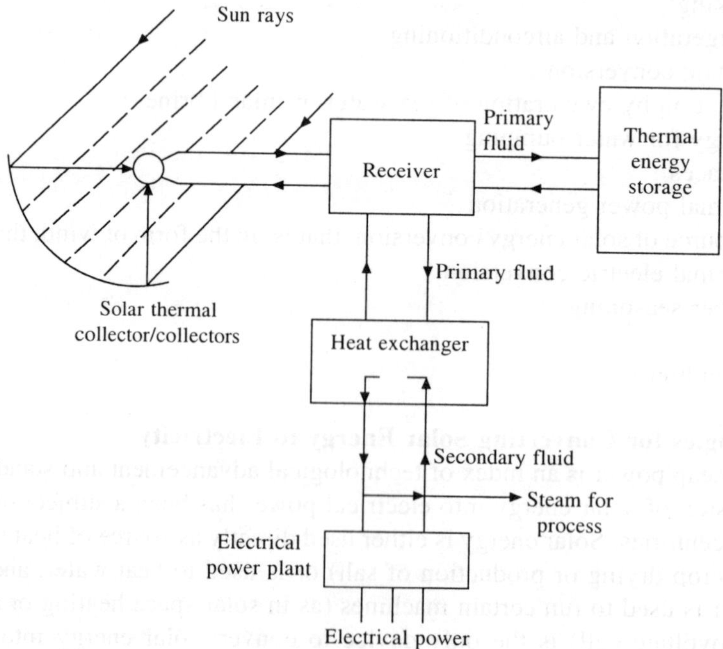


**Fig. 1.7  Schematic of a solarthermal electrical power plant**

exchanger in high temperature cycle. Refrigerants like R-11, R-113 and R-114 are used as secondary fluid in low temperature cycle where as water is used in medium temperature cycle.

### 1.7.9  Solar Energy Collection Systems

Solar energy available in nature is in dilute form with low power density (kw/m$^2$). Hence we go for its collection by using collectors covering a large area on the ground.

#### *Solar Energy Collectors*

A solar energy collector is a device to collect solar radiation and transfer the energy to a passing fluid in contact with it. Utilization of solar energy requires solar collectors. They can be classified into two general types:

   (i)  Flat plate collector or non-focusing type
  (ii)  Concentrating (focusing) type collector

#### *Flat Plate Solar Collector (Non-focusing type)*

Non-focusing types of collectors have been made to produce temperatures upto 100°C. Their operation is dependent on the properties of certain materials (glass for example) which transmit energy in the visible and near visible wave lengths (0.39 to 0.78 $\mu$m) but are opaque to infrared radiation (more than 0.78 $\mu$m). Figure 1.8 gives the schematic diagram of a flat plate solar collector. The suface to be heated is coated with an absorbing black paint and is placed behind one, two or three sheets of glass in a box totally insulated at the back as well as the sides. The incident sun rays heat the surface and the protective glass shield prevents the infrared radiation emitted by the hot surface from escaping back into space. Hence the incoming solar energy is entrapped inside the box and working fluid gets heated recovering from the surface. This effect is known as the green house effect. Wavelength of entrapped radiation is increased which is unable to escape fully. Tubes carrying water or air to be heated are usually brazed or welded to the hot surface. The warm water or air issuing from there can be utilized as a heating medium. The system has the advantage that the collector does not have to be rotated to face the sun. The system has collection efficiency of about 65 percent. Solar collectors of this type are in extensive use in

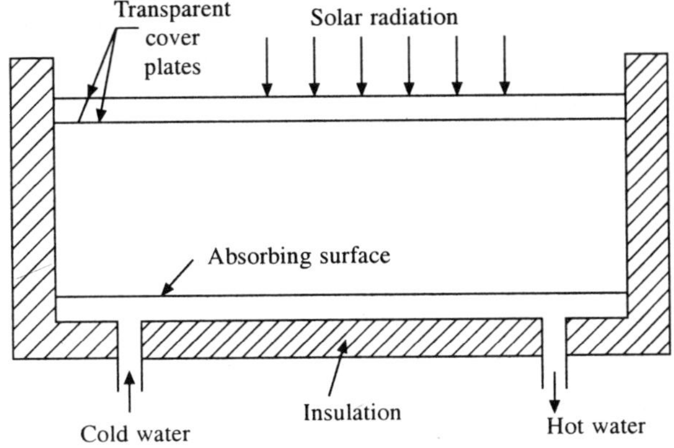


**Fig. 1.8   Flat plate solar collector**

Australia and India. In addition to utilizing solar energy to heat water, efforts are being made to use the energy to dry vegetable and industrial products, distil mineral water, refrigeration and conditioning purposes.

Solar collectors should must face south to receiv more radiation, if they are located in northern hemisphere. They should be inclined at an optimum angle depending upon energy demand and period. The demand may be high for space heating in the winter months of December, January and February. On the other hand, if solar energy is to be used to run an absorption refrigeration plant, the duty would be highest in months of April, May and June owing high ambient temperatures. In such cases, it would obviously be desirable to use a tilt greater than the latitude for a winter application and the reverse for a summer application. The usual practice is to recommend values of $(\phi + 10°)$ or $(\phi + 15°)$ for the former, $(\phi - 10°)$ or $(\phi - 15°)$ for the latter and $(\phi + 5 \text{ to } 10°)$ for throughout year application, where $\phi$ is the latitude angle for the place at which collector is to be located.

### *Focsing Type Collector*

The focusing collector (Fig. 1.9) is parabolic or semi-cylindrical with a highly reflecting front surface. The reflecting surface may be obtained by polishing aluminium on the front surface or by using plastic aluminized plates, an extremely highly polished aluminium sheet or other material to cover the front surface. In America, aluminized plates are easily available and are stuck to a main surface which is parabolic or semi-cylindrical. It is not easy to obtain such plates in India, so polished aluminium sheets may be used. At the focal point of the mirror there is a tube carrying fluid to be heated. To absorb the concentrated sun rays, the tube is coated black at the portion facing the mirror. This system has the disadvantage that the tube and mirror must rotate to face the sun at all times. If properly designed, the system can obtain extremely high temperature nearly 300°C – 350°C by using a hemispherical or paraboloidal surface.

Since Solar energy incident on the earth surface is 1 kw/m$^2$ to develop 1 kw of useful power, we require 15 m$^2$ area considering boiler and engine having overall efficiency of 15%.

### 1.7.10  Will the Sun Last?

The sun has always been a source of wonder and awe. The energy consumed by the whole world in one year comes from all conventional sources, is produced by the sun only in 3 minutes. So there is no energy crisis but only a crisis of ignorance, selfishness and fear. For the ancients, sun was the God of Light. For the modern man it is ball of fire, burning the same way as kerosene burns in a lamp. But it is not difficult to show that had this been the case, it would never have lasted till today.

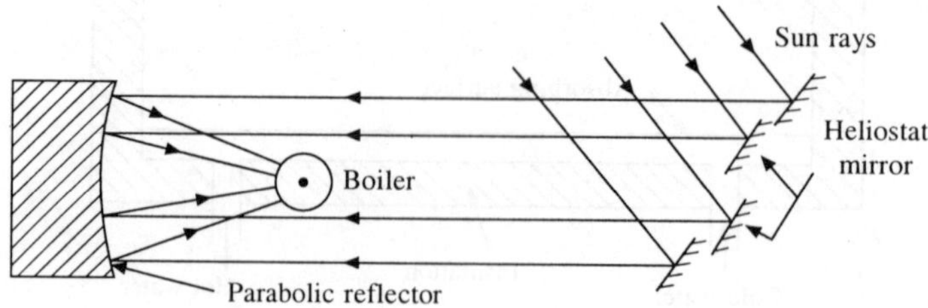


**Fig. 1.9  Focusing type collector**

We all know that the sun is hot. But the first attempt to calculate the temperature was made by Arthur Eddington, the famous British astronomer. He found that the temperature at the core was about 2 million degrees centigrade. At this temperature, matter cannot exist in the normal state. It exists in a state called plasma, that is, as a hot soup of electrons, protons and neutrons.

What is the source of the sun's tremendous energy? This was a riddle for a long time. Many theories were put forward and none were satisfactory. Lord Kelvin (1824 –1907) and Helmholtz (1821–94) simultaneously proposed a theory based on the equivalence mechanical energy and heat. The mass of the sun is so enormous that it has a tendency to collapse under its own weight. The energy produced during this collapse is converted into heat. They calculated that this process can provide enough to last for two hundred million years.

When Einstein proposed his equation $E = MC^2$, it became known that matter can be converted into energy; and also that the energy stored within the atomic nucleus exceeds, by a factor of million, the energy liberated in ordinary chemical reactions. The Austrian physicist FG Houstermans and the British astronomer R Atkinson concluded that at the temperature existing in the solar interior, thermonuclear reactions between hydrogen nuclei and other light elements can produce sufficient energy in the sun.

In 1937 M Bethe of the USA and C Von Veizsasker of Germany proposed a theory known as the carbon cycle, wherein the carbon atom works like a nuclear catalyst and binds together four protons into an alpha particle. The alpha particle is then released and the carbon reverts to its original state. The cycle is then repeated. These reactions can take place only in the hot interior of the sun. The American physicist, Charles Critchtied propounded another theory known as the H-H reaction. In this reaction, two protons collide and stick together to form a deuterium nucleus. Then though a series of nuclear reactions deuterium nuclei fuse together to form a particle. Since about 60% of the solar material is hydrogen, the reaction can take place on a large scale.

The sun can radiate energy till the supply of hydrogen lasts. The amount of hydrogen is enormous, but it is not inexhaustible. Calculations show that it can last for about five billion years. What will happen then?

Charles Crithfield and George Gamow suggested an answer to this problem. According to them, a spherical shell will gradually expand and the luminosity of the sun will increase tenfold. The oceans of the earth will start boiling and life would become extinct on the earth. After this last effort, the sun will slowly shrink into a cold dense body. However, there is no, reason for panic as we still have five billion years to go.

## 1.8    FOSSIL FUELS—Detailed discussion is given in chapter 14.
###     STEAM POWER PLANT—Detailed discussion is given in chapter 8 and 9

## 1.9    NUCLEAR ENERGY

Nuclear energy means controlled fission of heavier unstable atoms such as $U^{235}$. The $U^{232}$ and artificial elements $P_u^{239}$ liberate large amount of heat energy. This enormous release of energy from a relatively small mass of nuclear fuels make this source of energy of great interest. The energy, released by the complete fission of one kg of $U^{235}$ is equal to the heat energy obtained by complete combustion of 4500 tonnes = $4.5 \times 10^6$ kg of high graded coal. However, there are some difficulties in the use of nuclear energy, namely, high capital cost of nuclear power plants, limited availability of raw materials, and difficulties associated with disposal of radioactive wastes. The detailed discussion follows in chapter 15.

## 1.10 HYDRAULIC POWER

Water power potential created by rivers, flowing down from hills into valleys and finally into the sea, is actually an indirect evidence of solar energy. Large quantity of raw water deposited in rivers and on mountains can be used to generate electricity. The total world production of electrical power from hydraulic power sources amounted to $690 \times 10^{12}$ kw, in 1960, out of which $166 \times 10^{12}$ kwh came from India. But in 1963, the world and Indian productions had gone upto $980 \times 10^{12}$ kwh and $22.9 \times 10^{10}$ kwh respectively, while in 1971, the estimated corresponding productions were $100 \times 10^{12}$ and $54 \times 10^{12}$ kwh. Modern hydroelectric power generation stations use difference in elevation ranging from 20 m to as high as 1,600 m for power generation. Suitable hydraulic turbines have been designed even for small head but, large discharge. Final power available from a hydraulic power plant $= \dfrac{\omega Q H}{1000} \times \eta$ overall. where $Q$ is discharge rate available at a head H. Hydro-electric power is only about 2.57% of the total world utilization of energy. India's potential for hydro-electric energy is about $41 \times 10^6$ kw of power while the world's potential is $3000 \times 10^6$ kw. Disadvantages are: (i) high initial cost of dam building and storage and erection (ii) big units lead to flooding of fertile land. However, hydro-electric power is cheaper than thermal power.

## 1.11 WIND POWER

Wind power is actually an indirect evidence of solar energy which causes large wind flows. About 1–2% of the total solar radiation that reaches the earth is converted in the atmosphere into the energy of wind. When the air is in motion it is, called wind. Wind velocity 21–72 kmph is required to generate electrical power. Wind results from the differential heating of the earth and the atmosphere by the sun. This uneven heating of the earth from the equator to the poles and over the oceans and continents. The air circulates from cold to warm areas producing winds. The wind resource is concentrated in certain regions and can vary a great deal with time and location. The wind speed generally increases with height. During day time the air over land gets heated up more than that over the ocean, river, ponds and lakes. The hot air over the land rises up and the heavier air from large water bodies rushes in to take its position which results in winds. During night time reverse happens because air cools on land at a faster rate compared to that on water bodies.

It has been estimated that roughly 10 million MW of energy is continuously available in the earth's wind. The utilization of some of this energy through various mechanical conversion devices has played a decessive role in the economic development of many villages of the countries where winds are strong and steady. Wind power is proportional to (diameter)$^3$ and (speed)$^3$. Bigger units produce more energy. 10 kmph winds produce 8 times more power than 5 kmph wind. It is estimated that $20 \times 10^6$ MW of wind power can be tapped.

*Advantages of Wind Power*

Wind is one of the most eco-friendly, clean and safe energy because it has no pollution. No fuel is required to run the wind project since wind is freely available. It has very low operation and maintenance cost in the range of 20–30 paise/unit in comparison with other often used back-up sources. Wind projects appear to be feasible for mechanical milling, grinding grain into flour, lifting of water from wells in villages and other applications. In similar future wind would be a conventional resource. Windmills cannot drive cars, but are suitable devices in remote villages where electricity will be costly due to long distribution distances.

Pollution sowing per year from a typical 200 kw wind electric generator is given below:

| | | |
|---|---|---|
| Average yearly output | — | 400 000 kwh |
| Substitution of coal | — | 120–200 Tonnes |
| Sulphur dioxide ($SO_2$) | — | 2–3.2 Tonnes |
| Nitrogen oxide ($NO_2$) | — | 1.2–204 Tonnes |
| Carbon dioxide ($CO_2$) | — | 300–500 Tonnes |
| Slag and fly ash | — | 16–28 Tonnes |
| Particulates | — | 160–280 kg |

### 1.11.1  Wind Power Potential in India

India is endowed with substantial wind resources because of its unique geographical location exposed to both southwest and northwest monsoon winds. The westerly winds of the southwest provide bulk of wind potential with the total installed capacity of 1,080 MW. The total potential for wind power in India as assessed by Ministry of Non-conventional Energy Sources (MNES) and Indian Institute of Tropical Metrology (IITM) is about 45,000 MW. India had 4,051 installed wind turbines with a capacity to produce 1022.075 MW by 1999 March-end. There was an increase in wind turbines by 380 and capacity 118.270 MW on March 31, 2000.

In India, there were 30 wind electric generator (WEG) supplier in the 90's. Today there are just half a dozen active WEG manufacturers. Some companies do maintain the machines they had supplied.The leaders among the active companies are: Vestas RRB, Suzlon Energy Ltd, NEPC India, Enercon India Ltd, Wescare (Das-Langerway), Asian Wind Turbine Ltd. and Pioneer-Wincon. They have been listed with installed capacity in Table 1.1

Among Indian states, Tamil Nadu still maintains its unassailable leading position in wind power due to well conceived policies and the facilities offered by the state government. It had first private sector wind farm set up its in 1990, 81 percent of the installed capacity of the wind farms has been contributed by Tamil Nadu. The single largest area of private wind farms of the country is existing only in Tamil Nadu with an installed capacity over 394.715 MW and this is next only to the cluster of wind farms existing at California in the US The total wind power potential in the state is estimated to be 20,000 MW. States like Gujarat, Maharashtra, Rajasthan, Karnataka and Andhra Pradesh are now supporting the wind power companies and investors with liberal policy initiatives.

The world is depending more and more on the power of wind to energize its development. The growth rate has been spectular and is said to be next only to infotech industry. The installed wind power capacity in the world is estimated at 13,932 MW. India stands fifth in the world ranking with installed capacity of 1035 MW as shown in the Table 1.2.

**Table 1.1   Leading Wind Electric Generator Companies in India**

| Name of the company | No. WEGS 31-3-1999 | Installed MW 31-3-1999 | No. WEGS 31-3-2000 | Total MW 31-3-2000 | Shares as on 31-3-2000 |
|---|---|---|---|---|---|
| NEPC | 1,334 | 326.19 | 135 | 330.865 | 29.02 |
| Vestas RRB | 682 | 145.245 | 775 | 174.420 | 15.30 |
| Aban kenetch | 231 | 94.710 | 231 | 94.710 | 8.31 |
| AMTL–Wind world | 327 | 82.440 | 330 | 83.190 | 7.30 |
| Suzlon–Sudwind | 106 | 36.940 | 214 | 74.740 | 6.55 |
| Bhel–Nordex | 239 | 55.800 | 263 | 61.800 | 5.42 |
| Enercon | 124 | 28.520 | 223 | 55.360 | 4.86 |
| Das–Langerway | 192 | 46.470 | 206 | 49.970 | 4.38 |
| TTG–Husummer | 130 | 32.500 | 134 | 33.500 | 2.94 |
| Ned wind–Windia | 59 | 30.250 | 59 | 30.250 | 2.65 |
| Others | 627 | 143.010 | 645 | 151.400 | 13.28 |

**Table 1.2   The Top Tens**

| Countries | MW in 1998 | Installed MW 1999 | Total MW in 1999 |
|---|---|---|---|
| Germany | 2874 | 1568 | 4442 |
| US | 2141 | 477 | 2445 |
| Spain | 880 | 932 | 1812 |
| Denmark | 1420 | 325 | 1736 |
| India | 992 | 43 | 1035 |
| Netherlands | 379 | 54 | 433 |
| UK | 338 | 24 | 362 |
| Italy | 197 | 80 | 277 |
| China | 200 | 62 | 262 |
| Sweden | 176 | 44 | 220 |

In Asia, India is far ahead of its neighbours with 1,145 MW. China comes next with 262 MW, Japan 68 and others 11 MW.

The rest of the world has a total installed capacity of 151 MW.

### 1.11.2   Types of Wind Turbines

Modern power generating wind turbines are generally classified according to the following:

1. Axis of rotation
2. Number of blades
3. Up wind or down wind
4. Power regulation

The axis of the wind turbine can be designed for either horizontal or vertical mounting. Both Horizontal Axis Wind Turbines (HAWT) (Fig. 1.11) and vertical Axis Wind Turbines (VAWT) (Fig. 1.10) are suitable for electricity power generation but former is more popular for 15 kw-3

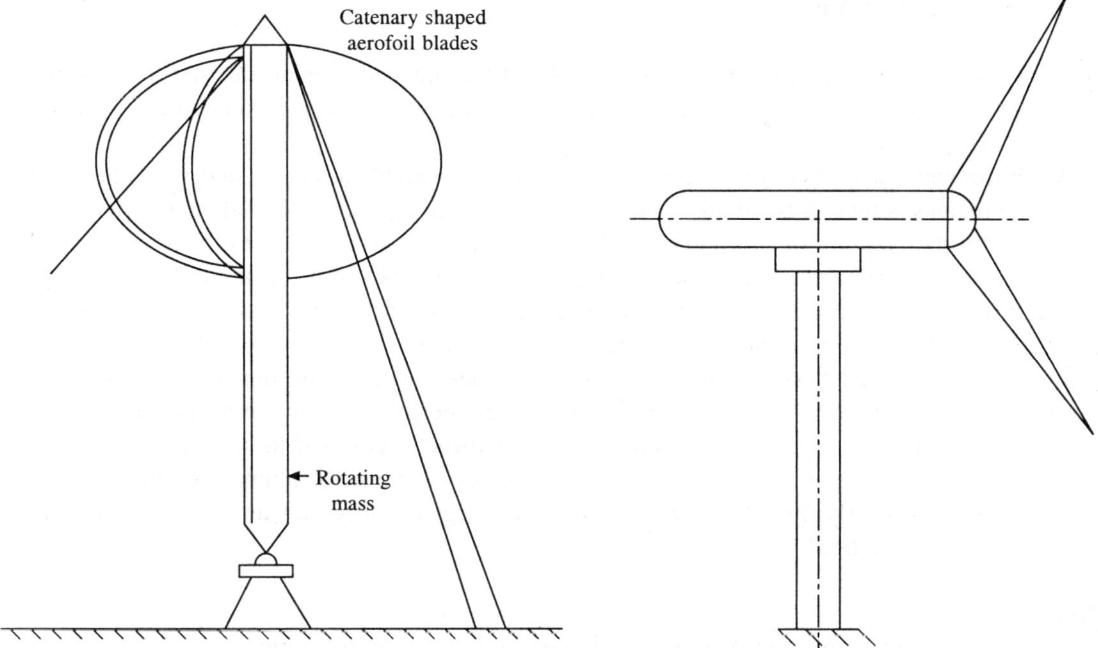



Fig. 1.10  **Vertical Axis Wind Turbine (VAWT)**  Fig. 1.11  **Horizontal Axis Wind Turbine (HAWT), Two Bladed**

MW. The propeller type horizontal axis wind turbine has a central shaft with a hub and a propeller (wheel). The shaft is mounted on two bearings. The propeller (wheel) has a few blades with aerofoil design. The wind passes through the propeller and gives a circumferential force (torque) and axial thrust. The torque is responsible for converting wind power to rotary mechanical power. The HAWT has one, two or three blades rotor revolving on a horizontal axis and the blades move on a vertical plane.

High rotational speed and low solidity are preferable for generating huge amounts of electricity. Large HAWT have been manufactured with one, two and three blades. Two and three blades have proved most popular due to smooth power output and balanced gyroscopic forces. There is no need to teeter the rotor, allowing the use of a simple rigid hub.

If blades are mounted down wind they can be coned outwards in a down-wind direction to give adequate blade to tower clearance with a small and less costly nacelle. In the upwind configuration the gravitational moment of the blades about the tower is counterbalanced by the wind loading. The main advantage of down wind turbine is that it allows the use of free yaw system. It also allows the blade to deflect away from the tower when loaded.

For optimum performance of power regulation a blade must be twisted with a large pitch angle near the root which gradually decreases as the tip is approached. In some designs of wind turbine the pitch angle is fixed while in another the pitch angle is varied by a pitch change mechanism after being placed in the blade root.

## 1.12   TIDAL POWER

Tide is the periodic rise and fall of the water level of the sea. Tides occur due to the attraction of sea water by the moon. These tides can be used to produce electrical power which is known as tidal power.

When the water level is above the mean sea level it is called flood tide and when the level is below the mean sea level it is called ebb tide (Fig. 1.12) 1 to 2 percent of total energy demand can be met with tidal power. For a country like India which is surrounded by sea, tidal power will be of much use. In 1966, France had used turbines to generate 245 MW of power both when tides rises and when it falls. For harnessing tidal energy, a dam is constructed in such a way that a basin gets separated from the sea and a difference in the water level is obtained between the sea and basin. The constructed basin is filled during high tides and emptied during low tide passing through sluices and turbines respectively. The potential energy stored in the basin is used to drive the turbine which in turn generates electricity as it is directly coupled to alternator.

Tidal power is superior to conventional hydro power as the hydroplants are known for their large seasonal and yearly fluctuation in the output of energy because they are entirely independent on nature cycle of rainfall.

### Disadvantages
   (i)   These power plants can be developed if natural sites are available.
  (ii)   The supply of power is not continuous as it depends upon the timing of tides.
 (iii)   The navigation is obstructed.

## 1.13   GEOTHERMAL ENERGY

According to various theories, the earth has a molten core. The centre of the earth is solid and metallic. Outside this, there is a molten core (iron + nickel) surrounded by a mantle of rock 3,400 km thick in layers. Then there is a thin skin, 35 km thick. The core and base of the mantle are very hot, temperature ranging from 3000°C to 4000°C. Almost 80 percent of the earth's heat comes from the slow decay of radioactive isotopes of Uranium and thorium within the earth. Such unstable particle gradually become stable (uranium into lead) emitting radiation and heat. By appropriate technology, we can harness this energy. Since coal, oil, gas and nuclear materials are depletable, geothemal energy assumes great importance as an alternative. It is reliable and plentiful. It is almost inexhaustable. No combustion is involved and the cost is low. An estimate states that

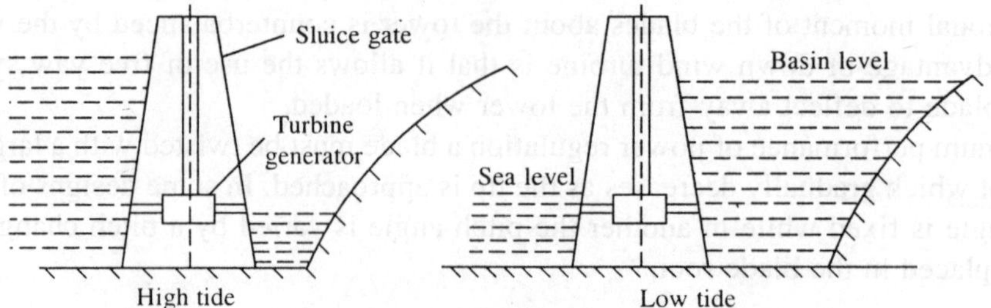


**Fig. 1.12   Schematic of a Simple Tidal Energy Conversion Plant**

in the US; G.T. systems can produce 300,000 times the present annual consumption. At least in 80 countries, there are geological conditions permitting us to use this energy. Byproducts include chemical mineral and gaseous products. Desalination of sea water is carried out in Japan with geothemal energy. In Ireland, 60 percent of homes are heated with this energy. There are 250 hot springs in India giving wet steam under 140°C. Generating plants can be installed to use this energy. At present about 1500 MW of power is being generated in the world from geothermal source. Presently this source contributes only about 1/1600 the part of the world's electrical needs.

The available geothermal energy is classified in four categories: (i) dry steam systems (ii) wet steam system (iii) hot dry rock systems, (iv) magmatic (molten rock) chamber systems. The diagramatic sketch of the dry steam system is shown in Fig. 1.13.

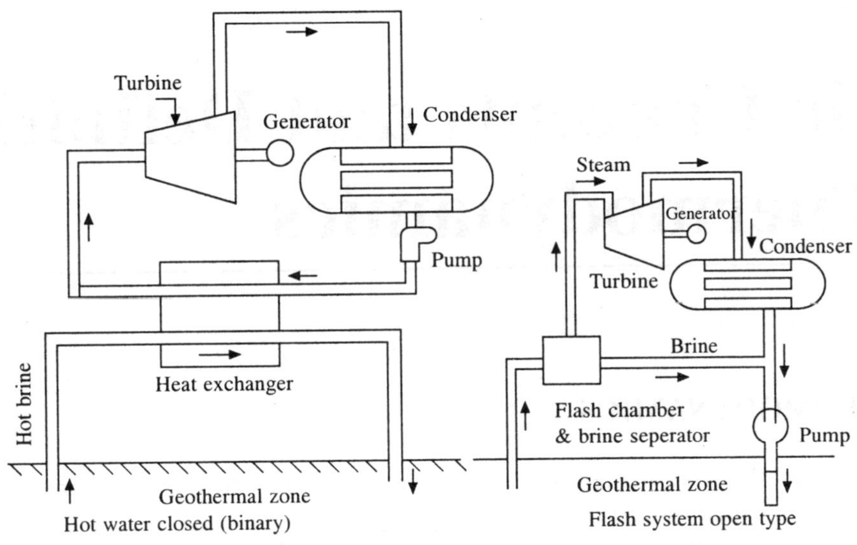


**Fig. 1.13**

It has been concluded that energy demands of the world are increasing at faster rate while the available energy sources are depleting. It is essential, therefore, that new and as yet untapped sources are utilized. Each type of system has got its limitations and problems. In the subsequent chapters conversion systems will be discussed.

## EXERCISES

1.1. Write down in brief about energy scenario in the world and India.
1.2. What are the sources of energy?
1.3. Write down the details of direct solar energy. How it is useful for human beings?
1.4. Define the followings:
    Latitude angle, longitude angle, declination angle, azimuth angle and zenith angle.
1.5. Why we go for solar energy collection? What are the collection devices?
1.6. Write down the details about wind power generation and its advantage.
1.7. What is the contribution of India in wind power generation?
1.8. What do you mean by tidal energy and geothermal energy? How these are tapped?

# 2

# Basic Concepts and Definitions of Thermodynamics

## 2.1 THERMODYNAMICS

Thermodynamics, the word is made of thermo, meaning hot or heat, and dynamics, meaning the study of matter in motion—power, or powerful.

The word thermodynamics means the study of heat related to matter in motion. It is the branch of science which deals with energy and its transfer from one form to another and the relationship between them. Due to energy transfer the physical properties of working substances change. As a matter of fact, thermodynamics deals with the transformation of energy of all kinds from one form to another, but mainly of two forms, that is, heat and work. It is the science that provides a theoretical foundation for understanding of thermal science.

The subject was developed mainly by Carnot, Mayer, Clausius, Joule, Kelvin, Maxwell, Planck, Earatheodory and Gibbs.

Based on observations of common experience and extensive experiemental work, this science has been formulated into four basic laws. They are zeroth law, first law, second law and third law of thermodynamics. These laws govern the principles of energy conversion especially from heat energy to work energy and vice-versa.

There is a wide range of application of thermodynamics in the engineering field. It is used in the design of thermal prime movers, such as steam engines, steam turbines, internal combustion engines and gas turbines. It is used in fuel cells, thermoelectric and thermionic generators. It is used in refrigeration and air conditioning, jet propulsion and compressors.

## 2.2  MACROSCOPIC AND MICROSCOPIC VIEW POINT

The study of thermodynamics has been divided into two groups based on macroscopic and microscopic approach.

Classical thermodynamics — based on macroscopic approach
Statistical thermodynamics — based on microscopic approach

Macroscopic view means a total view. The view analyses the system or equipment as a whole. We study the bulk nature of the system. We are not concerned with the action of individual molecule and its behaviour and no special assumptions are necessary to describe the system. The values of the properties of the system are their average values. Therefore, we can consider the matter as continuous, or the whole thing as a continuum. Let us consider a sample of a gas in a closed container. The pressure of the gas is the average value of the pressure exterted by millions of individual molecules. Similarly the temperature of this gas is the average value of translational kinetic energies of millions of individual molecules. These properties like pressure and temperature can be measured very easily. The changes in the properties can be felt by our senses. The analysis of macroscopic system requires simple mathematical formulae. Microscopic view considers that the system is made up of a very large number of particles known as atoms or molecules as the case may be. In mircoscopic analysis 1 cm$^3$ of a monoatomic gas requires the number of equation which would be equal to 6 times the number of atoms, specifying 3 location co-ordinates and 3 velocity components for each atom. We can just imagine the possibilities. Under these situations we take help of digital computer with the concept of kinetic theory of gases. Microscopic approach can be used in the systems that have high vacuum and region of outer space. The properties like velocities, momentum, impulse, kinetic energy and force of impact, which describe the molecule, cannot be felt by our senses.

However, the behaviour of the system under both these views is same and the results given are compatible. In short, the gross values of microscopic analysis are identical with those given by macroscopic analysis.

In this book, macroscopic view has been adopted which is direct and simple.

## 2.3  CONTINUUM

In macroscopic thermodynamic analysis, continuum is used. We consider the matter as continuous rather than consisting of discrete particles. Generally in engineering field, we consider pressures and temperatures of large number of molecules in the system. The substance in such system is considered as continuous. Such a continuous substance is known as continuum. The concept of continuum is not applicable when number of molecules in the system are very small such as systems under high vaccuum.

## 2.4  THERMODYNAMIC EQUILIBRIUM

When a system is isolated from its surroundings, then there would no change in macroscopic property and the system is said to be in thermodynamic equilibrium. To attain a state of thermodynamic equilibrium the following three types of equilibrium states must be achieved:

   (i) Mechanical equilibrium—If there is no unbalanced force within the system and also between the system and surroundings, the system is said to be in mechanical equilibrium. Unbalanced forces exists in the system or between system and surrounding till mechanical equilibrium is attained.

  (ii) Thermal equilibrium — If there is uniformity of temperature in the system, then the system is in thermal equilibrium. That is, the system involves no temperature differentials, which constitute the driving potentials for heat flow.

 (iii) Chemical equilibrium — When there is no chemical reaction taking place in any part of a system, then the system is in chemical equilibrium.

So if a system is in thermodynamic equilibrium, it must be in mechanical, thermal and chemical equilibrium simultaneously. Thermodynamic properties are used only for thermodynamic equilibrium states and are used as co-ordinates to describe a system.

## 2.5  SYSTEM

The term system is defined as a prescribed region of space or a collection of matter that is completely enclosed by an envelope called the boundary. The boundary separates the system from the rest of the universe called surrounding. The system boundary may be real or imaginary, it may be fixed or moving. A transfer of mass or energy to or from a system is said to occur only if it crosses the system boundary.

   There are three types of systems:

   (i) Closed system (ii) Open system (iii) isolated system.

   (i) A closed system has a constant mass and no mass can come in or go out of it. It has exchange of only energy with the surrounding on its boundary. This is also called a non flow system.

     In Fig. 2.1 (a) a rigid conducting cylinder is filled with gas and heated. The gas is an example of a closed system. The boundary (the inner surface of the cylinder) is real and fixed. The cylinder wall forms a part of the surrounding. We see that there is no mass transfer but energy (in the form of heat) crosses its boundary. If the cylinder is not rigid the system expands. It is still a closed system but then it has a moving boundary. Figure 2.1 (b) shows a similar system with work (energy) being transferred across its boundary by means of a paddle wheel.

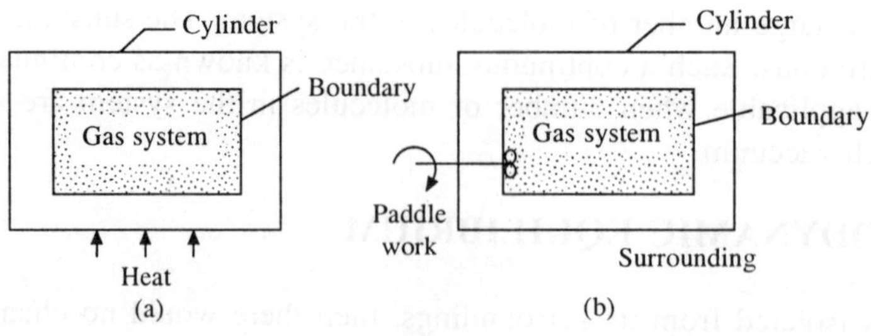


**Fig. 2.1**

(ii) An open system has exchange of mass and energy both with the surrounding. Thus mass may flow in or flow out simultaneously. Similarly energy can flow in or flow out simultaneously.

Boiler is an example of an open system. In a boiler while mass (water) goes in and mass (steam) goes out, energy (heat) is added to the system from outside (see Fig. 2.2 (a)). While studying flow processes, generally attention is focussed on a fixed region rather than on a fixed mass. Such a fixed region as space is called a control volume. This control volume is then treated as a system. Both mass and energy may cross the boundary of the control volume.

An open system may be further classified as (a) steady flow open system and (b) unsteady flow open system.

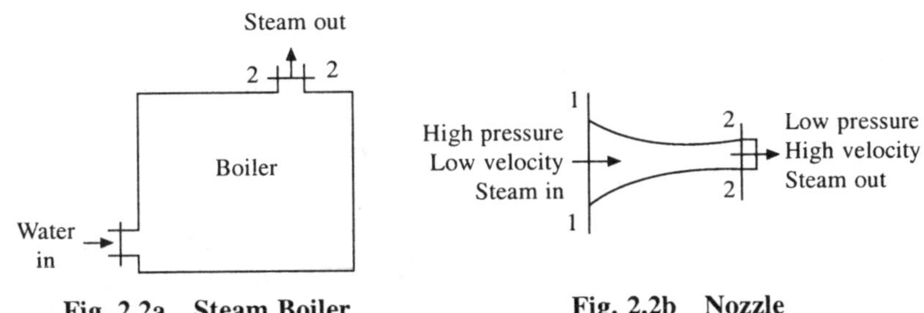


**Fig. 2.2a   Steam Boiler**          **Fig. 2.2b   Nozzle**

(iii) In an isolated system, no exchange of mass or energy with the surrounding takes place. In real life there is no such system. However, if the whole universe is taken as a system it behaves as an isolated system because then nothing is left with which it can exchange mass or energy. For example, gas enclosed in an insulated box.

## 2.6  PROPERTY

Properties of a system are its measurable characteristics describing the system. A gaseous system is described by its temperature, pressure, volume and density. which are measurable. These are therefore its properties. There are many other properties used to identify the condition of a system, for example, internal energy, entropy, velocity, surface area, height and electrical potential. The properties are used as thermodynamic coordinates to identify the condition or state of a system. The value of a property depends only on the state of the system and not on the manner in which the state is reached. So, it is a point function. Any quantity whose change is fixed by only the end states, that is, independent of the process of change is a property.

Properties are of two types (1) intensive and (2) extensive.

1. Intensive property — If the value of a property does not depend on the mass (or extent) of the system it is called an intensive property. Pressure, temperature, velocity, height, concentration, and specific volume are a few examples of intensive properties. The intensive properties are the driving potentials that can cause change in the condition of a system.

2. Extensive property—If the value of a property depends on the mass of the system, it is called an extensive property. Volume, internal energy and mass are a few examples of extensive properties. It is important to note that specific values of these are intensive whereas, volume is an extensive property.

## 2.7   STATE

The thermodynamic condition of a system is called its state and is described by the properties of the system. For example, the state of a gas system is described by its properties like pressure, temperature, and volume. In Fig. 2.3 (a) two different states $A$ and $B$ of a gas system have been shown as fixed by different values of pressure and specific volumes. To fix the state of a system, at least two independent intensive properties should be known.

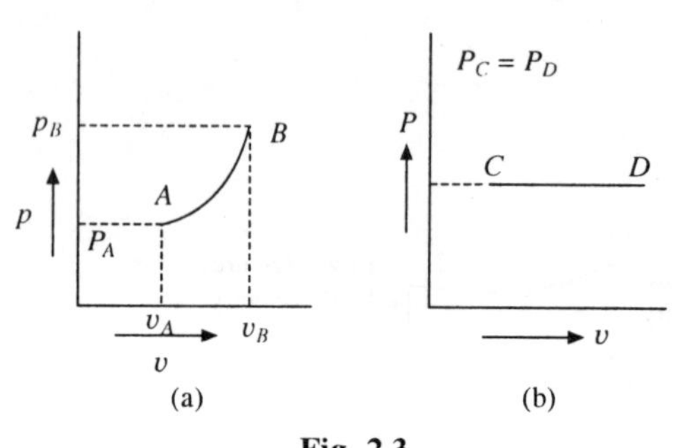

Fig. 2.3

## 2.8   PATH

The path of a change of state is the line joining the series of states through which the system passes. In the above figure $AB$ is the path for the change of state from $A$ to $B$.

## 2.9   PROCESS

A process is said to occur when a system changes from one state to another. In the above figure process AB has been shown to take place when the state changes from A to B along path AB.

If a process occurs according to some fixed law, it has a specific name. For example a process is said to be a constant pressure process if the pressure remains constant as the state changes. In Fig. 2.3 (b) CD is a constant pressure process. There are similarly other processes which follow specific laws, for example, constant volume process and constant temperature process are given as below with pressure ($P$) and volume (V) relation:

(i) Constant pressure process or isobaric process — the pressure is kept constant during the process $P = C$.

(ii) Constant volume process or isochoric process — the volume is kept constant during the process $V = C$.

(iii) Constant temperature process or isothermal process or hyperbolic process — temperature is kept constant during the process. $Pv = C$.

(iv) Constant entropy process or isentropic process or reversible adiabatic process — the entropy is kept constant during the process. $Pv^{\gamma} = C$.

Fig. 2.4 shows some of the processes on $P - v$ diagram.

A process is said to be cyclic if the process starts from one state and after passing through a path returns to the same state.

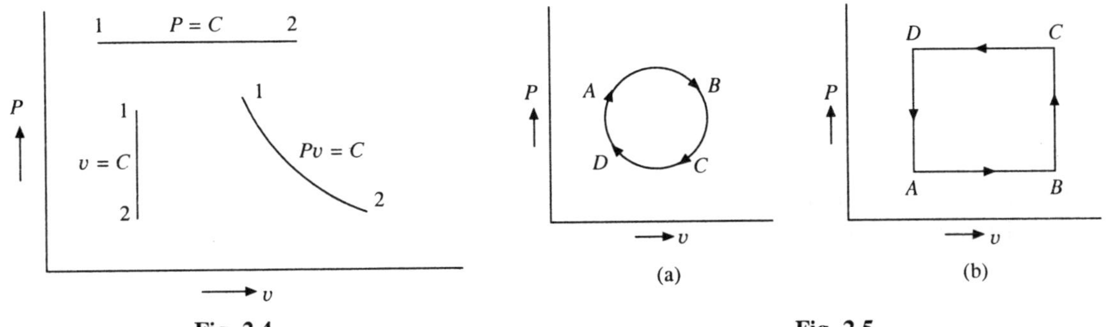

**Fig. 2.4**                    **Fig. 2.5**

In Fig. 2.5(a), a cyclic process *ABCA* has been shown which starts from state *A* and ends at the same point. Some times a number of processes together form a cyclic process. In Fig. 2.5(b), *ABCDA* is a cyclic process of two constant pressure processes *AB* and *CD* and two constant volume processes *BC* and *DA*.

## 2.10  TEMPERATURE AND TEMPERATURE SCALE

To understand the concept of temperature, we should think about the human sense of feeling. By feeling we mean whether it is hot or cold? The decision is made as a result of natural reaction to the sense of feeling. This sense of feeling is the satisfactory method to determine the hotness or coldness of a body.

Temperature scale is referred to a thermometer for measuring inequality of temperature. To construct a thermometer, two reference points are chosen, namely, ice point and steam point.

These are easily reproducible. For getting ice point, consider a glass tube with one end open and other closed, containing mercury at a pressure of one atmosph. Put the closed end of mercury tube in a container containing a mixture of ice and water. Note the mercury column and scratch on the glass which indicates ice point ($\theta_{ice}$). To get another point a system of water boiling at a pressure of one atmosphere is taken. Put the closed end of mercury tube in this container and mark the mercury column by scratching on glass tube which indicates steam point ($\theta_{steam}$).

Many attempts have been made to lay drown scale of temperature. Generally two scales are used, namely, Celsius and Fahrenheit. The celsius scale (often referred to as the centigrade scale) is named after Anders Celsius (1701–44) and Fahrennheit scale is named after Daniel Gabriel Fahrenheit (1686–1736). The lower fixed point (ice point) is the temperature at which pure ice melts. The upper fixed point (stream point) is the temperature at which pure water boils. These are also known as freezing point and boiling point respectively. Of interest, these are considered as 0°C and 100°C in the Celsius scale and 32°F and 212°F in the Fahrenheit scale. To derive the relation between these two scales the following equation may be used:

$$\frac{C - \theta_{ice}}{\theta_{Steam} - \theta_{ice}} = \frac{F - \theta_{ice}}{\theta_{Steam} - \theta_{ice}} \tag{2.1}$$

or,
$$\frac{C - 0}{100 - 0} = \frac{F - 32}{212 - 32}$$

or,

$$\frac{C}{100} = \frac{F - 32}{180}$$

or

$$\frac{C}{5} = \frac{F - 32}{9} \tag{2.2}$$

Likewise, we can get relations between different scales as given below:

1. To convert Celcius to Fahrenheit

$$F = \frac{9}{5}C + 32 \tag{2.3}$$

2. To convert Fahrenheit to Celcius

$$C = \frac{5}{9}(F - 32) \tag{2.4}$$

3. To convert Celcius to Kelvin

$$k = C + 273 \tag{2.5}$$

4. To convert Fahrenheit to Rankine

$$R = F + 460 \tag{2.6}$$

5. To convert Kelvin to Rankine

$$R = \frac{9}{5}k \tag{2.7}$$

6. To convert Rankine to Kelvin

$$K = \frac{5}{9}R \tag{2.8}$$

**Comparison of Temperature Scales**

| Scale | Fahrenheit °F | Centigrade °C | Kelvin K | Rankine °R |
|---|---|---|---|---|
| (i) Stem Point | 212 | 100 | 373.15 | 671.67 |
| (ii) Triple Point | 32.02 | 0.01 | 273.16 | 491.69 |
| (iii) Ice Point | 32 | 0 | +273.15 | 491.67 |
| (iv) Absolute zero | −459.67 | −273.15 | 0 | 0 |

The temperature indicated by a temperature scale is dependen t on thermometric property of the fluid used. As we know triple point of water $T_{dp}$ is a fixed point (273.16 k), the relation between absolute temperature $T$ and the thermometric property $M$ can be written as:

$$T/T_{tp} = M/M_{tp} \text{ or, } T = T_{tp} \times M/M_{tp} \tag{2.9}$$

where $M_{tp}$ is the value of thermometric property at triple point.

# 2.11  DEVICES USED FOR THE MEASUREMENT OF TEMPERATURE

1. Fluid thermometers

   (a) Mercury in glass-thermometers.
   (b) Beckman thermometer –It can measure temperature rise of usually 6°C with an accuracy of 0.01°C.
   (c) The constant volume gas thermometer.

2. Temperature gauges using fluids:

   (a) Temperature gauge. .

3. Dimetallic strip method
4. Pyrometers

   (a) Thermocouples — measured due to Peltier and Thomson effect. The thermocouples used are:

   | | |
   |---|---|
   | Chromel (+) | Alumel (–) |
   | Iron (+) | Constantan (–) |
   | Copper (+) | Constantan (–) |
   | Chromel (+) | Copel (–) |

   (b) The resistance thermometer — the equation used:
   $$R = R_0 (1 + \alpha t)$$
   where $R_0$ = original resistance
   $\alpha$ = co-efficient of increase of resistance with temperature $t$

   (c) The optical Pyrometer

   (i) The radiation Pyrometer.
   (ii) The fusion Pyrometer.
   (iii) Thermal Paints.

## 2.11.1  The Constant Volume Gas Thermometer

The Fig. 2.6 illustrates a constant volume gas thermometer which consists of a small bulb $A$ in which a small amount of gas (normally hydrogen) is collected. It is connected to a glass tube $B$. Glass tube $B$ is connected to manometer $C$, consisting a U-bend-flexible plastic tube $D$ to get the arm $C$ up and down, thus maintaining the meniscus in the other arm to a fixed mark $F$. To get the difference in column a fixed scale $E$ is located. Arm $C$ of the manometer is open to the atmosphere.

The pressure of the gas in the bulb can be

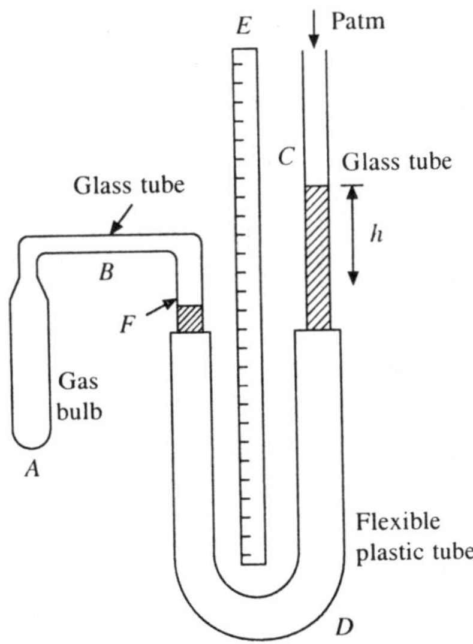


**Fig. 2.6  Constant Volume Gas Thermometer**

obtained by determining the difference in mercury level h and, to this, add barometric height to give the absolute pressure.

It is assumed in this thermometer that equal changes of pressure of the gas in the bulb are produced by equal changes of temperature and thus linear relationship between pressure and temperature of the gas in the bulb is achieved.

Now, for measuring the temperature, the gas bulb is first placed in a constant temperature bath maintained at triple point of water and arm $C$ of manometer is lowered to maintain the mercury level at fixed mark $F$. Let the pressure reading taken at this temperature $T_{tp}$ be $P_{tp}$. Then the bulbs is placed in a location where the temperature $T$ is to be measured. Again, in a similar fashion, the meniscus is brought to fixed mark $F$. Assuming the new pressure reading is $P$. Then by using Eq. (2.9), temperature $T$ is given by

$$T = 273.16 \frac{P}{P_{tp}} \tag{2.10}$$

This thermometer can be used over a wide range of temperatures. At very temperatures, helium gas is used which has lower condensation temperature and is substitute for hydrogen. Nitrogen is the substitute for higher temperature measurement. Glass can be replaced by platinum or platinum-iridium alloy to avoid melting, for higher temperature measurement.

## 2.12 ZEROTH LAW OF THERMODYNAMICS

We know from experience that when a hot body is brought into contact with a cold body, heat is transferred from the hot body to cold body until both bodies attain the same temperature. At that instant of time, the heat transfer stops and the two bodies are said to have reached thermal equilibrium. The equality of temperature is the only requirement for thermal equilibrium.

This law gives the idea of temperature. It states that: if two bodies, isolated from the environment, are in thermal equilibrium with a third body then the two bodies will be in thermal equilibrium with each other. The zeroth law is illustrated in Fig. 2.7. Let $A$ and $B$ be two bodies insolated from each oher. Let $C$ be the third body which acts as a thermometer containing mercury in a glass tube. When the thermometer is steady, it is assumed that the mercury, the glass tube and the body whose temperature is being measured, are all at the same temperature and hence, are in thermal equilibrium. If $A$ and $B$ are in thermal equilibrium with C separately and simultaneously its means showing same temperature reading on C. It should be noted that bodies $A$ and $B$ need not be brought into actual contact with each other. The body (thermometer) is brought into contact with $A$ and $B$, in succession.

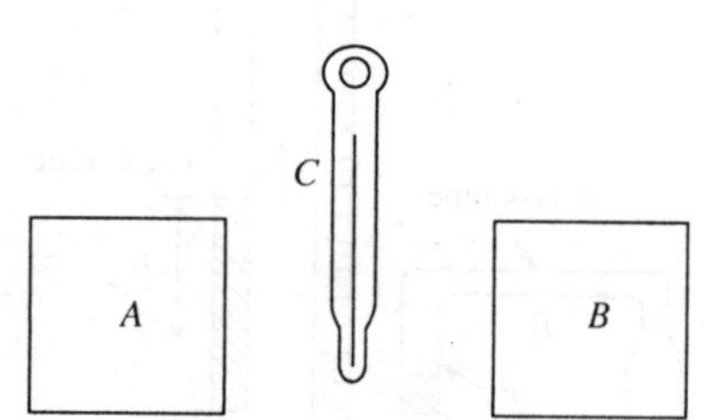

**Fig. 2.7  Establishing zeroth law of thermodynamics**

This law was one of the last thermodynamic laws to be developed. It was introduced by RH Fowler and EA Guggenheium in 1939. It should have been stated before all the laws as the concept of temperature is essential at the begining. It is, therefore, considered that this law should be designated numerically before the other laws and hence called zeroth law (law number zero).

Such an obvious fact of thermal equilibrium as stated as Zeroth Law of thermodynamics cannot be concluded from the other laws of thermodynamics because it serves as a basis for the validity of temperature measurement. By replacing the third body with a thermometer, the Zeroth law can be restated as:

Two bodies are said to be in thermal equilibrium if both have the same temperature reading even if they are not in contact.

## 2.13  HEAT AND WORK

For transfer of energy between a system and the surrounding some driving factor is essential. The driving factors may be temperature difference, pressure difference, difference of velocity and difference of electrical potential between the system and the surrounding. Depending upon what driving factor responsible for the flow of energy, we recognize two forms of energy—heat and work.

*Heat:* Heat is energy transferred without transfer of mass across the boundary of a system because of temperature difference between system and surrounding. The sole reason for the flow of this energy is temperature difference.

In the Fig. (2.8) (a) if $T_2 > T_1$ heat flows from the surrounding to the system. If $T_1 > T_2$ heat flows from system to surrounding on similar lines  the definition of work is given as.

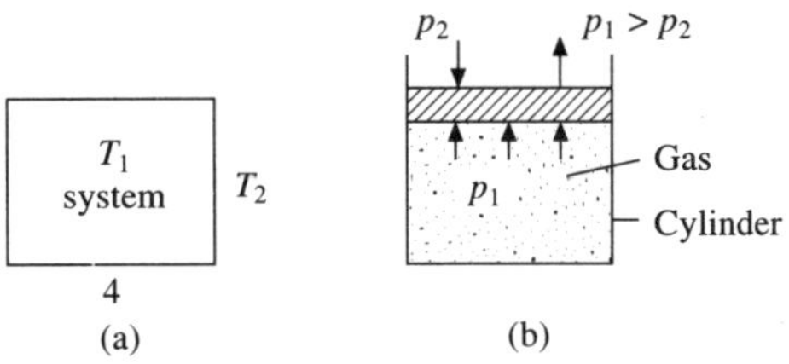

$p_2$ $\qquad$ $p_1 > p_2$

$T_1$ system $\quad T_2 \qquad$ Gas

$p_1$ $\quad$ Cylinder

4

(a) $\qquad$ (b)

Fig. 2.8

*Work:* Work is energy transferred without transfer of mass across the boundary of a system because of an intensive progress in the difference other than temperature that exists between the system and surrounding.

Thus work may be done because of pressure difference and the electrical potential difference between the system and surrounding.

In Fig. 2.8(b) work is done by a high pressure gas as it expands because of pressure difference between gas and atmospheres and forces the piston up. This is mechanical work in which physical movement of point of application of force is necessary.

Work is also defined as thermodynamic work which flows from a system (and to another) during a given operation if the sole effect external to the system could be the rise of a weight.

*Types of Work*
(i) *pdv* work or displacement work (ii) flow work (iii) paddle wheel work or stirring work (iv) electrical work, and (v) elastic work.

Some important points about heat and work:

1. Heat and work are transitional in nature. These terms are used as long as the energy is being transferred to or from a system. Once the energy is transferred it is no longer heat or work. These terms are similar to rain. Rain is the term used when it is raining. Once it has rained it is called water and not rain.

2. Both heat and work are boundary phenomena. They are realised only when they cross the boundary of the system. Inside the system, they have no meaning.

3. According to convention, heat added to a system is positive and heat subtracted from it is negative. Work done by a system (work energy following out) is positive and work done on the system is negative. Letters $Q$ and $W$ will be used to denote heat and work for a system.

4. Work can be converted entirely into heat but heat cannot be converted into work entirely. This is an important difference between these two forms of energy.

5. Heat and work are not properties because they appear only when a change of state occurs. They cannot be used to represent the state of a system.

6. The values of heat and work during a process depend not only on the end states but also on the path. So they are path functions and are different for different paths. In Fig. 2.9 (b) different amounts of heat and work are transferred while the processes follow different paths like $1A2$, $1\,B2$ or $1C2$.

Thus
$$Q_{1A_2} \neq Q_{1B_2} \neq Q_{1C_2}$$

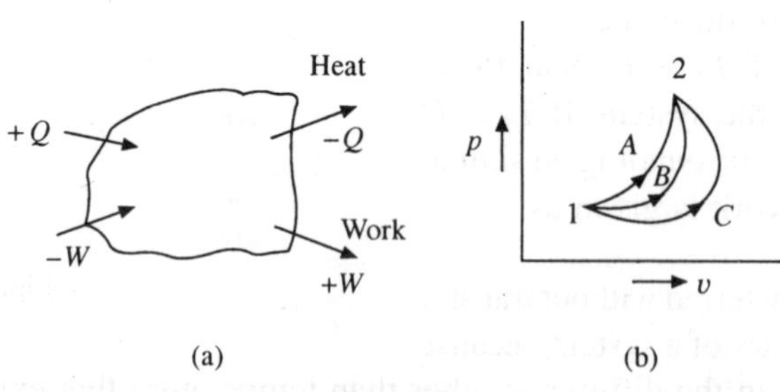


**Fig. 2.9**

and
$$W_{1A_2} \neq W_{1B_2} \neq W_{1C_2}$$

7. Heat transfer takes place because of temperature difference only. All other energy transfers may be termed as work transfers.

## 2.14 INTERNAL ENERGY

Internal energy is the intrinsic energy associated with a system due to its constituents and configuration and motion of its molecules, atoms and subatomic particles. The internal energy of 1 kg coal is different from the internal energy of 1 kg petrol because of difference of constituents. Again one kg of hydrogen gas at one bar pressure and 20°C has different amount of internal energy than 1 kg of the same gas at 2 bar pressure and 20°C as the configuration of molecules is changed. In fact the absolute value of internal energy of any system is not known. Hence the change in internal energy in a process is determined. Letter $U$ will be used to denote internal energy of a system and $u$ for 1 kg of it, that is, specific internal energy.

# 2.15 THERMODYNAMIC EQUILIBRIUM, QUASI STATIC PROCESS AND REVERSIBLE PROCESS

### *Thermodynamic equilibrium*

A system is said to be in a state of thermodynamic equilibrium if every point of the system is in the same state. Thus a system in thermodynamic equilibrium is incapable of spontaneous change when it is isolated.

*Quasi-Static Process* — Quasi means almost. A process which is almost stationary or proceeds with extreme slowness is called a quasi-static process. A quasi-static process is one during which the system deviates from one equilibrium state to another by only infinitesimal amount throughout the entire process. In other words a process closely passing through a series of equilibrium states is known as a quasi-static process.

As an example, let us consider the pressure $P_1$ of a gas acting on one face of a frictionless piston fitted in a cylinder and a pressure $P_2$ on the opposite face. If the two pressures are equal, the piston is in equilibrium. If now pressure $P_1$ be infinitesimally larger than $P_2$ the gas on the side of $P_1$ will undergo an infinitesimal expansion as the piston moves although the gas is in almost equilibrium state. The process followed during this phenomena is an example of a quasi-static process. As shown in the Fig. 2.10, 1-2 represents a quasi-static process on a $p$-$v$ plane. At the successive states from 1, such as $a, b, c, d, e$. the system is in very nearly thermodynamic equilibrium. The quasi-static processes are very slow.

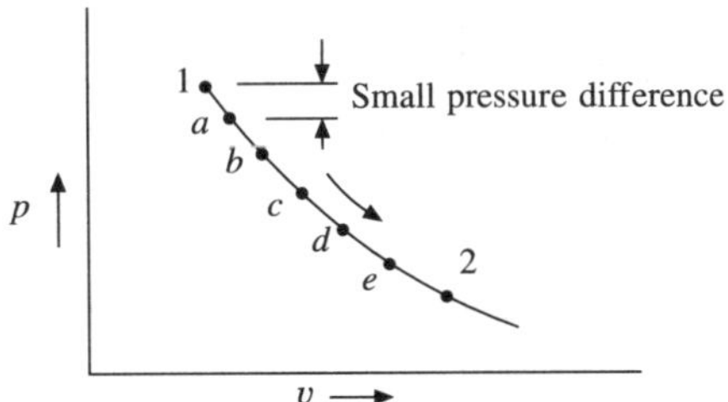


**Fig. 2.10    Quasi-static Process**

### *Reversible process*

Let us again consider a high pressure gas entrapped in a non-conducting cylinder behind a non-conducting piston (Fig. 2.11). There is no friction between piston and cylinder and the piston moves out from position 1 to 2 as the gas expands slowly passing through various quasi-static states.

Work is given out by the gaseous system during its expansion. In this process the gas loses its internal energy and does an equivalent amount of work as there is no energy input, that is, $W = dU = U_1 - U_2$. The surrounding gets this work $W$.

If now force is applied on the piston from outside in opposite direction and it is moved inside so that the process exactly goes along the same path in reverse direction, that is, 2 to 1, work $W$ is transferred to the gas and it is stored there in the form of internal energy; thus restoring

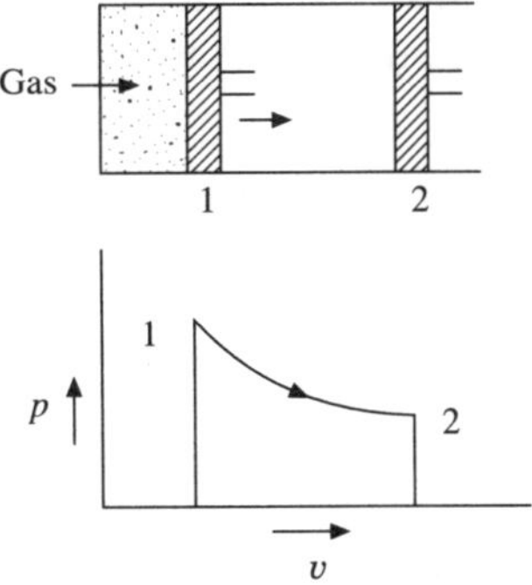


**Fig. 2.11    Gas Entrapped in a Frictionless Cylinder and Piston**

the original state of the gas at 1. The surrounding is also restored to its original state. We say that the process has been completely reversed.

Thus, A process is called reversible if the system and the surrounding can be completely restored to their respective initial states after the process has occurred. Alternatively, "a reversible process may be defined as a process which when undone will leave no evidence of the events in the surrounding."

It is important to understand that simple return of the piston to its initial position does not ensure the above process to be reversible. It should be noted that a process is not reversible if it involves (i) friction (ii) transfer of heat between system and surrounding due to finite temperature difference (iii) free expansion to a lower pressure (iv) and rapid expansion or compression.

Actually speaking no process is truely reversible and in nature all processes are irreversible, but some processes may approach reversibly to a close approximation. A few examples of nearly reversible process are:

  (i) Frictionless relative motion.
 (ii) Expansion and compression of a spring.
(iii) Frictionless expansion or compression of a fluid.

## 2.16 WORK DONE DURING A NON-FLOW REVERSIBLE PROCESS (Pdv WORK)

In Fig. 2.12 a non-flow process is shown occuring in a cylinder fitted with a frictionless piston. A high pressure gas which forms the system is entrapped between cylinder and piston. If the piston is allowed to move slowly the gas expands and work is done by it (the gas). The variation of gas pressure and displacement of piston is shown in the figure.

At any time the force on piston is $P \times A$, where $A$ is the area of cross-section of piston and $p$ the pressure acting on the piston at that instant. For a small displacement $dl$ of the piston the work transferred during the process 1–2 will be given by

$$(p \cdot A) \times dl = P (A \, dl) = pdv, \quad (2.11)$$

Where $dv$ is the small change of volume of gas corresponding to displacement $dl$ of piston.

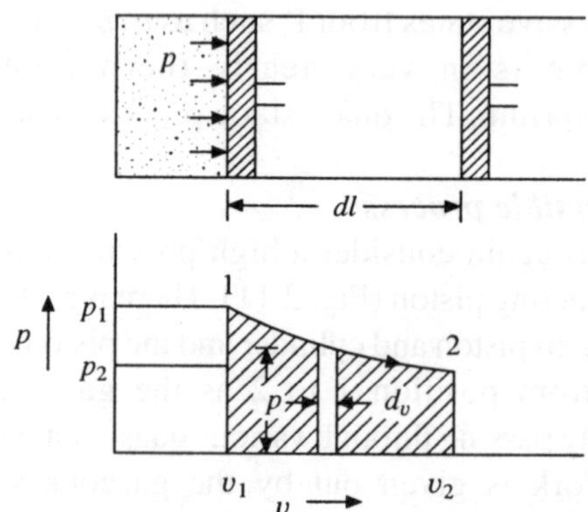

Fig. 2.12   **Non-flow expansion process.**

We see from the above figure that $\int_1^2 pdv$ is the shaded area under the curve 1–2 that represents the path of the process. If the path 1–2 changes, the work done also changes. It is thus clear that to know the work done in a process, the knowledge of the process is essential. It is to be noted that during expansion of the gas work is done by the gas and the work goes out of it. So it is positive. On the other hand if the gas is compressed from state 2 to 1 as the piston moves from position 2 to 1 the work done is given by $\int_1^2 p\,dv$ and this work is done on the system (gas) and is therefore negative.

It must be remembered that $\int_1^2 p\,dv$ gives the work done during a reversible non-flow process when $p$ is the pressure acting on the moving boundary.

In case of a free expansion $p\,dv \neq 0$ but work done is zero because there is no moving boundary.

Similarly incase of a paddle work done on a system the $\int p\,dv = 0$ but $w \neq 0$. The explanation of these lies in the fact that $\int p\,dv$ gives the work done in a non-flow reversible process only. The previous two cases were examples of irreversible processes.

## 2.17  FLOW WORK OR FLOW ENERGY

Let us consider the processes shown in the Fig. 2.13.

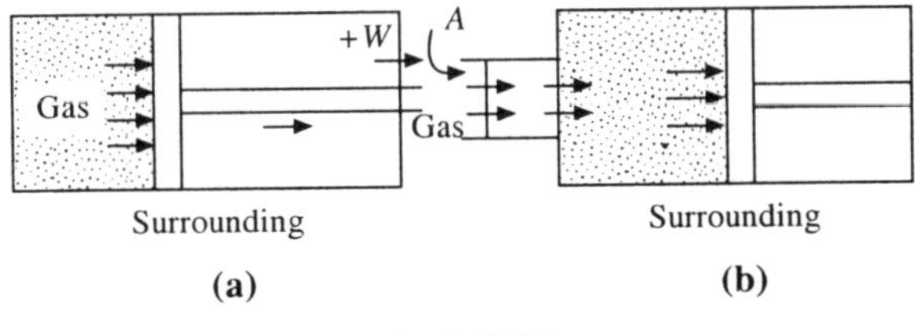


**Fig. 2.13(b)**

In the Fig. 2.13 (a) the gas pushes a frictionless piston and does work. This process is a non flow process and does not involve any mass flow from outside. Such work is called shaft work or $p\,dv$ work. If there be no heat exchange with the surrounding the work is done at the cost of the internal energy which decreases.

Now consider the process occuring in Fig. 2.13(b). Here the cylinder has an opening at the end and as the high pressure gas enters the cylinder from the surrounding, it applies pressure on the piston. The piston moves and work flows out. This work involves mass flow and work is done without any loss of internal energy of the gas. In fact the energy used to accomplish work is transmitted from a pump somewhere in the surrounding which forces the gas into the system. This work which accompanies mass flow across the system boundary is called flow work.

For a volume $A \cdot \Delta L$ flowing into the system the work done $W = p \times A \cdot \Delta L = P\Delta V$. For unit mass, flow work $W = p \times v$, where $v$ is the specific volume of the gas.

Thus if a fluid at a pressure $p$ enters a system every unit mass of it brings with it an energy equal to $pv$ together with its internal energy, kinetic energy, and potential energy without any change in its state (Fig. 2.14). This is flow energy.

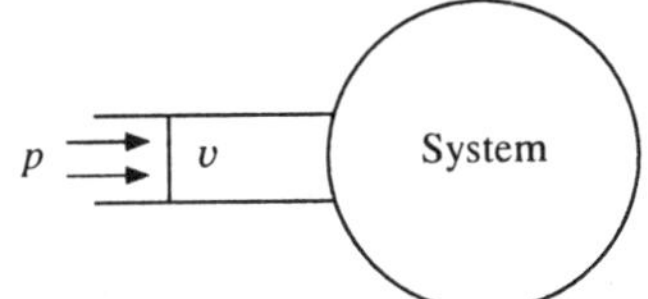


**Fig. 2.14**

## 2.18   ENTHALPY

While dealing with different processes, we often come acorss cases when the flow energy accompanies internal energy. Hence we use a composite term known as enthalpy for the sum of flow energy and internal energy.

Then for unit mass enthalpy,    $h = u + pv$    (2.12)

For a system of mass $m\,kg$, $mh + mpv$

or    $H = U + m\,pv$    (2.13)

## 2.19   TOTAL ENERGY

Total energy associated with a system is the sum of all forms of energy which are either internal energy, flow energy, kinetic energy or potential energy.

Thus, total energy content of a system of mass $m$ is given by total energy

$$= I.E. + F.E. + K.E. + P.E.$$    (2.14)

or,    $$E = U + m\,pv + mV^2/2g_c + mZg/g_c$$    (2.15)

for 1 kg mass

$$e = u + pv + V^2/2g_c + Z\,g/g_c$$

or,    $$e = h + v^2/2g_c + Z\,g/g_c$$    (2.16)

Note
(1)  All these energy terms must be expressed in the same unit.
(2)  $g = 9.81$ m/sec$^2$, $g_c = 9.81$ kg m/kg $f$-sec$^2$ in mks unit and $= 1$ kg m/N $-$ sec$^2$ in SI unit.

## 2.20   WORK DONE IN A FLOW PROCESS

In the absence of $KE$ and $PE$ the work done in the steady flow process can be determined as under. In the Fig. 2.15(a) a piston is shown fitting in a cylinder.

Let, unit mass of fluid at pressure $p_1$ and specific volume $v_1$ enter the cylinder through an inlet valve at $A$ and push the piston down to $B$ with valve $D$ closed (Fig. 2.15(b). The valve $A$ then closes and the piston moves as gas expands to a lower pressure at $C$. At $C$ the exhaust valve $D$

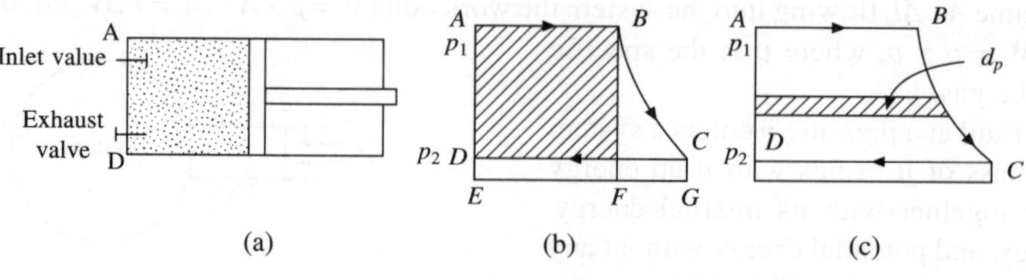


**Fig. 2.15**

opens and the flow pressure gas is exhausted at constant pressure along *CD*. Let these processes occur reversibly. For unit mass of fluid, total work done = areas $(ABFE + BCGF - CDEG) = p_1 v_1$ $+ \int_1^2 p\, dv - p_2 v_2$. The above area comes to be *ABCD* which is nonthing but the intergration of $v\, dp$ between pressure limits $p_2$ to $p_1$ (see Fig. 2.15(c)).

i.e.
$$W = \int_{p_2}^{p_1} v\, dp$$

or,
$$W = \int_{p_1}^{p_2} v\, dp \tag{2.17}$$

In general we write $W = -\int v\, dp$. In this example a single piston cylinder arrangement does not have a continuous flow of mass and energy. But if a number of such cylinders be connected by a common intake and exhaust pipe the system will work as a steady flow process.

The above relation gives work done in any reversible steady flow process where *PE* and *KE* are negligible.

## 2.21  UNITS AND DIMENSIONS

The progress of science and technology is related with the systems of units, dimensions, measurement and standardisation. Energy science and technology involves several physical quantities. Each quantity has certain basic/derived dimensions in terms of length ($L$), mass ($M$) and time ($T$) A measurement is expressed in terms of a number and its units, for example, 10 kg, 15 m$^3$. Each measurement simultaneously identifies a particular dimension, a measured number and relevant units. Standardisation is an important aspect of measurements. Measurements are carried out in terms of standard units acceptable to everyone. Standard units are specified by standard organisations recognised by various nations.

The science and technology involves measurements of several quantities. During 1950s, several systems of measurements and units prevailed in different countries. This made the energy calculations quite complex, difficult and time consuming. In 1960, the international systems of units or system international of units (SI) was introduced. In SI units all forms of energy are measured in terms of the same unit joule (J). This has made energy calculations very simple and easy.

In the present text, the SI system of units has been used. The basic units in this system are given in Table 2.1

The dimensions of all others quantities are derived from these basic units which are given in Table 2.2.

Some SI derived units have been mentioned in terms of base units in Table 2.3.

### 2.21.1  Force
The force acting on a body is derived from Newton's second law of motion and is defined as the force which will accelerate 1 kg of mass with an acceleration of 1 metre per second.

**Table 2.1  Basic Units**

| Quantity | Unit | Symbol |
|---|---|---|
| Length ($L$) | metre | m |
| mass ($M$) | kilogramme | kg |
| Time (t) | second | S |
| Temperature (T) | kelvin | K |
| Plane angle | radian | rad |
| Amount of substance | mole | mol |
| Electric current | ampre | A |
| Luminous intensity | candela | Cd |
| Solid angle | steradian | Sr |

**Table 2.2  Derived Units**

| Quantity | Unit | Symbol |
|---|---|---|
| Force | Newton | N |
| Pressure | Pascal | Pa or N/m$^2$ |
| Energy, heat or work | Joule | J = Nm |
| Power | Watt | W = J/s |
| Torque | Newton-metre | Nm = J |
| Specific heat | Joule per kilogramme per degree Kelvin | J/kg k |
| Dynamic viscosity | Newton second | N$_S$/m$^2$ |
| Thermal conductivity | Watt-per metre per degree Kelvin | W/mk = J/Sm k |
| Frequency | hertz | H$_z$ = C/S |

**Table 2.3  Derived SI Units in Terms of Base Units**

| Quantity | Symbol | Expression in terms of other unit | Expression in terms of SI base units |
|---|---|---|---|
| Force | N | – | kgm/S$^2$ |
| Pressure, stress | Pa | N/m$^2$ | kg/mS$^2$ |
| Energy, work, heat | J | N-m | kgm$^2$/S$^2$ |
| Power | W | J/S | kg m$^2$/S$^3$ |
| Temperature | °C | – | K |

Newton's second law gives us the relation

$F \alpha\, m \times a$, where $F$ is force acting on mass $m$ and producing an acceleration a

or,
$$F = \frac{ma}{g_c} \tag{2.18}$$

Where $g_c$ = constant of proportionality

As one Newton produces acceleration of 1 m/s$^2$ in 1 kg of mass, we have

$$1 \text{ Newton} = \frac{1}{g_c} \times 1 \text{ kg} \times 1 \text{ m/s}^2$$

$$\therefore \qquad g_c = 1$$

When $g_c = 1$ the system is said to be coherent or consistent and the product of units of mass and acceleration becomes the unit of force.

$$\therefore \qquad 1 \ N = 1 \text{ kg m/s}^2$$

Note: SI units derived from propor nouns are denoted by the first letters of the noun written in capital, for example, $K$ for Kelvin and $J$ for Joule, others utilize lowercase letters.

The metric unit of force is kilogram force (kgf). One kgf is the weight of one kilogram mass (kgm), under a standard gravitational acceleration of 9.80665 m/s$^2$. Generally we use 9.81 m/s$^2$.

Thus, writing the Newton's second law of motion as

Force $$F = \frac{mg}{g_c} = \text{weight}$$

Where $a = g$ = acceleration due to gravity, we have

$$1 \text{ kg} f = \frac{1 \text{ kgm} \times 9.80665}{g_c} \text{ m/s}^2$$

So that $g_c = 9.80665$ kgm m/kgf s$^2$.

If, local value of $g$ is numerically the same as $g_c$, then the weight of 1 kgm becomes equal to 1 kgf.

So, if the mass is allowed to fall freely under the action of standard gravitational force, it is accelerated at the rate of 9.80665 m/s$^2$ (9.81 m/s$^2$) and we have

$$\text{Force} = 1 \text{ kg} \times 9.81 \text{ m/s}^2 = 9.81 \text{ N}$$

It follows that the weight of 1 kg mass (1 kgf) equals 9.81 Newtons

that is, $$1 \text{ kgf} = 9.81 \text{ N}$$

### 2.21.2 Density ($\rho$)

Density is defined as mass (not weight) per unit volume. It is also known as mass density. Its unit is kilogram per cubic metre (kg/m$^3$). Density of water is 1,000 kg/m$^3$ or 1 tonne/m$^3$. The reciprocal of density is called specific volume and is defined as volume occupied by unit mass of a substance. Its unit is cubic metre per kilogram (m$^3$/kg).

$$v = m^3/\text{kg} \quad \text{and} \quad v = \frac{1}{\rho}$$

### 2.21.3 Specific Weight ($\omega$)

It is defined as weight per unit volume. It is also known as weight density. It is expressed in kgf/m$^3$ in MKS units and N/m$^3$ in SI units and depends upon both the density of the substance and gravitational acceleration g. Let $W$ represents the weight corresponding to certain mass $m$; then

$$W = \frac{1}{g_c} mg \qquad (2.19)$$

Dividing both sides of this equation by volume, we get

$$\frac{W}{v} = \frac{1}{g_c} \frac{mg}{v}$$

or,

$$\omega = \frac{mg}{v} \ \text{N/m}^3 \ (g_c = 1 \text{ in SI unit}) \tag{2.20}$$

### 2.21.4  Pressure (P)

The pressure at a point is defined as the normal force per unit area of the bounding surface. Thus, $F$ is the force normal to area $A$, then

$$P = \frac{F}{A} \tag{2.21}$$

The unit of pressure in the SI system is Pascal ($P_a$), which is the force of one netwon acting on an area of $1 \ m^2$.

$$1 \ P_a = 1 \ \text{N/m}^2$$

The unit of Pascal is very small. So we often use kilopascal (kPa) and megapascal (MPa).

$$1 \ \text{kPa} = 10^3 \text{Pa}, \ 1 \ \text{mPa} = 10^6 \ \text{Pa}.$$

Standard atmospheric pressure is 101325 Pa = 101.325 kPa = $1.01325 \times 10^5$ Pa. The unit of bar for pressure is extensively used. Standard atmospheric pressure = 1.01325 bar.

$$1 \ \text{bar} = 10^5 \ \text{Pa} = 10^5 \text{N/m}^2.$$

1 kgf/cm² is also called technical atmosphere and denoted by ata.

Mostly, instruments indicate pressure relative to the atmospheric pressure whereas the pressure of a system is its pressure above zero, or relative to perfect vacuum. The pressure relative to the atmosphere is called gauge pressure. The pressure relative to a perfect vacuum is called absolute pressure. Barometer measures atmospheric pressure. The pressure of a fluid is measured by instruments known as gauges. Gauges which measure pressure greater than atmospheric pressure and called pressure gauges such as Bourdon Pressure Gauge. This gauge is used to measure the pressure of steam in a boiler. Gauges which measure pressure less than atmospheric pressure are called vacuum gauges. A vacuum gauge is used in a condenser to measure the difference between the atmospheric pressure and the pressure inside the condenser. This differences is called vacuum.

Thus following equations can be written as

$$P_{abs} = P_{atm} + P \text{ gauge}$$

$$P_{abs} = P_{atm} - P_{vacuum}$$

Fig. 2.16 shows this relationship. The datum for absolute pressure is a perfect vacuum, datum for gauge pressure is atmospheric pressure.

Another useful unit of pressure is,

$$1 \ t_{or} = 1 \ \text{mm of Hg} = 133.32 \ \text{N/m}^2 = 133.32 \ \text{Pa}$$

Conversion factors for pressure have been shown in the Table 2.4.

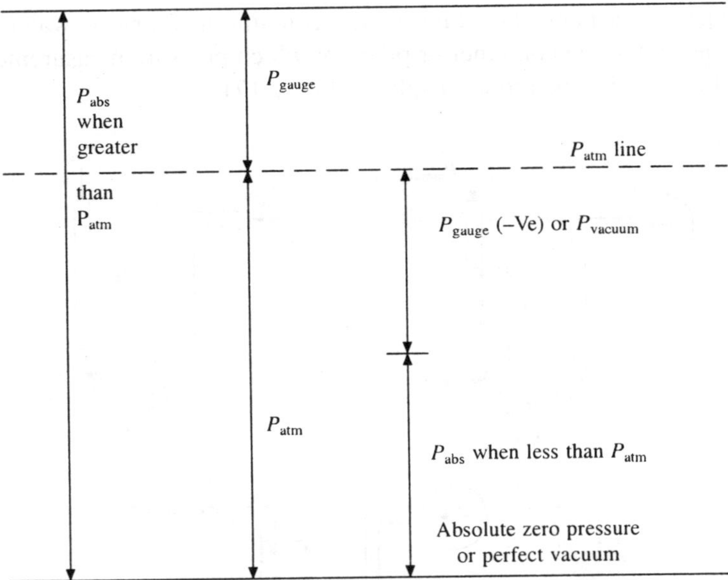


**Fig. 2.16    Pressure Relationship**

**Table 2.4    Conversion Factors for Pressure**

| | atm | mm H$_2$O at (21°C) | mm Hg at (21°C) torr | N/m$^2$ or Pa | kgf/cm$^2$ ata | dyne/cm$^2$ | bar |
|---|---|---|---|---|---|---|---|
| 1 atm | 1 | 10332.3 | 760 | $1.01325 \times 10^5$ | 1.03323 | $1.01325 \times 10^6$ | 1.01325 |
| mm H$_2$O at 21°C | $96.7838 \times 10^{-6}$ | 1 | 0.073556 | 9.80665 | $10^{-4}$ | 98.0665 | $98.0665 \times 10^{-6}$ |
| 1 mm Hg at 21°C) | $1.31578 \times 10^{-3}$ | 13.5951 | 1 | $1.333223 \times 10^2$ | $10.3595 \times 10^{-3}$ | $1.333223 \times 10^3$ | $1.333223 \times 10^{-3}$ |
| N/m$^2$ or Pa | $0.986923 \times 10^{-5}$ | $10197.2 \times 10^{-5}$ | $750.062 \times 10^{-5}$ | 1 | $10.1972 \times 10^{-6}$ | 10 | $10^{-5}$ |
| kgf/cm$^2$ or ata | 0.967838 | 10000 | 735.559 | $0.980665 \times 10^5$ | 1 | $0.980665 \times 10^6$ | 0.980665 |
| dyne/cm$^2$ | $0.986923 \times 10^{-6}$ | $10197.2 \times 10^{-6}$ | $750.062 \times 10^{-6}$ | 0.1 | $1.01972 \times 10^{-6}$ | 1 | $10^{-6}$ |
| bar | 0.986923 | 10197.2 | 750.062 | $10^5$ | 1.01972 | $10^6$ | 1 |

## 2.22    MANOMETER AND PRESSURE GAUGE

Manometer utilizes a column of liquid to measure the pressure, the height of the column indicates the magnitude. If mercury is used as manometric fluid, it is called mercury manometer, and when water is used, the instrument is called water manometer. Generally these two types of fluids are used in manometers. Barometer which measures atmospheric pressure using mercury, is a manometer-type instrument.

$U$-tube manometer is a bent glass tube which contains mercury or water. One end of this manometer is connected to the container or pipe for which pressure measurements is to be made and the other end is usually open to atmosphere (Fig. 2.17).

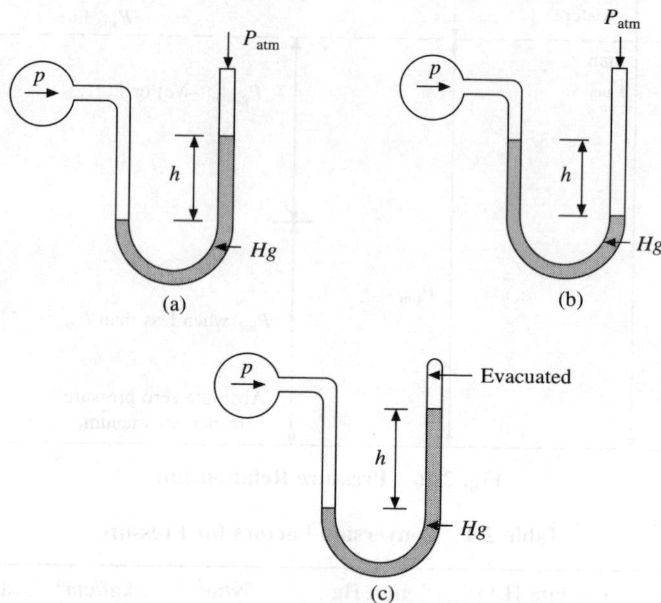


**Fig. 2.17** **(a) Open $U$-tube monometer indicating gauge pressure; (b) Open $U$-tube manometer indicating vacuum, and (c) Closed $U$-tube manometer indicating absolute pressure**

Fig. 2.17(a) Shows a $U$-tube manometer. One limb of the $U$-tube is open to the atmosphere and the other is connected to a vessel pressure higher than atmospheric pressure. Suppose the manometer reads $h$ mm, then absolute pressure of the gas in the vessel $(P)$ = atmospheric pressure + manometer reading

or, $$P = P_{atm} + h \tag{2.22}$$

Fig. 2.17(b) shows a $U$-tube manometers measuring pressure less than atmospheric pressure. One end of the U tube is open to atmosphere and the other end is connected to a vessel that has pressure less than that of the atmosphere.

In this case, $$P = P_{atm} - h \tag{2.23}$$

Fig. 2.17 (c) shows a closed $U$-tube indicating absolute pressure. If $P$ is the atmospheric pressure, then this gauge is known as barometer.

If $h$ is the difference in the heights of the liquid columns in two limbs of the U-tube, $\rho$ the density of the fluid and $g$ the acceleration due to gravity, then from elementary principle of hydrostatics, the pressure $P$ can be measured as

$$= \rho g h \text{ N/m}^2 \tag{2.24}$$

The manometer is an accurate, sensitive and simple device but is limited to fairly small pressure differentials. Due to inertia and friction of the fluid, it is not suitable for fluctuating pressure unless the rate of change of pressure is small. Manometers using water as the measuring fluid are particularly useful for measuring very small pressures.

### 2.22.1 The Bourdon Pressure Gauge

Mercury manometer becomes quite long if the value of gauge pressure approximately 0.13 MN/·m² above atmospheric is to be measured. Above this value the Bourdon Pressure Gauge is the suitable device which is shown in Fig. 2.18.

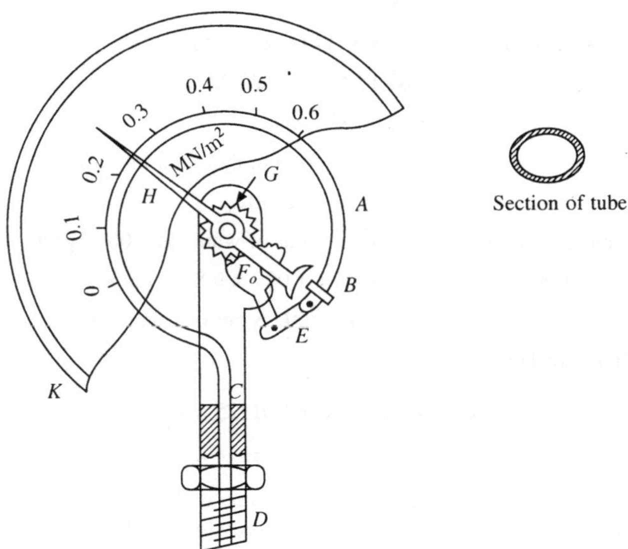


**Fig. 2.18   The Bourden Pressure Gauge**

The basic element of this type of pressure gauge is a tube *A* which of elliptical cross-section and is bent into an arc of a circle as shown. End *B* is conneccted to a link *E* and end *C* is bent down and fixed in a connecting union *D*. When fluid under pressure enters the tube, its cross-section tends to become circular and end *B* moves. This change of cross-section causes the tube to become straight. The movement of end *B* is transferred to the pinion *G* through link *E* and quadrant gear *F*. The guadrant gear engages with the pinion gear on to which a pointer *H* is attached. The pointer rotates over a colibrated pressure scale *J*. The whole mechanism is enclosed in a cylindrical case *K*. Such gauges can be obtained to measure pressure below, as well as above, atmospheric pressure.

### Solved Problems

2.1  A temperature scale of a certain thermometer is given by the relation $t = a \ln p + b$, where $a$ and $b$ are constants and $P$ is thermometric property of the fluid in the thermometer. If at the ice point and steam point, thermometric properties are found to be 1.5 and 7.5 respectively, what will be the temperature corresponding to the thermometric property of 3.5 on celsius scale.

(Nov. 87, P.U) Poona Univ.

**Soln.**

$$t = a \ln p + b \qquad\qquad (1)$$

For celsius scale, steam point = 100 and ice point = 0

So,

$$100 = a \ln 7.5 + b \qquad\qquad (ii)$$

and

$$0 = a \ln 1.5 + b \qquad\qquad (iii)$$

Equating equations (ii) and (iii), we get

$$a = 62.15$$

Putting the value of $a$ in equation (ii),

$$100 = 62.15 \ln 7.5 + b$$

$$\therefore \qquad b = -25.22$$

By putting the values of $a$ and $b$ in equation (1) for $p = 3.5$,

$$t = 62.15 \ln 3.5 - 25.22$$

$$= 52.63°C \quad Ans$$

**2.2**  Convert 73 cm of mercury pressure into water column and then into a bar.

**Soln.**

$$(\rho gh)_{mercury} = (\rho gh)_{water}$$

Taking density of mercury as 13,600 kg/m$^3$ and that of water as 1,000 kg/m$^3$

$$13,600 \times 9.81 \times 73 \times 10^{-2} = 1,000 \times 9.81 \times hw \times 10^{-2}$$

$$\therefore \qquad h_w = 992.8 \text{ cm of water} \quad Ans.$$

1 atmosphere = 76 cm of Hg = 1.01325 bar

$$\therefore \qquad 73 \text{ cm of mercury} = 1.01325 \times \frac{73}{76}$$

$$= 0.97325 \text{ bar} \quad Ans.$$

**2.3**  A pressure gauge fitted on a steam boiler records a pressure of 30 bar. The vacuum gauge on the condenser indicates the vacuum of 0.95 bar. If the atmospheric pressure is 1 bar, calculate

(a)    absolute pressure in the steam boiler
(b)    absolute pressure in the condenser

**Soln.**    (a)                    For steam boiler

$$P_{abs.} = P_{atm.} + P_{gauge}$$

$$= 1 + 30 = 31 \text{ bar} \quad Ans.$$

(B)                    For condenser

$$P_{abs} = P_{atm} - P_{vacuum}$$

$$= 1 - 0.95 = 0.05 \text{ bar} \quad Ans.$$

**2.4**  A barometer reads 76 cm of Hg. What would be the absolute pressure of (a) a pressure gauge connected to a steam main line leading to inlet of steam turbine reads 28 bar and (b) a vacuum gauge connected to exhaust line of the same turbine reads equivalent to 910 cm of water column. Express the absolute pressure in both cases in kPa.

**Soln:**                    Barometric pressure $P = \rho gh$

$$= 13600 \times 9.81 \times 76 \times 10^{-2}$$

$$= 101396 \text{ Pa}$$

$$= 101.396 \text{ kPa}$$

(a) Absolute pressure of steam in steam mainline

$$= P_{\text{barometric}} + P_{\text{gauge}}$$

$$= 101.396 + 28 \times 10^5 \times 10^{-3}$$

$$= 2901.396 \text{ kPa} \quad \text{Ans}$$

(b) Absolute pressure of steam in exhaust line

$$= P_{\text{barometric}} - P_{\text{gauge}}$$

$$= 101.396 - (1000 \times 9.81 \times 910 \times 10^{-2} \times 10^{-3})$$

$$= 12.125 \text{ kPa} \quad \text{Ans}$$

2.5  A barometer of mountain hiker reads 955 mbars at the begining of a hiking trip and 820 mbars at the end. Neglecting the effect of altitude on local gravitational acceleration, determine the vertical distance climbed. Assume average air density = 1.2 kg/m$^3$ and $g$ = 9.7 m/s$^2$.  (Dec. 93 P.U.) Poona Univ.

*Soln.* Refer to Fig. 2.19
Pressure difference between beginning and end of the hiking trip

$$\Delta p = h \rho g, \quad h = \text{vertical distance climbed}$$

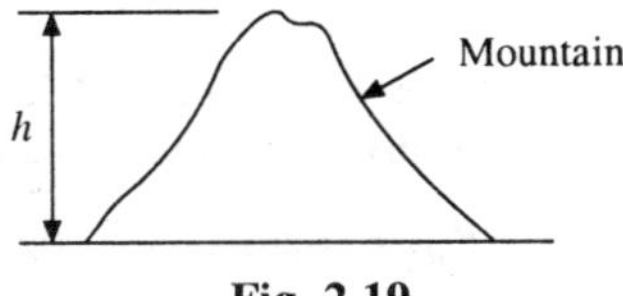


$$\therefore \qquad h = \frac{\Delta P}{\rho g}$$

**Fig. 2.19**

$$= \frac{(955 - 820) \times 10^{-3} \times 10^5}{1.2 \times 9.7}$$

$$= 1159.79 \text{ m}$$

$$= 1.159 \quad \text{km.} \quad Ans.$$

2.6  A pressure gauge is attached to a closed tank which contains some gas. The gauge reads 1.3 bar when the barometer reads 75 cm of Hg. If the barometer changes to 72 cm of Hg, what will be the reading on the gauge for same absolute pressure?  (Dec. 93 P.U.) Poona Univ.

*Soln.*
$$P_{\text{abs}} = 1.3 + \frac{75}{76} \times 1.01325$$

$$= 2.3 \text{ bar}$$

For same absolute pressure

$$2.3 = P_{\text{gauge}} + \frac{72}{76} \times 1.01325$$

$$\therefore \qquad P_{\text{gauge}} = 1.34 \text{ bar} \qquad Ans.$$

2.7  Air tank has three manometers connected to it, the fluids in them are oil (sp. gravity = 0.8), water and mercury (sp. gravity = 13.6). If the absolute pressure in the tank is 1.2 bar and the manometer reads 760 mm of Hg. Estimate the height of fluid in each manometer.

(Nov. 90, P.U.) Poona Univ.

*Soln.* Absolute pressure in the tank

$$= P_{\text{atm}} + \text{Pressure due to column of the fluid}$$

If $h_0$ is the column of oil in manometer,

$$1.2 \times 10^5 = 1.01325 \times 10^5 + h_0 \times 0.8 \times 1000 \times 9.81$$

$$\because \qquad [760 \text{ mm of Hg} = 1.01325 \text{ bar}]$$

$$\therefore \qquad h_0 = 2.379\text{m} \quad Ans.$$

Similarly water column,

$$hw = \frac{1.2 \times 10^5 - 1.01325 \times 10^5}{1000 \times 9.8}$$

$$= 1.90 \ m \quad \text{A\textsc{ns}}$$

and mercury column,

$$hm = \frac{1.2 \times 10^5 - 1.01325 \times 10^5}{13.6 \times 1000 \times 9.81}$$

$$= 0.1399 \text{ m} \quad Ans.$$

2.8  A U-tube manometer with one arm open to atmosphere is used to measure the pressure of steam flowing through a pipe. Mercury is used as a manometric fluid. The height of the column of mercury in the open end is 10 cm greater than the mercury level in the arm connected to the steam pipe. Steam condenses in the manometer on the steam side. The column of condensate is 3.5 cm in height. The atmospheric pressure is 760 mm of mercury. What is the absolute pressure of steam in the pipe in bar?

(May 91. P.U.) Poona Univ.

*Soln.* Refer to Fig. 2.20
Applying pressure on a common level $A - A$

$$P_{abs} + (\rho gh)_{c.steam} = (\rho gh)_{mercury} + P_{atm.}$$

$$\therefore P_{abs.} = (\rho gh)_{mercury} - (\rho gh)_{c.steam} + P_{astm}$$

$$= 13600 \times 9.81 \times 10 \times 10^{-2}$$

$$- 1000 \times 9.81 \times 3.5 \times 10^{-2}$$

$$+ 1.01325 \times 10^5$$

$$[\because 760 \text{ mm of Hg} = 1.01325 \times 10^5 \text{N/m}^2]$$

$$= 114323.25 \text{ N/m}^2$$

$$= 1.143 \text{ bar} \quad Ans.$$

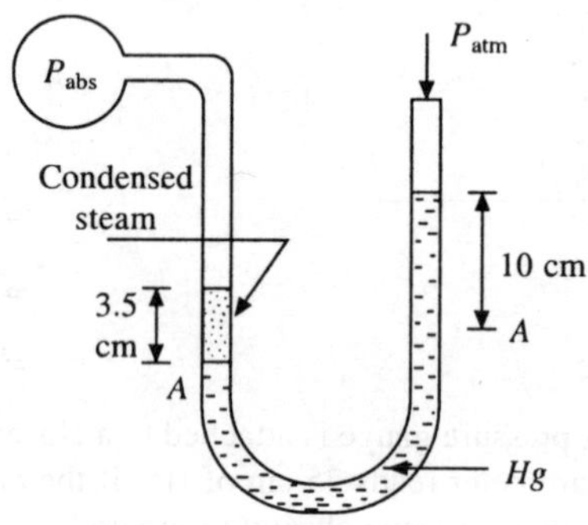


**Fig. 2.20**

2.9  A *U*-tube manometer with one arm open to atmosphere is used to measure the pressure of kerosene vapour passing through a pipe. Mercury used gives the height of the column in the arm open to atmosphere greater than 15 cm the mercury level in the arm connected to kerosene pipe. There is condensation of kerosene vapour on the mercury column of 5.1 cm height. If the atmospheric pressure is 75.5 cm of mercury, find the absolute pressure of vapour in kPa. Assume specific gravity of kerosene is 0.8 and that of mercury is 13.6.

(May 94, P.U.) Poona Univ.

*Soln.* Refer to Fig. 2.21
Applying pressure at common level $A - A$

$$P_{abs} + (\rho gh)_{kerosene} = (\rho gh)_{mercury} + P_{atm.}$$

$$\therefore \quad P_{abs} = (\rho gh)_{mercury} - (\rho gh)_{kerosene} + P_{atm}$$

$$= 13.6 \times 1000 \times 9.81 \times 15 \times 10^{-2}$$

$$- 0.8 \times 1000 \times 9.81 \times 5.1 \times 10^{-2}$$

$$+ \frac{75.5}{76} \times 1.01325 \times 10^5$$

$$= 120270.54 \quad \text{N/m}^2 = Pa$$

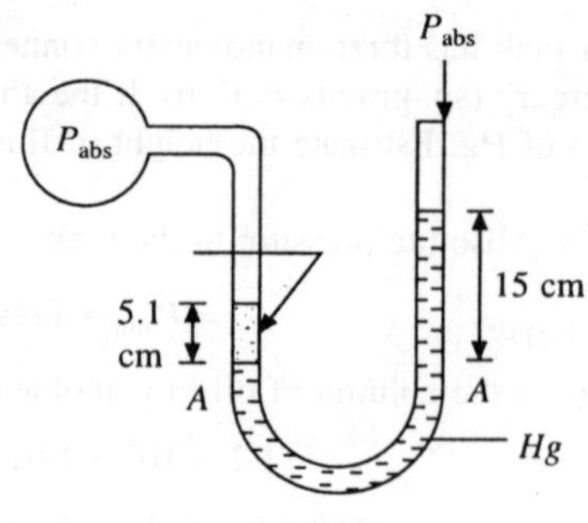


**Fig. 2.21**

$$= 120.27 \text{ kPa} \quad \text{Ans}$$

2.10 The following data is as shown in the figure below.

Pressure at $A = 400$ kPa

Find pressure at $B$.

*Soln.* Equating the pressure in both arms above the common liquid level $x - x$

Refer to Fig. 2.22

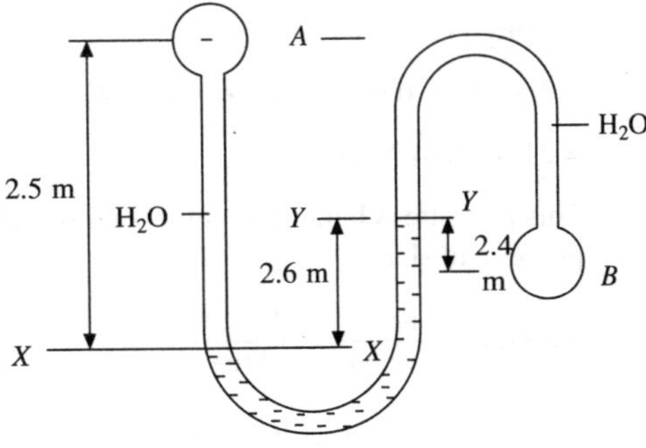


**Fig. 2.22**

or,

$$400 + 1000 \times 9.81 \times 2.5 \times 16^{-3}$$

$$- 13600 \times 9.81 \times 0.6 \times 10^{-3}$$

$$+ 1000 \times 9.81 \times 0.4 \times 10^{-3} - P_B = 0$$

$$\therefore \quad P_B = 348.39 \text{ kPa} \quad Ans.$$

2.11 A vertical composite liquid column with its upper end exposed to atmospheric pressure, comprise 45 cm of Hg (sp.gr. 13.6), 65 cm of water and 80 cm of oil (sp.gr. 0.8). Calculate the pressure in bar,

(i)     At the bottom of the column.

(ii)    At the interface of oil and water

(iii)   At the interface of water and Hg.

Assume atmospheric pressure as 100 kPa and $g = 9.81$ m/s$^2$

*Soln.* Refer to Fig. 2.23

$$P_{atm} = 100 \text{ kPa} = \frac{100 \times 10^3}{10^5} = 1 \text{ bar}$$

(i)   Pressure at the bottom

$$= P_{atm} + P_{oil} + P_{H_{20}} + P_{Hg}$$

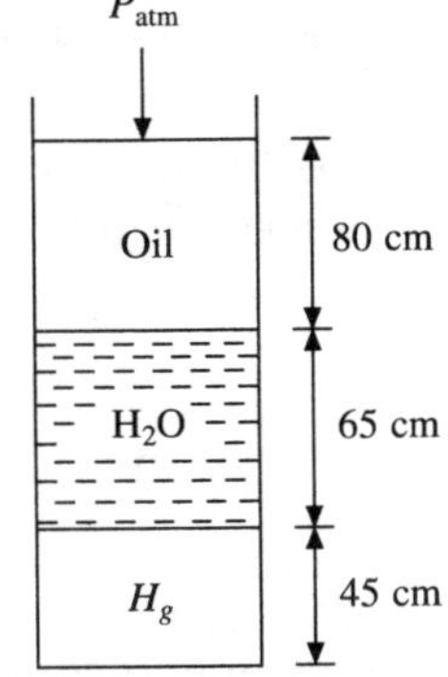


**Fig. 2.23**

$$= 1 + \frac{0.8 \times 1000 \times 9.81 \times 0.8}{10^5} + \frac{1000 \times 9.81 \times 0.65}{10^5} + \frac{13.6 \times 1000 \times 9.11 \times 0.45}{10^5}$$

$$= 1 + 0.0627 + 0.0637 + 0.6003 = 1.7267 \text{ bar} \quad Ans.$$

(ii)  Pressure at the interface of oil and water

$$= P_{atm} + P_{oil}$$

$$= 1 + 0.0627 = 1.0627 \text{ bar} \quad Ans.$$

(iii) Pressure at the interface of water and Hg

$$= P_{atm} + P_{oil} + P_{H_2O}$$

$$= 1 + 0.0627 + 0.0637 = 1.1264. \text{ bar} \quad Ans$$

2.12  The cylinder and tubing shown in Fig. 2.24. Contain oil of density $0.902 \times 10^3$ kg/m³ for gauge reading of 2 bar. What is the total weight of piston and slab placed on it?

*Soln.*

At level $x - x$, pressure value will be same.
Total pressure at this level

$= $ Gauge press. $ + (\rho gh)_{oil}$

$= 2 \times 10^5 + 0.902 \times 10^3 \times 9.81 \times 2$

$= 217697.24$ N/m²

This pressure is balanced by the total weight of piston and slab.

$$\text{Pressure} = \frac{\text{Force}}{\text{Area}} = \frac{\text{Weight}}{\text{Area}}$$

$\therefore$   weight $= \frac{\pi}{4} 2^2 \times 217697.24$

$= 683916.05$ N $= 683.916$ KN   *Ans.*

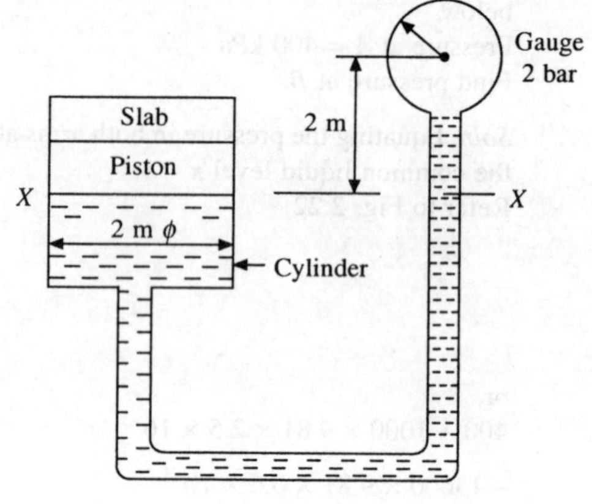


**Fig. 2.24**

2.13  The piston shown in Fig. 2.25 is held in equilibrium by pressure of gas flowing through the pipe. The piston has a mass of 21 kg. $P_1 = 600$ kPa, $P_2 = 170$ kPa. Determine pressure in the gas $P_3$.

*Soln.* The total downward force acting will be due to self weight of piston, pressure $P_1$ and $P_2$.

Force due to mass $= 21 \times 9.81 = 206.01$ N $= 0.20601$ KN

Force acting on 10 cm diametre piston due to pressure $P_1 = 600 \times \frac{\pi}{4} (0.10)^2 = 4.7123$ KN

Force acting on 20 cm diametre piston due to pressure

$P_2 = 170 \times \frac{\pi}{4} (0.2^2 - 0.1^2) = 4.0055$ KN

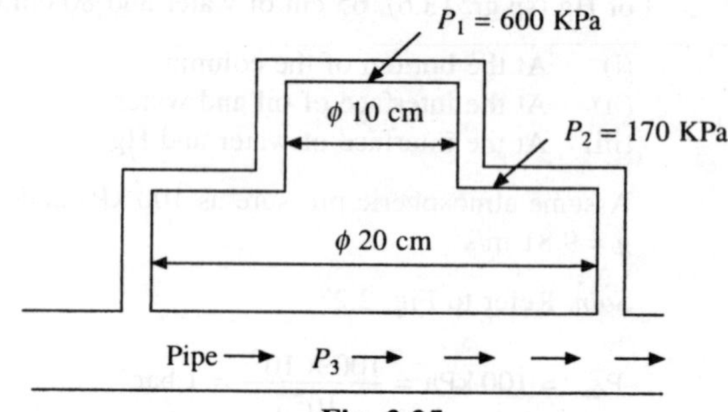


**Fig. 2.25**

Total downward load $= 0.20601 + 4.7123 + 4.0055 = 8.9238$ KN

This is balanced by $P_3$ at 20 cm dia. Piston

$$P_3 = 8.9238/\frac{\pi}{4} (0.2)^2 = 284.053 \text{ kPa}   \textit{Ans.}$$

## EXERCISES

Answer the following questions

2.1  What is thermodynamics?

2.2  What is meant by thermodynamic equilibrium?

2.3  Explain microscopic and macroscopic point of view.

2.4  What are forms of energy sources?

2.5  Explain with simple sketches. Thermodynamic systems – closed, open, adiabatic, isolated.

2.6  Explain thermodynamic properties.

2.7  What is meant by reversible and irreversible process? Give two examples of reversible process.

8.8 State the conditions to be satisfied for a process to be reversible.

2.9 Give the thermodynamic definition of work.

2.10 Differentiate between absolute and gauge pressure.

2.11 Define pressure. Explain with the help of a diagram showing relation between gauge pressure, vacuum pressure and atmospheric pressure.

2.12 State and explain Zeroth Law of Thermodynamics. Why it is called so?

2.13 Name the devices to measure temperature.

2.14 What are the devices to measure pressure? Explain the Bourdan Pressure Gauge.

2.15 A thermocouple with test junction at $t°C$ on gas thermometer scale and reference junction at ice point gives the emf as,

$$\varepsilon = 0.20t - 5 \times 10^{-4}t^2 \text{ mv}$$

The millivoltmeter is calibrated at ice and steam points. What will be reading on this thermometer where the gas thermometer reads 60°C                                                                                      *Ans.* (68°C)

2.16 Two celsius thermometers $A$ and $B$ with temperature readings $ta$ and $t_b$ agree at *ice* point (0°C) and steam point (100°C) but elsewhere are related by $t_a = l + mt_b + nt_b^2$ where $l$, $m$ and $n$ are constants. When they are immersed in an oil both $A$ shows 51°C while $B$ 50°C. Determine $t_b$ when $t_a$ is 25°C.

*Ans.* (24.26°C)

2.17 Convert followig pressure readings into kPa assuming barometric pressure of 750 mm of Hg.

   (i)  55 cm of Hg
  (ii)  3.3 atm
 (iii)  3.2 m of $H_2O$
 (iv)  4.6 bar

2.18 A glass tube contains 40 cm column of Hg and 80 cm column of water, tube being open to atmosphere from the top. Find the gauge pressure at the base of the tube in KN/m².     *Ans.* (61.48 KN/m²)

2.19 A pressure gauge reads 2.5 MPa and the barometer reads 98 kPa. Calculate the absolute pressure in MPa.

*Ans.* (2.598 MPa)

2.20 A vacuum gauge on the condenser reads 640 mm of mercury. What is the absolute pressure in the condenser if the barometer reads 760 mm of mercury.     *Ans.* (16 kPa)

2.21 The pressure in a constant gas thermometer is measured as 32 mm of Hg above the atmospheric pressure at triple point of water, determine the temperature in °C. When pressure is 76 mm of Hg above atmospheric pressure. Barometer reads 752 mm of Hg.

2.22 A U-tube manometer contains a liquid having a density of 800 kg/m³. When the manometer is connected to a gas pipe, the elevation in the open arm is 30 cm higher than the level in the column connected to the pipe. Find the absolute pressure of the gas in bar if the barometric pressure is 75 cm of Hg.

2.23 A U-tube manometer with one arm open to atmosphere is used to measure the pressure of steam flowing through a pipe using mercury as manometric fluid. The level of mercury column to the pipe is 125 mm below the level of mercury in the open arm. The column of condensate collected is 35 mm. Taking $g = 9.81$ m/s², density of mercury = 13560 kg/m³ and the atoms pressure as 760 mm of mercury, estimate the absolute pressure of steam in the pipe in bar.

2.24 Two liquids of different densities $\rho_1 = 1500$ kg/m³ and $\rho_2 = 500$ kg/m³ are poured together into a 100 litre tank so that tank is completely filled. The resulting density of fluid is 800 kg/m³. Find the amounts of the liquid and weight of the mixture. Local $g = 9.675$ m/s².

*Ans.* ($m_1 = 45$ kg, $m_2 = 35$ kg, $W = 774N$)

2.25 At sea level the value of gravitational acceleration '$g$' is 9.806 m/s². If '$g$' decreases 0.003 m/s² per 1000 meters ascent, find the height in kilometers above sea level at a point for which value of '$g$' = 9.616 m/s²

*Ans.* (63.33 km)

# 3

# The First Law of
# Thermodynamics

## 3.1 INTRODUCTION

There are number of energies existing in the universe, such as potential energy, mechanical energy, electrical energy, heat energy and several, ohers. Whether we are on land, in mines, in space or any where in this universe, we require these energies for survival. All these various forms of energies are mutually interchangeable. The units which convert one form of energy into another work on principles of thermodynamics or laws of thermodynamics. Although these laws do not have mathematical powers, but are deduced from experimental observations and are based on logical reasonings.

## 3.2 JOULE'S EXPERIMENT

Prior to James P Joule (1818–90) heat was considered as an invisible fluid flowing from a body of higher calorie to a body of lower calorie and it was known as caloric theory of heat. It was Joule who first established that heat is a form of energy. During 1840–49, he conducted several experiemnts for the formulation of the first law of thermodynamics as shown in Fig. 3.1. Work was transferred to the measured mass of water kept in an adiabatic vessel by means of a paddle wheel driven by the falling of a weight. The rise in temperature of the water was recorded by a thermometer. Let the initial temperature of water in the vessel be $T_1$. The paddle wheel work increased the water temperature to $T_2$. The cover was then removed and bath was allowed to return to its original temperature $T_1$.

The system thus executed a cycle (Fig. 3.2), which consisted of a definite amount of work input $W_{1-2}$ into the system followed by the transfer of an amount of work input $W_{1-2}$ into transfer of an amount of heat $Q_{2-1}$ from the system.

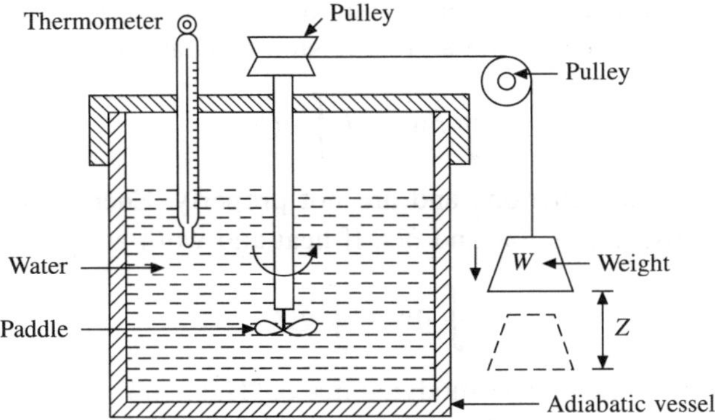


**Fig. 3.1   The Joule's experiment**

Joule repeated this experiment for different weight falling through different distances and measured each temperature rise.

Joule used mercury also as a fluid system and later a solid system of metal blocks which obsorbed work by friction when rubbed against each other. Other experiemnts involved the supplying of work in an electric current.

In each and every case Joule found that work and heat have a constant relation. This constant of propertionality is called Joule's constant or the mechanical equivalent of heat ($J$).

$$\frac{\text{Work}}{\text{Heat}} = \text{Constant (J)}$$

In the Fig. 3.2 if the cycle involves several work and heat transfers, the result is the same and can be written as

$$\sum W_{\text{cycle}} = J \sum Q_{\text{cycle}} \qquad (3.1(a))$$

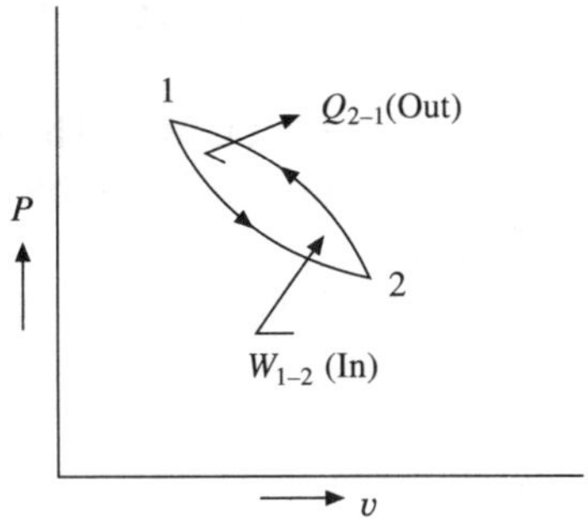

In the SI System, work is measured in Newton-metre (Nm) and heat in Joules (J) and relation is 1 Nm) = 1 Joule and hence Joule's constant is unity, that is, $J = 1$

$$\therefore \qquad \sum W_{\text{cycle}} = \sum Q_{\text{cycle}}$$



**Fig. 3.2**

This can be also represented in the form

$$\oint dw = \oint dQ \qquad\qquad 3.1\ (b)$$

This equation deals with the first law of thermodynamics for a closed system undergoing a cycle.

## 3.3   FIRST LAW OF THERMODYNAMICS

We know the law of conservation of energy according to which energy can neither be created nor destroyed but can only be converted from one form to another. This general law when applied to a system dealing with work and heat is called First Law of Thermodynamics and is stated as under:

"If a system undergoes a cyclic change of state then the algebraic sum of the work delivered to the surrounding is proportional to the algebraic sum of heat taken from the surrounding",

$$\oint \delta w = J \oint \delta Q$$

Both heat and work are path functions and are not property. Energy is designated by inexact differential mathematically, that is, $\delta w$ and $\delta Q$. If heat and work are expressed in the same unit,

$$\oint \delta w = \oint \delta Q$$

Another statement of First Law of Thermodynamics is "Heat and work are mutually convertible but since energy can neither be created nor destroyed the total energy associated with an energy conversion remains constant."

The First Law is also stated on basis of the concept of "perpetual motion machine of the first kind". Perpetual motion machine of the first kind is an imaginary machine which gives out work continuously without taking energy from anywhere thus violating the first law. Obviously such machine *a* can not be made. So, we state First Law of Thermodynamics as follows:

"Perpetual motion machine of the first kind, abbreviated to PMM I, is an impossibility."

It is important to note that there is no theoretical proof of First Law of Thermodynamics but it is found to hold invariably.

## 3.4  INTERNAL ENERGY IS A PROPERTY

Proof on the basis of First Law of Thermodynamics. Let us consider that a closed system undergoes a change of state from 1 to 2 as shown in Fig. 3.3. along path $A$ and then from 2 to 1 along either of two paths $B$ or $C$.

We thus have two cyclic processess $1 - A - 2 - B - 1$ and $1 - A - 2 - C - 1$.

Now, applying First law of Thermodynamics to the cyclic process 1–A–2–B–1, we have

$$\oint (\delta Q - \delta W) = 0$$

or, 
$$\int_{1A}^{2} (\delta Q - \delta W) + \int_{1B}^{2} (\delta Q - \delta W) = 0$$

$$(3.2)$$

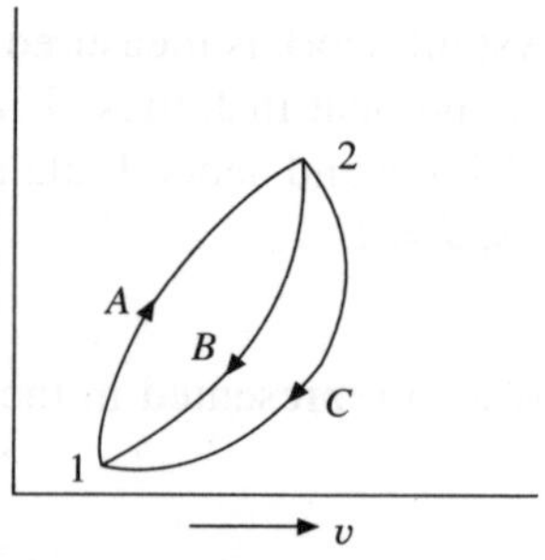

Similarly for the cyclic process 1–A–2–C–1 we have,

$$\oint (\delta Q - \delta W) = 0$$

or, 
$$\int_{1A}^{2} (\delta Q - \delta W) + \int_{1C}^{2} (\delta Q - \delta W) = 0 \qquad (3.3)$$

**Fig. 3.3**

Subtracting eq. 3.3 from Eq. 3.2 we, get,

$$\int_{2B}^{1} (\delta Q - \delta W) = \int_{2C}^{1} (\delta Q - \delta W) \tag{3.4}$$

We, thus see that when the system changes from prescribed state 2 to 1 along path $B$ or $C$ the integral of $(\delta Q - \delta W)$ remains same. It depends only on the end states 1 and 2 and therefore, it represents a change in a property of a system as it satisfies the criteria of a property.

So we have
$$(\delta Q - \delta W) = dE \tag{3.5}$$

Where $\delta Q$ = Elemental heat transfer to the system

$\quad\quad \delta W$ = Elemental work done by the system

$\quad\quad dE$ = change of energy in the system

Notation $dE$ is used here as against $\delta E$ since $dE$ is an exact differential. The symbol $E$ indicates the total energy stored in a system through the two modes, namely, macroscopic energy mode and microscopic energy mode. The macroscopic energy mode includes the macroscopic potential energy and kinetic energy of a system. If a fluid element of mass $m$ moving with a velocity $v$ and elevated from an arbitrary datum $Z$, macroscopic potential energy and kinetic energy can be written as

$$E_P = mgz/g_c \tag{3.6}$$

and
$$E_k = mv^2/2g_c \tag{3.7}$$

respectively.
$$(g_c = 1 \text{ in S.I. units})$$

The microscopic energy mode refers to the energy stored in the molecular and atomic structure of the system, which is called the molecular internal energy or simply internal energy, generally denoted by the symbol $U$.

In an ideal gas there are no intermolecular forces of attraction and repulsion and the internal energy depends only on temperature. Thus

$$U = f(T) \text{ only for an ideal gas.}$$

A system can possess other forms of energies also like magnetic energy, electrical energy and surface (tension) energy. In the absence of these forms, the total energy $E$ of a system is given by

$$E = E_k + E_p + U \tag{3.8}$$

It should be noted that in the absence of motion and gravity $E_k = 0$ and $E_P = 0$

$$\therefore \quad\quad\quad\quad\quad\quad E = U$$

Therefore Eq. 3.5 becomes

$$dQ - dW = dU \tag{3.9}$$

As from Eq. 3.4, $(dQ - dW)$ is independent of path and depends upon end states only, internal energy $U$ which is equal to $\int (dQ - dW)$ is also independent of path and hence a property of the system.

So internal energy can be defined as a form of stored energy in a system in the absence of magnetism, electricity, capillarity, surface tension, motion and gravity. Eq. 3.9 indicates the another way of stating First Law of Thermodynamics mathematically and it is written in terms of total energy as given below:

$$Q = \Delta U + W \text{ (for non flow process, that is, closed system)}$$

## 3.5 APPLICATIONS OF FIRST LAW OF THERMODYNAMICS

### 3.5.1 Application of First Law on Closed Systems (Non Flow Process)

(i) Constant volume, isochoric or isometric process. Let us consider a fluid in a rigid container (fixed volume). The state of of the system can be changed by either adding heat or by doing work on it by means of a paddle wheel as shown in Figs. 3.4 (a) and 3.4 (b). The change of state has been shown on a $p - v$ plot by a vertical line in Fig. 3.4 (c).

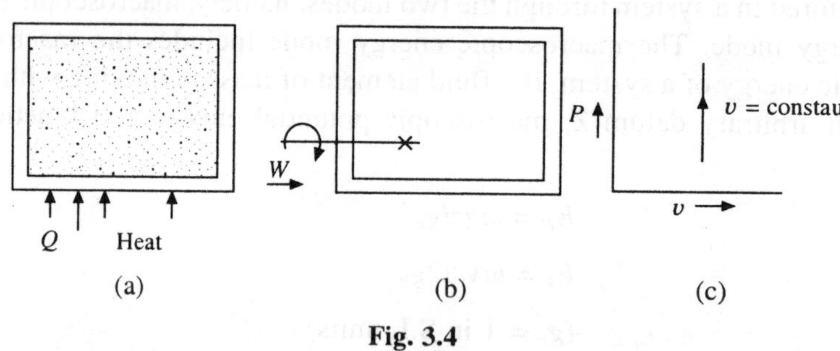


**Fig. 3.4**

In the two processes, namely, heating or doing paddle work the state of the system changes from the same initial state 1 to the same final state 2. In the first, there is only heat addition and no work done as there is no change of volume $(dv = 0)$. Applying the First Law of Thermodynamics we have

$$\delta Q - 0 = dU$$

or,
$$\delta Q = dU \tag{3.10}$$

or,
$$Q_{1-2} = U_2 - U_1$$

In the second case there is no heat transfer but work (irreversible frictional work) is added by paddle wheel. Applying the First Law to this process, we get

$$0 + \delta W = dU$$

or,
$$\delta W = dU \tag{3.11}$$

or,
$$W_{1-2} = U_2 - U_1$$

We find that for the same end states the values of work and heat depend on the process. In the first case work is zero and in the second heat is zero. So, we conclude that heat and work are not properties.

### *Definition of Specific Heat at Constant Volume*

We know that specific heat of a substance is defined as the amount of heat required to raise the temperature of unit mass by one degree. In the above case of heating process heat added = mass × specific heat × rise in temperature.

or,
$$\delta Q = mx \text{ specific heat} \times \delta T$$

or,
$$\text{specific heat} = \delta Q/m\,\delta T$$

For gaseous system denoting the specific heat for constant volume heating by $Cv$ we have

$$Cv = \delta Q/m\delta T = dU/mdT \; [\because \delta Q = dU]$$

or,
$$Cv = dU/dT \tag{3.12}$$

where $dT$ is the differential change in temperature

or,
$$dU = Cv\, dT \tag{3.13}$$

or,
$$C_v = \left(\frac{\partial u}{\partial T}\right)_v \tag{3.14}$$

(ii) Constant pressure or isobaric process

Let us consider a gas to be entrapped between a cylinder and a friction less piston (Fig. 3.5).

The system is, say in state 1 and if heat is added to it the gas extands and the piston rises up while the pressure remains constant. If the process is accomplished slowly it approaches a reversible process. The change of state from 1 to 2 with process 1–2 has been shown on a $p$–$v$ plot in Fig. 3.5 (c).

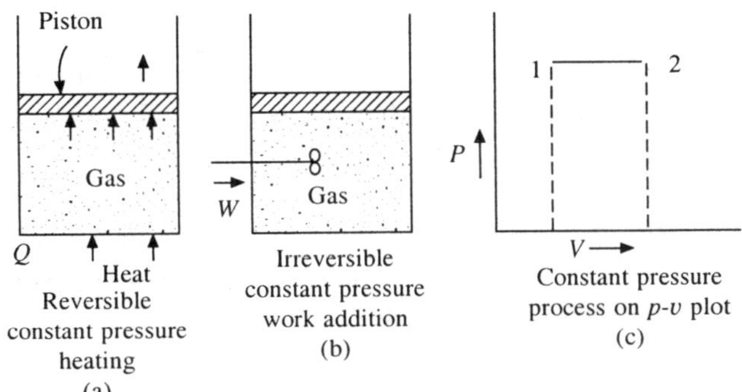


**Fig. 3.5**

Applying the First Law to this non-flow process we have

$$\delta Q - \delta W = dU$$

or,
$$\delta Q - pdV = dU$$

or,
$$\delta Q = dU + pdV$$

$$= m(du + pdv)$$

$$= m.d(u + pv) \text{ since } p \text{ is constant}$$

or, $$\delta Q = m \cdot dh \qquad (3.15)$$

If $C_p$ indicates the specific heat of the gas for constant pressure process then according to definition of specific heat we have for the above process

$$\delta Q = m\, Cp\, dT \qquad (3.16)$$

From Eqs. 3.15 and 3.16

$$mcp\, dT = mdh$$

or, $$cp = \partial h / \partial T \qquad (3.17)$$

or, $$C_p = \left(\frac{\partial h}{\partial T}\right)_p \qquad (3.18)$$

Integrating between 1 and 2

We have $$h_2 - h_1 = C_p\,(T_2 - T_1) \qquad (3.19)$$

The above change of state of gas could be obtained by addition of work by an irreversible paddle wheel process.

In this case, applying First law

$$\delta Q + \delta W = dU$$

or, $$\delta W = dU \qquad (3.20)$$

or, $$W_{1-2} = U_2 - U_1 \qquad (3.21)$$

(iii) Constant internal energy process

This process can be achieved in two ways (1) if in a process the amount of heat added is equal to the amount of work done by it, (2) there is neither any heat transfer nor any work done during the process.

First Case: In the figure shown above the cylinder head is conducting but the cylinder walls and the piston are non-conducting. If a heat source (Fig. 3.6a) is brought near the cylinder head and the heat transferred is so regulated that the amount of heat added is just equal to the amount of work done, that is, $dQ = dW$, then from First law

$$\delta Q - \delta W = dU$$

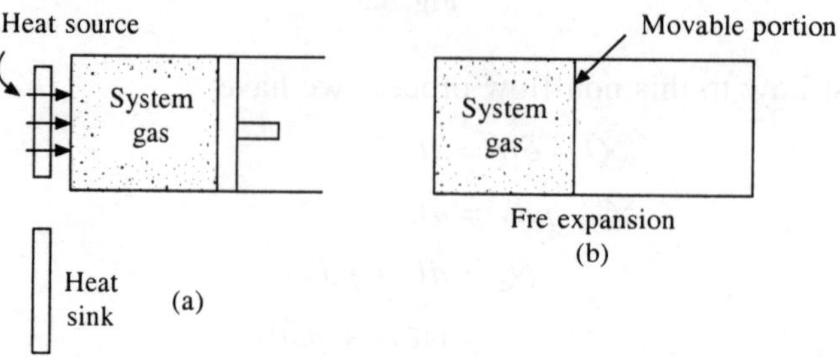


Fig. 3.6

or,
$$du = 0$$

That is the internal energy remains constant.

The process can be a reversible one if the piston returns back to addition of equal amount of work that has been done in the forward process and by rejecting equal amount of heat to a sink brought in place of the source of heat.

Second Case: Another way of achieving a constant internal energy process is to let the gas expand freely without any addition of heat as shown in Fig. 3.6(b).

Here the gas is allowed to expand freely inside the system boundary by means of a movable partition. As the work is not transferred across the system boundary, this work cannot be taken into account. The cylinder wall is nonconducting and so no heat transfer takes place. Thus we have $dQ$ and $dW$ both equal to zero and so from first law $0 - 0 = dU$, that is, the internal energy of the system remains constant. It is to be noted that this free expansion process is an irreversible process.

(iv) Isothermal process

In the Fig. 3.7(a) heat is added to the gas and work is done by it due to its expansion in such a way that its temperature remains constant. For such a process the plot of pressure versus volume is given by a curve. The first law applies as usual to such a process and

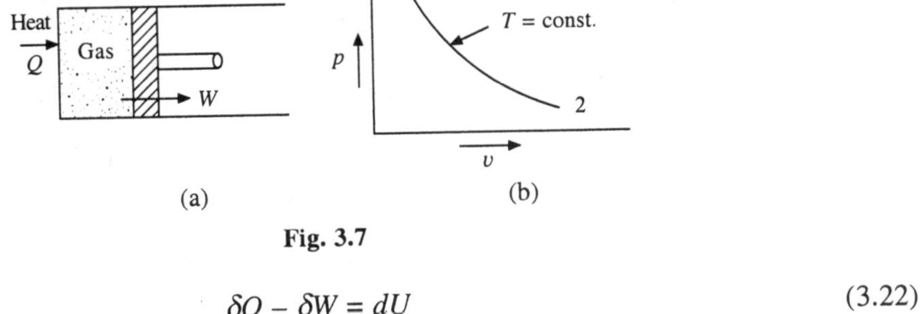


(a)                    (b)

**Fig. 3.7**

$$\delta Q - \delta W = dU \qquad (3.22)$$

(v) Adiabatic Process

It is a process in which there is no heat transfer between system and the surroundings.

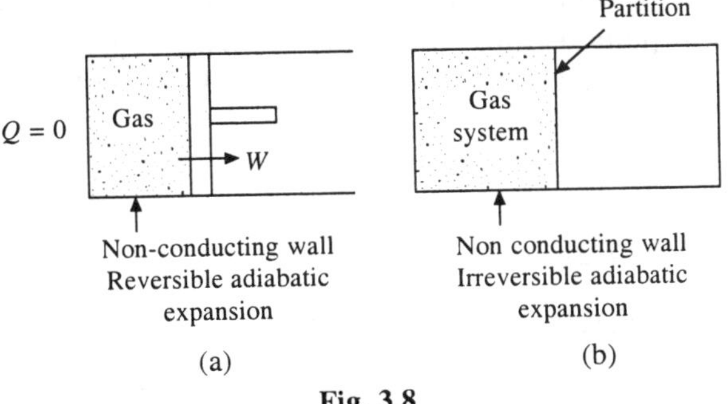


**Fig. 3.8**

If a gas expands reversibly in a non-conducting cylinder work is done while heat transfer is zero (refer to Fig. 3.8(a)).

Applying the First Law we get

$$0 - \delta W = dU$$

or,
$$\delta W = - dU$$

or,
$$W_{1-2} = U_1 - U_2 \tag{3.23}$$

Thus, if work is done by the gas the internal energy will decrease but if work is done on it the internal energy will increase. A free expansion of a gas in a non-conducting cylinder is an example of an irreversible adiabatic process (Fig. 3.8b)

Here,
$$\delta Q = 0$$

$$\delta W = 0$$

and so
$$dU = 0$$

### 3.5.2  Application of First Law to Steady Flow Systems

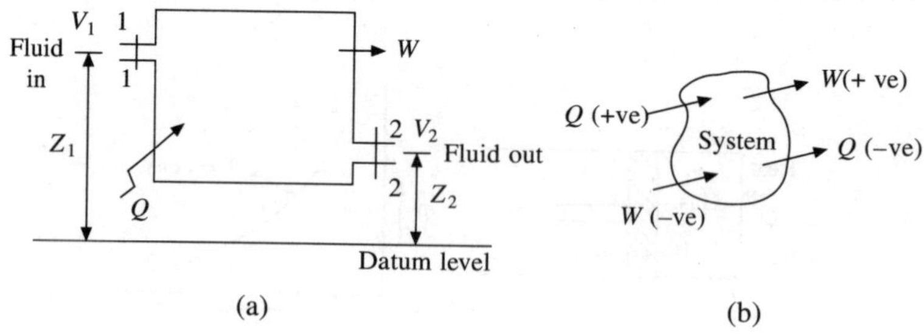


**Fig. 3.9**

Let us consider a steady flow open system with the working fluid entering at point 1 and leaving at point 2 as shown in Fig. 3.9(a).

Let $m$ kg mass of fluid enter and leave per unit time. Also let $Q$ be the amount of heat entering the system per unit time and $W$ the amount of work done by the system.

Now, we can apply the First Law of thermodynamics to the control volume bounded by say two imaginary planes 1–1 and 2–2 passing through points 1 and 2 respectively.

With $m$ kg of fluid at point 1, different forms of energy enter the system, namely, internal energy, flow energy, kinetic energy and potential energy. Using suffix 1 for the energies entering at point 1 are

$$\text{I.E.} = mu_1$$

$$\text{K.E.} = mV_1^2/(2g_c)$$

$$\text{P.E.} = (m\,Z_1)g/g_c$$

$$\text{F.E.} = mp_1v_1 \quad \text{Where } V_1, v_1 \text{ and } Z_1 \text{ indicate velocity, specific volume and height}$$

above datum at inlet.

Thus, the total amount of energy entering the system with mass $m$

$$= m\left[u_1 + P_1v_1 + V_1^2/(2gc) + Z_1 \times \frac{g}{g_c}\right]$$

Similarly, if suffix 2 be used to indicate the point 2, the amount of energy leaving the system with mass $m$ is

$$m[u_2 + p_2v_2 + V_2^2/(2g_c) + (Z_2)\times(g/gc)]$$

Taking into account the total heat added $Q$ and the work done $W$ and as per their sign conventions mentioned in Fig. 3.9(b), we have

$$Q = W + m\left\{u_2 + p_2v_2 + V_2^2/2gc + Z_2 \times \frac{g}{g_c}\right\}$$

$$-m \times \left\{u_1 + p_1v_1 + V_1^2/(2g_c) + Z_1g/(g_c)\right\}$$

or $\qquad Q - W = m\left\{(h_2 - h_1) + (V_2^2 - V_1^1)/(2g_c) + (Z_2 - Z_1)g/g_c\right\}$ $\qquad\qquad$ (3.24)

$Q \qquad\qquad (h = u + pv),\ g_c = 1$ in S.I. System

This equation is called steady flow energy equation (SFEE)

or, $\qquad\qquad Q - W = m[\Delta h + \Delta K \cdot E + \Delta P \cdot E]$

(i) *Continuity Equation*

We know that in the steady flow process the mass entering per second, $m$ is equal to the mass leaving per second and it remains constant.

Or, $m_1 = m_2 = m$ at the two sections namely,

1 – 1 and 2 – 2 (Fig. 3.10)

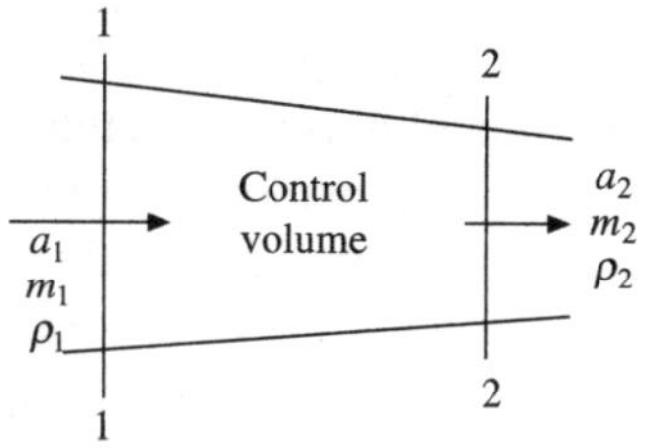

But mass rate of flow,

$$m = \frac{\text{Velocity} \times \text{area of cross-section}}{\text{specific volume}}$$

**Fig. 3.10**

or, $\qquad\qquad V_1\,a_1/v_1 = V_2\,a_2/v_2 \quad$ or, $\quad a_1\,V_1\,\rho_1 = a_2V_2\,\rho_2$ $\qquad\qquad$ (3.25)

Where $a_1\,a_2$ and $\rho_1,\ \rho_2$ are the inlet and exit areas and fluid densities respectively.

Eq. (3.25) is called continuity equation.

(ii) *Throttling Process*

Let us consider the flow of a gas through a horizontal insulated pipe fitted with a valve (Fig. 3.11). Let the valve be only partially open. On flowing a gas through the pipe it can be experimentally found that the pressure of the gas falls down markedly as it passes through the valve although there is no change in the velocity of the gas. If we take the portion of the pipe between sections 1 and 2 as control volume, we find that the velocities at sections 1 and 2 are same but the pressure at section 2 is much less than the pressure at section 1. Neither there is heat transfer as the pipe

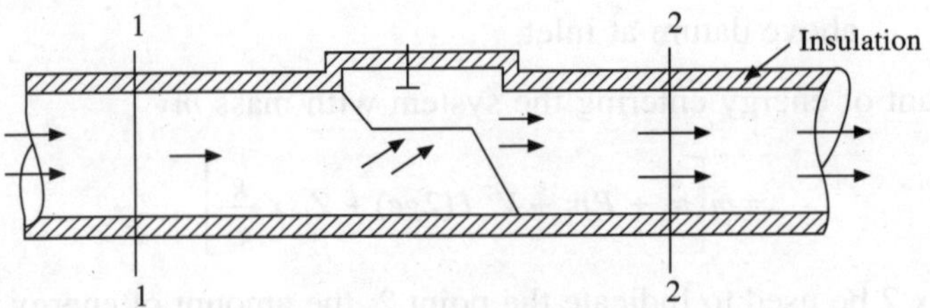

**Fig. 3.11. Throttling Process**

is insulated nor there is any work done. When the pipe is horizontal then the potential energy at section 1 is equal to the potential energy at section 2. Therefore, applying the steady flow energy Eq. 3.24 to the control volume we have,

$$Q - W = m\left\{(h_2 - h_1) + \frac{(V_2^2 - V_1^2)}{2g_c} + \frac{(Z_2 - Z_1)g}{g_c}\right\}$$

$$Q = 0, \ W = 0, \ Z_2 = Z_1$$

$$\therefore \quad h_2 - h_1 = 0 \quad \text{or,} \quad h_2 = h_1 \tag{3.25}$$

This means that the enthalpies before the valve and after the valve is the same. The above process is an example of a throttling process. The flow is irreversible as the gas flows through the valve under viscous resistance and across large pressure differences.

To provide restriction across fluid-flow another experiment was conducted and named as Joule–Thomson Porous Plug Experiment (Fig. 3.12). In this experiment a simple porous plug which acts to drop the pressure of a high stream pressure gas, has been provided instead of a valve. The porous plug is completely insulated from the surrounding ensuring no heat transfer in the process. It is also evident that no work is done. The gas expands through a minute aperture (Porous plug, narrow throat of valve or crack). Due to fall in pressure the gas should come out with large velocity but due to friction between the gas and restricting material the kinetic energy is converted into heat which warms up the gas to its initial temperature. The net enthalpy of gas remains constant. After ensuring a steady flow through a porous plug and applying Eq. 3.24 the result may be $h_1 = h_2$.

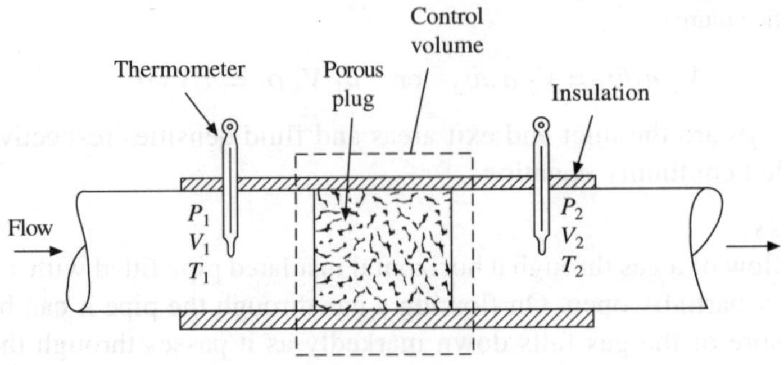

**Fig. 3.12 Porous Plug Experiment**

If we reduce the pressure step by step after the plug and trace a line of constant enthalpy on a temperature pressure diagram (Fig. 3.13) we

would see that $\left(\dfrac{\partial t}{\partial p}\right)_h$ is a constant. This constant
is called the Joule-Thomson co-efficient. For carbon dioxide and steam this co-efficient has a small positive value within the usual range of condition whereas for hydrogen and liquid water, it is negative at room temperature and pressure. Joule-Thomson co-efficient is measured to determine the relations between properties of substances.

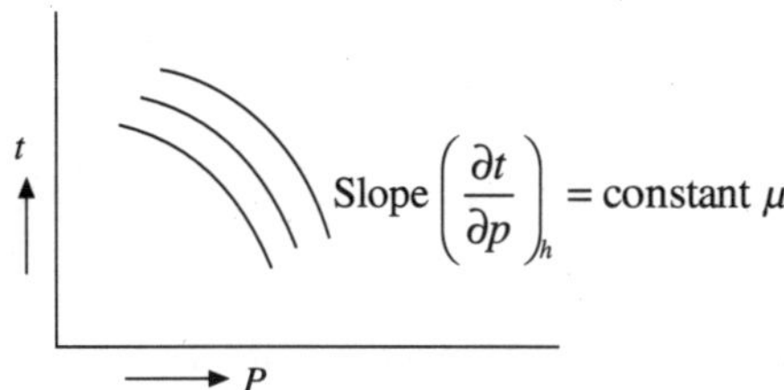


**Fig. 3.13   Joule-Thomson Co-efficient**

(iii) *Nozzle*

It is a device through which a high pressure fluid is passed to convert its enthalpy into kinetic energy as a result of the fall in pressure of the fluid. It is an insulated passage of varying cross-section and produces a jet of high velocity as shown in Fig. 3.14. There is no heat or work transfer.

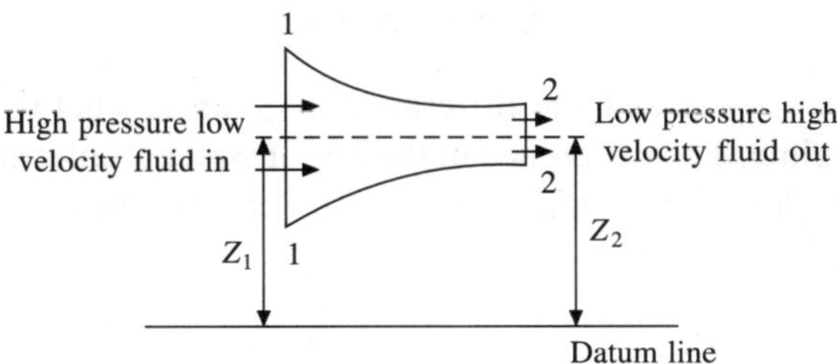


**Fig. 3.14   Nozzle**

Applying the steady flow energy equation and assuming $Z_1 = Z_2$, we have

$$Q - W = m[(h_2 - h_1) + (V_2^2 - V_1^2)/2g_c + (Z_2 - Z_1) \times (g/g_c)]$$

or,   $$(V_2^2 - V_1^2)/(2g_c) = h_1 - h_2 \qquad (3.26)$$

or,   $$V_2^2 = V_1^2 2g_c (h_1 - h_2)$$

or,   $$V_2^2 = 2g_c (h_1 - h_2) + V_1^2$$

or,   $$V_2 = \sqrt{2g_c(h_1 - h_2) + V_1^2} \qquad (3.27)$$

From continuity equation

$$a_1 V_1/v_1 = a_2 V_2/v_2$$

or,   $$V_1 = V_2 \times (a_2/a_1) \times \frac{v_1}{v_2} \qquad (3.28)$$

Putting the value of $V_1$ into Eq. 3.26 we have

$$V_2^2[1 - \{(a_2/a_1) \times v_1/v_2\}^2]/2g_c = h_1 - h_2$$

or,
$$V_2 = \frac{2g_c(h_1 - h_2)}{1 - (a_2 \cdot v_1 / a_1 v_2)^2} \qquad (3.29)$$

If $v_1$ is negligible, then from Eq. 3.26

$$V_2 = \sqrt{2g_c(h_1 - h_2)} \qquad (3.30)$$

### (iv) *Boiler*

A boiler is a device to produce steam from water by the application of heat which is obtained by combustion of some fuel.

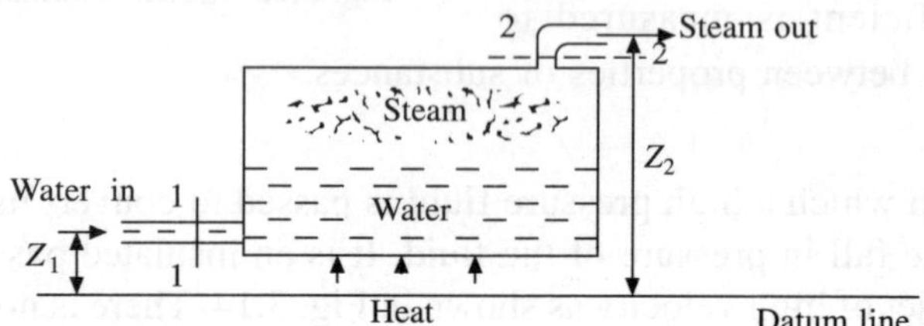


**Fig. 3.15   Boiler**

No work is done in the boiler and the difference in $z_1$ and $z_2$ between the inlet and exit is generally small and negligible. The difference in the KE of water at inlet and KE of steam at outlet is small and so can be neglected. On application of the energy equation on the control volume between section $1 - 1$ and $2 - 2$ we have

$$Q - W = m[(h_2 - h_1) + \{(Z_2 - Z_1)\} \times (g / g_c) + (V_2^2 - V_1^2)/(2g_c)]$$

or,
$$Q = m(h_2 - h_1) = H_2 - H_1 \qquad (3.31)$$

Thus the increase in enthalpy is equal to the amount of heat added.

### (v) *Turbine*

It is a device to produce mechanical work at the cost of enthalpy of a working fluid which passes through it. The fluid at high pressure enters the nozzles which are located adjacent to the wheel (Fig. 3.16).

Along the periphery of the wheel there are curved plates called blades or buckets. The fluid first passes through the nozzle (or nozzles) and attains a high velocity. This high velocity fluid issued from nozzle impinges on the blades and after flowing over them it is discharged out. The change of momentum of the striking fluid causes the wheel to rotate. As the wheel rotates news blades take the position of old ones and the rotation continues giving out work.

The general energy equation may be applied to the turbine (nozzle and wheel together) between sections $1 - 1$ and $2 - 2$.

The potential energy difference between the fluid at outlet and inlet $(Z_2 - Z_1)$ is usually small and may be neglected. For $m$ kg of mass flow we, therefore, have

$$Q - W = m[(h_2 - h_1) + (V_2^2 - V_1^2)/(2g_c) + \{(Z_2 - Z_1)\} \times (g / g_c)]$$

or,
$$W = Q + m\{(h_1 - h_2) + (V_1^2 - V_2^2)/(2g_c)\} \qquad (3.32)$$

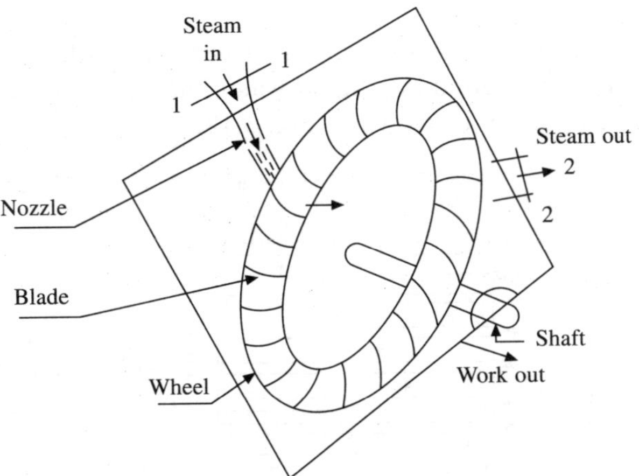


**Fig. 3.16    Steam Turbine**

$Q$ is generally, negative. It is the heat lost from the casing of turbine to the surrounding.

(vi) *Reciprocating Compressor*

A reciprocating compressor is a device where low pressure gas is compressed at a high pressure (Fig. 3.17).

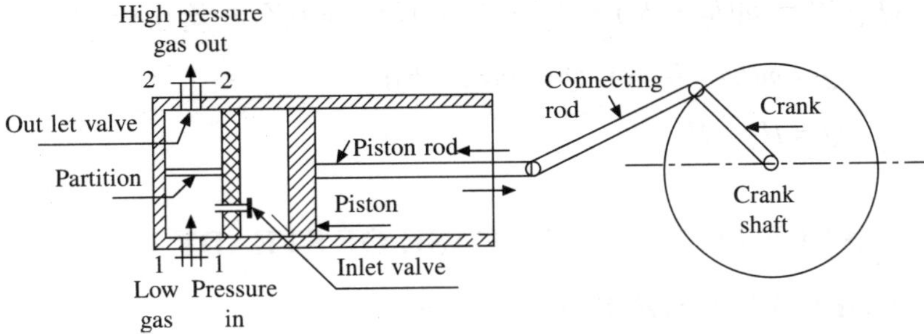


**Fig. 3.17    Reciprocating Compressor**

A piston is made to move to and fro by means of a motor inside a cylinder. In the cylinder end which is hollow, there are two valves fitted on either side of the partition as shown. As the piston moves out partial vacuum is created between cylinder head and the piston. The low pressure gas which is at higher pressure than the space inside the cylinder forces through the inlet valve and fills the cylinder. The valve at outlet so far closed. During the return stroke of the piston the gas is compressed and is forced out through the outlet valve to go to the receiver.

Here, $\Delta KE$ and $\Delta PE$ are negligible. Due to compression temperature of the system rises and so cooling is done by means of water. This heat is rejected. Work done is negative. The energy equation therefore applied is as under

$$-Q + W = m[(h_2 - h_1) + (V_2^2 - V_1^2)/(2g_c) + \{(Z_2 - Z_1)\} \times (g/g_c)]$$

or,     $$W = Q + m\,(h_2 - h_1) \tag{3.33}$$

### (iii) *Condenser*

It is used to condense the steam by a flow of cooling water. The cooling water takes up the latent heat from steam and steam is converted into liquid water known as condensate. The arrangement of flow of cooling water and steam is shown in Fig. 3.18.

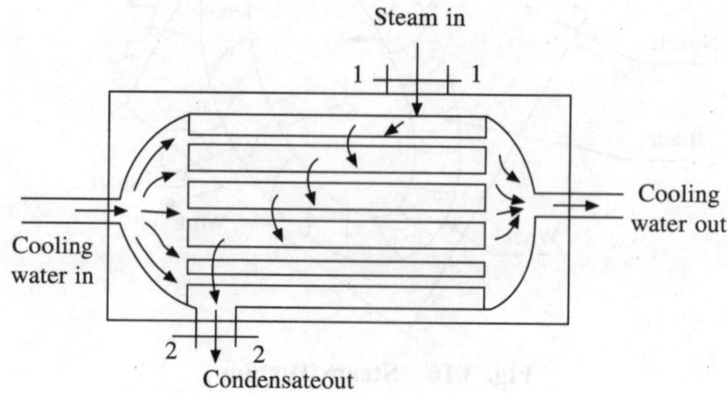


**Fig. 3.18  Steam Condenser**

Here heat is rejected from steam and so it is negative.

There is no work transfer. The $\Delta KE$ and $\Delta PE$ are negligible.

Therefore, applying the energy equation we have

$$-Q - W = m[(h_2 - h_1) + (V_2^2 - V_1^2)/(2g_c) + \{(Z_2 - Z_1)\} \times (g/g_c)]$$

or,

$$-Q = m(h_2 - h_1) \text{ or, } Q = m(h_1 - h_2) \tag{3.34}$$

or,

$$Q = H_1 - H_2$$

## 3.6  SIGNIFICANCE OF $\int pdv$ IN CASE OF STEADY FLOW PROCESS AND NON-FLOW PROCESS

Refer to Fig. 3.19 the energy equation for a steady flow process on unit mass basis is as under

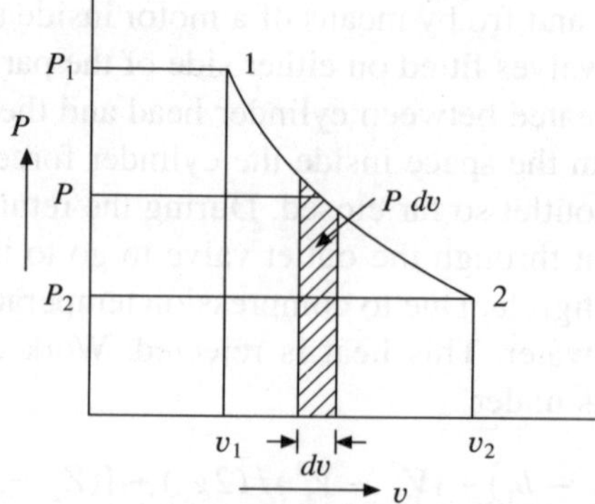


**Fig. 3.19**

$$\frac{V_1^2}{2} + Z_1 g + u_1 + p_1 v_1 + q = \frac{V_2^2}{2} + Z_2 g + u_2 + p_2 v_2 + w$$

Suffix 1 and 2 denote conditions at entrance and exit respectively.

or,
$$q = \Delta u + \Delta pv + \Delta KE + \Delta pE + w$$

In differential form

$$\delta q = du + d(pv) + d(K \cdot E) + d(P \cdot E) + \delta \omega$$

But for any reversible process

$$\delta q = du + Pdv$$

$$\therefore \quad du + Pdv = du + d\,(Pv) + d\,(K \cdot E) + d\,(P \cdot E) + \delta w$$

$$\therefore \quad Pdv = d(P\,v) + d(K \cdot E) + d(P \cdot E) + \delta w$$

$$\int_1^2 Pdv = \Delta(Pv) + \Delta(K \cdot E) + (P \cdot E) + w \tag{3.35}$$

that is, $\int_1^2 Pdv$ for a steady flow process is the sum of change in flow work, plus change in kinetic

energy, as well as change in potential energy and shaft work. $\int_1^2 Pdv$ is the area under the curve

$1 - 2$ represents shaft work when the pressure and volume change from $P_1$ to $P_2$ and $v_1$ to $v_2$

respectively. For a non-flow reversible process.

that is,
$$\int_1^2 Pdv = {}_1W_2 \tag{3.36}$$

# 3.7  SIGNIFICANCE OF $-\int vdp$ IN CASE OF REVERSIBLE STEADY FLOW PROCESS

Refer to Fig. 3.20. The energy equation for a steady flow process for unit mass can be written as

$$\left(\frac{V_1^2}{2} + Z_1 g + h_1\right) + q$$

$$= \left(\frac{V_1^2}{2} + Z_2 g + h_2\right) + w$$

or,  $q - w = \Delta h + \Delta P \cdot E + \Delta K \cdot E$

In differential

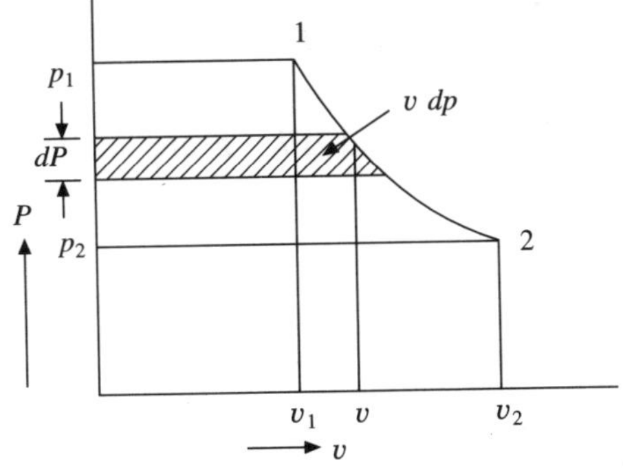


**Fig. 3.20**

$$\delta q = \delta w + dh + dP \cdot E + d K \cdot E \tag{3.37}$$

But From First law of Thermodynamics

$$\delta q = \delta w + du$$

and as $\delta w = Pdv$ for a reversible process

$$\delta q = Pdv + du \tag{3.38}$$

As we know, $h = u + Pv$

$\therefore$

$$dh = du + Pdv + vdp \tag{3.39}$$

Substituting values of $\delta q$ and $dh$ from Eqs. (3.38) and (3.39) in Eq. (3.37), we get

$$Pdv + du = \delta w + du + Pdv + vdp + dP \cdot E + d K \cdot E.$$

Cancelling $Pdv$ and $du$ on both sides, and rearranging

$$-\int vdP = \int \delta w + \Delta P \cdot E + \Delta K \cdot E$$

or,

$$-\int vdP = w + \Delta P \cdot E + \Delta K \cdot E \tag{3.40}$$

which is to say that, in a reversible steady flow process $-\int vdp$ equals the shaft work plus changes in kinetic energy and potential energy.

(i)   If $\Delta P \cdot E$ is negligible

$$-\int v \, dp = w + \Delta K \cdot E$$

(ii)   If both $\Delta P \cdot E$ and $\Delta K \cdot E$ are negligible, then

$$-\int v \, dp = w$$

(iii)   If $w$ and $\Delta P \cdot E = 0$, then

$$-\int v \, dp = \Delta K \cdot E$$

Generally $\Delta P \cdot E$ and $\Delta K \cdot E$ are negligible, hence

$$-\int v \, dp = w_{\text{shaft}}$$

Thus in a reversible steady flow process $-\int v \, dp$ equals shaft work when changes in kinetic energy and potential energy are neglected. on $P - V$ diagram this represents the area behind the curve.

If kinetic energy and potential energy are not negligible then the expression is modified to account for these energies. The expression $-\int v\,dp$ applies to all open systems including rotary machines with or without valves such as a turbine or a centrifugal or an axial-flow compressor as represented in Fig. 3.21 by section of increasing and decreasing areas respectively. In the case of reciprocating compressors, the velocities are truly negligible, whereas the same in the case of centrifugal and axial compressors and turbines may be significant.

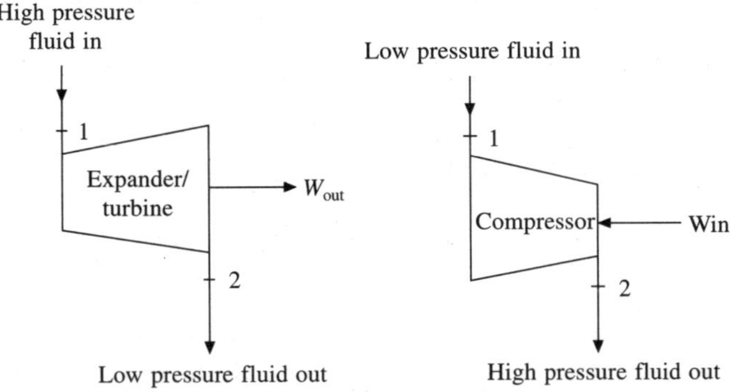

**Fig. 3.21    Representations of expander/turbine and compressor**

## 3.8    COMPARISON OF SFEE WITH EULER AND BERNOULLI EQUATIONS

The steady flow energy Eq. (3.24) for unit mass can be written as

$$\delta q = (h_1 - h_1) + \frac{V_2^2 - V_1^2}{2} + (Z_2 - Z_1)g + \delta w$$

In the differential form the SFEE becomes

$$dq = dh + V\,dV + g\,dz + \delta w \tag{3.41}$$

Now, $h = u + Pv$ and $\delta q = du + P\,dv$ for any reversible process. Eq. (3.41) can be written as

$$du + P\,dv = d(u + Pv) + V\,dV + g\,dz + \delta w$$

or, $$du + P\,dv = du + P\,dv + vdp + VdV + gdz + \delta w$$

for an incompressible fluid ($v$ = constant), $\delta w = Pdv = 0$

$\therefore$ $$v\,dp + V\,dV + g\,dz = 0 \tag{3.42}$$

This is Euler equation. If we integrate between two sections 1 and 2 of the pipe

$$\int_1^2 vdp + \int_1^2 VdV + \int_1^2 gdz = 0$$

or, $$v(p_2 - p_1) + \frac{V_2^2}{2} - \frac{V_1^2}{2} + g\,(Z_2 - Z_1) = 0 \tag{3.43}$$

or,
$$\frac{P_1}{\rho} + \frac{V_1^2}{2} + Z_1 g = \frac{P_2}{\rho} + \frac{V_2^2}{2} + Z_2 g \quad \left[ \because v = \frac{1}{\rho} \right] \tag{3.44}$$

or,
$$\frac{P}{\rho} + \frac{V^2}{2} + Z g = \text{constant} \tag{3.45}$$

This is Bernoulli equation which is valid for an inviscid incompressible fluid and can be written as

$$\Delta \left( Pv + \frac{V^2}{2} + gZ \right) = 0 \tag{3.46}$$

Eq. (3.24) or Eq. (3.41) can be written with ($h = u + pv$) as

$$Q - W = \Delta \left( u + Pv + \frac{V^2}{2} + gZ \right) \tag{3.47}$$

Eqs. (3.46) and (3.47) have several terms in common. Bernoulli equation is a limiting case of SFEE because it is restricted to frictionless incompressible fluid whereas, SFEE is valid for viscous compressible fluid.

## 3.9 THE UNSTEADY FLOW PROCESS

During a steady flow process, no changes occur within the control volume. In unsteady flow or transient-flow processes, the rates at which mass and energy enter the control volume may not be the same as the rate of flow of mass and energy go out of the control volume. For instance, filling of a storage tank with a fluid, charging of rigid vessels from a supply line, evacuating gas cylinders driving a gas turbine with pressurized air, inflating tubes or balloons and cooking with a pressure cooker, are all processes, involving transient flow. In unsteady flow processes, changes start and stop after some time rather than continuing. Therefore, for transient flow processes we have to deal with changes that occur over some time interval. Such processes can be analyzed by the control volume technique.

Consider the tank-filling process shown in Fig. 3.22. In this system a rigid tank is connected

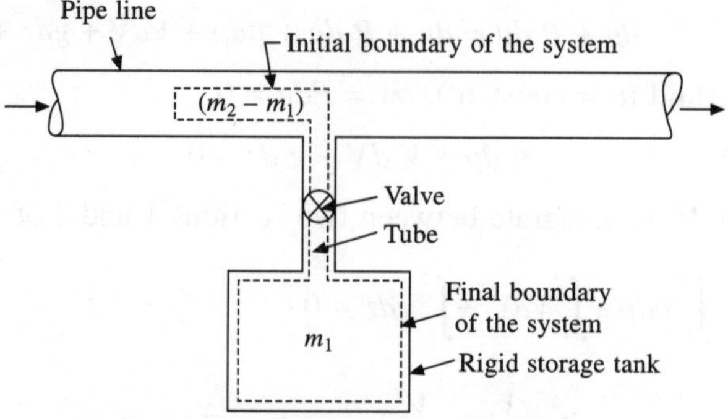


**Fig. 3.22   Unsteady Flow Process (Closed System)**

through a valve to a higher pressure supply pipe. When the valve is opened, the rigid tank is filled from the supply pipe until the tank pressure is equal to the supply pressure. The subscripts "1" and "2" refer to the initial and final conditions in the storage tank respectively. $P$ indicates the conditions of gas in the pipe line. Analysis of this type of problem can be made either by considering closed system or a control volume approach.

## (i) Closed System Analysis

Assume a closed system of gas in the pipeline and the tube which would eventually enter the tank as shown in the figure.

Energy stored already in the tank before filling it

$$E_1 = m_1 u_1 + (m_2 - m_1)\left(\frac{V_p^2}{2} + u_p\right) \quad [P \cdot E \text{ is almost neglisibe}]$$

where $(m_2 - m_1)$ is the mass of gas to be entered the tank.
Energy of gas after filling

$$E_2 = m_2 u_2$$

$\therefore$ Change in stored energy

$$\Delta E = E_2 - E_1 = m_2 u_2 - \left[m_1 u_1 + (m_2 - m_1)\left\{\frac{V_p^2}{2} + u_p\right\}\right] \qquad (3.48)$$

After filling, the gas is in rest in the tank
So $K \cdot E$ is neglected.
Now, work applied to introduce the gas mass $(m_2 - m_1)$ entering the storage tank at its volume and pressure equal to $P_p$,

$$W = P_p \times \Delta V_p = P_p(m_2 - m_1)v_p$$

where $v_p$ = sp. volume of gas in the pipeline

$\therefore$ Using the first law for the process

$$Q = \Delta E + W$$

$$= m_2 u_2 - \left[m_1 u_1 + (m_2 - m_1)\left\{\frac{V_p^2}{2} + u_p\right\}\right] - P_p(m_2 - m_1)v_p$$

$$= m_2 u_2 - m_1 u_1 - (m_2 - u_1)\left[\frac{v_p^2}{2} + (u_p + P_p v_p)\right]$$

$$= m_2 u_2 - m_1 u_1 - (m_2 - u_1)\left[\frac{v_p^2}{2} + h_p\right] \qquad (3.49)$$

For steady flow process $m_1 = m_2 = m$ and $\Delta E = 0$

## (ii) Control Volume Analysis

For an unsteady flow process undergoing in *a* control volume during time interval $\Delta t$, the principle of conservation of energy can be written as

$$
\begin{bmatrix}
\text{total energy} \\
\text{crossing the} \\
\text{boundary as heat} \\
\text{and work during } \Delta t
\end{bmatrix}
+
\begin{bmatrix}
\text{total energy transported} \\
\text{by mass into the} \\
\text{control volume} \\
\text{during } \Delta t
\end{bmatrix}
=
\begin{bmatrix}
\text{change of energy} \\
\text{of the control} \\
\text{volume during} \\
\Delta t
\end{bmatrix}
$$

$$
\therefore \qquad Q - W + (m_2 - m_1)\left( h_p + \frac{V_p^2}{2} \right) = m_2 u_2 - m_1 u_1
$$

In this case there is no work interaction. Being the gas in motion in the pipeline it indicates enthalpy rather than internal energy.

$$
Q = m_2 u_2 - m_1 u_1 - (m_2 - m_1)\left( h_p + \frac{V_p^2}{2} \right) \tag{3.50}
$$

The result is similar to Eq. (3.49) using closed system analysis.


## Solved Problems

**3.1**  A closed system undergoes a thermodynamic cycle *ABCDA*. The heat transfers per minute during processes *AB*, *BC* and *CD* are — 500 kJ, 10,000 kJ and — 1000 kJ respectively. The work transfers per second during processes AB, BC, CD and DA are — 10,000, zero, 17,000 and — 1000 kJ respectively. Find the rate of heat transfer during the process *CD* and net rate of work input in kW.

*Soln.*  From first law of thermodynamics $\oint \delta Q = \oint \delta w$,

Let $Q_{CD}$ represents the heat transfer during the processes CD. Then

$$
\oint \delta Q = -500 + 10{,}000 - 1000 + Q_{CD}
$$

$$
= 8500 + Q_{CD} \text{ kJ/minute}
$$

$$
\oint \delta w = -10{,}000 + 0.0 + 17{,}000 - 1000 \text{ kJ/sec.}
$$

$$
= 6000 \times 60 \text{ kJ/min}
$$

$$
= 360000 \text{ kJ/min}
$$

$$
\therefore \qquad 8500 + Q_{CD} = 360000
$$

or, $\qquad Q_{CD} = 351500 \text{ kJ/min}$   *Ans*

Rate of work done/sec = 6000 kJ/sec

$$
= 6000 \text{ kW}   \textit{Ans}
$$

**3.2**  The internal energy of a certain substance is given by the following equation

$$u = 3.56 \, pv + 84$$

where $u$ is in kJ/kg, $p$ is in $kPa$, $v$ is= in m$^3$/kg. A system composed of 3 kg of this substance expands from an initial pressure of 500 kPa and a volume of 0.22 m$^2$ to a final pressure of 100 $kPa$ in a process in which pressure and volume are related by $pv^{1.2} = c$.

   (i)  If the expansion is quasi-static, find $Q$, $\Delta u$ and $w$ for the process.

  (ii)  In another process the same system expands according to the same pressure-volume relation as in case (i) and from the same initial state and same final state as in case (i), but the heat transfer for this case is 30 kJ. Find the work transfer for this process.

 (iii)  Explain the difference in work transfer in case (i) and (ii).

*Soln.*  Given that, $u = 3.56 \, pv + 84$

(i)     Change in internal energy $\Delta u = u_2 - u_1$

i.e.     $u_2 - u_1 = 3.56 \, p_2 v_2 + 84 - 3.56 \, p_1 v_1 - 84$

$$= 3.56 \, (p_2 v_2 - p_1 v_1) \tag{i}$$

$$\because \quad p_1 v_1^{1.2} = p_2 v_2^{1.2} \qquad \therefore \quad v_2 = v_1 \left( \frac{p_1}{p_2} \right)^{\frac{1}{1.2}}$$

or     $$v_2 = 0.22 \left( \frac{500}{100} \right)^{\frac{1}{1.2}} = 0.841 \text{ m}^3$$

Now, from eqn. (i)

$$\Delta U = 3.56(100 \times 0.841 - 500 \times 0.22) \text{ kJ}$$

$$= -92.204 \text{ kJ} \quad Ans.$$

For a quasi static process   $$w = \int p \, dv$$

$$= \frac{p_2 v_2 - p_1 v_1}{1 - n}$$

or,     $$w = \frac{100 \times 0.841 - 500 \times 0.22}{1 - 1.2} = 129.5 \text{ kJ} \quad Ans$$

$\therefore$     $$Q = \Delta U + W = -92.204 + 129.5 = 37.296 \text{ kJ} \quad Ans$$

(ii)    Here $Q = 30$ kJ. Since the end states are same as in case (i)

$$W = Q - \Delta U = 30 - (-92.204) = 122.204 \text{ kJ} \quad Ans$$

(iii)   The work in case (ii) is not equal to $\int p \, dv$. So the process is not quasi-static.

**3.3**  The internal energy of a certain system is a function of temperature only and is given by $U = 30 + 0.25$ $t$ Joule. When this system executes a certain process, the work done by it per degree rise in temperature is given by $\delta w / dt = 1.068$ kJ/degree C. Find the heat transferred when the temperature changes from 100°C to 300°C, where $t$ denotes temperature in °C.

*Soln.*  Given     $\delta w / dt = 1.068$ kJ/°C $= 1068$ J/°C

Again     $U = 30 + 0.25 \, t$

$\therefore$     $dU / dt = 0.25$

But $\qquad \delta Q - \delta W = dU$

or $\qquad \delta Q/dt - \delta W/dt = dU/dt$

or, $\qquad \delta Q/dt = \delta W/dt + dU/dt = 1068 + 0.25 = 1068 \cdot 25 \; J/°C$

$$Q = 1068.25 \, dt = \int_{100}^{300} 1068.25 \, dt = [1068.25 \, t]_{100}^{300}$$

$$= 1068 \cdot 25 \times 200 = 213650 \; J$$

$$= 213.65 \; kJ \quad Ans.$$

**3.4** In a non-flow reversible process, the pressure and volume are related by

$$p = v^2 + 10/V \text{ where } p \text{ is in N/m}^2 \text{ abs}$$

$$\text{and } V \text{ is volume in m}^3.$$

During the the process the volume changes from 1.5 m$^3$ to 4.5 m$^3$ and heat added is 3,000 Joule, determine the change in internal energy.

*Soln.*  The work done during the process is given by

$$W = \int_{v_1}^{v_2} p \, dv$$

Now pressure $\qquad p = v^2 + 10/v \; \text{N/m}^2 \; abs$

$$\therefore \qquad W = \int_{v_1}^{v_2} (v^2 + 10/v) \, dv \quad \text{Joule}$$

$$= [v^3/3 + 10 \ln v]_{1.5}^{4.5} \quad \text{Joule}$$

$$= [1/3 \, (4.5^3 - 1.5^3) + 10 \ln (4.5/1.5)] \; \text{Joule}$$

$$= [1/3 \times (91.125 - 3.375) + 10 \ln 3] \; \text{Joule}$$

$$= [29.25 + 10.986] \; \text{Joule}$$

$$= 40.236 \; \text{Joule}$$

But according to the First Law of Thermodynamics

$$Q - W = \Delta U$$

or, $\qquad 300 - 40.236 = 2959 \cdot 764 \; J \quad Ans$

**3.5** The specific heat at constant pressure of certain system is a function of temperature only and may be expressed as $cp = 0.5 + 10/(T + 100)$ Joule/°C where $T$ is the temperature of the system in °C.
The system is heated while it is maintained under a pressure of 1 atmosphere until its volume increases from 2000 c.c. to 2400 c.c. and temperature 0°C to 100°C

(i)  How much heat is added to the system?
(ii)  How much does the internal energy of the system increase?

Assume the mass of system as one kg.

*Soln.*  We know that in a reversible process at constant pressure, $\delta Q = dH = m \, dh$

Also $\qquad Cp = dh/dT \text{ or, } dh = cp \, dT$

or,
$$dh = [0.5 + 10/(T + 100)]dT$$

or,
$$\Delta h = [0.5T + 10 \ln (T + 100)]_0^{100}$$

$$= 0.5 \times 100 + 10 [\ln (200/100)]$$

mass = 1 kg.

or
$$Q_{1-2} = \Delta h = 50 + 10 \times 0.6931 = 56.931 \text{ Joule } \textit{Ans.}$$

Again
$$w_{1-2} = \int_{v_1}^{v_2} p\,dv = p(v_2 - v_1)$$

$$= 1.01325 \times 10^5 (2400 - 2000)/10^6 \text{ Nm}$$

$$= 40.53 \text{ Nm} = 40.53 \text{ Joule } \textit{Ans.}$$

From First law

$$Q - W = \Delta U$$

Increase in *I.E.* = 56.931 – 40.53 = 16.401 Joule  *Ans.*

**3.6** When a system is taken from state "*a*" to state "*b*" as shown in the Fig. 3.23 along the path *acb*, 20 k. Joules of heat flows into the system and the system does 8 k Joules of work.

(i) How much heat flows into the system along path *adb* if the work done by the system is 2.5 k joules?

(ii) The system is returned from state "*b*" to state 'a' along the curved path *ba*. The work done on the system is 5 k joules. Does the system absorb or give out heat and how much?

(iii) If the value of internal energy at state "*d*" with respect to state 'a' is 10 k Joules, find the heat absorbed in the processes $a - d$ and $d - b$.

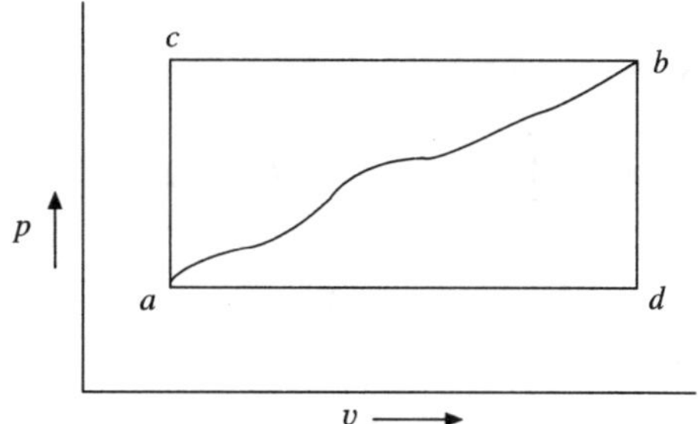


**Fig. 3.23**

*Sol.* Refer to Fig. 3.23

(i)
$$U_b - U_a = Q_{acb} - W_{ac-b} = 20 - 8 = + 12 \text{ k Joules}$$

$$Q_{adb} = (U_b - U_a) + W_{adb} = 12 + 2.5 = 14.5 \text{ Joules } \textit{Ans.}$$

(ii)
$$Q_{ba} = (U_a - U_b) + W_{ba} = - (U_b - U_a) + W_{ba}$$

$$= -12 + (-5) = -17 \text{ k Joules } \textit{Ans.}$$

The system gives out heat

(iii)
$$Q_{db} = (U_b - U_d) + W_{db} = (U_b - U_a) - (U_d - U_a) + W_{db}$$

$$= 12 - 10 + W\,db$$

But
$$W_{d\,b} = 0 \text{ as } db \text{ is a constant volume}$$

Process

$$Q_{db} = 2 + 0 = 2\ k \text{ Joules}$$

Again
$$Q_{adb} = (U_b - U_a) + W_{adb}$$

or,
$$Q_{ad} + Q_{db} = (U_b - U_a) + W_{adb}$$

or,
$$Q_{ad} + 2 = 12 + 2.5$$

or,
$$Q_{ad} = 14.5 - 2 = 12.5 \text{ k Joules} \quad Ans.$$

**3.7**  A closed system of constant volume experience a temperature rise of 20°C when a certain process occurs. The heat transfer in the process is 18 kJ. The specific heat at constant volume for the pure substance comprising the system is 1.2 kJ/kg °C and the system contains 2 kg of this substance. Determine the internal energy and work change what type of work is this?

*Sol.*
$$\Delta U = mCv\Delta T = 2 \times 1.2 \times 20 = 48 \text{ kJ} \quad Ans.$$

$$Q = 18 \text{ kJ}$$

$\therefore$ From first law of thermodynamics

$$Q - W = \Delta U$$

or,
$$18 - W = 48 \text{ or, } W = 18 - 48 = -30 \text{ kJ} \quad Ans.$$

This work is done on the system
The system is of constant volume
The *pdv* work is zero. Therefore,
This work done must be an irreversible work like one by a paddle wheel

**3.8**  Prove the following relations:

(a) $\quad -\int v\, dp = Q - \Delta H$ (for steady flow processes) when changes in KE and PE are neglected

(b) $\quad \int p\, dv = Q - \Delta U$ (for non-flow processes) when changes in KE and PE are neglected.

(c) $\quad -\int v\, dp - \int p\, dv = -\Delta pv$

*Soln.*  (a)  Steady flow energy equation for unit mass can be written as

$$\frac{V_1^2}{2} + z_1 g + h_1 + q = \frac{V_2^2}{2} + z_2 g + h_2 + w$$

If changes in k.E. & P.E. are neglected,

$$h_1 + q = h_2 + w$$

or,
$$q = (h_2 - h_1) + w$$

or,
$$q = \Delta h + w \text{ for per unit mass}$$

or,
$$m \cdot q = m \cdot \Delta h + mw$$

$\therefore$
$$Q = \Delta H + w$$

But for steady flow process $W = -\int v\, dp$

$\therefore$
$$Q - \Delta H = -\int v\, dp \quad \text{proved}$$

(b)    For the non-flow process, from the First Law

$$Q = \Delta U + W$$

But

$$W = \int p\, dv$$

$\therefore$

$$Q = \Delta U + \int p\, dv$$

$\therefore$

$$\int p\, dv = Q - \Delta U \quad \text{proved}$$

(c)    For steady flow work, $W = -\int v\, dp = Q - \Delta H$

or,

$$-\int v\, dp = Q - (\Delta U + \Delta pv)$$

and for non-flow work, $W = \int p\, dv = Q - \Delta U$

$\therefore$

$$-\int v\, dp - \int p\, dv = (Q - \Delta U - \Delta pv) - (Q - \Delta U)$$

$$= -\Delta pv \quad \text{proved}$$

It is important to note that only difference between steady flow and non-flow is the flow work. Flow work is required in steady flow process as mass enters and leaves the system. Flow work is absent in non-flow process as masm does not enter or leave the system.

$\therefore$ For steady-flow process

$$Q = \Delta H + W = \Delta U + \Delta Pv + W$$

and for non flow process

$$Q = \Delta U + W$$

**3.9** The work and heat transfer per degree temperature change for a system executing a steady non-flow process are given by

$$\frac{\delta w}{dT} = \frac{1}{8}\,\text{kJ/k} \quad \text{and} \quad \frac{\delta Q}{dT} = 0.4\,\text{kJ/k}$$

Determine the change in internal energy of the system when the temperature increases from 55°C to 250°C.

*Soln.*

$$W = \int_{55}^{250} \frac{dT}{8} = \frac{1}{8}(250 - 55) = 24.375\,\text{kJ}$$

and

$$Q = \int_{55}^{250} 0.4\,dT = 0.4(250 - 55) = 78\,\text{kJ}$$

Applying the First Law to the process

$$Q = \Delta U + W$$

or, $$\Delta U = Q - W$$

$$= 78 - 24.375$$

$$= 53.625 \text{ kJ}$$

This indicates that the internal energy of the system is increased.

**3.9a**  A thin partition divides a rigid closed insulated vessel into two compartments. One compartment contains sulphuric acid and the other contains water. If the partition is punctured and the temperature rises to $T_2$ from an initial temperature of $T_1$ find the change in internal energy of the vessel and its contents.

*Soln.*  Since the vessel is insulated there is no heat transfer. Also, as it is rigid, the work done is zero. Therefore, applying the First Law, we have

$$\delta Q - \delta W = dU$$

or, $$O - O = dU$$

that is, $$dU = 0$$

So, the change in IE is zero.

**3.10**  A piston cylinder arrangement contains a fluid which is continuously agitated for 1 minute by a stirrer passing through the cylinder wall. The piston is frictionless and faces atmospheric pressure from outer side. The stirring motor has 1 Hp. If the diameter of the cylinder is 0.15 m, find the net work done in kW on the fluid when (i) the piston is stationary (ii) the piston moves 0.50 m outwards. Assume no heat transfer.

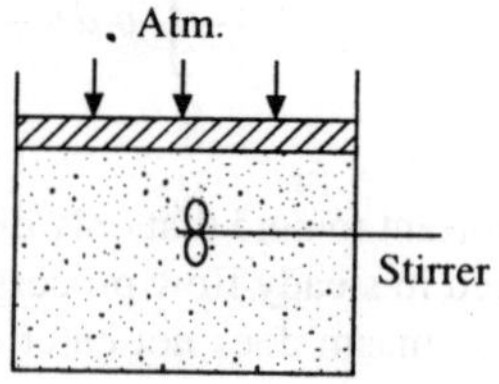

*Sol.*  Refer to Fig. 3.24    1 H.P. = 736 Watts
Work input by stirrer



**Fig. 3.24**

(i) There is no work done by the system as the piston is fixed
The only work done on the fluid due to stirring for 1 minute

$$= \text{work done by stirrer} = 736 \times 60 \text{ J}$$

$$= 44.16 \text{ kJ}$$

(ii) In this case the piston moves out against atmospheric pressure doing work done

$$WD = \text{force} \times \text{displacement} = 1.01325 \times 10^5 \times \pi/4 \times (0.15)^2 \times 0.50$$

$$= 895.280 \text{ N} - \text{m} = 895.280 \text{ J}$$

$$= 8.9528 \text{ kJ}$$

Net work done on the fluid

$$= 44.16 - 8.9528 = 35.207 \text{ kJ} \quad \textit{Ans.}$$

**3.11**  100 kJ of heat is rejected per kg of fluid during a reversible steady flow process. During the process, pressure and volume change from $p_1 = 3.6$ bar to $p_2 = 1.2$ bar

and $$v_1 = 0.0568 \text{ m}^3 \text{ to } 0.142 \text{ m}^3$$

the relation between the pressure and volume during the process is given by $pv^n = c$

Find the change in enthalpy. Take

$$\Delta PE = 0 \text{ and } \Delta KE = 0$$

*Soln.* The energy equation for steady flow reversible process is

$$Q - W = \Delta H + \Delta KE + \Delta PE$$

Here, given

$$\Delta KE = 0 \text{ and } \Delta PE = 0$$

$$\therefore \qquad Q = W + \Delta H \qquad\qquad (1)$$

But work done per kg for steady flow process is given by

$$W = -\int_{p_1}^{p_2} v \, dp$$

Also given $pv^n = C$ $\therefore$ $v = (c/p)^{\frac{1}{n}}$

$$-\int_{p_1}^{p_2} v \, dp = -\int (c/p)^{\frac{1}{n}} \, dp = c$$

$$= -c^{1/n}\left[\int_{p_1}^{p_2} p^{-1/n} \, dp\right] = -c^{\,1/n}\left[\frac{p^{1-1/n}}{1-{1/n}}\right]_{p_1}^{p_2}$$

$$= -n/n - 1(c)^{1/n}[p^{(n-1)n}]_{p_1}^{p_2} = (n/n - 1)c^{1/n}[p_2^{n-1/n} - p_1^{n-1/n}]$$

$$= -n/n - 1[c^{1/n}p_2^{n-1/n} - c^{1/n}p_1^{n-1/n} = -n/n - 1 \times (p_2 v_2^n)^{1/n}$$

$$X_{p_2}^{n-1/n} - (p_1 v_1^n)^{1/n}(p_1)^{n-1/n}]$$

$$\therefore \qquad = pv^n = p_1 v_1^n = p_2 v_2^n = c$$

$$= -n/n - 1[p_2^{1/n}v_2 p_2^{n-1/n} - p_1^{1/n}v_1 p_1^{(n-1/n)})$$

$$= -n/n - 1[p_2 v_2 - p_1 v_1] = n/n - 1(p_1 v_1 - p_2 v_2) \qquad\qquad (2)$$

To find $n$, as $[p_1 v_1^n = p_2 v_2^n]$

$$\therefore \qquad\qquad p_1/p_2 = (v_2/v_1)^n$$

or, $\qquad\qquad (3.6/1.2) = (0.142/0.0568)^n$ or, $3 = (2.5)^n$

or, $\qquad\qquad \ln 3 = n \ln 2.5$

or, $\qquad\qquad n = \ln 3/\ln 2.5 = 1.09861/0.91629$

$$= 1.2$$

$$\therefore \qquad W = (1.2/0.2) \times 10^5 \times (3.6 \times 0.\,0568 - 1.2 \times 0.142)$$

$$= 6 \times 10^5(0.2044 - 0.1704) = 6 \times 10^5 \times 0.034$$

$$= 20400$$

$$= 20{\cdot}4 \text{ kJ}$$

From equation (1)

$$-100 = 20.4 + \Delta h$$

or,
$$\Delta h = -79.6 \text{ kJ}$$

The −ve sign indicates that the enthalpy of the fluid decreases.

**3.12** Steam enters a nozzle at a velocity of 40 m/sec and enthalpy 681.4 kJ/kg. It leaves the nozzle at a velocity of 700 m/sec. The mass flow rate through the nozzle is 1,500 kg/hr. The heat lost from the nozzle is 3,000 kJ/hr.

Determine the final enthalpy of steam and the nozzle exit area if the specific volume at exist is 1.24 m³/kg.

*Soln.* As nothing is mentioned about the elevation of inlet and exit points, we can assume $z_1 = z_2$. The work done = 0

Rate of mass flow/sec = 1,500/3,600 = 5/12 = 0.417 kg/sec

Rate of heat transfer/kg mass = −3,000/1,500

$$= -2 \text{ kJ/kg}$$

Applying steady flow energy equation we have

$$Q - W = m\left[(h_2 - h_1) + \frac{V_2^2 - V_1^2}{2g_c} + \{(z_2 - z_1)\} \times g / g_c\right]$$

or,
$$Q = m(h_2 - h_1) + m \times \{(V_2^2 - V_1^2)/2g_c\}$$

or,
$$Q/m = (h_2 - h_1) + (V_2^2 - V_1^2)/2g_c$$

or, for 1 kg of flow rate

$$Q = (h_2 - h_1) + (V_2^2 - V_1^2)/(2g_c)$$

or,
$$-2 = h_2 - 681.4 + \frac{700^2 - 40^2}{2 \times 1000}$$

or,
$$h_2 = 681.4 + \frac{(40^2 - 700^2)}{2 \times 1000} - 2$$

$$= 681.4 - 244.2 - 2 = 435.2 \quad \text{kJ/kg} \quad Ans.$$

Let the area at the exit be $a_2$ then mass flow rate/sec

$$\frac{a_1 V_1}{v_1} = \frac{a_2 V_2}{v_2}$$

or,
$$0.417 = a_2 \times 700/1.24 \quad \text{or} \quad a_2 = 0.417 \times 1.24/700 \text{ m}^2$$

or,
$$a_2 = (0.417 \times 1.24/700) \times 10^4 \text{ cm}^2 = 7.387 \text{ cm}^2 \quad Ans.$$

**3.13** Air is passed through a nozzle to expand from an initial pressure of 3 bar, and initial temperature of 150°C to a final pressure of 1 bar and at 36°C. If the initial velocity is 90 m/sec, find the final velocity. Neglect the change in elevation given $cp$ for air = 1005 Joules/kg k

*Soln.* We know that $h = cpT$

$$h_1 = C_p T_1 = 1005 \times (273 + 150) = 4.25 \times 10^5 \text{J/kg}$$
$$h_2 = cpT_2 = 1005 \times (273 + 36) = 3.10 \times 10^5 \text{ J/kg}$$

Applying the steady flow energy equation

$$Q / kg - W / kg = [(h_2 - h_1) + (V_2^2 - V_1^2) / (2g_c) + (Z_2 - Z_1) \times (g / g_c)]$$

Flow in nozzle is adiabatic and it does not produce work

that is, $$\qquad Q = 0, \; W = 0$$

or, $$\qquad h_2 - h_1 = (V_1^2 - V_2^2)/(2g_c)$$

or, $$\qquad V_1^2 - V_2^2 = 2 \times (3.10 \times 10^5 - 4.25 \times 10^5)$$
$$= -2 \times 1.155 \times 10^5$$

or, $$\qquad V_2^2 = V_1^2 + 2 \times 1.155 \times 10^5$$
$$= 8100 + 2.31 \times 10^5$$
$$= 239100$$

or, $$\qquad V_2 = 488.978 \text{ m/s} \quad Ans.$$

**3.14** Two kg of a fluid per second enters the control volume of a steady flow system at 1 bar and leaves at 10 bar. The inlet and exit velocities are 20 m/s and 40 m/s respectively. During the process, $72 \times 10^3$ kJ of heat is transferred per hour from the control volume. The fall in enthalpy of the fluid is 15 kJ/kg. Calculate the power developed in KW and H.P. both.

*Soln.* $$\Delta z = 0, \; (h_2 - h_1) = -15 \times 10^3 \text{J/kg}$$

$$Q/\text{kg-sec} = \frac{-72 \times 10^3 \times 1000}{2 \times 3600}$$

$$= -10{,}000 \text{ J/kg sec}$$

Now applying the steady flow energy equation, we get

$$Q - W = m[(h_2 - h_1) + (V_2^2 - V_1^2) / (2g_c) + (z_2 - z_1) \times g / g_c]$$

or, $$W = Q - m[(h_2 - h_1) + (V_2^2 - V_1^2) / (2g_c)]$$

or, $$W/kg = Q/kg + (V_1^2 - V_2^2) / (2g_c) - (h_2 - h_1)$$
$$= -10{,}000 + (20^2 - 40^2)/2 \times 1 + 15 \times 10^3$$
$$= 4400 \text{ J/s}$$

or, total work done/sec $= + 4400 \times 2 = + 8800$ J/S

Developed power $= 2 \times 4400$ J/s $= 2 \times 4400$ $w = 8.8$ kW  *Ans.*

H.P. (Metric) $= 8800/736 = 11.957$ H.P.  *Ans.*

**3.15** A steam turbine receives a flow of 1 kg/sec of steam from the boiler at a velocity of 4,000 m/min, specific enthalpy of 700 kJ/kg and an elevation of 5 m. The steam leaves the turbine at a velocity of 8,000 m/min specific enthalpy of 500 kJ/kg and an elevation of 2 m. Heat losses from the turbine take place of the rate of 31.67 kJ/min. Calculate the power developed by the turbine.

*Soln.*  We have for steady flow $Q - W = m[(h_2 - h_1) + (V_2^2 - V_1^2)/(2g_c) + \{(z_2 - z_1)\}(g/g_c)]$

Here $m = 1$ kg/sec

$$Q = -31.67/60 = -0.527/\text{kg}$$

$$V_2 = 8000/60 = 133.33 \text{ m/sec}$$

$$V_1 = 4000/60 = 66.66 \text{ m/sec}$$

$$h_2 - h_1 = (500 - 700) = -200 \text{ kJ/kg}$$

Putting these values in energy equation we get

$$-0.527 - W = 1\left[-200 + \frac{133.33^2 - 66.66^2}{2 \times 1000} + \frac{(2 - 5) \times 9.81}{1000}\right]$$

or,  $\qquad -W = 0.527 - 200 + 6.667 - 0.029$

or,  $\qquad + W = 192.835$ kW

Power developed = 192.835 kW  *Ans.*

**3.16**  Two kg of a gas is passed through an insulated and opened valve such that its pressure falls from 20 bar abs. to 1.5 bar abs. In the process the internal energy falls by 0.16 kJ. If the initial volume of the gas is 0.44 m$^3$, find the final volume.

*Soln.*  This is the case of a throttling process.
The energy equation reduces to

$$h_1 = h_2 \tag{1}$$

or,  $\qquad u_1 + p_1 v_1 = u_2 + p_2 v_2$

or,  $\qquad p_2 v_2 = (u_1 - u_2) + p_1 v_1$

or,  $\qquad p_2 v_2 = (u_1 - u_2) + p_1 v_1 \tag{2}$

Now,  $\qquad u_1 - u_2 = 0.16/2 = 0.08$ kJ/kg

$$p_1 = 20 \times 10^5 \text{ N/m}^2$$

$$p_2 = 1.5 \times 10^5 \text{ N/m}^2$$

$$v_1 = 0.44/2 = 0.22 \text{ m}^3/\text{kg}$$

From (2) we have

$$1.5 \times 10^5 \times v_2 = (0.08) + 20 \times 10^5 \times 0.22$$

or,  $\qquad v_2 = \dfrac{0.08 + 20 \times 10^5 \times 0.22}{1.5 \times 10^5} = 2.933 \text{ m}^3/\text{kg}$

Total final volume = $2\cdot933 \times 2 = 5.866$ m$^3$  *Ans.*

**3.17**  Steam enters a steam condenser at the rate of 3,600 kg per hour. The inlet and exit specific enthalpies of steam and condensate are respectively 605 kJ/kg and 32.2 kJ/kg. If 264.6 m$^3$ of cooling water at a specific enthalpy of 21.1 kJ/kg is passed through the condenser, find the final specific enthalpy of the cooling water.

*Sol.*  Heat lost by steam = heat gained by water or, fall of enthalpy of steam = rise of enthalpy of water. Rate of steam flow/sec = 3,600/3,600 = 1 kg/sec mass rate of cooling water circulated/sec

$$= \frac{264.600 \times 10^3}{3,600} = 264,600/3,600 \text{ kg}$$

$$= 73.5 \text{ kg/sec}$$

or, $\qquad 1 \times (h_{s1} - h_{s2}) = 73.5 \times (h_{w2} - h_{w_1})$

$$1 \times (605 - 32.2) = 73.5 \times (h_{w2} - 21.1)$$

or, $\qquad h_{w2} - 21.1 = 7.793$

or, $\qquad h_{w2} = 28.896 \text{ kJ/kg} \quad Ans.$

**3.18** In an air compressor air enters at a velocity of 5 m/sec, pressure 1 bar and specific volume 0.5 m³/kg. The corresponding exit conditions are velocity 7.5 m/sec, pressure 7 bar and specific volume 0.15 m³/kg. The mass rate of flow of air is 0.5 kg per second. The internal energy of the air leaving the compressor is 40 kJ/kg greater than that at inlet point and during the process the system loses 3,600 kJ/min of energy dissipated as heat to the cooling water and surroundings.
Find (a) the shaft work input the air in kW. (b) Ratio of inlet to outlet areas.

*Soln.* Heat lost to cooling water and surrounding = 3,600 kJ/min

$$= 3,600 \text{ kJ/min}$$

$$= \text{kJ/sec} = 60/0.5 = 120 \text{ kJ/kg}$$

Applying the steady flow energy equation for unit mass and neglecting P.E. we get

$$Q - W = 1 \times [(h_2 - h_1) + (V_2^2 - V_1^2)/(2g_c) + (z_2 - z_1) \times g/g_c]$$

or, $\qquad -120 - W = (u_2 - u_1) + (p_2 v_2 - p_1 v_1) + (V_2^2 - V_1^2)/(2g_c)$

or, $\qquad W = -120 - 40 - (7 \times 0.15 - 1 \times 0.5) \times 10^5 \times 10^{-3} - (7.5^2 - 5^2)/2 \times 1000$

or, $\qquad W = -120 - 40 - 55 - 0.0156 = -215.0156 \text{ kJ/kg}$

Total work/sec $= -215.0156 \times 0.5 \text{ kJ/sec} = -107.5078 \text{ kW}$

–ve sign indicates that work is added to the system
shaft work input to the Air = 107.5078 kW $\quad Ans.$
(b) From continuity equation

$$a_1 V_1/v_1 = a_2 V_2/v_2$$

or, $\qquad a_1/a_2 = v_1/v_2 \times V_2/V_1 = (0.5/0.15) \times (7.5/5) = 5 \quad Ans.$

**3.19** Water enters in a boiler at specific enthalpy of 40 kJ/kg and leaves in the form of steam at specific enthalpy of 700 kJ/kg. The exit point is 3 m above the inlet. Heat added by burning fuel is 680 kJ/kg while some heat is lost from boiler to the surrounding by radiation. Find the loss of heat/kg. Neglect any KE change.

*Soln.* Applying the general energy equation for 1 kg

$$Q - W = 1 \times [(h_2 - h_1) + (V_2^2 - V_1^2)/(2g_c) + (z_2 - z_1) \times \{g/g_c\}]$$

or, $\qquad (680 - Q \text{ lost}) = (700 - 40) + 0 + 3 \times 9.81/1000 = 660.029$

or, $\qquad Q \text{ lost} = 680 - 660.029 = 19.971 \text{ kJ} \quad Ans.$

## EXERCISES

### Objective Questions

3.1  For each of the following statements indicate whether it is true or false.

   (i)   For a closed system $\int p\ dv$ for a gas from one state to another does not depend on the path for reversible processes.

   (ii)  Work is always given by $\int p\ dv$

   (iii) The First Law of Thermodynamics requires that the total energy of any system be conserved within the system.

   (iv)  The energy of an isolated system must be a constant.

   (v)   If stirring work is done on an ideal gas in a closed system during a constant volume process, then $dQ \neq Cv\ dT$

   (vi)  The reversible work done in an open flow system is given by $-\int v\ dp$

   (vii) If hydrogen and oxygen in a combustible mixture in an insulated closed vessel forms a system and a minute spark causes then to combine the internal energy of the system will not change.

   (viii) For an isothermal process of an ideal gas the change in internal energy is same as the change in enthalpy between the same given states.

   (ix)  A system has exchange of heat and work with the surrounding but the heat received is equal to heat rejected and work done on the system is equal to work done by the system. This is an isolated system.

   (x)   According to first law of thermodynamics for any process heat added must be equal to work done by it i.e. $dQ = dW$

   (xi)  Specific volume is an intensive property.

   (xii) Internal energy is a point function and therefore, a property.

   (xiii) Work is a path function and therefore, it is not a property.

   (xiv) Heat is energy contained in a body above absolute zero temperature.

   (xv)  Concept of temperature comes from Zeroth Law of Thermodynamics.

   (xvi) Concept of internal energy comes from the First Law of Thermodynamics

   (xvii) During throttling process enthalpy remains constant

   (xviii) In a beaker containing cold water a glass of hot water is powred. We can say that heat is added to the beaker.

   (xix) If the cyclic integral of differential of a thermodynamic variable is zero it must be a property.

   (xx)  If a thermodynamic variable has exact differential it is a property.

   (xxi) Energy contained in the hot exhaust gases from an automobile is heat.

   (xxii) The thermodynamic state of system can not be changed without energy transfer in the form of either heat or work.

| Ans | i | ii | iii | iv | v | vi | vii | viii | ix | x | xi | xii |
|-----|---|----|----|----|----|----|----|----|----|----|----|----|
|     | F | F | F | T | T | T | T | T | F | F | T | T |

| | xiii | xiv | xv | xvi | xvii | xviii | xix | xx | xxi | xxii |
|---|----|----|----|----|----|----|----|----|----|----|
|   | T | F | T | T | T | F | T | T | F | F |

3.2  For a fluid passing through a nozzle the inlet and outlet parameters are as follows

|        | Enthalpy    | Velocity  | Area    | Specific volume |
|--------|-------------|-----------|---------|-----------------|
| Inlet  | 3000 kJ/kg  | 60 m/sec  | 0.1m$^2$ | 0.185 m$^3$/kg  |
| Outlet | 2770 kJ/kg  | –         | –       | 0.495 m$^3$/kg  |

Find the outlet velocity, rate of flow of fluid and exit area of the nozzle. The nozzle is horizontal and there is no heat loss.

(680.5 m/sec, 32.43 kg/sec, 0.236 m$^2$)

3.3  A vertical cylinder 0.2 m bore has its piston initially 0.3 m above bottom pisition. The piston is loaded with 1000 Newton weight and can move frictionlessly. The atmospheric pressure is 1 bar ($10^5$N/m$^2$). Inside the cylinder there is a gas at 293 k. The gas is heated with heater of 1 watt for an hour. Assuming no heat loss to the surrounding, calculate the change in internal energy of the gas and the distance moved by the piston.

For gas given $C_p$ = 820 J/kg k

$$R = 189 \text{ J/kg k} \qquad (2780 \text{ J, } 0.82 \text{ m})$$

3.4  A fluid undergoes a non-flow friction less process from initial volume of 6 m$^3$ to a final volume of 2 m$^3$. The pressure volume law for the process is $p = (14.7 + 0.5v)$ bar when V is in m$^{3.}$ The heat lost by the system is 210 kJ/kg. Find the work done by the system and the change in internal energy.

(–2403.22 kJ/kg, 2193.9 kJ/kg)

3.5  The properties of a fluid in a steady flow process at inlet and exit are respectively, pressure $6 \times 10^5$ N/m$^2$, $1.25 \times 10^5$ N/m$^2$

Velocity 300 m/s,    200 m/s

Internal energy 2,000 kJ, 1,500 kJ

specific volume 0.3 m$^3$/kg 1.2 m$^3$/kg

The mass rate of flow is 4 kg/s. If there be no change of potential energy and work done by the machine be 515 kJ/kg, find the heat lost to the surrounding per kg of gas. (40 kJ/kg)

3.6  A system completes a reversible cyclic process in the following manner. First it executes an adiabatic process. It then undergoes a constant volume process during which 357 kJ of heat is added to it. The system is then brought to  its initial state by a constant pressure process during which 84 kJ of work is done on it and it rejects 378 kJ of heat. Determine the value of adiabatic work done and internal energy at all end states if the initial value is 840 kJ.

(63 kJ, $U_1$ = 840 kJ

$U_2$ = 1197 kJ, $U_3$ = 903 kJ)

3.7  In a gas turbine the gas enters at 6.867 bar, 860°C and 164 m/sec. The exit takes place at 1.962 bar, 640° and 330 m/sec. Find the power output of the turbine in kW if the rate of mass flow is 2kg/sec.

Assume $C_p$ = 1.13 kJ/kgk

(410.64 kW)

3.8  A refrigerator uses 2 kwh of electrical energy in cooling the food inside and the internal energy of the system decreases by 10,000 kJ as the temperature drops. Find the amount of heat transferred during the process (–17,200 kJ).

3.9  A gas flows steadily through a passage of varying cross-sectional area. The flow is frictionless and adiabatic. At a particular section the gas flows with a velocity of 600 m/s, the pressure is 2 bar and temperature 400 k. If at another section the relocity is 300 m/s, find the changes in pressure, temperature and internal energy per kg of mass flow. Assume,

$$c_p = 1.25 \text{ kJ/kg K and } C_v = 0.84 \text{ kJ/kg}$$

(381 k, 2.144 bar, 90.7 KJ/kg)

3.10  Feed water at the rate of 44.44 kg/sec enters a boiler at a specific enthalpy of 840 kJ/kg. The steam leaves at a specific enthalpy of 3,360 kJ/kg. 4.166 kg of coal with a calorific value of 31,080 kJ/kg is burnt per second. 13 kg of air at enthalpy of 67 kJ/kg is supplied per kg of fuel burnt. 14 kg of flue gases per kg of fuel burnt leave the boiler at enthalpy of 294 kJ/kg. If the changes in kinetic and potential energies are negligible, find the heat lost from the boiler by radiator and other methods.          (1,014 kJ/kg of coal).

3.11  A fluid is connfined in a cylinder by a springloaded, frictionless piston so that the pressure in the fluid is a linear function of the volume ($p = a + bv$), The internal energy of the fluid is given by the following equation

$$U = 40 + 3.25 \, Pv$$

Where $U$ is in kJ, $P$ in $kPa$, and $v$ in $m^3$. If the fluid changes from an initial state of 170 $kPa$, 0.03 m$^3$ to a final state of 400 $kPa$, 0.06 m$^3$ with no work other than that done on the Piston, find the magnitude and direction of heat transfer. [May 98, Mumbai Univ.]

3.12  A fluid system undergoes a non-flow frictionless process from $V_1 = 0.015$ m$^3$ to $V_2 = 0.5$m$^3$. The relation between pressure and volume is given by the expression $p = 1.5/V + 2$ bar. Where $V$ is in m$^3$. During this process the system rejects 22.4 kJ of heat. Find the change in enthalpy. (Ans $\Delta H = 142.39$ kJ)

3.13  5 kg of air is heated reversibly in a constant volume process till its pressure becomes three times the initial pressure. The initial temperature is 100°C. Determine (1) final temperature (2) change in internal energy (3) change in enthalpy and (4) heat transfer during the process. $R = 0.289$ kJ/kgk, $C_v = 0.75$ kJ/ kg K. [Ans. (1) 1119 K (2) 2665.95 kJ, (3) 3744.92 k (4) 2665.95 kJ]

3.14  Air flows in a steady flow system at a rate of 0.5 kg/s. Air enters at 6m/s velocity, 1 bar pressure and volume 0.85 m$^3$/kg and it leaves at 3m/s velocity, 7 bar pressure and volume 0.17 m$^3$/kg. The internal energy of air leaving is 80 kJ/ kg greater than that of entering. Cooling water in system jacket absorbs heat from air at a rate of 40 kW. Determine the rate of shaft work transfer. If the pipe diameter of air entering the system is 30 cm. What would be the exit diameter? [May 97, Mumbai Univ.]

3.15  0.5 kg of air at 177°C expands adiabatically to three times its original volume and its temperature falls to 17°C during the expansion. The work done during the process is 57.40 kJ. Calculate $Cp$ and $C_V$.
[Ans. $C_p = 1.002$ kJ/kgk, $C_v = 0.716$ kJ/kgk]

3.16  A room is fitted with two fans each consuming 0.2 kW power. There are three lamps in the room each consuming 200 W. Ventilation air enters the room with enthalpy of 85 kJ/kg and leaves the room with enthalpy of 60 kJ/kg. The rate of air flow is 100 kg/hr. There are five persons in the room and heat generated by each person is 600 kJ/hr. Determine the rate at which the heat is to be removed by a room cooler, so that steady state is maintained in the room. [Ans. 2.52 kW]

# 4

# The Second Law of Thermodynamics

## 4.1 LIMITATIONS OF THE FIRST LAW AND INTRODUCTION TO THE SECOND LAW

According to the First Law, work and heat are mutually convertible and during any process the sum of all forms of energy remain constant. While nobody can deny the above fact. It is a common experience that while work is very easily converted into heat (by rubbing both hands we generate heat) but heat cannot be easily converted into work. Not only that, work is easily converted into an equal amount of heat, that is, a 100% conversion but one cannot convert all heat into work in a continuous manner. That is, a system can not go on taking heat and converting all this into work. In other words, in a cyclic process all heat supplied cannot be converted into work. In reality, any system can convert, in a cyclic process, only part of the heat supplied to it into work while the rest of heat is rejected. Had this restriction not been then? One could take heat from the ocean (which is so large) and convert it continuously into work thereby avoiding the need any other source of energy.

Although, there is no theoretical basis to deny complete conversion of heat into work, all attempts have failed to work in a cycle and continuously convert all heat into work.

This experience of man has led him to believe that there is a natural law which restricts the continuous conversion of heat into work. This is called the Second Law of Thermodynamics which denies the possibility of ever completely converting into work till the heat supplied to a system operating on a cycle exist, no matter how perfectly designed a machine may be.

## 4.2  STATEMENTS OF SECOND LAW OF THERMODYNAMICS.

A statement of Second Law by Kelvin-Planck is as follows: "It is impossible to construct an engine which will work in a complete cycle and reduce no other effect except the rising of a weight (the production of work) and the exchange of heat with a single reservoir."

Thus, according to this law an engine $E$ of Fig. 4.1 working on cyclic process (which means that there is no change in the internal energy for the complete cycle), cannot simply receive an amount of heat say $Q$ from a heat source and deliver work $W$, such that $W$ is equal to $Q$. Such an engine would satisfy the First Law of Thermodynamics but violate the Second Law.

An engine which obeys both first and second laws and can work continuously is shown in Fig. 4.2. Thus, a practical engine must reject some of the heat which it receives. In Fig. 4.2 we see that the engine receives $Q_H$ amount of heat from a heat source at a higher temperature $T_H$ and rejects a smaller amount of heat $Q_L$ to a body at a lower temperature $T_L$ (called sink). The rest of the heat, that is, $Q_H - Q_L$ is converted into work.

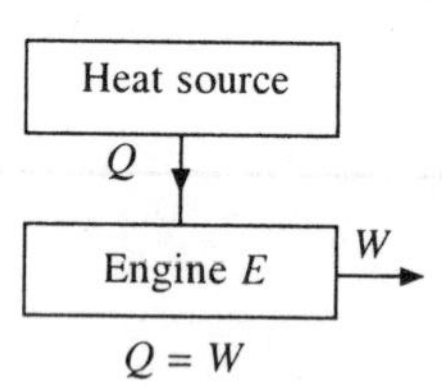


**Fig. 4.1  An impossible engine**

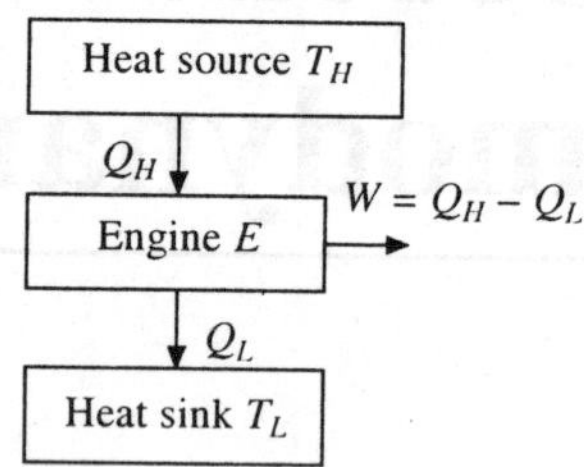


**Fig. 4.2  A possible engine**

Thus the First Law is not opposed to the complete conversion of work into heat or heat into work. But the Second Law oppooses the complete conversion of heat into work while it does not put any restriction to complete conversion of work into heat. Thus, the Second Law makes direction of the process an important aspect.

Another statement of the Second Law the given by Clausius. It concerns the heat pumps or refrigerators instead of heat engines. An engine is a device to convert heat into work where as a heat pump is a device to pump (transfer) heat from a lower temperature to a higher while operating in a cycle. "It is impossible to construct a heat pump which will work in a complete cycle and produce no effect except to transfer heat from cold reservoir to a hot reservoir." In other words it is stated as "heat itself cannot flow from a colder to a hotter body."

In fact a heat pump can transfer heat from a lower temperature to a higher one, only when it gets external help, that is, work is added to it. Fig. 4.3 shows a heat pump which is taking heat $Q_L$ from a heat sink at lower temperature $T_L$ and is supplying heat $Q_H$ to a heat source at a higher temperature $T$ while work is added to it (the heat pump). The Clausius's statement says that addition of work to the heat pump is a must for it to work in a complete cycle.

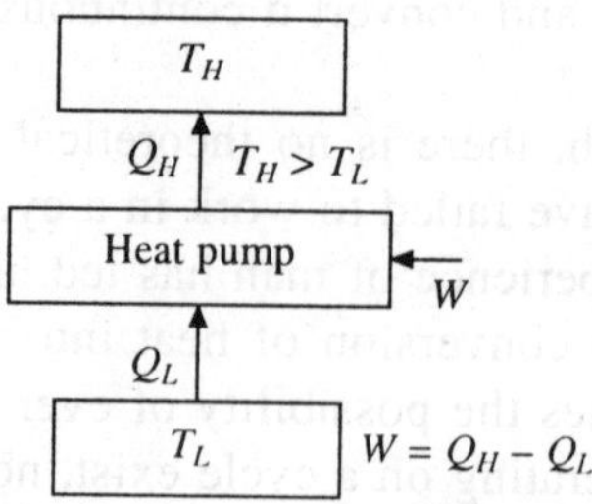


**Fig. 4.3  Heat Pump**

### 4.2.1  Equivalence of Kelvin Planck Statement and Clausius Statement

At the first glance it appears that the two statements are quite different. But in fact they amount to the same thing because violation of one statement violates the other.

### (A) *Violation of Kelvin-Planck statement results into violation of Clausius statement*

Let us consider a system which consists both of a heat engine and a heat pump working between the same temperatures $T_1$ and $T_2$ as shown in Fig. 4.4(a).

Let us suppose that the engine $E$ violates the Kelvin-Planck statement but pump $P$ obeys Clausius statement. That is, the engine $E$ takes heat $Q_H$ from the source at $T_H$ and without rejecting any heat does work $W$ equal to $Q_H$ thus violating the Second Law (Kelvin-Planck statement). The heat pump is connected to the engine and it receives work $W (= Q_H)$ from the engine, takes heat $Q_L$ from sink at temperature $T_L$ and rejects the total energy $(Q_H + Q_L)$ to the source at temperature $T_H$, thus obeying the Clausius statement (Fig. 4.4a).

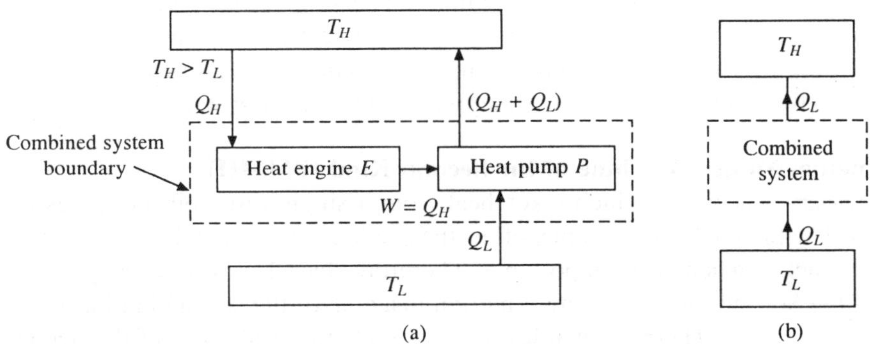


**Fig. 4.4**  **(a) Heat engine $E$ violating $K$-$P$ statement. Heat pump $P$ obeying Clausius statement; (b) Combined system violating Clausius statement**

Now if we consider the entire system (Fig. 4.4b) consisting of the engine and the pump, we find that source $T_H$ is receiving a net amount of heat $(Q_H + Q_L) - Q_H$, that is, $Q_L$ while the same amount is taken from sink at $T_L$. Thus the whole system transfers heat $Q_L$ from a lower temperature to a higher temperature without taking any work; clearly this is a violation of the Clausius statement. Thus violation of Planck's statement amounts to the violation of Clausius statement.

### (B) *Violation of clausius statement results into violation of Kelvin-Planck statement*

Consider a heat pump removing heat $Q_L$ from a low temperature reservoir at temperature $T_L$ and delivering it to a high temperature reservoir at temperature $T_H$ in a cyclic process (Fig. 4.5a) without any external aid (work). This violates Clausius statement of the second Law.

Let a heat engine operate between these same reservoirs and let heat $Q_H$ be supplied to the heat engine. The engine rejects heat $Q_L$ to a low temperature reservoir and produces work $(W = Q_H - Q_L)$.

Now if we consider the whole combined system (Fig. 4.5b) consisting of both the devices, we find that combination receives heat $(Q_H - Q_L)$ from the high temperature reservoir and converts it completely to work $(W = Q_H - Q_L)$ without rejecting any heat to the low temperature reservoir. Thus, it leads to a violation of the Kelvin-Planck statement also.

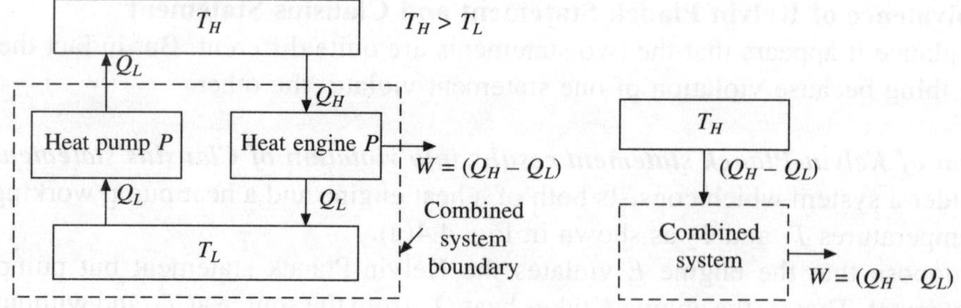

**Fig. 4.5** **(a) Heat pump *P* violating Clausius statement. Heat engine *E* obeying *K-P* statement; (b) Combined system violating *K-P* statement**

Hence, both the Kelvin-Planck and Clausius statements of the Second Law of thermodynamics are equivalent. This indicates that if a heat pump transfers heat from sink to source reservoir then it is also possible that a heat engine which will receive heat from a single reservoir and convert it completely to work. Likewise, if a heat engine receives heat from a single reservoir and converts it completely to work, then it is also possible to have a refrigerator without consuming any work.

### 4.2.2 Perpetual Motion Machine of the Second Kind (PMMII)

It is an imaginary machinery which takes heat from a single reservoir and gives an equivalent amount of work without leaving any other effect thus violating the Second Law of Thermodynamics.

Obviously such a machine is impossible. Therefore, based on this concept there is another statement of the Second Law: "Perpetual motion machine of the second kind is an imposibility."

Like the First Law of Thermodynamics, there is no theoretical proof of the Second Law also. But it has been found to hold good invariably and all attempts to disprove it have failed. This is the greatest proof of it.

## 4.3 EFFICIENCY AND COP

Thermal efficiency of a heat engine is a measure of its ability to convert heat into work. It is the ratio of work done by it to the heat supplied to it. From Fig. 4.2, we have

$$\eta_{th} = \frac{W}{Q_H} = \frac{(Q_H - Q_L)}{Q_H} = 1 - \frac{Q_L}{Q_H} \tag{4.1}$$

According to the Second Law $W$ is not equal to $Q_H$. So the efficiency cannot be 100%. Even though engineers constantly strive to increase the efficiency of heat engines. However, the thermal efficiencies of these work producing devices (prime movers) are very low. Usually, the spark ignition automobile engines have a thermal efficiency of about 20%. The thermal efficiency of diesel engine is about 30% and that of large gas-turbine plants is also about the same. Steam power plants operate at about 40% efficiency.

For heat pump or refrigerator a term known as co-efficient of performance (COP) is used as a measure of its performance. COP and efficiency, both terms have the same meaning. The term COP used to distinguishing heat pump and refrigerator from prime movers that have efficiency term. Very often COP > 1, and hence the term co-efficient is used rather than efficiency. If the value of COP of refrigerator or heat pump is more, that means they are more efficient.

As discussed earlier, a heat pump has its duty to transfer heat from a lower temperature to a higher one. Here there can be two cases. Let us have two examples. Let us suppose that a house is to be airconditioned. In winter its temperature is to be raised and so a heat pump is used to pump heat from the surroundings (which are at a lower temperature) to the room (which is at a higher temperature). The heat pump will have to be supplied with work from outside for its working. One would like to reject as much heat to the room as possible at the cost of small amount of work as possible. So the interest lies in heating, that is, the heat rejected to the room. For efficiency we write as output to the input. Likewise, we write output as the purpose (aim, that is, heating of room by rejecting heat from heat pump). Therefore, the COP of such a device is the ratio of heat rejected to work supplied. With reference to Fig. 4.3, we can write

COP (Heat Pump)

$$\xi_{H.P.} = \frac{Q_H}{W} = \frac{Q_H}{(Q_H - Q_L)} \tag{4.2}$$

Let us take the case of the same room in summer. In summer the room will require cooling. So a heat pump will be needed to pump out heat from the room (which is at a lower temperature) to the surrounding (which is at a higher temperature). Here the interest lies in the amount of heat abstracted from the room and not in the heat rejected. Therefore, for such a device called refrigerator the C.O.P is the ratio of heat abstracted to the work supplied. Thus with reference to Fig. 4.3 again, we can write COP (refrigerator) as

$$\xi_R = \frac{Q_L}{W} = \frac{Q_L}{(Q_H - Q_L)} \tag{4.3}$$

By subtracting and adding $Q_L$ in numerator part of Eq. 4.2, we get

$$\xi_{H.P} = \frac{Q_H - Q_L + Q_L}{(Q_H - Q_L)} = \frac{(Q_H - Q_L)}{(Q_H - Q_L)} + \frac{Q_L}{(Q_H - Q_L)}$$

$$\xi_{H.P} = 1 + \xi_R \tag{4.4}$$

∴    COP (heat pump) = 1 + COP (refrigerator)

So COP of a heat pump is always greater than unity. Most of the present day heat pumps have COPs between 2 and 3. It is important to note that a heat pump and a refrigerator are the same. The different names given depend on the purpose for which they are used, that is, heating or cooling.

### 4.3.1   Energy Efficiency Rating (EER)
In modern terminology, the performance of refrigerators and air-conditioners (basically refrigerators whose refrigerated space is very large, like a room or building instead of household refrigerator space) is expressed in terms of energy efficiency rating (EER) which is the amount of heat removed from the cold space for 1 Wh (watt-hour) of electricity consumed. A unit which removes 1 kwh (= 3600 kJ) of heat from the cooled space for each kwh of electric power it consumes (COP = 1) will have an EER of 3.412. Therefore, the relation between EER and COP is as follows:

$$EER = 3.412 \, COP_R \qquad (4.5)$$

Thus, most air-conditioners with COPs between 2.3 and 3.5 will have EERs between 8 and 12.

## 4.4 REVERSIBLE AND IRREVERSIBLE PROCESSES

We observe that thermal efficiency of a heat engine is never equal to 100%. But we are always eager to achieve maximum possible efficiency. The concept of reversible process (see section 2.15) are considered to be ideal processes.

A reversible process in a system is defined as a process which after taking place can be reversed and in doing so leaves no change in either system or surroundings, that is, system and surroundings both come back to their initial states. No external energy is required to return a system to its original state in a reversible process. In addition, the interactions between the system and the surroundings are also equal and opposite in direction. A reversible process thus leaves no history of the process after it is reversed.

A process which is not reversible, is an irreversible process.

The expansion of a gas after removing infinitesimal weights slowly from the piston one by one (Fig. 4.6a) is an illustration of a process tending to reversible. The gas can be brought back by compression after putting the weights slowly back on to the piston. The work done by the gas on the surroundings in raising the weights during expansion is returned to the gas during compression when the weights are lowered.

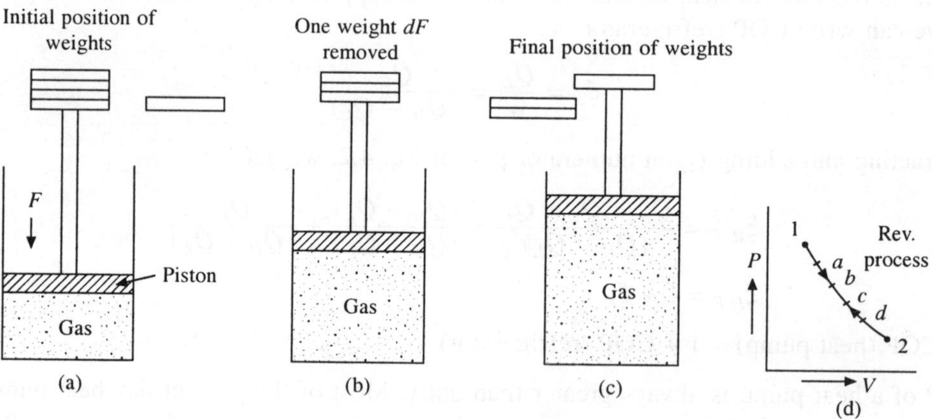


**Fig. 4.6  Reversible process**

Being the ideal ones, reversible processes do not occur in nature. This means, all the processes occuring in nature are irreversible. Then normally, it is asked about why such type of fictitious processes need to be considered in our study? Reason behind this is as follows:

(i) They are easy to analyse since a system passes through a series of equilibrium states $a$, $b$, $c$ and $d$. (Fig. 4.6(d) during a reversible process.

(ii) They serve as idealized models to which actual processes can be compared.

An illustration of an irreversible process is provided in Fig. 4.7. Let us consider a gas at high pressure entrapped in a piston-cylinder arrangement in which piston is locked by a pin. As the piston removes, a non-equilibrium process occurs. There is no equilibrium between gas pressure $P$ gas and external pressure $P_{ext}$ due to sudden expansion taking place. Now, the piston comes to an equilibrium position as shown in Fig. 4.7(b). In this case, some work has been developed which is equal to $\int P_{ext}\,dv$. Now, to bring the gas back to its initial state, a step of two processes are required, involving first compression to initial pressure ($P_{gas}$) followed by cooling at constant pressure to the position of initial temperature and volume as shown in Fig. 4.7(d). Since $P_{ext} <$ $P_{gas}$, the work input in the reverse process $\int P_{gas}\,dv$ is greater than the work output $\int P_{ext}\,dv$ in the forward process by the amount

$$-\int P_{gas}\,dv + \int P_{ext}\,dv$$

There are many processes which are totally irreversible. A hot cup of tea cools by virtue of heat transfer to the cooler surrounding but once it is cooled, it can never be heated by addition of heat from the cooler surroundings.

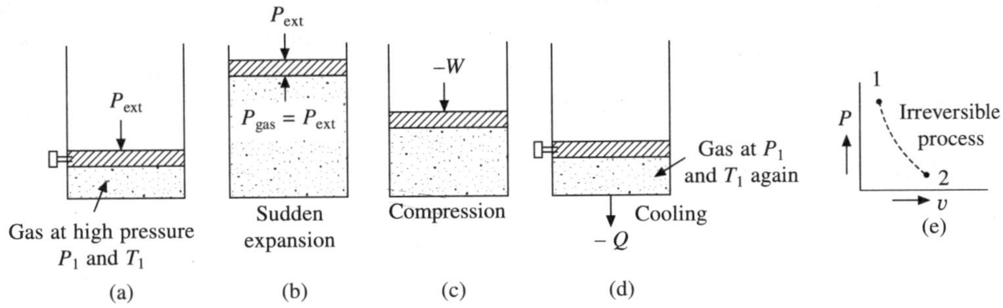


**Fig. 4.7  Irreversible Process**

Once fuel burns with air, and forms exhaust gages, it is impossible to re-convert the exhaust gages into the original fuel and air. Such processes are known as totally irreversible.

The factors that make processes irreversible are listed below:

(i) Friction
(ii) Unrestrained expansion
(iii) Mixing of substances
(iv) Inelastic deformation of solids
(v) Non-quasi equilibrium changes
(vi) Combustion
(vii) Flow of electricity through a resistor
(viii) Heat transfer over a finite temperature difference

Few examples of reversible processes (ideal) are listed below:

(i) Frictionless adiabatic expansion or compression

(ii)  Frictionless isothermal expansion or compression

(iii)  Condensation and boiling of liquids

The irreversibility of a system is classified as internal and external irreversiblity. Internal irreversibility is associated with friction due to rubbing of layers of fluid and turbulence, temperature variations within the fluid, combustion and diffusion of fluid. External irrevessibility is associated with friction at bearing and between the atmosphere and rotating members.

Thermodynamic system boundaries are so selected at the time of analysing, that possibility of irreversibility is excluded. Therefore, the system boundary can be subjected to reversible processes. For a process to be perfectly reversible, it must be internally and externally reversible as well as it must be mechanically and thermally reversible.

## 4.5  PERPETUAL MOTION MACHINE OF THE THIRD KIND (PMM III)

As discussed above that friction makes a process irreversible and can be demonstrated by the Second Law of Thermodynamics. As a natural phenomenon friction is always present in moving devices and may be reduced by suitable lubrication but can never be eliminated. If complete elimination is possible, a movable device could be kept in motion in continual motion without violating either of the two laws of thermodynamics. Thus a movable device (flywheel, bearing) keeps on moving in a continual motion in the complete absence of friction is known as perpetual motion machine of the third kind. Likewise PMM I and PMM II, PMM III is an impossible device as complete elimination of friction is not possible.

## 4.6  CARNOT CYCLE

In 1824 Sadi Carnot, a French engineer introduced an imaginary ideal cycle which consisted of two reversible isothermal process and two isentropic (adiabatic reversible) processes. This cycle was called Carnot cycle after his name. The operations of the cycle could be done either as a (i) vapour cycle or as a (ii) gas cycle.

A cycle is referred to as a vapour cycle when the processes involve change of phase of the working substance from liquid to vapour and from vapour to liquid. And a cycle, in which the working substance remains a gas throughout, is referred to as a gas cycle.

The working fluid is supposed to be entrapped in a cylinder that has a frictionless piston. The walls of the cylinder and piston are supposed to be perfect insulators. The cylinder head is supposed to act as a perfect conductor or as perfect insulation when desired. There is supposed to be available a large heat source at temperature $T_H$ and another larger heat sink at temperature $T_L$ used to heat and cool the fluid when desired. Carnot cycle for gas is given in detail in this section.

The processes occuring are shown in Fig. 4.8. The thermodynamic processes are also shown on a $p\text{-}v$ plot and $T\text{-}S$ plot corresponding to piston movements.

*Process 1–2:*  The working fluid is initially in state 1 with the piston at position 1. The heat source is in contact with the cylinder end which is acting as a perfect heat conductor. The heat passes from hot body at temperature $T_H$ to the fluid which is equal to temperature $T_L$. The fluid expands slowly and work is done such that the temperature of the fluid remains constant. This process

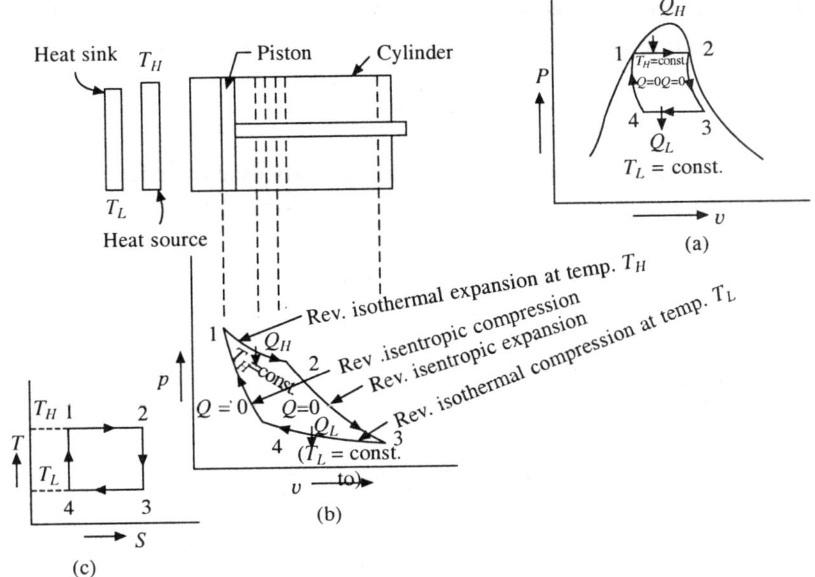



**Fig. 4.8**  (a) Carnot cycle for vapour; (b) Carnot cycle for gas

continues upto position 2 of the piston. During this period the process 1–2 is followed which is shown on the $p$-$v$ plot with temperaute $T_H$ constant. Work is done by the fluid which is equal to heat supplied

$$Q_H = P_1v_1 \ln v_2/v_1 = mRT_H \ln r \qquad (4.6)$$

where $r = v_2/v_1$

*Process* 2–3:   The hot body (heat source) is removed from the cylinder end which, now acts as perfect heat insulator. So, no heat flows to or from the fluid. The fluid expands reversibly and adiabatically such that the piston moves from position 2 to position 3. The isentropfic process is shown on the $p$-$v$ plot as 2–3. During this process the temperature of the fluid falls from $T_H$ to $T_L$. Here also work is done by the fluid

$$= \frac{P_2v_2 - P_3v_3}{\gamma - 1} \qquad (4.7)$$

*Process* 3–4:   The cold body (heat sink) at temperature $T_L$ is now brought in contact with the cylinder end which is now a conductor. Heat is transferred from the fluid isothermally and reversibly at temperature $T_L$ to the heat sink. The piston returns back from position 3 to 4. The process is extremely slow and is shown on the $p$-$v$ plot by curve 3–4. In this case gas is compressed and work is done on it. Heat is rejected during this process

$$Q_L = P_3v_3 \ln v_3/v_4 = mRT_L \ln r \qquad (4.8)$$

$$[\because \quad v_3/v_4 = v_2/v_1 = r]$$

*Process* 4–1:   The cold body is again removed and the cylinder end behaves as an insulator. The piston moves from position 4 and back to 1. The process is shown by curve 4–1. Work is done on the fluid. The temperautre of the fluid rises to $T_H$ again. Work done on the

$$\text{fluid} = \frac{P_4 v_4 - P_1 v_1}{\gamma - 1} \tag{4.9}$$

Thus a cycle is completed 1–2–3–4–1 in which all the processes are reversible. So the cycle is reversible and the engine working on this cycle is a reversible engine.

Heat is added during isothermal expansion process 1–2 and rejected during isothermal compression process 3–4. During processes 2–3 and 4–1 there is no heat transfer as these are adiabatic processes.

$\therefore$   Net work done = Heat supplied – Heat rejected

$$= Q_H - Q_L = mR(T_H - T_L) \ln r$$

$$\text{Thermal efficiency of Carnot cycle} = \frac{\text{work done per cycle}}{\text{heat supplied per cycle}}$$

$$= \frac{mR (T_H - T_L) \ln r}{mRT_H \ln r} = \frac{T_H - T_L}{T_H} = 1 - \frac{T_L}{T_H}$$

or,
$$\eta_{\text{carnot}} = \left( 1 - \frac{T_L}{T_H} \right) \tag{4.10}$$

Note that isothermal process 1–2 and adiabatic process 2–3 are completed during one stroke of the piston and cycle is completed during two strokes of the piston or 360° of crank rotation.

### *Important Observations*

1. From above equation, thermal efficiency of Carnot engine depends upon the absolute unit (kelvin scale) of soure temperature ($T_H$) and sink temperature ($T_L$). Carnot efficiency is independent of nature of working fluid, that is, if any fluid is used as working fluid, the thermal efficiency will be the same.

   The efficiency can be increased either by increasing the source temperature ($T_H$) or decreasing sink temperature ($T_L$) or both. Maximum value of the Carnot efficiency is limited as mentioned below:

   Due to metallurgical limit of boiler-tube, piston, cylinder and turbine blades maximum temperature of working fluid should not exceed about 650°C. Temperature at which heat is rejected from sink should be more than atmospheric temperature. In India, an average atmospheric temperature is considered to be 25°C. Let us consider that the heat rejection temperature is about 40°C (more than 25°C for better heat transfer). So maximum value of Carnot efficiency would be about

$$\eta_{\text{carnot}} = 1 - \frac{(40 + 273)}{(650 + 273)} = 1 - \frac{313}{923} = 0.660 \tag{4.11}$$

   that is, Carnot efficiency should not exceed 66%.

2. Carnot efficiency, however, gives the maximum value of efficiency which can be achieved by any engine operating between the same temperature limits.

3. The construction of Carnot engine using carnot cycle is more or less impossible. This cycle serves as a guideline in design and development of different heat engines or heat pumps indicating ultimate value of efficiency under same temperature limits.

### 4.6.1  Impractibility of the Carnot Cyle

(i) All the four processes are reversible (ideal) which are almost impossible to achieve because friction cannot be eliminated completely, occuring between the layers of the working fluid.

(ii) Isothermal processes of heat supplied (process 1–2) and heat rejected (process 3–4) need very slow motion of the piston to allow time for heat transfer so that temperature remains constant.

(iii) Adiabatic processes of expansion (process 2–3) and compression (process 4–1) need very high speed of the piston to avoid heat transfer from the system to the surroundings due to short interval of time.

In the Carnot cycle, as discussed earlier, isothermal and adiabatic processes take place in the same stroke which means that for part of the stroke the piston should move very slowly and for the remaining part very fast. We know that this variation of speed in the same stroke is not possible in actual practice.

(iv) $p$-$v$ diagram of this cycle is extended in vertical as well as horizontal position directing maximum pressure and volume of the engine. Even though, enclosed area under $p$-$v$ diagram which represents the work output of the engine is very small.

(v) It is not practicable to have an interchangable cylinder head.

(vi) It is impossible to completely eliminate the heat losses due to conduction and, radiation.

(vii) In vapour power cycle superheating of steam (occurs due to supply of additional heat to dry steam) takes place at constant pressure. If Carnot cycle works on vapour power cycle, heating of dry steam has to take place at constant temperature rather than constant pressure which is not possible. It means Carnot vapour cycle will work within a limit for dry steam which produces less work compared to superheated steam.

### 4.6.2  The Reversed Carnot Cycle

As mentioned in section 4.6, the Carnot heat engine cycle (Fig. 4.8) is a totally reversible cycle. When all the four processes are reversed in direction it becomes the Carnot refrigeration cycle. While reversing both heat and work interactions are reversed. This is, $Q_L$ is absorbed from the low temperature reservoir and $Q_H$ is rejected to high temperature reservoir by requiring a work input to achieve this.

The $p$-$v$ diagram of the reversed Carnot cycle is shown in Fig. 4.9. This cycle is same as that of the Carnot cycle, shown in Fig. 4.8, except that the direction of the processes are reversed.

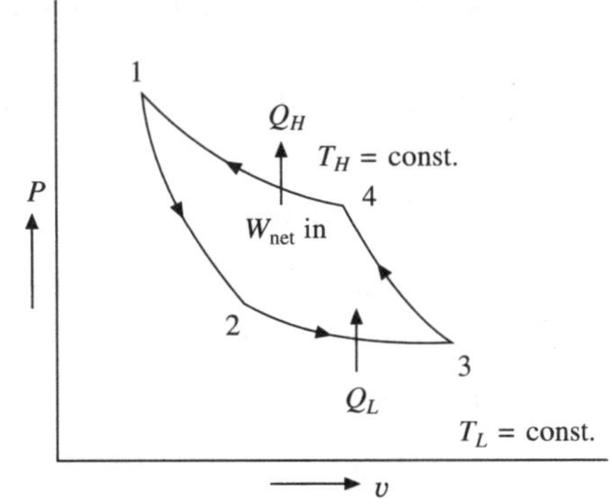


**Fig. 4.9  $p$-$v$ diagram of the reversed Carnot cycle**

## 4.7   TWO REVERSIBLE ADIABATIC PATHS CAN NOT INTERSECT EACH OTHER

This statement can be proved by contradiction. Assuming that two reversible adiabatic processes $AC$ and $BC$ intersect each other at point $C$. Let a reversible isothermal process $AB$ intersects adiabatics $AC$ and $BC$ at point $A$ and $B$ (Fig. 4.10). These processes $AC$, $CB$ and $BA$ constitute a reversible cycle. An engine operating on this cycle receives heat during process $AB$ from a single reservoir and produces work which is equal to area under diagram $ABC$. This engine does not reject any heat to any reservoir. This violates Kelvin-Planck statement of the Second Law of Thermodynamics. Hence the assumption made that two revesrsible adiabatic processes intersect each other is not correct. Therefore, it is proved that two reversible adiabic processes cannot intersect each other.

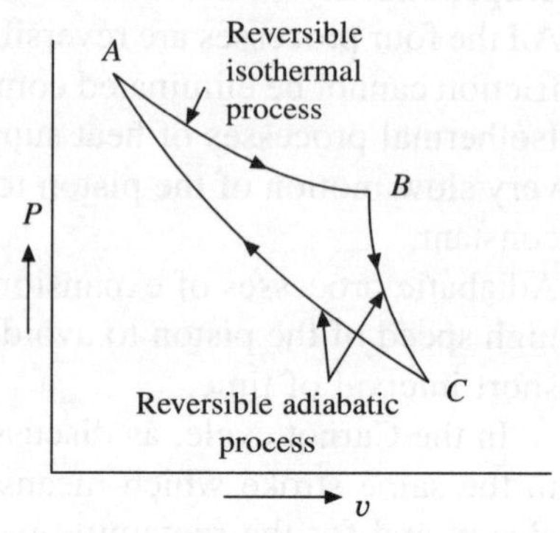


**Fig. 4.10**

## 4.8   ANY REVERSIBLE PROCESS CAN BE REPLACED BY TWO REVERSIBLE ADIABATIC PROCESSES AND ONE ISOTHERMAL PROCESS

Consider any reversible process $CD$ on a $p$-$v$ diagram as shown in Fig. 4.11. The state $D$ can be reached by means of processes $CA$, $AB$ and $BD$ in which $CA$ and $BD$ are adiabatic processes and AB is isothermal process. $AB$ is performed in such a way that when it is plotted on $p$-$v$ diagram, area $ACO$ is equal to the area $BDO$, so that the sum of the areas under the curve $CA$, $AB$ and $BD$ is equal to the area under the curve $CD$.

As we know area under $p$-$v$ diagram gives work.

$$\therefore \qquad W_{CD} = W_{CABD}$$

Applying the First Law of Thermodynamics for these processes $CD$ and $CABD$, we can write

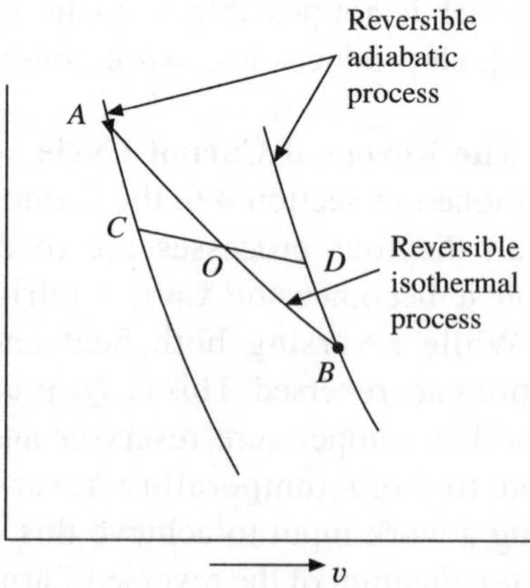

$$Q_{CD} = (U_D - U_C) + W_{CD} \qquad (4.12)$$

$$Q_{CABD} = (U_D - U_C) + W_{CABD}$$

$$= (U_D - U_C) + W_{CD} \quad [\because \quad W_{CD} = W_{CABD}] \qquad (4.13)$$



**Fig. 4.11**

Comparing Eqs. (4.12) and (4.13), we can write

$$Q_{CD} = Q_{CABD}$$

$$= Q_{CA} + Q_{AB} + Q_{BD} \tag{4.14}$$

But heat transfer during two adiabatic processes $CA$ and $BD$ is zero, that is, $Q_{CA} = Q_{BD} = 0$.

$$\therefore \qquad Q_{CD} = Q_{AB} \tag{4.15}$$

The Eq. (4.15) means that, a reversible process can be replaced by two adiabatic processes and one isothermal process if heat transfer during reversible process is equal to heat transfer during isothermal process.

## 4.9   COROLLARIES OF THE SECOND LAW OF THERMODYNAMICS

### 4.9.1   Corollary-1 Carnot's Theorem

No engine can be more efficient than a reversible engine working between the same limits of temperature.

*Proof*: Let us take two engines $I$ and $R$ working between the same heat source and heat sink at temperature $T_H$ and $T_L$ respectively. Engine $I$ is an irreversible engine and $R$, a revessible engine as shown in Fig. 4.12.

Now, if the corollary is not true, let the engine $I$ be more efficient than engine $R$. Let both the engines receive the same amount, of heat $Q_H$ from the heat source. Let $W_I$ and $W_R$ represent the work done by the two engines.

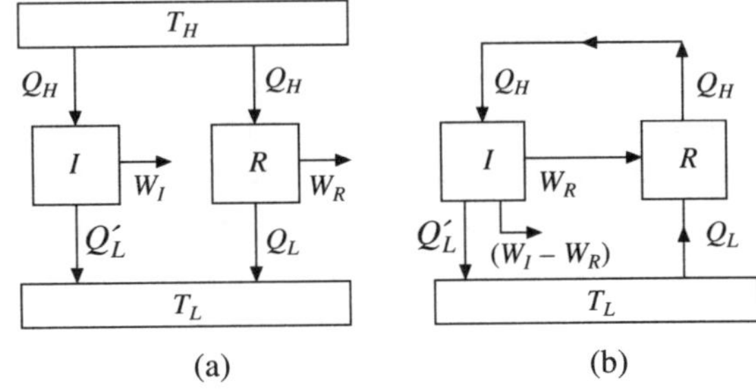


**Fig. 4.12**

Then

$$(\eta_{\text{th}})_I > (\eta_{\text{th}})_R \tag{4.16}$$

or,

$$W_I/Q_H > W_R/Q_H$$

or,

$$W_I > W_R \tag{4.17}$$

The engine $R$ is a reversible engine and it can be reversed. It receives heat $Q_H$ rejects heat $Q_L$ and does work $W_R = Q_H - Q_L$. If the engine is reversed it becomes heat pump. It will take heat $Q_L$ from the sink at temperature $T_L$ and reject heat $Q_H$ to source at $T_H$ while work equal to $W_R$ will have to be added to it.

Now if the engine $I$ and the reversed engine $R$ are connected (Fig. 4.12b) in such a way that the heat rejected by $R$ is passed to $I$ and work required by $R$ is supplied by $I$ (as $W_I > W_R$) the two engines together serve as one engine which transacts heat with only one body (the sink here) and gives a net amount of work equal to $(W_I - W_R)$. So the system produces a net work while exchanging heat with only one reservoir, the heat sink at $T_L$. Thus it forms a perpetual motion machine of the second type. But such a machine is not possible according to the Second Law. Therefore, the very assumption that engine $I$ is more efficient than engine $R$ is wrong.

### 4.9.2   Corollary 2

All reversible engines working between the same limits of temperature have the same efficiency.

The proof is done in a way similar to that of corollary 1. Let there be two reversible engines $R_1$ and $R_2$ working between temperature $T_H$ and $T_L$ and if say $R_1$ is more efficient than $R_2$, then taking the same amount $Q_H$ from the heat source at $T_H$, $R_1$ will produce more work than that of $R_2$ (Fig. 4.13(a)). Now if $R_2$ is reversed and connected to $R_1$, the two together form a system which exchanges heat with only the heat sink at $T_L$ and gives a net amount of work (Fig. 4.13b). This forms a perpetual motion machine of the second type and therefore $R_1$ cannot move efficiently than $R_2$. Similarly $R_2$ cannot move efficiently than $R_1$ and for that matter all reversible engines between $T_H$ and $T_L$ must have the same efficiency.

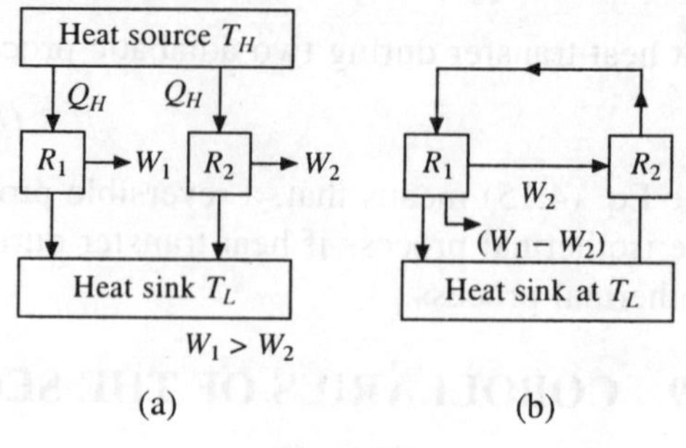


**Fig. 4.13**

### 4.9.3   Corollary 3

It is possible to define a temperature scale which is independent of the thermometic substance. We know different thermometers. They use the variation in some of their material property with temperature. For example a mercury in glass thermometer uses the property of expansion of mercury with temperature. A resistance thermometer uses the variation of resistance of platinum wire with temperature and so on.

From the Second Law of Thermodynamics we know that the efficiency of a reversible engine depends only on the temperatures of the heat source and heat sink and not on the properties of the materials of the reversible engine. Thus if one temperature is known (that of heat source or heat sink) a reversible engine can be used to find the temperature of the second. The reversible engine will therefore, serve as a thermometer and will give a temperature scale which does not depend on any material property.

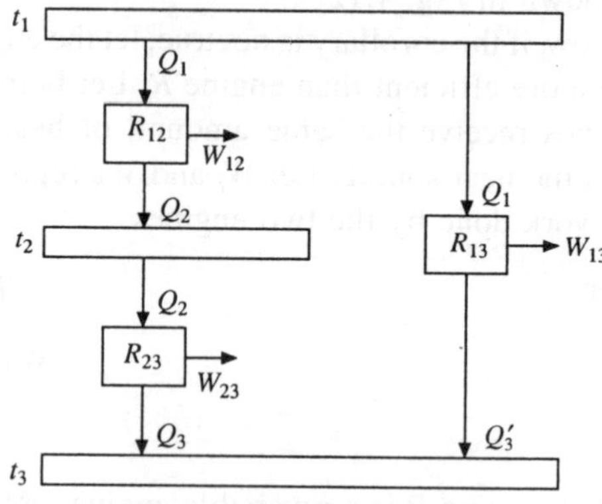


**Fig. 4.14**

Let there be three reversible engines $R_{12}$, $R_{23}$ and $R_{13}$ working respectively between reservoirs at temperatue $t_1$ and $t_2$, $t_2$ and $t_3$ and $t_1$ and $t_3$. Let engine $R_{12}$ receive heat $Q_1$ at temperature $t_1$ and reject heat $Q_2$ at temperature $t_2$. Similarly let engine $R_{23}$ receive heat $Q_2$ at temperature $t_2$ and reject heat $Q_3$ at temperature $t_3$ and engine $R_{13}$ receive heat $Q_1$ at temperature $t_1$ and reject heat $Q_3'$, at temperature $t_3$.

The efficiency of a reversible engine $R_{12}$ will depend only on temperature $t_1$ and $t_2$. So it is a function of temperatures $t_1$ and $t_2$ as shown in Fig. 4.15.

Let $\eta_{12} = f(t_1, t_2)$ where $f$ indicates some function

But
$$\eta_{12} = 1 - Q_2/Q_1 \qquad (4.18)$$

$\therefore$
$$Q_2/Q_1 = \phi(t_1, t_2)$$

where $\phi$ is another function on similar lines, we can write

$$Q_2/Q_3 = \phi(t_2, t_3) \qquad (4.19)$$

and
$$Q_1/Q_3' = \phi(t_1, t_3) \qquad (4.20)$$

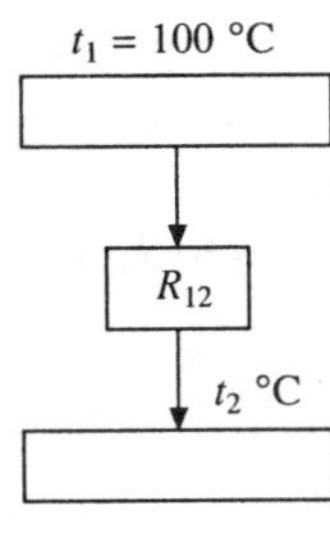


**Fig. 4.15**

Since engines $R_{12}$ and $R_{23}$ together form a reversible engine working between reservoirs 1 and 3, the efficiency of the combination must be equal to the efficiency of engine $R_{13}$. But since the heat taken is the same namely $Q_1$, the heat rejected also must be same, that is, $Q_3 = Q_3'$.

Therefore we have

$$Q_1/Q_3 = \phi(t_1, t_3) \qquad (4.21)$$

Dividing Eq. 4.21 by Eq. 4.19, we get

$$Q_1/Q_2 = \frac{\phi(t_1, t_3)}{\phi(t_2, t_3)} \qquad (4.22)$$

or, from Eqs. 4.22 and 4.18

$$\phi(t_1, t_2) = \frac{\phi(t_1, t_3)}{\phi(t_2, t_3)} \qquad (4.23)$$

Eq. 4.23 can be true only if $t_3$ is cancelled from the right hand side giving

$$\phi(t_1, t_2) = \psi(t_1)/\psi(t_2) \qquad (4.24)$$

where, $\psi$ denotes another function

or,
$$Q_1/Q_2 = \psi(t_1)/\psi(t_2) \qquad (4.25)$$

the form of function will depend on the chosen scale of temperature. Kelvin chose the scale of temperature and denoted it by $T$, thus

$$\psi(t) = T$$

which gives
$$Q_1/Q_2 = T_1/T_2 \qquad (4.26)$$

The efficiency of a reversible engine $R_{12}$ will then become

$$\eta = 1 - T_1/T_2 \qquad (4.27)$$

The temperature scale defined by Kelvin is called Thermodynamic or absolute scale. We know that in centigrade scale two reference temperatures are used, namely, melting point of ice and boiling point of water. The interval between there is 100°C. In case of absolute scale, also the difference between the ice point and the boiling point is taken as 100° but in this case the ice point is not taken as 0 (in case of centigrade scale this temperature has been taken arbitrarily equal to

0). To determine the temperature in absolute scale corresponding to 0°C, the efficiency of a nearly reversible engine working between 100°C and 0°C was found very carefully. The value was obtained as 0.268.

If 0°C corresponds to $T_0$ K, 100°C will correspond to $(100 + T_0)$ K. The efficiency would then be given as

$$0.268 = 1 - T_0/T_{100} \text{ or, } 1 - T_0/(100 + T_0)$$

or,
$$0.268 = (100 + T_0 - T_0)/100 + T_0)$$

$$= 100/(100 + T_0)$$

or,
$$100 + T_0 = 100/0.268 = 373$$

or,
$$T_0 = 273$$

Thus if $T$ is used to denote the temperature in absolute units and $t$ in 0°C units we have

$$T = (t + 273) \tag{4.28}$$

### 4.9.4   Corollary 4 – Clausius Theorem

For any reversible cycle process the cyclic integral of $\delta Q/T$ is equal to 0, that is $\oint_R \dfrac{\delta Q}{T} = 0$.

*Proof:*   Consider that a system is passing through a reversible cyclic process *ABCDA* which is divided into a number of small Carnot cycles by drawing adiabatic and isothermal lines in such a way that the sum of heat transferred during the set of isothermals is equal to the amount of heat transferred in the original reversible cycle (see section 4.8) as shown in Fig. 4.16(a). It is observed from the figure that inside boundaries of the Carnot cycle cancels out, that is, 2–3 is adiabatic expansion while 5–6 is adiabatic compression. So, the process 2–3 is cancelled out with regard to energy transfer by 5–6. The number of Carnot cycles are assumed to be infinitely large so that the isothermals which complete the cycles are approximated to original reversible cycle *ABCDA*. One of the strips as Carnot cycles from Fig. 4.16(a) is shown in Fig. 4.16(b) for ready reference.

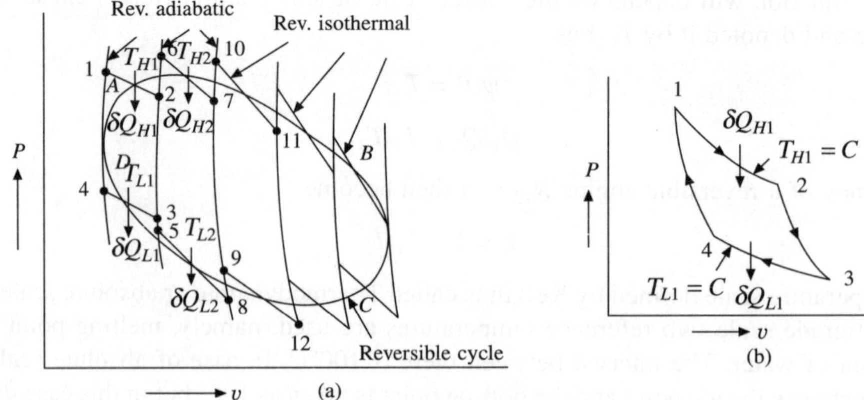


**Fig. 4.16   Clausius theorem**

Carnot cycle 1-2-3-4-1 receives heat $\delta Q_{H_1}$ at temperature $T_{H_1}$ during reversible isothermal process 1–2 and rejects heat $\delta Q_{L_1}$ to $T_{L_1}$ during reversible isothermal process 3–4. Processes 2–3 and 4–1 are reversible adiabatic processes indicating expansion and compression action respectively. Referring to section 4. 9. 3, we can write for the cycle 1–2–3–4–1 as follows:

$$\frac{\delta Q_{H_1}}{T_{H_1}} = \frac{\delta Q_{L_1}}{T_{L_1}} \tag{4.29}$$

Using the sign convention as, heat added ($\delta Q_{H_1}$) is +ve and heat rejected ($\delta Q_{L_1}$) is –ve.

$$\left(\frac{\delta Q_{H_1}}{T_{H_1}}\right) + \left(\frac{\delta Q_{L_1}}{T_{L_1}}\right) = 0 \tag{4.30}$$

Similarly for cycle 5–6–7–8–5

$$\left(\frac{\delta Q_{H_2}}{T_{H_2}}\right) + \left(\frac{\delta Q_{L_2}}{T_{L_2}}\right) = 0$$

If we take all number of cycles then for whole (complete) original cycle we can write

$$\left(\frac{\delta Q_{H_1}}{T_{H_1}} + \frac{\delta Q_{L_1}}{T_{L_1}}\right) + \left(\frac{\delta Q_{H_2}}{T_{H_2}} + \frac{\delta Q_{L_2}}{T_{L_2}}\right) + \ldots = 0 \tag{4.31}$$

or,
$$\oint_R \frac{\delta Q}{T} = 0 \tag{4.32}$$

### 4.9.5  Corollary 5 — Clausius Inequality

If a system executes a complete cyclic process, the cyclic integral of $\delta Q/T$ is less than zero, or in the limit, is equal to 0.

*Proof:*  According to Clausius theorem,

$$\oint_R \frac{\delta Q}{T} = 0$$

According to Carnot's theorem, we know that efficiency of reversible engine is more than that of an irreversible engine under same temperature limit, that is,

$$\eta_R > \eta_I$$

or,
$$\eta_I < \eta_R$$

or,
$$\left[1 - \frac{\delta Q_L}{\delta Q_H}\right]_I < \left[1 - \frac{\delta Q_L}{\delta Q_H}\right]_R \tag{4.33}$$

According to section 4.9.3, for reversible engine

$$\frac{\delta Q_L}{\delta Q_H} = \frac{T_L}{T_H}$$

$$\therefore \qquad \left[1 - \frac{\delta Q_L}{\delta Q_H}\right]_I < \left[1 - \frac{T_L}{T_H}\right] \text{ [For example } 1 - 0.6 < 1 - 0.4] \qquad (4.34)$$

$$\text{or,} \qquad \left(\frac{\delta Q_L}{\delta Q_H}\right)_I > \frac{T_L}{T_H} \quad [\because \ 0.6 > 0.4]$$

$$\text{or,} \qquad \left(\frac{\delta Q_H}{\delta Q_L}\right)_I < \frac{T_H}{T_L} \left[\because \ \frac{1}{0.6} < \frac{1}{0.4}\right]$$

$$\text{or,} \qquad \left(\frac{\delta Q_H}{T_H}\right)_I - \left(\frac{\delta Q_L}{T_L}\right)_I < 0 \qquad (4.35)$$

We know that, heat added ($\delta Q_H$) should be positive and heat rejected ($\delta Q_L$) should be negative.

$$\therefore \qquad \left(\frac{\delta Q_H}{T_H}\right)_I + \left(\frac{\delta Q_L}{T_L}\right)_I < 0 \quad \text{ and so on} \qquad (4.36)$$

Getting sum of all for the complete original irreversible cycle, we get

$$\left[\left(\frac{\delta Q_{H_1}}{T_{H_1}}\right)_I + \left(\frac{\delta Q_{L_1}}{T_{L_1}}\right)_I\right] + \left[\left(\frac{\delta Q_{H_2}}{T_{H_2}}\right)_I + \left(\frac{\delta Q_{L_2}}{T_{L_2}}\right)_I\right] + \ldots < 0$$

$$\text{or,} \qquad \oint \frac{\delta Q}{T} \leq 0 \quad \text{ for an irreversible cycle} \qquad (4.37)$$

From Clausius theorem $\oint_R \dfrac{\delta Q}{T} = 0$ (see Eq. 4.32) for a reversible cycle.

Combining results for reversible and irreversible cycle, we may write

$$\oint \frac{\delta Q}{T} \leq 0 \qquad (4.38)$$

This expression is known as Clausius inequality. It implies whether any cyclic process is reversible or irreversible or impossible.

*Note:*    According to Clausius inequality

(i)   $\oint \dfrac{\delta Q}{T} = 0$ for a reversible cycle

(ii)   $\oint \dfrac{\delta Q}{T} < 0$ for an irreversible cycle

(iii)   $\oint \dfrac{\delta Q}{T} > 0$ for an impossible result

(iv)   All temperatures must be in $k$

(v)   Heat received by the engine is +ve and −ve for heat rejected.

## 4.10    ENTROPY

In corollary 4, Carnot theorem says that the cyclic integral of $\dfrac{\delta Q}{T}$ for a reversible cycle is 0. This facts suggests that the quantity $\dfrac{\delta Q}{T}$ of a reversible clycle is a point function and is, therefore a property of the system. This property is called entropy. At this, property depends upon the mass of the system and hence it is extensive property. Like internal energy the absolute value of entropy of a system is not known. Traditionally entropy of a system at 0°C is assumed to be 0. For a reversible process from state 1 to state 2 the change in entropy of a system is given by

$$\Delta s = s_2 - s_1 = \int_1^2 \frac{\delta Q}{T} \tag{4.39}$$

in differential form

$$ds = \left(\frac{\delta Q}{T}\right)_R \tag{4.40}$$

### 4.10.1    Entropy is a Property

In getting the proof, we have to prove that the change of entropy does not depend upon path but it depends upon end states, that is, change in entropy is a point function. Consider a closed system passing through a reversible cyclic process 1 A2 B1.

Applying Clausius theorem

$$\oint_R \frac{\delta Q}{T} = 0$$

Hence for reversible cyclic process 1A 2B1.

$$\int_{1(A)}^2 \frac{\delta Q}{T} + \int_{2(B)}^1 \frac{\delta Q}{T} = 0 \tag{4.41}$$

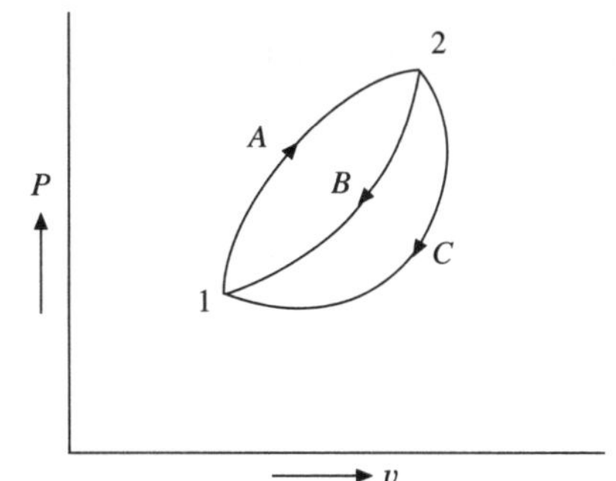

Fig. 4.17    **Entropy is a point funtion**

Now consider another reversible cyclic process 1A2C1.

$$\int_{1(A)}^2 \frac{\delta Q}{T} + \int_{2(C)}^1 \frac{\delta Q}{T} = 0 \tag{4.42}$$

Subtracting Eq. (4.42) from Eq. (4.41), we get

$$\int_{2(B)}^1 \frac{\delta Q}{T} = \int_{2(C)}^1 \frac{\delta Q}{T} = ds$$

or,

$$\int_{2(B)}^1 \frac{\delta Q}{T} = \int_{2(C)}^1 \frac{\delta Q}{T} = ds \tag{4.43}$$

The magnitude of $\dfrac{\delta Q}{T}$ is same for the paths $B$ and $C$ and it does not depend upon the path. So, it depends upon the end states 1 and 2, it is a point function and hence a property of the system.

### 4.10.2 Entropy of an Isolated System Increases or in the Limit Remains Constant

In getting the proof, let us consider a system undergoing a cyclic change 1–2–1. Let the system complete the cycle in two different ways. In one, the cycle consists of two reversible processes $A$ and $B$. In other, it consists of a reversible process $A$ and an irreversible process $C$ (Fig. 4.18).

There are two cycles. The cycle 1A 2B1 is a reversible cycle while the cycle 1a2c1 is irreversible because of the irreversible process 2c1.

We have proved for the reversible cycle 1A2B1.

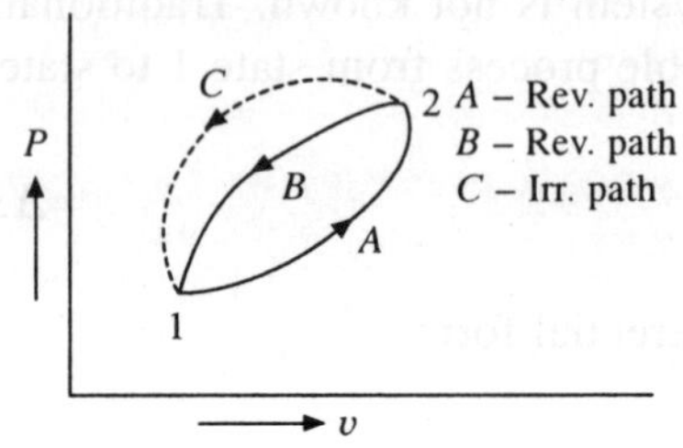


**Fig. 4.18**

$$\oint_R \frac{\delta Q}{T} = \int_{1(A)}^{2} \left( \frac{\delta Q}{T} \right) + \int_{2(B)}^{1} \left( \frac{\delta Q}{T} \right) = 0 \qquad (4.44)$$

Applying inequality of Clausius for irreversible cycle 1A2C1, we can write

$$\oint_I \left( \frac{\delta Q}{T} \right) = \int_{1(A)}^{2} \left( \frac{\delta Q}{T} \right) + \int_{2(C)}^{1} \frac{\delta Q}{T} < 0 \qquad (4.45)$$

Substracting Eq. (4.45) from Eq. (4.44), we get

$$\int_{2(B)}^{1} \left( \frac{\delta Q}{T} \right) - \int_{2(C)}^{1} \left( \frac{\delta Q}{T} \right) \geq 0$$

$$\therefore \qquad \int_{2(B)}^{1} \left( \frac{\delta Q}{T} \right) \geq \int_{2(C)}^{1} \left( \frac{\delta Q}{T} \right)$$

Differentiating the above equation, we can write

$$\left( \frac{\delta Q}{T} \right)_{\text{along } 2B1} \geq \left( \frac{\delta Q}{T} \right)_{\text{along } 2C1}$$

But

$$\left( \frac{\delta Q}{T} \right)_{\text{along } 2B1} = ds \text{ (for reversible process)}$$

$$ds \geq \frac{\delta Q}{T} \qquad (4.46)$$

This relation is commonly known as the principle of increase of entropy. Equality sign holds for a reversible process and the inequality for irreversible process.

Eq. (4.46) can be written as

$$\Delta s = s_2 - s_1 \geq \int_1^2 \left( \frac{\delta Q}{T} \right)$$

where $s_2 - s_1 = \dfrac{\delta Q}{T}$ valid for reversible process and $s_2 - s_1 > \dfrac{\delta Q}{T}$ valid for irreversible process.

Eq. 4.46 holds good when $\delta Q = 0$, or $\delta Q < 0$ or $\delta Q > 0$. It means this equation is valid for isolated system also because according to definition of isolated system there should be no heat and work transfer with the surrounding (environment) and the total energy remains constant, that is, $\delta Q = 0$.

$$\therefore \qquad \Delta s_{\text{isolated}} \geq 0 \qquad\qquad (4.47)$$

Therefore, for an isolated system if the entropy-change remains constant (as $\delta Q = 0$), it must go through reversible process and if entropy-change increases, it would go through irreversible process. This criteria is used to decide whether a process is reversible or irreversible. Eq. (4.47) is stated, therefore, as follows: "change of entropy of an isolated system increases in all real processes (irreversible processes) and is conserved in reversible processes." This statement was given by Planck and considered to be one of the most important statements of the Second Law of Thermodynamics.

Equation (4.47) gives the idea for the principle of an increase of entropy. Following is the principle of increase of entropy, the entropy of the universe increases due to natural processes (irreversible processes), that is,

$$\Delta s_{\text{(universe)}} = \Delta s_{\text{(system)}} + \Delta s_{\text{(surrounding)}} > 0 \qquad\qquad (4.41)$$

For reversible processes

$$\Delta s_{\text{(system)}} = \left( \frac{\delta Q}{T} \right)_R$$

and

$$\Delta s_{\text{(surrounding)}} = -\left( \frac{\delta Q}{T} \right)_R$$

$$\therefore \qquad \Delta s \text{ (universe)} = 0 \text{ for reversible process} \qquad\qquad (4.42)$$

Thus, we obtain from Eqs. (4.41) and (4.42) that

$$\Delta s_{\text{(universe)}} \geq 0 \qquad\qquad (4.43)$$

In nature all real processes are irreversible and therefore Eq. (4.43) can be stated as "Entropy of the universe increases in all irreversible processes and is conserved in reversible processes."

While increasing the entropy of the universe, when it reaches the maximum value then it is called dead state. There is no known mechanism by which entropy can be destroyed.

### 4.10.3 Physical Interpretation of Entropy

Thermodynamics is the study of three $E's$ namely energy, equilibrium and entropy. Like the property internal energy which was derived from the First Law of Thermodynamics, entropy also had been discovered from the Second law of Thermodynamics by Clausius in 1865. This word "entropy" is taken from the Greek word "tropee" which means transformation. This terms has assumed great significance in the energy, technology and ecology.

According to the Second Law of Thermodynamics an engine operating on a cycle can convert only a portion of heat supplied to it into work. This portion of heat is called the available energy. As per our earlier discussion, if a reversible engine operates between a heat source at $T$ and a heat sink at $T_0$ its efficiency is given by $\left(1 - \dfrac{T_0}{T}\right)$. But efficiency is equal to work done divided by heat supplied. Therefore, the maximum work done is $\delta Q\left(1 - \dfrac{T_0}{T}\right)$. In other words,

$$\text{available energy} = \delta Q\left(1 - \frac{T_0}{T}\right) \tag{4.44}$$

and

$$\text{unavailable energy} = \delta Q - \delta Q\left(1 - \frac{T_0}{T}\right)$$

$$= \delta Q \times \frac{T_0}{T}$$

$$= T_0 \times \frac{\delta Q}{T} \tag{4.45}$$

But

$$\left(\frac{\delta Q}{T}\right)_R = ds$$

$\therefore$

$$\text{unavailable energy} = T_0\, ds \tag{4.46}$$

This indicates that increase in entropy is associated with an increase in unavailable energy. Thus entropy may be regarded as the degradation of energy or an index of unavailable of energy.

In a heat engine cycle, the maximum portion of input thermal energy which can be converted to work is called the available energy (AE) or exergy. The portion of thermal energy to be necessarily rejected to surroundings (environment) is called the unavailabe energy (UE) or anergy. The names "Exergy" and "Anergy" were introduced by Rant around 1955 and are shown in Fig. 4.19.

The term exergy comes from the Greek words ex and ergon, meaning from and work: the exergy of a system can be increased if work is done on it. Exergy is the quality of energy which gives useful work from the given energy.

From microscopic considerations the entropy is defined as a measure of disorder or molecular disorder or molecular randomness. As the system becomes more disordered, the positions of the molecules become less predictable and the entropy increases. Thus, the entropy of a substance is the lowest in the solid phase and the highest in the gas phase. The gas molecules possess considerable amount of kinetic energy. As we know due to heating of solid or gas, its molecular disorder will increase and hence entropy will increase. Consider a metal ball resting at the top of an inclined rolling platform as shown in Fig. 4.20. When the ball is allowed to roll on the platform, it will keep on rolling till its whole potential energy is converted into kinetic energy and heat energy

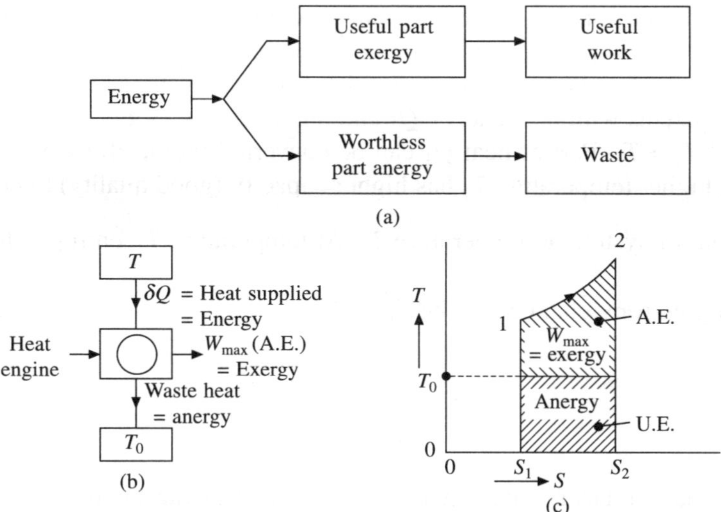

**Fig. 4.19   Concept of available energy, unavailable energy and entropy**

(developed due to friction occuring between ball and platform). The generated heat, received by the molecules of the ball will cause to increase the molecular disorder and therefore, entropy will increase and will reach the minimum value when the ball comes to rest position.

Consider another example for gages. Let two gases $N_2$ and $H_2$ be separated by a barrier in an isolated box that has the same volume and temperature as shown in Fig. 4.21. After removal the barrier, molecules of both gases get more space to move randomly and hence collisions take place between the molecules of the same gas as well as both the gases. After some time equilibrium will be attained.

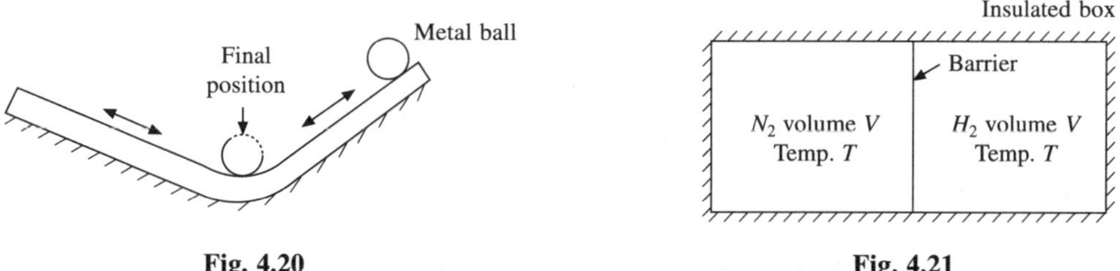

**Fig. 4.20**                    **Fig. 4.21**

After removing the barrier each gas has occupied double the previous volume while the temperature and internal energy are same. But we find that the entropy increases due to increase in randomness. As the  mixing process is irreversible (see section 4.4) and irreversible process is always associated with increase in entropy as per the Second Law, hence entropy increases. When the system comes to equilibrium state after mixing, the distribution of the molecules of each gas in the system is changed. This means, entropy has to do something with the distribution of molecules.

It is to be noted that entropy deals with the quality of energy. Let us consider $E_1 = 100$ kJ of heat energy due to solar radiation of sea water and same quantity of heat $E_2 = 100$ kJ generated by burning of fuel in air at considerably high temperature 1,000°C. Part of heat $E_2$ can be converted into work but not part of heat $E_1$ can be converted into work. This means, available quantity of energy is not important than quality of energy. The quantity of energy in sea water and in the surroundings is abundant but this energy is of no use as its quality is very poor.

Consider two systems with heat energy $Q_1$ and temperatures $T_1$ and $T_2$ respectively, above the surroundings. Let $T_1 > T_2$. Part of heat $Q_1$ can be converted into useful work (mechanical).

The system at higher temperature $T_1$ has higher capacity (good quality) to convert heat energy into work than that of system at temperature $T_2$. At temperature $T_1$, entropy change will be $\dfrac{Q_1}{T_1}$.

At temperature $T_2$, entropy change will be $\dfrac{Q_1}{T_2}$

As
$$T_1 > T_2,$$
$$\frac{Q_1}{T_1} < \frac{Q_1}{T_2}$$

Hence, it follows that for a same given quantity of heat energy the smaller the value of its entropy $\dfrac{Q_1}{T_1}$ then the greater is its capacity to convert into work.

Consider a hot body at temperature $T_1$ which rejects heat $Q$ (Fig. 4.22). Therefore, the loss of entropy for hot body is $\dfrac{Q_1}{T_1}$. Now let this heat $Q$ is received by cold body at a lower temperature $T_2$. Therefore gain of entropy for cold body is $\dfrac{Q}{T_2}$.

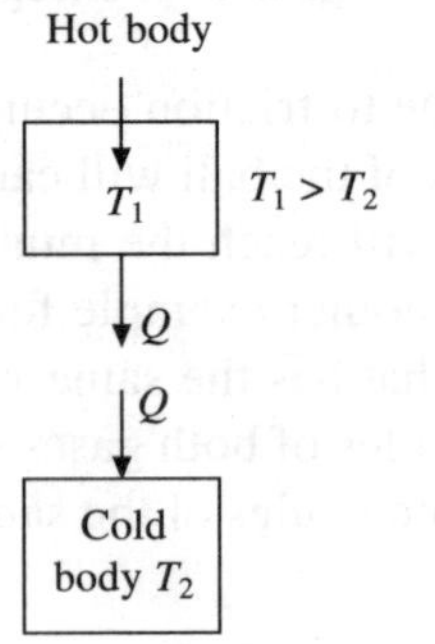

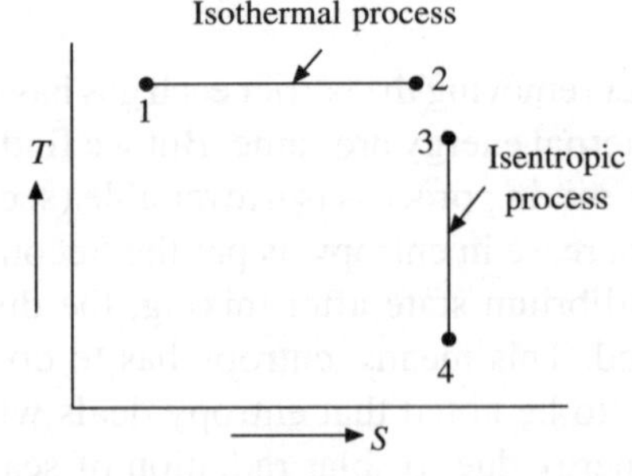


**Fig. 4.22**

As $T_1 > T_2$, the gain of entropy is greater than loss of entropy. Hence as temperature falls, during a heat transfer process entropy increases. Heat transfer from body at high temperature to body at low temperature is a natural process. Hence natural heat transfer process results in increase in entropy.

## 4.11 TEMPERATURE ENTROPY DIAGRAM

Some processes are very well defined on a plot with temperature and entropy as co-ordinates as shown in Fig. 4.23. An isothermal process can be shown on such a plot by a horizontal line as temperature remains constant. Similarly an reversible adiabatic process can be shown by a vertical line as entropy remains constant during this.

We have seen earlier in article 2.16 that in case of a non-flow process shown on $p$-$v$ plot,



**Fig. 4.23**

the work done is found by the area under the curve that represents the process and is given by $\int p\,dv$. Similarly, as $ds = \dfrac{\delta Q}{T}$ or, $\delta Q = T\,ds$, for a reversible process, the heat transferred can be found by the area under the curve (area $a - 1 - 2 - b$) representing the process on $T$-$S$ plane as shown in Fig. 4.24 and given by

$$\int_{1R}^{2} T\,ds = Q_{1-2} \qquad (4.47)$$

Since, in thermodynamics we are dealing with conversion of heat into work or work into heat and therefore, generally $p$-$v$ and $T$-$S$ diagrams are drawn.

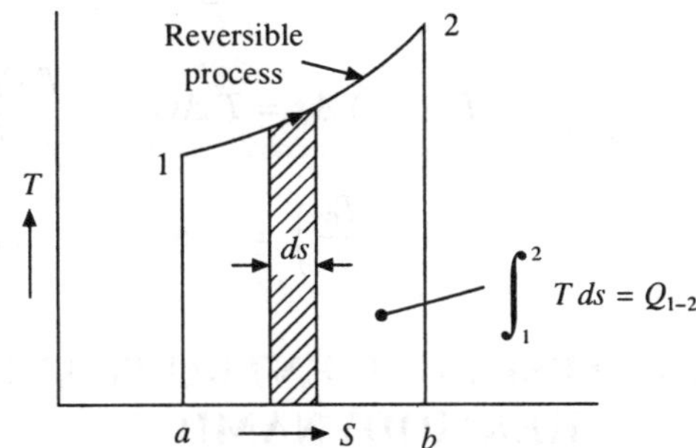


**Fig. 4.24   Reversible process on a $T$–$S$ plane**

## 4.12   CARNOT CYCLE ON $T$-$S$ DIAGRAM AND AVAILABLE ENERGY

The Carnot cycle described in section 4.6 can be represented on a $T$-$S$ diagram more clearly. In Fig. 4.25(a) and 4.25(b) a Carnot cycle has been shown on $p$-$v$ and corresponding $T$-$S$ planes respectively. Processes 1–2 and 3–4 are reversible isothermal and process 2–3 and 4–1 are reversible adiabatic. The heat added during isothermal process 1–2 is given by

$$Q = \int_{1}^{2} \delta Q = \int_{1}^{2} T\,ds = T(s_2 - s_1)$$

$$= T\,\Delta s = \text{area } 1\text{–}2\text{–}b\text{–}a$$

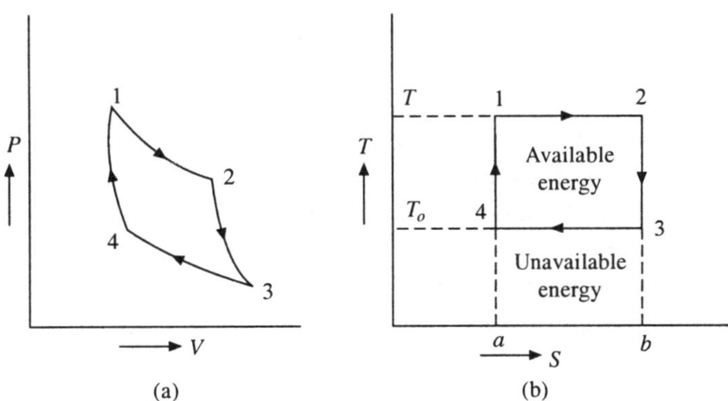


**Fig. 4.25**

Heat rejected during process 3–4

$$= \int_{3}^{4} \delta Q = \int_{3}^{4} T_0\,ds = T_0(s_4 - s_3)$$

$T_0\Delta s = \text{area } 3 - 4 - a - b = $ unavailable energy.

There is no heat addition or rejection during reversible adiabatic process 2–3 or 4–1.
Net heat added = area 1 – 2 – 3 – 4.

$$= \text{work done during the cycle} = \oint \delta Q = \oint \delta W$$

$$= (T - T_0)\,\Delta s = T\,\Delta s \left(1 - \frac{T_0}{T}\right) \tag{4.48}$$

$$= Q\left(1 - \frac{T_0}{T}\right) = \text{Available energy} \tag{4.49}$$

## 4.13   ABSOLUTE ENTROPY AND THE THIRD LAW OF THERMODYNAMICS

The Third Law of Thermodynamics, often referred to as Nernst law, provides the basis for the calculation of absolute entropies of substances. The issential ideas for the law were put forth by WH Nernst (1864–1941) and Max Planck (1858–1947) in 1906 developed from quantum statistical mechanics therories, for which novel won the 1920 Nobel prize in chemistry.

Like enthalpy which has been discussed earlier, we require a common zero point reference state from which to measure the entropies of all the substances. Enthalpy and internal energy have no physically well-defined absolute zero values. Even at absolute zero temperature it can be shown that the enthalpy and internal energy are not generally zero. Entropy, on the other hand, does have an absolute zero point dictated by a state of absolutely perfect molecular order. This is the postulate called the Third Law of Thermodynamics. By making measurements of heat capacities of different crystalline forms of substances, and then calculating entropies at 0 K, Nernst and Planck found that the entropy values were the same. These results led to the Third Law Statement:

"The entropy of all prefect crystalline solid is zero at absolute zero temperature."

The absolute zero temperature, 0 K or 0 R, exists in the thermodynamic temperature scale. This corresponds to –273.15°C and –459.67°F on the Celcius and Fahrenheit scales respectively. In practice, one can never achieve absolute zero temperature as it is impossible but temperatures very close to it have been attained due to perfect insulation.

For unattainability of absolute zero temperature, let us consider a series of reversible engines extending from a source temperature $T_1$ to a lower temperature (Fig. 4.26).

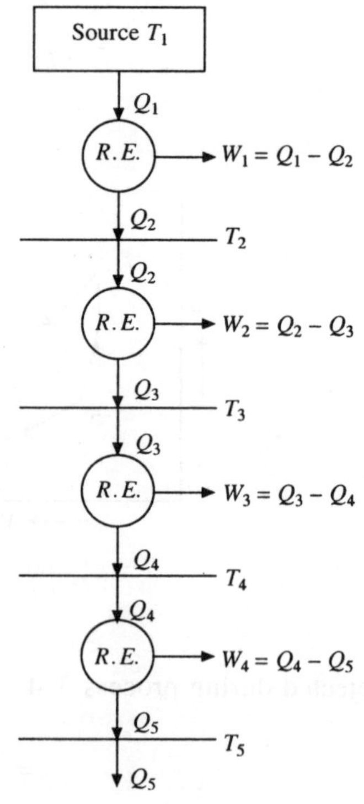


**Fig. 4.26**

Assuming that they are working in series between equal temperature intervals

that is,
$$T_1 - T_2 = T_2 - T_3 = T_3 - T_4 = \ldots$$

and producing equal quantities of work

that is,
$$W_1 = W_2 = W_3 = \ldots$$

According to the First Law of Thermodynamics, total work output of all the engines ($W_1 + W_2 + W_3 + \ldots$) will be equal to $Q_1$. It will be true only when heat rejected from the last engine should be zero. But by the Second Law, however, the operation of a cyclic heat engine with zero heat rejection can not be achieved, although it may be approached as a limit. Zero heat rejection indicates to use sink maintained at zero temperature which is not possible to attain without a violation of the Second Law. It means any value which can be attained will be more than absolute zero temperature. Thus it can be stated according to third law of thermodynamics as follows: It is impossible to reduce any system to absolute zero by any process of idealization. This is called Fowler-Guggenheim statement of the Third Law. This law is not an extension of Second Law but behaves as an independent law.

## 4.14 COMPARISON OF FIRST AND SECOND LAW OF THERMODYNAMICS

| First Law | Second Law |
|---|---|
| 1. According to the First Law heat and work are of same quality indicating 100% efficiency of a cyclic engine. | 1. Work is considered to be a high grade energy whereas heat as a low grade energy. |
| 2. Results in the definition of the extensive propety; internal energy. | 2. Results in the definition of the extensive property; entropy. |
| 3. States that energy of an isolated system can be neither created nor destroyed. | 3. States that entropy of an isolated system can not be destroyed, but it can be created. |
| 4. Energy of the universe is constant. | 4. The entropy of the universe increases towards a maximum. |
| 5. Energy is conserved in every real process. | 5. Energy is degraded in every real process. |
| 6. For the possibility of real processes, no directional implication are given | 6. It is a natural law giving directional possibility for real processes. |
| 7. In symbols<br><br>(a) cyclic process $\oint \delta Q = \oint dw$<br>(b) non-cyclic process $Q - W = \Delta E$ | 7. $\oint \dfrac{\delta Q}{T} \leq 0$ |
| 8. Concludes that the construction of PMM I is not possible | 8. Concludes that the construction of PMM II is not possible. |

Note: Change of entropy during different thermodynamic processes has been given in the next chapter.

## Solved Problems

**4.1**  A freezer is to be maintained at a temperature of 238 k when the ambient temperature is 306 k. In order to maintain the freezer box at 238 k it is necessary to remove heat from it at the rate of 2,460 J/s. What is the maximum possible co-efficient of performance of the freezer and what is the minimum power that must be supplied to the freezer?

*Soln.*  The maximum co-efficient of performance is obtained when the freezer operates reversibly

Max. *COP* as a refrigerator

$$= Q_2/Q_1 - Q_2 = T_2/T_1 - T_2 = 238/306 - 238 = 3.5 \quad Ans.$$

But   $Q_2 = 2460$ J/s and $Q_1 - Q_2 = W$

$$2460/W = 3.5 \text{ or, } W = 2460/3.5 = 702.8 \text{ J/s}$$

$$= 702.8 \ W \quad Ans.$$

**4.2**  An inventor claims to have developed a cyclic engine which exchanges heat with reservoirs at 130°C and –40°C. It receives only 2,100 kJ/min of heat and develops 17.66 kW. Is his claim feasible?

*Soln.*  The maximum efficiency of a heat engine can be equal to that of a Carnot engine

$$\eta_{max} = T_1 - T_2/T_1$$

$$= \frac{(273 + 130) - (273 - 40)}{(273 + 130)}$$

$$= 433 - 239/403 = 0.422 \text{ or, } 42.2\%$$

The claimed efficiency = work output/heat input

$$= \frac{17.66}{2100/60} = 0.5045$$

$$= 50.45\%$$

The claimed efficiency is even greater than the Carnot engine efficiency which is not possible. Therefore the claim is not feasible.

**4.3**  A heat engine works between hot and cold reservoirs at 556 k and 278 k. The engine receives 278 kJ/s of heat. The following results were reported.

(i)   70 kJ/s are rejected
(ii)   139 kJ/s are rejected
(iii)   208 kJ/s are rejected

Indicate which of the results show a reversible cycle, irreversible cycle or impossible cycle.

We know that $\oint dQ/T = 0$ for a reversible cycle

$\oint dQ/T < 0$ for an irreversible cycle and $\oint dQ/T > 0$ is impossible.

Now taking the three cases we have

Since
$$\oint dQ/T = \int_{\substack{heat \\ add.}} dQ/T + \int_{\substack{heat \\ reject.}} dQ/T$$

$$\therefore \qquad \text{for case (i)} \quad \oint dQ/T = 278/556 - 70/278 = 0.248$$

This is positive so case (i) is impossible

Case (ii) $$\oint dQ/T = 278/556 - 139/278 = 0$$

$\therefore$ The cycle is reversible

Case (iii) $$\oint dQ/T = 278/556 - 208/278 = -0.248$$

This is negative

$\therefore$ The cycle is irreversible

**4.4** A cyclic process abcd is completed by a system as follows.

(i) Constant pressure process *ab* during which 252 kJ of heat is added and 48.3 kJ of work is done by the system
(ii) Adiabatic expansion process be during which 78.75 kJ of work is delivered by the system.
(iii) Constant volume process *cd* during which 145.95 kJ of heat is rejected and (ii) adiabatic process *d a* which brings back the system to its original state *a*. Determine the thermal efficiency of the cycle and the work done during *d a*.

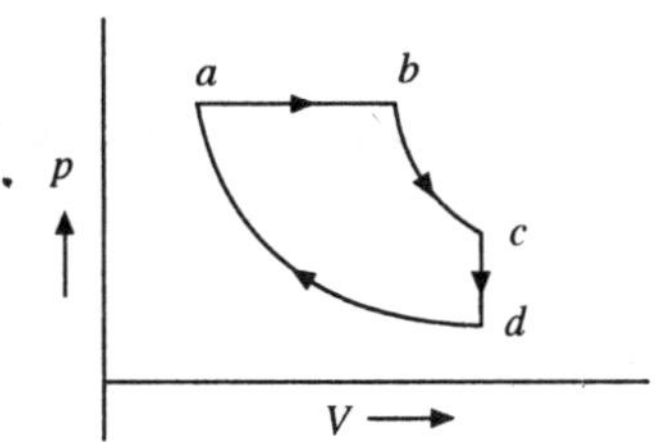


**Fig. 4.27**

*Soln.* Refer to Fig. 4.27.
For process *a-b*

$$Q = 252 \text{ kJ and } W = 48.3 \text{ kJ}$$

Applying first law to this process

$$252 - 48.3 = U_b - U_a = 203.7 \text{ kJ}$$

For process *b-c*,

$$Q = 0, \ W = 78.75 \text{ kJ}, \ U_c - U_b = -78.75 \text{ kJ}$$

For process *c-d*,

$$Q = 145.95 \text{ kJ}, \ W = 0$$

$$U_d - U_a = -145.95 \text{ kJ}$$

Now as *abcd* is a cyclic process

$$U_b - U_a + U_c - U_b + U_d - U_c + U_a - U_d = 0$$

Therefore,

$$203.7 - 78.75 - 145.95 + U_a - U_d = 0$$

or,$\qquad\qquad\qquad U_a - U_d = 78.75 + 145.95 - 203.7 = 21 \text{ kJ}$

$\therefore \qquad\qquad\qquad 0 - W_{da} = U_a - U_d$ or, $W = -21$ kJ.  *Ans.*

Heat added during the heating process $ab = 252$ kJ

Net work done during the cycle

$$= \text{Heat supplied} - \text{heat rejected}$$

$$= 252 - 145.95 = 106.05 \text{ kJ}$$

Thermal efficiency $= 106.05/252$

$$= 0.4208 = 42.08\% \quad Ans.$$

**4.5**  A reversible engine is supplied with heat from two constant temperature sources at 627°C and 327°C and rejects energy to a constant temperature reservoir at 27°C. Assuming that the engine executes a number of complete cycle while developing 42 kW and rejecting 1680 kJ of heat per minute, calculate the heat supplied by each source and the efficiency of the engine.

*Soln.*  Refer to Fig. 4.28
The arrangement of the system is shown in the figure $W.D$/min by engine $= 42 \times 60 = 2,520$ kJ/min

Total heat supplied $= WD +$ heat rejected

$$= 2,520 + 1,680$$

$$= 4,200 \text{ kJ/min}$$

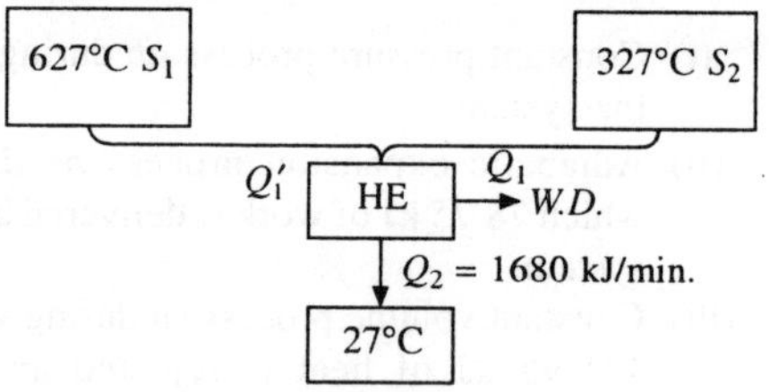

627°C $S_1$    327°C $S_2$

$Q_1'$   HE   $Q_1$   W.D.

$Q_2 = 1680$ kJ/min.

27°C

**Fig. 4.28**

Now, let $Q_1'$ unit of heat be supplied by source $S_1$ and $4,200 - Q_1'$ by source $S_2$. As the processes are reversible the total entropy change must be zero

or,$\qquad \{Q_1'/(627 + 273)\} + (4200 - Q_1')/(327 + 273) - 1680/(27 + 273) = 0$

or,$\qquad\qquad Q_1'/900 + (4200 - Q_1')/600 - 1680/300 = 0$

$$\frac{2Q_1' + 12600 - 3Q_1' - 10080}{1800} = 0$$

or,$\qquad\qquad\qquad -Q_1' = -2520$

or,$\qquad\qquad\qquad Q_1' = 2520 \text{ kJ/min}$

Heat supplied by source $S_2$

$$= 4200 - 2520 = 1680 \text{ kJ/min}$$

Efficiency,$\qquad\qquad \eta = 2520/4200 = 0.60$ or, 60%  *Ans.*

**4.6.**  Water in a constant pressure container is agitated by a paddle unit, the temperature rises from 60°C to 100°C, compute the change in entropy.

*Soln.*  The padde agitation is an irreversible process. The change in entropy will, however, depend only on the end states and not on the process. Therefore, the change in entropy will be the same in the irreversible process as would have been during a reversible heating from 60°C to 100°C.

$\therefore$   Here   $\Delta s$ irr, $= \Delta s$ rev $= \int \delta Q/T = \int m\,cp\,\delta T/T$

or,   $s_2 - s_1 = m\,cp \int_1^2 dT/T = m\,cp \ln T_2/T_1$

For 1 kg of water, $s_2 - s_1 = 1 \times 4.187 \times \ln (273 + 100)/(273 + 60)$

$$= 4.187 \ln 373/333 = 0.474 \text{ kJ/kg k}\quad Ans.$$

**4.7** Two Carnot engines combined in series operate between temperatures of 906 k and 586 k. What should be the intermediate temperatures so that both the engines produce equal work.

*Soln.*  Refer to Fig. 4.29.

Denoting the given temperatures by $T_1$ and $T_3$ and the intermediate temperature by $T_2$, the efficiencies of the engines will be

$$\eta_1 = W/Q_1 = (Q_1 - Q_2)/Q_1$$

$$= (T_1 - T_2)/T_1$$

$$\eta_1 = \frac{W}{Q_1} = \frac{W}{W + Q_2} = (T_1 - T_2)/T_1 \quad (1)$$

and  $\eta_2 = W/Q_2 = (Q_2 - Q_3)/Q_2$

$$= (T_2 - T_3)/T_2 \qquad (2)$$

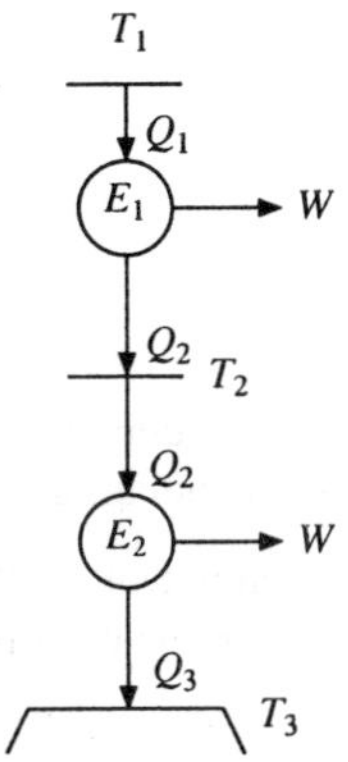

Now from equation (1)

$$W = Q_2(T_1 - T_2)/T_2 \qquad (3)$$

and from equation (2)

$$W = Q_2(T_2 - T_3)/T_3 \qquad (4)$$

Again from 3 and 4 we get

$$Q_2(T_1 - T_2)/T_2 = Q_2(T_2 - T_3)/T_3$$

or,   $T_1 T_3 - T_2 T_3 = T_2^2 - T_2 T_3$

or,   $T_2 = \sqrt{T_1 T_3}$

The intermediate temperature

$$T_2 = \sqrt{906 \times 586} = 728.63 \text{ k} \qquad \text{Ans.}$$

**Fig. 4.29**

**4.8.** A heat engine operates between a limited source and a limited sink of masses $m$ and specific heat $cp$. If the temperatures of the source and sink be denoted by $T_1$ and $T_2$ respectively, find the maximum work that can be produced by the engine.

*Soln.*  The source and sink are limited, so as the engine draws heat from the source and rejects heat to the sink, the temperature $T_1$ of the source will go on falling and that of sink will go on rising till they attain a common value given by say $T_i$.

The engine will then stop. For maximum work the engine must be a reversible one and then $\oint \delta Q/T = 0$

or,   $\delta Q/T_1 + \delta Q_2/T_2 = 0$

$$\text{or,} \qquad mcp \int_{T_1}^{T_i} dT_1 / T_1 + mcp \int_{T_2}^{T_i} dT_2 / T_2 = 0$$

$$\text{or,} \qquad m\,cp \ln T_i / T_1 + m\,cp \ln T_i / T_2 = 0$$

$$\text{or,} \qquad \ln T_i/T_1 + \ln T_i/T_2 = 0$$

$$\text{or,} \qquad \ln (T_i^2 / T_1 T_2) = 0 = \ln 1$$

$$\text{or,} \qquad T_i^2 = T_1 T_2 \quad \text{or,} \quad T_i = \sqrt{T_1 T_2}$$

$$W = Q_1 - Q_2$$

$$= m\,cp\,(T_1 - T_i) - mcp\,(T_i - T_2)$$

$$= m\,cp\,(T_1 + T_2 - 2T_i)$$

$$= m\,cp\,(T_1 + T_2 - 2\sqrt{T_1 T_2}\,)$$

$$= m\,cp\,(\sqrt{T_1} - \sqrt{T_2}\,)^2 \qquad Ans.$$

**4.9.** A gear box of an automobile receives 147 kW and delivers 140 kW when operating in a steady state at 100°C. The surroundings are at 10°C. Find the entropy increase of the gear box and also of the surroundings.

*Soln.* There is a difference of 7 kW in the input and output of the gear box which will be converted into heat. This will increase the entropy of the gear box.

$$\text{Rate of heat generation} = 7 \text{ kW} = 7 \text{ kJ/sec}$$

Increase in entropy = 7/(273 + 100) = 0.0187 kJ/k sec. The heat will be lost to the surrounding at 10°C. Increase of entropy of the surroundings.

$$= 7/(273 + 10) = 7/283 = 0.0247 \text{ kJ/k sec.} \quad Ans.$$

**4.10.** Two kg of ice at 0°C is mixed with 10 kg of water at 26.5°C. The system is open to atmosphere which is at 1 bar pressure. Find the change of entropy for adiabatic mixing of the two phases of water. sp. heat of water = 4.187 kJ/kg k and latent heat of fusion = 336 kJ/kg.

*Soln.* Let $t$ be the final temperature of the mixture. Then heat gained = heat lost

$$\text{or,} \qquad 2 \times 336 + 2 \times 4.187 \, (t - 0) = 10 \times 4.187 \, (26.5 - t)$$

$$\text{or,} \qquad 8.374t + 672 = 1109.55 - 41.87t$$

$$\text{or,} \qquad 50.244t = 437.55 \text{ or, } t = 8.70°C$$

Now, change of entropy of ice at 0°C to water at 8.70°C

$$= 2 \times 336/(0 + 273) + 2 \times 4.187 \ln [8.70 + 273)/(0 + 273)]$$

$$= 2.461 + 0.262 = 2.723 \text{ kJ/k}$$

Change in entropy of water from 26.5°C to 8.70°C

$$= m\,cp \ln [(8.70 + 273)/(26.5 + 273)]$$

$$= 10 \times 4.187 \times \ln (281.73/299.5)$$

$$= -2.560 \text{ kJ/k}$$

Total change of entropy of the whole system = (2.748) + (–2.560)

$$= 0.188 \text{ kJ/k} \quad Ans.$$

**4.11.** The compressor of a refrigerator is run by a 2.205 kW electric motor. The refrigerator works between 1°C and 45°C. If the amount of heat absorbed by the refrigerator at low temperaure is 1,800 k cal/hr, find the amount of heat rejected per hour and the irreversibility in k cal/hr.

*Soln.* 2.205 KW motor run for 1 hour will give $W = 2.205 \times 860 = 1896$ k cal/hr. [Q  1 KW = 860 k cal/hr]

$$Q_2 = 1800 \text{ k cal/hr}$$

Heat rejected at higher temperature

$$Q_1 = 1896 + 1800 = 3696 \text{ k cal/hr} \quad Ans.$$

In the irreversible cycle. If $I$ be the irreversibility, we have $dQ/T + I = 0$

$$Q_1/T_1 + Q_2/T_2 + I = 0$$

or $\qquad I = -Q_2/T_2 + Q_1/T_1 = 3696/(45 + 273) - 1800/(1 + 273)$

$$= 3696/318 - 1800/274 = 5.050 \text{ k cal/hr} \quad Ans.$$

**4.12.** A domestic food freezer maintains a temperature of – 15°C. The ambient air temperature is 30°C. If heat leaks into the freezer at the continuous rate of 1.75 kJ/s, what is the minimum power required to pump this heat out continuously? [Nov. 98, May 99, Mumbai univ.]

*Soln.* The refrigerator cycle removes heat from the freezer at the same rate at which heat leaks into it (Refer to Fig. 4.30).
For minimum power requirement

$$\frac{Q_2}{T_2} = \frac{Q_1}{T_1}$$

$\therefore \qquad Q_1 = Q_2 \dfrac{T_1}{T_2}$

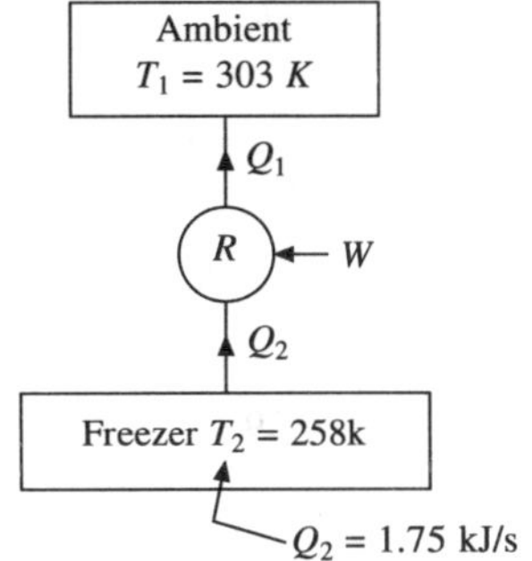


**Fig. 4.30**

$$= 1.75 \times \frac{303}{258} = 2.055 \text{ kJ/s}$$

$\therefore \qquad W = (Q_1 - Q_2)$

$$= (2.055 - 1.75) = 0.305 \text{ kJ/s}$$

$$= 0.305 \text{ kW} \quad Ans.$$

$$\text{C.O.P.} = \frac{Q_2}{W} = \frac{1.75}{0.305} = 5.73 \quad Ans.$$

**4.13.** An ice plant working on the principle of Carnot refrigerator produces 20 tonnes of ice per day. The ice is formed from water at 0°C and maintained at 0°C. Heat is rejected to atmosphere at 27°C. The ice plant is run by a carnot engine which absorbs heat from a source which is maintained at 227°C by burning fuel of $CV = 21,000$ kJ/kg. Find the power developed by the engine and the fuel consumed per hour. Latent heat of fusion = 336 kJ/kg. [Nov, 99, Mumbai Univ.]

*Soln.* Refer to Fig. 4.31.

$$\text{C.O.P.}_{(C.R)} = \frac{0 + 273}{(27 + 273) - (0 + 273)} = 10.11$$

Refrigerating effect (RE)

$$= \frac{336 \times 20 \times 10^3}{24 \times 60 \times 60} = 77.78 \text{ kW}$$

$$\therefore \quad \text{Power required } W = \frac{\text{RE}}{\text{COP}} = \frac{77.78}{10.11}$$

$$= 7.69 \text{ kW}$$

$$\eta_{\text{C.E.}} = 1 - \frac{(27 + 273)}{(227 + 273)} = 0.40$$

or, $\quad 0.40 = \dfrac{W}{Q} = \dfrac{7.69}{Q}$

$$\therefore \quad Q = 19.225 \text{ kW} = \frac{\text{kJ}}{\text{s}}$$

$$Q = m \times CV$$

$$19.225 = m \times 21000 \quad \therefore \ m = 9.15 \times 10^{-4} \text{ kg/s} = 3.29 \text{ kg/hr} \quad \textit{Ans.}$$

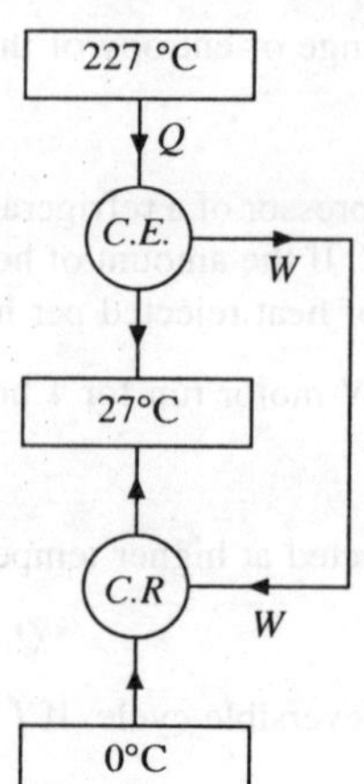


**Fig. 4.31**

**4.14.** A reversible heat engine works between 150°C and 27°C. The work output of this engine is supplied to a reversible refrigerator working between 0°C and 27°C. Find the energy removed as heat from the cold body by refrigerator? Also find heat rejected to the surrounding at 27°C by both the units. Assume 100 kJ heat is supplied to the engine from the reservoir at 150°C    [Nov. 95, Mumbai Univ.]

*Soln.* Refer to Fig. 4.32

$$\eta_E = \frac{W}{Q_1} = 1 - \frac{Q_2}{Q_1}$$

$$= 1 - \frac{(27 + 273)}{(150 + 273)} = 0.29$$

or, $\quad \dfrac{W}{Q_1} = 0.29$

$\therefore \qquad W = 29 \text{ kJ}$

$\because \qquad Q_1 = W + Q_2$

or, $\qquad Q_2 = Q_1 - W$

$$= 100 - 29$$

$$= 71 \text{ kJ}$$

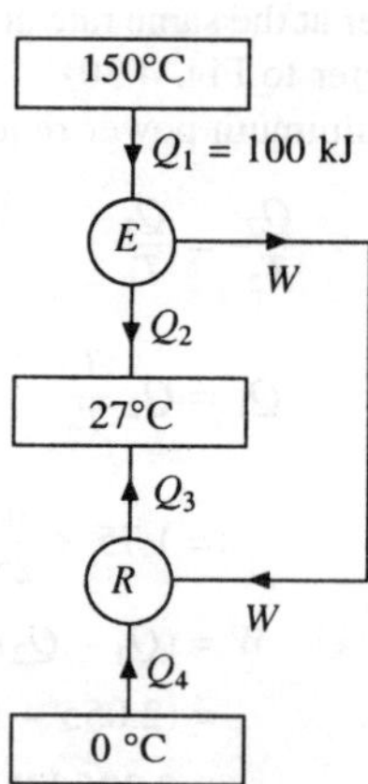


**Fig. 3.1   The Joule's experiment**

$$\text{C.O.P}_{(R)} \text{ i.e. } \xi_R = \frac{0 + 273}{(27 + 273) - (0 + 273)}$$

$$= 10.11 = \frac{Q_4}{W}$$

$$\therefore \qquad Q_4 = 10.11 \times W = 10.11 \times 29 = 293.19 \text{ kJ}$$

$\therefore \quad$ Heat removed from cold body = 293.19 kJ    *Ans.*

Now, $\qquad W + Q_3 = Q_4 \quad \therefore \ Q_3 = Q_4 - W$

$$= 293.19 - 29 = 264.19 \text{ kJ}$$

$$\therefore \qquad Q_2 + Q_3 = (71 + 264.19) = 335.19 \text{ kJ}$$

$\therefore$ Heat rejected to the surrounding at

$$27°C = 335.19 \text{ kJ} \quad Ans.$$

**4.15.** Three Carnot engines $E_1$, $E_2$, $E_3$ operate in series between two heat reservoirs which are at temperatures of 1,000 k and 300 k. Calculate intermediate temperatures if amount of work produced by these engines is in the proportions of 5:4:3.

*Soln.* Refer to Fig. 4.33.
From carnot theorem,

we have $\qquad \dfrac{Q_1 - Q_2}{Q_1} = \dfrac{W_1}{Q_1} = \dfrac{T_1 - T_2}{T_1}$ $\qquad$ (1)

and $\qquad \dfrac{Q_2 - Q_3}{Q_2} = \dfrac{W_2}{Q_2} = \dfrac{T_2 - T_3}{T_2}$ $\qquad$ (2)

From Eqs. (1) and (2), we get

$$\frac{W_1}{W_2} = \frac{(T_1 - T_2)}{(T_2 - T_3)} \times \frac{Q_1}{Q_2} \times \frac{T_2}{T_1}$$

$$= \frac{T_1 - T_2}{T_2 - T_3} \left[ \text{as } \frac{Q_1}{T_1} = \frac{Q_2}{T_2} \right]$$

So, for Carnot engines operating in series, the temperature difference between the reservoirs, is proportional to the work ratios.

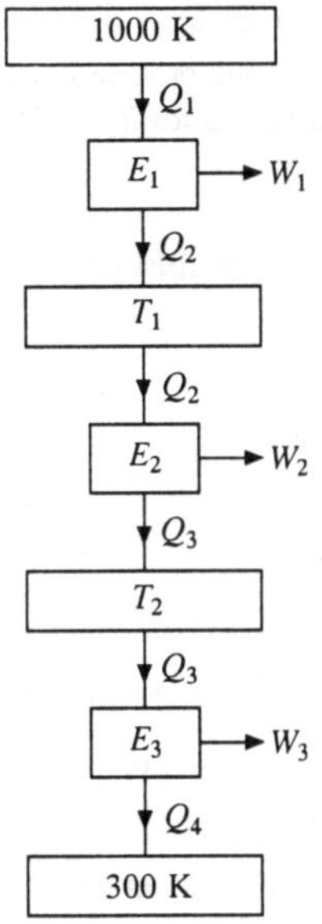

$$\therefore \qquad 1000 - T_1 : T_1 - T_2 : T_2 - 300 = 5 : 4 : 3$$

$$\therefore \qquad \frac{1000 - T_1}{T_1 - T_2} = \frac{5}{4} \qquad (3)$$

and $$\frac{T_1 - T_2}{T_2 - 300} = \frac{4}{3} \qquad (4)$$

From Eq. (3), we get

$$4000 - 4T_1 = 5T_1 - 5T_2$$

or, $$4000 - 9T_1 = -5T_2 \qquad (5)$$

Similarly from Eq. (4)

$$3T_1 - 3T_2 = 4T_2 - 1200$$

or, $$3T_1 = 7T_2 - 1200$$

or, $$9T_1 = 21T_2 - 3600 \qquad (6)$$

Adding Eq. (5) and Eq. (6), we get

$$4000 - 9T_1 + 9T_1 = -5T_2 + 21T_2 - 3600$$

$$\therefore \qquad 16T_2 = 7600$$

$$T_2 = 475 \text{ k} \quad Ans.$$



**Fig. 4.33**

Substuting $T_2 = 475$ in Eq. (5)

$$\therefore \qquad 4000 - 9T_1 = (-5) \times 475$$

or

$$9T_1 = 4000 + 2375$$

$$\therefore \qquad T_1 = 708.333 \text{ k} \quad Ans.$$

**4.16.** (a)  In a certain reversible process the rate of heat transfer to the system per unit temperature rise is given by $dQ/dT = 1.05$ kJ/K. Find the increase in entropy of the system if its temperature rises from 300°K to 400°K. Also find the change in specific entropy if mass of system is 2 kg.

(b)  In a second process between the same end states the temperature rise is obtained by stirring accompained by a heat addition half as great as in case (a). Find the increase in entropy in the case.

*Soln.* $\qquad \delta Q/dT = 1.05$, or, $\delta Q = 1.05 \, dT$

(a)  for reversible process,

$$ds = 1.05 \, dT/T$$

or,

$$s = 1.05 \int dT/T = 1.05 \ln T_2/T_1$$

$$= 1.05 \ln 400/300$$

or, $\qquad 1.05 \times 0.2876 = 0.3019$ kJ/k  *Ans.*

Change in specific entropy $= 0.3019/2 = 0.15095$ kJ/k kg  *Ans.*

(b)  Between same end states the change of entropy will not depend on the process. Therefore in this case also the increase of entropy is the same, namely, 0.3019 kJ/k  *Ans.*

**4.17.**  A Carnot engine operates between the temperature limits of 946 k and 622 k. If 210 kJ are absorbed from the source at 946 k, calculate:

(a)  The work performed or available energy
(b)  The heat rejected or unavailable energy and the change of entropy of the universse.

*Soln.*  (a) carnot efficiency $= (T_1 - T_2)/T_1 = (946 - 622)/946 = 0.343$

But $\eta =$ work output/heat input $= W/Q_1$

Work output $= \eta \times$ heat input $= 0.343 \times 210 = 72.03$ kJ  *Ans.*

(b)  $\qquad W = Q_1 - Q_2$ or, $Q_2 = Q_1 - W = 210 - 72.03 = 137.97$ kJ  *Ans.*

(c)  Entropy change will take place during isothermal expansion and compression.

$$\Delta s_{\text{isothermal expn.}} = \Delta s_{\text{system}} + \Delta s_{\text{surrounding}}$$

$$= (210/946) + (-210/946) = 0$$

$$\Delta s_{\text{isothermal comp.}} = \Delta s_{\text{system}} + \Delta s_{\text{surrounded}}$$

$$= - 137.97/622 + 137.97/622 = 0$$

$$s_{\text{universe}} = 0 + 0 = 0 \quad Ans.$$

**4.18.**  One kg of a certain gas is expanded reversibly and adiabatically from its initial volume of 0.0624 m$^3$ and initial temperature of 555 k to a final volume of 0.1872 m$^3$ and final temperature of 388 k. What is the change of entropy in the process?

In a second case the same gas is expanded from the same initial state into an evacuated space to same final volume of 0.1872 m$^3$ but to a final temperature of 540k. Find the change of entropy in the second case

take $$Cv = 1.26 \text{ kJ/kg k}$$

*Soln.* Case (i) In this case as the process is adiabatic there is no heat transfer and, therefore, $\Delta s = \int \delta Q/T = 0$

Case (ii) This process is irreversible and to we cannot find the change of entropy directly. But the change of value of a property depends only on the end states. Therefore we can find the value of $\Delta s$ between the same end states but along reversible paths.

For example, we can first expand the gas adiabatically and reversibility to the final volume of 0.1872 m$^3$ and temperature 388 k (state 2). From then the gas may go along a constant volume process from 388 k to 540 k as shown in figure (2′ to 2). Refer to Fig. 4.34.

$\therefore \quad \Delta s$ (irrev) $(1 - 2) = \Delta s$ (rev $(1 - 2')$) + $\Delta s$ (rev $(2' -2)$)

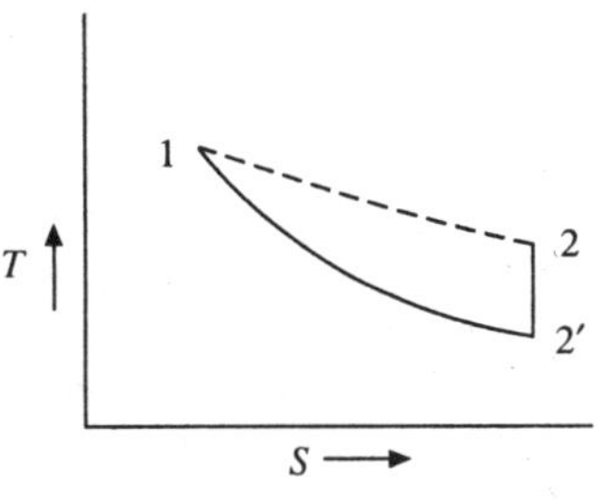

$$= \int_{388}^{540} m\, cv\, dT/T + 0$$

$$= 1 \times 1.26 \ln dT/T$$

$$= 1.26 \ln (540/388)$$

$$= 0.4165 \text{ kJ}$$

$\therefore \qquad \Delta s_{\text{irriv}} = 0.4165 \text{ kJ} \quad Ans.$

**Fig. 4.34**

# EXERCISES

## Objective Questions

4.1. State True or False

(a) In an adiabatic process either reveresible or irreversible change of entropy must be zero because $dQ = 0$ — *F*.

(b) Entropy of an isolated system must decrease — *F*

(c) If a system undergoes an irreversible change from a initial equilibrium state 1 to a final equilibrium statê *f*, the entropy change of the surroundings must be less (algebraically) than it would be if the system changed for *i* to *f* reversibly — *F*.

(d) If a system undergoes a process under which it s entropy does not change, it is necessarily true that the process is reversible and adiabatic. — *F*

(e) Heat is always given by the integral $\int T ds$. — *F*

(f) For any process the Second law of Modynamics requires that the entropy change of the system be zero or +ve —*F*

(g) The second law says that all heat can not be converted into work. For a cyclic process the First Law says that work is equal to heat. First law violates second law — *F*

(h) A system is changed from a given initial equilibrium state to the same final equilibrium state by two different processes, one reversible and one irreversible, than the following is true

$$\Delta s_{\text{irr}} = \Delta s_{\text{rev}} \text{ --- } T$$

(i) The cyclic integral of $dQ/T$ for a system exchang heat end work with its surrounding cannot be +ve — *T*

(j)  No engine can be as efficient as a heat engine working between the same limits of temperature — $F$.

(k)  Entropy change between two end states can be calculated only if the process is reversible. — $F$

(l)  Heat can never flow from lower to a higher temperature — $F$

(m)  Cop of a heat pump is always greater than one — $T$.

4.2.  Calculate the minimum work required to manufacture 5 kg of ice from water initially at 273.15 k. Assume that the surroundings are at 298 k. The latent heat of fusion of water at 273.15 k is 338.7 kJ/kg. (154 kJ)

4.3.  A modern nuclear power plant generates 750,000 kW for which the reactor temperature is 586 k. Necessary heat rejection is done to a river with a water temperature of 293 k. (a) what is the maximum possible thermal efficiency of the plant and what is the minimum amount of heat that must be discarded to the river? (b) If the actual thermal efficiency of the plant is 60% of the maximum how much heat must be discarded into the river, and what will be the temperature rise of the river if it has a flow rate of 165 m³/ s? (50%, 750,000 kW, 1750, 000 kW 2.54 k

4.4.  A Carnot engine rejects heat to a cooling pond at 27°C. The heat rejected to the pond is 840 kW/min. If the efficiency of the engine is 30%. Find the power of the engine and temperature of the engine.

(6 kW, 155.5°C)

4.5.  An engineer claims to have developed a refrigerator which takes in 10500 kJ of heat at a temperature of –23°C and rejects 12474 kJ to the atmosphere at 25°C. It receives 1965, k Joules of work for its working. Is his claim valid. Find the maximum possible *cop.* of the refrigerator and the claimed CoP. Is the claim valid? (5.319, 5.208, No)

4.6.  A non-flow reversible process is shown in Fig. 4.35 a $T$-$S$ plane by a straight line. The initial temperature is 473 k and initial entropy 0.945 kJ/k. If 559.65 kJ of heat is rejected by the system during the process and the final temperature is 373 k, find the final entropy. (2.268 kJ/k)

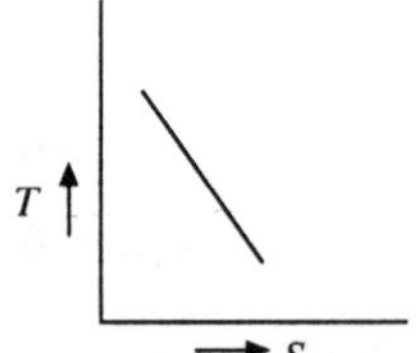

**Fig. 4.35**

4.7.  An electric motor of 10 kW is subjected to a breaking test for one hour. The heat generated due to friction is transferred to the surroundings at 20°C. What is the increase in entropy?(122.808 kJ/k)

4.8.  A piece of red hot iron is plunged suddenly into a mass of 20 kg water. The initial temperature of water was 20°C and final temperature becomes 55°C. Find the change of entropy water (12.217/k).

4.9.  420 kJ of heat from a source (a reservoir at 727°C is supplied to 2 kg of gas initially at 1.96 bar and 77°C in a closed tank. $C_v = 0.861$ kJ/kg k for the gas. Find the loss in available energy due to heat transfer. Take the temperature of the surroundings as 27°C. (147 kJ)

4.10.  A mass of $m$ kg of a liquid at a temperature $T_1$ is mixed with an equal mass of the same liquid at temperature $T_2$. The system is thermally insulated. Show that the entropy change of the universe must be positive and is given by

$$2\, m\, cp \, \log \{(T_1 + T_2)/(2\sqrt{T_1 \cdot T_2})\}$$

4.11.  A reversible engine operates between source at 927°C and two sinks, one at 127°C and another at 27°C. The heat rejected to both sinks is same, what is the efficiency of the engine? (71.5%).

4.12.  A reversible heat engine operates between two reservoirs at temperature of 650°C and 35°C. The engine drives a reversible refrigerator which operates between temperature of 35°C and –20°C. The heat transfer to the engine is 1000 kJ/min and net work output of the combined device is 180 kJ/min. Estimate the refrigerating effect in kW and net heat transfer to the reservoir at 35°C. What is the power developed by the engine in kW.  [Nov. 87, Poona Univ. R.E. = 37.284 kW, 50.95 kW, 11.105 kW]

4.13.  A Carnot cycle receives heat at 527°C causing increase in entropy equal to 5 kJ/kg k. The engine delivers 2,000 kJ/kg of work. Determine the efficiency of the cycle and lowest temperature in the cycle.

[May 91, Poona Univ. 50%, 400 k]

4.14. A refrigerator storage is supplied 20,400 kg of fish at 26.7°C. For preserving fish has to be cooled to – 95°C. The cooling takes place in 10 hours. The specific heat of fish which is 2.93 above freezing point of fish and 1.256 below freezing point of fish which is –2.2°C. The latent heat of freezing is 232 kJ/kg. What is the capacity of the plant in tons of refrigeration for cooling the fish? What would be the ideal Carnot cycle COP between the same temperature range? If the actual COP Is 1/3 that of ideal, find the power required to run the plant.

# 5

# The Perfect Gas

## 5.1 THE PERFECT GAS

A simple gas in a space contains a number of molecules that have velocities and are forever changing because of collisions. Each molecule consists of one or more atoms and have associated with it as a quantity of energy. There is existence of forces of attraction or repulsion between the molecules. As the molecules move about in space they collide with each other and interchange energy. Whether or not a collision occurs will depend on the distance between molecules and the law of attraction between molecules. The gas can be made indefinitely rare to minimize the effect of such collisions on the observed properties of the gas. By so doing a limiting condition is approached by real gases where the behaviour is unaffected by the presence of other molecules. Since all gases can approach this limit, it is considered to be an ideal or perfect state. In this perfect state the molecules behave as elastic spheres in a conservative mechanical system.

A perfect or ideal gas can be proposed as a standard of comparison for real gases. This idealized gas will exhibit two characteristics:

(i) All force interactions between molecules will be absent.
(ii) The aggregate size of all molecules will always be negligible in comparison with the size of the confining space.

In strict sense there is no perfect gas but all common gases at low pressures and high temperatures behave as perfect gases.

## 5.2 LAWS OF THE PERFECT GAS

At pressures which are low relative to the critical pressure and temperatures which are high compared to the critical temperatures, the *p-v-t* relation for a gas is governed by the following statements known as the Perfect Gas Laws.

*Boyle's Law*—When the temperature of a given mass of gas is constant, the product of absolute pressure and specific volume is constant and is mathematically given as

$$(Pv)_T = \text{Constant or, } v \, \alpha \, \frac{1}{P} \tag{5.1}$$

Between two states 1 and 2 at constant temperature

$$P_1 v_1 = P_2 v_2 \tag{5.2}$$

*Charles' Law*—It may be stated in two ways:

(a) If pressure remains constant, the volume of a given mass of gas varies directly as the absolute temperature.

that is, $\qquad v \, \alpha (T)_P = \text{constant or, } \left( \dfrac{v}{T} \right)_P = \text{constant}$ $\tag{5.3}$

Between two states 1 and 2 at constant pressure

$$\frac{v_1}{T_1} = \frac{v_2}{T_2} \tag{5.4}$$

(b) If the volume remains constant, the pressure of a given mass of gas varies directly with temperature

that is, $\qquad P \, \alpha (T)_v = \text{constant} \quad \text{or, } \left( \dfrac{P}{T} \right)_v = \text{constant}$ $\tag{5.5}$

between two states 1 and 2 at constant volume

$$\frac{P_1}{T_1} = \frac{P_2}{T_2} \tag{5.6}$$

## 5.3 CHARACTERISTIC GAS EQUATION OR EQUATION OF STATE FOR PERFECT GASES

For a gas that obeys the Boyle's and Charles' laws,

as $v \, \alpha \dfrac{1}{p}$ at $T = \text{constant}$

and $v \, \alpha \, T$ at $p = \text{constant}$

$\therefore \qquad\qquad v \, \alpha \dfrac{1}{p}$ when $T$ and $p$ both vary

or, $\qquad\qquad \dfrac{pv}{T} = \text{constant} = R \text{ or, } pv = RT$ $\tag{5.7(a)}$

Where, $R$ is a dimensional constant called characteristic gas constant.

Between two states 1 and 2 we have $\dfrac{p_1 v_1}{T_1} = \dfrac{p_2 v_2}{T_2} = R$ $\tag{5.7(b)}$

The value of this constant $R$ will be different for different gases and its unit will be kJ/kg k. In terms of total volume V of a given mass $m$ of a gas as, $V = m \cdot v$ we have

$$pV = mRT \tag{5.8}$$

Eqs. 5.7 or 5.8 are called characteristic gas equations or equations of state for a perfect gas.

## 5.4   AVOGADRO'S HYPOTHESIS AND UNIVERSAL GAS CONSTANT

*Avogadro's Hypothesis*—Equal volumes of perfect gases under the same conditions of temperature and pressure contain equal number of molecules.

From the above hypothesis it follows that the densities of perfect gases at the same temperature and pressure must bear the same relationship to each other as their molecular weights (M)

For two gases,
$$\frac{\rho_1}{\rho_2} = \frac{M_1}{M_2} \tag{5.9}$$

or,
$$\frac{v_2}{v_1} = \frac{M_1}{M_2} \tag{5.10}$$

Applying Eq. 5.7(b) to Eq. 5.10, we have

$$\frac{M_1}{M_2} = \frac{v_2}{v_1} = \frac{\left(R_2 \dfrac{T}{P}\right)}{\left(R_1 \dfrac{T}{P}\right)} = \frac{R_2}{R_1}$$

or,
$$R_1 M_1 = R_2 M_2 = R_0 \tag{5.11}$$

where $R_0$ is known as universal gas constant and is given by the product of characteristic gas constant and the molecular weight of the particular gas. Values of $R_0$ have been experimentally determined. Its value is 848 kg m/kg-mole $k$ or 8.314 KJ/kg mole $k$ (in SI units)

Following table gives the values of $R_0$ and $R$ of some of the gases and confirms the universality of $R_0$.

**Table 5.1**

| Gas | M | R | $M \times R = R_0$ | |
|---|---|---|---|---|
| Air | 29.00 | 29.27 kgm/kg K. | $29 \times 29.27 = 848$ kg m/kg mole °k. | |
| | | 0.287 KJ/kg k. (in SI units) | $29 \times 0.287 = 8.323$ KJ/kg mole k. (in SI units) | |
| $H_2$ | 2.016 | 4.124 kJ/kg k | $2.016 \times 4.124 = 8.313$ kJ/kg mole k. | $= 8.314$ kJ/kg mole k. |
| $N_2$ | 28.016 | 0.296 kJ/kg k | $28.016 \times 0.296 = 8.292$ kJ/kg mole k. | |
| $O_2$ | 32 | 0.259 kJ/kg k | $32 \times 0.259 = 8.288$ kJ /kg mole k. | |

## 5.5  JOULE'S LAW

This law states that the internal energy of perfect gas depends on its absolute temperature only that is, $u = f_1(T)$. If the temperature of a perfect gas changes from $T_1$ to $T_2$, the change in internal energy $(u_2 - u_1)$ remains the same; no matter how $p$ and $v$ might have changed.

Again since enthalpy $h = u + pv$ or, $h = u + RT$ where, $R$ is constant; it is seen that $h$ is a function of $T$ only

or,
$$h = f_2(T)$$

## 5.6  RELATION OF SPECIFIC HEATS OF GASES WITH GAS CONSTANT

We have known in Chapter three that any gas has two specific heats known as specific heat at constant pressure and specific heat at constant volume and are denoted by $C_p$ and $C_v$ respectively.

We had from Eqs. 3.13 and 3.18 that $C_v = \dfrac{du}{dT}$ and $C_p = \dfrac{dh}{dT}$ or we can write $du = C_v\, dT$ and $dh = C_p\, dT$

But,
$$h = u + pv$$

$\therefore$ differentiating this equation we get,

$$dh = du + d\,(pv\,\cdot)\quad \text{or,}\quad dh = du + d\,(RT) = du + R\,dT$$

Substituting the values of $dh$ and $du$, we get

$$C_p\, dT = C_v\, dT + R\, dT$$

or,
$$C_P - C_v = R \tag{5.12}$$

The ratio of two specific heats is denoted by $\gamma$* (or some times by $k$)

$\therefore$
$$\gamma = \frac{C_p}{C_v} \tag{5.13}$$

From Eq. 5.12

$$C_v\gamma - C_v = \text{or,}\quad C_v = \frac{R}{(\gamma - 1)} \tag{5.14}$$

But
$$C_p = \gamma\, C_v$$

$\therefore$
$$C_p = \frac{\gamma}{(\gamma - 1)} \times R \tag{5.15}$$

If we multiply Eq. 5.12 by molecular weight of the gas, we get

$$M\, C_P - MC_v = MR$$

or,
$$C_p^{-} - C_v^{-} = R_0 \tag{5.16}$$

where, $C_p^{-}$ and $C_v^{-}$ are molar specific heats at constant pressure and at constant volume.

------

*($\gamma$ gamma)

Note: An approximate value of $\gamma$ can be found from the equation given below if the number of atoms in a molecule of the gas be known.

$$\gamma = \frac{i+2}{i}$$

where $i$ = number of degrees of freedom of the molecules of gas. Thus for monoatomic gases, as

$$i = 3 \quad \therefore \quad \gamma = \frac{3+2}{3} \quad \text{(for Ar, He)}$$

or, $$\gamma = \frac{5}{3} = 1.667$$

for a diatomic gas $$i = 5, \gamma = \frac{7}{5} = 1.4$$

(for $O_2$, $H_2$, $N_2$)

for a triatomic gas $$i = 6, \gamma = \frac{8}{6} = 1.333$$

($H_2O$, $CO_2$)
For air $\gamma = 1.4$

## 5.7 REVERSIBLE NON-FLOW PROCESSES OF A PERFECT GAS

For a reversible non-flow process, we have

$$dq = du + p\,dv = T\,ds$$

or, $$ds = \frac{du}{T} + \frac{p\,dv}{T}$$

But, $du = C_v dT$ and $pv = RT$ i.e. $\dfrac{p}{T} = \dfrac{R}{v}$

$$\therefore \qquad ds = \frac{C_v dT}{T} + \frac{R\,dv}{v} \qquad\qquad (5.17)$$

Again as $T = \dfrac{pv}{R}$ and $C_p - C_v = R$

$\therefore$ From 5.17,
$$ds = C_v \frac{d\left(\dfrac{pv}{R}\right)}{\left(\dfrac{pv}{R}\right)} + (C_p - C_v)\frac{dv}{v}$$

or, $$ds = C_v \frac{(p\,dv + v\,dp)}{R} \times \frac{R}{pv} + C_p \frac{dv}{v} - C_v \frac{dv}{v}$$

or, $$ds = C_v \frac{p\,dv}{pv} + C_v \frac{v\,dp}{pv} + C_p \frac{dv}{v} - C_v \frac{dv}{v}$$

or, $$= C_v \frac{dv}{v} + C_v \frac{dp}{p} + C_p \frac{dv}{v} - C_v \frac{dv}{v}$$

or,
$$ds = C_p \frac{dv}{v} + C_v \frac{dp}{p} \qquad (5.18(a))$$

Again putting $C_v = (C_p - R)$ in the above equation we have

$$ds = C_p \frac{dv}{v} + (C_p - R) \frac{dp}{p} \qquad (5.18(b))$$

we know $pv = RT$
differentiating, we get $p\, dv + v\, dp = R\, dT$
dividing this equation by $pv = RT$, we get

$$\frac{p\, dv + v\, dp}{pv} = \frac{R\, dT}{RT}$$

or,
$$\frac{dv}{v} + \frac{dp}{p} = \frac{dT}{T} \qquad (5.19)$$

Putting value of $\dfrac{dv}{v}$ in Eq. 5.18(b)

We get,
$$ds = C_p \frac{dT}{T} - C_p \frac{dp}{p} + cp \frac{dp}{p} - R \frac{dp}{p}$$

or,
$$ds = C_p \frac{dT}{T} - R \frac{dp}{p} \qquad (5.20)$$

For the change of state from 1 to 2 the increase in entropy can be obtained by integrating Eqs. 5.17, 5.18 and 5.20 in different forms

Thus   (i) $s_2 - s_1 = C_v \ln \dfrac{T_2}{T_1} + R \ln \dfrac{v_2}{v_1}$

(ii) $s_2 - s_1 = C_p \ln \dfrac{V_2}{V_1} + C_v \ln \dfrac{P_2}{P_1}$

(iii) $s_2 - s_1 = C_p \ln \dfrac{T_2}{T_1} R \ln \dfrac{P_2}{P_1}$

Depending on the process, suitable equations can be used for finding $(s_2 - s_1)$.

### 5.7.1  Calculation of Entropy Change for Different Processes

Now, let us take the different non-flow processes one by one and find out work done, heat transfer and change of entropy per kg of gas between two states 1 and 2.

(i)  Constant volume process or, ($v$ = constant)

For 1 kg of gas,

$$W_{1-2} = \int_1^2 pdv = 0 \quad \text{as} \quad dv = 0$$

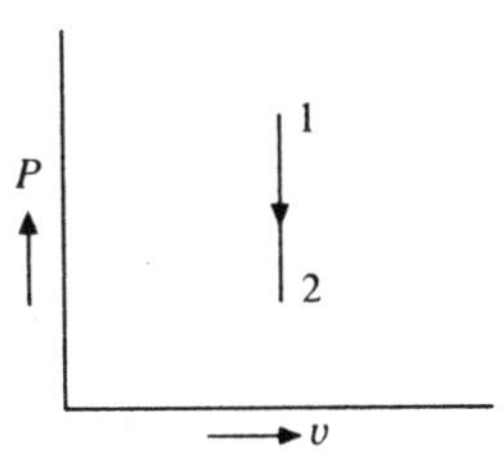

Fig. 5.1

$$q_{1-2} = \int_1^2 cv\, dT = C_v\,(T_2 - T_1) = u_2 - u_1 = u_{1-2}$$

$$s_2 - s_1 = \int_1^2 \frac{dq}{T} = \int_1^2 Cv\frac{dT}{T} = C_v \ln\frac{T_2}{T_1} \tag{5.21}$$

**(ii) Constant pressure process ($p$ = constant)**

For 1 kg of gas,

$$u_{1-2} = C_v\,(T_2 - T_1)$$

$$W_{1-2} = \int_1^2 pdv = p(v_2 - v_1) = R(T_2 - T_1)$$

$$q_{1-2} = \int_1^2 C_P dT = C_p\,(T_2 - T_1) = h_2 - h_1 = h_{1-2}$$

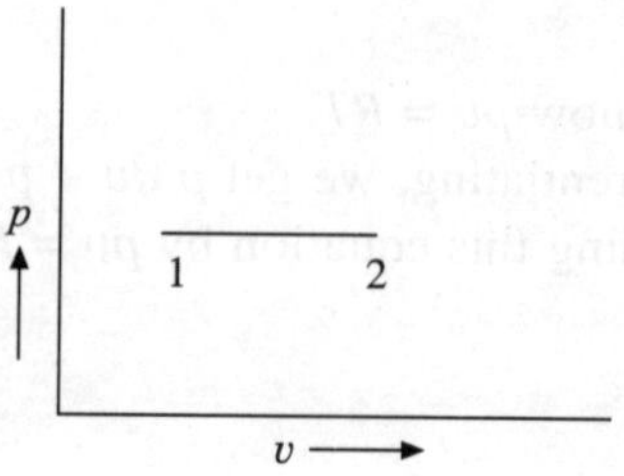

$$s_2 - s_1 = \int_1^2 \frac{dq}{T} = \int_1^2 C_p\frac{dT}{T} = C_p \ln\frac{T_2}{T_1} \tag{5.22}$$

**Fig. 5.2**

**(iii) Polytropic process**

A polytropic process is a general process and is expressed by $pv^n = c$(constant) where index $n$ is a constant.

Thus,

$$p_1 v_1^n = p_2 v_2^n = C \text{ But } \frac{p_1 v_1}{T_1} = \frac{p_2 v_2}{T_2}$$

$\therefore \qquad T_1 v_1^{n-1} = T_2 v_2^{n-1}$

or

$$\frac{T_2}{T_1} = \left(\frac{v_1}{v_2}\right)^{n-1}; \text{ Also } \frac{T_2}{T_1} = \left(\frac{p_2}{p_1}\right)^{\frac{n-1}{n}}$$

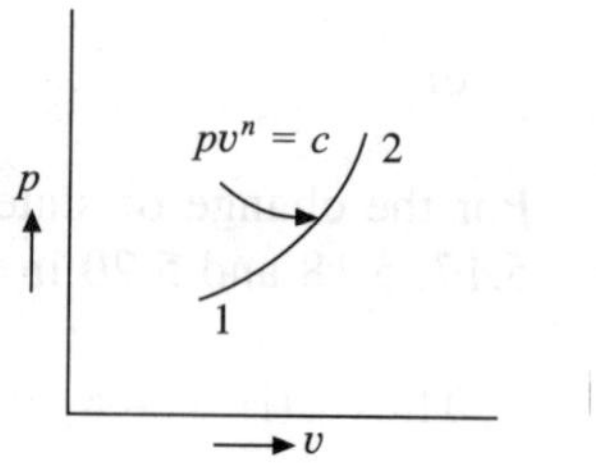

Work done/kg $= W_{1-2} = \int_1^2 pdv = \frac{Cdv}{v^n} = C\int_1^2 \frac{dv}{v^n} = C\int_1^2 v^{-n} dv$

**Fig. 5.3**

$$= C\left[\frac{v^{1-n}}{1-n}\right]_1^2 = \frac{C}{n-n}(v_2^{1-n} - v_1^{1-n})$$

$$= \frac{1}{1-n}\left[Cv_2^{1-n} - Cv_1^{1-n}\right] = \frac{1}{n-1}[p_2 v_2^n\, v_2^{1-n} - p_1 v_1^n\, v_1^{1-n}]$$

$$= \frac{1}{1-n}(p_2 v_2 - p_1 v_1)\frac{p_1 v_1 - p_2 v_2}{n-1} = \frac{R(T_1 - T_2)}{(n-1)} \text{ in work units} \tag{5.23}$$

Again heat transfer per kg is given by

$$q_{1-2} = W_{1-2} + \Delta u_{1-2} = \frac{R(T_1 - T_2)}{(n-1)} + C_v\,(T_2 - T_1) = \frac{R(T_1 - T_2)}{(n-1)} - C_v\,(T_1 - T_2)$$

$$= (T_1 - T_2)\left[\frac{R}{(n-1)} - C_v\right] = (T_1 - T_2)\left[\frac{Cv(\gamma - 1)}{(n-1)} - Cv\right].$$

$$\left(\because R = C_p - C_v \text{ and } \gamma = \frac{C_p}{C_v}\right)$$

$$\therefore \qquad = \frac{(T_1 - T_2)}{n - 1} [C_v(\gamma - 1) - C_v(n - 1)]$$

$$= \frac{T_1 - T_2}{n - 1} [C_v(\gamma - 1 - n + 1) = C_v\frac{\gamma - n}{n - 1} \times (T_1 - T_2)$$

$$= C_v\frac{\gamma - n}{1 - n}(T_2 - T_1) = C_n(T_2 - T_1) \qquad (5.24)$$

where, $C_n$ is called polytropic specific heat and is given by $C_n = C_v = \dfrac{(\gamma - n)}{1 - n}$ and change of

entropy per kg

$$s_2 - s_1 = \int_1^2 \frac{dq}{T} = C_n \int_1^2 \frac{dT}{T} = C_n(T_2 - T_1) = C_v\frac{(\gamma - n)}{1 - n} \ln \frac{T_2}{T_1} \qquad (5.25$$

**(iv) Isothermal process ($T$ = constant)**
For perfect gas as $pv = RT$. Therefore as $T$ is constant, $pv$ is constant.

Let us say, $p_1v_1 = p_2v_2 = C$. For this process $\Delta u = 0$ as internal energy is a function of temperature only for a perfect gas and here $T_1 = T_2$.

Then $$W_{1-2} = \int_1^2 p\, dv = \int_1^2 \frac{C}{v} dv$$

$$= C \ln \frac{v_2}{v_1} = p_1v_1 \ln \frac{v_2}{v_1} = p_2v_2 \ln \frac{p_1}{p_2} = RT_1 \ln \frac{v_2}{v_1} \qquad (5.26)$$

$\therefore$ For 1 kg of gas, $$u_{1-2} = C_v(T_2 - T_1) = 0$$

$$q_{1-2} = w_{1-2}$$

$$= RT_1 \ln \frac{v_2}{v_1}$$

$$s_2 - s_1 = \int_1^2 \frac{dq}{T} = \frac{1}{T} \int_1^2 dq \ (\because T = \text{constant})$$

$$= T_1 = T_2$$

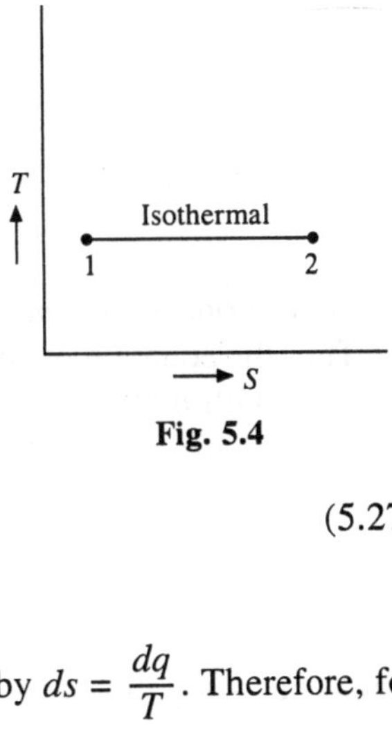

$$= \frac{1}{T} \times q_{1-2} = \frac{RT_1}{T_1} \ln \frac{v_2}{v_1} = R \ln \frac{v_2}{v_1} = R \ln \frac{p_1}{p_2} \qquad (5.27)$$

**Fig. 5.4**

**(v) Adiabatic reversible or isentropic process**

For a reversible process we know that the change in entropy is given by $ds = \dfrac{dq}{T}$. Therefore, for

an adiabatic reversible process as $dq = 0$. We have $ds = \dfrac{0}{T} = 0$. This means that the entropy change is zero. That is why a reversible adiabatic process is called an isentropic process.

Equation of an adiabatic reversible process:

We know that for a non-flow reversible process

$$dq = du + dw = du + p\,dv = C_v\,dT + p\,dv \tag{5.28}$$

$\therefore$ for an adiabatic reversible process, $\quad du + p\,dv = 0$

or, $\qquad\qquad\qquad\qquad\qquad C_v\,dT + p\,dv = 0$

Again $pv = RT$ for a perfect gas

$\therefore$ differentiating this equation we get,

$$p\,dv + v\,dp = R\,dT \text{ or, } dT = \frac{p\,dv + v\,dp}{R}$$

Now, putting this value of $dT$ we get for an adiabatic reversible process

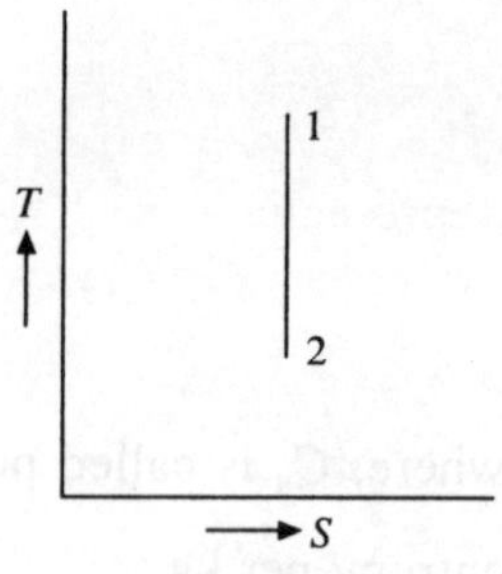

**Fig. 5.5**

$$\frac{C_v}{R}\,(p\,dv + v\,dp) + p\,dv = 0$$

or, $\qquad\qquad\qquad C_v(p\,dv + v\,dp) + R\,p\,dv = 0$

But $\qquad\qquad\qquad\qquad\qquad\qquad C_p - C_v = R$

$\therefore \qquad\qquad\qquad C_v(p\,dv + v\,dp) + (C_p - C_v)\,p\,dv = 0$

or, $\qquad\qquad\qquad C_v(p\,dv + v\,dp) + (C_p - C_v)\,p\,dv = 0$

or, $\qquad\qquad\qquad\qquad C_v\,v\,dp + C_P\,p\,dv = 0$

Dividing throughout by $C_v\,Pv$ we get

$$\frac{dp}{p} + \frac{C_p}{C_v}\frac{dv}{v} = 0 \quad \text{or,} \quad \frac{dp}{p} + \gamma\frac{dv}{v} = 0$$

Integrating this equation, we get

$$\ln P + \gamma \ln v = 0 \text{ constant}$$

or, $pv^\gamma = $ constant or, $pv^\gamma = $ constant

Thus the equation $pv^\gamma = C$ represents an isentropic process. Thus this is of the form of polytropic process with $n = \gamma$.

Putting $n = \gamma$ in the expression for work done for a polytropic process, we get work done per kg of gas during an isencropic process is given by

$$W_{1-2} = \frac{R(T_1 - T_2)}{\gamma - 1} = \frac{p_1v_1 - p_2v_2}{\gamma - 1} \tag{5.29}$$

There is no heat transfer, that is, $q = 0$ and so $s_2 - s_1 = 0$
From the First Law since $q_{1-2} = u_{1-2} + W_{1-2}$
and $q = 0$

$$\therefore \qquad u_{1-2} = -W_{1-2} = \frac{p_2v_2 - p_1v_1}{(\gamma - 1)} = \frac{R(T_2 - T_1)}{(\gamma - 1)} \tag{5.30}$$

## 5.8 DIFFERENT NON-FLOW REVERSIBLE PROCESSES AS SPECIAL CASE OF THE POLYTROPIC PROCESS:

The polytropic process is represented by $pv^n = C$. The different non-flow processes can be represented by $pv^n = C$ with specific value of index $n$. Figs. 5.6(a) and 5.6(b) show the different processes on the $p$–$v$ and $T$–$S$ plots with values of index $n$.

(a) Constant volume process
    Polytropic equation is $pv^n = C$

$$\therefore \qquad v = \left(\frac{c}{p}\right)^{\frac{1}{n}} . \text{ Now if } n = \infty, v = \text{constant}$$

$\therefore$ For constant volume process index $n = \infty$

(b) Constant pressure process

$$pv^n = C$$

If $n = 0$ then $pv^0 = C$ or, $p = $ constant
$\therefore$ For constant pressure process, index $n = 0$

(c) Isothermal process

$$pv^n = C$$

If $n = 1$ then $pv = C$ which represents isothermal process. Therefore index $n$ for this process $= 1$.

(d) Isentropic process
    We know that isentropic process is given by $pv^\gamma = C$. Thus index $n$ for isentropic process is $\gamma$

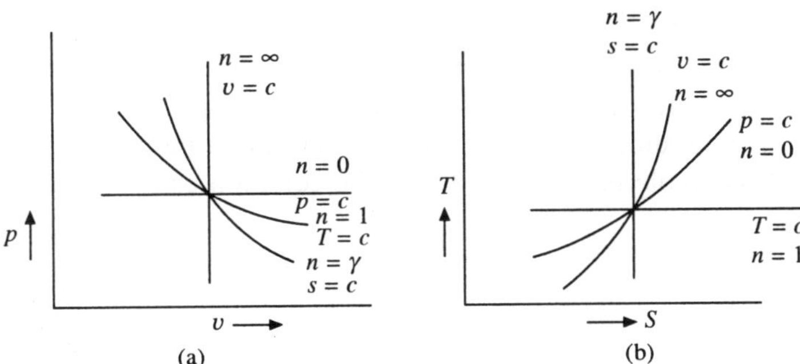


**Fig. 5.6**

## 5.9 FREE EXPANSION AND THROTTLING PROCCESSES OF A PERFECT GAS

*Free expansion process:* In this process a gas expands freely without any resistance from a higher pressure to a lower pressure without any heat transfer. Thus $q = 0$ and $W = 0$

$\therefore$ from the First Law $du = 0$ which means that temperature remains constant. This process is irreversible because after free expansion the gas cannot be compressed without doing work, that is, freely. The change in entropy per kg is given by

$$s_2 - s_1 = R \ln \frac{v_2}{v_1} \tag{5.31}$$

*Throttling process:* When a gas expands and flows through minute openings resulting into decrease in velocity the process is called throttling process. During the process $h_1 = h_2$. For ideal gases it means a constant temperature. This process is also irreversible because during the process kinetic energy is converted into heat due to wall friction. The change in entropy per kg of gas is given by

$$s_2 - s_1 = R \ln \frac{p_1}{p_2} \tag{5.32}$$

## 5.10  VANDER WAALS EQUATION FOR REAL GASES

We have known that there is no perfect gas in the strict sence. But most of the real gases at low pressure and high temperature approximately behave as perfect gases. For example, in case of air at normal atmospheric temperatures the error in the approximation to the true equation of state is of the order of 1% at 20 atmospheres and 0.1% at atmospheric pressure, and at higher temperatures the error becomes still less.

However, Vander Waals in 1879 has proposed an equation of state for a real gas. This is expressed as

$$\left( p + \frac{a}{v^2} \right) = \frac{RT}{v - b} \tag{5.33}$$

or,
$$P = \frac{RT}{v - b} - \frac{a}{v^2} \tag{5.34}$$

This equation represents, in a qualitative way, for both liquid and vapour phase conditions.

The constants $a$ and $b$ are specific constants and depend upon the type of fluid considered. $v$ represents the volume per unit mass and $P$, $R$ and $T$ are the usual terms.

If the volume of one mole is considered then the equation can be written as

$$\left( p + \frac{a}{\bar{v}^2} \right)(\bar{v} - b) = R_0 T \tag{5.35}$$

where $p = $ N/m$^2$, $\bar{v} = $ m$^3$/kg made, $T = K$ and $R_0 = 8314.4$ Nm/kg mole K.

The co-efficient "$a$" was introduced to account for the existence of mutual attraction between the molecules. The term $a/v^2$ is called the force of cohesion. The co-efficient "$b$" was introduced to account for the volumes of the molecules and is known as co-volume. The force exerted by molecules on walls is reduced due to attractive forces. This reduction in pressure is known as internal pressure which according to Vander Waals is proportional to square of density and is equal to $a/v^2$. Hence the term $- a/v^2$ occurs in Eq. 5.34.

At atmospheric pressure and temperature the volume occupied by molecules is only about

1000th of the volume that of a room. But volume occupied by the molecules becomes a greater part of the total volume as pressure increases. This is taken account by replacing $v$ in the ideal gas relation with the quantity $v - b$, where $b$ represents the volume occupied by the gas molecules per unit mass. The value of $b$ is of the same order of magnitude as the volume of the liquified gas.

### 5.10.1  Determination of constants of Vander Waals equation

Equation 5.34 of Vander Waals can be written as

$$pv^3 - (pb + RT)\,v^2 + av - ab = 0 \tag{5.36}$$

It will give three roots of $v$ of which only one needs to be real. For low temperatures, three positive real roots exist for a certain range of pressure. As the temperature increases the three real roots approach one another and at the critical isotherm $T_c$ at the critical state on the $p$–$v$ plane (Fig. 5.7) where the three real roots of the Vander Waals equation coincide, not only has a zero slope, but also its slope changes at the critical state (point of inflection), so that the first and second derivatives of $P$ with respect to $v$ at $T = T_c$ are each equal to zero. Eq. 5.34 could be rewritten as $P = \dfrac{RT}{v - b} - \dfrac{a}{v^2}$

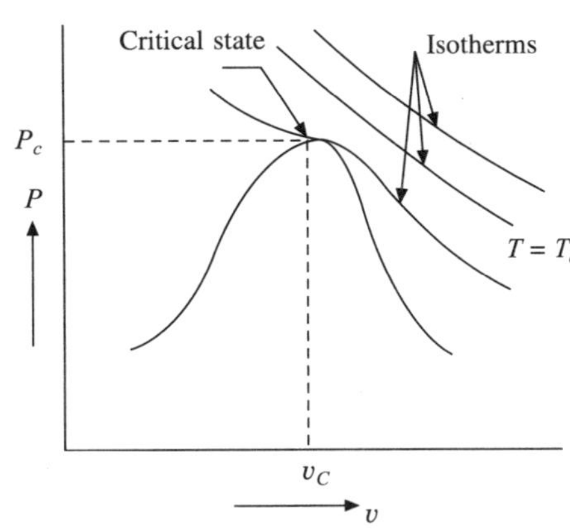


**Fig. 5.7**

$\therefore$
$$\left(\frac{\partial P}{\partial v}\right)_{T = T_c} = \frac{RT_c}{(v_c - b)^2} + \frac{2a}{v_c 3} = 0$$

or,
$$RT_c = \frac{2a}{v_c 3}\,(v_c - b)^2 \tag{5.37}$$

$$\left(\frac{\partial^2 P}{\partial v^2}\right)_{T = T_c} = \frac{2RT_c}{(v_c - b)^3} + \frac{6a}{v_c 4} = 0$$

or,
$$RT_c = \frac{3a}{v_c 4}\,(v_c - b)^3 \tag{5.38}$$

From Eqn. 5.36 and 5.37 by equating, we get

$$\frac{2a}{v_c 3}\,(v_c - b)^2 = \frac{3a}{v_c 4}\,(v_c - b)^3$$

or,
$$2v_c = 3\,(v_c - b) \quad \text{or,} \quad \boxed{v_c - 3b} \tag{5.39}$$

Substituting this value in Eq. 5.38, we get

$$\boxed{T_c = \frac{8a}{27\,Rb}} \tag{5.40}$$

Now substuting Eq. 5.39 and 5.40 in Eqn. 5.34, we get

$$P_c = \frac{a}{27\,b^2} \tag{5.41}$$

The constants for Vander Waals equation can be obtained if the critical constants for the particular gas are known. As $P_c$ and $T_c$ can be measured more easily and more accurately than $v_c$, $a$ and $b$ are expressed in terms of $P_c$ and $T_c$.

From Eqs. 5.39 and 5.41

$$a = 3P_c\,v_c^2 = \frac{9}{8}\,RT_c v_c = \frac{27}{64}\frac{R^2 T_c^2}{P_c^2} \tag{5.42}$$

$$b = \frac{v_c}{3} = \frac{RT_c}{8\,P_c} \tag{5.43}$$

and

$$R = \frac{8}{3}\frac{P_c v_c}{T_c} \tag{5.44}$$

Substuting these values of $a$, $b$ and $R$ in terms of critical properties in Vaals Waals equation of state, we get

$$\left(P + \frac{a}{v^2}\right)(v - b) = RT$$

or,

$$\left(P + \frac{3\,p_c v_c^2}{v^2}\right)\left(v - \frac{v_c}{3}\right) = \frac{8}{3}\frac{P_c v_c}{T_c}\,T$$

or,

$$\left(\frac{p}{p_c} + \frac{3\,v_c^2}{v^2}\right)\left(\frac{v}{v_c} - \frac{1}{3}\right) = \frac{8}{3}\frac{T}{T_c}$$

using the reduced parameters $\left(P_r = \dfrac{p}{p_c},\, v_r = \dfrac{v}{v_c},\, T_r = \dfrac{T}{T_c}\right)$,

we get

$$\left(P_r + \frac{3}{v_r^2}\right)\left(v_r - \frac{1}{3}\right) = \frac{8}{3}\,T_r \tag{5.45}$$

### 5.10.2 Limitations of Vander Waals equation

The Vander Waals equation is not followed closely under actual conditions and thereby its validity fails due to following reasons.

(i) The values of constants $a$ and $b$ are varying with change in temperature. Therefore, the equation does not give correct results.

(ii) The Vander Waals equation can be compared with ideal gas equation considering the value of $pv/RT$ at critical point of the gas.

$$\frac{P_c v_c}{RT_c} = \frac{3}{8} \text{ (for Vander Waals equation)}$$

and
$$\frac{P_c v_c}{RT_c} = 1 \text{ (for ideal gas equation)}$$

The above equation indicates that $\frac{P_c v_c}{RT_c}$ is 3/8 for Vander Waals gas whereas experiments shows that this value varies from 0.2 to 0.3 for most of the gases.

So in the critical region Vander Waals equation is not accurate.

However, Vander Waals equation was used for a long time for most of the gases as it is found to be valid for real gases above critical point.

**Table 5.1**

| Gas | $a$<br>KN/m$^4$/(kg mole)$^2$ | $b$<br>m$^3$/kg mole | Compressibility factor<br>$Z_c = P_c v_c/RT_c$ |
|---|---|---|---|
| Air | 135.8 | 0.0365 | 0.284 |
| $O_2$ | 138.0 | 0.0318 | 0.290 |
| $N_2$ | 136.7 | 0.0386 | 0.291 |
| $H_2O$ | 551.7 | 0.0304 | 0.230 |
| $CO_2$ | 365.6 | 0.0428 | 0.276 |
| $NH_3$ | 424.6 | 0.0373 | 0.242 |
| He | 3.42 | 0.0235 | 0.300 |

# 5.11 COMPRESSIBILITY FACTOR AND COMPRESSIBILITY CHART

$Pv = RT$ is referred to perfect gas only for the condition when pressure approaches 0. At high pressure and low temperature real gas departs from ideal gas. The perfect gas equation can be modified by applying a correction factor Z, called compressibility factor, to indicate this deviation

$$pv = Z\,RT$$

For the ideal gas $Z = 1$. For real gas $z$ is determined from expeimental data.

At all pressure-temperature conditions the value of compressibility factor of a perfect gas is unity. For any gas compressibility approaches unity as the pressure is reduced because the gas acts more like a perfect gas as the pressure is lowered.

As we know $P_r = P/P_c$, $T_r = T/T_c$

and
$$v_r = v/v_c = v/RTC/Pc \quad \text{as} \quad v_c = \frac{RT_c}{P_c}$$

Therefore, the compressibility factor for any fluid is a function of only two properties, that is, $T_r$, $P_r$ because the third property $v_r$ is dependent on above two. Hence,

$$Z = f(T_r, P_r)$$

using equation $Z = \frac{Pv}{RT}$ for compressibility factor for real gases and substuting $v = v_r v_c$, $P = P_r P_c$ and $T = T_r T_c$, we get

$$Z = \frac{(P_r P_c)(v_r v_c)}{R\,(T_r T_c)} = \frac{P_c v_c}{RT_c} \times \frac{P_r v_r}{T_r} = Z_c\,\frac{P_r v_r}{T_r}$$

or,
$$Z = Z_c\,\frac{P_r v_r}{T_r} \tag{5.46}$$

where $Z_c$ is the compressibility factor at critical point of the gas and $P_r$, $v_r$ and $T_r$ are non-dimensional. The above equation suggests that compressibility factor is also a universal function of $p_r$ and $T_r$ for all gases that have the same critical compressibilities because the law of corresponding states indicates that $v_r$ is a universal function of $P_r$ and $T_r$.

Compressibility chart is drawn which represents $Z$ on Y-axis and $P_r$ on X-axis. The curves, $Z$ versus $P_r$ for different $T_r$ (taking $T_r$ as a parameter) are drawn as shown in Fig. 5.8.

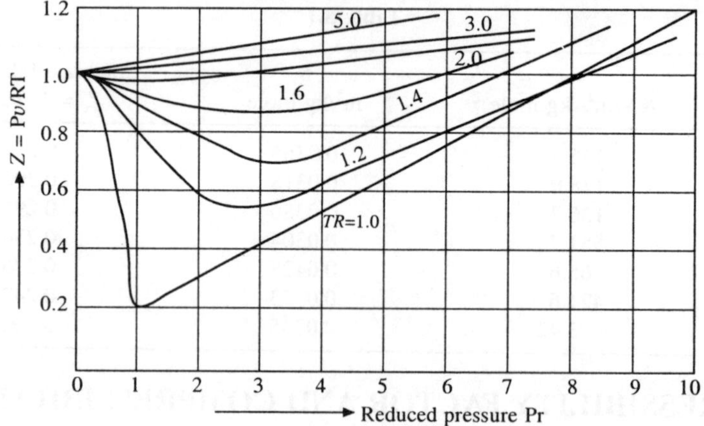


**Fig. 5.8   Generalised compressibility chart**

This chart is known as generalized compressibility chart as it can be used for any gas at any condition provided its properties ($P_c$, $v_c$, $T_c$) at critical point are known. This chart mention about deviation from ideal gas behaviour giving an accuracy of ± 5%. A gas having compressibility factor less than unity is more compressible than a perfect gas.

From this chart it can be observed clearly that both temperature and pressure deviate from ideal behaviour. This chart gives best results for the regions well removed from the critical state for all gases.

By using reduced pressure and temperature for all gases a generalized chart has been developed by Nelson and Obert. The chart is used for solving the problems.

## 5.12   BEATTIE-BRIDGEMAN EQUATION

Beattie-Bridgeman has proposed an equation of state, in 1928, which gives good results and thereby widely used. It is based on five experimentally determined constants.

$$P = \frac{R_0 T\,(1 - \epsilon)}{v^2}\,(v + B) - \frac{A}{v^2} \tag{5.47}$$

where $A = A_0\left(1 - \dfrac{a}{v}\right)$, $B = B_0\left(1 - \dfrac{b}{v}\right)$, $\in = \dfrac{c}{vT^3}$ and $A_0$, $a$, $B_0$, $b$ and $c$ are constants whose values are different for different gases.

This equation of state predicts properties accurately, within 2%, in regions where the density is less than 0.8 times the critical density.

## 5.13 VIRIAL EQUATION OF STATE

Kammerlingh has introduced the equation of state, in 1901, which is known as virial equation of state. It represents an expansion of the $pv$ product in infinite series form as given under

$$Pv = RT\left(1 + \frac{B}{v} + \frac{c}{v^2} + \frac{D}{v^3} + \ldots\right)$$

or,

$$\frac{Pv}{RT} = \left(1 + \frac{B}{v} + \frac{c}{v^2} + \frac{D}{v^3} + \ldots\right) \tag{5.48}$$

In this equation, the $P$-$V$-$T$ relation is expressed in terms of the infinite series in powers of density. The co-efficients $B$, $C$, $D$ each of which is a function of temperature only, are called the second, third and fourth virial co-efficients. They state the deviation of the real gases in terms of the force of attraction between the molcules. The word virial arises from the Latin word for force. Based on an assumed law of force between the molecules, theoretical deviations of the equation of state usually lead to a virial equation of state.

The virial equation of state is applicable only to gases of low or medium densities. As the pressure approaches to zero value, all the virial co-efficients will vanish and the equation will reduce to the ideal gas equation of state. The P-V-T behaviour of a substance can be represented accurately with the virial equation of state over a wider range by including a sufficient number of terms.

## 5.14 *P-V* DIAGRAM FOR REAL SUBSTANCE

A pure (real) substance is a homogeneous substance which retains its chemical composition even though there may be change of phases. It may exist in one or more phases.

The saturated liquid line and saturated vapour line incline towards each other forming saturation or vapour line. The two lines meet at the critical state solid region is situated to the left of saturated solid line. Between the saturated solid line and saturated liquid line solid-liquid mixture ($S + L$) region exists. Between the two saturated liquid lines is the compressed liquid region. Between the saturated liquid and saturated vapour lines the liquid-vapour mixture region ($L + v$) exists within the vapour dome. To the right of the saturated vapour line is the vapour region. The triple point is a point-on the $P$-$v$ diagram, where all the three phases, solid, liquid and gas exist in equilibrium. At a pressure below triple point line, the substance can not exist in the solid phase, and the substance, when heated transforms from solid to vapour (known as sublimation) by absorbing talent heat of sublimation from the surroundings. Therefore, the region below the triple point line is the solid vapour ($S + v$) mixture region (Figs. 5.9 and 5.10).

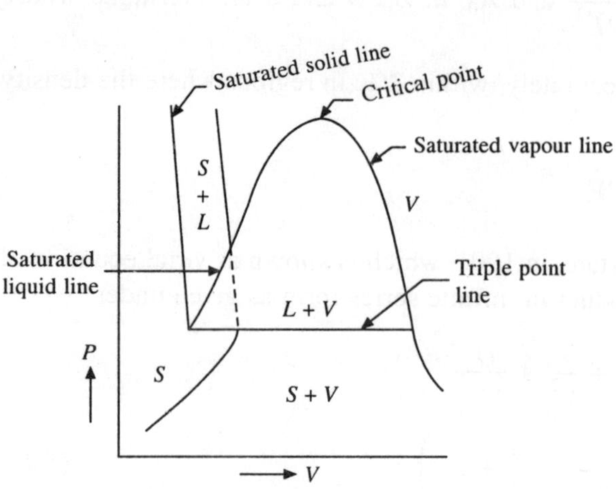

Fig. 5.9   *p-v* diagram of water whose volume decreases on melting

Fig. 5.10   *p-v* diagram of a pure substance other than water whose volume increases on melting

In most power cycles the working fluid is liquid and vapour and, therefore, Figs. 5.9 and 5.10 may be omitted and Fig. 5.11 may be referred.

If the vapour at state *A* is compressed slowly and isothermally, the pressure will rise until there is saturated vapour at point *B*. At constant pressure and temperature condensation takes place if compression is continued after *B*. At any point between *B* and *C*, the liquid and vapour are in equilibrium. Since a very large increase in pressure is required, line CD is almost vertical. *ABCD* is a typical isotherm of a pure substance on a *P-v* diagram. Some isotherms have been drawn in Fig. 5.11. As the temperature increases the liquid vapour line *BC* goes on decreasing and becomes

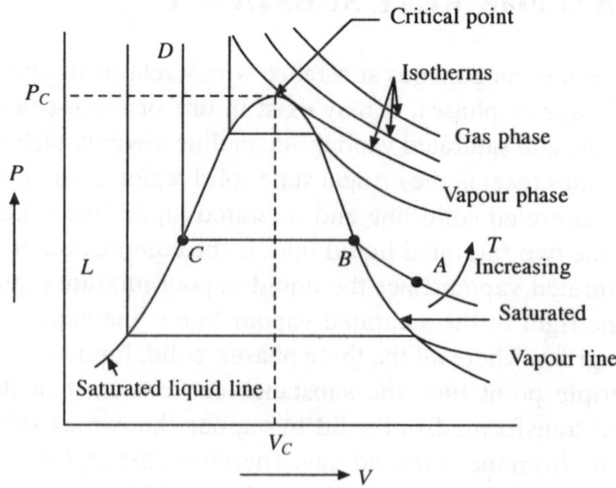

Fig. 5.11   Saturation curve on *p-v* diagram

zero at critical point. Below the critical point only a saturated liquid absorbs latent heat of vaporization on heating and becomes saturated vapour at a constant pressure and temperature in the liquid vapour transition zone and reverse occurs on cooling. However, above critical point a liquid suddenly flashes into vapour on heating and a vapour changes into liquid on cooling. There is no distinct zone from liquid to vapour and vice versa. The isotherms passing through the critical point is called critical isotherm and the corresponding temperature is known as the critical temperature ($t_c$). The pressure and volume at the critical point are called critical pressure ($P_c$) and critical volume ($v_c$) respectively.

For water
$$P_c = 221.2 \text{ bar}$$
$$t_c = 374.15°C$$
$$v_c = 0.00317 \text{ m}^3/\text{kg}$$

## 5.15  *P-T* DIAGRAM FOR REAL SUBSTANCE

In Figs. 5.9 and 5.10 on the *P-V* plane we have seen the changes in state of a pure substance on slow heating at constant pressures. If these state changes are plotted on *P-T* co-ordinates, the diagram as shown in Fig. 5.12 will be obtained.

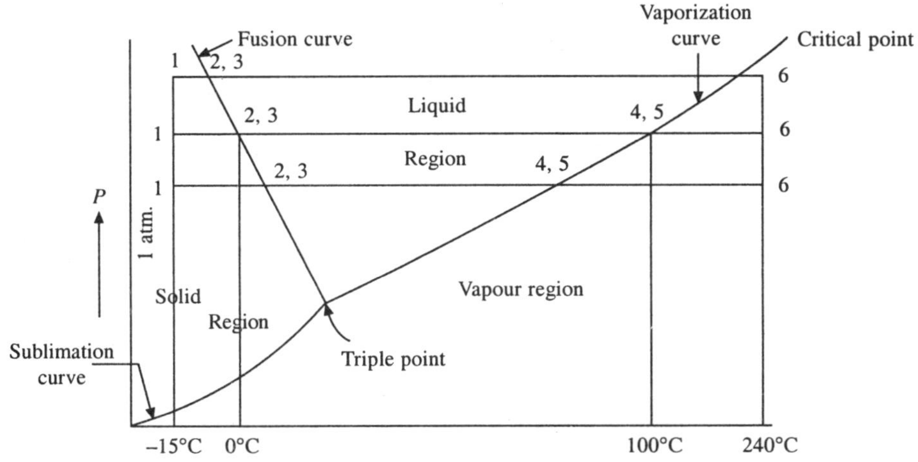


**Fig. 5.12   Phase equilibrium diagram on *P–T* plane**

Let us consider ice at −15°C is heated till it converts completely into steam at 240°C at the constant pressure of 1 atm. So different zones can be achieved as given below.

    1–2 = solid (ice) heating

    2–3 = melting of ice at 0°C

    3–4 = liquid heating

    4–5 = vapourization of water at 100°C

and     5–6 = heating in the vapour phase.

when cooling takes place the process will be reversed from state 6 to state 1. The curve passing through the 4–5 points is called vapourization curve is obtained when vapour pressure of solid is plotted at different temperatures. All these three curves meet at triple point.

For all substances the slope of the sublimation and vapourization curves are positive. The slope of the fusion curve for most substances is positive, but for water, it is negative.

The triple point of water is at 4.58 mm Hg (0.006114 bar) and 273.16 K, whereas that of $CO_2$ is at 3885 mm Hg (about 5 atm) and 216.55K. So when solid $CO_2$ (dry ice) is exposed to atmosphere it gets converted into vapour directly by absorbing latent heat from atmosphere. The atmosphere gets cooled and the process is known as sublimation of ice.

## 5.16   *P-V-T* SURFACE FOR REAL SUBSTANCE

Three dimensional *P-V-T* surfaces have been drawn in Figs. 5.13 and 5.14 to show the relationships between pressure, specific volume and temperature. Fig. 5.13 shows for water which expands upon freezing whereas Fig. 5.14 shows for substances like solid $CO_2$ which contracts upon freezing.

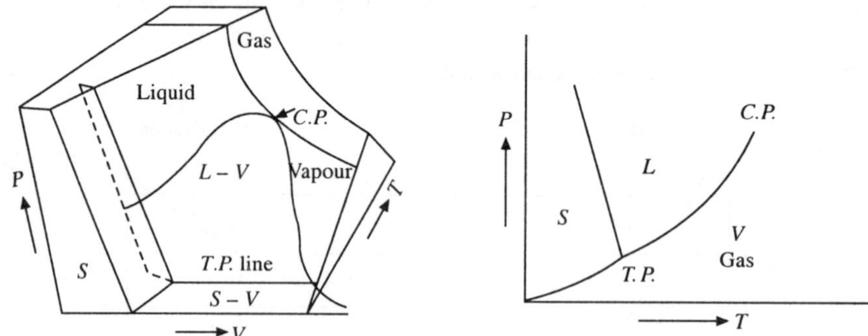


**Fig. 5.13   *P-V-T* surface for water which expands on freezing.**

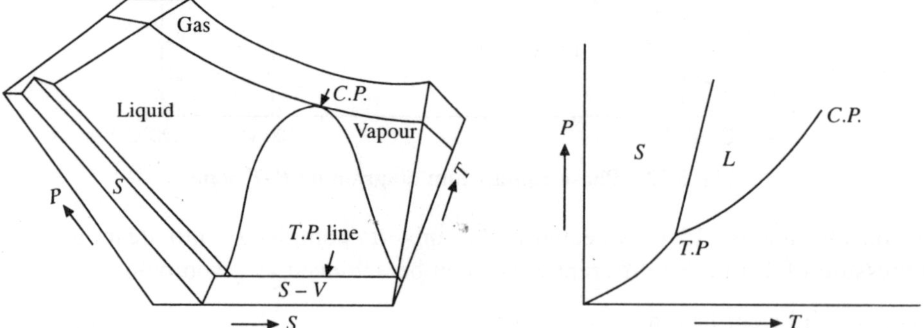


**Fig. 5.14   *P-V-T* surface of a substance (solid $CO_2$) which contracts on freezing.**

An equilibrium state of the substance is represented by any point on the *P-V-T* surface. The triple point line when projected on the *P-T* plane becomes a point. The critical isotherm has a point of inflection at the critical point.

## 5.17  RELATIONS FOR PERFECT GASES SUMMARIZED

$$pv = RT \; ; \frac{p_1 v_1}{T_1} = \frac{p_2 v_2}{T_2}$$

$$C_v = \frac{du}{dT} \; ; du = C_v dT$$

$$C_p = \frac{dh}{dT} \; ; dh = C_p dT$$

$$C_p - C_v = R \; ; \frac{C_p}{C_v} = Y$$

$$C_p = \frac{\gamma R}{\gamma - 1} \; ; C_v = \frac{R}{\gamma - 1}$$

$$T\,ds = C_v\,dT + p\,dv; \quad T\,ds = C_p\,dT - v\,dp$$

$$ds = C_v\,\frac{dT}{T} + R\,\frac{dv}{v}; \quad ds = C_p\,\frac{dT}{T} - R\,\frac{dp}{p}; \quad ds = C_p\,\frac{dv}{v} + C_v\,\frac{dp}{p}$$

$$Pdv + vdp = RdT; \quad \frac{dv}{v} + \frac{dp}{p} = \frac{dT}{T}$$

## 5.18  FIRST AND SECOND LAW COMBINED

From the Second Law, we got

$$\delta Q_{\text{rev}} = T\,ds$$

and from the Second Law, for a closed non-flow system,

$$\delta Q = du + p\,dv$$

$$T\,ds = du + p\,dv \tag{5.49}$$

$$T\,ds = dH - v\,dp \tag{5.50}$$

$$[\because H = U + PV, \; dH = du + p\,dv + v\,dp\,]$$

Eqs. (5.49) and (5.50) are related to the properties of the system, we get useful equations after combining the First and Second laws which are given below

(i) $\delta Q = dE + \delta w$ – It holds good for reversible as well as irreversible and any system
(ii) $\delta Q = du + \delta w$ – It holds good for any process undergone by a closed system
(iii) $\delta Q = du + pdv$ – It holds good for a closed system when only $p\,dv$ work is present.
(iv) Eq. (5.49) – It is true for a reversible process.
Eq. (5.50) – It holds good for any process reversible and irreversible undergoing in a closed system.

**Table 5.1   Consolidated formulae for heating and expansion of perfect gases in a closed system**

| S.N. | Types of expansion | Equation | $P$-$V$-$T$ Relation | Work done $W$ | Change in Internal Energy ($\Delta U$) | Heat added or substracted $Q$ | Enthalpy change ($H_2 - H_1$) | Entropy change ($S_2 - S_1$) |
|---|---|---|---|---|---|---|---|---|
| 1 | 2 | 3 | 4 | 5 | 6 | $7 = 5 + 6$ | 8 | 9 |
| 1. | Constant | $pv^\circ = C$ | $\dfrac{v_1}{T_1} = \dfrac{T_1}{T_2}$ | $p(v_2 - v_1)$ | $m\,C_v(T_2 - T_1)$ | $m\,C_p(T_2 - T_1)$ | $m\,C_p(T_2 - T_1)$ | $= m\,C_v \ln \dfrac{v_2}{v_1}$ |
| | Pressure | i.e. $p = C$ | | $= mR(T_2 - T_1)$ | | | | $= m\,C_p \ln \dfrac{T_2}{T_1}$ |
| | (Isobaric) | | | | | | | |
| 2. | Constant | $pv^\infty = C$ | $\dfrac{p_1}{T_1} = \dfrac{p_2}{T_2}$ | zero | $m\,C_v(T_2 - T_1)$ | $m\,C_v(T_2 - T_1)$ | $m\,C_p(T_2 - T_1)$ | $m\,C_v \ln \dfrac{T_2}{T_1}$ |
| | Volume | i.e. $v = C$ | | | | | | $= m\,C_v \ln \dfrac{p_2}{p_1}$ |
| | (Isochoric) | | | | | | | |
| 3. | Constant | $pv = C$ | $\dfrac{p_1}{p_2} = \dfrac{v_2}{v_1}$ | $P_1v_1 \ln \dfrac{v_2}{v_1}$ | zero | $P_1v_1 \ln \dfrac{v_2}{v_1}$ | zero | $m\,R \ln \dfrac{v_2}{v_1}$ |
| | temp. or | i.e. $T = C$ | | $= P_1v_1 \ln \dfrac{p_1}{p_2}$ | | $= P_1v_1 \ln \dfrac{p_1}{p_2}$ | | $= mR \ln \dfrac{p_1}{p_2}$ |
| | Isothermal | | | $= mRT \ln \dfrac{v_2}{v_1}$ | | $= mRT \ln \dfrac{v_2}{v_1}$ | | |
| 4. | Reversible | $pv^\gamma = C$ | $\dfrac{T_2}{T_1} = \left(\dfrac{p_2}{p_1}\right)^{\frac{\gamma-1}{\gamma}}$ | $\dfrac{P_1v_1 - P_2v_2}{\gamma - 1}$ | $m\,C_v\,(T_2 - T_1)$ | Zero | $m\,C_p(T_2 - T_1)$ | Zero |
| | adiabatic or | | $= \left(\dfrac{v_1}{v_2}\right)^{\gamma-1}$ | or | | | | |
| | Isentriopic | | | $\dfrac{mR(T_1 - T_2)}{(\gamma - 1)}$ | | | | |

5. General

expansion or

polytropic

expansion

$$pv^n = C$$

$$\frac{T_2}{T_1} = \left(\frac{p_2}{p_1}\right)^{\frac{n-1}{n}}$$

$$= \left(\frac{v_1}{v_2}\right)^{n-1}$$

$$\frac{p_1 v_1 - p_2 v_2}{n - 1}$$

$$mC_v(T_2 - T_1)$$

$$\frac{p_1 v_1 - p_2 v_2}{(n - 1)} + mC_p(T_2 - T_1)$$

$$m C_v(T_2 - T_1)$$

or

$$m C_v(T_2 - T_1)$$

or $\dfrac{\gamma - n}{\gamma - 1} \times \text{W. D}$

$$m C_n \ln \frac{T_2}{T_1}$$

where

$$C_n = C_v \frac{(\gamma - n)}{(1 - n)}$$

## Solved Problems

**5.1** The characteristic equation of a certain gas is $pV = mRT$. Calculate the specific heat of the gas at constant pressure if the specific heat of the gas at constant volume is 0.651 and the molecular weight is 32.

If 0.8 kg of this gas occupies a volume of 0.6 m³ at a pressure of 2 bar, calculate the internal energy of the gas if the value of internal energy at –40°C be assumed zero.

*Soln.*   $R = \dfrac{R_0}{M} = \dfrac{8.314}{32} = 0.260$ kJ/kg k

Again $C_p = C_v + R$  $\therefore$  $C_p = 0.651 + 0.260 = 0.911$ *Ans.*

Given, $pV = mRT$

$\therefore$   $2 \times 10^5 \times 10^{-3} \times 0.6 = 0.8 \times 0.260 \times T$ or, $T = 576.92$ k

Now,   $-40°C = -40 + 273 = 233$ k

$\therefore$   $\Delta U = m\, C_v\, [576.92 - 233] = 0.8 \times 0.651 \times 343.92$

$= 179.11$ kJ *Ans.*

**5.2** Two kg mole of $N_2$ is contained in a vessel of volume 5 m³ at 373 k. (i) Evaluate the mass, the pressure and the specific volume of gas (ii) If the ratio of specific heat is 1.4, calculate the value of $C_p$ and $C_v$. (iii) If the gas is cooled to 30°C what is the final pressure? (iv) Also find the inverse in specific value of internal energy, enthalpy and entropy. Also find the heat transfer.

*Soln.*   (i) Molecular weight of $N_2 = 28$

$\therefore$ mass of gas = 2 mole = $28 \times 2 = 56$ kg *Ans.*

$R = \dfrac{R_0}{M} = \dfrac{8.314}{28} = 0.296$ kJ/kg k.

Again as $pV = mRT$ $\therefore$  $P = \dfrac{mRT}{V} = \dfrac{56 \times 0.296 \times (273 + 100)}{5}$ N/m²

$= \dfrac{56 \times 0.296 \times 373}{5 \times 10^5} = 0.0123$ bar *Ans.*

Specific volume $v = \dfrac{V}{m} = \dfrac{5}{56} = 0.089$ m³/kg. *Ans.*

(ii)   Given $\dfrac{C_p}{C_v} = 1.4$ $\therefore$ $C_p = 1.4\, C_v$

Also since, $C_p - C_v = R$

$\therefore$   $1.4\, C_v - C_v = R$  or,  $C_v = \dfrac{0.2961}{0.4} = 0.740$ kJ/kg k

$C_p = 1.4\, C_v = 0.740 \times 1.4 = 1.036$ kJ/kg k.

(iii) We know $\dfrac{p_1 v_1}{T_1} = \dfrac{p_2 v_2}{T_2}$ But $V_1 = V_2$ as vessel has fixed volume.

Here $T_1 = 100 + 273 = 373$, $T_2 = 273 + 30 = 303$ k

$\therefore$   $P_2 = \dfrac{P_1 \times T_2}{T_1} = \dfrac{0.0123 \times 303}{373} = 0.0099$ bar *Ans.*

(iv)
$$\Delta u = u_2 - u_1 = C_v(T_2 - T_1) = 0.740 \ (303 - 373)$$
$$= 0.740 \times (-70) = -51.8 \text{ kJ/kg } Ans.$$
$$\Delta h = h_2 - h_1 = C_p(T_2 - T_1) = 1.036 \ (-70) = -72.52 \text{ kJ/kg } Ans.$$

For constant volume process change in entropy per kg is given by, $\Delta s = C_v \ \ln \dfrac{T_2}{T_1} = 0.740 \ \ln \dfrac{303}{373}$

$= -0.1538$ kJ/kg k. *Ans.*

From first law $q = W + \Delta u = \Delta u$ as $W = 0$ (constant volume process)

$\therefore$ $\qquad\qquad\qquad\qquad q = -51.8$ kJ/kg.

$\therefore$ $\qquad\qquad$ Total heat transfer $Q = m \times q = -56 \times 51.8 = -2900.8$ kJ *Ans.*

The – ve sign indicates that heat is rejected by the gas.

**5.3** An insulated vessel of capacity 1.136 $m^3$ is filled with a gas at a pressure of 12.36 bar abs. The gas is stirred by a paddle till the pressure increases to 20.60 bar abs. Find the change in enthalpy, work input, heat transfer and change in entropy per kg of gas.
Assume $C_p = 0.920$ kJ/kg k, $C_v = 0.659$ kJ/kg k.

*Soln.* $\qquad\qquad C_p - C_v = R \ \therefore \ R = (C_p - C_v) = (0.920 - 0.659) = 0.261$ kJ/kg k

$$h_2 = u_2 + p_2v_2, \ h_1 = u_1 + p_1v_1$$

$\therefore$ $\qquad\qquad \Delta H = m(u_2 - u_1) + (p_2 - p_1) \ V \ [\because \ V_1 = V_2 = V]$

$$= mC_v(T_2 - T_1) + (p_2 - p_1) \ XV = C_v \ (mT_2 - mT_1) + (p_2 - p_1)V$$

$$= C_v \left( \frac{P_2 V_2}{R} - \frac{P_1 V_1}{R} \right) + (p_2 - p_1) \ V \ (\because \ pV = mRT)$$

$$= \frac{C_v}{R} \times V(p_2 - p_1) + (p_2 - p_1) \ V$$

$$= \frac{0.659}{0.261} \times 1.136(20.60 - 12.36) \times 10^5 + (20.60 - 12.36) \times 10^5 \times 1.136$$

$$= 3299535.9 \text{ J} = 3.299 \text{ MJ } Ans.$$

From the First Law $Q = W + \Delta U$
Here $Q = 0$ as vessel is insulated

$\therefore$ $\quad 0 = W + \Delta u$ or, $W = -\Delta U = -\dfrac{C_v}{R} (P_2 - P_1) \ V = \dfrac{0.659}{0.261} (20.60 - 12.36) \times 10^2 \times 1.136$

$$= -2363.47 \text{ kJ } Ans.$$

$$\Delta s = C_v \ \ln \frac{T_2}{T_1} = 0.659 \ \ln \frac{P_2}{P_1} = 0.659 \ \ln \frac{20.60}{12.36} = 0.337 \text{ KJ/kg k. } Ans.$$

**5.4** 3 kg of an ideal gas is expanded from a pressure of 6.86 bar abs. and volume 1.5 $m^3$ to a pressure 1.37 bar abs. and volume 4.5 $m^3$. The specific heat at constant volume for the gas is 1.05 KJ/kg k and decrease in internal energy is 525 KJ. calculate, (i) gas constant, (ii) decrease in enthalpy and (iii) initial and final temperatures.
$\hfill$ (Patna University 1989)

*Soln.* $\ m = 3$ kg, $p_1 = 6.86$ bar, $V_1 = 1.5 \ m^3$, $p_2 = 1.37$ bar

$V_2 = 4.5 \ m^3$, $C_v = 1.05$ kJ/kg k, $u_1 - u_2 = 525$ kJ.

Now,
$$U_1 - U_2 = m\, C_v(T_1 - T_2) \text{ or, } 525 = 3 \times 1.05\,(T_1 - T_2)$$

$$\therefore \quad (T_1 - T_2) = \frac{525}{3 \times 1.05} = 166.667 \text{ kJ} \tag{1}$$

Again
$$\frac{p_1 V_1}{T_1} = \frac{p_2 V_2}{T_2} \therefore \frac{T_1}{T_2} = \frac{p_1 V_1}{p_2 V_2} = \frac{6.86 \times 1.5}{1.37 \times 4.5} = 1.669$$

or
$$T_1 = 1.669\, T_2 \tag{2}$$

From (1) and (2), $166.667 = (1.669\, T_2 - T_2)$ or, $T_2 = \dfrac{166.667}{0.669}$

or,
$$T_2 = 249.12 \text{ k } Ans.$$

$T_1 = 1.669\, T_2 = 415.78 \text{ k. } Ans.$

$$P_1 V_1 = mRT_1 \text{ or, } R = \frac{P_1 V_1}{mT_1} = \frac{6.86 \times 10^2 \times 1.5}{3 \times 415.78} = 0.8249 \text{ kJ/kg k. } Ans.$$

We know, $C_p - C_v = R$   or,   $C_p = R + C_v$

$$\therefore \quad C_p = 0.8249 + 1.05 = 1.8749 \text{ kJ/kg k.}$$

$\therefore \quad$ decrease in enthalpy $= m(h_1 - h_2) = m\, C_p(T_1 - T_2)$

$$= 3 \times 1.8749\,(416.65 - 250) = 3 \times 1.8749 \times 166.65$$

$$= 937.356 \text{ kJ } Ans.$$

**5.5**  5 m$^3$ of a gas at 7.84 bar and 180°C are heated at constant pressure until the volume is doubled. $C_p = 1.00$ and $C_v = 0.709$. Determine the change in internal energy and the external work done during the process.

*Soln.*  Here, $V_1 = 5$ m$^3$, $T_1 = 180 + 273 = 453$ k.

$P_1 = 7.84$ bar $= P_2$, $V_2 = 10$ m$^3$

$\therefore$ from $\dfrac{P_1 V_1}{T_1} = \dfrac{P_2 V_2}{T_2}$ as $P_1 = P_2$, $\dfrac{V_1}{T_1} = \dfrac{V_2}{T_2}$

or
$$T_2 = \frac{T_1 V_2}{V_1} = \frac{453 \times 10}{5} = 906 \text{ k.}$$

Again as $pV = mRT \therefore m = \dfrac{pV}{RT}$

But $R = (C_p - C_v) = (1.00 - 0.709) = 0.291$ kJ/kg k.

$$\therefore \quad m = \frac{7.84 \times 10^2 \times 5}{0.291 \times 453} = 29.73 \text{ kg}$$

$\therefore$ Heat added during the process

$$Q_{1-2} = m\, C_p(T_2 - T_1) = 29.73 \times 1.00\,(906 - 453) = 13467.69 \text{ kJ } Ans.$$

External work done during the process

$$W_{1-2} = P_1(V2 - V_1) = 7.84 \times 5 = 3920 \text{ kJ }  Ans.$$

From First law, $\Delta U = Q - W$

$$\therefore \qquad U_2 - U_1 = 13467.69 - 3920 = 9547.69 \text{ kJ } Ans.$$

**5.6** 0.5 kg of air initially at 1 bar and 156°C is compressed isothermally till the volume is reduced to 0.14 m³. Determine the work done, heat exchange, change in internal energy and entropy.

*Soln* We know $pV = mRT$

$$\therefore \qquad V_1 = \frac{mRT_1}{P_1} = \frac{0.5 \times 0.287 \times (273 + 156) \times 10^3}{10^5} = 0.615 \text{ m}^3$$

(i) Work done/kg $= p_1 v_1 \ln \dfrac{v_2}{v_1}$

$$\therefore \qquad \text{Total work done} = \text{mass} \times p_1 v_1 \ln \frac{v_2}{v_1}$$

$$= p_1 \times V_1 \ln \frac{v_2}{v_1} = 10^5 \times 0.615 \ln \frac{0.14}{0.615} \text{ Joules} = -91 \text{ kJ}$$

−ve sign indicates that work is added to the system. For isothermal process of a perfect gas change in internal energy is zero.

$$\therefore \qquad \text{Work done} = \text{Heat transfer} = -91 \text{ kJ. Heat is lost from the system.}$$

$$\text{Change in entropy} = m \times (s_2 - s_1) = mR \ln \frac{v_2}{v_1} = 0.5 \times 0.287 \ln \frac{0.14}{0.615} = -0.212 \text{ kJ/kg k } Ans.$$

**5.7** Air is compressed in a reversible steady flow polytropic process from 1.89 bar and 40°C to 10.1 bar following the law $Pv^{1.25} = $ constant. Neglecting any change in kinetic and potential energies, find the work done and heat transfer per kg of air.

*Soln.*
$$\frac{T_2}{T_1} = \left(\frac{P_2}{P_1}\right)^{\frac{n-1}{n}} = \left(\frac{10.1}{1.89}\right)^{\frac{0.25}{1.25}} = 1.398$$

or $\qquad T_2 = T_1 \times 1.398 = (273 + 40) \times 1.398 = 437.57 \text{ k} = 164.57°C$

Work done during a steady flow process $= -\displaystyle\int v \, dp$

$$\therefore \qquad \text{Work} = -\int_1^2 v \, dp = \frac{nR}{n-1}(T_1 - T_2)$$

In a reversible flow polytropic process $pv^n = C$

$$\text{Work done} = -\int_1^2 v \, dp = -\int_1^2 \left(\frac{C}{P}\right)^{\frac{1}{n}} dp = -C^{\frac{1}{n}} \int_1^2 p^{-\frac{1}{n}} dp$$

$$= -C^{\frac{1}{n}} \left[\frac{p^{1-\frac{1}{n}}}{1-\frac{1}{n}}\right]_1^2 = \frac{-n}{n-1} \times C^{\frac{1}{n}} \times \left(p_2^{\frac{n-1}{n}} - p_1^{\frac{n-1}{n}}\right) = -\frac{n}{n-1} \times \left[C^{\frac{1}{n}} p_2^{\frac{n-1}{n}} - C^{\frac{1}{n}} p_1^{\frac{n-1}{n}}\right]$$

$$= -\frac{n}{n-1} \left[p_2^{\frac{1}{n}} \times v_2 \times p_2^{\frac{n-1}{n}} - p_1^{\frac{1}{n}} \times v_1 \times p_1^{\frac{n-1}{n}}\right]$$

$$= -\frac{n}{n-1} [p_2 v_2 - p_1 v_1] = -\frac{n}{n-1}[RT_2 - RT_1] = \frac{+nR}{n-1}(T_1 - T_2)$$

$$\therefore \qquad \text{Work} = \frac{1.25}{0.25} \times 0.287 \times (40 - 223) = -262.6 \text{ kJ/kg } Ans.$$

(in heat units)

Again in the absence of $k\,E$ and $P\,E$ change,

$$q = W + (h_2 - h_1) = -262.6 + C_p\,(T_2 - T_1)$$

$$= -262.6 + 1.005(164.57 - 40) = -137.40 \text{ kJ/kg } Ans.$$

**5.8**  Two kg of air at 6.86 bar abs. and 90°C pass through a reversible non flow polytropic process represented by $pV^{1.1}$ = constant till the pressure falls to 1.37 bar. Find (a) the final temperature, specific volume and change in entropy (b) work and heat transfer (c) what will be the answers if the process was irreversible and adiabatic between the same end states?

*Soln.*  (a) We know $pV = mRT$

$$\therefore \qquad \text{Specific volume } v_1 = \frac{RT_1}{p_1} = \frac{0.287 \times (90 + 273)}{6.86 \times 10^5 \times 10^{-3}} = 0.151 \text{ m}^3/\text{kg}$$

As the process is polytropic

$$\therefore \qquad p_1 v_1^{1.1} = p_2 v_2^{1.1} \text{ or, } v_2 = v_1 \times \left(\frac{P_1}{P_2}\right)^{\frac{1}{1.1}} = 0.151 \times \left(\frac{6.86}{1.37}\right)^{\frac{1}{1.1}}$$

$$= 0.151 \times (5)^{\frac{1}{1.1}} = 0.656 \text{ m}^3/\text{kg } Ans.$$

$$T_2 = \frac{p_2 v_2}{R} = \frac{1.37 \times 10^2 \times 0.656}{0.287} = 313°\text{K } Ans.$$

Change in entropy for one kg is given by

$$\Delta s = C_v \ln \frac{T_2}{T_1} + R \ln \frac{v_2}{v_1} = 0.714 \ln \frac{313}{363} + 0.287 \ln \frac{0.656}{0.151}$$

$$= -0.105 + 0.421 = 0.316 \text{ kJ/kg°k}$$

$$\therefore \qquad \Delta s \text{ for 2 kg} = 2 \times 0.316 = 0.632 \text{ kJ/k } Ans.$$

(b) Work done for 1 kg of air  $W = R \dfrac{(T_1 - T_2)}{(n - 1)}$

$$= 0.287 \times \frac{(363 - 314)}{1.1 - 1} = 140.63 \text{ kJ/kg.}$$

$$\therefore \qquad \text{Total work done} = 140.63 \times 2 = 281.26 \text{ kJ } Ans.$$

(c)  As the end states are same the changes in properties do not vary as they are independent of path and depend on end states only.

$$\therefore T_2 = 314 \text{ k, } v_s = 0.656 \text{ m}^3/\text{kg}$$

$$\Delta S = 0.632 \text{ kJ/k } Ans.$$

But, heat transfer in this case is zero as the process is adiabatic, $\Delta u$ will not change

As $$q = W_{1-2} + \Delta u$$

$$\therefore \qquad W_{1-2} = -\Delta u = 34.986 \text{ kJ/kg}$$

$$\therefore \qquad \text{Total work done in this case} = 2 \times 34.986 = 69.972 \text{ kJ } Ans.$$

**5.9** 0.5 kg of air at 1 bar and 300 k is compressed isentropically to six times its original pressure. Find the final temperature and the work done. If the air is then cooled at constant pressure to the original temperature what amount of heat is rejected and what further work is done?

If instead of these two processes the final state is obtained by only a reversible isothermal compression what is the work done and heat transfer in this case?

*Soln.* Refer to Fig. 5.15

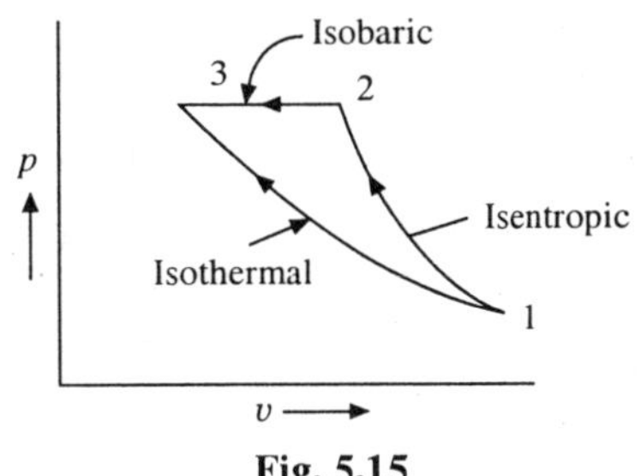



**Fig. 5.15**

$$\frac{T_2}{T_1} = \left(\frac{P_2}{P_1}\right)^{\frac{\gamma-1}{\gamma}} = \left(\frac{6}{1}\right)^{\frac{0.4}{1.4}} \quad \text{or,} \quad T_2 = 300 \times 1.67 = 501 \text{ k } Ans.$$

Process 1–2 is isentropic, that is, adiabatic reversible.

$$\therefore \quad Q_{1-2} = 0 \text{ and } W_{1-2} = -\Delta U_{1-2} = -m \times C_v(T_2 - T_1)$$

$$= -0.5 \times 0.714 \, (501 - 300)$$

$$= -71.757 \text{ kJ } Ans.$$

For isobaric process 2–3, heat rejected $Q_{2-3} = mC_p(T_2 - T_3)$

$$= 0.5 \times 1.009 \, (501 - 300) = 101.404 \text{ kJ } \quad Ans.$$

At this range of temperature $C_{p\,air} = 1.009$ kJ/kg k
Work of compression during 2–3

$$W_{2-3} = mp_2(v_2 - v_2) = mR(T_3 - T_2)$$

$$= 0.5 \times 0.287 \times (300 - 501) = -28.843 \text{ kJ. } Ans.$$

For isothermal compression 1–3

$$W_{1-3} = m\, P_1 V_1 \ln \frac{v_3}{v_1} = m \times RT_1 \ln \frac{p_1}{p_3}$$

$$= 0.5 \times 0.287 \times 300 \ln \frac{1}{6} = -77.135 \text{ kJ } \quad Ans.$$

For isothermal process 1–3, $\Delta U = 0$

$$\therefore \quad\quad\quad Q_{1-3} = W_{1-3} = -77.135 \text{ kJ } Ans.$$

**5.10** (a) Show that the approximate change of entropy during a polytropic process equals the quantity of heat transferred divided by the mean absolute temperature.

(b) One kg of air at a pressure of 1 bar and temperature 20°C is compressed to a pressure of 8 bar according to $pV^{1.25}$ = constant. Find the approximate change of entropy and the correct change of entropy.

*Soln.* Refer to Fig. 5.16

(a) The process is shown in the figure by curve 1–2 on a $T$ - $S$ plot.

The area under the curve 1-2-*a*-*b* gives the heat transfer we see that the area under the straight line 1–2 (dotted shown) is approximately the same on the curvature of the curve 1–2 is small.

$$\therefore \; Q_{approx.} = \text{area 1 2 3 (triangle)} + 3\,2\,a\,b$$

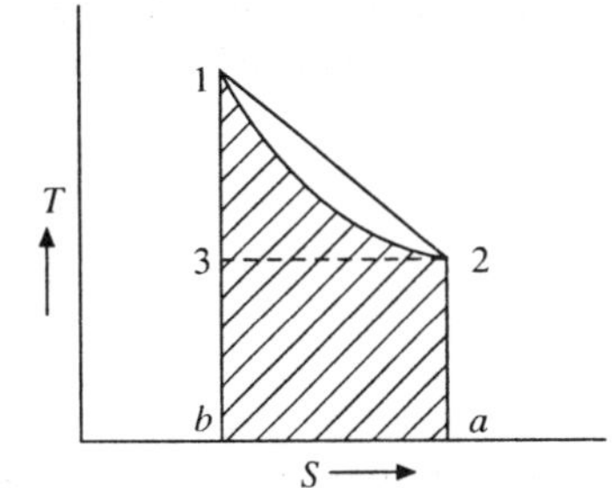



**Fig. 5.16**

$$= \frac{1}{2}(S_2 - S_1)(T_1 - T_2) + T_2(S_2 - S_1)$$

$$= (S_2 - S_1)\left\{T_2 + \frac{T_1 - T_2}{2}\right\} = (S_2 - S_1) \times \frac{T_1 + T_2}{2}$$

or

$$(S_2 - S_1) \times \text{Mean temperature}$$

or,

$$S_2 - S_1 = \frac{Q_{approx}}{\text{mean temp.}} \quad \text{Proved}$$

(b)

$$R = C_p - C_v = 1.005 - 0.714 = 0.291$$

or,

$$R = 0.291 \text{ kJ/kg k}$$

$$v_1 = \frac{RT_1}{p_1} = \frac{0.291 \times 293}{1 \times 10^2} = 0.852 \text{ m}^3/\text{kg}$$

$$T_2 = T_1 \left(\frac{P_2}{P_1}\right)^{\frac{n-1}{n}} = 293 \times (8)^{\frac{0.25}{1.25}} = 449°\text{K}$$

$$v_2 = v_1 \left(\frac{p_1}{p_2}\right)^{\frac{1}{n}} - = 0.948 \times \left(\frac{1}{8}\right)^{\frac{1}{1.25}} = 0.18 \text{ m}^3$$

$\therefore$ Heat transferred during polytropic process/kg

$$= \frac{\gamma - n}{\gamma - n} \times \frac{p_1 v_1 - p_2 v_2}{n - 1} = \frac{1.43 - 1.25}{1.43 - 1} \times \left(\frac{1 \times 0.852 - 8 \times 0.18}{0.25}\right) \times 10^2 = -98.45 \text{ kJ}$$

$\therefore$ Correct change of entropy per kg

$$= S_2 - S_1 = C_v \frac{\gamma - n}{n} \ln \frac{p_1}{p_2} = 0.714 \times \frac{1.43 - 1.25}{1.25} \ln \frac{1}{8}$$

$$= -0.2137 \text{ kJ/kg k} = \text{correct value}$$

$$T_{mean} = \frac{293 + 499}{2} = 371$$

$$(s_2 - s_1)_{approx.} = -\frac{98.45}{371} = -0.2653 \text{ kJ/°k}$$

**5.11**  In a throttling process the air expands from a pressure of 7 bar and temperature of 93°C to a pressure of 3.5 bar. Determine the change in entropy.

*Soln*  The process is irreversible. So, change of entropy for this process cannot be directly calculated. However, since change of entropy depends only on end states its value can be calculated for any reversible process between the same end states. $T_2 = T_1$ during the throttling as $h_2 = h_1$ and for a perfect gas $h = f(T)$ only.

$\therefore$

$$\frac{V_2}{V_1} = \frac{P_1}{P_2} = \frac{7.0}{3.5} = 2.0$$

and

$$s_2 - s_1 = C_v \ln \frac{T_2}{T_1} + R \ln \frac{V_2}{V_1}$$

$$= 0 + 0.287 \ln 2.0 = 0.199 \text{ kJ/kg k} \quad \text{Ans.}$$

**5.12**  Critical pressure and critical volume of helium are $0.23 \times 10^6$ N/m$^2$ and $58 \times 10^{-6}$ m$^3$/mole respectively calcultate Vander Waals constants $a$ and $b$.

*Soln.* Given $\qquad P_c = .0.23 \times 10^6$ N/m$^2$, $V_c = 58 \times 10^{-6}$ m$^2$/mole

$$v_c = 3b \text{ and } P_c = \frac{a}{27b^2}$$

$$b = \frac{v_c}{3} = \frac{58 \times 10^{-6}}{3} = 19.67 \times 10^{-6} \text{ m}^2/\text{mole} \tag{i}$$

$$a = 27\, b^2\, P_c \tag{ii}$$

Substituting the values of $b$ and $P_c$ in equation (ii), we get

$$a = 27\,(19.67 \times 10^{-6})^2 \times 0.23 \times 10^6$$

$$= 2.32 \times 10^{-3} \text{ N–m}^4/\text{mole}^2$$

**5.13** 10 kg of air at 100°C is stored in a rigid cylinder of volume 0.05 cubic meters. Calculate the pressure using Vander Waals equation of state. The properties of air at critical point are $P_c = 38.467$ bar, $T_c = 137.24$ k and $v_c = 0.093$ m$^3$/kg mole. $\qquad$ [Nov. 95, Mumbai University]

*Soln.* Vander Waals equation is written as

$$\left( P = \frac{a}{v^2} \right)(\bar{v} - b) = R_0 T, \qquad \bar{v} = \text{molal volume}$$

$R_0$ = Universal gas const.

$a = 3\, P_c\, v_c^2, \ b = v_c/3$

$$a = 3\, P_c\, v_c^2$$

$$= 3 \times 38.467 \times 10^5 \times 10^{-3} \times 0.093^2$$

$$= 99.81 \text{ kN-m}^4/(\text{kg-mole})^2$$

$$b = v_c/3 = 0.093/3 = 0.031 \text{ m}^3/\text{kg mole}$$

Pressure $\qquad\qquad P = \dfrac{R_0 T}{\bar{v} - b} - \dfrac{a}{\bar{v}^2};$

molal volume $\qquad\qquad \bar{v} = \dfrac{0.059 \times 29}{10} = 0.145 \text{ m}^3/\text{kg mole}$

or, $\qquad\qquad P = \dfrac{8314.4 \times (100 + 273)}{0.145 - 0.031} - \dfrac{99.81 \times 10^3}{(0.145)^2}$

$$= 22456928 \text{ N/m}^2 = 224.569 \text{ bar } Ans.$$

Now, using ideal gas equation, we get

$$R = \frac{R_0}{M} = \frac{8314.4}{29} = 286.703 \text{ N-m/kg-k}$$

using ideal gas equation

$$P = \frac{mRT}{v} = \frac{10 \times 286.703\,(100 + 273)}{0.05} = 21388044 \text{ N/m}^2 = 213.88 \text{ bar}$$

**5.14** The specific volume of $CO_2$ at 100°C is 1 m$^3$/kg. (i) Determine the pressure exerted by $CO_2$ using Vander Waals equation (ii) Compare the result obtained if $CO_2$ is treated as ideal gas. Take $a = 362\,850$ Nm$^4$/(kg - mole)$^2$, $b = 0.0422$ m$^3$/kg mole $\qquad$ [May 98, Mumbai University]

*Soln.*  molal volume $\bar{v} = 1 \times 44 = 44$ m³/kg mde

$$R_0 = 8314.4 \text{ J/kg mole } k$$

From Vander Waals equation

$$P = \frac{R_0 T}{\bar{v} - b} - \frac{a}{\bar{v}^2} = \frac{8314.4\,(100 + 273)}{44 - 0.0422} - \frac{362850}{44^2}$$

$$= 70363.679 \text{ N/m}^2 = 0.7036 \text{ bar Ans.}$$

Now, treating $CO_2$ as ideal gas

$$P = \frac{RT}{v}; \qquad R = \frac{R_0}{M} = \frac{8314.4}{44} = 188.96 \text{ J/kg k}$$

$$= \frac{188.96\,(100 + 273)}{1} = 70482.08 \text{ N/m}^2$$

$$= 0.7048 \text{ bar} \quad Ans.$$

**5.15**  Determine the specific volume of steam at a pressure of 20 MPa and temperature 600°C

(i) using ideal gas equation
(ii) using vander waals equation.

The constants in Vander Waals equation are
$a = 5.538 \times 10^5$ Nm⁴/(kg-mole)² $b = 0.0305$ m³/mg mole

*Soln.*  (i) Using ideal gas equation $pv = RT$ for unit mass

$$\therefore \qquad v = \frac{RT}{P} = \frac{8314.4/18 \times (600 + 273)}{20 \times 10^6} = 0.02016 \text{ m}^3/\text{kg} \quad \text{Ans.}$$

(ii) Using Vander Waals equation

$$\left(P + \frac{a}{\bar{v}^2}\right)(\bar{v} - b) = R_0 T$$

$$\left(20 \times 10^6 + \frac{5.538 \times 10^5}{\bar{v}^2}\right)(\bar{v} - 0.0305) = 8314.4(600 + 273)$$

Solving this equation by trial and error, we get

$$\bar{v} = 0.3222$$

$$\therefore \qquad \text{Specific volume} = \frac{\bar{v}}{M} = \frac{0.3222}{18} = 0.0179 \text{ m}^3/\text{kg} \quad Ans.$$

## EXERCISES

5.1  Objective Questions (say true or false)

  (i)   All ideal gases have the same molar heat capacity at constant pressure.
  (ii)  The molar heat capacity at constant volume of an ideal gas is independent of temperature.
  (iii) The molar capacity at constant pressure of an ideal gas is independent of temperature.
  (iv)  Specific heat of an ideal gas is a function of temperature only.
  (v)   Specific enthalpy of an ideal gas is a function of temperature only.
  (vi)  In true sense there is no ideal gas and therefore the ideal gas equations are useless.

(vii)   The work done during a polytropic process between same and states is the both for a perfect gas and a real gas.

(viii)  If a system undergoes a process during which its entropy does not change, it is necessarily true that the process is reversible and adiabatic.

(ix)   If stirring work is done on an ideal gas in a closed system during a constant volume process, then $dQ \neq C_v\, dT$

| Ans. | (i) | (ii) | (iii) | (iv) | (v) | (vi) | (vii) | (viii) | (ix) |
|---|---|---|---|---|---|---|---|---|---|
|  | F | F | T | T | T | F | F | F | T |

5.2.   For each statement three answers are given. Write down the correct answer for each:

(a)   In a reversible steady flow compression (neglecting KE, PE and heat energy the work supplied is equal to drop in enthalpy/increase in internal energy/increase in entropy.

(b)   For a constant volume process the exponent $n$ in the relation $pv^n$ = constant is zero/infinity/one.

(c)   In a cyclic process the change in internal energy of a closed system is positive/negative/zero.

(d)   In a constant volume process the work done is positive. The process must be reverisble/irreversible/ may be any of two.

(e)   For a reversible adiabatic process volume/entropy/temperature must remain constant.

Ans. (a) drop in enthalpy (b) infinity (c) zero (d) irreversible (e) entropy


# NUMERICAL QUESTIONS.

5.3   Calculate the volume occupied by 3 kg of hydrogen gas having the gas constant $R = 4.127$ kJ/kg k at a pressure of 1.96 bar and temperature of 25°C *Ans.* 1881.8 m$^3$.

5.4   An oxygen cylinder 30 cm in diameter and 120 cms long contains the gas at 17.16 bar *abs.* and 300 k. After utilising some of the oxygen the pressure in the tank was 13.74 bar *abs.* and the temperature was 293 k. Calculate the percentage of oxygen used                                      *Ans.* 18.05 percent

5.5   If oxygen at N.T.P. has a specific volume of 0.7 m$^3$ and ratio of specific heats at constant pressure to that at constant volume be 1.4 find the values of the specific heats.

5.6   A perfect gas is cooled from 473 k and 20.6 bar *abs.* to a temperature of 333 k. If the process takes place at a constant volume of 0.135 m$^3$, find the final pressure, decrease of internal energy, work done, heat rejected and change of entropy.
(Given $C_p = 1.008$ kJ/kg $k$; $C_v = 0.756$ kJ/kg k)

(*Ans.* 14.51 bar, 233 kJ, 233 kJ, zero, –0.14)

5.7   14.8 kg of a perfect gas at 7.85 bar and 180°C are heated at constant pressure until its volume is doubled. If the gas constant $R$ is equal to 0.292 kJ/kgk and $\gamma = 1.4$, determine the change in internal energy and external work done.

(*Ans.* 4754.4 kJ, 1962.67 kJ)

5.8   500 litres of air is compressed isothermally till its volume is halved. The initial pressure is 14.7 bar and the entropy decreases by 1.05 kJ/k during the process. Find the initial temperature, mass of gas and heat transferred.

(*Ans.* 15°C, 5.62 kg, –302.4 kJ)

5.9   An air compressor comprsses 800 m$^3$ of air per minute from one atmospheric pressure to 4.5 atmosphere. Law of compression is $pV^{1.25}$ = constant. If only 88 percent of work supplied to compressor is used in compression process, find the power of compressor.

(*Ans.* 2082.88 kW)

5.10   2 cu.m of air at 423 k and 4 bar is expanded to one bar reversibly and adiabatically and then heated isobarically by the addition of 140 kJ. If the same final condition is reached by a polytropic process determine the index of expansion.

(*Ans.* $n = 1.305$)

5.11   3.54 litres of air at a pressure of 30.5 bar is expanded isentropically till the absolute temperature is halved. Find the final volume and pressure of air. Also find the work done and change in internal energy.

*(Ans.* 20 lit, 2.7 bar abs., 13.49 kJ, −13.44 kJ)

5.12   One kg. of air at 305 k is expanded in a steady flow reversible process according to the law $p_v^{1.25} = C$ until the pressure falls by 50 percent. If the initial velocity of air be 305 m/sec. and final velocity be 122 m/sec. compute $\Delta h$, $\Delta s$, $\Delta w$ and $q$.

*Ans.* (39.9 kJ, 0.06 kJ/k, 49.22 kJ and 16.8 kJ)

5.13   15 kg of air at 250°C is stored in a rigid cylinder of volume 0.07 cubic meters. Calculate the pressure using vander waals equation of state. The properties of air at critical point are

$$P_c = 38.467 \text{ bar}, \ T_c = 132.4 \text{ k}$$

and
$$v_c = 0.093 \text{ m}^3/\text{kg mole}$$

[May 97, Mumbai University]

5.14   Calculate "$a$" for dry air, given that $T_c = 132$ k, $P_c = 37.2$ atmosphere and $R$ per mole $= 82.07$ cm$^2$ atm k.

5.15   Calculate the pressure of $CO_2$ at 100°C and having a mass of 20 kg and a volume 1200 litres. The constants $a$ and $b$ of vander waals equation are

$$a = 3.6235 \ \frac{\text{N}-\text{litre}^2}{(\text{gm-mole})^2}, \ b = 0.6423 \ \frac{\text{litre}}{\text{gm-mole}}$$

[May 96, Mumbai University]

5.16   Compute from Vander Waals equation; the pressure exerted by 20 kg by hydrogen at 200°C if specific volume is 1 m$^3$/kg. $a = 25.105 \times 10^3$ Nm$^4$/kg mole$^2$, $b = 0.0262$ m$^3$/kg mole.

[Nov. 96, Mumbai University]

5.17   The pressure and temperature of nitrogen contained in a 29 m$^3$ vessel are 21 MPa and 210 k respectively. Calculate the mass of nitrogen using.

 (i) The ideal gas equation of state
 (ii) The generalized compressibility chart given for hydrogen $P_c = 3.4$ MPa, $T_c = 126.2$ k

**6**

# The Gas Power Cycle

## 6.1  INTRODUCTION

In the indirect energy conversion systems, or conventional systems, the chemical energy of fuel is first converted into heat through combustion process and then the heat is converted into work. Examples of such systems are internal combustion engines, gas turbines or steam power plants. In these systems the working fluid of some of them execute a cyclic process. The essential elements of a thermodynamic power cycle are as under:

(i)  A working fluid which undergoes a change of state as it receives or rejects heat and produces work.

(ii)  A heat source to supply heat to the fluid and a heat sink to which heat may be rejected.

(iii)  An arrangement is essential to get the work out of the working fluid. This may be done by a piston-cylinder or a turbine or any other arrangement.

In this chapter we shall limit our analysis of the power plants which use gas as the working fluid and execute a cyclic process for examples a gas turbine or an internal combustion engine.

## 6.2  TERMS USED IN PISTON CYLINDER ARRANGEMENT

Let us take the case of a petrol engine and consider its working. A piston is fitted closely inside a cylinder as shown in Fig. 6.1. The two extreme positions of the piston are known as inner dead centre (I.D.C.) position and outer dead centre (O.D.C.) position in a horizontal engine and top dead centre and bottom dead centres (T.D.C. and B.D.C.) in vertical engines. The space between the cylinder head and the I.D.C. is called clearance volume. The volume between the I.D.C. and O.D.C. is called stroke volume or swept volume. An inlet valve and an exhaust valve are fitted on the cylinder head to admit the working fluid and to discharge it. These are one way valves and so they allow only unidirectional flow of fluid. The piston is connected to the crank by a rod called

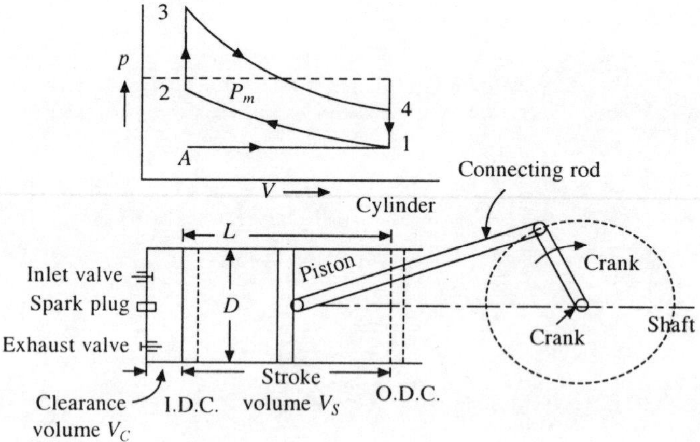

**Fig. 6.1**

the connecting rod. The crank is keyed to the shaft. Any linear motion of the piston is converted into rotary motion of the shaft by means of the connecting rod and the crank.

*Working*: A mixture of air and petrol is prepared outside the cylinder in a unit called carburettor (not shown here). First to start the engine the crank shaft is rotated by means of a handle in the direction shown by the arrow. This results into the movement of the piston from its I.D.C. towards the O.D.C. A partial vacuum is created in the space between the cylinder head and the piston and the inlet valve opens and admits the air petrol mixture. This continues till the piston reaches the O.D.C. when the inlet valve closes. The exhaust valve was kept closed during this period. This stroke of the piston is called suction stroke. After reaching the O.D.C. the piston returns towards the I.D.C. with both valves closed and compresses the air petrol mixture which rises in temperature and pressure. This stroke is called compression stroke. When the piston reaches the I.D.C. a spark is provided by means of a spark plug and the fuel is ignited. As the fuel burns and gases are produced the pressure inside the cylinder increases tremendously which forces the piston towards the O.D.C. and work is produced. This is called the working or expansion stroke. When the piston reaches the O.D.C. it returns due to inertia of a flywheel mounted on the crank shaft. The exhaust valve now opens and the burnt gases are exhausted out. This completes the cycle. In the next cycle a fresh air-fuel mixture will be taken in and the process continues.

The change of state of the working fluid is shown on the $p$-$V$ plot. Process $A$-1 indicates the suction at constant pressure, 1-2 the adiabatic compression, 2–3 the combustion of fuel and sudden rise of pressure, 3-4 the adiabatic expansion of gases and 4-1-A the exhaust. Stroke volume $= \dfrac{\pi}{4} D^2 \times L$ where $D$ is cylinder diameter and $L$ stroke length.

We see that the total volume of the cylinder = clearance volume + swept volume

$$= V_c + V_s \qquad (6.1)$$

Compression ratio: The ratio of the cylinder volume to the clearance volume is called compression ratio and is denoted by $r$

Thus,
$$r = \frac{V_c + V_s}{V_c} = 1 + \frac{V_s}{V_c} \qquad (6.2)$$

*Mean Effective Pressure* (mep): We know that during a cyclic process the net work done is given by the enclosed area on the $p$-$v$ plot. It is noted that the pressure is changing during the cyclic process. We can think of an average pressure which if exerted on the piston would give the same work done in only one stroke as the total work done in the entire cycle. This average pressure is called mean effective pressure. Thus, mean effective pressure is defined as an imaginary constant pressure which, if exerted on the piston during one stroke would develop the same network for the cycle as was obtained with variable pressure.

Thus mean effective pressure, $\quad p_m = \dfrac{\text{Work done/cycle}}{\text{Stroke volume}} \qquad (6.3)$

or, $\qquad\qquad$ work done per cycle = $p_m \times$ stroke volume

or, $\qquad\qquad W_{\text{cycle}} = p_m \, AL$ when $A$ = area of cross section of piston

$$= \frac{\pi}{4} D^2$$

or, $\qquad\qquad$ Work done per minute = $p_m \, L \, A \, N$, $N$ being the rpm

Thus, the indicated power of the engine

$$\text{IP} = \frac{p_m \, L \, A \, N}{60000} \, kW \qquad (6.4)$$

## 6.3.  AIR STANDARD CYCLES

The actual analysis of an internal combustion engine is extremely complex due to the many variables that must be considered. The actual cycle deals with a mixture of air and fuel. The air standard analysis of an internal combustion cycle was developed to give a simplified analysis of the performance for use as a guide to the engineer. It also shows the influence of such factors as the compression ratio on the efficiency. Air is assumed to be the working medium, thus removing the effect of different media used in different cycles (petrol and air in otto cycle, and diesel and air in diesel cycle). The specific heat of air is assumed to be constant and the air is assumed to follow the perfect gas laws. The actual cycle is replaced by a cycle made up of reversible processes which approximate as closely as possible the actual cycle. In actual cycle the heating of media is done by combustion of fuel and cooling by rejection of hot exhaust gases. In air standard cycle the heating is supposed to be done by bringing a hot source in contact with the cylinder head and cooling by bringing a cold sink in contact with the cylinder head. An important difference between the two cycles is that in an actual cycle in every cycle fresh air and fuel are admitted. Then the cycle is an open cycle, whereas in case of air standard cycle the same air goes on repeating the cycles, that is, there is no exhaust or suction. Thus it forms closed cycles.

The thermal efficiency of air standard cycle is given by,

$$\text{Thermal efficiency} = \frac{\text{Work done}}{\text{Heat supplied}} = \frac{\text{Heat supplied} - \text{Heat rejected}}{\text{heat supplied}}$$

## 6.4   THE CARNOT CYCLE AIR STANDARD EFFICIENCY

As discussed earlier in Chapter 4 the Carnot cycle is an ideal cycle, however, on which no engine really works. It helps in estimating the maximum possible efficiency, that can be obtained in any heat engine working under the given temperature limits. The cycle (Fig. 6.2) comprises two isothermal and two isentropic processes. We have already realised the efficiency of a Carnot cycle, however, the same can be found in terms of the compression ratio

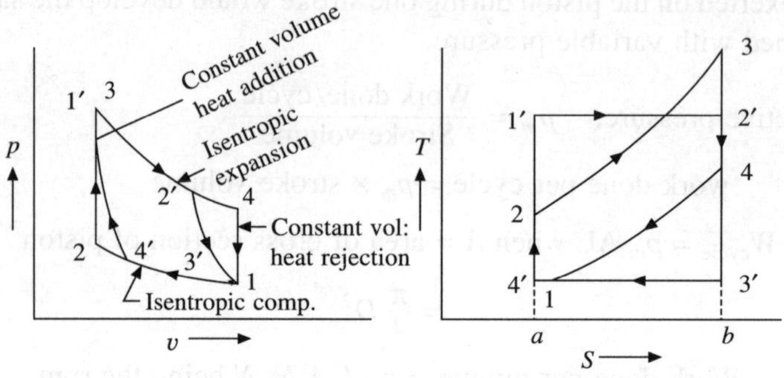



**Fig. 6.2   Carnot cycle**

Heat added/cycle (process 1′–2′)

$$Q_1 = \text{Area } 1' - 2' - b - a = T_{1'}(S_{2'} - S_{1'})$$

Heat rejected/Cycle (Process 3′–4′)

$$Q_2 = \text{Area } 3' - 4' - a - b = T_{3'}(S_{3'} - S_{4'})$$

$\therefore$

$$\eta \text{ of Carnot cycle} = \frac{T_{1'}(S_{2'} - S_{1'}) - T_{3'}(S_{3'} - S_{4'})}{T_{1'}(S_{2'} - S_{1'})}$$

$$= 1 - \frac{T_{3'}}{T_{1'}} \quad (\because S_{2'} - S_{1'} = S_{3'} - S_{4'})$$

$$= 1 - \frac{T_{3'}}{T_{2'}} \quad (\because T_{2'} = T_{1'})$$

$$= 1 - \left(\frac{V_{2'}}{V_{3'}}\right)^{\gamma - 1}$$

$$\left[ \because \text{ for isentropic process } 2'-3', \ \frac{T_{3'}}{T_{2'}} = \left(\frac{v_{2'}}{v_{3'}}\right)^{\gamma - 1} \right]$$

or,

$$\eta = 1 - \frac{1}{r^{\gamma - 1}} \tag{6.5}$$

where $r = \dfrac{v_{3'}}{v_{2'}} = \dfrac{v_{4'}}{v_{1'}}$; the adiabatic compression or expansion ratio

## 6.5  THE OTTO CYCLE OR CONSTANT VOLUME CYCLE

The petrol engines work on the Otto cycle. To find the air standard efficiency of Otto cycle we require air as the working fluid. This was invented by a German engineer Nikolaus August Otto (1832–91) in 1876.

The different processes of the cycle have been shown on $p$-$v$ and $T$-$S$ planes in Fig. 6.3.

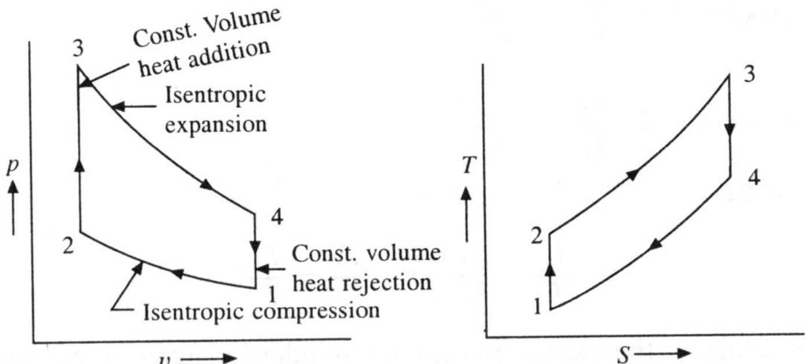


**Fig. 6.3  Otto cycle**

The various processes are:

Process 1–2: Insentropic compression of air from pressure $p_1$ to $p_2$. Work is done on the system and the temperature of gas increases. The volume of gas decreases from $V_1$ to $V_2$ $(= V_C$, the clearance volume). No heat is added or rejected. This is compression stroke.

Process 2–3: Constant volume heating of air takes place. Heat is added from a hot source (hot reservoir) to the gas reversibly by bringing it in contact with a cylinder head. The pressure rises from $p_2$ to $p_3$ while volume remains constant so there is no work done during the process.

Process 3–4. Isentropic expansion of hot air takes place. While the volume of air increases its temperature and pressure fall. The state of air is indicated by $p_4$, $V_4$ and $T_4$. During this process work is done by air but there is no heat addition or rejection.

Process 4–1: This is a constant volume heat rejection process during which heat is rejected to the cold sink by bringing it in contact with the cylinder head. There is no work done during this process as there is no change of volume.

The cycle is thus completed as the air has returned to its initial state $p_1$, $v_1$, $T_1$.

Now heat added per kg of air (during process 2–3)

$$= C_v(T_3 - T_2)$$

Similarly heat rejected per kg of air (during process 4–1) $= C_v(T_4 - T_1)$. There is no heat addition or rejection during process 1–2 and 3–4.

$\therefore$  Net heat added/kg $= C_v(T_3 - T_2) - C_v(T_4 - T_1)$

We know that in a cyclic process: $Q_{Net} = W_{net}$

$\therefore$  Net work done per kg during the cycle $= C_v(T_3 - T_2) - C_v(T_4 - T_1)$

$\therefore$ Air standard efficiency

$$\eta = \frac{C_v(T_3 - T_2) - C_v(T_4 - T_1)}{C_v(T_3 - T_2)} = 1 - \frac{T_4 - T_1}{T_3 - T_2} = 1 - \frac{T_1\left(\dfrac{T_4}{T_1} - 1\right)}{T_2\left(\dfrac{T_3}{T_2} - 1\right)}$$

But for isentropic process 3–4, $\dfrac{T_3}{T_4} = \left(\dfrac{v_4}{v_3}\right)^{\gamma-1}$

and for isentropic process 1–2, $\dfrac{T_2}{T_1} = \left(\dfrac{v_1}{v_2}\right)^{\gamma-1}$

But $v_1 = v_4$ and $v_2 = v_3$

$$\therefore \qquad \frac{T_2}{T_1} = \frac{T_3}{T_4} \quad \text{or,} \quad \frac{T_3}{T_2} = \frac{T_4}{T_1}$$

Now, putting these values in the expression for air standard efficiency we get,

$$\eta = 1 - \frac{T_1}{T_2} = 1 - \frac{1}{\left(\dfrac{v_1}{v_2}\right)^{\gamma-1}} = 1 - \frac{1}{(r)^{\gamma-1}} \tag{6.6}$$

It is obvious from the above equation that the air standard efficiency of the otto cycle depends upon the compression ratio only and it increase with increase in compression ratio Fig. 6.4. The compression ratio in an actual engine lies between 5 and 10 and depends upon the type of fuel used.

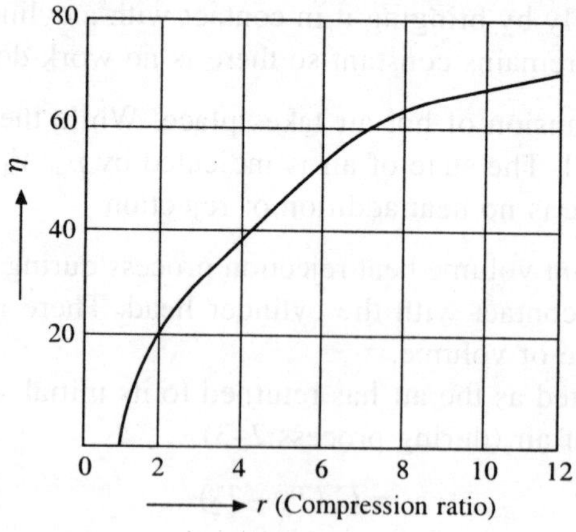


**Fig. 6.4**

*Mean Effective Pressure for Otto Cycle*
Mean effective pressure is defined as the constant net pressure acting on the piston that will produce the same amount of work as produced by the actual varying pressure produces during the cycle.

$$\text{m.e.p} = P_m = \frac{\text{work done per cycle (N–m)}}{\text{Stroke volume (m}^3\text{)}}$$

i.e. $P_m(v_1 - v_2) =$ Area under process 3–4 – Area under process 1–2.

$$\therefore \qquad P_m(v_1 - v_2) = \frac{P_3 v_3 - P_4 v_4}{\gamma - 1} - \frac{P_2 v_2 - P_1 v_1}{\gamma - 1}$$

Dividing both sides by $v_2$

$$P_m(r - 1) = \frac{P_3 - P_4 r}{\gamma - 1} - \frac{P_2 - P_1 r}{\gamma - 1}$$

$$\therefore \qquad P_m = \frac{1}{r - 1}\left[\frac{(P_3 - P_4 r) - (P_2 - P_1 r)}{\gamma - 1}\right] \qquad (6.7)$$

Same value of $P_m$ we get from following equation

$$P_m = P_1 r\frac{(\alpha - 1)}{(\gamma - 1)}\left[\frac{r^{\gamma-1} - 1}{r - 1}\right] \qquad (6.8)$$

Where $\alpha =$ explosion ratio or pressure ratio $= \dfrac{P_3}{P_2} = \dfrac{T_3}{T_2}$. The Eq. (6.8) is only valid for reversible adiabatic expansion and compression and not for polypopic process because in polytropic process some heat transfer occurs during expansion and compression.

If the processes are polytropic then the actual area should be calculated and then divided by the stroke volume to obtain mean effective pressure and $\gamma$ should be replaced by $n$.

## 6.6   THE DIESEL CYCLE

This cycle differs from the otto cycle in one respect. The heat addition is done at constant pressure instead of constant volume. Original diesel engine was invented by Rudolf Christian Karl Diesel (1858–1913), France, in 1892. In actual cycle diesel is used as fuel and its combustion gives heat. In air standard cycle, however, air will be the medium which will be heated by heat source and are cooled by a heat sink.

Following are the processes of the cycle (refer to Fig. 6.5):

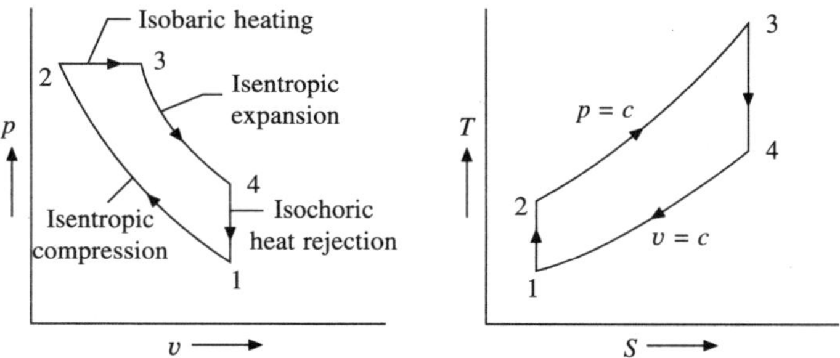


**Fig. 6.5   Diesel Cycle**

(i) Isentropic compression process 1–2 during which work is added to the system.

(ii) Constant pressure heat addition process 2–3. Heat is added at the same time air expands and does work.

(iii) Isentropic expansion process 3–4. During this process work is done by the system although there is no heat transfer.

(iv) Constant volume cooling process 4–1. During this process work done is zero but heat is rejected by the system.

Net work done during the cycle = Net heat added.

$\therefore$　　　　Work done/kg = Heat added per kg of air – heat rejected/kg.

$$= C_p(T_3 - T_2) - C_v(T_4 - T_1)$$

$\therefore$　　　　$$\text{Efficiency} = \frac{\text{Heat added} - \text{Heat rejected}}{\text{Heat added}}$$

$$= \frac{C_p(T_3 - T_2) - C_v(T_4 - T_1)}{C_p(T_3 - T_2)}$$

$$= 1 - \frac{C_v}{C_p} \frac{T_4 - T_1}{T_3 - T_2} = 1 - \frac{1}{\gamma}\left(\frac{T_4 - T_1}{T_3 - T_2}\right)$$

$$= 1 - \frac{1}{\gamma} \times \frac{T_1\left[\dfrac{T_4}{T_1} - 1\right]}{T_2\left[\dfrac{T_3}{T_2} - 1\right]} \tag{6.9}$$

Let,　　　　　　　　$$\frac{v_1}{v_2} = r = \text{Compression ratio}$$

and　　　　　　$$\frac{v_3}{v_2} = r_c = \text{cut off ratio (This name, because heat}$$

$$\text{addition is cut off at } v_3$$

Now, for isentropic process 1–2

$$\frac{T_2}{T_1} = \left(\frac{v_1}{v_2}\right)^{\gamma-1} \quad \text{or,} \quad \frac{T_2}{T_1} = r^{\gamma-1}$$

Similarly from process 3–4,　　$$\frac{T_4}{T_3} = \left(\frac{v_3}{v_4}\right)^{\gamma-1}$$

$$\frac{T_4}{T_3} \times \frac{T_2}{T_1} = \left(\frac{v_3}{v_4}\right)^{\gamma-1} \times \left(\frac{v_1}{v_2}\right)^{\gamma-1} = \left(\frac{v_3}{v_2}\right)^{\gamma-1} \quad (\because v_1 = v_4)$$

or,　　　　　　　$$\frac{T_4}{T_1} = \frac{T_3}{T_2} \times \left(\frac{v_3}{v_2}\right)^{\gamma-1}$$

But from constant pressure process 2–3

$$\frac{T_3}{T_2} = \frac{v_3}{v_2} = r_c$$

$$\frac{T_4}{T_1} = \left(\frac{v_3}{v_2}\right) \times \left(\frac{v_3}{v_2}\right)^{\gamma-1} = \left(\frac{v_3}{v_2}\right)^{\gamma}$$

or,

$$\frac{T_4}{T_1} = r_c^{\gamma}$$

Putting the values of $\dfrac{T_1}{T_2}, \dfrac{T_3}{T_2}$ and $\dfrac{T_4}{T_1}$ into Eq. 6.9, we get

$$\eta = 1 - \frac{1}{\gamma} \times \frac{1}{r^{(\gamma-1)}} \left[\frac{r_c^{\gamma}-1}{r_c-1}\right] \tag{6.10}$$

or,

$$\eta = 1 - \frac{1}{r^{\gamma-1}} \left[\frac{r_c^{\gamma}-1}{\gamma(r_c-1)}\right] \tag{6.11}$$

It is noticed from Eqs. 6.6 and 6.11 that the thermal efficiency of the air standard diesel cycle differs from the thermal efficiency of the air standard otto cycle only by the bracketed term which is always greater than unity as given under

| $r_c$ | = | 3 | 2.5 | 2 | 1.5 |
|---|---|---|---|---|---|
| $\dfrac{r_c^{\gamma}-1}{\gamma(r_c-1)}$ | = | 1.31 | 1.24 | 1.17 | 1.092 |

Therefore, the efficiency of the diesel cycle is less than that of the otto cycle for the same compression ratio.

*Mean Effective Pressure for Diesel Cycle*
The mean effective pressure of a cycle is defined as the work done per cycle divided by the stroke volume

$$P_m = \frac{P_2(v_3 - v_2) + \dfrac{(P_3 v_3 - P_4 v_4)}{\gamma - 1} - \dfrac{(P_2 v_2 - P_1 v_1)}{\gamma - 1}}{v_S} \tag{6.12}$$

$$= \frac{P_2(v_3 - v_2) + \dfrac{(P_3 v_3 - P_4 v_4)}{\gamma - 1} - \dfrac{(P_2 v_2 - P_1 v_1)}{\gamma - 1}}{v_1 - v_2}$$

Dividing both sides by $v_2$, we get

$$P_m = \frac{P_2(r_c - 1) + \dfrac{P_3 r_c - P_4 r}{\gamma - 1} - \dfrac{(P_2 - P_1 r)}{\gamma - 1}}{r - 1} \tag{6.13}$$

$$\left[\because \text{cut off ratio } r_c = \frac{v_3}{v_2}, r = \frac{v_4}{v_2} = \frac{v_1}{v_2}\right]$$

The Eq. (6.12) can be expressed in terms of $P_1$, $r_c$ and $r$ as all these three parameters are generally known.

$$P_2 = P_1 \left( \frac{v_1}{v_2} \right)^{\gamma} = P_1 (r)^{\gamma}$$

$$P_3 = P_2 = P_1 (r)^{\gamma}$$

$$P_4 = P_3 \left( \frac{v_3}{v_4} \right)^{\gamma} = P_3 \left( \frac{1}{r} \right)^{\gamma} = P_3 \left( \frac{r_c}{r} \right)^{\gamma} = P_1 (r)^{\gamma} \cdot \left( \frac{r_c}{r} \right)^{\gamma} = P_1 r_c^{\gamma}$$

$$v_1 = v_s + v_c = v_s + \frac{v_s}{r-1} = v_s \left( \frac{r}{r-1} \right)$$

$$v_2 = v_c \cdot r_c = \frac{v_s \cdot r_c}{r-1}$$

$$v_4 = v_1 = v_s \left( \frac{r}{r-1} \right)$$

Putting the above values of $P_2$, $P_3$, $P_4$, $v_1$, $v_2$, $v_3$ and $v_4$ into Eq. (6.12), we get

$$P_m = \frac{1}{v_s} \left[ P_1 r^{\gamma} \left\{ \frac{v_s r_c}{r-1} - \frac{v_s}{r-1} \right\} \right]$$

$$+ \left\{ \frac{P_1 r^{\gamma} [v_s r_c /(r-1)] - P_1 r_c^{\gamma} \cdot [v_s r/(r-1)]}{\gamma - 1} \right\}$$

$$- \left\{ \frac{P_1 r^{\gamma} [v_s /(r-1)] - P_1 [v_s r/(r-1)]}{\gamma - 1} \right\} \Bigg]$$

$$= \frac{P_1}{r-1} \left[ r^{\gamma} (r_c - 1) + \left\{ \frac{r^{\gamma} r_c - r_c^{\gamma} r}{\gamma - 1} \right\} - \left\{ \frac{r^{\gamma} - r}{\gamma - 1} \right\} \right]$$

$$= \frac{P_1}{(r-1)(\gamma-1)} \left[ r^{\gamma} (r_c - 1)(\gamma - 1) + r^{\gamma} \cdot r_c - r r_c^{\gamma} - r^{\gamma} + r \right]$$

$$= \frac{P_1}{(r-1)(\gamma-1)} [r^{\gamma} \cdot \gamma \cdot (r_c - 1) - r(r_c^{\gamma} - 1)] \tag{6.14}$$

## 6.7 DUAL CYCLE OR COMPOSITE CYCLE

As we have seen that in case of diesel cycle efficiency increases as cut-off ratio $r_c$ decreases. Therefore, it is desirable to keep the value of $r_c$ as small as possible to increase the efficiency. This could be done in a dual cycle by the combustion of fuel taking place partly at constant volume and partly at constant pressure. Hence, it is also referred to as a mixed cycle (Fig. 6.6)

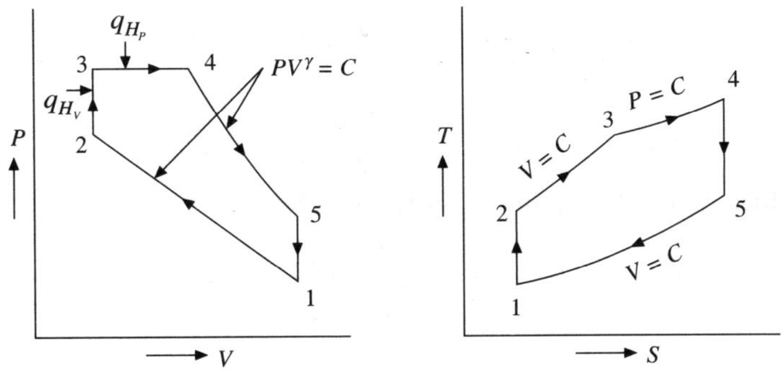

**Fig. 6.6   Dual cycle, *P-V* and *T-S* diagram**

The following 5 processes are involved in this cycle given as under

process 1–2 = isentropic compression, $PV^\gamma = C$
process 2–3 = heat addition at constant volume, $V = C$
process 3–4 = heat addition at constant pressure, $P = C$
Process 4–1 = isentropic expansion during which work is obtained by the system, $PV^\gamma = C$
process 5–1 = heat rejection at constant volume, $V = C$

Assuming one kg of working fluid
Total heat supplied during both processes

that is,
$$q_H = q_{H_v} + q_{H_p}$$
$$= C_v(T_3 - T_2) + C_p(T_4 - T_3)$$

and heat rejected

$$q_L = C_v(T_5 - T_1)$$

Hence, the thermal efficiency of air standard dual cycle is

$$\eta = 1 - \frac{q_L}{q_H} = 1 - \frac{T_5 - T_1}{(T_3 - T_2) + \gamma(T_4 - T_3)} \tag{6.15}$$

compression ratio, $r = \dfrac{v_1}{v_2}$, pressure ratio, $\alpha = \dfrac{P_3}{P_2}$ cutt off ratio, $r_c = \dfrac{v_4}{v_3}$, expansion ratio,

$r_e = \dfrac{r}{r_c} = \dfrac{V_5}{V_4}$ considering process 1–2, $T_2 = T_1 \times r^{\gamma-1}$. considering process 2–3, $(v_2 = v_3)$

$$\frac{P_2 v_2}{T_2} = \frac{P_3 v_3}{T_3}$$

$\therefore$
$$T_3 = T_2 \cdot \frac{P_3}{P_2} = T_1 \times r^{\gamma-1} \times \alpha$$

Considering process 3–4, $(P_3 = P_4)$

$$\frac{P_3 v_3}{T_3} = \frac{P_4 v_4}{T_4} \quad \therefore \quad T_4 = T_3 \cdot \frac{v_4}{v_3} = T_3 \times r_c = T_1 r^{\gamma-1} \times \alpha \times r_c$$

Considering process 4–5

$$T_5 = T_4 \left(\frac{v_4}{v_5}\right)^{\gamma-1} = T_1 \times r^{\gamma-1} \times \alpha \times r_c \times r_c \frac{\gamma-1}{r^{\gamma-1}}$$

$$= T_1 \times \alpha \times r_c^\gamma$$

Substituting the values of $T_2$, $T_3$, $T_4$ and $T_5$ in Eq. (6.15)

$$\eta = 1 - \frac{T_1 \times \alpha \times r_c^\gamma - T_1}{[T_1 r^{\gamma-1}\alpha - T_1 r^{\gamma-1}] + \gamma[T_1 r^{\gamma-1}\alpha r_c - T_1 r^{\gamma-1}\alpha]}$$

$$= 1 - \frac{1}{r^{\gamma-1}}\left[\frac{\alpha r_c^\gamma - 1}{(\alpha - 1) + \gamma\alpha(r_c - 1)}\right] \tag{6.16}$$

When $\alpha = 1$, $\eta_{dual} = \eta_{Diesel}$
When $r_c = 1$, $\eta_{dual} = \eta_{Otto}$

It can be observed from Eq. (6.16) that the efficiency of the dual cycle can be increased by increasing the value of $\alpha$ (more heat is added during constant volume process) and decreasing the value of $r_c$ (less heat is added during constant pressure process) keeping total heat added constant.

$$\text{Mean effective pressure for dual cycle} = \frac{\text{Work done}}{\text{Swept volume}}$$

$$P_m = \frac{P_1 r^\gamma\left[\alpha\gamma(r_c - 1) + (\alpha - 1) - \dfrac{1}{r^{\gamma-1}}(\alpha r_{c-1}^\gamma)\right]}{(\gamma - 1)(r - 1)} \tag{6.17}$$

*Comparison Between Otto, Diesel and Dual Cycles*
With regard to thermal efficiency of these three cycles for the same compression ratio and same heat input, $P$-$v$ and $T$-$S$ diagrams are given in Fig. 6.7 (a) and 6.7 (b). It has been persumed that mechanical and thermal stresses remain unchanged. The air standard thermal efficiency of a cycle is given by:

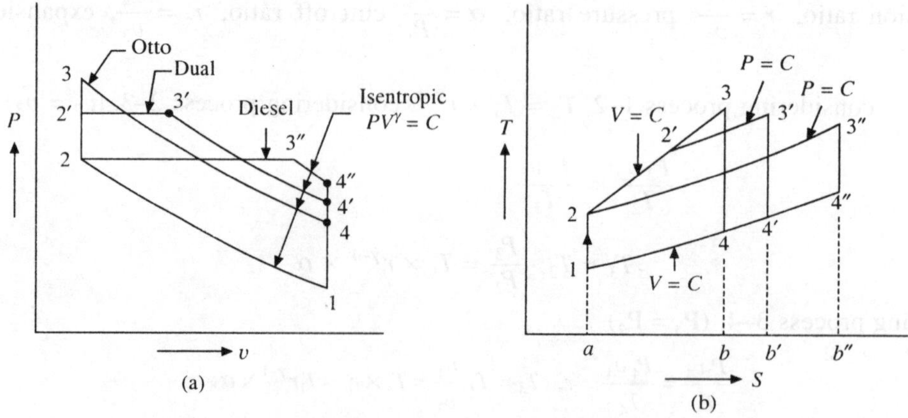


**Fig. 6.7  Comparison of Otto, Diesel and Dual Cycles**

$$\eta_{th} = 1 - \frac{Q_r}{Q_s}, \text{ where } Q_s = \text{Heat supplied and } Q_r = \text{Heat rejected.}$$

For same heat input ($Q_s$), the efficiency would be maximum when rejected heat ($Q_r$) would be less.

Amount of rejected heat from $T - s$ are given as under:

For Otto cycle, $\qquad\qquad Q_r = \text{Area } 1 - a - b - 4$

For Dual cycle, $\qquad\qquad Q_{r'} = \text{Area } 1 - a - b' - 4'$

and for Diesel cycle $Q_{r''} = \text{Area } 1 - a - b'' - 4''$.

As $Q_r < Q_{r'} < Q_{r''}$, so under the given condition we can say that otto cycle is more efficient as compared to diesel cycle and dual cycle lies in between two.

## 6.8  THE BRAYTON OR JOULE CYCLE

The air standard Brayton cycle also called the Joule cycle is a theoretical cycle on which run the gas turbines. A Brayton cycle employs an air compressor, a combustion chamber and a gas turbine. Air is compressed in the compressor and then goes to the combustion chamber. For open cycles the fuel is burnt in the combustion chamber at constant pressure and then the gas (here air) passes to the gas turbine after which it is exhausted. The air standard cycle will, however, be a closed one and the combustion chamber is replaced by a heat exchanger where the air gets the heat. Again after leaving the gas turbine the air passes through a heat exchanger where it rejects heat and cools down before going back to the compressor.

In the air standard cycle, the compression and the expansion process in the compressor and turbine are assumed isentropic. Heat addition to air and heat rejection from it is at constant pressure. The cycle is shown in Fig. 6.8.

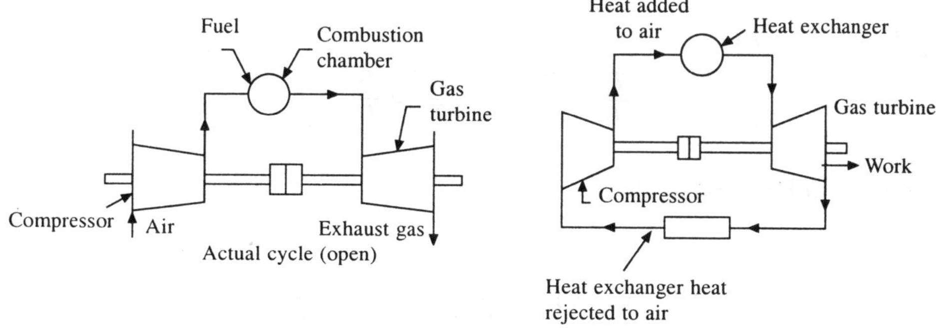


**Fig. 6.8  Air standard cycle (closed)**

$P$-$V$ and $T$-$S$ diagrams are shown in Fig. 6.9.

Heat added during the constant pressure heat addition process 2–3 per kg of air = $C_p (T_3 - T_2)$

Heat rejected during constant pressure heat rejection process 4–1 per kg of air = $C_p(T_4 - T_1)$

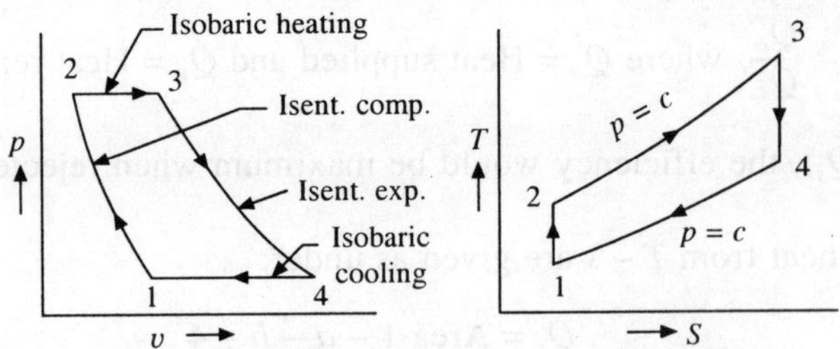

**Fig. 6.9   Brayton cycle, *P-v* and *T-S* diagram**

$\therefore$        Net work done/kg = Net heat added/kg

$$= C_p(T_3 - T_2) - C_p(T_4 - T_1)$$

Air standard efficiency      $\eta = \dfrac{\text{Work done}}{\text{Heat supplied}}$

$$= \dfrac{C_p(T_3 - T_2) - C_p(T_4 - T_1)}{C_p(T_3 - T_2)}$$

$$= 1 - \dfrac{T_4 - T_1}{T_3 - T_2} = 1 - \dfrac{T_1\left(\dfrac{T_4}{T_1} - 1\right)}{T_2\left(\dfrac{T_3}{T_2} - 1\right)} \tag{6.18}$$

From isentropic compression process $1 - 2$, $\dfrac{T_2}{T_1} = \left(\dfrac{p_2}{p_1}\right)^{\frac{\gamma-1}{\gamma}}$

Similarly from isentropic expansion process 3–4, $\dfrac{T_3}{T_4} = \left(\dfrac{P_3}{P_4}\right)^{\frac{\gamma-1}{\gamma}}$

Again $p_2 = P_3$ and $p_1 = p_4$

$\therefore$          $\dfrac{T_2}{T_1} = \dfrac{T_3}{T_4}$   or,   $\dfrac{T_4}{T_1} = \dfrac{T_3}{T_2}$

Putting $\dfrac{T_4}{T_1} = \dfrac{T_3}{T_2}$ in Eq. 6.10 we get,

$$\eta = 1 - \dfrac{T_1}{T_2} = 1 - \left(\dfrac{P_1}{P_2}\right)^{\frac{\gamma-1}{\gamma}} = 1 - \left(\dfrac{1}{\alpha}\right)^{\frac{\gamma-1}{\gamma}} \tag{6.19}$$

Where $\alpha = P_2/P_1 = $ pressure ratio

But           $p_1 v_1^{\gamma} = P_2 v_2^{\gamma}$   $\therefore$   $P_2/P_1 = (v_1/v_2)^{\gamma}$   or,   $\alpha = r^{\gamma}$

Putting this value in Eq. (6.19), we get

$$\eta = 1 - (1/r)^{\gamma-1} \qquad\qquad (6.20)$$

Thus for the same compression ratio the efficiency of Brayton cycle is same as that of otto cycle.

## 6.9  THE ATKINSON CYCLE

This cycle is named after J Atkinson, a British Engineer, who carried out some work on the gas turbine in 1885. Atkinson gas engine is peculiar as it has short and long stroke obtained from a special link mechanism. The short stroke is meant for compression and long stroke, for expansion process. However, this cycle has also been used as a gasturbine cycle.

The Atkinson cycle for gas engine shown in Fig. 6.10 as a $P$-$v$ and $T$-$S$ diagram, comprises the following four processes:

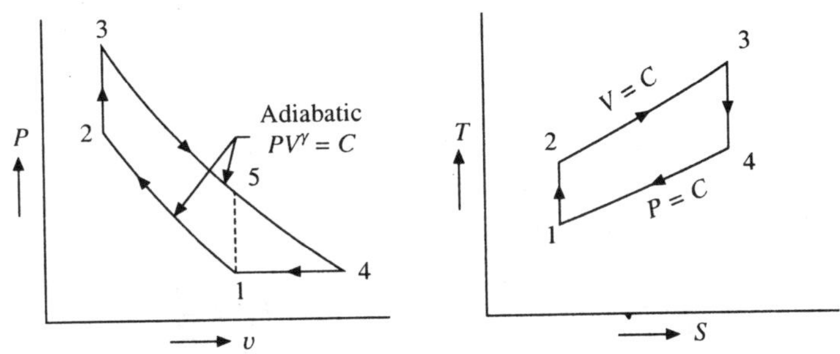


**Fig. 6.10**  **Atkinson cycle, $P$-$V$ and $T$-$S$ diagram**

Process 1–2 : Reversible adiabatic (isentropic) compression of air from pressure $P_1$ to increased pressure $P_2$ and with compression ratio $r = \dfrac{v_1}{v_2}$.

Process 2–3: Heat supply at constant volume from an external source. So pressure rises from $P_2$ to $P_3$ suddenly and the ratio $\dfrac{P_3}{P_2} = \alpha$ is called explosion ratio.

Process 3–4: Reversible adiabatic (isentropic) expansion of air up to the lowest pressure $P_4 = P_1$ with expansion ratio $r_e = v_4/v_3$.

Process 4–1: Heat rejection at constant pressure to cool sink. This process completes the cycle and returns the air to its original state.

No heat is received or rejected during adiabatic operations.
The amount of heat received

$$Q_s = mC_v(T_3 - T_2)$$

The amount of heat rejected

$$Q_r = mC_p (T_4 - T_1)$$

$\therefore$ Work done $= Q_s - Q_r = mC_v(T_3 - T_2) - {}_mC_p (T_4 - T_1)$ and the thermal efficiency of the cycle will be:

$$\eta_{th} = \frac{\text{work done}}{\text{Heat supplied or received}}$$

$$= \frac{m[C_v(T_3 - T_2) - C_P(T_4 - T_1)]}{mC_v(T_3 - T_2)}$$

$$= 1 - \gamma\left[\frac{(T_4 - T_1)}{T_3 - T_2}\right] \quad \left(\because \frac{C_p}{C_v} = \gamma\right) \tag{6.21}$$

Let us convert all the temperatures in Eq. (6.21) in terms of highest operating temperature $T_3$ to derive the expression of thermal efficiency.

From constant volume process 2–3

$$\frac{T_3}{T_2} = \frac{P_3}{P_2} = \alpha \quad \therefore T_2 = \frac{T_3}{\alpha}$$

From isentropic compression process 1–2

$$\frac{T_2}{T_1} = \left(\frac{v_1}{v_2}\right)^{\gamma-1} \quad \therefore \ T_1 = \frac{T_2}{r^{\gamma-1}} = \frac{T_3}{\alpha r^{\gamma-1}}$$

From isentropic expansion process 3–4

$$\frac{T_3}{T_4} = \left(\frac{v_4}{v_3}\right)^{\gamma-1} = (r_e)^{\gamma-1} \quad \therefore \ T_4 = \frac{T_3}{r_e^{\gamma-1}}$$

From constant pressure process 4–1

$$\frac{T_4}{T_1} = \frac{v_4}{v_1} \quad \therefore \ T_4 = T_1\left(\frac{v_4}{v_1}\right)$$

$$\frac{v_4}{v_1} = \frac{v_4}{v_3} \times \frac{v_3}{v_1} = \frac{v_4}{v_3} \times \frac{v_2}{v_1} \quad (\because v_2 = v_3)$$

$$= \frac{v_4}{v_3} \div \frac{v_1}{v_2} = \frac{r_e}{r}$$

$$\therefore \qquad T_4 = T_1\left(\frac{r_e}{r}\right) = \frac{T_3}{\alpha r^{\gamma-1}}\left(\frac{r_e}{r}\right)$$

Substituting the above values in Eq. (6.21), we get

$$\eta_{th} = 1 - \gamma\left[\frac{\dfrac{T_3}{\alpha r^{\gamma-1}}\left(\dfrac{r_e}{r}\right) - \dfrac{T_3}{\alpha r^{\gamma-1}}}{T_3 - \dfrac{T_3}{\alpha}}\right]$$

$$= 1 - \gamma\left[\frac{\dfrac{r_e}{r\alpha\, r^{\gamma-1}} - \dfrac{1}{\alpha r^{\gamma-1}}}{1 - \dfrac{1}{\alpha}}\right]$$

$$= 1 - \gamma \left[ \frac{(r_e - r)\,\alpha}{r\alpha \quad r^{\gamma-1}(\alpha - 1)} \right]$$

$$= 1 - \frac{\gamma}{r^{\gamma-1}} \left[ \frac{(r_e - r)}{r\,(\alpha - 1)} \right]$$

$$= 1 - \frac{\gamma}{r^{\gamma-1}} \left[ \frac{\dfrac{r_e}{r} - \dfrac{r}{r}}{(\alpha - 1)} \right]$$

$$= 1 - \frac{\gamma}{r^{\gamma-1}} \left[ \frac{(r_e/r) - 1}{(\alpha - 1)} \right] \tag{6.22}$$

This cycle gives more work as compared to Otto cycle 1–2–3–5–1 by an area 1–4–5 but practically unsuitable due to the complex drive mechanism and bigger cylinder dimension. Equation 6.22 is derived in terms of three ratios, namely, compression ratio $\gamma \left( = \dfrac{v_1}{v_2} \right)$ pressure or explosion ratio $\alpha \left( = \dfrac{P_3}{P_2} \right)$ and expansion ratio $r_e \left( = \dfrac{v_4}{v_3} \right)$. Adiabatic index $\gamma = \dfrac{C_P}{C_V} = 1.4$ for air.

## 6.10  THE STIRLING CYCLE

This cycle is named after Stirling brothers—Dr Robert Stirling and James Stirling who brought out an air engine in 1815 working on a reversible cycle. In 1845, one such engine was used in a Dundee foundry. Heat was supplied by a furnace through a heating surface to air in a closed system. The heat transfer into the air was slow and, therefore, heater surface burned out putting the engine out of use.

As we have discussed earlier that no cycle can have efficiency greater than that of the Carnot cycle working between given temperature limits $T_H$ and $T_L$. Stirling cycle has equal efficiency to that of Carnot cycle. It is superior to the carnot cycle as it has higher work ratio.

The Stirling cycle is represented on $P$-$v$ and $T$-$S$ diagrams along with schematic diagram of the engine in Fig. 6.11. The cycle consists of two isothermals and two constant volume processes.

Following is the sequence of operation:

1. The air is compressed isothermally from state 1 to state 2 in cylinder no. 1.

2. The air at state 2 is passed into the regenerator made of matrix of wire gauge. The air passing through the matrix gets heated from lower temperature ($T_L$) to highest temperature ($T_H$) at constant volume. The heat transfer taking place from matrix to air is considered thermodynamically reversible because in each element of the matrix heat is transferred due to an infinitesimal difference in temperature.

3. The air expands from state 3 to 4, isothermally in the cylinder no. 2.

4. The air coming out from cylinder no. 2 which is at highest temperature $T_H$ and at state 4, enters the regenerator again for cooling through wire matrix with the help of a cooling media

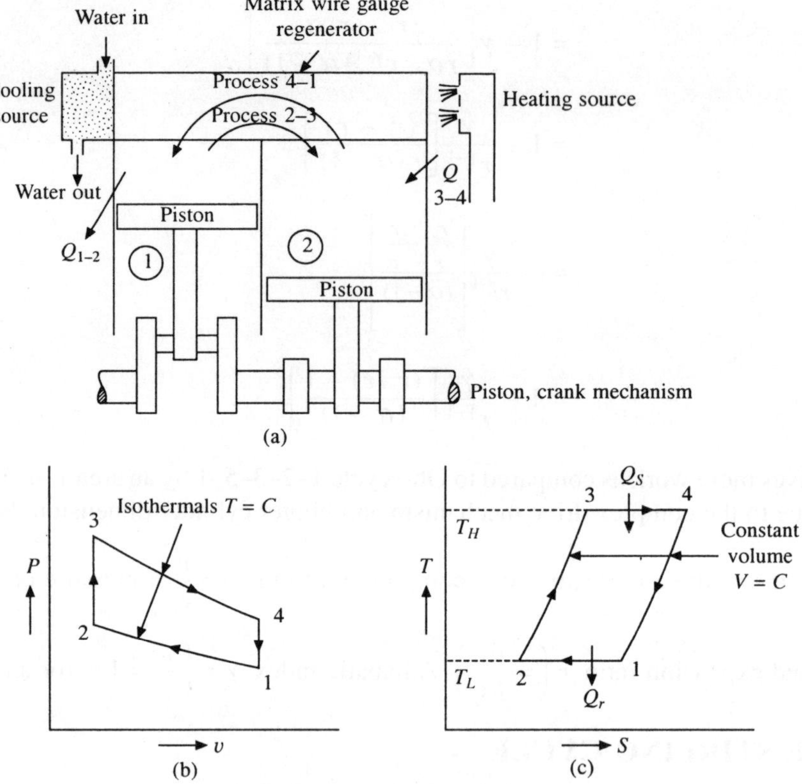



**Fig. 6.11    Stirling Engine**

(water). The heat transfer during the process 4–1 is also considered thermodynamically reversible because of the very small temperature difference in any section between the air (gas) and the matrix.

A temperature gradient from lowest to highest temperature along the matrix is maintained by heating one end by a furnace and cooling the other by keeping a continuous flow of water through the tubes. The regenerative processes 2–3 and 4–1 take place at constant volume and is internal to the cycle. There is a revival of interest in the Stirling cycle with its process of regeneration from time to time.

Air-standard efficiency of this cycle is given as under:

Heat supplied ($Q_s$) during the process 3–4 isothermally from external source

$$= P_3 v_3 \ln\left(\frac{v_4}{v_3}\right) = RT_H \ln\left(\frac{v_4}{v_3}\right) = RT_H \ln r \text{ per kg of air}$$

where $r$ = compression ratio $= \dfrac{v_1}{v_2} = \dfrac{v_4}{v_3}$  Heat rejected ($Q_r$) during the process 1–2 isothermally to external sink

$$= P_1 v_1 \ln\left(\frac{v_1}{v_2}\right) = RT_L \ln r \text{ per kg of air}$$

$\therefore$ work produced by one kg of air

$$W = Q_S - Q_r = R(T_H - T_L) \ln r$$

Hence the air standard efficiency (thermal efficiency) of the cycle will be:

$$\eta_{th} = \frac{\text{Work output}}{\text{Heat input}} = \frac{R(T_H - T_L)\ln r}{R\, T_H \ln r} = \frac{T_H - T_L}{T_H} \qquad (6.23)$$

Thus, the efficiency of Stirling cycle is equal to that of Carnot cycle, both working within same temperature limits. It is true only when the efficiency of the regenerator is 100%. This means that all heat stored in the regenerator during heat rejection is available to air during the process of heat addition. But in actual practice 10 to 20% heat losses occur in the regenerator and the efficiency of the regenerator lies between 80 to 90%.

The cycle has not become very popular even though its ideal efficiency is sufficiently high, because the regenerator volume has to very large compared with the engine size.

## 6.11   THE ERICSSON CYCLE

This cycle is so named after John Ericsson (1803–89), a Swedish engineer. He eventually, settled in America and developed hot air engines there in 1853.

The Ericsson cycle is represented on $P$-$v$ and $T$-$S$ diagrams in Fig. 6.12. It will be seen that it is composed of two isothermal processes and two constant pressure processes. The Ericsson cycle is similar to the Stirling cycle except that the two isothermals are connected by constant pressure processes. Since these two constant pressure processes are bound by same temperature limits, hence regeneration is possible in this cycle also. Heat rejected to the air during the process 4–1 is reutilized during heat addition process 2–3. With the incorporation of regenerator, this cycle also becomes thermodynamically reversible and hence efficiency is same as that of Carnot cycle working between same temperature limits.

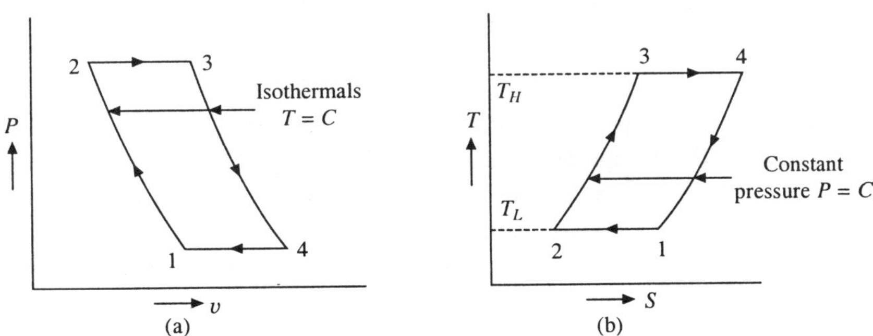


**Fig. 6.12   The Ericsson cycle**

Rejected heat in process 4–1 is reutilized in process 2–3, hence $mC_P(T_3 - T_2) = m_{cp}(T_4 - T_1)$ working between temperature limits $T_H$ and $T_L$. And hence efficiency could be written as

$$\eta_{th} = \frac{T_H - T_L}{T_H} \qquad (6.24)$$

## 6.12   GAS CYCLE FOR REFRIGERATION

In the previous articles gas power cycles have been discussed. If these cycles are reversed, they behave as refrigeration cycle according to second law of thermodynamics. In the gas power cycle energy is received at high temperature, rejected at low temperature and work is obtained from the cycle. In the refrigeration cycle, however, energy is received at low temperature, rejected at high temperature and work (or heat) is required to perform the cycle. Due to the transfer of energy from low to high temperature, the refrigerator is sometimes called as a heat pump. Refrigeration may be defined as the process by which the temperature of a given space or a substance is lower than the atmosphere or surroundings. Depending upon the temperature achieved lower than the surroundings, different ranges of refrigeration are given below:

(i)  Cooling – temperature drop is small
(ii)  Refrigeration – upto – 150°C (123 k)
(iii)  Cryogenics – from –150°C (123 k) to –273°C

The conventional Carnot cycle for power generation is already discussed in chapter 4. A reversed Carnot cycle is shown on $P$–$v$ and $T$–$S$ plane in Fig. 6.13 which is used for refrigeration.

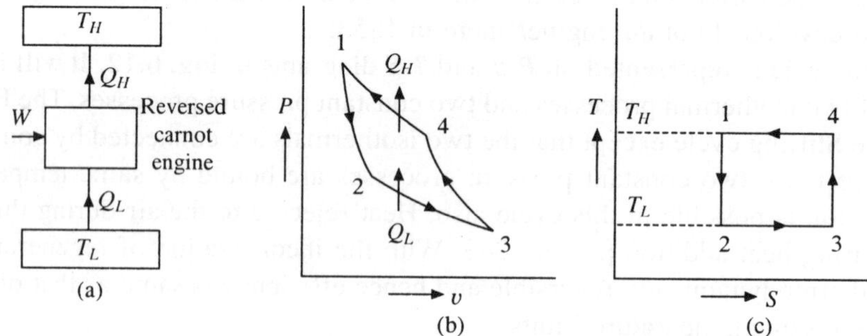


**Fig. 6.13   Reversed Carnot Engine (Refrigeration)**

The sequence of operation is given below:

(i)  Process 1–2    Isentropic expansion of air contained in the clearance space of the cylinder. The temperature falls from $T_1(=T_H)$ to $T_2$ $(=T_L)$.

(ii)  Process 2–3    Isothermal expansion during which heat is absorbed at temperature $T_L$ from the space being enclosed.

(iii)  Process 3–4    Isentropic compression of air by the addition of external work. The temperature rises from $T_L$ to $T_H$.

(iv)  Process 4–1    Isothermal compression during which heat is liberated at temperature $T_H$ to the another body.

Heat absorbed by the reversed Carnot engine

$$Q_L = T_L(S_3 - S_2) = T_L(S_4 - S_1)$$

Heat rejected from the reversed Carnot engine

$$Q_H = T_H(S_4 - S_1)$$

Work input $W = Q_H - Q_L = (T_H - T_L)(S_4 - S_1)$

Refrigerating effect $= Q_L = T_L(S_4 - S_1)$

The COP of the Carnot refrigerator

$$= \frac{\text{Refrigerating effect}}{\text{Work input}} = \frac{Q_L}{W} = \frac{T_L(S_4 - S_1)}{T_H - T_L}$$

$$= \frac{T_L}{(T_H - T_L)} \tag{6.25}$$

and COP of the Carnot heat pump

$$= \frac{\text{Heat rejected by engine at higher temperature}}{\text{Work input}}$$

$$= \frac{Q_H}{W} = \frac{T_H(S_4 - S_1)}{(T_H - T_L)(S_4 - S_1)} = \frac{T_H}{(T_H - T_L)} \tag{6.26}$$

The temperatures in the above equations should be taken in Kelvin scale.

Like Carnot engine, reversed Carnot engine also cannot be used for refrigeration purposes as the isentropic process requires high speed whereas the isothermal process requires very low speed. The variation of the speed during a stroke is not practical.

In practice reversed Stirling, reversed Ericsson or irreversible Joule Thomson (JT) expansion cycles are used for refrigeration and that to specially at very low temperature. The refrigerators working on very low temperature (cryogenics range) are known as cryocoolers (cryogenic refrigerators).

## 6.13  CRYOCOOLERS

Cryocoolers are the small cryogenic refrigerators achieving low temperatures (cryogenic range). They may be classified on the following basis:

   (i)  The function they perform (the delivery of liquid cryogens, the separation of mixture of gases and maintenance space at cryogenic temperature),
  (ii)  Their refrigerating capacity, and
 (iii)  The temperature they reach

The following gases which are used in cryocoolers and the lowest temperature achieved are mentioned, against their name:

   (i)  LNG – Liquified natural gas (120 K or –153°C)
  (ii)  $LO_2$ – Liquified oxygen (83 k or – 190°C)
 (iii)  $LN_2$ – Liquified nitrogen (77 k or – 196°C)
  (iv)  $LH_2$ – Liquified hydrogen (20 k or – 235°C)
   (v)  LHe – Liquified helium (4k or – 269°C)

Evans-Perkins compression cycle is being currently used by domestic refrigerators. The multistage or cascade (with several different cryogenic fluids as mentioned above) versions of this type of cycle are limited in the temperature range of natural gas liquefaction. To achieve cryogenic temperatures, other cycles are required such as Ericsson, Stirling or Joule Thomson (J T) isenthalpic expansion.

Application of cryocoolers will be given later in this chapter.

### 6.13.1   The Stirling and Ericsson Cycle for Cryorefrigeration

In reversed ideal Carnot cycle working between cryogenic temperature limits requires very large and technologically unrealistic operation pressure and, therefore, it is impossible to use it. The possible reversed Stirling and Ericsson cycles are represented on $T$-$S$ plot alongwith reversed Carnot cycle in Fig. 6.14. The cooling power (refrigerating effect) as mentioned below in process 1–2 and the mechanical energy involved (represented by the cycle area) remain unchanged. It is thus possible to use Stirling and Ericsson cycle which give same optimal theoretical thermodynamic efficiency. It is important to notice that contrary to isentropic transformations (Carnot cycle), isochoric (Stirling cycle) and isobaric (Ericsson cycle) ones lead to heat transfer (area under processes 2–3 and 4–1). In the $T$-$S$ diagram the isochors and isobars lines for a perfect gas are parallel (with respective slopes $1/C_v$ and $1/C_p$). Consequently the calorific energy to be provided to the cycle gas during process 2–3 is equal to the calorific energy to be extracted during process 4–1. It is thus possible to imagine a storage device which alternatively stores and rejects this energy during each cycle which is nothing but the regenerator.

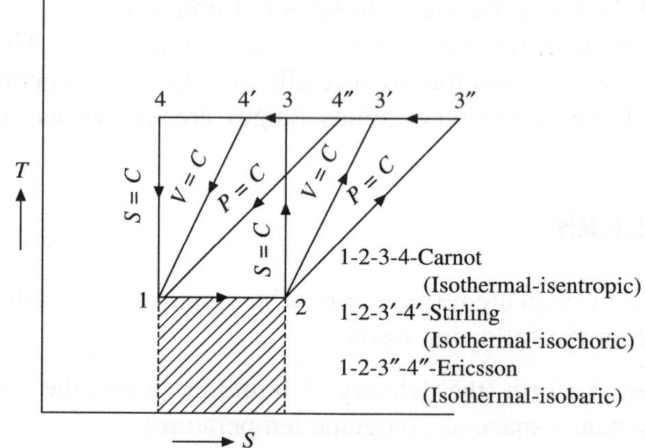


**Fig. 6.14   Ideal cycles for cryocoolers**

The regenerator is generally made of a porous material with a large heat capacity to increase the overall thermal resistance. In the temperature range 300 k–80 k, metallic gauze discs (stainless steel or phosphorous bronze) are generally used (wire diameter and opening of the order of 30 to 100 $\mu$m). In the temperature range 80 k–10 k achieved in two stage refrigerators, local shot (diameter of the order of 200 $\mu$m) is commonly used since this material exhibits a rather large heat capacity at low temperature. At low temperature, the volume heat capacity of common regenerative materials decreases rapidly whereas that of helium, the unique potential cycle fluid remaining in the gaseous state in this temperature range, increases strongly. The regeneration is therefore, inefficient. This is the reason that Stirling or Ericsson type cryocoolers have rather limited operation, that is, above 10 k. The practical operation of cryocoolers based on Stirling or Ericsson leads to the use of moving parts like expander and compressor in the cold temperature to generate cycle fluid volumetry and displacement.

As far as operation in the cycle is concerned, the heat transfer gas, typically helium, is pumped back and forth by a compressor. As the compressor piston depresses, it pushes the gas through the regenerator (to absorb much of the heat from the gas). The gas expands, cools and then flows through heat exchangers that remove heat, cooling the equipment. As the compressor piston retracts, the gas is drawn back through the system. When it passes over the regenerator material, some of the absorbed heat is released into the gas to moderate its temperature. The cycle is repeated at a high rate, often as fast as 30 times per second.

### 6.13.2  The Joule-Thomson Isenthalpic Expansion Cycle

This is irreversible process cycle in which isenthalpic expansion takes place by a simple pressure drop in a calibrated orifice, a capillary or an adjustable needle valve. In getting the cooling effect by such an expansion, the cycle gas has to be precooled at an appropriate level of temperature which is below the so called inversion temperature (Fig. 6.15).

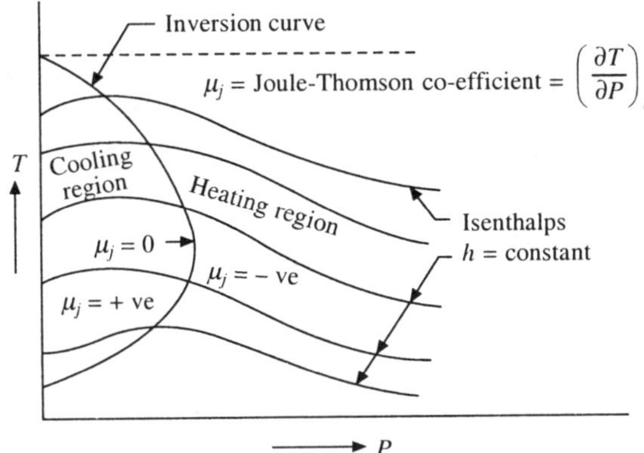


**Fig. 6.15   Typical Inversion Temperature Curve**

For main cryogenic fluid, inversion temperature with indication of optimal high pressure range for the cooling effect during expansion is given in Table 6.1

**Table 6.1   Inverson Temperature and Optimal Pressure**

| Fluid | Inversion temp. (K) | Optimal pressure (MPa) |
|---|---|---|
| Nitrogen | 621 | 20–60 |
| Hydrogen | 205 | 10–30 |
| Helium | 51 | 1.2–2.5 |

It has been observed that for nitrogen or oxygen or air no precooling below room temperature is required for isenthalpic expansion. Helium is the only substance which remains fluid at temperature below 14 k. It is most difficult among all gases to liquefy helium being its maximum inversion temperature 40 k and boiling point 4.2 k. Therefore, to achieve cooling by isenthalpic expansion, helium must be previously precooled below 40 k. A typical *J-T* cycle diagram and *T-S* plot are given in Fig. 6.16. A counter flow heat exchanger is provided to obtain a liquid vapour mixture after expansion at state 3. The cooling effect is then produced by the latent heat of vaporization (process 3–4) of the liquid phase. The operation at liquid-vapour equilibrium allows an extremely good temperature stability of the cold sink (as long as the expansion pressure is well controlled).

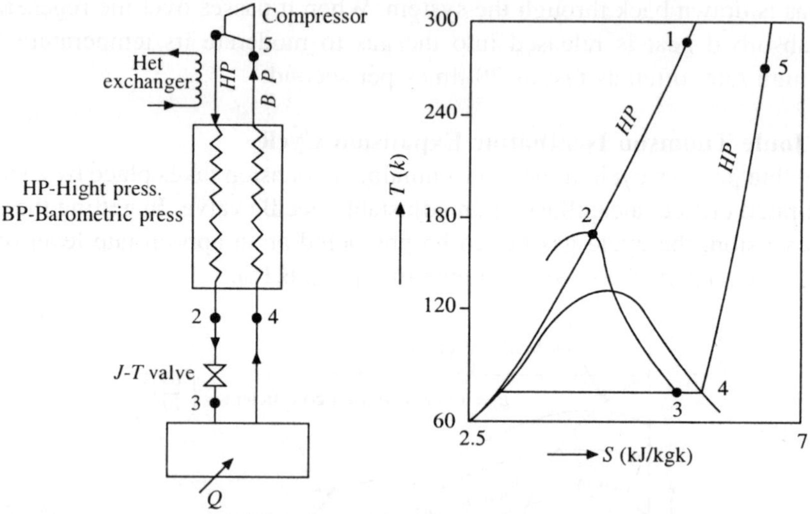


**Fig. 6.16   J.T. expansion cycle**

The main theoretical drawback of the J-T expansion is its irreversible nature which does not allow for an optimal thermodynamic efficiency. However, J-T coolers offer a simple and easily miniaturizable technology with no moving part in the cold region and have long been used in the collins style helium liquefiers.

### 6.13.3   Regenerator Material for Cryocoolers

Regenerator is the most important and integral part of cryocoolers which decides their efficiencies. It must be capable of absorbing large amounts of heat without changing its material temperature. Although the regenerator is only 2–3 inches in length and about a half an inch in diameter, it maintains a constant low temperature. According to recent developments about its material, following achievents/suggestions are given below:

The Ames Lab Scientists have developed a set of 36 alloys to serve as regenerator materials for cryocoolers that reach temperatures of 10–16 k. Erbium, a rare-earth metal, is the main component of the alloys. Differing amounts of other metals control the magnetic ordering temperature and therefore the heat capacity of each alloy. This gives manufacturers flexibility in choosing the best alloy for a specific use. As we know that at temperature close to absolute zero (0 kelvin), normal materials have a negligible heat capacity. A small energy input will disrupts the equilibrium, causing the temperature to rise. As suggested by scientist Vitalij Pecharsky (a professor of material science and engineering at Iowa State University), erbium-based materials prevent that from happening because they have an outstanding heat capacity.

Most 10–60 k cryocoolers use lead regenerators. Being the soft metal it can be used in the form of small spheres. The erbium-based alloys can absorb 25–175 percent more heat than lead in that temperature range and are much harder than lead, making them versatile to be used in a variety of forms, as mentioned by Gschneidner. According to him "A cryocooler using an erbium-based regenerator would be more efficient because it could cool the load to a lower temperature, or it could cool a bigger load without needing additional power." Unlike lead, the erbium-based alloys pose no health or environmental hazards. However, lead is cheaper than erbium. Moreover, for cooling the instruments erbium-based alloys enable the scientists to switch to cryocoolers rather than the use of liquid helium.

Research for other alloys that can be used in cryocoolers continues. Scientists have applied for a patent on the alloys. Atlas Scientific, a California-based company that designs and builds cryocoolers, recently signed two licencing agreements to use the materials and is gearing up to conduct tests of the alloys using a pulse-tube cryocooler (a new concept in this field, which will be discussed hereafter in this chapter). Additionally, the University of Victoria in British Columbia is testing the alloys for a Gifford-McMahon cryocooler.

Ames Laboratory operated by the Department of Energy by Iowa State University conducts research into various areas of national concern, including energy resources, high speed computer design, environmental cleanup and restoration, and the synthesis and study of new materials.

### 6.13.4   Applications of Cryocoolers

Cryocoolers are required in many technology areas. Few applications are listed below:

(i) Missile launching—Cryogenic fluids (liquid $O_2$, liquid $H_2$) play an important role in successful missile launching. Cryogenic fluids provide considerable high specific impulse (basic requirement) compared with conventional hydrocarbon fuels. The specific impulse is directly proportional to the temperature and inversely proportional to the molecular weight of the combustion products inside the chamber. Hence $H_2$ proves to be an effective fuel. Liquefaction of these gases is done by cryocoolers.

(ii) Semen preservation—Bull semen preservation is almost entirely a cryogenic operation. The frozen semen refrigerated by liquid nitrogen (–196°C), is marketed by organizations.

   Another possibility is the preservation of human sperm by freezing for "test tube" babies. This could mean the survival of humanity in the event of a large-scale warfare.

(iii) Computers—There is a great demand for very small and compact units of computers for storage purposes. Computers which should work at such a speed that the solution by human efforts is not feasible. The human brain has $10^{10}$ switching elements while the power consumption is only 60 watts. A cryotson is a tiny superconducting device used for the purpose and with its use it is possible to keep the power consumption very low as an electric current flowing through a superconductor does not meet with any resistance at all and so does not consume any power. The new computer with cryostons can accomodate $10^{10}$ components per cubic inch—a density far greater than that of the human brain.

(iv) Cryogenic for underground power lines—It is a demand of tomorrow to use underground cryogenic power lines to meet soaring power demands of the world because the cryogenic transmission power lines can transmit incredible amounts of power into dense city centres. It has been observed that a power cable supplying 435 kV was able to transmit 3500 MW power with the same cable insulation when the cable was chilled to – 220°C by liquid nitrogen.

(v) Defence purpose—Infra-red detectors have improved sensitivity and extension of range when operating at cryogenic temperatures. Cryogenically cooled infra-red detectors are already extensively used as the eye of the missile guidance system or a infra-red camera which can take photographs of enemies during war in the dark.

(vi) In the recent years large super conducting magnets have been developed for a variety of applications including fusion reactor, energy storage, electric power transmission, high energy Physics experiment and magnetic levitated trains.

(vii) In a nuclear power plant, krypton and xenon are among the waste products. A number of problems are being faced for their removal and disposal. One kg of activated charcoal in an absorption bed at cryogenic temperature removes as much radioactive krypton and xenon from the gas stream of the reactor as two tonnes of the same charcoal at room temperature. The radioactive elements can then be held in the cold charcoal until it is safe to release them.

(viii) Cryocoolers are helpful in separation of
   (a)  gaseous mixture such as constituents of the atmospheric gas.
   (b)  $H_2$ from petrolium refinery gases.
   (c)  $H_2$ and CO from coke oven and coal-water gas reactors, and
   (d)  He from natural gas.

### 6.13.5  Pulse Tube Cooler: A New Concept

Pulse tube cryocooler is a technical evolution of the existing system. It has attracted much interest because of its potential for high reliability and no moving parts at low temperatures.

Different types of refrigerators (producing cryogenic temperature) such as stirling refrigerator Gifford Mc Mahon refrigerator and pulse tube refrigerators are most commonly used. The concept of pulse tube refigeration was originally given by Gifford and Longsworth in 1964, is how commonly known as the basic pulse tube (BPT) refrigerator. Mikulin have modified and improved the original concept in 1984 by introducing an orifice and a buffer volume connected to the warm end of the pulse tube through an impedance and is called an orifice pulse tube refrigerator. In the last few years, interest has been grown due to its merits of mechanical simplicity, high relability and no moving parts at the cold end, resulting in low vibration levels. The pulse tube refrigerators are classified based on different parameters. The classification based on arrangement of components is (a) conventional or linear pulse tube refigerator (b) U-tube refrigerator (c) annular pulse tube refrigerator and (d) co-axial pulse tube refrigerator.

Basically all components of a classical regerative cooler are present, that is, a compressor with a distribution valve or a pressure oscillator to generate a cyclic pressure wave and a regenerator. A linear pulse tube refrigerator, working on air as working fluid, has been disigned, fabricated and tested by Gawali BS and Sane NK at Walchand College of Engineering, Sangali (India) in the year 2000.

The schematic diagram of experimental set-up is given in Fig. 6.17. A compressure which works as pressure oscillator, is connected to the pulse tube refrigerator through an rotary valve driven by variable speed DC motor. For regenerator, a stainless steel tube of 1 mm wall thickness and 13 mm inner diameter has been used. This tube is tested for 80,100 and 120 mm length with bronze and phosphor bronze as the wire screen material. Better results have been obtained by using 120 mm length with phospher bronze screen and hence considered to be constant parameter.

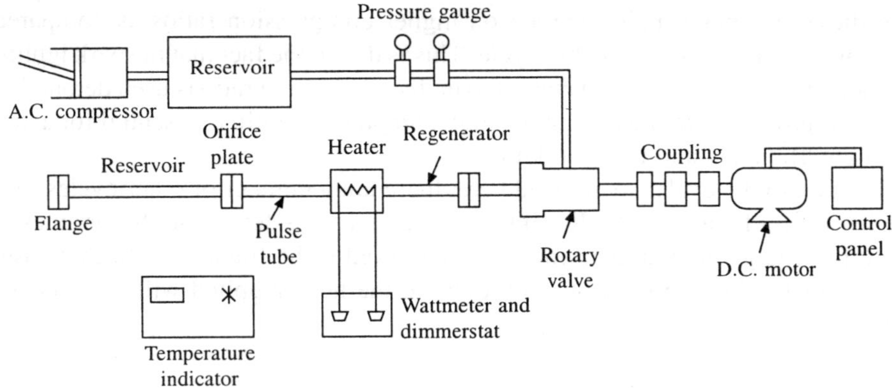


**Fig. 6.17   Schematic of a pulse tube cooler**

The pulse tube made of seamless stainless steel with 13 mm inner diameter, 1 mm thickness and 400 mm length is used for experimentation. Regenerator covers 120 mm from one side, while remaining is occupied by pulse tube including cold end and hot end heat exchangers. Water is used for cooling the hot end heat exchanger. A resistance heater is also provided for cooling power measurements. At room temperature the buffer volume is connected to the pulse tube through flanges with orifice disk in between. Minimum temperature is achieved around 207 k at pulse rate of 8.3 bar and mesh equal to 100 as a result.

A three-year research project on miniature pulse tube cryocoolers has started in January 2000 at ILK Dresden, Germany. The aim of this project is the development of cryocoolers with a cooling power larger than 1 W at 80 k and a minimum temperature less than 70 k. The electric input power requirement for these cryocoolers should be less than 100 W. Two different cold heads, one of them made in the conventional way by use of metallic materials, the second one made of non metallic and non-magnetic materials, should be combined with a SL200 double piston linear motor compressors produced by the German Company AIM, Helilbronn for miniature stirling cryocoolers.

A pulse tube liquefier prototype was completed for NASA/Johnson as part of a program for developing the technology of liquefying oxygen on Mars. NASA expects to have a satellite sent to Mars in the year 2007 that would convert the carbon dioxide atmosphere of Mars into oxygen and then be liquefied using the pulse tube liquefiers technology. After two years, a sufficient quantity of liquid oxygen would be collected to free rockets for lifting off of Mars and returning to Earth with Mars rock samples. Extensive measurements with the liquefier prototype were completed and the results were used to update the models of pulse tube refrigerators.

*Some Important Points on Cycles*

1. Relative efficiency — The thermal efficiency of an ideal aircycle is called the air standard

efficiency. Another terms called relative efficiency or efficiency ratio is used which takes into account the effect of friction and other losses.

$$\text{Relative efficiency} = \frac{\text{Actual thermal efficiency}}{\text{Air standard efficiency}}$$

2. The diesel engines usually operate on higher compression ratios as compared to spark ignition engines working on Otto cycle. This is due to the fact that in spark ignition engines air fuel mixture is compressed and if too high compression ratio is used detonation becomes a serious problem. On the other hand high compression ratio is essential for a diesel engine for high efficiency and to avoid knock.

3. In addition to the cycles mentioned here, there are a number of other cycles, for example, Cayley cycle, Lenoir cycle, Reitlinger cycle and Crossley cycle. Sometimes a fictitious cycle may be given and efficiency may be asked to be found out. In that case the usual method of finding the ratio of work done to the heat supplied will give the air standard efficiency.

## Solved Problems

**6.1** With the help of *T-S* diagram prove that for the same compression ratio Otto cycle is more efficient than diesel cycle.

*Soln.*   Refer to Fig. 6.18

For the same compression ratio starting with the same initial point 1. The state points after isentropic compression will also be the same (say 2) for the two cycles Now, for Otto cycle the heating process is constant volume (Line 23′) whereas for diesel cycle it is constant pressure (line 23). As state points 3′ and 3 are different the isentropic expansions will be different as 3′ 4′ and 3 4. The final constant volume heat rejection processes will be along the same curve i.e. 4′ 1 will lie on 4 1.

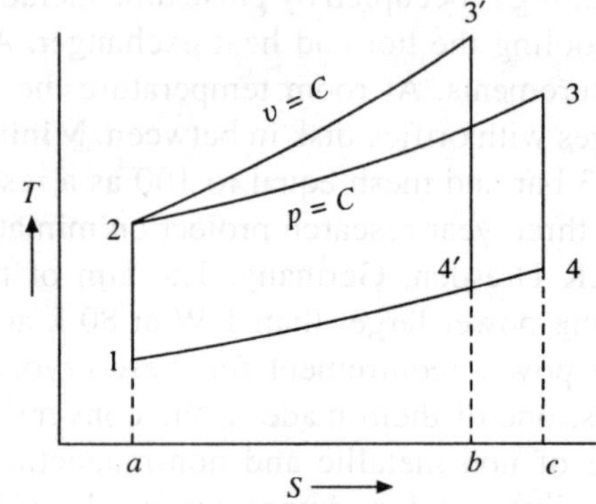


**Fig. 6.18**

Let us extend 3′ 4′ to b, 3 4 to C and 21 to a. To make the comparison we shall assume that equal amount of heat is added in the two cases.

Now we know that constant volume curves are steeper than constant pressure curves. The heat added in the case of Otto cycle will be given by area a 2 3′ ba2 and in case of diesel cycle by area a 2 3 C a and they will be equal as assumed. However, the heat rejected in case of otto cycle will be given by area a14′ ba and in case of diesel cycle by area a1 4ca. Thus we see that heat rejected in case of diesel cycle is more which show that there will be less net work done. Therefore its efficiency must be less than that of otto cycle.

**6.2** The highest pressure in an otto cycle is 19 bar with a compression ratio of 5 and a minimum pressure of 1 bar. Find the air standard efficiency and mean effective pressure.

*Soln.*   Air standard efficiency. (Refer to Fig. 6.19)

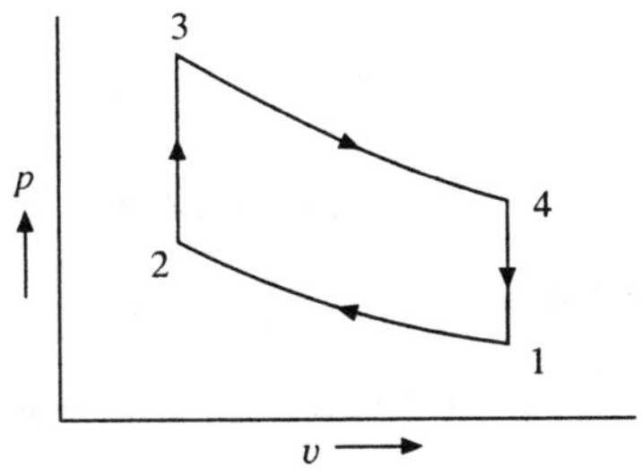

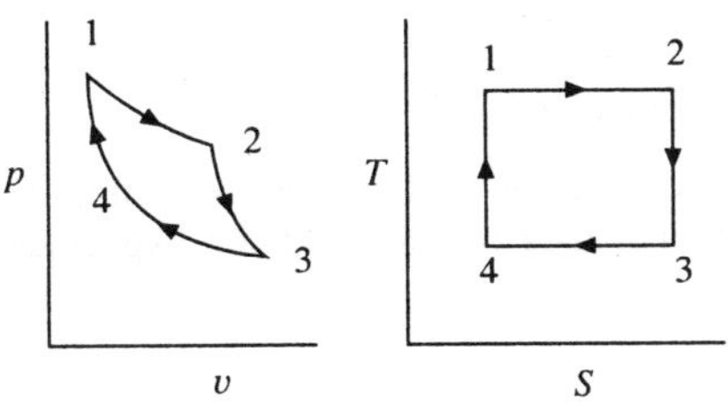

Fig. 6.19

$$= 1 - \frac{1}{r^{\gamma-1}} \quad \text{(for air } \gamma = 1.4\text{)}$$

$$= 1 - (1/5)^{0.4} = 1 - 0.525 = 0.475 \text{ or, } 47.5\%$$

Since $\qquad p_4/p_3 = (V_3/V_4)^\gamma \quad \therefore \quad p_4 = 19 \times (1/5)^{1.4} = 19/9.5 = 2 \text{ bar}$

Again $\qquad p_2/p_1 = (V_1/V_2)^\gamma \text{ or, } p_2 = 1 \times (5)^{1.4} = 9.5 \text{ bar}$

Work done/kg = work done expansion process − work done in compression process

$$= \frac{p_3 v_3 - p_4 v_4}{\gamma - 1} - \frac{p_2 v_2 - p_1 v_1}{\gamma - 1}$$

Also since $\qquad r = \dfrac{v_4}{v_3} = \dfrac{v_1}{v_2} = 5$

$$\text{work done/kg} = \left[ \frac{19 \times V_3 - 2 \times 5V_3}{0.4} - \frac{9.5\, V_3 - 5V_3}{0.04} \right] \times 10^5 \text{ Nm}$$

$$= \frac{4.5\, v_3}{0.4} \times 10^5 \text{ Nm}$$

$$\text{m.e.p} = \frac{\text{W.D.}}{\text{Stroke volume}} = \frac{4.5\, v_3}{0.4} \times \frac{10^5}{4v_3}$$

$$= 281250 \text{ N/m}^2 = 2.8125 \text{ bar} \quad Ans.$$

**6.3** In a Carnot cycle the minimum pressure is 1 bar and the corresponding specific volume is 0.825m³/kg. The pressure at end of isothermal compression is 3.5 bar and at-end of isentropic compression is 10.5 bar. Find $\eta$, mep. and power output. Engine speed is 3 cycles/S, mass of air in a kg.

*Soln.* Refer to Fig. 6.20

Fig. 6.20

$$p_3 \, v_3 = RT_3 \quad \therefore \quad T_3 = \frac{p_3 \, v_3}{R}$$

$$= \frac{10^5 \times 0.85}{287} = 288 \text{ K}$$

Again from isothermal process 3–4

$$v_4 = \frac{p_3 \, v_3}{p_4} = (1.0/3.5) \times 0.825 = 0.236 \text{ m}^3/\text{kg}$$

From isentropic process 4–1

$$v_1 = v_4 \times (p_1/p_4)^{\frac{1}{\gamma}} = 0.236 \times (3.5/10.5)^{1/1.4} = 0.108 \text{ m}^3/\text{kg}$$

$$\text{stroke volume} = v_3 - v_1 = 0.825 - 0.108 = 0.717 \text{ m}^3/\text{kg}$$

Now from process 4–1

$$\frac{T_1}{T_4} = (v_4/v_1)^{\gamma-1} \quad \therefore \; T_1 = 288 \times \left(\frac{0.236}{0.108}\right)^{0.4} = 395 \text{ k}$$

Knowing the two temperature we can find the efficiency.

Thus (i)
$$\eta = \frac{T_1 - T_3}{T_1} = \frac{395 - 288}{295} = 0.271 = 27.1\%$$

Work done in a complete cycle = Net heat added in complete cycle/kg

$$RT_1 \ln \frac{v_2}{v_1} - RT_3 \ln \frac{v_3}{v_4} \qquad \left[\because \frac{v_2}{v_1} = \frac{v_3}{v_4} = r\right]$$

$$= R \ln \frac{v_2}{v_1} \, (T_1 - T_3)$$

Compression ratio $r = v_3/v_4$ $\quad \therefore \quad$ Work done $= R \ln \dfrac{v_3}{v_4}(T_1 - T_3)$

$$= R \ln \frac{p_4}{p_3} \, (T_1 - T_3)$$

$$= 0.287 \times \ln 3.5 \times 107 = 38.47 \text{ kJ}$$

$$\text{m.e.p} = \frac{\text{W.D.}}{V_S} = \frac{38.47}{0.717} = 53.65 \text{ kN/m}^2$$

W.D/Cycle = 38.47 kJ

∴ Power output for 3 cycles = 38.47 × 3 = 115.41 kJ   *Ans.*

**6.4**  A diesel engine has compression ratio of 14. The fuel is cut off at 0.08 of stroke. The relative efficiency is 52%. Find the mass of the fuel of calorific value 42,000 kJ/kg which would be required per kW per hour (U.P.S.C. 1992)

*Soln.*  Refer to Fig. 6.21.

The diesel cycle is shown in the figure on a *p–V* plot compression ratio $r = \dfrac{v_1}{v_2} = 14$

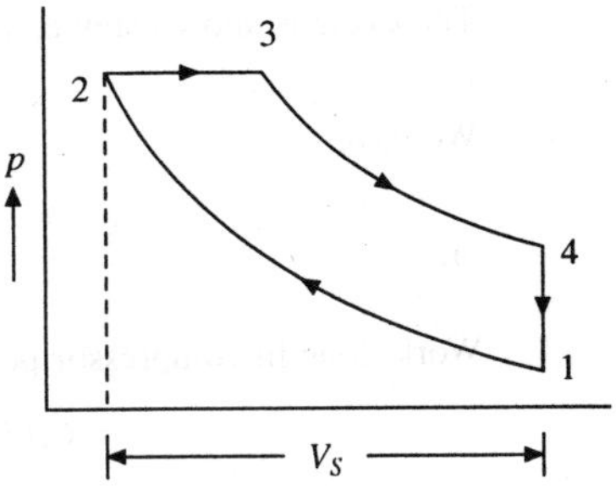

Also                                 $v_3 - v_2 = 0.08 \times v_S$

or,                                  $v_3 - v_2 = 0.08(v_1 - v_2)$ dividing by $v_2$,

$$\frac{v_3}{v_2} - 1 = 0.08\left(\frac{v_1}{v_2} - 1\right)$$

$$\frac{v_3}{v_2} = 0.08(14 - 1) + 1$$

or                                   $r_c = 2.04$

The air standard efficiency of diesel engine

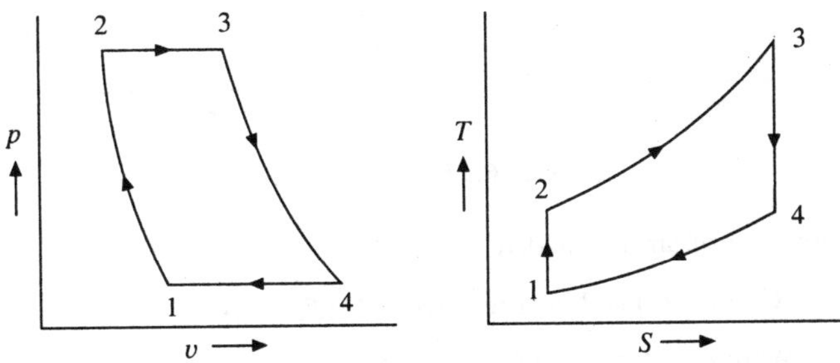

**Fig. 6.21**

$$= 1 - (1/r)^{\gamma-1} \times \frac{1}{\gamma}\left[\frac{r_c^{\gamma} - 1}{r_C - 1}\right]$$

$$\text{A.S.E.} = 1 - (1/14)^{1.4-1} \times 1/1.4\left[\frac{2.4^{1.4} - 1}{2.4 - 1}\right]$$

$$= 0.581 = 58.1\%$$

But relative efficiency $= \dfrac{\text{Actual thermal efficiency}}{\text{Air standard efficiency}}$

Actual thermal efficiency $= 0.52 \times 0.581 = 0.302$

Again actual thermal efficiency $= \dfrac{\text{Actual work done}}{\text{Heat added}}$

If $m$ be the amount of fuel required per kW hour, then heat added per kW hour $= 42,000 \times m = 42,000m$
Also per kW hour work done $= 3,600$ kJ

$$0.302 = \frac{3,600}{42,000 \times m}$$

or,                     $m = \dfrac{3,600}{0.302 \times 42000} = 0.283$   *Ans.*

**6.5**  Air at 15°C and atmospheric pressure is taken in an open cycle gas turbine power plant. In the compressor the pressure rises to five times. The compressed air is then heated to 800°C and then expanded in the turbine to the atmospheric pressure. Find the power developed by per kg of fuel and the air standard efficiency.

*Soln.*  Refer to Fig. 6.22

**Fig. 6.22**

The cycle is shown above. The flow here is steady flow.

We have $\qquad T_2/T_1 = (r_p)^{\frac{\gamma-1}{\gamma}}$ where $r_p$ is pressure ratio

or, $\qquad T_2 = 238 \times (5)^{\dfrac{0.4}{1.4}} = 456\text{k}$

Work done in compressor per kg of air $= h_2 - h_1$

$$= C_p(T_2 - T_1) = 1.005(456 - 288) = 168.84 \text{ kJ/kg}$$

Again in $\qquad T_3/T_4 = (r_p)^{\frac{\gamma-1}{\gamma}}$

or, $\qquad T_4 = \dfrac{(800 + 273)}{(5)^{\frac{0.4}{1.4}}} = 677.5 \text{ k}$

Turbine work per kg of air $= h_3 - h_4$

$$= C_p(T_3 - T_4) = 1.005 \times (1073 - 677.5) = 397.47 \text{ kJ/kg}$$

Net work done $= 397.47 - 168.84 = 228.63 \text{ kJ/kg}$

Heat added/kg $= cp(T_3 - T_2) = 1.005(1073 - 456) = 620.08 \text{ kJ/kg}$

Air standard efficiency $= W$ net/heat added $= \dfrac{228.63}{620.08} = 0.3687 = 36.87\%$

Output power per kg $= 228.63$ kW *Ans.*

**6.6** In an ideal Otto cycle, if $T_3$ and $T_1$ represent the maximum and minimum temperature respectively, show that for maximum work done in the cycle

$$T_2 = T_4 = \sqrt{T_1 T_3}$$

Where $T_2$ and $T_4$ are the temperature at the end of compression and expansion

*Soln.* Ref to Fig. 6.23

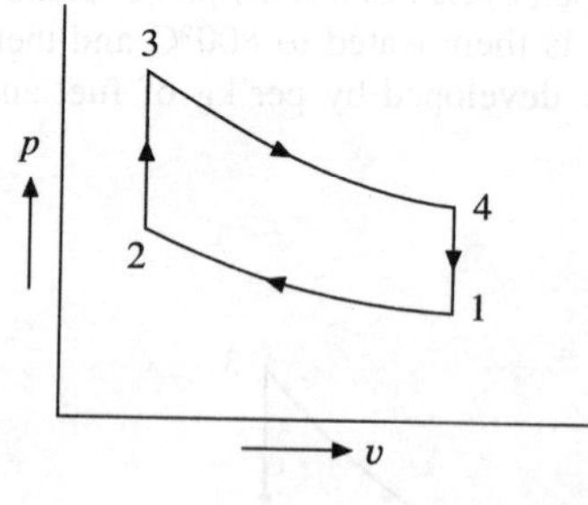
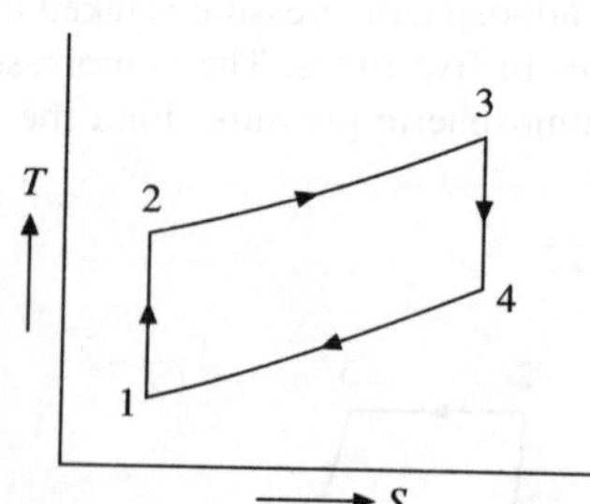

**Fig. 6.23**

We have for the cycle for 1 kg of air heat added

$$= C_v(T_3 - T_2) \text{ and heat rejected} = C_v(T_4 - T_1)$$

$$\text{Work done/kg} = C_v(T_3 - T_2) - C_v(T_4 - T_1)$$

$$= C_v[(T_3 + T_1) - (T_2 + T_4)]$$

$$v_1/v_2 = v_4/v_3 = \text{compression ratio } r$$

from process 1–2 $\qquad T_2/T_1 = (v_1/v_2)^{\gamma-1} = r^{\gamma-1}$

from process 3–4 $\qquad T_3/T_4 = (v_4/v_3)^{\gamma-1} = r^{\gamma-1}$

$$T_2/T_1 = T_3/T_4 \text{ or, } T_2 = T_1 T_3/T_4$$

Substituting the value of $T_2$ in the expression work done we have

$$W/\text{kg} = C_v\left[(T_1 + T_3) - \left(\frac{T_1 T_3}{T_4} + T_4\right)\right]$$

$T_1$ and $T_3$ are fixed, therefore, to get the maximum work, the differential of this with respect to $T_4$ will be zero

$$\frac{d_w}{d_{T_4}}\ 0, \quad \text{or,} \quad 0 - T_1 T_3/T_4^2 + 1 = 0$$

or $\qquad T_1 T_3/T_4^2 = 1 \text{ or, } T_4 = \sqrt{T_1 \cdot T_3}$

Also since, $\qquad T_2/T_1 = T_3/T_4, \ T_2 = T_1 T_3/T_4$

$$T_2 = \frac{T_1 T_3}{\sqrt{T_1 T_3}} = \sqrt{T_1 T_3}$$

Thus $\qquad T_2 = T_4 = \sqrt{T_1 \cdot T_3} \text{ Proved}$

**6.7** The bore and stroke of an engine cylinder are 18 cm and 32 cm respectively. The clearance volume is $0.00254 \text{ m}^3$. If the engine works on Otto cycle, find the compression ratio and the air standard efficiency.

*Soln.*  Ref. to Fig. 6.24

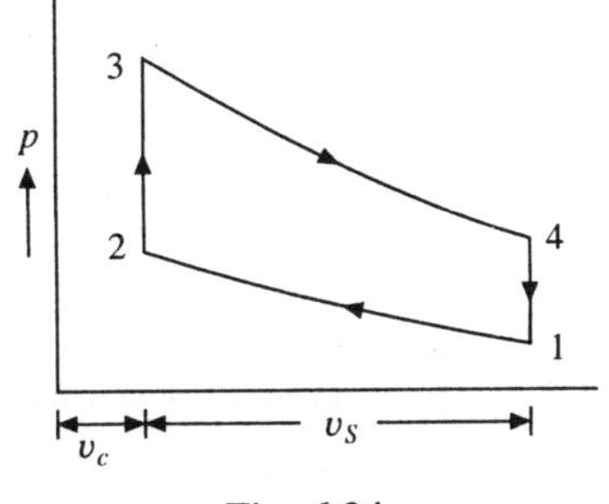

$$v_c = v_2 = v_3 = \text{clearance volume} = 0.00254 \text{ m}^3$$

$$\text{stroke volume} = v_1 - v_2 = \pi/4\ D^2 \times L$$

$$= \pi/4 \times 0.18^2 \times 0.30 = 0.00763 \text{ m}^3$$

$$\text{cylinder volume} = v_c + v_s = 0.00254 + 0.00763 \text{ m}^3$$

$$= 0.01017 \text{ m}^3$$

$$\text{Compression ratio } r = \frac{0.01017}{0.00254} = 4.00 \quad Ans.$$

**Fig. 6.24**

Hence the air standard efficiency is given by

$$= 1 - (1/r)^{\gamma-1}$$

$$= 1 - (1/4)^{0.4} = 1 - 0.574 = 0.426$$

$$= 42.6\% \ Ans.$$

**6.8** In a certain air standard cycle the following processes occur:

(i) Compression from 40°C, 1 bar and 0.01 m$^3$ through a compression ratio of 6, following the law $pv^{1.25} = C$.

(ii) Heat added at constant volume till the pressure reaches 50 bar.

(iii) Expansion of air following $pv^{1.25} = C$ to the initial volume. Show the cycle on *p-v* plot and find the work done per cycle and the mean effective pressure.

*Soln.* Refer to Fig. 6.25

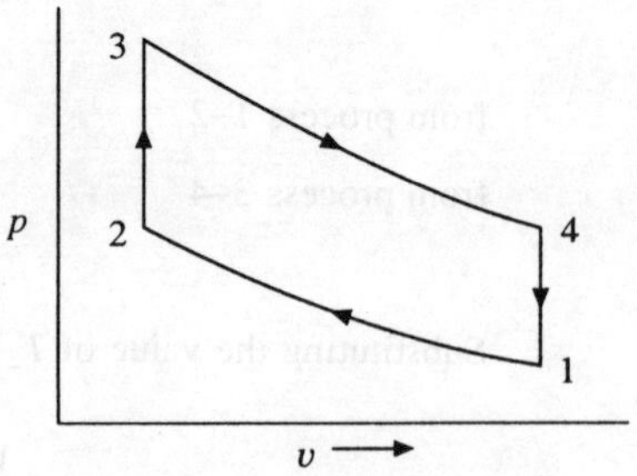

$$v_1 = 0.01 \text{ m}^3, \ v_2 = 0.01/6 = 0.00166 \text{ m}^3$$

Stroke volume $v_s = v_1 - v_2 = 0.01 = 0.00166$

$$= 0.00834 \text{ m}^3$$

From process 1–2

**Fig. 6.25**

$$p_2/p_1 = (v_1/v_2)^n = (6)^{1.25} = 9.4$$

$$p_2 = 9.4 \times 1 = 9.4 \text{ bar}$$

from process 3–4

$$P_4/P_3 = (v_3/v_4)^n \text{ or, } p_4 = \frac{50}{6^{1-25}} = 5.32 \text{ bar}$$

Thus we know each state point and so the *P-V* plot can be completed.

work done/cycle = area $1 - 2 - 3 - 4$

= work done during expansion work done during compression

$$= \frac{P_3 v_3 - P_4 v_4}{n - 1} - \frac{P_2 v_2 - P_1 v_1}{n - 1}$$

$$= \frac{(50 \times 10^5 \times 0.00166 - 5.32 \times 10^5 \times 0.01)}{1.25 - 1} - \frac{(9.4 \times 10^5 \times 0.00166 - 0.01 \times 10^4)}{1.25 - 1}$$

$$= \frac{2445.7}{0.25} = 9782.8 \text{ Nm} = 9.7828 \text{ kJ} \quad Ans.$$

$$\text{Mean effective pressure} = \frac{\text{W.D./cycle}}{\text{stroke volume}}$$

$$= \frac{9782.8}{0.00834} = 11.73 \times 10^5 \text{ N/m}^2 = 11.73 \text{ bar} \quad Ans$$

**6.9** The following particulars refer to a 2-stroke diesel engine: bore = 10 cm, stroke = 15 cm, piston speed = 300 m/min, torque developed = 58 Nm, piston speed = 300 m/min, torque developed = 58 Nm, mechanical efficiency = 80%, indicated thermal efficiency = 40%, calorific value of fuel used = 44000 kJ/kg. Determine: (a) indicated power (b) indicated mean effective pressure, and (c) fuel consumption per kW hour on brake power basis.

*Soln.* Given $D$ = 10 cm, $L$ = 15 cm = 0.15 m, Piston speed = $2LN$ = 300 m/min, $T$ = 58 Nm. Type of engine = 2 stroke diesel engine, $C \cdot V$ = 44000 kJ/kg, $\eta_{mech}$ = 80%, $\eta_{idi.th.}$ = 40%

$$\text{Piston speed} = 2LN = 300 \quad \therefore \quad N = \frac{300}{2L} = \frac{300}{2 \times 0.15} = 1000 \text{ r.p.m}$$

$$\eta_{mech} = \frac{\text{Brake power}}{\text{Indicated power}}$$

Now, $\qquad$ Brake Power (BP) $= \dfrac{2\pi NT}{60 \times 1000} = \dfrac{2\pi \times 1000 \times 58}{60 \times 1000} = 6.073 \text{ kW}$

$\therefore \qquad$ Indicated power (IP) $= \dfrac{\text{BP.}}{\eta_{mech}} = \dfrac{6.073}{0.8} = 7.5921 \text{ kW}$

$$\text{IP/cylinder} = \frac{P_{m(act)} \times L \times A \times N}{60 \times 1000} \text{ kW}$$

$$7.5921 = \frac{P_{m(\text{act})} \times 0.15 \times \pi(0 \cdot 1)^2 \times 1000}{60 \times 1000} \text{ kW}$$

$\therefore$ Indicated mean effective pressure

$$P_{m(\text{act})} = \frac{7.5921 \times 60 \times 1000}{0.15 \times \pi \times 0.01 \times 1000} = 386666.67 \text{ N/m}^2 = 3.87 \text{ bar}$$

$$\text{Indicated thermal efficiency} = \frac{IP \times 60 \times 60}{mf \times c \cdot V}$$

$$\text{Fuel consumption per hour, } mf = \frac{7.5921 \times 3600}{0.4 \times 44000} = 1.5529 \text{ kg/hr}$$

$\therefore$ Fuel consumption  per kw hour on brake power

$$\text{basis} = \frac{mf}{B.P.} = \frac{1.5529}{6.073} = 0.2557 \text{ kg}$$

**6.10**  The compression ratio and expansion ratio of an oil engine working on the dual cycle are 9 and 5 respectively. The initial pressure and temperature of the air are 1 bar and 30°C. The heat liberated at constant pressure is twice the heat liberated at constant volume. The expansion and compression follow the law $PV^{1.25}$ = constant. Determine:

(i) Pressure and temperatures at all salient points
(ii) Mean effective pressure of the cycle
(iii) Efficiency of the cycle
(iv) Power of the engine if working cycles per second are 8.
Assume: cylinder bore = 250 mm and stroke length = 400 mm.

(Mumbai Univ. winter 2000, summer 2001)

*Soln.*  Refer to Fig. 6.26

Given $\qquad\qquad r = \dfrac{v_1}{v_2} = 9$

$$r_e = \frac{v_5}{v_4} = 5$$

$P_1 = 1$ bar, $T_1 = 30 + 273 = 303$ K, $n = 1.25$

$$q_{H_p} = 2 \times q_{H_v}$$

$D = 250$ mm, $L = 400$ mm

$$\frac{T_2}{T_1} = \left(\frac{P_2}{P_1}\right)^{\frac{n-1}{n}} = \left(\frac{v_1}{v_2}\right) = (9)^{1.25-1}$$

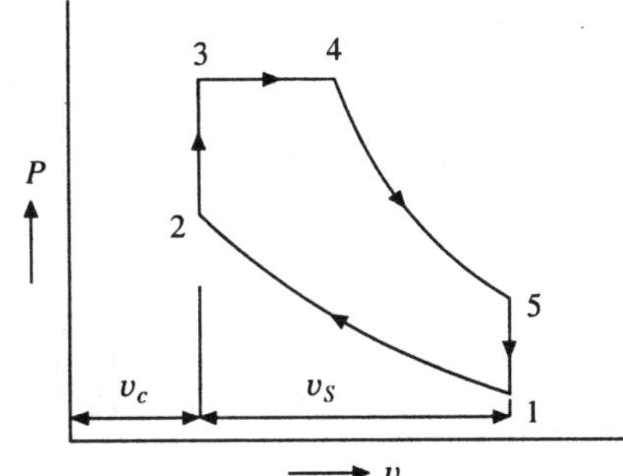


**Fig. 6.26**

$\therefore \qquad T_2 = 303 \times 9^{0.25} = 524.81 \text{k}$

$$P_2 v_2^n = P_1 v_1^n$$

$\therefore \qquad P_2 = P_1\, r^n = 1 \times 9^{1.25} = 15.58 \text{ bar}$

For process 2–3

$$\frac{P_2\, v_2}{T_2} = \frac{P_3\, v_3}{T_3} \qquad \because\ v_2 = v_3$$

$$\therefore \qquad P_3 = P_2 \frac{T_3}{T_2} \qquad\qquad\qquad (i)$$

Given
$$C_p(T_4 - T_3) = 2 \times C_v(T_3 - T_2) \qquad\qquad\qquad (ii)$$

for process 3–4

$$\frac{P_3 v_3}{T_3} = \frac{P_4 v_4}{T_4} \quad \because \; P_3 = P_4$$

$$\therefore \qquad \frac{T_4}{T_3} = \frac{v_4}{v_3} = r_c = \frac{v_1/v_3}{v_1/v_4} = \frac{v_1/v_2}{v_1/v_4} = \frac{v_1/v_2}{v_5/v_4} = \frac{r}{r_e}$$

or,
$$\frac{T_4}{T_3} = \frac{9}{5} = 1.8 = \frac{v_4}{v_3} = r_c$$

$$\therefore \qquad T_4 = 1.8 \; T_3$$

Putting the value in Eq. (ii)

$$1.005 \,(1.8 \, T_3 - T_3) = 2 \times 0.7(T_3 - 524.81)$$

or,
$$0.804 \, T_3 = 1.4 \, T_3 - 734.73$$

$$\therefore \qquad T_3 = 1232.76 \text{ k}$$

Putting in Eq. (i)

$$P_3 = 15.58 \, \frac{1232.76}{524.81} = 36.59 \text{ bar}$$

$$P_3 = P_4 = 36.59 \text{ bar}$$

$$T_4 = 1.8 \, T_3 = 1.8 \times 1232.76 = 2218.96 \text{ k}$$

$$\frac{T_5}{T_4} = \left(\frac{v_4}{v_5}\right)^{n-1} = \left(\frac{1}{5}\right)^{1.25-1}$$

$$\therefore \qquad T_5 = 2218.96 \left(\frac{1}{5}\right)^{0.25} = 1483.90 \text{ k}$$

$$\left(\frac{P_5}{P_4}\right)^{\frac{n-1}{n}} = \left(\frac{v_4}{v_5}\right)^{n-1}$$

$$\therefore \qquad P_5 = P_4 \left(\frac{v_4}{v_5}\right)^{n} = 36.59 \left(\frac{1}{5}\right)^{1.25} = 4.89 \text{ bar}$$

(ii) Mean effective pressure

$$P_m = \frac{P_1 r^n \left[\alpha n(r_c - 1) + (\alpha - 1) - \dfrac{1}{r^{n-1}}(\alpha r_c^n - 1)\right]}{(n-1)(r-1)}$$

here $P_1 = 1$ bar, $r = 9$, $n = 1.25$, $\alpha = \dfrac{P_3}{P_2} = \dfrac{36.59}{15.58} = 2.348$

$$r_c = \frac{v_4}{v_3} = \frac{T_4}{T_3} = 1.8$$

$$\therefore \quad P_m = \frac{1 \times 9^{1.25}\left[2.348 \times 1.25(1.8-1) + (2.348-1) - \dfrac{1}{9^{1.25-1}}(2.348 \times 1.8^{1.25}-1)\right]}{(1.25-1)(9-1)}$$

$$= \frac{15.588[2.348 + 1.348 - 2.248]}{0.25 \times 8} = 11.29 \text{ bar} \quad Ans$$

(iii) Work done per cycle $= P_m\, v_s$

$$\text{Stroke volume, } v_s = \frac{\pi}{4}D^2 L = \frac{\pi}{4} \times (0.25)^2 \times 0.4 = 0.01963 \text{ m}^3$$

$\therefore$ Work done per cycle

$$= 11.29 \times 10^5 \times 10^{-3} \times 0.01963 = 22.16 \text{ kJ/cycle}$$

Heat supplied per cycle = mqs

$$\text{mass, } m = \frac{P_1 v_1}{RT_1}$$

$$v_1 = v_s + v_c = \frac{r}{r-1}\, v_s = \frac{9}{9-1} \times 0.01963 = 0.02208 \text{ m}^3$$

$$\therefore \qquad m = \frac{1 \times 10^5 \times 0.02208}{287 \times 303} = 0.02539 \text{ kg/cycle}$$

$\therefore$ Heat supplied per cycle

$$= 0.02539\,[C_v(T_3 - T_2) + C_p\,(T_4 - T_3)]$$

$$= 0.02539\,[0.7(1232.76 - 524.81) + 1.005 \times (2218.96 - 1232.76)]$$

$$= 0.02539\,[495.565 + 991.131] = 37.74 \text{ kJ/cycle}$$

$$\therefore \quad \text{Efficiency of the cycle } = \frac{\text{work done}}{\text{Heat supplied}}$$

$$= \frac{22.16}{37.74} \times 100 = 58.71\% \quad Ans.$$

(iv) Power of the engine in 8 cycles/second

$$= 22.16 \times 8$$

$$= 177.28 \, \frac{\text{kJ}}{\text{cycle}} \times \frac{\text{cycle}}{\text{second}} = 177.28 \text{ kW} \quad Ans$$

**6.11** A four stroke limited pressure cycle engine or dual fuel cycle engine operates on 13 litres of air at 1 bar and 27°C per cycle. The addition of heat at constant volume is adjusted for a maximum pressure in the cycle of 70 bar. The heat addition at constant pressure continues for 6% of the stroke. Calculate.

  (i)   Pressure ratio and cut off ratio
  (ii)  Heat added per cycle
 (iii)  Heat rejected per cycle
 (iv)  Work done per cycle
  (v)  Thermal efficiency
 (vi)  Indicated power developed if engine runs at 1200 r.p.m.
Assume $C_p = 1.005$ kJ/kg k, $\gamma = 1.4$, compression ratio = 17.

*Soln.*   Refer to Fig. 6.27

$\qquad$ *Given* $P_1 = 1$ bar, $T_1 = 27 + 273 = 300$k

$$P_3 = 70 \text{ bar}$$

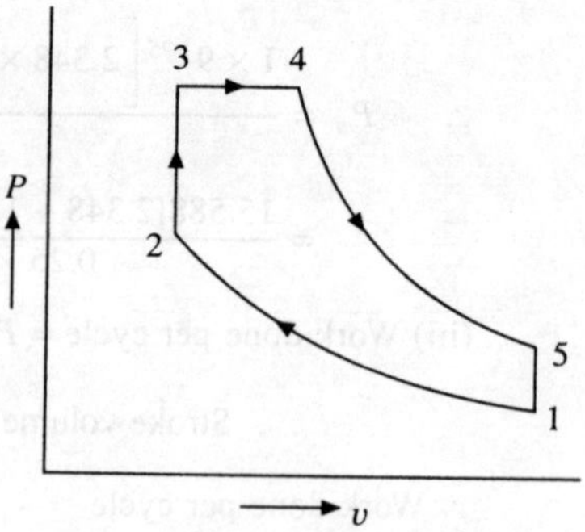

$$\text{Stroke volume} = v_1 - v_2 = 13 \times 1000 \text{ cm}^3$$

$$= 0.013 \text{ m}^3$$

$$\text{compression ratio } r = \frac{v_1}{v_2} = 17$$

$\therefore \qquad\qquad\qquad \dfrac{v_1 - v_2}{v_2} = 17 - 1$

**Fig. 6.27**

or, $\qquad\qquad\qquad \dfrac{0.013}{v_2} = 16 \quad \therefore v_2 = 8.125 \times 10^{-4} \text{ m}^3$

$$v_1 = v_5 = 0.013 + 0.0008125 = 0.0138125 \text{ m}^3$$

For process 1–2

$$\frac{P_2}{P_1} = \left(\frac{v_1}{v_2}\right)^{\gamma}$$

$\therefore \qquad\qquad\qquad P_2 = 1 \times (17)^{1.4} = 52.79 \text{ bar}$

$\qquad\qquad\qquad\quad P_3 = 70 \text{ bar}$

$\therefore \qquad\qquad$ explosion ratio = pressure ratio $\alpha = \dfrac{P_3}{P_2}$

$$\alpha = \frac{70}{52.79} = 1.326$$

$$\frac{T_2}{T_1} = \left(\frac{v_1}{v_2}\right)^{\gamma - 1}$$

$\therefore \qquad\qquad\qquad T_2 = 303(17)^{1.4-1} = 941.07 \text{ k}$

$\qquad\qquad\qquad\quad T_3 = \dfrac{P_3}{P_2} \cdot T_2 = \dfrac{70}{52.79} \times 941.07 = 1247.85 \text{ k}$

And $\qquad\qquad\qquad v_4 = v_3 + \dfrac{6}{100}(v_1 - v_2)$

$$= 0.0008125 + 0.06 \times 0.013 = 0.00159 \text{ m}^3$$

$\therefore \qquad\qquad$ cut off ratio $r_c = \dfrac{v_4}{v_3} = \dfrac{0.00159}{0.0008125} = 1.956$

and $\qquad\qquad\qquad T_4 = T_3 \dfrac{v_4}{v_3} = 1247.85 \times 1.956 = 2440.79 \text{ k}$

For process 4–5

$$\frac{T_5}{T_4} = \left(\frac{v_4}{v_5}\right)^{\gamma - 1} \quad \therefore T_5 = 2440.79\left(\frac{0.00159}{0.0138125}\right)^{1.4-1}$$

$$= 1027.96 \text{ k}$$

Mass flow rate/cycle

$$m = \frac{1 \times 10^5 \times 0.0138125}{287 \times 300} = 0.01604 \text{ kg}$$

Heat added per cycle $= mC_v(T_3 - T_2) + mC_p(T_4 - T_3)$

$$= 0.01604 \times \frac{1.005}{1.4}(1247.85 - 941.01) + 0.01604 \times 1.005\,(2440.79 - 1247.85)$$

$$= 22.76 \text{ kJ/cycle}$$

Heat rejected per cycle $= mC_v(T_5 - T_1)$

$$= 0.01604 \times \frac{1.005}{1.4}\,(1027.96 - 300) = 8.38 \text{ kJ/cycle}$$

work done/cycle = Heat added – Heat rejected

$$= 22.76 - 8.38 = 14.38 \text{ kJ/cycle}$$

$\therefore$ Thermal efficiency $\eta = \dfrac{\text{work done/cycle}}{\text{Heat added/cycle}} = \dfrac{14.38}{22.76} \times 100 = 63.18\%$    *Ans*

Engine r.p.m = 1200

$\therefore$ working cycle for four stroke engine

$$= \frac{1200}{60 \times 2} = 10 \text{ cycles/sec.}$$

$\therefore$      Indicated power = work done/cycle $\times$ No. of cycles/sec.

$$= 14.38 \times 10 = 140.38 \text{ kJ/sec} = 140.38 \text{ kW}    \textit{Ans}$$

**6.12** Obtain an expression for the specific work done by an engine working on the otto cycle in terms of the maximum and minimum temperature of the cycle, the compression ratio $R$ and constant of working fluid.

Hence show that the compression ratio for maximum specific work output is given by $R = \left(\dfrac{T_{\max}}{T_{\min}}\right)^{\frac{1}{2(\gamma-1)}}$

(Mumbai Univ. winter 1999, summer 2000).

*Soln.* Refer to Fig. 6.28
work done per kg of fluid

$$W = C_v(T_3 - T_2) - C_v\,(T_4 - T_1)$$

But      $T_2 = T_1 R^{\gamma-1}$   and   $T_3 = T_4 R^{\gamma-1}$

$\therefore$      $W = C_v\left[T_3 - T_1 R^{\gamma-1} - \dfrac{T_3}{R^{\gamma-1}} + T_1\right]$

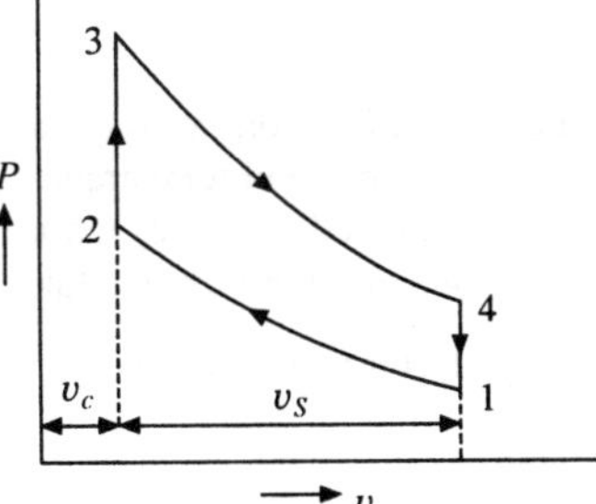

**Fig. 6.28**

This expression is a function of $R$ when $T_3$ and $T_1$ are fixed. The value of $W$ will be maximum when

$$\frac{dW}{dR} = 0 \quad \therefore \quad \frac{dw}{dR} = -T_1(\gamma - 1)R^{\gamma-2} - T_3(1 - \gamma)R^{-\gamma} = 0$$

or,      $T_3 R^{-\gamma} = T_1 R^{\gamma-2}$   or,   $\dfrac{T_3}{T_1} = R^{2(\gamma-1)}$

$\therefore$      $R = \left(\dfrac{T_3}{T_1}\right)^{1/2(\gamma-1)}$   for air $\gamma = 1.4$ $\therefore$ $R = \left(\dfrac{T_3}{T_1}\right)^{1.25}$

**6.13** In a Brayton cycle, prove that the optimum pressure ratio $r_p$ for maximum network done between the temperatures $T_1$ and $T_3$, where $T_3$ is the maximum and $T_1$ is minimum temperature obtained, is given by

$$r_p = \left(\frac{T_2}{T_1}\right)^{\gamma} / \{2(\gamma - 1)\}$$

*Soln.* Refer to Fig. 6.29

We have
$$\frac{T_3}{T_4} = \frac{T_2}{T_1} = \left(\frac{P_2}{P_1}\right)^{(\gamma-1)/\gamma} = r_p^{(\gamma-1)/\gamma} = y$$

Let $T_3/T_1 = Z$

Therefore, net work done/kg

$$W = (h_3 - h_4) - (h_2 - h_1)$$
$$= C_p(T_3 - T_4) - C_p(T_2 - T_1)$$
$$= C_p T_3\left(1 - \frac{1}{y}\right) - C_p T_1(y - 1)$$

or,
$$\frac{W}{C_p T_1} = \frac{T_3}{T_1}\left(1 - \frac{1}{y}\right) - (y - 1) = z - \frac{z}{y} - y + 1$$

$$d\left\{\frac{W}{C_p T_1}\right\} = 0 \text{ for maximum work done}$$

$\dfrac{T_3}{T_1}$ is a constant, only $y$ is a variable

Therefore,
$$0 = \frac{z}{y^2} - 1 + 0 = 0$$

or,
$$y = \sqrt{z}$$

i.e.
$$r_p^{(\gamma - 1)/\gamma} = \left(\frac{T_3}{T_1}\right)^{1/2}$$

or,
$$r_p = \left(\frac{T_3}{T_1}\right)^{\gamma/\{2(\gamma - 1)\}} \qquad \text{Proved}$$

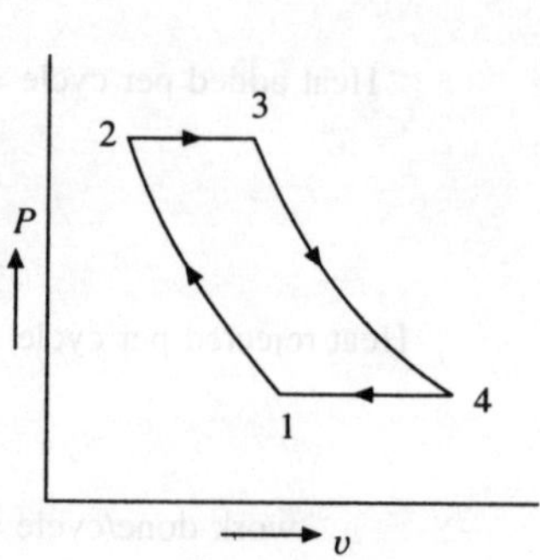

**Fig. 6.29**

**6.14** In a Brayton cycle the air enters the compressor at 1 bar and 25°C. The pressure leaving the compressor at 3 bar and temperature at turbine inlet is 650°C. Determine per kg of air (a) cycle efficiency (b) heat supplied to air (c) work available (d) heat rejected in the cooler at the shaft and (e) temperature of air leaving the turbine. Take $\gamma = 1.4$ and $C_p = 1.005$ kJ /10k

*Soln.* Refer to Fig. 6.30

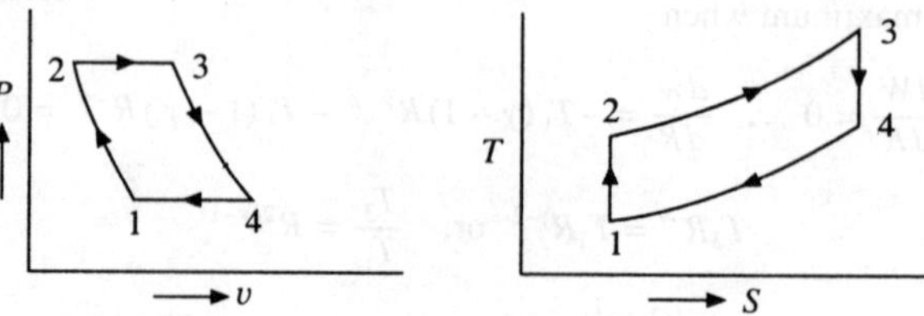

**Fig. 6.30**

Given
$$P_1 = 1 \text{ bar}, \ T_1 = 25°C = 25 + 273 = 298 \text{ k}$$
$$P_2 = 3 \text{ bar}, \ T_3 = 650°C = 650 + 273 = 923 \text{ k}, \ m = 1 \text{ kg}$$

(a)  Cycle efficiency,     $\eta = 1 - \left(\dfrac{P_1}{P_2}\right)^{\frac{\gamma-1}{r}} = 1 - \left(\dfrac{1}{3}\right)^{\frac{1.4-1}{1.4}} = 26.9\%$

(b)  Considering process 1–2

$$T_2 = T_1 \times \left(\frac{P_2}{P_1}\right)^{\frac{\gamma-1}{\gamma}} = 298 \,(3)^{\frac{0.4}{1.4}} = 407 \text{ k}$$

Heat supplied to air $Q_1 = C_p(T_3 - T_2) = 1.005(923 - 407) = 515$ kJ/kg

(c)          Work out put $= \eta \times$ Heat input $= 0.269 \times 515 = 138.5$ kJ /kg

(d)          Work output $= Q_1 - Q_2$

$\therefore$          $Q_2 = 515 - 138.5 = 376.5$ kJ/kg

(e)
$$T_4 = \frac{T_3}{\left(\dfrac{P_2}{P_1}\right)^{\frac{\gamma-1}{\gamma}}} = \frac{923}{(3)^{\frac{0.4}{1.4}}} = 678 \text{k}$$

**6.15**  At the begining of adiabatic compression in an Atkinson cycle the temperature, volume and pressure are 20°C, 20 litre and 100 kN/m² respectively. After adiabatic compression volume reaches to 4 litres. The maximum pressure of the cycle is 2 MN/m². Determine

  (i)  The thermal efficiency
 (ii)  The heat received
(iii)  The heat rejected
 (iv)  The net work
  (v)  The work ratio
 (vi)  The mean effective pressure
(vii)  The Carnot efficiency within the cycle temperature limits. Take $\gamma = 1.4$, $R = 0.287$ kJ/kgk

*Soln.*  Refer to Fig. 6.31

(i)
$$T_1 = 20 + 273 = 293 \text{ k}$$
$$v_1 = 20 \text{ litres}, \ v_2 = 4 \text{ litres}$$
$$P_1 = 100 \text{ kN/m}^2$$
$$\frac{T_2}{T_1} = \left(\frac{v_1}{v_2}\right)^{(\gamma-1)}$$

Fig. 6.31

$\therefore$          Heat received $= mC_v(T_3 - T_2)$

$$= 0.0237 \times 0.7175 \,(1172 - 557.77)$$
$$= 10.44 \text{ kJ} \quad Ans$$

(iii)          Heat rejected $= mC_p(T_4 - T_1), \ c_p = \gamma C_v$

$$= 0.0237 \times 1.4 \times 0.7175(497.97 - 293)$$

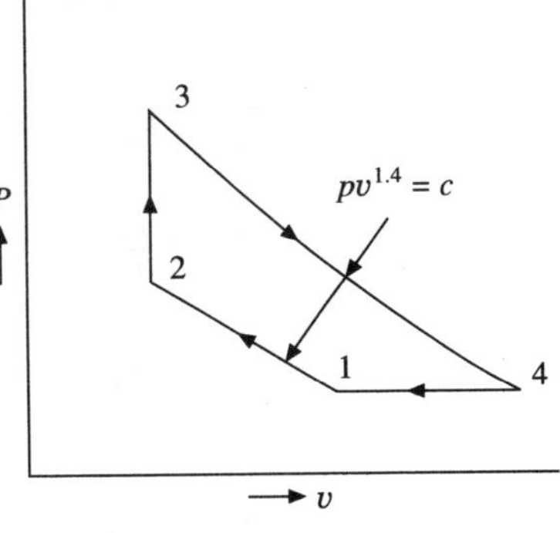


**Fig. 6.31**

(iv)           The net work = Heat received − Heat rejected

$$= 10.44 - 4.87 = 5.57 \text{ kJ} \quad Ans.$$

(v)      Work ratio $= \dfrac{\text{Net work}}{\text{Positive work done}} = \dfrac{5.57}{\dfrac{(P_3 v_3 - P_4 v_4)}{(\gamma - 1)}}$

$$= \dfrac{5.57}{\dfrac{(2000 \times 0.004 - 100 \times 0.0339)}{1.4 - 1}} = 0.483 \quad Ans$$

(vi)   Mean effective pressure $= \dfrac{\text{Network}}{\text{Stroke volume}}$

$$= \dfrac{\text{Net work}}{(v_4 - v_2)} = \dfrac{5.57}{(0.0339 - 0.004)} = 186.28 \text{ kN/m}^2 \quad Ans.$$

(vii)          $\eta_{\text{carnot}} = \dfrac{T_3 - T_1}{T_3} = \dfrac{1172 - 293}{1172} = 0.75 = 75\% \quad Ans.$

**6.16**  An ideal double acting hot air working on Stirling cycle, has, at the begining of isothermal compression a temperature, volume and pressure of 30°C, 0.125 m³ and 1 bar. Compression ratio is equal to 2. The maximum temperature of the cycle is 550°C. Speed of the engine is equal to 50 rpm. Regenerative efficiency is equal to 90%. Heat supplied is equal to 2,000 kJ. Determine

$$\therefore \qquad T_2 = T_1\left(\frac{v_1}{v_2}\right)^{(\gamma-1)} = 293\left(\frac{20}{4}\right)^{1.4-1} = 557.77 \text{ k}$$

compression ratio $\qquad r = \dfrac{v_1}{v_2} = \dfrac{20}{4} = 5$

$$\frac{P_2}{P_1} = \left(\frac{v_1}{v_2}\right)^{\gamma} \quad \therefore P_2 = 100\left(\frac{20}{4}\right)^{1.4} = 951.82 \text{ kN/m}^2$$

$$P_3 = 2 \ MN/m^2 = 2000 \text{ kN/m}^2$$

$\therefore$  pressure ratio or explosion ratio $\alpha = \dfrac{P_3}{P_2} = \dfrac{2000}{951.82} = 2.10$

$$\frac{P_2}{T_2} = \frac{P_3}{T_3} \quad \therefore T_3 = \frac{P_3}{P_2} \cdot T_2 = \frac{2000}{951.82} \times 557.77 = 1172 \text{ k}$$

$$\frac{T_4}{T_3} = \left(\frac{P_4}{P_3}\right)^{\frac{\gamma-1}{\gamma}} \quad \therefore T_4 = T_3\left(\frac{P_1}{P_3}\right)^{\frac{\gamma-1}{\gamma}} = 1172\left(\frac{100}{2000}\right)^{\frac{1.4-1}{1.4}} = 497.97 \text{ k}$$

$$\frac{T_3}{T_4} = \left(\frac{v_4}{v_3}\right)^{\gamma-1} = (r_e)^{\gamma-1} \text{ or, } r_e = \left(\frac{T_3}{T_4}\right)^{\frac{1}{\gamma-1}} = \left(\frac{1172}{497.97}\right)^{\frac{1}{1.4-1}} = 8.49$$

$$8.49 = \frac{v_4}{v_3} = \frac{v_4}{v_2} \quad \therefore v_4 = 8.49 \times 0.004 = 0.0339 \text{ m}^3$$

$\therefore$ Thermal efficiency $\qquad \eta_{\text{th}} = 1 - \dfrac{\gamma(T_4 - T_1)}{(T_3 - T_2)}$

$$= 1 - \frac{1.4(497.97 - 293)}{(1172 - 557.77)} = 1 - 0.467 = 0.533 = 53.3\% \quad Ans.$$

By using second formula

$$\eta_{th} = 1 - \frac{\gamma}{r^{\gamma-1}}\left[\frac{\left(\dfrac{r_e}{r}\right)-1}{\alpha-1}\right] = 1 - \frac{1.4}{5^{1.4-1}}\left[\frac{\dfrac{8.49}{5}-1}{2.10-1}\right]$$

$$= 1 - 0.466 = 0.534 = 53.4\% \quad Ans$$

(ii) $\qquad\qquad P_1 v_1 = mRT_1$

$\therefore \qquad\qquad m = \dfrac{P_1 v_1}{RT_1} = \dfrac{100 \times 0.02}{0.287 \times 293} = 0.0237 \text{ kg}$

$$C_v = \frac{R}{(\gamma-1)} = \frac{0.287}{1.4-1} = 0.7175$$

$\therefore \qquad$ Mean effective pressure $= \dfrac{\text{Total work done}}{\text{Stroke volume}}$

$$= \frac{0.0995 \times 105.43}{0.0417} \text{ KN/m}^2$$

$$= 251.56 \text{ kN/m}^2 = 2.5156 \text{ bar} \quad Ans.$$

(iv) $\qquad$ Indicated power $= 9.95 \times 105.43$ kJ/minute

$$= 1049.0285 \text{ kJ/minute} = 17.48 \text{ kW} \quad Ans.$$

**6.17** In an ideal Ericsson cycle, using air, the initial conditions of temperature, volume and pressures are recorded as 25°C, 0.07 m$^3$ and 100 kN/m$^2$ respectively. Compression ratio is equal to 5. After compression the volume of air is doubled during the constant pressure process. Determine:

  (i) The maximum temperature
  (ii) The net work done
  (iii) The ideal thermal efficiency
  (iv) The thermal efficiency if the process of regeneration is not included.
Take $R = 0.287$ kJ/kgk, $C_p = 1.006$ kJ/kgk.

*Soln.* Refer to Fig. 6.12.

(i) $\qquad\qquad T_1 = 25 + 273 = 298$ k

$$\frac{v_3}{T_3} = \frac{v_2}{T_2} \text{ As } T_2 = T_1$$

$\therefore \qquad\qquad T_3 = T_2\dfrac{v_3}{v_2} = 298 \times 2 = 596 \text{ k} = 323°\text{C}$

$\therefore \qquad$ Maximum temperature $= 323°\text{C} \quad Ans.$

(ii) $\qquad\qquad m = \dfrac{P_1 v_1}{RT_1} = \dfrac{100 \times 0.07}{0.287 \times 298} = 0.0818 \text{ kg}$

  (i) Work done/kg of air
  (ii) thermal efficiency
  (iii) mean effective pressure
  (iv) indicated power

*Soln.* Refer to Fig. 6.12

(i) Minimum temperature of the cycle

$$T_1 = T_L = 30 + 273 = 293 \text{ k}$$

Maximum temperature of the cycle

$$T_3 = T_H = 550 + 273 = 823 \text{ k}$$

Work done/kg of air $= R\,(T_H - T_L) \ln r$

$$= 0.287(823 - 293) \times \ln 2 = 105.43 \text{ kJ} \quad Ans$$

(ii) Heat supplied per kg of air

$$= R\,T_4 \ln r + (1 - \eta_g)\,C_v(T_4 - T_1)$$

$$= 0.287 \times 823 \ln 2 + (1 - 0.9) \times 0.7(823 - 293) = 200.82 \text{ kJ}$$

∴ Thermal efficiency of the cycle

$$\eta_{th} = \frac{105.43}{200.82} = 0.5249 = 52.49\% \quad Ans$$

(iii) Mass of air used/minute $= \dfrac{2000}{200.82} = 9.95$ kg

Being the engine double acting, net working cycle/minute $= 2 \times 50 = 100$

∴ Mass of air per working cycle $= \dfrac{9.95}{100}\, 0.0995$ kg $\quad P_1 v_1 = mrT_1$

∴

$$v_1 = \frac{0.0995 \times 0.287 \times 293}{1 \times 10^5 \times 10^{-3}} = 0.0836 \text{ m}^3$$

compression ratio

$$r = \frac{v_1}{v_2} = 2$$

∴

$$v_2 = \frac{0.0836}{2} = 0.0417 \text{ m}^3$$

∴ Stroke volume $= v_1 - v_2 = 0.0836 - 0.0417 = 0.0417$ m$^3$

Net work done, $W$

$$= mR(T_3 - T_2) + mRT_3 \ln \frac{v_4}{v_3} - mR(T_4 - T_1) - mRT_1 \ln \frac{v_1}{v_2}$$

$$= mR \ln \frac{v_1}{v_2} \times (T_3 - T_1) \text{ Since } T_3 = T_4 \text{ and } T_1 = T_2$$

also

$$\frac{v_4}{v_3} = \frac{v_1}{v_2}$$

∴

$$W = 0.0818 \times 0.287 \ln 5 \times (596 - 298) = 11.259 \text{ kJ}$$

(iii) Ideal thermal efficiency

$$\eta_{th} = \frac{T_H - T_L}{T_H} = \frac{(596 - 298)}{596} = 0.50 = 50\% \quad Ans.$$

(iv)  Thermal efficiency $= 1 - \dfrac{\text{Heat rejected}}{\text{Heat received}}$

$$= 1 - \frac{\left[ mC_p (T_4 - T_1) + mRT_1 \ln \dfrac{v_1}{v_2} \right]}{\left[ mC_p (T_3 - T_2) + mRT_3 \ln \dfrac{v_4}{v_3} \right]}$$

$$= 1 - \frac{\left[ C_p (T_4 - T_1) + RT_1 \ln \dfrac{v_1}{v_2} \right]}{\left[ C_p (T_3 - T_2) + RT_3 \ln \dfrac{v_4}{v_3} \right]}$$

$$= 1 - \frac{[1.006 \times 298 + 0.287 \times 298 \ln 5]}{[1.006 \times 298 + 0.287 \times 596 \ln 5]}$$

$$= 1 - \frac{437.436}{575.085} = 0.2393 = 23.93\% \quad Ans$$

## EXERCISES

6.1  Objective questions
   (i)  For the same power, a two-stroke cycle engine occupies the floor area which as compared to four-stroke cycle is (a) more (b) less (c) equal (d) non of the above
   (ii)  The flywheel used in two-stroke cycle engine as compared to four stroke cycle engine is
(a) heavy in weight (b) same in weight
(c) light in weight (d) non of the above
   (iii)  The process, in which the violent sound pulsations within the cylinder of an IC engine are produced, is known as
(a) supercharging (b) scavenging (c) Polymerisation (d) detonation
   (iv)  The ratio of shaft power to the energy supplied by the fuel is called
(a) volumetric efficiency (b) relative efficiency
(c) mechanical efficiency (d) indicated thermal efficiency
   (v)  Morse test is used to determine
(a) indicated power for multicylinder engines
(b) Shaft power
(c) mean effective pressure
(d) temperature of the exhaust gases
   (vi)  The efficiency of a Carnot engine is given as 0.75. If the cycle direction is reversed, what will be the value of COP of reversed Carnot cycle
(a) 0.75 (b) 1.33 (c) 0.33 (d) 0.25
   (vii)  A gas turbine works on
(a) Rankine cycle (b) Carnot cycle (c) Otto cycle
(d) Brayton cycle (e) Ericsson cycle
   (viii)  Brayton cycle cannot be used in reciprocating engines even for same adiabatic compression ratio and work output because
(a) Otto cycle is more efficient
(b) Brayton cycle is less efficient
(c) Brayton cycle is for slow speed engines
(d) Brayton cycle requires large air-fuel ratio

(e) large volume of low pressure air can not be efficiently handled in reciprocating engines

(ix) The charge in a diesel engine consists of
   (a) Air + diesel + lubricating oil
   (b) Air + diesel
   (c) Air + lubricating oil
   (d) Diesel + lubricating oil
   (e) Air

(x) In a 2-stroke engine we get one power stroke in
   (a) 180° of crank rotation
   (b) 270° of crank rotation
   (c) 360° of crank rotation
   (d) 540° of crank rotation
   (e) 720° of crank rotation

(xi) A carburettor is used to supply
   (a) Diesel + air + lubricating oil
   (b) petrol + air + lubricating oil
   (c) Petrol + lubricating oil
   (d) Air + lubricating oil
   (e) Petrol + air

(xii) By applying choke in an automobile we get
   (a) stoichiometric mixture
   (b) chemically correct mixture
   (c) Lean mixture
   (d) rich mixture
   (e) weak mixture

(xiii) In case of a four cylinder diesel engine the heat carried away by exhaust gases will be
   (a) 5 to 7 percent (b) 7 to 10 percent
   (c) 10 to 20 percent (d) 35 to 50 percent
   (e) 20 to 35 percent

(xv) The thermal efficiency of theoretical Otto cycle
   (a) increases with increase in compression ratio
   (b) increases with increase in isetropic index $\gamma$
   (c) as independent of the compression ratio
   (d) follows all the above

(xvi) The work output of theoretical Otto cycle
   (a) increases with increase in compression ratio
   (b) increases with increase in pressure ratio
   (c) increases with increase in adiabatic index $\gamma$
   (d) follows all the above

(xvii) In air standard Diesel cycle, at fixed compression ratio and fixed value of adiabatic index ($\gamma$)
   (a) thermal efficiency increases with increase in cut-off ratio
   (b) thermal efficiency decreases with increase in cut-off ratio
   (c) thermal efficiency remains same with increase in to cut-off ratio
   (d) non of the above

**Answers**

|  |  |  |  |  |
|---|---|---|---|---|
| (i) b | (ii) c | (iii) d | (iv) d | (v) a |
| (vi) c | (vii) d | (viii) e | (ix) e | (x) c |
| (xi) e | (xii) d | (xiii) d | (xiv) e | (xv) d |
| (xvi) d | (xvii) a | | | |

## NUMERICAL PROBLEMS

6.2  In an Otto cycle the upper and lower limits for absolute temperatures are $T_3$ and $T_1$ respectively. Show that for maximum work, the compression ratio is given by

$$r = (T_3/T_1)^{1.25}$$

6.3  In a Brayton cycle, prove that the optimum pressure ratio $r_p$ for maximum net work done between the same temperatures $T_1$ and $T_3$, when $T_3$ is the maximum temperature obtained and $T_1$ is the minimum temperature is given by

$$r_p = (T_3/T_1) \ \gamma/2(\gamma - 1)$$

6.4  In an air standard cycle the air at the beginning is at 17°C, 1 bar and 0.856 m³/kg. The isentropic compression ratio is 15 and 838 kJ/kg heat is added at constant pressure. Find pressure and temperature at each point of the cycle, thermal efficiency and power output of an ideal engine working on this cycle and consuming 0.1 kg of air per second ($\eta$ = 60.4, 51.18. KW)

6.5  For an ideal Otto engine working on air, the temperature at the end of isentropic compression is 452°C and at the end of expansion 1347°C. If the compression ratio be 7.5, find the work done in a cycle and efficiency (980 kJ/kg, 47.6%)

6.6  An air standard engine works on diesel cycle. The compression ratio is 16 and cut off ratio 2. The total cylinder volume is 0.001416 m³. At the beginning of compression, pressure is 1 bar and temperature is 4.5°C, calculate the thermal efficiency. (53.2%)

6.7  What will be the source and sink temperatures of a Carnot engine whose efficiency is 20% and the efficiency becomes 40%. When the temperature of sink is reduced by 60°C (27°C,–33°C).

6.8  In an air standard cycle working on diesel cycle the pressure at the beginning of compression is 1 bar abs and at the end it is 21 bar abs. Find the compression ratio. If the cut off takes place at ane twentieth of the stroke, find the efficiency (8.71,54.9%)

6.9  A gas engine works on otto cycle. The clearance volume is 0.001470m³ and the stroke and diameter are both 0.178 m what is the air standard efficiency of the cycle? (0.43)

6.10  The initial pressure and tempeature of the air used in an oil engine working on theoretical diesel cycle are 1 bar and 30°C. cut off takes place at 10% of stroke. If the cylinder bore and stroke are 20 cm and 30 cm respectively. Find power of the engine if there be 200 cycles per minute (31.14 kW)

6.11  A Brayton cycle operates with inlet temperatures of 300 k and 923 k at inlet to the compressor and turbine respectively. The pressure ratio is 6. Find the specific net work. (170.16 kJ/kg).

6.12  A gas turbine engine working on an air-standard Brayton cycle, operates between the temperature limits of 300 k and 1200 k and pressure limits of 101 kPa and 505kPa. Determine (a) the thermal efficiency of the cycle, (b) the compressor work, (c) the turbine work, and (d) the air-flow rate required for 2.0 kW of net power output.         [*Ans* (a) 38.86%, (b) – 175.94 kJ/kg, (c) 444.33 kJ/kg, (d) 13.41 kg/hr]

6.13  The compression ratio of a Dual cycle is 10. The pressure and temperature at a begining of the cycle are 1 bar and 27°C. The maximum pressure of the cycle is limited to 70 bar and heat supplied is limited to 675 kJ/kg of air. Find the thermal efficiency of the cycle. [*Ans.* 59.5%]

6.14  An ideal dual cycle has a compression ratio of 15. Air is at 97 kPa, 0.084 m³ and 28°C at the begining of compression.' The maximum pressure and temperature of the cycle are 6.2 MPa and 1320°C respectively. Determine (a) the net work done, (b) the thermal efficiency and (c) the mean effective pressure of the cycle.         [*Ans.* (a) 36.58 kJ, (b) 65.34%, (c) 466.58 kPa].

6.15  In an ideal dual combustion cycle the temperature, volume and pressure at the begining of the adiabatic compression are 24°C, 0.05 m³ and 93 kN/m² respectively. The volume ratio of the adiabatic compression is 9:1. The constant volume heat addition pressure ratio is 1.5 : 1 and the constant pressure heat addition volume ratio is 2 : 1. Determine (a) the thermal efficiency, (b) the net work done, (c) the work ratio, (d) the mean effective pressure and (e) the carnot cycle efficiency within the cycle temperature limits.         [*Ans.* (a) 48.3%, (b) 35 *kJ*, (c) 0.685, (d) 788 kN/m² (e) 84.8%]

6.16 In an Atkinson cycle the volume ratios of the adiabatic expansion and the adiabatic compression are 9:1 and 5:1 respectively. At the begining of the adiabatic compression the pressure and temperature of the gas are 96.5 kN/m$^2$ and 27°C respectively. Determine, (a) the thermal efficiency, (b) the network done/kg of gas, (c) the work ratio, (d) the mean effective pressure, (e) the carnot cycle efficiency within the cycle temperature limits.

[*Ans.* (a) 52%, (b) 260 kJ, (c) 0.487 (d) 73.4 kN/m$^2$, (e) 75.8%]

6.17 An ideal Stirling cycle with regeneration, using air as the working fluid, has air at 110 kPa, 0.05 m$^3$ and 30°C at the begining of compression. The minimum volume of the cycle is 0.005 m$^3$. The maximum temperature of the cycle is 700°C. Determine (a) the net work done, (b) the ideal thermal efficiency, and (c) the thermal efficiency if the process of regeneration was not included.

[*Ans.* (a) 27.98 kJ, (b) 68.85%, (c) 39.4 %]

6.18 An ideal Ericsson cycle, using air and including regeneration, has air at 100 kPa, 20°C and 0.08 m$^3$ at the begining of isothermal compression. The volume becomes 1/5 th of the initial value at the end of compression. After isothermal compression, the volume becomes double during the constant pressure process. Determine (a) the maximum temperature, (b) the net work done, (c) the ideal thermal efficiency, and (d) the thermal efficiency if the process of regeneration was not included.

[*Ans.* (a) 586.3k, (b) 12.86 kJ, (c) 50% (d) 23.84%]

6.19 An ideal Ericsson cycle using helium as the working fluid operates between the temperature limits of 290 k and 1600 k and pressure limits of 101 kPa and 800 kPa. Assuming a mass flow rate of 5 kg/s, determine (a) the thermal efficiency of the cycle (b) the heat transfer rate in the regenerator, and (c) the power delivered

{*Ans.* (a) 81.88%, (b) 33999 kW, (c) 28.15 MW]

# 7

# Properties of Steam and Processes

## 7.1  INTRODUCTION:

A "pure substance" is defined as a homogeneous substance which retains its chemical composition even though there may be change of phase. Water which is one of the pure substances is an important working fluid. It exists in all the three phases i.e., solid (ice), liquid (water) and gas (steam). In Chapter 6, we discussed the power cycles using gas as a working substance which does not change its phase. In the present chapter formation of steam and properties will be discussed.

## 7.2  PHASE TRANSFORMATION AT CONSTANT PRESSURE

The transformation of 1 kg of ice into 1 kg of superheated steam has been shown in the Fig. 7.1 on enthalpy-temperature diagram. At constant pressure change in enthalpy is equal to heat transferred.

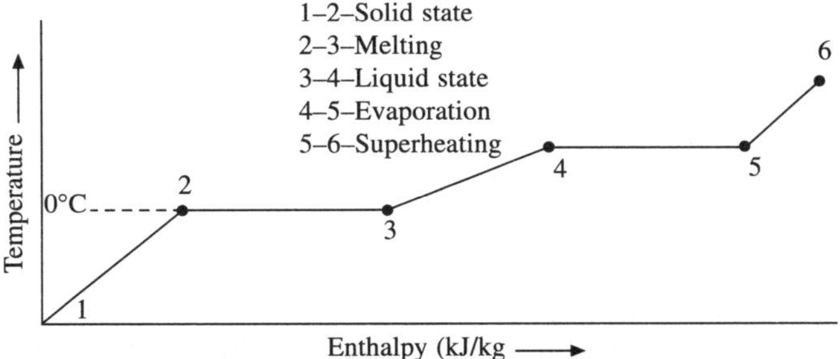


**Fig. 7.1   Phase transformation**

It is assumed that ice is available at atmospheric pressure and at a temperature much below 0°C. As heat is added, temperature of ice increases till the meeting point (0°C at atmospheric pressure), is attained. On further addition of heat, ice starts melting at 0°C taking latent heat of melting, on further addition of heat, temperature of water starts increasing till the saturation temperature (boiling point). Boiling starts, temperature remains constant till entire 1 kg of water is completely converted into steam. The heat required to completely evaporate water at saturation temperature is called latent heat of evaporation. Further addition of heat increases the temperature of the steam. This process is called superheating.

Similar transformation curve can be drawn for other substances. The amount of heat required to raise the temperature of ice to the melting point

$$= h_i = c_{pi}(t_i - t_o) \tag{7.1}$$

where $t_o$ = initial temperatures of ice, $t_i$ = melting point of ice and $c_{pi}$ = specific heat of solid ice.

$$\text{Total heat required for melting of ice} = c_{pi}\,(t_i - t_o) + h_{if} \tag{7.2}$$

where $h_{if}$ = enthalpy of melting = latent heat of fusion. Addition of heat to water from latent point to saturated liquid at saturation temperature called sensible enthalpy

$$= h_f = 1 \times C_{pw}(t_s - t_i) \tag{7.3}$$

where $C_{PW}$ = specific heat of water and $t_s$ = saturation temperature.

$$\text{Latent heat of evaporation} = h_{fg} \tag{7.4}$$

On further heating, superheated steam is generated. Total heat added from initial temperature to final temperature of superheated steam ($t_{\text{sup}}$) can be given by:

$$h_{\text{sup}} = C_{pi}\,(t_i - t_o) + h_{if} + h_f + h_{fg} + C_{ps}\,(t_{\text{sup}} - t_s) \tag{7.5}$$

Where $C_{ps}$ = specific heat of superheated steam and $t_s$ = saturation temperature. Average value of $C_{ps}$ = 2.0934 kJ/kgk

## 7.3 EFFECT OF PRESSURE ON SATURATION TEMPERATURE

The boiling temperature of water increases with increase of pressure. The boiling temperature of water at a particular pressure is called saturation temperature and corresponding pressure is known as saturation pressure. The saturation temperature at every pressure is uniquely fixed. The variation of saturation temperature with saturation pressure is shown in the Fig. 7.2.

*Critical Temperature* The critical temperature of water is defined as the temperature at which

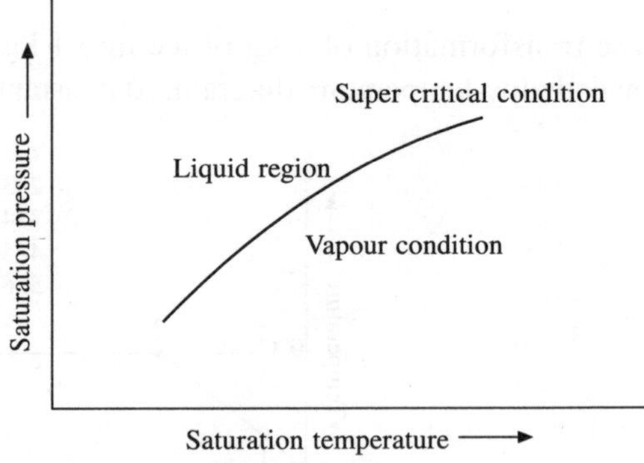


**Fig. 7.2 Effect of pressure on saturation temperature**

water evaporates into steam without taking any latent heat. Above this temperature it is impossible to liquify the steam. The saturation pressure corresponding to critical temperature is called critical pressure. The critical temperature and critical pressure of water are 374.15°C and 221.20 bar respectively. Above critical point, all differences in the properties of liquid and vapour disappear.

*Triple Point*   At very low pressure melting point and boiling point coincide together and all the three states solid, liquid and gas remain in equilibrium. This temperature is called triple point. For water at 0.006114 bar, the triple point is 273.16°K.

## 7.4  GENERATION OF STEAM

The most important vapour used in engineering is steam. It may be either dry saturated, wet or superheated. It does not behave as a perfect gas rather real gas in saturated and superheated condition. For studying the various properties of steam, let us consider 1 kg of water at 0°C (datum temperature of enthalpy) contained in a cylinder fitted with a frictionless piston exerting a constant pressure ($p$) in the cylinder all the time.

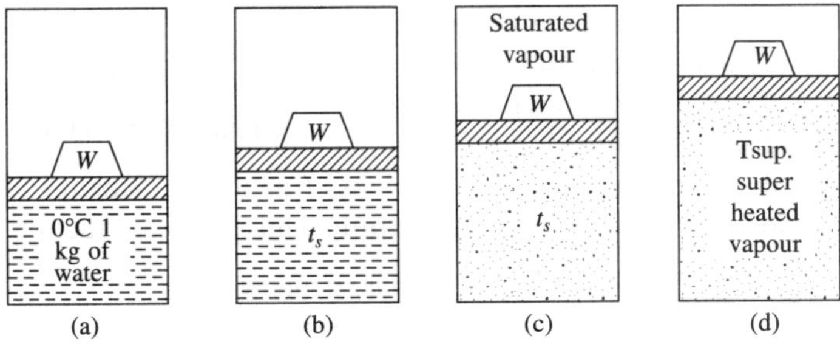


**Fig. 7.3   Generation of steam**

As heat is supplied, the temperature of water continues to increase until saturation temperature is attained. This condition is shown in Fig. 7.3(b). The heat supplied in this process is given by

$$h_f = C_{pw}(t_s - o) \tag{7.6}$$

where $C_{pw}$ = Specific heat of water at constant pressure. This heat is called enthalpy of saturated liquid. Further addition of heat is used to change the phase from liquid to vapour. This is continued until entire liquid is converted into vapour.

Now the total heat required to convert entire 1 kg of water at 0°C into steam at $t_s$°C is given by

$$h_g = h_f + h_{fg} \tag{7.7}$$

*hfg* is called latent enthalpy or enthalpy of evaporation and $h_g$ is called enthalpy of saturated vapour.

Further addition of heat will increase the temperature of steam above the saturation temperature. This can be continued to any desired temperature. One such condition has been shown in Fig. 7.3(d). The quantity of heat during superheating is given by

$$C_{ps}(t_{sup} - t_s)$$

and total heat of superheated steam

$$= h_{\text{sup}} = h_g + C_{ps}(t_{\text{sup}} - t_s) \tag{7.8}$$

$(t_{\text{sup}} - t_s)$ is called degree of superheat

## 7.5 QUALITY OF STEAM

The steam may exist in three conditions namely, wet, dry-saturated or superheated. Steam generated in the presence of water with which it is in a state of equilibrium is known as wet steam. The water present in the wet steam is saturated liquid at saturation temperature. The steam which does not contain any liquid particle is called dry saturated steam.

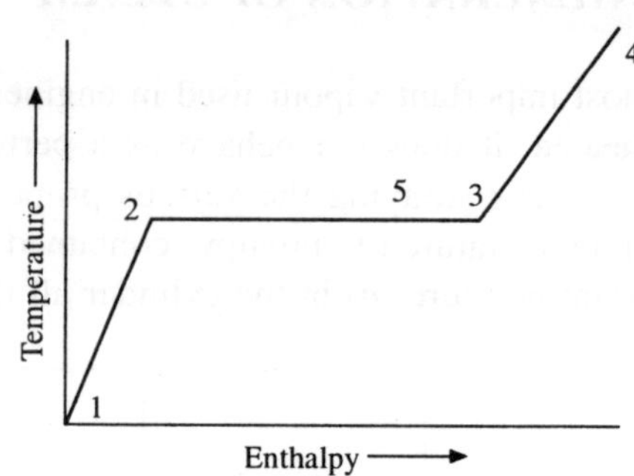


**Fig. 7.4    Condition of steam**

The steam ordinarily produced in any boiling vessel is always wet steam as it contains liquid particles in a finely divided liquid droplet etc. Entire liquid has not taken latent heat. The state of wet steam will be somewhere in between, saturated liquid state (2) and dry vapour state (3). Let us say state 5. The relative amounts of vapour phase and saturated liquid phase determine the quality of steam. The quality or "dryness fraction" of steam is defined as the ratio of the mass of dry vapour to the total mass of mixture. Thus dryness fraction =

$$x = \frac{\text{mass of dry vapour in the mixture}}{\text{mass of the mixture}} \tag{7.9}$$

For example if 1 kg of steam contains 0.95 kg of dry vapour then dryness fraction of steam is 0.95. $(1 - x)$ represents wetness of the steam.

The heat required to produce 1 kg of wet steam is given by:

$$h = h_f + xh_{fg} \tag{7.10}$$

If the temperature of the dry steam is increased more than the saturation temperature, steam becomes superheated. The state 4 in the Fig. 7.4, represents condition of superheated steam.

## 7.6 THERMODYNAMIC PROPERTIES OF STEAM

In the study of various engineering processes and vapour power cycles, thermodynamic properties of steam like specific volume, enthalpy, entropy, internal energy and saturation temperature must be known. The values of these properties are determined either experimentally or otherwise. These properties are available in the form of table and chart. The properties of saturated liquid, saturated vapour and superheated are available. The properties of wet steam is calculated by knowing dryness fraction and pressure of the steam. The calculation of various properties are given below:

1. *Specific Volume*

Let $v_f$ = specific volume of saturated liquid

$v_{fg}$ = specific volume of evaporation

$v_g$ = specific volume of dry steam, then $v_g = v_f + v_{fg}$

Specific volume of wet steam is given by:

$$v = v_f + xv_{fg} = v_f + x(v_g - v_f) = (1 - x)\, v_f + xv_g$$

but the specific volume of saturated liquid is very very small compared to specific volume of saturated vapour.

Therefore $\qquad\qquad\qquad\qquad\qquad\qquad v = xv_g$ $\qquad\qquad\qquad\qquad\qquad\qquad$ (7.11)

Specific volume of superheated steam is determined approximately by the relation:

$$v_{sup} = v_g\, \frac{T_{sup}}{T_s} \qquad\qquad (7.12)$$

because superheated steam behaves as a perfect gas approximately.

2. *Internal Energy of Steam*: Internal energy is the difference of enthalpy and flow work. Therefore internal energy of saturated liquid is given by:

$$u_f = h_f - P \cdot v_f \qquad\qquad (7.13)$$

Internal energy of wet steam is:

$$u = (h_f + xh_{fg}) - pxv_g \qquad\qquad (7.14)$$

Internal energy of dry saturated steam

$$= u_g = h_g - pv_g \qquad\qquad (7.15)$$

Internal energy of superheated steam

$$= u_{sup} = h_{sup} - pv_{sup} = h_g + C_p(t_{sup} - t_s) - p\, v_{sup} \qquad\qquad (7.16)$$

3. *Entropy of Steam*

From definition of change of entropy $ds = \dfrac{dQ}{T}$

Entropy of saturated liquid

$$ds = -\int_{273+0}^{273+t_s} C_{pw}\, \frac{dT}{T} = C_{pw}\, \ln \frac{273 + t_s}{273}$$

$$S_f = C_{pw}\, \ln \frac{T_s}{273} \qquad\qquad (7.17)$$

0°C (273°K) has been taken as datum.

Latent entropy or entropy of evaporation is $S_{fg} = \dfrac{hf_g}{T_s}$ because during evaporation $T_s$ remains constant.

$$\text{Entropy of wet steam} = S = S_f + x\, S_{fg} \tag{7.18}$$

$$\text{Entropy of superheated steam} = S_{\text{sup}} = S_g + \int_{T_s}^{T_{\text{sup}}} C_{p_s}\, dT$$

or,

$$S_{\text{sup}} = S_g + C_{p_s} \ln \frac{T_{\text{sup}}}{T_s} \tag{7.19}$$

## 7.7   STEAM TABLE

The generation of steam at different pressures has been studied experimentally and various properties of steam have been obtained at different conditions. The properties have been listed in tables called steam tables. The steam tables are available for

1. Saturated water and steam—on pressure basis
2. Saturated water and steam—on temperature basis
3. Superheated steam—on pressure and temperature basis for enthalpy, entropy and specific volume
4. Supercritical steam—on pressure and temperature basis above 221.2 bar and 374.15°C for enthalpy, entropy and specific volume.

### 7.7.1   Some Important Points Regarding Steam Tables

(a) The steam table gives values for 1 kg of water and 1 kg of steam
(b) The steam table gives values of properties from the triple point of water to the critical point of steam.
(c) For getting values of thermodynamic properties either saturation pressure or saturation temperature need to be known. Pressure based steam table (i.e. extreme left pressure column is placed) is used when pressure-value is known, similarly temperature based steam table is used when temperature-value is known.
(d) At low pressure the volume of saturated liquid is very small as compared to the volume of dry steam and usually the specific volume of liquid is neglected. But at very high pressures the volume of liquid is comparable and should not be neglected.
(e) The specific enthalpy and specific entropy at 0°C are both taken as zero and measurements are made from 0°C onwards.
(f) In computing the properties for wet steam it should be noted that only $h_{fg}$ and $S_{fg}$ are affected by dryness fraction but $h_f$ and $S_f$ are not affected by dryness fraction. This means that for steam with dryness fraction $x$,

$$h_{fg} = h_f + x\, hfg$$

$$Sg = S_f + x\, Sfg$$

(g) The values of properties for superheated steam can be taken from the table for superheated steam or can be calculated by application of ideal gas equation.

To compute the thermodynamic properties different types of empirical equations have been used which are supposed to be complex and therefore could be readily used whenever required as given in tabular form

**Property Table**

| Property | Wet steam | Dry steam | Super heated steam |
|---|---|---|---|
| Volume | $(1 - x)\, v_f + x v_g$ | $v_g$ | $v_g\, \dfrac{T_{\text{sup}}}{T_s}$ |
| Enthalpy | $h_f + x\, h_{fg}$ | $h_f + h_{fg} = h_g$ | $h_g + C_{p_s}\, (t_{\text{sup}} - t_s)$ |
| Entropy | $S_f + x\, S_{fg}$ | $S_f + S_{fg} = S_g$ | $S_g + C_{p_s}\, \ln\, \dfrac{T_{\text{sup}}}{T_s}$ |

A skeleton of steam table has been shown on next page.

## 7.8   STEAM PROPERTY CHARTS

In addition to steam table, properties can be represented graphically on chart. The most convenients charts are the temperature entropy $(T - S)$ and enthalpy - entropy $(h$-$s)$ charts which were first prepared by Mollier. These two charts are discussed below:

(a) Temperature—Entropy chart. $(T - S$ diagram) is useful in the sense that in any process on $T - S$ diagram, heat added and rejected can be represented. The isothermal and isentropic processes will be horizontal and vertical lines respectively. Heat added or rejected will be equal to area under the process line, two ordinates and entropy axis $\left( \text{i.e. } Q = \displaystyle\int T\, ds \right)$.

Such a diagram has been shown in Fig. 7.5. Consider 1 kg. of steam at an absolute pressure $p_1$

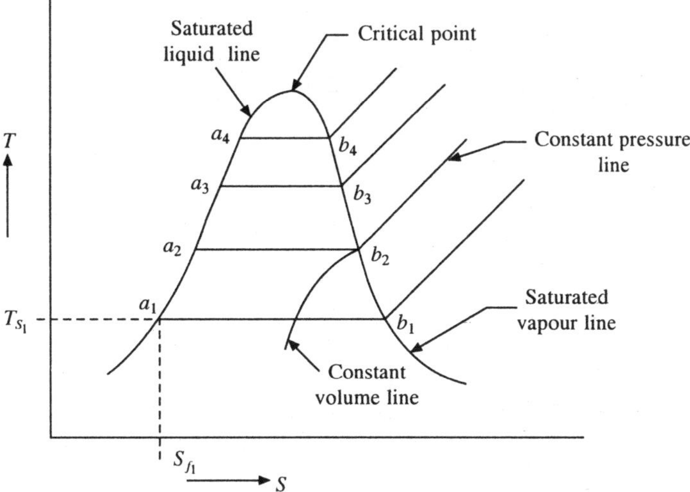


**Fig. 7.5   Temperature entropy diagram**

**Saturated water and steam**

| Absolute Pressure bar | Temperature °C | Specific Volume m³/kg | | Specific internal energy | | Specific Enthalpy kJ/kg | | | Specific Entropy kJ/kg k | | |
|---|---|---|---|---|---|---|---|---|---|---|---|
| | | Water $v_f$ | steam $v_g$ | Water $u_f$ | Steam $u_g$ | Water $h_f$ | Evaporation $h_{fg}$ | Steam $h_g$ | Water $S_f$ | Evaporation $S_{fg}$ | Steam $S_g$ |
| 1.0 | 99.63 | 0.001043 | 1.6937 | 417.39 | 2506.63 | 417.5 | 2256.9 | 2676.0 | 1.303 | 6.057 | 7.360 |
| 15.0 | 198.3 | 0.001154 | 0.13167 | 842.86 | 2592.39 | 844.6 | 1945.3 | 2789.9 | 2.314 | 4.126 | 6.441 |

Superheated steam

Pressure = 20 bar, $t_s$ = 212.4, $v_g$ = 0.0995, $u_g$ = 2598.2, $h_g$ = 2797.2, $S_g$ = 6.337

| Temp.°C | Sp. Volume m³/kg | Enthalpy kJ/kg | Entropy kJ/kg k |
|---|---|---|---|
| 250 | 0.1115 | 2902.4 | 6.545 |
| 300 | 0.1255 | 3025.0 | 6.770 |

and saturation temperature $T_{s1}$, entropy of saturated liquid can be taken from steam table. On $T$-$S$ diagram corresponding to temperature $T_{s1}$ and entropy $S_{f1}$ a point $a_1$, can be represented. Similarly corresponding to pressures $p_2$, $p_3$ other points $a_2$, $a_3$, $a_4$ . . . can be represented on the $T$-$S$ diagram. The locus of all these points represent *saturated liquid line* on $T$-$S$ diagram. Similarly corresponding to pressures $p_1$, $p_2$ . . . $p_n$, $S_{g1}$, $S_{g2}$ . . . can be represented on $T$-$S$ diagram by the points $b_1$, $b_2$, $b_3$ . . . Locus of all these points represent *saturated vapor line*. From the diagram, we find that the gap between the saturated liquid line and saturated vapour line goes on decreasing with increase of pressure and finally both the lines meet at a point called critical point where latent entropy and latent enthalpy become zero. Constant pressure line in the two phase region is also a constant temperature line. But in superheated region entropy increases with increase of temperature and therefore constant pressure line is inclined in superheated region.

(b) Enthalpy-Entropy chart (Mollier chart) Most of the thermodynamic systems deal with flow of stem in steady condition where change in enthalpy is encountered. Therefore the most convenient method of computing change in enthalpy is the enthalpy-entropy chart. The enthalpy and entropy of the saturated liquid and vapour can be found from steam tables at various pressures. These two lines can be drawn on $h$-$s$ diagram. On this chart lines of constant pressure, constant volume, constant temperature are also shown. Saturated liquid region is not required for solving engineering problems and therefore only a part of chart near saturated vapour region and superheat region is shown.

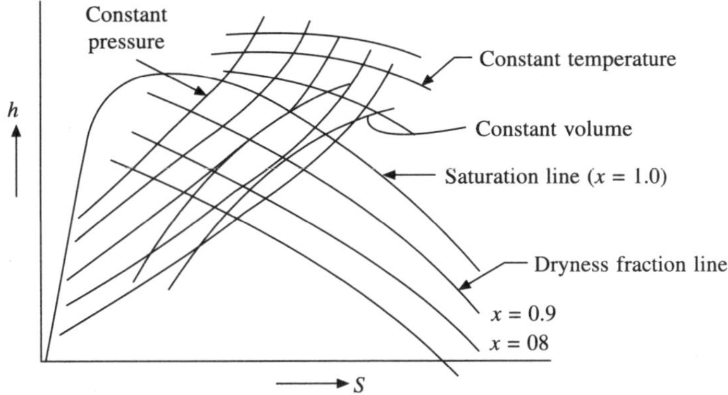


**Fig. 7.6 $h$-$s$ diagram**

## 7.9 THERMODYNAMIC PROCESSES

The basic equations derived from first and second law of thermodynamics can be used for gases as well as for vapours, but ideal gas equation $pv = RT$ is not valid for steam. The basic energy equations for non-flow and flow processes are:

$$dQ = du + p\, dv \text{ for non-flow processes} \tag{7.20}$$

$$dQ = dh - v\, dp \text{ for flow processes} \tag{7.21}$$

The various thermodynamic processes are:

## (a) Constant Volume Heating or Cooling

The constant volume heating process represented on $p$-$v$ and $T$-$S$ diagrams has been shown in the Fig. 7.7.

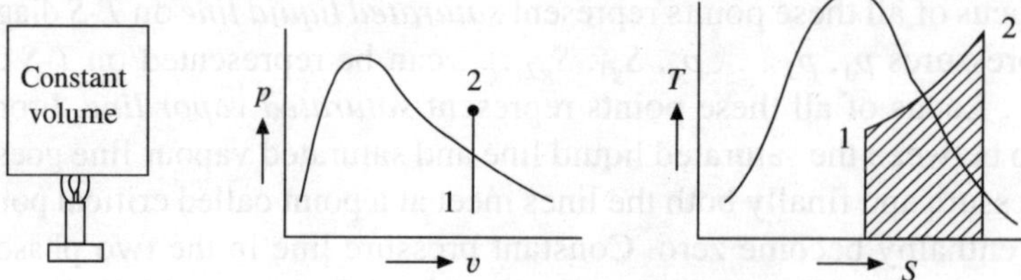

**Fig. 7.7    Constant volume heating**

It is assumed that initially the steam is in wet condition. At the end of heating, steam is superheated to a pressure $p_2$. As the mass of the steam during the process remains constant,

$$\frac{v}{x_1 v_{g1}} = \frac{v}{v_{\text{sup}_2}}, \quad \text{but} \quad v_{\text{sup}_2} = v_{g2}\frac{T_{\text{sup}_2}}{T_{S2}} = x_1 v_{g1}$$

or
$$\frac{T_{\text{sup}2}}{T_{S2}} = \frac{x_1 v_{g1}}{v_{g2}} \tag{7.22}$$

The temperature $T_{\text{sup}2}$ can be found out. The change in other properties can be determined. If the steam is still wet even after heating, then

$$x_1 v_{g1} = x_2 v_{g2} \quad \text{or,} \quad x_2 = x_1 \frac{v_{g1}}{v_{g2}}$$

The work done = 0 and heat transferred

$$=Q = u_2 - u_1 = (h_2 - p_2 x_2 v_{\text{sup}2}) - (h_1 + p_1 x_1 v_{g1}) \tag{7.23}$$

The above analysis can also be used for constant volume cooling also.

## (b) Constant Pressure Heating or Cooling

Constant Pressure heating process has been shown in the Fig. 7.8. An example of this process occurs in the generation of steam in the boiler.

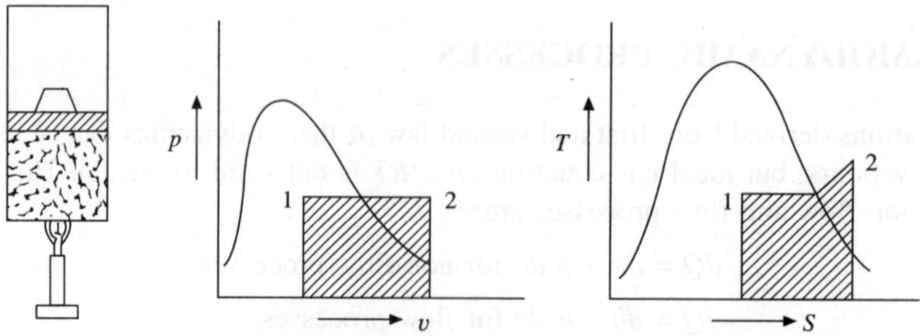

**Fig. 7.8    Constant pressure heating**

$$\text{Heat added } = Q = u + \int_1^2 pvd = (u_2 - u_1) + P(v_2 - v_1)$$

$$= (u_2 + P_2v_2) - (u_1 + P_1v_1) = h_2 - h_1 \tag{7.24}$$

Thus heat added or rejected at constant pressure heating is change in enthalpy.

Work done is given by
$$W = \int_{V_1}^{V_2} pdv = p(v_2 - v_1) \tag{7.25}$$

*(c) Constant Temperature or Isothermal Expansion*
Constant temperature process is also a constant pressure process in the wet region during evaporation and condensation. Once the steam becomes super heated, it behaves like a gas and constant temperature process in superheated region is hyperbolic ($pv$ = constant). During expansion in superheated region, pressure decreases and volume increases.

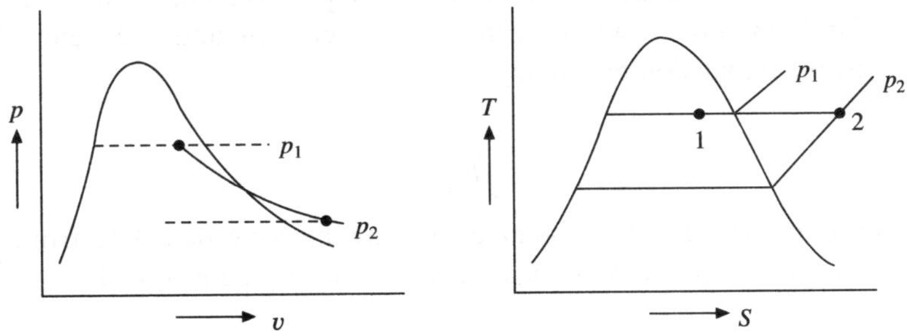

Fig. 7.9   Isothermal expansion

$$\text{Heat added } = Q = \Delta u + \int_{V_1}^{V_2} pdv = (u_2 - u_1) + P_1v_1 \ln \frac{v_2}{v_1} \tag{7.26}$$

In the wet region, constant temperature process is not a hyperbolic process.

*(d) Isentropic Expansion*
The isentropic expansion process has been shown in Fig. 7.10 on *T-S* and *h-s* planes.

The process is a vertical line on *T-S* and *h-s* planes. Entropy remains constant during the process. $S_1 = S_2$ is used to locate the state 2. If the process is non-flow and reversible adiabatic then

$$Q = \Delta u + \int_1^2 p\,dv = \Delta u + W$$

but $Q = 0$ Therefore
$$W = -\Delta u = u_1 - u_2 \tag{7.27}$$

That is work done is equal to decrease in internal energy. If the process is reversible adiabatic steady flow, then by applying steady flow energy equation we get:

$$Q + h_1 = W + h_2 \quad \text{or,} \quad 0 + h_1 = W + h_2$$

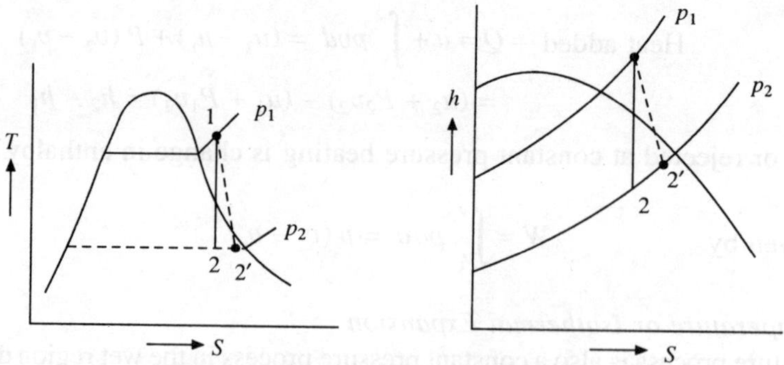

**Fig. 7.10  Isentropic Expansion**

or,
$$W = h_1 - h_2 \tag{7.28}$$

That is, in reversible adiabatic expansion, work done is equal to enthalpy drop. If the process is adiabatic but irreversible then there will be generation of entropy and end state will be on 2'. In such situation, isentropic expansion efficiency is defined as

$$\eta_{\text{isen}} = \frac{h_1 - h_2'}{h_1 - h_2} \tag{7.29}$$

On Mollier diagram end state 2 can be located by drawing a vertical line from state 1 to the pressure $p_2$. Therefore on $h$-$s$ diagram end state 2 can be easily determined.

### (e) Throttling Process

Throttling process is an isenthalpic process. If the steam is throttled from higher pressure to lower pressure, enthalpy remains constant, entropy increases, pressure decreases and quality of steam improves.

The process has been shown in Fig. 7.11 on $T$-$S$ and $h$-$s$ planes. The end state can be located by equating enthalpy at the two states.

$$h_1 = h_2 \tag{7.30}$$

The common applications of throttling processes are:

(a)  throttling of steam in throttled governed engines and steam turbines,

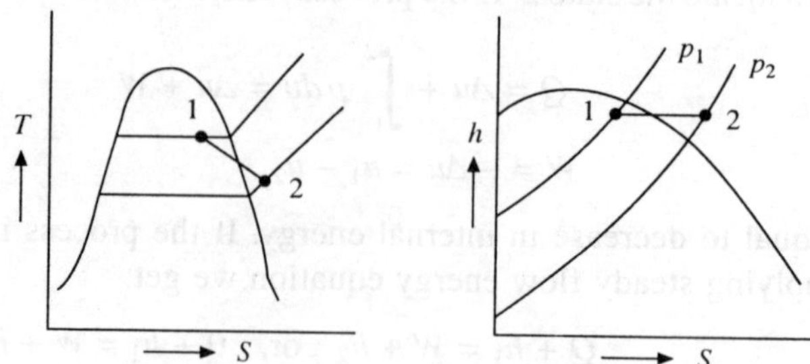

**Fig. 7.11  Throttling process**

(b)  determination of dryness fraction of steam by throttling calorimeter and

(c)  expansion of light pressure refrigerant in throttle value or capillary tube. Throttling process is a horizontal line on $h$-$s$ diagram.

## 7.10  METHODS OF DETERMINING THE DRYNESS FRACTION OF STEAM

There are three methods used for determining the dryness fraction of steam as given below

(a)  Separating calorimeter

(b)  Throttling calorimeter

(c)  Combined separating and throttling calorimeter

**Separating Calorimeter**

The apparatus used is shown in Fig. 7.12. The supply steam is admitted to the calorimeter from main pipe through a sampling tube. The incoming steam strikes the baffle plate and direction of steam is completely reversed. Therefore, water particles being heavier are separated and collected at the bottom of the vessel. The quantity of separated water is noted from the indicator attached with inner vessel. The dry steam passes down through the annular space between inner and outer vessel. A barrel calorimeter rests on a weighing platform. This calorimeter receives the dry steam in the form of condensate which quantity is obtained by weighing machine.

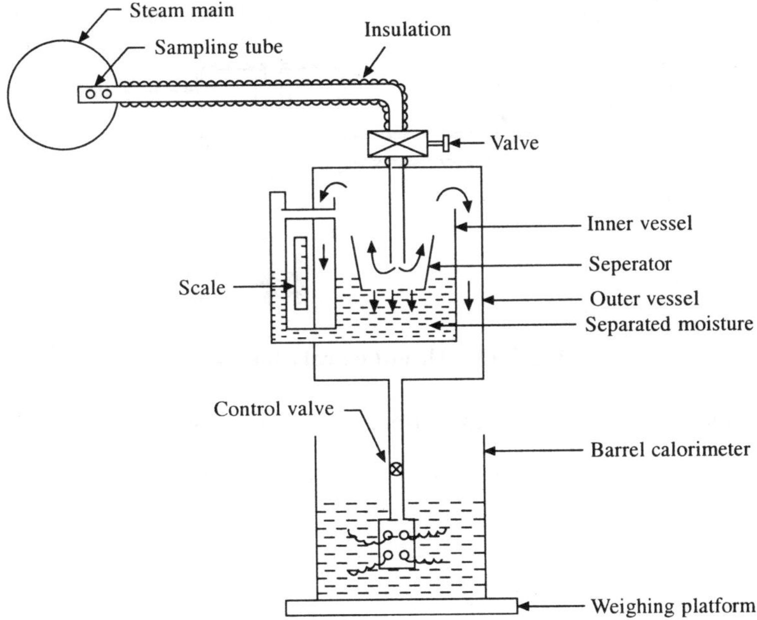


**Fig. 7.12  Separating calorimeter**

Let   $m_w$ = mass of moisture collected in inner vessel in kg

$m_s$ = mass of steam condensed in barrel calorimeter in kg

$$x = \text{dryness fraction of steam}$$

$\therefore$ 

$$\text{dryness fraction of steam} \quad x = \frac{m_s}{(m_s + m_w)}$$

**Limitation**

(a) The size of the water particles is very small and may be carried away with the steam at high dryness fraction. Therefore, this calorimeter can not be used for high dryness fraction of steam

(b) Complete separation of water is not possible

(c) Method is approximate

**Throttling calorimeter**

The sample of the wet steam enters through the sampling tube as shown in Fig. 7.13. The sample of the steam is throttled. A pressure gauge and a mercury manometer are provided to measure pressure before and after throttling respectively. The steam after throttling must be superheated. With the help of mercury thermometer, temperature of throttled steam can be measured. During throttling, the enthalpy of steam remains constant. The enthalpy of superheated steam is known if its pressure and temperature are known. If wet steam is throttled to become superheated, then the dryness fraction can be evaluated by equating before and after throttling. The properties of steam are noted from the steam table.

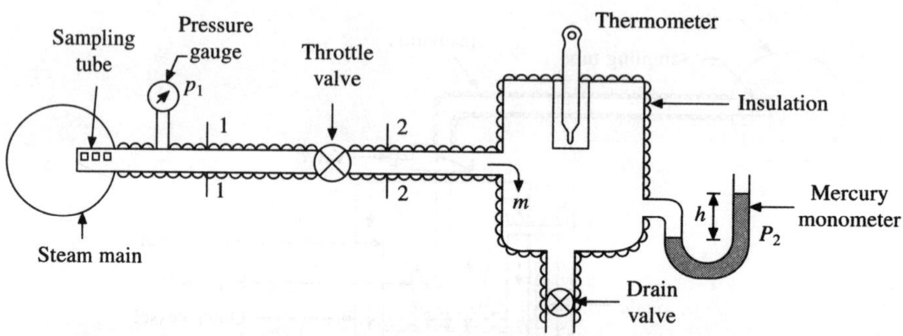


**Fig. 7.13    Throttling calorimeter**

Equating enthalpies before and after throttling, we get $h_1 = h_2$

$$h_{f_1} + x_1 h_{fg_1} = h_{g2} + c_{p_s}(t_{\text{sup}} - t_s)$$

$\therefore$

$$x_1 = \frac{h_{g2} + c_{p_s}(t_{\text{sup}} - t_s)}{h_{fg_1}}$$

**Limitation:**

(a) For the measurement of minimum value of dryness fraction the steam should be superheated by minimum 5°C.

(b) This method can not be used for low dryness fraction

## Combined Separating and Throttling Calorimeter

The arrangement is shown in Fig. 7.14. It uses both calorimeters. Steam from main pipe is collected through sampling tube. First it enters into separating calorimeter where direction is

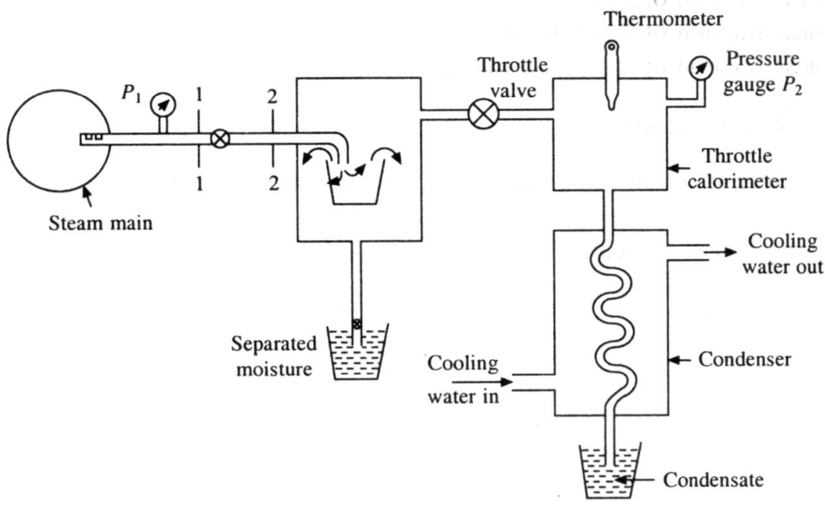


**Fig. 7.14   Combined separating and throttling calorimeter**

reversed and thereby heavy water particles are collected and separated at the bottom of the separator. The steam with considerable dryness fraction leaves the separator and enters the throttling calorimeter where it is collected and taken to superheated region. The process is shown on $h$-$s$ diagram in Fig. 7.15. The process 1–2 represents separation of moisture at constant pressure. The dryness fraction $x_1$ at point 1 is improved to $x_2$ at point 2. process 2–3 represents throttling.

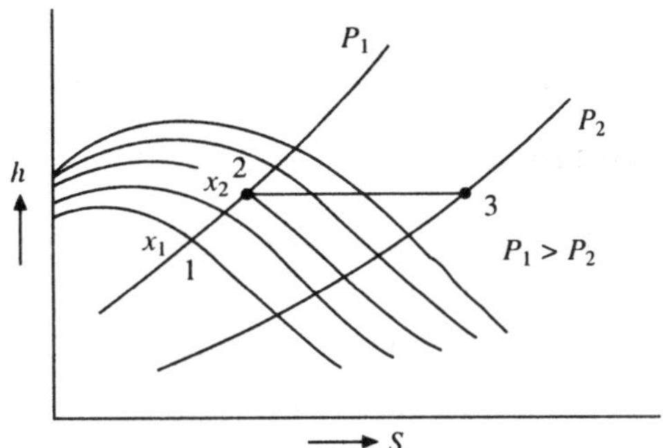


**Fig. 7.15   Representation of separating and throttling processes**

## Calculation of Dryness Fraction

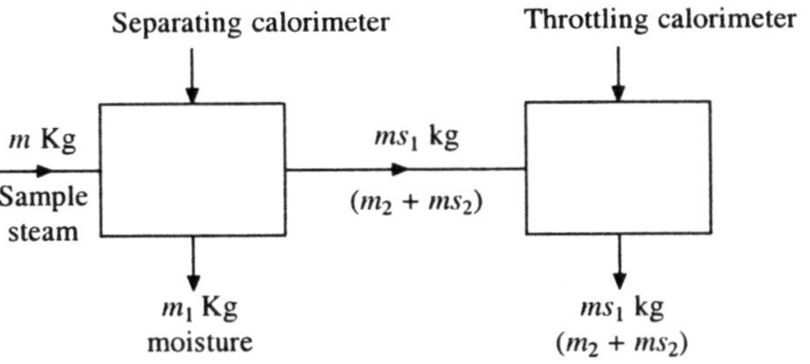


**Fig. 7.16   Line diagram for calculation**

Let  $m$ = mass of steam passing through sampling tube in kg

    $m_1$ = mass of moisture separated in separator in kg

    $m_{s_1}$ = mass of steam coming from separating calorimeter in kg

    $X$ = dryness fraction of steam in steam main

    $x_1$ = dryness fraction of steam leaving the separating calorimeter

    $x_2$ = dryness fraction of steam leaving the throttling calorimeter

The dryness fraction of steam after separating calorimeter is given by $x_1 = \dfrac{m_{s_1}}{m}$

It should be noted that the sample $m_{s_1}$ contains moisture $m_2$ and dry steam $m_{s_2}$

$\therefore$   The dryness fraction of steam after throttling is given by $x_2 = \dfrac{m_{s_2}}{m_{s_1}}$

The total dryness fraction of steam will be the ratio of final dry steam to total initial steam given as

$$X = \frac{m_{s_2}}{m} = \frac{m_{s_2}}{m_{s_1}} \times \frac{m_{s_1}}{m} = x_2 \cdot x_1$$

$\therefore$
$$\boxed{X = x_1 x_2}$$

Thus the dryness fraction of steam main is the product of dryness fractions given by separating and throttling calorimeter respectively.

### Solved Problems

**7.1** Evaluate enthalpy, internal energy, volume and entropy of 1 kg of steam having dryness fraction 0.85 and pressure of 20 bar.

*Soln.*  From steam table at 20 bar, properties are

$t_s$ = 212.4°C, $v_f$ = 0.00177 m³/kg, $v_g$ = 0.099549 m³/kg

$h_f$ = 908.6 kJ/kg, $hfg$ = 1888.7 kJ/kg, $hg$ = 2797.2 kJ/kg

$S_f$ = 2.447 kJ/kg k, $S_{fg}$ = 3.890 kJ/kg k, $S_g$ = 6.337 kJ/kg k

$$\text{Enthalpy} = h_f + x h_{fg}$$
$$= 908.6 + 0.85 \times 1888.7 = 2513.995 \text{ kJ/kg}\quad Ans.$$

$$\text{Volume} = (1 - x)\, v_f + x v_g \simeq x v g$$
$$= (1 - 0.85) \times 0.00177 + 0.85 \times 0.099549$$
$$= 0.0848 \text{ m}^3\text{/kg}\quad Ans.$$

$$\text{Internal energy} = h - Pv$$
$$= 2{,}513.995 - 20 \times 10^5 \times 10^{-3} \times 0.0848$$
$$= 2{,}344.395 \text{ kJ/kg}\quad Ans.$$

$$\text{Entropy} = s_f + x\, s_{fg}$$

$$= 2.447 + 0.85 \times 3.890$$

$$= 5.753 \text{ kJ/kg k} \quad Ans.$$

**7.2** Evaluate the same properties as mentioned in just previous problem at the condition of pressure 20 bar and steam is superheated by 50°C. Assume $C_{ps} = 2.1$ kJ/kg k

*Soln.*
$$t_{\text{sup}} = (212.4 + 50) = 262.4°C,\ T_{\text{sup}} = (262.4 + 273) = 535.4 \text{ k}$$

$$\text{Enthalpy} = h_g + C_{ps}\,(t_{\text{sup}} - t_s),\ \text{given } (t_{\text{sup}} - t_s) = (T_{\text{sup}} - T_s) = 50$$

$$= 2{,}797.2 + 2.1\,(50)$$

$$= 2{,}902.2 \text{ kJ/kg} \quad Ans.$$

*Note* If value of $C_{ps}$ is not given, value of enthalpy can be taken from superheated steam table at 20 bar and 262.4°C.

$$\text{Volume} = v_g\,\frac{T_{\text{sup}}}{T_s} = 0.099549\,\frac{(262.4 + 273)}{(212.4 + 273)}$$

$$= 0.1098 \text{ m}^3\text{/kg} \quad Ans.$$

$$\text{Internal energy} = h - Pv$$

$$= 2902.2 - 20 \times 10^5 \times 10^{-3} \times 0.1098$$

$$= 2682.6 \text{ kJ/kg}$$

$$\text{Entropy} = S_g + C_{ps} \ln \frac{T_{\text{sup}}}{T_s}$$

$$= 6.337 + 2.1 \ln \frac{(262.4 + 273)}{(212.4 + 273)}$$

$$= 6.542 \text{ kJ/kg k} \quad Ans.$$

**7.3** Determine the state of the steam i.e. whether it is wet, dry or superheated in the following cases.

(i) Pressure 8 bar and specific volume 0.196 m$^3$/kg
(ii) Pressure 14 bar and temperature 225°C
(iii) Pressure 25 bar and 2,700 kJ/kg of heat is required to generate steam from water at 0°C.

*Soln.*   (i) The given value of specific volume = 0.196 m$^3$/kg is less than that of taken from steam table at 8 bar i.e. 0.24026 m$^3$/kg.

$$\therefore \qquad \text{dryness fraction } x = \frac{0.196}{0.24026} = 0.815 \quad Ans.$$

(ii) The saturated temperature at 14 bar is 195°C which is less than temperature of given sample. Hence sample is superheated and degree of superheat = (225 − 195) = 30°C  *Ans.*

(iii) At pressure 25 bar, $h_g = 2809.9$ kJ/kg and for sample enthalpy is less than it
$\therefore$ enthalpy of wet steam = $h_f + x\, h_{fg}$

$$2700 = 961.9 + x\,1839$$

$$\therefore \qquad x = 0.945 \quad Ans.$$

**7.4** In a turbine steam expands from 20 MPa, 550°C to 0.005 MPa isentropically. Evaluate the work done per kg of steam.

*Soln.*

$$20 \text{ MPa} = 20 \times 10^6 \text{ pa}$$
$$= 20 \times 10^6 \text{ N/m}^2 = 200 \text{ bar}$$
$$0.005 \text{ MPa} = 0.005 \times 10^6 \text{ pa}$$
$$= 0.05 \text{ bar}$$

Isentropic expansion process 1 – 2 is shown here, Fig. 7.17.

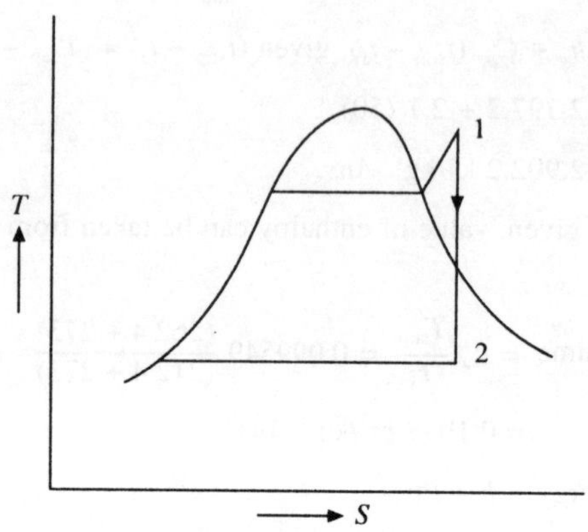

**Fig. 7.17**

$$S_1 = S_2$$

$$S_1 = S_g + C_{ps} \ln \frac{T_{\text{sup}}}{T_s}$$

Value of $C_{ps}$ is not given. However, average value of $C_{ps} = 2.0934$ kJ/kg k. From steam table saturation temperature $t_s = 365.7°C$ at 200 bar which is less than 550°C. Therefore steam is superheated. So directly superheated steam table can be used for

$$S_1 = 6.325 \text{ kJ/kg k and } h_1 = 3388.3 \text{ kJ/kg at 0.05 bar}$$

$$S_2 = Sf_2 + xSfg_2$$

$$6.325 = 0.476 + x\,7.920$$

$$x = 0.738$$

$$\therefore \qquad h_2 = hf_2 + xhfg_2$$

$$= 137.8 + 0.738 \times 2423.8 = 1926.56$$

$$\therefore \qquad \text{Work done per kg} = (h_1 - h_2)$$

$$= (3388.3 - 1926.56)$$

$$= 1{,}461.74 \text{ kJ} \quad \textit{Ans.}$$

**7.5** Steam at 10 bar and 0.9 dryness fraction is available. Find the final dryness fraction of steam in each of the following two cases

(i)   170 kJ of heat is removed per kg of steam at constant pressure

(ii)  steam expands isentropically to a pressure 0.5 bar in a turbine in a flow process. The turbine develops 300 kJ of work per kg of steam.         [Nov. 96 Mumbai Univ.]

Refer to Fig. 7.18

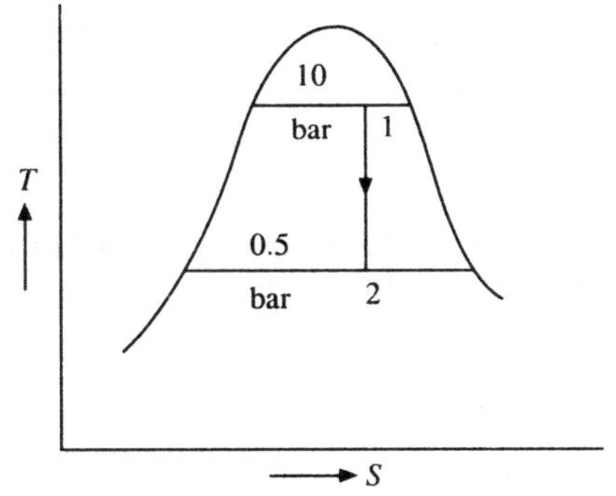


**Fig. 7.18**

*Soln.* (i) At 10 bar and 0.9 dryness, heat available $= h_f + x_1 hf_g = 762.6 + 0.9 \times 2{,}013.6$

$$= 2574.84 \text{ kJ/kg}$$

If 170 kJ heat is removed, then rest heat $= 2{,}574.84 - 170 = 2{,}404.84$ kJ/kg

$\therefore$ $\qquad 2{,}404.84 = 762.6 + x_2 \times 2013.6$

$\therefore$ $\qquad x_2 = 0.815$ *Ans.*

(ii) $\qquad h_1 = 2{,}574.84$ kJ/kg

$\qquad 300 = 2{,}574.84 - (340.6 + x_2 \times 2{,}305.4)$ at 0.5 bar

$\therefore$ $\qquad x_2 = 0.83$ *Ans.*

**7.6** Find the specific enthalpy and specific entropy of steam when the pressure is 1 MPa and specific volume is 0.16 m³/kg. If 5 kg of the steam at the above state is heated at constant pressure to a temperature of 250°C; calculate the heat transfer during the process. [Nov. 95, Mumbai Univ.]

*Soln.* $\qquad 1 \text{ MPa} = 1 \times 10^6 \text{ Pa} = 10 \text{ bar}$

At 10 bar from steam table specific volume $v_g = 0.19430$ m³/kg but given value (0.16 m³/kg) is less, therefore steam is wet

$\therefore$ $\qquad x = 0.16/0.19430 = 0.823$

So for wet steam, specific enthalpy $\qquad = h_f + x\, h_{fg}$

$$= 762.6 + 0.823 \times 2{,}013.6$$

$$= 2{,}419.79 \text{ kJ/kg} \quad Ans.$$

and specific entropy $= S_f + x\, S_{fg}$

$$= 2.138 + 0.823 \times 4.445$$

$$= 5.79 \text{ kJ/kg k} \quad Ans.$$

Initial heat (enthalpy) of 5 kg of steam at

$$10 \text{ bar} = 5 \times 2{,}419.79 = 1{,}2098.95 \text{ kJ}$$

Final heat (enthalpy) of same quantity of steam after heating to 250°C at 10 bar

$$= 5 \times 2{,}943 = 14{,}715 \text{ kJ}$$

$\therefore$  Heat-transfer during-the heating process = 14,715 − 12098.95 = 2,616.05 kJ   *Ans.*

**7.7** A pressure cooker contains 2 kg of dry and saturated steam at 3 bar. Find the amount of heat which must be rejected so as to reduce the dryness fraction to 0.6. What is the temperature and pressure at this new state?

[May 97, Nov. 97 Mumbai Univ.]

*Soln.* At 3 bar for dry saturated steam $v_g = 0.60553$ m³/kg, $h_g = 2,724.7$ kJ/kg, $t_s = 133.5$°C While rejecting heat from cooker volume of steam at 0.6 dry also will be same. Initial volume

$$= 2 \times 0.60553 = 1.211 \text{ m}^3$$

Initial volume = final volume

$$1.211 = 2\,[(1 - x)\,v_{f2} + x\,v_{g2}] \simeq 2 \times 0.6\,v_{g2}$$

$\therefore$ 
$$v_{g2} = 1.211/2 \times 0.6 = 1.009 \text{ m}^3/\text{kg}$$

Corresponding to 1.009 m³/kg from steam table pressure $P_2 = 1.75$ bar and temperature $t_s = 116$°C   *Ans.*

Internal energy of steam/kg at initial state

$$u_1 = h_g - pv_g$$

$$= 2724 - 3 \times 10^5 \times 10^{-3} \times 0.60553 = 2542.341 \text{ kJ/kg}$$

Internal energy at final per kg at 1.75 bar

$$u_2 = h_2 - p_2 v_2$$

$$= (486.95 + 0.6 \times 2213.25) - 1.75 \times 10^5 \times 10^{-3} \times 0.6 \times 1.009$$

$$= 1,814.9 - 105.945 = 1708.955 \text{ kJ/kg}$$

Pressure cooker is a closed system

$\therefore$ 
$$dQ = du + dw$$

$$= du + p\,dv\ [p\,dv = 0, \text{ since no change in volume}]$$

or, 
$$dQ = du$$

Therefore heat transfer at constant volume

$$(u_1 - u_2) = (2542.341 - 1708.955)$$

$$= 2543.04 \text{ kJ/kg}$$

$\therefore$ 
$$\text{Total heat transfer} = 2(u_1 - u_2)$$

$$= 2 \times 2543.04$$

$$= 1668.17 \text{ kJ}\quad Ans.$$

**7.8** In a test on combined separating and throttling calorimeter the following data were obtained
Pressure of steam sample = 15 bar
Pressure of steam at exit = 1 bar
Temperature of steam at exit = 150°C
Discharge of water separating calorimeter = 0.2 kg/min.
Discharge collected at exit = 10 kg/min
Calculate the dryness fraction of steam in the main.

[June 94 Mumbai Univ.]

*Soln.* Dryness fraction obtained separating calorimeter $x_1 = \dfrac{10}{10 + 0.2} = 0.98$

Dryness fraction obtained throttling calorimeter $x_2$ can be calculated as

$$hf_1 + xhfg_1 = hg_2 + C_{ps}(t_{sup} - t_s)$$

$$\underbrace{844.6 + x_2\,1945.3}_{\text{at 15 bar}} = \underbrace{2675.4 + 2.0934\,(150 - 99.63)}_{\text{at 1 bar}}$$

$$\therefore \qquad x_2 = 0.995$$

$\therefore$  Dryness fraction of steam in the main

$$x = x_1 x_2$$

$$= 0.980 \times 0.995 = 0.975 \quad Ans.$$

**7.9** A spherical vessel of 0.9 m³ capacity contains steam at 8 bar and 0.9 dryness fraction. Steam is blown off until the pressure drops to 4 bar. The valve is then closed and steam is allowed to cool until the pressure falls to 3 bar. Assuming that enthalpy of steam in the vessel remains constant during blowing off period, determine:

(i)  the mass of steam blown off
(ii)  the dryness fraction of steam in the vessel after cooling
(iii)  the heat lost per kg of steam during cooling                [May 98, Mumbai Univ.]

*Soln.*  At 8 bar from steam table

$$hf_1 = 720.9 \text{ kJ/kg},\ hfg_1 = 2046.5 \text{ kJ/kg},\ vg_1 = 0.24026 \text{ m}^3/\text{kg},\ ts_1 = 170.4°\text{C}.$$

$$hf_1 = hf_1 + x_1 hfg_1$$

$$= 720.9 + 0.9 \times 2046.5 = 2562.75 \text{ kJ/kg}$$

$$v_1 = x_1 vg_1 = 0.9 \times 0.24026 = 0.2162 \text{ m}^3/\text{kg}$$

$$\therefore \qquad \text{mass of initial steam} = \frac{0.9}{0.2162} = 4.162 \text{ kg}$$

As the total heat remains constant during blowing off period, we have

$$2562.75 = hf_2 + x_2 hfg_2$$

At 4 bar from steam table

$$hf_2 = 604.7 \text{ kJ/kg},\ hfg_2 = 2132.9 \text{ kJ/kg},\ vg_2 = 0.46220 \text{ m}^3/\text{kg},\ ts_2 = 143.6°\text{C}$$

$$\therefore \qquad 2562.75 = 604.7 + x_2 \times 2132.9$$

$$\therefore \qquad x_2 = 0.918$$

$$\therefore \qquad v_2 = x_2 vg_2 = 0.918 \times 0.46220 = 0.424 \text{ m}^3/\text{kg}$$

$$\therefore \qquad \text{Mass of steam after blowing off} = \frac{0.9}{0.424} = 2.122 \text{ kg}$$

(i)  $\therefore$  Mass of steam blown off $= (4.162 - 2.122) = 2.04$ kg   *Ans.*

At pressure 3 bar from steam table

$$hf_3 = 561.4 \text{ kJ/kg},\ hfg_3 = 2163.2 \text{ kJ/kg},\ vg_3 = 0.60553 \text{ m}^3/\text{kg},\ ts_3 = 133.5°\text{C}$$

(ii)  The quality $x_3$ of steam after cooling is given by $0.9 = m_2 \times x_3 \times vg_3$

$$\therefore \qquad x_3 = 0.9/2.122 \times 0.60553 = 0.70 \quad Ans.$$

(iii)   The heat/kg after cooling is given by

$$h_3 = hf_3 + x_3 hf_3 = 561.4 + 0.70 \times 2163.2 = 2075.64 \text{ kJ/kg}$$

As the cooling is at constant volume, the heat lost is equal to the difference of internal energies.

$$U_2 = h_2 - P_2 x_2 v g_2$$

$$= 2562.75 - 4 \times 10^5 \times 10^{-3} \times 0.918 \times 0.4622$$

$$= 2393.030 \text{ kJ/kg}$$

and

$$U_3 = h_3 - P_3 x_3 v g_3$$

$$= 2075.64 - 3 \times 10^5 \times 10^{-3} \times 0.70 \times 0.60553 = 1948.478 \text{ kJ/kg}$$

$\therefore$   Total heat lost during cooling per kg of steam

$$= (U_2 - U_3)$$

$$= (2393.030 - 1948.478) = 444.552 \text{ kJ}   \textit{Ans.}$$

10.  Followings were the observations on an experiment with a separating and throttling calorimeter
Pressure in the main = 17.5 bar
Pressure after throttling = 13.6 cm of water
Barometric pressure = 75 cm of Hg
Temperature of steam after throttling = 110°C. Amount of water collected in the separator in 5 minutes
= 120 C.C. Steam condensed after throttling in 10 minutes = 5 Litre. Find the quality of steam in the main.

*Soln.*   Absolute pressure after throttling $= 75 + \dfrac{1000 \times 13.6}{13,600} = 75 + 1 = 76$ cm of Hg

$$= 1.01325 \text{ bar}$$

During same interval of time (i.e. 10 minutes) water collected in the separator

$$m_1 = 120 \times \frac{10}{5} = 240 \text{ C.C}$$

Steam condensed after throttling in 10 minutes

$$ms_2 = 5 \text{ litre} = 5,000 \text{ C.C.}$$

dryness fraction obtained by separating calorimeter

$$x_1 = \frac{5000}{240 + 5,000} = 0.954$$

dryness fraction obtained by throttling calorimeter $x_2$ can be calculated as

$$\underbrace{hf_2 + x_2 hfg_2}_{\text{at 17.5 bar}} = \underbrace{hg_3 + C_{ps}(t_{\text{sup}} - t_s)}_{\text{at 1.01325 bar}}$$

$$878.2 + x_2 \times 1915.9 = 2676 + 2.0934(110 - 100)$$

$\therefore$

$$x_2 = 0.949$$

$\therefore$   Dryness fraction of steam in the main

$$x = x_1 x_2$$

$$= 0.954 \times 0.949 = 0.905   \textit{Ans.}$$

## EXERCISES

7.1  Objective Type Questions
   (i)  The critical point is the point at which
      (a)  melting and boiling temperature became equal
      (b)  the change of volume accompanying evaporation is maximum
      (c)  the change of volume accompanying evaporation is zero
      (d)  non of the above
  (ii)  For steam
      (a)  the critical temperature is 221.2°C and critical pressure is 374.15 bar
      (b)  the critical temperature is 374.15°C and critical pressure is 221.2 bar
      (c)  the critical temperature is 221.2°C and pressure is 221.2 bar
      (d)  the critical temperature is 374.15°C and critical pressure is 37.4.15 bar.
 (iii)  Which one of the following is a pure
      (a)  mixture of water and oil
      (b)  hot coffee
      (c)  mixture of ammonia and carbon-dioxide
      (d)  mixture of oxygen and nitrogen
  (iv)  The locus of saturated liquid line and saturated vapour line meets at
      (a)  boiling point
      (b)  critical point
      (c)  ice point
      (d)  triple point
   (v)  At a low pressure
      (a)  boiling point rises markedly and melting point drops slightly
      (b)  boiling point rises slightly and melting point drops markedly
      (c)  boiling points drops markedly and melting point rises slightly
      (d)  boiling point drops slightly and melting point rises markedly
  (vi)  with the increase in pressure
      (a)  the boiling point of water decreases and enthalpy of evaporation increases
      (b)  the boiling point of water increases and enthalpy of evaporation decreases
      (c)  both the boiling point of water and the enthalpy of evaporation decreases
      (d)  both the boiling point of water and the enthalpy of evaporation increases.
 (vii)  The temperature at which the melting and boling boiling temperature become equal, is equal to
      (a)  273.16 K     (b)  647.3 K     (c)  373.16 K     (d)  173.16 K
(viii)  At the critical point, the specific volume of water is equal to
      (a)  1 $m^3$/kg     (b)  0.00317 $m^3$/kg     (c)  zero     (d)  0.00622 $m^3$/kg
  (ix)  The enthalpy of dry saturated steam . . . with the increase in pressure
      (a)  decreases     (b)  increases     (c)  remains constant
   (x)  Dryness fraction of wet steam is given by
      (a)  ratio of mass of dry steam to the mass of suspended liquid water
      (b)  ratio of mass of suspended liquid water to mass of dry steam
      (c)  ratio of mass of dry steam to the sum of mass of suspended liquid water and dry steam
      (d)  ratio of sum of mass of suspended liquid water and dry steam to the mass of dry steam.

### Answers

| | | | |
|---|---|---|---|
| (i)  c | (ii)  b | (iii)  b | (iv)  b |
| (v)  c | (vi)  b | (vii)  a | (viii)  b |
| (ix)  a | (x)  c | | |

7.2  Definc a pure substance. Give three examples of pure substances

7.3 Explain how water and mixutre of its different phases are pure substances.

7.4 Explain "triple point of water" and show it on *P-T* diagram.

7.5 Explain "ciitical point of water vapour" and show it on *p-h* diagram.

7.6 Draw *P-V* and *T-S* diagram for water starting from its liquid phase to superheated phase.

77. Draw *h-s* diagram of water and show important properties of water on it.

7.8 What is meant by saturation temperature and saturation pressure?

7.9 Define the following terms

    (i) sensible heat of water

    (ii) latent heat of vaporisation

    (iii) total heat of steam

    (iv) dryness fraction of steam; and

    (v) volume of superheated steam

7.10 Describe methods of finiding dryness fraction of steam

7.11 Describe a combined seperating and throttling calorimeter and find the expression for finding dryness fraction of steam.

7.12 Determine the quantity of heat required to generate 5 kg of steam at a pressure of 8 bar absolute, from water at a temperature of 40°C

    (i) When dryness fraction of steam is 0.98

    (ii) Superheated up to 300°C. Assume mean specific heat of superheated steam is 2.3 kJ/kg k

    [MSBTES-2000]                  *Ans*. 12797.35 kJ, 15715.67 kJ

7.13 State the conditions of steam in following cases:

    (i) At a pressure of 10 bar, total heat is 2646 kJ/kg.

    (ii) At a pressure of 15 bar, temperature is 197.39°C.

    (iii) At a pressure of 20 bar, temperature is 220°C.

    (iv) At a pressure of 7 bar, specific volume is 0.25 m$^3$/kg

    (v) At a pressure of 8 bar, total heat is 2800 kJ/kg         [A.M.I.E 1993]

7.14 (a) Explain the terms: (i) critical temperature (ii) critical pressure and (iii) critical density

    (b) The pressure of steam in a condenses is 0.12 bar absolute and the dryness fraction is 0.88. How many heat units must be abstracted from the steam in order to condensate? (a) 1 kg, (b) 1 m$^3$.

                     [A.M.I.E 1992]

7.15 A one litre closed vessel contains water at its critical conditions. This vessel is cooled until its pressure drops to 1 MPa. Calculate the mass of  water in the vessel and the find dryness fraction.

                    (*Ans*. 0.31696 kg, 0.0105]

7.16 One kg of water at a pressure of 500 KN/m$^2$ and temperature 30°C is heated at constant pressure to raise its temperature to 300°C. Show the process on *p-v* diagram and calculate

    (i) increase in enthalpy

    (ii) increase in internal energy

    (iii) increase in specific volume, and

    (iv) increase in entropy.

7.17 A closed vessel of 1.2 m$^3$ capacity contains steam at 300 kN/m$^2$ and 0.85 dry steam at 1.0 MN/m$^2$ and 0.96 dry is supplied to the vessel until the pressure inside the vessel becomes 500 KN/m$^2$. Calculate the mass of steam supplied to the closed vessel and the final dryness fraction of the steam in the vessel.

7.18 A certain quantity of steam in a closed vessel of volume 0.14 m$^3$ exerts a pressure of 1.0 MN/m$^2$ at 250°C. If the vessel is cooled so that the pressure falls to 0.36 MN/m$^2$, calculate:

    (i) final quality of steam;

    (ii) final temperature;

(iii)  change in internal energy; and
(iv)  change of entropy

7.19  Steam at 1.15 MN/m$^2$ and 0.88 dry is contained in a closed vessel of 0.80 m$^3$ volume. Certain quantity of steam is blown-off until its pressure reduced to 400 KN/m$^2$. Later, the remaining steam is cooled in the vessel until its pressure become 300 KN/m$^2$. Assuming that specific entropy of steam remained constant during blown-off process, calculate:

  (i)  the mass of steam blown-off
  (ii)  quality of steam in the vessel after cooling; and
(iii)  heat transfered per kg of steam during cooling.

7.20  In an experiment, it was found that the steam enters a throttling calorimeter at a pressure of 12.25 bar. After throttling, the pressure and temperature was measured as 1.013 bar and 115°C respectively. Estimate the dryness fraction of steam.  *[Ans.*0.96]

7.21  During a test in a combined seperating and throttling calorimeter, the following data were obtained

  (i)  Pressure of steam in the boiler = 1.5 MN/m$^2$
  (ii)  Pressure of steam after throttling part of the calorimeter = 100 KN/m$^2$
(iii)  tamperature of steam after the throttling = 120°C
(iv)  Water collected in the separating calorimeter = 0.25 kg
  (v)  Steam condensed after throttling = 3.5 kg calculate dryness fraction of steam.  *[Ans.* 0.899]

7.22  In a combined seperating and throtting calorimeter, the following observations were made. Total quantity of steam = 23.4 kg; water drained from seperator = 1.2 kg; steam pressure before throttling = 8.25 bar; temperature of steam on leaving = 111.4°C; steam pressure on leaving = 1.01325 bar. Find the dryness fraction of steam on entry. Specific heat of superheated steam is 2 kJ/kg K.  *[Ans.* 0.914]

# 8

# Steam Power Cycles

## 8.1 INTRODUCTION

Mechanical energy is one of the most desirable energy forms in that it can be converted into thermal energy with an efficiency of 100 percent and it can also be converted into electrical energy with very high-conversion efficiency. Unfortunately, the production of mechanical energy from thermal energy is not as easy. Almost all of the mechanical energy produced today is produced from conversion of thermal energy in some sort of a heat engine. The operation of all heat engines can usually be approximated by an ideal thermodynamic power cycle of some kind. A thermodynamic cycle is composed of a saries of thermodynamic processes that returns the working fluid to its initial state. The power producing cycles may utilize either a gas or a vapour. Internal combustion engines and gas turbines utilize air or mixture of gases as working substance whereas external combustion engines like steam engines and steam turbines use water vapour (steam) as working substance. The thermadynamic cycles which use vapour as the working substance are called *Vapour power cycles*. The most commonly used vapour is steam but mercury and other liquid metals in vapour form are also sometimes used. The following are the advantages of vapour power cycles over gas power cycles: (a) The external combustion systems using vapour as the working fluid is less polluting than internal combustion systems using gas. The primary pollutants from I.C. engines are unburnt hydrocarbons, carbon monoxide and oxides of nitrogen. In vapour power system carbon monoxide is reduced to a minimum value and oxides of nitrogen are completely absent due to combustion at low temperature, (b) Vapour power systems can utilize less expensive and low graded fuels like coal, residual fuel oil, organic refuse or any combustible fuel whereas gas power cycles of I.C. engines and gas turbines use most refined fuel like petrol and diesel, (c) Vapour power cycles can utilizie nuclear energy also as the energy source, (d) A number of different working fluids can also be used in vapour power cycles, (e) Vapour power cycles utilize a working substance which does not contact the fuel, (f) Isothermal heat addition and rejection are easy to be executed in vapour power cycle.

In gas cycles the working fluid remains in one phase wheras in vapour power cycles the working fluid undergoes a change from vapour to liquid and back again to vapour. In this chapter we shall discuss vapour power cycles using steam as the working substance. The complete vapour power cycles consists of a device of generating steam (called boiler), an expanding device (steam turbine or engine), heat rejection heat exchanger (called condenser).

## 8.2  CARNOT CYCLE*

Carnot cycle is an ideal cycle, efficiency of which is independent of the working substance $\left(\eta_{\text{carnot}} = 1 - \dfrac{T_2}{T_1}\right)$. The Carnot cycle using vapour as a working substance is shown in the Fig. 8.1. It consists of two reversible isothermal and two reversible adiabatic (Isentropic) processes. Isothermal heat addition and heat rejection processes using vapour can be accomplished with high speed whereas isothermal processes using gases are very slow and impracticable. Isothermal heat addition (Process 4-1) is an evaporation process therefore it is isobaric also. Similarly isothermal heat rejection is condensation which is also isobaric.

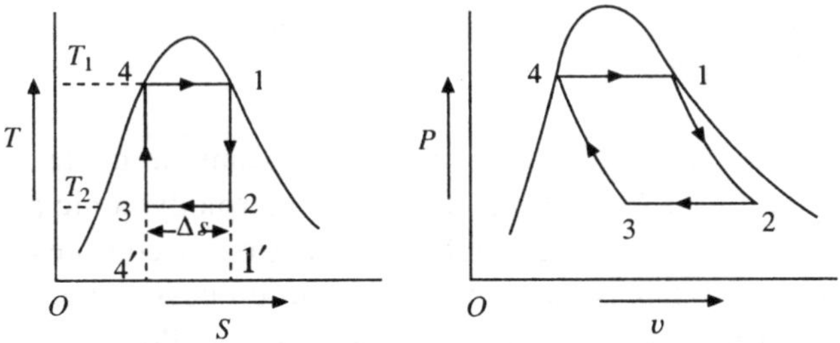

**Fig. 8.1 Carnot cycle on *p-v* and *T-S* planes**

In the Fig. 8.1, the different processes of Carnot cycle are:

(a) Isothermal (Isobaric also) heat addition (Process 4-1).
(b) Isentropic expansion of steam in an expander (Process 1-2).
(c) Isothermal heat rejection in the condenser (Process 2-3).
(d) Isentropic compression of a mixture of vapour and liquid (Process 3-4).

Considering 1 kg of working fluid circulated through the cycle, heat added = $Q$ = area (4-1–1'-4'-4) on *T-S* plane = $T_1$, $\Delta S$. Where $\Delta S$ = change in entropy

Heat rejected = $Q_2$ = area (2-1'-4'-3-2) on *T-S* plane = $T_2 \Delta S$.

$$\text{Net work done} = W = Q_1 - Q_2 = T_1 \, \Delta S - T_2 \cdot \Delta S = (T_1 - T_2) \cdot \Delta S.$$

$$\text{Thermal efficiency of the cycle} = \frac{\text{Work done}}{\text{Heat added}}$$

$$= \frac{W}{Q_1} = \frac{Q_1 - Q_2}{Q_1} = 1 - \frac{Q_2}{Q_1} = 1 - \frac{T_2 \cdot \Delta S}{T_1 \cdot \Delta S} = 1 - \frac{T_2}{T_1} \qquad (8.1)$$

---

*The principle of the reversible cycle was put forward by French Engineer N.L. Sadi Carnot in 1824.

### Limitation of Carnot Cycle Using Vapour as Working Substance

Though the carnot cycle is the most efficient cycle, it is not taken as a standard cycle for reference for steam power plants due to following reasons:

1. During isothermal condensation, the condensation process has to be stopped at state 3, before the saturated liquid line is reached. This is due to the fact that after isentropic compression, isothermal heat addition can take place. But in this situation the device which increases the pressure has to handle a mixture of liquid and vapour (water and steam). Therefore it is neither a compressor nor a pump. This type of device is very difficult to design which handles a two phase fluid.

2. Isentropic compression of a vapour requires more work due to its high specific volume thereby reducing the work ratio.

3. Isothermal heat addition after the saturated vapour line is very difficult to achieve as it involves heat addition at the same time expansion of steam.

4. Specific steam consumption is high and accordingly for a given output the plant size is bigger.

Therefore due to the reasons mentioned earlier, Carnot cycle is not adopted as a standard cycle for vapour power systems.

## 8.3 RANKINE CYCLE*

The simplest of the vapour power cycles in use is the Rankine cycle. It is composed of four components shown in the Fig. 8.2. and one of the four thermodynamic processes occurs in each component. Process 3–4 is a reversible adiablatic (isentropic) compression, process that takes place in the pump. Process 4-1 is isobaric heat addition process in the boiler which consists of sensible heating latent heating and some-times superheating also. Process 1-2 is isentropic expansion in the turbine or in the engine. Process 2–3 is isobaric (isothermal also) reversible heat rejection

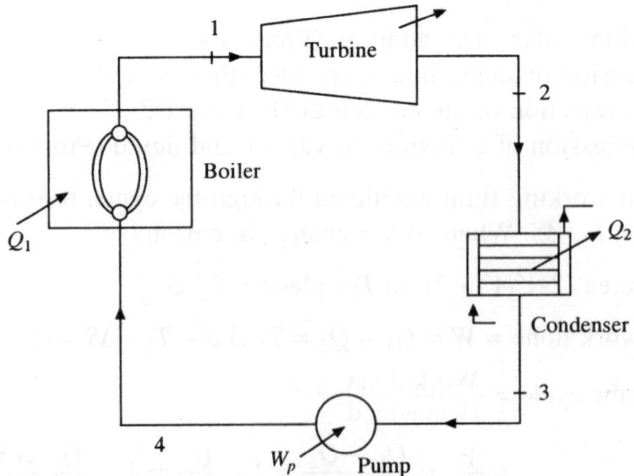


**Fig. 8.2   The Rankine cycle components and system**

---

*This was suggested by William John M. Rankine (1820–1872), a Professor at Glasgow University.

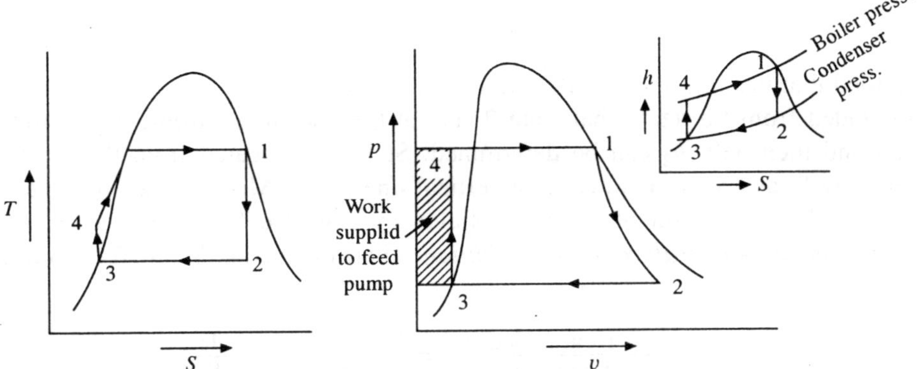



**Fig. 8.3   Rankine Cycle on $T$-$S$, $p$-$v$ and $h$-$s$ diagram**

in the cendenser. It is observed with reference to the Fig. 8.3, that the heat transfer at the source (boiler) occurs not at a constant temperature, but at temperatures varying from $T_4$ to $T_1$. This should be contrasted with the carnot cycle in which a source at constant temperature transfers heat reversibly to the working substance. It is, therefore, to be expected that the Rankine cycle efficiency is lower than that of a Carnot cycle operating between the same high and low temperature limits.

In order to analyse the cycle and determine its efficiency, consider a steady flow conditions at all the states and 1 kg of steam is circulating through the cycle. The heat supplied by the boiler per kg of steam generated

$$= Q_1 = (h_1 - h_4) = (h_1 - h_3) - (h_4 - h_3) = (h_1 - h_3) - W_p$$

Where $W_P = (h_4 - h_3)$ is called pump work per kg of steam.
Heat rejected into the condenser $= Q_2 = (h_2 - h_3)$

$$\text{Net work done per kg of steam} = Q_1 - Q_2 = (h_1 - h_3) - W_P - (h_2 - h_3)$$

$$= (h_1 - h_2) - W_P = W_T = W_T - W_P$$

Where $W_T$ = Turbine Work = $(h_1 - h_2)$ = isentropic enthalpy drop during expansion.

$$\text{Rankine efficiency} = \eta_R = \frac{\text{Net work done}}{\text{Heat supplied}} = \frac{W}{Q_1} = \frac{(h_1 - h_2) - W_p}{(h_1 - h_3) - W_p}$$

or,
$$\eta_R = \frac{(h_1 - h_2) - W_p}{(h_1 - h_3) - W_p} \qquad (8.2)$$

The pump work can also be expressed as $W_p = (p_1 - p_b) \, v_3$ $\qquad (8.3)$

Where $p_1$ = boiler pressure, $p_b$ = back pressure and $v_3$ = specific volume of saturated liquid at back pressure. Therefore Rankine efficiency can be expressed as:

$$\eta_R = \frac{(h_1 - h_2) - (p_1 - p_b)v_3}{(h_1 - h_3) - (p_1 - p_b)v_3} \qquad (8.4)$$

Generally pump work $(W_p)$ is very very small as compared to turbine work $(h_1 - h_2)$ and heat added $(h_1 - h_3)$ therefore it can be fairly neglected. Then Rankine efficiency without pump work

$$\eta_R = \frac{h_1 - h_2}{h_1 - h_3} \tag{8.5}$$

It is understood that heating in the boiler starts from state 3 itself.

It is concluded from Eq. (8.5) that state 2 (i.e. at the end of isentropic expansion) must be known then and then only $h_2$ can be determined. State 2 is located from the steam table by equating entropy $S_1$ and $S_2$ or by drawing a vertical line on the Mollier chart from state 1 to the back pressure. The expression of thermal efficicnecy can also be developed by introducing *thermodynamic mean temperature of heat addition*. Thermodynamic Mean Temperature of heat addition

$$T_m = \frac{\text{Heat added}}{\text{Change in entropy}} = \frac{h_1 - h_4}{S_1 - S_4} = \frac{h_1 - h_4}{S_1 - S_3} \tag{8.6}$$

If pump work is neglected then

$$T_m = \frac{h_1 - h_3}{S_1 - S_3} \tag{8.7}$$

Therefore, Rankine efficiency becomes $= \eta_R = 1 - \dfrac{T_2}{T_m}$

or,
$$\eta_R = 1 - \left(\frac{S_1 - S_3}{h_1 - h_3}\right) T_2 \tag{8.8}$$

**Specific Steam Consumption (S.S.C.)**

It is defined as the steam consumption (kg/s) to produce unit Power (kW). Mathematically

$$\text{S.S.C.} = \frac{\text{Mass flow rate per hour}}{\text{Net power output}} = \frac{\text{kg/s}}{\text{KW}} = \frac{\text{kg}}{\text{KWS}} = \frac{3600}{(h_1 - h_2)} \text{ kg/kWhr}$$

$(h_1 - h_2)$ KJ work is obtained from 1 kg of steam.

$\because$ $\qquad\qquad\qquad\qquad$ 1 kW hr = 3600 kJ

Therefore, $\qquad\qquad\qquad$ $\text{S.S.C.} = \dfrac{3600}{(h_1 - h_2)}$ kg/kW hr $\tag{8.9}$

In case of steam power plant, the specific steam consumption is an indicator of the relative size of the plant.

Work ratio $(W_r)$: It is the ratio of net work done to the turbine work

$$W_r = \frac{(h_1 - h_2) - W_p}{(h_1 - h_2)} \tag{8.10}$$

*Relative Efficiency* or *Efficiency Ratio*: It is defined as the ratio of thermal efficiency (either on I.P. basis or B.P basis) to the corresponding Rankine efficiency.

$$\text{Relative efficiency} = \frac{\text{Thermal efficiency}}{\text{Rankine efficiency}} \tag{8.11}$$

## 8.4   RANKINE ENGINE

When only the prime mover is considered (say steam engine) and it receives steam at a higher pressure and rejects heat after expansion at a lower pressure, then the efficiency is referred to the engine alone instead of the cycle. This efficiency is known as Rankine engine efficiency. The $p$-$v$ diagram of the Rankine engine, neglecting clearance, is shown in the Fig. 8.4. If the expansion is isentropic, the work done = $W = (h_1 - h_2)$
Efficiency of Rankine engine

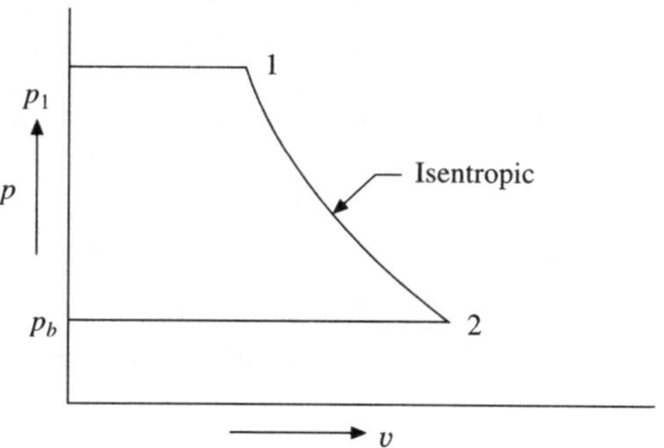


$$\eta_R = \frac{h_1 - h_2}{h_1 - h_{f2}} \qquad (8.12)$$

**Fig. 8.4   Rankine engine cycle**

Where $h_{f2}$ = enthalpy of saturated liquid corresponding to back pressure.

If the pump work is neglected then Rankine cycle efficiency is equal to the Rankine engine efficiency.

---

**Example 8.1**   Dry saturated steam at 15 bar absolute is supplied to a steam turbine. The exhaust takes place at 1.1 bar.

Determine: (a) Rankine efficiency neglecting pump work (b) pump work (c) work ratio, (d) steam consumption per KWhr if efficiency ratio is 0.65 (e) thermodynamic mean temperature of heat addition (f) Carnot efficiency for a given pressure limit using steam as a working substance. By introducing a jet condenser the exhaust pressure is reduced to 0.2 bar, find the percentage increase in Rankine efficiency, percentage decrease in steam consumption and percentage increase in moisture content.

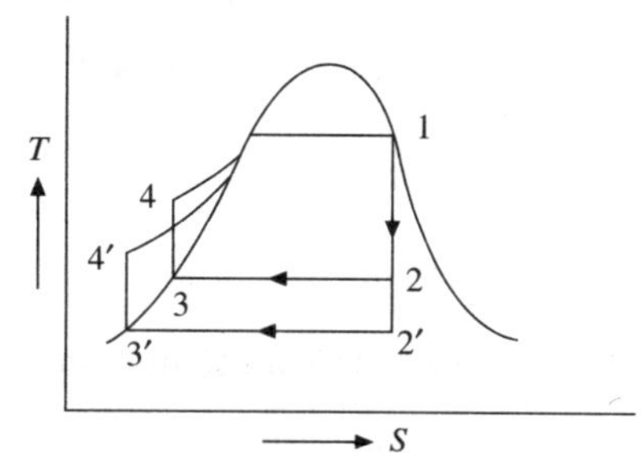


*Solution*   Ref. to Fig. 8.5

**Fig. 8.5**

$p_1 = 15$ bar abs., $p_b = 1.1$ bar abs. From steam table at 15 bar, enthalpy of steam = $h_1 = h_g = 2789.9$ kJ/kg, $t_1 = 198.3°C$, $S_1 = S_{g1} = 6.441$ kJ/kg K°

To determine the quality of steam at state 2:

$$S_1 = S_2$$

or,

$$S_1 = S_{f2} + X_2 S_{fg2}$$

or,

$$6.441 = 1.333 + X_2\, 5.995 \quad \therefore \quad x_2 = 0.852$$

$$h_2 = h_{f_2} + x_2\, h_{fg_2} = 428.8 + \times 2{,}250.8 = 2{,}346.4 \text{ kj/kg}$$

$$h_3 = h_{f_3} = 428.8 \text{ kJ/kg (from steam table).}$$

(a) Rankine efficiency, neglecting pump work

$$\eta_R = \frac{h_1 - h_2}{h_1 - h_{f_3}} = \frac{2789.9 - 2346.4}{2789.9 - 428.8} = 0.1878 = 18.78 \text{ percentage.}\quad Ans.$$

(b) Pump work per kg of steam $= W_p = (p_1 - p_b)v_3$, sp. volume of

$$\text{water} = \frac{1}{\rho} = \frac{1}{1000}\ \text{m}^3/\text{kg}$$

$$\therefore\ W_p = \frac{1}{1000} \times \frac{(p_1 - p_2) \times 10^5}{1000} = \frac{\text{m}^3}{\text{kg}} \times \frac{\text{kN}}{\text{m}^2} = \frac{(p_1 - p_2)}{10}\ \frac{\text{kJ}}{\text{kg}} \quad \text{when } p_1 \ \& \ p_2 \text{ are in bar.}$$

$$= \frac{(15 - 1.1)}{10} = 1.39 \text{ kJ/kg}$$

(c) Work ratio $= W_r = \dfrac{\text{net work}}{\text{turbine work}} = \dfrac{(h_1 - h_2) - W_p}{h_1 - h_2}$

$$= \frac{(2789.9 - 2346.4) - 1.39}{(2789.9 - 2346.4)} = \frac{442.11}{443.5} = 0.9968 \quad Ans.$$

(d) Efficiency ratio $= \dfrac{\text{Indicated thermal efficiency}}{\text{Rankine efficiency}}$

$\therefore$ Indicated thermal efficiency $=$ Efficiency ratio $\times$ Rankine efficiency

$$= 0.65 \times 0.1878 = 0.122$$

Indicated thermal $\eta = \dfrac{3600}{m_s(h_1 - h_3)}$   or,   $0.122 = \dfrac{3600}{m_s(2789.9 - 428.8)}$

$\therefore\ m_s = 12.49$ kg/hr kW   [$\rm Q$   1 kJ $= 3600$ kW hr]

(e) Thermodynamic mean temperature of heat addition

$$T_m = \frac{h_1 - h_4}{S_1 - S_3} = \frac{h_1 - (h_3 + W_p)}{S_1 - S_3}$$

$$= \frac{2789.9 - (428.8 + 1.39)}{(6.441 - 1.333)} = 461.96 \text{ K}\quad Ans.$$

(f) Carnot efficiency $= \eta_C = 1 - \dfrac{T_2}{T_1} = 1 - \dfrac{(273 + 102.3)}{(273 + 198.3)} = 20.36$ per cent   *Ans.*

If the exhaust pressure is reduced to 0.2 bar then new cycle is $1 - 2' - 3' - 4' - 1$

$$S_1 = S_2' \text{ or, } 6.441 = 0.832 + x_2' \times 7.077 \therefore\ x_2' = 0.792$$

$$h_2' = h_{f_2'} + x_2' h_{f_{g_2}} = 251.5 + 0.792 \times 2358.4 = 2119.35 \text{ kJ/kg.}$$

Rankine efficiency $\eta_R = \dfrac{h_1 - h_2'}{h_1 - h_3'} = \dfrac{2789.9 - 2119.35}{2789.9 - 251.5} = 0.264 = 26.4$ percent

Percentage increase in efficiency $= \dfrac{(0.264 - 0.1878) \times 100}{0.1878} = 40.5$ percent

New indicated thermal efficiency $= 0.65 \times 0.264 = 0.171$

$$\eta_{I.T} = \frac{3600}{m_s\,(h_1 - h_3')} \quad \text{or,} \quad 0.171 = \frac{3600}{m_s\,(2789.9 - 251.5)}$$

or, $\qquad\qquad m_s = 8.29$ kg/hr. kW.

Percentage decrease in steam consumption $= \dfrac{12.49 - 8.29}{12.49} \times 100 = 33.62$ percent

Percentage of moisture in the first case $= (1 - x_2) \times 100 = 1 - 0.852 = 14.8$ percent

Percentage of moisture in the exhaust steam in the second case

$$= (1 - x_2') \times 100 = (1 - 0.792) \times 100 = 20.8 \text{ percent}$$

$\therefore$ Percentage increase in moisture content $= \dfrac{(20.8 - 14.8)}{14.8} \times 100 = 40.54$  *Ans.*

**Example 8.2**  A steam power plant uses steam at a pressure of 50 bar and temperature 500°C and exhausted into a condenser where a pressure of 0.05 bar is maintained. The mass flow rate of steam is 150 kg/sec. Determines (a) the Rankine cycle efficiency, (b) Rankine engine efficiency, (c) power developed, (d) specific steam consumption, (e) the heat rejected into the consenser per hour, (f) Carnot efficiency.

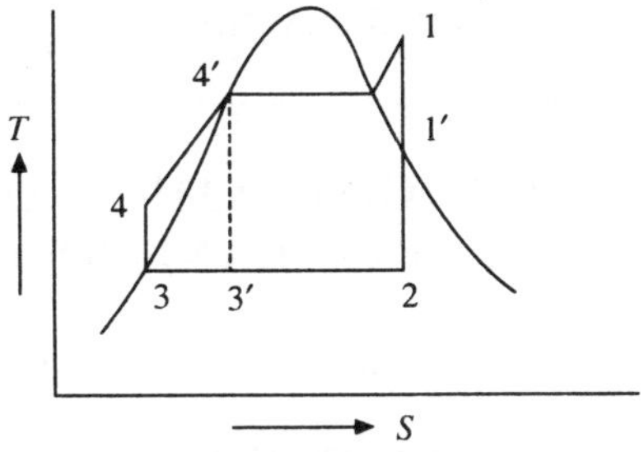

*Solution*  $p_1 = 50$ bar, $t_1 = 500°C$ $p_b = 0.05$ bar, $m_s = 150$ kg/sec.
Ref. to Fig. 8.6
From steam table

**Fig. 8.6**

$$h_1 = 3433.7 \text{ kJ/kg } S_1 = 6.977 \text{ kJ/kg k}$$

at pressure 50 bar and tem. 500°C

$$S_1 = S_2 = S_{f_2} + x_2 S_{fg_2} , 6.977 = 0.476 + x_2 \times 7.920$$

$\therefore \qquad\qquad x_2 = 0.82$ $h_2 = h_{f_2} + x_2 h_{fg_2} = 137.8 + 0.82 \times 2423.8$

$$= 2125.316 \text{ kJ/kg}$$

$$v_3 = v_{f_2} = 1.005 \times 10^{-3} \text{ m}^3/\text{kg}$$

(a)   Rankine cycle efficiency $= \eta_R = \dfrac{(h_1 - h_2) - W_p}{(h_1 - h_3) - W_p}$

$$= \frac{(3433.7 - 2125.316) - (50 - 0.05)/10}{(3433.7 - 137.8) - (50 - 0.05)/10}$$

$$= \frac{1308.384 - 4.995}{3295.9 - 4.995} \times 100 = 39.6 \text{ percent}\quad Ans.$$

(b)   Rankine engine efficiency $= \dfrac{h_1 - h_2}{h_1 - h_3} = \dfrac{1308.384}{3295.9} = 0.3969 = 39.69 \text{ percent}\quad Ans.$

*Note*   Rankine engine efficiency is almost equal to Rankine efficiency. Therefore in Rankine cycle efficiency, pump work is neglected.

(c)   Power developed $= m_s \times$ work done per kg $= 150\,(h_1 - h_2)$

$$= 150 \times 1308.384 = 196257.6 \text{ kW} = 196.257 \text{ MW}\quad Ans$$

(d)   $\qquad\qquad$ S.S.C. $= \dfrac{3600}{(h_1 - h_2)} = \dfrac{3600}{1308.384} = 2.751 \text{ kg/hr. kW}\quad Ans.$

(e)   Heat rejected into the condenser $= Q_2 = m_s(h_2 - h_3) = 150 \times (2125.316 - 137.8)$

$$= 298127.4 \text{ kJ/s}\quad Ans.$$

(f)   Carnot efficiency $\eta_C = 1 - \dfrac{T_2}{T_1} = 1 - \dfrac{(273 + 32.9)}{(273 + 263.9)} = 0.43 = 43 \text{ percent}\quad Ans.$

## 8.5   THERMODYNAMIC VARIABLES AFFECTING EFFICIENCY AND OUTPUT OF RANKINE CYCLE

The important thermodynamic variables affecting the rankine efficiency are:

(i)   Steam pressure at inlet to turbine or engine, called pressure at throttle conditon.

(ii)   Steam temperature or degree of superheat at inlet to tubrine or engine, called temperature at throttle condition.

(iii)   Steam pressure at exhaust or condenser pressure (also called back pressure).

*(i) Effect of Pressure at Throttle condition.*
Figure 8.7 shows the two cycles with the same minimum temperature but a different throttle pressure at the inlet. For both the cases exhaust pressure is assumed constant.

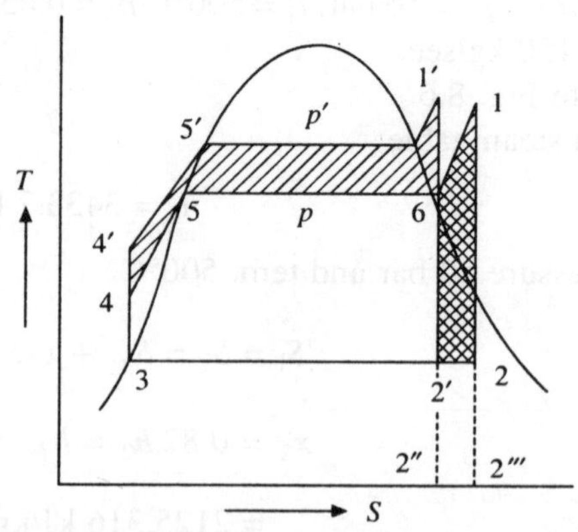


**Fig. 8.7**   **Effect of Admission pressure at inlet to turbine**

Cycle 1-2-3-4-5-1 is for throttle pressure $P$ and $1'$-$2'$-$3'$-$4'$-$5'$-1 is for higher throttle pressure $p'$. From the figure we observe that at higher pressure $p'$, work done is reduced by the hetched area (1-2-$2'$-6-1) but increased by the area (4-$4'$-$5'$-$1'$-6-5-4). Both the areas are almost equal but at higher pressure heat rejected is less by the amount of area $2'''$ $2''$ $2'$ 2. Therefore the efficiency is increased.

However, the cycle efficiency does not increase continuously with boiler pressure upto the ciritical pressure. This is due to that latent heat decreases at higher pressure. Therefore the fraction of heat added at constant temperature to total heat decreases which finally results, after certain stage in lowering the average temperature of heat addition.

The increase in pressure increases the cost of the boiler, pipe line and turbine at the same time. We also notice that increase in pressure results in decrease in dryness fraction or increase in pressure results in decrease in dryness fraction or increase in moisture content. The variation of cycle efficiency and S.S.C. has been shown in the Fig. 8.8.

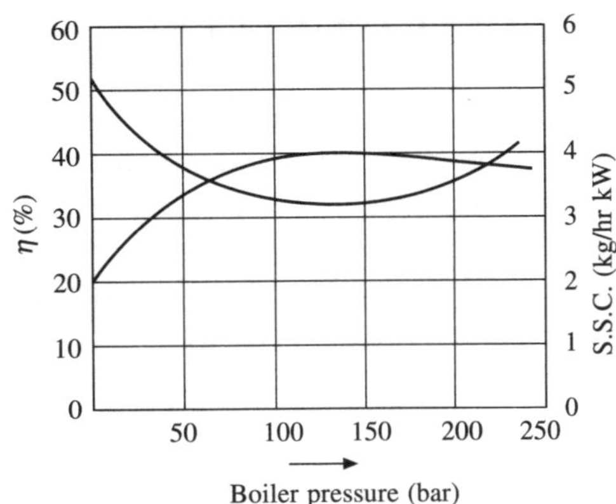


Fig. 8.8 Variation of Rankine efficiency witn boiler pressure

### (ii) Effect of Temperature

Figure 8.9 shows the effect of supper heating of steam before entering the turbine. It is noted that increasing the initial temperature from $T_1$ to $T_1'$, the work done increases by the amount of shaded area 1-$1'$-$2'$-2-1. The heat supplied is also increased by the amount of area 2-$2'$-$2'''$-$2''$-2. However the ratio of extra work done and extra heat supplied is more than the ratio for rest of the cycle, therefore, the net efficiency of the cycle increases with increase in degree of super heat. This can also be inter-pretated that degree of superheat increases the average temperature of heat addition. The followings are the advantages of using super heated steam initially (Fig. 8.10).

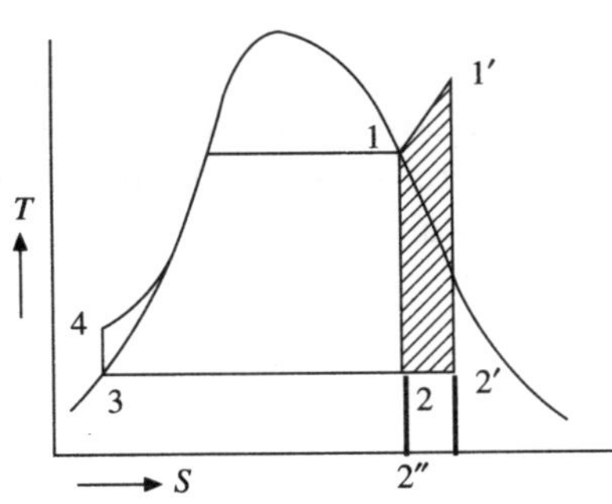


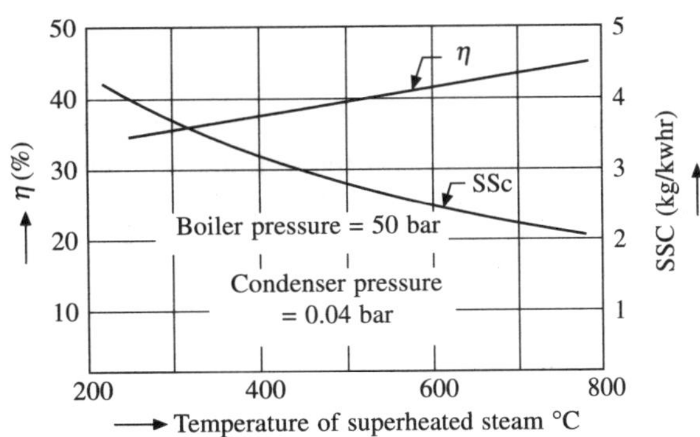


Fig. 8.9 Rankine cycle with super heat

Fig. 8.10 Effect of super heat on Rankine cycle efficiency and SSC

*(iii) Effect of condenser pressure*

By decreasing the condenser pressure, heat rejected temperature is decreased Fig. 8.11. The work done is increased at the same time heat added is also increased. On the whole the efficiency of the Rankine cycle increases. The disadvantage is moisture content in the exhaust steam is also increased which causes corrosion and errosion of the blades.

## 8.6 MODIFIED RANKINE CYCLE (STEAM ENGINE CYCLE)

If the steam is expanded upto the back pressure ($p_b$) in the steam engine cylinder (upto state 2'),

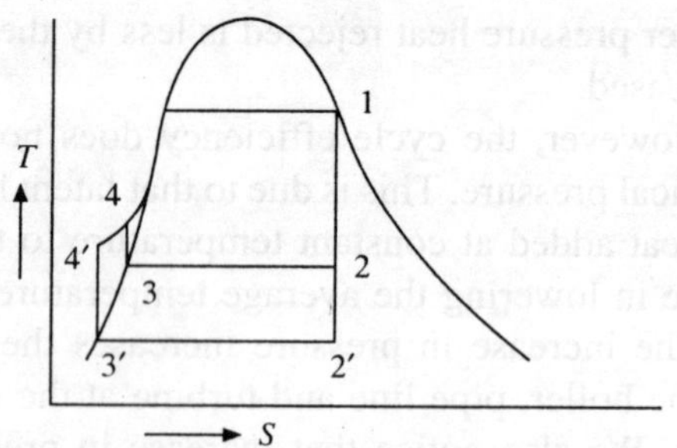

Fig. 8.11   Effect of condenser pressure

the stroke length would be very large at the same time the work which is obtained at the tail end (shaded area) of the stroke is not even sufficient to overcome the frictional resistance between cylinders and piston. Therefore in actual practice, the expansion is terminated at some suitable pressure (called release pressure) at state 2 and then steam is released at constant volume upto back pressure. In the Fig. 8.12, Process 1-2 is isentropic expansion similar to simple Rankine

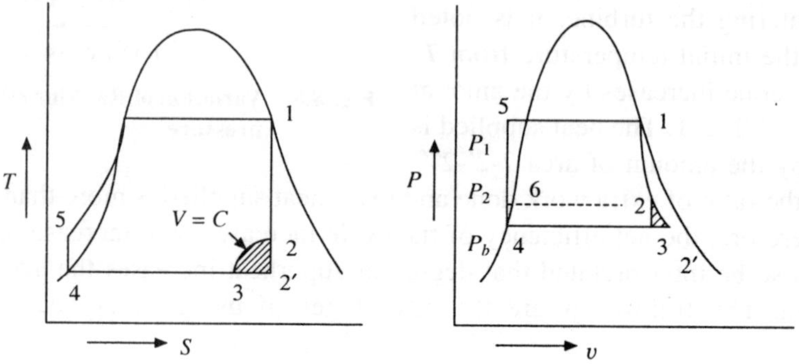

Fig. 8.12   Modified Rankine Cycle

Cycle and Process 2-3 represents the release of steam at constant volume. In modified Rankine cycle, there is loss of work due to incomplete expansion but at the cost of this small work stroke length and hence size of the cylinder is considerably reduced. Therefore steam engine works on modified Rankine Cycle whereas steam turbine works on simple Rankine Cycle. Neglecting the pump work, heat added per kg of steam

$$Q_1 = (h_1 - h_4) = (h_1 - h_{f3}) \tag{8.12}$$

$W$ = work done = $(h_1 - h_2)$ + area (2-3-4-6-2) on $p$-$v$ diagram if clearance volume is neglected, then work done

$$W = (h_1 - h_2) + (p_2 - p_b) v_2 \tag{8.13}$$

Efficiency of modified Rankine cycle

$$\eta = \frac{(h_1 - h_2) + (p_2 - p_b)\,v_2}{(h_1 - h_{f3})} \tag{8.14}$$

**Example 8.3**  Super heated steam at 10 bar abs. and 300°C admitted into the cylinder of a steam engine expands isentropically to a pressure of 0.7 bar. The pressure then falls at constant volume to a back pressure of 0.28 bar. Determine: (a) Modified Rankine Cycle efficiency (b) steam consumption per kW hour (c) mean effective pressure (d) heat removed in the condenser per kg of steam (e) loss of work due to incomplete expansion (f) if the cylinder diameter and strokes are 30 cm and 58 cm. respectively, what would be the new stroke if the steam is allowed to expand without any restriction upto the condenser pressure.

*Solution:* Ref. Fig. 8.13(a), (b)

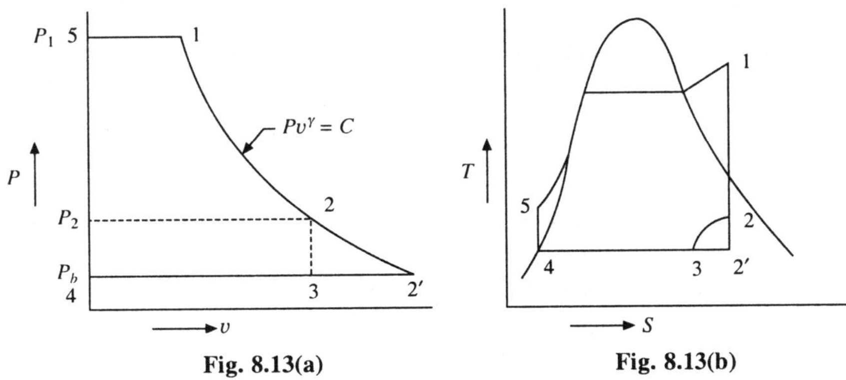



**Fig. 8.13(a)**　　　　　**Fig. 8.13(b)**

From steam table, at $p_1$ = 10 bar and $t_1$ = 300°C

$$h_1 = 3052.1 \text{ kJ/kg}, \; S_1 = 7.125 \text{ kJ/kg k.}$$

At 0.7 bar, $h_{f_2}$ = 376.8, $h_{fg}$ = 2283.3, $S_f$ = 1.192 $S_{fg}$ = 6.228, $v_g$ = 2.364

At 0.28 bar, $h_f$ = 282.7 $h_{fg}$ = 2340, $S_f$ = 0.925, $S_{fg}$ = 6.868

$$v_f = 0.001021, \; v_g = 5.5784$$

Equating entropy at states 1 and 2

$$S_1 = S_2 \therefore 7.125 = 1.192 + x_2 \times 6.228 \therefore x_2 = 0.95$$

$$h_2 = h_{f_2} + h_{fg2} = 376.8 + 0.95 \times 2283.3 = 2545.93$$

(a)  Work done = $W = (h_1 - h_2) - (p_2 - p_b)/10$

$$= (3052.1 - 2545.93) - (10 - 0.28)/10 = (506.17 - 0.972)$$

$$= 505.19 \text{ kJ/kg}$$

Heat supplied $Q_1 = h_1 - h_4 = h_1 - h_{f_3} = 3052.1 - 282.7 = 2769.4$ kJ/kg

Modified Rankine Cycle efficiency $= \dfrac{505.19}{2769.4} = 0.1827 = 18.27$ percent

(b)  Steam consumption per kW hour $= \dfrac{3600}{\text{Work per kg}} = \dfrac{3600}{505.19} = 7.12$ kg/hr kW   *Ans.*

(c)  Mean effective pressure $= \dfrac{\text{Work done}}{\text{Swept volume}}$

$$= \dfrac{505.19}{v_2} = \dfrac{505.19}{0.95 \times 2.364} = 224.948 \text{ N/m}^2 \quad Ans.$$

(d)  heat removed in the condenser $= h_3 - h_4$ but $h_3$ has not been *foundout*. Heat rejected can also be determined as $=$

heat added $-$ Work done $= (2769.4 - 505.19) = 2264.21$ kJ/kg   *Ans.*

(e)  If the expansion process is allowed upto the back pressure then $W = h_1 - h_2'$

Now $$S_1 = S_2' = S_{f_2} + x_2' \, S_{fg_2'}$$

or $$7.125 = 0.925 + x_2' \times 6.868, \quad \therefore \quad x_2' = 0.902$$

$$h_2' = h_{f_2'} + x_2' h_{fg_2'} = 282.7 + 0.902 \times 2340 = 2393.38 \text{ kJ/kg}$$

Work done with complete expansion $= h_1 - h_2' = (3052.1 - 2393.38) = 658.72$

Loss of work due to incomplete expansion

$$= (h_1 - h_2') - (h_1 - h_2) = (658.72 - 506.17) = 152.55 \text{ kJ/kg}$$

(f)  Volume of cylinder $= \dfrac{\pi}{4} D^2 L = \dfrac{\pi}{4}(0.30)^2 (0.58) = 0.041 \text{ m}^3$.

$$\text{Mass of steam at state } 2 = \dfrac{v_{2cy}}{v_2} = \dfrac{0.041}{0.95 \times 2.364} = 0.01825 \text{ kg/stroke}$$

Volume at state $2' = $ mass $\times$ sp. volume at state $2'$

$$= 0.01825 \times 0.902 \times 5.5784 = 0.0918 \text{ m}^3$$

If $L$ is the new stroke length, then $0.0918 = \dfrac{\pi}{4} D^2 L$.

$\therefore$ $$L = \dfrac{4 \times 0.0918}{\pi \times (0.30)^2} = 1.298 \text{ m} \quad Ans.$$

*Note*: This problem illustrates the advantage of adopting the modified Rankine cycle for the steam engine. In modified cycle stroke length is 58 cm and in Rankine cycle it increases to 129.8 cm.

## 8.7 DESIRABLE PROPERTIES OF AN IDEAL FLUID USED IN A VAPOUR POWER CYCLE

Though water is extensively used in vapour power cycle, its thermodynamic properties are not really ideal for obtaining high thermal efficiency. From thermodynamic point of view, the following properties are desirable.

1. The critical temperature should be well above the useful temperature set by the available material for use in the steam power plant. This avoids superheating and thermodynamically all the heat should be added at high temperature. Unfortunately critical temperature for water is 374.15°C whereas mercury has critical temperature of 1460°C.
2. The saturation pressure at critical temperature (called critical pressure) should be moderate to reduce the cost of pressure vessel. Water has very high critical pressure (221.2 bar) whereas mercury has low (20.58 bar).
3. Fluid must have small specific heat of liquid. This will render the sensible heat negligible in comparison to heat added at constant temperature (boiling). The Rankine cycle will then approach the carnot cycle.
4. High enthalpy of evaporation and low specific volume (high density) reduce the plant size.
5. The saturated vapour line should be very steep, so that the fluid after expansion has the high dryness fraction without recourse to superheating.
6. The saturation pressure at the exhaust temperature should be slightly higher than the atmospheric pressure. This will avoid the air leakage into the condenser.

## 8.8 REHEAT CYCLE

Efficiency of Rankine cycle can be improved with the help of reheat cycle as well as by regenerative cycle. In practice reheat and regeneration, both are used to improve the overall cycle efficiency. Efficiency of ordinary Rankine cycle can be improved by increasing the pressure and temperature of the steam entering the turbine. As the initial pressure increases, the expansion ratio of the turbine increases and steam becomes quite wet at the end of the expansion. This is undesirable because moisture content causes erosion of the blades and increases the frictional losses. The main purpose of reheat cycle is to increase the dryness fraction of the steam entering L.P. stages as dryness fraction below 88 percent is undesirable.

In reheat cycle, the steam is extracted from suitable point in the turbine after expansion in the H.P. turbine and is reheated with the help of flue gases in the boiler furnace (Fig. 8.14). The following are the advantages of reheat cycle:

(i) Wet steam causes erosion and corrosion to the blades of the turbine. Reheated steam eliminates these problems.
(ii) The output of the turbine is increased.
(iii) Thermal efficiency of the cycle is increased. A gain of 4 to 7 percent of thermal efficiency over an equivalent non-reheat cycle is possible.
(iv) Final dryness fraction of steam is improved.
(v) Nozzle and blade efficiencies are increased.
(vi) Specific steam consumption is decreased.

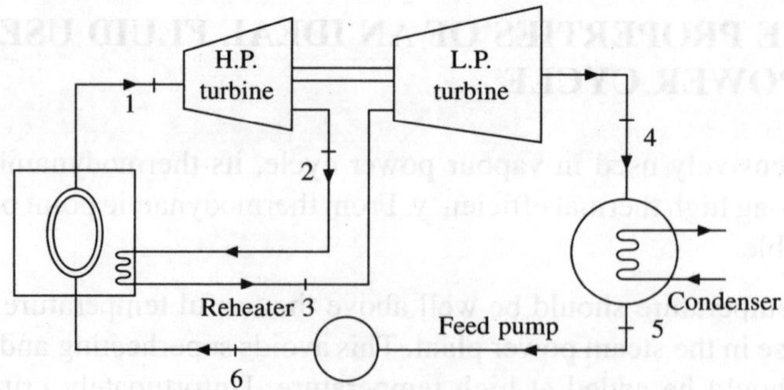



**Fig. 8.14   Reheat cycle equipments**

Reheating is generally employed when the pressures are high say above 100 bar. At unnecessary low reheat pressures the reheat cycle efficiency may be less than the Rankine cycle efficiency since above temperature during reheating will then below. The reheat pressure is generally kept within 40 percent of the initial pressure. The reheat cycle equipments have been shown in the Fig. 8.14 and $T$-$S$ and his diagrams have been shown in the Fig. 8.15. Referring to Fig. 8.15, the total heat added per kg of

$$\text{steam} = Q_1 = (h_1 - h_5) + (h_3 - h_2) - W_p$$

$$\text{Work done} = W = (h_1 - h_2) + (h_3 - h_4) - W_p$$

Where $W_P$ = Pump work = $h_6 - h_5$

$$\text{Efficiency of reheat cycle} = \frac{W}{Q_1} = \frac{(h_1 - h_2) + (h_3 - h_4) - W_p}{(h_1 - h_5) + (h_3 - h_2) - W_p} \tag{8.15}$$

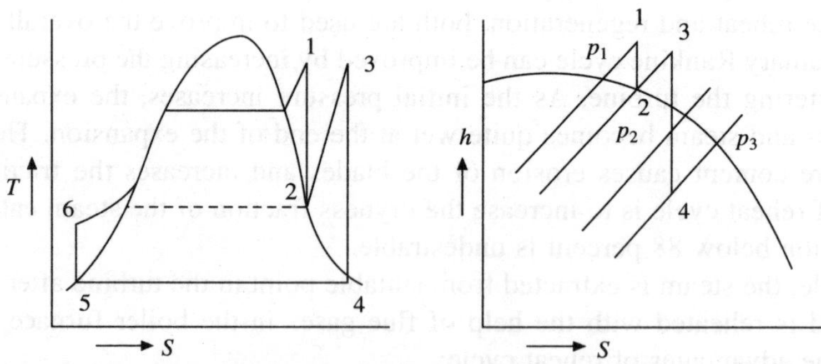



**Fig. 8.15   Reheat cycle**

**Example 8.4**  In a reheat cycle steam at 90 bar absolute and 480°C is supplied to a steam turbine. The steam is reheated to its original temperature passing it through a reheater at 12 bar absolute. The expansion after reheating takes place to condenser pressure of 0.07 bar absolute. Find the efficiency of the reheat cycle and workout for the flow rate of 1 kg/sec. Neglect the pressure loss in the system and assume isentropic expansion in both the stages.

*Solution*:   $p_1 = 90$ bar, $t_1 = 480°C$, $p_2 = 12$ bar, $t_2 = 480°C$, $p_b = 0.07$ bar
Ref. to Fig. 8.16(a), (b).

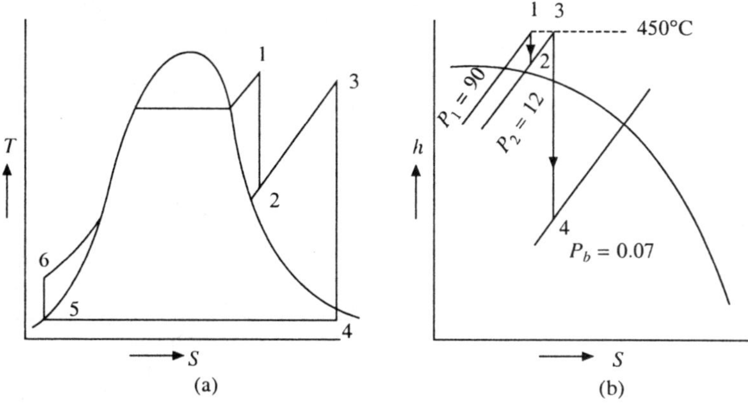


**Fig. 8.16(a), (b)**

The cycle has been shown on *T-S* and *h-s* diagrams. The problem can be solved easily by Mollier Chart. The state 1 is located on 90 bar pressure and teperature of 480°C, state 2 is located by drawing a vertical line from 1 up to pressure 12 bar. state 3 is on 12 bar and temperature 480°C. Similarly state 4 is located. The various values of enthalpy are:

$$h_1 = 3315 \text{ kJ/kg } h_2 = 2790 \text{ kJ/kg } h_3 = 3440 \text{ kJ/kg}$$

$$h_4 = 2400 \text{ kJ/kg } h_{f_5} = 163.4 \text{ kJ/kg (from steam table)}$$

Neglecting pump work, work done per kg of

$$\text{steam} = W = (h_1 - h_2) + (h_3 - h_4) = (3315 - 2790) + (3440 - 2400) = 1565 \text{ kJ/kg}$$

$$\text{Heat added} = Q_1 = (h_1 - h_{f_5}) + (h_3 - h_2) = (3315 - 163.4) + (3440 - 2790)$$

$$= 3801.16 \text{ kJ/kg.}$$

$$\text{Eficiency of reheat cycle} = \frac{W}{Q_1} = \frac{1565}{3801.16} = 0.4117$$

$$= 41.17 \text{ percent } Ans.$$

Power generated = 1 kg/s × 1565 kJ/kg = 1565 kW   *Ans.*

**Example 8.5** In a reheat cycle steam enters the H.P. turbine at 100 bar and 500°C. The expansion is continued to a pressure of 8.5 bar with isentropic efficiency of 80 percent. There is a pressure drop of 1.5 bar in the reheater and then steam enters the L.P. turbine at 7 bar and 500°C in which expansion is continued to a back pressure of 0.04 bar with isentropic efficiency of 85 percent. Determine (a) thermal efficiency and (b) S.S.C

*Solution*:   $p_1 = 100$ bar $t_1 = 500°C$
Refer to Fig. 8.17(a), (b)

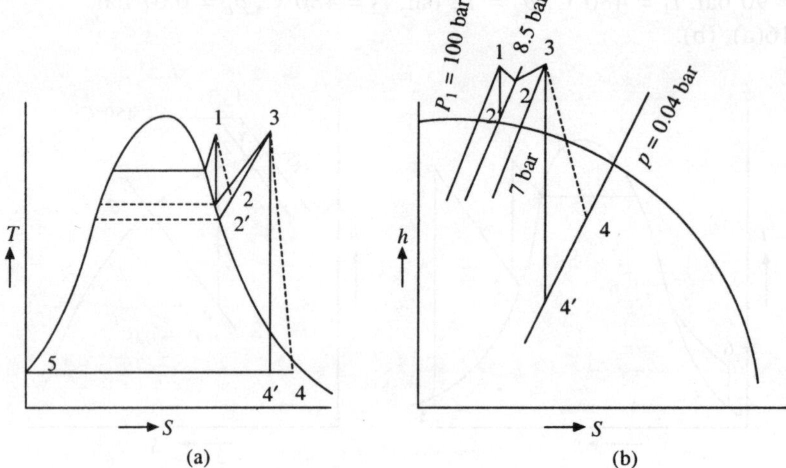

**Fig. 8.17(a), (b)**

From Mollier chart $h_1 = 3420$ kJ/kg  $h_2' = 2740$

$$\eta_{\text{isen}} = \frac{h_1 - h_2}{h_1 - h_2'} \qquad \therefore \quad h_2 = h_1 - \eta_{\text{isen}} \times (h_1 - h_2')$$

$$= 3420 - 0.8\,(3420 - 2740) = 2876 \text{ kJ/kg}$$

$$h_3 = 3490 \text{ kg/kg}, \; h_4' = 2375 \text{ kJ/kg}$$

$$h_4 = h_3 - \eta_{\text{isen}} \times (h_3 - h_4') = 3490 - 0.85\,(3490 - 2375) = 2542.25 \text{ kJ/kg}$$

$$h_{f_4} = h_5 = 121.4 \quad \text{(from steam table)}$$

Work done per kg $= W = (h_1 - h_2) + (h_3 - h_4)$

$$= (3420 - 2876) + (3490 - 2542.25) = 1491.75 \text{ kJ/kg}$$

Heat supplied $= Q_1 = (h_1 - h_{f_4}) + (h_3 - h_2)$

$$= (3420 - 121.4) + (3490 - 2876) = 3912.6$$

(a) Thermal efficiency $= \eta = \dfrac{W}{Q_1} = \dfrac{1491.75}{3912.6} = 0.3812$

$$= 38.12 \text{ per cent} \quad \textit{Ans.}$$

(b) S.S.C $= \dfrac{3600}{W} = \dfrac{3600}{1491.75} = 2.41$ kg/hr. kW  *Ans.*

## 8.9  REGENERATIVE CYCLE

The efficiency of Rankine cycle is less than the Carnot cycle because in the former all the heat is not added at the highest temperature of the cycle as in the latter. The sensible heating of feed water from condensing temperature to evaporation temperature is not isothermal which reduces the mean temperature of heat addition and hence decreases the Rankine cycle efficiency. If the external supply of heat could be eliminated during this sensible heating by reversibly transferring heat from some part of the cycle, where the fluid is at the same temperature as the feed, all the heat supplied from an external source would be at the maximum temperature of the cycle and the carnot efficiency could be achieved. An ideal arrangement to achieve this is represented diagramatically and on *T-S* diagram neglecting pump work in Fig. 8.18. In the figure the steam enters the turbine dry saturated at temperature $T_1$ and expands to temperature $T_2$ non-isentropically. The condensate from the condenser is pumped back through an annular space in the turbine casing and flows in a direction opposite to that of expanding steam and heated to a temperature slightly smaller than $T_1$ before it enters into the boiler. At all the points the temperature difference is infinitismal between the condensing steam and the water and therefore process is reversible. Such heating is known as regenerative heating as steam is used to heat itself. Due to this loss of heat, the expansion in the turbine is no more adiabatic and it follows the curved path 1-2. The heat gained by feed water is area (4-5-6-3-4) and heat lost by expanding steam is area 1-7-8-2-1 i.e. area (4-3'-3-4) = area (1-2'-2-1) adding area (1-2-3'-4-1) both hand sides. We get area (1-2-3-4) = area (1-2'-3'-4-1) or work = $(T_1 - T_2)\, \Delta S$.

Extarnally heat supplied = $T_1 \times \Delta S$.

$$\therefore \text{ Efficiency of regenerative cycle } = \frac{\text{Work done}}{\text{heat supplied}}$$

$$= \frac{(T_1 - T_2)\, \Delta S}{T_1 \times \Delta S} = \frac{T_1 - T_2}{T_1} = \text{Carnot efficiency} \tag{8.16}$$

Therefore the ideal regenerative cycle has the same efficiency as carnot. However the ideal regenerative cycle carnot be achieved in practice due to the following reasons.

(1) It is impracticable to design a turbine which would operate as a heat exchanger for feed heating as well as expander for steam.

(2) The steam expanding in turbine will have impractically low dryness fraction which will cause errosion to the blades of the turbine.

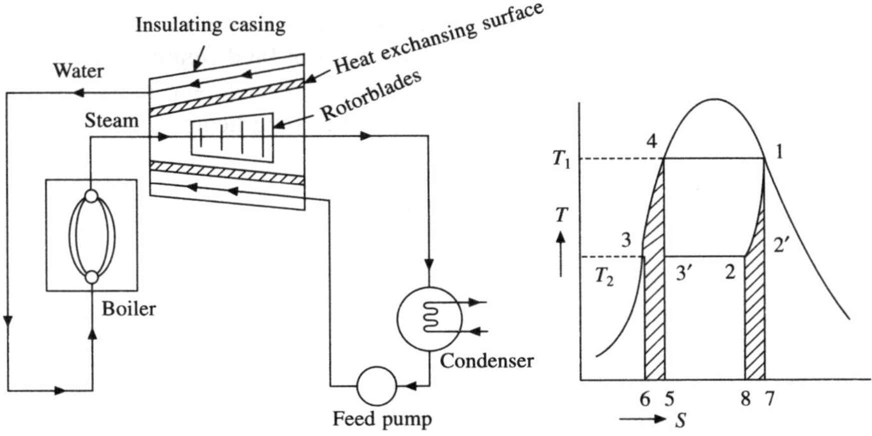


**Fig. 8.18  Ideal regenerative cycle**

*Practical Regenerative Cycle*:  In actual practice the advantage of regenerative cycle is taken by extracting (bleeding) certain portion of expanding steam at many suitable pressures to reheat the feed water so that the dryness fraction of the remaining part is not greatly reduced (Fig. 8.19).

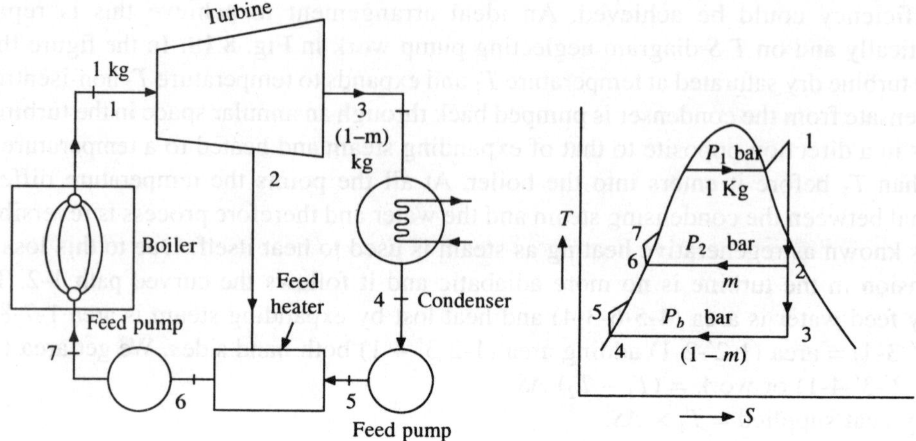


**Fig. 8.19    A simple regenerative cycle with single feed water heater.**

Consider 1 kg of steam flowing from the boiler to the turbine. After expansion of 1 kg of steam to state 2, m kg of the steam is bled off and taken to a feed heater. The remaining (1-m) kg of steam is expanded to condenser pressure and leaves the turbine at state 3. After condensation to state 4. condensate water enters to the heater maintained at bleeding pressure. (1-*m*) kg. of condensate mixed with m kg of bled steam and the mixture comes out at stage 6. Latent heat of bled steam has been used to heat the feed water up to saturation temperature corresponding to bleeding pressure. Then 1 kg of heated condensate is pumped to the boiler where heating starts from state 7. This cycle is not ideal cycle as mixing in feed water heater is not reversible. However with infinite number of bleedings, the mixing process becomes reversible and theoretically carnot efficiency can be attained. The mass of bled steam (m) can be determined by energy balance and mass balance equations applied to feed heater (Fig. 8.20). This gives

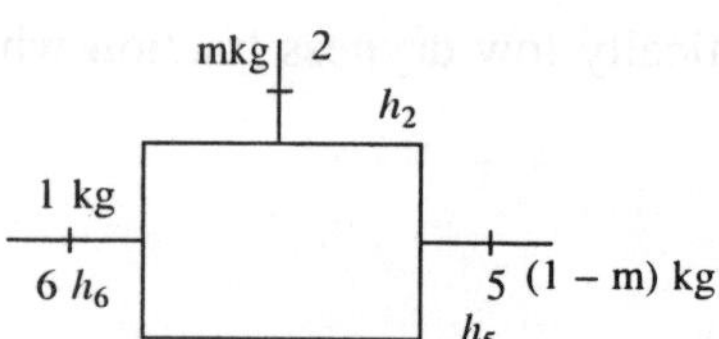


**Fig. 8.20    Feed heater**

$$mh_2 + (1\text{-}m)\, h_5 = 1 \times h_6$$

$h_5 = h_4$ if pump work is neglected. Then

$$m\, h_4 + (1 - m)h_4 = 1 \times h_6$$

knowing $h_2$, $h_4$ and $h_6$, m can be determined.

Total work done = $W = 1(h_1 - h_2) + (1 - m)\,(h_2 - h_3)$

Heat supplied = $Q_1 = (h_1 - h_6)$ neglecting pump work.

The efficiency of regenerative cycle is

$$= \frac{1(h_1 - h_2) + (1 - m)(h_2 - h_3)}{(h_1 - h_6)} \tag{8.17}$$

If pump work is considered,

$$W_p = (1 - m)(P_2 - P_b)/10 + 1(P_1 - P_2)/10 \text{ kJ/kg}$$

In actual practice 3 to 8 feed water heaters are used and temperature rise in each feed heater is about 10 to 15°C.

*Advantages of Regenerative Cycle*:   The following are the main advantages of the regenerative cycle:
1. Average temperature of heat addition to the cycle is increased due to high temperature of feed water entering the boiler.
2. The thermal stresses in the boiler are reduced due to hotter feed water.
3. Irreversibility in the feed water heating process in the boiler is minimised due to which efficiency of the cycle increases.
4. Less amount of steam is passed through the L.P. stages, so blade height will be less.
5. The condenser capacity is reduced.
6. The hotter feed prevents the condensation of sulphur dioxide gases on economiser.

*Disadvantages*
1. Cost of the plant increases.
2. Work done per kg of steam is reduced due to which boiler capacity is increased for a given output.

---

**Example 8.8**   A steam turbine plant equipped with a single regenerative feed heater of contact type operates under the following conditions:
Initial steam pressure = 16.5 bar initial degree of superheat = 93°C, bleeding pressure = 2 bar abs, exhaust pressure = 0.05 bar. Determine the followings (a) thermal efficiency, (b) steam consumption in kg/kW hr and (c) steam condensed per kW hr. It is assumed that expansion is isentropic. Determine the above quantities for non-regenerative cycle also.

*Solution*:   Ref. to Fig. 8.21 (a), (b).
From steam table initial temp = 202.9 + 93 = 295.9°C. At pressure 16.5 bar and temperature 295.9°C, from Mollier chart $h_1 = 3025$ kJ/kg, $h_2 = 2600$ kJ/kg, $h_3 = 2090$ kJ/kg, $h_4 = h_{f_4} = 137.8$ kJ/kg (from table) $h_6 = 504.7$ kJ/kg (from table)
Neglecting pump work, $h_4 = h_5$, $h_6 = h_7$.
Energy balance equation for the feed heater

gives:
$$m \times h_2 + (1 - m) h_4 = 1 \times h_6$$

or,
$$m = \frac{h_6 - h_4}{h_2 - h_4} = \frac{504.7 - 137.8}{2600 - 137.8} = 0.149 \text{ kg}$$

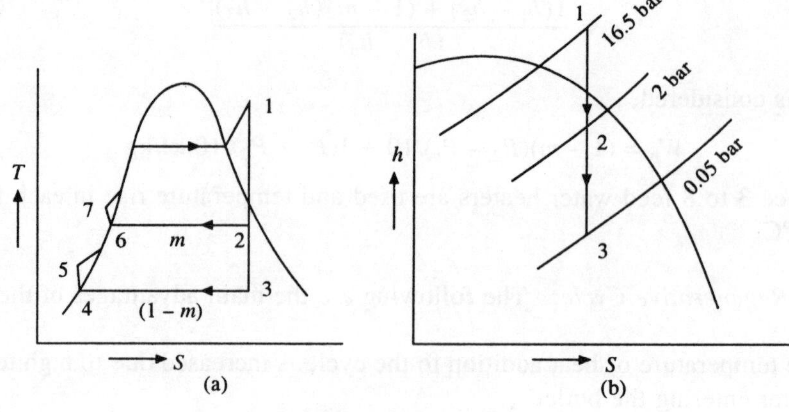



**Fig. 8.21 (a), (b)**

Neglecting pump work, work done in the cycle = W

$$= (h_1 - h_2) + (1 - m)(h_2 - h_3) = (3025 - 2600) + (1 - 0.149)(2600 - 2090)$$

$$= 859.01 \text{ kJ/kg}$$

Heat supplied $= Q_1 = h_1 - h_6 = (3025 - 504.7) = 2520.3$ kJ/kg

Efficiency of regenerative cycle $= \dfrac{859.01}{2520.3} = 0.340 = 34.0$ percent   *Ans.*

Steam consumption per kW hr $= \dfrac{3600}{\text{work done per kg of steam}}$

$$= \dfrac{3600}{859.01} = 4.19 \text{ kg/kW hr.}   \textit{Ans.}$$

Steam condensed in the condenser per kW hr

$$= (1 - m) \times 4.19 = 3.56 \text{ kg/kW hr}   \textit{Ans.}$$

For non-regenerative cycle

$$\text{Work done} = h_1 - h_3 = (3025 - 2090) = 935 \text{ kJ/kg}$$

$$\text{Heat supplied} = h_1 - h_4 = (3025 - 137.8) = 2887.2 \text{ kJ/kg}$$

$$\text{Efficiency} = \dfrac{935}{2887.2} = 0.323 = 32.3 \text{ percent}   \textit{Ans.}$$

$$\text{S.S.C.} = \dfrac{3600}{935} = 3.85 \text{ kg/kW hr}   \textit{Ans.}$$

Steam condensed without regeneration = 3.85 kg/kW hr   *Ans.*

### Types of feed water heaters

The feed water heaters are of two types:

   (i)  Open or direct contact type feed water heaters

  (ii)  closed or indirect type feed water heaters.

In an open or direct contact type heater, the extracted steam is allowed to mix with feed water and both leave the heater at a common temperature (Fig. 8.19). Open feed water heaters are simple in construction, have low cost and high transfer heat capacity. The disadvantage is that each heater in the plant needs a pump to deal with large amount of hot feed water. For, this reason, the system is usually not employed in practice.

In case of closed or indirect type feed water heater, the fluids are kept separate and are not allowed to mix together (Fig. 8.22).

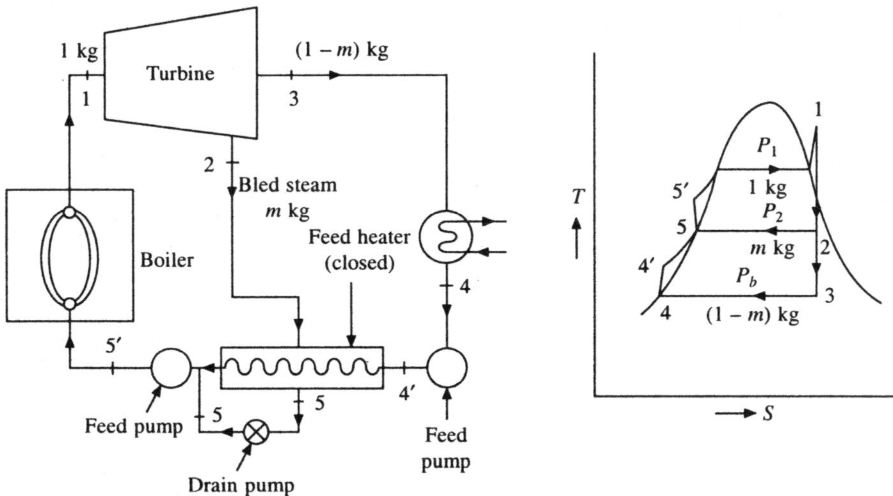


**Fig. 8.22   Indirect type feed heating with single feed heater and drain pump**

In this method the condensate of bled steam from closed feed water heaters is extracted by the drain pumps and discharged into the feed pipes of the next higher pressure feed water heater. This system also suffers from the same disadvantage that the feed pump has to deal with hot feed water. The only advantage in that the pump has to deal with condensed steam only. This method is also called as *drain pump method*.

Under indirect heating the second method is named as *all drains in hot-well method* (Fig. 8.23). In this method the condensate of all the feed water heaters and the condensate of the condenser are all discharged into the hot well. The mixed condensate (feed water) from the hot well is pumped into feed water heaters for feed water heating purposes. This arrangement may be used in case of break down of other complicated arrangements.

The third system of indirect heating is cascade method (Fig. 8.24). In this method the condensate of the high pressure (H.P.) heater is discharged to the next lower pressure feed water heater and the total condensate of the low pressure (L.P.) feed water heater and also the condensate of the condenser is discharged to a hot well. The combined condensate from the hot-well is circulated to feed water heaters with the help of feed pump.

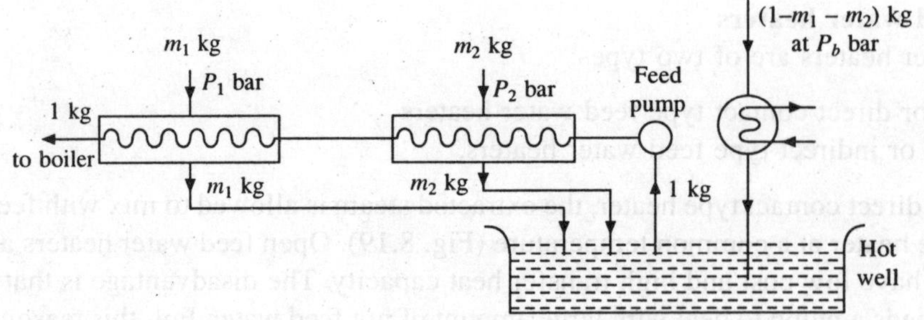

**Fig. 8.23 All drains in hot-well method**

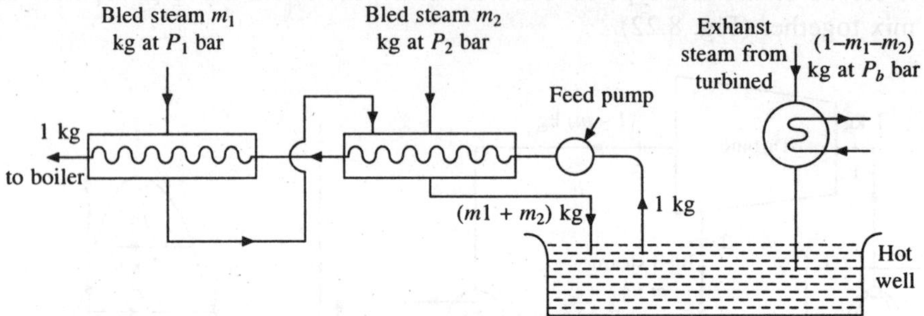

**Fig. 8.24 Cascade method**

The system can use both closed type and open type feed waters. In case of cascade system only one feed pump is used, thus there is a sowing in cost of feed pump.

Closed feed water heaters are costly but their main advantage is that a single feed pump is needed regardless of number of feed water heaters in the system.

Complete carnotization of Rankine cycle plant is not possible with a finite number of heaters. Through the efficiency increases with the increased number of feed water but the net gain in efficiency with each additional heater keeps on decreasing. The cycle efficiency is maximum when the total enthalpy rise of feed water from the condenser temperature to the boiler saturation temperature is divided equally between the feed water heater and the economiser in a single bleed cycle.

The effect of number of heaters used in a plant on its cycle efficiency employing various boiler pressures is shown in Fig. 8.25. The percentage gain in efficiency against the feed water temperatures is shown in Fig. 8.26.

**Efficiency of Direct Contact Heating Cycle**

The arrangement consisting two feed water heaters with $T$-$S$ and $h$-$s$ diagram is shown in Fig. 8.27. The processes 5-5′, 6-6′ and 7-7′ represent pump proceses and processes 5′-6 and 6′-7 represent for feed water heating in heaters. Process 1-2-3-4 represent isentropic expansion during bleeding of steam from turbine.

Work done per kg of steam supplied by the turbine is given by

$$W_T = (h_1 - h_2) + (1 - m_1)(h_2 - h_3) + (1 - m_1 - m_2)(h_3 - h_4) \text{ kJ} \qquad (8.18)$$

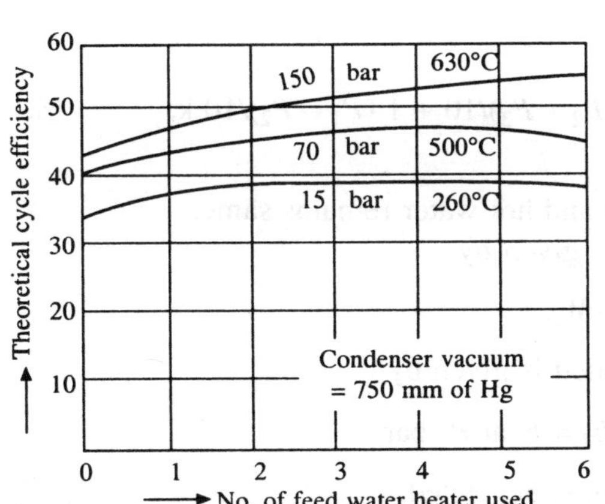

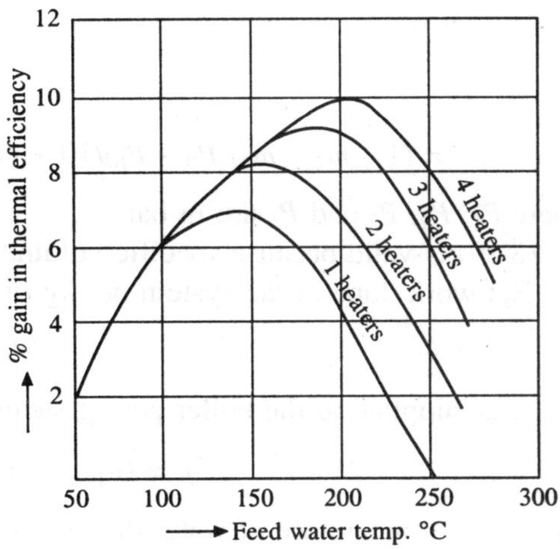

**Fig. 8.25    Effect of number of feed water heaters on cycle efficiency**

**Fig. 8.26    Gain in thermal efficiency $V_s$ feed water temperature**

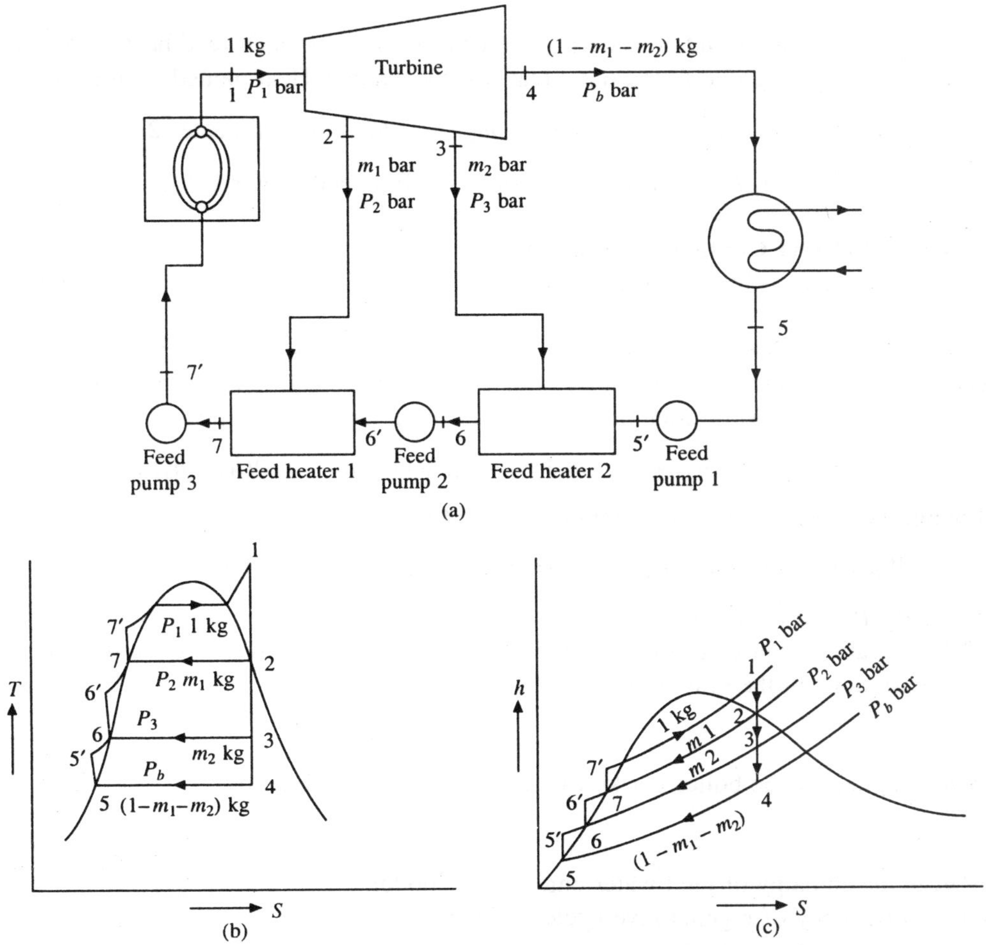

**Fig. 8.27    Regenerative cycle with two direct contact feed water heaters**

Total pump work per kg of steam supplied given by

$$= (1 - m_1 - m_2)(P_3 - P_b)/10 + (1 - m_1)(P_2 - P_3)/10 + 1\,(P_1 - P_2)/10 \text{ kJ} \qquad (8.19)$$

where $P_1$, $P_2$, $P_3$ and $P_b$ are in bar

Eq. (8.19) is valid because specific volume of cold and hot water remains same.

$\therefore$  Net work done in the system per kg of steam is given by

$$W_N = W_T - W_P$$

The heat supplied to the boiler per kg steam generated is given by

$$q_s = (h_1 - h_7'), \ h_7' \simeq h_7 = h_f \text{ at } P_2 \text{ bar}$$

The rise in temperature of water due to pump action is neglected.

Therefore, efficiency of regenerative cycle is given by

$$\eta = \frac{W_N}{q_s} = \frac{W_T - W_P}{h_1 - h_7)} \qquad (8.20)$$

Neglecting the heat losses and considering the heat balance for each feed heater, the amount of bled steam $m_1$ and $m_2$ can be found out. Heat lost by steam = Heat gained by the water

$\therefore$ $\qquad\qquad m_2(h_3 - h_6) = (1 - m_1 - m_2)(h_6 - h_5)$ For feed heater 2

and $\qquad\qquad m_1(h_2 - h_7) = (1 - m_1)(h_7 - h_6)$ For feed heater 1

**Efficiency of Indirect Contact Heating Cycle**

The arrangement consisting two feed water heaters with T-S and h-s diagram is shown in Fig. 8.28.

The processes 5-5′, 8-8′ and 9-9′ represent pump processes and processes 5′-6 and 6-7 represent for feed water heating in heaters. *Processes* 1-2-3-4 represent isentropic expansion during bleeding of steam from turbine. Work done per kg of steam supplied from the turbine is given by

$$W_T = (h_1 - h_2) + (1 - m_1)(h_2 - h_3) + (1 - m_1 - m_2)(h_3 - h_4) \text{ kJ} \qquad (8.21)$$

Total pump work per kg of steam supplied is given by

$$W_P = (1 - m_1 - m_2)(P_1 - P_b)/10 + (1 - m_1)(P_1 - P_3)/10 + (P_1 - P_2)/10 \text{ kJ} \qquad (8.22)$$

where $P_1$, $P_2$, $P_3$ and $P_b$ are in bar

Eq. (8.22) is valid because specific volume of cold water and hot water remains same.

$\therefore$  Net work done in the system per kg of steam is given by

$$W_N = W_T - W_P$$

The heat supplied by the boiler per kg of steam generated is given by

$$q_s = (h_1 - h_b)$$

where $h_b$ is the enthalpy of feed water entering the boiler.

Therefore, efficiency of regenerative cycle is given by

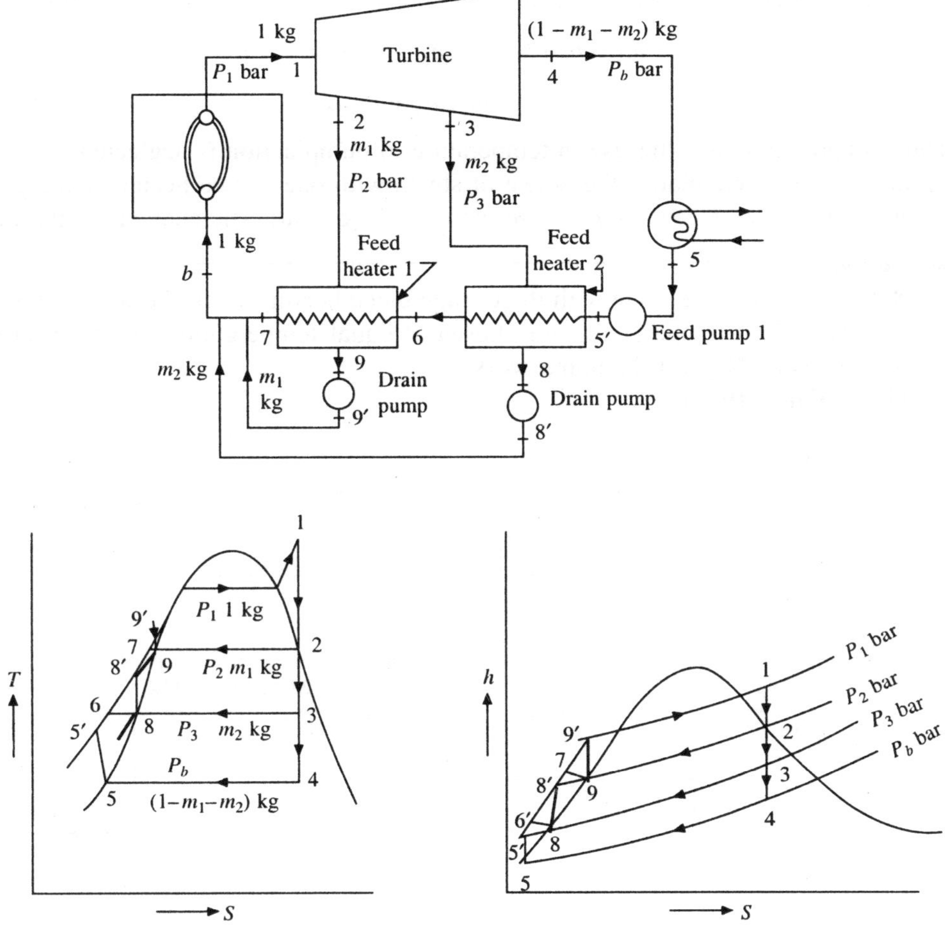

**Fig. 8.28   Regenerative cycle with two indirect contact feed water heaters**

$$\eta = \frac{W_N}{q_s} = \frac{W_T - W_P}{(h_1 - h_b)} \tag{8.23}$$

Neglecting the heat losses and considering the heat balance for each feed heater, we can find the values of $m_1$ and $m_2$ required to be bled from steam turbine.

$$m_2(h_3 - h_8) = (1 - m_1 - m_2)(h_8 - h_5) \text{ for feed heater 2} \tag{8.24}$$

and $\qquad m_1(h_2 - h_9) = (1 - m_1 - m_2)(h_9 - h_8) \text{ for feed heater 1} \tag{8.25}$

Eqs. (8.24) and (8.25) are valid when the rise in temperature of water due to pump action is neglected as it is very small and it is considered that the temperature of feed water going out of heat exchanger is equal to the saturation temperature of steam leaving the heat exchanger.

Therefore, on h-s diagram processes 6-8 and 9.7 are horizontal lines. i.e. $h_6 = h_8$, $h_9 = h_7$ and $h_5' = h_5$.

The condition of the feed water entering to the boiler is given by following expression

$$(1 - m_1 - m_2) \, h_9 + m_2 h_8 + m_1 h_9 = 1 \times h_b$$

$$\therefore \qquad h_b = (1 - m_2) \, h_9 + m_2 h_8 \qquad\qquad (8.26)$$

In the above expression also, the rise in temperature to pump action is neglected.

In actual thermal power plants, the reheat of steam with one or two heaters and regeneration with one or two heaters are generally used to increase the overall efficiency of the thermal plant.

---

**Example 8.9** A regenerative cycle with three stage bleed heating works between 30 bar, 450°C and 0.04 bar. The bleed temperatures are chosen at equal temperature ranges. Determine the efficiency of the cycle. Neglect the pump work.
(Mumbai Univ. Winter 1989)

*Solution* Ref. to Fig. 8.29(a), (b)
The figure shows the layout and *h-s* diagram. process 1-2-3-4-5 indicates insentropic expansion. At points 2, 3 and 4 bleeding of steam takes place.

The temperature of steam from steam table at 0.040 bar = 28.98°C

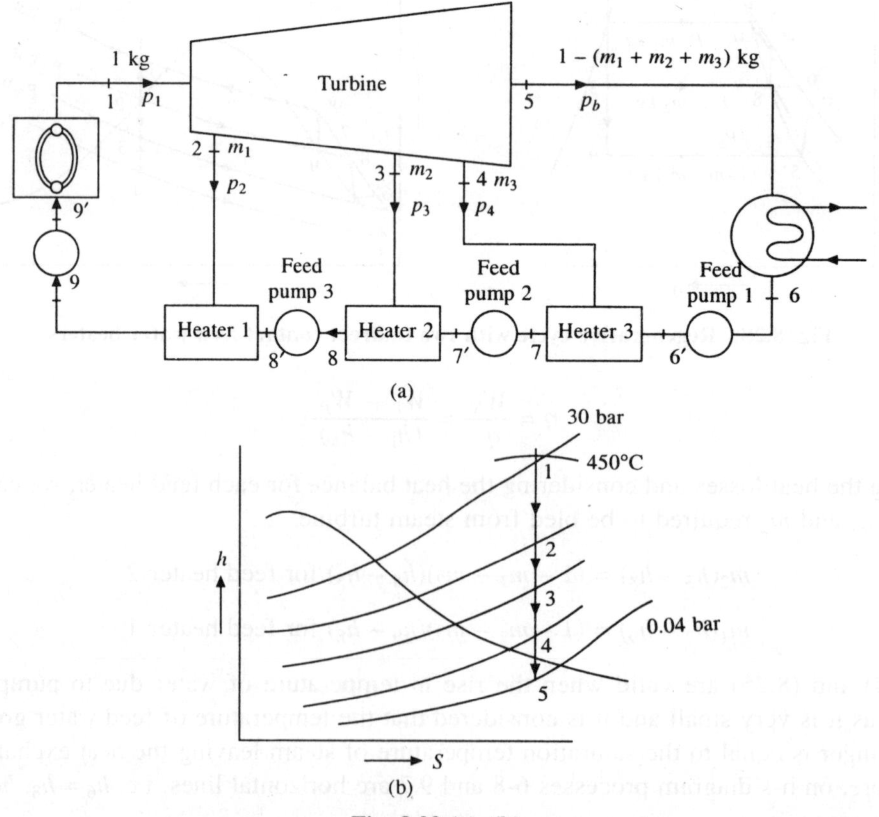


**Fig. 8.29 (a), (b)**

The maximum temperature = 450°C

∴   Temperature range = 450 – 28.98 = 421.02°C

∴ Equal temperature differences $= \dfrac{421.02}{4} = 105.25°C$

Bleedings are taking place at equal interval of temperature. Hence the temperatures at which bleed steam takes place, are

$$450 - 105.25 = 344.75°C$$

$$344.75 - 105.25 = 239.5°C$$

$$239.5 - 105.25 = 134.25°C$$

By coming down from point 1 on $h$-$s$ diagram the points 2, 3, 4 can be located at the meeting points by temperatures lines of 344.75, 239.5 & 134.25°C. The enthalpies of steam from Mollier chart at points 1, 2, 3, 4 and 5 are as follow

$$h_1 = 3340 \text{ kJ/kg}, \ h_2 = 3120 \text{ kJ/kg}, \ h_3 = 2935 \text{ kJ/kg}$$

$$h_4 = 2720 \text{ kJ/kg}, \ h_5 = 2120 \text{ kJ/kg}$$

pressure values at these points are read from $h$-$s$ chart

$$P_2 = 14.5 \text{ bar}, \ P_3 = 6.5 \text{ bar}, \ P_4 = 2.6 \text{ bar}$$

From steam table at these pressure values

$$h_{f_1} = 1008.3 \text{ kJ/kg (at 30 bar)}, \ h_{f_2} = 837.5, \ h_{f_3} = 678.75$$

$$h_{f_4} = 540.9 \ \text{ and } \ h_{f_5} = 121.4, \ h_{f_5} = h_6$$

Pump work is neglected

Energy balance for heaters 3, 2, 1 are as follow

$$m_3(h_4 - h_{f_4}) = [1 - (m_1 + m_2 + m_3)](h_{f_4} - h_{f_5})$$

$$m_2(h_3 - h_{f_3}) = [1 - (m_1 + m_2)](h_{f_3} - h_{f_4})$$

$$m_1(h_2 - h_{f_2}) = (1 - m_1)(h_{f_2} - h_{f_3})$$

After substituting the values

$$m_3(2720 - 540.9) = [1 - (m_1 + m_2 + m_3)](540.9 - 121.4)$$

∴ $\qquad\qquad m_3 = [1 - (m_1 + m_2 + m_3)] \times 0.1925$ $\qquad\qquad$ (i)

$$m_2(2935 - 678.75) = [1 - (m_1 + m_2)] (678.75 - 540.9)$$

∴ $\qquad\qquad m_2 = [1 - (m_1 + m_2)] \times 0.061$ $\qquad\qquad$ (ii)

$$m_1(3120 - 837.5) = (1 - m_1)(837.5 - 678.75)$$

∴ $\qquad\qquad m_1 = (1 - m_1) \times 0.0695$ $\qquad\qquad$ (iii)

or, $\qquad\qquad m_1 = 0.064 \text{ kg}$

Solving equations (i) and (ii)

$$m_2 = 0.0538 \text{ kg}, \ m_3 = 0.1424 \text{ kg}$$

work done per kg of steam

$$= 1 \ (h_1 - h_2) + (1 - m_2)(h_2 - h_3) + (1 - m_1 - m_2)(h_3 - h_4)$$
$$+ (1 - m_1 - m_2 - m_3)(h_4 - h_5)$$
$$= 1 \ (3340 - 3120) + (1 - 0.0538)(3120 - 2935)$$
$$+ (1 - 0.064 - 0.0538)(2935 - 2720)$$
$$+ (1 - 0.064 - 0.0538 - 0.1424)(2720 - 2120)$$
$$= 220 + 175.04 + 189.67 + 443.88$$
$$= 1028.59 \text{ kJ/kg}$$

Heat supplied per kg of steam $= (h_1 - h_{f_2})$

$$= (3340 - 837.5) = 2502.5 \text{ kJ/kg}$$

$$\therefore \qquad \eta_{\text{th}} = \frac{1028.59}{2502.5} \times 100 = 41.10\% \quad Ans$$

**Example 8.10**   In a regenerative feed heating cycle, the steam enters the turbine at 25 bar and 25°C. The condenser pressure is 0.05 bar. The steam is bled off for feed heating for a closed heater at 3.5 bar and for an open heater at 0.7 bar. The condensate of the closed heater is discharged into feed water of the low pressure open heater. Calculate the thermal efficiency of the cycle. Neglect the pump work. Also, determine the corresponding efficiency of the Rankine cycle.

*Solution*   Ref. to Fig. 8.30 (a), (b)
From steam tables

$$h_1 = 2879.45 \text{ kJ/kg}$$
$$h_9 = h_{10} = h_{11} = 584.3 \text{ kJ/kg}$$
$$h_7 = h_8 = 376.8 \text{ kJ/kg}$$
$$h_5 = h_6 = 137.8 \text{ kJ/kg}$$

and
$$S_1 = 6.408 \text{ kJ/kgk}$$

Considering the isentropic expansion process
1-2-3-4

$$S_1 = S_4 = S_3 = S_2$$
$$S_4 = S_f + x_4 S_{fg} \text{ at 0.05 bar}$$
$$6.408 = 0.476 + x_4 \times 7.920$$

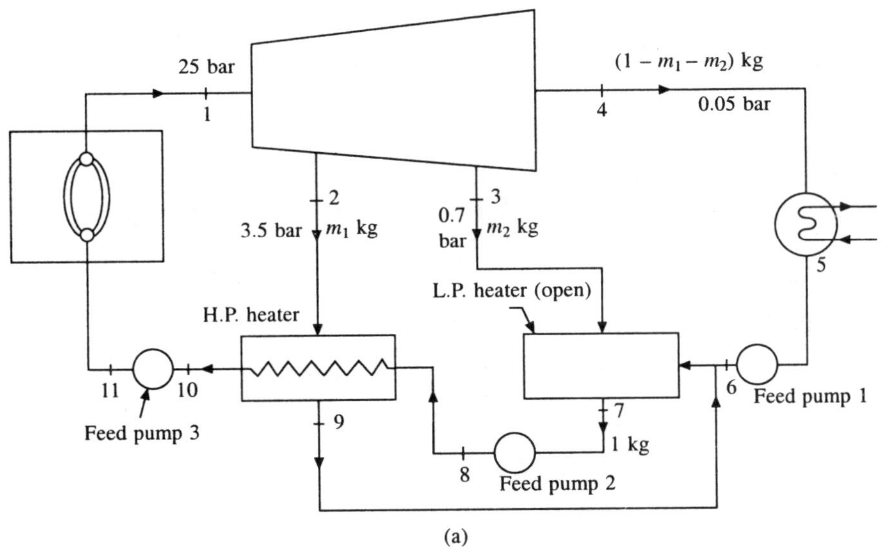



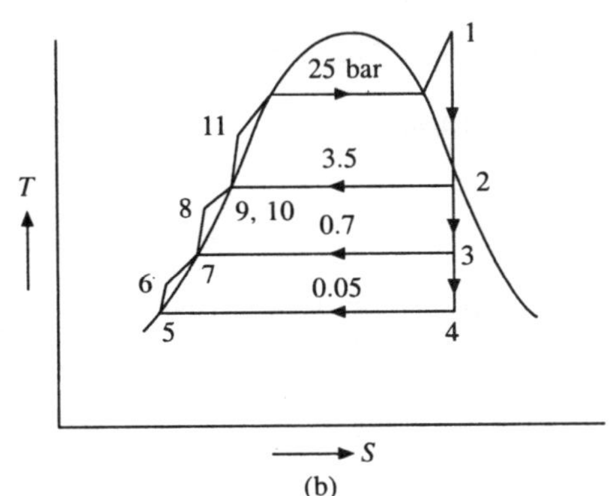



**Fig. 8.30**

$$\therefore x_4 = 0.748,\ h_4 = 137.8 + 0.748 \times 2423.8 = 1950.80 \text{ kJ/kg}$$

$$S_3 = S_f + x_3\, S_{fg} \text{ at } 0.7 \text{ bar}$$

$$6.408 = 1.192 + x_3 \times 6.288$$

$$\therefore x_3 = 0.829,\ h_3 = 376.8 + 0.829 \times 2283.3 = 2269.65 \text{ kJ/kg.}$$

$$S_2 = S_f + x_2 S_{fg} \text{ at } 3.5 \text{ bar}$$

$$6.408 = 1.727 + x_2 \times 5.212$$

$$\therefore x_2 = 0.898,\ h_2 = 584.3 + 0.898 \times 2147.3 = 2512.57 \text{ kJ/kg}$$

Energy balance for H.P. heater

$$m_1 \times (h_2 - h_9) = (1)(h_{10} - h_8)$$

or, $m_1(2512.57 - 584.3) = (1)(584.3 - 376.8)$

or, $$1928.27\ m_1 = 207.5$$

$\therefore\quad m_1 = 0.1076$ kg/kg of steam

Energy balance for L.P. heater

$$m_2 h_3 + m_1 h_9 + (1 - m_1 - m_2)\ h_6 = 1 \times h_7$$

or, $m_2 \times 2269.65 + 0.1076 \times 548.3 + (1 - 0.1076 - m_2) \times 137.8 = 376.8$

or, $2269.65\ m_2 + 62.87 + (0.8924 - m_2)\ 137.8 = 376.8$

$\therefore\ m_2 = 0.0895$ kg/kg of steam

$\therefore\quad$ Turbine work $W_T = (h_1 - h_2) + (1 - m_1)(h_2 - h_3) + (1 - m_1 - m_2)(h_3 - h_4)$

$$= (2879.45 - 2512.57) + (1 - 0.1076)(2512.57 - 2269.65)$$

$$+ (1 - 0.1076 - 0.0895)(2269.65 - 1950.80)$$

$$= 366.88 + 216.78 + 256.60$$

$$= 839.66 \text{ kJ/kg}$$

Heat supplied $q_s = (h_1 - h_{11})$

$$= (2879.45 - 584.3)$$

$$= 2295.15 \text{ kJ/kg}$$

$$\therefore \eta_{th} = \frac{W_T}{q_s} = \frac{839.66}{2295.15} \times 100 = 36.58\% \quad Ans.$$

$$\eta_{\text{Rankine}} = \frac{h_1 - h_4}{h_1 - h_5} = \frac{2879.45 - 1950.80}{2879.45 - 137.8} \times 100$$

$$= 33.87\% \quad Ans.$$

## 8.10  BINARY VAPOUR CYCLE

At high temperatures few better fluids are used which are (i) diphenyl ether $(C_6H_5)_2$ O, (ii) aluminium bromide $AlBr_3$ and (iii) liquid metals like mercury, sodium, potassium and so on. Among these, only mercury has actually been used in practice. Except this most organic substances decompose at high temperature. Aluminium bromide is a possiblity and yet to be considered.

Carnot cycle efficiency will increase with an increase in initial temperature or with the decrease in exit temperature. As at pressure of 12 bar, the saturation temperatures for water, aluminium bromide and mercury are 187°C, 482.5°C and 560°C respectively. The highest cyclic temperature consistent with the best available material for use in power plant is about 550–600°C. Therefore, mercury is a better working fluid in the high temperature range, because its vaporisation pressure is relatively low. Its critical pressure and temperature are 1080 bar and 1,460°C respectively. Where as, water has critical pressure and temperature as 221.2 bar and 374.15°C respectively. For achieving high temperature at high pressure of steam creates many difficulties in design, operation and control.

Mercury fulfills the requirement to produce mercury vapour at high temperature with low pressure which avoid the difficulties connected with high pressures.

But in the low temperature range, mercury is unsuitable because its saturation pressure becomes excedingly low, and it would be in practical to maintain such a high vacuum in the condenser. At 30°C, the saturation pressure of mercury is only $2.7 \times 10^{-4}$ cm of $H_g$ and 3.17 cm of $H_g$ that for water. Its specific volume at such a low pressure is very large, and it would be difficult to accommodate such a large volume flow. For this reason, to take advantage of the beneficial features of mercury in the high temperature range and water in low temperature range, mercury vapour leaving the mercury turbine is condensed at a higher temperature and pressure, and the heat released during the condensation of mercury is utilised in evaporating water to form steam to operate a conventional turbine.

The schematic diagram for mercury-steam binary vapour cycle is shown in Fig. 8.31(a) and the corresponding cycle on *T-S* diagram is shown in Fig. 8.31(b).

First experimental mercury turbine of 1800 kW capacity was installed at Dutch-point conn in 1923 (Denmark). Number of difficulties were experienced during its operation. The recent mercury steam installation is schiller station of 40 MW capacity at New Hampshire installed in 1949 which was developed by General Electric of the U.S.A.

Over and above, mercury is expensive, limited in supply and highly toxic. Because of the low latent heat (263 kJ/kg) over a wide range of desirable condensation temperature, therefore, several kg (7 to 8 kg times) must be circulated per kg of water evaporated, are needed. Capital costs are high. Although the New Hampshire plant has now been dismantaled due to higher overall cost.

The mercury plants are preferred where acute shortage of important facilities exists such as deficiency of condenser cooling water, insufficient steam-boilers and demand for electrical generating capacity.

With an advancement of critical and supercritical boilers, the binary-vapour plants are rarely used now a days as modern thermal power plants give equally high thermal efficiencies.

## Calculation for Cycle Efficiency

The mercury cycle 1-2-3-4-4'-1 is named as topping cycle and steam cycle *a-b-c-d-e-a* as bottoming cycle. The mercury leaves the condenser as saturated liquid and steam leaves as the saturated vapour (*e – f*). The condensed mercury liquid is pumped back to its boiler with the help of mercury pump. The evaporated steam is superheated in the boiler (*f-a*), after being sent to economiser (*d-e*. It is then expanded isentropically in steam turbine to point *b*, finally the steam is condensed in the steam condenser up to point *C* and pumped to the steam boiler.

Let 1 kg of steam be evaporated in the mercury condenser and it requires $m_{hg}$ of mercury vapour. Energy balance for mercury condenser can be written as

$$m_{hg}(h_2 - h_3) = 1 \ (h_f - h_e)$$

or,
$$m_{hg} = \frac{(h_k - h_f)}{(h_2 - h_3)} \qquad (8.27)$$

By neglecting pump work.

$$m_{hg} \times x_2 h_{fg2} = 1 \times (h_f - h_e) = 1 \times h_{fg}$$

Net work done per kg of steam

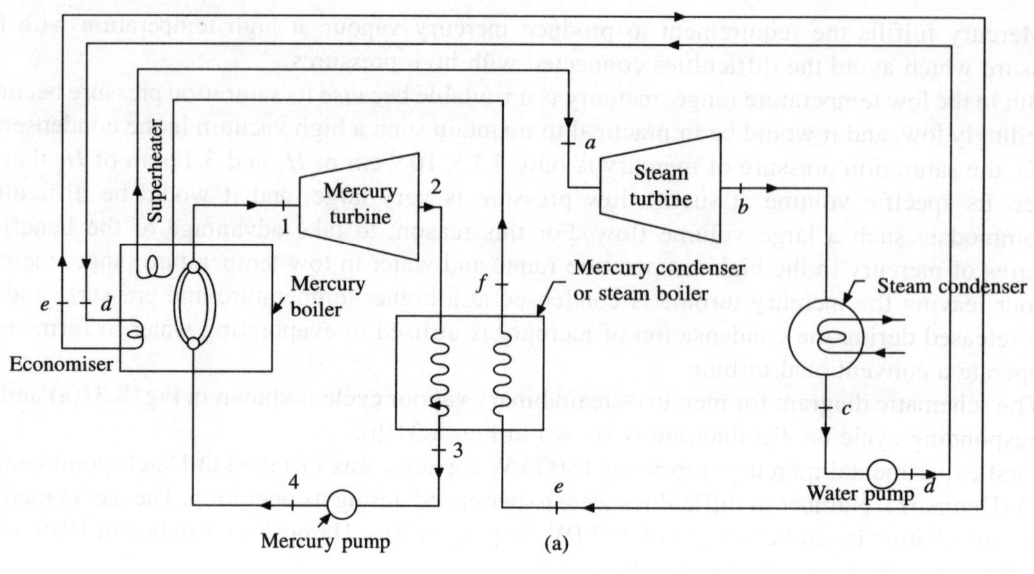



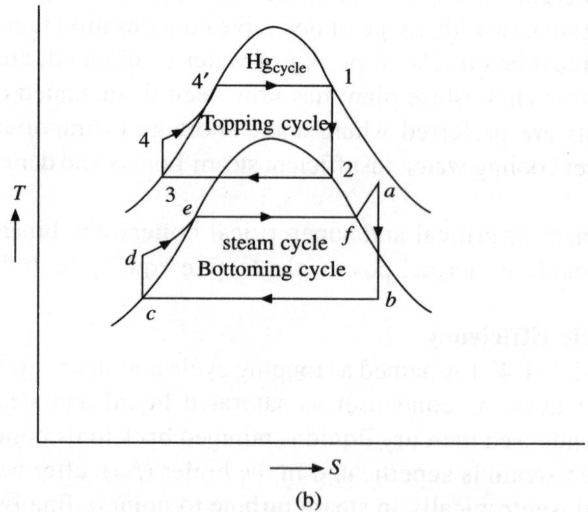



**Fig. 8.31  Binary vapour cycle, layout and *T-S* diagram**

$W_{net}$ = (Mercury turbine work) – (mercury pump work) + (Steam turbine work) – (steam pump work)

$$= m_{hg}(h_1 - h_2) - m_{hg}(h_4 - h_3) + 1\,(h_a - h_b) + 1\,(h_d - h_c)$$

If the pump work is neglected

$$W_{net} = m_{hg}(h_1 - h_2) + (h_a - h_b)$$

Heat supplied per kg of working fluid

$$q_s = m_{hg}(h_1 - h_4) + (h_a - h_f) + (h_e - h_d)$$

Heat rejected per kg of working fluid (steam)

$$q_r = (h_b - h_c)$$

$\therefore$ cycle efficiency, $\quad \eta_{\text{cycle}} = \dfrac{q_s - q_r}{q_s} = \dfrac{W_T - W_P}{q_s} = \dfrac{W_{\text{net}}}{q_s}$ $\qquad$ (8.28)

Specific steam consumption rate (S.S.C) $= \dfrac{3600}{W_{\text{net}}}$ kg/kW hr $\qquad$ (8.29)

like the above cycle, another coupled cycle (tertiary cycle) has been shown in Fig. 8.32. Three cycles have been coupled in series namely sodium cycle, mercury cycle and steam cycle.

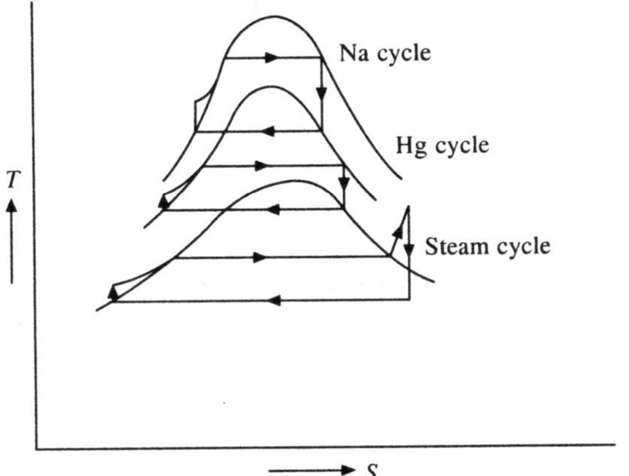


**Fig. 8.32   Coupled cycles**

## 8.11   THERMODYNAMICS OF RANKINE-RANKINE COUPLED CYCLES

When two cycles are combined with each other, the heat lost by toppling cycle is absorbed by bottoming cycle (Fig. 8.33), as in the mercury-steam binary cycle.

Let $\eta_1$ and $\eta_2$ be the efficiencies of the topping and bottoming cycles respectively and $\eta$ be the overall efficiency of the combined cycle.

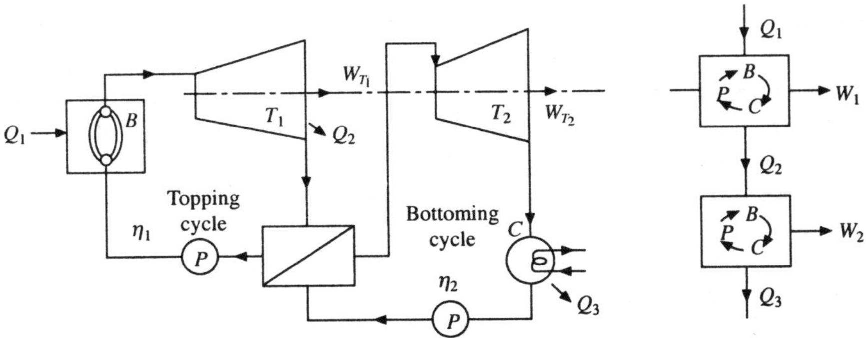


**Fig. 8.33   Two vapour power cycles coupled in series**

$$\eta_1 = 1 - \frac{Q_2}{Q_1} \quad \text{and} \quad \eta_2 = 1 - \frac{Q_3}{Q_2}$$

or, $Q_2 = (1 - \eta_1)Q_1$ and $Q_3 = (1 - \eta_2)Q_2$

Now,

$$\eta = 1 - \frac{Q_3}{Q_1} = 1 - \frac{(1 - \eta_2)Q_2}{Q_1}$$

$$= 1 - \frac{(1 - \eta_2)(1 - \eta_1)Q_1}{Q_1}$$

$$\eta = 1 - (1 - \eta_1)(1 - \eta_2) \tag{8.30}$$

For $n$ cycles coupled in series

$$\eta = 1 - (1 - \eta_1)(1 - \eta_2) \quad (1 - \eta_n)$$

or,

$$1 - \eta = \sum_{i=1}^{n} (1 - \eta_i)$$

For two cycles coupled in series Eq. (8.30) simplified to

$$\eta = 1 - (1 - \eta_1 - \eta_2 + \eta_1\eta_2)$$

$$= 1 - 1 + \eta_1 + \eta_2 - \eta_1\eta_2$$

$$\eta = \eta_1 + \eta_2 - \eta_1\eta_2 \tag{8.31}$$

Let

$$\eta_1 = 0.50 \text{ and } \eta_2 = 0.40$$

$$\therefore \quad \eta = 0.5 + 0.4 - 0.5 \times 0.4 = 0.70$$

So it is concluded that by combining two cycles in series, even if individual efficiencies are low, it is possible to achieve a fairly high combined efficiency which can not be attained by a single cycle.

It is almost impossible to achieve such a high efficiency in a single cycle.

Therefore, there would be fuel economy by improving cycle efficiency using multifluid coupled cycles. This leads to bulk power generation.

---

**Example 8.11**   A binary power plant uses mercury and water as working fluids. The $H_g$ cycle works between temperatures of 556°C and 222°C. Dry saturated mercury vapour from mercury boiler is used in $H_g$ turbine. The steam cycle works between pressure limits of 20 bar and 0.07 bar. The steam is superheated in a superheater to 400°C using mercury boiler the gas before entering into the steam turbine. The temperature of the feed water is raised to saturation temperature corresponding to steam generator pressure in an economiser placed in mercury boiler flue gas path and is then evaporated into steam in the mercury condenser cum steam generator. Draw the line diagram equipment and determine

(i)  Mass of $H_g$ required per kg of steam
(ii) Work out put from $H_g$ turbine per kg of steam generated in mercury condenser

(iii)  Work done by steam turbine per kg of steam, and

(iv)  Theoretical overall efficiency of cycle.

Neglect feed pumps work for both $H_g$ and steam pumps and take the following properties for $H_g$.

| Saturated temperature °C | Specific enthalpies kJ/kg | | | Specific entropy kJ/kg k | |
|---|---|---|---|---|---|
| | $h_f$ | $h_{fg}$ | $h_g$ | $s_f$ | $s_g$ |
| 556°C | 75.6 | 290 | 365.6 | 0.153 | 0.511 |
| 222°C | 29.20 | 302 | 331.6 | 0.079 | 0.704 |

(Mumbai Univ. Winter 1997)

*Solution*

For line diagram follow the Fig. 8.31(a) and for *T-S* diagram, Fig. 8.34.

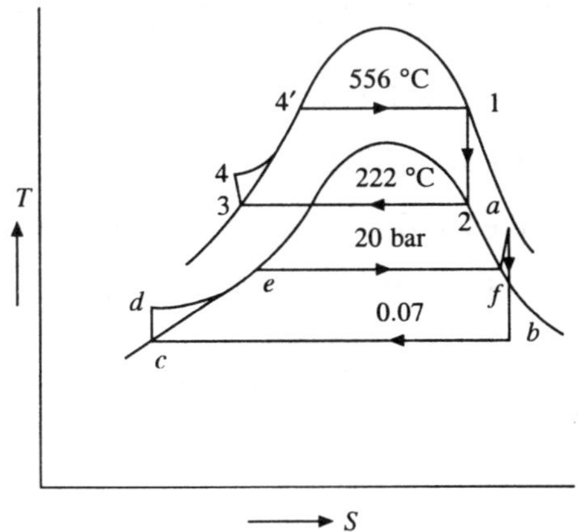


**Fig. 8.34  *T-S* diagram**

From steam Table

| | $t_s$ | $h_f$ | $h_{fg}$ | $h_g$ | $s_f$ | $s_{fg}$ | $s_g$ |
|---|---|---|---|---|---|---|---|
| 20 bar | 212.4 | 908.6 | 1888.7 | 2797.2 | 2.447 | 3.890 | 6.337 |
| 0.07 bar | 39.03 | 163.4 | 2409.2 | 2572.6 | 0.559 | 7.718 | 8.277 |

From $H_g$ Table

$$S_2 = S_1 = 0.511 \text{ kJ/kg k}$$

$$S_2 = S_f + x_2 S_{fg2}$$

or,
$$0.511 = 0.079 + x_2 (0.704 - 0.079)$$

$$\therefore \qquad x_2 = 0.6912$$

$$h_2 = 29.2 + 0.6912 \times 302 = 237.94 \text{ kJ/kg}$$

(i) **Mass flow rate of mercury**

Heat lost by mercury = Heat gained by Steam

$$m_{hg}(h_2 - h_3) = (h_f - h_e)$$

$$\therefore \qquad m_{hg} = \frac{(h_f - h_e)}{(h_2 - h_3)} = \frac{2797.2 - 908.6}{237.94 - 29.2} = 9.047 \text{ kJ/kg of steam}$$

(ii) $\therefore$ work done per kg of steam generated in mercury turbine

$$= m_{hg}(h_1 - h_2) = 9.047 \,(365.6 - 237.94) = 1154.94 \text{ kJ/kg} \quad Ans.$$

**Condition of steam at point "b"**

$$S_a = S_b$$

$$S_a = S_g + C_{ps} \ln \frac{T_{sup}}{T_s} = 6.337 + 2.1 \ln \frac{400 + 273}{212.4 + 273} = 7.023 \text{ kJ/kg k}$$

$$S_b = 0.559 + x_b \times 7.718$$

or, $\qquad 7.023 = 0.559 + x_b \times 7.718 \quad \therefore \quad x_b = 0.837$

$$\therefore \qquad h_b = 163.4 + 0.837 \times 2409.2 = 2179.9 \text{ kJ/kg}$$

From superheated steam table at 20 bar and 400°C

$$h_a = 3248.7 \text{ kJ/kg}$$

(iii)

$\therefore$ Work done per kg of steam $= (h_a - h_b)$

$$= (3248.7 - 2179.9)$$

$$= 1068.8 \text{ kJ/kg} \quad Ans.$$

$\therefore$ Total work done $= 1154.94 + 1068.8$

$$= 2223.74 \text{ kJ/kg of steam}$$

Heat supplied per kg of steam

$$= \text{Heat given to } H_g \text{ in combustion chamber}$$

$$+ \text{ Heat given to feed water in economiser}$$

$$+ \text{ Heat supplied to superheat the steam}$$

$$= m_{hg}(h_1 - h_3) + 1(h_e - h_c) + 1 \times C_{ps} (T_{sup} - T_s)$$

$$= 9.047 \,(365.6 - 29.2) + 2 \,(908.6 - 163.4) + 1 \times 2.1 \,(400 - 212.4)$$

$$= 4182.57 \text{ kJ/kg}$$

(iv)  $\therefore$  Overall efficiency

$$\eta_{\text{overall}} = \frac{\text{Work done}}{\text{Heat supplied}} = \frac{2223.74}{4182.57} \times 100$$

$$= 53.16\% \quad Ans.$$

## 8.12  COMBINED REHEAT REGENERATIVE CYCLE

As we have seen already that reheating is adopted when the vaporization pressure is high and, advantageous to get drier steam in the low pressure stages. Regeneration is adopted for increasing the thermal efficiency of the cycle. Thus it will be advisable to execute a cycle adopting both reheating and regeneration. A modern steam power plant is equipped with both. Figure 8.35(a), (b) indicate diagram and $T$-$S$ plot for such a system. At point 2, $m_1$ kg of steam is extracted (bled

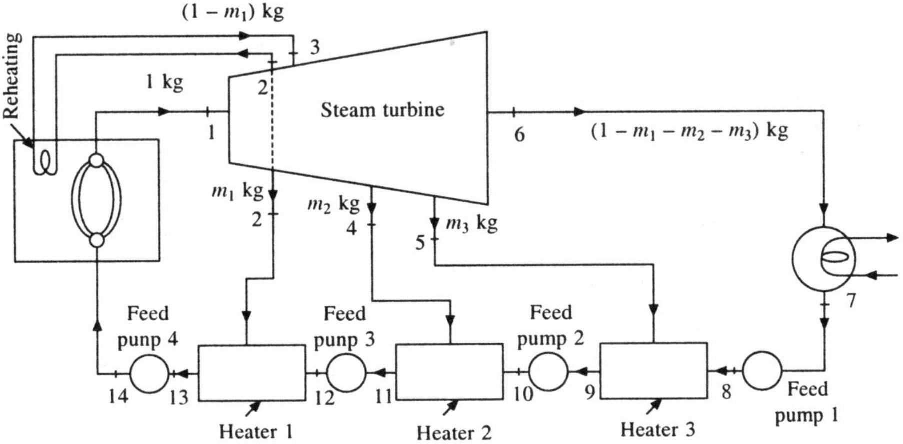


**Fig. 8.35(a)  Reheat-regenerative cycle**

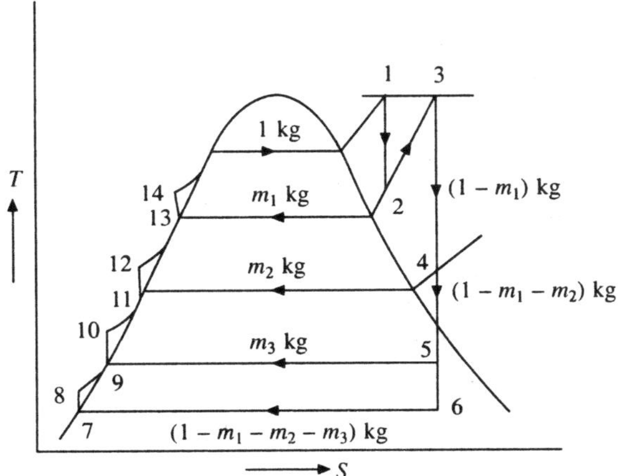


**Fig. 8.35(b)  $T$-$S$ diagram**

off) from turbine and supplied to open heater number 1. Remaining amount of steam i.e. $(1 - m_1)$ kg enters the boiler for reheating and then comes back to turbine. After this, at point 4 and 5 $m_2$ and $m_3$ kg of steam is extracted (bled off) and supplied to heaters 2 and 3 respectively. Remaining steam $(1 - m_1 - m_2 - m_2)$ kg of steam expands to condenser pressure i.e. point 6, and further condenses to water down to point 7.

Total turbine work

$$W_T = (h_1 - h_2) + (1 - m_1)(h_3 - h_4) + (1 - m_1 - m_2)(h_4 - h_5)$$
$$+ (1 - m_1 - m_2 - m_3)(h_5 - h_6) \text{ kJ/kg of steam}$$

Total pump work

$$W_P = (1 - m_1 - m_2 - m_3)(h_8 - h_7) + (1 - m_1 - m_2)(h_{10} - h_9)$$
$$+ (1 - m_1)(h_{12} - h_{11}) + (h_{14} - h_{13}) \text{ kJ/kg of steam}$$
$$= (1 - m_1 - m_2 - m_3)(P_5 - P_6)/10 + (1 - m_1 - m_2)(P_4 - P_5)/10$$
$$+ (1 - m_1)(P_2 - P_4)/10 + (P_1 - P_2)/10 \text{ kJ/kg of steam}$$

Since specific volume of feed water rarely changes with increase in temperature. Heat supplied to the cycle is given by

$$q_s = (h_1 - h_{14}) + (1 - m_1)(h_3 - h_2) \text{ kJ/kg of steam}$$

Heat rejected in condenser

$$q_r = (1 - m_1 - m_2 - m_3)(h_7 - h_6) \text{ kJ/kg of steam}$$

The energy balances of heaters 1, 2 and 3 give
Heat lost by steam = Heat gained by feed water

$$m_1(h_2 - h_{13}) = (1 - m_1)(h_{13} - h_{12})$$
$$m_2(h_4 - h_{11}) = (1 - m_1 - m_2)(h_{11} - h_{10})$$
$$m_3(h_5 - h_9) = (1 - m_1 - m_2 - m_3)(h_9 - h_8)$$

from which $m_1$, $m_2$ and $m_3$ can be calculated
Thus the thermal efficiency of the cycle is given by

$$\eta_{\text{cycle}} = \frac{W_{\text{net}}}{q_s} = \frac{W_T - W_p}{q_s} \tag{8.32}$$

---

**Example 8.12**  A steam power plant equipped with regenerative as well as reheat arrangement is supplied with steam to the H.P. turbine at 80 bar and 470°C. For feed heating a part of steam is extracted at 7 bar and the remainder of steam is reheated to 350°C in a reheater and then expanded in L.P. turbine down to 0.035 bar. Determine

   (i)  Amount of steam bled off for feed heating
  (ii)  Amount of steam in L.P. turbine
 (iii)  Heat supplied in boiler and reheater

(iv) Out put of turbine

(v) Cycle efficiency

(Mumbai Univ. Summer 1998, winter 2000)

*Solution*

Refer to Fig. 8.36 (a), (b). From Mollier chart enthalpies values at state points 1, 2, 3, & 4 are taken as mentioned below

$$h_1 = 3330 \text{ kJ/kg}, h_2 = 2725 \text{ kJ/kg},$$

$$h_3 = 3165 \text{ kJ/kg}, h_u = 2245 \text{ kJ/kg}.$$

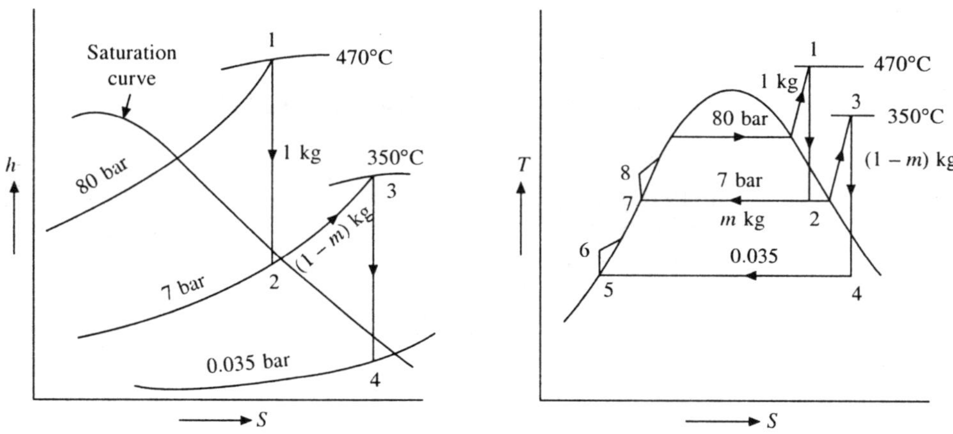


**Fig. 8.36(a), (b)   h-s and T-S diagram**

Neglecting pump work i.e. $h_5 = h_6$, $h_7 = h_8$



From steam table $h_5 = 111.8$ kJ/kg, $h_7 = 697.1$ kJ/kg
energy balance of heater

$$mh_2 + (1 - m)\, h_5 = 1 \times h_7$$

or, Heat lost by steam = Heat gained by feed water

$$m(h_2 - h_7) = (1 - m)(h_7 - h_5)$$

or, $m(2725 - 697.1) = (1 - m)(697.1 - 111.8)$
or, $2027.9\, m = 585.3 - 585.3$ m
$\therefore m = 0.2239$ kg/kg of steam

Amount of steam bled off for feed heating = 0.2239 kg per kg of steam  *Ans.*
Amount of steam in L.P. turbine

$$= 1 - m$$

$$= (1 - 0.2239) = 0.7761 \text{ kg/kg of steam}  \textit{Ans.}$$

Heat supplied in boiler and reheater

$$= (h_1 - h_7) + (h_3 - h_2)(1 - m)$$

$$= (3330 - 697.1) + (3165 - 2725)(1 - 0.2239)$$

$$= 2632.90 + 341.48 = 2974.38 \text{ kJ/kg of steam}  \textit{Ans.}$$

Output of turbine

$$= (h_1 - h_2) + (h_3 - h_4)(1 - m)$$

$$= (3330 - 2725) + (3165 - 2245)(1 - 0.2239)$$

$$= 1319.012 \text{ kJ/kg}$$

$$\therefore \text{ cycle efficiency } \eta_{\text{cycle}} = \frac{1319.012}{2974.38} \times 100$$

$$= 44.34\%  \textit{Ans.}$$

**Example 8.13**   A steam turbine equipped with single stage reheating and regenerative feed heating operates under the following conditions:
Initial steam pressure = 100 bar
Initial steam temperature = 530°C
Pressure at which steam is reheated in reheater = 20 bar
Temperature after reheater = 510°C
Feed temperature at outlet from low pressure feed heater = 170.4°C
Exhaust pressure of steam from low pressure turbine stage = 0.055 bar
steam bled off before reheated = 4% of steam entering turbine
Total steam bled off at all the different stages of extraction including the one before heater = 21% of steam entering the turbine
Isentropic efficiency of H.P. turbine stage = 83%
Isentropic efficiency of I.P. and L.P. stages = 88%
Efficiency of alternator = 0.98%
Neglecting heat losses and pressure losses due to reheating, determine

  (a)  The heat rate in kJ/kW hr
  (b)  The total volume rate of flow of steam passing through the exhaust in m³/sec if the electrical output is 1,00,000 kW.

(Mumbai Univ. Winter 1998)

Solution
Refer to Fig. 8.37 (a), (b)

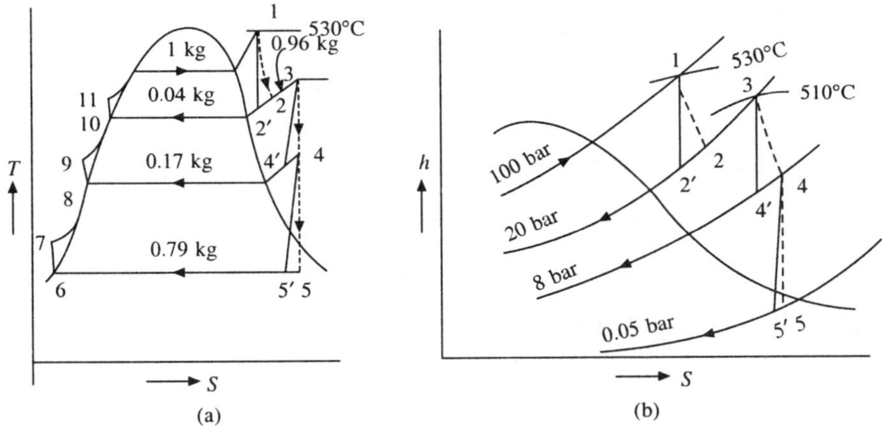

**Fig. 8.37**  *T-S* and *h-s* diagram

At temperature 170.4°C, from steam table saturation pressure is 8 bar. Let 1 kg be entering the steam turbine. 4% i.e. 0.04 kg is bled off from H.P. turbine and remaining amount i.e. 0.96 kg supplied to reheater. Total amount of bled off steam is 21% of entering. So remaining amount i.e. (0.21 – 0.04) = 0.17 kg bled off from I.P. turbine at 8 bar. According to the conditions of steam state points are shown in *T-S* and *h-s* diagrams. Amount of exhaust steam from L.P. turbine = 1 – (0.04 + 0.17) = 0.79 kg/kg of steam entering.

From Mollier chart

$$h_1 = 3450 \text{ kJ/kg}, \ h_2' = 2970 \text{ kJ/kg}, \ h_3 = 3500 \text{ kJ/kg},$$

$$h_5' = 2295 \text{ kJ/kg}$$

$$\eta_{\text{H.P.}} = \frac{h_1 - h_2}{h_1 - h_2'}$$

or, $$0.83 = \frac{3450 - h_2}{3450 - 2970}$$

or, $$h_2 = 3051.6 \text{ kJ/kg}$$

Knowing this value locate the state point 2 at 20 bar.

$$\eta_{\text{I.P.}} = \frac{h_3 - h_4}{h_3 - h_4'}$$

or, $$0.88 = \frac{3500 - h_4}{3500 - 3200}$$

or, $$h_4 = 3236 \text{ kJ/kg}$$

knowing this value locate the state point 4 at 8 bar.

$$\eta_{\text{L.P.}} = \frac{h_4 - h_5}{h_4 - h_5'}$$

or, $$0.88 = \frac{3236 - h_5}{3236 - 2295}$$

or, $$h_5 = 2407.92 \text{ kJ/kg}$$

Neglecting pump work. Total turbine work

$$W_T = (h_1 - h_2) + 0.96\,(h_3 - h_4) + 0.79\,(h_4 - h_5)$$

$$= (3450 - 3051.6) + 0.96\,(3500 - 3236) + 0.79\,(3236 - 2407.92)$$

$$= 398.40 + 253.44 + 654.18 = 1306.02 \text{ kJ/kg}$$

Heat supplied

$$q_s = (h_1 - h_{11}) + 0.96\,(h_3 - h_2)$$

$$[\because \quad h_{10} = h_{11} = 908.6 \text{ kJ/kg from steam table}]$$

$$= (3450 - 908.6) + 0.96\,(3500 - 3051.6)$$

$$= 2971.86 \text{ kJ/kg}$$

$$\text{cycle efficiency} = \frac{W_T}{q_s} = \frac{1306.02}{2971.86} \times 100 = 43.94\%$$

Specific steam consumption

$$\text{S.S.C.} = \frac{3600}{1306.02} = 2.756 \text{ kg/kW hr}$$

Heat rate = S.S.C. $\times\, q_s$

$$= 2.756 \times 2971.86 = 8190.44 \text{ kJ/kW hr} \quad Ans.$$

$$\text{Turbine output} = \frac{1,00,000}{0.98} = 102040.82 \text{ kW}$$

$$\text{mass flow rate } m = \frac{102040.82}{1306.02} = 78.12 \text{ kg/s}$$

$\therefore$   Mass of exhaust steam through L.P. turbine

$$= 0.79 \times 781.2 = 61.71 \text{ kg/s}$$

At state point 5, specific volume of steam from mollier chart = 27.5 m$^3$/kg

$\therefore$   Total volume flow rate $= 61.71 \times 27.5$

$$= 1697.025 \text{ m}^3\text{/s} \quad Ans.$$

## EXERCISES

### A. Objective Questions

8.1 (i)    In a thermal power plant thermodynamic cycle used is
     (a) Erricson
     (b) Brayton
     (c) Joule
     (d) Carnot
     (e) Rankine

(ii)    The efficiency of a Rankine cycle may be expected to
     (a) Increase with decreasing temperature of heat rejection
     (b) Decrease with decreasing temperature of heat rejection
     (c) Decrease with increasing temperature of heat rejection
     (d) Increase with increasing exhaust pressure
     (e) None of the above.

(iii)    A Rankine cycle working with saturated differs thermodynamically from a carnot cycle only in the fact that
     (a) The carnot cycle can not be used for vapour
     (b) Steam is not the working substance for carnot cycle
     (c) There are more than two sources of heat in Rankine cycle
     (d) Heat is supplied to the water at temperatures below the maximum temperature of the cycle
     (c) Non of the above

(iv)    Bleeding of steam
     (a) Increases the thermodynamic efficiency
     (b) Reduces the temperature stresses in the boiler
     (c) Prevents the condensation of $SO_2$ gases on economiser
     (d) All of the above
     (e) None of the above

(v)    Bleeding of steam
     (a) Increases the cost of plant
     (b) Reduces the work done per kg of steam
     (c) Both (a) and (b)
     (d) None of the above

(vi)    Reheating
     (a) Increases thermal efficiency
     (b) Increases work done per kg of steam
     (c) Increases dryness fraction of steam
     (d) Reduces blade erosion
     (e) All of the above
     (f) None of the above

(vii)    The efficiency of steam turbines may be improved by
     (a) Reheating of steam
     (b) Regenerative feed heating
     (c) Binary vapour plants
     (d) Any of the above methods
     (e) None of the above methods

### Answers

| | | | |
|---|---|---|---|
| (i)   e | (ii)  a | (iii) d | (iv) d |
| (v)   c | (vi)  e | (vii) d | |

## B. Theoretical Problems

8.1  Explain why the Rankine cycle rather than carnot cycle is used as a standard of reference for the performance of steam plants?

8.2  What are the difficulties experienced in the practical use of carnot cycle in steam power plants?

8.3  What should be the properties of the working fluid so that Rankine cycle approaches the carnot cycle?

8.4  Sketch a schematic diagram of a steam power plant and explain the various processes of Rankine cycle on $T$-$S$ and $p$-$v$ diagrams.

8.5  Represent the modified Rankine cycle on $T$-$S$ and $p$-$v$ diagrams. Explain why modified Rankine cycle rather than Rankine cycle is used in steam engine.

8.6  Discuss with the help of $T$-$S$ diagram, the effect of the variation of the following variables on efficiency and power output:

(a)  Inlet pressure,

(b)  initial temperature with inlet pressure maintaining constant and

(c)  condenser pressure.

8.7  What are the advantages of reheat cycle over Rankine cycle?

8.8  State briefly the advantages of a regenerative feed heating in steam power cycle.

8.9  What are the advantages of using superheated steam in a steam power plant?

8.10  What are the difficulties encountered in use of ideal regenerative feed heating system?

## C. Numerical Problems

8.1  A steam engine operates on ideal carnot cycle using dry steam at 17.5 bar abs. The exhaust takes place at 0.07 bar into a condenser. Assuming that the expansion and compression are isentropic and liquid enters the boiler as saturated liquid, find:

(a) efficiency of operating cycle and (b) power developed by the engine if steam consumption is 20 kg/minute.  *(Ans.* = 35.1 percent 235.8 kW).

8.2  Saturated steam at 18 bar abs. enters the turbine of a steam power plant and expands to a condenser pressure of 0.8 bar. Determine the carnot and Rankine efficiencies.  *(Ans.* 23.7 percent, 21.7 percent)

8.3  A big pressure turbine working on the Rankine cycle uses superheated steam at 270°C and condenser pressure of 0.5 bar. The turbine efficiency is 85 percent. Determine the maximum pressure at which steam is to be supplied to the turbine if the moisture in the exhaust not to exceed 7 percent

*(Ans.* 33 bar).

8.4  Dry saturated steam at 15 bar absolute is supplied to a steam turbine. The exhaust takes place at 1.1 bar. Determine the followings.

(a) Rankine efficiency, (b) steam consumption per kW. hour if the efficiency ratio is 0.65, (d) carnot efficiency for the given pressure limit using steam as a working fluid; (d) If the exhaust pressure is reduced to 0.2 bar abs. by introducing a jet condenser, find the percentage increase in Rankine efficiency and percentage decrease in specific steam consumption.

*(Ans.* (a) 18.8 percent (b) 12.43 kg/kW. hr (c) 20.3 percent (d) Percentage increase in $\eta$ = 28.6 percent, Percentage decrease in S.S.C. = 34.4 percent).

8.5  Steam at a pressure of 10 bar abs. is supplied to a prime mover and the exhaust takes place at 0.2 bar. Find the Rankine efficiency in the following cases:

(a) steam is dry saturated at inlet, (b) steam is 0.9 dry at inlet and (c) steam has degree of superheat by 50°C at inlet. Neglect pump work.  *(Ans.,* 24.6 percent, 24 percent, 24.9)

8.6  In a steam power plant, steam is supplied to the turbine at 10 bar and 50°C degree of superheat. Exhaust takes place at 0.08 bar. Determine:

(a) heat supplied, (b) heat rejected to condenser, (c) pump work, (d) net work, (e) work ratio, (f) Rankine cycle efficiency and (g) S.S.C. Assume isentropic expansion of steam

8.7  An ideal reheat cycle utilizes steam as the working fluid. Steam at 100 bar, 400°C is expanded in the high pressure turbine to 16 bar. After this, it is reheated to 380°C and is then expanded in the low pressure turbine to a pressure of 0.6 bar. Determine the thermal efficiency of the cycle  *(Ans.* 32.2 percent).

8.8  In an ideal reheat cycle, steam at 60 bar, 400°C is supplied from a high pressure boiler to a high pressure steam turbine. After it is expanded in the turbine to 10 bar, the steam is reheated under constant pressure to 400°C. The steam is then expanded in low pressure turbine to 0.05 bar. Assuming isentropic expansion and compression in the turbines and in the feed pump, determine per kg of steam (a) the work done by the H.P. and L.P. turbines, (b) pump work, (c) overall thermal efficiency and (d) S.S.C.
   (*Ans.* (a) H.P. turbine work = 420.83 kJ/kg. L.P. turbine work = 991.98 kJ/kg (b) 6.07 kJ/kg, (c) $\eta_{ev}$ = 40 percent (d) SSC = 2.554 kg/kW. hr)

8.9  A reheat cycle operating between 30 bar and 0.04 bar has a superheat and reheat temperature of 450°C. The expansion in the H.P. turbine takes place till the steam is dry saturated and then the reheat takes place. Neglecting pump work, determine the ideal cycle efficiency.                    (*Ans* 38.7 percent).

8.10  In a steam power plant, steam is admitted to H.P. turbine at 84 bar and 370°C. After expansion in the H.P. turbine to 10.5 bar, the steam is removed and reheated to 315°C, upon the completing expansion in L.P. turbine, the steam is exhausted to 0.04 bar. Find the efficiency of the cycle. Find the efficiency of the cycle with and without reheating. In each case you may assume that isentropic efficiency of H.P. turbine is 80 percent and that of L.P. turbine is 75 percent. Comment on the values obtained in the two cases.  (*Ans.* 31.6 percent, 31.8 percent)

8.11  A regenerative cycle utilizes steam as the working . . . steam is supplied to the turbine at 42 bar, 430°C and the condenser pressure is 0.07 bar. After expansion in the turbine to 4.2 bar, same of the steam is extracted from the turbine for heating the feed water from the condenser in an open heater. The pressure in the boiler is 42 bar and state of fluid leaving the heater is saturated liquid water at 4.2 bar. Assuming isentropic heat drop in the turbine and pumps, compute the efficiency of the cycle and the total work done by the pumps per kg of steam.                    (*Ans.* 39 percent, 5.2 kJ/kg).

8.12  An ideal reheat regenerative cycle operates with steam as the working fluid. Steam enters the turbine at 80 bar, 450°C. Part of steam is extracted from the turbine at 5 bar for heating the feed water in an open type feed heater, while the remainder is reheated by means of the flue gases of the boiler. The reheated steam enters the turbine at 5 bar, 400°C and expands to a pressure of 0.1 bar. Determine the thermal efficiency of the cycle.                    (*Ans.* 40.3 percent)

8.13  In a regenerative feed heating cycle, steam enters the turbine at 20 bar and 350°C. Bleeding takes place at 4 bar. The blede steam just comes out saturated. This bled steam heats the condensate in an open heater to its saturation temperature at bleeding pressure. The rest of the steam in the turbine expands to a condenser pressure of 0.1 bar. Assuming the turbine efficiency to be the same before and after bleeding, find: (a) the turbine efficiency and the steam quality at the exit of the last stage:
   (b) the mass flow of bled steam per kg of steam flow at turbine inlet, (c) power output per kg of steam flow per second; and (d) overall cycle efficiency.
   (*Ans.* (a) $\eta_T$ 79.8 percent, $x$ = 0.904, (b) 0.162 kg/kg (c) 717.7 kW (d) $\eta$ = 28.3 percent)

8.14  A steam engine receives dry steam at 20 bar and exhaust takes place at 1.2 bar. The release pressure is 3 bar. Determine (a) efficiency of modified Rankine cycle, (b) theoretical loss of work per kg of steam due to incomplete expansion and (c) Rankine efficiency.
   (*Ans.* 18.5 percent, 85.41 kJ/kg, 22.8 percent)

8.15  A reciprocating steam engine working on modified Rankine cycle takes steam at 10 bar abs. and 240°C and expands to a release pressure of 1.2 bar abs. and exhausts at 0.2 bar. The steam consumption of the engine is 10.8 kg/kW hr. Find: (a) the indicated thermal efficiency, (b) efficiency of modified Rankine cycle and (c) Carnot cycle efficiency between the same pressure limit.
   (*Ans.* 12.39 percent, 21.5 percent, 26.3 percent)

8.16  In a modified Rankine cycle, steam is supplied at a pressure of 12 bar abs. and 200°C. It expands is entropically to 2 bar abs and then released at constant volume to exhaust pressure of 1 bar abs. using steam table only, determine:
   (a) Work done per kg of steam, (b) efficiency of modified Rankine cycle and (c) heat rejected to condenser per kg of steam.                    (*Ans.* 402.7 kJ/kg, 16.76 percent, 1957.84 kJ/kg.).

8.17  A mercury cycle is superposed on the steam cycle operating between the boiler outlet condition of 40 bar, 400°C and the condenser temperature of 40°C. The heat released by mercury condensing at 0.2 bar is used to impart the latent heat of vaporization to the water in the steam cycle. Mercury enters the mercury turbine as saturated vapour at 10 bar. Compute (a) kg of mercury circulated per kg of water, and (b) the efficiency of the combined cycle. The property, values of saturated mercury are given below

| $P$ (bar) | $t(°C)$ | $h_f$ | $h_g$ | $S_f$ | $S_g$ | $v_f$ | $v_g$ |
|---|---|---|---|---|---|---|---|
| | | (kJ/kg) | | (kJ/kg k) | | (m³/kg) | |
| 10 | 515.5 | 72.23 | 363.0 | 0.1478 | 0.5167 | $80.9 \times 10^{-6}$ | 0.0333 |
| 0.2 | 277.3 | 38.35 | 336.55 | 0.0967 | 0.6385 | $77.4 \times 10^{-6}$ | 1.163 |

8.18   A binary vapour plant uses mercury between a temperaute range of 540°C and 205°C. The mercury is dry and saturated at the higher temperature limit. The steam cycle works between 17.35 bar supply pressure and 73.66 mm of $H_g$ of vacuum in condenser. The supply temperature of steam is 370°C. The feed water is raised to 200°C in the economiser and is evaporated to dry steam in the mercury condenser and is superheated by the gases. Assuming isentropic expansion in mercury and steam turbines and using the property values given in the following tables, find:

(a)   Mass of mercury circulated per kg of steam generated
(b)   The work done by mercury and steam per kg separately
(c)   The ideal efficiency of the plant

| Temperature | $h_f$ | $h_{fg}$ | $h_g$ | $S_f$ | $S_{fg}$ | $S_g$ |
|---|---|---|---|---|---|---|
| °C | | (kJ/kg) | | | (kJ/kg k) | |
| 540 | 72.6 | 294.1 | 366.7 | 0.151 | 0.421 | 0.572 |
| 205 | 29.0 | 301.5 | 330.5 | 0.08 | 0.6235 | 0.7035 |

8.19   A binary vapour cycle operates on mercury and steam. Saturated mercury vapour at 4.5 bar is supplied to the mercury turbine, from which it exhausts at 0.04 bar. The mercury condenser generates saturated steam at 15 bar which is expanded in a steam turbine to 0.04 bar. (a) Find the overall efficiency of the cycle (b) If 50,000 kg/hr of steam flows through the steam turbine, what is the flow through the mercury turbine? (c) Assuming that all processes are reversible, what is the useful work done in the binary vapour cycle for the specified steam flow? (d) If the steam leaving the mercury condenser is superheated to a temperature of 300°C in a superheater located in the mercury boiler, and if the internal efficiencies of the mercury and steam turbines are 0.85 and 0.87 respectively. Calculate the overall efficiency of the cycle. The properties of saturated mercury are given below.

(*Ans.* (a) 53%, (b) $59.35 \times 10^4$ kg/hr (c) 28.5 mW (d) 46.2%)

| $P$ (bar) | $t(°C)$ | $h_f$ | $h_g$ | $S_f$ | $S_g$ | $v_f$ | $v_g$ |
|---|---|---|---|---|---|---|---|
| | | (kJ/kg) | | (kJ/kg k) | | (m³/kg) | |
| 4.5 | 450 | 62.93 | 355.98 | 0.1352 | 0.5397 | $79.9 \times 10^{-6}$ | 0.068 |
| 0.04 | 216.9 | 29.98 | 329.85 | 0.0808 | 0.6925 | $76.5 \times 10^{-6}$ | 5.178 |

8.20   Steam at 17 bar and 208°C is supplied to a steam turbine. Two feed heaters are used at 3.4 bar and 0.6 bar respectively. The exhaust pressure of the steam is 0.06 bar. The stage (or isentropic) efficiency between 17 bar and 3.4 bar is 70% and is 65% in the other two stages. Assuming that the temperature of feed water is raised to that of bled steam temperature and neglecting pump work and other losses, determine (a) the mass of bled steam at each heater per kg of steam supplied to the turbine, (b) work done per kg of steam supplied to the turbine, and (c) the thermal efficiency of the cycle.

8.21 In a reheat-regenerative steam cycle, steam enters the H.P. turbine at 80 bar, 480°C and expands to 7 bar. The steam is then reheated to 440°C before entering the L.P. turbine, where it expands to the condenser pressure of 0.08 bar. Steam extracted from h.p. turbine at 20 bar is fed to a closed feed water heater through a trap. Steam extracted from the L.P. turbine at 3 bar is also fed into the open feed-water heater, from which the total flow is pumped into the steam generator. The net power output of the cycle is 100 MW. Assuming ideal process, determine (a) the cycle efficiency (b) the rate of steam generation in kg/hr.

$\qquad$ (*Ans.* (a) 43%, (b) $2.8 \times 10^5$ kg/hr)

8.22 Steam at 20 bar and 300°C is supplied to a steam turbine. The steam expands isentropically in the H.P. turbine to 8 bar and then it is reheated to its original temperature at constant pressure. The reheated steam expands isentropically in I.P. turbine to 1 bar. Enough steam from exit of I.P. turbine is removed for feed heating and remaining steam expands isentropically in L.P. turbine to the condenser pressure of 0.035 bar. The condensate from feed heater is pumped ahead into the feed water line. Find the efficiency of the cycle and compare it with the efficiency of the Rankine cycle working between the same pressure limits.

# 9

# Boiler

## 9.1 INTRODUCTION

Steam is the most important vapour used as a working substance in steam engines and turbines working on vapour power cycles. The equipment used for producing steam is called steam generators or boiler. Technically speaking, all steam generators are not boilers but all boilers are steam generators. The American society of Mechanical Engineers (ASME) gives the following definition of steam generator: "A combination of apparatus for producing, furnishing or recovering heat together with the apparatus for transferring the heat so made available to the fluid being heated and vaporised."

A steam generator covers the whole unit encompassing furnace, superheater, reheater, boiler or evaporator, economiser, air preheater and water-wall tubes along with various auxiliaries such as pulverizers, stokers, dust collectors, burners, fans, ash handling equipment, and chimney or stack. The boiler or evaporator is that part of steam generator which consists only of the containing vessel and convection heating surfaces where phase change (or boiling) occurs from liquid (water) to vapour (steam), essentially at constant pressure and temperature. Minimum capacity of boiler should not be less than 22.4 litres. However, the term "boiler" is traditionally used to mean the whole steam generator.

The boiler consists of a drum called shell in which water is contained. The thermal energy released due to combustion of fuel (may be solid, liquid or gaseous) in the combustion chamber is used to evaporate water and to produce steam at a desired pressure and temperature. The steam thus generated is used for:

(a) Power generation: Mechanical work or electrical power may be generated by expanding steam in steam engine or steam turbine.

(b) Heating: The steam is utilised for heating the residential and industrial buildings, in winter season and for producing hot water for hot water supply.

(c) Industrial Processing: Steam is used in textile industries sugar mills and other chemical industries.

## 9.2  ESSENTIALS OF A GOOD BOILER

We may down the following requisites of a good boiler for efficient generation of steam.

1. It should be capable of producing maximum amount of steam with minimum fuel consumption with minimum attention and minimum initial and operating cost.
2. It should occupy less space and be safe in working.
3. It should be capable of quick starting and should be able to meet any rapid fluctuation of load.
4. It should be strong enough against temperature stresses and strains.
5. All the components and parts should be easily accessible for inspection and repair purposes.
6. It should have a well designed combustion chamber in order to have easy ignition and complete and efficient combustion of fuel in the furnace to reduce fuel consumption.
7. It should be equipped with test gauges, safety valves and other mountings for control and safety.
8. It should have a constant and thorough circulation of water throughout the boiler.
9. The construction of all the components should be such that they may be easily transported.
10. It should have its heating surfaces, as near as possible, at right angle to currents of heated gases for better heat transfer.
11. It should have minimum frictional power loss for the flow of hot gases, water and steam.
12. There should be no deposition of mud and other forgeign particles on the heated surfaces.
13. It should have only minimum joints and those too should be away from direct flame.
14. It should confirm to the safety regulation, as laid down in the "Boiler Act".

## 9.3  CLASSIFICATION OF BOILERS

Boilers are mainly classified according to the followings.

1. Relative Position of hot gases and water.
   (a) Fire tube boiler. In fire tube boilers, hot gases pass through tubes which are surrounded with water. There may be single tube as in the case of Lancashire boiler or there may be a bank of tubes as in a locomotive boiler, and cochran boiler. Due to their simplicity and because of small capacity requirements of individual users, these boilers are generally used in many industries like paper sugar and chemical industries producing process steam.
   (b) Water tube boiler: The tubes contain water and the hot gases produced by combustion of fuel flow outside. A bank of water tubes are connected through the header to the boiler drum. Steam is generated inside the tube and collected in the drum. The boiler drum is not exposed to hot gases at all. Large capacity boilers are water tube boilers. Examples are:
      (i) Babcock and Wilcox boiler which has usually straight but inclined tubes which connect the inlet and outlet headers.
      (ii) Sterling boiler which has multi bent tubes to connect the drum and headers.
2. Pressure of Stream. Boilers generating steam at a pressure above 80 bar are called high pressure boilers. The high pressure boilers are Babcock and Wilcox, Lamont, Velox and Benson etc. The Boilers producing steam at pressure below 80 bar are called low pressure boilers. Example are: Cochran, Cornish, Lancashire and Locomotive boilers.

3.  Method of Firing.

    (a)  Internally fired boiler. The furnace region in which combustion takes place is located inside the boiler shell and is completely surrounded by water cooled surfaces. This method of firing is used in Lancashire, Locomotive and scotch boilers.

    (b)  Externally fired boiler. The combustion chamber is provided outside the boiler. This method of firing has the advantage that furnace region is simple and can be constructed separately. Example of such system is Babcock and Wilcox boiler.

4.  Method of Circulation of Water.

    (a)  Natural circulation. Low pressure boilers generally use natural circulation of water where water circulation takes place by natural convection currents set up due to density difference.

    (b)  Forced circulation: At high steam pressure the density difference between steam and water becomes very low due to which natural circulation is slowed down. In such cases a pump is used to force water through the tubes. Some times forced circulation is also adopted in low pressure boiler for rapid evaporation.

5.  Position and Number of drums. Single or multidrum may be positioned longitudinally or crosswise.

6.  Nature of service to be performed.

    (a)  Stationary or Land boilers. The boilers which are stationed at a particular place are called stationary boilers. All the thermal power plants have stationary boilers.

    (b)  Mobile boilers: The boilers which move with the service to be performed are mobile boilers. Examples are locomotive and marin boilers.

7.  Number of gas passes. The flue gases may flow through a single pass or through a multi passs.

8.  Nature of draught.

    (a)  Natural or chimney draught. When the air enters the combustion chamber by natural pressure difference to combustion furnace and hot gases after transferring thermal energy to water comes out through the chimney without any fan or blower, we call it natural draught.

    (b)  Artificial draught: When air is forced over the furnace bed by artificial fan, the draught is called forced draught. Power plants boilers use artificial draught system.

9.  Fuel used: This classification is according to the type of fuel used by the boiler. Electrical energy or nuclear energy may also be used for generating steam.

10.  Material of construction of boiler shell: Depending upon the material of the construction of the boiler shell. In this group boilers are either cast iron, boiler or steel boiler. Power generating boilers are steel boilers.

## 9.4   BOILERS PARTS

The followings are the details of a boiler.

1.  Shell. It is the main component of a boiler. It is a container usually cylindrical in shape. It contains water and steam at high pressure. In once through boiler, boiler shell is completely removed. The steam generated in the tubes is directly sent to engine or steam turbine.

2.  Furnace. It is the combustion space where combustion of fuel takes place. It may have a grate to burn coal or a burner to atomise and burn liquid fuel. Sufficient space is provided for efficient combustion.

3. Water flow path. Water flow path is the path through which water flows and receives heat from the hot gases of combustion. The frictional loss should be minimum and treated water is used for generating steam in order to eliminate corrosion problem.

4. Gas Flow Path. Heat gas path is arranged in both the boilers, fire tube and water tube boilers such that maximum heat exchange takes place from hot gases to water. The boiler efficiency largely depends upon gas flow and water flow path.

5. Steam Path: In most of the boilers steam is taken from the top of the boiler shell to eliminate water particles being carried with the steam. In order to reduce water particles being carried with the steam, steam is generally taken through steam separater in case of large boiler.

6. Mountings. For safety and control of the boiler, certain fittings are mounted on the body of the boiler. These are called mountings. Important mountings are: (a) water level indicator (b) safety valve (c) pressure gauge (d) blow off cock (e) fusible plug (f) steam stop valve and (g) feed check valve.

7. Accessories. To increase the efficiency of a boiler, certain components are attached to the boiler. These fittings are called accessories. Important accessories are air preheater, economiser, superheater, feed pump, steam injector, antipriming pipe, steam seperator, steam strap.

8. Draught Equipment. The equipment required to supply the required amount of air, for proper combustion of fuel is called draught equipment. Draught may be achieved by either providing a tall chimney or by forced fan or steam jet.

## 9.5   DISTINCTION BETWEEN FIRE TUBE AND WATER TUBE BOILERS

The following are the differences between the fire tube and water tube boilers:

| *Fire Tube Boiler* | *Water Tube Boiler* |
|---|---|
| 1. Hot gases (products of Combustion) flow inside flows the tubes, the water remains outside, within the shell. | 1. Hot gases flow outside the tube, while water inside. |
| 2. The boiler shell is large to accommodate the tubes also and has to withstand the large steam pressure, hence <br>(a) thickness of the shell will be larger <br>(b) Construction of shell is difficult and costlier. <br>(c) Operation of the larger vessel at high pressure is more risky. | 2. The largest part is the drum which is of smaller size. Therefore only smaller drum is subjected to larger pressure. |
| 3. This boiler can not be used at very high pressure. | 3. Suitable for very high pressure also (about 200 bar). |
| 4. Maximum generating capacity of these boilers is about 9000 kg/hr. | 4. Generating capacity is much larger. |
| 5. Difficult to transport because of large size of the shell. | 5. Since the various parts can be separated easily and hence easy to transport. |
| 6. Compact and occupies less space. | 6. Large and occupies more space. |
| 7. It is internally fired boiler and therefore combustion space can not be altered according to load fluctuation. | 7. It is externally fired and therefore furnace area can be easily altered to meet the fuel requirement. |

*(Contd.)*

| *Fire Tube Boiler* | *Water Tube Boiler* |
| --- | --- |
| 8. Slow rate of steam generation renders the fire tube boilers un-suitable for steam power plants. | 8. Fast steam generation rate and hence suitable for steam power plants. |
| 9. Failure in feed water supply for some time does not cause damage to boiler as it contains large quantity of water. | 9. Failure in feed water supply even for short period is liable to make the boiler over heated. |
| 10. Due to less combustion space combustion efficiency is low. | 10. Due to large combustion space, combustion efficiency is very high. |
| 11. Heating surfaces are not so effective. | 11. Heating surfaces are more effective because hot gases flow at right angle to the direction of flow of water. |
| 12. Initial cost is low. | 12. Initial cost is high. |

## 9.6  BOILER MOUNTINGS

Fittings and devices which are necessary for the safety and control are known as boiler mountings. The mountings form an integral part of the boiler and are mounted directly over the boiler shell. Fittings which are essential from the safety point of view are as follows:

(1)  Water level indicaters (2) Safety valves (3) Fusible plug.

Control fittings are: (1) Pressure gauge (2) Steam stop Valve (3) Feed check valve (4) Blow off cock (5) Man hole and mud box. Brief description of the mountings is given below:

1. Water level Indicators: There are two such indicators mounted on a boiler to ascertain the level of water inside the boiler shell. If the water level falls below a certain minimum value, the operator should make an arrangements either to supply feed water or to stop firing the fuel.

2. Safety Valves: There are two safety valves on a boiler shell so that at least one of them is working at a time. The function of the safety valves is not to allow steam pressure to rise more than the design value. Automatically they open if steam pressure becomes more than the design value and some steam leaks out producing a hooking sound. They are mounted on the top of the boiler shell.

3. Feed Check Valve: It is a one way valve and is used to control the supply of feed water to the valve. It allows the feed water to enter into the boiler shell but prevents any high pressure water of boiler from escaping through it.

4. Fusible Plug. Each boiler furnace is provided with one fusible plug. It is used to protect the boiler from over heating due to low water level in the shell. Fusible plug is made of a fusible alloy, melting point of which is much less than that of the boiler shell. If the boiler gets over heated, fusible plug melts and water from the boiler shell comes down over the fire grate and extinguishes the fire. It is located at the bottom of the shell.

5. Pressure Gauge: It is an instrument which records the inside pressure of the boiler. The pressure recorded is gauge pressure.

6. Steam Stop Valve. It is used to regulate the flow of steam from the boiler through the steam pipe to the steam engine or steam turbine.

7. Blow off Cock: This valve is used to empty the boiler shell for cleaning the mud and other sediments collected in the boiler shell for long operation of the boiler. It is opened after taking shut down of the boiler.

8. Man hole and Mud Box: Man hole is a large opening through which a man can enter into the boiler shell for cleaning and inspection purposes. Mud box is replaced at the bottom of the boiler to collect mud. Blow off cock is connected to this mud box.

## 9.7 BOILER ACCESSORIES

Fittings or devices which are provided to increase the thermal efficiency of the boiler and help in smooth working of the plant are called accessories. Boiler accessories are: 1. Economiser 2. Air Preheater 3. Superheater 4. Feed pump 5. Steam injector 6. Anti Priming pipe 7. Steam seperator, and 8. Steam trap.

1. Economiser: The greatest item of heat loss in boiler plant is that of the heat carried away by the flue gases through the chimney slack. Some of the heat carried away by flue gases is used to heat up the feed water. Hot gases are allowed to pass over the various tubes in which feed water is circulated. This feed water gets slightly heated before it is fed into the boiler shell. A saving of 10 percent (5 to 6°C rise in temperature of feed water) is often achieved by the use of economiser.

2. Air preheater: The function of air preheater is similar to economiser. It recovers some portion of waste heat from the flue gases coming out of the economiser and uses it to heat the air before it passes into the furnace. Hot air is used for combustion in the furnace increasing the efficiency of combustion. By using air preheater the overall efficiency will increase by about 10%.

   There are two types of air preheaters

   (a) recuperative and (b) regenerative.

   In a recuperative air preheater, the heat from the flue gases is transferred to air through a metallic medium. In a regenerative air preheater, air and flue gases are made to pass alternatively through the matrix. When the hot gases pass through the matrix it receives heat from the hot gases and transfers it to the cold air.

3. Superheater: A superheater is one of the most important accessories of the boiler. The function of the superheater is to increase the temperature of saturated steam without increasing its pressure. Super heated steam increases the efficiency of the power plant at the same time problem of corrosion and erosion of the turbine blades are minimised. Superheaters are the tubes in which dry or wet steam is allowed to pass for superheating by the hot gases of the furnace.

4. Feed pump: The pressure inside a boiler is very high. Therefore, the pump required to pump feed water into the boiler shell is called feed pump. Small quantity of feed water or make up water can also be fed to the boiler shell by steam injector.

   Generally, the pressure of feed water is 20% more than that in boiler.

A feed pump may be of centrifugal type or reciprocating type. But a double acting reciprocating pump is commonly used as a feed pump these days. The reciprocating pumps are run by the steam from the same boiler in which water is to be fed. These pumps may be classified as simplex, duplex and triplex pumps according to the number of pump cylinders. The common type of pump used is a duplex feed pump for supplying feed water continuously.

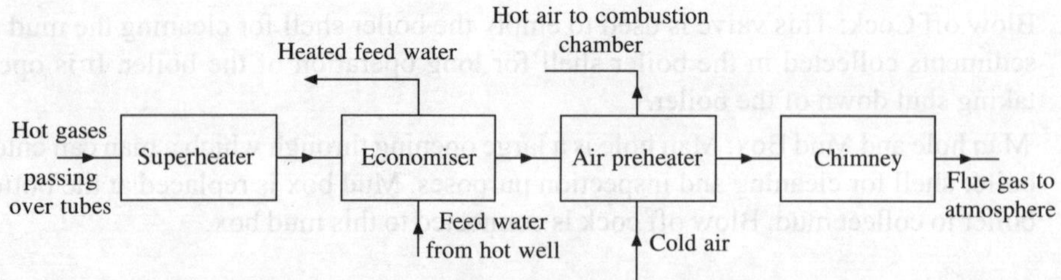



**Fig. 9.1    Position of economiser and air preheater**

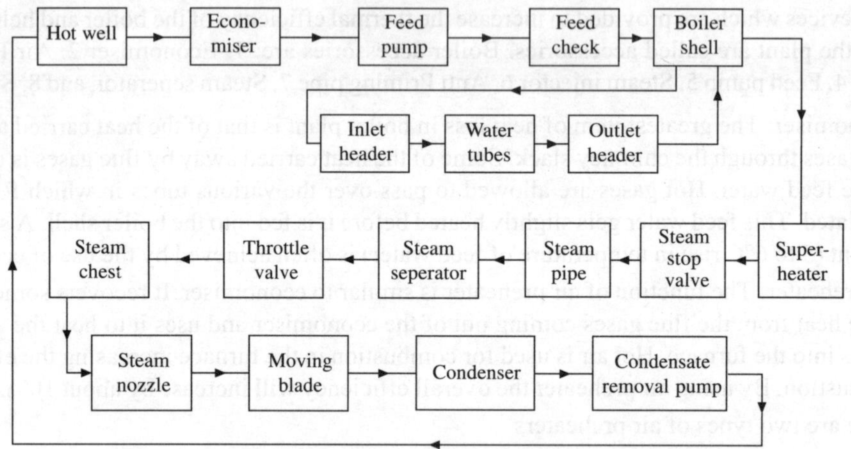



**Fig. 9.2    Water flow and steam flow path**

5. **Steam Injector:** When the feed pump fails due to any reasons, then steam injector, injects or delivers feed water to the boiler drum, by the use of steam of same boiler of vertical and locomotive type. It consists of three jets—steam jet, suction jet, combining and delivery jet.

The steam expands in the steam nozzle where its pressure drops but its velocity increases. As the steam passes across space between steam and suction nozzles a vacuum is developed in the suction chamber. The water is drawn into the suction chamber from the feed tank. The high speed steam jet forces the water along with the steam into the combining and delivery jet. Here the steam is condensed and the delivery jet receives the water and the condensate. The delivery jet is so designed that a considerable amount of kinetic energy of the jet changes into pressure energy which is sufficient to force the water in the boiler. In the begining water and steam escapes out through the overflow value as long as sufficient pressure is not developed in the discharge jet. There will be no overflow when the steam and water are in proper ratio. For the injector to act properly there is a definite relation between the quantity of steam and water entering the injector. When the ratio of steam to water is greater, the pressure of steam is lower. This arrangement is shown in Fig. 9.3. The injector though limited in use to small boilers has the following advantages:

    (i)  occupies minimum space (ii) comparatively cheap in first cost and every day maintenance

    (iii)  since it supplies hot feed water, the boiler parts are free from thermal stresses.

When the requirement of steam is suddenly increased, the boiler has to supply and raise steam production which leads to carry water particles with it. Before being supplied to heat engine or steam turbine, water particles have to be removed and for that we require following devices.

6. An antipriming pipe is a cast iron box which is fitted in the steam space of the boiler shell and under the mounting block on which the steam stop valve is to be bolted. When the steam with water particles passes through the perforations made in the upper half of the antipriming pipe, the heavier water particles separate out and are collected at the bottom of the pipe. The water thus collected is later on drained to the boiler through the holes which are made at the ends of the pipe.

7. Steam separator: It is also known as drier. It also serves the same purpose as that of anti priming pipe. It is located in between boiler and steam engine or steam turbine and placed nearer to latter one to supply steam without water particles resulting due to partial condensation of steam in connecting pipe. The steam is led to pass through a chamber in which it strikes the baffle plate, gets deflected and made to change its direction. The water particles are thrown out because of greater inertia. The water collected at the bottom of the chamber is drained out at regular intervals. The arrangement is shown in Fig. 9.4.

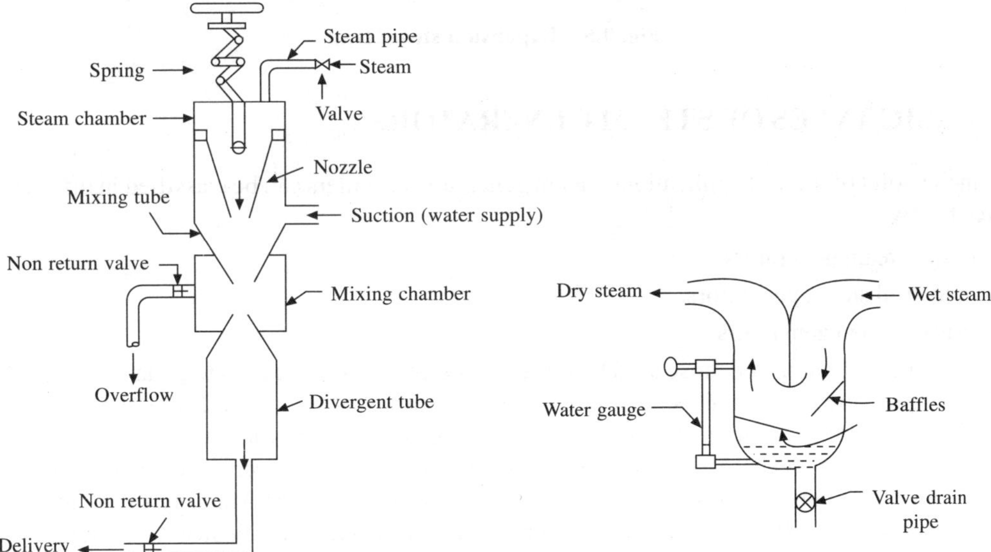


**Fig. 9.3   Steam injector**

**Fig. 9.4   Steam separator**

8. Steam trap: The function of steam trap is to drain away automatically the condensed steam from the steam pipes, steam jackets and steam separators without permitting any steam to escape. The steam condensed in the steam pipe flows to the steam trap by gravity. The steam traps are classified as follows:

   1. Expansion or thermostatic type

   2. Bucket or float type

Figure 9.5 shows an expansion steam trap. A hollow spring tube $B$ of nickel steel contains a liquid which becomes a gas at temperatures above the lowest temperature that the steam is likely to have. The water enters at $A$, and, being lower in temperature than the steam, it causes the tube B to contract and the valve is opened. When all the water is discharged and steam enters the trap, the increased temperature converts the liquid in the tube B into the gas, the curved tube tends to straighten itself and consequently the valve V is closed. By adjusting the screws the trap can be arranged to discharge either continuously or intermittently. The end $C$ of the tube is held against the adjusting screw by the spring, one end of which presses against the lugs $L$ fixed to the casing.

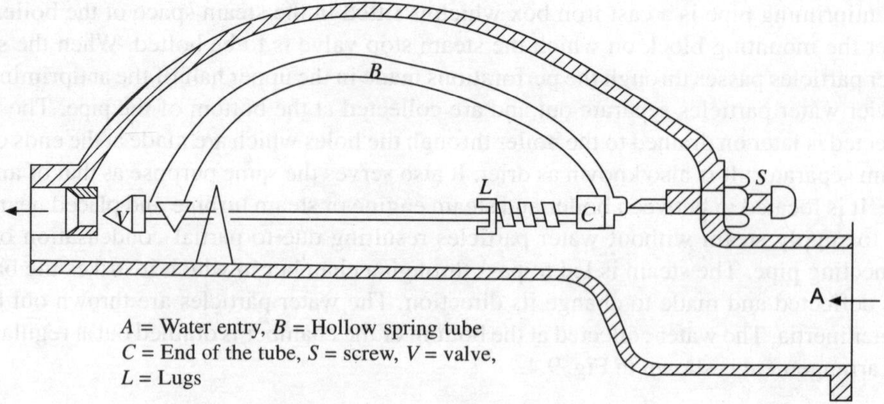



**Fig. 9.5   Expansion steam trap**

## 9.8   BASIC TYPES OF STEAM GENERATORS

From the point of view of applications, steam generators or boilers can be classified in different ways as given below:

- (a)  utility steam generators
- (b)  industrial steam generators
- (c)  marine steam generators

utility steam generators are those used by utilities for electric-power generating plants. Depending on whether the pressure of steam is below or above the critical presssures (221.2 bar), they can be either subcritical or supercritical units. The subcritical steam generators are water tube-drum type and they usually operate at between 130 and 180 bar steam pressure. The supercritical steam generators are drumless once-through type and operate at 240 bar presssure or higher. Majority of the utility steam generators are of the 170–180 bar water tube-drum variety. Modern utility steam generators produce power ranges from 125 to 1,300 MW. Pulverized coal is mainly the fuel used.

Industrial steam generators are those used in process industry like sugar, paper, jut and so on, and institutions like hospital, commercial and resendential building complexes. They are smaller in size. They can be pulverized coal fired. They operate at pressures ranging from 5 to 105 bar. Normally, they do not produce superheated steam. They supply wet or saturated steam, and sometimes even only hot water.

Marine steam generators are used in many marine ships and ocean liners driven by steam turbines. They are usually oil-fired. They produce superheated steam at about 60–65 bar and 540°C.

## 9.9   DESCRIPTION OF SOME IMPORTANT BOILERS

1. *Cochran Boiler:* It is a vertical multi-tube fire tube low pressure boiler commonly used for small capacity steam generation. It is made in different sizes for evaporating capacities ranging from 150 to 3,000 kg per hour and working pressure upto 20 bar. The Fig. 9.6 shows the different components of the boiler. It consists of a cylinderical shell surmounted by a hemispherical crown plate. The fire box is also of the hemispherical form pressed from a single plate in a special hydraulic machinery. The grate is placed at the bottom of the furnace and the ash pit is located

below the grate as shown in the figure. The coal is fed into the grate through the fire door and ash formed is collected in the ash pit located just below the fire grate and then it is removed manually. The furnace and the combustion chamber are connected through a short pipe. The back of the combustion chamber is lined with fire bricks. The hot gases from the combustion chamber flow through the horizontal tubes immersed in water. The tubes are generally 6.25 cm external diameter and 165 to 170 in number. The gases passing through the fire tubes transfer a large portion of the heat to water by convection The flue gases after transferring maximum heat to water leaves through the chimney to the atmosphere. The spherical top and the spherical shape of the combustion space is for least material for the plate to have maximum volume. The hemispherical crown of the shell gives maximum strength to withstand high pressure and spherical shape of the fire box is to absorb maximum heat due to radiation from the furnace.

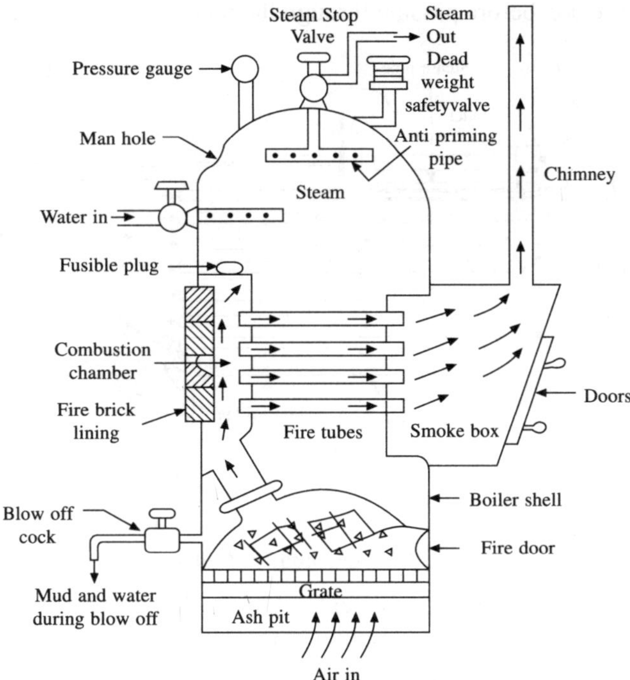


**Fig. 9.6  Cochran boiler**

Coal or oil can be used as fuel in this boiler. If coal is used as fuel, no grate is provided but the bottom of the furnace is lined with fire bricks. Oil burners are fitted at suitable locations below the fire door.

A man-hole near the top of the crown of the shell is provided for cleaning. In addition to this, a number of hand holes are provided around the outer shell at the bottom for cleaning purposes. The smoke box is provided with the doors for cleaning the interior of the fire tubes. The air flows through the grate due to draught produced by means of the chimney. A damper is provided near the base of the chimney to control the discharge of the flue gases through the chimney thereby, it controls the flow of air into the furnace. The outstanding features of this boiler are:

1. It is very compact and requires less floor area.
2. Any type of fuel can be used with this boiler.
3. It is most suitable for small capacity plant.

4.  It gives about 70 percent thermal efficiency with coal firing and 75 percent with oil.

5.  It is provided with all the boiler mountings but accessories like superheater, economiser and air preheaters are not provided.

2. *Locomotive boiler:* Locomotive boiler is an internally fired multi-tube fire tube boiler. The arrangement of the various parts is shown in the Fig. 9.7. In this boiler, the shell or drum axis is horizontal and fire box is separate. It is attached with the main drum on the one end. On the other end of the drum is the smoke box. The hot gases of combustion from fire box pass to the smoke box through the tubes immersed in water. The steam is collected on the top of the shell. A large door is provided for access to the smoke box and tubes for cleaning.

It may be noted that in the fire box the grate shopes downwards towards the front. This helps in deflecting the products of combustion causing them to some in contact more thoroughly with the whole heating surface of the fire box before passing through the tubes.

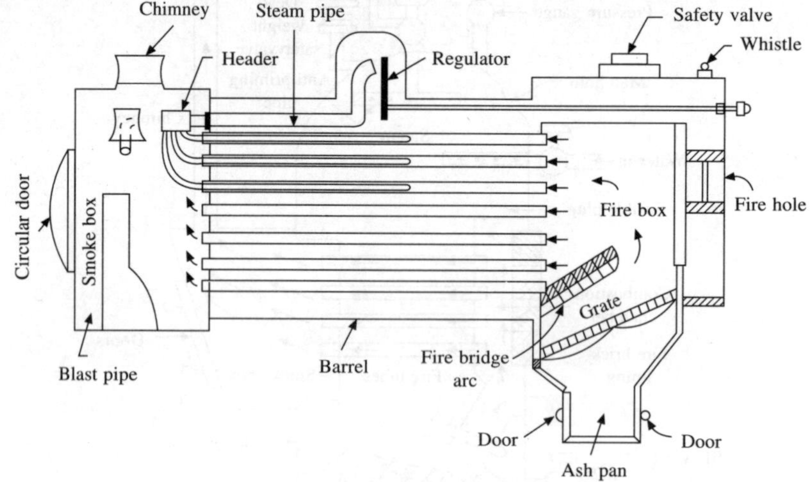


**Fig. 9.7   Locomotive boiler**

When the regulator is opened, the wet steam enters and goes to inlet headers of the superheater and after passing through the superheater tubes it returns to the outlet header of the superheater, from which it is led to the engines.

The boiler draught is provided by expanding the exhaust steam of steam engine in the nozzles. Vacuum created the outlet of the nozzle draws the hot gases through the fire tubes. Finally at the top of the smoke box, a diverging section is provided to increase the pressure of the hot gases more than atmosphere to exhaust into it. A blower is also provided for use when necessary steam supply to engine is shut off.

3. *Babcock and Wilcox Boiler:* Babcock and Wilcox boiler is a water tube high pressure boiler. This is used for large generation rate at high pressure. The boiler has three main components; steam and water drum, water tubes and furnace.

The water and steam drum is made of steel plates welded or riveted together. The end cover plates are removable. Just below the drum, water tubes are arranged. These tubes are connected with each other and with the drum by vertical passages called downtake and uptake headers. The downtake header is joined at the bottom to mud box where blow off cock is connected. The water from drum is carried through the down take header to the tubes. As water flows through the tubes, it is converted into steam resulting in decrease in density. Low density steam is collected in the drum through uptake headers. Circulation of water is natural due to density difference. A super heater is placed between the drum and water tubes. During the first turn of the hot gases, the gases are passed over the superheater tubes and super heated steam is taken through steam stop valve. When the steam is being raised from cold boiler the superheater is filled with water to the drum. Water level, which is essential to prevent the overheating of superheater tubes. The super heater remains flooded with water until steam reaches the working pressure. Once the rated pressure in the boiler is reached then the water from the superheater is drained and steam is fed to it for superheating purpose. Economiser and air preheaters are provided in the boiler to increase its thermal efficiency. Smaller diameters tubes are used for generating steam at high pressure. The different mountings have been shown in the Fig. 9.8.

4. *Scotch Marine Fire Tube Boiler:* This is the most common type of marine fire tube boiler for steaming capacities up to about 7000 kg per hour and a pressure of about 20 bar. It is a self contained, horizontal, multitubular, internally fired, natural circulation fire tube boiler. It provides a large area of heating surface for the space occupied. It consists of a cylindrical water shell, housing one or more cylindrical shell having diameter ranges from 2.5 to 3.5 metres. Fig. 9.9 shows the details. The shell is made of two plates rivietted by butt joint or welded steel. The shell may contain one to four furnaces. The furnaces are corrugated and are made of steel. They are about 1 m to 1.2 m in diameter. The front end of the furnace is attached to a flanged opening in the front head end. The rear end is connected to the front wall of the combustion chamber into which the furnace opens. The furnace is entirely surrounded by water and therefore does not require a fire brick lining. A thin ash plate rests below the grate at the bottom of the ashpit to facilitate the removal of the ashes at regular intervals.

There are a number of fire tubes passing from front plate of the internal combustion chamber to the front plate of the shell. All the fire tubes are immersed in water. There is a provision for uptake passage from the boiler front to the stack. Three manholes are placed on the front end, one near the top, and one between the centre and each side furnace. Various mountings and accessories are provided in the boiler.

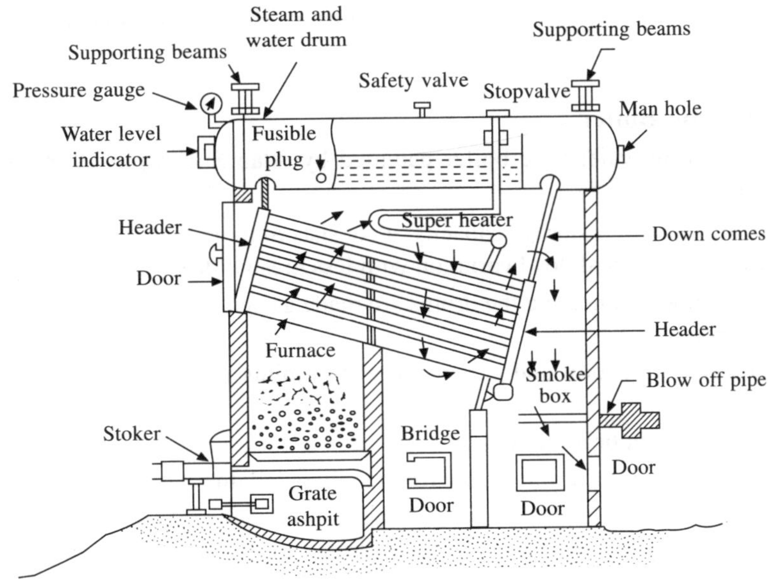


**Fig. 9.8  Bobcock Wilcox boiler**

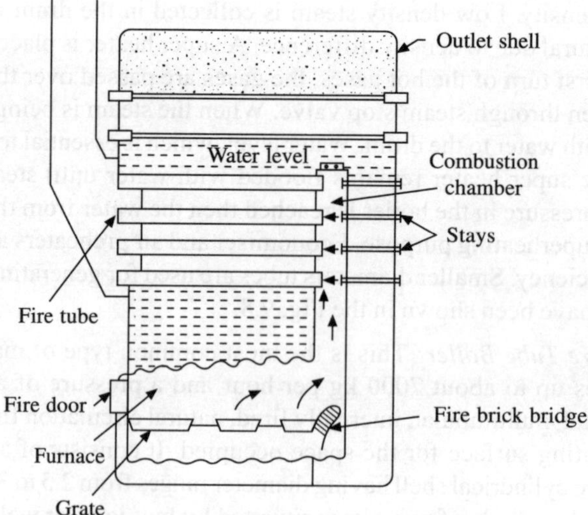


**Fig. 9.9    Scotch Marine fire tube boiler**

When this type of boiler is adopted for stationary work and its combustion chamber is lined with refractory, it is called "dry back" type boiler. The hot gases caused by burning of fuel on the grate are deflected upwards by the fire brick bridge and pass to the combustion chamber. From the combustion chamber the gases enter the fire tubes and travel to the front portion of the shell. After this they enter the uptake and are discharged through the stack. During their travel from the grate to the stack, the hot gases give heat to the water and steam is produced.

The water circulation is natural, caused by convection currents. The advantages of this boiler are (i) compactness (ii) Minimum amount of brick work (iii) low head room. Disadvantages are (i) During starting, circulation difficulties are experienced due to the large diameter of the shell (ii) there is a practical limit to capacity and pressure (iii) cleaning and inspection are difficult and (iv) fuel charge requirements are difficult to meet.

## 5. Stirling Bent Tube Boiler

This multitube, multidrum water tube boiler patented by Alan Stirling in 1988 has the following distinguishing features:

(i)  For more steam and water storage capacity, it consists of more than two steam drums in number thereby it can meet the peak load variation with less presure drop.

(ii)  The number of tubes connecting the drums are made as bent tubes rather than straight tubes for the following reasons:

    (a)  due to free contraction and expansion at the joints near the drum thermal stresses are minimised.

    (b)  tubes replacement becomes easier

    (c)  the tubes can enter the drums in approximately radial direction.

Figure 9.10 shows the principle of working of a stirling boiler with two water drums (mud drum) and three steam drums connected in the series. The water drums are 10 cm to 25 cm larger in diameter than steam drums. The steam drums are usually suspended in the steel structures and these are connected by the bent tubes to the waterdrum. The steam drum contains water and steam and all are connected in series by the tubes above and below the water lines called as steam and water equalising tubes.

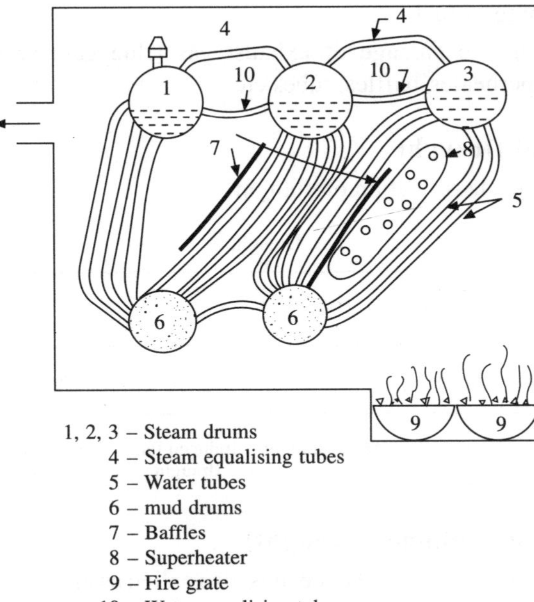


1, 2, 3 – Steam drums
4 – Steam equalising tubes
5 – Water tubes
6 – mud drums
7 – Baffles
8 – Superheater
9 – Fire grate
10 – Water equalising tubes

**Fig. 9.10   Stirling bent tube boiler**

Steam equalising tubes are used to equalise the pressure and water equalising tubes will maintain same level of water in the steam drums. Feed water is first fed into drum-1 which then passes down to the water drum through the rear bank of tubes. The mud particles are collected in this drum. Water from here is circulated to remaining upper drums 2 and 3. The flue gases from the fire grate rise above and they pass over the bent tubes. The flue gases are deflected by the baffles for their proper distribution and finally they are discharged through the flue gas outlet. The wet steam generated in the drums is passed through the superheater tubes to superheat the steam as and when required.

The stirling boilers can use solid, liquid or gaseous fuels economically. These are lighter in construction and flexible in operation. These boilers can generate steam up to a pressure of 50 bar and 450°C temperature up to a capacity of 200,000 kg/hr.

## 9.10   BOILER DRAUGHT(DRAFT)

### 9.10.1   Introduction

We have discussed about the formation of steam. As we know the rate of formation or generation of steam depends upon the rate at which the fuel is burnt. Burning of fuel depends upon area of grate and rate of air supply to it. So finally formation of steam depends mainly due to a pressure difference by virtue of which air should flow into the furnace and go out into atmosphere through chimney. So this created and maintained pressure difference above and below the fire grate for the supply of fresh air continuously and to exhaust the products of combustion from the combustion chamber of boiler is known as boiler draught.

Amount of boiler draught further depends upon the followings

(i) nature of fuel used and depth on the grate

(ii) design of combustion chamber

(iii) method of burning of fuel

(iv) the resistance in the passage of exhaust gas (flue gas) created due to air preheater, economiser, superheater, baffles, tubes etc.

### 9.10.2   Classification of Draught

The following chart gives the classification of draught.

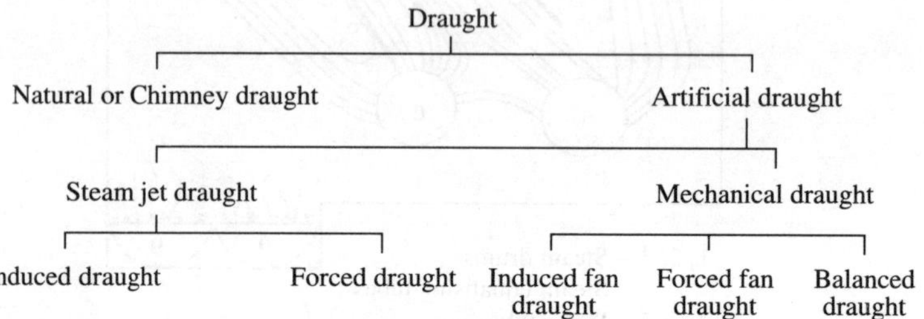


### 9.10.3   Natural Draught (Chimney Draught)

This is produced with the help of chimney, hence, it is also called as chimney draught. The difference in static pressure by the chimney is produced due to difference in weight of column of hot gases inside the chimney and the weight of equal column of cold air outside the chimney. Since the density of cold air outside is more than the density of hot gases in the chimney, it follows that pressure just outside the level of chimney base would be more than the pressure inside the chimney at its base. This pressure difference is responsible for the flow of air from the surroundings to the furnace grate.

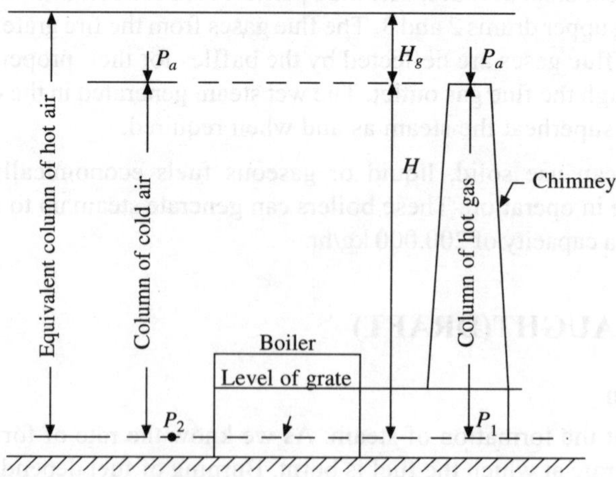


**Fig. 9.11   Chimney draught**

Let    $P_a$ = Atmospheric pressure, N/m$^2$

$\omega_a$ = Specific weight of atmospheric air = $\rho_a g$, N/m$^3$

$\omega_g =$ Average specific weight of hot gas

$T_N =$ Temperature corresponding to $0°C = 273$ k (at N.T.P. Condition)

$T_a =$ Temperature of atmospheric air, k

$T_g =$ Mean temperature of hot gases inside the chimney, k

$V_N =$ Volume of air at N.T.P Condition, $m^3$

$H =$ Height of chimney measured above from grate level, $m$

$m_a =$ Mass of air supplied per kg of fuel

$(m_a + 1) =$ Mass of flue gases, kg/kg of fuel

$\rho_w =$ Density of water, 1000 kg/$m^3$

The pressure at the grate level on the chimney side

$P_1 = P_a$ i.e. atmospheric pressure on the top of the chimney + pressure due to column of hot gases of height $H$

or ,
$$P_1 = P_a + w_g H \tag{9.1}$$

pressure on the grate level outside the chimney

$P_2 = P_a +$ column due to cold atmospheric air

$$= P_a + w_a H \tag{9.2}$$

$\therefore$ Net pressure difference causing the flow through the combustion chamber

$$= \Delta_p = P_2 - P_1$$
$$= H(w_a - w_g), \text{ N/m}^2$$

$\therefore$ Natural draught $\Delta p = H(w_a - w_g)$ $\tag{9.3}$

Natural draught has very small value and therefore it is measured in mm of water column indicated by the manometer which is placed at the base of the chimney shown in Fig. 9.12.

1 mm of water head $= 1 \times 10^{-3}$ m of water

head $= 9.806 \text{N/m}^2$

The U-tube of manometer is filled with water. Due to static pressure difference between the pressures exerted on two open legs of U-tube, the level in the tube is altered to balance the pressures. The difference in level of water column (mm) between the two legs measures the draught as $h_w$. It may be noted that the draught gauge measures only the static pressure difference at the same level. It does not measure the absolute pressure. Hence, if the draught gauge is connected to the top of the chimney it will read zero, and also at the grate level in ash pit, it will read zero in case of the natural draught system.

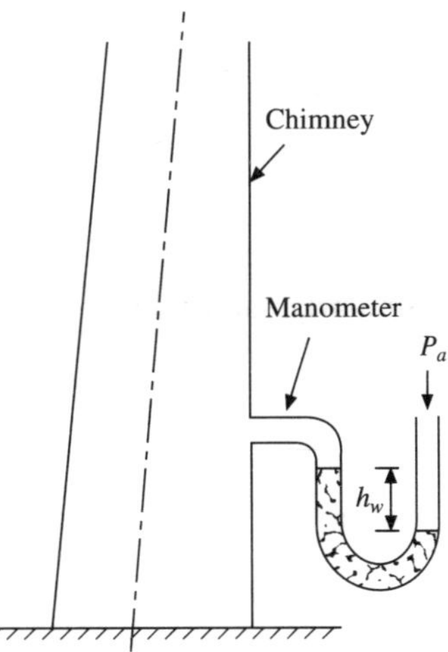


**Fig. 9.12   Natural draught in mm of water colum**

### 9.10.4 Determination of Height and Diameter of Chimney

We know that,  $\qquad C + O_2 = CO_2$   (9.4)

$$1 \text{ vol.} + 1 \text{vol.} = 1 \text{ vol.}$$

and

$$2\,H_2 + O_2 = 2\,H_2O$$
$$2 \text{ vol.} + 1 \text{ vol.} = 2 \text{ vol.}$$   (9.5)

From above two equations it is observed that one volume of $O_2$ from air used for combustion of carbon in fuel, produces one volume of gaseous products as $CO_2$; but one volume of $O_2$ from air produces two volumes of steam ($H_2O$) in the gaseous products. Since the hydrogen content in solid or liquid fuel is quite small, so steam formed is neglisible compared to other constituents. $N_2$ and other constituents do not take part in combustion.

Therefore, it may be assumed that volume of air required for combustion is approximately same as the products of combustion.

$\therefore$ Volume of 1 kg of flue gases at 0°C (273 k) = volume of 1 kg of air at 0°C (273 k)

$$v_N = \frac{RT_N}{P_N} = \frac{287 \times 273}{1.01325 \times 10^5} \text{ at N.T.P. Conditions}$$

$$= 0.7734 \text{ m}^3/\text{kg}$$

The pressure difference being very small, having order of maximum of 25 mm of water column, the pressure can be treated as constant at furnace and chimney base for the purpose of the calculation of volumes at higher temperature.

By applying charle's law (at constant pressure) volume of air outside the chimney

$$V_a = V_N \frac{T_a}{T_N} = 0.7734\, m_a \times \frac{T_a}{273}$$   (9.6)

Its density,  $\qquad \rho_a = \dfrac{\text{mass}}{\text{volume}} = \dfrac{m_a}{v_a}$

$$= \frac{273 \times m_a}{0.7734\, T_a\, m_a} = \frac{353}{T_a}$$   (9.7)

Similarly, same volume of hot gases should be occupied inside the chimney at temperature $T_g$,

$$V_g = 0.7734\, ma \, \frac{T_g}{273}$$   (9.8)

Its density,  $\qquad \rho_g = \dfrac{\text{mass}}{\text{volume}} = \dfrac{(m_a + 1)}{V_g}$

$$= \frac{(m_a + 1)}{0.7734\, ma \times \dfrac{T_g}{273}} = \left(\frac{m_a + 1}{m_a}\right) \frac{353}{T_g}$$   (9.9)

So, by using Eq. 9.3, we get

$$\text{natural draught } \Delta p = H(\rho_a \times g - \rho_g \times g) \text{ N/m}^2$$

$$= Hg\,(\rho_a - \rho_g)$$

$$= Hg\left[\frac{353}{T_a} - \left(\frac{m_a + 1}{m_a}\right)\frac{353}{T_g}\right]$$

$$= 353Hg\left[\frac{353}{T_a} - \frac{(m_a + 1)}{m_a}\frac{1}{T_g}\right] \text{ N/m}^2 \tag{9.10}$$

If natural draught is expressed in terms of water column $h_w$, then

$$h_w = \frac{\Delta P}{\rho_w g} = \frac{353Hg\left[\dfrac{1}{T_a} - \dfrac{(m_a + 1)}{m_a}\dfrac{1}{T_g}\right]}{1000 \times g} \; m \text{ of water}$$

or,

$$h_w = \frac{1000 \times 353Hg\left[\dfrac{1}{T_a} - \dfrac{(m_a + 1)}{m_a}\dfrac{1}{T_g}\right]}{1000 \times g} \; \text{mm of water}$$

or,

$$h_w = 353H\left[\frac{1}{T_a} - \frac{(m_a + 1)}{m_a}\frac{1}{T_g}\right] \text{ mm of water} \tag{9.11}$$

If the boiler draught is represented in terms of meter of hot gas $H_g$, then

$$\Delta p = w_g H_g$$

$$= \rho_g \times g \times H_g$$

$$= \frac{353}{T_g}\frac{(m_a + 1)}{m_a} Hg \times g, \text{ N/m}^2$$

then

$$H_g = \frac{\Delta p}{\dfrac{353}{Tg}\dfrac{(m_a + 1)}{m_a} \times g}$$

By taking $\Delta p$ from Eq. (9.10)

$$= \frac{353Hg\left[\dfrac{1}{T_a} - \dfrac{(m_a + 1)}{m_a}\dfrac{1}{T_g}\right]}{\dfrac{353}{T_g}\dfrac{(m_a + 1)}{m_a}\,g}$$

or,

$$H_g = H\left[\frac{m_a}{(m_a + 1)} \times \frac{T_g}{T_a} - 1\right] m \text{ of hot gas column} \tag{9.12}$$

*Note*: Eqs. (9.10) and (9.11) only represent the theoretical value of the draught and is known as static draught. The actual value of draught is less than the theoretical value due to the following reasons:

   (i)  The effect of frictional resistance offered to the passage of air flow through baffles, bends of tubes, chimney, grate etc.

  (ii)  Weather condition

 (iii)  Construction material of chimney i.e. brick machinery, steel, reinforced concrete, reinforced plastic etc.

## Diameter of Chimney

The theoretical velocity of hot gas flow across the chimney is given by

$$V_g = \sqrt{2g(H_g - h_f)}$$

Where $h_f$ = head loss due to friction in the chimney ($m$ of hot gases)

or,

$$V_g = \sqrt{2g}\,\sqrt{H_g}\,\sqrt{1 - \frac{h_f}{H_g}} = k\sqrt{H_g} \tag{9.13}$$

where

$$k = \sqrt{2g\left(1 - \frac{hf}{H_g}\right)}$$

which depends upon the nature of the chimney

$$k = 0.825 \text{ for brick chimney}$$
$$= 1.1 \text{ for steel chimney}$$

The rate of flow of gases flowing through any cross section of the chimney is given by

$$m_g = \rho_g A\, V_g \tag{9.14}$$

$\therefore$

$$A = \frac{m_g}{\rho_g V_g} = \frac{\pi}{4} D^2$$

$\therefore$

$$D = 1.128\sqrt{\frac{m_g}{\rho_g V_g}} \tag{9.15}$$

Where $D$ is the diameter at any cross section of the chimney.

### 9.10.5  Condition for Maximum Discharge Through the Chimney

We have already discussed that the height of hot gas column producing draught is

$$H_g = H\left[\frac{m_a}{(m_a + 1)}\frac{T_g}{T_a} - 1\right]$$

and the velocity of hot gases (flue gases) through the chimney,

$$V_g = \sqrt{2 \cdot g \cdot H_g}\,, \text{ when } hf = 0$$

$$= \sqrt{2g \cdot H \left[ \frac{m_a}{(m_a+1)} \frac{T_g}{T_a} - 1 \right]}$$

We know that density of hot gas is inversely proportional to its temperature, i.e.

$$\rho_g \propto \frac{1}{T_g} \quad \text{or,} \quad \rho_g = \frac{C}{T_g}$$

where $C$ is a constant of proportionality

By using Eq. (9.14), mass rate of flow of

$$\text{hot gas } mg = \frac{C}{T_g} \times A \sqrt{2gH \left[ \frac{m_a}{(m_a+1)} \frac{T_g}{T_a} - 1 \right]}$$

Substituting $C \cdot A = C_1$, another constant, we have

$$m_g = \frac{C}{T_g} \sqrt{2gH \left[ \left( \frac{m_a}{(m_a+1)} \right) \frac{T_g}{T_a} - 1 \right]}$$

Again substituting $C_1 \sqrt{2gH} = C_2$, another constant,

We have

$$m_g = \frac{C_2}{T_g} \sqrt{\frac{m_a}{(m_a+1)} \times \frac{T_g}{T_a} - 1}$$

$$= C_2 \sqrt{\frac{m_a}{(m_a+1)} \times \frac{T_g}{T_a T_g} - \frac{1}{T_g^2}}$$

$$= C_2 \left[ \frac{m_a}{(m_a+1)} \cdot \frac{1}{T_a T_g} - \frac{1}{T_g^2} \right]^{1/2}$$

For given value of $T_a$, $m_a$ and height of chimney $H$, $m_g$ is maximum, when

$$= \left[ \frac{m_a}{(m_a+1)} \cdot \frac{1}{T_a T_g} - \frac{1}{T_g^2} \right] \text{ is maximum i.e.}$$

$$= \frac{d}{dT_g} \left[ \frac{m_a}{(m_a+1)} \cdot \frac{1}{T_a T_g} - \frac{1}{T_g^2} \right] = 0$$

or,
$$= \frac{m_a}{(m_a+1)} \frac{1}{T_a} \cdot \frac{(-1)}{T_g^2} + \frac{2}{T_g^3} = 0$$

or,
$$= \frac{m_a}{(m_a+1)} \cdot \frac{1}{T_a} \cdot \frac{2}{T_g} \quad \text{or,} \quad T_g = \frac{2(m_a+1)}{m_g} \cdot T_a \qquad (9.16)$$

It shows that for maximum discharge the temperature of flue gases is slightly greater than twice the atmospheric temperature.

The draught in mm of water column for maximum discharge is given by

$$h_w = 353\,H\left[\frac{1}{T_a} - \frac{(m_a+1)}{m_a}\frac{1}{T_g}\right]$$

or,

$$h_w = 353\,H\left[\frac{1}{T_a} - \frac{(m_a+1)}{m_a}\cdot\frac{m_a}{2(m_a+1)}\cdot\frac{1}{T_a}\right]$$

$$= 353\,H\left[\frac{1}{T_a} - \frac{1}{2T_a}\right] = \frac{353}{2}H\cdot\frac{1}{T_a}$$

$$= 176\cdot 5\,\frac{H}{T_a}$$

or,

$$hw = 176\cdot 5\,\frac{H}{T_a} \tag{9.17}$$

### 9.10.6  Height of Hot Gas Column Under Maximum Discharge

The height of hot gas column under the maximum discharge condition can be calculated by substituting the value of $T_g$ from Eq. (9.16) in Eq. (9.12), we get,

$$H_g = H\left[\frac{m_a}{(m_a+1)}\times\frac{2(m_a+1)}{m_a}\,T_a\cdot\frac{1}{T_a} - 1\right]$$

or,     $H_g = H(2-1) = H$

or,     $H_g = H$ $\tag{9.18}$

It shows that under the maximum discharge condition of chimney, the drought produced in terms of the hot gas column equals to the height of chimney.

### 9.10.7  Efficiency of a Chimney

With a given height of the chimney, a minimum temperature value is required to produce a given draught. The temperature of the flue gases leaving a chimney, in case of natural draught, is higher than that of flue gases leaving it in case of artificial draught system. Therefore, the heat carried away by the flue gases is more in case of natural draught.

Hence chimney efficiency may be defined as the ratio of the energy required to produce the artificial draught, expressed in metres head or J/kg of flue gas, to the mechanical equivalent of extra heat carried away per kg of flue gases due to the natural draught. This leads to a reduction in a boiler efficiency.

Let   $T_{g1} =$ Absolute temperature of the flue gases leaving the chimney in the natural draught system, $k$

$\quad\ T_{g2} =$ Absolute temperature of the flue gases leaving the chimney in the artificial draught system, $k$

$\quad\ C_{Pg} =$ Specific heat of flue gases in kJ/kg k. Its value may be considered as 1.005 kJ/kgK

$\therefore$ The extra heat carried away by 1 kg of flue gas due to higher temperature required to produce the natural draught

$$= 1 \times C_{Pg} \left( T_{g1} - T_{g2} \right) \text{ kJ} \tag{9.19}$$

The draught pressure produced by the natural draught system in metre of hot gases

$$H_g = H \left[ \frac{m_a}{\left( m_a + 1 \right)} \times \frac{T_g}{T_a} - 1 \right]$$

The maximum energy this lead would give to 1 kg of flue gas

$$m \cdot g \cdot H_g = 1 \times 9.81 \times H \left[ \frac{m_a}{\left( m_a + 1 \right)} \times \frac{T_g}{T_a} - 1 \right] Nm$$

$$= \frac{9.81 H}{1000} \left[ \frac{m_a}{\left( m_a + 1 \right)} \times \frac{T_g}{T_a} - 1 \right] \text{kJ} \tag{9.20}$$

$\therefore$ Efficiency of the chimney

$$= \frac{\dfrac{9.81 \times H}{1000} \left[ \dfrac{m_a}{\left( m_a + 1 \right)} \times \dfrac{T_g}{T_a} - 1 \right]}{C_{Pg} \left( T_{g_1} - T_{g_2} \right)} \tag{9.21}$$

### 9.10.8 Draught Losses

The actual or available draught is less than the theoretical draught obtained by calculations, because of the draught losses viz:

1. The head loss due to frictional resistance offered by the flues, gas passages and fuel bed. Fuel bed resistance depends upon the size of the fuel, bed thickness and combustion rate.
2. The head loss due to bends in gas flow circuit
3. The head loss due to frictional resistance offered by economiser, air preheater, superheater etc.
4. Head equivalent to velocity of gas flow

The draught losses in some boilers and their equipments are given below.

| | |
|---|---|
| Lancashire or Locomotive boiler | = 6 to 8 mm of water |
| Babcock and Wilcox boiler | = 5 to 10 mm of water |
| Stirling boiler | = 12 mm of water |
| Air preheater | = 25 to 50 mm of water |
| Economiser | = 15 to 40 mm of water |
| Superheater | = 4 mm of water |

The total drought loss in a chimney is 20% of the total drought produced by it. Hence, available drought is about 80% of the total static drought.

### 9.10.9  Advantages of Natural Draught

1. Easy to construct
2. Long life of chimney
3. For producing the draught no power is required
4. Maintenance is not required

## Disadvantages

1. Poor efficiency
2. Tall chimney is required
3. Decreases with increase in outside temperature
4. No flexibility to create more draught to take peak loads.

### 9.10.10 Artificial Draught

Natural draught is dependent on the atmospheric conditions and it becomes less when the air temperature outside is high. In bigger power plants, the draught of the order of 25–350 mm of $H_2O$ column is required. For producing this much draught, the chimney height has to be increased considerably, which is neither convenient nor economical. Therefore it becomes necessary to provide some mechanical means to meet the larger draught requirements and the draught so produced called arti icial draught. The artificial draught used in modern power plants must be independent of the climatic conditions. It is provided either by a steam jet called steam jet draught or by a fan called fan draught. For small installations and in locomotives, steam jet draught is used, while mechanical draught is used in central power stations.

### 9.10.11 Mechanical Draught

**Forced draught (Forced fan draught)**

In a forced draught system, a fan or blower is provided as shown in Fig. 9.13 located outside and before the grate which forces the air in the combustion chamber. In the combustion chamber combustion of air and fuel takes place and hot gases are generated.

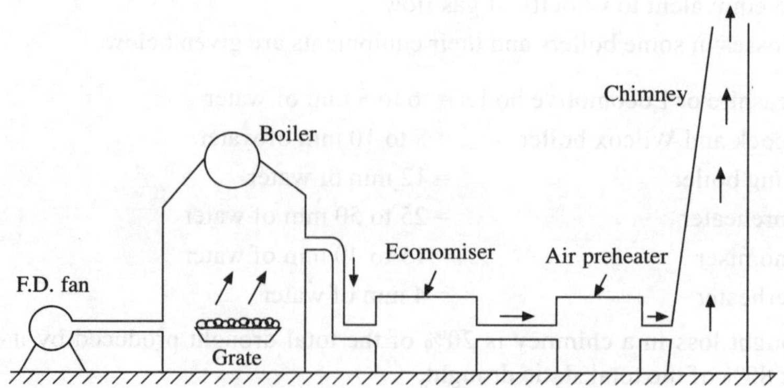


**Fig. 9.13   Forced draught**

These gases are then forced to pass through the air preheater, economiser and then they are exhausted after recovering heat. The pressure created due to fan or blower ranges from 2.5 cm to about 7.5 cm of water. This draught system is known as positive draught system because the pressure of air throughout the system is above the atmospheric pressure.

## Induced draught (Induced fan draught)

If the fan or blower is placed at or near the base of the chimney then the system is known as induced draught system. The fan draws the flue gases from the furnace so that the pressure above the furnace is reduced below the atmospheric pressure as shown in the Fig. 9.14. This induces the surrounding air to flow through the grate. This is similar in action to natural draught system. It is usually employed when the plant uses the economiser and the air preheater.

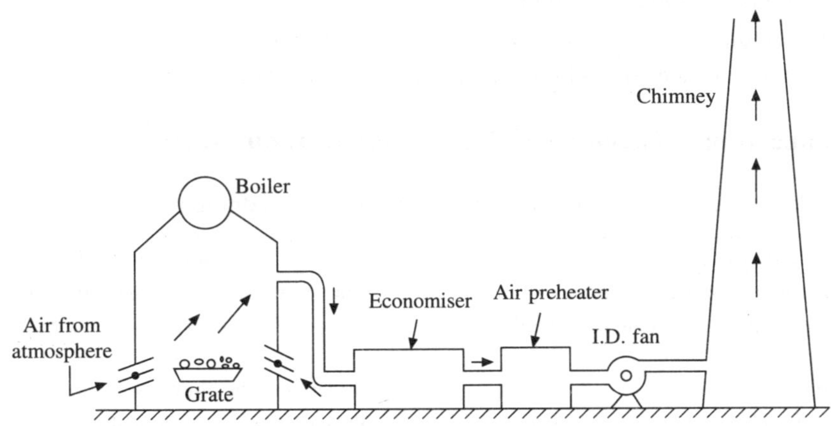


**Fig. 9.14   Induced draught**

## Balanced draught

The use of the combination of I.D. and F.D. fan is mostly preferred instead of forced or induced draught alone. If forced draught alone is used then the furnace can not be opened for firing for inspection because the high pressure gases inside the furnace will try to blow out and there is every chance of blowing out of the fire completely and the furnace may stop. If induced draught fan alone is used, then also furnace can not be opened either for firing or for inspection because the cold air will try to rush into the furnace which reduces the effective draught. To overcome these difficulties balanced draught is used. In this case I.D. fan and F.D fan are provided.

## 9.10.11.1   Advantages and Disadvantages of Mechanical Draught

In modern Power plant mechanical draught is generally used. It has following advantages and disadvantages over the natural draught.

## Advantages

1. It is more economical.
2. It reduces the necessary height of the chimney
3. It allows to burn low grade fuels

4. It increases the evaporative capacity of the boiler due to increased quantity of fuel burnt per m$^3$ of grate area.

5. It is better in control

6. The flow of air through the grate and furnace is uniform

7. The air flow can be regulated according to changing requirements.

8. It reduces the amount of smoke

9. It increases efficiency of the plant

10. It reduces the fuel consumption. About 15% of fuel is saved for the same amount of work

## Disadvantage

1. Initial installation cost is high

2. The running cost is high due to electrical power required to operate fans.

3. The maintenance cost is high due to increased wear and repairs.

### 9.10.12   Comparison Between Forced Draught and Induced Draught

The following table gives the comparison between forced draught and induced draught

| *Forced draught* | *Induced draught* |
|---|---|
| 1. The fan is placed before the fire grate. | The fan is placed after the fire grate. |
| 2. The air is forced into the combustion chamber and <br><br> from atmosphere. | Hot gases are sucked from combustion chamber <br><br> forced into the chimney. |
| 3. Pressure remains above atmospheric inside the furnace. | Pressure remains below atmospheric inside the furnace. |
| 4. Forced draught fan does not require water cooled bearings since it handless the cold air. | Induced draught fan requires water cooled bearings since it handles hot gages. |
| 5. Since the forced draught fan handles cold air only which has less volume, so power required to run it will be less. | Power required to run induced draught fan is more since it handles more volume of hot gas at high temperature. |
| 6. The flow of air through grate and furnace is more uniform. | The flow of air through grate and furnace is less uniform. |
| 7. With forced draught the leakages are outwards, therefore, the care must be taken while opening the fire doors in order to avoid the danger of blow out during the period of fan operation | Such type of problem is not faced with induced draught since all leakages are inwards, but, due to air filteration the available draught is reduced. |

### 9.10.13   Regulations For Chimney Height (With Reference To Ibr)

1 A plan of the boiler house and chimney is to be approved by the chief inspector of Indian Boiler Regulations (IBR).

2 Chimney fabrication drawings are to be approved before fabrication.

3 A plan of 1m = 1 cm on map showing the exact position of furnace, flues, chimney, duly certified by the chartered engineer/owner should be submitted.

4. Substantial structure certificate for the chimney standing that the chimney is self supported/ supported by wire ropes and withstand a wind load of 200 bar is to be submitted.

5. Chimney height minimum required as per IBR is 100 feet = 30.48 m from the firing floor of the boiler.

6. An undertaking for the use of specific fuel is to be submitted by the owner.

7. An undertaking for the height of chimney is to be submitted by the owner.

8. No objection certificate (NOC) is to be obtained from Air force department.

9. Chimney should be painted with Red and White strips of equal width in total 7 strips.

10. Anti Collision lights are to be fitted for sighting in the night and poor visibility.

11. Following details are to be submitted

    (a) Height of own factory building/boiler house
    (b) Height of tallest building within a radius of 300 m around ground level.
    (c) Capacity of boiler.
    (d) Working pressure
    (e) Type of fuel
    (f) Quantity of fuel/hour
    (g) Type of chimney-self supported/supported by wire ropes.
    (h) Material of construction.

## 9.10.14 Steam Jet Draught

It is a simple and cheap method of producing artificial draught. It uses the exhaust steam of non-condensing steam. It may be induced draught or forced draught depending upon where the steam jet to produce draught is located. It is mostly used in locomotive boilers, where the exhaust steam from the engine cylinder is discharged through a blast pipe placed at the smoke box and below the chimney. An induced steam jet draught is shown in Fig. 9.15. When locomotive is stationary, steam from boiler/ non-condensing engine is passed through the nozzle which is located near the smoke box to create draught. It induces a flow of gases through the tubes, ash pit, grate and flues creating a pressure below that of the atmospheric in the smoke box. When the locomotive is running, due to the motion, the air enters through the damper and forces its way through the grate and flues to the smoke box;

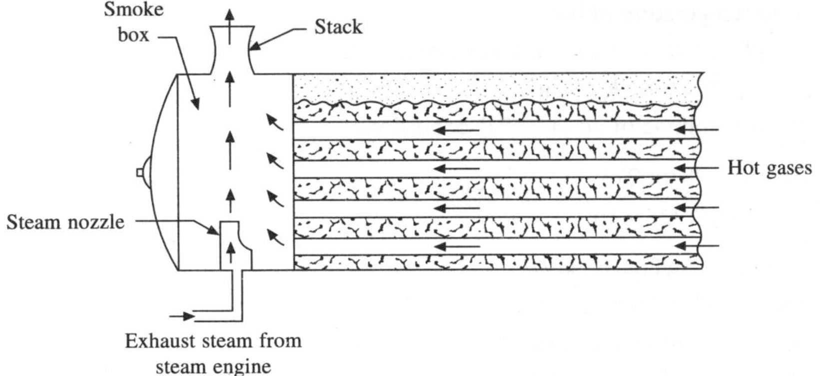


**Fig. 9.15   Steam jet draught of induced type**

but besides this, the exhaust steam from the cylinder can be led to the nozzle in the smoke box, and induced draught created.

In Forced type steam jet draught system, steam from the boiler is throttled by a valve to a pressure of about 1.5 to 2.5 bar gauge and passed through the nozzles projecting in a diffuser pipe. The steam thus emerges out with greater velocity and drags the column of air along with it thus allowing fresh outside air to force through the fuel bed, furnace and flues to the chimney. Forced draught developed by steam jet is also called turbine draught.

If the nozzles are periodically inspected and renewed when eroded, the steam used to creat the draught should not exceed 5% of the output of the boiler.

### 9.10.14.1 Advantages of Steam Jet Draught

(a) It is very simple and economical

(b) It occupies very little or no useful space.

(c) Cheap and plenty of low grade fuels can be used with this system.

(d) Its maintenance cost is nil.

(e) It requires little attention

(f) Draught is automatically adjusted to suit the requirements of the boiler with thin system.

### *Disadvantage of steam jet draught*

(a) It can not be started until high pressure steam is available.

(b) The forced draught is not favoured because the specific heat of the products of combustion increases resulting in more heat loss to flue gases. Moreover increased water column is likely to cause more corrosion especially with fuels rich in sulphur contents.

### 9.10.15 Power Required to Drive a Draught Fan

Forced fan handles air and induced fan handles gas. Volume of gas which is at high temperature, is more. Hence power required by induced fan is more than that by forced fan. Let

$T_a$ = Temperature of atmospheric air

$m_a$ = mass of air supplied per kg of fuel

$T_g$ = Mean temperature of hot gas

$h_w$ = Draught required in mm of water column

$= gh_w(\text{N/m}^2) = 9.81\ h_w(\text{N/m}^2)$

$v_N$ = Specific volume of air at N.T.P. conditions

$$= \frac{RT}{P} = \frac{287 \times 273}{1.01325 \times 10^5} = 0.7734\ \text{m}^3\text{/kg}$$

$m_f$ = Amount of fuel burnt, kg/s

$\eta$ = Mechanical efficiency of the fan

*(i) Power required for forced draught fan*

Mass of air to be handled $= m_a \times m_f$ (kg/s) by applying charle's law (at constant pressure), Specific volume of atmospheric air $(v_a)$:

$$\therefore \qquad \frac{P_N v_N}{T_N} = \frac{P_a v_a}{T_a} \qquad\qquad (\because P_N = P_a)$$

$$\therefore \qquad v_a = v_N \frac{T_a}{T_N} = 0.7734 \frac{T_a}{273} \ \text{m}^3/\text{kg}$$

Total volume of air to be handled by the fan per second:

$$V = (ma \times m_f) \times 0.7734 \frac{T_a}{273} \ \text{m}^3/\text{s}$$

Fan Power, 
$$P_1 = h_w \times v, \frac{N}{m^2} \times \frac{m^3}{s} = \frac{N-m}{s} = \frac{J}{s} = \text{Watt}$$

$$= 9.81 \times N_w \times (m_a \times m_f)\, 0.7734 \frac{T_a}{273}, = \text{Watt} \qquad\qquad (9.22)$$

$$\therefore \qquad \text{Motor Power} = \frac{P_1}{\eta} \qquad\qquad (9.23)$$

(ii) Power required for an induced draught fan

Mass of hot gas to be handled $= (m_a + 1)m_f$, kg/s

Specific volume of hot gases $= 0.7734 \dfrac{T_g}{273}\left(\dfrac{m_a}{m_a + 1}\right),$ m$^3$/kg $\qquad\qquad$ [Refer Eq. 9.9]

Total volume to be handled

$$V = (m_a + 1)\, m_f \times 0.7734 \ \frac{T_g}{273}\left(\frac{m_a}{m_a + 1}\right) \ \text{m}^3/\text{s}$$

$$= m_a \cdot m_f \times 0.7734 \frac{T_g}{273}$$

$$\therefore \text{Fan Power,} \qquad P_2 = h_w \times V$$

$$= 9.81 \times h_w \times (m_a \times m_f)\, 0.7734 \frac{T_g}{273}, \ \text{watt} \qquad\qquad (9.24)$$

$$\therefore \text{Motor Power} \qquad = \frac{P_2}{\eta} \qquad\qquad (9.25)$$

Now, 
$$\frac{\text{Power required to drive induced draught fan}}{\text{Power required to drive induced draught fan}} = \frac{T_g}{T_a} \qquad\qquad (9.26)$$

Since, $T_g$ is nearly 50 to 100% more than $T_a$, induced draught fan requires nearly 50 to 100% more power for the same draught developed.

## 9.11  BOILER PERFORMANCE

### 9.11.1  Introduction

Steam is generated into the boiler at a certain pressure and temperature by using certain grade of fuel located at particular heating surface area. The economic generation of steam by a boiler, therefore, indicates its capacity or its performance. Two important terms called "equivalent evaporation" and "boiler efficiency" are introduced to measure the performance of a boiler.

### 9.11.2  Evaporation Rate

The amount of water evaporated into steam by the boiler in kg per hour at full load is known as evaporative capacity of the boiler. The evaporation rate is expressed on the basis of fuel burnt or grate area or furnace volume with units kg/kg of fuel or $kg/m^2$ hr or $kg/m^3$/hr. It is required to design the grate, furnace, heating surface and other necessary parts of the boiler.

### 9.11.3  Equivalent Evaporation

The performance of an individual boiler may be judged by finding its thermal efficiency. Its value may be used for comparing the performance of various boilers. However, if different boilers are to be compared with respect to the evaporating capacities then the term equivalent evaporation is more meaningful. Different boilers generate steam at different rate at different pressure and of different quality. For comparison purpose if we can imagine that

1  Feed water is available at 100°C

2  Evaporation of water takes place at the same temperature and at standard atmospheric pressure of 1.01325 bar.

3  Steam produced is just dry saturated.

Then under above mentioned conditions, the evaporation rate in kg/hr is called equivalent evaporation. Thus equivalent evaporation may be defined as the rate of generation of steam which would be achieved if the feed water was supplied at 100°C and converts into dry steam at 100°C at standard atmospheric pressure of 1.01325, it requires latent heat of evaporation as 2,256.9 kJ/kg. It is generally writen as "from and at 100°C" for water evaporation.

Let  $m_w$ = mass of water actually evaporated into steam per hour per kg of fuel burnt, kg/kg of fuel

$\quad h_s$ = enthalpy of steam, kJ/kg

$\quad h_f$ = enthalpy of feed water, kJ/kg

$\quad \therefore$ Total heat absorbed by the water at actual conditions = $m_w(h_s - h_f)$

Rate of heat absorbed by water at standard condition = $m_e \times h_{fg}$

$\quad$ Where $m_e$ = equivalent evaporation

$\qquad h_{fg}$ = latent enthalpy of evaporation at standard atmospheric pressure

$\qquad\quad$ = 2256.9 kJ/kg

Then equating the two equations

$$m_e \times h_{fg} = m_w(h_s - h_f)$$

or,

$$m_e = m_w \frac{\left(h_s - h_f\right)}{h_{fg}} = \frac{m_w\left(h_s - h_f\right)}{2256.9} \tag{9.27}$$

The expression $\dfrac{\left(h_s - h_f\right)}{h_{fg}}$ is always greater than unity and is called factor of evaporation or generation factor. The factor of evaporation may be defined as the ratio of heat absorbed by 1 kg of feed water under working conditions to that absorbed by 1 kg of water from and at 100°C (standard conditions).

## 9.11.4  Boiler Efficiency or Thermal Efficiency

Boiler efficiency is defined as rates of actual heat i.e. energy utilised by feed water in converting it into steam in the boiler to the energy released by complete combustion of fuel during the same time. Thus

$$\text{Boiler efficiency} = \frac{\text{Energy absorbed by feed water}}{\text{Energy released by combustion of fuel}}$$

or, $$\eta_{\text{boiler}} = \frac{m_w\left(h_s - h_f\right)}{m_f \times C \cdot V} \tag{9.28}$$

where $m_w$ = mass of water supplied to be converted into steam or mass of steam generated per hour

$h_s$ = enthalpy of steam, kJ/kg

$h_f$ = enthalpy of feed water supplied, kJ/Kg

$m_f$ = rate of fuel consumed, kg/hr

$C.V$ = Calorific value of fuel, kJ/kg

If a boiler consisting of an economiser and superheater, is considered to be a single unit, then the efficiency is termed as overall efficiency of the boiler and it takes into account the efficiency of all three elements i.e. boiler, economiser and superheater.

## 9.11.4.1  Factors Affecting Boiler Efficiency

There are two main categories which affect the boiler efficiency namely: fixed factors and variable factors.

(a)  Fixed factors:
   (i)  Boiler design: shape and size of furnace, arrangement of flue gas and water circulation.
   (ii)  heat recovery equipments; economiser, superheater, air preheater
   (iii)  properties of fuel used
   (iv)  Rate of firing
(b)  Variable factors
   (i)  Humidity and temperature of atmospheric air
   (ii)  excess air fluctuations
   (iii)  Fuel condition at the time of firing
Boiler efficiency is expressed in other form also as

$$= \frac{\text{Fuel heating value/kg} - \text{Losses/kg}}{\text{Fuel heating value/kg}}$$

## 9.11.5  Economiser Efficiency

Feed water gets heated by absorbing extra heat from flue gases being carried by them. The efficiency of economiser is defined as heat absorbed by feed water to that of lost by flue gases.

$$\text{Economiser efficiency} = \frac{\text{heat absorbed by feed water in the economiser}}{\text{heat in flue gases entering the economiser}}$$

$$\text{or} \qquad \eta_{\text{economiser}} = \frac{m_w \, C_{P_w} \, w_2 \, w_1}{m_g \, C_{P_g} \left( t_g = t_a \right)} \tag{9.29}$$

where  $m_w$ = mass of water circulated through economiser

$m_g$ = mass of flue gases circulated through economiser

$t_{w2}$ and $t_{w1}$ = temperature of water leaving and entering the economiser respectively.

$t_a$ = temperature of atmospheric air being supplied to the boiler

$c_{Pw}$ and $c_{Pg}$ = specific heat of water and gases supplied to economiser

$t_g$ = temperature of gas entering the economiser

## 9.11.6  Boiler Power

It is used to measure the capacity of a boiler. According to A.S.M.E boiler power is equivalent to an evaporation of 21.296 kg of water per hour from and at 100°C into dry saturated steam.

$$\text{Boiler power} = \frac{\text{Equivalent evaporation from and at } 100°\text{C per hour}}{21.296}$$

## 9.11.7  Sources of Heat Losses in a Boiler

As we know that heat utilised in the boiler system is always less than heat liberated due to burning of the fuel. The difference between heat liberated in the furnace and heat utilised is known as losses in the boiler system. These losses are categorised as follows:

1.  Heat carried away by flue gases through the chimney-

Heat lost to dry flue gases per kg of fuel = $m_g \times c_{pg} (t_g - t_a)$ (9.30)

where  $m_g$ = Mass of gases formed per kg of fuel

$c_{pg}$ = Mean specific heat of flue gases, kJ/kg k

$t_g$ = Temperature of the flue gases leaving the chimney, °C

$t_\alpha$ = temperature of boiler room, °C

2.  Heat lost due to presence of hydrogen and moisture in the fuel

The reaction $2H_2 + O_2 = 2H_2O$ takes place due to hydrogen present in the fuel.

4 kg Hydrogen liberates 36 kg $H_2O$ (steam)

∴  $H_2$ kg Hydrogen liberates $9H_2$ kg (steam)

Considering, $m$ = mass of moisture present in the fuel, kg

$h_f$ = specific enthalpy of water at boiler room temperature, kJ/kg

$c_{ps}$ = Specific heat of superheated steam, kJ/kg k

Total amount of steam produced due to hydrogen and moisture present in the fuel = $(9H_2 + m)$. Assuming that the steam is in superheated state in the flue gases at atmospheric pressure and at flue gas temperature.

$\therefore$ Heat lost to steam in the flue gases per kg of fuel burnt

$$= (9\,H_2 + m) \times [h_{\text{sup}} - hf]$$
$$= (9\,H_2 + m) \times [2676 + c_{ps}\,(t_g - 100) - h_f]\ kJ \qquad (9.31)$$

where $\qquad 2676 = h_g$ at 1.01325 bar (std. atm. press.)

$\qquad\qquad 100 = t_s$ (saturation temperature at 1.01325 bar)

3. Heat lost due to incomplete combustion of carbon to carbon monoxide

$$C + O_2 = CO_2$$
$$2C + O_2 = 2CO$$

From above two chemical equations, mass of carbon burnt to $CO = \dfrac{CO \times C}{CO_2 + CO}$ $\qquad (9.32)$

where $\quad CO_2 = $ Percentage by volume in flue gas

$\qquad CO = $ Percentage by volume in flue gas

$\qquad C = $ Fraction of carbon in 1 kg of fuel

$\therefore$ Heat lost due to incomplete combustion of carbon per kg of fuel

$$= \dfrac{CO \times C}{CO_2 + CO} \times C \times V\ kJ/kg\ of\ fuel \qquad (9.33)$$

where C.V. = calorific value of carbon burnt to CO.

4. Heat lost due to unburnt fuel which is collected on the ashpit

$$\text{Heat lost} = m_1 \times C \cdot V \qquad (9.34)$$

where $\quad m_1 = $ mass of unburnt fuel per kg of fuel

$\qquad C.\,V = $ Calorific value of carbon

5. Heat lost due to radiation and convection from furnace to outside

The loss is calculated by substracting the above losses from heat liberated due to burining of fuel.

## 9.11.8  Boiler Plant

The black diagram of a boiler plant is shown in Fig. 9.16 which is to be used while considering performance trial of boiler. Steam and water path along with flue gas have been shown in this figure while in Fig. 9.17 details are given for former one.

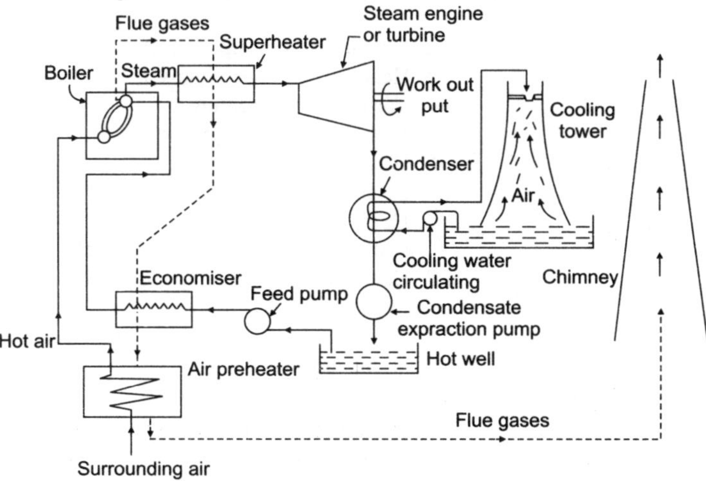


**Fig. 9.16   Block diagram of a boiler plant**

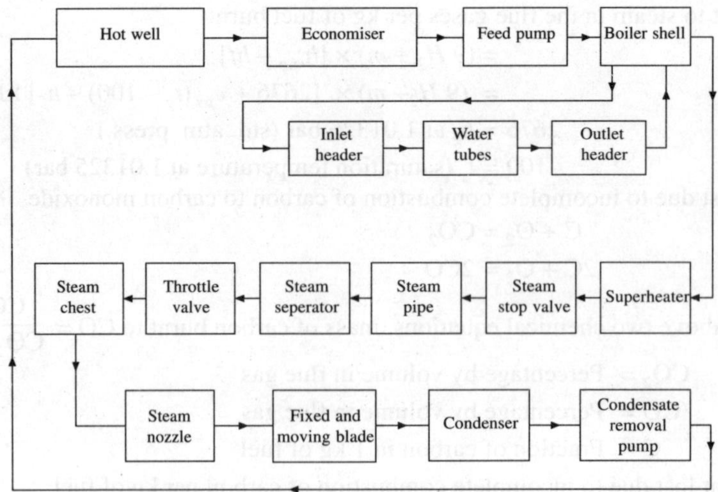

**Fig. 9.17    Water flow and steam flow path**

## 9.11.9    Boiler Trial And Heat Balance Sheet

Boiler trial is conducted for 4 to 8 hours of running to see the detail information regarding pressure, temperature, rate of feed water supply, thermal, efficiency of the plant and steam generating capacity during the test period. After a steady run of atleast 2 hours readings are taken every 10–15 minutes.

The heat balance sheet represents an account of heat energy supplied in fuel and the break up of heat energy utilised by the various components of the boiler plant. The heat balance sheet is usually drawn either on the basis of kJ/min or on the basis of heat energy per kg of fuel burnt or on the percentage basis.

The heat balance sheet for a boiler trial per kg of fuel is drawn as below.

**Table 9.1**

| Heat supplied | KJ | % | Heat utilised | KJ | % |
|---|---|---|---|---|---|
| Heat supplied by fuel | $Q = mf \times C \cdot V$ | 100% | 1. Heat utilised in generating steam | $Q_1$ | ... |
| | | | 2. Heat utilised in superheater | $Q_2$ | ... |
| | | | 3. Heat utilised in economiser for feed water heating | $Q_3$ | ... |
| | | | 4. Heat utilised in air preheater | $Q_4$ | ... |
| | | | 5. Heat carried away by the gases through chimney | $Q_5$ | ... |
| | | | 6. Heat lost due to presence of hydrogen and moisture in the fuel | $Q_6$ | ... |
| | | | 7. Heat lost due to incomplete combustion of carbon to carbon monoxide | $Q_7$ | ... |
| | | | 8. Heat lost due to unburnt fuel collected on the ashpit | $Q_8$ | ... |
| | | | 9. Heat lost due to radiation and convection by difference | $Q-(Q_1 + Q_4 + Q_3 + Q_4 + Q_5 + Q_6 + Q_7 + Q_8)$ | |
| Total | Q | 100% | Total | Q | 100% |

## Solved Problems

9.1 Find the minimum temperature of the flue gases required to produce a draught of 1.5 cm of water by a chimney of 30 m height, when air-fuel ratio required in the combustion is 20. Atmospheric air temperatureis 27°C . [Poona Univ. 1993]

*Soln.* Given

$$h_w = 15\text{mm}, \ H = 30 \text{ m}, \ \frac{\text{Air}}{\text{Fuel}} = \frac{20}{1},$$

$$T_a = 27°\text{C} = 27 + 273 = 300 \text{ k}, \ T_g = ?$$

$\dfrac{\text{Air}}{\text{Fuel}} = \dfrac{20}{1}$, Therefore, mass of air $m_a = 20$ kg and mass of flue gases $(m_a + 1) = 20 + 1 = 21$ kg/kg of

We know that,

$$hw = 353 \, H \left[ \frac{1}{T_a} - \frac{(m_a + 1)}{m_a} \frac{1}{T_g} \right] \text{ mm of water}$$

or,

$$15 = 353 \times 30 \left[ \frac{1}{300} - \frac{21}{20} \frac{1}{T_g} \right]$$

or,

$$15 = 10590 \left[ \frac{1}{300} - \frac{21}{20} \frac{1}{T_g} \right]$$

or,

$$15 = \frac{10590}{300} - 10590 \times \frac{21}{20} \frac{1}{T_g}$$

or,

$$15 = 35.3 - \frac{11119.5}{T_g}$$

or,

$$\frac{11119.5}{T_g} = 35.3 - 15 = 20.3$$

$$T_g = \frac{11119.5}{20.3} = 547.75 \text{ k}$$

$$= 274.75°\text{C}$$

∴ Minimum temperature of flue gas = 274.75°C *Ans.*

9.2 How much air is used/kg of coal burnt in a boiler having chimney of 35 m height to create a draught of 20 mm of water, when the temperature of flue gas in the chimney is 370°c and the boiler house tempeature is 34°C. Does this chimney satisfy the condition of maximum discharge?

[Poona Univ. 1987]

*Soln.* Given,

$$H = 35 \text{ m}, \ h_w = 20 \text{ mm}, \ T_g = 370°\text{C}$$

$$= 370 + 273 = 643 \text{ k}, \ T_a = 34°\text{C} = 307 \text{ k}$$

$$m_a = ?$$

We know that,

$$h_w = 353 \, H \left[ \frac{1}{T_a} - \frac{(m_a + 1)}{m_a} \frac{1}{T_g} \right] \text{mm of water}$$

$$20 = 353 \times 35 \left[ \frac{1}{307} - \frac{(m_a + 1)}{m_a} \frac{1}{643} \right]$$

$$\therefore \qquad m_a = 18.86 \text{ kg/kg of coal}$$

For maximum discharge

$$T_g = 2\frac{(m_a + 1)}{m_a} \times Ta$$

$$= 2\frac{(m_a + 1)}{m_a} \times 307$$

$$= 646.55 \text{ } k = 373.55°C$$

where as actual $T_g = 643 \text{ } k < 646.55 \text{ } k$. Therefore, condition is satisfied.

9.3  Find the draught produced in mm of water by a chimney of 40 m height. Let the mass of flue gas be 20 kg/kg of fuel burnt in the combustion chamber. The temperature of flue gases and the ambient air are 270°C and 23°C. Assuming the diameter of the chimney is 150 cm and 30% of the theoretical draught is lost in friction, find the mass of the flue gases passing through the chimney per minute.
[Poona Univ. 1986}

*Soln:* Given $\qquad hw = ?, H = 40 \text{ } m, (m_a + 1) = 20 \text{ kg/kg of fuel}$

$$T_g = 270 + 273 = 543 \text{ } k, T_a = 23 + 273 = 296 \text{ } k, D = 1.5 \text{ } m$$

We know that,

$$h_w = 353 \text{ } H \left[ \frac{1}{T_a} - \frac{(m_a + 1)}{m_a} \times \frac{1}{T_g} \right]$$

$$= 353 \times 40 \times \left[ \frac{1}{296} - \frac{20}{19} \times \frac{1}{543} \right]$$

$$= 20.33 \text{ mm of water}$$

Mass of flue gases discharged

$$m_g = \rho_g A V_g, \frac{\text{kg}}{m^3} \times m^2 \times \frac{m}{S} = \frac{\text{kg}}{S}$$

Area, $\qquad A = \frac{\pi}{4} D^2 = \frac{\pi}{4} \times 1.5^2 = 1.767 \text{ m}^2$

$$\rho_g = \frac{(m_a + 1)}{m_a} \frac{353}{T_g} \qquad\qquad \text{[Ref. Eq. 9.9]}$$

$$= \frac{20}{19} \frac{353}{543} = 0.684 \text{ kg/m}^3$$

Velocity of gas $\qquad Vg = \sqrt{2gH_g}, Hg = \text{Actual draught}$

To find $H_g$, we know that

$$Hg = H \left[ \frac{m_a}{(m_a + 1)} \frac{T_g}{T_a} - 1 \right] \qquad\qquad \text{(Eq. 9.12)}$$

$$= 40 \left[ \frac{19}{20} \times \frac{543}{296} - 1 \right] = 29.709 \text{ m}$$

Since 30% of draught is lost in friction only, so remaining 70% of the draught is available.

$$\therefore \text{ Actual avilable draught} = 0.70 \times 29.709$$

$$H_g = 20.796 \text{ m of gas column}$$

$$\therefore \qquad V_g = \sqrt{2gH_g}$$

$$= \sqrt{2 \times 9.81 \times 20.796}$$

$$= 20.199 \text{ m/s}^2$$

∴ Mass of flue gases discharged

$$m_g = \rho_g A V_g$$

$$V_g = 0.684 \times 1.767 \times 20.199$$

$$= 24.41 \text{ kg/s}$$

$$= 1464.784 \text{ kg/minute } Ans.$$

9.4  In a boiler trial following data were noted,

Height of chimney = 25 m.

Temperature of flue gases = 320°C.

Ambient temperature = 20°C.

Mass of air = 17 kg/kg of coal.

Atmospheric pressure = 720 mm of Hg Calculate $h_w$.

   *Soln:* Given, $\qquad H = 25$ m, $T_g = 320°C$, $T_a = 22°C$

$$ma = 17 \text{ kg/kg of coal, Atm. press.} = 720 \text{ mm of Hg}$$

$$h_w = ?$$

We know that

$$h_w = 353\, H \left[ \frac{1}{T_a} - \frac{(m_a + 1)}{m_a} \cdot \frac{1}{T_g} \right] \text{ mm of water}$$

[*Note*: In this equation 353 comes only when pressure is standard atmospheric i.e. 760 mm of Hg]

$$h_w = 353 \times \frac{720}{760} \times 25 \left[ \frac{1}{(22 + 273)} - \frac{(17 + 1)}{17} \times \frac{1}{(320 + 273)} \right]$$

$$= 13.41 \text{ mm of water } Ans.$$

9.5.  Determine the height of chimney required for a boiler having an average coal consumption of 1500 kg/ hour and producing 15 kg of dry flue gases per kg of coal fired. The pressure drop, in mm of water, at various stages in a boiler flue passage are

  (i)  fuel bed 4.9 mm

  (ii)  boiler flue 7.8 mm

  (iii)  boiler tube passage 2.7 mm

  (iv)  through bends, dampers etc. 0.6 mm

Actual draught may be taken as 25% greater than the theoretical draught. The flue gases have a mean temperature of 277°C and the ambient temperature is 27°C. Does the chimney fulfill the condition of maximum discharge?

   *Soln.* Given, $\qquad m_f = 1500$ kg/hr, $(m_a + 1) = 15$ kg/kg of coal

$$T_g = 277°C = 550\, k, T_a = 27 + 273 = 300\, k, H = ?$$

Actual draught required to overcome the losses

$$= 4.9 + 7.8 + 2.7 + 0.6 = 16 \text{ mm of water}$$

Actual draught = 1.25 × Theoretical draught?

$\therefore$ Theoretical draught $(h_w) = \dfrac{16}{1.25} = 12.8$ mm of water

Now, we know that

$$h_w = 353\, H \left[ \frac{1}{T_a} - \frac{(m_a + 1)}{m_a} \cdot \frac{1}{T_g} \right] \text{ mm of water}$$

or,

$$12.8 = 353\, H \left[ \frac{1}{300} - \frac{15}{14} \cdot \frac{1}{550} \right]$$

or,

$$12.8 = 0.489\, H$$

$\therefore$

$$H = 26.17 \text{ m}$$

Chimney height = 26.17 m *Ans.*

For maximum discharge

$$T_g = 2\frac{(m_a + 1)}{m_a} \times Ta$$

$$= 2 \left( \frac{15}{14} \right) \times 300 = 642.85 \text{ k}$$

Where as actual $\quad T_g = 550$ k $< 642.85$ k.

Even

$$h_w = \frac{176.5\, H}{T_a} \text{ for maximum discharge}$$

$$= \frac{176.5 \times 26.17}{300} = 15.39 \text{ mm of water}$$

$\therefore$ Theoretical draught $12.8 < 15.39$ mm of water

Both conditions indicate that chimney fulfills the condition.

9.6 Estimate the height of a chimney to produce a static draught of 20 mm of water if the mean temperature of the hot gases is 250°C and ambient temperature is 20°C. Assume the density of air and hot gases as 1.293 kg/m$^3$ and 1.34 kg/m$^3$ respectively at N.T.P. Also estimate the mass of flue gases formed per kg of fuel. Barometer reads 760 mm of *Hg*.  [Poona Univ. 1993]

*Soln.*  Given $h_w = 20$ mm of water, $Tg = 250°C = 523$ k,

$$T_a = 20°C = 293 \ k, \ \rho_a = 1.293 \text{ kg/m}^3, \ \rho_g = 1.34 \text{ kg/m}^3$$

$$H = ?, \text{ mass of flue gas} = ?$$

From gas equation

$$PV = mRT$$

or,

$$P = \frac{m}{V} RT = \rho\, RT$$

or,

$$\rho T = \text{constant at a given pressure?}$$

$\therefore$

$$\rho_N T_N = \rho_a T_a \ (N \text{ indicates for N.T.P. condition})$$

$\therefore$

$$\rho_a = \rho_N \frac{T_N}{T_a} = \frac{1.293 \times 273}{293}$$

$$= 1.204 \text{ kg/m}^3 \text{ at atmospheric temperature } Ta$$

Now, density of hot gas at $T_g$;

$$\rho_g = \frac{1.34 \times 273}{523} = 0.699 \text{ kg/m}^3$$

Total draught produced $P = Hg\,(\rho_a - \rho_g)\text{ N/m}^2$

or, $\qquad \rho_w \times g \times h_w \times 10^{-3} = Hg(1.204 - 0.699)$

or, $\quad 1000 \times g \times 20 \times 10^{-3} = 0.505\,Hg\text{ N/m}^2$

$\therefore \qquad\qquad\qquad H = \dfrac{20}{0.503} = 39.6\text{ m } Ans.$

Now, $\qquad\qquad h_w = 353\,H\left[\dfrac{1}{T_a} - \dfrac{(m_a + 1)}{m_a} \cdot \dfrac{1}{T_g}\right]$

or, $\qquad\qquad 20 = 353 \times 39.6\left[\dfrac{1}{293} - \dfrac{(m_a + 1)}{m_a} \cdot \dfrac{1}{523}\right]$

or, $\qquad\qquad m_a = 27.24\text{ kg/kg of fuel}$

$\therefore$ Mass of gases formed per kg of fuel

$$= (m_a + 1)$$
$$= (27.24 + 1)$$
$$= 28.24 \text{ kg } Ans$$

**9.7** Find the height of chimney necessary to produce a draught of 30 mm of water column. The atmospheric air temperature is 27°C and gas temperature in chimney is 217°C. Air fuel ratio is 13.5. What will be the Power required if induced draught fan is used for producing the above draught? fuel consumption is 1500 kg/hr. [Mumbai Univ. 1994]

*Soln.* Given $\qquad\qquad h_w = 30\text{ mm of water, } T_a = 27°C = 300\text{ k,}$

$$T_g = 217° C = 490\text{ k,} \frac{13.5}{1} \cdot H = ?,\, m_f = 1500\text{ kg/hr}$$

we know that, $\qquad h_w = 353 \cdot H\left[\dfrac{1}{T_a} - \dfrac{(m_a + 1)}{m_a} \cdot \dfrac{1}{T_g}\right]$

or, $\qquad\qquad 30 = 353 \cdot H\left[\dfrac{1}{300} - \dfrac{14.5}{13.5} \times \dfrac{1}{490}\right]$

$\therefore \qquad\qquad\qquad H = 74.46\ m\ Ans.$

Using Eq. 9.24, we get

$\qquad\qquad$ Fan power $= h_w \times V$

$$= \frac{9.81 \times h_w \times (m_a \times m_f)\,0.7734}{1000} \frac{T_g}{273}\text{ kw}$$

$$= \frac{9.81 \times 30 \times 13.5 \times 1500}{1000 \qquad 3600} \times 0.7734 \times \frac{490}{273}\text{ kw}$$

$$= 2.29\text{ kw } Ans$$

**9.8** Following data relate to a boiler trial.

Mean temperature of flue gases = 550 k

Temperature of cold air = 300 k

Air supplied per kg of fuel = 18

Fuel consumption = 1800 kg/hr

Draught required = 100 mm of water

Mechanical efficiency of fan = 80%

Calculate motor power for forced and induced draught fans allowing 10% of leakage of air in both the cases. Assume, specific volume of air at N.T.P. as 0.7734 m$^3$/kg.

*Soln.*   Given $T_g$ = 550 k, $T_a$ = 300 k, $m_a$ = 18 kg/kg of

$$\text{fuel, } m_f = \frac{1800}{3600} = 0.5 \text{ kg/s}$$

$$\text{leakage factor} = 1.1$$

$$\eta = 0.8$$

using Eq. 9.23, we get

$$\text{Motor power for forced fan} = \frac{9.81 \times 100 \times 18 \times 0.5 \times 0.7734 \times 300 \times 1.1}{1000 \times 0.8 \times 273}$$

$$= 10.317 \text{ kW } Ans$$

Similarly by using Eq. 9.25, we get for motor power for induced fan

$$= \frac{9.81 \times 100 \times 18 \times 0.5 \times 0.7734 \times 300 \times 1.1}{1000 \times 0.8 \times 273}$$

$$= 18.914 \text{ Kw } Ans$$

**9.9**  A draught of 60 mm of water column is produced by using F.D fan. The ambient temperature and flue gas temperature are 30°C and 320°C respectively. The fuel consumption is 1500 kg/hr and air-fuel ratio is 15:1. Mechanical efficiency of fan is 88% and transmission efficiency of fan driving gear is 90%. Find the power of motor required to drive the fan. What will be the percentage change in power if I.D. fan is used instead of F.D. fan to produce the same draught? Assume I.D. fan efficiency 85% and same efficiency of driving gear.

*Soln.*   Given, $h_w$ = 60 mm of water, $T_a$ = 30°C = 303 k,

$$T_g = 320°C = 593 \; k, \quad \frac{A}{F} = \frac{15}{1} \quad m_f = \frac{1500}{3600} = 0.4167 \text{ kg/s}$$

$$\eta_{\text{mech}} = 88\%, \; \eta_{\text{mech}}. = 90\%, \text{ F.D. fan Power} = ?$$

Motor power for forced fan

$$= \frac{9.81 \times 60 \times 15 \times 0.4167 \times 0.7734 \times 303}{1000 \times 0.88 \times 0.90 \times 273}$$

$$= 3.987 \text{ kw } Ans$$

$$\text{Motor Power for induced fan} = \frac{9.81 \times 60 \times 15 \times 0.4167 \times 0.7734 \times 593}{1000 \times 0.85 \times 0.90 \times 273} \quad 8.079 \text{ kW } Ans.$$

For producing same draught, percentage change in power

$$= \frac{8.079 - 3.987}{3.987} \times 100$$

$$= 102.63\% \; Ans$$

**9.10**  A boiler plant uses forced draught fan. It delivers atmospheric air at 16 m/s and the draught lost through the grate is 30 mm of water column. The oil burnt is 9000 kg/hr and the air supplied is 18 kg per kg of oil. Ambient air is at 27°C and at 1 bar. Find the shaft power required to drive the fan if its mechanical efficiency is 81%. [Mumbai Univ. 1995]

*Soln.* Given velocity of air $V = 16$ m/s,

$$h_w = 30 \text{ mm of water,}$$
$$m_f = 9000 \text{ kg/hr,}$$
$$m_a = 18 \text{ kg/kg of oil,}$$
$$T_a = 27°C = 300 \text{ k,}$$
$$P_a = 1 \text{ bar, shaft power } = ?$$
$$\eta_{mech} = 81\%$$

Velocity head imparted to air by fan $= \dfrac{V^2}{2g}$

$$= \dfrac{16^2}{2 \times 9.81} = 13.04 \text{ m}$$

pressure head equivalent to velocity head at N.T.P. $= \rho g h$

$$= 1.293 \times 9.81 \times 13.04 \qquad \text{(at N.T.P, } \rho = 1.293 \text{ kg/m}^3)$$
$$= 165.40 \text{ N/m}^2$$

Draught lost across the grate $= 30$ mm of water

$$= 30 \times 9.81$$
$$= 294.3 \text{ N/m}^2$$

∴ Total draught required to be created by forced draught fan, $\quad h = 165.40 + 294.3$

$$= 459.7 \text{ N/m}^2$$

Mass of air handled by fan $= \dfrac{9000}{3600} \times 18$

$$= 45 \text{ kg/s}$$

Now, $\qquad PV = m \, RT$

$$1 \times 10^5 \times V = 45 \times 287 \times (27 + 273)$$

∴ $\qquad$ Rate of volume $V = 38.745 \text{ m}^3\text{/s}$

$$\text{Shaft power } = \dfrac{h \cdot V}{\eta} = \dfrac{459.7 \times 38.745}{0.81 \times 1000} \text{ kW}$$

$$= 21.988 \text{ kW } Ans.$$

**9.11** Fuel consumption of a 125 MW steam power plant is 0.36 kg/kw. hr. Air pressure at entry to forced draught fan is 1.013 bar and air temperature is 30°C. The total pressure drop in air and gas path is 290 mm of water. Determine the power consumption of forced draught fan if fan efficiency is 80%. Assuming the properties of air and flue gases to be same, estimate the power consumption of induced draught fan for the above plant. The temperature of flue gas at entry to the fan is 250°C and fan efficiency is 74%. Air fuel ratio is 17. [Mumbai Univ. 1992]

*Soln.* Given, $\qquad m_f = \dfrac{0.36 \times 125 \times 10^3}{3600} = 12.5 \text{ kg/s}$

$$P_a = 1.013 \text{ bar}, \ T_a = 30°C = 303 \text{ k}, \ h_w = 290 \text{ mm}$$

of water, F.D. fan power = ? $\eta_{if} = 80\%$, $T_g = 250°C$

$$= 523 \text{ k}, \ \eta_{if} = 74\%$$

Mass of air handled by forced fan = 12.5 × 17, kg/s

Now,
$$P_a V_a = m_a R_a T_a$$

$$V_a = \frac{212.5 \times 287 \times 300}{1.013 \times 10^5} = 180.61 \text{ m}^3/s$$

Rate of volume flow of air = 180.61 m³/s

Power required for forced fan

$$= \frac{290 \times 9.81 \times 180.61}{1000 \times 0.80} \text{ kw}$$

$$= 642.27 \text{ kw } Ans$$

Power required for induced fan

$$= \frac{290 \times 9.81 \times 212.5 \times 0.7734 \times 523}{1000 \times 273 \times 0.74}$$

$$= 1210.42 \text{ kw } Ans.$$

9.12   The equivalent evaporation of a boiler from and at 100°C is 10.4 kg of steam per kg of fuel. The calorific value of the fuel is 29800 kJ/kg. Determine the efficiency of the boiler. If the boiler produces 15 tons of steam per hour at 2.0 MN/m² from feed water at 40°C and the fuel consumption is 1650 kg/hr. Determine the condition of steam produced.

*Soln.* Given,
$$m_e = 10.4 \text{ kg/kg of fuel,}$$
$$C \cdot V = 29800 \text{ kJ/kg}$$
$$m_w = 15000 \text{ kg/hr,}$$
$$\text{Pressure} = 2 \text{ MN/m}^2,$$
$$\text{Feed water temp.} = 40°C, \ m_f = 1650 \text{ kg/hr}, \ x = ?$$

Equivalent evaporation $m_e = m_w \ \dfrac{(h_s - h_f)}{2256.9}$

or,
$$10.4 = m_w \frac{1500 \times \left(h_s - 40 \times 4.187\right)}{1650 \ 2256.9}$$

specific heat of water = 4.187 kJ/kgk

$h_f = 4.187 \times (40 - 0)$ or $h_f$ value can be directly taken from steam table at 40°C.

From above equation
$$h_s = 2749.373 \text{ kJ/kg}$$

Now,
$$\eta_{\text{boiler}} = \frac{m_w (h_s - h_f)}{m_f \times C \cdot V}$$

$$= \frac{15000(2749.373 - 40 \times 4.187)}{1650 \times 29800} \times 100$$

$$= 78.76\% \; Ans.$$
$$h_s = h_f + x \, h_{fg}$$

At pressure $2 \text{ MN/m}^2 = 2 \times 10^6 \times 10^{-5} = 20$ bar,

from steam table, $h_f = 908.6$ kJ/kg and $h_{fg} = 1888.7$ kJ/ kg

$\therefore \qquad 2749 \cdot 373 = 908.6 + x \, 1888.7$

or, $\qquad\qquad\qquad x = 0.974$

condition of steam produced with dryness fraction = 0.974 *Ans.*

9.13  A boiler delivers steam at the rate of 6000 kg/hr. pressure of steam generated is 800 kN/m$^2$ and dryness fraction 0.98. Feed water supplied is at 40°C. If the efficiency of the boiler is 75%, determine the rate of coal consumption of calorific value as 31000 kJ/kg. What is the equivalent evaporation from this boiler? If the superheater is used then the temperature leaving the boiler is 250°C, what is the efficiency of the boiler then? Assume $c_p$ for superheated steam = 2.27 kJ/kg k.

*Soln.*

Given $m_w = 6000$ kg/hr, pressure = 800 kN/m$^2$, $x = 0.98$,

Feed water temp. = 40°C, $\eta_{\text{boiler}} = 75\%$, $mf = ?$

C.V = 31000 kJ/kg, $m_e = ?$ $t_{\text{sup}} = 250$°C, $\eta_{\text{boiler}} = ?$

$c_{ps} = 2.27$ kJ/kgk.

$$\eta_{\text{boiler}} = \frac{m_w (h_s - h_f)}{m_f \times C \cdot V}, \text{ pressure } 800 \text{ kN/m}^2 = 800 \times 10^3 \times 10^{-5} = 8 \text{ bar}$$

or $\qquad\qquad 0.75 = \dfrac{6000\,[(720.9 + 0.98 \times 2046.5) - 40 \times 4.187]}{m_f \times 31000}$

$$m_f = 660.384 \text{ kg/hr}$$

$\therefore$ Rate of coal consumption = 660.384 kg/hr Ans.

$$m_e = \frac{6000\,[(720.9 + 0.98 \times 2046.5 - 40 \times 4.187)]}{660.384 \times 2256.9}$$

$$= 10.301 \text{ kg/kg of fuel burnt}$$

$\therefore$ Equivalent evaporation = 10.301 kg/kg of fuel burnt *Ans.*

$$\eta_{\text{boiler}} = \frac{m_w\,[h_{\text{sup.}} - h_f]}{m_f \times C \cdot V} = \frac{m_w[\{h_g + c_{p_s}(t_{\text{sup.}} - t_s)\} - h_f]}{m_f \times C \cdot V}$$

$$= \frac{6000\,[\{720.9 + 2.27\,(250 - 170.4)\} - 40 \times 4.187)]}{660.384 \times 31000} \times 100$$

$$= 81.49\% \; Ans.$$

9.14  During boiler trial on fire tube boiler, following readings were noted,

Duration of trial = 24 bar

Amount of water evaporated = 20 tonnes.

Feed water temperature = 50°C.

Condition of steam = 0.92.

Absolute pressure = 15 bar.

Coal burnt = 6 tonnes.

Grate area $= 4\text{m}^2$,

Heating surface $= 3200\text{m}^2$.

C.V. of coal $= 20000$ kJ/kg

Calculate

1. Mass of coal burnt per $\text{m}^2$ of grate area per hr.

2. Equivalent of evaporation from & at 100°C in kJ/kg of coal

3. Thermal efficiency

4. Equivalent evaporation/$\text{m}^2$ of heating surface per hr.

$$\text{Given} \qquad m_w = \frac{20000}{24} = 833.33 \text{ kg/hr}$$

$$m_f = \frac{6000}{24} = 250 \text{ kg/hr}$$

Feed water temp. $= 50$°C, $x = 0.92$, pressure $= 15$ bar,

$$A = 4\text{m}^2, \text{ heating surface} = 3200 \text{ m}^2$$

$$C \cdot V = 20000 \text{ kJ/kg}$$

1. Mass of coal burnt per $\text{m}^2$ of grate area

$$\text{per hour} = \frac{250}{4} = 62.5 \text{ kg } Ans.$$

2. Equivalent evaporation in kg/kg of coal

$$m_e = \frac{m_w\left(h_s - h_f\right)}{2256.9}$$

$$= \frac{20000[(844.6 + 0.92 \times 1945.3) - 50 \times 4.187]}{6000 \times 2256.9}$$

$$= 3.581 \text{ kg/kg of coal}$$

3.
$$\eta_{th} = \frac{m_w\left(h_s - h_f\right)}{mf \times C \cdot V} \quad \frac{20000[(844.6 + 0.92 \times 1945.3) - 50 \times 4.187]}{6000 \times 2000}$$

$$= 0.4343 \times 100 = 43.43\% \text{ Ans.}$$

4. Equivalent evaporation/$m^2$ of heating surface per hour

$$= \frac{833.33}{3200} \quad \frac{[(844.6 + 0.92 \times 1945.3) - 50 \times 4.187]}{2256.9}$$

$$= 0.279803 \text{ kg/hr. } m^2 \text{ Ans.}$$

9.15 Following particulars refer to a steam power plant consisting of a boiler, superheater and an economiser steam pressure = 15 bar, mass of steam generated = 5000 kg/hr, mass of coal used = 700 kg/hr, calorirific value of coal = 30,000 kJ/kg, dryness fraction of steam leaving the boiler = 0.97, temperature of superheated steam leaving the superheater = 300°C, temperature of faced water entering the economiser = 40°C, temperature of feed water leaving the economiser = 100°C. Determine

(i) Equivalent evaporation

(ii) Boiler efficiency with economiser and superheater [Poona Univ. 1992]

*Soln.*

Given pressure = 15 bar,

$$m_w = 5000 \text{ kg/hr,}$$
$$m_f = 700 \text{ kg/hr,}$$
$$\text{C.V} = 30{,}000 \text{ kJ/Kg,}$$

$$x = 0.97, \, t_{\text{sup}} = 300°C, \, t_{w_1} = 40°C, \, t_{w_2} = 100°C, \, m_e = ?, \, \eta_{boiler} = ?$$

At 15 bar,
$$h_s = h_f + x h_{fg}$$
$$= 845 + 0.97 \times 1945 = 2731.65 \text{ kJ/kg}$$
$$h_w = 100 \times 4.187 = 418.7 \text{ kJ/kg}$$

Equivalent evaporation $m_e = \dfrac{5000}{7000} \dfrac{(2731.65 \times 418.7]}{2256.9}$

$$= 7.317 \text{ kg/kg of coal burnt } Ans.$$

$$\eta_{\text{boiler}} = \dfrac{5000}{7000} \dfrac{(2731.65 \times 418.7]}{30{,}000} \times 100$$

$$= 55\% \, Ans.$$

Boiler efficiency with economiser

$$= \dfrac{5000[2731.65 - 40 \times 418.7] \times 100}{700 \times 30{,}000}$$

$$= 61\%$$

Boiler efficiency with superheater

$$= \dfrac{5000[3038.9 - 418.7]}{700 \times 30{,}000} \times 100 = 62.38\% \, Ans.$$

Where $h_{\text{sup}} = 3038.9$ kJ/kg from steam table.

9.16 In a boiler trial, following readings were noted, Mean temperature of feed water = 25°C. Mean boiler working pressure = 1.2 MPa. Mean dryness fraction = 0.95. Mass of coal burnt/hr = 250 kg. C.V of coal burnt = 32400 kJ/kg. Mass of water supplied to the boiler in 7 hours and 14 minutes = 16500 kg. Mass of water in the boiler at the end of test was less than that at the commencement by 1000 kg calculate

  (i) Actual evaporation/kg of coal

  (ii) Equivalent evaporation from and at 100°C

(iii) Thermal efficiency of boiler.

*Soln.* up to the prescribed limit water would be filled into the boiler befor firing. The quantity of water supplied during the test period of 7 hours and 14 minutes = 16500 kg and at the end of the test water in the drum was less by 1000 kg.

(i) ∴ Total mass of steam generated in 7 hrs and 14 minutes = 16500 + 1000 = 17500 kg

$$\text{Mass of steam generated/hr} = \dfrac{17500}{7 + \dfrac{14}{60}} = 2419.354 \text{ kg}$$

Actual evaporation or steam generated per kg of coal

$$= \frac{2419.354}{250} = 9.678 \text{ kg} = m_w \; Ans.$$

(ii) Equivalent evaporation $m_e = \dfrac{m_w \left( h_s - h_f \right)}{2256.9}$, $h_s$ at 12 bar from steam table

or,
$$m_e = \frac{9.678[(798.4 + 0.95 \times 1984.3) \times 25 \times 4.187]}{2256.9},$$

$$= 11.058 \text{ kg/kg of coal} \quad Ans$$

(iii)  Boiler thermal efficiency

$$\eta_{th} = \frac{m_s (h_s - h_f)}{m_f \times C \cdot V}$$

$$= \frac{9.678[(798.4 + 0.95 \times 1984.3) - 25 \times 4.187]}{32400},$$

$$= 0.7702 \times 100 = 77.02\% \; Ans.$$

9.17  Following data relate to a trial on boiler using economiser, air preheater and superheater: Condition of steam at exit of boiler = 20 bar, 0.96 dry. Temperature of steam at exit of superheater = 300°C steam evaporation rate/kg of fuel = 12 kg. Room temperature = 25°C. Temperature of feed water at exit of economiser = 50°C. Temperature of air at exit of airpreheater = 70°C. The temperature of flue gases at inlet to superheater, economiser, airpreheater and exit of airpreheater are 650°C, 430°C, 300°C and 180°C respectively. Assume that air supplied as 19 kg/kg of fuel of calorific value of 45,000 kJ/kg, find

   (i)  Equivalent evaporation with and without economiser,
  (ii)  Thermal efficiency of the boiler with and without economiser,
 (iii)  Thermal efficiency of superheater, economiser and air preheater, and
  (iv)  Draw up heat balance sheet, Assume specific heat of flue gases and air as 1.05 and 1.01 kJ/kgk respectively.

*Soln.* Refer Fig. 9.27

From steam table at 20 bar and 300°C

$$h_{sup} = 3025 \text{ kJ/kg}$$

and $h_s$ for 0.96 dry steam

i.e.
$$h_s = h_f + x h_{fg} = 908.6 + 0.96 \times 1888.7$$
$$= 2721.75 \text{ kJ/kg}$$

(i) Equivalent evaporation without economiser

$$m_e = \frac{m_s (h_{sub} - h_f)}{2256.9}$$

$$= \frac{12 (3025 - 50 \times 4.187)}{2256.9}$$

$$= 14.97 \text{ kg/kg of fuel} \; Ans.$$

Equivalent evaporation with economiser

$$m_e = \frac{12\,(3025 \times 25 \times 4.187)}{2256.9}$$

$$= 15.52 \text{ kg/kg of fuel } Ans.$$

(ii) Thermal efficiency of boiler without economiser

$$\eta_{th} = \frac{12\,(3025 - 50 \times 4.187)}{45000} \times 100$$

$$= 75.07\% \ Ans.$$

9.18  Equivalent evaporations from and at 100°C for a boler when it uses coal and oil fuel are 8 kg/kg of fuel and 16 kg/kg of fuel respectively. How many litres of oil of specific gravity 0.9 are equivalent to one tonne of coal? Find out the calorific value of oil fuel if calorific value of coal is 24000 kJ/kg of coal. Calculate the efficiency of boiler assuming boiler efficiency to be the same for both the fuels.  [Poona Univ. 1993]

*Soln.* Assuming *o* for oil and *c* for coal

Given  $me_o = 16$ kg/kg, $me_c = 8$ kg/kg

Sp. gravity of oil = 0.9, $mc = 1$ tonne = 1000 kg

$$C \cdot V_c = 24000 \text{ kJ/kg}$$

$$\eta_{\text{boiler}} = \frac{m_w (h_s - h_f)}{m_c \times C \cdot V_c} = \frac{m_w (h_s - h_f)}{m_o \times C \cdot V_c}$$

Equivalent evaporation

$$m_e = \frac{m_w (h_s - h_f)}{m_f \times 2256.9}$$

$$\therefore \qquad \eta_{\text{boiler}} = \frac{m_e \times 2256.9}{C \cdot V}$$

$$\therefore \qquad \frac{m_{eo} \times 2256.9}{C \cdot V_o} = \frac{m_{ec} \times 2256.9}{C \cdot V_c}$$

$$\text{or,} \qquad \frac{16 \times 2256.9}{C \cdot V_o} = \frac{8 \times 2256.9}{24000}$$

$$\therefore \qquad C \cdot V_o = 48000 \text{ kJ/kg } Ans$$

$$\eta_{\text{Boiler}} = \frac{16 \times 2256.9}{48000} \times 100 = 75.23\% \ Ans.$$

Also, heat supplied by oil = heat supplied by coal

$$m_o \times C \cdot V_o = m_c \times C \cdot V_c$$

$$\text{or,} \qquad m_o \times 48000 = 1000 \times 24000$$

$$\therefore \qquad m_o = 500 \text{ kg}$$

Density of oil = $0.9 \times 1000 = \dfrac{\text{mass}}{\text{volume}}$

$$\text{or,} \qquad 0.9 \times 1000 = \frac{500}{\text{volume}}$$

$$\therefore \qquad \text{volume of oil used} = \frac{500 \times 1000}{0.9 \times 1000} = 555.55 \text{ litres } Ans. \qquad [\because 1 \text{ m}^3 = 1000 \text{ litres}]$$

9.19  The following data were recorded during a trial on steam boiler: pressure of steam = 12 bar, mass of feed water = 4500 kg/hr, temperature of feed water = 75°C, Dryness fraction of steam = 0.96; coal used = 490 kg/hr, calorific value of coal = 35700 kJ/kg, ash and unburnt coal collected in ash pit = 2% by mass, moisture in coal = 4% by mass, calorific value of ash and unburnt coal = 13500 kJ/kg mass of dry flue gases = 18.57 kg/kg of coal, temperature of flue gases = 300°C boiler house temperature = 16°C, specific heat of flue gases = 0.97 kJ/kg K. Draw the heat balance sheet of the boiler per kg of coal.

*Soln.*

From steam table at 12 bar

$$h_s = h_f + x h_{fg}$$
$$= 798.4 + 0.96 \times 1984.3$$
$$= 2703.328 \text{ kJ/kg}$$

(a) Heat utilised to generate steam per kg of coal

$$= \frac{4500}{490}(h_s - h_f) = \frac{4500}{490}(2703.328 - 75 \times 4.187)$$
$$= 21942.579 \text{ kJ}$$

(b) Heat carried away by dry flue gas

$$= m_g \times c_{pg}(t_g - t_a)$$
$$= 18.57 \times 0.97(300 - 16) = 5115.663 \text{ kJ}$$

(c)  Heat carried away by moisture in the coal per kg of coal

$$= m\,[2676 + c_{ps}(t_g - 100) - h_f]$$
$$= 0.04[2676 + 2.093(300 - 100) - 75 \times 4.187]$$
$$= 111.223 \text{ kJ}$$

(d)  Heat lost in ash and unburnt coal $= 0.02 \times 13500 = 270 \text{ kJ}$

(e)  As the moisture and ash and unburnt coal in the coal are 0.04 kg and 0.02 kg per kg of coal respectively, therefore heat supplied by 1 kg of coal

$$= (1 - 0.04 - 0.02) \times 35700 = 33558 \text{ kJ}$$

(f)  unaccounted heat loss = 33558 − (21942.579 + 5115.663 + 111.223 + 270)

$$= 33558 - 27439.465 \text{ kJ} = 6118.535 \text{ kJ}$$

### Heat balance sheet

| Heat supplied | kJ | % | Heat utilised | kJ | % |
|---|---|---|---|---|---|
| Heat supplied by coal | 33558 | 100 | 1. Heat utilised to generate steam | 21942.579 | 65.38 |
| | | | 2. Heat carried away by dry flue gas | 5115.663 | 15.24 |
| | | | 3. Heat carried away by moisture in the coal | 111.223 | 0.33 |
| | | | 4. Heat lost in ash and unburnt coal | 270 | 0.80 |
| | | | 5. Heat loss due to radiation, convection i.e. unaccounted loss (by difference) | 6118.535 | 18.24 |
| Total | 33558 | 100% | Total | 33558 | 100% |

9.20 Following particulars were observed during a boiler trial . Duration of trial $= 24$ hours, evaporation capacity $= 12500$ kg, Dryness fraction $= 0.90$, Economiser water temperature of outlet $= 120°C$, calorific value of coal used $= 30{,}000$ kJ/kg, $\dfrac{A}{F} = \dfrac{22}{1}$, coal consumed $= 1600$ kg, pressure $= 7.5$ bar (gauge), Hot $F1$ well temperature $= 40°C$. By providing superheating - steam is superheated to $228°C$. Coal lost in ash pit $= 11$ kg, temperature of flue gas $= 280°C$, ambient temperature $= 30°C$. Draw a heat balance on kJ/ hr basis and on % basis to show the performance of each unit.

*Soln.*

(a) Heat produced due to burning of coal

$$= \frac{(1600 - 11)}{24} \times 30{,}000 = 1986250 \text{ kJ/hr}$$

(b) Heat lost due to coal in ash pit

$$= \frac{11}{24} \times 30{,}000 = 13750 \text{ kJ/hr}$$

(c) Steam pressure $= 7.5$ bar (gauge)?

Absolute steam pressure $= (7.5 + 1.01325) = 8.51$ bar

From steam table at 8.51 bar

| $t_s$ | $h_f$ | $h_{fg}$ | $h_g$ |
|---|---|---|---|
| 172.9°C | 732.05 kJ/kg | 2037.8 kJ/kg | 2769.85 kJ/kg |

$h_s = h_f + x h_{fg} = 172.9 + 0.90 \times 2037.8 = 2566.07$ kJ/kg

$h_f = 120 \times 4.187 = 502.44$ kJ/kg Heat utilised in generating steam

$$= \frac{12500}{24} (h_s - h_f)$$

$$= \frac{12500}{24} (566.07 - 502.44)$$

$$= 1074800.4 \text{ kJ/hr}$$

(d) Heat utilised by economiser

$$= \frac{12500}{24} \times 4.187(120 - 40)$$

$$= 174458.33 \text{ kJ/hr}$$

(e) Heat utilised by superheater 12500

$$= \frac{12500}{24} (h_{\text{sup}} - h_s)$$

$$= \frac{12500}{24} (2898.9 - 2566.07) \; [h_{\text{sup}} = 2898.9 \text{ kJ/kg from steam table}]$$

$$= 173348.96 \text{ kJ/hr}$$

(f) Heat carried away by flue gas $(1600 - 11)$

$$= (22 + 1) \times 1.005 (280 - 30) \times \frac{(1600 - 11)}{24}$$

$$= 382601.41 \text{ kJ/hsr} \quad [c_{pg} = 1.005 \text{kJ/kgk}]$$

(g) Heat unaccounted i.e. radiation & convection

$$\text{loss (by difference)} = 1986250 - (13750 + 1074800 + 1\,174458.33$$
$$+\ 173348.96 + 382601.41)$$
$$= 1986250 - 1818958.7 = 167291.3 \text{ kJ/hr}$$

### Heat balance sheet on kJ/hr basis

| Heat supplied | kJ/hr | % | Heat utilised | kJ/hr | % |
|---|---|---|---|---|---|
| Heat supplied by coal | 1986250 | 100 | 1. Heat utilised in generating steam | 1074800.4 | 54.11 |
| | | | 1 Heat utilised by economiser | 174458.33 | 8.78 |
| | | | 2 Heat utilised by superheater | 173348.9 | 6 8.73 |
| | | | 3 Heat lost in ash pit | 13750 | 0.70 |
| | | | 4 Heat carried away by flue gas | 382601.41 | 19.26 |
| | | | 5 Heat unaccounted | 167291.3 | 8.42 |
| Total | 1986250 | 100% | Total1 | 986250 | 100% |

9.21 Following particulars were observed during boiler trial:

Temperature of feed water = 75°C

Quantity of feed water supplied = 4000 kg

Steam pressure = 15 bar (gauge)

Coal fired per hour = 390 kg

Barometer reading = 750 mm of H$g$

Higher calorific value of coal = 40000 kJ/kg

Moisture in coal = 5.3% by weight

Throttling calorimeter temperature at outlet = 120°C

Pressure of steam after throttling = 50 mm of Hg

Specific heat of superheated steam = 2.1 kJ/kgk.

Partial pressure of steam in flue gases = 0.08 bar

Temperature of flue gases leaving the boiler = 370°C

Boiler room temperature = 28°C

Percentage composition of dry coal was as follows

$C$ = 85%, H2 = 4%, Ash 5% and rest other volatile matters. The analysis of flue gas by volume was as follows

$CO_2$ = 12%, CO = 1.8%, $O_2$ = 8% and N2 = 78.2%. Assume specific heat of dry flue gases as 0.95 kJ/ kg k. Heating surface of boiler = 25 m2. Draw up the heat balance sheet for the boiler per kg of coal fired and calculate the boiler efficiency and equivalent evaporation per kg of fuel and per m$^2$ of heating surface per hour.

*Soln.*

Given $m_w$ = 4000 kg, $t_w$ = 75°C, $P_1$ = 15 bar (gauge) $m_f$ = 130 kg/hr, C.V. = 40,000 kJ/kg, $t_g$ = 370°C,

$c_{p_g}$ = 2.1 kJ/kgk, $t_a$ = 28°C, $t_{sup}$ = 120°C, $P_2$ = 0.08 bar, $c_{p_g}$ = 0.95 kJ/kgk, $A$ = 25 m$^2$

As we know in throttling calorimeter enthalpy remains constant. ? Enthalpy before throttling = Enthalpy afterthrottling pressure before throttling

$$\text{pressure befor throttling} = 15 + \left(15 + \frac{750}{760} \times 1.01325\right) = 16 \text{ bar at dryness fraction } x.$$

pressure after throttling $= \times 1.01325$

$$760 = 1.067 \text{ bar at } 120°C$$

$$\underbrace{h_f + xh_{fg}}_{at\ 16\ bar} = \underbrace{h_g}_{at\ 1.067\ bar} + c_{ps}(t_{\text{sup}} - t_s),\ t_s \text{ at } 1.067 \text{ bar} = 101.26°C$$

$$858.5 + x\,1933.2 = 2677.8 + 2.1(120 - 101.26)$$

$$x = 0.961$$

In order to prepare heat balance sheet, the following heats are required to be calculated

(a)  Heat utilised in generating steam per kg of coal $= \dfrac{4000}{390} [(858.5 + 0.961 \times 1933.2) - 75 \times 4.187]$

$$= 24638.771 \text{ kJ/kg of coal}$$

(b)  Heat supplied per kg of coal fired $= (1 - 0.053) \times 40{,}000 = 37880 \text{ kJ/kg}$

(c)  Heat carried away by dry flue gases

Firstly mass of dry flue gases ($mg$) per kg of coal should be calculated with the help of dry flue gas analysis given

| Constituent | Volume in $1m^3$ of flue gas | Molecular mass | Proportional mass | Mass of constituent in kg per kg of flue gas |
|---|---|---|---|---|
| | (a) | (b) | $c = a \times b$ | $d = \dfrac{c}{\Sigma c}$ |
| $CO_2$ | 0.12 | 44 | 5.28 | $\dfrac{5.28}{30.24} = 0.1746$ |
| $O_2$ | 0.08 | 32 | 2.56 | $\dfrac{2.56}{30.24} = 0.0847$ |
| $CO$ | 0.018 | 28 | 0.504 | $\dfrac{0.504}{30.24} = 0.0167$ |
| $N_2$ | 0.782 | 28 | 21.896 | $\dfrac{21.896}{30.24} = 0.7240$ |
| | 1.000 | | $\Sigma c = 30.24$ | 1.000 |

mass of carbon in kg of flue gas

$$= \frac{3}{11} CO_2 + \frac{3}{7} CO = \frac{3}{11} \times 0.1746 + \frac{3}{7} \times 0.0167$$

$$= 0.05477 \text{ kg}$$

$\therefore$  Mass of dry flue gases per kg of coal burnt

$$m_g = \frac{\text{mass of carbon in 1 kg of coal}}{\text{mass of carbon in 1 kg of flue gas}}$$

$$= \frac{0.85}{0.05477} = 15.519 \text{ kg/kg of flue gas}$$

$\therefore$ Heat carried away by dry flue gas per kg of coal

$$= m_g \times c_{p_g} (t_g - t_a)$$
$$= 15.519 \times 0.95 (370 - 28) = 5042.123 \text{ kJ}$$

(d) Heat carried away by steam formed due to moisture and hydrogen present in the coal per kg of coal.

$$= \text{mass of steam formed specific enthalpy}$$
$$= (9H_2 + m) [h_{sup} - h_f]$$

where, $h_{sup}$ = specific enthalpy of superheated steam at partial pressure of 0.08 bar, kJ/kg

and $\qquad h_f$ = specific enthalpy of water at boiler room temp. kJ/kg

$\therefore$ Heat carried in steam (vapour)

$$= (9 \times 0.04 + 0.053)[ h_g + C_{p_s} (t_g - t_s) - h_f]$$
$$= 0.413[2577.1 + 2.1 (370 - 41.54) - 4.187 \times 28]$$
$$= 1300.79 \text{ kJ}$$

(e) Heat lost by radiation, convection etc. (by difference)

$$= 37880 - (24638.771 + 5042.123 + 1300.79)$$
$$= 6898.316 \text{ kJ}$$

### Heat balance sheet per kg of coal

| Heat supplied | KJ | % | Heat utilised | kJ | % |
|---|---|---|---|---|---|
| Heat supplied by coal | 37880 | 100 | 1. Heat utilised in generating steam | 24638.771 | 65.04 |
| | | | 2. Heat carried away by dry flue gas | 5042.123 | 13.31 |
| | | | 3. Heat carried away to moisture & hydrogen in coal | 1300.79 | 3.43 |
| | | | 4. Heat unaccounted (by difference) | 6898.316 | 18.22 |
| Total | 37880 | 100% | Total | 37880 | 100 |

$$\text{Boiler efficiency} = \frac{m_w (h_s - h_f)}{m_f \times C.V}$$

$$= \frac{24638.771}{37880} \times 100 = 65.04 \% \quad Ans.$$

Equivalent evaporation per kg of coal

$$= \frac{m_w (h_s - h_f)}{2256.9} = \frac{24638.771}{2256.9} = 10.91 \text{ kg} \quad Ans.$$

Equivalent evaporation per $m^2$ per hour

$$= \frac{\text{Equivalent evaporation Per kg of fuel}}{\text{Heating Surface area}}$$

$$= \frac{10.91}{25} = 0.4364 \, kg \quad Ans$$

# EXERCISES

**Objective Questions**

9.1(i)  Within the boiler, the temperature of steam is highest in

    (a)  water drum                        (b)  water tubes

    (c)  water walls                        (d)  super heater

(ii)  If in a boiler, the circulation of water is by convection currents which are set up during the heating of water, then the boiler is known as

    (a)  forced circulation boiler         (b)  natural circulation boiler

    (c)  internally fired boiler            (d)  externally fired boiler

(iii)  If in a boiler, the circulation of water is by pumps, then the boiler is known as

    (a)  forced circulation boiler         (b)  natural circulation boiler

    (c)  internally fired boiler            (d)  externally fired boiler

(iv)  If in a boiler, the combustion takes place outside the region of boiling water, the boiler is known as

    (a)  forced circulation boiler         (b)  natural circulation boiler

    (c)  internally fired boiler            (d)  externally fired boiler

(v)  Which of the following is a high pressure boiler?

    (a)  La-Mont boiler                (b)  Benson boiler

    (c)  Loeffer boiler                 (d)  all of the above

(vi)  The rate of flow of steam in case of water-tube boilers as compared to fire-tube boilers is

    (a)  less                            (b)  more

    (c)  same                          (d)  none of the above.

(vii)  Boilers using Bagasse as fuel use

    (a)  spreader stokers              (b)  underfeed stokers

    (c)  travelling grate stroker        (d)  pulverized fuel burners

    (e)  Chain grate stokers

(viii)  The fitting mounted on the boiler, whose function is to put off the fire in the furnace, when level of water falls to an unsafe limit is called

    (a)  safety valve                 (b)  stop valve

    (c)  fusible plug                 (d)  blow off cock

(ix)  The fitting mounted on the boiler, whose function is to control the flow of steam from boiler to the main steam pipe and to shutt off steam completely when required, is called

    (a)  Safety valve         (b)  stop valve

    (c)  fusible valve         (d)  blow off cock

(x)  Which of the following are boiler mountings?

    (a)  economiser         (b)  fusible plug

    (c)  super heater         (d)  injector

(xi)  If a steam injector lifts the water but fails to force it into the boiler, the trouble may not be due to

    (a)  combining tube clogged         (b)  insufficient steam pressure

    (c)  dry steam         (d)  leak in water pipe

    (e)  any of the above

(xii)  Safety valve generally used on high pressure boiler is

    (a)  dead weight type         (b)  lever type

    (c)  combined lever and spring type         (d)  spring type

(xiii)  With reference to a safety valve the blow down pressure is the pressure at which

    (a)  valve opens         (b)  valve closes after opening

    (c)  valve spring pressure is adjusted         (d)  non of the above

(xiv)  Choose the correct statement

    (a)  size of the boiler tubes is specified by inside diameter

    (b)  an economiser decreases the steam raising capacity of a boiler

    (c)  the basic purpose of a drum in boiler is to serve as storage of steam

    (d)  injector is used for pumping into the boiler and also for heating the feed water

(xv)  Which safety valve of the following should be used for portable boilers?

    (a)  dead weight safety valve         (b)  spring loaded safety valve

    (c)  lever safety valve         (d)  high steam and low water safety valve

(xvi)  The function of junction valve is to

    (a)  regulate the supply of water(b) regulate the flow of steam

    (c)  Put off the fire in the furnace (d) to indicate the level of water

(xvii)  The device, which separates the suspended water particles from steam which is on its way from the boiler to the engine, is called

    (a)  steam seperator         (b)  steam trap

    (c)  injector         (d)  economiser

## Answers

| | | | | |
|---|---|---|---|---|
| (i)  d | (ii) b | (iii)  a | (iv)  d | (v)  d |
| (vi)  b | (vii) a | (viii)  c | (ix)  b | (x)  b |
| (xi)  c | (xii) d | (xiii)  b | (xiv)  d | (xv)  b |
| (xvi)  b | (xvii) b | | | |

**9.2  Objective type questions**

(i)  Boiler draught is defined as

    (a)  the device which produces vacuum in the boiler

    (b)  sum of pressures above and below the fire grate

      (c)  small pressure difference which causes flow of gases to take place inside the boiler

      (d)  non of the above.

(ii)  The draught may be produced by a

      (a)  mechanical fan           (b)  chimney

      (c)  steam jet           (d)  all of these

(iii)  Draught produced by a chimney is known as

      (a)  induced draught           (b)  forced draught

      (c)  natural draught           (d)  none of the above

(iv)  Stack refers to a chimney which is made of

      (a)  masonry structure           (b)  metal

      (c)  concrete structure           (d)  none of the above

(v)  The chimney draught varies with

      (a)  climatic conditions           (b)  temperature of furnace gases

      (c)  height of chimney           (d)  all of these

(vi)  The amount of draught produced by a brick chimney as compared to the draught produced by a steel chimney is

      (a)  more           (b)  less

      (c)  same           (d)  none of the above.

(vii)  The draught in locomotive boilers is produced by

      (a)  chimney           (b)  centrifugal fan

      (c)  steam jet           (d)  none of these

(viii)  The condition for maximum discharge through a chimney is that height of chimney should be

      (a)  equal to the height of column of hot gases producing draught

      (b)  two times the height of column of hot gases producing draught

      (c)  half the height of column of hot gases producing draught

      (d)  non of the above

(ix)  The mechanical draught produces . . . . . . draught than natural draught

      (a)  more           (b)  less

(x)  The mechanical draught . . . . . . the amount of smoke.

      (a)  increases           (b)  decreases

      (c)  does not effect

(xi)  The efficiency of the plant. . . . . . with the mechanical draught

      (a)  increases           (b)  decreases

      (c)  does not effect

(xii)  The type of draught, which is more economical when draught required is more than 40 mm, is

      (a)  natural draught           (b)  artificial draught

      (c)  both (a) and (b)           (d)  none of the above

(xiii) The air pressure at the fuel bed is reduced below that of atmosphere by means of a fan placed at or near the bottom of the chimney to produce a draught. Such a draught is called

    (a) natural draught            (b) induced draught

    (c) forced draught            (d) balanced draught

(xiv) If a fan is installed near the base of a boiler to force air through fuel and another fan is installed near the base of the chimney to suck the gasfrom boiler, the type of draught produced is called

    (a) forced draught            (b) induced draught

    (c) balanced draught         (d) none of the above

(xv) For the same draught developed, the forced draught fan as compared to induced draught fan requires

    (a) more power            (b) less power

    (c) same power            (d) none of the above

(xvi) The density of air at 0°C is equal to

    (a) 1.293 kg/m3            (b) 12.93 kg/m3

    (c) 129.3 kg/m3            (d) 0.1293 kg/m3

(xvii) For maximum discharge of flue gases through a chimney, the draught ($hw$) in mm of water is equal to

    (a) $353 \dfrac{H}{T_a}$            (b) $176.5 \dfrac{H}{T_a}$

    (c) $\dfrac{T_a}{353H}$            (d) $\dfrac{T_a}{176.5H}$

Where $H$ = Height of chimney in meter

$Ta$ = Absolute temperature of outside air

(xviii) The volume per unit mass of air at 0°C is equal to

    (a) 7.734 m3/kg            (b) 0.7734m3/kg

    (c) 0.07734 m3/kg        (d) 1.293 m3/kg

(xix) The power required in kw to run a fan, which produces draught is equal to

    (a) $\dfrac{h_w \, V \cdot \eta}{60000}$            (b) $\dfrac{9.81 h_w}{\eta \times 60000}$

    (c) $\dfrac{9.81 \, hw \, V}{60000}$         (d) $\dfrac{h_w \, \eta}{V \times 60000}$

Where $V$ = volume of gas flowing through fan in m³/min

$\eta$ = Efficiency of fan

$h_w$ = Draught produced by fan in mm of water column

(xx) The velocity of flue gases through the chimney under a static draught of $Hg$ meters is given by

    (a) $4.429 \, Hg$            (b) $4.429 \sqrt{H_g}$

    (c) $(4.429 \, H_g)^2$         (d) $4.429 \, (H_g)^2$

## Answers

|       |   |        |   |         |   |        |   |       |    |
|-------|---|--------|---|---------|---|--------|---|-------|----|
| (i)   | c | (ii)   | d | (iii)   | c | (iv)   | b | (v)   | d  |
| (vi)  | b | (vii)  | c | (viii)  | a | (ix)   | a | (x)   | b  |
| (xi)  | a | (xii)  | b | (xiii)  | b | (xiv)  | c | (xv)  | b  |
| (xvi) | a | (xvii) | b | (xviii) | b | (xix)  | b | (xx)  | b. |

## Theoretical Problems

9.3  State how the boilers are classified.

9.4  Explain the difference between a fire tube and a water tube boiler. State which type of boiler is used for power generation and why?

9.5  Sketch and decribe a Cochran boiler? What are its special features?

9.6  What are the factors involved in the selection of a boiler?

9.7  Sketch and describe the working of a locomotive boiler. Show the positions of fusible plug, blow off cock, feed check valve and superheater. Mention the function of each.

9.8  How draught is created in a locomotive boiler?

9.9  Discuss the chief advantages of water tube boiler over fire tube boiler.

9.10  Explain why the superheated tubes are flooded with water at the starting of the boiler?

9.11  What are the high pressure boilers? How do they differ in construction and working principle from ordinary boilers?

9.12  Sketch and describe a Babcock and Wilcox boiler.

9.13  What are the essential features of a good boiler?

9.14  What is the basic difference between a mounting and an accessory of a boiler? Name six boiler mountings and five boiler accessories.

9.15  Describe the functions and locations of the following:

(a) steam stop valve (b) safety valve (c) feed check valve (d) fusible plug (e) water level indicator and (f) pressure gauge.

9.16  List down all the mountings in the two divisions of safety fittings and control fittings and mention about the numbers of each particular in boiler setting.

9.17  What is the function of a boiler chimney and explain why there is no chimney in the case of a locomotive boiler?

9.18  What is the function of blow off cock? Where it is located?

9.19  With the help of heat sketch, explain an injector for feeding water to the boiler drum. Why it is not used for large capacity boilers? Explain its location in boiler installation.

9.20  What do you understand by "Boiler draught"? What are various methods for producing draught?

9.21  What are the functions of a boiler chimney? Why chimney is not provided to locomotive boiler?

9.22  Explain the terms forced draught, induced draught and balanced draught.

9.23  Prove that the draught produced, in mm of water by a chimney, is given by

$$h = 353\,H\left[\frac{1}{T_a} - \frac{(m_a + 1)}{m_a} \times \frac{1}{T_g}\right]$$

9.24  Prove that the discharge will be maximum for a given height of the chimney, when

$$T_g = 2T_a \frac{(m_a + 1)}{m_a}$$

9.25 Show that under maximum discharge conditions, the draught produced in terms of hot gas column is equal to the height of chimney.

9.26 Define chimney efficiency and derive the expression for the same.

9.27 What do you understand by steam jet draught? What are its advantages and disadvantages?

9.28 Deduce an expression for the power required to drive (a) a forced draught fan, and (b) an induced draught fan.

9.29 Define equivalent evaporation from and at $100°$ C and the boiler efficiency and discuss their practical importance.

9.30 What are the factors affecting boiler efficiency.

9.31 Define economiser efficiency.

9.32 Give a schematic sketch of a boiler plant. What are the observations to be recorded during a boiler trial?

9.33 What are the sources of heat loss in a boiler plant? What are the methods used to reduce these losses?

9.34 What are the uses of conducting a trial on a boiler?

9.35 A 30 m high chimney is used to produce a natural draught of 15 mm of water. The temperature of hot gases in the chimney is 287°C. If the temperature of outside air is 27°C; find the mass of air used per kg of fuel (*Ans*. 15.6 kg/kg of fuel)

9.36 A boiler is equipped with a 25 m high chimney. The ambient temperature is 26°C and the temperature of the flue gases in the chimney is 310°C. The quantity of air supplied to the boiler is 19 kg per kg of fuel burnt. Determine

  (a)  the theoretical draught produced in cm of water

  (b)  the velocity of the flue gases in the chimney if 55% of the theoretical draught is lost in friction at the grate and Passage. (*Ans*. 1.358 cm of water, 13.716 m/s)

9.37 A boiler has a chimney of 35 m height. It produces a draught of 20 mm of water column. The mean temperature of flue gases is 370°C and the boiler house temperature is 34°C. Estimate the air-fuel ratio. Does this boiler satisfy the condition for maximum discharge? Also fined the height of hot gas column under maximum discharge condition. (Poona Univ. 1992)

[*Ans*. nearly fulfills, 35 m of hot gas]

9.38 A chimney is 30 m high which produces a draught of 18 mm of water when the flue gas temperature is 350°C and cold air temperature is 20°C. Determine the amount of air supplied/kg of fuel, the draught produced in metres of hot gas column and the condition for maximum discharge.

(*Ans*. 14.85 kg/kg of fuel, 29.76 m, 625.46 k)

9.39 A steam power plant developing 1700 kW has a chimney of 50 m height. The coal consumption is 3 kg/ kw hr. The amount of air supplied is 18 kg per kg of coal burnt. Assuming that the air temperature is 27°C and discharge through the chimney is maximum, determine the diameter of the chimney.

(*Ans*. 1.364 m)

9.40 In a steam power plant a chimney of 40 m height produces a natural draught equivalent to 32 m column of hot gases. The mass of flue gas produced per kg of fuel burnt is 20 kg and the ambient air temperature is 30°C. Calculate

(a) temperature of the flue gases leaving the chimney

(b) extra heat carried away by the flue gases if the flue gas temperature could be reduced to 125°C by artificial draught which is the limit imposed due to condensation of water vapour. The mean specific heat of the flue gases is 1.05 kJ/kgk.

(c) the efficiency of the chimney.

(d) percentage of heat of the fuel used for creating the draught.

(e) temperature of the chimney gases and the corresponding draught for maximum discharge condition. Take calorific value of fuel. = 30000 kJ/kg.

(Ans. 301.1°C, 184.905 kJ, 0.1697%, 12.32%, 364.89c, 40 m of gas column)

9.41 A boiler using 1500 kg/hr of fuel requires 20 kg of air per kg of fuel burnt. Temperature of outside air is 27°C and of hot gases is 327°C. Draught required is 70 mm of water. Find the motor power of I.D. and F.D. fans if their efficiencies are 80%.

(Ans. 12.158 kW, 6.079 kW)

9.42 A forced draught fan delivers air at 10m/s against a draught of 25 mm of water across the grate. Air supplied is 13 kg per kg of fuel when the fuel burnt is 9000 kg/hr. Find the fan power required if its mechanical efficiency is 80%. Assume the atmospheric pressure and temperatures to be 10N/cm2 and 27°C respectively. (*Ans.* 10.598 kW)

9.43 Find the total power consumed by forced and induced draught fans when balanced draught is used in a boiler. The pressure loss before and after the furnace are 50 and 100 cms of water respectively. The boiler uses 3,00,000 kg/hr of fuel with *A/F* ratio of 18.75 and outside air is 27°C and the flue gases are at 170°C.F.D and I.D. fan efficiencies are 0.7. . [Mumbai Univ. Nov. 1997]

(Ans 9305.15 + 13740.61 = 23045.76 kW)

9.44 A boiler evaporates 8 kg of water per kg of coal fired from feed water at 46° c when working at 10 bar absolute. Determine the equivalent evaporation from and at 100°C per kg of coal fired when the steam produced is (i) 0.92 dry (ii) dry saturated and (iii) superheated to 250°C.

[Ans. 8.594 kg/kg of coal, 9.166 kg, 9.687 kg]

9.45 A boiler is supplied with 15,900 kg of feed water at 15°C during a trial period of 8 hours 20 minutes. The mass of water at the end of trial was 680 kg less than that at the commencement of trial. The steam is produced at 12.5 bar and 0.95 dry. The coal fired has a calorific value of 30000 kJ/kg and the coal consumed is 275 kg/hr Calculate

(i) Actual evaporation per kg of coal.

(ii) Equivalent evaporation per kg of coal.

(iii) Thermal efficiency of the boiler.

[Poona Univ. 1993]

[Ans. (i) 7.235 kg/kg of coal (ii) 8.4 kg/kg (iii) 63.3%]

9.46 A boiler generates steam at 18 bar and 325° C. Feed water is supplied at 41.45°C. The thermal efficiency of the boiler is 80%. It uses furnace oil of C.V. 45500 kJ/kg. The steam is supplied to a turbine producing 500 kW power, having a specific steam consumption of 10 kg/hr kw. Calculate the furnace oil consumption in kg/hr and equivalent evaporation.

[Poona Univ. 1992]

[Ans. 395.72 kg/hr, 6382.55 kg/hr]

9.47 The following readings were noted during are hour trial on a boiler:

$$\text{coal burnt} = 750\,\text{kg}$$
$$\text{water fed} = 7000\,\text{kg}$$
$$\text{C.V of coal} = 33500\,\text{kJ/kg}$$

Feed water temperature entering the economiser = 35 °C feed water temperature leaving the economiser = 90 °C. Boiler working pressure is 10 bar and steam dryness 0.96. Temperature of steam leaving the superheater is 250 °C. Determine the thermal efficiency of the plant and also calculate the enthalpy received by feed water in various components as % of total enthalpy increase. [Poona Univ . 1990]

[*Ans.* ECO. 8.24%, Boiler 82.91%, Superheater = 8.58% $\eta$th= 77.91%]

9.48 A steam generating plant consists of a boiler , superheater and economiser . The temperature of water entering and leaving the economiser are 40 °C and 1 10°C. The pressure and quality of the steam leaving the boiler are 16 bar and 0.99 drynness. The temperature of the steam leaving the superheater is 300°C. The steam generating rate is 8 kg/kg of coal. Assuming H.C.V of coal is 33500 kJ/kg, find the (i) Efficiency of the plant (ii) Efficiency of the boiler alone (iii) Increase in ef ficiency due to economiser alone (heat gained in the economiser by water as a percentage of heat generated). (iv) increase in efficiency due to superheater alone. (Heat gained by steam as a percentage of heat generated).

[*Ans.* (i) 68.4%, 55.2%, 7%, 6.15%]

9.49 The following results were obtained in a boiler trial. Feed water per hour = 700 kg at 27 °C steam produced at 8 bar , 0.97 dry , coal used = 100 kg/hr of C.V = 25000 kJ/kg, ash and unburnt coal collected from beneath the grate bars = 7.5 kg/hr of C.V. of 2000 kJ/kg, mass of flue gases produced per kg of fuel = 17.3 kg, flue gas temperature = 327°C, temperature of air in the room = 16°C, specific heat of flue gases = 1.025 kJ/kg k. Draw the ener gy balance on minute basis. What is the boiler efficiency? [Poona Univ . 1986]

[*Ans.* 72.6%]

9.50 The following observations were made during a boiler trial: Weight of feed water = 1520 kg/hr, temperature of feed water = 30°C, steam pressure = 8.5 bar, steam temperature = 172.9°C, dryness fraction of steam = 0.95, coal burnt = 200 kg/hr, calorific value of coal = 27200 kJ/kg, ash and unburnt coal collected = 16 kg/hr, calorific value of ash and unburnt coal = 2720 kJ/kg, weight of flue gases = 17.3 kg/kg of coal, temperature of flue gases = 330°C, boiler room temperature = 18°C, $Cp$ for gases = 1.006 kJ/kg °C. Calculate the boiler efficiency . [*Ans.* 70%]

9.51 Compare the thermal efficiency of two boilers for which the data are given below:

|  | Boiler 1 | Boiler 2 |
| --- | --- | --- |
| Steam pressure | 14 bar | 14 bar |
| Steam produced per kg of coal fired | 10 kg | 14 kg |
| Quality of steam | 0.9 dry | Superheated to 240°C |
| Feed water temperature | 27°C | 27°C |
| Calorific value of fuel | 34000 kJ/kg | 46000 kJ/kg |

Specific heat of feed water is 4.187 kJ/kgk and specific heat of steam is 2.1 kJ/kgk.

[*Ans.* 7.29% (Boiler 1), 79.5% (Boiler 2)]

9.52 The following data refer to a boiler trial; coal used per hour = 6100 kg having a moisture content of 2% and gross calorific value (dry) of 35400 kJ/kg, the ultimate analysis of dry coal being $c = 84\%$, $O_2 = 4\%$, $H_2 = 4\%$, ash = 8% by mass. The dry flue gas analysis was, by volume $CO_2 = 9.5\%$, $O_2 = 10.48\%$ and 80.02%. Temperature of flue gas was = 300°C, boiler house temperature = 30°C, mean specific heat of dry flue gas = 1.005 kJ/kgk and of steam in flue gas = 2.005 kJ/kgk, mass of steam generated 55000 kg/ hour , dry saturated at 20 bar from feed water at 50°C. Draw up a heat balance per kg of coal above boiler house temperature, given that air contains 23.1% $O_2$ by mass. Assume partial pressure of water vapour in flue gas = 0.07 bar .

[*Ans.* mass of dry flue gas = 22 kg, total $H_2O = 0.3804$ kg, mass of air = 21.44 kg, Air for correct combustion = 10.9 kg, heat supplied in raising steam = 24600 kJ, (68.01%), heat in dry flue gas = 31 10 kJ(8.6%), Excess = 2860 kJ(7.91%), in vapour = 1 134 kJ (3.13%), radiation etc. 4466 kJ (12.35%)]

# 10

# Air Compressor

## 10.1  INTRODUCTION AND GENERAL CLASSIFICATION

Purpose of a fan, blower and compressor is to supply air at a higher pressure value. Depending on the discharge pressure value these are categorized. In a fan there is a appreciable change in density of air and thus flow is considered as incompressible like water. The pressure rise in the ring of fan blade is less than 0.05 bar. If the effect of compression of air is substantial and taken into the consideration of its design and applying the laws of thermodynamics, the unit is called a blower or a compressor. In a blower discharge pressure of air reaches upto 2.5 bar.

Mechanically, air compressors are classified into two ways (i) on the basis of mechanism-reciprocating or rotary, (ii) on the basis of flow-positive displacement and non-positive displacement. The classification on the basis of displacement is shown as follows:

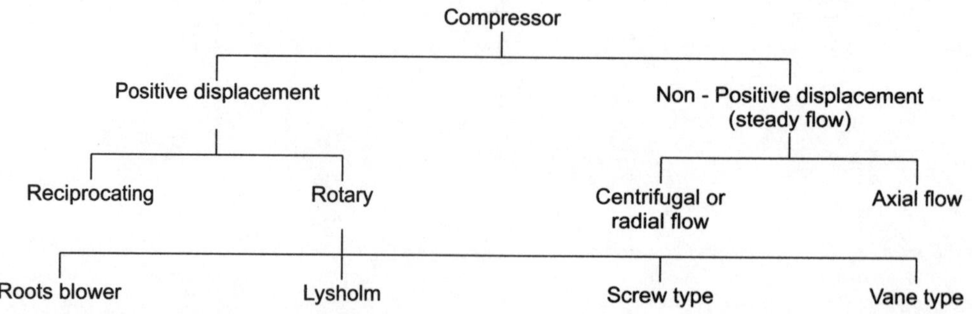


In positive displacement type compressors there is a positive admission and delivery of air preventing undesired reversal of flow due to the arrangement of valves. They have intermittent operation, subjecting the fluid to non-flow processes and work is transferred by the virtue of a hydrostatic force on the moving boundary formed by the piston (in the case of reciprocating compensator). In rotary positive displacement compressors the change in pressure is by backflow of air (in the case of roots blower) or by both squeezing action and back flow of air (vane type).

In non-flow or steady flow type, viz., centrifugal compressors have no means to prevent the reversal of flow. The air is subject to flow processes and the work is transferred by virtue of the change of momentum (dynamic effect) of a stream of air flowing at a high speed over blades, impellor or vanes attached to a rotor.

However, positive displacement compressors can also be regarded as open system supplying air steadily although the internal processes are intermittent as in a closed system.

## 10.2  RECIPROCATING COMPRESSOR

Reciprocating air compressors are suitable for delivering small quantities of air at high pressures. It works on the principle of low discharge with high pressure. The maximum value of air delivered is about 30 m$^3$ per minute with pressure as high as 1000 bar. Due to less inertia to reciprocating parts it is also called a slow speed machine. The compressed air is used for the following applications:

1.  to fill air to automobile tires, bicycles.
2.  to drive air motors which are used in mines
3.  to operate drills, hammers, air brakes for locomotives and railways, water pumps and paint sprays.
4.  to start large diesel engines.
5.  to cool large buildings and aircraft.
6.  to clean workshop machines, generators.
7.  to operate pneumatic drills, lift, cranes.
8.  to start and supercharge of internal combustion engines.
9.  to inject fuel in air injection Diesel engines.
10.  to operate blast furnaces, baseemer, converters used in steel plants.
11.  to supply compressed air in gas turbine plant.

### 10.2.1  Working of Reciprocating Compressor

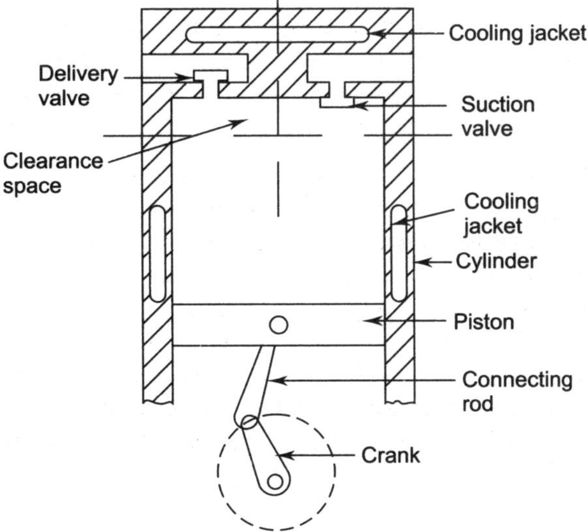


**Fig. 10.1   Single stage compressor**

A single stage compressor is shown in Fig. 10.1 with its parts like cylinder, piston, connecting rod, crank, inlet and outlet valves.

Atmospheric air is sucked inside the cylinder through suction valve during downward movement of piston. It is continued till piston reaches bottom dead centre (*BDC*) position. Soon after suction valve closes and piston starts moving in upward direction and thereby air is compressed inside the cylinder having value more than delivery side. Then delivery valve opens and air is delivered during the remaining upward motion of the piston to the receiver. Then high pressure is left in small quantity in clearance space of the cylinder, which starts expanding during downward movement of piston. Pressure inside the cylinder falls below atmospheric valve and hence suction valve opens, as a result atmospheric air is sucked and the cycle is repeated. The suction, compression and delivery of the air take place within two strokes of the piston or one revolution of the crank.

### 10.2.2   Air compressor Terminology

1. **Single acting compressor**  In this type of compressor, the suction, compression and delivery of the air take place on one side of the Piston while crankshaft completes one revolution.

2. **Double acting compressor**  In this type of compressor, the suction, compression and delivery of the air take place on both sides of the piston. Such compressor would have two delivery strokes per revolution of the crank shaft.

3. **Single stage compressor**  In this type of compressor, compression of air from initial pressure to final pressure is done in one cylinder only.

4. **Multi stage compressor**  In this type of compressor, compression of air from initial pressure to final pressure is done in more than one cylinder.

5. **Compression ratio or pressure ratio**  It is the ratio of absolute discharge pressure to the absolute inlet pressure having value more than one.

6. **Free air delivered (FAD)**  It is the actual volume delivered at the stated pressure and reduced to intake pressure and temperature. It is expressed in cubic meter per minute.

7. **Displacement of the compressor**  It is swept volume achieved by first stage piston in cubic meter per minute.

8. **Compressor capacity**  It is the actual free air delivered by the cylinder per cycle or per minute. It is always less than the displacement of the stroke.

9. **Volumetric efficiency**  It is the ratio of the actual capacity of the compressor to the piston displacement of the compressor.

### 10.2.3   Work in reciprocating compressor

(a) **Without clearance**

Fig. 10.2 shows the theoretical *p-v* diagram for a single stage and single acting air compressor without clearance.

Process 4-1 represents suction of air at pressure $P_1$.

Process 1-2 represents compression of air polytropically

Process 2-3 represents the discharge of air to the receiver at pressure $P_2$.

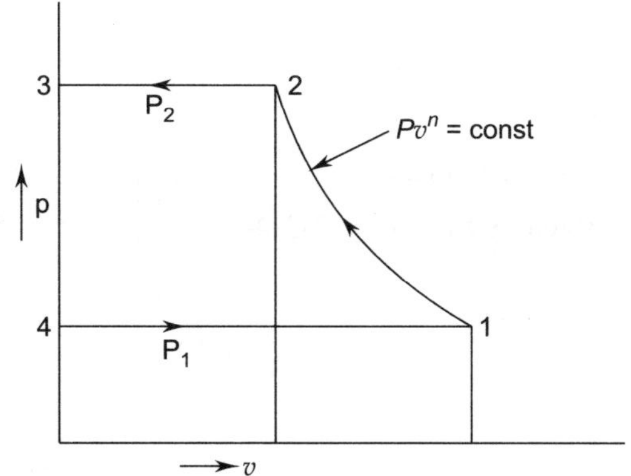


**Fig. 10.2   Theoretical *p-v* diagram**

Area enclosed in the diagram 4-1-2-3-4 represents the work done by cycle.

Work required,
$$W = p_2 v_2 + \frac{p_2 v_2 - p_1 v_1}{n-1} - p_1 v_1$$

$$= (p_2 v_2 - p_1 v_1) + \frac{p_2 v_2 - p_1 v_1}{n-1} = (p_2 v_2 - p_1 v_1)\left(1 + \frac{1}{n-1}\right)$$

$$= \frac{n}{n-1}(p_2 v_2 - p_1 v_1) = \frac{n}{n-1} mR(T_2 - T_1)\ \text{Nm/cycle} \qquad (10.1)$$

$$= \frac{n}{n-1} p_1 v_1 \left(\frac{p_2 v_2}{p_1 v_1} - 1\right) = \frac{n}{n-1} p_1 v_1 \left[\left(\frac{p_2}{p_1}\right)^{\frac{n-1}{n}} - 1\right] \qquad (10.2)$$

$$= \frac{n}{n-1} mRT_1 \left[\left(\frac{p_2}{p_1}\right)^{\frac{n-1}{n}} - 1\right]\ \text{Nm/cycle} \qquad (10.3)$$

for per kg of air *i.e.* $m = 1$ kg, the above equation reduces to

$$W = \frac{n}{n-1} RT \left[\left(\frac{p_2}{p_1}\right)^{\frac{n-1}{n}} - 1\right]\ \text{Nm/kg} \qquad (10.4)$$

## (b)  **With clearance**

Fig. 10.3 shows the theoretical p-v diagram for a single acting air compressor with clearance.

A certain clearance space is provided between top dead centre (*TDC*) position and cylinder cover to prevent the piston from striking the cover. The volume thus lift unswept is known as clearance volume $(v_c)$. The volume of air drawn in at the end of suction stroke is $(v_a)$ whereas, $v_s$ is the swept volume.

Process 4-1 represents suction

Process 1-2 represents polytropic compression

Process 2-3 represents delivery

Process 3-4 represents polytropic expansion

Work required per cycle is given by the area 1-2-3-4-1 on p-$v$ diagram.

$$\therefore \quad W = \text{Area } 1\text{-}2\text{-}5\text{-}6\text{-}1 - \text{Area } 3\text{-}4\text{-}5\text{-}6\text{-}3$$

$$= \frac{n}{n-1} p_1 v_1 \left[ \left( \frac{p_2}{p_1} \right)^{\frac{n-1}{n}} - 1 \right] - \frac{n}{n-1} \left[ \left( \frac{p_3}{p_4} \right)^{\frac{n-1}{n}} - 1 \right]$$

As $p_3 = p_2$ and $p_4 = p_1$

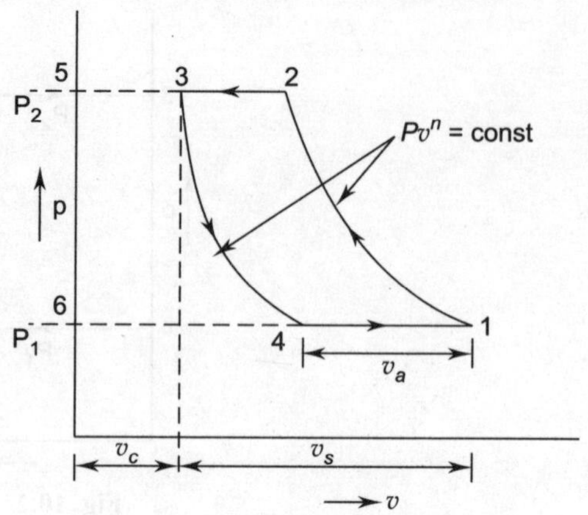

Fig. 10.3  Theoretical $p$-$v$ diagram

$$\therefore \quad W = \frac{n}{n-1} p_1 \left( v_1 - v_4 \right) \left[ \left( \frac{p_2}{p_1} \right)^{\frac{n-1}{n}} - 1 \right] \tag{10.5}$$

$$= \frac{n}{n-1} p_1 v_a \left[ \left( \frac{p_2}{p_1} \right)^{\frac{n-1}{n}} - 1 \right] \tag{10.6}$$

where, actual volume $\qquad v_a = (v_1 - v_4)$

$$\therefore \qquad W = \frac{n}{n-1} m_1 R T_1 \left[ \left( \frac{p_2}{p_1} \right)^{\frac{n-1}{n}} - 1 \right] \text{ Nm/cycle} \tag{10.7}$$

for per kg of air *i.e.* $m = 1$ kg, the above equation reduces to

$$W = \frac{n}{n-1} R T_1 \left[ \left( \frac{p_2}{p_1} \right)^{\frac{n-1}{n}} - 1 \right] \text{ Nm/kg} \tag{10.8}$$

Eqns (10.4) and (10.8) clearly suggest that clearance volume does not affect the work of compression per kg of air.

### 10.2.4  Volumetric efficiency

As mentioned in 10.2.2, volumetric efficiency

$$\eta_v = \frac{\text{Actual volume}}{\text{Swept or stroke volume}} = \frac{v_a}{v_s} = \frac{v_1 - v_4}{v_1 - v_3}$$

clearance volume is always represented in terms of percentage of stroke volume

*i.e.* $\qquad c = \frac{v_c}{v_s} \times 100 = \text{clearance ratio}$

Actual volume $\qquad v_a = v_1 - v_4$

$\therefore \qquad p_3 v_3^n = p_4 v_4^n$

or,
$$\frac{v_4}{v_3} = \left(\frac{p_3}{p_4}\right)^{1/n} = \left(\frac{p_2}{p_1}\right)^{1/n}$$

$\therefore$
$$v_4 = v_3 \left(\frac{p_2}{p_1}\right)^{1/n}$$

$\therefore$
$$v_a = v_1 - v_3 \left(\frac{p_2}{p_1}\right)^{1/n}$$

$$= (v_s + v_c) - v_c \left(\frac{p_2}{p_1}\right)^{1/n} \qquad\qquad [\because \quad v_1 = v_s + v_c]$$

$$= v_s - v_c \left[\left(\frac{p_2}{p_1}\right)^{1/n} - 1\right]$$

$\therefore$
$$\eta_v = \frac{v_a}{v_s} = \frac{v_s - v_c \left[\left(\dfrac{p_2}{p_1}\right)^{1/n} - 1\right]}{v_s}$$

$$= 1 - \frac{v_c}{v_s} \left[\left(\frac{p_2}{p_1}\right)^{1/n} - 1\right]$$

$$= 1 + c - c \left(\frac{p_2}{p_1}\right)^{1/n} \qquad\qquad (10.9)$$

The ratio $\left(\dfrac{p_2}{p_1}\right)$ is called the pressure ratio. Thus the volumetric efficiency depends upon the pressure ratio and clearance ratio. Volumetric efficiency becomes unity when there is no clearance.

The methods of improving the volumetric efficiency include the following:

(i) Providing clearance as small as possible

(ii) Maintaining low pressure ratio

(iii) Cooling during compression

(iv) Reducing pressure drops at the valves by designing a light weight valve mechanism, minimizing value overlaps and choosing suitable lubricating oils.

## 10.2.5  Isothermal  efficiency

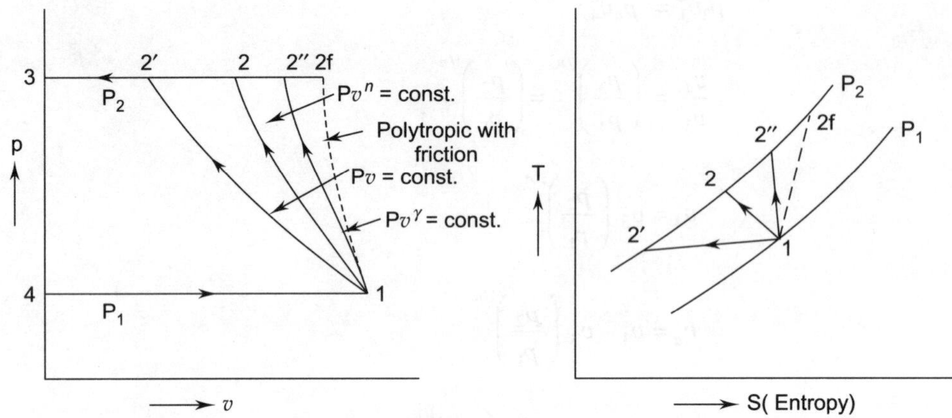


**Fig. 10.4   Isothermal, polytropic and isentropic compression on $p$-$v$ and T-s diagram**

As we know area under $p$-$v$ diagram gives work. Therefore Fig. 10.4 suggests that isothermal process requires least work as compared with polytropic, isentropic and compression (polytropic) with friction $(1 - 2f)$.

Work required per kg of air when compression is polytropic, is given as follows

$$W_a = \frac{n}{n-1} RT_1 \left[ \left( \frac{p_2}{p_1} \right)^{\frac{n-1}{n}} - 1 \right] \text{ Nm/kg} \tag{10.10}$$

Area $1 - 2' - 3 - 4 - 1$ on $p$-$v$ diagram gives work required for isothermal compression which is given as

$$W_i = p_1 v_1 \ln \frac{v_1}{v_2'} + p_2 v_2' - p_1 v_1$$

According to the isothermal condition

$$p_1 v_1 = p_2 v_2' = \text{constant}$$

$\therefore$
$$W_i = p_1 v_1 \ln \frac{v_1}{v_2} = mRT_1 \ln \left( \frac{p_2}{p_1} \right) \text{ Nm/cycle}$$

for per kg of air, *i.e.*  $m = 1$ kg

$$W_i = RT_1 \ln \left( \frac{p_2}{p_1} \right) \text{ Nm/Kg} \tag{10.11}$$

Therefore saving in the work after shifting from polytropic with cooling (process $1 - 2$) to isothermal (process $1 - 2'$) compression would be

$$\Delta W = W_a - W_i$$

where $a$ and $i$ represent actual and isothermal

$$\therefore \quad \Delta W = \frac{n}{n-1} RT_1 \left[ \left( \frac{p_2}{p_1} \right)^{\frac{n-1}{n}} - 1 \right] - RT_1 \ln \left( \frac{p_2}{p_1} \right)$$

$$= RT_1 \left[ \frac{n}{n-1} \left\{ \left( \frac{p_2}{p_1} \right)^{\frac{n-1}{n}} - 1 \right\} - \ln \left( \frac{p_2}{p_1} \right) \right] \qquad (10.12)$$

Isothermal efficiency of the compressor is the ratio of isothermal work to the actual work with polytropic compression.

$$\eta_i = \frac{W_i}{W_a} = \frac{RT_1 \ln \left( \dfrac{p_2}{p_1} \right)}{\dfrac{n}{n-1} RT_1 \left[ \left( \dfrac{p_2}{p_1} \right)^{\frac{n-1}{n}} - 1 \right]}$$

$$= \frac{\ln \left( \dfrac{p_2}{p_1} \right)}{\dfrac{n}{n-1} \cdot \left[ \left( \dfrac{p_2}{p_1} \right)^{\frac{n-1}{n}} - 1 \right]} \qquad (10.13)$$

For isentropic compression, $n = \gamma$ and for isothermal compression, $n = 1$. Higher the value of $\gamma$ for a substance, more the work required and higher the discharge temperature. To reduce the work of compression and also to lower the discharge temperature, it is necessary to resort to cooling during compression. Process 1-2″ represents compression with isentropic condition having, $n = \gamma$. It means after neglecting process of polytropic with friction $(1 - 2f)$, the actual value of compression should lie between $n = 1$ and $n = \gamma = 1.4$ for air. With proper cooling of compressor the practical value of n comes around 1.125 for air.

The term polytropic compression is used in two different senses. It may mean either (a) reversible but non-adiabatic compression, in which heat is removed during the process or, (b) irreversible but adiabatic compression, in which there is friction but not heat transfer, or (c) both.

Reciprocating compressors may approach case (a) with the cooling of the cylinder, provided the velocities are small. Thus n will be less than $\gamma$.

Centrifugal compressors approach case (b), *i.e.* friction effects are considerable but the flow is nearly adiabatic and n will be greater than $\gamma$. We know that the isothermal compression process is the best but it would be extremely slow and is not possible to achieve in practice. Actual compression processes are nearly adiabatic. We can reduce the work of compression to some extent by cooling the work of compressor cylinder and achieve a process of the type (a) viz., polytropic with cooling such as $1 - 2$ in Fig. 10.4. Such a process will have work less than that of an adiabatic process and more than that of an isothermal process. If the value of $\gamma$ for a gas is very high such as 1.4 for air and 1.3 for ammonia and/or the pressure ratio is also high, the compressor cylinders are cooled by water-jacketing. If the value of $\gamma$ is not so high as in the case of fluorocarbons, cooling by air through natural convection is found satisfactory.

Adiabatic efficiency can be defined as

$$\eta_a = \frac{\text{Work done on compressor with polytropic compression}}{\text{Actual work on compressor with adiabatic compression}}$$

$$= \frac{\dfrac{n}{n-1} nRT_1 \left[\left(\dfrac{p_2}{p_1}\right)^{\frac{n-1}{n}} - 1\right]}{\dfrac{\gamma}{\gamma-1} nRT_1 \left[\left(\dfrac{p_2}{p_1}\right)^{\frac{\gamma-1}{\gamma}} - 1\right]}$$

$$= \frac{\dfrac{n}{n-1} \left[\left(\dfrac{p_2}{p_1}\right)^{\frac{n-1}{n}} - 1\right]}{\dfrac{\gamma}{\gamma-1} \left[\left(\dfrac{p_2}{p_1}\right)^{\frac{\gamma-1}{\gamma}} - 1\right]} \tag{10.14}$$

Adiabatic efficiency is the most commonly used term. As we know, the minimum work of compression, with cooling, is isothermal work. Often isothermal efficiency is, therefore, used to express the performance of reciprocating compressors which are invariably cooled.

The minimum work of compression, without cooling, is isentropic work. Adiabatic efficiency is, therefore, used to express the performance of centrifugal compressors in which it is not possible to arrange cooling during compression.

### 10.2.6  Methods for improving isothermal efficiency

As we have seen that during compression of air, temperature $T_2$ increases and the practical values of index of compression n ranges from one (isothermal) to $\gamma$ (isentropic) *i.e.* 1.4 for air. Thus method adopted to keep the value of n nearer to one requires following criteria:

1. **Spray injection.** This is the old method to spray water into the compressor cylinder at the end of compression stroke with the object of cooling the air. This system has some disadvantages.
   (a)  A special gear is required for injection
   (b)  Before using the air the water mixed with it should be separated.
   (c)  There is a possibility to damage cylinder walls and valves because injected water may interferes with cylinder lubrication.
2. **Water jacketing.** During compression of air, heart developed is transferred from cylinder to water jacket. This method is commonly used in all types of reciprocating compressor.
3. **Intercooling.** When speed of the compressor and pressure ratio required are high, water jacketing is not much effective with single stage compressor. Inter-cooling is used in addition to water

jacketing by dividing the compression into two or more stages. The air is compressed in first stage is cooled in a heat exchanger known as inter-cooler to its original temperature before it is taken to the second stage.

4. **External fins.** For small capacity air compressor effective cooling can be achieved with the use of fins on the external surface of the compressor.

5. **Suitable cylinder dimensions.** To have effective heat transfer from cylinder, it should have large diameter and short stroke. But this has the limitation since increased diameter increases clearance volume and hence reduces volumetric efficiency.

### 10.2.7 Effect of Clearance on Work of Compression

Referring the Fig. 10.3, it is considered that law of compression and expansion follow $pv^n = c$. However, in practice the index of expansion and compression may not be the same. Thus the effect of clearance is to reduce the volume of air actually sucked in per working cycle.

Moreover, the effect of the clearance volume on the work of compression is due to the different values of the exponents of the compression and expansion processes. If the exponent of compression is $n_1$ and of expansion is $n_2$, the net work is given by

$$W = \int_1^2 v\,dp + \int_4^3 v\,dp$$

$$= -\frac{n_1}{n_1 - 1} p_1 v_1 \left[ \left( \frac{p_2}{p_1} \right)^{\frac{n_1 - 1}{n_1}} - 1 \right] + \frac{n_2}{n_2 - 1} p_4 v_4 \left[ \left( \frac{p_3}{p_4} \right)^{\frac{n_2 - 1}{n_2}} - 1 \right] \tag{10.15}$$

It two exponents are equal *i.e.* $n_1 = n_2 = n$ and as $p_2 = p_3$ and $p_1 = p_4$

$$W = \frac{n}{n - 1} p_1 (v_1 - v_4) \left[ \left( \frac{p_2}{p_1} \right)^{\frac{n - 1}{n}} - 1 \right] \tag{10.16}$$

$$= \frac{n}{n - 1} mRT_1 \left[ \left( \frac{p_2}{p_1} \right)^{\frac{n - 1}{n}} - 1 \right] \tag{10.17}$$

where $(v_1 - v_4) = v_s$ = volume of air sucked and $m$ is the mass of air passing through the compressor.

Thus the work is only proportional to the suction volume. The clearance air merely acts like spring, alternately expanding and contracting. In practice, however, a large clearance volume results in a low volumetric efficiency and hence large cylinder dimensions, increased contact area between the piston and cylinder and so, increased friction and work.

### 10.2.8 Free Air Delivery (F.A.D)

As defined earlier, *F.A.D* is the volume of air delivered by an air compressor when reduced to atmospheric conditions of pressure and temperature. From continuity equation,

Mass flow rate at the inlet of compressor = Mass flow rate at the outlet of compressor

Since $pv = \dot{m}\,RT$ and assuming clearance

$$\frac{pv}{RT} = \dot{m} \text{ (mass flow rate)}$$

$$\therefore \quad \frac{P_f v_f}{T_f} = \frac{P_1\left(v_1 - v_4\right)}{T_1} = \frac{P_2\left(v_2 - v_3\right)}{T_2} \tag{10.18}$$

If clearance is neglected then,

$$\frac{P_f v_f}{T_f} = \frac{P_1 v_1}{T_1} = \frac{P_2 v_2}{T_2} \tag{10.19}$$

where $p_f$, $v_f$ and $T_f$ are free air conditions and $v_f$ will be *F.A.D.* For convenience, $p_f = 101.325\ kPa$ and $T_f = 288\ K$ is assumed, unless stated otherwise. F.A.D is used in technical literature as an index of compressor capacity.

### 10.2.9  Multi-Stage Compression

For the requirement of compressed air with high pressure, it is essential to create a large pressure ratio in the compressor. This can be done in one cylinder or more than one cylinder in series. The air after compression in the first cylinder and before entering the next cylinder is cooled at constant pressure by passing it through an intercooler.

It is obvious from Eqn. (10.9) that volumetric efficiency of a reciprocating compressor is a function of clearance ratio c, pressure ratio $\left(\dfrac{p_2}{p_1}\right)$ and index of compression. The volumetric efficiency of a compressor with a fixed clearance decreases with an increase in pressure ratio.

This is also obvious from Fig. 10.5 that when pressure ratio increased from $\dfrac{p_2}{p_1}$ to $\dfrac{p_3}{p_2}$, the effective volume sucked in reduces from $(v_1 - v_4)$ to $(v_1 - v_7)$.

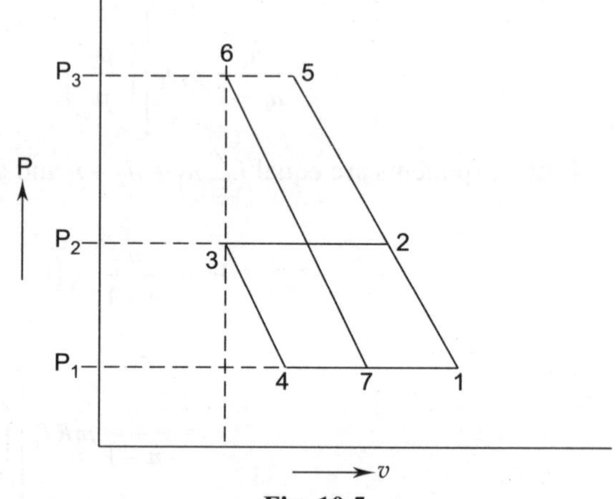

**Fig. 10.5**

Thus it is observed that increase in pressure ratio tends to decrease the volumetric efficiency and at a certain limit it will even approach zero which means no delivery. Therefore, before reaching to this extreme condition, the compression is split into two or more stages rather than simply increase the size of the cylinder to obtain the desired flow. The following arrangement is normally used with the regard to the number of stages:

1. Single stage compression – delivery pressure upto 5.5 bar
2. Two stage compression – delivery pressure from 5.5 to 34.5 bar
3. Three stage compression – delivery pressure from 34.5 to 82.5 bar
4. Four stage compression – delivery pressure above 82.5 bar.

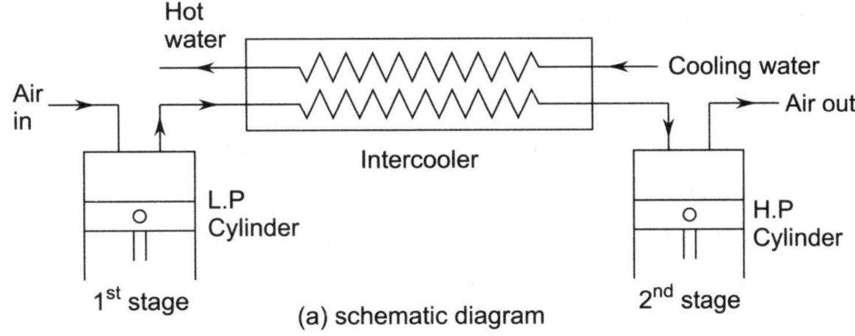



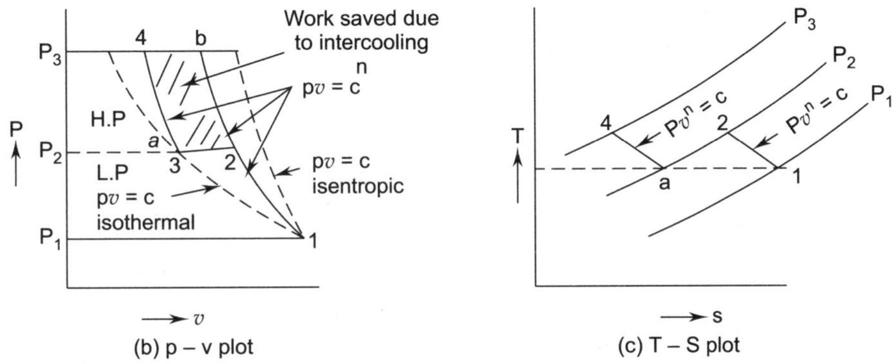



**Fig. 10.6   Two stage reciprocating air compressor with intercooling**

Fig. 10.6 shows the schematic diagram of a two stage reciprocating air compressor along with $p$-$v$ and $T$-$S$ plot. The air is first taken into the low pressure ($L.P$) cylinder. After compression to some desired intermediate pressure $P_2$, the air from the $L.P$ cylinder at condition 2 is passed through intercooler. The condition of air leaving the intercooler is shown by the point 3 where its temperature is reduced from $T_2$ to $T_3$ at same pressure of $p_2$. If the air is cooled to the original temperature $T_1$, the cooling is said to be perfect intercooling. However, if the cooling is not to the original temperature, it is said to be imperfect intercooling. It is possible to cool the air in the intercooler up to initial temperature $T_1$ by properly regulating the supply of cooling water in the intercooler. Finally the air is compressed from condition 3 in the $H.P.$ cylinder and is discharged to the receiver.

For percent cooling, the points 1 and a would lie on the same isothermal curve. Without intercooling the compression would have followed the path 1-2-$b$. The shaded area 2-$a$-4-$b$ represents the saving in work input due to intercooling. Normally, 20% saving is achieved for high pressure rises. Both the cylinders ($L.P.$ and $H.P.$) are mounted on the same shaft and are driven by a prime mover, mostly, electric motor.

### Assumptions in two stage compression with intercooler

1. Effect of clearance is neglected
2. There is no pressure drop in the intercooler
3. The compression in both the cylinders is polytropic *i.e.* follows $pv^n = c$.
4. The suction and delivery of air takes place at constant pressure.

### Advantages of multi-stage compressor

1.  Work required for compression is reduced at same delivery pressure.
2.  Maximum temperature is reduced resulting in little chance of lubrication troubles.
3.  It gives more uniform torque, hence a smaller flywheel is required.
4.  Leakage loss is reduced.
5.  Volumetric efficiency is improved for given pressure ratio.

### 10.2.10   Work done by a two stage reciprocating air compressor with intercooling

Refer Fig. 10.6 for a two stage reciprocating air compressor with intercooling between *H.P.* and *L.P.* cylinders.

**Case 1:** Incomplete intercooling

Work required per cycle to compress air in L.P. cylinder is

$$W_1 = \frac{n}{n-1}\, p_1 v_1 \left[ \left( \frac{p_2}{p_1} \right)^{\frac{n-1}{n}} - 1 \right]$$

Similarly, compression work in *H.P.* cylinder

$$W_2 = \frac{n}{n-1}\, p_2 v_2 \left[ \left( \frac{p_3}{p_2} \right)^{\frac{n-1}{n}} - 1 \right]$$

Total work required,    $W = W_1 + W_2$

$$\therefore \quad W = \frac{n}{n-1}\, p_1 v_1 \left[ \left( \frac{p_2}{p_1} \right)^{\frac{n-1}{n}} - 1 \right] + \frac{n}{n-1}\, p_2 v_2 \left[ \left( \frac{p_3}{p_2} \right)^{\frac{n-1}{n}} - 1 \right]$$

$$W = \frac{n}{n-1}\left[ p_1 v_1 \left\{ \left( \frac{p_2}{p_1} \right)^{\frac{n-1}{n}} - 1 \right\} + p_2 v_2 \left\{ \left( \frac{p_3}{p_2} \right)^{\frac{n-1}{n}} - 1 \right\} \right] \tag{10.20}$$

**Case 2: Complete intercooling**

for complete intercooling $p_1 v_1 = p_2 v_2 = pv = c$, thus above equation reduces to

$$W = \frac{n}{n-1}\, p_1 v_1 \left[ \left( \frac{p_2}{p_1} \right)^{\frac{n-1}{n}} + \left( \frac{p_3}{p_2} \right)^{\frac{n-1}{n}} - 2 \right] \tag{10.21}$$

or,

$$W = \frac{n}{n-1}\, mRT_1 \left[ \left( \frac{p_2}{p_1} \right)^{\frac{n-1}{n}} + \left( \frac{p_3}{p_2} \right)^{\frac{n-1}{n}} - 2 \right] \tag{10.22}$$

Power required to drive a two stage compressor would be, $P = \dfrac{W \times Nw}{60}$ watts

where $\quad Nw$ = No. of working strokes per minute

$Nw$ = N for single acting compressor

$Nw$ = 2$N$ for double acting compressor

$N$ = r.p.m.

## 10.2.11 Minimum work required for a two stage reciprocating compressor (single acting)

Referring Eqn (10.22), written as

$$W = \frac{n}{n-1} \, p_1 v_1 \left[ \left( \frac{p_2}{p_1} \right)^{\frac{n-1}{n}} + \left( \frac{p_3}{p_2} \right)^{\frac{n-1}{n}} - 2 \right]$$

If intake pressure $P_1$ and delivering pressure $P_3$ are fixed, then least value of intercooler pressure $P_2$ is obtained by differentiating the above equation with respect to $P_2$ for which work required to drive the compressor is minimum.

$$\therefore \qquad \frac{dw}{dp_2} = 0$$

$$\therefore \quad \frac{d}{dp_2} \left[ \frac{n}{n-1} \, p_1 v_1 \left\{ \left( \frac{p_2}{p_1} \right)^{\frac{n-1}{n}} + \left( \frac{p_3}{p_2} \right)^{\frac{n-1}{n}} - 2 \right\} \right] = 0$$

Let $\qquad \dfrac{n-1}{n} = \text{constant} = x$

$$\therefore \qquad \frac{1}{x} \frac{d}{dp_2} \left[ p_1 v_1 \left\{ \left( \frac{p_2}{p_1} \right)^{x} + \left( \frac{p_3}{p_2} \right)^{x} - 2 \right\} \right] = 0$$

$$\therefore \frac{1}{x} p_1 v_1 \left[ \left( \frac{1}{p_1} \right)^{x} \times x \times (p_2)^{x-1} + (p_3)^{x} (-x)(p_2)^{-x-1} \right] = 0$$

$$\therefore \qquad (p_1)^{-x} (p_2)^{x-1} = (p_3)^{x} (p_2)^{-x-1}$$

or, $\qquad (p_2)^{x-1} (p_2)^{x+1} = (p_3)^{x} (p_1)^{x}$

or, $\qquad (p_2)^{2x} = (p_3)^{x} (p_1)^{x}$

$$\therefore \qquad p_2 = \sqrt{p_3 p_1} \ \text{condition for perfect intercooling} \quad (10.23)$$

In other words

$$\frac{p_2}{p_1} = \frac{p_3}{p_2} = \left( \frac{p_3}{p_1} \right)^{\frac{1}{2}} \tag{10.24}$$

Putting this condition in Eqn. (10.22), we get

$$W = \frac{n}{n-1} p_1 v_1 \left[ \left( \frac{p_2}{p_1} \right)^{\frac{n-1}{n}} + \left( \frac{p_2}{p_1} \right)^{\frac{n-1}{n}} - 2 \right]$$

$$= 2 \times \frac{n}{n-1} p_1 v_1 \left[ \left( \frac{p_2}{p_1} \right)^{\frac{n-1}{n}} - 1 \right]$$

$\therefore \qquad W = 2 \times \text{work for each stage}$

Similarly work required for 3 stage compressor

$$W = 3 \times \text{work of each stage}$$

Thus for a compressor with $q$ stages,

Minimum work required, $\qquad W = q \times \dfrac{n}{n-1} p_1 v_1 \left[ \left( \dfrac{p_{q+1}}{p_1} \right)^{\frac{n-1}{qn}} - 1 \right]$ \hfill (10.25)

If index of polytropic compression are $n_1$ and $n_2$ in *L.P.* and *H.P.* cylinders respectively, as shown in Fig. 10.7, work required could be written as

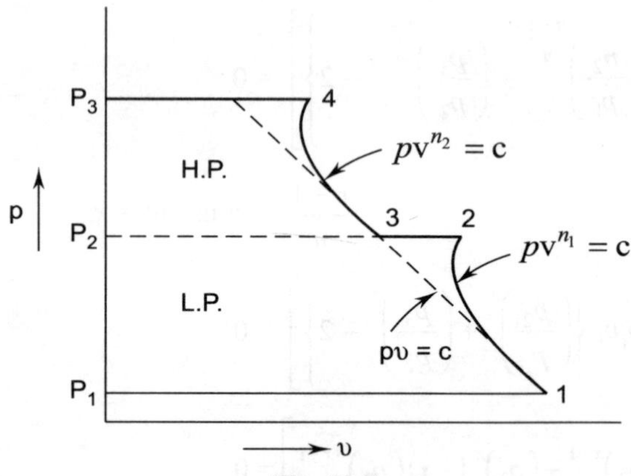


**Fig. 10.7**

$$W = \frac{n_1}{n_1 - 1} RT_1 \left[ \left( \frac{p_2}{p_1} \right)^{\frac{n_1 - 1}{n_1}} - 1 \right] + \frac{n_2}{n_2 - 1} RT_3 \left[ \left( \frac{p_3}{p_2} \right)^{\frac{n_2 - 1}{n_2}} - 1 \right] \qquad (10.26)$$

Assuming $\qquad \dfrac{n_1 - 1}{n_1} = x$ and $\dfrac{n_2 - 1}{n_2} = y$

$$\therefore \qquad W = \frac{1}{x} RT_1 \left[ \left( \frac{p_2}{p_1} \right)^x - 1 \right] + \frac{1}{y} RT_3 \left[ \left( \frac{p_3}{p_2} \right)^y - 1 \right]$$

$$\frac{dW}{dp_2} = 0 \text{ for minimum work condition}$$

$$\frac{d\left[ \frac{1}{x} RT_1 \left\{ \left( \frac{p_2}{p_1} \right)^x - 1 \right\} + \frac{1}{x} RT_3 \left\{ \left( \frac{p_3}{p_2} \right)^y - 1 \right\} \right]}{dp_2} = 0$$

$$\therefore \quad \frac{1}{x} T_1(x)(p_1)^{-x}(p_2)^{x-1} + \frac{1}{y} T_3(p_3)^y(-y)(p_2)^{-y-1} = 0$$

or, $\qquad T_1(p_2)^{x-1}(p_1)^{-x} - T_3(p_3)^y(p_2)^{-(y+1)} = 0$

or, $\qquad T_1(p_2)^{x-1}(p_2)^{y+1} - (p_1)^x T_3(p_3)^y = 0$

or, $\qquad (p_2)^{x-1+y+1} = (p_1)^x (p_3)^y \dfrac{T_3}{T_1}$

or, $\qquad p_2^{x+y} = \dfrac{T_3}{T_1} p_1{}^x p_3{}^y$

$$\therefore \qquad p_2 = \left( \frac{T_3}{T_1} p_1{}^x \, y \, p_3{}^y \right)^{\frac{1}{x+y}} \text{ condition for minimum work}$$

$$(10.27)$$

If $\qquad\qquad\qquad n_1 = n_2 = n, \text{ then } x = y$

Therefore, Eqn (10.27) reduces to

$$p_2 = \left[ \frac{T_3}{T_1} \left( p_1 p_3 \right)^x \right]^{1/2x}$$

$$= \left( p_1 p_3 \right)^{1/2} \left( \frac{T_3}{T_1} \right)^{1/2x}$$

or, $\qquad\qquad\qquad p_2 = \sqrt{\left( p_1 p_3 \right) \left( \frac{T_3}{T_1} \right)^{1/2x}}$ $\qquad\qquad (10.28)$

If intercooling is perfect, then $T_3 = T_1$

$$\therefore \qquad\qquad\qquad p_2 = \sqrt{p_1 p_3} \text{ as achieved in Eqn. (10.23)}$$

## 10.2.12 Cylinders dimensions of a multistage compressor

Applying continuity equation for each cylinder with steady state flow,

$$V_{a_1} \rho_1 = V_{a_2} \rho_2 = V_{a_3} \rho_3 = \text{constant} \qquad\qquad (10.29)$$

Where $va_1$, $va_2$ and $va_3$ are the actual volume of air per stroke during suction stroke during respective cylinders.

$$\therefore \qquad pV = mRT$$

or,

$$\frac{m}{V} = \frac{P}{RT}$$

or,

$$\rho = \frac{P}{RT}$$

$$\therefore \qquad V_{a1}\frac{P_1}{RT_1} = V_{a2}\frac{P_2}{RT_2} = V_{a3}\frac{P_3}{RT_3} = \cdots$$

If perfect intercooling takes place

*i.e.*

$$T_1 = T_2 = T_3$$

$$\therefore \qquad V_{a1}p_1 = V_{a2}p_2 = V_{a3}p_3 = \cdots$$

$$\because \qquad \eta_v = \frac{V_a}{V_s}, \; V_s = \text{stroke volume}$$

$$\therefore \qquad V_{s1}\,\eta_{v1}p_1 = V_{s2}\,\eta_{v2}\,p_2 = V_{s3}\,\eta_{v3}\,p_3$$

$$\therefore \qquad \frac{\pi}{4}\,D_1^2\,L_1\,\eta_{v1}p_1 = \frac{\pi}{4}\,D_2^2\,L_2\,\eta_{v2}\,p_2 = \frac{\pi}{4}\,D_3^2\,L_3\,\eta_{v3}\,p_3 = \cdots$$

Generally stroke length for all cylinders is the same

*i.e.*

$$L_1 = L_2 = L_3, \, D = \text{cylinder diameter}$$

$$\therefore \qquad D_1^2\,\eta_{v1}\,p_1 = D_2^2\,\eta_{v2}\,p_2 = D_2^3\,\eta_{v3}\,p_3 = \cdots$$

If all cylinders have same clearance as the percentage of stroke

*i.e.*

$$\eta_{v1} = \eta_{v2} = \eta_{v3}$$

$$\therefore \qquad D_1^2\,p_1 = D_2^2\,p_2 = D_3^2\,p_3 \cdots \tag{10.30}$$

## 10.2.13  Heat rejected in reciprocating compressor

Heat produced during compression of air is rejected through cylinder wall fins and in intercooler. Considering per kg of air, heat rejected $q = q_1 + q_2$

where $\quad q_1 = \dfrac{c_v\,(\gamma - n)\,(T_2 - T_1)}{n - 1}$ heat rejected during polytropic compression.

and $\quad q_2 = c_p(T_2 - T_3)$ heat rejected to intercooler

where $\quad \gamma = $ adiabatic index

$\qquad\qquad n = $ polytropic index

$\qquad c_p, c_v = $ specific heat of air at constant pressure and volume respectively

$\qquad\qquad T_1 = $ Inlet air temperature

$\qquad\qquad T_2 = $ Intercooler temperature

$\qquad\qquad T_3 = $ Discharge temperature of air

## 10.2.14  Reciprocating air motor

Air operated motors are suitable forms of power source at places where there are safety requirements to be met as in mining applications. Pneumatic breakers, rammers, vibrators and rivetors form a range of hand tools which have wide applications in constructional work. The operation of such tools is possible with air-operated design.

The cycle in air motor is reverse of reciprocating air compressor. The indicator diagram for a reciprocating air motor is shown in Fig. 10.8.

The high pressure air from supply main is applied to the cylinder of air motor through the valves at pressure $P_1$. The air expands polytropically from 1 to 2. The exhaust valve opens and the air is released at constant volume from 2 to 3 to pressure $p_3 = p_b$ which is the back pressure. Then the air is exhausted along the line 3-4 at constant pressure. At point 4 the exhaust valve closes and the polytropic compression of trapped air takes place along 4 to 5. At point 5 the admission valve opens and the pressure in the cylinder rises rapidly to the inlet pressure $p_1$. Again air is supplied at constant pressure up to the point of cut-off 1.

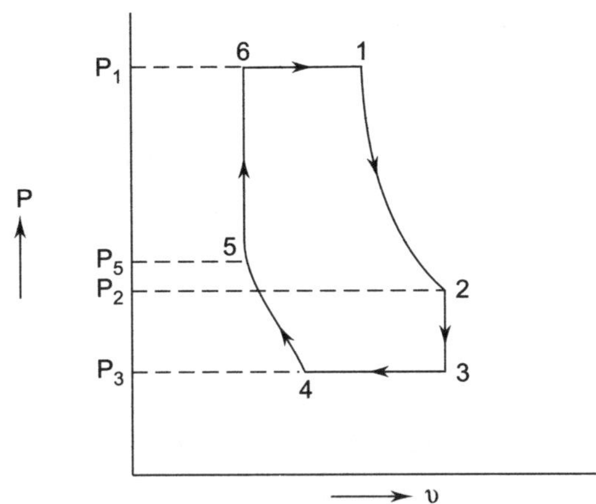

**Fig. 10.8  Hypothetical indicator diagram of an air motor**

The work done is given by

$$w = p_1\left(v_1 - v_6\right) + \frac{\left(p_1 v_1 - p_2 v_2\right)}{n-1} - p_3\left(v_3 - v_4\right) - \frac{\left(p_5 v_5 - p_4 v_4\right)}{n-1} \quad (10.31)$$

## 10.3  POSITIVE DISPLACEMENT ROTARY COMPRESSORS

These types of machines are smaller than reciprocating machines due to continuous rotary action. The compression is adiabatic and these machines are high speed machines and are not generally cooled. Rotary machines can handle larger flow rates of air and are used as superchargers in internal combustion engines and as vacuum pumps. There are many types of positive displacement rotary compressors, the chief among them being roots blower and vane type blower.

(a) **Roots blower.** The arrangement of this blower is shown in Fig. 10.9. It consists of two rotors, rotating in opposite directions. One of the rotors is connected to the drive and the second rotor is gear driven from the first. The lobes of the rotors are of epicycloids, hypocycloid or involute profiles because this ensures correct mating. The high pressure side of the compressor is sealed from low pressure side for all angular positions of the rotors. A small clearance is maintained between the surface to prevent wear. The leakage through the clearance increases with increasing pressure ratio and reduces the efficiency of the compressor.

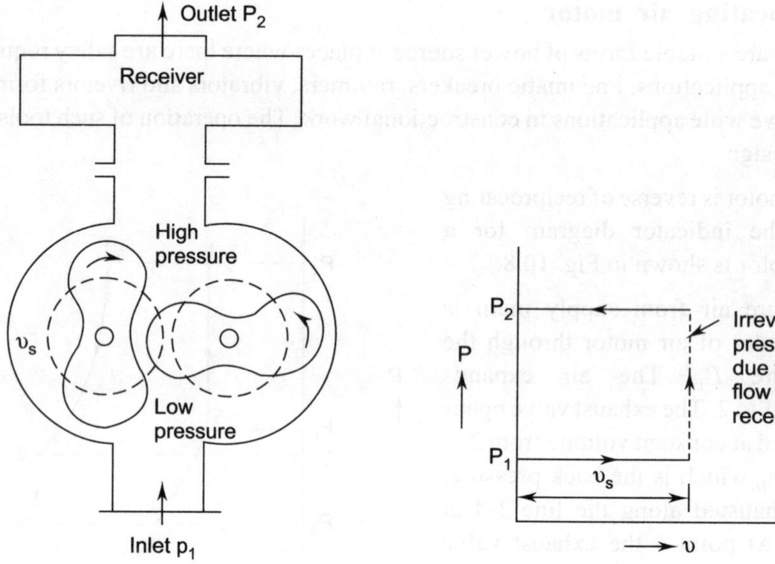



**Fig. 10.9   Roots blower**          **Fig. 10.10   $p$-$v$ diagram for roots blower**

Since in roots blower the flow area does not decrease from entry to exit, theoretically the air should not undergo any pressure rise. During rotation volume of air $v$ at atmospheric pressure $p_1$ is entrapped between the left hand rotor and casing. This air is positively displaced at constant pressure without change in volume until the space opens to high pressure region. At this instant some high pressure air will rush back from the receiver and mix irreversibly with blower air $v$ and compress it until the pressure is equalized. Assuming the receiver to be of infinite size, the equalized pressure would be the receiver pressure $p_2$. This happens four times with a two lobe rotor and six times with a three lobe rotor. The free air delivery for two lobe rotor is thus $4 \times v$ per revolution.

work done per revolution,

$$w = (p_2 - p_1)v \times 4 \text{ Nm for two lobes rotor} \tag{10.32}$$

If $v_s$ is the volume compressed per minute

$$w = (p_2 - p_1)\, v_s$$

which is greater than work required to drive a reversible adiabatic compressor which is equal to

$$w = \frac{\gamma}{\gamma - 1} p_1 v_s \left\{ \left( \frac{p_2}{p_1} \right)^{\frac{\gamma-1}{\gamma}} - 1 \right\}$$

$$\text{Roots efficiency} = \frac{\text{work done isentropically}}{\text{Actual work done}} = \frac{w_i}{w_a}$$

$$= \frac{\dfrac{\gamma}{\gamma - 1} p_1 v_s \left\{ (r_p)^{\frac{\gamma-1}{\gamma}} - 1 \right\}}{p_1 v_s (r_p - 1)}$$

(where $r_p = \dfrac{p_2}{p_1}$ = pressure ratio)

$$\text{or, Roots efficiency} \quad = \frac{c_p}{R}\left\{\frac{\left(r_p\right)^{\frac{\gamma-1}{\gamma}}-1}{\left(r_p-1\right)}\right\} \qquad \because \quad \frac{\gamma}{\gamma-1}=\frac{c_p}{R} \quad (10.33)$$

At low pressure ratios efficiencies up to 80% are achieved while under high pressure ratios, say 2 : 1, the efficiency reached is about 50 to 60% only. The poorer efficiency at high pressure ratios is due to the losses involved in the operating principle.

The roots blower is designed to supply the air from $0.15$ m$^3$/min to about $1500$ m$^3$/minute and the pressure ratios in the order to 1 to 3.6 for single stage machines. The maximum rotational speed used is 12500 r.p.m.

(b)  **Vane type blower.** The arrangement of this blower is shown in Fig. 10.11. It consists of a rotor located eccentrically in a cylindrical outer casing. A set of spring loaded vanes (non-metallic fibre or carbon) mounted in slots of rotor and rotating concentrically in the outer casing. Like roots blower a volume of air $v$ at atmospheric pressure $p_1$ is entrapped between the vanes due to rotation. As the rotation takes place the trapped air first expands from $b$ to $a$ but is then compressed to d when it comes in communication with receiver pressure $p_2$. The back flow from the

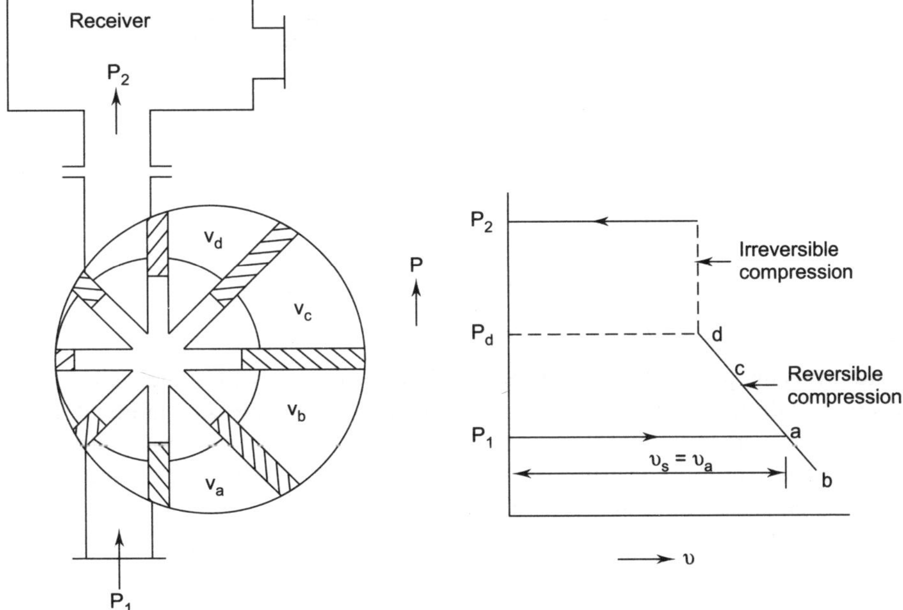


**Fig. 10.11   Vane type blower**

receiver takes place and the pressure equalizes by irreversible compression to $p_2$. This air is then displayed in the receiver. About half the pressure rise takes place by internal reversible compression and half by back flow irreversible compression.

The $p$-$v$ diagram for the compression process is shown in Fig. 10.11. The work done per revolution with $N$ vanes is given as under:

$$W = N \frac{\gamma}{\gamma - 1} p_1 v_1 \left[ \left( \frac{p_d}{p_1} \right)^{\frac{\gamma - 1}{\gamma}} - 1 \right] + N \left( p_2 - p_d \right) v_d \qquad (10.34)$$

For a given air flow and given pressure ratio the vane blower needs less work input than a roots blower. The vane blowers are commonly used for capacities upto 150 $m^3$/min. and pressure ratios up to 8.5:1. The blower speed is limited to 3000 r.p.m.

The vane and root blowers are being superceded by centrifugal compressors (blower) in respect of their use in supercharging aero engines. This is because the centrifugal blowers are comparatively much more efficient. Centrifugal blowers can be more easily fitted into the design of aero-engines, and can be driven at much higher speeds and are much more efficient.

## 10.3(a)   TURBINE VS COMPRESSOR BLADES

An essential difference between turbines and compressors is that in former the passage between the blades is converging and the flow is accelerating, whereas in the latter the passage between the blades is diverging and hence the flow is diffusing or decelerating. The blades profile with passage area is shown in Fig. 10.12.

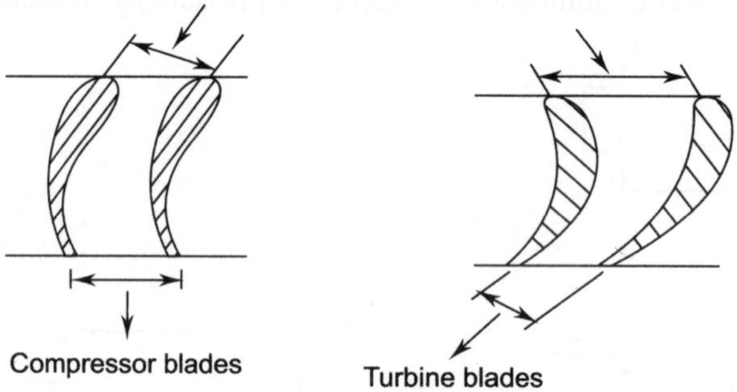


**Fig. 10.12    Blade profile with passage area**

A diffusing flow is less stable than a converging flow.

In too rigid diffusion, particularly in curves, the fluid breaks away from the walls of the diverging passage, reverses its direction and flows back in the direction of the pressure gradient. For this reason blade shape and profile are much more important for a compressor than for a reaction turbine. Whereas turbine blades may have profile consisting of circular arcs and straight line, the compressor blades must be of aerofoil section based on the aerodynamic theory.

The pressure ratio per stage of axial flow compressor is limited to 1.2 to keep gas velocity below the sonic value and to avoid drop in isentropic efficiency.

## 10.4   NON POSITIVE DISPLACEMENT OF STEADY-FLOW ROTARY COMPRESSORS

The two types of non positive or dynamic compressor or steady flow compressors are centrifugal and axial flow. The centrifugal compressor is like a radial flow turbine whereas an axial flow compressor is like an axial flow reaction turbine.

## 10.5   CENTRIFUGAL COMPRESSOR

A centrifugal compressor like a pump is a head or pressure producing device. The contribution of the centrifugal energy in the total change in the energy level is significant. These days centrifugal compressors are used in small power units because in small sizes axial flow compressors can not maintain their high efficiency. It is also used in gas turbines for vehicle application and also as blowers and as supercharger for I.C. engine. The centrifugal compressors are used to supply large quantities of air but at a lower pressure ratio. It (Fig. 10.13) consists of a rotating impeller, diffuser and casing. The impeller rotating at high speed (up to 20,000 to 30,000 r.p.m) consists of a disc on which radial blades are attached. The diffuser is the other important part of the compressor which surrounds the impeller and provides diverging passages for flow thus increasing the air pressure. The air coming out from the diffuser is collected in the casing and taken out from the outlet of the compressor.

The air enters the eye of the compressor at low velocity and atmospheric pressure. The air moves radially outward passing through the impeller and is guided by the impeller vanes. The impeller increases the momentum of the air flowing through it, causing a rise in pressure and temperature of the air. The air leaving the impeller enters the diffuser where its velocity is reduced by providing more cross-sectional area for the air flow. Part of the kinetic energy of the air is converted into pressure energy and pressure of the air is further increased.

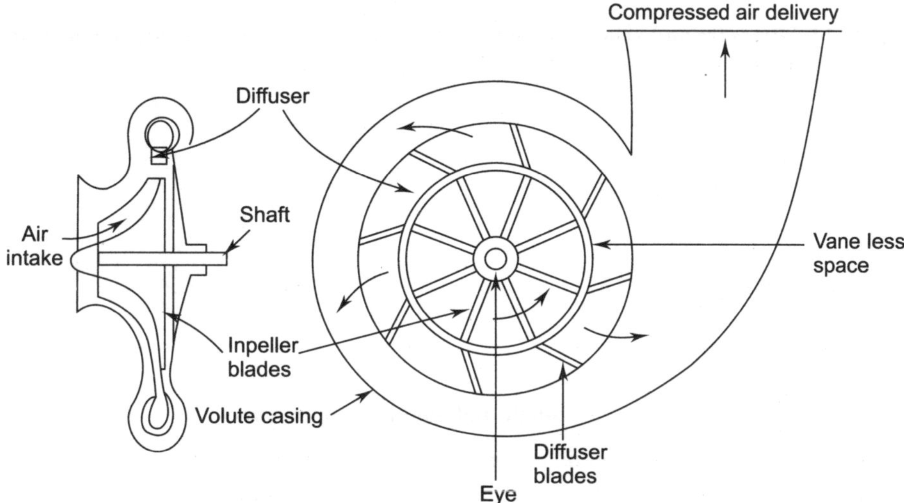


**Fig. 10.13   Centrifugal compressor**

Nearly half of the total pressure rise is achieved in the impeller and remaining half in the diffuser. A pressure ratio of 4 can be achieved with single stage centrifugal compressor. For higher pressure ratios, multistage compressors are used. In multistage compressors, the outlet of the first stage is passed to the second stage and so on. A pressure ratio of 12 : 1 is possible with multi-stage centrifugal compressors.

The change of pressure and velocity of air passing through the impeller and diffuser are shown in the Fig. 10.14

The impellers which are generally used are of two types, single eye type and double eyed type. In single eye type, the air enters into the compressor from one side only and in double eyed type, the air enters from both sides as shown in Fig. 10.15.

The impeller having entries from both ends is subjected to equal in axial forces in opposite directions, which is an advantage over single eye impeller.

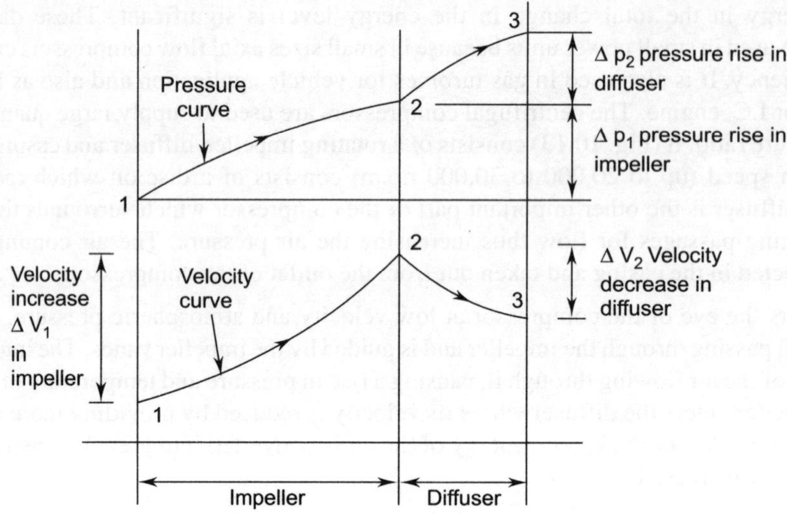



**Fig. 10.14   Pressure and velocity variation of air passing through impeller and diffuser**

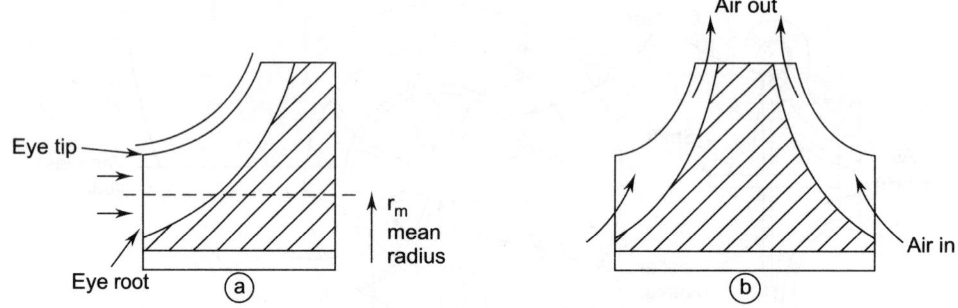



**Fig. 10.15   Single and double eyed impellers**

As we have seen that part of the kinetic energy of air is converted into pressure energy and pressure of the air is further increased. Finally the air at a high pressure is delivered to the receiver.

Let $\quad p_1$ = Initial pressure of air

$\qquad v_1$ = Initial volume of air

$\qquad T_1$ = Initial temperature of air

$p_2, v_2, T_2$ = corresponding values at the final condition

$\qquad m$ = mass of air compressed per minute

Work required to run the compressor for all three processes are listed below:

(a)  Isothermal compression

$$w = p_1 v_1 \ln \left( \frac{v_1}{v_2} \right) = mRT_1 \ln \left( \frac{v_1}{v_2} \right) \qquad (10.35)$$

(b)  Polytropic compression

$$w = \frac{n}{n-1} p_1 v_1 \left[ \left( \frac{p_2}{p_1} \right)^{\frac{n-1}{n}} - 1 \right] = \frac{n}{n-1} mRT_1 \left[ \left( \frac{p_2}{p_1} \right)^{\frac{n-1}{n}} - 1 \right] \quad (10.36)$$

(c)  Adiabatic compression

$$w = \frac{\gamma}{\gamma-1} p_1 v_1 \left[ \left( \frac{p_2}{p_1} \right)^{\frac{\gamma-1}{\gamma}} - 1 \right] = \frac{\gamma}{\gamma-1} mRT_1 \left[ \left( \frac{p_2}{p_1} \right)^{\frac{\gamma-1}{\gamma}} - 1 \right]$$

$$= mc_p (T_2 - T_1) \qquad \left[ \because \ \frac{T_2}{T_1} = \left( \frac{p_2}{p_1} \right)^{\frac{\gamma-1}{\gamma}} \right] \quad (10.37)$$

## 10.5.1  Velocity vector diagram for centrifugal compressor

The following notations are used in the analysis of centrifugal compressor:

Let,   $u_1$ = Linear velocity of the moving blades at inlet

    $V_1$ = Absolute velocity of the air entering the blade

    $V_{f1}$ = Velocity of low at inlet

    $\beta_1$ = Inlet angle to the rotor or impeller. This is the angle which the relative velocity $V_{r1}$ makes with the direction of motion of blade

$m_2, V_2, V_{f2}, V_{r2}, \beta_2 =$               corresponding values at oulet

    $m$ = Mass of air compressed in kg/s

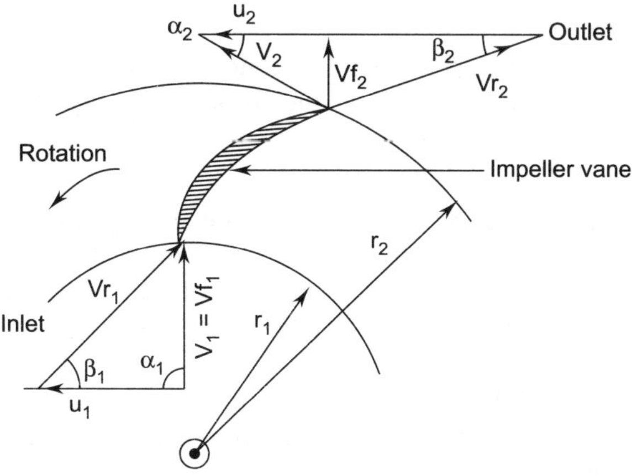


**Fig. 10.16   Velocity diagram for centrifugal compressor**

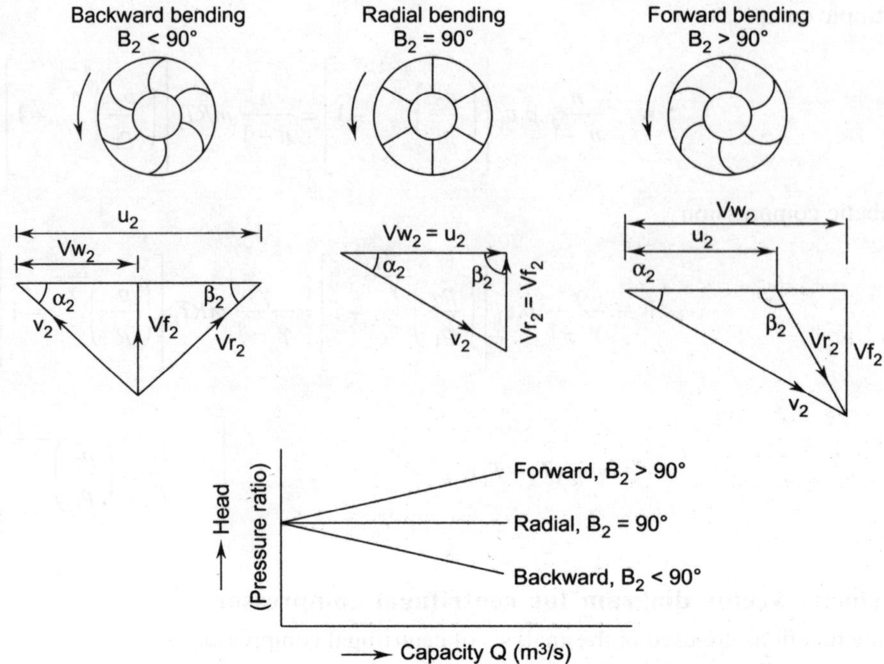

Fig. 10.17   Characteristics of backward, radial and forward curved vanes

The velocity diagram at the inlet and outlet of the impeller are shown in Fig. 10.16. It is assumed that the entry of the air is axial, therefore the whirl component at the inlet $(V_{w_1})$ is zero and therefore $V_1 = V_{f_1}$. The blade must be parallel to the relative velocity of air at inlet or outlet to avoid shock at entry and exit and therefore $\beta_1$ and $\beta_2$ are the impeller blade angles at the inlet and outlet. Velocity diagrams when the blades are backward curved, radial and forward curved are shown in Fig. 10.17. In all the cases, inlet velocity diagram remains same because it is assumed that absolute inlet velocity at inlet is axial as mentioned above. It may be seen from Fig. 10.17 that for backward-curved vanes $(\beta_2 < 90°)$ the tangential component $(V_{w_2})$ is much reduced and consequently for a given impeller speed, the impeller will have a lower energy transfer.

In the case of forward curved vanes $(\beta_2 > 90°)$, $V_{w2}$ is increased and consequently the energy transfer for forward curved vanes is maximum. However, the absolute velocity at impeller outlet $(v_2)$ is also increased. The high value of $V_2$ is not desirable as its conversion into static pressure cannot be very efficiently carried out in the diffuser section.

Normally, backward vanes with $\beta_2$ between $20 - 25°$ are employed except in the case where high head is the major consideration. Sometimes compromise is made between the low energy transfer (backward curved vanes) and high outlet velocity (forward curved vanes) by using radial vanes. Moreover, the radial vanes can be manufactured easily and are free from complex bending stresses.

## 10.5.2   Width of impellor blades

Width or height of the impeller blades of a centrifugal compressor is calculated considering the constant mass flow through the system.

Let          $b_1$ = width or height of the impeller blade at inlet

$D_1 = 2r_1$ = Diameter of the impeller

$$V_{f1} = \text{velocity of flow at inlet}$$

$$v_{s_1} = \text{specific volume of air at inlet}$$

$$b_2, D_2, V_{f_2}, V_{s_2} = \text{corresponding values at outlet}$$

$$m = \text{mass of air flowing through the impeller}$$

$$= \frac{\text{Circumferential area of flow} \times \text{Radial velocity of air}}{\text{Specific volume of air}}$$

$$= \frac{(\pi D_1 - nt) \times b_1 \times v_{f_1}}{v_{s_1}} \tag{10.38}$$

where $n = $ no. of blades or impeller

$t = $ thickness of the blade

If no. of blades and thickness are not considered,

then
$$m = \frac{\pi D_1 \, b_1 \, V_{f_1}}{v_{s_1}} \tag{10.39}$$

Similarly, at the outlet of impeller

$$m = \frac{\pi D_2 \, b_2 \, V_{f_2}}{v_{s_2}}$$

As the mass of air flowing through the impeller is constant, therefore

$$\frac{\pi D_1 \, b_1 \, v_{f_1}}{v_{s_1}} = \frac{\pi D_2 \, b_2 \, V_{f_2}}{v_{s_2}} \tag{10.40}$$

### 10.5.3  Work done on impeller

By Newton's law, the rate of change of angular momentum of air or gas stream is equal to torque applied to the body casing. Thus, for flow rate of Q m$^3$/s

$$\text{Torque} = \rho Q(Vw_2 \, r_2 - Vw_1 \, r_1) \; N\text{-}m \tag{10.41}$$

For ideal case, let it be assumed that the impeller is radial vaned, and there is no frictional loss, no heat transfer and the gas leaves the impeller with a tangential velocity $V_{w_2}$.

For the inlet velocity vector diagram, it has been assumed that air enters the impeller eye in an axial direction, *i.e.* $V_{w_1} = 0$ $\qquad (\because \; \alpha_1 = 90^*)$

and
$$V_1 = V_{f1}$$

$\therefore \qquad \text{Torque} = \rho Q(Vw_2 \, r_2)$

$\therefore \qquad \text{Energy transfer} = \text{torque} \times \text{radian/s}$

$$= \rho Q \, Vw_2 \, r_2 \times \omega$$

$$= \rho Q \, Vw_2 \, u_2 \qquad (\because \; u_2 = r_2 w) \tag{10.42}$$

blade velocity at impeller outlet

Energy transfer per unit mass is called the change of head of fluid and is designated by H.

This is also called Euler work

$$H = Vw_2 \, u_2$$

Power required to run the compressor

$$= \frac{\rho Q V w_2\, u_2}{1000}\ \text{kW}$$

By considering radial bending, $V_2 = V w_2$

$\therefore$ 
$$\text{Work/kg} = V w_2\, u_2 = u_2^2$$

Assuming the heat transfer during the flow of air through the impeller is zero ad considering the steady flow at the inlet and outlet of the impeller

$$h_1 + \frac{V_1^2}{2} - W = h_2 + \frac{V_2^2}{2} - q, \quad [\because \quad q = 0)$$

$\therefore$ 
$$\text{Work/Kg}\ W = \left( h_1 + \frac{V_1^2}{2} \right) - \left( h_2 + \frac{V_2^2}{2} \right)$$

Also, 
$$h_1 + \frac{V_1^2}{2} = c_p \left( T_1 + \frac{V_1^2}{2 c_p} \right) \tag{10.43}$$

Where $T_1$ is the static temperature at inlet. The stagnation or the total temperature at inlet

$$T_{01} = T_1 + \frac{V_1^2}{2 c_p}$$

$\therefore$ 
$$\text{Work/Kg} = c_p (T_{01} - T_{02}) \tag{10.44}$$

While for high compression ratio, where $c_p$ is not constant $\left[ h_1 + \dfrac{V_1^2}{2} \right]$ can be considered as stagnation enthalpy $h_{01}$ in which

$$\text{Work/Kg} = (h_1 - h_2)$$
$$= cp(T_1 - T_2) \tag{10.45}$$

Combining Eqn. (10.44) and (10.45). Neglecting the negative sign for compression

$$\text{Work/Kg} = c_p\, T_{01} \left( \frac{T_{o2}}{T_{ol}} - 1 \right) \tag{10.46}$$

But 
$$T_{01} = \left( \frac{P_{ol}}{P_1} \right)^{\frac{\gamma - 1}{\gamma}} \times T_1$$

Where 
$$T_1 = \text{static temp at } 1, p_1 = \text{static press at } 1$$

and 
$$T_{02} = T_2 \left( \frac{P_{o2}}{P_2} \right)^{\frac{\gamma - 1}{\gamma}}$$

$$\therefore \qquad \text{Work/kg} = c_p\, T_{01} \left[ \left( \frac{P_{o2}}{P_1} \right)^{\frac{\gamma-1}{\gamma}} - 1 \right]$$

$$= c_p\, T_{01} \left[ \left( r_{op} \right)^{\frac{\gamma-1}{\gamma}} - 1 \right] \tag{10.47}$$

In most practical problems, $V_1 = V_2$

Then Eqn. (10.46) and (10.47) are reduced to

$$W = c_p (T_2 - T_1)$$

or, $\qquad\qquad W = c_p\, T_1 \left[ \left( r_p \right)^{\frac{\gamma-1}{\gamma}} - 1 \right] \tag{10.48}$

$$u_2^2 = c_p\, T_{01} \left[ \left( \frac{P_{o2}}{P_{o1}} \right)^{\frac{\gamma-1}{\gamma}} - 1 \right]$$

$$\therefore \qquad \frac{P_{o2}}{P_{o1}} = \left[ \frac{u_2^2}{c_p T_{o1}} + 1 \right]^{\frac{\gamma}{\gamma-1}}$$

if $\qquad\qquad V_1 \simeq V_2$

$$u_2^2 = c_p T_1 \left[ \left( \frac{p_2}{p_1} \right)^{\frac{\gamma-1}{\gamma}} - 1 \right]$$

Thus $\qquad\qquad \frac{p_2}{p_1} = \left[ \frac{u_2^2}{c_p T_1} + 1 \right]^{\frac{\gamma}{\gamma-1}} \tag{10.49}$

Thus the power input to the compressor as per Eqn. (10.48) depends upon the following factors

(a) mass flow of air through the compressor

(b) Total temperature at the inlet of the compressor

(c) Total press. ratio of the compressor which depends upon the square of the impeller tip velocity in the above analysis $p_{01}, p_{02}, T_{01}$ and $T_{02}$ are total head pressures and temperatures at the inlet and outlet if the impeller and $p_1, p_2, T_1$ and $T_2$ are the static pressures and temperatures at the inlet and outlet of the compressor.

### 10.5.4   Slip factor, ($\sigma$)

As per Fig. 10.17, under ideal condition $Vw_2 = u_2$. But due to the inertia of the fluid entrapped between impeller vanes it tends to lag behind the movement of impeller. This phenomenon is known as slip. Therefore, $\beta_2 < 90°$ for radial straight vanes and $Vw_2 < u_2$. Thus slip factor is defined as the ratio of outlet whirl velocity to the blade velocity at outlet. *i.e.* $\sigma = Vw_2/u_2$ (10.50)

## 10.6  AXIAL FLOW COMPRESSOR

In axial flow compressors the flow of fluid is essentially parallel to the axis of the compressor. The blades are arranged as in a reaction turbine, the fixed blades on an outer casing (called stator) while moving blades on a central drum (called rotor) driven by a driven shaft. The reduction in volume may be obtained by flaring the stator or by flaring the rotor. This keeps the axial velocity (flow velocity) constant.

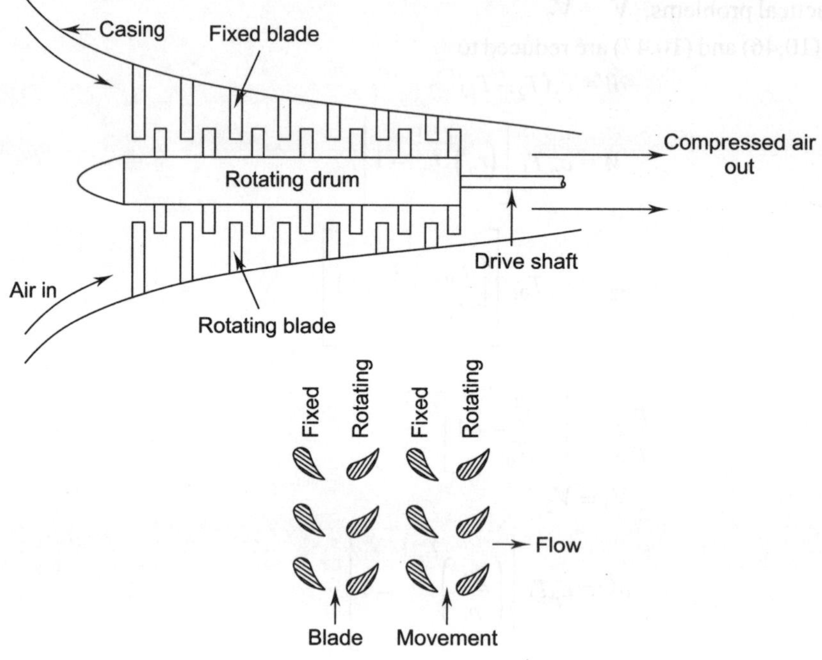


**Fig. 10.18    Axial flow compressor**

One ring of moving blades followed by a ring of fixed blades constitute a stage. The air flows axially through the moving and fixed blades in turn. In the moving blades the enthalpy and pressure rise at the expense of reduction in the relative velocity of the fluid. This is accomplished by providing diffuser passage between blades however, the absolute velocity of the fluid increases due to the work input to the moving blades through the rotor shaft. This increase in kinetic energy is further converted into pressure energy in the fixed blades also guide the fluid at the correct angle into the next stage so that entry is without shock. There is equal pressure and temperature rise in moving and fixed blades and the axial velocity is usually kept constant throughout. Pressure rise per stage is small (say pressure ratio about 1.2). Pressure ratios upto 10 : 1 are obtained by multi-staging. The number of stages used vary from 5 to 14. It is arranged to have work input in each stage.

Each stage is made similar with regard to air velocity and blade angles.

The construction of rotor and casing is similar to a reaction turbine. The rotor is built of discs of steel or light alloy and the blades are fitted into tee-shaped or dove-tailed slots in the periphery of the disc. The stator blades are arranged on to a ring which is turn fixed to the casing.

## 10.6.1  Velocity vector diagram for axial compressor

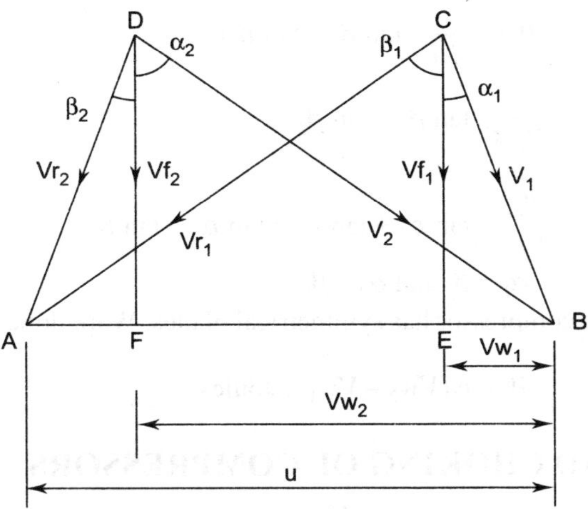



**Fig. 10.19   Combined velocity diagram for axial flow compressor**

Fig. 10.1p9 shows the combined velocity diagram for one stage of rotor blades of an axial flow compressor. All angles are measured from the axial direction and the blade velocity $u$ is taken to be the same at blade entry and exit. This is because the air enters and leaves the blades at almost equal radii. Air approaches the rotating blade with absolute velocity $V_1$ and at an angle $\alpha_1$. The relative velocity $V_{r_1}$ obtained by the vectorial addition of absolute velocity $V_1$ and blade velocity $u$, has the inclination $\beta_1$ with the axial direction. Due to diffusion in the diverging passages formed by rotating blades, there is some pressure rise. This is at the expense of relative velocity and so the relative velocity decreases from $Vr_1$ to $Vr_2$. Since work is being done on the air by rotating blades, the air would ultimately leave the rotating blades with increased absolute velocity $V_2$. The air then enters the stator blades and the diffusion and deacceleration take place in the diverging passage of the stator blades.

One stage consists of one row of rotor and one row of stator blades. There is increase in enthalpy, and therefore of pressure, in both rows. Therefore, degree of reaction comes in picture, which is defined as

$$R \text{ (Degree of reaction)} = \frac{\text{pressure rise in rotor blades}}{\text{pressure rise in the stage}}$$

$$R = \frac{Vr_1^2 - Vr_2^2}{2u\left(Vw_2 - Vw_1\right)} \tag{10.51}$$

From velocity triangles at inlet and outlet, we have

$$Vw_1 = u - V_f \tan \beta_1, \quad V_{w2} = u - V_f \tan \beta_2$$

$$\therefore \qquad Vw_2 - Vw_1 = V_f (\tan \beta_1 - \tan \beta_2) \text{ and}$$

$$Vr_2^2 = V_f^2 + (V_f \tan \beta_2)^2, \quad V_{r_1}^2 = V_f^2 + (V_f \tan \beta_1)^2$$

$$\therefore \qquad Vr_1^2 - Vr_2^2 = V_f^2 (\tan^2 \beta_1 - \tan^2 \beta_2)$$

$$\therefore \qquad R = \frac{V_f^2 \left(\tan^2 \beta_1 - \tan^2 \beta_2\right)}{2u V_f \left(\tan \beta_1 - \tan \beta_2\right)} = \frac{V_f \left(\tan \beta_1 + \tan \beta_2\right)}{2u} \tag{10.52}$$

The value of degree of reaction is taken as 0.5.

$$\therefore \qquad 0.5 = \frac{V_f}{2u}\,(\tan\beta_1 + \tan\beta_2)$$

$$\text{or,} \qquad \frac{u}{V_f} = \tan\beta_1 + \tan\beta_2$$

$$\text{But} \qquad \frac{u}{V_f} = \tan\alpha_1 + \tan\beta_1 = \tan\alpha_2 + \tan\beta_2$$

$$\therefore \qquad \alpha_1 = \beta_2 \text{ and } \alpha_2 = \beta_1$$

So for 50% reaction, the compressor has symmetrical blades. Work done by the compressor is given by

$$W = m(Vw_2 - Vw_1)u \text{ Joules} \tag{10.53}$$

## 10.7   SURGING AND CHOKING OF COMPRESSORS

The theoretically delivery pressure ratio vs. mass flow in a dynamic compressor at constant speed is shown in Fig. 10.20. When the delivery valve is shut the pressure developed is given by point '*a*'. This pressure is equal to the centrifugal pressure head produced by the action of the impeller on the air entrapped between the vanes. As the delivery valve is opened, flow starts and the pressure ratio increases with the flow because of the contribution of diffuser to pressure rise. At some point '*b*' the pressure ratio reaches a maximum value and the efficiency is maximum. Increasing the mass flow further reduces the pressure ratio. From the points '*a*' to '*b*' where the characteristics has a positive slope, the flow is unstable due to the phenomenon of surging.

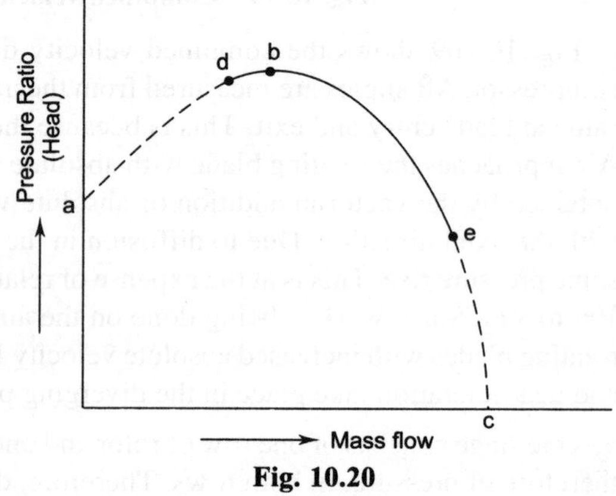


**Fig. 10.20**

This phenomenon can be explained by taking some point '*d*' between the points '*a*' and '*b*'. At this point decrease in mass flow is accompanied by fall in delivery pressure. Now if the pressure in the pipe line downstream of the compressor does not fall off quickly the air will have a tendency to reverse its direction and flow upstream to compressor in the direction of the pressure gradient. When this occurs after certain time the pressure downstream of the compressor falls and hence the compressor picks up again the air flow is back to its normal direction this to, and fro motion of the air will cause pulsation, i.e. noise and vibration and this phenomenon is known as surging.

When the mass flow is increased beyond '*b*' the pressure ratio falls and the efficiency also falls because the air angles become widely different from the vane angles causing breakaway of the air. Finally, point '*c*' is reached where the pressure ratio is unity and efficiency zero. All the power is absorbed in overcoming internal

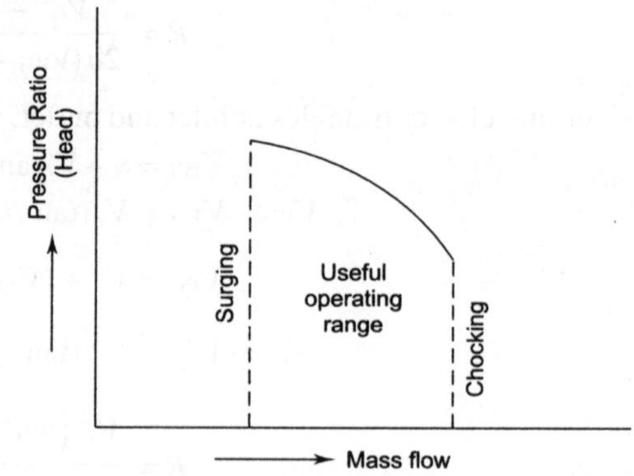


**Fig. 10.21**

friction resistance. The point '*e*' on the curve '*bc*' represents the maximum mass flow which is practically possible in a compressor and is known as choking mass flow. Chocking means fixed mass flow rate regardless of pressure ratio *i.e.* characteristic becomes vertical as shown in Fig. 10.21.

The surging causes overheating and stress reversal in the vanes or blades and damages the machine. The region between the points '*a*' and '*b*' is known as surging region and it is objectionable to operate the machine in this region. It is not possible to get the curve '*ab*' and '*ce*' due to surging and chocking of the compressor. The compressor has to work in the region '*be*'.

## 10.8 DIFFERENCE BETWEEN CENTRIFUGAL AND AXIAL FLOW COMPRESSOR

| *Properties* | *Centrifugal compressor* | *Axial flow compressor* |
|---|---|---|
| 1. Direction of flow | flow is in a plane perpendicular to the axis of compressor | Flow is in a plane parallel to the axis of compressor |
| 2. Pressure ratio | For single stage 4 : 1 | For single stage 1.2 : 1 |
| 3. Efficiency | Isentropic efficiency = 70 – 80% with multistage | Isentropic efficiency = 85 – 90% with multistage |
| 4. Starting torque | Low starting torque | High starting torque |
| 5. Frontal area | Large | Less for same mass flow rate |
| 6. Manufacturing & running cost | Low | High |
| 7. Suitability for multi-scope | Not suitable | Suitable |
| 8. Applications | Suitable for supercharging I.C. Engines, aircraft, refrigerants and industrial gases. | Suitable for large gas turbines. |
| 9. Efficiency vs speed curve | More flat Fig. 10.22 | Less flat comparatively Fig. 10.22 |

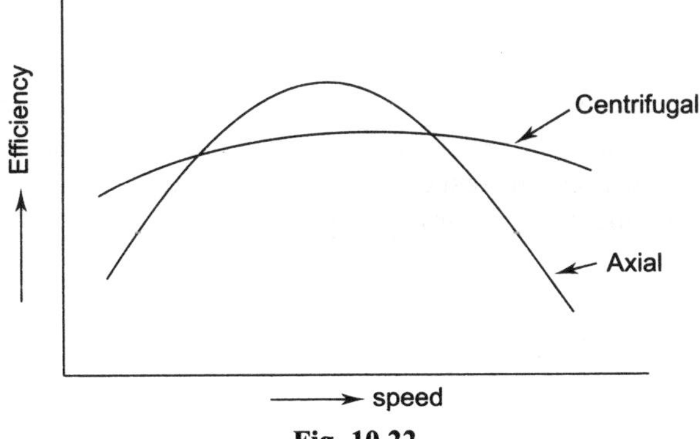


**Fig. 10.22**

## 10.9 STATIC AND TOTAL HEAD (OR STAGNATION) QUANTITIES IN ROTARY COMPRESSORS

A very high velocity of air is encountered in rotary compressor and therefore kinetic energy has to be considered while stagnation quantities come in picture.

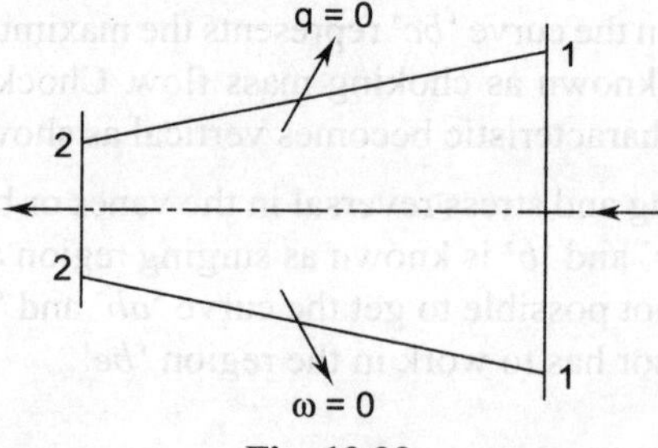

**Fig. 10.23**

Let,     $T$ = Static temperature

$T_o$ = Stagnation or total temperature

$h$ = Static enthalpy

$h_o$ = Stagnation or total enthalpy

$p$ = Static pressure

$p_o$ = Stagnation or total head pressure

$V$ = Velocity of air

Considering a horizontal passage of varying area with no external heat transfer and work transfer as shown in Fig. 10.23. For 1 kg of air flow, by the steady flow energy equation (S.F.E.E),

$$h_1 + \frac{V_1^2}{2} = h_2 + \frac{V_2^2}{2} \qquad (\because \quad \Delta P.E = 0)$$

or,

$$C_p T_1 + \frac{V_1^2}{2} = C_p T_2 + \frac{V_2^2}{2}$$

or,

$$C_p T_1 + \frac{V^2}{2} = \text{constant} \qquad (10.54)$$

The static temperature $T$ is the actual temperature recorded by a moving thermometer alongwith air with same speed. If the air is brought to rest under reversible adiabatic conditions, and without external work transfer, the total kinetic energy of the air is converted into internal energy, increasing the temperature and pressure of the air. The new temperature and pressure of the air are called the total head or stagnation temperature and pressure respectively. The stagnation temperature is the temperature recorded by stationary thermometer when air is brought to rest.

$$C_p T + \frac{V^2}{2} = C_p T_o$$

or,

$$(T_o - T) = \frac{V^2}{2c_p} = \text{(Dynamic temperature)} \qquad (10.55)$$

Similarly,

$$(h_o - h) = \frac{V^2}{2} \qquad (10.56)$$

The total head or stagnation enthalpy remains constant in a moving system in the absence of heat and work transfer.

Similarly,
$$\frac{p_o}{p} = \left(\frac{T_o}{T}\right)^{\frac{\gamma}{\gamma-1}} \qquad (10.57)$$

## 10.10 ISENTROPIC EFFICIENCY IN ROTARY COMPRESSOR

There is a difference between reciprocating and rotary compressor while calculating thermodynamic efficiency of both. For less work required for both compressor, we need isothermal efficiency reciprocating compressor whereas, adiabatic efficiency is considered in rotary compressor. The reason is that in reciprocating compressors due to slow speed, cooling of cylinder and itnerstage cooling

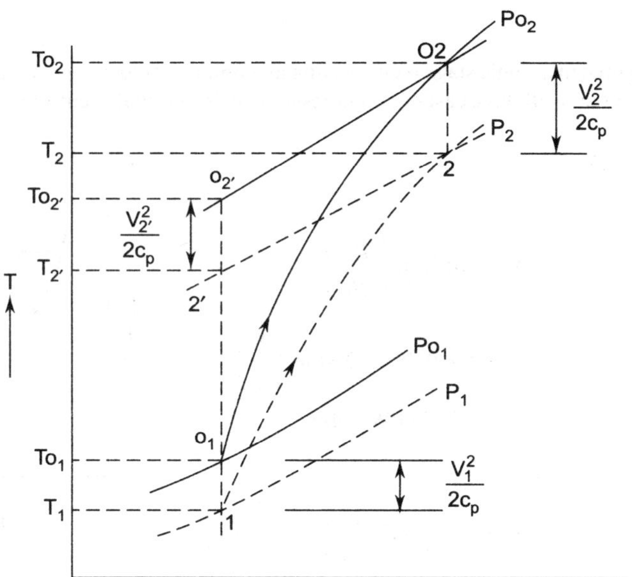


Fig. 10.24   Stagnation and static quantities on T-S diagram

the ideal isothermal compression is approached. In uncooled rotary compressors due to high speed there is friction and eddying which causes internal reheating and, therefore, the index of compression is always more than $\gamma$. Moreover, due to shock at entry and turbulence caused in air in rotary compressor, there is an increase in enthalpy of the air without increase of pressure, therefore, the actual temperature of air coming out from the compressor is more than the temperature of air if it is compressed isentropically. The actually work required for the same increase in pressure ratio is more due to irreversibilities. The actual and isentropic compression with stagnation and static quantities are given in Fig. 10.24.

The isentropic efficiency $\eta_{\text{isen}}$ is defined as the ratio of isentropic work to actual adiabatic work.

*i.e.*
$$\eta_{\text{isen}} = \frac{\text{work to compress isentropically}}{\text{Actual adiabatic work to compress}} = \frac{h'_{o2} - h_{o1}}{h_{o2} - h_{o1}} \qquad (10.58)$$

Assuming, $V_2 = V'_2$, the $\eta_{\text{isen}}$ is given in terms of temperature as

$$\eta_{\text{isen}} = \frac{c_p\left(T_{o2'} - T_{o1}\right)}{c_p\left(T_{o2} - T_{o1}\right)} = \frac{T_{o_{2'}} - T_{o1}}{T_{o_2} - T_{o1}} \qquad (10.59)$$

*i.e.* isentropic efficiency is also defined as the ratio of isentropic temperature rise to actual temperature rise. Eqn. (10.59) can be written as

$$T_{o2} - T_{o1} = \frac{1}{\eta_{isen}} \left( T_{o2'} - T_{o1} \right) = \frac{T_{o1}}{\eta_{isen}} \left( \frac{T_{o2'}}{T_{o1}} - 1 \right)$$

$$= \frac{T_{o1}}{\eta_{isen}} \left[ \left( \frac{p_{o2}}{p_{o1}} \right)^{\frac{\gamma-1}{\gamma}} - 1 \right] \tag{10.60}$$

## Solved Problems

**10.1** Find the work required in a single stage reciprocating air compressor to compress air at the rate of 1 kg/s from 1 bar and 300 k to 8 bar in all three cases of compression i.e. isothermal, isentropic and polytropic with index as 1.25.

*Soln.* Refer Fig. 10.25

**Isothermal**

$$w = mRT_1 \ln \left( \frac{p_2}{p_1} \right)$$

$$= 1 \times 0.287 \times 300 \ln \left( \frac{8}{1} \right)$$

$$= 179.03 \text{ kW} \quad \textit{Ans.}$$

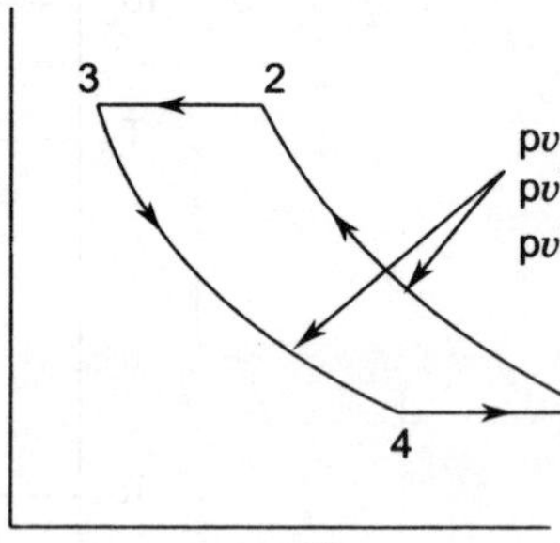

**Isentropic**

$$w = \frac{\gamma}{\gamma - 1} mRT_1 \left[ \left( \frac{p_2}{p_1} \right)^{\frac{\gamma-1}{\gamma}} - 1 \right]$$

$$= \frac{1.4}{1.4 - 1} \times 1 \times 0.287 \times 300 \left\{ \left( \frac{8}{1} \right)^{\frac{1.4-1}{1.4}} - 1 \right\}$$

$$= 244.52 \text{ kW} \quad \textit{Ans.}$$

**Polytropic**

$$w = \frac{n}{n - 1} mRT_1 \left[ \left( \frac{p_2}{p_1} \right)^{\frac{n-1}{n}} - 1 \right]$$

$$= \frac{1.25}{1.25 - 1} \times 1 \times 0.287 \times 300 \left\{ \left( \frac{8}{1} \right)^{\frac{1.25-1}{125}} - 1 \right\}$$

$$= 222.02 \text{ kW} \quad \textit{Ans.}$$

**Fig. 10.25**

**10.2**  A single stage, single acting reciprocating air compressor with 0.3 m bore and 0.4 m stroke length runs at 400 r.pm. The suction pressure is 1 bar at 300 k and the delivery pressure is 5 bar. Find the power required to run it, if compression is isothermal, compression follows the law $pv^{1.3} = C$ and compression is reversible adiabatic. Also find the isothermal efficiency.

*Soln.* Refer Fig. 10.26

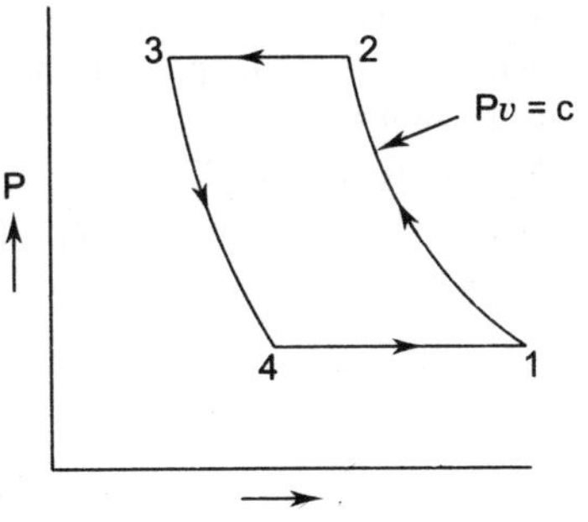

Volume of air compressed per min $= \dfrac{\pi}{4} D^2 LN$

$$= \frac{\pi}{4} \times 0.3^2 \times 0.4 \times 400$$

$$= 11.31 \text{ m}^3/\text{min}$$

Isothermal compression work

$$w = p_1 v_1 \ln\left(\frac{p_2}{p_1}\right)$$

$$= 1 \times 10^5 \, \frac{11.31}{60} \, \ln\left(\frac{5}{1}\right)$$

$$= 30337.9 \text{ W} = 30.34 \text{ kW} \quad \textit{Ans.}$$

Polytropic work of compression

$$w = \frac{n}{n-1} p_1 v_1 \left\{ \left(\frac{p_2}{p_1}\right)^{\frac{n-1}{n}} - 1 \right\}$$

$$= \frac{1.3}{1.3-1} \times 1 \times 10^5 \times \frac{11.31}{60} \left\{ \left(\frac{5}{1}\right)^{\frac{1.3-1}{1.3}} - 1 \right\}$$

$$= 36739.2 \text{ W} = 36.74 \text{ kW} \quad \textit{Ans.}$$

$$\eta_{\text{isothermal}} = \frac{\text{Isothermal work}}{\text{Actual work}} \times 100 = \frac{30.34}{36.74} \times 100 = 82.58\% \quad \textit{Ans.}$$

Isentropic work of compression

$$w = \frac{\gamma}{\gamma-1} p_1 v_1 \left\{ \left(\frac{p_2}{p_1}\right)^{\frac{\gamma-1}{\gamma}} - 1 \right\}$$

$$= \frac{1.4}{1.4-1} \times 1 \times 10^5 \times \frac{11.31}{60} \left\{ \left(\frac{5}{1}\right)^{\frac{1.4-1}{1.4}} - 1 \right\}$$

$$= 38517.5 \text{ W} = 38.52 \text{ kW} \quad \textit{Ans.}$$

$$\eta_{\text{adiabatic}} = \frac{36.74}{38.52} \times 100 = 78.76\% \quad \textit{Ans.}$$

**10.3** A single stage double acting air compressor delivers air at 7.5 bar. The pressure and temperature at the end of suction stroke are 1 bar and 25°C. It delivers 2.2 $m^3$ of free air per minute when compressor is running at 310 r.p.m. The clearance volume is 5% of stroke volume. The pressure and temperature of ambient air are 1.03 bar and 20°C. Determine (i) Volumetric efficiency (ii) Diameter and stroke of the cylinder if $L = 1.5\,D$ and (iii) *I.P* of compressor and *B.P* if $\eta_{mech} = 85\%$. Take, index of compression = 1.25 and index of expansion = 1.3

(Mumbai Univ. summer 2003)

*Soln.* Refer Fig. 10.27

Given

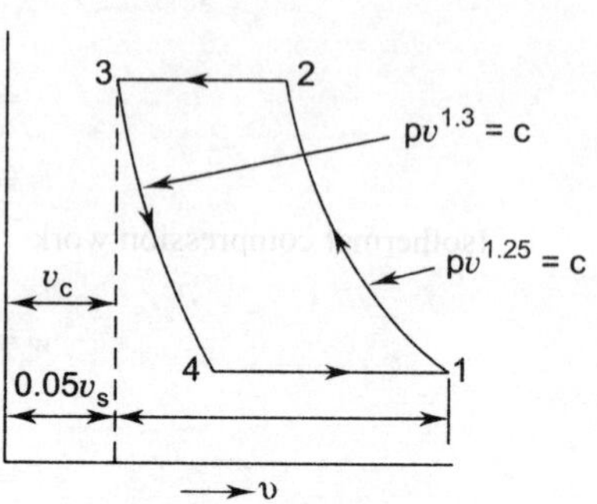


**Fig. 10.27**

$$T_1 = 25 + 273 = 298\ K,\ P_1 = 1\ \text{bar},$$
$$P_2 = 7.5\ \text{bar}$$
$$v_{\text{free air}} = 2.2\ \text{m3/min.}$$
$$N = 310\ \text{r.p.m},\ v_c = 0.05\ v_s$$
$$p_f = 1.03\ \text{bar},\ T_f = 20 + 273 = 293\ \text{k}$$
$$\eta_{mech} = 85\%,\ L = 1.5\ D$$

(i) $\quad \eta_{vol} = \dfrac{\text{Actual volume}}{\text{Stroke volume}} = \dfrac{v_a}{v_s} = \dfrac{v_1 - v_4}{v_1 - v_3}$

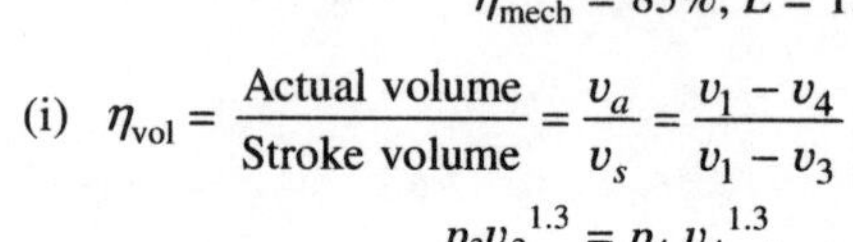

$$p_3 v_3^{1.3} = p_4\, v_4^{1.3}, \qquad \because\ v_3 = v_c$$

$\therefore \qquad v_4 = v_c \left(\dfrac{p_2}{p_1}\right)^{1/1.3} = v_c (7.5)^{0.769} = 4.71\ v_c$

But $\qquad v_c = 0.05\ v_s$

$\therefore \qquad v_4 = 4.71 \times 0.05\ v_s = 0.2355\ v_s$

$\therefore \qquad \eta_{vol} = \dfrac{v_1 - v_4}{v_s} = \dfrac{(v_s + v_c) - 0.02355\ v_s}{v_s}$

$$= 0.814 \times 100 = 81.4\%\quad Ans.$$

(ii) The volume of air delivered at suction condition-Reffering Eqn. (10.18) as

$$\dfrac{p_f v_f}{T_f} = \dfrac{p_1 (v_1 - v_4)}{T_1},\ \text{where } p_f,\ v_f \text{ and } T_f \text{ are free air conditions}$$

$\therefore \qquad (v_1 - v_4) = \dfrac{p_f \times v_f \times T_1}{p_1 \times T_1} = \dfrac{1.03 \times 2.2 \times 298}{1 \times 293}$

$$= 2.305\ m^3/\text{min}$$

The volume delivered per minute

$\because \qquad \eta_{vol} = \dfrac{v_a}{v_s + 2N} = \dfrac{(v_1 - v_4)}{v_s \times 2N}$ (Being the compressor double acting)

$\therefore \qquad v_s = \dfrac{(v_1 - v_4)}{\eta_{vol} \times 2N} = \dfrac{2.305}{0.814 \times 2 \times 310} = 0.00456\ m^3$

$$= 4560\ cm^3$$

$\therefore \qquad \dfrac{\pi}{4}\ D^2 L = 4560$

or, $\qquad \dfrac{\pi}{4}\ D^2 \times 1.5\ D = 4560$

$\therefore \qquad$ Diameter, $D = 15.70$ cm and $L = 23.55$ cm   *Ans.*

(iii)  Indicated power (*I.P*) and Break power (*B.P*)

work required per cycle for double acting

$$w = 2\left[\frac{n_1}{n_1 - 1} p_1 v_1 \left\{\left(\frac{p_2}{p_1}\right)^{\frac{n-1}{n}} - 1\right\} - \frac{n_2}{n_2 - 1} p_4 v_4 \left\{\left(\frac{p_3}{p_4}\right)^{\frac{n_2-1}{n_2}} - 1\right\}\right]$$

$\because \qquad p_2 = p_3$ and $p_4 = p_1$, $v_1 = v_c + v_s = 0.05\,v_s + v_s = 1.05\,v_s$ and $v_4 = 0.2355\,v_s$

$$w = 2 \times 1 \times 10^5 \times 0.00456 \left[\frac{1.25}{1.25 - 1} \times 1.05 \left\{(7.5)^{0.25/1.25} - 1\right\}\right.$$

$$\left. - \frac{1.3}{1.3 - 1} \times 0.2355 \left\{(7.5)^{0.3/1.3} - 1\right\}\right]$$

$$= 912[(5 \times 1.05 \times 1.496) - 1.0205 \times 1.9]$$

$$= 912\,(7.854 - 1.622) = 5683.584 \text{ Nm/cycle}$$

$\therefore \qquad \text{I.P} = \dfrac{w.D/\text{cycle} \times r.p.m}{60 \times 1000} = \dfrac{5683.584 \times 310}{60 \times 1000}$

$$= 29.365 \text{ kW}$$

$\because \qquad \eta_{\text{mech}} = \dfrac{I.P}{B.P} \quad \therefore \quad B.P = \dfrac{29.365}{0.85} = 34.582 \text{ kW}$   *Ans.*

**10.4**  The stroke volume of a compressor is 0.2 m³/min. Air is taken at 98.5 KN/m² and 20°C. After 85 seconds, the pressure in the receiver is 982 KN/m² and temperature is 50°C. The capacity of the receiver is 1.48 $m^3$. If initial pressure in the receiver is 101 KN/$m^2$ and temperature is 16°C, calculate volumetric efficiency of the compressor. (Mumbai University Summer 2005)

*Soln.*

Initial mass of air in receiver

$$m_1 = \frac{p_1 v_1}{RT_1} = \frac{101 \times 1.48}{0.287\,(16 + 273)} = 1.8055 \text{ kg}$$

Final mass of air in receiver

$$m_2 = \frac{p_2 v_2}{RT_2} = \frac{982 \times 1.48}{0.287\,(50 + 273)} = 15.6780 \text{ kg}$$

$\therefore$  Actual mass of air supplied by compressor to receiver

$$= \frac{m_2 - m_1}{85} \times 60 = \frac{15.6780 - 1.8055}{85} \times 60$$

$$= 9.79 \text{ kg/min}$$

$\therefore$  Actual volume of this mass at suction condition

$$v_a = \frac{mRT}{P} = \frac{9.79 \times 0.287\,(20 + 273)}{98.5} = 8.35 \ m^3/\text{min}$$

$\therefore \qquad \eta_V = \dfrac{\text{Actual volume}}{\text{Stroke volume}} = \dfrac{8.35}{10.2} \times 100 = 81.86\%$   *Ans.*

**10.5** A single acting compressor draws air at pressure of 1 bar and delivers at 5 bar. It has clearance volume 4% of stroke volume. During overhauling its clearance volume was changed and now it was found that compressor delivers air at pressure 8 bar for same initial condition of intake. The speed and power remains same. The law of compression and expansion is $pv1.25 = c$. Determine percentage change in mass delivered and new clearance ratio.

(Mumbai University Summer 2006).

*Soln.* Refer Fig. 10.28

Given
$$p_1 = 1 \text{ bar}, p_2 = 5 \text{ bar}, v_3 = v_c = 0.04\, v_s$$

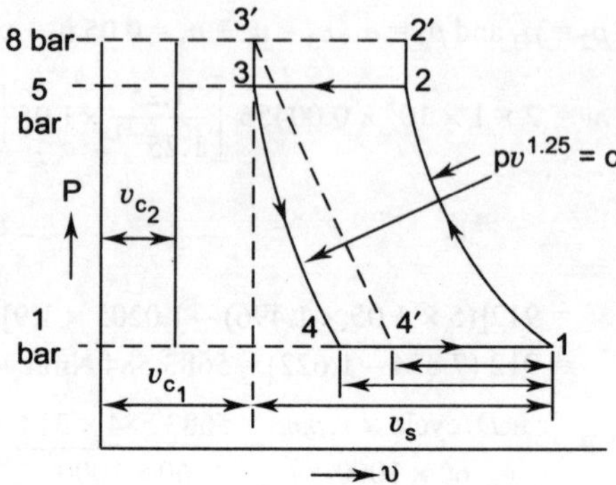


**Fig. 10.28**

Cylinder volume
$$(v_1 - v_4) = v_{a1}$$
$$(v_1 - v_{4'}) = v_{a2}$$

Stroke volume in both cases remains same which is $v_s$. Initiating the clearance volume is $v_{c_1}$ and after changing, it changes to $(v_{c_1} + v_{c_2})$ where $vc_2$ is increased clearance volume.

Initially, work done per cycle,

$$w_1 = \frac{n}{n-1} mRT_1 \left[ \left( \frac{p_2}{p_1} \right)^{\frac{n-1}{n}} - 1 \right]$$

After overhauling, work done per cycle

$$w_2 = \frac{n}{n-1} m_2RT_2 \left[ \left( \frac{p_2'}{p_1} \right)^{\frac{n-1}{n}} - 1 \right]$$

∵ Power remains same after over hauling *i.e.* $w_1 = w_2$

∴
$$\frac{m_2}{m_1} = \frac{\left[ \left( \dfrac{p_2}{p_1} \right)^{\frac{n-1}{n}} - 1 \right]}{\left[ \left( \dfrac{p_2'}{p_1} \right)^{\frac{n-1}{n}} - 1 \right]} = \frac{\left[ \left( \dfrac{5}{1} \right)^{\frac{0.25}{1.25}} - 1 \right]}{\left[ \left( \dfrac{8}{1} \right)^{\frac{0.25}{1.25}} - 1 \right]} = \frac{0.379}{0.515} = 0.734$$

Initially $m_1$, delivered by the cycle

$$m_1 = \frac{p_1 v_{a1}}{RT_1} = \frac{p_1 \times \eta_v \times v_s}{RT_1} = \frac{p_1 \times \left[1 + c_1 - c_1\left(\dfrac{p_2}{p_1}\right)^{1/n}\right] v_s}{RT_1}$$

$$= \frac{p_1\left[v_s + v_{c_1} - v_{c_1}\left(\dfrac{p_2}{p_1}\right)^{1/n}\right]}{RT_1} = \frac{p_1 v_s}{RT_1}\left[1 - \frac{v_{c_1}}{v_s}\left\{\left(\frac{p_2}{p_1}\right)^{1/n} - 1\right\}\right]$$

Similarly,

$$\text{mass } m_2 = \frac{p_1 v_{a2}}{RT_1} = \frac{p_1\left(u_1 - v'_4\right)}{RT_1}$$

$$= \frac{p_1}{RT_1}\left[\left\{v_s - \left(v_{c_1} + v_{c_2}\right)\right\}\left\{\left(\frac{p'_2}{p_1}\right)^{1/n} - 1\right\}\right]$$

$$= \frac{p_1 v_s}{RT_1}\left[1 - \left\{\frac{v_{c_1}}{v_s} + \frac{v_{c_2}}{v_s}\right\}\left\{\left(\frac{p'_2}{p_1}\right)^{1/n} - 1\right\}\right]$$

$$\therefore \quad \frac{m_2}{m_1} = \frac{\left[1 - \left\{\dfrac{v_{c_1}}{v_s} + \dfrac{v_{c_2}}{v_s}\right\}\left\{\left(\dfrac{p'_2}{p_1}\right)^{1/n} - 1\right\}\right]}{\left[1 - \dfrac{v_{c_1}}{v_s}\left\{\left(\dfrac{p_2}{p_1}\right)^{1/n} - 1\right\}\right]} = 0.734$$

Assuming
$$\frac{v_{c_1}}{v_s} = \frac{c_1}{100} \text{ and } \frac{v_{c_2}}{v_s} = \frac{c_2}{100}$$

$$\therefore \quad \frac{m_2}{m_1} = \frac{\left[1 - \left\{\dfrac{c_2}{100} + \dfrac{c_1}{100}\right\}\left\{\left(\dfrac{p'_2}{p_1}\right)^{1/n} - 1\right\}\right]}{1 - \dfrac{c_1}{100}\left\{\left(\dfrac{p_2}{p_1}\right)^{1/n} - 1\right\}} = 0.734$$

given,
$$c_1 = 4, \quad \frac{p_2}{p_1} = 5 \text{ and } \frac{p'_2}{p_1} = 8$$

$$\therefore \quad \frac{\left[1-\left\{\frac{4}{100}+\frac{c_2}{100}\right\}\left\{(7)^{1/1.25}-1\right\}\right]}{1-\frac{4}{100}\left\{(5)^{1/1.25}-1\right\}} = 0.734$$

or, $$\frac{100-\left(c_2+4\right)\times 3.74}{100-4(2.62)} = 0.734$$

$$\therefore \qquad c_2 = 5.16$$

$$\frac{c_2}{100} = \frac{v_{c2}}{v_s} = \frac{5.16}{100}$$

$\therefore \qquad$ Increase in clearance volume = 5.16% of stroke

$\therefore \quad$ New clearance volume $\qquad = vc_1 + vc_2 = 4 + 5.16 = 9.16\%$ of stroke   *Ans.*

Percentage increase in volume due to overhauling

$$= \frac{5.16}{4} \times 100 \; = 129\%$$

Percentage decrease in mass $\qquad = \dfrac{m_1 - m_2}{m_1} \times 100$

$$= \left(1-\frac{m_2}{m_1}\right)\times 100 \; = (1-0.734)\times 100 = 26.6\% \quad \textit{Ans.}$$

**10.6** Two stage compressor, compresses air from 1 bar and 20°C to 42 bar. If $pv^{1.3}$ = constant and intercooling is complete to 20°C. Find per kg of air

(i) Work done (ii) Mass of water necessary for abstracting the heat in the intercooler if temperature rise of cooling water is 25°C. Take $R = 287$ J/kg-k, $c_p = 1$ kJ/kgk of air. (Mumbai University Summer 07)

*Soln.* Refer Fig. 10.29

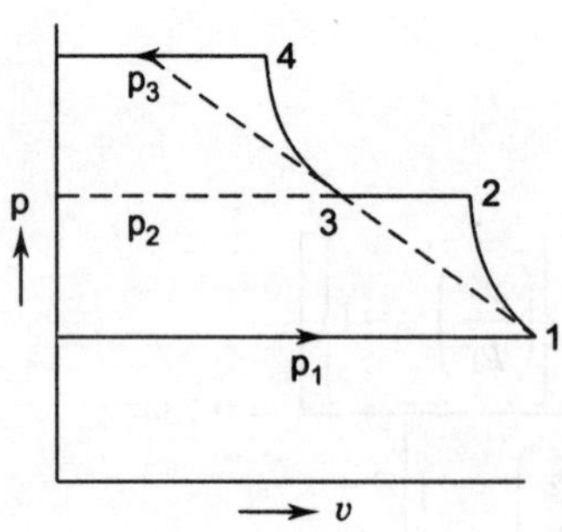

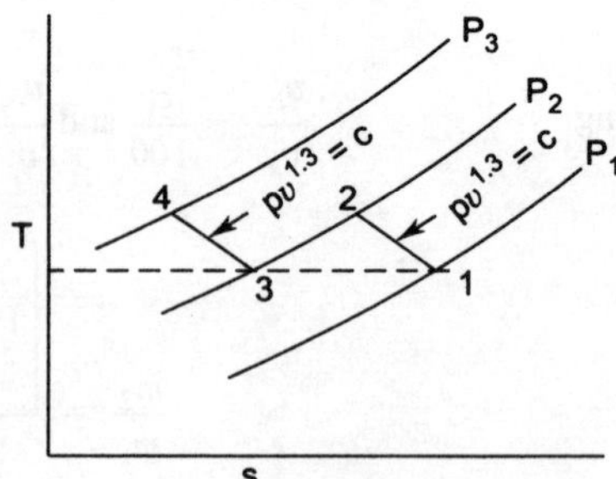


**Fig. 10.29**

Given, $\qquad\qquad P_1 = 1$ bar, $T_1 = 20 + 273 = 293$ K

$\qquad\qquad\qquad\qquad P_3 = 42$ bar, $n = 1.3$

Temperature rise in cooling water = 25°C = $(T_2 - T_3)$

For perfect intercooling, intermediate of intercooler pressure $p_2 = \sqrt{p_1 p_3} = \sqrt{1\times 42} = 6.48$ bar

Initial volume

$$v_1 = \frac{mRT_1}{p_1} = \frac{1 \times 287 \times 293}{1 \times 10^3} = 0.84 \ \text{m}^3/\text{kg of air}$$

Work required

$$W = \frac{n}{n-1} p_1 v_1 \left[ \left( \frac{p_2}{p_1} \right)^{\frac{n-1}{n}} + \left( \frac{p_3}{p_2} \right)^{\frac{n-1}{n}} - 2 \right]$$

$$= \frac{1.3}{1.3-1} \times 1 \times 10^5 \times 0.84 \left[ \left( \frac{6.48}{1} \right)^{\frac{1.3-1}{1.3}} + \left( \frac{42}{6.48} \right)^{\frac{1.3-1}{1.3}} - 2 \right]$$

$$= 364000 \times (1.539 + 1.539 - 2)$$

$$= 392392 \ \text{N-m} = 392.392 \ \text{kJ} \quad Ans.$$

$$\frac{T_2}{T_1} = \left( \frac{p_2}{p_1} \right)^{\frac{n-1}{n}} \quad \therefore \quad T_2 = 293 \left( \frac{6.48}{1} \right)^{\frac{1.3-1}{1.3}} = 450.92 \ K$$

Heat gained by water = Heat lost by air

$$m_w \, Cp_w \, (\Delta t)_w = m_a \, Cp_a \, (\Delta t)_a$$

$$m_w \times 4.187 \times 25 = 1 \times 1 \times (450.92 - 293)$$

$\therefore \qquad m_w = 1.508 \ \text{kg water} \quad Ans.$

**10.7** A single acting two stage air compressor with complete intercooling delivers 6 kg/min of air at 16 bar. Assume intake at 1 bar and 15°C. The compression and expansion follow the law $pv^{1.3}$ = constant. Calculate (i) Power required (ii) Isothermal efficiency (iii) Free air delivery (iv) If clearance ratio for L.P. and H.P. are 0.04 and 0.06; calculate volumetric efficiency and swept volume for each cylinder. Speed of compressor is 420 r.p.m

*Soln.* Refer Fig. 10.30

Given, $\qquad\qquad m = 6$ kg/min, $p_1 = 1$ bar, $T_1 = 15°C$, $pv^{1.3} = C$, $p_3 = 16$ bar

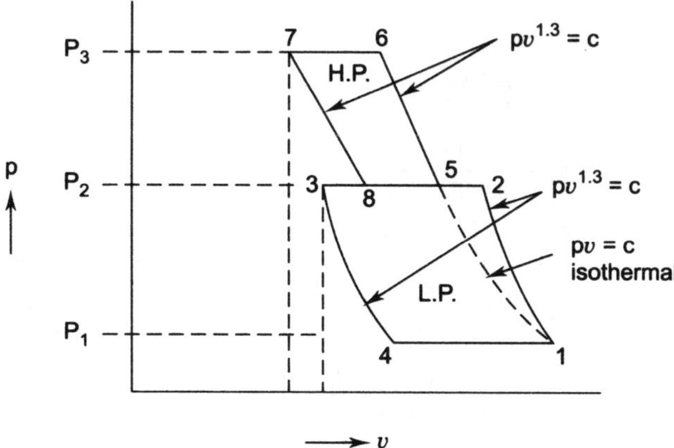


**Fig. 10.30**

For complete intercooling

$$p_2 = \sqrt{p_1 p_3} = \sqrt{1 \times 16} = 4 \text{ bar}$$

**(i) Power required**

$$w = \frac{2n}{n-1} \times mRT_1 \left[ \left( \frac{p_2}{p_1} \right)^{\frac{n-1}{n}} - 1 \right]$$

$$= \frac{2 \times 1.3}{1.3 - 1} \times \frac{6}{60} \times 0.287 \,(15 + 273) \left[ (4)^{0.3/1.3} - 1 \right] \text{kJ/s}$$

$$= 27 \text{ kW} \quad Ans.$$

**(ii) Isothermal work,**
$$W_i = mRT_1 \ln \frac{p_3}{p_1}$$

$$= \frac{6}{60} \times 0.287 \times 288 \ln 16 = 22.92 \text{ kW}$$

$$\therefore \qquad \eta_{\text{iso}} = \frac{\text{Isothermal work } W_i}{\text{Actual work, } W} = \frac{22.92}{27} \times 100 = 84.9\% \quad Ans.$$

**(iii)**
$$FAD = \frac{mRT_1}{P_1} = \frac{6 \times 0.287 \times 28.8}{1 \times 10^5 \times 10^{-3}} = 4.96 \text{ m}^3/\text{min} \quad Ans.$$

**(iv) L.P. Cylinder**

$$\eta_v = 1 - C \left[ \left( \frac{P_2}{P_1} \right)^{\frac{1}{n}} - 1 \right]$$

$$= 1 - 0.04 \left[ (4)^{1/1.3} - 1 \right] = 0.942 = 94.2\% \quad Ans.$$

**H.P Cylinder**

$$\eta_v = 1 - 0.06 \left[ (4)^{1/1.3} - 1 \right] = 0.887 = 88.7\% \quad Ans.$$

**Swept volume**
L.P Cylinder

$$\eta_v = \frac{v_a}{v_s \times N}$$

$$\therefore \qquad v_s = \frac{4.96}{0.942 \times 420} = 0.01278 \text{ m}^3$$

$\therefore$ clearance volume $\quad v_c = 0.04 \times 0.01278 = 0.00051 \text{ m}^3 \quad Ans.$
H.P. cylinder

$$v_s = \frac{4.96}{0.887 \times 420} \times \frac{1}{r_p}, \text{ where } r_p = \frac{p_2}{p_1} = \frac{p_3}{p_2}$$

$$= \frac{4.96}{0.887 \times 420} \times \frac{1}{4} = 0.00333 \ m^3$$

$\therefore$ clearance volume, $\quad v_c = 0.06 \times 0.00333$

$$= 0.002 \ m^3 \quad Ans.$$

**10.8** Find the percentage saving in work by compressing air in two stages from 1 bar to 7 bar instead of one stage. Assuming compression index 1.35 in both cases and optimum pressure and complete intercooling in two stage compressor.                    (Mumbai University Winter 2003)

*Soln.* Refer Fig. 10.31

Given            $p_1 = 1$ bar, $p_3 = 7$ bar, $n = 1.35$

For two stage compression with perfect intercooling, intermediate pressure

$$p_2 = \sqrt{p_1 \times p_3}$$

$$= \sqrt{1 \times 7} = 2.6458 \text{ bar}$$

Work of compression in single stage

$$w_1 = \frac{n}{n-1} p_1 v_1 \left[ \left( \frac{p_3}{p_2} \right)^{\frac{n-1}{n}} - 1 \right]$$

$$= \frac{1.35}{1.35-1} \times p_1 v_1 \left[ \left( \frac{7}{1} \right)^{\frac{1.35-1}{1.35}} - 1 \right]$$

$$= 2.5308\, p_1 v_1$$

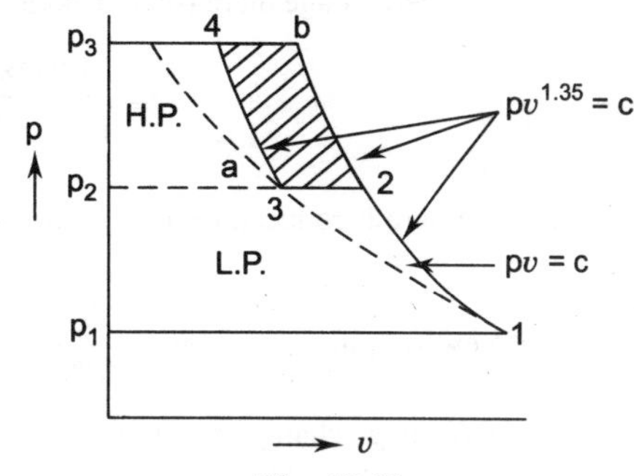


**Fig. 10.31**

Work of compression in two stages for same delivery pressure of 7 bar *i.e.*

$$w_2 = \frac{n}{n-1} p_1 v_1 \left[ \left( \frac{p_2}{p_1} \right)^{\frac{n-1}{n}} - 1 \right]$$

$$= 2 \times \frac{1.35}{1.35-1} p_1 v_1 \left[ \left( \frac{2.6458}{2} \right)^{\frac{1.35-1}{1.35}} - 1 \right]$$

$$= 2.2133\, p_1 v_1$$

$\therefore$    Percentage saving in work $= \dfrac{w_1 - w_2}{w_1} \times 100$

$$= \frac{2.5308 - 2.2133}{2.5308} \times 100$$

$$= 12.54\% \quad \textit{Ans.}$$

**10.9** A reciprocating air compressor takes in air at 40°C and 1.033 bar in day time.

(i)  Find the % increase of mass output in night, if the night temperature is 30°C less.

(ii) If the compressor is shifted to a place, where barometer reads 70 cm of Hg. Find percentage change in output. Assume no change in temperature.   (Mumbai University Summer 2004)

*Soln.*

Let $v$ be the handling capacity of compressor

(i)  Mass of air at 1.033 bar and 40°C,

$$m = \frac{pv}{RT} = \frac{1.033 \times 10^5 \times v}{287(40 + 273)} = 1.1499\, v$$

Night temperature,    $T_1 = 40 - 30 = 10°C = 283\ K$

∴    Mass to be delivered at night

$$m_1 = \frac{pv}{RT_1} = \frac{1.033 \times 10^5 \times v}{287 \times 287} = 1.2718\, v$$

∴    Percentage increase of mass at night

$$= \left(\frac{1.2718v - 1.1499v}{1.499v}\right) \times 100 = 10.67\%\quad Ans.$$

(ii)  Pressure is reduced to $p_1 = 70$ cm of $Hg = \dfrac{70}{70} \times 1.01325\ = 0.933$ bar

New output,        $m_1 = \dfrac{p_1 v}{RT_1} = \dfrac{0.933 \times 10^5 \times v}{287 \times (40 + 273)} = 1.0386\, v$

percentage change in output

$$= \frac{m_1 - m}{m} \times 100 = \frac{(1.0386v - 1.1499v)}{1.499v} \times 100$$

$$= -9.67\%\quad Ans.$$

Negative sign shows that output decreases.

**10.10**  A two stage single acting air compressor having capacity 5 m³/min measured under free air conditions of 1.01325 bar and 15°C. The pressure during suction is 0.98 bar. The temperature at the start of compression in each stage is 27°C. The delivery pressure is 15 bar. The clearance volume of *L.P.* cylinder is 5% of the stroke volume. The index of compression and expansion is 1.3 and speed is 150 r.p.m. The intercooler pressure is such that the work is shared equally between the two cylinders. Determine:

(i)  The indicated power

(ii)  Diameter and stroke of L.P. cylinder if bore is equal to the stroke.

(Mumbai University Winter 2004)

*Soln.* Refer Fig. 10.32

Given, at free air conditions

$p_f = 1.01325$ bar,

$T_f = 15°C = 288\ K$,

$P_1 = 0.98$ bar,

$T_1 = 27°C = 300\ K$

$c = \dfrac{v_c}{v_s} = 5\%, n = 1.3$,

$N = 150$ r.p.m, $v_f = 5$ m³/min, $p_3 = 15$ bar

(i)  Indicated power

Since the work is shared equally in both cylinders

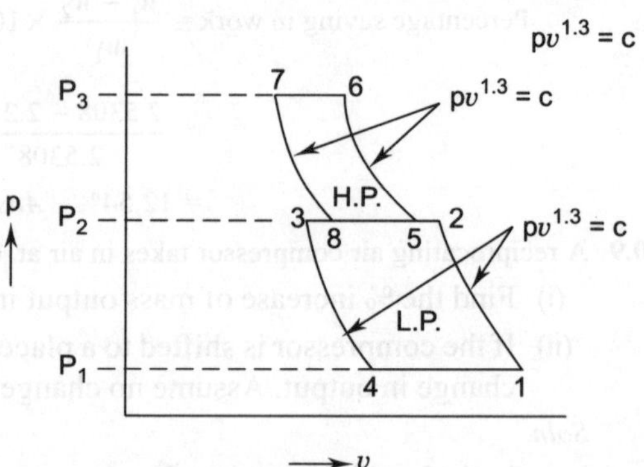


**Fig. 10.32**

$$\therefore \qquad P_2 = \sqrt{P_1 \times P_3} = \sqrt{0.98 \times 15} = 3.834 \text{ bar}$$

Mass of air handled per minute at free air conditions

$$m = \frac{p_f\, v_f}{RT_f} = \frac{(1.01325 \times 10^5) \times 5}{287 \times 288} = 6.129 \text{ kg/min}$$

$$\text{I.P} = 2\,\frac{n}{n-1}\, mRT_1 \left[\left(\frac{P_2}{P_1}\right)^{\frac{n-1}{n}} - 1\right]$$

$$= 2 \times \frac{1.3}{1.3-1} \times 0.287 \times 300 \left[\left(\frac{3.834}{0.98}\right)^{\frac{1.3-1}{1.3}} - 1\right]$$

$$= 28.2 \text{ kW Ans}$$

(ii)  Diameter and stroke of L.P. cylinder

Given $L = D$, Let $(v_1 - v_4)$ = volume of air handled at suction conditions.

$$p_1\,(v_1 - v_4) = mRT_1$$

$$(v_1 - v_4) = \frac{mRT_1}{P_1} = \frac{6.129 \times 287 \times 300}{0.98 \times 10^5} = 5.384 \text{ m}^3/\text{min}$$

volumetric efficiency at suction condition

$$\eta_v = 1 + c - c\left(\frac{P_2}{P_1}\right)^{\frac{1}{n}}$$

$$= 1 + 0.05 - 0.05\left(\frac{3.834}{0.98}\right)^{\frac{1}{1.3}}$$

$$= 0.9072$$

$$\therefore \qquad v_1 = \eta_v\,(v_1 - v_4)$$

$$= 0.9072 \times 5.384 = 4.88 \text{ m}^3/\text{min}$$

$$\text{stroke volume} = \frac{4.88}{N} = \frac{4.88}{150} = 0.0325 \text{ m}^3$$

But

$$v_1 = \frac{\pi}{4}\,D^2.L = \frac{\pi}{4}\,D^2 \times D$$

or,

$$0.0325 = \frac{\pi}{4}\,D^3$$

$$\therefore \qquad D = 0.3458 \text{ m} = 34.58 \text{ cm}$$

and stroke, $\qquad L = 34.58 \text{ cm} \quad Ans$

**10.11**  A three stage single acting reciprocating compressor has perfect intercooling. The pressure and temperature at the end of suction stroke in L.P. Cylinder is 1.013 bar and 15°C respectively. If 8.4 m3 of free air is delivered by the compressor at 70 bar per minute and the work done is minimum calculate- (i) L.P and I.P delivery pressures (ii) Ratio of cylinder volumes and (iii) Total indicated power.

Neglect clearance and assume $n = 1.2$. $\hfill$ (Mumbai University Winter 2003)

*Soln.* Refer Fig. 10.33

Given $\qquad p_1 = 1.013$ bar, $T_1 = 15 + 273 = 288$ k

$\qquad\qquad p_4 = 70$ bar.

$$\text{I.P(3 stage)} = \frac{3n}{n-1}\, p_1 v_1 \left[\left(\frac{p_4}{p_1}\right)^{\frac{n-1}{3n}} - 1\right]$$

$$= \frac{3 \times 1.2}{1.2 - 1} \times 1.013 \times 10^5 \times \frac{8.4}{60}\left\{\left(\frac{70}{1.013}\right)^{\frac{0.20}{3 \times 1.2}} - 1\right\}$$

$$= 67.64 \text{ kW} \quad Ans.$$

Now, $\qquad\qquad \dfrac{p_2}{p_1} = \dfrac{p_3}{p_2} = \dfrac{p_n}{p_3} = K$

and $\qquad\qquad \left(\dfrac{p_{x+1}}{p_1}\right)^{1/x} = K$

where $\qquad\qquad x = $ no. of stage

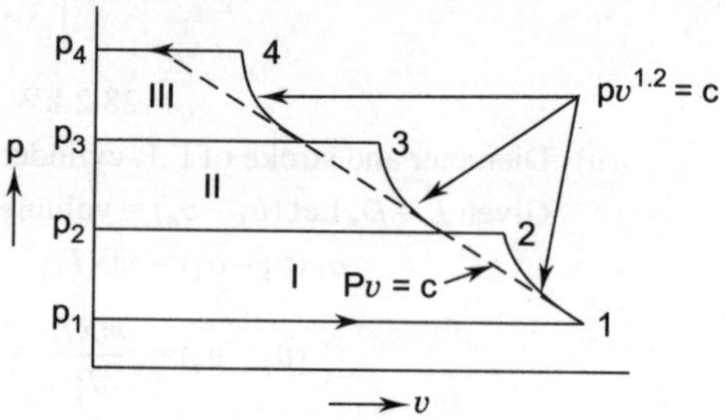

or, $\qquad\qquad \left(\dfrac{70}{1.013}\right)^{1/3} = 4.103$

$\therefore \qquad\qquad p_2 = 4.103\, p_1 = 4.156$ bar

$\qquad\qquad p_3 = 4.103\, p_2 = 17.05$ bar $\quad Ans.$

**Fig. 10.33**

**Cylinder volumes**

Let $v_1 : v_2 : v_3$ be ratio of cylinder volumes for all three stages.

For minimum work

$$p_1 v_1 = p_2 v_2 = p_3 v_3$$

$\therefore \qquad\qquad \dfrac{v_1}{v_2} = \dfrac{p_2}{p_1} = \dfrac{4.56}{1.013} = 4.103$

$\therefore \qquad\qquad v_1 = 4.103\, v_2$

Similarly

$\qquad\qquad \dfrac{v_2}{v_3} = \dfrac{p_3}{p_2} = \dfrac{17.05}{4.156} = \dfrac{4.103}{1}$

$\therefore \qquad\qquad v_2 = 4.103\, v_3$

Assuming $\qquad v_3 = 1 \ m^3$, then ratio becomes

$$v_1 : v_2 : v_3 = 4.103 \times 4.103 : 4.103 : 1$$

$$= 16.83 : 4.103 : 1 \quad Ans.$$

**10.12** A multistage compressor takes in air at 1 bar, 25°C and compresses it to 20 bar. Assume perfect intercooling and that the amount of energy rejected at intercooler is equal to the energy rejected during compression due to cooling of cylinder. The compressor runs at 900 r.p.m and delivers 3 kg per minute of air. All pistons have a stroke of 170 min. Calculate (i) the index of compression (ii) the number of stages if the temperature at the end of compression is not to exceed 400 k (iii) the temperature and the pressure at the end of each stage (iv) the total power input to the compressor if $\eta_{\text{mech}} = 80\%$ for each stage and (v) cylinder dimensions if $\eta_v = 80\%$ for each stage.

(Mumbai university winter 2005)

*Soln.* Given

$$N = 900 \text{ r.p.m}$$
$$m = 3 \text{ kg/min}$$
$$L = 170 \text{ mm}$$
$$p_1 = 1 \text{ bar}, \ T_1 = 25°C = 298 \text{ K}$$
$$p_{x+1} = 20 \text{ bar if } x = \text{no. of stages}$$

(i) **Index of compression**

Heat rejected at intercooler = Heat rejected during compression

$$C_v(T_2 - T_1) = C_v \left( \frac{\gamma - n}{n - 1} \right) (T_2 - T_1)$$

or,
$$\frac{c_p}{c_v} = \frac{\gamma - n}{n - 1} \quad \text{or, } n = \frac{2\gamma}{\gamma + 1} = \frac{2 \times 1.4}{1.4 + 1} = \frac{7}{6} = 1.167$$

∴
$$\frac{n}{n - 1} = \frac{\frac{7}{6}}{\frac{7}{6} - 1} = 7 \quad Ans.$$

(ii) **No. of stages**

Let $p_i$ be the intermediate pressure, then

$$\left( \frac{p_i}{p_1} \right) = \left( \frac{T_2}{T_1} \right)^{\frac{n}{n-1}} = \left( \frac{400}{298} \right)^7$$

But
$$\frac{p_i}{p_1} = \left( \frac{p_2}{p_1} \right)^{1/x} \quad \text{for minimum work}$$

$$\left( \frac{p_{x+1}}{p_1} \right)^{1/x} = \left( \frac{400}{298} \right)^7, \text{ where } x = \text{no. of stages}$$

or,
$$\left( \frac{20}{1} \right)^{1/x} = 7.85$$

or,
$$(20)^{\frac{1}{x}} = 7.85$$

or,
$$\frac{1}{x} \ln 20 = \ln 7.85$$

$$\frac{1}{x} \times 2.995 = 2.06$$

∴
$$x = 1.453 \ i.e. \ 2 \text{ stages} \quad Ans.$$

(iii)
$$\frac{T_i}{T_1} = \left( \frac{p_{x+1}}{p_1} \right)^{\frac{n-1}{2n}} = \left( \frac{20}{1} \right)^{\frac{1}{2 \times 7}} = 20^{\frac{1}{14}}$$

∴
$$T_i = T_1 \times 20^{1/14} = 298 \times 20^{1/14} = 369 \, k$$

which is less than given value of $400 \, K$, therefore satisfactory $p_i = \sqrt{20 \times 1} = 4.475$ bar

(iv)  Minimum work

$$w = \frac{2n}{n-1}\, mRT_1 \left[ \left( \frac{p_2}{p_1} \right)^{\frac{n-1}{2n}} - 1 \right]$$

$$= 2 \times 7 \times \frac{3}{60}\, 0.286 \times 298 \left[ (20)^{1/14} - 1 \right] = 14.28 \text{ kW}$$

$\therefore$   Input power to the compressor $= \dfrac{14.28}{\eta_{mech}} = \dfrac{14.28}{0.8}$

$$= 17.85 \text{ kw} \quad Ans.$$

(v)  Volume of air filled in L.P cylinder

$$v_s = \frac{mRT_1}{p_1 \times \eta_v} = \frac{3 \times 287 \times 298}{1 \times 10^5 \times 0.8} = 3.21 \text{ m}^3/\text{min}$$

or,   $\dfrac{\pi}{4}\, D^2 \times 1.7 \times 900 = 3.21$

$\therefore$   $D = 0.164$ m $= 16.4$ cm (cylinder dia.)   *Ans.*

Volume of air handled by H.P cylinder at pressure pi

*i.e.*   $v' = \dfrac{p_1 v_1}{p_i} = \dfrac{p_1 v_s}{p_i} = \dfrac{1 \times 3.21}{4.475} = 0.718 \text{ m}^3/\text{min}$

Since the stroke is same for both cylinders

$\therefore$   diameter of H.P cylinder

$$\frac{\pi}{4}\, D'^2 \times 0.17 \times 900 = 0.718$$

$\therefore$   $D' = 0.0772\ m = 7.72$ cm   *Ans.*

**10.13**  A rotary air compressor working between 1 bar and 2.5 bar has internal and external diameter of impellers of as 30 cm and 60 cm respectively. The vane angle at inlet and outlet are 30° and 45° respectively. If the air enters the impellers at 15 m/s, find (i) speed of the impeller in r.p.m and (ii) work done by the compressor per kg of air.

*Soln.* Refer Fig. 10.34

Given   $p_1 = 1$ bar

$p_2 = 2.5$ bar

$D_1 = 30$ cm

$D_2 = 60$ cm

$\beta_1 = 30°,\ \beta_2 = 45°,\ v_1 = 15$ m/s

From inlet velocity triangle

$$\tan \beta_1 = \frac{v_1}{u_1} \quad \therefore\ u_1 = \frac{15}{\tan 30°} = 25.98 \text{ m/s}$$

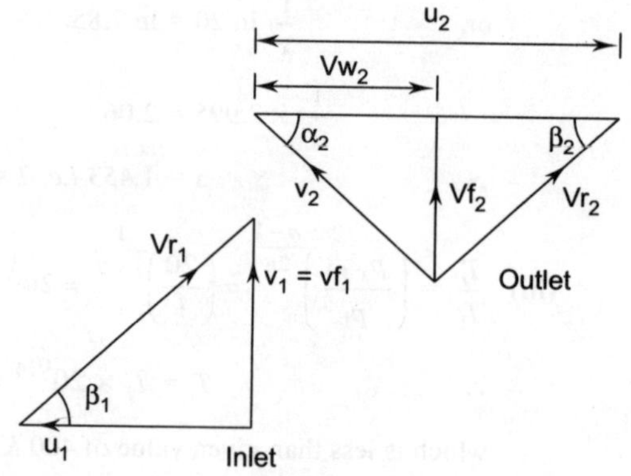


**Fig. 10.34**

$$u_1 = \frac{\pi D_1 N}{60} \quad \therefore \quad r.p.m, \ N = \frac{60 \times 25.98}{\pi \times 0.30} = 1653.93 \ \text{r.p.m} \quad Ans.$$

From velocity triangle at outlet, we know that

$$\frac{u_2}{u_1} = \frac{D_2}{D_1} \quad \therefore \quad u_2 = \frac{D_2}{D_1} \times u_1$$

$$= \frac{0.60}{0.30} \times 25.98 = 51.96 \ \text{m/s}$$

$$v_{f1} = v_{f2} = v_1 = 15 \ \text{m/s}$$

$$\therefore \qquad \tan \beta_2 = \frac{V_{f2}}{(u_2 - v_{w2})} \quad \text{or,} \quad (u_2 - v_{w2}) = \frac{V_{f2}}{\tan \beta_2} = \frac{15}{\tan 45°} = 15 \ \text{m/s}$$

$$\therefore \qquad v_{w2} = u_2 - (u_2 - v_{w2}) = 51.96 - 15 = 36.96 \ \text{m/s}$$

$$\therefore \quad \text{Work done by the compressor} = m \times v_{w2} \, u_2 = 1 \times 36.96 \times 51.96$$

$$= 1920.44 \ \text{Watt} \quad Ans.$$

**10.14** A centrifugal compressor delivers 50 kg of air per min. at pressure of 2 bar and 97°C. The intake pressure and temperature of air is 1 bar and 15°C. If no heat is lost to the surrounding, find:

    (i)  index of compression

    (ii)  power required if compression is isothermal, take $R = 287$ J/kg-k for air.

(Mumbai university summer 2007)

*Soln.* Given, $\qquad m = 50$ kg/min, $p_2 = 2$ bar, $T_2 = 97°C$, $p_1 = 1$ bar, $T_1 = 15°C$

$$\frac{T_2}{T_1} = \left(\frac{p_2}{p_1}\right)^{\frac{n-1}{n}}$$

$$\ln \frac{T_2}{T_1} = \frac{n-1}{n} \ln\left(\frac{p_2}{p_1}\right)$$

$$\ln \frac{(97 + 273)}{(15 + 273)} = \frac{n-1}{n} \ln\left(\frac{2}{1}\right)$$

or, $\qquad\qquad 0.2505 = \frac{n-1}{n} \times 0.693$

$\therefore \qquad\qquad n = 1.567 \quad Ans.$

Power required for isothermal compression

$$w = RT_1 \ln \frac{p_2}{p_1}$$

$$= 0.287 \times (15 + 273) \ln\left(\frac{2}{1}\right) = 57.29 \ \text{kJ/kg}$$

$$\therefore \quad \text{Total power produced} \quad = \frac{50}{60} \times 57.29 = 47.74 \ \text{kW} \quad Ans.$$

**10.15** The centrifugal compressor running at 10,000 r.p.m delivers 660 m³/min of free air. The air is compressed from 1 bar and 20°C to a pressure ratio of 4 with an isentropic efficiency of 82%. Blades are radial at outlet of impeller and flow velocity of 62 m/s may be assumed throughout constant. The outer radius of impeller is twice the inner and the slip factor is assumed as 0.9. The blade area co-efficient may be assumed 0.9 at inlet. Calculate (i) Final temperature of air (ii) Theoretical (iii) Impeller diameter at inlet and outlet (iv) Breadth of impeller at inlet (v) Impeller blade angle at inlet (vi) Diffuser blade angle at inlet.

(Mumbai University summer 2006)

*Soln.* Given, $N = 10,000$ r.p.m, $p_{o1} = 1.5$ bar, $T_{o1} = 20°C$, $m = 660$ m$^3$/min, $r_p = 4$, $\eta_{isen} = 82\%$, $v_f = 62$ m/s, $r_2 = 2 \times r_1$, $\phi = 0.9$, $Ka = 0.9$

Temperature rise across the compressor

$$\frac{\Delta T_o}{T_{o1}} = \left\{ \left( r_p \right)^{\frac{\gamma-1}{\gamma}} - 1 \right\}$$

$$\therefore \qquad \Delta T_o = 293 \left\{ (4)^{\frac{1.4-1}{1.4}} - 1 \right\} = 142.38 \, K$$

$\therefore$ Actual temperature rise $= \dfrac{142.38}{\eta_{isen}} = \dfrac{142.38}{0.82} = 173.64 \, K$

$\therefore$ Final temperature $= 173.64 + 273 = 446.64$ K   *Ans.*

Mass flow rate $\qquad m = \dfrac{PV}{RT} = \dfrac{1 \times 10^5 \times 660}{287 \times 293} = 13.08$ Kg/s   *Ans.*

Work required to run the compressor

$$w = mc_p \, \Delta T_o = 13.08 \times 1.005 \times 173.64$$
$$= 2285.45 \text{ kW}$$

For radial blades, energy transfer is given by

$$E = w = \phi_w \, \phi_s \, v^2 = cp \, \Delta To$$

$$v_2 = \sqrt{\frac{c_p \Delta T_o}{\phi_w \phi_s}} = \sqrt{\frac{1.005 \times 100 \times 173.64}{1.04 \times 0.9}} = 136.62 \text{ m/s}$$

$$\frac{\pi D_2 N}{60} = 136.62$$

$\therefore \qquad\qquad D_2 = 0.26$ m or, 26.09 cm   *Ans.*

and $\qquad\qquad D_1 = \dfrac{D_2}{2} = \dfrac{26.09}{2} = 13.045$ cm

Volume flow rate $\qquad = \pi D_1 b_1 \, k_a \, v_{f1} = \dfrac{660}{60}$

$\therefore \qquad\qquad b_1 = \dfrac{\dfrac{660}{60}}{\pi \times 13.045 \times 0.9 \times 62} = 0.48$ m

$$\tan \beta_1 = \frac{v_{f1}}{u_1} \quad \because \quad \frac{v_2}{v_1} = \frac{D_2}{D_1}, \quad u_1 = u_2 \frac{D_1}{D_2} = 136.62 = 68.31 \text{ m/s}$$

$$= \frac{62}{68.31} = 0.90$$

$\therefore \qquad\qquad \beta_1 = 42.22°$   *Ans.*

$$\tan \alpha_2 = \frac{Vf_2}{Vw_2} = \frac{V_{f1}}{\phi u_2} = \frac{6.2}{0.9 \times 136.62} = 0.5042$$

$$\therefore \qquad \alpha_2 = 26.75°$$

*i.e.* Diffuse blade angle at inlet, $\alpha_2 = 26.75°$  *Ans.*

**10.16** A single inlet type centrifugal compressor handles 6 kg/s air. Ambient air conditions are 1 bar and 20°C. Compressor runs at 22000 r.p.m with isentropic efficiency of 82%. Air is compressed in compressor from 1 bar static pressure to 4.2 bar total pressure. Air enters impeller eye with velocity of 150 m/s with no pre whirl. Assuming ratio of impeller speed to tip speed is 0.9. Find

  (i)  Rise in total temperature during compression ($\Delta KE = 0$)

 (ii)  Tip diameter of impeller

(iii)  Power required

 (iv)  Eye diameter if hub diameter = 10 cm          (Mumbai university summer 2005)

*Soln.* Given, $m = 6$ kg, $p_{o1} = 1$ bar, $T_{o1} = 20°C = 293\ K$, $N = 22000$ r.p.m, $\phi = 0.9$, $D_i = 10$ cm, $\eta_{isen} = 82\%$, $p_{o2} = 4.2$ bar, $v_1 = 150$ m/s, $v_{w1} = 0$, *i.e.* $v_1 = v_{f1}$

Temperature drop equivalent to K.E at inlet

$$T_{o1} - T_1 = \frac{V_1^2}{2c_p}$$

$$293 - T_1 = \frac{150^2}{2 \times 1000} \qquad \text{(assume } c_p = 1 \text{ kJ/kgK)}$$

$$\therefore \qquad T_1 = 281.75\ K$$

$$\frac{T'_{o2}}{T_{o1}} = \left(\frac{P_{o2}}{P_{o1}}\right)^{\frac{\gamma-1}{\gamma}}$$

$$\therefore \qquad T'_{o2} = 293 \left(\frac{4.2}{1}\right)^{\frac{1.4-1}{1.4}} = 441.49\ K$$

$$\eta_{isen} = \frac{T'_{o2} - T_{o1}}{T_{o2} - T_{o1}}$$

$$\text{or,} \qquad 0.82 = \frac{441.49 - 293}{T_{o2} - 293}$$

$$\therefore \qquad T_{o2} = 474.08\ K$$

Rise in total temperature $\quad = T_{o2} - T_{o1} = (474.08 - 293)$

$$= 181.08\ K \quad Ans.$$

Work, per kg of air

$$W = c_p(T_{o2} - T_{o1})$$
$$= 1(181.08) = 181.08 \text{ kJ/kg}$$

Also, $\qquad w = v_w \times u = \psi_s\, u_2^2$

$$\therefore \qquad u_2^2 = \frac{181.08}{0.9} = 201.2$$

$$\therefore \qquad u_2 = 14.18 \text{ m/s}$$

Also,
$$u_2 = \frac{\pi D_2 N}{60}$$

or,
$$14.18 = \frac{\pi D_o \times 22000}{60}$$

$\therefore$   $D_o = 0.0123$ m $= 1.23$ cm   *Ans.*

Power
$$= \dot{m} \times W$$
$$= 6 \times 181.08 = 1086.48 \ K$$

Also,
$$\left(\frac{T_{o1}}{T_1}\right)^{\frac{\gamma}{\gamma-1}} = \frac{P_{o1}}{P_1}$$

$$\left(\frac{293}{281.75}\right)^{\frac{1.4}{1.4-1}} = \frac{1}{p_1}$$

$\therefore$   $p_1 = 0.872$ bar

$$R = c_p\left(1 - \frac{1}{r}\right) = 1\left(1 - \frac{1}{1.4}\right) = 0.286 \text{ kJ/kgk}$$

Specific value of air at inlet

$$v_1 = \frac{RT_1}{p_1} = \frac{0.286 \times 281.75}{0.872 \times 10^5} = 9.24 \times 10^{-4} \ m^3/kg$$

mass flow rate,

$$m = \frac{\text{Area of flow at inlet} \times V_{f1}}{V_1}$$

$$6 = \frac{A \times 150}{9.24 \times 10^{-4}}$$

$$A = 3.696 \times 10^{-5} \ m^2$$

$$A = \frac{\pi}{4}\left(D_o^2 - D_i^2\right)$$

$$= \frac{\pi}{4}\left(D_o^2 - 0.1^2\right)$$

$$3.696 \times 10^{-5} = \frac{\pi}{4}\left(D_o^2 - 0.1^2\right)$$

$\therefore$   $D_o = 0.10$ m   *Ans.*

**10.17**  An axial flow compressor draws air at 20°C and delivers it at 50°C. Assuming 50% degree of reaction, calculate velocity of flow of blade velocity is 100 m/s, work factor is 0.85. Take $c_p = 1 \ kj/kg \ K$. Assume $\alpha = 10°$ and $\beta = 40°$. Find the number of stages.                    (Mumbai university winter 2003)

*Soln.* Given

$$p_1 = 40°\text{C} = 293 \ K, \ T_2 = 50°\text{C} = 323 \ K$$

Degree of reaction,   $R = 50\%, \ u = 100$ m/s

work factor   $\psi = 0.85, \ cp = 1 \text{ kJ/kg } K, \ \alpha_1 = 10°, \ \beta_1 = 40°$

please refer Fig. 10.19

For $\qquad R = 50\%$

$$\beta_1 = \alpha_2 \text{ and } \alpha_1 = \beta_2$$

From $\Delta$ $CBE$ and $BDF$

$$\tan \alpha_1 = \frac{Vw_1}{V_{f1}} \quad \text{or,} \quad V_{f1} = \frac{Vw_1}{\tan \alpha_1}$$

and $\qquad \tan \alpha_2 = \frac{Vw_2}{V_{f2}} \quad \text{or,} \quad V_{f2} = \frac{Vw_2}{\tan \alpha_2}$

$\because \qquad V_{f1} = V_{f2} \quad \therefore \quad \frac{Vw_1}{\tan \alpha_1} = \frac{Vw_2}{\tan \alpha_2} = \left( \frac{u_1 - Vw_1}{\tan \alpha_2} \right)$

$$5.671\, V_{w1} = (100 - v_{w1}) \times 1.191$$

$\therefore \qquad V_{w1} = 17.35 \text{ m/s}$

$\therefore \qquad V_{f1} = V_{f2} = V_f = \dfrac{17.35}{\tan 10°} = 98.39 \text{ m/s} \quad Ans.$

$\therefore \qquad \Delta V_w = V_{w2} - V_{w1} = (u - V_{w1}) - V_{w1} = u - 2V_{w1}$

$\qquad\qquad\qquad = 100 - 2 \times 17.35 = 65.3 \text{ m/s}$

Theoretical work done, $\quad w' = u \times \Delta V_w = 100 \times 65.3 = 6530 \text{ } N\text{-}m/\text{kg}$

Actual work done, $\qquad w = \text{Theoretical work done} \times \text{work factor}$

$\qquad\qquad\qquad = 6530 \times 0.85 = 5550.5 \text{ } N\text{-}m/\text{kg}$

$$W = \frac{\gamma}{\gamma - 1} RT_1 \left[ \left( r_p \right)^{\frac{\gamma - 1}{\gamma}} - 1 \right]$$

$$5550.5 = \frac{1.4}{1.4 - 1} \times 287 \times 293 \left[ \left( r_p \right)^{\frac{1.4 - 1}{1.4}} - 1 \right]$$

$\therefore \qquad r_p \text{ (pressure ratio)} = 1.067$

If $K$ is the number of stages, then

$$(r_p)^K = \left( \frac{T_2}{T_1} \right)^{\frac{\gamma}{\gamma - 1}}$$

$$(1.067)^K = \left( \frac{323}{293} \right)^{\frac{1.4}{1.4 - 1}}$$

$\therefore \qquad K = 6.265 \text{ say 6 stages} \quad Ans.$

**10.18** Air enters a single stages centrifugal compressor at a pressure of 1 bar and temperature 15°C, velocity at inlet is 60 m/s. Impeller diameter at inlet = 0.6 m, impeller diameter at outlet = 12 m, r.p.m = 3000, velocity of flow is constant, blade angle at outlet = 90°. Determine: (i) Blade angle at inlet (ii) Total pressure at outlet (iii) Static temperature and pressure at exit (iv) Angle made by absolute velocity at exit (v) Blade width at exit. Assume mass flow rate = 7 kg/s. $\qquad$ (Mumbai univ. winter 2007)

*Soln.* Given, $P_1 = 1$ bar, $T_1 = 15°C = 288\,K$, $V_1 = 60$ m/s, $D_1 = 0.6\,m$, $D_2 = 1.2$ m, $N = 3000$ r.p.m, $v_{f1} = v_{f2}$, $\beta_2 = 90°$, $m = 7$ kg/s.  $\because$  $\beta_2 = 90°$  $\therefore$  Radial bending

Refer Fig. 20.24 and 20.35.

$$u_1 = \frac{\pi D_1 N}{60} = \frac{\pi \times 0.6 \times 3000}{60}$$

$$= 94.24 \text{ m/s}$$

$$u_2 = \frac{\pi D_2 N}{60}$$

$$= \frac{\pi \times 1.2 \times 3000}{60}$$

$$= 188.49 \text{ m/s}$$

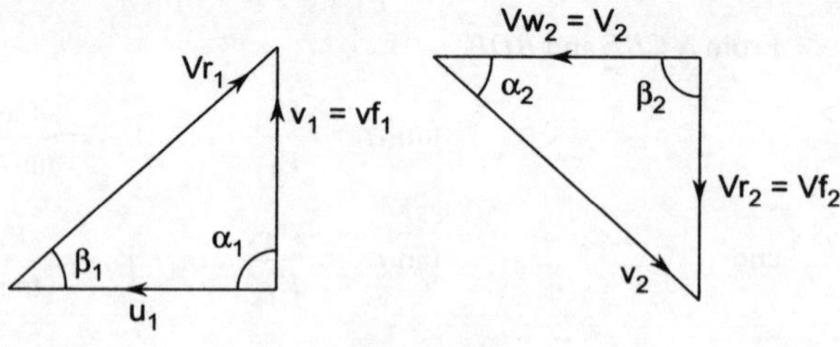


**Fig. 10.35**

$$V_{r1} = \sqrt{V_1^2 + u_1^2} = \sqrt{60^2 + 94.24^2} = 111.71 \text{ m/s}, \tan\beta_1 = \frac{u_1}{v_1} = \frac{60}{94.24} = 0.6366$$

$\therefore$  $\beta_1 = 32.48°$  *Ans.*

$$T_{o1} = T_1 + \frac{V_2^2}{2c_p} = 288 + \frac{60^2}{2 \times 1.005 \times 1000} = 289.79\,K$$

$$\left(\frac{p_{o1}}{p_1}\right)^{\frac{\gamma-1}{\gamma}} = \frac{T_{o1}}{T_1} \quad \therefore \quad p_{o1} = p_1 \times \left(\frac{T_{o1}}{T_1}\right)^{\frac{\gamma}{\gamma-1}}$$

or,  $$p_{o1} = 1\left(\frac{288.79}{288}\right)^{\frac{1.4}{1.4-1}} = 1.0219 \text{ bar}$$

$$\frac{p_{o2}}{p_1} = \left(\frac{u_2^2}{c_p T_{o1}} + 1\right)^{\frac{\gamma}{\gamma-1}} = \left(\frac{188.49^2}{1.005 \times 289.79 \times 1000} + 1\right)^{\frac{1.4}{1.4-1}}$$

$$= 1.496$$

$\therefore$  $$p_{o2} = 1.0496 \times 1.0219 = 1.528 \text{ bar} \quad Ans.$$

$$\frac{p_2}{p_1} = \left(\frac{u_2^2}{c_p T_1} + 1\right)^{\frac{\gamma}{\gamma-1}} = \left(\frac{188.49^2}{1.005 \times 1000 \times 188} + 1\right)^{\frac{1.4}{1.4-1}}$$

$$= 1.827$$

$\therefore$  $$p_2 = 1.827 \times 1 = 1.827 \text{ bar} \quad Ans.$$

Work done per *kg*  $$= c_p(T_2 - T_1)$$

$\therefore$  $$u_2^2 = c_p(T_2 - T_1)$$

$$(188.492)^2 = 1.005 \times 1000\,(T_2 - 288)$$

$\therefore$  $$T_2 = 323.35\,K \quad Ans.$$

Mass flow rate,  $$m = \pi D_1 \beta_1 V_{f1}$$
$\because$  $$p_1 = \rho_1 R T_1$$

$$\therefore \qquad \rho_1 = \frac{p_1}{RT_1} = \frac{1 \times 10^5 \times 10^3}{0.287 \times 288} = 1.209 \text{ Kg/m}^3$$

$$\therefore \qquad \frac{Q}{\rho_1} = \pi D_1\, \beta_1\, V_{f1} = m$$

$$\frac{7}{1.209} = \pi \times 0.6 \times b_1 \times 60$$

$$\therefore \qquad b_1 = 0.06\, m = 6.01 \text{ cm} \quad Ans.$$

$$\tan \alpha_2 = \frac{Vf_2}{u_2} = \frac{60}{188.49} = 0.3183$$

$$\therefore \qquad \alpha_2 = 17.65° \quad Ans.$$

**10.19** An axial flow compressor having 8 stages and with 50% reaction design compresses air in the ratio of 4 : 1. The air enters the compressor at 20°C and flows through it with a constant speed of 90 m/s. The rotating blades of the compressor rotate with a mean speed of 180 m/s. Isentropic efficiency of the compressor may be taken as 82%. Calculate:

  (i) Work done by the machine      (ii) Blade angles

Assume        $\gamma = 1.4$, $c_p = 1.005$ kJ/kg k     (Mumbai University, winter 2004)

*Soln.* Given,    stages $K = 8$, $R = 50\%$, $r_p = 4$

$$T_1 = 20°C = 293\ K,\ V_1 = 90 \text{ m/s},\ u = 180 \text{ m/s}$$

$$\eta_{isen} = 82\%,\ c_p = 1.005 \text{ kJ/kg } k,\ \gamma = 1.4$$

please refer Fig. 10.19

(i) Work done by the machine/kg of air

$$w = \left(\frac{\gamma}{\gamma - 1}\right) RT_1 \left[ (r_p)^{\frac{\gamma - 1}{\gamma}} - 1 \right]$$

$$= \frac{1.4}{1.4 - 1} \times 287 \times 293 \left[ (4)^{\frac{1.4 - 1}{1.4}} - 1 \right]$$

$$= 143028.44 \text{ Nm/kg of air}$$

$$\eta_{isen} = \frac{\text{Isentropic work, } W_i}{\text{Actual work, } W}$$

$\therefore$ Actual work done   $w = \dfrac{w_i}{\eta_{isen}} = \dfrac{143028.44}{0.82} = 174424.93$ N-m/kg of air *Ans.*

(ii) Blade angles

work done per stage    $= \dfrac{174424.93}{8} = 21803.11$ N-m/kg of air

For 50% reaction turbine

$$\beta_1 = \alpha_2 \text{ and } \alpha_1 = \beta_2$$

Let      $AE = FB = x$

From $\Delta\, CBE$ and $\Delta\, BDF$

$$V_{f1} = x \cot \alpha_1$$
$$90 = x \cot \alpha_1$$
$$DF = V_{f2} = 90 = (u - x) \cot \alpha_2$$

or, $\qquad\qquad 90 = (180 - x) \cot \alpha_2$

Work done per stage $\qquad = u \times \Delta V_w = u(Vw_2 - Vw_1)$

$$21803.11 = 180\,(Vw_2 - Vw_1)$$

or, $\qquad\qquad (Vw_2 - Vw_1) = 121.128$ m/s

But $\qquad\qquad Vw_2 - Vw_1 = u - 2x = 180 - 2x$

or, $\qquad\qquad 121.128 = 180 - 2x$

$\therefore \qquad\qquad x = 29.436$ m/s

$\therefore \qquad\qquad 90 = x \cot \alpha_1$

$$90 = 29.436 \cot \alpha_1$$

$\therefore \qquad\qquad \alpha_1 = 18.11° \quad$ *Ans.*

Also $\qquad\qquad 90 = (180 - x) \cot \alpha_2$

or, $\qquad\qquad 90 = (180 - 29.436) \cot \alpha_2$

$$90 = 150.564 \cot \alpha_2$$

$\therefore \qquad\qquad \alpha_2 = 59.13° \quad$ *Ans.*

**10.20** Overall stagnation isentropic efficiency of an axial flow compressor is 90% with overall stagnation pressure ratio of 4.5. The inlet stagnation pressure and temperature are 1 bar and 330 K. The mean blade speed is 200 m/s. The degree of reaction is 0.5 at the mean radius with relative air angles of 30° and 10° at rotor inlet and outlet respectively. the work done factor is 0.92. Calculate the stagnation temperature. If the hub-tip ratio is 0.45, mass flow rate is 25 kg/s, calculate the blade height in the first stage.

*Soln.* Given, $\eta_{\text{isen}} = 0.90$, $r_p = 4.5 = \dfrac{P_{o2}}{P_{o1}}$, $p_{o1} = 1$ bar, $T_{o1} = 330\,k$, $u = 200$ m/s, $R = 0.5$, $\alpha_1 = 10°$, $\alpha_2 = 30°$

Refer Fig. 10.36

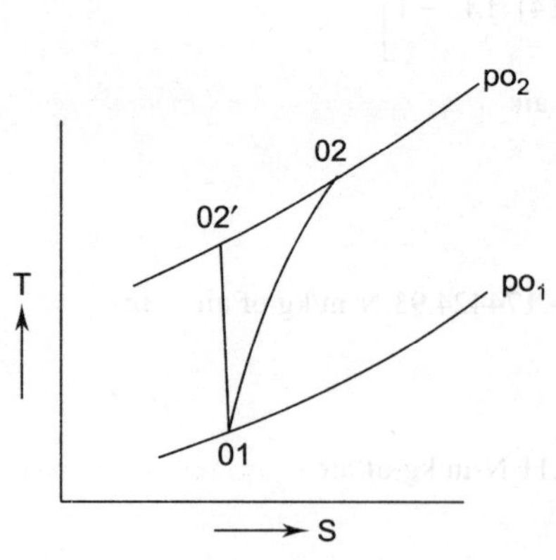
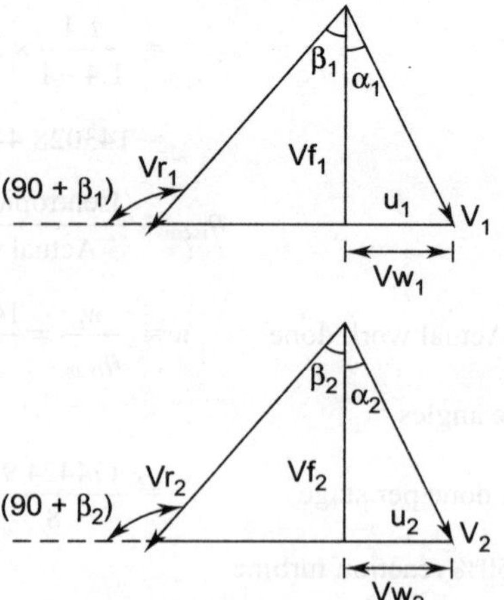




**Fig. 10.36**

$$\text{Degree of reaction} \quad = \frac{\text{Enthalpy increase in rotor balde}}{\text{Enthalpy increase in both}}$$

$= 0.5$ *i.e.* 50% reaction m/c, the inlet and outlet velocity diagrams are similar. Hence

$$\alpha_1 = \beta_2 = 10° \text{ and } \alpha_2 = \beta_1 = 30°$$

Due to isentropic compression, the temperature at the end

$$\frac{T_{o2'}}{T_{o2}} = \left(\frac{P_{o2}}{P_{o2}}\right)^{\frac{\gamma-1}{\gamma}}$$

$$\therefore \qquad T_{o2'} = 330(4.5)^{(1.4-1)/1.4}$$

$$T_{o2'} = 507.15 \ K$$

$$\eta_{\text{isen}} = \frac{T_{o2'} - T_{o1}}{T_{o2} - T_{o1}} = \frac{507.16 - 330}{T_{o2} - 330} = 0.90$$

or, $\qquad T_{o2} = 526.84 \ K$

$$\eta_{\text{std.polytropic}} = \frac{\ln\left(\frac{P_{o2}}{P_{o1}}\right)^{\frac{(\gamma-1)}{\gamma}}}{\ln\left(\frac{T_{o2}}{T_{o1}}\right)} = \frac{\ln(4.5)^{\frac{1.4-1}{1.4}}}{\ln\left(\frac{526}{330}\right)}$$

or, $\qquad 0.9217 = 92.17\%$ *Ans.*

**From velocity triangles**

$$u_1 = u_2 = u \text{ mean blade speed}$$

$$\tan \alpha_1 + \tan \beta_1 = \frac{u_1}{v_{f1}}$$

$$\tan 10° + \tan 30° = \frac{200}{v_{f1}}$$

$$\therefore \qquad V_{f1} = 265.36 \text{ m/s} = v_{f2}$$

$$V_{w2} = V_{f2} \tan \alpha_2$$

$$= 265.36 \tan 30° = 153.20 \text{ m/s}$$

$$V_{w1} = V_{f1} \tan \alpha_1 = 265.36 \tan 10° = 46.79 \text{ m/s}$$

$$\text{Work done factor} \qquad \lambda = \frac{\text{work consumed per stage}}{\text{work supplied}}$$

$$= \frac{c_p\left(T_{o2'} - T_{o1}\right)}{\left(V_{w2} - V_{w1}\right)u}$$

$\therefore \quad$ Work consumed per stage $= 0.92(153.20 - 46.79) \times 200$

$$= 19579.44$$

$$= 19.57 \text{ kJ/kg}$$

Total work consumed by the compressor

$$= c_p(T_{o2} - T_{o1}) = 1.005(526.84 - 330)$$
$$= 197.82 \text{ kJ/kg}$$

$\therefore$ Number of stages $= \dfrac{197.82}{19.57} = 10.1 \simeq 10$ stages   *Ans.*

The absolute velocity $V_1$ at exit from the guide vanes and approaching to moving blades of first stage is

$$\cos \alpha_1 = \frac{V_{f1}}{V_1} \quad \therefore \quad V_1 = \frac{265.36}{\cos 10°} = 269.45 \text{ m/s}$$

And the temperature

$$T_1 = T_{o1} - \frac{V_1^2}{2c_p} = 330 - \frac{269.45^2}{2 \times 1.005 \times 1000} = 293.87 \ K$$

Assuming reversible flow through the guide vanes put head of the first stage

$$\left(\frac{p_1}{P_{o1}}\right) = \left(\frac{T_1}{T_{o1}}\right)^{\frac{\gamma}{\gamma - 1}} \quad \text{or,} \quad \left(\frac{p_1}{1}\right) = \left(\frac{293.87}{330}\right)^{\frac{1.4}{1.4 - 1}}$$

$\therefore \qquad\qquad\qquad p_1 = 0.667 \text{ bar}$

Hence the density of air approaching to first stage

$$p_1 V_1 = mRT_1$$

or, $\qquad$ density $\rho_1 = \dfrac{m}{V_1} = \left(\dfrac{p_1}{RT_1}\right)$

$$= \frac{0.667 \times 10^5 \times 10^{-3}}{0.287 \times 293.87}$$
$$= 0.79 \text{ kg/m}^3$$

From the continuity eqn.

$$\dot{m} = \rho_1 A_1 V_{f1} = 25 \text{ kg/s}$$

or, $\qquad 0.79 \times \pi r_t [1 - (0.45)^2] \times 265.36 = 25$

or, $\qquad r_t = 0.2181 \ m = 21.81 \text{ cm}$

But $\qquad \dfrac{r_r}{r_t} = 0.45$

$\therefore \qquad\qquad r_r = 21.81 \times 0.45 = 9.81 \text{ cm}$

Hence the height of the blade in the first stage is

$$l = r_t - r_r = 21.81 - 9.81 = 12 \text{ cm} \quad Ans.$$

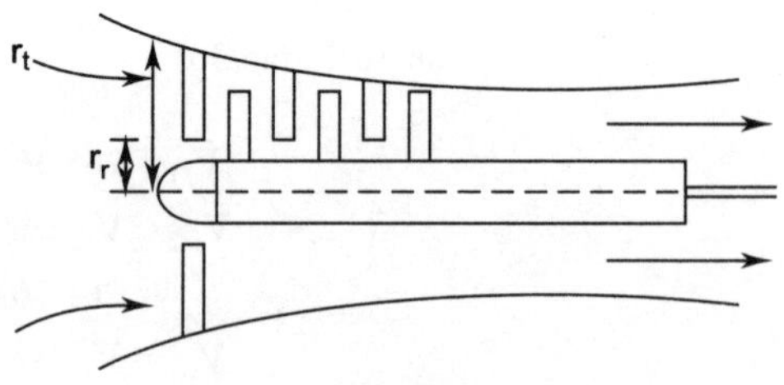


**Fig. 10.37**

# EXERCISES

10.1 Objective questions

(i) Power required to run air compressor is minimum of compression follows

    (a) adiabatic law, $pv^\gamma = c$          (b) isothermal law, $pv = c$

    (c) $pv^{1.1} = \text{constant}$          (d) $pv^{1.25} = \text{constant}$

(ii) Isothermal compression is required, but it necessitates
    (a) high running of machine
    (b) very slow running of machine
    (c) running of machine at constant speed
    (d) none of the above

(iii) The practical values of index of compression (*i.e. n*) for a compressor in the range of
    (a) 1.4 to 1.6    (b) 1.3 to 1.4    (c) 1.2 to 1.4    (d) 1.0 to 1.3

(iv) Power required to run the compressor is maximum if compression follows the process of
    (a) isothermal    (b) isentropic    (c) polytropic    (d) adiabatic

(v) The clearance volume in reciprocating air compressors is provided to
    (a) reduce the work done per kg of air delivered
    (b) increase the volumetric efficiency of the compressor
    (c) accommodate valves in the head of compressor
    (d) create turbulence in the air to be delivered

(vi) With increase in clearance volume, the ideal work of compression of 1 kg of air
    (a) increases
    (b) decreases
    (c) remains same
    (d) first increases and then decreases

(vii) The clearance ratio for a single stage compressor lies between
    (a) 15% to 20%    (b) 20% to 30%    (c) 4% to 10%    (d) 1% to 2%

(viii) The maximum theoretical volume of air that can be delivered by the compressor is called
    (a) free air delivery
    (b) compressor capacity
    (c) swept volume
    (d) none of these

(ix) The actual volume of air sucked by the compressor during the suction stroke is
    (a) less than swept volume
    (b) more than swept volume
    (c) equal to swept volume
    (d) equal to cylinder volume

(x) The ratio of initial volume to the final volume during compression is known as
    (a) compressor capacity
    (b) compression ratio
    (c) compressor efficiency
    (d) mean effective pressure

(xi) Volumetric efficiency of reciprocating compressor is defined as the ratio of actual volume to
    (a) air sucked in
    (b) free air delivered
    (c) compressed air delivered
    (d) swept volume

(xii) Mechanical efficiency of reciprocating air compressor is expressed as
    (a) $\dfrac{B.P}{I.P}$    (b) $\dfrac{I.P}{B.P}$    (c) $\dfrac{F.P}{B.P}$    (d) $\dfrac{F.P}{I.P}$

(xiii) Free air condition means the air at
    (a) 15°C and 1.01325 bar with relative humidity 60%
    (b) standard atmospheric conditions and 0°C
    (c) atmospheric conditions at any specific location
    (d) 20°C and 1.01325 bar and relative humidity 36%

(xiv) The compressor capacity is equal to
    (a) Volume of the air sucked
    (b) Volume of air delivered
    (c) sum of (a) and (b)
    (d) none of the above

(xv) With the increase of pressure ratio, the volumetric efficiency of an air compressor
    (a) increases
    (b) decreases
    (c) constant
    (d) first increases then decreases

(xvi)  The interstage pressure for minimum work in a two stage reciprocating compressor is
  (a)  half the delivery pressure
  (b)  twice the suction pressure
  (c)  square root of the product of suction and delivery pressure
  (d)  square root of delivery pressure

(xii)  Free air delivered ($F.A.D$) is the volume of delivered air converted to
  (a)  atmospheric condition
  (b)  $NPT$
  (c)  interstage pressure
  (d)  all

(xiii)  In centrifugal air compressor the pressure developed depends on
  (a)  impellor velocity
  (b)  inlet temperature
  (c)  compression index
  (d)  all of the above

(xiv)  In a centrifugal compressor the pressure ratio is increased by
  (a)  increasing the speed of impeller keeping its diameter fixed
  (b)  increasing the diameter of the impeller keeping its speed constant.
  (c)  reducing inlet temperature, keeping impeller diameter and speed fixed
  (d)  all of the above

(xx)  The efficiency of vane type air compressor as compared to roots air compressor, it pressure ratio exceed 1.4 is
  (a)  more
  (b)  less
  (c)  same
  (d)  may be more or less

(xxi)  In case of air engines of air motors
  (a)  intake is at higher pressure and discharge at lower pressure
  (b)  intake is at lower pressure and discharge at higher pressure
  (c)  intake and discharge is at same pressure
  (d)  none of the above

## Answer

|         |     |          |     |          |     |          |     |
|---------|-----|----------|-----|----------|-----|----------|-----|
| (i)     | (b) | (ii)     | (b) | (iii)    | (d) | (iv)     | (b) |
| (v)     | (c) | (vi)     | (c) | (vii)    | (c) | (viii)   | (c) |
| (ix)    | (a) | (x)      | (b) | (xi)     | (d) | (xii)    | (b) |
| (xiii)  | (d) | (xiv)    | (b) | (xv)     | (b) | (xvi)    | (c) |
| (xvii)  | (a) | (xviii)  | (d) | (xix)    | (d) | (xx)     | (a) |
| (xxi)   | (a) |          |     |          |     |          |     |

10.2  Enumerate the applications of compressed air.

10.3  Classify the various types of compressors.

10.4  Show that cylinder clearance does not affect the theoretical work required to compress and deliver 1 kg of air, provided that the suction and delivery pressure remain same and indices of compression and expansion have the same value.

10.5  Prove that the work done/kg of air in a compressor is given by

$$W = RT_1 \frac{n}{n-1}\left[\left(r_p\right)^{\frac{n-1}{n}} - 1\right], \text{ where } r_p = \text{pressure ratio}$$

10.6  Prove that the volumetric efficiency of a single stage compressor is given by

$$\eta_v = 1 + c - c\left(\frac{p_2}{p_1}\right)^{1/n}, \text{ where } c = \frac{v_c}{v_s}$$

10.7   State the conditions which lower the volumetric efficiency.

10.8   What factors limit the delivery pressure in a reciprocating compressor.

10.9   What are the advantages of multi stage compression over a single stage compression for the same pressure ratio? Why inter-cooling is necessary in multi-stage compression.

10.10   Prove that the minimum work required per cycle with perfect intercooling is given by

$$W = \frac{xn}{n-1} p_1 v_1 \left[ \left( \frac{p_{x+1}}{p_1} \right)^{\frac{n-1}{xn}} - 1 \right]$$

where,                                        $x$ = no. of stages.

10.11   Explain the effect of intake pressure and temperature of air on the output of a compressor.

10.12   (a)  What is the difference between centrifugal and axial flow compressor?

 (b)  Deduce the expression for roots efficiency.

10.13   (a)  Define the following (i) ship (ii) ship factor (iii) degree of reaction

 (b)  Obtain an expression for the work done by a centrifugal compressor.

10.14   What you mean by surging and choking in a compressor?

## Numerical Problems

10.15   Atmospheric air at 1.013 bar and 21°C is taken into a single stage compressor having zero clearance. It is compressed according to the law $pv^{1.2}$ = constant to the constant discharge pressure of 4.052 bar. The discharge is taken through a regulating valve into a closed vessel of 2.8 m³ capacity. Here the initial conditions were 1.013 bar and 21°C and after charging for 4.2 min. Were 3.4 bar and 27°C, Calculate

 (a)  the volume of air taken per minute of measured at atmospheric conditions, and

 (b)  the indicated power to drive the compressor.

*[Ans.* 1.5259 m³/min, 4.01769 kW]

10.16   A single stage single acting air compressor takes in air at 1 bar and compresses it to 10 bar and delivers at a rate of 0.05 m³/s. If the compressor and expansion follow the law $pv^{1.3} = c$, find the power required if the clearance volume is 6% of the swept volume and the swept volume is 14,500 c.c. Also find the speed of compressor and the volumetric efficiency.

*[Ans.* 23 kW, 293 r.p.m, 77%]

10.17   A simple stage single acting air compressor running at 1000 r.p.m delivers air at 25 bar. For air the induction and free air conditions can be taken as 1.01325 bar and 15°C, and the free air delivery 0.25 m³/min. The clearance volume is 3% of the swept volume and the stroke/bore ratio is 1.2 : 1. Calculate the bore and stroke and the volumetric efficiency of this machine. Take the index of compression as 1.3. Calculate also the indicated power and the isothermal efficiency.

*[Ans.* 73.2 mm, 87.84 mm, 67.6%]

10.18   A single acting reciprocating compressor with a bore and stroke of 0.4 m runs at 5 r.p.s. The clearance volume is 1750 cm³. The expansion and compression follow the law $pv^{1.3} = c$. The pressure and temperature at the end of suction are 1 bar and 300 K. Calculate the indicated power of compressor and volumetric efficiency if the delivery pressure is 7 bar.

*[Ans.* 51 kW, 87%]

10.19   A single stage double acting air compressor delivers air at 7 bar pressure. The amount of free air delivered is 2 m³ at 300 r.p.m. The pressure and temperature at the end of suction stroke are 1 bar and 27°C. The ambient conditions are 1.01325 bar 20°C. The clearance is 5% of stroke. Determine (i) Brake power required to run the compressor if the $\eta_{mech}$ = 80% (ii) The diameter and stroke of the cylinder, if both are equal. Assume compression and re-expansion follow the law $pv^{1.3}$ = constant.

*[Ans. B.P* = 10.8 KW, $L = D$ = 17.95 cm]

10.20   A single stage double acting air compressor is driven by 39 KW electric motor with a transmission efficiency of 95%. Air enters the cylinder at 0.99 bar and 289 $K$. The index of compression is 1.2. The delivery pressure is 5.9 bar. Find the size of the cylinder for a piston speed of 2.5 m/s and a shaft speed of 1.67 r.p.s. Neglect clearance.

$$[Ans.\ D = 0.32\ m,\ L = 0.75\ m]$$

10.21   A single cylinder, single acting air compressor running at 300 r.p.m is driven by 23 KW. The mechanical efficiency of the drive between motor and compressor is 87%. The air inlet conditions are 1.01325 bar and 15°C and the delivery pressure is 8 bar. Calculate the free air delivery, volumetric efficiency and the bore and stroke of the compressor. Assume that the index of compression and expansion, $n = 1.3$, that the clearance volume is 7% of the swept volume and that the stroke.

$$[Ans.\ 4.47\ m^3/min,\ 73\%,\ 296\ mm]$$

10.22   A two stage air compressor consists of 3 cylinders having the same bore and stroke. The delivery pressure is 7 bar and the free air delivery is 4.2 m$^3$/min. Air is drawn in at 1.01325 bar, 15°C and intercooler cools the air to 38°C. The index of compression is 1.3 for all the three cylinders. Neglect clearance, calculate

   (i)   The intermediate pressure                      (ii)   The power required to drive the compressor

 (iii)   The isothermal efficiency.

$$[Ans.\ (i)\ 2.19\ bar,\ (ii)\ 16.3\ KW,\ 84.5\%]$$

10.23   A two stage compressor, compress air from 15°C and 100 $KPa$ to 6000 $KPa$. The air is cooled in the intercooler to 30°C and the intermediate pressure is steady at 733 $KPa$. The low pressure cylinder is 10 cm in diameter and the stroke for both cylinders is 11.25 cm. Assuming a compression law of $pv^{1.35}$ = constant and that the volume of air at atmospheric condition drawn in per stroke is equal to the low pressure cylinder swept volume. Find the power of the compressor when running at 250 r.p.m. Find also the diameter of the high pressure cylinder.

$$[Ans.\ 2.0425\ kW,\ D = 3.78\ cm]$$

10.24   A three stage compressor delivers air at 70 bar from an atmospheric pressure of 1 bar and 30°C. Assuming the intercooling is complete and maximum efficiency conditions, find for one kg of air per second.

   (i)   Intermediate pressure                         (ii)   Power required

 (iii)   Heat rejected in each intercooler

 (iv)   Compare the power required if the compressor is single stage.

$$[Ans.\ 4.11\ bar,\ 16.9\ bar,\ 369\ kW]$$

10.25   A single acting two stage compressor with complete intercooling delivers 10 kg/min of air at 16 bar. Suction occurs at 1 bar and 15°C. The expansion and compression processes are reversible polytropic with $n = 1.25$. Find

   (i)   The power required                          (ii)   The thermal efficiency

 (iii)   The free air delivery                         (iv)   Heat transferred in intercooler

  (v)   If the clearance ratio for $L.P$ and $H.P$ cylinders are 0.04 and 0.06 respectively. Calculate the swept and clearance volume for each cylinder. The speed of the compressor is 400 r.p.m.

$$[Ans.\ (i)\ 44\ KW\ (ii)\ 86.8\%\ (iii)\ 8.266\ m^3/min\ (iv)\ 15.41\ kW$$
$$(v)\ 0.0225\ m^3,\ 0.0009\ m^3,\ 0.00588\ m^3,\ 0.000353\ m^3]$$

10.26   A three stage single acting compressor delivers 2.25 m$^3$/min of free air from 1 bar to 64 bar. Calculate the power required to operate the compressor, if $n = 1.3$. The mean piston speed is 140 m/min. Find the piston area, neglecting clearance.

$$[Ans.\ 18\ KW,\ 321\ cm^2,\ 80\ cm^2,\ 20\ cm^2].$$

10.27   An air compressor works between the pressures of 1 bar and 125 bar in four stages. At each stage, the air is cooled down to 308 $K$. The index of compression is 1.23. Assuming optimum interstage pressures, estimate the isothermal efficiency of the compressor referred to the intake temperature of 298 $K$.

$$[Ans.\ 87\%]$$

10.28  A multistage air compressor is to be designed to elivate the pressure from 1 bar to 120 bar such that stage pressure ratio will not exceed 4. Determine (i) no. of stage (ii) exact stage pressure ratio (iii) intermediate pressure

[*Ans.* (i) 4 (ii) 3.31 (iii) 36.25 bar, 10.95 bar, 3.31 bar]

10.29  A centrifugal compressure is designed to have a positive ratio of 3.50 : 1. The inlet eye of the compressor impeller is 30 cm in diameter. The axial velocity at inlet is 130 m/s and the mass flow is 10 kg/s. The velocity in the delivery duct is 115 m/s. The tip speed of the impeller is 450 m/s and runs at 16000 r.p.m. head isentropic efficiency of 78% and pressure co-efficient of 0.72. The ambient conditions are 1.01325 and 15°C. Calculate the (i) static pressure ratio (ii) static pressure and temperature at inlet and outlet and compressor (iii) work of compressor per kg of air and (iv) The theoretical power required.

[*Ans.* (i) 4.21 (ii) 0.917 bar, 279.6 *K*, 3.86 bar (iii) 180.29 kJ/kg of air (iv) 18

10.30  A centrifugal compressor delivers 54 kg of air per minute at a pressure of 200 Kpa, when compressing from 100 kPa and 15°C. If the temperature of the air delivered is 97°C and no heat is added to the air from external sources during compression, determine the efficiency of the compressor relative to ideal adiabatic compression and determine the power absorbed.

[*Ans.* 76.8%, 74.058 KW]

10.31  A centrifugal compressor running at 9000 r.p.m delivers 600 m$^3$/min of free air. Air is compressed from 1 bar and 20°C to a pressure ratio of 4 with an isentropic efficiency of 0.82. Blades are radial at outlet of impeller and the flow velocity of 62 m/s may be assumed throughout constant. The outer radius of impeller is twice the inner and the slip factor may be assumed as 0.9. The blade are co-efficient assumed as 0.9 at the inlet, calculate (i) Final temperature of air (ii) Theoretical power (iii) Impellers diameter at inlet and outlet (iv) Breadth of the impeller at inlet (v) Impeller blade angle of inlet (vi) Diffuse blade angle of inlet.

[*Ans.* (i) 466.85 K (ii) 2077.7 KW (iii) 46.745 cm, 94.9 cm (iv) 12.2 cm (v) 15.7° (vi) 8.9°]

10.32  Two compensators, one reciprocating and one centrifugal, both are taking in 1 Kg/s of air at 1 bar and 10°C and deliver at 3.5 bar. The isothermal efficiency of the reciprocating compensator and the isentropic efficiency of the centrifugal compressor are each 75%. Calculate theoretical power required to drive each compressor and the temperature of the air delivered by the centrifugal compressor. Estimate the peripheral speed of a straight vaned impeller used in the centrifugal compressor, pre-whirl being negligible.

[*Ans.* 135.7 kW, 163.5 kW, 172.7°C, 404.4 m/s]

10.33  In an axial flow compressor, the overall stagnation pressure ratio achieved is 4 with overall stagnation isentropic efficiency of 88%. The inlet stagnation pressure and temperature are 1 bar and 320 *K*. The mean blade speed is 190 m/s. The degree of reaction is 0.5 at the mean radius with relative air angles of 10° and 30° at rotor inlet and outlet respectively. The work done factor is 0.9. Calculate: (i) Stagnation polytropic efficiency (ii) Number of stages (iii) Inlet temperature and pressure (iv) Bode height in the first stage if the hub-tip ratio is 4, mass flow rate is 20 Kg/s.

[*Ans.* (i) 88.4% (ii) 11 (iii) 287.39 K, 0.6864 bar (iv) 11.4 cm]

10.34  The first stages of an axial flow compressor are to be designed for an axial velocity of 120 m/s, an air flow is 23 Kg/s and for ambient conditions of 1.01325 bar and 10°C. At the mean radius both moving and fixed blade sections are to be suitable for relative air angles of 50° at inlet and 30° at outlet. A row of guide vanes giving an outlet air angle of 30° is to be fitted before the stage. Taking the ratio of hub to tip diameter as 0.68, assuming free vortex conditions and ignoring and effects, draw the velocity triangles for mean and tip positions, indicating on them the air velocities. Calculate (i) The compressor speed in r.p.s (ii) The theoretical temperature rise across the stage (iii) the number of similar stages which would theoretically be required to give a total head delivery pressure of 8.1 bar, if the stage isentropic efficiency is 85%.

[*Ans.* $D_{\text{lip}}/D_m = 1.1905$, $U = 212.3$ m/s. At tip $V_{w1} = 58.2$ m/s, $V_{w2} = 120.1$ m/s, $\beta_1 = 54.4°$, $\beta_2 = 47.9°$, $\alpha_1 = 25.9°$, $v_1 = 0.802$ m$^3$/kg, $D_m = 0.507$ m, $N = 133.3$ r.p.s. $\Delta T = 15.65°$, no. of stages = 13]

11

# Refrigeration and Air Conditioning

## 11.1  FUNDAMENTALS

According to the natural phenomena, heat always passes downhill, from a warm body to a cooler one until both sides are at the same temperature. A body is cooled when heat is removed from it. The word 'refrigerate' means to chill or freeze a substances i.e. to lower its temperature by removing some of its heat from a substance and rejecting the heat so removed to the atmosphere which is at a higher temperature level. The carrier substance used to carry the heat is called a 'refrigerant'.

This process has been very popular right from the advancement, industrial growth and growing standard of living of human beings. Today it becomes most essential requirements as per our needs. The equipment employed to maintain the system at a low temperature is termed as refrigerating system and the system which is kept at lower temperature is called refrigerated system. Refrigeration is generally produced in one of the following ways:

  (i)  By melting of solid (ice)

  (ii)  By sublimation of a solid (dry ice)

  (iii)  By evaporation of a liquid

The earliest method of refrigeration was melting of ice which is still used. Ice melts at 0°C. So when it melts in a warmer space at 0°C, heat flows into the ice and the space is cooled or refrigerated. Due to latent heat of fusion of ice, which is being supplied by the surrounding or space, it changes its phase from solid to liquid. Likewise dry ice or solid carbondioxide when exposes to atmosphere conditions, it directly converts from solid to vapour which is called sublimation. So dry ice is very suitable for low temperature refrigeration.

**Important refrigeration applications**

1. Food preservation by freezing
2. Ice making
3. Transportation of food products from one place to other
4. Medical and surgical aids
5. Air conditioning for comfort feeling
6. Industrial applications
7. Controlling humidity and temperature in the process industry

**Refrigeration systems**

1. Ice refrigeration system
2. Air refrigeration system
3. Vapour compression refrigeration system
4. Vapour absorption refrigeration system
5. Other special refrigeration system
    (i) Steam jet refrigeration system
    (ii) Cascade refrigeration system
    (iii) Absorption refrigeration system
    (iv) Vortex tube refrigeration system
    (v) Thermoelectric refrigeration

## 11.1.1  Ice  Refrigeration

This method was the only means to produce cooling effect in early days. Ice refrigeration has not become absolute till today as it has many applications for which other method of refrigeration are not suitable.

In ice refrigeration method, the temperature of a space is decreased by the use of ice naturally harvested or artificially manufactured. The ice absorbs 335 kJ of heat when one kg of it melts at atmospheric pressure. The loss of heat by the space results in lowering down its temperature. A simple ice refrigerator is shown in Fig. 11.1 where ice is kept in a tray with proper leaffing in the space to be cooled in this refrigerator. The heat from the space and the products stored there in goes to the ice by convective currents set-up in the air of the refrigerated space. The ice is placed near the tip of the refrigeration to ensure proper circulation of air. This method is known as direct contract method because ice comes in

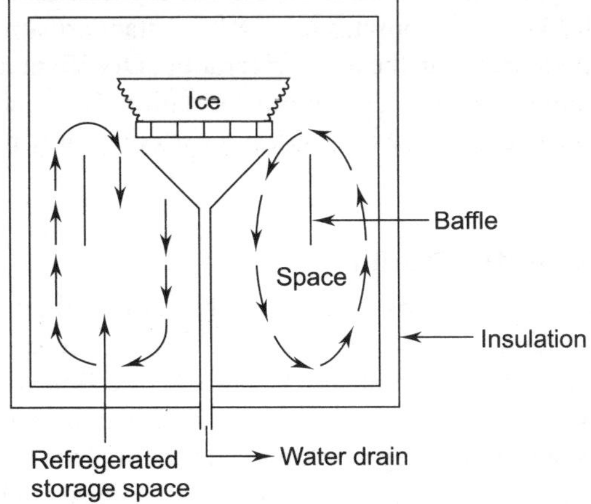


**Fig. 11.1    Ice refrigerator**

direct contact with the space to be refrigerated. The ice refrigeration method may be made continuous by making refrigeration as shown in Fig. 11.2 where secondary refrigerant usually brine is continuously chilled in separate compartment by ice and circulated through the space to be cooled.

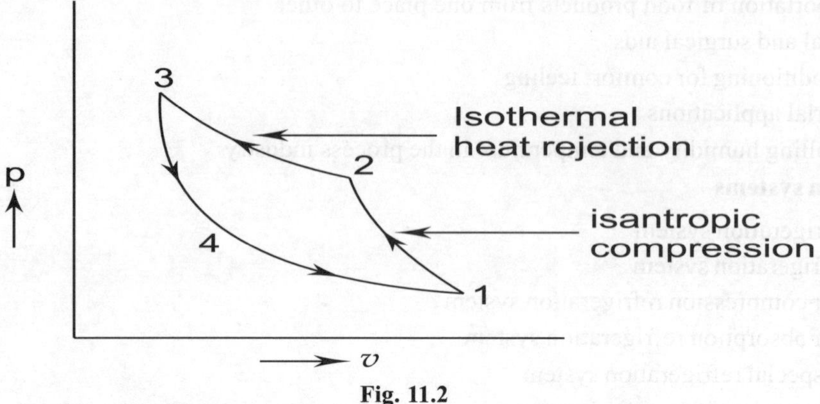


**Fig. 11.2**

**Advantages of ice refrigeration are:**

1. It prevents dehydration of fresh fruit, vegetables, fish and poultry
2. It preserves the appearance of eatable products and imparts them eye appeal.

**The disadvantages are**

1. Ice is fed to the refrigerator off and on
2. There is a problem of disposal of water
3. The control on rate of refrigeration is impossible.
4. There is a limit of temperatures of space to be refrigerated.

## 11.1.2   Dry ice Refrigeration

In this method solid carbondioxide (commonly known as dry ice) in place of ice, is used. Dry ice has the characteristics of changing from solid phase to vapour state directly without entering into the liquid phase. During change of phase from solid to vapour the dry ice absorbs heat equivalent to its latent heat of evaporation (605.5 KJ/kg). It evaporates at $-78°C$ at standard atmosphere pressure. So this method can produce lower temperatures than the ice refrigeration. Dry ice method of refrigeration is exclusively used to preserve perishable food stuff during transportation from one place to another especially in air transportation. Dry ice is placed on top or at sides of cartoons carrying eatables and thus the articles are frozen.

## 11.1.3   Evaporative  Refrigeration

When a liquid evaporates it absorbs heat from the surroundings equivalent to its latent heat of evaporation, thereby the temperature of the surroundings is decreased. A drop of sprit placed on our plam of hand produces cold because it absorbs heat from the skin when it evaporates. The evaporation of moisture from the skin of a human body helps to keep it cool. Another common application of this principle is the desert bag used to keep drinking water cool. This bag consists of tightly woven fabric bag fitted with drinking water. The bag is not water proof, consequently some water sweeps through and surface of the bag remains moist. Under desert condition which are usually both hot and dry, moisture on the surface evaporate rapidly. The same principle is used in evaporative, condenser. Thermodynamically, the evaporative cooling is the adiabatic exchange of heat between air and water spray or wetted surface.

## 11.1.4  Air Expansion Refrigeration

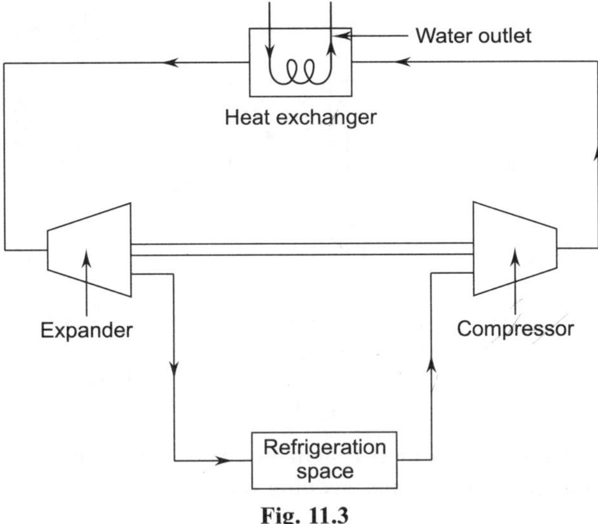


**Fig. 11.3**

The temperature of air and other such gases can be lowered down by expanding them adiabatically. The cooling effect is produced in this method, shown in Fig. 11.3, without changing the phase of the gases.

This cooling effect is governed by the well known thermodynamic relation $\dfrac{T_2}{T_1} = \left(\dfrac{p_2}{p_1}\right)^{\frac{\gamma-1}{\gamma}}$ for air $\gamma = 1.4$,

so the relation becomes $\dfrac{T_2}{T_1} = \left(\dfrac{p_2}{p_1}\right)^{\frac{2}{7}}$.

If air from at morphemic temperature of 15°C or 285 $K$ is compressed, adiabatically to a pressure of five

atmospheres, its temperature will become $T_2 = T_1 \left(\dfrac{p_2}{p_1}\right)^{\frac{2}{7}} = 28 \left(\dfrac{5}{1}\right)^{\frac{2}{7}} = 460.8\ K$. This compressed air can

be cooled down in a suitable heat exchanger without pressure loss to its origin at temperature of 15°C. Now if the cooled air at 15°C and at 5 atmospheric pressure is expanded again in a suitable device to 1

atmospheric pressure, then $T_1 = \left(\dfrac{p_1}{p_2}\right)^{\frac{\gamma-1}{\gamma}}$. $T_2 = 288 \left(\dfrac{1}{5}\right)^{\frac{2}{7}} = 180\ K = -93°C$ which is very low

temperature.

### 11.1.5 Cooling by Throttling

During a throttling process there is no change in temperature. Actual gases, however, diverge, sufficiently from perfect gases and they show a substantial change, usually decrease, in temperature after being throttled.

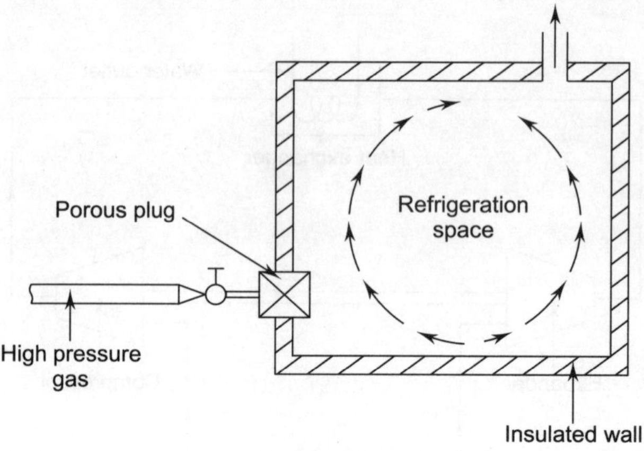


**Fig. 11.4**

The Fig. 11.4 shows a simple refrigeration system based upon this principle where air or some other gas at high pressure is throttled through a porous plug into a space.

### 11.1.6 Steam Jet Refrigeration

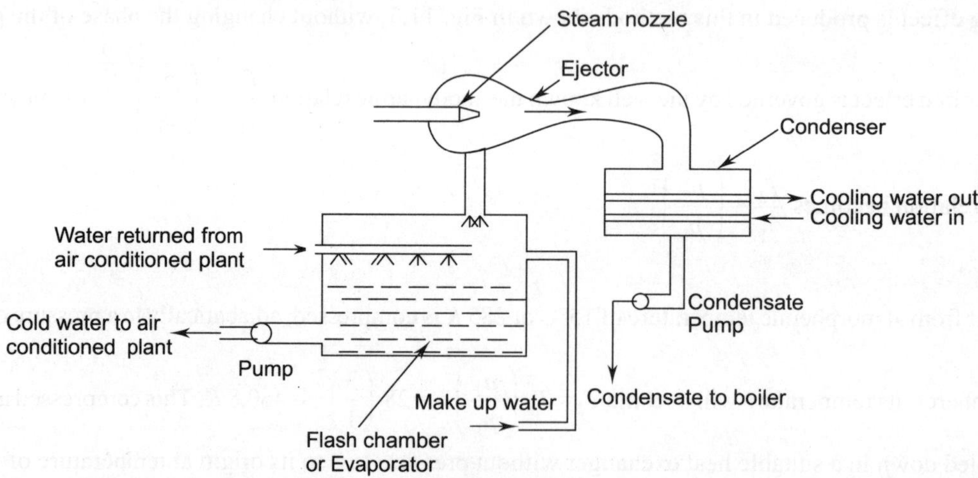


**Fig. 11.5**

Fig. 11.5 shows the schematic diagram of steam jet refrigeration. Boiling temperature of water can considerably be decreased by reducing the pressure on it ordinarily, water boils at 100°C when it is under standard atmospheric pressure. By suitably reducing the pressure to 6.5 cm of water, the water can boil and evaporate at 10°C. If the pressure is further reduced to **5 cm** of water it may boil even **at 6°C.**

The *above principle of boiling* the *water at low temperature by reducing its pressure is adopted in steam jet refrigeration. The steam required for this process of refrigeration may* either be live steam from the boiler or exhaust or waste steam from other industrial process.

## 11.1.7  Liquid Gas Refrigeration

Cooling can be accomplished by the evaporation of liquid gases. Liquid gases like nitrogen and carbon dioxide are used for cooling in the **transportation vehicles.** When the liquid gases evaporate, they lower the temperature of the space. The gas after taking the heat of evaporation is released to the atmosphere. This method shown in Fig. 11.6 of refrigeration is becoming popular in cooling the vehicles carrying perishable articles. As the gases used for this purpose are to be non-toxic so common gases available in liquid state are liquid nitrogen and liquid carbon dioxide

| | |
|---|---|
| Liquid oxygen | $= -181.8°C$ |
| Liquid nitrogen has boiling point | $= -195.6°C$ |
| Liquid neon | $= -245.9°C$ |
| Liquid hydrogen | $= -252.6°C$ |
| Liquid helium | $= -272°C$ |

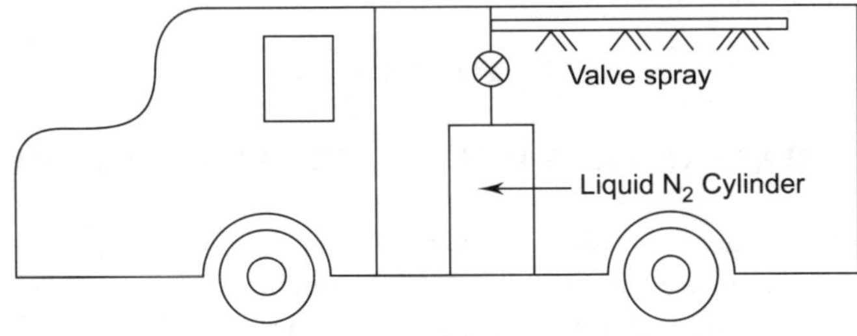


**Fig. 11.6**

## 11.1.8  Vapour Compression Refrigeration

It is the most common method of refrigeration which is usually termed as mechanical refrigeration. In this method shown in Fig. 11.7, the liquid when evaporates absorbs heat is employed. The only speciality of this method is that same refrigerant is used again and again in a cycle. The refrigerant continuous changing from liquid to vapour state when absorbing heat and from vapour to liquid state when gives out heat.

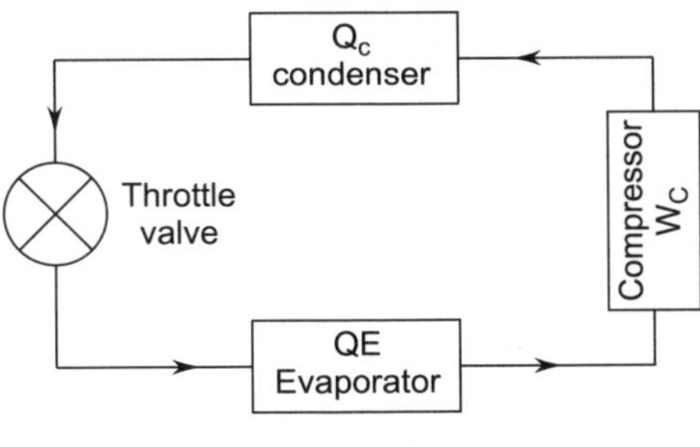


**Fig. 11.7**

### 11.1.9  Vapour Absorption System

In this system shown in Fig. 11.8, heat energy is used to produce refrigeration unlike mechanical energy used in vapour compression system. The vapour absorption system uses two substances which have greater affinity for each other but which can be separated easily by the application of heat. The system finds its application in both domestic and commercial installations

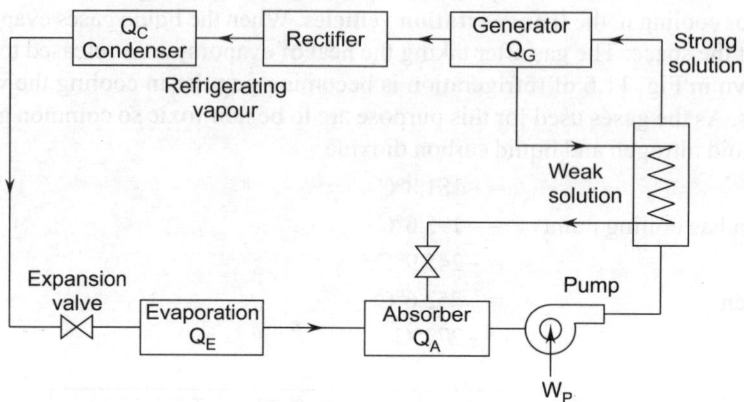


**Fig. 11.8**

## 11.2  STANDARD RATING OF A REFRIGERATION MACHINE

The rating of a refrigeration machine is obtained by refrigerating effect or amount of heat extracted in a given time from a body. The power or capacity of mechanical equipment is generally given in H.P. Similarly the capacities of electrical equipments are generally given in KW for small units and in M.W for power plants. A unit used in the field of refrigeration and air conditioning system is known as Ton of refrigeration. The rate of removal of heat in cooling operation was expressed in terms of ponds, kilograms or tons of ice required per unit time usually hour or day.

A ton of refrigeration is defined as the quantity of heat required to be removed from one metric ton of water at 0°C within 24 hours to produce ice at 0°C because some cooling effect will be given by melting the same quality of ice.

$$1 \text{ Ton of refrigeration} = \frac{2240 \times 144}{24 \times 60} \text{ BTU/hr}$$

$$= 224 \text{ BTU/min}$$

$$= \frac{224}{3.968} = 56 \text{ Kcal/min}$$

∵   In U.S.A. one ton is taken as 2000 lb

∵        $$1 \text{ Ton of refrigeration} = \frac{2000 \times 144}{24 \times 60} \text{ BTU/min}$$

$$= 200 \text{ BTU/min}$$

$$= \frac{200}{3.968} = 50.4 \, \text{Kcal/min}$$

$$\text{Ton in metric unit} = \frac{1000 \times 80}{24 \times 60} = 55.4 \, \text{Kcal/min}$$

In Russia 1 Ton of refrigeration is defined as 50 Kcal/min.

In M.K.S, 1 Ton of refrigeration (TR)

$$= 50 \, \text{Kcal/min}$$

$$1 \, \text{TR} = 50 \, \text{Kcal/min} = 211 \, \text{KJ/min} = 3.5167 \, \text{KW}$$

(It should be noted that the word 'ton' in 'ton of refrigeration' should be spelled only as 'ton', as it represents the heat unit as distinguished from the weight unit spelled as 'tonne').

## 11.3   HEAT ENGINE, REFRIGERATION AND HEAT PUMP

Systems which have thermodynamic importance are mainly divided into two groups viz.

1. Work developing systems– examples are heat engine
2. Work absorbing systems– examples are compressors, refrigerators and heat pumps. For the purpose of refrigeration and air conditioning, we are mostly interested in refrigerators and heat pumps. All the three systems are shown in Fig. 11.9.

Source and sink contain infinite energy at constant temperature. Source temperature is always higher than the sink temperature.

The performance of any system depends upon what we get and what we pay. In case of heat engine we are interested to get maximum amount of work with minimum amount of energy and therefore we define efficiency, $\eta = \dfrac{W}{Q}$

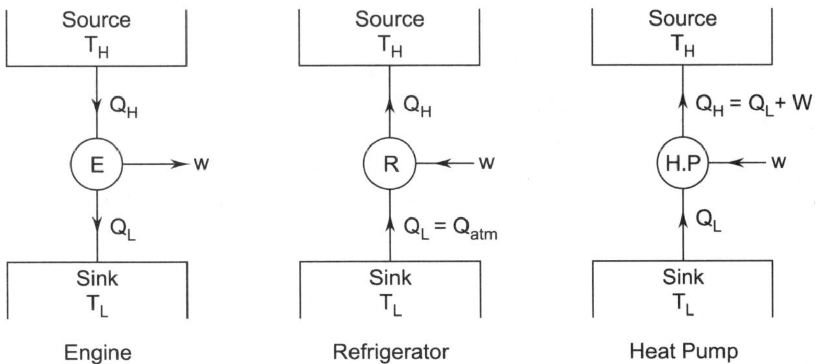


**Fig. 11.9   Engine, refrigerator and heat pump**

In case of refrigerator we are interested to maintain temperature $T_L < T_H$, where $T_H$ is atmospheric temperature and for doing so most economically, the maximum $Q_L$ must be taken from sink with minimum expense of $W$. So the performance of refrigerator is taken into the account by the ratio $\dfrac{Q_L}{W}$ .

$$\therefore \qquad \text{C.O.P.} = \xi_R = \frac{Q_L}{W}.$$

This is known as theoretical C.O.P.

$$\text{Relative C.O.P.} = \frac{\text{Actual C.O.P.}}{\text{Theoretical C.O.P.}}$$

The cycle with is used for refrigerator is also used for heat pump. Only difference is, we are interested to deliver the quantity of heat to the source instead of absorbing the heat from the sink. If more quantity of heat is absorbed from the sink, then more quantity of heat is delivered to the source.

Then we define Energy Performance Ratio for heat pump.

$$\text{E.P.R.} = \frac{Q_L + W}{W} = 1 + \frac{Q_L}{W}$$

$$\therefore \qquad \text{E.P.R.} = 1 + \text{C.O.P} \qquad\qquad (11.1)$$

**Please refer article 4.3 and 4.3.1 for C.O.P and E.E.R.**

Refrigerating efficiency $\qquad \eta = \dfrac{\text{C.O.P. of the cycle}}{\text{C.O.P. of carnot cycle}} = \dfrac{\xi}{\xi_{\text{carnot}}}$

Heating efficiency, $\qquad \eta_n = \dfrac{\text{C.O.P. of heat pump of cycle}}{\text{C.O.P. of heat pump of carnot cycle}}$

Thermal efficiency of a heat engine is of the order 30%

*i.e.* $\qquad\qquad \eta_{\text{th}} = \dfrac{Q_H - Q_L}{Q_H} = 0.3$

$\therefore$   carnot C.O.P, for refrigerator,

$$\xi_c = \frac{Q_L}{Q_H - Q_L} = \frac{1 - \eta_{\text{th}}}{\eta_{\text{th}}} = 2.33$$

carnot C.O.P. for heat pump $\xi_h = \dfrac{Q_H}{Q_H - Q_L} = \dfrac{1}{\eta_{\text{th}}} = 3.33$

For vapour compression system $\xi_c$ is of the order of 2 for air conditioning application, 1 for domestic refrigerators and 3 for water coolers. For air cycle refrigeration system $\xi_c = 1$ and for vapour absorption system, it is well below unity.

Fig. 11.10 shows the comparison between heat engine, heat pump and refrigerating machine with temperature level, where $T_o$ represents temperature of cooled space as lowest temperature $T_L$, $T_a$ represents ambient temperature and $T_H$ represents the source temperature which is highest one acting as heat source for heat engine.

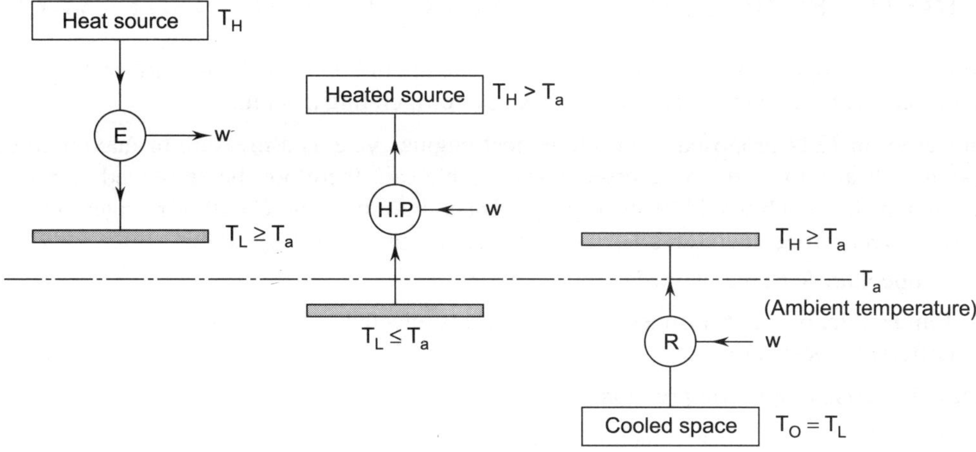


**Fig. 11.10   Comparison of heat engine, heat pump and refrigerating
machine with temperature level**

As for the heat pump, there is no difference in the cycle of operation between a refrigerating machine and a heat pump. The same machine can be utilized either

(i) to absorb heat from a cold body (a cooled space) at temperature To and rejected it to the surroundings at temperature $T_H \geq T_a$, or

(ii) to absorb heat from the surroundings at temperature $T_o \leq T_a$ and reject it to a hot body (a heated space) at temperature $T_H$.

## 11.4   SELECTION OF OPERATING TEMPERATURES

The selection of cooled space temperature To depends on the particular of refrigeration as shown in Table 1.1

**Table 11.1   Refrigeration temperature requirements of common applications**

| No. | Applications | Refrigeration temp. $t_o$, °C |
|---|---|---|
| 1. | Summer air conditioning | 0 to 10 |
| 2. | Cold storages | − 10 to 2 |
| 3. | Domestic refrigerators | − 15 |
| 4. | Frozen foods | − 35 |
| 5. | Freeze drying | − 35 to − 45 |
| 6. | Ice plant | brine temp. − 7 to − 10 |

In domestic refrigerator, temperature in the freezer is around − 15°C. Usual temperature in its cabinet ranges from 7°C to 10°C depending upon opening and closing period of the door. In blood bank recommended temperature ranges from 4°C to 6°C whereas − 20°C temperature is required for storing human tissue like eye, kidney.

## 11.5   IDEAL REFRIGERATING CYCLE (REVERSED CARNOT CYCLE)

We know that a reversible refrigeration cycle has the maximum C.O.P. We know, further that a reversible heat engine can be reversed in operation to work as a refrigerating machine.

Sadi Carnot in 1824 proposed a reversible heat engine cycle as a measure of maximum possible conversion of heat into work. A reversed Carnot cycle can, therefore, be employed as a reversible refrigeration cycle, which would be measuring of maximum possible C.O.P of refrigerating machine operating between two temperatures $T_H$ of heat rejection and $T_o$ of refrigeration.

$T_O$ = Temperature to be maintained in refrigeration

$T_H$ = Temperature of the atmosphere to which heat is rejected
     Different process are

Process 1 – 2  Isentropic compression

Process 2 – 3  Isothermal heat rejection to hot reservoir

Process 3 – 4  Isentropic expansion

Process 4 – 1  Isothermal heat absorption at cold rervervior to

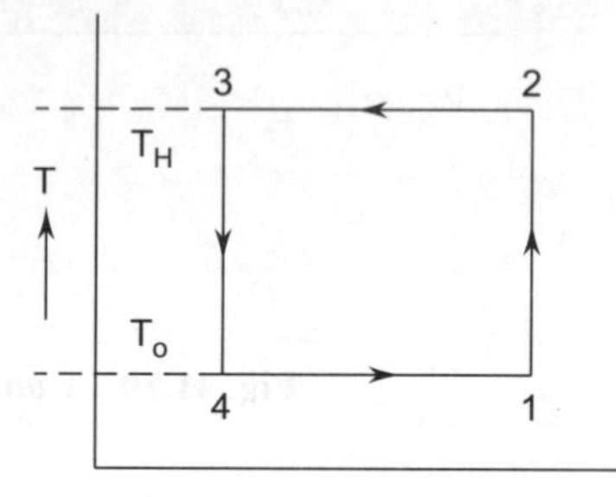

Fig. 11.10(a)

$$\xi = \text{C.O.P.} = \frac{Q}{W} = \frac{T_o\left(S_1 - S_4\right)}{\left(T_H - T_o\right)\left(S_1 - S_4\right)} = \frac{T_o}{T_H - T_o} = \frac{1}{\dfrac{T_H}{T_o} - 1}$$

$$\text{C.O.P of refrigeration} = \frac{1}{\dfrac{T_H}{T_o} - 1} = \xi_c$$

$$\text{E.P.R of heat pump} = \frac{T_H}{T_H - T_o} = \frac{1}{1 - \dfrac{T_o}{T_H}} = \xi_h$$

Carnot C.O.P. depends upon operating temperatures $T_o$ and $T_k$ and not upon the working substance (refrigerant). For cooling the lowest possible refrigerator temperature is $T_o = 0$ (absolute zero at which $\xi = 0$)

The highest possible refrigeration temperature is $T_o = T_H$ i.e. the refrigeration temperature is equal to the surroundings at which $\xi = \infty$. Thus Carnot C.O.P for cooling varies between 0 and $\infty$.

For heating, $T_k$ is the temperature of heat absorption from the surroundings and $T_k$ is heating temp. Theoretically the C.O.P for heating various between 1 and $\infty$.

It, may, therefore, be noted that to obtain maximum possible C.O.P in any application.

1.  The cold body temperature $T_o$ should be as high as possible and
2.  The hot body temperature $T_k$ should be as low as possible.

The lower the refrigeration temperature required and higher the temperature of the surroundings the larger is the power consumption of the refrigerating machine.

**Temperature limitation**

$$\text{C.O.P.} = \frac{T_o}{T_H - T_o} = \frac{1}{\dfrac{T_H}{T_o} - 1}$$

C.O.P. can be increased either by decreasing $T_H$ or by increasing $T_o$. The value of $T_o$ has more pronounced effect upon C.O.P. than $T_H$ as shown in Fig. 11.11. If the Control over the temperatures is only decoding factor thus if $T_o = T_k$ then C.O.P. $= \infty$, but in actual practice certain temperature requirements are always imposed upon the refrigerating system. For example if the refrigeration system has to maintain a cold room at 0°C and reject heat to the temperature at 25°C thin these temperatures are the limitations with which the cycle must abide.

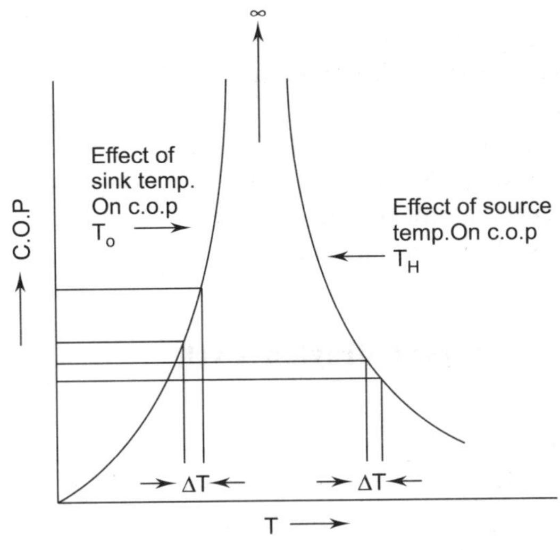


**Fig. 11.11**

During the heat rejection process, the temperature of the refrigerant must be higher than 25°C and during the heat absorption refrigerant temperature must be below 0°C.

## 11.6 AIR REFRIGERATION SYSTEM

Air refrigeration system is one of the earliest systems of cooling which uses air as a working substance. In olden days this system was universly employed due to its cheapness and availability of air in abundance as a natural gift and having quite safe properties. Due to the fact that air does not change its phase through out the cycle, it has a low heat carrying capacity and thus large volume of air is required to produce refrigeration. This results in low C.O.P and high operating coat.

With the advancement in air flight and was the idea of applying air conditioning system to aeroplane air refrigeration system is gaining further research and development. As high pressure air is already available in the aeroplane, it has been applied recently to aircraft systems where with low equipment weight, it can utilize a portion of air from supercharger.

The air refrigeration can be divided into a closed cycle and open cycle. The difference between the closed system and open system is only that the air after used in refrigerator is thrown into the atmosphere in stead of recirculating in closed system.

There are two basic cycles on which are refrigeration system works.

(a) Reversed Carnot cycle and

(b) Reversed Brayton or Joule or Bell Coleman cycle.

Although the reversed Carnot cycle as shown in Fig. 11.12 is the most efficient between fixed temperature limits and useful as a criterion of perfection it has its inherent draw backs when a gas is used as a refrigerant and can be seen from its *p-v* diagram. These are:

1. Extreme pressure and large volumes are developed since the pressure rise takes place during isentropic compression as well as isothermal heat rejection process.

2. Isothermal heat transfer process with a gas *i.e.,* a medium with finite specific heat which are imperable to achieve in practice.

3. The *p-v* diagram of the cycle is working with a gas is so narrow that even small irreversible of the individual process cause significant increase in work done which forms a large portion of net work of the cycle.

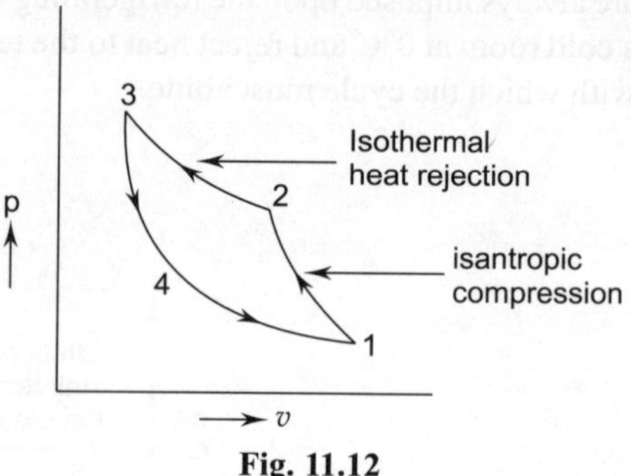


**Fig. 11.12**

## 11.6.1   Bell coleman or Reversed Brayton Cycle

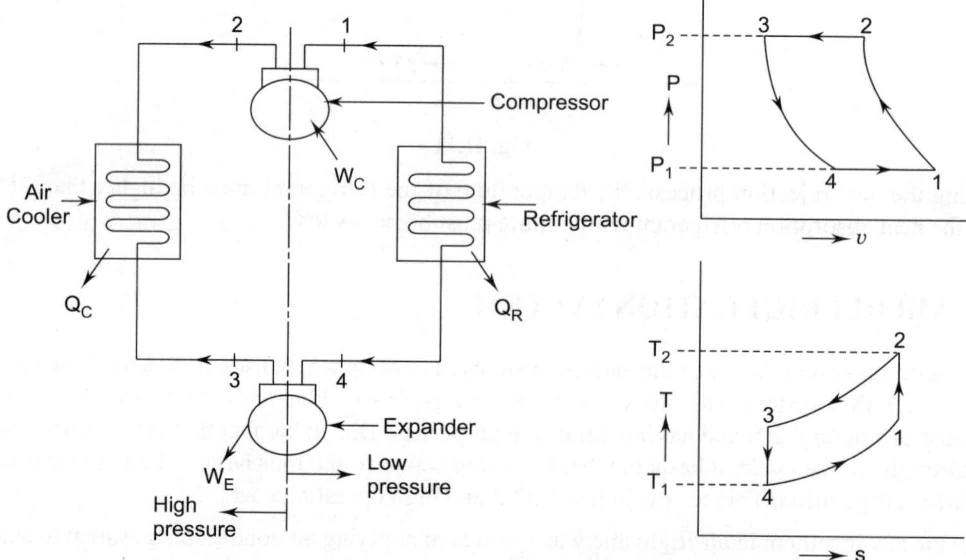


**Fig. 11.13   Bell coleman cycle with p.v and T-S diagram**

The Bell-coleman refrigerator as shown in Fig. 11.13 working on air cycle was developed by Bell-coleman and Light foot. It was used for preventing meat in ships in earlier times. It consists of four major components, a compressor, a cooler, an expander and a refrigerator. It consists of four processes.

Process $1 - 2$   Isentropic compression

Process $2 - 3$   Constant pressure cooling (Isobaric cooling)

Process $3 - 4$   Isentropic expansion

Process $4 - 1$   Constant pressure expansion (Isobaric expansion)

The hot return air from the refrigerator is drawn in to the compressor during suction stroke. The air gets heated up in a constant entropy process. The hot air is then cooled in a cooler, where it rejects heat to the surrounding. The cooled air is then expanded in the expander in an isentropic expansion process. The air is finally admitted to the refrigerator where it cools the application. In doing so it picks up heat from the application. The hot return air then again enters the compressor and the cycle repeats itself.

Heat rejected by air per kg $\qquad - q_C = c_p(T_2 - T_3)$

Heat absorbed by air per kg $\qquad = q_R = c_p(T_1 - T_4)$

$\therefore$ Net work done during cycle per kg of air

$$= q_C - q_R = c_p(T_2 - T_3) - c_p(T_1 - T_4) = W_C - W_E$$

$\therefore \qquad\qquad$ C.O.P. $= \dfrac{\text{Heat absorbed}}{\text{Network input to machine}}$

$$= \dfrac{c_p\left(T_1 - T_4\right)}{c_p\left(T_2 - T_3\right) - c_p\left(T_1 - T_4\right)} = \dfrac{\left(T_1 - T_4\right)}{\left(T_2 - T_3\right) - \left(T_1 - T_4\right)} \tag{11.2}$$

$\therefore \qquad\qquad$ C.O.P. $= \dfrac{T_4\left(\dfrac{T_1}{T_4} - 1\right)}{T_3\left(\dfrac{T_2}{T_3} - 1\right) - T_4\left(\dfrac{T_1}{T_4} - 1\right)} \tag{11.3}$

For isentropic compression $\dfrac{T_2}{T_1} = \left(\dfrac{p_2}{p_1}\right)^{\frac{\gamma-1}{\gamma}}$

For isentropic expansion $\dfrac{T_3}{T_4} = \left(\dfrac{p_3}{p_4}\right)^{\frac{\gamma-1}{\gamma}}$

$\because \qquad\qquad p_2 = p_3 \quad \text{and} \quad p_1 = p_4$

$\therefore \qquad\qquad \dfrac{T_2}{T_1} = \dfrac{T_3}{T_4} \quad \text{or,} \quad \dfrac{T_1}{T_4} = \dfrac{T_2}{T_3}$

$\therefore$ Eqn. 11.3 reduces to

$$\text{C.O.P.} = \dfrac{T_4}{T_3 - T_4} = \dfrac{1}{\dfrac{T_3}{T_4} - 1} = \dfrac{1}{\left(\dfrac{p_3}{p_4}\right)^{\frac{\gamma-1}{\gamma}} - 1} = \dfrac{1}{\left(\dfrac{p_2}{p_1}\right)^{\frac{\gamma-1}{\gamma}} - 1} \tag{11.4}$$

$\therefore \qquad\qquad$ C.O.P. $= \dfrac{1}{\left(r_p\right)^{\frac{\gamma-1}{\gamma}}} \tag{11.5}$

where $\qquad\qquad r_p = \dfrac{p_3}{p_4} = \dfrac{p_2}{p_1} = \text{pressure ratio}$

The above equation of C.O.P is only valid under the following ideal conditions.

   (i)  The compression and expansion follow the isentropic law

   (ii)  There is perfect intercooling in the heat exchanger

   (iii)  There are no pressure losses in the system

In actual practice, the compression and expansion follow polytropic laws and perfect intercooling is not possible. The pressure losses in the heat exchanger and refrigerator are inevitable.

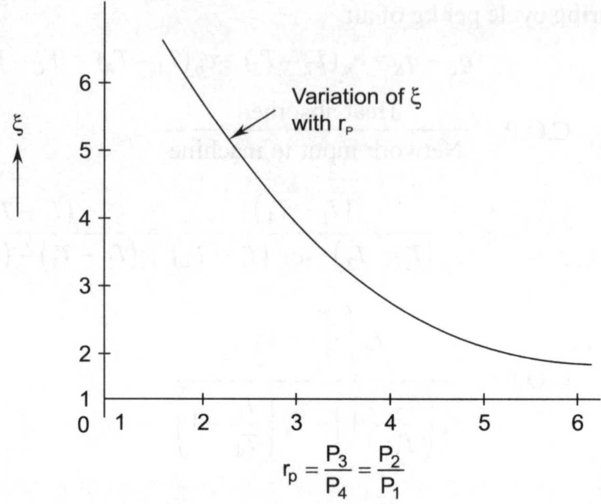


$$r_p = \frac{P_3}{P_4} = \frac{P_2}{P_1}$$

**Fig. 11.14**

**Table 11.2   Variation of $\xi$ with $r_p$**

| $r_p$ | 1 | 2 | 3 | 4 | 5 | 6 |
|---|---|---|---|---|---|---|
| $\xi$ | $\infty$ | 4.56 | 2.71 | 2.05 | 1.72 | 1.50 |

As we observe that C.O.P. ($\xi$) increases as pressure ratio ($r_p$) decreases. From Eqn. (11.4) it may be noted that C.O.P approaches infinity as the temperature difference ($T_3 - T_4$) approaches zero. In such a case the value of source (refrigerator) temperature $T_4$ may be greater and there may not be any refrigeration. The pressure ratio should be such that the value of $T_4 < T_1$ and the minimum pressure ratio is given by

$$r_p = \left( \frac{T_2}{T_1} \right)^{\frac{\gamma}{(\gamma - 1)}} \tag{11.6}$$

At this pressure ratio the refrigeration effect is zero. Higher is the range of pressure ratio, larger is the temperature difference available for refrigeration and smaller is the mass flow per KW of refrigeration. Hence there must be a compromise between the COP value and the rate of circulation. Larger the mass flow, larger are the losses and deviation from the ideal value of C.O.P.

Though, theoretically this is the most efficient cycle its COP is very low, about $\frac{1}{2}$ to $\frac{3}{4}$, which is nearly $\frac{1}{10}$ th of vapour compression cycle. This is because of low specific heat and poor conductivity of air.

The important disadvantages of Bell-coleman cycle may be listed as

(i)  Les efficient and low value of C.O.P

(ii) Due to poor conductivity of air the size of machine is bulky.

### C.O.P. of Bell-coleman cycle working on Actual law of compression and expansion

As we have seen that in actual practice perfect isentropic process is not possible and so assuming compression and expansion to be polytropic i.e. $pv^n$ = constant as shown in Fig. 11.15.

Work required by the compressor per kg of air

$$W_C = \frac{n}{n-1}\,(p_2 v_2 - p_1 v_1)$$

$$= \frac{nR}{n-1}\,(T_2 - T_1)$$

Work delivered by the expander per kg of air

$$W_E = \frac{n}{n-1}\,(p_3 v_3 - p_4 v_4)$$

$$= \frac{nR}{n-1}\,(T_3 - T_4)$$

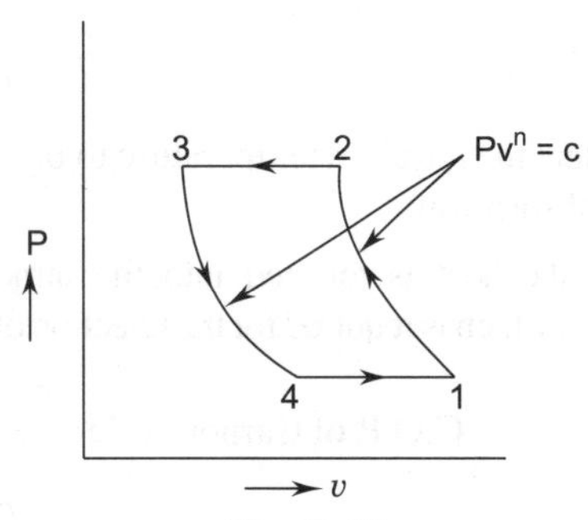


**Fig. 11.15**

$\therefore$    Net work required,

$$W_{\text{Net}} = W_C - W_E = \frac{nR}{n-1}\,[(T_2 - T_1) - (T_3 - T_4)]$$

Refrigerating effect per kg of air

$$= c_p (T_1 - T_4)$$

$\therefore$

$$\text{C.O.P.} = \frac{c_p\,(T_1 - T_4)}{\dfrac{n}{n-1}\,R\left[(T_2 - T_1) - (T_3 - T_4)\right]}$$

$$= \frac{c_p\,(T_1 - T_4)}{\dfrac{n}{n-1}\,R\,(T_1 - T_4)\left[\dfrac{T_2 - T_3}{T_1 - T_4} - 1\right]}$$

$$= \frac{c_p}{\dfrac{n}{n-1}\,R\left(\dfrac{T_2 - T_3}{T_1 - T_4} - 1\right)} \tag{11.7}$$

$\because$          $c_p - c_V = R$

or,
$$1 - \frac{1}{\gamma} = \frac{R}{c_p}$$

or,
$$\frac{\gamma - 1}{\gamma} = \frac{R}{c_p} \qquad \therefore \quad R = \frac{c_p\,(\gamma - 1)}{\gamma}$$

$\therefore$
$$\text{C.O.P.} = \cfrac{1}{\cfrac{n}{n-1} \cdot \cfrac{\gamma - 1}{\gamma} \left( \cfrac{T_2 - T_3}{T_1 - T_4} - 1 \right)}$$

but
$$\frac{T_2}{T_1} = \frac{T_3}{T_4} = \frac{T_2 - T_3}{T_1 - T_4}$$

$\therefore$
$$\text{C.O.P.} = \cfrac{1}{\cfrac{n}{n-1}\left(\cfrac{\gamma - 1}{\gamma}\right)\left[\cfrac{T_2}{T_1} - 1\right]} = \cfrac{1}{\cfrac{n}{n-1}\left(\cfrac{\gamma - 1}{\gamma}\right)\left[\cfrac{T_3}{T_4} - 1\right]} \qquad (11.8)$$

Here the required temperature to be maintained in the refrigerator is $T_1$ and $T_4 < T_1$ which is required for the absorption.

If the heat is rejected into the atmosphere which is at a temperature $T_3$ in Bell-coleman cycle and $T_2 > T_3$ which is required for the rejection of heat by the fluid of atmosphere. If it is carnot cycle, $T_4 = T_1$ and $T_3 = T_2$

$\therefore$
$$\text{C.O.P. of Carnot cycle} = \cfrac{1}{\cfrac{T_2}{T_1} - 1} = \frac{T_1}{T_2 - T_1}$$

### 11.6.2  Actual Air Cycle

The actual gas cycle will have significant pressure drops in heat exchangers and irreversibilities of the compressor and expander as shown in Fig. 11.16. The actual gas cycle is represented as 1-2-3-4-1 whereas ideal cycle is 1-2-3′-4″-1. There are also other losses associated with heat gain from surroundings and pressure drops at the compressor and expander. Hence, we define compressor efficiency and expander efficiency as follows

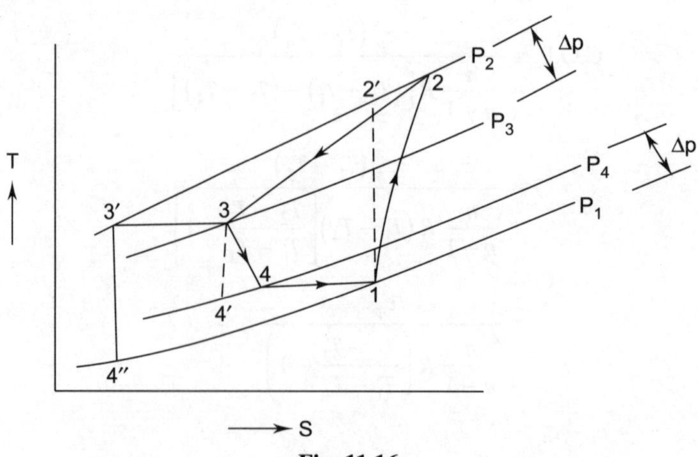


**Fig. 11.16**

$$\eta_c = \frac{c_p\left(T_{2'} - T_1\right)}{c_p\left(T_2 - T_1\right)} \tag{11.9}$$

and

$$\eta_{\text{exp.}} = \frac{c_p\left(T_3 - T_4\right)}{c_p\left(T_3 - T_{4'}\right)} \tag{11.10}$$

### 11.6.3   Necessity of Aircraft Refrigeration

At first sight it seems surprising as to why the refrigeration system is needed at all in the aircraft, when the higher we go, the cooler, we find, is actually true. To understand this let us consider the reasons due to which the cabin conditions are not in conformity with the physiological requirements of human beings such as temperature, humidity and atmospheric pressure.

The temperature of the cabin goes up due to various reasons such as

    (a)  heat release from occupants        (b)  ramming of air

    (c)  solar radiator                  (d)  control device

    (e)  air resistance (skin friction)

Fig. 11.17 shows the variation of power required according to the altitude of the plane.

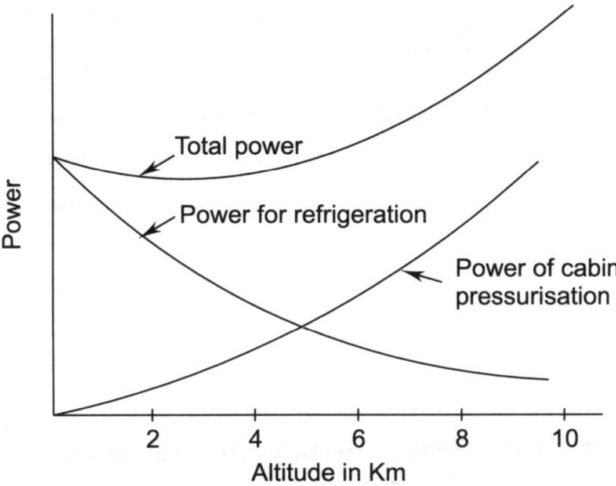


**Fig. 11.17   Variation of power for pressurization of cabin and refrigeration with altitude.**

### 11.6.4   Application to Aircraft Refrigeration

The gas cycle is exclusively used in air conditioning systems of military and commercial aircrafts, though refrigerated cargo aircrafts generally use dry ice. It is more appropriate to call it air cycle refrigeration since only air has found application as a working substance in this cycle. The C.O.P. of this cycle is lower than that of vapour compression cycle. Nevertheless, the air cycle continues to be favoured for aircraft refrigeration because of its many advantages which are:

   1.  Small amounts of leakage are tolerable with air as the refrigerant.

   2.  Since air is used as cooling medium no cost of refrigerant is involved and hence quite cheap.

   3.  Cabin pressurization and air conditioning can be combined into one operation.

4. Initial compression of the air is obtained by the ram effect due to high kinetic energy of the ambient air relative to the aircraft.

5. Air cycle in its simplest form with an open system requires only one heat exchange.

6. It is light weight per ton of refrigeration as compared to other refrigeration system.

7. No additional vibrational problem is involved which otherwise would have resulted from the use of vapour compression system.

## 11.6.5  Simple Aircraft Refrigeration Cycle With Ram Compression

The initial compression of the ambient air due to ram effect and a simple aircraft refrigeration cycle are shown in Fig. 11.18. The ram effect is shown by the line 1–2. point 2′ denotes the state after isentropic compression to pressure $p_{2'}$ and temperature $T_{2'}$ so that we have from the energy equation $h_2 = h_{2'} = c_p T_1 + \dfrac{V^2}{2}$

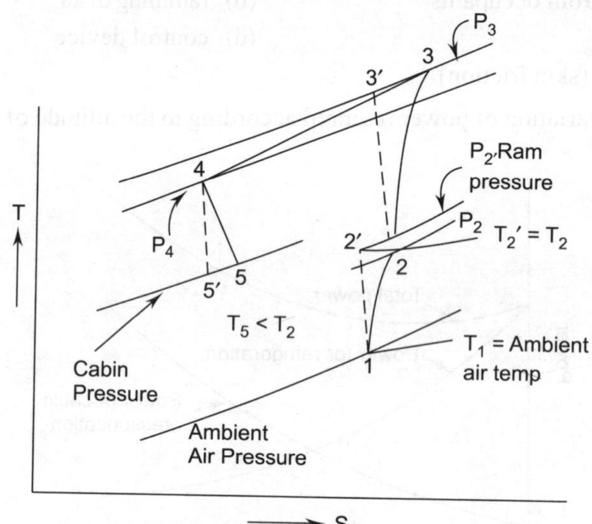


**Fig. 11.18   Simple Aircraft refrigeration cycle**

$$h_2 = h_1 + \frac{V^2}{2}$$

where $V$ is the velocity of aircraft.

$$C_p T_2 = C_p T_{2'} = C_p T_1 + \frac{V^2}{2} \quad \text{or,} \quad T_2 = T_{2'} = T_1 + \frac{V^2}{2C_p}$$

The above relation can be simplified to

$$\frac{T_{2'}}{T_1} = 1 + \frac{V^2}{2 T_1 C_p} = 1 + \frac{V^2}{2 T_1 \dfrac{\gamma}{\gamma - 1} R} = 1 + \frac{\gamma - 1}{2} \frac{V^2}{a^2} \tag{11.11}$$

where

$$a = \text{velocity of sound } a = \sqrt{\gamma R T_1}$$

or
$$\boxed{\frac{T_{2'}}{T_1} = 1 + \frac{\gamma - 1}{2} \cdot M^2}$$
where  M = Mach number.  (11.12)

The temperature $T_2$ is the stagnation temperatures of the ambient air. The stagnation pressure after isentropic compression $p_2'$ is given by the equation

$$\frac{p_{o'}}{p_1} = \left(\frac{T_{2'}}{T_1}\right)^{\frac{\gamma}{\gamma - 1}}.$$

The irreversible compression in the ram, however, results in air reaching point 2 instead of point 2' i.e. at the same stagnation temperature but at a reduced stagnation pressure $p_2$ which is obtained from the knowledge of the ram efficiency $\eta_r$ defined by

$$\eta_r = \frac{\text{Actual pressure recovery}}{\text{Ideal pressure recovery}}$$

$$\boxed{\eta_r = \frac{p_2 - p_1}{p_{2'} - p_1}}$$
(11.13)

The rest of the cycle is also shown. The temperature after the cooler process (process 3-4), $T_y$ is shown to be higher than the stagnation temperature $T_2$ of the ambient air. It implies that the air cannot be cooled to a temperature below $T_2$ unless some method is devised to extract the kinetic energy of the ambient air. Another point to note is the pressure after expansion, $p_5$ which is slightly above the cabin pressure and is also higher than the ambient air pressure. Refrigerating effect produced is

$$Q_R = C_p(T_i - T_5), \quad \text{where} \quad T_i = \text{inside temperature of cabin.}$$

of cabin. The network of the cycle is the difference of work for process 2–3 of the compressor and process. 4–5 of the expander plus the ram work $C_p(T_2 - T_1)$ which is derived directly from the engine. The ambient air temperature varies with the altitude of the flight of the aircraft. Generally, the temperature drops by 0.64°C per 100 m of height from the sea level temperature.

### 11.6.6  Refrigeration Systems for Aircrafts

The development of high speed jet aircraft and missile has necessitated the use of a compact and simple refrigeration system capable of high capacity with a reduced pay load. The cooling requirements per unit volume of the space are quite heavy due to compactness of all equipments and engine. The refrigeration system required for an ordinary passenger aeroplane is about 8 tons and that of jet fighter of 20 tons.

The sources contributing heat load to compartment may be grouped into the following two categories:

1. External heat source
2. Internal heat source

### External heat sources

1. Solar heat which finds its way into the passenger cabin through the area exposed to the sun.
2. The aeroplane has to fly up in the sky where the atmospheric pressure decreases. The decrease in pressure will cause a breathing problem to the passengers inside. As such it is necessary to compress the ambient air from the ambient pressure to the required cabin pressure which consequently results in increase of temperature.

3. When the aeroplane moves with a high speed through air, due to skin friction, the air gets compressed and heated up thus adding heat load through the surface of the aeroplane to the cabin.

## Internal sources

1. Aeroplane load which means heat given off by the passengers and pilot. A normal man gives out about 420 kJ of heat per hour when at rest.

2. The engine part adds good amount of heat which is transmitted to the cabin by conduction, convection and radiation.

3. The various items of electrical and electronic equipment also generate heat which should also be taken into account while designing a refrigeration system for an aircraft.

## 11.6.7  Simple Cooling System

The system is shown in Fig. 11.19

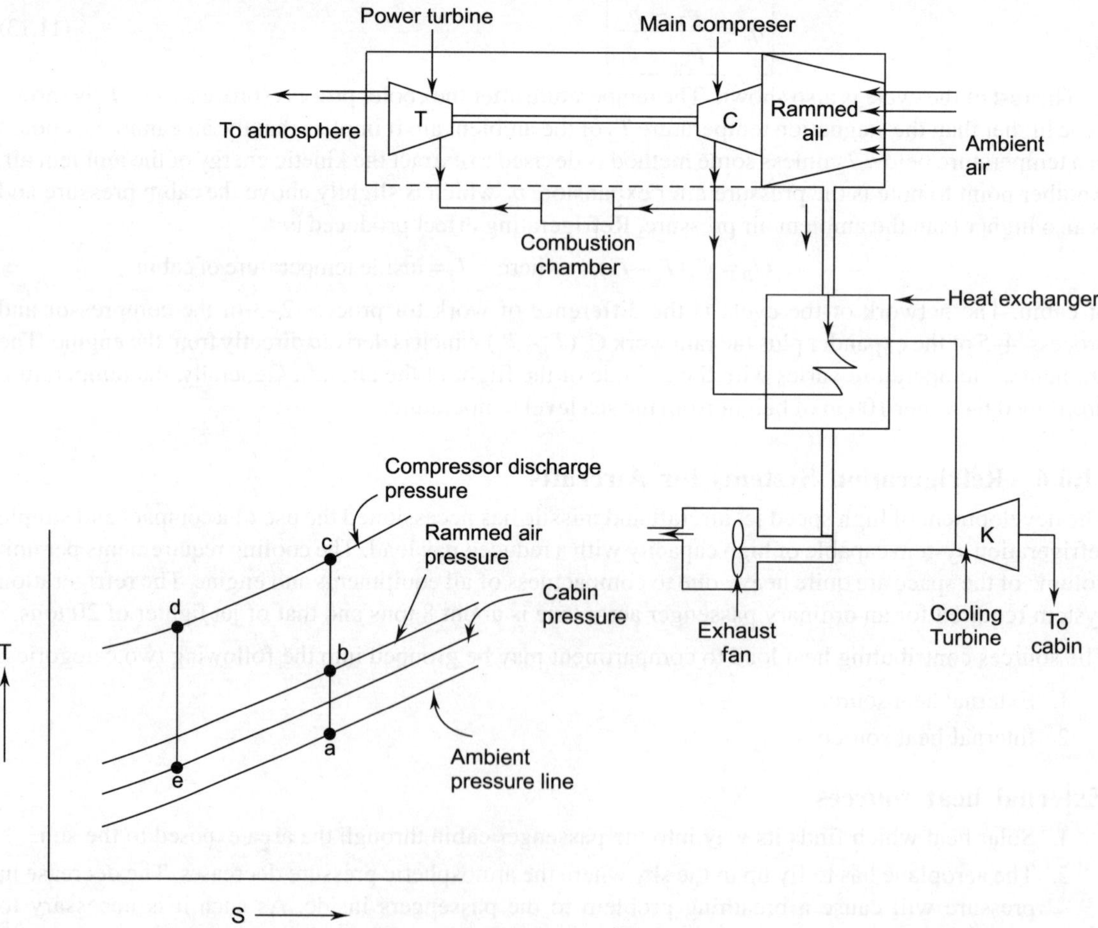



**Fig. 11.19   Simple cooilng cycle**

The main components of this system are air to air heat exchanger or cooler and a cooling turbine. The ambient air at a velocity equal to the plane speed in which the unit is installed, enters the inlet section which is designed as a diffuser. The air is slowed down and a part of the kinetic energy of the air stream is converted into pressure. This type of compression is called *ram compression*. This increase in pressure of this ramming action is shown by line ab on *T-S* diagram. Further compression of air to the desired pressure takes place in the main compressor $C$ and the process is shown by $bc$. Now a part of this high pressure air is taken off for refrigeration system as shown. This high pressure and high temperature air is first cooled in the heat exchanger or cooled (process $c - d$) where cooling medium is the **ramming** air. Further cooling of air takes place in the cooling turbine by the process of expansion with work, extraction to the desired cabin pressure and temperature as shown by the line $de$. The work obtained from the cooling turbine is utilized to drive the exhaust fan which draws rammed air over the heat exchanger.

## 11.6.8  Boot Strap Air Cycle Refrigeration System

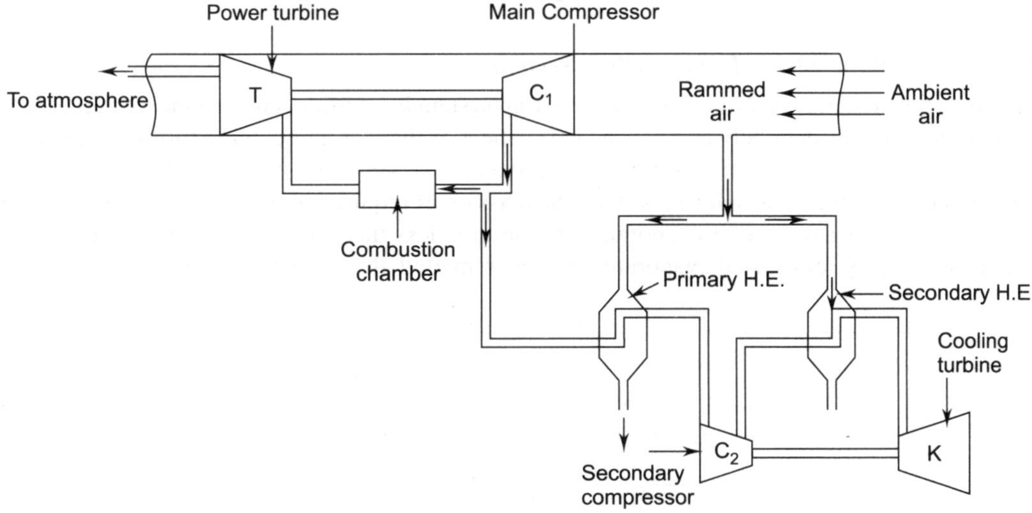


**Fig. 11.20   Boot strap air cycle refrigeration system**

If in an air cycle refrigeration system the pressure of air is raised to a pressure higher than the main compressor pressure by using a secondary compressor, the system is known as boot strap system shown in Fig. 11.20. The power required to drive for secondary compressor is taken from the cooling turbine. Like simple cooling cycle ambient air as enters the inlet section, gets compressed due to ramming action and the process is shown in Fig. 11.21 by $a$-$b$. Further compression in the main compressor $C_1$ and the process is represented by $bc$. A part of this high pressure air is taken for the refrigeration system which first enters the primary heat exchanger where it gives out some of its heat to the ramed air which is used as cooling medium.

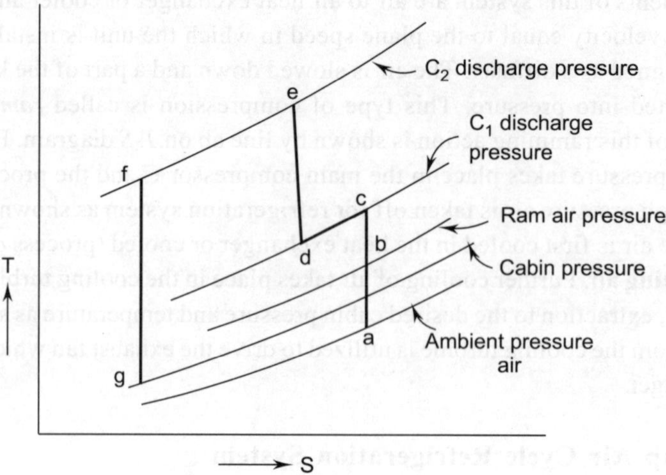



**Fig. 11.21    T-S diagram of boot strap air cycle**

## 11.6.8.1    *The Boot Strap Evaporative System*

This system shown in Fig. 11.22 is similar to boostrap system in addition, an evaporative cooler located in the high pressure air duct between the secondary heat exchanger and cooling turbine. Any suitable evaporant may be used as a heat sink for evaporative cooler. The mass of air required in evaporative boostrap system for given $T.R$ will be less than the mass of air required in the bootstrap system as outlet temperature of cooling turbine in evaporative boostrap is less than the outlet temperature of cooling turbine in boostrap system. Usual evaporants are water, methyl alcohol and ammonia.

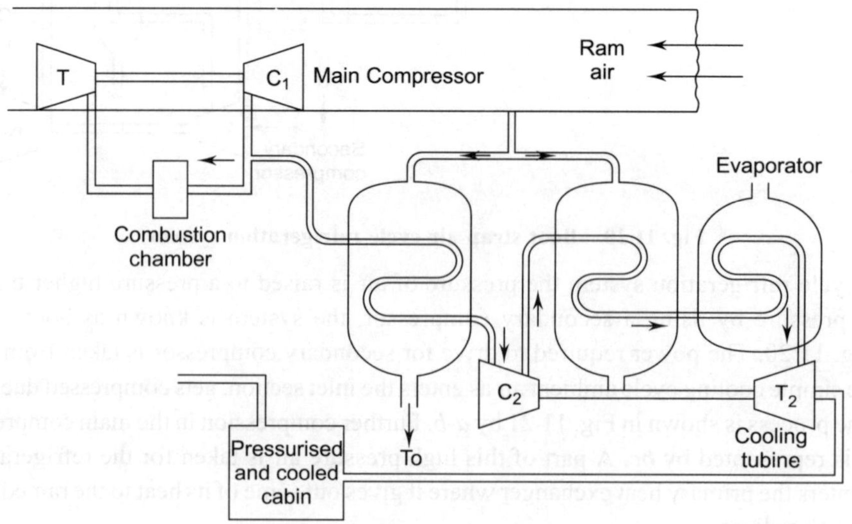



**Fig. 11.22    Boot strap evaporative system**

Processes are

1—2 → Ramming action

2—3 → Compression in the main compressor

3—4 → Cooling in primary heat exchanger

4—5 → Compression in the secondary compressor

5—6 → Cooling in the secondary heat exchanger

6—7 → Cooling in the evaporation

7—8 → Expansion in the cooling turbine

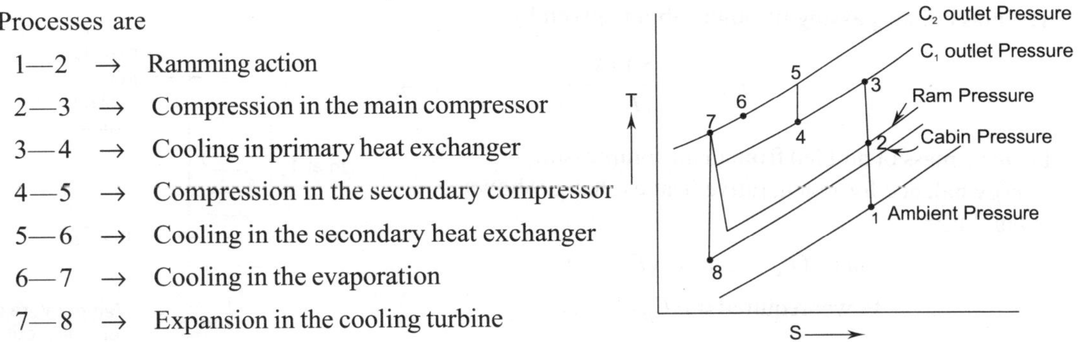


**Fig. 11.23  T-S diagram of Boot Strap Evaporative system**

## 11.6.9  Regenerative Cooling System

If the cooling turbine discharge temperature of a simple system is too high then the regenerative system is used. This is suitable for ground cooling. This system shown in Fig. 11.24 consists of a primary air-to-air heat exchange a regenerative heat exchanger and a cooling turbine. High pressure high temperature air coming from compressor is cooled first in the primary heat exchanger then in the regenerative heat exchanger and finally in the cooling turbine. A part of cold air from the turbine is recirculated in the regenerative heat exchanger. This type of cooling system is generally preferred for supersonic airplanes.

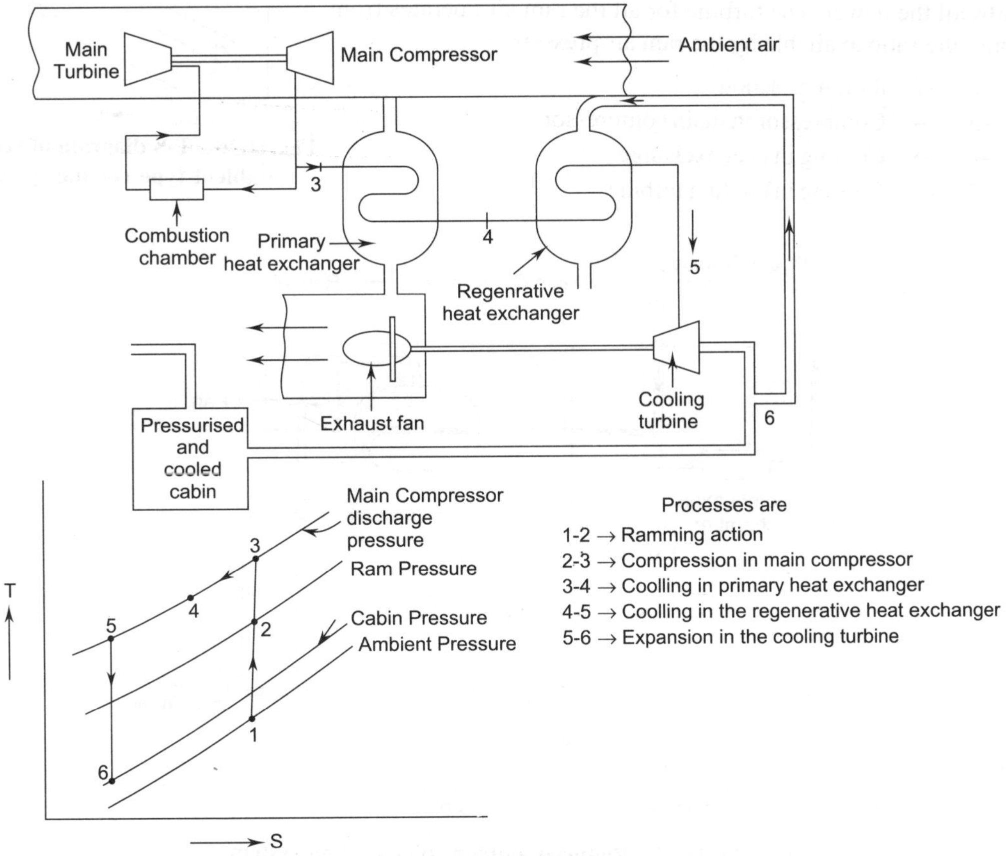

**Fig. 11.24  Regenerative cooling system**

The mass of air passing through cabin is given by

$$m_1 = \frac{50\,TR}{c_p\left(T_i - T_b\right)} \qquad (11.14)$$

Let $m$ be mass of air bled from main compressor.
Energy balance for regenerative heat exchange shown
in Fig. 11.25

$$m\,C_p\left(T_4 - T_5\right) = m_2\left(T_7 - T_6\right)$$

$$\text{Power required} = mC_p\left(T_3 - T_2\right)$$

$$\text{C.O.P.} = \frac{50\,TR}{m\,c_p\left(T_3 - T_2\right)} \qquad (11.15)$$

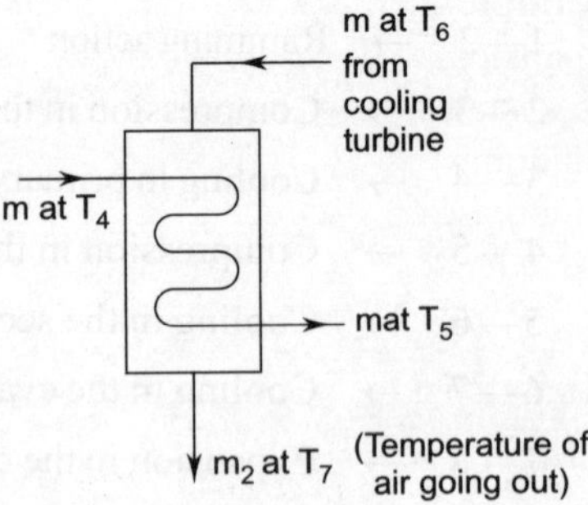


**Fig. 11.25**

## 11.6.10   Reduced Ambient Type Cooling System

This system is used for exceptionally high speed aircraft when the
ram air temperature is too high. In this system, there are two expansion
turbines. One in cabin air stream and other in cooling air system.
Both turbines are connected to the shaft driving the fan which
absorbs all the power. The turbine for all the ram air operates from
the pressure ratio available by the ram air pressure.

1—2   $\rightarrow$   Ramming action
2—3   $\rightarrow$   Compressor in main compressor
3—4   $\rightarrow$   Cooling in heat exchange
4—5   $\rightarrow$   Cooling in cooling turbine

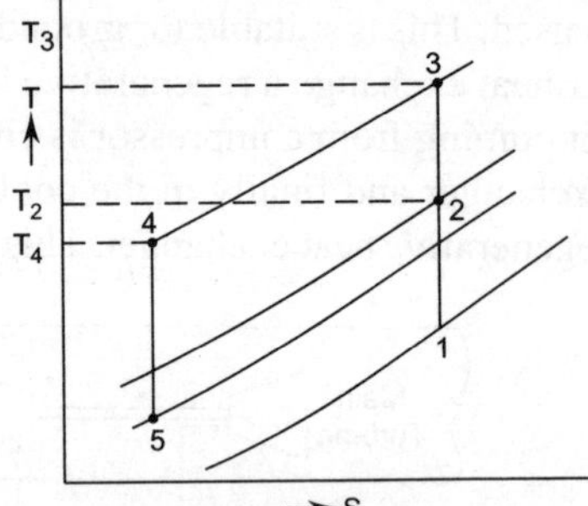

**Fig. 11.26    T-S diagram of reduced
ambient type cooling system**

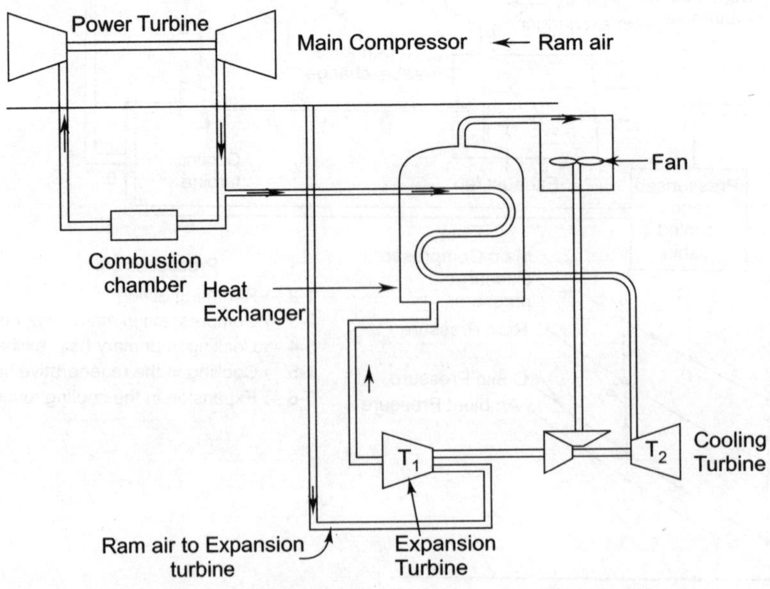


**Fig. 11.27    Reduced ambient type cooling system**

**Example 11.1**  Bell-Coleman cycle works between 1 and 6 bar pressure limit. The compression and expansion indices are 1.25 and 1.3 respectively. Obtain COP, heat rejected in the cooler and tonnage of the system for an air flow rate of 0.5 kg/sec. Neglect clearance volume and take temperature at the beginning of compression and expansion to be 7°C and 37°C respectively. Find the bore of expander and compressor cylinders if the unit runs at 250 rpm and stroke of both the cylinders is equal to the bore of expander. Assume double acting and $\eta_{vol} = 0.88$.

*Solution*   Given   $p_2 = 6$ bar, $p_1 = 1$ bar, $\dot{m} = 0.5$ kg/sec

$$= 0.5 \times 60 = 30 \text{ kg/min}$$

$$T_1 = 273 + 7 = 280°K, \ T_3 = 273 + 37 = 310 \text{ K}$$

$$\frac{T_2}{T_1} = \left(\frac{p_2}{p_1}\right)^{\frac{n_1-1}{n_1}} \quad \text{or} \quad T_2 = T_1 \left(\frac{p_2}{p_1}\right)^{\frac{n_1-1}{n_1}}$$

$$= 280 \left(\frac{6}{1}\right)^{\frac{1.25-1}{1.25}} = 400.67 \, K$$

$$\frac{T_4}{T_3} = \left(\frac{p_1}{p_3}\right)^{\frac{n_2-1}{n_2}} = \left(\frac{1}{6}\right)^{\frac{1.3-1}{1.3}}$$

$$\therefore \qquad T_4 = 310 \left(\frac{1}{6}\right)^{\frac{1.3-1}{1.3}} = 205.02 \text{ K}.$$

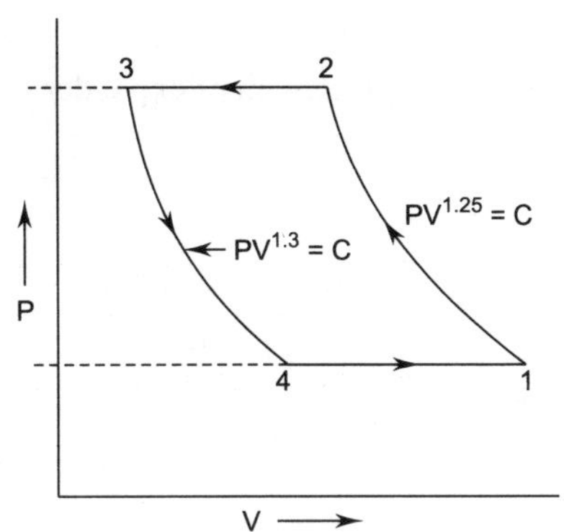


**Fig. 11.28**

Refrigerating effect per kg $= q = C_p(T_1 - T_4) = 1 \times (280 - 205.02) = 74.98$ KJ/kg

$$\text{Net work required} = W = \frac{n_1}{n_1 - 1}[p_2 v_2 - p_1 v_1] - \frac{n_2}{n_2 - 1}[p_3 v_3 - p_4 v_4]$$

$$= \frac{n_1}{n_1 - 1} R(T_2 - T_1) = \frac{n_2}{n_2 - 1} R(T_3 - T_4)$$

$$= \frac{1.25 \times 0.287}{1.25 - 1}(400.67 - 280) - \frac{1.3}{1.3 - 1} \times 0.287(310 - 205.02)$$

$$= 42.6 \text{ kJ/kg}$$

$$\therefore \qquad \text{COP} = \frac{q_o}{W} = \frac{74.96}{42.6} = 1.76 \quad Ans.$$

$\therefore$  Heat rejected in the cooler $= Q_k = \dot{m}q_k = \dot{m}C_p(T_2 - T_3) = 30 \times 1 (400.67 - 310)$

$$= 2720 \text{ kJ/min} \quad Ans.$$

$$\text{Capacity} = \frac{\dot{m} q_o}{211} TR = \frac{30 \times 74.98}{211}$$

$$= 10.66 \, TR \quad Ans.$$

Piston displacement for compressor

$$V_1 = \frac{\dot{m}\,RT_1}{p_1} = \frac{30 \times 0.287 \times 280 \times 1000}{1 \times 10^5} = 24.63\ m^3/\text{min},\ R = 0.287\,\text{kJ/kgK}$$

$$V_1 = \frac{\pi}{4}\cdot D_c^2\,L_c N \times 2\eta_{\text{vol}} \quad \text{or,} \quad 24.63 = \frac{\pi}{4}\cdot D_c^2 \times L_c \times 250 \times 2 \times 0.88$$

$$\therefore \qquad D_c^2 = \frac{0.0627}{0.88\,L_c} = \frac{0.07125}{L_c}$$

Piston displacement of expander

$$V_4 = \frac{\dot{m}R\,T_4}{p_4} = \frac{\pi}{4}\,D_E^2 \times L_E \times N \times 2 \times \eta_{\text{vol}}$$

or, $$\frac{30 \times 287 \times 205.02}{1 \times 10^5} = \frac{\pi}{4}\cdot D_E^2 \times D_E \times 250 \times 2 \times 0.88$$

or, $$\qquad D_E = 0.3710\,\text{cm}\quad Ans.$$

and $$\qquad L_E = L_C = 0.3710\,\text{cm}\quad Ans.$$

$$\therefore \qquad D_C = \sqrt{\frac{0.07125}{0.3710}} = 0.438\,\text{cm}\quad Ans.$$

**Example 11.2**   An air refrigerator working on Bell Coleman cycle takes air into the compressor at 1 bar and $-5°C$. It is compressed in the compressor to 5 bar and cooled to 25°C at the same pressure. It is further expanded in the expander to 1 bar and discharged to take the cooling load. Isentropic efficiencies of compressor and expander are 85% and 90.7% respectively. Find:

(a)  refrigerator capacity of the system if air circulation is 40 kg/min

(b)  power required for the compressor

(c)  C.O.P of the system

(d)  Heat rejected into the cooler.

*Solution*   $\quad p_2 = 5\,\text{bar},\ p_1 = 1\,\text{bar}$

$\qquad\qquad T_1 = -5 + 273 = 268\,\text{K},\ T_3 = 25 + 273$

$\qquad\qquad = 298\,K,\ \dot{m} = 40\,\text{kg/min}.$

$$\frac{T_{2'}}{T_1} = \left(\frac{p_k}{p_o}\right)^{\frac{\gamma-1}{\gamma}} \quad \therefore \quad T_{2'} = T_1\left(\frac{p_2}{p_1}\right)^{\frac{\gamma-1}{\gamma}}$$

$$= 268\left(\frac{5}{1}\right)^{\frac{1.4-1}{1.4}} = 424.65\,\text{K}$$

$$\eta_C = \frac{T_{2'} - T_1}{T_2 - T_1} \quad \therefore \quad \text{or,}\ \ 0.085 = \frac{424.65 - 268}{T_2 - 268}$$

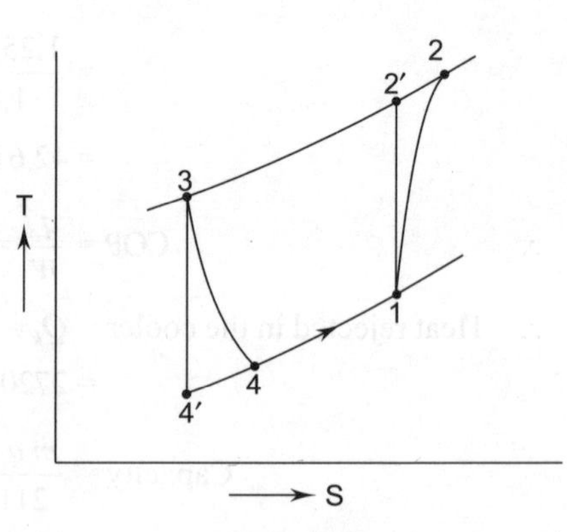


**Fig. 11.29**

$$\therefore \qquad T_2 = \textbf{452.3 K} \quad \text{Now} \quad \frac{T_{4'}}{T_3} = \left(\frac{p_4}{p_3}\right)^{\frac{\gamma-1}{\gamma}} \qquad \therefore \quad T_{4'} = T_3 \left(\frac{p_4}{p_3}\right)^{\frac{\gamma-1}{\gamma}}$$

$$= 298 \left(\frac{1}{5}\right)^{\frac{0.4}{1.4}} = \textbf{188 K}$$

$$\eta_E = \frac{T_3 - T_4}{T_3 - T_{4'}}, \quad \text{or,} \quad 0.95 = \frac{298 - T_4}{298 - 188} \quad \therefore \quad T_4 = 199\,\text{K}$$

Refrigerating effect per kg $= q_o = c_p(T_1 - T_4) = 1(268 - 199)$
$$= 69\,\text{kJ/kg}$$

Net work required per kg $= w_c = C_p(T_2 - T_1) - C_p(T_3 - T_4)$
$$= 1[(452.3 - 268) - (298 - 199)] = 85.3\,\text{kJ/kg}$$

(a)  Capacity $= \dfrac{\dot{m}\,q_o}{211}\,\text{TR} = \dfrac{40 \times 69}{211} = \textbf{13 TR}$

(b)  Compressor power $= m.w = \dfrac{40}{60} \times 85.3\,\text{kW} = \textbf{56.86}\,\text{kW}$

(c)  C.O.P. $= \dfrac{R.E.}{\text{Work}} = \dfrac{q_o}{W_C} = \dfrac{69}{85.3} = 0.808 \quad Ans.$

(d)  Heat rejected into the cooler $= Q_k = \dot{m}\,q_k = \dot{m}\,C_p(T_2 - T_3)$
$$= 40 \times 1(452.3 - 298) = 6172\,\text{kJ/min} \quad Ans.$$

**Example 11.3**  A dense air refrigeration used to produce 10 ton of refrigeration. The expander and compressor cylinders are double acting. The pressure limit for compression and expression cylinders is 4 bar and 16 bar. The compressor sucks in air at 4°C and discharges air at 20°C to expansion cylinders after the air cooler. Mechanical efficiency of the compressor and expander cylinder drive is 85%. For 250 rpm and 25 cm stroke, determine

(a)  power required to drive the unit

(b)  bore of compressor and expander.

Assume isentropic compression and expansion with $\gamma = 1.4$.

*Solution*    $p_2 = 16\,\text{bar},\, p_1 = 4\,\text{bar}$

$T_1 = 4 + 273 = 277\,\text{K},\, T_3 = 20 + 273$

$\qquad = 293\,\text{K},\, TR = 10,\, \eta_{\text{mech}} = 0.85$

for both.    $N = 250\,rpm,$

$\qquad L = 25\,\text{cm} = 0.25\,m$

$$\frac{T_2}{T_1} = \left(\frac{p_2}{p_1}\right)^{\frac{\gamma-1}{\gamma}}$$

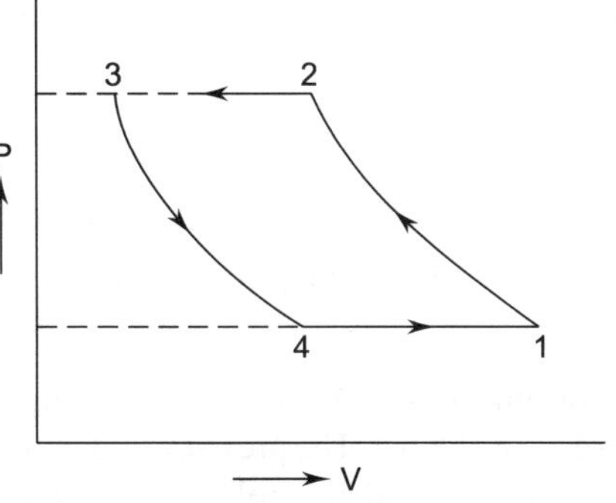


**Fig. 11.30**

or
$$T_2 = T_1 \left(\frac{p_2}{p_1}\right)^{\frac{\gamma-1}{\gamma}} = 277 \left(\frac{16}{4}\right)^{\frac{0.4}{1.4}} = 411.62 \, k$$

$$\frac{T_4}{T_3} = \left(\frac{p_4}{p_3}\right)^{\frac{\gamma-1}{\gamma}} \quad \therefore \quad T_4 = T_3 \left(\frac{p_4}{p_3}\right)^{\frac{\gamma-1}{\gamma}} = 293 \left(\frac{4}{16}\right)^{\frac{0.4}{1.4}} = 197.17 \, k$$

Refrigerating effect per kg $= q_o = C_p(T_1 - T_4) = 1(277 - 197.17) = 79.83 \, \text{kJ/kg}$

Rate of mass flow $= \dfrac{TR \times 211}{q_o} = \dfrac{10 \times 211}{79.83} = \mathbf{26.42} \, \text{kg/min}$

Work done per kg $= \dfrac{\gamma}{\gamma-1} R \dfrac{[T_2 - T_1]}{\eta_{\text{mech}}} - \dfrac{\gamma}{\gamma-1} R(T_3 - T_4) \, \eta_{\text{mech}}$

$$= C_p \frac{(T_2 - T_1)}{\eta_{\text{mech}}} - C_p(T_3 - T_4) \, \eta_{\text{mech}} = 1\left(\frac{411.62 - 277}{0.85}\right)$$

$- 1(293 - 197/17) \times 0.85 = 77.27 \, \text{kJ/kg}$

Compressor power $= \dot{m}W = \dfrac{26.43}{60} = 77.27 = 34.03 \, \text{kW}$ *Ans.*

(b)
$$V_1 = \frac{\dot{m} \, RT_1}{p_1} = \frac{26.43 \times 0.287 \times 277}{4 \times 10^5} = 5.26 \, m^3/min$$

$$V_1 = \frac{\pi}{4} D_C^2 \, L_C N \times 2 \text{ for double acting}$$

or,
$$5.26 = \frac{\pi}{4} D_C^2 \times 0.25 \times 250 \times 2; \quad D_C = 0.231 \, \text{m} = 23.1 \, \text{cm}$$

$$\therefore \quad \text{compressor diameter} = 23.1 \, \text{cm}$$

Similarly
$$V_4 = \frac{\dot{m} \, RT_4}{p_4} = \frac{26.43 \times 0.287 \times 197.17}{4 \times 10^5} = 3.74 \, m^3/min$$

$\therefore$
$$V_4 = \frac{\pi}{4} D_E^2 \times L_E N \times 2$$

$\therefore$
$$3.74 = \frac{\pi}{4} D_E^2 \times 0.25 \times 250 \times 2$$

$\therefore$
$$D_E = 0.195 \, m = 19.5 \, \text{cm}$$

$\therefore$ Diameter of expander cylinder $= 19.5 \, \text{cm}$ *Ans.*

**Example 11.4** A 5 ton refrigerating machine operating on Bell-Coleman cycle has an upper limit of pressure of 5 bar. The pressure and temperature at the start of compression are 1 bar and 17°C respectively. The compressed air cooled at constant pressure to a temperature of 40°C enters the expansion cylinder. Assuming both expansion and compression power to be adiabatic with $\gamma = 1.4$, determine

  (i)  C.O.P

 (ii)  quantity of air in circulation per minute

(iii)  piston displacement of compressor and expander

(iv)  bore of compressor and expander cylinders. The unit runs at 250 rpm and is double acting stroke = 20 cm

 (v)  H.P. required to drive the unit.

*Solution*
$$p_1 = 1 \text{ bar}, t_1 = 17°C.$$
$$T_1 = 273 + 17 = 290 \text{ K}$$
$$T_3 = 273 + 40 = 313 \text{ K}.$$

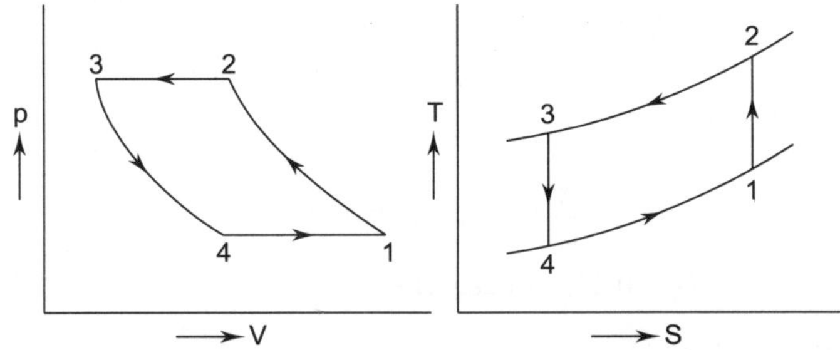

**Fig. 11.31**

Now
$$\frac{T_2}{T_1} = \left(\frac{p_2}{p_1}\right)^{\frac{\gamma-1}{\gamma}} = \left(\frac{5}{1}\right)^{\frac{1.4-1}{1.4}} = 1.585$$

$\therefore$
$$T_2 = T_1 \times 1.585 = (290)\,1.585 = \mathbf{459.65\,K}$$

$$\frac{T_3}{T_4} = \frac{T_2}{T_1} = 1.585, \quad T_4 = \frac{T_3}{1.585} = \frac{313}{1.585} = 197.5 \text{ K}.$$

 (i)  C.O.P of the machine $= \dfrac{T_1 - T_4}{\left(T_2 - T_3\right) - \left(T_1 - T_4\right)} = \dfrac{1}{\dfrac{T_2 - T_3}{T_1 - T_4} - 1}$

$$= \frac{1}{\dfrac{T_3}{T_4} - 1} = \frac{T_4}{T_3 - T_4} = \frac{197.5}{313 - 197.5} = 1.715 \quad Ans.$$

(ii)  Refrigerating effect per kg of air $= C_p(T_1 - T_4) = 1(290 - 197.5)$

$$= 92.5 \text{ kJ/kg}$$

Hence mass of air required per minute = 92.5 kJ/kg

$$= \dot{m} = \frac{5 \times 211}{92.5} = \mathbf{11.4\ kg/min}$$

Piston displacement of compressor is given by the volume corresponding to point 1.

$$v_1 = \frac{mRT_1}{p_1} = \frac{11.3 \times 2927 \times 290}{1 \times 10^4} = \textbf{9.60}\ \text{m}^3/\text{min}$$

Now $\qquad 2 \times \dfrac{\pi}{4} D_C^2\ \text{L.N} = 9.60$ since compressor is double acting

or, $\quad 2 \times \dfrac{\pi}{4} D_C^2 \times 0.20 \times 250 = 9.60 \quad$ or, $\quad D_C = 0.35\ \text{m} = 35\ \text{cm}$

$\therefore \qquad\qquad\qquad D_C = 35\ \text{cm} \quad Ans.$

Piston displacement of expansion cylinder is given by the volume corresponding to point 4.

$$v_4 = \frac{mRT_4}{p_4} = \frac{11.3 \times 29.27 \times 197.5}{10^4 \times 1} = 6.55\ \text{m3/min}$$

Now $\qquad 2 \times \dfrac{\pi}{4} D_e^2 \times L N = 6.55$

$$= 2 \times \frac{\pi}{4} D_e^2 \times 0.2 \times 250 = 6.55$$

or $\qquad\qquad\qquad D_e = 0.289\ m = 28.9\ \text{cm}$

(v) Work done $= \dfrac{R.E.}{C.O.P.} = \dfrac{5 \times 211}{1.715} = 615.16\ \text{kJ/min}$

$$\text{power} = \frac{615.16}{60} = 10.25\ \text{kW} \quad Ans.$$

**Example 11.5**   A refrigerator unit working on Bell-Coleman cycle takes air from cold chamber at $-10°C$ and compresses it from 1 bar to 6.5 bar with index of compression being 1.2. The compressed air is cooled to a temperature 5°C above the of 25°C before it is expanded in the expander where the index of expansion is 1.35. Determine

(i) C.O.P.

(ii) Quantity of air circulated per minute for production of 2000 kg of ice per day at 0°C from water at 20°C.

(iii) Capacity of plant in ton of refrigeration, assume $C_p = 1$ kJ/kgk for air

*Solution* $\quad p_1 = 1$ bar, $T_1 = 273 - 10 = 263$ K,

$\qquad\qquad p_2 = 6.5$ bar $n_1 = 1.2$, $T_3 = (273 + 25 + 5)$

$\qquad\qquad = 303$ K, $n_2 = 1.35$.

Net work $= W = W_C - W_e$

$$= \frac{n_1}{n_1 - 1}[p_2 v_2 - p_1 v_1] - \frac{n_2}{n_2 - 1}[p_3 v_3 - p_4 v_4]$$

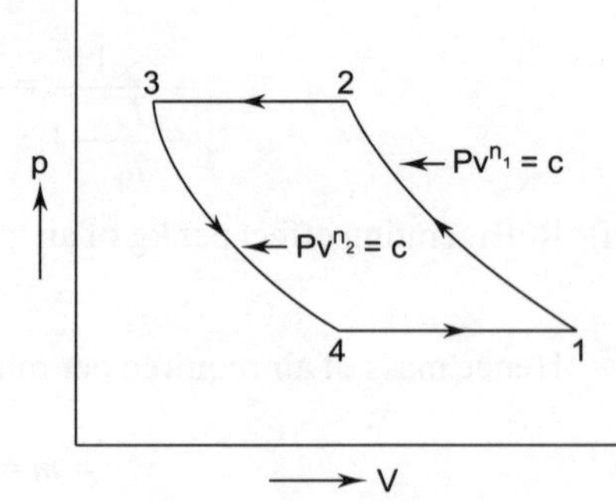



**Fig. 11.32**

$$= \frac{n_1}{n_1 - 1} RT_1 \left[ \left( \frac{p_2}{p_1} \right)^{\frac{n_1 - 1}{n_1}} - 1 \right] - \frac{n_2}{n_2 - 1} RT_4 \left[ \left( \frac{p_3}{p_4} \right)^{\frac{n_2 - 1}{n}} - 1 \right]$$

$$\frac{T_4}{T_3} = \left( \frac{p_4}{p_3} \right)^{\frac{n_2 - 1}{n_2}} \quad \text{or,} \quad T_4 = T_3 \left( \frac{p_4}{p_3} \right)^{\frac{n_2 - 1}{n_2}} = \frac{303}{(6.5)^{\frac{1.35 - 1}{1.35}}} = 186.5 \text{ k}$$

$$R = 29.27 \text{ kgm/kgk} = 29.27 \times \frac{29.81}{1000} = 0.287 \text{ KJ/kgK}$$

$$\therefore \qquad W = \frac{1.25}{(1.2 - 1)} \times 0.287 \times 263 \left[ (6.5)^{\frac{1.2 - 1}{1.2}} - 1 \right] - \frac{1.35}{1.35 - 1} \times 0.287 \times 186.5$$

$$\left[ (6.5)^{\frac{0.35}{1.35}} - 1 \right] = 165.75 - 129.03 = 36.72 \text{ kJ/kg}$$

Refrigerating effect per kg $= q_o = C_p (T_1 - T_4) = 1(263 - 186.5)$
$$= 76.5 \text{ kJ/kg}$$

(i) C.O.P. $= \dfrac{q_o}{W} = \dfrac{76.5}{36.72} = 2.08 \quad$ *Ans.*

Total refrigeration effect $Q_o = \dfrac{2000}{24 \times 60} \left[ 4.2(20 - 0) + 336 \right]$
$$= 583.33 \text{ kJ/min}$$

Air circulation per minute $= \dfrac{Q_o}{q_o} = \dfrac{583.33}{76.5} = 7.625 \text{ kg/mim}$

(iii) Tonnage $= \dfrac{583.33}{211} = 2.764 \quad$ *Ans.*

$$\because \quad 1 \, TR = \mathbf{211} \text{ kJ/min}$$

**Example 11.6** In a Bell Coleman cycle, air is drawn into compressor at 1 bar and $-7°C$ and is compressed to 5.5 bar isentropically. The air is then cooled to $18°C$. It is then expanded in an expander by following $pv^{1.25} = c$ to 1 bar and discharged into a refrigerating chamber. Find the C.O.P of the plant. If mass of air circulated through the cycle is 50 kg per minute; find the refrigerating capacity of the plant. Show the cycle on $p - v$ and $T - S$ diagram.

*Solution* Refer Fig. 11.31, for process 1-2, $pv\gamma = c$ and for process 3-4, $pv^{1.2} = C$.

Given $\qquad\qquad\qquad p_1 = 1 \text{ bar}, T_1 = -7 + 273 = 266 \text{ k}, p_2 = 5.5 \text{ bar},$

$$\dot{m} = 50 \text{ kg/min}, T_3 = 18 + 273 = 291 \text{ k}$$

$$\frac{T_2}{T_1} = \left(\frac{p_2}{p_1}\right)^{\frac{\gamma-1}{\gamma}} \qquad \therefore \quad T_2 = T_1\left(\frac{p_2}{p_1}\right)^{\frac{\gamma-1}{\gamma}}$$

$$= 266\left(\frac{5.5}{1}\right)^{\frac{1.4-1}{1.4}} = 432.92\,k$$

$$\frac{T_3}{T_4} = \left(\frac{p_3}{p_4}\right)^{\frac{n-1}{n}} \qquad \therefore \quad T_4 = \frac{T_3}{\left(\dfrac{p_3}{p_4}\right)^{\frac{n-1}{n}}} = \frac{291}{\left(\dfrac{5.5}{1}\right)^{\frac{1.2-1}{1.2}}} = 219\,k$$

$$\therefore \qquad \text{Net power required,}\ W = mR\left[\frac{\gamma}{\gamma-1}\left(T_2 - T_1\right) - \frac{n}{n-1}\left(T_3 - T_4\right)\right]$$

$$= \frac{50}{60} \times 0.287\left[\frac{1.4}{1.4-1}\left(432.92 - 266\right) - \frac{1.2}{1.2-1}\left(291 - 219\right)\right]$$

$$= 36.472\,KW \quad Ans.$$

$$\text{Refrigerating effect} = mc_p(T_1 - T_4)$$

$$= \frac{50}{60} \times 1\,(266 - 219) = 39.167\,KW$$

$$\text{Refrigerating capacity} = \frac{39.167 \times 60}{211} = 11.13\,TR \quad Ans.$$

$$\text{C.O.P.} = \frac{39.167 \times 60}{360.472} = 1.073 \quad Ans.$$

**Example 11.7**   The speed of an aircraft flying at an altitude of 8000 m, where the ambient air is at 0.341 bar pressure and 263 k temperature, is 900 km/hr. The compression ratio of the air compressor is 5. The cabin pressure is 1.01325 bar and the temperature is 27°C. Determine the power requirement of the aircraft for pressurization (excluding the ram work) additional power required for refrigeration and refrigerating capacity on the basis of 1 kg/sec flow of air.

(a)  Determine the same if the following are to be accounted

   Compressor efficiency,   $\eta_c = 0.82$

   Expander efficiency        $\eta_e = 0.77$

   Heat exchanger effectiveness

   $$\xi = 0.8$$

   Ram efficiency $= \eta_r = 0.84$

*Solution*        Speed of aircraft $V = \dfrac{900 \times 1000}{3600}$

$$= 250\ \text{m/sec}$$

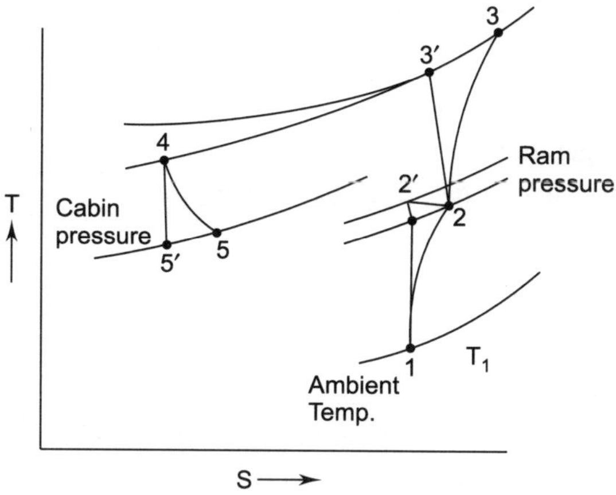


**Fig. 11.33**

Stagnation temperature

$$T_2 = T_1 + \frac{V^2}{2 \times C_p} = 263 + \frac{(250)^2}{2 \times 1 \times 10^3} = 263 + 31.25 = 294.25$$

Stagnation pressure

$$= p_{2'} = p_1 \left(\frac{T_2}{T_1}\right)^{\frac{\gamma}{\gamma-1}} = 0.341 \left(\frac{294.25}{263}\right)^{\frac{1.4}{0.4}}$$

Power requirement for pressurization,

$$p_{2'} = \mathbf{0.504}\ \text{bar}$$

$$(\text{excluding ram work}) = W_1 = m\, C_p\, T_2 \left\{\left(\frac{p_{cabin}}{p_{2'}}\right)^{\frac{\gamma-1}{\gamma}} - 1\right\}$$

$$= 1 \times 1 \times 294.25 \left[\left(\frac{1.01325}{0.504}\right)^{\frac{1.4-1}{1.4}} - 1\right] = 64.798\ \text{kW}$$

Compressor discharge temperature

$$= T_3 = T_2 \left(\frac{p_3}{p_{2'}}\right)^{\frac{\gamma-1}{\gamma}} = 294.25\,(5)^{\frac{1.4-1}{1.4}} = 466\ \text{k}$$

$$\text{Expander exit temperature} = T_5 = \frac{T_4}{\left(\frac{p_4}{p_5}\right)^{\frac{\gamma-1}{\gamma}}} = \frac{294.25}{\left(\frac{5 \times 0.504}{1.01325}\right)} = 0.286$$

$$= 226.6\ \text{k}$$

Power required for refrigeration (excluding ram work)

$$= W_2 = W_c - W_e = m\,C_p[(T_3 - T_2) - (T_4 - T_5)]$$
$$= 1 \times 1[(466.0 - 294.25) - (294.25 - 226.6)] = 104.1\,\text{kW}$$

Additional power for refrigeration

$$W_2 - W_1 = (104.1 - 64.798) = 39.302\,\text{kW}$$

Refrigerating capacity

$$Q_o = m\,q_o = mc_p\,(T_{\text{cabin}} - T_5)$$
$$= 1 \times 1(300 - 226.6) = 73.4\,\text{kW}$$

(b)   Ram pressure $= p_2 = p_1 + \eta_r\,(p_{2'} - p_1)$
$$= 0.341 + 0.84\,(0.504 - 0.341) = 0.478\,\text{bar}$$

Power requirement for pressurization (excluding ram work)

$$= W_1 = \frac{1 \times 1 \times 294.25}{0.82}\left\{\left(\frac{1.01325}{0.478}\right)^{0.286} - 1\right\}.$$

$$= 86.4\,\text{kW}$$

Temperature after actual compression

$$= T_3 = T_2 + \frac{T_{3'} - T_2}{\eta_C} = 294.25 + \frac{466 - 294.25}{0.82} = 503.7\,\text{k}$$

Effectiveness of heat exchanger

$$= \xi = \frac{T_3 - T_4}{T_3 - T_2} = \frac{\text{Actual cooling}}{\text{Maximum possible cooling}}$$

or,    $$T_4 = T_3 - \xi(T_3 - T_2) = 503.7 - 0.8\,(503.7 - 294.25) = 336.14\,\text{k}$$

Temperature after expansion $= T_{5'} = \dfrac{336}{\left(\dfrac{5 \times 0.504}{1 - 0.1325}\right)^{0.286}}$

$$= 258.9\,\text{k}$$

Actual temp after expansion $= T_5 = T_4 - \eta_T(T_4 - T_{5'})$
$$= 336 - 0.77(336 - 258.9) = 276.7\,\text{k}$$

Power required for refrigeration (excluding ram work)
$$= W_2 = m\,c_p[(T_3 - T_2) - (T_4 - T_5)] = 1 \times 1$$
$$[503.4 - 294.25) - (336.14 - 276.7)] = 149.71\,\text{kW}$$
$$\text{Additional power} = W_2 - W_1 = 149.71 - 86.4 = 63.31\,\text{kW}$$

**Example 11.8**   An air plane working on boot strap evaporating system has 20 *TR* capacity with speed of 1.3 Mach. The ambient conditions are $-10°C$ and 0.5 bar. The pressure of the air bled-off the main compressor is 4 bar and this is further compressed in the secondary compressor to 5 bar. The internal efficiency of the main compressor is 90% and that of secondary compressor is 80%. The internal efficiency of cooling turbine is 80%. The heat exchanger effectiveness of the primary ram air heat exchanger is 0.4 and that of secondary also 0.4. The air is further cooled in the evaporator to 100°C. Ram efficiency may be assumed as 90%. Assume $\gamma = 1.4$ and $c_p = 1$ kJ/kgk and cabin to be maintained at 20°C. Determine

  (i)   mass flow rate bled off the compressor
  (ii)  main compressor power used for refrigerating system

*Solution*   Given  $T_1 = -10 + 273 = 263$ k

$$M = 1.3 = 20\ TR$$
$$\eta_{c_1} = 0.90$$
$$\eta_{c_2} = 0.80$$
$$\eta_T = 0.80$$
$$\xi_1 = \xi_2 = 0.4$$
$$T_7 = 100 + 273 = 373\ \text{k}$$
$$\eta_{\text{ram}} = 0.90$$
$$\gamma = 1.4,\ C_p = 1\ \text{KJ/kgk}$$
$$T_9 = 25 + 273 = 298\ \text{k}$$

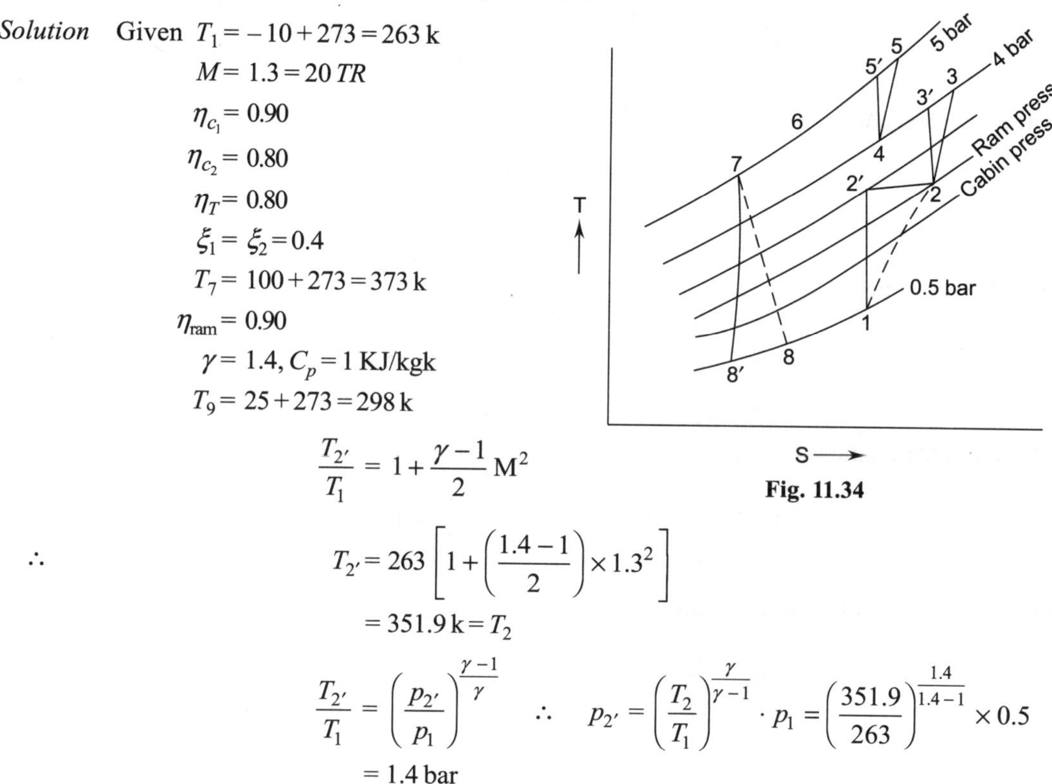


**Fig. 11.34**

$$\frac{T_{2'}}{T_1} = 1 + \frac{\gamma - 1}{2}\,\text{M}^2$$

$\therefore$

$$T_{2'} = 263\left[1 + \left(\frac{1.4 - 1}{2}\right) \times 1.3^2\right]$$

$$= 351.9\ \text{k} = T_2$$

$$\frac{T_{2'}}{T_1} = \left(\frac{p_{2'}}{p_1}\right)^{\frac{\gamma-1}{\gamma}} \qquad \therefore \quad p_{2'} = \left(\frac{T_2}{T_1}\right)^{\frac{\gamma}{\gamma-1}} \cdot p_1 = \left(\frac{351.9}{263}\right)^{\frac{1.4}{1.4-1}} \times 0.5$$

$$= 1.4\ \text{bar}$$

Now, for ramming action

$$\eta_{\text{ram}} = \frac{p_2 - p_1}{p_{2'} - p_1}$$

or,

$$0.90 = \frac{p_2 - 0.5}{(1.4 - 0.5)} \qquad \therefore \quad p_2 = 1.31\ \text{bar}$$

$$\frac{T_3}{T_2} = \left(\frac{p_3}{p_2}\right)^{\frac{\gamma-1}{\gamma}} \qquad \therefore \quad T_3 = 351.9\left(\frac{4}{1.31}\right)^{\frac{0.4}{1.4}} = 484\ \text{k}$$

Now, for first compressor

$$\eta_{c_1} = \frac{T_{3'} - T_2}{T_3 - T_2} = \frac{484 - 351.9}{T_3 - 351.9} = 0.90$$

$\therefore$

$$T_3 = 498.7\ \text{k}$$

Now, for first heat exchanger

$$\xi_1 = \frac{T_3 - T_4}{T_3 - T_2} = \frac{498.7 - T_4}{498.7 - 351.9} = 0.4$$

$$\therefore \qquad T_4 = 440\,\text{k}$$

$$\frac{T_{5'}}{T_4} = \left(\frac{p_{5'}}{p_4}\right)^{\frac{\gamma-1}{\gamma}} \qquad \therefore \quad T_{5'} = 440\left(\frac{5}{4}\right)^{\frac{0.4}{1.4}} = 469\,\text{k}$$

Now, for second compressor

$$\eta_{c_2} = \frac{T_{5'} - T_4}{T_5 - T_4} = \frac{469 - 440}{T_5 - 440} = 0.80$$

$$\therefore \qquad T_5 = 476.3\,\text{k}$$

Now, for second heat exchanger

$$\xi_2 = \frac{T_5 - T_6}{T_5 - T_2} = \frac{476.3 - T_6}{476.3 - 351.9} = 0.4$$

$$\therefore \qquad T_6 = 427\,\text{k} \quad \therefore \quad T_7 = 373\,\text{k}$$

The cabin is maintained at 1 bar

$$\frac{T_{8'}}{T_7} = \left(\frac{p_{8'}}{p_7}\right)^{\frac{\gamma-1}{\gamma}} \qquad \therefore \quad T_{8'} = 373\left(\frac{1}{5}\right)^{\frac{0.4}{1.4}} = 235.5\,\text{k}$$

Now, for expansion in cooling turbine

$$\eta_T = \frac{T_7 - T_8}{T_7 - T_{8'}} = \frac{373 - T_8}{373 - 235.5} = 0.80$$

$$\therefore \qquad T_8 = 263\,\text{k} = -10°\text{C}$$

$$\text{Mass flow rate } \dot{m} = \frac{20 \times 211}{c_p\left(T_1 - T_8\right)} = \frac{20 \times 211}{1\left[25 - (-10)\right]}$$

$$= 120.57\,\text{kg/min} \quad \textit{Ans.}$$

Main compressor power required for refrigeration system

$$= \dot{m}\,(T_3 - T_2)$$

$$= 120.57 \times 1(498.7 - 351.9)$$

$$= 17699.67\,\text{kJ/min} = 295\,\text{kW} \quad \textit{Ans.}$$

**Example 11.9** A boot strap cooling system of 12 *TR* capacity is used in an aeroplane. The ambient air temperature and pressure are 20°C and 0.85 bar respectively. The pressure of air increases from 0.85 bar to 1 bar due to ramming action of air. The pressure of air discharged from the main compressor is 3 bar. The discharge pressure of air from auxiliary compressor is 4 bar. Isentropic efficiency of each compressor is 85% and isentropic efficiency of turbine is 80%. 50% of the enthalpy of air discharged from the main compressor is removed in the first heat exchanger and 30% of the enthalpy of air discharged from the auxiliary compressor is removed in the second heat exchanger. Assuming ramming action to be isentropic, the required cabin pressure is 0.9 bar and the temperature of air having the cabin as 20°C. Find

   (i) Power required to operate the system

  (ii) C.O.P, state the assumptions made.

[Poona university summer 1998]

*Solution*   Given

Refrigerating capacity = 12 *TR*

$$T_1 = 20 + 273 = 293 \text{ k}, p_1 = 0.85 \text{ bar}$$
$$p_2 = 1.0 \text{ bar}, p_3 = 3 \text{ bar}, p_6 = 4 \text{ bar},$$
$$\eta_c = 0.85, \eta_T = 0.80, p_{\text{cabin}} = 0.9 \text{ bar}$$

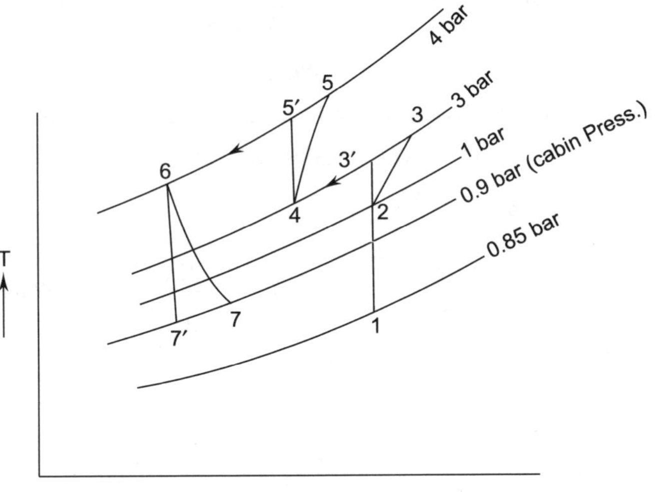



**Fig. 11.35**

$$\frac{T_2}{T_1} = \left(\frac{p_2}{p_1}\right)^{\frac{\gamma-1}{\gamma}}$$

∴

$$T_2 = 293 \left(\frac{1}{0.85}\right)^{\frac{1.4-1}{1.4}} = 306.92 \text{ k}$$

$$\frac{T_{3'}}{T_2} = \left(\frac{p_{3'}}{p_2}\right)^{\frac{\gamma-1}{\gamma}} \qquad \therefore \quad T_{3'} = 306.92 \left(\frac{3}{1}\right)^{\frac{1.4-1}{1.4}} = 420 \text{ k}$$

For first compressor

$$\eta_{c_1} = \frac{T_{3'} - T_2}{T_3 - T_2} = \frac{420 - 306.92}{T_3 - 306.92} = 0.85$$

∴

$$T_3 = 439.95 \text{ k} = 166.95°\text{C}$$

Since 50% of total heat is removed in the first heat exchanger, therefore

$$t_4 = (1 - 0.5) \times 166.95 = 83.475°C$$

$$\therefore \quad T_4 = 83.475 + 273 = 356.475 \, k$$

$$\frac{T_{5'}}{T_4} = \left(\frac{p_{5'}}{p_4}\right)^{\frac{\gamma - 1}{\gamma}}$$

$$\therefore \quad T_{5'} = 356.475 \left(\frac{4}{3}\right)^{\frac{1.4 - 1}{1.4}} = 387 \, k$$

For second compressor

$$\eta_{c_2} = \frac{T_{5'} - T_4}{T_5 - T_4} = \frac{387 - 356.475}{T_5 - 356.475} = 0.85$$

$$\therefore \quad T_5 = 392.38 \, k = 119.386°C$$

Since 30% of energy (enthalpy) is lost in second heat exchanger, therefore,

$$t_6 = (1 - 0.3) \times 119.386 = 83.57°C$$

$$\therefore \quad T_6 = 356.57 \, k$$

$$\frac{T_6}{T_{7'}} = \left(\frac{p_6}{p_{7'}}\right)^{\frac{\gamma - 1}{\gamma}} \qquad \therefore \quad T_{7'} = \frac{356.57}{\left(\dfrac{4}{0.9}\right)^{\frac{1.4 - 1}{1.4}}} = 232.84 \, k$$

For expansion in cooling turbine

$$\eta_T = \frac{T_6 - T_7}{T_6 - T_{7'}} = \frac{356.57 - T_7}{356.57 - 232.84} = 0.80$$

$$\therefore \quad T_7 = 257.586 \, k = -15.414°C$$

Now, given cabin temperature $= 20°C = t_i$

$$\therefore \quad \text{Refrigerating effect}, \; q_o = c_p (t_i - t_7)$$

$$= 1[20 - (-15.414)] = 35.414 \, kJ/kg$$

$$\text{Work done} = C_p [(T_3 - T_2) + (T_5 - T_4) - (T_6 - T_7)]$$

$$= C_p (T_3 - T_2 + T_5 - T_4 - T_6 + T_7)$$

$$= 1(439.95 - 306.92 + 392.38 - 356.475 - 356.57 + 257.586) = 69.951 \, kJ/kg$$

$$\text{Mass flow rate} = \frac{12 \times 211}{69.951} = 36.196 \, kg/min$$

$$\therefore \quad \text{Power required} = \frac{36.196 \times 69.951}{60} = 42.20 \, kW \quad \textit{Ans.}$$

$$\therefore \quad \text{C.O.P.} = \frac{q_o}{\text{Work done}} = \frac{35.414}{69.951} = 0.506 \quad \textit{Ans.}$$

**Example 11.10**  A high altitude flight aircraft is flying at an altitude of 51,000 m with a speed of 1.2 Mach. The ambient atmospheric pressure and temp. are 0.1 bar and – 40°C. The cabin is pressurized to 0.7 bar and has to be maintained at 20°C. The main compressor pressuer ratio is 5 and air enters the cooling turbine at 40°C. The exit from the cooling turbine is at 0.75 bar. The cockpit cooling load is 10 TR. Assume internal efficiency of compressor is 85% and that of cooling turbine as 75%. Ram efficiency is 90%. Assume $\gamma = 1.4$ and $C_p = 1$ kJ/kg K. Determine

*Solution*

   (i)  Stagnation temp. and pressure of air entering and leaving the main compressor.

  (ii)  Mass flow rate of air

 (iii)  Ram air heat exchanger effectiveness

 (iv)  Volume handled by compressor and cooling turbine

  (v)  Net power delivered by engine to the refrigerating unit (pressurization + refrigeration)

 (vi)  COP of the system based on compressor work

 (vii)  Power for only pressurization (excluding ram work)

(viii)  Additional power for only refrigeration.

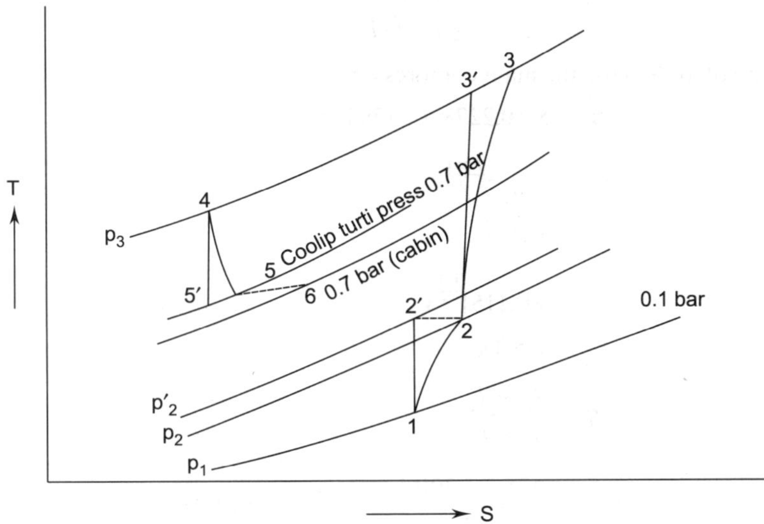


**Fig. 11.36**

 (i)  Process 1-2′ is the ramming process

$$\frac{T_{2'}}{T_1} = \frac{T_2}{T_1} = 1 + \frac{\gamma - 1}{2} M^2$$

$$= 1 + \frac{1.4 - 1}{2} \times 1.22 = 1.288$$

Therefore stagnation temp. of air after ramming and entering to the main compressor

$$T_{2'} = T_2 = T_1 \times 1.288$$

$$= (273 - 40) \times 1.288 = 300.1 \text{ k}$$

$$\frac{T_{2'}}{T_1} = \left(\frac{p_{2'}}{p_1}\right)^{\frac{\gamma-1}{\gamma}}$$

$$\therefore \qquad p_{2'} = \left(\frac{T_{2'}}{T_1}\right)^{\frac{\gamma}{\gamma-1}} \times p_1$$

$$= \left(\frac{300.1}{233}\right)^{\frac{1.4}{1.4-1}} \times 0.1 = 0.242 \text{ bar}$$

$$\eta_{Ram} = \frac{p_2 - p_1}{p_{2'} - p_1}$$

$$0.9 = \frac{p_2 - 0.1}{0.242 - 0.1} \qquad \therefore \quad p_2 = 0.2278 \text{ bar}$$

(ii)  Mass flow rate for 10 *TR*

$$\dot{m} = \frac{10 \times 211}{c_p \left(T_{cockpit} - T_5\right)}$$

The pressure of air leaving the main compressor

$$p_3 = 5 \times 0.2278 = 1.139 \text{ bar}$$

$$\frac{T_{3'}}{T_2} = \left(\frac{p_3}{p_2}\right)^{\frac{\gamma-1}{\gamma}}$$

$$\therefore \qquad T_{3'} = 300.1 \, (5)^{\frac{1.4-1}{1.4}}$$

$$= 475.3 \text{ k}$$

$$\eta_c = \frac{T_{3'} - T_2}{T_3 - T_2}$$

$$0.85 = \frac{475.3 - 300.1}{T_3 - 300.1} \qquad \therefore \quad T_3 = 506.02 \text{ k}$$

This air is cooled to 40°C i.e. $(273 + 40)$k in the ram air heat exchanger before entering the cooling turbine. i.e. $T_4 = (40 + 273) = 313$ k

$$\frac{T_{5'}}{T_4} = \left(\frac{p_{5'}}{p_4}\right)^{\frac{\gamma-1}{\gamma}}$$

$$\therefore \qquad T_{5'} = 313 \left(\frac{0.75}{1.139}\right)^{\frac{1.4-1}{1.4}}$$

$$= 277.77 \text{ k}$$

$$\eta_t = \frac{T_4 - T_{5'}}{T_4 - T_{5'}}$$

$$0.75 = \frac{313 - T_5}{313 - 277.77} \qquad \therefore \quad T_5 = 286.57 \, \text{k}$$

$$\therefore \qquad \dot{m} = \frac{TR \times 211}{c_p \left(T_{cockpit} - T_5\right)}$$

$$= \frac{10 \times 211}{1.0 \left\{(273 + 25) - (286.97)\right\}}$$

$$= 184.60 \, \text{kg/min} \quad \textit{Ans.}$$

(iii)  Ram air heat exchanger effectiveness

$$\xi = \frac{T_3 - T_4}{T_3 - T_2}$$

$$= \frac{(506.2 - 313)}{(506.2 - 300.1)} = 0.937 \quad \textit{Ans.}$$

(iv)  Volume handled by compressor

$$V_c = \frac{\dot{m} \, R \, T_2}{p_2}$$

$$= \frac{184.60 \times 287 \times 300.1}{0.2278 \times 10^5}$$

$$= 697.95 \, \text{m}^3/\text{min at entrance of compressor.}$$

Volume handled by cooling turbine

$$V_t = \frac{\dot{m} \, R \, T_4}{p_4}$$

$$= \frac{184.6 \times 287 \times 313}{1.139 \times 10^5}$$

$$= 145.59 \, \text{m}^3/\text{min at entrance of cooling turbine}$$

(v)  Net power delivered by the engine to the refrigeration unit. It is assumed that the compressor work done on bled off air from the compressor is the power delivered by the engine to the refrigeration unit. The cooling turbine power is not recoverable back to engine and is used up in refrigeration unit. Also ram work is again dissipated in exhaust air from the unit. Thus

$$W_{net} = \dot{m} \, C_p (T_3 - T_2)$$

$$= 184.60 \times 1.0(506.2 - 300.1)$$

$$= 38046.06 \, \text{kJ/min}$$

$$= \frac{38046.06}{60} = 634.10 \, \text{kW}$$

(vi)  C.O.P, $\xi = \dfrac{\dot{m}\, C_p \left(T_{\text{cockpit}} - T_5\right)}{\dot{m}\, C_p \left(T_3 - T_2\right)}$ , $T_{\text{cockpit}} = 25 + 273 = 298$

$$= \frac{(298 - 286.57)}{(506.2 - 300.1)}$$

$$= 0.055 \quad Ans.$$

(vii)  Power for pressurization (excluding ram work)

$$W = \frac{\dot{m}\, c_p\, T_2}{\eta_c} \left[ \left( \frac{p_{\text{cabin}}}{p_2} \right)^{\frac{\gamma - 1}{\gamma}} - 1 \right]$$

$$= \frac{184.6 \times 1 \times 300.1}{0.85} \left[ \left( \frac{0.7}{0.2378} \right)^{\frac{1.4 - 1}{1.4}} - 1 \right] = 24646 \text{ is kJ/min}$$

$$= 410.76\,\text{kW} \quad Ans.$$

Additional power required from engine only for refrigeration

$$W_{\text{ref}} = 634.10 - 410.76$$

$$= 223.34\,\text{kW} \quad Ans.$$

## 11.7   VAPOUR COMPRESSION REFRIGERATION SYSTEM

The invention of a refrigerant that undergoes a change of phase from liquid to vapour and vapour to liquid alternatively at low temperature range gave birth to the vapour compression refrigeration system as shown in Fig. 11.37. The cycle working on this system deals with all refrigerants used in refrigerators. The refrigerants used are ammonia, carbon dioxide, freons etc. This system is also used in water cooler, ice plants, air-conditioners and cold storages. The refrigerant is circulated through the system where it undergoes phase change. Refrigeration is produced by absorbing heat while the refrigerants evaporates. Vapour refrigerant is compressed in the compressor and therefore it is called vapour compression refrigeration. It is also called mechanical refrigeration. It consists of Evaporator, Compressor, Condenser, Receiver and or Pressure reducing mechanism like an expansion valve or capillary tube. This system is compact having very high COP and low running cost.

*p-h* and *T-S* diagram shown in Fig. 11.37 comprise following processes:

1—2  :  Isentropic compression of refrigerant vapour

2—3  :  Constant pressure heat rejection

3—4  :  Throttling process (or) constant enthalpy expansion

4—1  :  Constant pressure heat absorption or refrigeration process.

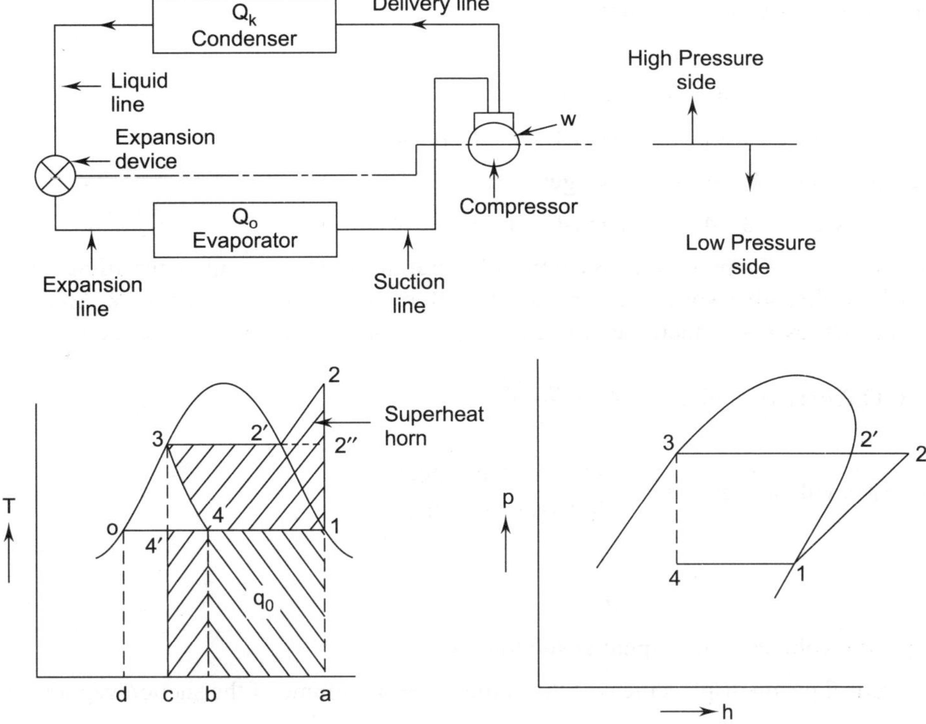



**Fig. 11.37   Vapour compression refrigeration system with p-h and T-S diagram**

Refrigerant liquid undergoes phase change during the constant pressure heat absorption process 4 to 1. This process is also known as refrigeration process and heat absorbed during this process is called refrigeration effect. The refrigerant vapour is then isentropically compressed from state points 1 to 2. The superheated refrigerant vapour is cooled during the constant pressure heat rejection process 2-3. During this process the refrigerant vapour is de-superheated (2-2′) and condensed into liquid.

The refrigerant liquid is then throttled to low pressure, that is expanded to low pressure during the process 3-4. The expansion device may be either thermostatic expansion valve (employed in large refrigeration system) or capillary tube. Throttling which is an expansion process from state 3 to 4, is an irreversible process. Irreversible process is shown by a dotted line assuming enthalpy remaining same instead of an isentropic. The network required is now equal to the compression work only as no work is obtained from the expansion process. The heat transfer in the condenser is unaffected, but the heat extracted in the evaporator is diminished. The rate of flow of working fluid is increased in this cycle as the heat extracted per unit mass of refrigerant is reduced. The refrigeration effect is a useful criterion for comparing cycles on the basis of size of plant required for a given duty. Refrigeration (Refrigerating) effect per kg of refrigerant $= q_o = (h_1 - h_4) = (h_1 - h_3) \because h_3 = h_4$

$$= \text{Area } 1 - a - b - 4 - 1$$

Heat rejected from condenser per kg of refrigerant

$$= q_k = h_2 - h_3 = \text{Area } 2 - a - c - 3 - 2$$

$$\text{Work done} = w = q_k - q_o = \text{Area } 1 - 2 - 3 - c - b - 4 - 1.$$

$$\text{Also work done } w = h_2 - h_1$$

Since process $3 - 4$ is throttling process

$$h_3 = h_4$$

or, $\qquad h_3 - h_o = h_4 - h_o$

or, $\qquad$ area $3 - c - d - o - 3 =$ Area $4 - b - d - o - 4$.

Subtracting area $o - 4' - c - d - o$, we get

$$\text{Area } o - 3 - 4' - o = \text{Area } 4' - 4 - b - c - 4'$$

Area $4' - 4 - b - c - 4'$ represents a loss of the refrigerating effect as a result of throttling. This area also represents a loss of positive work resulting from the failure to recover expansion work. Area $2 - 2' - 2''$ of superheat horn represents an increase of negative work as a result of dry compression.

$$\text{C.O.P. of refrigeration } \xi = \frac{q_o}{w} = \frac{h_1 - h_4}{h_2 - h_1} \tag{11.16}$$

$$\text{Refrigerant circulation rate, } m = \frac{\text{Refrigerating capacity}}{\text{Refrigeration effect}}$$

$$= \frac{Q_o}{q_o} \tag{11.17}$$

If, specific volume of the vapour at suction $= v_1$.

The theoretical piston displacement of the compressor or volume of the suction vapour $= mv_1$.

$$\text{Actual piston displacement of the compressor} = V_p = \frac{mv}{\eta_{\text{vol}}}$$

where $\qquad\qquad \eta_{\text{vol}} = $ volumetric efficiency of compressor

$$\text{Power consumption, } W = mw = m\,(h_2 - h_1)$$

$$\text{Hear rejected in the condenser} = Q_k = mq_k$$

$$= m\,(h_2 - h_3)$$

The theoretical COP of the vapour compression cycle is lower than that of the reversed carnot cycle. Nevertheless, it is closest to the carnot cycle as compared to other cycles and its COP approaches nearest to the Carnot value.

## Expansion cylinder Vs Throttle valve

With the reference to Fig. 11.37, we have seen that vapour compression cycle is similar to Carnot cycle except that the expander has been replaced by a throttle valve. Expansion process in the expander is represented by process 3-4' (isentropic). In throttling process the initial enthalpy equals the final enthalpy. This process shown by dotted line 3-4 is highly irreversible and thus the cycle is irreversible. However, cost of expander is saved by switching over to throttle valve but losing refrigerating effect from $(h_1 - h_{4'})$ to $(h_1 - h_4)$ and thereby less COP with same amount of compression.

COP with throttle valve

$$= \frac{\text{area } 1 - 4 - b - a - 1}{\text{area } 1 - 2 - 3 - 0 - d - a - 1} = \frac{h_1 - h_4}{h_2 - h_1} = \frac{h_1 - h_3}{h_2 - h_1} \tag{11.18}$$

COP with expansion cylinder

$$= \frac{\text{area } 1 - 4' - c - a - 1}{\text{area } 1 - 2 - 3 - 4' - c - a - 1} = \frac{h_1 - h_{4'}}{(h_2 - h_1) - (h_3 - h_{4'})} \tag{11.19}$$

where,    $(h_3 - h_{4'})$ = work of expansion in expansion cylinder.

However, with vapour compression system expansion cylinders are not used due to the following reasons:

1. More complicated, and involves more cost compared to gain in refrigerating effect and saving in work done but at the same time mechanical friction loss will come in picture in the cylinder. Hence substitution by a throttle valve would not significantly reduce the performance.
2. With throttle valve the flow of refrigerant can be considered easily which is not possible with expansion cylinder.

As the practical cycle becomes irreversible due to the introduction of throttle valve, COP is no longer independent of the working fluid as in the case of reversed Carnot cycle.

## 11.7.1  Determination of C.O.P. of Vapour Compression Refrigeration System from T-S Diagram

The following assumptions are made for drawing T-S diagram

1. The condition of the vapour leading the evaporator and entering the compressor is dry saturated.
2. The compression of vapour in the compressor is isentropic.
3. There is no pressure loss in the system.
4. There is no under cooling of the refrigerant in the condenser.
5. The work required to drive the system is equal to the difference between the heat rejected in the condenser and heat absorbed in the evaporator.
6. Transmission efficiency from motor to compressor is 100%.

**Case A:** When the vapour is dry and saturated at the end of compression.

Fig. 11.38 shows the cycle $1 - 2 - 3 - 4 - 1$ of operation. At point 1 the wet vapour is admitted into the compressor from the evaporator at low temperature $T_2$ and is compressed isentropically to point 2. The condition of the vapour coming out of the compressor is dry saturated at point 2. The condensation of the vapour starts from 2 to 3 taking place at constant pressure and temperature. During this process, refrigerant changes its phase from vapour to liquid. Liquid then enters the expansion valve. Dotted line $3 - 4$ represents throttling process. At point 4, the mixture of liquid and vapour enters the evaporator and absorbs its latent heat of evaporation from the space which to be cooled.

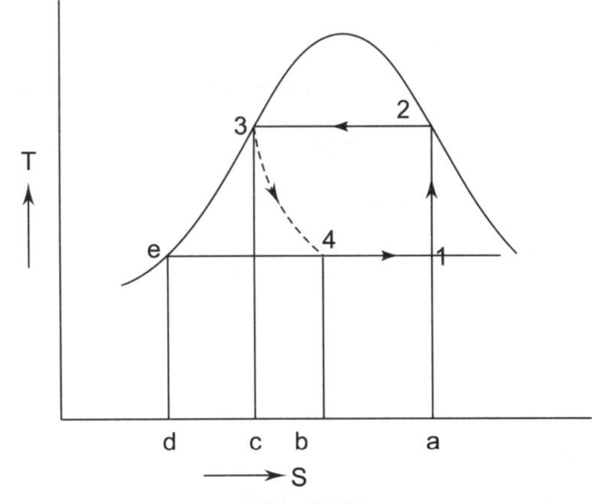

**Fig. 11.38**

Refrigerating effect per kg of refrigerant

$$q_o = \text{Area } 1 - 4 - b - a - 1 = h_1 - h_4 = h_1 - h_3$$

Work done on the compressor per kg of refrigerant

$$w = \text{Area } 1 - 2 - 3 - c - b - 4 - 1$$

$$= \text{heat at } 2 - \text{heat at } 1 = h_2 - h_1$$

$$\therefore \qquad \text{Theoretical C.O.P.} = \frac{q_o}{w} = \frac{h_1 - h_3}{h_2 - h_1} \qquad (11.20)$$

**Case B:** When the vapour is superheated at the end of compression.

During compression, when vapour becomes superheated additional refrigerating effect is achieved by the area $1 - 2 - 2' - e - a - 1$ shown in Fig. 11.39.

Refrigerating effect

$$q_o = \text{Area } 1 - a - b - 4 - 1$$

$$= (h_1 - h_4) = (h_1 - h_3)$$

Work done on the compressor,

$$w = \text{Area } 1 - 2 - 3 - c - b - 4 - 1.$$

$$= h_2 - h_1$$

$\therefore$ Theoretical

$$\text{C.O.P.} = \frac{q_o}{w} = \frac{(h_1 - h_3)}{(h_2 - h_1)} \qquad ...(11.21)$$

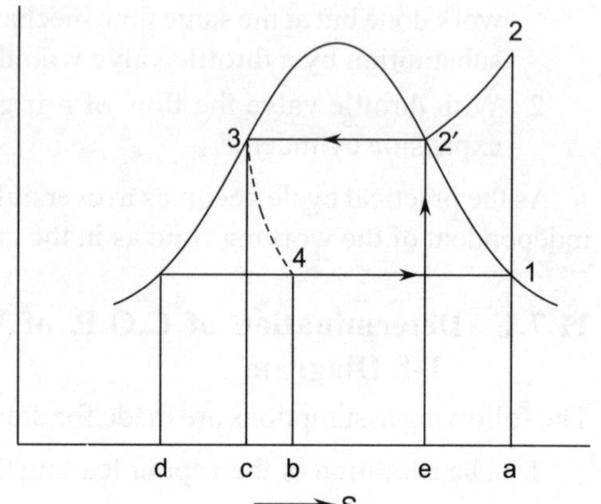

**Fig. 11.39**

**Case C:** When the vapour is wet at the end of compression

Refer Fig. 11.40 Refrigerating effect

$$q_o = \text{Area } 1 - a - b - 4 - 1$$

$$= (h_1 - h_4) = (h_1 - h_3)$$

Work done on the compressor

$$w = \text{Area } 1 - 2 - 3 - c - b - 4 - 1$$

$$= (h_2 - h_1)$$

$\therefore$ Theoretical

$$\text{C.O.P.} = \frac{q_o}{w} = \frac{(h_1 - h_3)}{(h_2 - h_1)} \qquad ...(11.22)$$

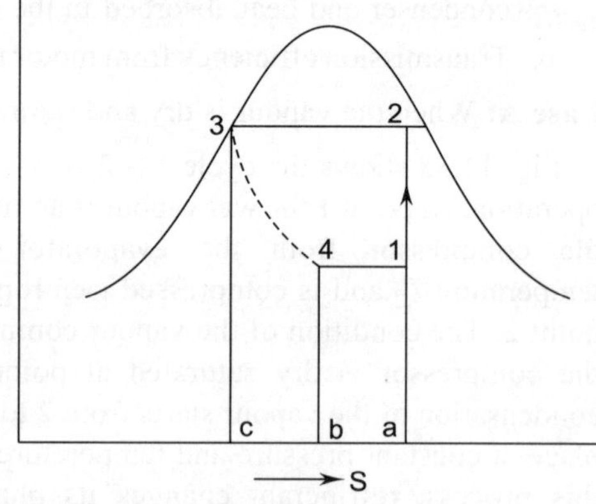

**Fig. 11.40**

## 11.7.2 Why Vapour Compression Cycle is Represented on p-h Diagram?

With the reference to p-h diagram drawn in Fig. 11.37, it is worth nothing that two of the processes are at constant pressure and one is at constant enthalpy. It, is therefore convenient to represent the vapour compression cycle on p-h diagram. Even though, the fourth process is an isentropic one, the p-h diagram is still convenient as the refrigerating effect and work done are solved numerically if enthalpies values at salient points are known.

## 11.7.3  Dry and wet Compression

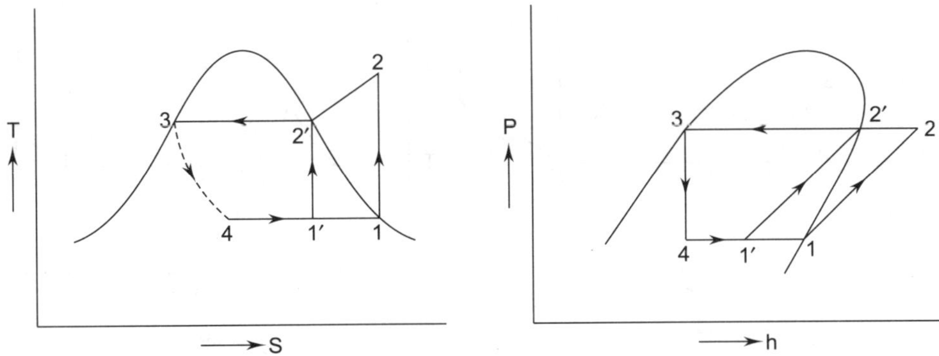

**Fig. 11.41  Dry and wet compression processes**

Process        $1-2$ = dry compression

Process        $1'-2'$ = wet compression

That is if the vapour enters the compressor in the dry or superheated condition the compression is called the dry compression. When the vapour contains liquid, entering the compressor, compression is called wet compression. Dry compression always results in the vapour being superheated at the end of compression as shown in Fig. 11.41. With a reciprocating compressor, wet compression is not found suitable due to the following reasons:

(i)  Liquid refrigerant when passes into the compressor it would tend to wash away the lubricating oil from its rubbing surfaces. Thus it would interfere with the lubrication. Further the oil would be carried to the evaporator where it might form a film on the tube surfaces and reduce the heat transfer.

(ii)  Liquid refrigerant may be trapped in the heat of the cylinder and may damage the compressor valves.

To ensure that no liquid refrigerant is carried into the compressor, the refrigerant at the entry of the compressor should not be only dry saturated but superheated. However, superheating is underable due to following reasons:

(i)  compressor work is increased due to increase in specific volume as a result capacity of compressor is reduced.

(ii)  Temperature of vapour refrigerant at the outlet of the compressor is increased and hence rejection of heat from condenser at moderate temperature would lead to departure from reversed carnot cycle.

The superheat is, therefore, kept to the minimum, so that COP may not be lowered, except marginally, in dry compression.

## 11.7.4  Subcooling or Undercooling of the Condensed Liquid

The refrigerant is said to be undercooled or subcooled when the liquid is cooled below the saturation temperature $T_1$ before throttling but at constant pressure. This undercooling is shown in Fig. 11.42($a$). In actual practice, the reduction in temperature during undercooling $(3-3')$ is hardly few degrees (maximum 5°C), therefore, for all calculation purposes, the point $3'$ is always taken on liquid saturation line as shown in Fig. 11.42($b$). Undercooling (subcooling) can be achieved by passing more cooling water in condenser. However, there is a limit of temperature (5°C) that can be achieved by this method and that depends on the temperature of condenser cooling water itself. When the liquid refrigerant is undercooled from 3 to 3' then throttling follows the line 3'4' instead of 34. Hence, extra refrigerating effect due to

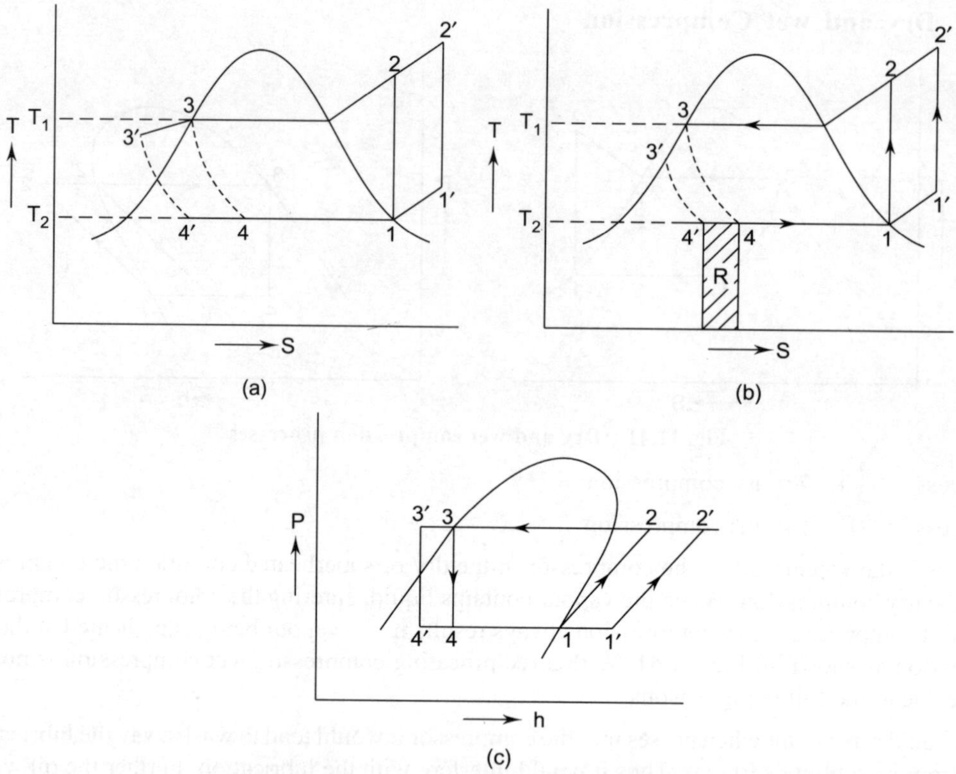

Fig. 11.42 **Undercooling of liquid refrigerant**

undercooling is increase by an area R. As no extra work $[(h_2 - h_1) \simeq (h_{2'} - h_{1'})]$ is required for undercooling, the net effect of undercooling is to increase COP of the cycle. Another advantages of this method are:

(i) Subcooling will ensure that throttling device receives liquid only.

(ii) Superheating before entry to compressor will ensure dry compression.

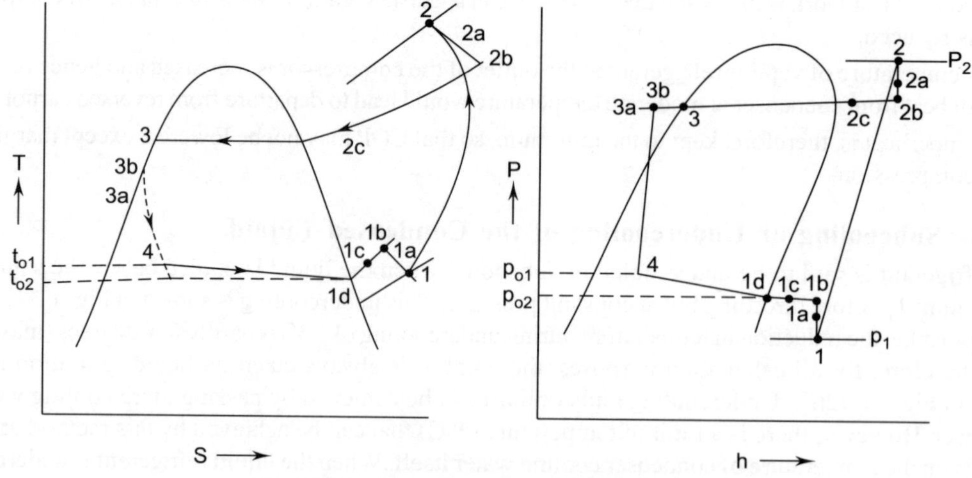

Fig. 11.43 **Actual vapour compression cycle**

Due to the flow of the refrigerant through the condenser, evaporator and piping, there will be drop in pressure. In a addition, there will be heat losses or gains depending upon the temperature difference between the refrigerant and the surroundings. Further compression will be polytropic with friction and heat transfer instead of isentropic. The actual vapour compression cycle may have some or all of the items of departure from the simple saturation cycle as discussed below:

1. Superheating of the vapour in the evaporator $1d - 1c$ (Fig. 11.43).
2. Heat gain and superheating of vapour in the suction line $1c - 1b$.
3. Pressure drop in the suction line $1b - 1a$.
4. Pressure drop due to wire drawing at the compressor suction valve $1a - 1$.
5. Polytropic compression with friction and heat transfer to the surroundings instead of isentropic compression, $1 - 2$.
6. Pressure drop at the compressor discharge valve $2 - 2a$.
7. Pressure drop in the delivery line $2a - 2b$
8. Heat loss and desuperheating of the vapour in the delivery line $2b - 2c$.
9. Pressure drop in the condensor $2b - 3$.
10. Subcooling of the liquid in the condensor or subcooler $3 - 3a$.
11. Heat gain in liquid line $3a - 3b$. The line $3 - 3a$ and $3a - 3b$ are along the saturated liquid line on T-S diagram as the constant pressure line in liquid region run close to it.
12. Pressure drop in the evaporator $4 - 1d$.

It may be noted that the pressure drop in the evaporator is large. This is due to cumulative effect of two factors. First the pressure drops in the evaporator due to friction. This is called frictional pressure drop. Secondly as evaporator proceeds, the volume increases ($m = \dfrac{A.V.}{v}$) and hence the velocity must also increase. The increase in K.E. comes from decreases in enthalpy and therefore, from a further pressure drop. This pressure drop is called momentum pressure drop.

In the condensor, the pressure drop is not significant, since the frictional drop is positive and the momentum pressure drop is negative. These might even be a pressure gain during condensation due to reduction in volume and K.E. and cobsequent conversion to enthalpy.

It may also be noted that due to the pressure drop in the evaporator from $-p_{o_1}$ to $p_{o_2}$, the temperature in the evaporator does not remain constant. It correspondingly changes from $t_{o_1}$ to $t_{o_2}$. Further due to various pressure drop, the capacity of the plant is decreased and unit power consumption is increased. Correspondingly C.O.P. of actual cycle is decreased.

## 11.7.6 Factors Affecting the Performance of Vapour Compression Refrigeration System

The effects of superheating and subcooling refrigerant has already been discussed. The remaining factors are listed below:

## Effect of Evaporator or Suction pressure

The original refrigeration cycle is represented by $1 - 2 - 3 - 4 - 1$ as shown in Fig. 11.44. Evaporator pressure is reduced by throttling in expansion valve. The new cycle is represented by $1' - 2' - 3' - 4' - 1$. From the figure it is clear that by reducing evaporator pressure refrigerating effect decreased from $(h_1 - h_4)$ to $(h_{1'} - h_{4'})$ and at the same time compression work increased from $(h_2 - h_1)$ to $(h_{2'} - h_{1'})$. Owing to both effects finally COP decreases. Therefore, the importance of maintaining the highest evaporator pressure consistent with requirement of temperature and load at the evaporator cannot be over emphasized.

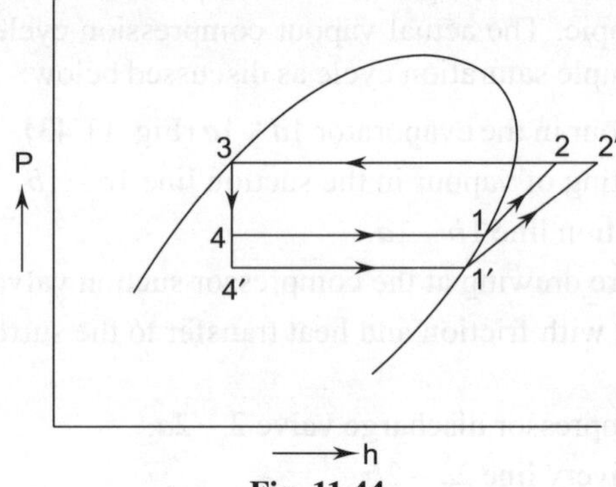

**Fig. 11.44**

The other effects of decreasing evaporator pressure are listed below:

(i) Volumetric efficiency decreases due to increase in pressure ratio since $\eta vol = 1 + c - c\left(r_p^{\frac{1}{n}}\right)$

(ii) Compressor size increases since it has to handle vapour refrigerant with increased specific volume at suction.

### Effect of Condenser pressure

The original refrigeration cycle is represented by $1 - 2 - 3 - 4 - 1$ as shown in Fig. 11.45. The new cycle is represented by $1 - 2' - 3' - 4' - 1$. From the figure it is clear that by increasing condenser pressure refrigerating effect reduced from $(h_1 - h_4)$ to $(h_1 - h_{4'})$ and compressor work increased from $(h_2 - h_1)$ to $(h_{2'} - h_1)$ as a result COP decreases.

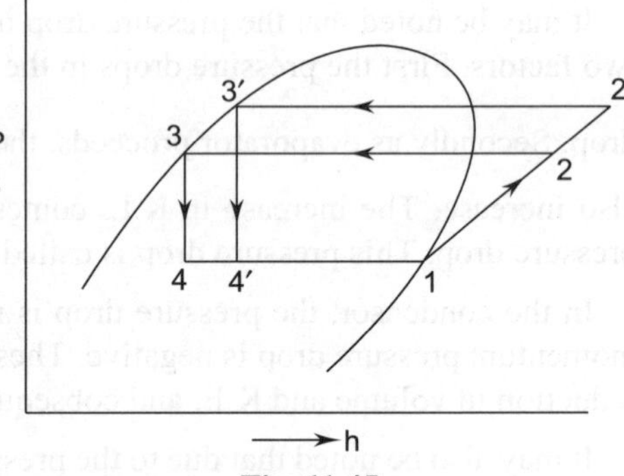

**Fig. 11.45**

As we know refrigerating device pumps heat from lower temperature to higher temperature. If this temperature difference is increased (corresponding to decrease in suction pressure and increase in delivery pressure) the work requirement for pumping the same amount of heat will be increased and hence the value of COP is decreased. Therefore the value of pressure difference between discharge pressure and suction pressure should be kept as low as possible.

## 11.7.7   Advantages of Vapour Compression System Over Air Refrigeration System

### Advantages:

1. Like reversed Carnot cycle, absorption and rejection of heat take place at constant temperature, hence it COP is quite high. C.O.P of the vapour cycle lies between 3 and 4 whereas the C.O.P of air cycle is always less than one.

2. Size of plant size is less as expander is eliminated.

3. The same refrigerant can be used over and over again.

4. The heat transfer co-efficient of refrigerant is quite high than that of air.

5. Running cost is around 20% of air refrigeration system.

## Disadvantages:

1. Initial investment cost is high.

2. Preventing leakage of refrigerant is a tedious task.

3. Refrigerant is more costly than freely available air.

### 11.7.8  Use of Liquid Vapour Regenerative Heat Exchanger

In this system as shown in Fig. 11.46 refrigerant vapour coming from evaporator is superheated in a regenerative heat exchanger with subsequent subcooling of the liquid from the condenser. Since the mass flow rate of the liquid and vapour is same, the heat transfer in heat exchanger may be written as

$$(h_{1'} - h_1) = (h_3 - h_{3'}) \tag{11.23}$$

Since specific heat of liquid refrigerant is more than that of vapour refrigerant, i.e. $C_{pl} > C_{pv}$

$$\therefore \qquad (t_i - t_o) > (t_k - t_{3'}) \tag{11.24}$$

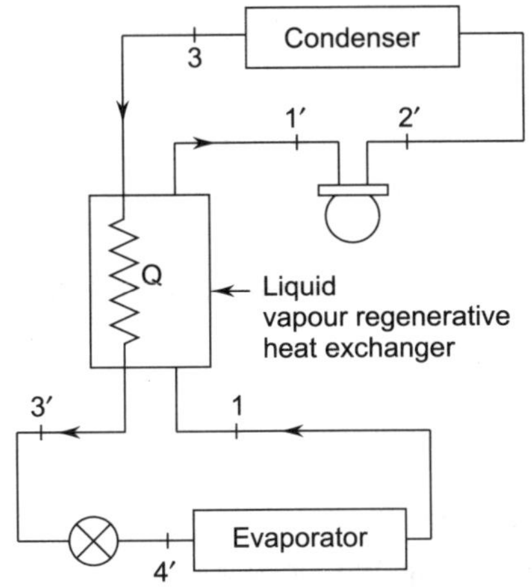


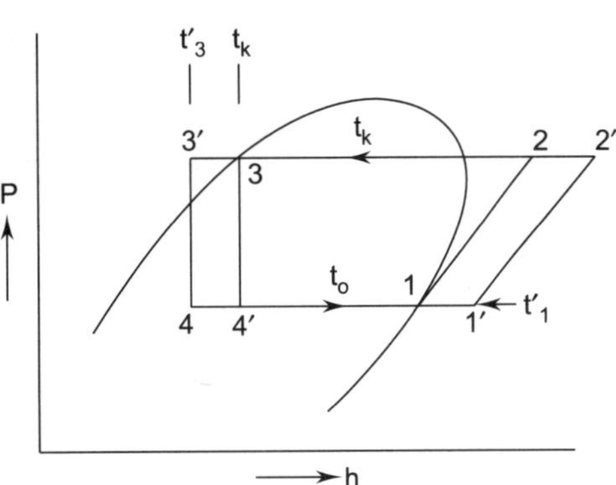


**Fig. 11.46**

The effects achieved are as follows:

(i)  refrigerating effect increases from $(h_1 - h_4)$ to $(h_{1'} - h_{4'})$

(ii)  compressor work per kg of refrigerant increases from $(h_2 - h_1)$ to $(h_{2'} - h_{1'})$

(iii)  refrigerating capacity

without heat exchanger $\quad q_o = \dot{m}\, q_o = \left(\dfrac{V_p\, \eta_{\text{vol}}}{v_1}\right)(h_1 - h_4)$

and with heat exchanger $\quad Q_{o'} = \dot{m}\, q_{o'} = \left(\dfrac{V_p\, \eta_{\text{vol}}}{v_{1'}}\right)(h_1 - h_{4'})$

$$\therefore \qquad \frac{Q_{o'}}{Q_o} = \frac{v_1}{v_{1'}}\frac{\left(h_1 - h_{4'}\right)}{\left(h_1 - h_4\right)} \tag{11.25}$$

$$\because \qquad (h_1 - h_4) < (h_1 - h_{4'}) \quad \text{and} \quad v_{1'} > v_1$$

$\therefore$  C.O.P. with heat exchanger

$$\xi' = \frac{h_1 - h_{4'}}{h_{2'} - h_{1'}} \quad \text{and} \quad \xi = \frac{h_1 - h_4}{h_2 - h_1} \tag{11.26}$$

In $\xi'$ expression both numerator and denominator increase whereas in $\xi$ expression both decrease. The net effect, whether positive, negative or zero depends on the refrigerant used and the operating conditions. It has been observed that the power per ton for $F_{12}$ is reduced but for $F_{22}$ and Ammonia increased. The additional advantage which we get that the refrigerant becomes superheat before entering into the compressor, which is generally provided to all domestic refrigerators. The enthalpy of refrigerant available for refrigeration is decreased due to undercooling of refrigerant.

**Example 11.11**  An ammonia refrigeration machine has to do an amount of refrigeration equal to the production of 20 tonnes of ice per 24 hours at $-3°C$ from water at $10°C$. If the temperature limits of the compressor are $27°C$ and $-12°C$, calculate the power of the compressor assuming the cycle is perfect one. Take the enthalpy of fusion of ice as 336 kJ/kg and its specific heat 2.1 kJ/kgK.

*Solution*  Given

$$T_H = 27 + 273 = 300 \text{ K}$$
$$T_L = -12 + 273 = 261 \text{ K}$$

$\therefore \qquad$ C.O.P. $\xi = \dfrac{T_L}{T_H - T_L} = \dfrac{261}{300 - 261} = 6.69$

Amount of heat required to be removed to produce 20 tonnes of ice at $-3°C$ from water at $10°C$ per 24 hr.

$$= \frac{20000}{24 \times 3600} [4.187(10-0) + 336 + 2.1\{0-(-3)\}]$$

$$= 88.95 \text{ KW}$$

$\because \qquad \xi = \dfrac{\text{Refrigerating effect}}{\text{Work supplied i.e., power required for compressor}}$

$\therefore$  Power required compressor $= \dfrac{88.95}{6.69} = 13.29 \text{ KW}$  *Ans.*

**Example 11.12**  Ice at $-5°C$ is formed from water at $30°C$. Temperature of brine is $-10°C$. Find the mass of ice formed per hour if 150 KW is required to drive the unit. Refrigeration machine works on reversible Carnot cycle. Take specific heat of ice as 2.1 kJ/kg, latent heat of ice as 335 kJ/kg, specific heat of water as 4.187 kJ/kgK.

*Solution*  Given,

$$T_L = -10 + 273 = 263 \text{ K}$$
$$T_H = 30 + 273 = 303 \text{ K}$$

$$\xi_{\text{carnot}} = \frac{263}{303 - 263} = 6.575$$

Amount of heat removed from water i.e.

$$Q_o = m[4.187(30-0)+335+2.1\{0-(-5)\}]=471.11\,m$$

$$\xi_{unit} = \frac{Q_o}{\omega} \quad \text{or,} \quad 6.575 = \frac{411.11\,m}{150}$$

$$\therefore \qquad m = 2.093 \text{ kg/s}$$

$$= 7536.45 \text{ kg/hr} \quad Ans.$$

**Example 11.13**  An ice plant produces 12 tons of ice per day at 0°C using water at 30°C. The plant operates on reversed carnot cycle between – 15°C and 28°C. If the actual COP is 50% of the ideal COP and overall electromechanical efficiency is 0.8, estimate the power rating of compressor motor. Latent heat of ice = 335 kJ/kg. Specific heat of water = 4.18 kJ/kgk.

$$\text{Refrigerating effect, } Q_o = \frac{12 \times 1000 \left[4.18\{30-0\}+335\right]}{24 \times 3600}$$

$$= 63.94 \text{ kJ/s}$$

$$\text{Theoretical COP} = \frac{-15+273}{(28+273)-(-15+273)} = \frac{258}{43} = 6$$

$$\text{Actual COP} = 0.5 \times 6 = 3 = \frac{Q_o}{\text{Power consumed}}$$

$$\therefore \qquad \text{Power consumed} = \frac{63.94}{3} = 21.31 \text{ kW}$$

$$\text{Power rating} = \frac{21.31}{8} = 26.64 \text{ kW} \quad Ans.$$

**Example 11.14**  A refrigerant F – 12 vapour compression system includes a liquid to vapour heat exchanger in the system. The heat exchanger cools saturated liquid coming out from the condenser from 32°C to 22°C with the help of vapour coming out from evaporator at – 12°C saturated. The compression is isentropic. Draw the line diagram of the components and find the following:

(i)  COP of the system

(ii)  If the compressor displacement is 1.2 m³/min, what is the refrigeration capacity of the system.

*Solution*

Given
$$t_{3'} = 22°C, t_3 = 32°C$$
$$t_1 = -12°C \text{ (saturated)}$$

From P – h chart of F – 12

$$h_{3'} = h_{4'} = 58 \text{ kJ/kg,}$$
$$h_3 = h_4 = 68 \text{ kJ/kg}$$
$$h_1 = 180 \text{ kJ/kg}$$
$$h_2 = 208 \text{kJ/kg}$$

Energy balance in heat exchanger

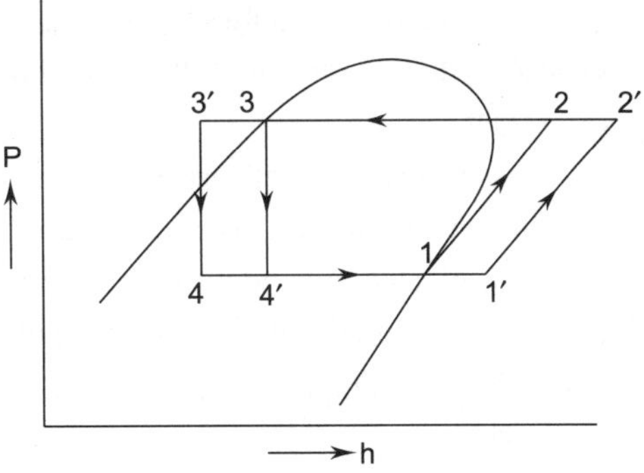


**Fig. 11.47**

Heat lost by liquid coming out from condensor = Heat gained by vapour coming out from evaporator

$$(h_3 - h_{3'}) = (h_{1'} - h_1)$$

or,

$$h_{1'} = (h_3 - h_{3'} + h_1)$$

$$= 68 - 58 + 180 = 190 \, \text{kJ/kg}$$

Locate point $2'$ on chart and get

$$h_{2'} = 218 \, \text{kJ/kg}$$

$$\therefore \qquad \text{C.O.P.} = \frac{h_1 - h_{4'}}{h_{2'} - h_{1'}} = \frac{180 - 58}{218 - 190} = 4.357 \quad Ans.$$

Mass flow rate of refrigerant

$$\dot{m} = \frac{1.2}{v_{s1'}} = \frac{1.2}{0.085 \, (\text{from table})} = 14.117 \, \text{kg/min}$$

$$\therefore \qquad \text{Refrigerating capacity} = \frac{\dot{m}\left(h_1 - h_{4'}\right)}{211}$$

$$= \frac{14.117 \left(180 - 58\right)}{211} = 8.16 \, TR \quad Ans.$$

If there is no heat exchanger

$$\xi = \frac{h_1 - h_4}{h_2 - h_1} = \frac{180 - 68}{208 - 180} = 4 \quad Ans.$$

$$\text{and refrigerating capacity} = \frac{\dot{m}\left(h_1 - h_4\right)}{211}$$

$$= \frac{114.117 \left(180 - 68\right)}{211} = 7.49 \, TR$$

*Note :* By applying heat exchanger C.O.P. and refrigerating capacity increased by $(4.357 - 4)$ and $(8.16 - 7.49)$.

**Example 11.15**   A freezer of 20 *TR* capacity has evaporating and condensing temperatures of $-30°C$ and $25°C$ respectively. The refrigerant R $-$ 12 is subcooled by 4°C before entering the expansion valve and superheated by 5°C before leaving the evaporator. The single acting compressor having six cylinders with stroke equal to the bore runs at 1000 rpm. Sketch the cycle on $p - h$ diagram and determine

(a)  C.O.P.

(b)  mass flow rate of refrigerant

(c)  the power required to drive the unit

(d)  heat rejected to condenser per minute

(e)  cylinder dimensions having 85% volumetric efficiency.

Specific heats of liquid and vapour refrigerant at working temperatures are 0.993 kJ/kgk and 0.6 kJ/kgk respectively.

*Solution*   From property table of F – 12

$$h_{1'} = 174.2 \text{ kJ/kg}, \; S_{1'} = 0.7171 \text{ kJ/kgk}$$

$$h_1 = h_{1'} + C_{pv}(t_1 - t_{1'})$$

$$= 174.2 + 0.6\{-25 - (-30)\} = 177.2 \text{ KJ/kg}$$

$$S_1 = S_{1'} + Cp_v \ln \frac{T_1}{T_{1'}} = 0.7171 + 0.6 \ln \frac{273 - 5}{273 - 30}$$

$$= 0.7286 \text{ kJ/kgk}$$

$$S_2 = S_{2'} + C_{pv} \ln \frac{T_2}{T_{2'}}$$

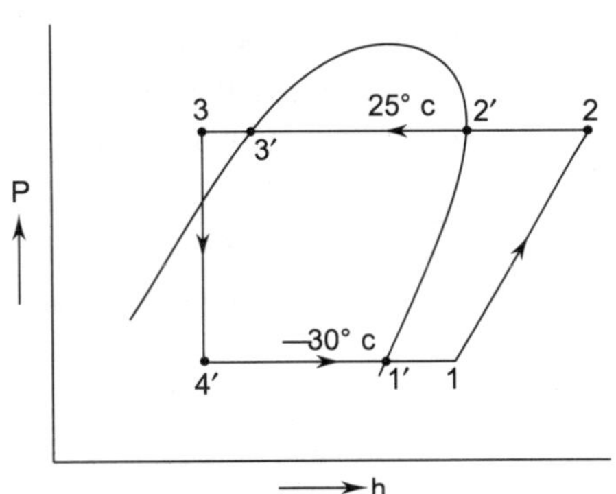


**Fig. 11.48**

or $\qquad S_1 = S_2 \quad \therefore \quad 0.7286 = 0.6869 + 0.6 \ln \dfrac{T_2}{273 + 25}$

$\therefore \qquad T_2 = 319.4\,°\text{K} = 46.4°\text{C}$

$$h_2 = h_{2'} + C_{pv}(t_2 - t_{2'}) = 197.7 + 0.6(46.4 - 25) = 211 \text{ kJ/kg}$$

$$h_3 = h_{3'} - C_{pl}(t_{3'} - t_3) = 59.7 - 0.993 \times 4 = 55.73 \text{ kJ/kg}.$$

(a) C.O.P. $= \dfrac{q_o}{w} = \dfrac{h_1 - h_4}{h_2 - h_1} = \dfrac{177.2 - 55.73}{211 - 177.2} = 3.59$   *Ans.*

(b) Rate of flow $\dot{m} = \dfrac{Q_o}{q_o} = \dfrac{TR \times 211}{q_o} = \dfrac{20 \times 211}{177.2 - 55.73} = 34.74 \text{ kg/min}$   *Ans.*

(c) Power $= \dot{m}\, w = \dfrac{34.74}{60}(h_2 - h_1) = \dfrac{34.74}{60}(211 - 177.2) = 19.56 \text{ kW}$   *Ans.*

(d) Heat rejected to condenser $= Q_k = \dot{m}\, q_k = \dot{m}(h_2 - h_3) = 34.74(211 - 55.73) = 5393 \text{ kJ/min}$   *Ans.*

(e) Piston displacement $V_p = \dot{m}\, v_1 = \dot{m}\, \dfrac{T_1}{T_{1'}}\, v_{1'} = \dot{m}\, \dfrac{273 - 25}{273 - 30} \times 0.1595 = 0.1626 \times \dot{m} = 0.1626 \times 34.74 \text{ m}^3/$

min $= 5.648 \text{ m}^3/\text{min}.$

There are six cylinders

$$5.648 \times 0.1626 \times 37.74 = 6 \times \frac{\pi}{4} D^2 L N \, \eta_{\text{vol.}} = \frac{\pi}{4} D_2 D \times 1000 \times 0.85$$

$$D = 0.1121 \text{ m} = 11.21 \text{ cm}$$

$\therefore \qquad D = L = 11.21 \text{ cm}$   *Ans.*

**Example 11.16**   An ammonia ice plant operates between a condenser temperature of 35°C and an evaporator temperature of – 15°C. It produces 12 ton ice per day from water at 28°C to ice at – 5°C. Assume simple saturated cycle, determine:

(a) capacity of the refrigeration plant in *TR*.

(b) mass flow rate of refrigerant.

(c)  the discharge temperature

(d)  the heat rejected in the condenser per minute

(e)  the compressor cylinder diameter and stroke if its volumetric efficiency is 0.65, $rpm = 1200$ and stroke bore ratio $\left(\dfrac{L}{D}\right) = 1.2$

(f)  Power of compressor motor if adiabatic efficiency of compressor is 0.85 and mechanical efficiency is 0.95.

(g)  theoretical and actual C.O.P.

*Solution*   Take sp. heat of ice $=1.94$ kJ/kg, latent heat $= 335$ kJ/kg

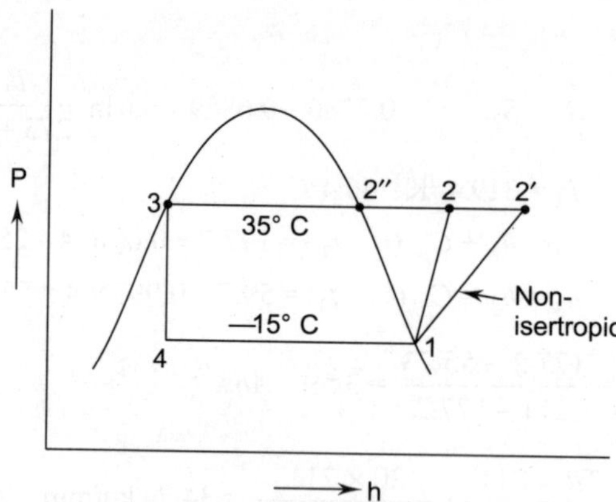


**Fig. 11.49**

Refrigerating capacity

(a)   $Q_o = \dfrac{12 \times 1000}{24}\,[4.18(28-0)+335+1.94(0-(-5))]$ kJ/hr

$\qquad = 230870$ kJ/hr $= \dfrac{230870}{60 \times 60} = 64.13$ kW

$\qquad = \dfrac{64.13 \times 60}{211} = 18.23\ TR$   *Ans.*

(b)   Refrigerating effect per kg $= q_o = h_1 - h_4 = h_1 - h_3$

$\qquad\qquad\qquad\qquad = 1426 - 347.5 = 1078.5$ kJ/kg.

Mass flow rate $\qquad \dfrac{Q_o}{q_o} = \dfrac{230870}{1078.5 \times 60}$ kg/min

$\qquad\qquad\qquad\qquad = \mathbf{3.5678}$ kg/min

(c)   $S_1 = S_2 = 5.549$ kJ/kgJ

Using chart $\qquad h_2 = 1685.4$ kJ/kg

$\qquad\qquad \eta_{\text{ediabatic}} = \dfrac{h_2 - h_1}{h_{2'} - h_1}$   or,   $0.85 = \dfrac{1685.4 - 1426}{h_{2'} - 1426}$

$\therefore \qquad\qquad h_{2'} = 1731.17$ kJ/kg

from chart $\qquad\qquad t_{2'} = $ discharge temp $= 129.92°$C   *Ans.*

(d)  Heat rejected to the condenser

$$\dot{m}\,(h_{2'}-h_3)=3.5678$$

$$(1731.17-347.5)=4936.7\,\text{kJ/min}\quad Ans.$$

(e)  Piston displacement $= V_p = \dfrac{\dot{m}\,v_1}{\eta_{\text{vol}}} = \dfrac{\pi}{4}D^2LN$

or,  $\dfrac{3.5678\times 0.509}{0.65} = \dfrac{\pi}{4}D^2(1.2\,D)\times 1200$

$\therefore\qquad\qquad D=0.135\,\text{m}=13.5\,\text{cm}$

$\qquad\qquad L=1.2\,D=1.2\times 3.5=0.162\,\text{m}\quad Ans.$

(f)  Power consumption $= \dfrac{\dot{m}\,(h_{2'}-h_1)}{\eta_{\text{Mech.}}} = \dfrac{3.5678\,(1731.17-1426)}{60\times 0.95} = 19.1\,\text{kW}\quad Ans.$

$\therefore$   Compressor power $= 19.1\,\text{kW}$

(g)  Theoretical   C.O.P. $= \dfrac{q_o}{\omega} = \dfrac{h_1-h_4}{h_{2'}-h_1} = \dfrac{1426-347.5}{1731.17-1426} = 3.53\quad Ans.$

Actual   C.O.P $= \dfrac{64.13}{19.1} = 3.36\quad Ans.$

**Example 11.17**   An R–12 reciprocating compressor with 5% clearance is to be designed for $8\,TR$ capacity at 5°C evaporator temperature and 40°C Condenser temperature. The compression index may be taken as 1.15. The number of cylinders may be selected as 2 and the mean piston speed as 3 m/sec. The stroke to bore ratio by fluorocarbon may be taken as 0.8. Pressure drops at suction and discharge. Valves may be assumed as 0.2 and 0.4 bar respectively. Determine:

(a)  power consumption of the compressor and C.O.P. of the cycle

(b)  volumetric efficiency of the compressor

(c)  bore, stroke and r.p.m of the compressor.

*Solution*      From table $p_o = 3.62$ bar

$\qquad\qquad P_k = 9.6$ bar

$\qquad\qquad p_s = 3.62-0.2 = 3.42$ bar

$\qquad\qquad p_d = 9.6+0.4 = 10.0$ bar

$\qquad\qquad h_1 = 189.7\,\text{kJ/kg},\; h_3 = 74.6\,\text{kg/kg}$

$\qquad\qquad q_o = h_1-h_4 = 189.7-74.6$

$\qquad\qquad\quad = 115.1\,\text{kJ/}k_q$

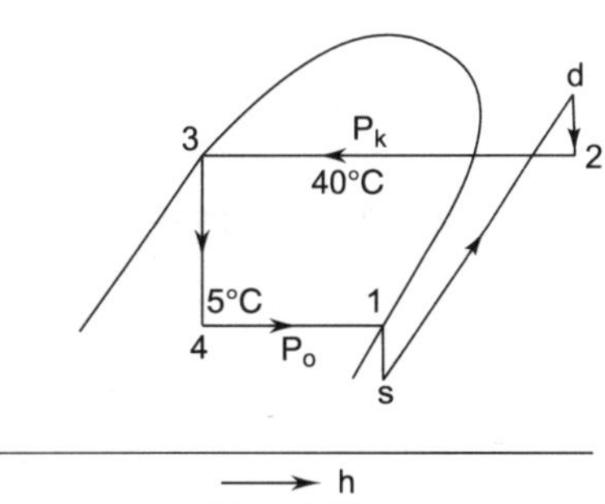


**Fig. 11.50**

$$\text{Specific work} = W = \frac{n}{n-1}p_s v_s\left[\left(\frac{p_d}{p_s}\right)^{\frac{n-1}{n}}-1\right]\frac{1}{1000}\ \text{kJ/kg}$$

$$= \frac{1.15}{1.15-1} \times (3.42 \times 105 \times 0.0525) \left[ \left( \frac{10}{3.42} \right)^{\frac{1.15-1}{1.5}} - 1 \right] \times \frac{1}{1000}$$

$$= 20.86 \text{ kJ/kg}$$

(a)  Rate of mass flow $= \dfrac{TR \times 211}{q_o} = \dfrac{8 \times 211}{115.1} = 14.67 \text{ kg/min}$

Power consumption $= \dot{m}w = \dfrac{14.67}{60} \times 20.86 = 5.1 \text{ kW}$

$$\text{C.O.P.} = \frac{q_o}{w} = \frac{115.1}{20.86} = 5.57 \quad Ans.$$

(b)  Volumetric efficiency

$$\eta_{\text{vol}} = (1 + C) \left( \frac{p_s}{p_1} \right)^{1/n} - C \left( \frac{p_d}{p_1} \right)^{1/n} - 0.01 \left( \frac{p_d}{p_s} \right)$$

allowing 1 per cent leakage per unit of compression ratio

$$\eta_v = (1 + 0.05) \left( \frac{3.42}{3.62} \right)^{1/1.15} - 0.05 \left( \frac{10}{3.62} \right)^{1/1.15} - 0.01 \left( \frac{10}{3.42} \right)$$

$$= 0.849 = 84.9\%$$

(c)  Piston displacement

$$V_p = \frac{\dot{m} V_s}{\eta_{\text{vol}}} = \frac{\pi}{4} D^2 L N$$

or,  $\left( \dfrac{14.67}{2} \right) \times \dfrac{0.0525}{0.849} = \dfrac{\pi}{4} D^2 \left( \dfrac{2LN}{2} \right) = \dfrac{\pi}{8} D^2 \times 3 \times 60$

Since piston speed,

$$2LN = 3 \times 60 \text{ m/min}$$

or,  $\qquad D^2 = \dfrac{8 \times 0.4535}{\pi \times 3 \times 60} \qquad \therefore \quad D = 0.08 \text{ m} \quad Ans.$

$$\frac{L}{D} = 0.8, L = 0.8 \, D = 0.8 \times 0.080 = 0.064 \text{ m} \quad Ans.$$

$\therefore \qquad$ Bore $= D = 8$ cm, stroke $= L = 6.4$ cm

$$\text{r.p.m.} = N = \frac{3 \times 60}{2L} = \frac{4 \times 60}{2 \times 0.064} = 1406 \text{ rpm} \quad Ans.$$

**Example 11.18** An R-12 refrigeration system works between pressure limits 1.83 bar and 9.63 bar respectively. The heat transfer from the condenser is found to be 72 kJ/min. The refrigerant vapour leaves the evaporator in the saturated state. The condensate leaves the condenser in just saturated state. The refrigerant flow rate through the system is found to be 0.5 kg/min. Obtain

  (a) C.O.P.

  (b) capacity of the plant and

  (c) energy input to the condenser.

*Solution*

  From refrigeration table

$$h_1 = 181 \text{ kJ/kg}, h_3 = 74.6 \text{ kJ/kg} = h_4$$

  Refrigerating effect

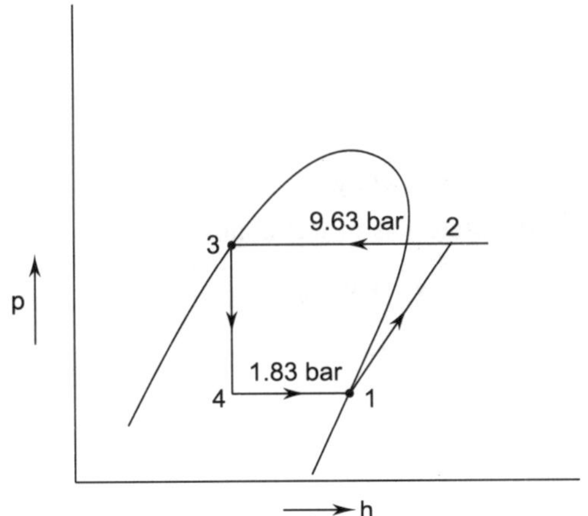


**Fig. 11.51**

$$= q_o = h_1 - h_4 = 181 - 74.6$$
$$= 106.4 \text{ kJ/kg}$$
$$Q_o = mq_o = 0.5 \times 106.4 = 53.2 \text{ kJ/min}$$
$$Q_k = 72 \text{ kJ/min}$$

$\therefore$    work input $= W = Q_k - Q_o = 72 - 53.2 = 18.8 \text{ kJ/min}$

$$C.O.P = \frac{Q_o}{W} = \frac{53.2}{18.8} = 2.83 \quad \textit{Ans.}$$

$$\text{Capacity} = \frac{Q_o}{211} = \frac{53.2}{211} = 0.252 \text{ TR} \quad \textit{Ans.}$$

**Example 11.19** A vapour compression refrigerator using carbondioxide as a refrigerant works between a pressure limits of 26.49 and 57.27 bar. After isentropic compression the temperature is 40°C and after condensation, the temperature is 10°C. Using the properties of $CO_2$ given below, estimate (a) the C.O.P (b) the mass flow rate of $CO_2$ in kg/min to produce a refrigerating effect of 5 ton, the corresponding compressor stroke volume. Assume volumetric efficiency of compressor as 0.8, find compressor cylinder diameter and stroke if it runs at 1200 rpm and $L = 1.2\,D$.

| Pressure bar | $ts$ (°C) | Enthalpy (kJ/kg) | | | Entropy (kJ/kgk) | | | Sp. heat (kJ/kgk) | | Vg m3/kg |
|---|---|---|---|---|---|---|---|---|---|---|
| | | $h_f$ | $h_{fg}$ | $h_g$ | $s_f$ | $s_{fg}$ | $s_g$ | $c_{pf}$ | $c_{pv}$ | |
| 57.27 | 20 | 144.11 | 155.51 | 299.62 | 0.523 | 0.5297 | 1.0527 | 2.889 | 2.135 | – |
| 26.49 | – 10 | 60.78 | 261.5 | 322.28 | 0.2381 | 0.9943 | 1.2324 | – | – | 0.014 |

*Sol.*

$$s_2 = s_{2'} + cp \ln \frac{T_2}{T_{2'}}$$

$$= 1.0527 + 2.135 \ln \frac{273 + 40}{273 + 20}$$

$$= 1.0527 + 0.14097$$

$$= 1.1936 \text{ kJ/kg}$$

which is less than $S_g$ at $p_o$, therefore state 1 is in wet condition

$$s_2 = s_1 = s_{f_1} + x_1\, s_{fg_1}$$

or, $\qquad 1.1936 = 0.238 + x_1 \times 0.9943$

∴ $\qquad x_1 = 0.9609$

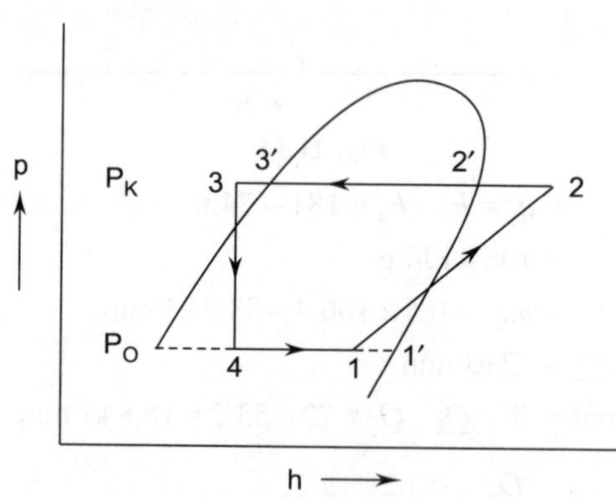


**Fig. 11.52**

$$h_1 = h_f + x_1\, s_{fg_1} = 60.78 + 0.960 \times 261.5 = 312.0 \text{ kJ/kg}$$

$$h_2 = h_{2'} + c_{pg}(t_2 - t_{2'}) = 299.62 + 2.135(40 - 20)$$

$$= 342.32 \text{ kJ/kg}$$

$$h_3 = h_{3'} - c_{pl}(t_{3'} - t_3) = 144.11 - 2.889(20 - 10) = 115.22 \text{ kJ/kg}$$

Refrigerating effect $\qquad = q_o = h_1 - h_4 = h_1 - h_3 = 312 - 115.22$

$$= 196.78 \text{ kJ/kg}$$

Compressor work $= w = h_2 - h_1 = 342.32 - 3/2 = 30.32 \text{ kJ/kg}$

$\therefore \quad \text{C.O.P} = \dfrac{q_0}{\omega} = \dfrac{196.8}{30.32} = 6.49$

Mass flow rate $= \dot{m} = \dfrac{Q_0}{q_0} = \dfrac{5 \times 211}{196.78} = 5.361 \text{ kg/min}$

Power required to run the compressor

$$= \frac{\dot{m} q_0}{60} = \frac{5.361 \times 30.32}{60} = 2.709 \text{ kw}$$

Piston displacement per minute $= \dfrac{\dot{m} x_1 \upsilon g_1}{\eta_{\text{vol}}} = \dfrac{5.361 \times 0.9609 \times 0.014}{0.8}$

*i.e.* $\qquad V_p = 0.09015 \text{ m}^3/\text{min}$

$$V_p = \frac{\pi}{4} D^2 LN = 0.09015$$

or, $\qquad 0.09015 = \dfrac{\pi}{4} D^2 \times 1.2 D \times 1200$

$\therefore \qquad D = 0.0430 \text{ m} = 4.3 \text{ cm}$

$\therefore \qquad L = 1.2 D = 5.16 \text{ cm} \quad Ans.$

**Example 11.20** A vapour compression refrigerating unit works between the temperature limits of 30°C and − 5°C as a condensing and evaporating temperatures respectively. The liquid temperature at the throttle valve entrance is 25°C and the vapour is 0.97 dry before leaving the evaporating coil. Determine (a) the condition of the refrigerant entering the evaporator (b) the theoretical C.O.P and (c) mass of cooling water used per kg of refrigerant for the condenser, if the temperature rise to be restricted to 20°C. The properties of the refrigerant are:

| Temp. °C | Enthalpy, kJ/kg | | Entropy, kJ/kgk | | Sp. heat, kJ/kgk | |
|---|---|---|---|---|---|---|
| | $h_f$ | $h_g$ | $s_f$ | $s_g$ | $c_{pl}$ | $c_{pv}$ |
| − 5 | 158.261 | 1431.885 | 0.63 | 5.4072 | – | – |
| 30 | 323.22 | 1465.38 | 1.2037 | 4.9839 | 5.024 | 3.35 |

*Sol.* Given, $\qquad t_3 = 30°\text{C}, t_4 = -5°\text{C}$

$\qquad\qquad\qquad t_{3'} = 25°\text{C}.$

$\qquad\qquad\qquad S_1 = S_2$

$$S_{f_1} + x_1 S_{fg_1} = S_{g_{2'}} + C_p \ln \frac{T_{\text{sup}}}{T_s}$$

or, $0.63 + 0.97(5.4072 - 0.63)$

$$= 4.9837 + 3.35 \ln \frac{T_{\text{sup}}}{(30 + 273)}$$

$$\therefore \qquad\qquad T_{\text{sup}} = 329.4\,\text{k} = 56.4°\text{C}$$

$$\therefore \qquad\qquad h_{3'} = h_4$$

$$\text{or,} \qquad h_3 - c_{pl}(T_3 - T_{3'}) = h_{f_4} + x_4\, h_{fg_4}$$

$$\text{or,} \qquad 323.22 - 4.024(303 - 298) = 158.261 + x_4(1431.885 - 158.261)$$

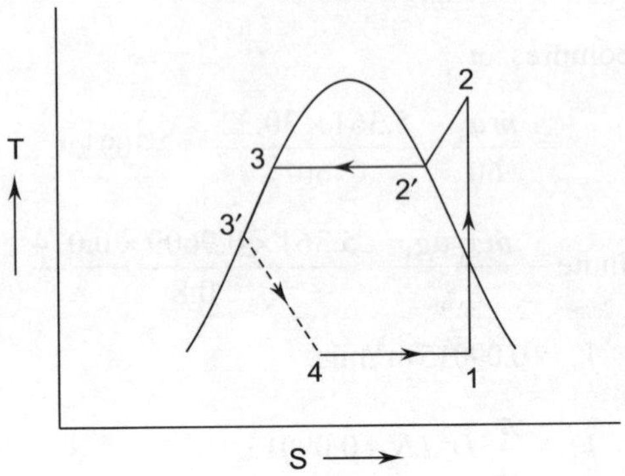

**Fig. 11.53**

$$\therefore \qquad x_4 = 0.1097 \;(\text{condition of the refrigerant entering the compressor}) \quad Ans.$$

$$h_1 = h_{f_1} + x_1\, h_{fg_1} = 158.261 + 0.97(1431.885 - 158.261)$$

$$= 1393.676 \text{ kJ/kg}$$

$$h_2 = h_{g_{2'}} + C_p v (T_{\text{sup}} - T_s)$$

$$= 1465.38 + 3.35(329.4 - 303) = 1553.82 \text{ kJ/kg}$$

$$h_4 = 158.261 + 0.1097(1431.885 - 158.261)$$

$$= 297.977 \text{ kJ/kg}$$

Refrigerating effect,
$$q_o = h_1 - h_4 = 1393.676 - 297.977$$
$$= 1095.699 \text{ kJ/kg}$$

work spent,
$$w = h_2 - h_1$$
$$= 1553.82 - 1393.676 = 160.144$$

$$\therefore \qquad \text{C.O.P} = \frac{q_0}{\omega} = \frac{1095.699}{160.144} = 6.841 \quad Ans.$$

Let
$$m_w = \text{mass of cooling water used per kg of refrigerant}$$

Heat gained by cooling water = Heat rejected from condenser
$$m_w \times 4.187 \times 20 = (h_2 - h_{3'})$$

$$\text{or,} \qquad m_w \times 4.187 \times 20 = 1553.82 - 297.977$$

$$\therefore \qquad m_w = 14.996 \text{ kg/kg of refrigerant} \quad Ans.$$

# 11.8  HOUSEHOLD REFRIGERATOR

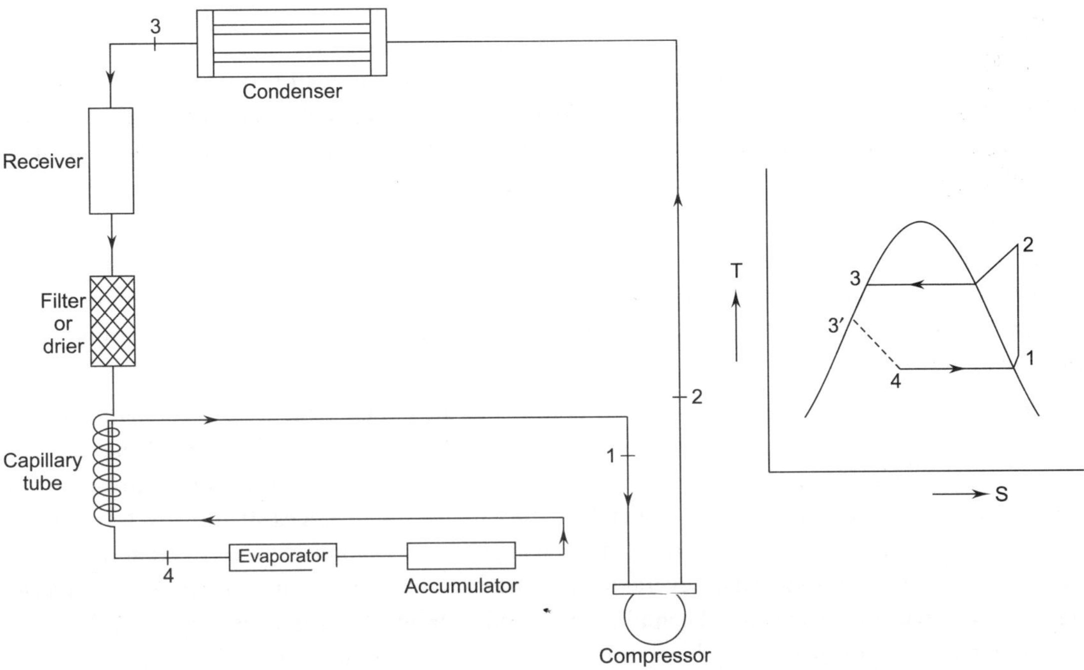


**Fig. 11.54   Schematic diagram of household refrigerator with *T-S* plot**

A typical household refrigerator works on the principle of vapour compression refrigeration cycle traditionally. Fig. 11.54 shows the schematic diagram with *T-S* plot. It contains the basic components like compressor, condenser, evaporator and capillary tube *i.e.* expansion device. The refrigerant vapour is compressed in the compressor and then stored in a receiver. Receiver acts as a storage device for the refrigerant. It also stores some additional refrigerant for overload and to compensate for leakage. When demand arises the refrigerant is first filtered and then expanded by a capillary tube. Here capillary tube acts as an expansion device. The temperature of refrigerant reduces during expansion and its volume increases. The refrigerant then enters the evaporator, where its temperature increases due to absorption of head from the application. The refrigerant evaporates and then passes through an accumulator. Accumulator prevents any liquid refrigerant droplets from entering the compressor. The vapour refrigerant then comes in indirect contact with capillary tube before entering the suction line of the compressor and is compressed again. This completes the cycle. When vapour comes in contact with capillary tube after accumulator, heat is transferred to convert it into superheated condition before entering the compressor. As a result refrigerating effect increases thereby C.O.P of refrigerator increases. The domestic refrigerator is shaped like a cabinet. The evaporator coils are wound over an ice box located at the topmost portion of the unit. The condenser which air cooled is located at the backside in the older models or as a side panel in newer refrigerators. The compressor is mounted at the base and is hermetically sealed compressor. Insulation is provided to the refrigerator cabinet to prevent loss of cool air to the surrounding. Freezer is the coldest part of the refrigerator cabinet with a temperature of about $-15°C$, while the refrigerant evaporates inside the evaporator tubes at around $-25°C$. Just below freezer there is a chiller tray. Further below are compartments with progressively higher temperatures. The bottommost part is the least cold portion which is meant for vegetables, milk, food etc. The cold air being heavier flows down from the freezer to the bottom of the refrigerator. The warm air being lighter rises from the vegetable compartment to the freezer gets cooled and flows down again. Thus, a natural convection current is set-up which maintains a temperature gradient between the top and the bottom of the refrigerator. Inside temperature of the cabinet varies from 7 to 12°C depending upon the rate of opening and closing of the door.

The controls used in the refrigerator are (1) starting relay (2) overload protector (3) thermostats (4) defrosting controls. The overload protector prevents burnout of the compressor motor in case of overload. Thermostat is used to set refrigerator temperature to a fixed value. Defrosting controls prevent formation of first on even rotor coils.

In the modern no-first refrigerators, the evaporator is located outside the freezer compartment. The cold air is made to flow by forced convection by a fan. $R_{12}$ or $F_{12}$ (dichloro difluro methane) has been used as a refrigerant in almost all refrigerators for a very long time. Now, according to Montreal protocol signed in Montreal (Canada) on 16th September 1987 under the auspices of the *UN* Environment programme//*UNEP* 1989//, a decision has been taken to phase out $R_{12}$ and a new refrigerant $R_{134a}$ (Tetra fluoro ethane) has been introduced in several major markets, including North America, South America, Asia and Australia. It is expected to be applied in selective applications in every region of the world. $R_{134a}$ has a major advantage over $R_{12}$ being non-flammable, zero ozone depletion potential (*ODP*) and less global warming potential (*GWP*). Comparison is given in Table 11.3 between both:

**Table 11.3  Difference between $R_{12}$ and $R_{134a}$.**

| Refrigerant | Normal boiling pt. | Freezing pt. | Molecular wt. | ODD | GWP | $\gamma = \dfrac{c_p}{c_v}$ | hfg kJ/kg |
|---|---|---|---|---|---|---|---|
| $R_{12}$ $(CF_2Cl_2)$ | $-29.8°C$ | $-136°C$ | 120.93 | 1.0 | 3.1 | 1.126 | 165.8 |
| $R_{134_a}$ $(CF_3CH_2F)$ | $-26.16°C$ | $-96.6°C$ | 102.03 | 0.0 | 0.27 | 1.102 | 222.5 |

$R_{134a}$ is now considered as eco-friendly refrigerant being implemented in almost all modern refrigerators.

Following is some technical data for a household refrigerator

1. Compressor power – 90 $W$ to 123 $W$
2. Capillary tube diameter – 0.82 min
3. Refrigerant charge – 160 – 190 grams
4. Power consumption – 2 to 3 kwh (165 litres)
$\qquad\qquad\qquad$ 3 to 4 kwh (286 litres)
5. Maximum running time = 40 to 60% of day time for small refrigerator and 60 to 80% for bigger refrigerators.

## 11.9  WATER COOLER

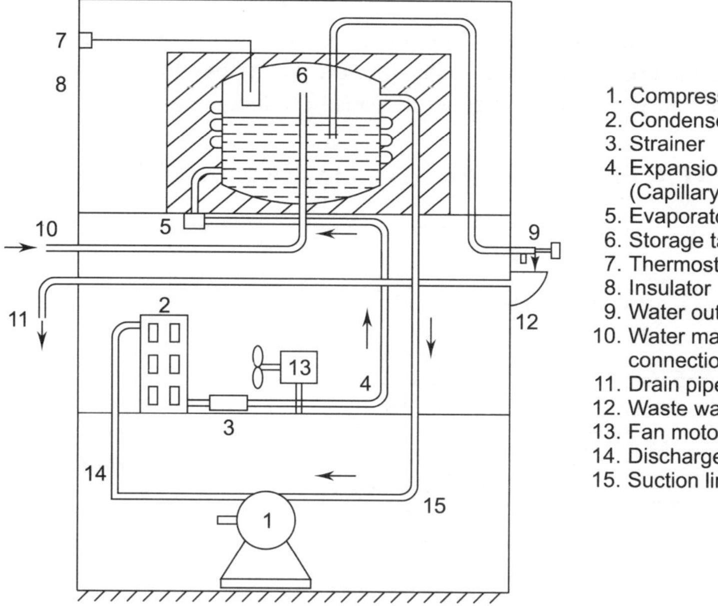



**Fig. 11.55  Water cooler**

This device shown in Fig. 11.55 uses the vapour compression system to chill water. These are commonly used in schools, offices, factories, stores and shops etc chilled water temperature ranges from 10°C to 20°C. Water coolers are classified as

1. Instantaneous type
2. Storage type (Fig. 11.55)
3. Instantaneous cum storage type

In the first type, there is instantaneous cooling as well as supply of the cold water and there is no float valve in the unit. The second type of water cooler has a facility of storage tank which stores the cold water and then it is supplied for consumption. A float valve is provided for proper storage of water and storing it as well as maintaining a continuous flow of cold water for a particular range of temperature.

In the refrigerant circuit $R_{134a}$ has been introduced in place of $R_{12}$ in modern water coolers. The vapour is compressed in the compressor, condensed in the condenser where it changes its phase to liquid. The condenser is cooled for forced air fan driven by electric motor. The condensed refrigerant is then expanded in a capillary tube and then passed to the evaporator. The evaporator coils are would over the water storage tank. The evaporator coil is a copper tube of 10 mm diameter and 10 to 15 m length. The refrigerant absorbs heat from the supplied water thereby cooling it. The refrigerant itself evaporates and is pushed into the accumulator which is usually a tube of about 30 mm diameter and 25 – 30 cm length. This accumulator is placed in such a position that at its outlet only refrigerant gas is pushed into the suction line of the compressor. This completes one cycle of the refrigeration cycle. The chilled water is supplied by a pipe to an orifice through a flow control valve. Whenever chilled water is required the user presses the knob which opens the valve causing the water to rush out in form of spray. A drain is provided at the back of the unit to remove spilled water. A purge line is used to remove any air bubbles from the storage tank. There is a thermostatic switch which is pre-set for certain temperature range (usually 10°C and 20°C) and installed in electric circuit in series with compressor. The electric supply is also cut off with an overload protector which operates when the motor is overloaded. The bimetallic overload gets heated and disconnects the power supply. Cold water is stored in water tank which is provided with insulation material all around it so that the cold water does not get heated from surrounding atmospheric. The entire unit is mounted on concrete pads to reduce vibrations.

## 11.10   MULTISTAGE OR COMPOUND COMPRESSION SYSTEM

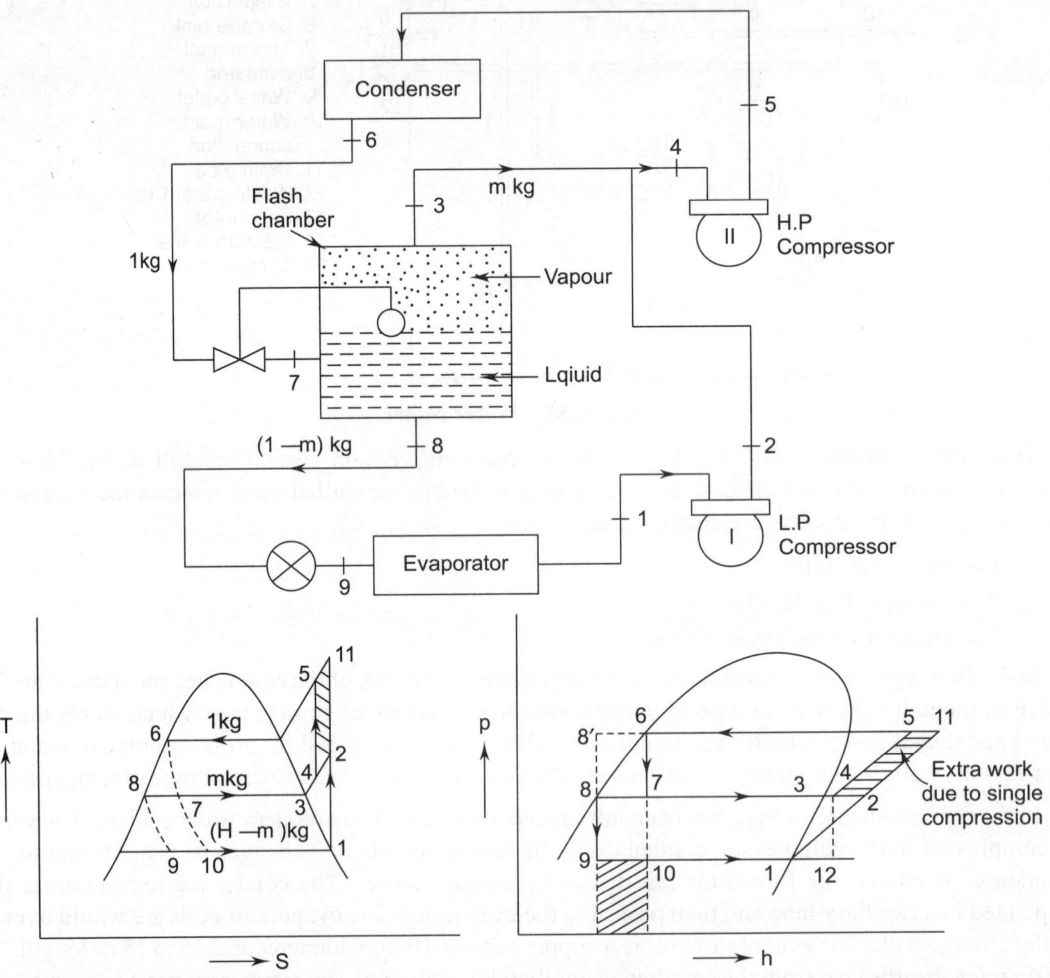


**Fig. 11.56   Schematic diagram of two stage compression with flash chamber, *T-S* plot and *p-h* plot**

Simple vapour compression system is a two pressure system. Systems with more than two pressures may arise either due to multistage compound compression for minimum work or due to feeding of a refrigerant to a multi evaporative systems. The whole idea of multi pressure system is to follow the compression very near the saturation curve. It is however, desirable to employ compound compression only when pressure ratio between the condenser and the evaporator is more than 4 or 5. This will happen either as a result of very high condensing temperature and or very low evaporating temperature. With increase in pressure ratio, volumetric efficiency of reciprocating compressor decreases and therefore, refrigerating capacity decreases due to reduction in the rate of flow. In order to increase the capacity and decrease the power consumption the multi pressure system is used. The two methods employed for cooling between stages are water intercooling and flash intercooling with flash gas removal.

Fig. 11.56 shows the schematic diagram for multistage compression with flash chamber and the corresponding cycles on *T-S* and *P-h* plot. Considering 1 kg of liquid refrigerant from condenser at 6 first expands to an intermediate pressure in flash chamber. From flash chamber the useless vapour, mass m kg is passed to the inlet of second stage compressor. The remaining liquid in the flash chamber, mass $(1 - m)$ kg is expanded in a second throttle valve to the final pressure and then passes to the evaporator and then is compressed in the first-stage compressor. In the simple system the flash vapour passes through the evaporator without doing any refrigeration, and work must be supplied to compress this vapour together with the useful vapour through the full range of pressure.

Process $6 - 7$ being a throttling process enthalpy remains same *i.e.* $h_6 = h_7$

$$h_6 = h_{f_7} + m h_{fg_7} \text{ or, } m = \frac{h_6 - h_{f_7}}{h_{fg_7}} \qquad \qquad ...(11.27)$$

Total work supplied $\qquad = w_{1-2} + w_{4-5} = (1 - m)\,(h_2 - h_1) + (h_5 - h_4)$ $\qquad$ ...(11.28)

Refrigerating effect $\qquad = (1 - m)\,(h_1 - h_8)$ $\qquad\qquad$ ...(11.29)

At the intermediate pressure, $(1 - m)$ kg of vapour at 2 is mixed adiabatically with m kg of flash vapour of enthalpy $h_{g_3}$. The resultant mixture at 4 is compressed in second stage compressor to pressure 5. By having enthalpy balance, condition of vapour at 4 can be calculated as

$$h_4 = (1 - m)\,h_{g_3} + m \times h_{g_3} \qquad\qquad ...(11.30)$$

Enthalpy rejected in condenser $= 1(h_5 - h_6)$ $\qquad\qquad$ ...(11.31)

In a system without a flash chamber, the liquid from the condenser expands straight to the evaporator pressure as shown by line 6-10. This is wasteful of energy as the vapour flashed at the intermediate pressure at 3 is also throttled to 12 at the evaporator pressure. A system with a flash chamber, thus, eliminates the undesirable throttling of the vapour generated at the intermediate pressure. Flash gas removal with multistage compression, therefore, results in power economy and is always desirable irrespective of any refrigerant.

## 11.11 VAPOUR ABSORPTION SYSTEM

Simple vapour absorptions system:

The vapour absorption refrigeration system differs from the vapour compression system only in the compression process. The function of the compressor in the vapour compression system is to continuously withdraw the vapour refrigerant from evaporator to Condenser. In the vapour absorption system the function of the compressor is accompanied in a three step process by the use of the absorber, pump and generator as given below:

1. Absorber: Absorption of the refrigerant vapour by its weak or poor solution in a suitable absorbent or adsorbent forming a strong or rich solution of the refrigerant in the absorbent/adsorbent.

2. Pump: pumping of the rich solution raising its pressure to condensers pressure.

3. Generator: Distribution of the vapour from the rich solution leaving the poor solution for recycling.

A simple vapour absorption system mainly consists of (1) Condenser (2) Expansion device (3) Evaporator as in vapour compression (4) Absorber (5) Pump (6) generator (7) Pressure reducing valve. Schematic diagram of this system is shown in Fig. 11.57.

Where,    $Q_o$ = Refrigerating effect

$Q_p$ = pump work in terms of heat unit

$Q_h$ = Heat added to generator

$Q_A$ = Heat rejected from absorber

$C$ = Condenser

$G$ = Generator

$E$ = Evaporator

$A$ = Absorber

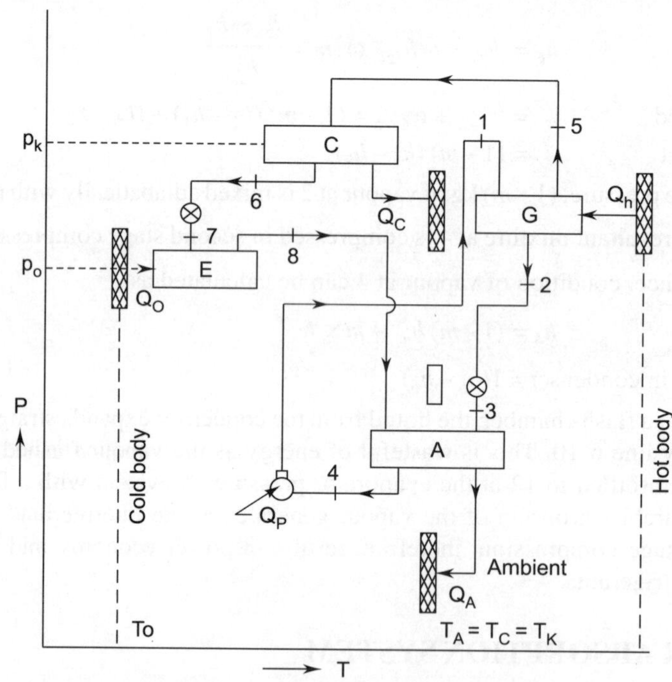


**Fig. 11.57    Vapour absorption system**

Energy balance of the system is $Q_o + Q_h + Q_p = Q_c + Q_A$

$$= Q_K \qquad \qquad ...(11.32)$$

Where $Q_K$ represents the net heat rejected to the environment

The pump work $Q_p = -\int vdp$ is very small compared to compressor work in the compression system

as the specific volume $v$ of the liquid is extremely small compared to that of the vapour ($V_f << V_g$). The energy consumptions is merely in the generator in the form of heat supplied $Q_k$.

Generally calculation for the vapour absorption system are based on a unit mass of refrigerant vapour distilled from the generator. All heat and mass flow quantities per unit mass of the vapour are then referred to as specific quantity. Thus if $D$ is the mass of the vapour distilled then the various specific heat quantities are

$$q_0 = Q_0/D$$

$q_p = Q_p/D$ and so the energy balance equations per unit mass of the vapour distilled is then

$$q_o + q_p + q_h = q_c + q_A = q_k \qquad \qquad ...(11.33)$$

C.O.P. is expressed as $\qquad = \dfrac{\text{refrigerating effect}}{\text{Energy supplied}} = \dfrac{q_0}{q_k + q_p}$

$$= \dfrac{q_0}{q_k - q_0} \qquad \qquad (11.34)$$

Neglecting the pump work which is very small.

$$\zeta = \dfrac{q_0}{q_h} \qquad \qquad (11.35)$$

## 11.11.1  Maximum  C.O.P

The vapour absorption system is a heat operated refrigerating machine. It may be considered as a combination of heat engines and a refrigerating machine. The energy supplied to the system is in the form of heat $Q_h$ at temperature $T_h$. The thermodynamics cycle compress a heat engine cycle operating between $T_h$ and $T_k$ and a refrigeration cycle between $T_o$ and $T_k$

$$\text{C.O.P} = \zeta = \dfrac{Q_0}{Q_h} = \dfrac{W}{Q_h} \cdot \dfrac{Q_o}{W} = \eta_{th} \times \zeta_c \qquad \qquad (11.36)$$

Which implies that the C.O.P of a heat operated refrigerating machine is equal to the product of the thermal efficiency of the heat engine, part of the cycle and C.O.P of the cooling of the refrigeration C.O.P. of heat operated refrigerating machine is maximum when each of the two terms has a maximum value which would be Carnot value.

$$\zeta_{max} = \eta_{th\ carnot} \times \zeta_{c\ carnot} = \left( \dfrac{T_h - T_k}{T_h} \right)\left( \dfrac{T_0}{T_k - T_0} \right) \qquad \qquad (11.37)$$

It can be seen from above expression that C.O.P depends on $T_h$, $T_k$ and $T_o$ and in order to be high, one should have:

1. $T_h$ for heat source as high as possible.
2. $T_k$ of heat sink as low as possible.
3. $T_o$ of refrigeration as high as possible.

$$\zeta_{max} = \left( 1 - \dfrac{T_k}{T_h} \right)\left( \dfrac{1}{\dfrac{T_k}{T_0} - 1} \right) \qquad \qquad (11.38)$$

It may be noted that in case the condensor and absorber temperatures are not the same and are equal to $T_c$ and $T_A$, then,

$$\zeta_{max} = \frac{T_h - T_A}{T_h} \times \frac{T_0}{T_c - T_0} \tag{11.39}$$

---

**Example 11.21**    In a vapour absorption refrigeration system, the refrigeration temperature is $-15\,c°$. The generator is operated by solar heat where the temperature reached is $110\,c°$, the temperature if heat is $55\,c°$. What is the maximum possible C.O.P of the system?

*Solution:* Given $\quad\quad\quad\quad\quad T_o = 273 - 15 = 258\,K,\ T_h = 273 + 110 = 383\,K\ T_k = 273 = 55 = 328\,K$

$$\zeta_{max} = \frac{T_h - T_k}{T_h} \times \frac{T_0}{T_k - T_0} = \frac{383 - 328}{383} \times \frac{258}{328 - 258}$$

$$= 0.0914 \times 3.69 = 0.34 \quad \textit{Ans.}$$

---

**Example 11.22**    In a vapour absorption system working on solar energy, the refrigeration temperature is $-5c°$. The generator is operated by solar heat where the temperature reached is $110\,c°$. The condenser temperature is $55\,c°$ and absorber temperature is $50\,c°$. What is the maximum possible C.O.P of the system? If refrigerating efficiency is 60%, find out the rate of heat added in the generator to produce 2 ton of refrigeration.

*Solution:* Given $\quad\quad\quad\quad\quad T_o = 273 - 5 = 268\,k,\ T_h = 273 + 100$

$$= 383\,k,\ T_c = 273 + 55 = 328\,k,\ T_A = 273 + 50$$

$$= 323\,k$$

Maximum possible $\quad\quad\quad$ C.O.P $= \eta_{carnot} \times \xi_{carnot}$

$$= \left(1 - \frac{T_A}{T_H}\right)\left(\frac{T_0}{T_k} - T_0\right)$$

$$= \left(1 - \frac{323}{383}\right)\left(\frac{268}{328 - 268}\right) = 0.7$$

C.O.P of absorption system $\quad = \xi_{carnot} \times$ refrigerating efficiency

$$= 0.7 \times 0.6 = 0.42$$

Now, $\quad\quad\quad\quad\quad\quad\quad \xi = \frac{Q_0}{Q_h}, \quad \therefore \quad Q_h = \frac{Q_0}{\xi} = \frac{2 \times 211}{0.43}\ \text{kJ/min}$

$$= \frac{2 \times 211}{0.42 \times 60}\ \text{kw} = 16.746\ \text{kw} \quad \textit{Ans.}$$

## 11.11.2    Actual vapour absorption cycle

Actual vapour absorption system is shown in Fig. 11.58

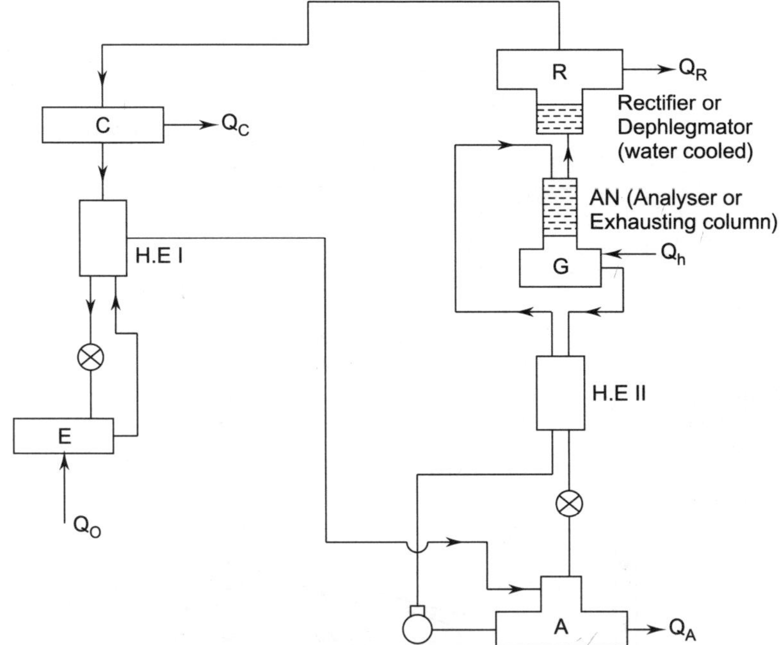


H.E.I: Liquid vapour heat exchange to sub-cool the liquid refrigerant from the condenser.
H.E.II: Liquid heat exchanger to reheat the liquid mixture from absorber.

**Fig. 11.58  Actual vapour absorption system**

Individual device has its own function actually as given below:

**Analyzer:** It is impracticable to attempt to remove all the water vapour from the ammonia and water vapour mixture but all except a very small fraction may be condensed out or absorbed by simple process of cooling. In the generator vapour, the ammonia vapour is super heated to a considerable greater degree than is the water vapour which would indicate that in cooling the water vapour would be first to condense. Only a negligible quantity of water vapour remains. When the mixture is cooled to temperature of about 10°C above the saturation temperature of ammonia. But the strong aqua ammonia from the absorber which is cool and aqua ammonia from the rectifier enter at the top of the analyzer and flow downward over trays into the generator. Large liquid surface area is exposed to vapour from generator and thus vapour cooled separating out water vapour as condensed water.

**Rectifier:** Purpose of rectifier is to further cool the vapour leaving the analyzer so that almost all water vapour is condensed leaving only dehydrated ammonia gas to pass to the condenser. Rectifier is generally water cooled.

## 11.11.3  Electrolux  Refrigerator

It is a three fluid system manufactured by Electrolux Company as shown in Fig. 11.59. It uses NH3 as refrigerant, water as absorbent and a lighter gas Hydrogen as a carrier to decrease the partial pressure of ammonia in the evaporator.

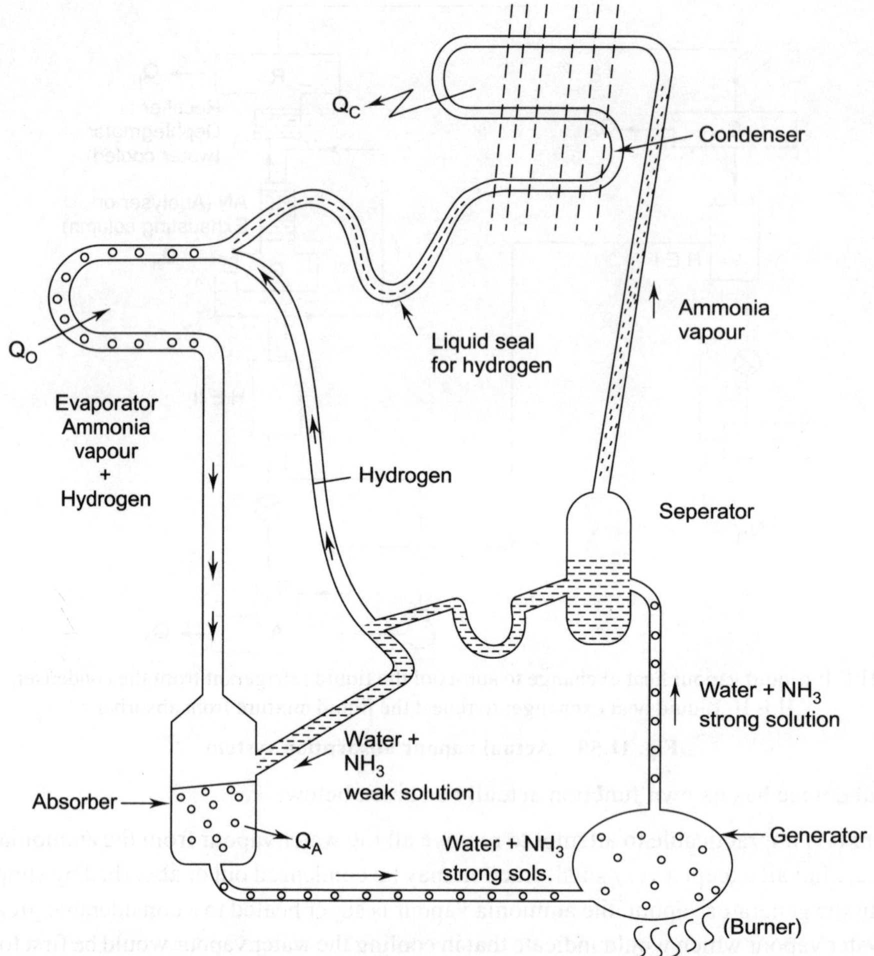


**Fig. 11.59  Electrolux refrigerator (Three fluid absorption system)**

The ammonia vapour in the condenser is condensed to liquid and flows to evaporator by gravity. The whole plant is charged to a pressure of about 14 bar in the evaporation the liquid ammonia meets an atmosphere of hydrogen of 12 bar. The partial pressure of ammonia falls by about 3 bar keeping the same total pressure and temperature falls about $-10°C$. The evaporation of ammonia produces refrigeration. It was developed by Munters and Baltzervon (Swedish engineers) at their undergraduate at Royal Institute of Technology, Strokholm in 1925 as commercially fabricated by Electrolux Company.

### 11.11.4  Thermodynamic analysis of the vapour absorption cycle

The thermodynamics analysis of the vapour absorption cycle is based on the following three equations which can be applied to any part of the system.

1.  Mass balance:

$$\sum m = 0 \tag{11.40}$$

2. Material balance (or partial mass balance)

$$\sum m_c = 0 \tag{11.41}$$

3. Energy balance:

$$\sum Q + \sum mh = 0 \tag{11.42}$$

Thus if mass handled by the pump is $F$ and the refrigerant vapour distilled is $D$ then by mass balance, the week solution returning to the absorber is $(F\text{-}D)$.

Also applying material balance to the generator-analyzer-dephlegmator, taking the refrigerant, say ammonia, under consideration, we have

$$FC_r' = (F-D)\,C_a' + DC_e'\ \text{In which}\ C_r' = C_{1a}\ C_a' = C_2$$

and
$$C_e'' = C_7$$

$$\therefore \qquad F\left(C_r' - C_a'\right) = D\left(C_e'' - C_a'\right) \tag{11.43}$$

$$\frac{F}{D} = f = \frac{C_e'' - C_a'}{C_r' - C_a'} \tag{11.44}$$

Also when rectifier is not used

$$f = \frac{C_e'' - C_a'}{C_r' - C_a'}$$

Where $f$ represents the quantity of rich solution handled by the pump per unit mass of the vapour distilled. This quality is termed as the specific rich solution circulation. The specific weak solution circulation is similarly given by

$$(f-1) = \frac{F - D}{F} = \frac{C_e'' - C_r'}{C_r' - C_a'} \tag{11.45}$$

## 11.11.5  Advantages of vapour absorption system over vapour compression system

1. As there is no moving part in the system, the operation is quite. Hence maintenance cost is low. For more capacity, motor power of pump is low than that of compressor.
2. Absorption system may be designed to use any readily vailable source of energy such that hot exhaust of engine or turbine, solar energy. This system can be used where electricity is not available.
3. Absorption system can be built in capacities well above 1000 tons.
4. At the same temperature of reduced load, the absorption system is equally efficient as at full load.
5. The steam which is bled from the turbine is used for winter heating, the same steam can be used for refrigeration in summer supplying to the generator.

**Comparison of vapour compression and vapour absorption system**

| S. No. | Vapour compression system | S. No. | Vapour absorption system |
|---|---|---|---|
| 1. | A high grade energy is supplied in the form of mechanical energy. | 1. | A low grade energy is supplied in the form of heat to the generator. |
| 2. | The system has more wear and tear, and noise due to moving parts of the compressor. | 2. | The system is very quite as it has minimum number of moving parts in only pump. |
| 3. | The system has poor performance at practical loads as energy supplied is low (Generally $\frac{1}{3}$ to $\frac{1}{4}$ of refrigerating effect). | 3. | The performance of the system is not affected by load variation. Energy supplied is about $1\frac{1}{2}$ times the refrigerating effect. |
| 4. | There are more chances of leakage of refrigerant from the system. | 4. | There is no compressor or any reciprocating component to cause leakage. |
| 5. | Liquid refrigerant traces from the evaporator to the compressor may damage the compressor | 5. | Liquid refrigerant traces from the evaporator constitute no danger. |
| 6. | C.O.P is higher than vapour absorption system. | 6. | C.O.P is lower than vapour compression system. |
| 7. | Charging of refrigerant is quite simple. | 7. | Charging of refrigerant is difficult. |
| 8. | Equipment less precise. | 8. | More precise equipment needed. |

## 11.12 REFRIGERANTS

**Definition:** A refrigerant is a medium of absorbing heat through phase change such as evaporation at lower temperature and pressure, of course with some exceptions where sensible energy transfer occurs.

Classifications:

1. Based on working principle: Under this heading the refrigerants are categorized as (1) Primary refrigerants and (2) Secondary refrigerants.

   The former passes through compression, cooling or condensation, expansion and evaporation or warming up during cyclic process. In case of phase change media such as ammonia, $R$-12, $R$-22, $CO_2$, $SO_2$ etc. Heat transfer is associated with phase change when suitable energy transfer takes place. A medium which does not undergo the cyclic process in refrigeration system, but is used only as a medium of heat transfer is often called a secondary refrigerant like water, brine solution of Nacl, Cacl$_2$ etc.

2. **Based on safety:** From safety considerations they come under three categories.

   (a) Safety refrigerant: They are non – toxics, non – flammable such as $R$-11, $R$-12, $R$-13, $R$-14, $R$-21, $R$-22, $R$-113, $R$-114, Methyle chloride, carbon dioxide and water.

   (b) Toxic and some what flammable refrigerant-They are: dichloro ethylene, methyl formaldehyde, sulphurdioxide, ammonia etc.

   (c) Very flammable refrigerant: They are butane, isobutane, propane, ethane, methane or ethylene.

3. Based on chemical composition. The refrigerant are grouped under the following categories.

   (i) Halo carbon compounds: They are obtained after replacing one or more hydrogen atoms in a hydrocarbon ethane or methane with halogens (chlorine, bromine or fluorine)

(ii)  Cyclic organic compound has been grouped in this class such as dichloro hexafluro cyclo butane.

(iii)  Azeotropes are mixtures of two or more refrigerants.

(iv)  Miscellaneous:

(v)  Oxygen and nitrogen compound

(vi)  Inorganic compound into group.

(a)  Cryogenic (in the range of 113 to 0 $K$) and $b$) non cryogenic (above 113 $K$)

## 11.12.1  Designation of refrigerant

The international designation of refrigerant uses Refrigerant or $R$ as the designation followed by certain numerals. Thus for a compound derived from a saturated hydro carbon denoted by the chemical formula

Complete designation is $R_{(m-1)(n+1)}P$

Hence for dichro tetra fluoro ethane ($CCl_2 F_4$)

$m = 2$, $n = 0$, $p = 4$, $q = 2$ we have $R_{114}$. In this manner

1. Dichlorodifluoro methane ($CCl_2 F_2$)-$R$-12

Monochloro difluoro methane ($CCl F_2$)-$R$-22

Trichloro monofluro methne ($CCl_3 F$)-$R$-11

The brominated refrigerant are denoted by putting an additional $B$ and a number to denote as how many chlorine atoms are replaced by bromine atoms. Thus $R$-13 $B_1$ is derived from $R$-13 with replacement of one chlorine atom by bromine atom $CF_3$ Br.

In case of common inorganic refrigerant, numerical designation have been given according to their molecular weight added to 700. Thus ammonia whose molecular weight is 17 is distinguished as $R$-717. Similarly water is designated as $R$-718 and carbon dioxide as $R$-744.

## 11.12.2  Selection of refrigerant

There is no working substances which could be called an ideal refrigerant. Different substances seem to satisfy different requirements and some times only partially. A refrigerant which is ideally suited in a particular application may be complete failure in the other. In general, a refrigerant may be required to satisfy requirement which may be classified as thermodynamic chemical and physical.

(a)  Primary: Those refrigerants that cool by taking latent heat: they directly take part in refrigeration. Primary refrigerant are $NH_3$, $CO_2$, $SO_2$, $CH_3Cl$, Freon.

(b)  Secondary: They absorb only sensible heat. Secondary refrigerants are air, water, brine.

## 11.12.3  Desirable properties of an ideal refrigerant

(a)  Thermodynamics properties.

(b)  Physical properties.

(c)  Safe working properties.

## Thermodynamics properties

1. Boiling point: low boiling point at atmosphere pressure for efficient refrigerants.

2. Freezing point: Very low freezing point is necessary for very low temperature in evaporator.

3. Evaporation and condenser pressure is always desirable to have positive pressure in evaporation and condenser, but the pressure should not be too high above atmosphere.

4. Critical temperature and pressure: The critical temperature should be higher than occurring in the condenser for easy condensation of the refrigerant vapour.

5. Latent heat of refrigerant: Latent heat of refrigerant at evaporator temperature is desirable because the refrigerating effect per kg of refrigerant will be high. Weight of refrigerant per ton will be reduced.

## Physical properties

1, Specific volume: Low specific volume of refrigerant at the solution into the compressor is always desirable because it reduces the size of the compressor.

2. Specific heat of liquid and vapour: Low specific heat of liquid refrigerant and high specific heat of vapour refrigerant are desirable because both tend to increase the refrigerating effect. Low specific heat of liquid tend to increase the sub cooling and high specific heat of vapour decreases the temperature of superheated vapour.

3. Thermal conductivity: High conductivities of both liquid and vapour states are desirable for high rate of heat transfer in evaporator and condenser.

4. Viscosity: Low viscosity of refrigerant are desirable for better heat transfer and low pumping cost.

5. Dielectric strength: The dielectric strength of a refrigerant is primarily important in hermetically sealed units in which electric motor is exposed to refrigerant.

## Safe working properties

1. Nontoxic
2. Low flammability
3. Less corrosive property
4. Chemical stability
5. No effect of stored product

## Other properties

1. Odour
2. Leak tendency
3. Cost and availability

## 11.12.4  Alternative refrigerants for refrigerators and air-conditioners

Chlorine based chemicals like chlorofluorocarbons (CFCS) and Hydro chlorofluorocarbons (HCFs) are being extensively used in refrigerators and air-conditioners as working refrigerants as well as blowing agents in foam. They possess most of the desirable characteristics, such as thermal and chemical stability, thermodynamic suitability, non-toxicity, non-flammability, material compatibility, low cost etc.

It is well known that the stratospheric ozone layer (at an altitude of 15 to 40 km) acts as a protective umbrella for the earth against the harmful ultra violete (*UV*) radiation from the sun. Recent studies have indicated that there has been significant depletion of the stratospheric ozone layer due to chain reaction between CFCs and ozone as given below:

$$CCl_2 F_2 \rightarrow CCl F_2 + Cl \tag{11.46}$$

$$Cl + O_3 \rightarrow ClO + O_2 \tag{11.47}$$

$$O + ClO \rightarrow Cl + O_2 \tag{11.48}$$

$$\overline{O + O_3 \rightarrow 2O_2} \tag{11.49}$$

Ozone is formed by the interaction of the UV radiation from the sun with oxygen ($O_2$). In this way $O_2$ and $O_3$ remain in photochemical equilibrium absorbing the *UV* rays. CFCs reach the stratosphere though the process of diffusion and are attacked by *UV* radiation. In this process chlorine gets liberated from the otherwise stable CFCs. This free Cl converts ozone into oxygen by shifting the equilibrium towards $O_2$ formation, and therefore depleting ozone layer. The total consequence of the stratospheric ozone layer depletion can be some health hazards on human and living creatures and other ecological environmental problems such as global warming, sea level rise, increased floods, drought and storms etc.

## Inter relationship of Montreal Protocols and Kyoto protocols

The Montreal protocol (*MP*) on substances that deplete the ozone layer was established in Montreal on 16th September 1987. A decision was taken to curtaile the production and phase out the use of CFCs on a global scale and has been signal by more-than hundred countries. India has becomes a party to the MP in 1992. As per MP ozone depletion substances (ODSs) have been phased out in developed countries by 1996 and have to be phased out in the developing countries by 2010. The protocol has been adjusted and amended in London, Copenhagen, Vienna and Montreal (1997).

The Kyoto protocol (*KP*) adopted at the 3rd conference on the framework convention of the Global climate change in Kyoto in 1997 has decided to put HFCs together with other five gases such as $CO_2$, $N_2O$, $CH_4$, PFCs and $SF_6$ in one basket of controlled substances. Although CFCs, HCFCs contribute to Global Warming (GW) the KP does not address these substances, since these are already controlled under Montreal protocol with specific phase out regimes.

The Montreal and Kyoto protocols are interconnected because of HFCs, which are alternatives to the ODS have been included in the basket of GHGS of Kyto protocol due to their high GWP. The MP adopted HFCs as viable alternatives primarily because of their zero ODP and very similar thermodynamic and thermo- physical characteristics as those of CFCs, HCFCs and other ODSs used in different applications. The ODS used as refrigerants are given in table 11.4.

### Table 11.4　Ozone depleting substances used as refrigerants

| Name | Formula | | ODP | GWP |
|---|---|---|---|---|
| CFC-11 | Trichlorofluoro methane | $CCl_3 F$ | 1.0 | 3800 |
| CFC-12 | Dichlorodifluoro methane | $CCl_2 F_2$ | 1.0 | 8100 |
| CFC-13 | Chlorofluoro methane | $CCl F_3$ | 1.0 | 11700 |
| CFC-14 | Tetrafluoro methane | $CCl F_2 CCl F_2$ | 1.0 | 11700 |
| HCFC-22 | Chlorodifluoro methane | $CHCl F_2$ | 0.05 | 1500 |

## Alternative refrigerants to CFC-12

In developing countries, CFC-12 is still used extensively in domestic and commercial refrigeration appliances like refrigerators, bottle coolers, water coolers etc. using small hermatic reciprocating compressors and mobile and transport refrigeration using open compressors. The Table 11.5 gives some of the alternatives refrigerants which have been extensively studied.

## Table 11.5    Alterative refrigerants to CFC-12

| Refrigerant | CFC-12 | HFC-152a | HFC-134a | MP66/39 | HC-290/600a | HC-600a Isobutane | HC-290 Propane |
|---|---|---|---|---|---|---|---|
| Formula | $CF_2Cl_2$ | $CH_3CHF_2$ | $CH_2FCF_3$ | HCFC/HFC blend | $C_3H_8$-$C_4H_{10}$ | $C_4H_{10}$ | $C_3H_6$ |
| Molecular weight | 120.93 | 66.05 | 102.03 | (dependent) | 51.12 | 58.13 | 44.1 |
| Critical temperature °C | 112.0 | 113.5 | 101.1 | 96.0 | 96.0 | 135.0 | 96.8 |
| Boil point (100 kPa) °C | – 29.8 | – 25.0 | – 26.16 | – 30.0 | – 30.0 | – 117.73 | – 42.1 |
| Density (kg/m³, –25°C) Sat. vapour Sat. liquid | 7.57 1472.0 | 3.25 1013.0 | 5.50 1371.0 | – – | 3.14 584.4 | 1.66 608.3 | 2.42 580.7 |
| Flammable limits (%in air) 25°C, 100 Kpa | None | 37 – 18.0 | None | None | 1.8 – 90 | 1.4 – 8.4 | 2.2 – 9.5 |
| TLV/OEL | 1000 | 1000 | 1000 | 1000 | S.a | 1000 | S.a |
| ODP (ozone depletion potential) | 1.0 | 0.0 | 0.0 | 0.036 | 0.0 | 0.0 | 0.0 |
| GWP (Global warming potential) | 3.1 | 0.03 | 0.27 | 0.25 | < 0.01 | < 0.01 | < 0.01 |

Amongst all HFCs with zero ODP value, HFC-134a (Tetra fluoro ethane) is considered to be the most eco-friendly potential substitute for CFC-12 according to many latest comparative thermodynamic and practical assessments. The initial apprehension on the energy penalty has been compressors with better efficiencies than CFC-12 based compressors. HFC-134a has a relatively high GWP. Several international chemical companies have invested in facilities to manufacture this non-flammable chemical. All major compressor manufactures are offering models for use with HFC-134a.

The volumetric capacity of HFC-134a is 9% below CFC-12 at the standard refrigerant rating conditions (– 25°C evaporator, 55°C condenser) used by compressor manufactures for performance measurements on calorimeters. The capacity loss is offset by increasing the displacement of the compressor. Efficiencies of compressors with the displacement change have been measured to be equivalent to CFC-12 compressors using similar viscosity oils. Recent energy consumption tests in refrigerator-freezers have shown that HFC-134a is approximately equal in performance to CFC-12 in optimized units. HFC-134a is benigh and shown to be compatible in sealed tube tests in the presence of metals, oil plastics and elastomers common refrigeration systems.

HFC-134a is not miscible with the naphthenic and alkaline benzene oils historically used with CFC-12. It demands a very stringent quality control. The polyol Ester (POE) based synthetic lubricant should be as dry as possible with moisture tolerance as low as 40 ppm. Therefore, special effort would be required to maintain as the moisture absorbing capacity of POE oils hundred times greater. Servicing of HFC-134a based refrigerators will also require an utmost care to prevent any moisture ingression into the system. HFC-134a cannot be considered a drop-in replacement for CFC-12, substantive modifications are required like (i) the compressor and system component design *e.g.* heat exchangers, lubricant, filter, dryer etc. (ii) use of process chemicals and lubricants used in the compressor manufacturers (iii) more restrictive capillary tubes, less refrigerant charge with new filter dryers (XH-9). Problems from capillary tube plugging due to sluge generation have been encountered by several manufacturers during developmental endurance testing of refrigerators containing HFC-134a. Both paraffinic motor winding insulation and incompatible fabrication process fluids have been identified as root causes for these problems. Observed problems have been resolved, but the application of HFC-134a is clearly more sensitive to contamination by foreign materials compared to CFC-12. Disciplined manufacturing cleanliness and control are essential for successful application of HFC-134a.

However, HFC-134a is the only alternative being used in mobile ACS all over the world, as there is no other well tested alternative to CFC-12. Most of all new generation cars made in India use HFC-134a.

# 11.13   AIR CONDITIONING

The science of air conditioning deals with supplying and maintaining a desirable internal atmospheric conditions irrespective of external conditions. The four important factors are involved in a complete air conditioning installation are:

1. Temperature control
2. Humidity control
3. Air movement and circulation
4. Air filtering, cleaning and purification.

The simultaneous control of these four factors within required limits when directed towards human comfort and health or when industrially directed towards conditions permitting the best product yield during manufacturing and storage can rightly be called air conditioning.

### 11.13.1   Classification of air conditioning system

The air conditioning systems are classified in several ways as discussed below:

1. classification as to major function
    (a) Comfort air conditioning system: The essential feature of comfort air conditioning system is to provide an environment which is comfortable to majority of the occupants. For example air conditioning system used in homes, offices, shops, restaurant, theatres, hospitals and schools etc. Comfort air conditioning may be classified into three groups.
        (i) Summer air conditioning: These systems control all the four atmosphere conditions for summer comfort. The major problems are to cool the air and to remove moisture front it. Cooling is accomplished by mechanical refrigeration. Removal of moisture (dehumidification) is accomplished as condensation of water vapour on cooling coil surfaces.

(ii) Winter air conditioning: The major problems of winter air conditioning are to heat and bring moisture content up to an acceptable level. Heating is accomplished by electrical heater or furnaces and boilers fired by gas, oil or coal. Humidifier may be of the simple pan type or spray type.

(iii) Year round air conditioning: This system assures the control of temperature and humidity of air in an enclosed space through the year when the atmospheric conditions are changing as per season.

(b) Industrial air conditioning: The purpose of these air conditioning system is to control atmospheric conditions primarily for the proper conduct of research and manufacturing process. Examples are air conditioning system used in textile mills, paper mills, machine part manufacturing plant, cool rooms, printing and photo processing plants.

(c) Classification as per equipment arrangement.

(i) Unitary system make use of air conditioners which are completely factor assembled. A single air conditioner may serve if building is a small one or the area may be divided into several zones, each zone being served by a conditioner of small to medium capacity. This type of system has the advantage of moderate cost and also that of flexibility of operation.

(ii) Central station system: When several rooms in the same building are intended for use which require air having approximately the same temperature and R.H, they can usually be air conditioned more economically from a control system than from a number of self conditioned units.

(iii) Combination systems: This type of system combines the feature of central station and unitary system. Heat energy is supplied in pipes to several units of air conditioner in the form of steam or hot water chilled water from the central refrigerating equipment is also piped to the air conditioner.

## 11.14  PSYCHROMETRIC TERMS

The study of atmospheric air, which is a mixture of dry air and water vapour, is important in the science of meterology and analysis of air conditioning systems and cooling towers. The science which deals with the behavior of moist air is given the name psychrometry or hygrometry.

Normally the atmosphere is not saturated with water vapour. Below saturation the vapour exists in the mixture as superheated vapour is at a very low pressure. At such low pressures (well below one atmosphere) the vapour behaves as a perfect gas and the properties of dry air-water vapour mixture (i.e. moist air) can be found using the Gi bb's-Dalton law as in the case of mixture of ideal gases.

Several special terms are used in psychrometry, which are defined below.

1. Moist air: The ordinary atmospheric air is the moist air. The atmospheric air is never dry; it contains a variable amount of water vapour. The mass of water vapour may be 1% to 3% of the mass of moist air. The maximum quantity of water vapour that can be present in air depends upon the temperature. The maximum quality of water vapour present in air at particular air temperature is known as the saturation capacity of air.

Though "dry air" does not exist and it is an useful concept which simplifies psychrometric calculations. The term dry air is used to denote air without water vapour whereas the terms "air" and "moist air" are used to denote natural mixture of dry air and water.

The composition of the atmospheric air excluding the variable water vapour is very nearly constant. If the molecular ratios are assumed to be fixed, the constituents of moist air other than the water vapour can be assumed to act as a single gas:

The moisture in the atmosphere is in superheated state and behaves as a perfect gas. The molecular weight of water vapour of moist air is 18.016 (18 for approximate calculations).

2. Dry bulb temperature (*dbt* or $t_{db}$): The temperature of air as registered by an ordinary thermometer is called the dry bulb temperature. It is the actual temperature of the air. It is not affected by the moisture in the air. When measuring the dry bulb temperature of the air, the bulb of the thermometer should be shaded to reduce the effects of direct radiation.

3. Dew point temperature (*dpt* or $t_{dp}$): It is defined as the temperature to which unsaturated air must be cooled to become saturated. The dew point of air is the saturation temperature corresponding to the partial pressure exerted by the water vapour. For any given volume of air the dew point temperature depends only on the mass of water vapour in the air. At dew point the condensation of moisture begins.

Consider atmospheric air at 1.01325 bar and temperature 30°C. The saturation pressure of water vapour corresponding to 30°C is 0.04246 bar. As the atmospheric air is generally far from saturation, the vapour pressure of water will be less than 0.0426 bar. Assume that the vapour pressure is 0.02339 bar. By Dalton's law of partial pressure

$$P = P_a + P_w \qquad (11.50)$$

where $\qquad\qquad P_w$ = partial pressure of water vapour

and $\qquad\qquad P_a$ = partial pressure of dry air

*i.e.* $\qquad\qquad P_a = P - P_W$

$$= 1.01325 - 0.02339 = 0.98986 \text{ bar.}$$

The saturation temperature corresponding to 0.02339 bar (water vapour partial pressure) is 20°C, hence the vapour in atmospheric air under these conditions has a superheat of (30 − 20) = 10°C. This state is indicated by point 1 on *T-S* diagram of Fig. 11.60. If this air is how cooled at constant pressure, condensation will start at point 2, called the dew point temperature. The dew point temperature corresponds to the saturation temperature of steam at the partial pressure of water vapour in the original mixture. Thus the dew point temperature of the air considered is 20°C.

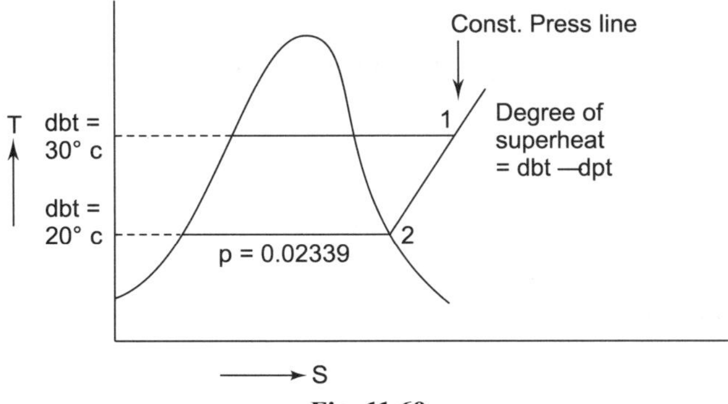


**Fig. 11.60**

For a given pressure, $t_{dp}$ depends only on the humidity ratio.

As long as the amount of water vapour in the air remains unchanged the dew point temperature of air remains same. However, if the amount of water vapour is increased, the partial pressure of water vapour will increase which will increase the dew point temperature; and vice-versa if the amount of water vapour is decreased, the partial pressure of water vapour will decrease which will cause decrease in the dew point temperature.

4.  Wet bulb temperature $(t_{wb})$: The temperature of air measured by a thermometer when its bulb is covered with wet cloth and is exposed to a current of air is known as wet bulb temperature. When the current of surrounding air passes over the wet cloth (silk wick) some of the water evaporates producing the cooling effect at the bulb. In this process heat is transferred from the air to the wick, evaporating water, and the equilibrium condition between the heat and mass transfer rates to and from the wetted sleeve is reached at which the wet bulb temperature is indicated. It is lower than the dry bulb temperature. Normally the air velocity around the wick is kept 5 to 10 m/s and the thermometer (set in sling psychrometer) is whirled for about one minute.

Wet bulb temperature depends on both the dry bulb temperature and relative humidity. The difference between the dry bulb temperature and wet bulb temperature is known as wet bulb depression (WBD) and it depends on the relative humidity of air. If the relative humidity is high, the rate of evaporation of water at the wick is low and hence the wet bulb depression is low. Similarly, if the relative humidity is low the rate of evaporation of water at the wick is high, and the wet bulb depression will be high. When the relative humidity of atmosphere equals to unity then there will be no evaporation of water (due to saturated condition of water vapour in air) and hence the dry bulb and wet bulb temperatures are same, which means that the wet bulb depression is zero. In this case dew point temperature is also equal to dry bulb temperature and wet bulb temperature.

Wet bulb temperature is an indirect measure of the moisture content of air. It is not a thermodynamic property due to various inaccuracies in the measurement and hence it can not be expressed in terms of $p_w$.

The following are some of the empirical relations for finding the value of partial pressure of water vapour for given dry bulb and wet bulb temperatures

Modified Apjohn's equation

$$p_w = p_{w,\,wbt} - (t_{db} - t_{wb})/1500 \tag{11.51}$$

Modified Ferrel's equation

$$p_w = (p_w)_{wbt} - 0.00066\, p_b\, (t_{db} - t_{wb}) \left[ 1 + \frac{t_{db}}{872.8} \right] \tag{11.52}$$

and the Carriers's equation

$$p_w = (p_w)_{wbt} - \frac{\left( p_d - p_{w,\,wbt} \right)\left( t_{db} - t_{wb} \right)}{\left( 15447 - 1.44\, t_{wb} \right)} \tag{11.53}$$

Here pressures are in bar

And
$$p_w = (p_w)wbt - \frac{\left[\left(p_b - p_{w,wbt}\right)\right]\left(t_{db} - t_{wb}\right) \times 1.8}{\left[2800 - 1.3\left(1.8\,tab + 32\right)\right]} \tag{11.54}$$

Here pressures are in mm of Hg

5. Saturated air: Saturated air is the moist air in which the partial pressure of the water vapour equals the saturation pressure of steam corresponding to the temperature of air. The objective 'saturated' in the saturated air does not refer to air but it refers to the water vapour which is dry and saturated. If saturated air is cooled, moisture separates and makes air misty or foggy. The dry bulb temperature of saturated air will be same as dew point temperature, and equals wet bulb temperature.

   Note that higher the air temperature more amount of water-vapour it can hold upto its saturation limit and higher will be the vapour pressure.

6. Absolute humidity: The mass of water vapour present in unit volume of dry air is known as absolute humidity. $\beta$ is expressed in gm/m$^3$ of dry air.

7. Specific humidity or humidity ratio ($\omega$): The water vapour in the air is called humidity. The specific humidity (or humidity ratio) or moist air is defined as the ratio of the mass of water vapour to the mass of dry air in a given volume of air water vapour mixture at same temperature and it is denoted by $w$. It is given in gram per kg of dry air.

   Let in a given mixture of volume v,
   $$m_a = \text{mass of dry air}$$
   $$m_w = \text{mass of water vapour}$$
   $$(\text{Total mass of mixture}) = (m_a + m_w)$$

then,
$$\omega = \frac{\text{mass of water vapour}}{\text{mass of dry air}} = \frac{m_w}{m_a} \tag{11.55}$$

It may be noted that the definition is given in terms of mass of dry air ma, instead of total mass of mixture, $(m_a + m_w)$, as the mass of the mixture of a given volume of air changes with humidity. Saturated air can also be defined as the condition of moist air that can exist in neutral equilibrium with a condensed phase presenting a flat face to it.

The masses $m_a$ and $m_w$ are given by the expression as
$$m_a = \frac{p_a V}{R_a T} \tag{11.56}$$

where $p_a$ is the partial pressure of dry air and $V$ is the volume of mixture and $R_a$ is gas constant for dry air

$$m_w = \frac{p_w V}{R_w T} \tag{11.57}$$

where $p_w$ = partial pressure water vapour and $R_w$ is gas constant for water vapour.

$\therefore$
$$\omega = \frac{p_w V}{R_w T} \cdot \frac{p_a V}{R_a T} = \frac{R_a}{R_w} \cdot \frac{p_w}{p_a} \tag{11.58}$$

But
$$R_a = \frac{\bar{R}}{M_a} \quad \text{and} \quad R_w = \frac{\bar{R}}{M_w}$$

where
$$\bar{R} = \text{universal gas const.}$$
$$M_a = \text{molecular weight of air}$$
$$M_w = \text{molecular weight of water vapour}$$

$$\therefore \quad \omega = \frac{M_w}{M_a} \times \frac{p_w}{p_a} = \frac{18}{29} \cdot \frac{p_w}{p_a} = 0.622 \frac{p_w}{p_a}$$

or,
$$\omega = 0.622 \frac{p_w}{p_b - p_w} \tag{11.59}$$

The masses of air and water vapour in terms of specific volumes are given by expressions as

$$m_a = \frac{V}{v_a} \text{ and } m_w = \frac{V}{v_w}$$

where $v_a$ and $v_w$ are the specific volumes of dry air and water vapour at the given mixture temperature and respective partial pressures.

$$\omega = \frac{v_a}{v_w} \tag{11.60}$$

8. **Degree of saturation ($\mu$):** The degree of saturation is defined as the ratio of mass of water vapour associated with unit mass of dry air to mass of water vapour associated with unit mass of dry air saturated at the same temperature and pressure, *i.e.* it is a ratio of specific humidity of moist air to the specific humidity of the saturated air at the same temperature. The degree of saturation is denoted by $\mu$.

$$\mu = \frac{\omega}{\omega_s} = \frac{\text{mass of water vapour associated with unit mass of dry air}}{\text{mass of water vapour associated with saturated unit mass of dry air at same temp. and pressure}}$$

$$\therefore \quad \mu = \frac{0.622\left(\dfrac{p_w}{p_b - p_w}\right)}{0.622\left(\dfrac{p_{ws}}{p_b - p_{ws}}\right)} = \frac{p_w}{p_{ws}}\left\{\dfrac{1 - \dfrac{p_{ws}}{p_b}}{1 - \dfrac{p_w}{p_b}}\right\} \tag{11.61}$$

where $p_{ws}$ is the partial pressure of water vapour when the air is fully saturated at the same temperature of air. This can be calculated from steam tables corresponding to the dry-bulb temperature of the air.

9. **Relative humidity ($\phi$):** Relative humidity is the term most widely used in air conditioning though thermodynamically it is not as directly related to the basic properties of the mixture as is the degree of saturation. The relative humidity of air is defined as the ratio of mole fraction of water vapour in moist air to the mole fraction of water vapour saturated at the same temperature and pressure.

$$\phi = \frac{\text{mass of water vapour in a given volume}}{\text{mass of water vapour in some volume if air is Saturated at the same temperature}}$$

$$= \frac{m_w}{m_{ws}} = \frac{p_w V/R_w T}{p_{ws} V/R_{ws} T} = \frac{p_w}{p_{ws}} \quad (\text{Assuming } R_w = R_{ws}) \tag{11.62}$$

The relative humidity can be defined as the ratio of the partial pressure of water vapour in a given volume of mixture to the partial pressure of water vapour when the same volume of mixture is saturated at the same temperature.

$$\mu = \frac{\omega}{\omega_s} = \phi \left[ \frac{1 - \dfrac{p_{ws}}{p_b}}{1 - \dfrac{p_w}{p_b}} \right] = \phi \left[ \frac{1 - \dfrac{p_{ws}}{p_b}}{1 - \dfrac{\phi p_{ws}}{p_b}} \right] = \phi \left[ \frac{p_b - p_{ws}}{p_b - \phi p_{ws}} \right] \tag{11.63}$$

$$\therefore \qquad \phi(p_b - p_{ws}) = \mu \cdot p_b - \mu \phi p_{ws}$$

$$\therefore \qquad \phi[p_b - p_{ws} + \mu p_{ws}] = \mu p_b$$

$$\therefore \qquad \phi = \frac{\mu p_b}{p_b - p_{ws} + \mu p_{ws}} = \frac{\mu}{1 - (1 - \mu)\dfrac{p_{ws}}{p_b}}$$

As $\qquad\qquad p_{ws} << p_b$

$$\therefore \qquad \phi \simeq \mu \tag{11.64}$$

At low temperatures, where the vapour pressure of water $P_{ws}$ is small compared to atmospheric pressure the numerical difference between f and $\mu$ is not great, but at temperatures over about 15°C it becomes appreciable. Of course, from their definitions, the values of two are identical for saturated air 1.0 and for dry air (zero), regardless of temperature.

In air conditioning practice, the percentage difference between relative humidity and degree of saturation is in the range 0.5% to 2% approximately.

The relative humidity $\phi$ can also be put in terms of $\mu$, $\omega$ and $\omega_s$

$$\phi = \frac{x_w}{x_{ws}} \qquad\qquad (x \text{ is the mole fraction})$$

$$= \frac{1 + \dfrac{0.622}{\omega_s}}{1 + \dfrac{0.622}{\omega}} = \mu \, \frac{0.622 + \omega_s}{0.622 + \omega} \tag{11.65}$$

## 11.15  PSYCHROMETRIC CHART

A chart which shows the inter relation of all the important properties as discussed above is known as psychrometric chart. This chart is constructed for a particular value of barometric pressure generally at standard sea level pressure i.e. 760 mm of Hg or 1.01325 bar. The pressure being fixed, specific humidity $\omega$ and dry bulb temperature t can be regarded as the only independent variables. The dry bulb temperature is taken along abscissa and specific humidity as ordinate to the right hand side of the chart. The chart is a representation of dry bulb temperature verses specific humidity and all other properties are shown by different lines on the chart shown in Fig. 11.61.

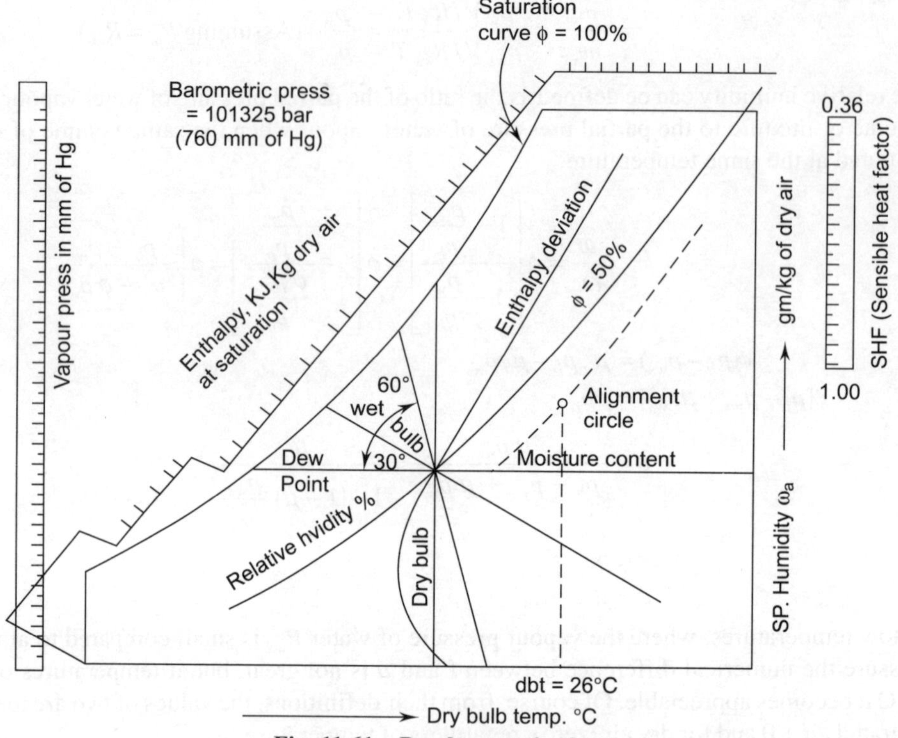



**Fig. 11.61    Psychrometric chart**

1. Sensible heat factor (SHF): The thermal properties of air can be separated into latent heat and sensible heat. The term sensible heat factor is defined as the ratio of sensible heat to the total heat.

$$\text{SHF} = \frac{\text{sensible heat}}{\text{total heat}} = \frac{\text{sensible heat}}{\text{sensible heat} + \text{latent heat}}$$

2. Enthalpy lines: Enthalpy of moist air per unit mass of dry air is given by

$$h = h_a + \omega h_w \tag{11.66}$$

$h_a$ depends on the dry bulb temperature of the air, $\omega h_w$ depends upon the specific humidity and pressure and temperature of the air. Hence lines of constant enthalpy can be drawn on psychrometric chart.

3. Volume lines Lines: of constant specific volume of the dry air in the mixture $V_a$ are also drawn on the psychrometric chart making 60° angle from horizontal. $V_a$ is a function of the dry bulb temperature $t$ and the partial pressure of air $(p_t - p_w)$, but $p_w$ is a function of $w$ and $p$ and hence for a given barometric pressure $p_t$, $V_a$ are a function of $t$ and $w$.

4. Relative humidity lines: The relative humidity is given by the expression

$$\phi = \frac{w(p_t - p_w)}{0.622\, p_{ws}} \tag{11.67}$$

$P_{ws}$ is the saturation pressure corresponding to the dry bulb temperature; $p_w$ is a function of specific humidity $w$ for a fixed barometric pressure. Therefore relative humidity $\phi$ is a function of $w$ and t. Hence curves of constant relative humidity are added to psychrometric chart. These curves are from 10 to 100% at an interval of 5%.

5. Wet bulb lines: Since humidity measurements are generally made in terms of wet and dry bulb temperature, lines of constant wet bulb are also included in psychrometric chart. These constant wet bulb lines are inclined at an angle of 30° from horizontal and follow very closely the lines of constant enthalpy.

6. Alignment circle: The chart has alignment circle having value of $\phi = 50\%$ and $t = 26°C$. The chart also has sensible heat factor scale on the right hand side.

---

**Example 11.21**   Obtain all the properties of moist air having DBT $= 36°C$ and RH $= 45\%$

*Sol.*

(i)
$$\phi = \frac{p_w}{p_{ws}}$$

∴
$$p_w = 0.45 \times 0.0597$$
$$= 0.0268 \text{ bar} \quad Ans.$$

∵   corresponding to 36°C, $p_{ws}$ from steam table $p_{ws} = 0.0597$ bar

(ii)  Humidity ratio
$$\omega = 0.622 \frac{0.0268}{(1.01325 - 0.0268)}$$

or,
$$\omega = 0.622 \frac{0.0268}{(1.01325 - 0.0268)}$$
$$= 0.01689 \text{ kg/kg of dry air} \quad Ans.$$

(iii)  Degree of saturation ($\mu$)

$$\mu = \frac{p_w}{p_{ws}} \left[ \frac{\left(1 - \dfrac{p_{ws}}{p_b}\right)}{\left(1 - \dfrac{p_w}{p_b}\right)} \right] = \frac{0.0268}{0.0597} \left[ \frac{\left(1 - \dfrac{0.0597}{1.01325}\right)}{\left(1 - \dfrac{0.0268}{1.01325}\right)} \right]$$

or,
$$\mu = 0.4339 = 43.39\% \quad Ans.$$

(iv)  Volume in m³/kg of dry air
$$(p_b - p_w)V_a = m_a R_a T_a, \quad m_a = 1 \text{ kg dry air}$$

∴
$$V_a = \frac{287 (273 + 36)}{(1.01325 - 0.0268) \times 10^5} \qquad\qquad [\because \quad R_a = 287 \text{ J/kgk}]$$
$$= 0.899 \text{ m}^3/\text{kg of dry air} \quad Ans.$$

(v)  Enthalpy per kg of dry air
$$h = 1.005t + w(2500 + 1.88t) \text{ kJ/kg of dry air}$$
$$= 1.005 \times 36 + 0.01689(2500 + 1.88 \times 36)$$
$$= 79.54 \text{ kg/kg of dry air} \quad Ans.$$

(vii)  Dew point temperature

This corresponds to the saturation temperature as read from steam tables for partial pressure $P_w$ of vapour,

$$p_w = 0.0268 \text{ bar}$$

(checked by chart, 0.k)

$$\therefore \quad t_{dp} = 21.28°C \quad Ans.$$

(viii)  Density

$$\rho_a = \frac{p_a}{R_a T} = \left(\frac{p_b - p_w}{R_a T}\right)$$

$$= \frac{(1.01325 - 0.0268) \times 10^5}{287 \times (36 + 273)}$$

$$= 1.1123 \text{ kg/m}^3$$

$$\rho_w = \frac{p_w}{R_v T} = \frac{0.0268 \times 10^5}{461 \times (36 + 273)}$$

$$= 0.0188 \text{ kg/m}^3$$

$\therefore \quad$ Density of moist air $\quad \rho = \rho_a + \rho_w$

$$= 1.1123 + 0.0188$$

$$= 1.1311 \text{ kg/m}^3 \quad Ans.$$

---

**Example 11.22**   Obtain all the properties of moist air having DBT = 38°C and dew point temperature = 18°C without using psychrometric chart.

*Sol.* Given,

Corresponding to $t_{dp}$, partial pressure of vapour from steam table

$$p_w = 0.0208 \text{ bar}$$

Corresponding to $t_{db}$, partial pressure of vapour (saturated) from steam table

$p_{ws}$ 0.0667 bar

(i)  Relative humidity $\quad \phi = \dfrac{p_w}{p_{ws}} = \dfrac{0.0208}{0.0667} \times 100$

$$= 31.18\% \quad Ans.$$

(ii)  Humidity ratio $\quad \omega = 0.622 \dfrac{p_w}{(p_b - p_w)}$

$$= 0.622 \frac{0.0208}{(1.01325 - 0.0208)}$$

$$= 0.013036 \text{ kg/kg of dry air} \quad Ans.$$

(iii)  Degree of saturation

$$\mu = \frac{p_w}{p_{ws}} \left[\frac{\left(1 - \dfrac{p_{ws}}{p_b}\right)}{\left(1 - \dfrac{p_w}{p_b}\right)}\right] = 0.2973 = 29.73\% \quad Ans.$$

(iv)  Volume in $m^3$/kg of dry air

$$(p_b - p_w)V_a = m_a R_a T_a \quad \because \quad m_a = 1 \text{ kg}$$
$$V_a = 0.9048 \text{ m}^3/\text{kg of dry air} \quad Ans.$$

(v)  Enthalpy per kg of dry air

$$h = 1.005t + w(2500 + 1.88t) \text{ kJ/kg of dry air}$$
$$= 75.71 \text{ kJ/kg of dry air} \quad Ans.$$

**Example 11.23**  A psychrometer reads 35°C DBT and 25°C WBT. The barometer reads 760 mm mercury. Calculate the relative humidity, specific humidity and dew point temperature of air. If the air is cooled to 15°C, what will be the amount of water vapour condensed?

*Sol.*

Given $\qquad$ DBT $= t_{db} = 35$°C, WBT $= t_{wb} = 25$°C, $p_b = 760$ mm of Hg $= 1.01325$ bar.

Corresponding to $t_{db}$, partial pressure of saturated vapour from steam table.

$$p_{ws} = 0.0563 \text{ bar}$$

and corresponding value at $t_{wb}$ i.e.

$$p_{ws,\,wbt} = 0.0317 \text{ bar}$$

Using the Carrier equation, partial pressure of water vapour

$$p_w = p_{w,\,wbt} - \frac{\left(p_b - p_{w,\,wbt}\right)\left(t_{ab} - t_{wb}\right)}{1547 - 1.44\, t_{wb}}$$

$$= 0.0317 - \frac{\left(1.01325 - 0.0317\right)\left(35 - 25\right)}{1547 - 1.44 \times 25}$$

$$= 0.0252 \text{ bar}$$

Corresponding to this pressure, temperature is called dew point temp. *i.e.* $t_{dp} = 21.1$°C  *Ans.*

Relative humidity, $\qquad$ $\phi = \dfrac{p_w}{p_{ws}} = \dfrac{0.0252}{0.0563} = 0.4476 = 44.76\%$  *Ans.*

Specific humidity, $\qquad$ $\omega = 0.622\, \dfrac{p_w}{p_b - p_w}$

$$= 0.622\, \frac{0.0252}{1.01325 - 0.052}$$

$$= 0.0157 \text{ kg/kg of dry air}$$
$$= 15.79 \text{ gm/kg of dry air} \quad Ans.$$

density of air

$$\rho_a = \frac{p_a}{R_a T} = \frac{\left(1.01325 - 0.0252\right) \times 10^5}{287\,\left(35 + 273\right)} = 0.11187 \text{ kg/m}^3$$

density of water vapour

$$\rho_w = \frac{p_w}{R_w T} = \frac{0.0252 \times 10^5}{461.88 \,(35 + 273)} \quad \left[ \because \quad R_w = \frac{\bar{R}}{M} = \frac{8314}{18} = 461.88 \, \text{J/kgk} \right]$$

$$= 0.0176 \, \text{kg/m}^3$$

Density of mixture
$$= \rho_a + \rho_w$$
$$= 0.11187 + 0.0176 = 0.12947 \, \text{kg/m}^3$$

Now, if air is cooled to 15°C of DBT at same value of relative humidity ($\phi$).

$$\therefore \qquad \phi' = \frac{p_{w'}}{p_{ws'}}, p_{ws'} \text{ at } 15°C = 0.01704 \, \text{bar}$$

$$\therefore \qquad p_{w'} = 0.4464 \times 0.01704 = 0.0076 \, \text{bar}$$

New specific humidity

$$\omega' = 0.622 \, \frac{p_{w'}}{p_b - p_{w'}} = 0.622 \, \frac{0.00760}{(1.01325 - 0.00760)}$$

$$= 0.0047 \, \text{kg/kg of dry air}$$

$\therefore$ amount of water vapour removed

$$= (\omega - \omega') = (0.01579 - 0.0047)$$
$$= 0.01109 \, \text{kg/kg of dry air}$$
$$= 11.09 \, \text{gm/kg of dry air Ans.}$$

---

**Example 11.24**  From a given sample of atmospheric air at 30°C and 80% RH, 5 grams of moisture per kg of dry air is removed. After removing the moisture the temperature of air reduces to 25°C. Determine (i) Relative humidity (ii) Dew point temperature. The barometer reads 760 mm of mercury.

*Sol.* Given $\qquad\qquad t_{db1} = 30°C, \phi = 80\%, m = 5 \, \text{gm/kg of dry air}, t_{db2} = 25°C, p_b = 760 \, \text{mm of Hg.}$

At $t_{db1}$, $\qquad\qquad p_{ws} = 0.04242 \, \text{bar}$

$$\therefore \qquad\qquad \phi = \frac{p_w}{p_{ws}} \text{ or, } 0.80 = \frac{p_w}{0.04242}$$

$$p_w = 0.0339 \, \text{bar}$$

$$\therefore \qquad\qquad \omega_1 = 0.622 \, \frac{0.0339}{1.01325 - 0.0339}$$

$$= 0.02153 \, \text{kg/kg of dry air}$$
$$= 21.53 \, \text{gm/kg of dry air}$$

Since 5 gm of moisture per kg of dry air is removed, so find value of specific humidity, $\omega' = (21.53 - 5)$

$$= 16.53 \, \text{gm/kg of dry air}$$
$$= 0.01653 \, \text{kg/kg of dry air}$$

At $\qquad\qquad t_{db} = 25°C, p_{ws'} = 0.03166 \, \text{bar}$

$$\omega' = 0.622 \, \frac{p_w}{p_b - p_{w'}}$$

or, $\qquad\qquad 0.01653 = 0.622 \dfrac{p_{w'}}{1.01325 - p_{w'}}$

$\therefore \qquad\qquad\qquad p_{w'} = 0.02623$ bar

$\therefore$  Relative humidity $\qquad = \dfrac{p_{w'}}{p_{ws'}} = \dfrac{0.02623}{0.03166} = 0.8284 = 82.84\%$  *Ans.*

Corresponding to $p_{ws'} = 0.02623$ bar, dew point temperature $t_{dp} = 21.8°C$  *Ans.*

---

**Example 11.25**  A sling psychrometer records dry bulb temperature and wet bulb temperature as 48°C and 34°C respectively. Calculate (i) the vapour pressure (ii) the relative humidity (iii) the specific humidity (iv) the degree of saturation (v) the vapour density of air (vi) the dew point temperature (vii) the enthalpy of the mixture, and (viii) the specific volume of moist air per kg of dry air. Assume atmospheric pressure as 1.01325 bar.

*Sol.* Given, $\qquad\qquad t_{db} = 48°C,\, t_{wb} = 34°C$

From steam table at $t_{db}$ of 48°C, $p_{ws} = 0.1118$ bar, sp. volume = 13.22 m³/kg and at $t_{wb}$ of 34°C, $p_{w,\,wbt} = 0.0532$ bar.

(i)  Using Carrier equation as

$$p_w = p_{w,\,wbt} - \frac{\left(p_b - p_{w,\,wbt}\right)\left(t_{db} - t_{wb}\right)}{\left(1547 - 1.44\, t_{wb}\right)}$$

$$= 0.0532 - \frac{\left(1.01325 - 0.0532\right)\left(48 - 34\right)}{1547 - 1.44 \times 34} = 0.0442 \text{ bar} \quad Ans.$$

(ii)  Relative humidity, $\qquad \phi = \dfrac{p_w}{p_{ws}} = \dfrac{0.0442}{0.118} = 0.3953 = 39.53\%$  *Ans.*

(iii)  Specific humidity, $\qquad \omega = 0.622 \dfrac{p_w}{p_b - p_w}$

$$= 0.622 \frac{0.0442}{0.01325 - 0.0442}$$

$$= 0.0283 \text{ kg/kg of dry air} \quad Ans.$$

(iv)  Degree of saturation

$$\mu = \phi \left[\frac{\left(1 - \dfrac{p_{ws}}{p_b}\right)}{\left(1 - \dfrac{p_w}{p_b}\right)}\right] = 0.3953 \left[\frac{\left(1 - \dfrac{0.1118}{1.01325}\right)}{\left(1 - \dfrac{0.0442}{1.01325}\right)}\right] = 0.3677 \quad Ans.$$

(v)  Vapour density of air $\qquad = \dfrac{0.3953}{13.22} = 0.0299 \text{ kg/m}^3$  *Ans.*

(vi)  From steam table, temperature corresponding to $p_w = 0.0442$ bar, $t_{dp} = 30.7°C$  *Ans.*

(vii)  Enthalpy of mixture can be calculated by any following method

    (a)  $h = c_{pa}\, t_{db} + w(2500 + cp_{w,\,sup} \times t_{db})$, kJ/kg of dry air

    (b)  $h = c_{pa}\, t_{db} + w(h_g) \times t_{db}$, kJ/kg of dry air

    (c)  $h = c_{pa}\, t_{db} + w[(h_g)_{dp} + cp_{w,\,sup}\,(t_{db} - t_{dp})]$, kJ/kg of dry air

    (d)  $h = c_{pa} \times t_{db} + w(h_g)_{dp}$, kJ/kg of dry air

where

$$cp_a = \text{sp heat of dry air} = 1.005 \text{ kJ/kgk}$$

$$cp_w = \text{sp. heat of water} = 4.187 \text{ kJ/kgk}$$

$$cp_{w,\,sup} = \text{sp. heat of superheated water vapour}$$

$$= 1.884 \text{ kJ/kgk}$$

$$h = cp_a \times t_{db} + w(h_g)_{dp}$$

$$= 1.005 \times 30.7 + 0.0283(2557.1)$$

$$= 103.219 \text{ kJ/kg of dry air} \quad Ans.$$

(viii)

$$R_a = \frac{\bar{R}}{M} = \frac{8.314}{29} = 0.2867 \text{ kJ/kgk}$$

$$v = \frac{mRT}{p_a} = \frac{1 \times 0.287 \times (273 + 48)}{(1.01325 - 0.0442) \times 10^5 \times 10^{-3}}$$

$$= 0.9506 \text{ m}^3/\text{kg of dry air}$$

It can be calculated by the following way also

$$R_w = \frac{\bar{R}}{M} = \frac{8.314}{18} = 0.4619 \text{ kJ/kgk}$$

$$v = \frac{mRT}{p_w} = \frac{0.0283 \times 0.4619\,(273 + 48)}{0.0442 \times 10^5 \times 10^{-3}}$$

$$= 0.9493 \text{ m}^3/\text{kg of dry air} \qquad \text{[values are almost same]}$$

## 11.16   PSYCHROMETRIC PROCESSES

In order to condition air to the conditions of human comfort, certain processes are to be carried out on the outside air available. The processes affecting the psychrometric properties of air are called psychrometric processes which are shown in Fig. 11.62.

They are

  (i)  Sensible heating – process OA

 (ii)  Sensible cooling – process OB

(iii)  Humidifying – process OC

(iv)  Dehumidifying – process OD

 (v)  Heating and humidifying – process OE

(vi)  Cooling and dehumidifying – process OF

(vii)  Cooling and humidifying – process OG

(viii)  Heating and dehumidifying – process OH

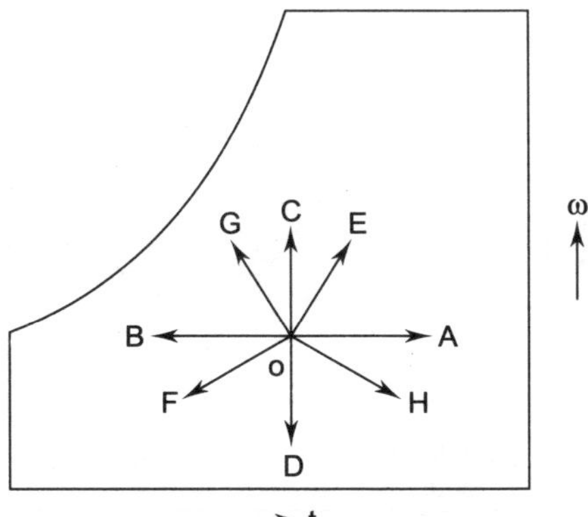


**Fig. 11.62  Psychrometric processes**

The first two processes viz. sensible heating and cooling involve only a change in the dry bulb temperature whereas the process of humidifying and dehumidifying involve a change in specific humidity.

1.  Mixing of air streams–Mixing of several streams in the process which is very frequently used in air conditioning. This mixing normally takes place without the addition or rejection of either heat or moisture, i.e. adiabatically and at constant total moisture content. Let us consider the adiabatic mixing of different quantities of air in two different states at constant pressure as shown in Fig. 11.63. Let subscripts 1 and 2 refer to the two streams of air and let $m_a$ refer to the mass of dry air in the stream. Then by moisture balance, we have for the specific humidity of the mixture

$$ma_3 w_3 = ma_1 w_1 + ma_2 w_2$$

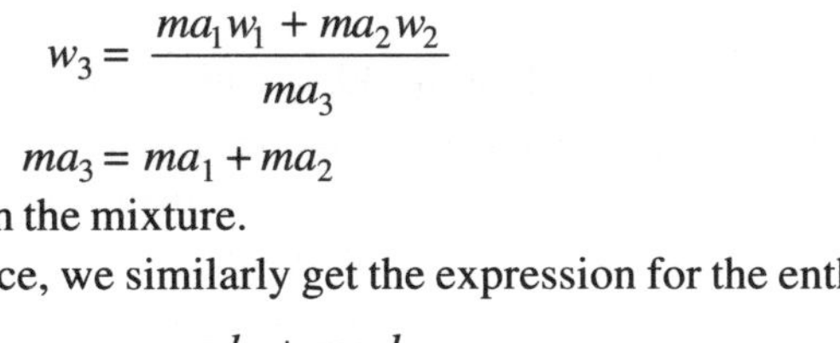

**Fig. 11.63**

or,
$$w_3 = \frac{ma_1 w_1 + ma_2 w_2}{ma_3} \tag{11.68}$$

where
$$ma_3 = ma_1 + ma_2$$

is the mass of dry air in the mixture.

Also, by energy balance, we similarly get the expression for the enthalpy of the mixture

$$h_3 = \frac{ma_1 h_1 + ma_2 h_2}{ma_3} \tag{11.69}$$

Enthalpy of moist air

$$h = h_a + w h_v \qquad \text{Gibb's law (11.70)}$$

where,    $h_a$ = enthalpy of dry air

$w h_v$ = enthalpy of associated water vapour part

$$h_a = c_{p_a} t_{db} = 1.005\,t,\,\text{kJ/kg}, \tag{11.71}$$

where    $c_{p_a}$ = sp. heat of dry air = 1.005 kJ/kgk

$t_{db}$ = dry bulb temp.

Taking reference state enthalpy as zero for a saturated liquid at 0°C, the enthalpy of the water vapour part, viz. at point A is expressed as

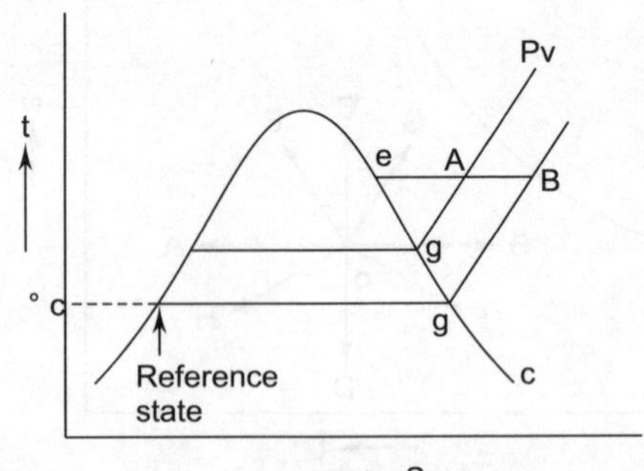


**Fig. 11.64**

$$h_v = h_A = Cp_w\, t_d + (h_{fg})_d + C_{pv}(t - t_d) \tag{11.72}$$

where  $Cp_w$ = sp heat of liquid water

$t_d$ = dew point temp.

$(h_{fg})_d$ = latent heat of vaporization of DPT

$Cp_v$ = sp. heat of superheated vapour

$Cp_w$ = 4.1868 kJ/kgk

$Cp_v$ = 1.88 kJ/kgk in the range of 0°C to 60°C.

Now,  $\quad h_A = h_B = (h_g)0°C + Cp_v(t-0)$, here $[t_d = 0°C]$ $\tag{11.73}$

Taking latent heat of vaporization of water at 0°C

as  $\quad$ 2500 kJ/kg, $hv = 2500 + 1.88t$ $\tag{11.74}$

By combining Eqs. (11.71) and (11.74) we have the enthalpy of moist air

$h = 1.005t + \omega(2500 + 1.88t)$, kJ/kg dry air $\tag{11.75}$

Eqn. 11.75 can also be written in the form.

*i.e.* multiply of moist air  $\quad h = (Cp_a + \omega Cpv)t + \omega(hf_g)0°C$

$$= cp_t + \omega(hf_g)°C \tag{11.76}$$

where  $\quad Cp = Cp_a + \omega Cpv$

$$= (1.005 + 1.88\omega) = \text{kJ/kg k of dry air}$$

The term Cpt governs the change in enthalpy of moist air with temperature at constant specific humidity and the term $w(hf_g)0°C$ governs the change in enthalpy with change in specific humidity *i.e.* due to the addition or removal of water vapour in air.

Since the second term 1.88 $w$ is very small compared to the first term 1.005, an approximated value of $C_p$ of 1.0216 kJ/kg d.a ($k$) may be taken for all practical purposes in air conditioning calculations. Substituting Eqn. 11.76 in Eqn. 11.69, we get

$$(c_p t_3 + h_{fg0}\omega_3) = \frac{ma_1}{ma_3}(c_p t_1 + h_{fg0}\omega_1) + \frac{ma_2}{ma_3}(c_p t_2 + h_{fg0}\omega_2)$$

After simplification

$$t_3 = \frac{ma_1 t_1 + ma_2 t_2}{ma_3} + \frac{h_{fg0}}{Cp}\left[\frac{ma_1}{ma_3}\omega_1 + \frac{ma_2\omega_2}{ma_3} - \omega_3\right] \tag{11.77}$$

The second term in the above expression being zero, we can write

$$t_3 \approx \frac{ma_1 t_1 + ma_2 t_2}{ma_3} \tag{11.78}$$

The sign of approximation has been used since an assumption has been made that the humid specific heat Cp is the same for the three streams of air.

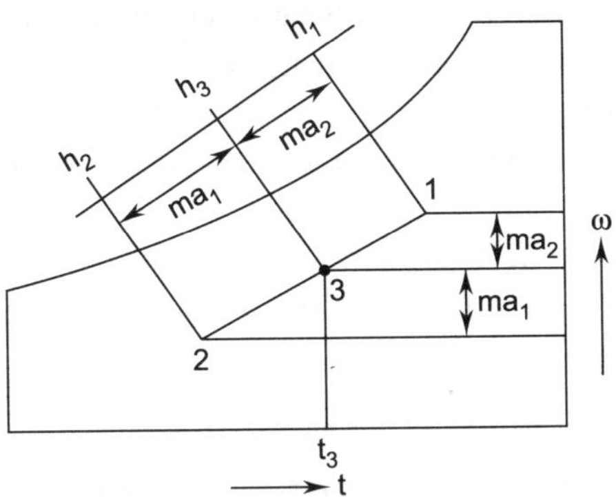

**Fig. 11.65  Mixing process on psychrometric chart.**

2. Sensible Heating – When air passes over a dry surface which is at a temperature greater than its (air) dry bulb temperature, it undergoes sensible heating. Thus the heating can be achieved by passing the air over heating coil like electric resistance heating coils or steam coils. During such a process, the relative humidity remains constant but the dry bulb temperature rises and approaches that of the surface. The extent to which it approaches the mean effective surface temperature of the coil is conveniently expressed in terms of the equipment by pass factor (B.F).

The by pass factor ($BF$) for the process is defined as the ratio of the difference between the mean surface temperature of the coil and leaving air temperature to the difference between the mean surface temperature and the entering air temperature. This in Fig. 11.66 air at temperature $t_1$ passes over a heating coil with an average surface temperature $t_3$ and leaves at temperature $t_2$. Then by pass factor is expressed as follows

$$BF = \frac{t_3 - t_2}{t_3 - t_1}$$

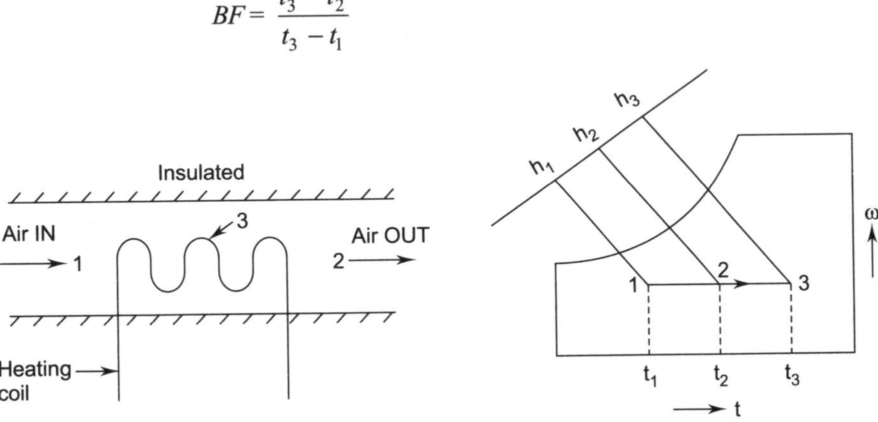

**Fig. 11.66  Sensible heating with psychrometric plot**

In terms of length on the chart, it is

$$BF = \frac{\text{length } 2\text{-}3}{\text{length } 1\text{-}3}.$$

The value of the by pass factor is a function of coil design and velocity. The head added to the air can be obtained directly from the entering and leaving enthalpies $(h_2 - h_1)$ or it can be obtained from the humid specific heat multiplied by the temperature difference $(t_2 - t_1)$.

In a complete air conditioning system the per-heating and reheating of air are among the familiar examples of sensible heating.

Note "By pass factor" can be considered to represent the fraction of air, which does not come into contact with coil surface.

Sensible heat added to air per kg of dry air

$$= h_2 - h_1 = (ha_2 + hv_2) - (ha_1 + hv_1)\ldots \tag{11.79}$$

Eqn. 11.79 can be evaluated as

$$(h_2 - h_1) = Cp_a(t_2 - t_1) + w_1\, Cpv(t_2 - t_1)$$

$$= (Cp_a + w_1 Cpv)\,(t_2 - t_1) = Cp_m(t_2 - t_1) \tag{11.80}$$

where $Cp_m = sp.$ heat of humid air or humid $sp.$ heat the value of Cpm is taken as 0.244 kcal/kg of dry air per 0°C or 1.0216 kJ/kg of dry air per K for calculation in psychrometry.

$$\therefore \quad (h_2 - h_1) = 1.0216(\Delta t) \text{ kJ/kg of dry air}$$

3. **Sensible cooling**

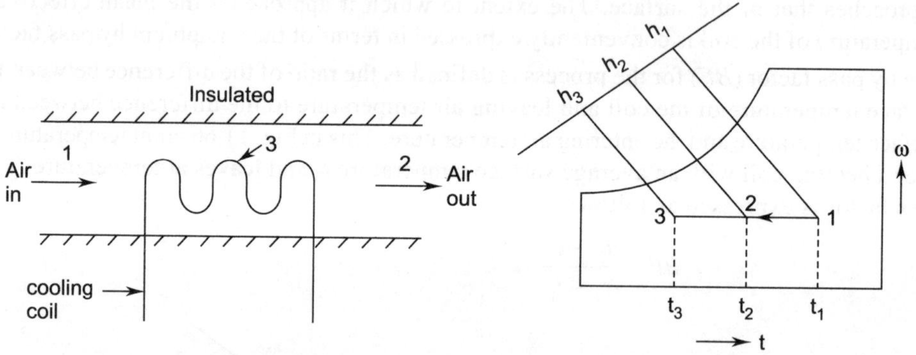


**Fig. 11.67    Sensible cooling with psychrometric plot**

Air undergoes sensible cooling whenever it passes over a surface that is at a temperature less than the dry bulb temperature of the air but greater than the dew point temperature. Thus sensible cooling can be achieved by passing the air over cooling coil like evaporating coil of the refrigerant cycle or secondary brine coil. During the process, the specific humidity remains constant and dry build temperature decreases, approaching the mean effective surface temperature. On a psychrometric chart (Fig. 11.67) the process will appear as a horizontal line (1-2), where point 3 represents the effective surface temperature. For this process

By pass factor $\qquad$ B.F $= \dfrac{t_2 - t_3}{t_1 - t_3}$ $\qquad$ (11.81)

Simple cooling can be done only up to the dew point temperature ($t_{dp}$). Cooling below this temperature will result in the condensation of moisture.

4. Cooling and Dehumidification

Whenever air is made to pass over a surface or though a spray of water that is at a temperature less than the dew point temperature of the air, condensation of the some of the water vapour in air will occur simultaneously with the sensible cooling process. Any air that comes into sufficient contact with the cooling surface will be reduced in temperature to the mean surface temperature along a path such as 1-2-3 in Fig. 11.68, with condensation and therefore dehumidification occurring between points 2 and 3. The air that does not contact the surface will be finally cooled by mixing with the portion that did, and the final state point will somewhere on the straight line connecting points 1 and 3. The actual path of air during the path will not be straight line shown but will be something similarly to the curved dashed line 1-4.

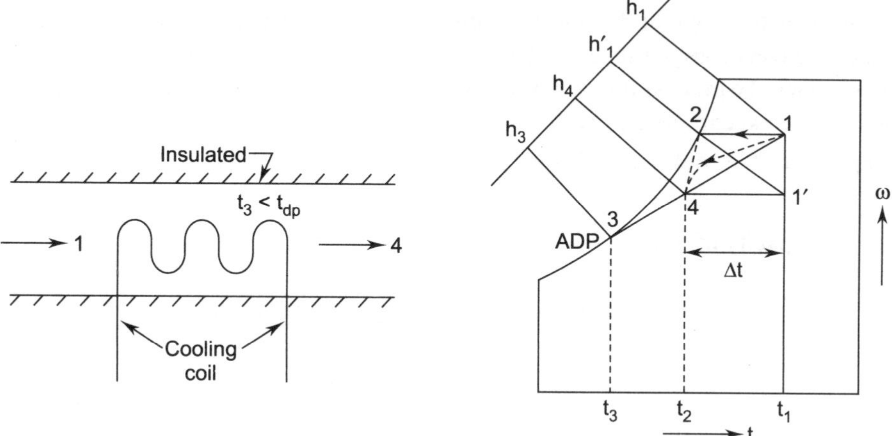

Fig. 11.68   **Cooling and dehumidification with psychrometric plot**

It will result from a continuous mixing of air which is connecting a particular part of the coil and air which is by passing it. It is convenient, however to analyse the problem with the straight line shown, and to assume that the final air state results from the mixing of air that has completely by passed the coil with air that has been cooled to the mean effective surface temperature. If there is enough contact between air and surface for all of the air to come to the mean surface temperature, the process is one of zero by pass. In any practical system, complete saturation is not obtained and final state will be a point such as 4 in the Fig. with an equivalent by pass factor equal to length 3-4. For processes involving condensation, length 3-1 the effective surface temperature, e.g. $t_3$ in Fig. 11.68 is called 'apparatus dew point' (ADP). The final state point of air passing through a cooling and dehumidifying apparatus is in effect a mixture condition that results from mixing the fraction of the air, which is equal to the equivalent by pass factor ($BF$) and is at initial state point and the remaining fraction which is equal to one minus by pass factor $(1 - BF)$ and is saturated at the apparatus Dew point ($ADP$).

Efficiency or effectiveness of cooling coil $= (1 - BF)$ \hfill (11.82)

Total heat removed from the air is given by

$$Q_t = h_1 - h_4 = (h_1 - h_{1'}) + (h_{1'} - h_4)$$
$$= Q_L + Q_s \hspace{2cm} (11.83)$$

where $\hspace{2cm} Q_L =$ Latent heat removed $(h_1 - h_{1'})$

$\hspace{3.5cm} Q_S =$ Sensible heat removed $(h_{1'} - h_4)$

The ratio $\dfrac{Q_s}{Q_t}$ is called sensible heat factor ($SHF$) or sensible heat ratio ($SHR$).

$$\therefore \hspace{2cm} SHF = \frac{Q_s}{Q_L + Q_s} \hspace{2cm} (11.84)$$

The ratio fixes the slope of the line $1 - 4$ on the psychrometric chart. Sensible heat factor slope lines are given on the psychrometric chart. If the initial condition and $SHF$ are known for the given processes, then the process line can be drawn through the given initial condition at a slope given by $SHF$ on the psychrometric chart.

The capacity of the cooling coil is tons of refrigeration is given by,

$$\text{Capacity of tons} = \frac{m_a \left( h_1 - h_4 \right)}{211} \hspace{2cm} (11.85)$$

where $\hspace{3cm} m_a =$ mass of air in kg/min passing over the coil.

Empirical relation for finding out S.H.F;

$$SHF = \frac{0.0204\, \Delta t}{0.0204\, \Delta t + 50\, \Delta \omega} \hspace{2cm} (11.86)$$

where, $\Delta \omega =$ change in specific humidity or humidity ratio in kg/kg of dry air

$$\Delta t = (t_1 - t_4)$$

5. **Heating and humidification**

   If air is passed through a humidifier which has heated water sprays instead of simply recirculated spray, the air is humidified and may be heated, cooled or unchanged in temperature. In such a process the air increases in specific humidity and the enthalpy, and the dry bulb temperature will increase or decrease according to the initial temperature of the air and that of the spray. If sufficient water is supplied relative to the mass flow of air, the air will approach saturation at water temperature. Examples of such processes are shown in the Fig. 11.69.

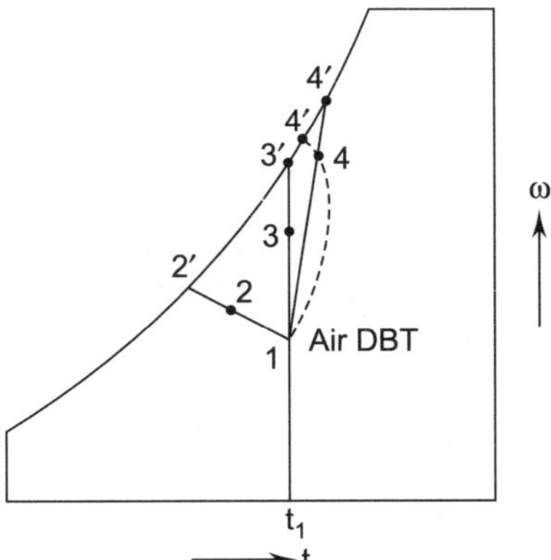


**Fig. 11.69   Heating and Humidification on Psychrometric plot**

process 1-2: denotes the cases in which the temperature of the heated spray water is less than the air *DBT*.

process 1-3: denotes the cases in which the temperature is equal to air *DBT*.

process 1-4: denotes the cases in which a spray temperature is greater than the air *DBT*.

As in the case of adiabatic saturation, the degree to which the process approaches saturation can be expressed in terms of the by pass factor or a saturating efficiency.

If the water rate relative to the air quantity is smaller, the water temperature will drop significantly during the process. The resultant process will be a curved line such as a dashed 1-4 where 4 represents the leaving water temperature.

Note-It is possible to accomplish heating and humidification by evaporation from an open pan of heated water, or by direct injection of heated water or steam. The latter is more common. The process line for is of little value because the process is essentially an instantaneous mixing or steam and the air. The final state point of the air can be found, however by making a humidity and enthalpy balance for the process. The solution of such a problem usually involves cut and try procedure.

6.  Heating and Dehumidification: (Adiabatic dehumidification) Heating and dehumidification can be simultaneously done if air is passed through a solid adsorbent surface or through a liquid adsorbent spray. The dehumidification is achieved because of the lower water vapour pressure at the surface of the adsorbent. Water vapour condenses out of air and the latent heat of condensation liberated heats the air supply. Hence due to condensation, specific humidity falls and this heat of condensation heats air increasing its dry bulb temperature. Hence this process is opposite of adiabatic saturation and follows the constant wet bulb temperature line as shown by line 1 –2 on thy psychrometric chart in the Fig. 11.70. The other line given in Fig. 11.70 shows deviation in actual practice. Some of the common solid adsorbents are silica get and activated alumina, and the liquid adsorbents are solutions of inorganic salts, or organic compounds like ethylene glycol.

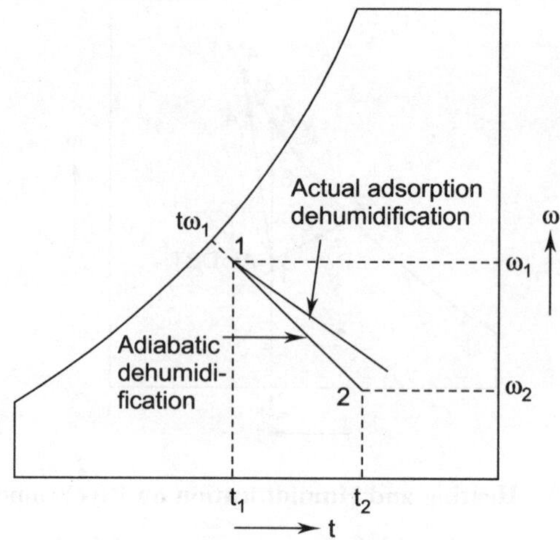


**Fig. 11.70   Heating and Dehumidification on Psychrometric plot**

# 11.17   ROOM (OR CONDITIONED SPACE) SENSIBLE HEAT FACTOR (RSHF)

The room sensible heat factor is the ratio of the room sensible heat to the sum of the room sensible heat
and the room latent heat. This ratio is expressed as

$$RSHF = \frac{RSH}{RSH + RLH} \qquad RSH = \text{Room sensible heat}$$

$$RLH = \text{Room latent heat}$$

$$= \frac{RSH}{RTH} \qquad RTH = \text{Room total heat}$$

The supply air to the conditional space must have the capacity to offset simultaneously both the
room sensible heat and the room latent heat loads. The room and supply air conditions to the space may
be plotted on psychrometric chart as shown in Fig. 11.71 and the points connected by a straight line 1-2.
This represents the psychrometric process of the supply air within the conditions space and is called
room sensible heat factor line. At the point where it intersects the saturation line is often called Room
*ADP*. The slope of the line explains the ratio of the sensible heat to the latent heat load within the room,
*i.e.* $\Delta h_s$ and $\Delta h_L$ respectively. Thus supply air having air bulb and wet bulb temperature, such that the
condition point falls on this line, will satisfy the requirements of the room with adequate supply of such
air. Thus the supply air with properties as at 1, 1*a*, 1*b* and 1*c* will satisfy the requirement, but only the
quantities of air will be different for different supply air points. Supply condition at 1 requires less air than
that at 1*a*. At 1*c*, it is highest of all four supply points.

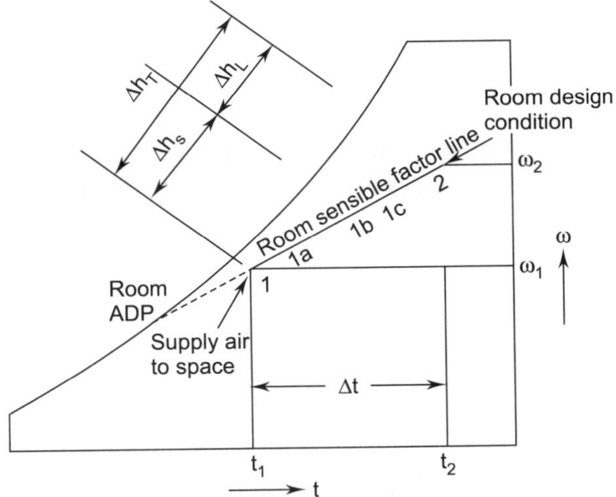



**Fig. 11.71   RSHF on Psychrometric plot**

The room sensible heat factor line can also be drawn from the calculated value of RSHF without knowing the supply condition which infact is normally required to be found. This method is illustrated.

**Step 1.** Draw a base line through the alignment circle (26°C *DBT* and 50% *RH* shown on psychrometric chart) and the calculated value of *RSHF* as shown on the sensible heat factor scale. This line is shown as 1-2.

**Step 2.** Draw the actual room sensible heat factor line through the room design conditions parallel to the base line step 1. This line is shown as 3-4. Thus the supply conditions of air falling on any point on the line 3-4 will offset the sensible heat and the latent heat loads on the room, provided adequate quantity of such supply air is supplied. Thus with property quantity of air, heat generation will be equal to the heat removal and Room Design Conditions will be maintained.

It is not necessary that SHF should be positive

(i)  If *SH* and *LH* both are positive (Gains)

$$SHF = \frac{SH}{SH + LH} \text{ is } + \text{ve}$$

(ii)  If *SH* and *LH* both are negative (Losses)

$$SHF = \frac{SH}{SH + LH} + \text{ve}$$

(iii)  If *SH* is negative (Loss) and *LH* is positive (Gain)

$$SHF = \frac{SH}{SH + LH} - \text{ve}$$

(iv)  If *SH* is positive (Gain) and *LH* is negative (Loss)

$$SHF = \frac{SH}{SH + LH} - \text{ve}$$

Thus the numerical value of $SHF = \dfrac{SH}{SH + LH}$ irrespective whether $SH$ is positive or negative or, $LH$ is positive or negative. But it will be positive when both $SH$ and $LH$ are either gains or losses. And it will be negative when one of them is positive and the other is negative. Negative $SHF$ lines can also be drawn on the chart where scale for positive values have already been given. Fig. 11.72 shows $SHF$ lines on psychrometric plot.

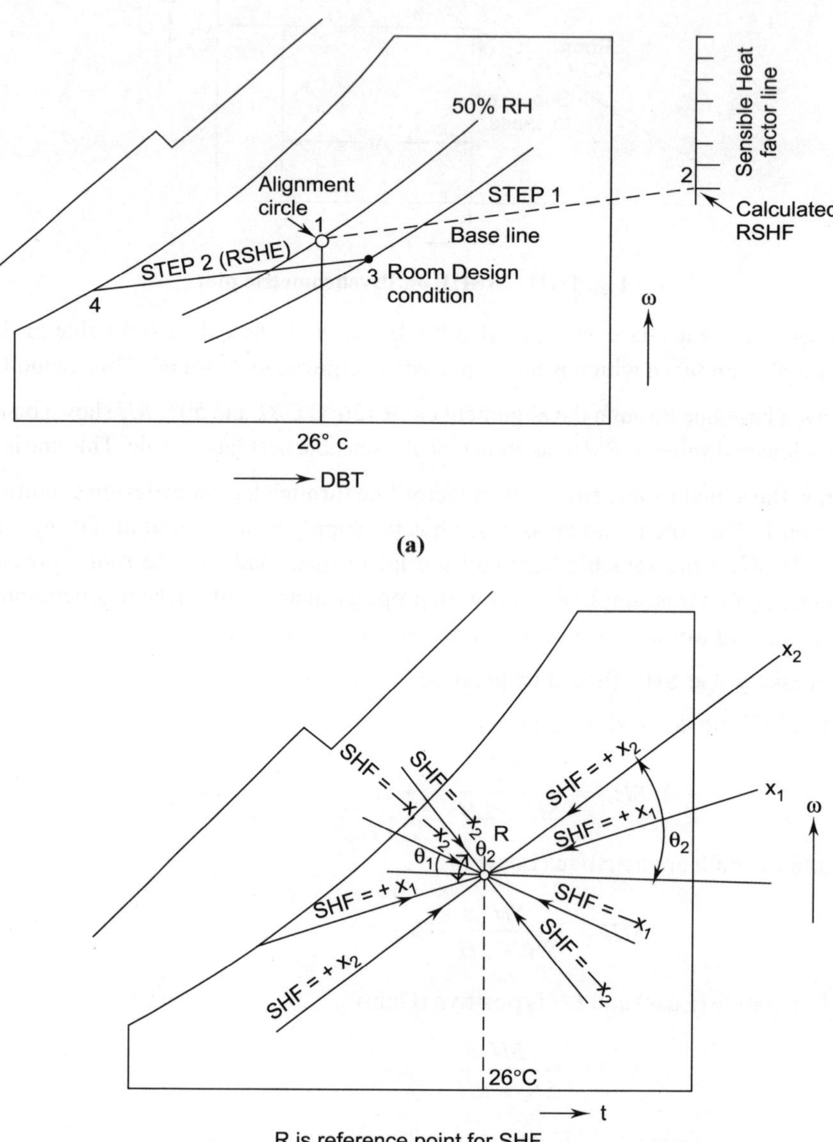


**Fig. 11.72   (a), (b) – SHF lines on psychrometric plot**

## 11.18  GRAND SENSIBLE HEAT FACTOR (GSHF)

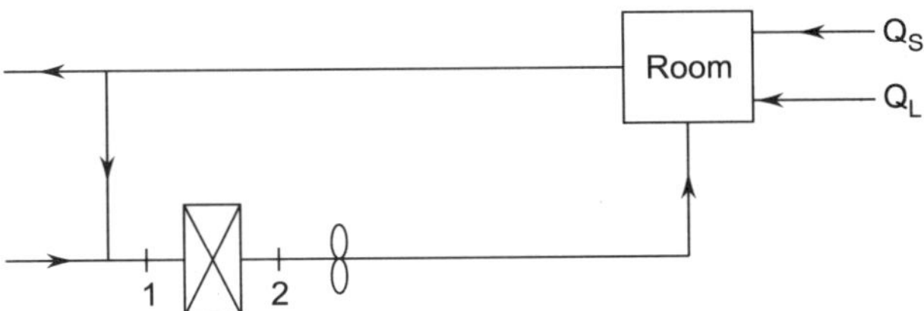

**Fig. 11.73  (a) GSHF**

The grand sensible heat factor is the ratio of total sensible heat to the grand total heat the cooling coil or the conditioning apparatus should handle. In practice, some outside air for ventilation and some infiltrated air from outside mixes with the room air. Thus, this outside air also adds the sensible and the latent heat load. Thus these mixture conditions reach the cooling coil or the conditioning apparatus get cooled and dehumidified and leave the apparatus (cooling coil) with decrease in temperature and moisture content. The fall in temperature and moisture content is determined by the total sensible heat and the latent heat required to be handled by the apparatus, if condition of the air leaving the apparatus is joined by a straight line 1-2, this gives the *GSHG* lines. This line 1-2 gives the psychrometric process undergone by air as it passes over the apparatus. Thus any point on this line will give the sensible heat, and the latent heat removed in a fixed proportion as given in Fig. 11.73 (b), the ratio of $\Delta h_s$ to $\Delta h_L$ or the ratio of $\Delta h_s$ to $\Delta h_T$ is the same for any point on this line.

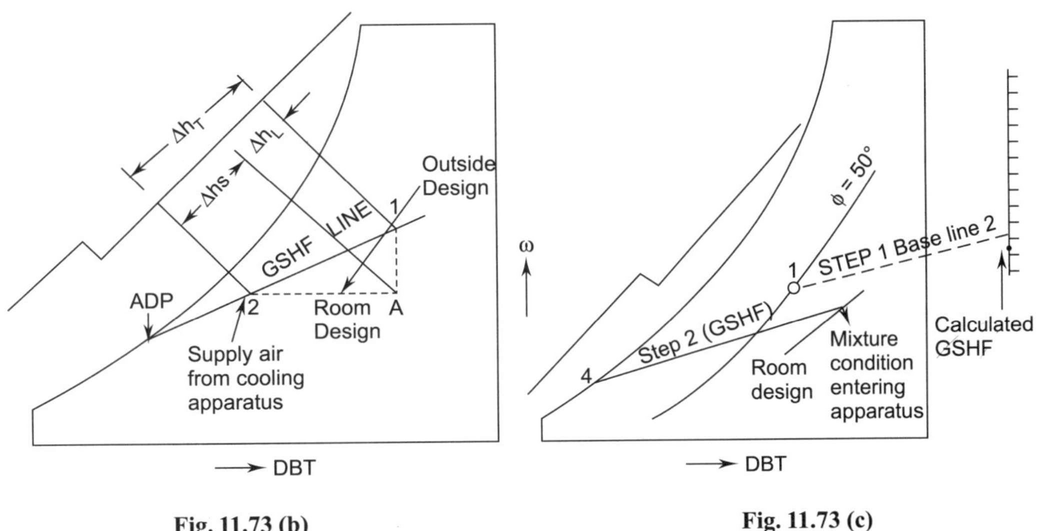

**Fig. 11.73 (b)**                    **Fig. 11.73 (c)**

If the grand sensible heat factor and the mixture condition entering the apparatus are given, the *GSHF* line can be drawn with the help of sensible heat factor scale in the same way as explained for *RSHF* line. This is given in Fig. 11.73 (c).

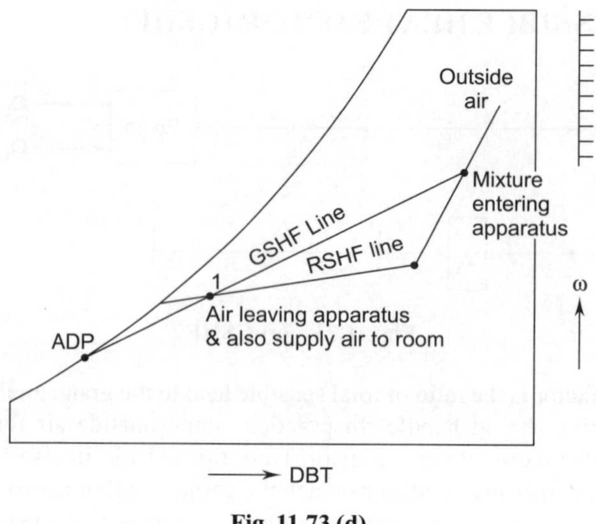



**Fig. 11.73 (d)**

Thus a little though reveals that a point that lies on *RSHF* line and the *GSHF* line is the ideal supply conditions to the room. This supply conditions takes proper care of the room sensible heat and the latent heats, and any outdoor heats reflected in the *GSHF* line for the apparatus. This is illustrated by point 1 in Fig. 11.73(d).

## 11.19 EFFECTIVE SURFACE TEMPERATURE

The surface temperature of the cooling or the heating coil varies throughout the surface as air passes over it. However, the effective surface temperature can be treated as that uniform surface temperature which would produce the same conditions of air leaving the apparatus as the non-uniform surface temperature obtained in actual operation. Referring the Fig. 11.74, conditioning the air through the apparatus is in principle the heat transfer between heating or cooling medium in the apparatus and the air passing through it. Therefore, there should be a common reference point. This point is the effective surface temperature of the apparatus. The actual heat transfers are relatively independent of each other but are quantitatively equal when referred to the effective surface temperature.

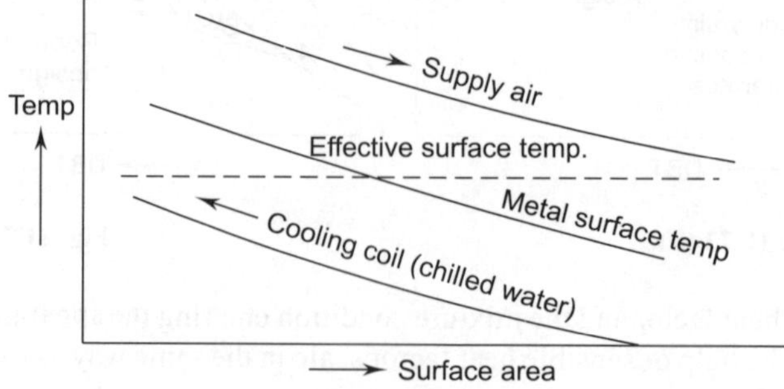



**Fig. 11.74  Effective surface temperature**

For process involving cooling and dehumidification, the effective surface temperature is at the point where the GSHF or the apparatus sensible heat factor line as it is sometimes called, cut the saturation or 100% relative humidity line on the psychrometric chart. As such, effective surface temperature is considered to be the dew point of the apparatus and hence in cooling and dehumidification processes, this term is used as Apparatus Dew Point (*ADP*). Thus for cooling and dehumidification the effective surface temperature falls on the saturation line. For other processes like evaporative cooling, sensible cooling or heating etc. the effective temperature will not necessarily on the saturation line.

Effective temperature in summer is taken as 21.7°C and the comfortable zone is *DBT* 25 ± 1°C, *RH* 50 ± 5% with air velocity at 0.4 m/s. In winter, the body acclimatizes to withstand lower temperature. Consequently, *DBT* 21°C at 50% *RH* with 0.15 to 0.2 m/s air velocity is quite comfortable.

## 11.20  COOLING AND HUMIDIFICATION

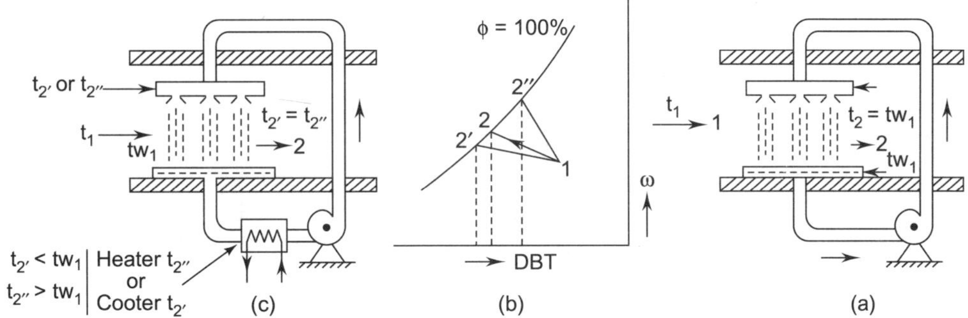


**Fig. 11.75    a, b, c-cooling and humidification with psy. plot**

If the air is washed through sprays of water maintained at a temperature higher than the dew point temperature of entering air, but lower than the dry bulb temperature of entering air, the air cools and gets humidified.

Case I No heat is supplied or rejected from the spray water if the same water is recirculated again and again. In this case the condition of adiabatic saturation will be reached. The temperature of spray water will reach the thermodynamic wet bulb temperature of the air entering the spray water. A condition of equilibrium is reached when the sensible heat drop in the entering air equals the heat of vaporization of water. This process is illustrated by Fig. 11.75(a) in schematic way and in process on the psychrometric chart is shown in Fig. 11.71(b) by line 1-2. This is a process along the constant wet bulb line.

Case II If the temperature of spray water is maintained at a temperature less than the wet bulb temperature of the entering air, by cooling the spray water by coolers before pumping it back to the spray nozzles as illustrated in Fig. 11.75(c) the process followed is shown on a psychrometric chart by line 1-2. Thus for ideal condition when humidifying efficiency is 100%, the condition of air at outlet shall be at temperature $t_{2'}$ and relative humidity 100%.

Case III If the temperature of the spray water is maintained constant at a temperature higher than the wet bulb temperature of the entering air but lower than the dry bulb temperature of air, by heating of water by heaters before pumping back to the spray nozzles the process followed is 1-2″ as shown in Fig. 11.75(b) and illustrated in Fig. 11.75(c). For the ideal condition when humidifying efficiency is 100%, the condition of air at outlet shall be at temperature $t_{2''}$ and relative humidity 100%.

## 11.21 HUMIDIFYING EFFICIENCY

In the previous discussion an ideal case of spray washing of the air has been considered. In practice the air will come out of the spray at 100%, relative humidity. And the extent to which humidification is affected depends on the velocity of air, the depth of the showers etc. Thus humidifying efficiency is defined as

$$\eta_{\text{humidifying}} = \frac{t_1 - t_3}{t_1 - t_2} = \frac{\omega_3 - \omega_1}{\omega_2 - \omega_1} \tag{11.87}$$

where $t_2 = t_{w_1}$ and $t_3$ is the actual dry bulb temperature of air at outlet.

For the other two cases the expression is given by

$$\eta_{\text{humidifying}} = \frac{t_1 - t_{3'}}{t_1 - t_{2'}} = \frac{\omega_{3''} - \omega_1}{\omega_{2''} - \omega_1}$$

when $t_{2''} > t_{w_1}$ (11.88)

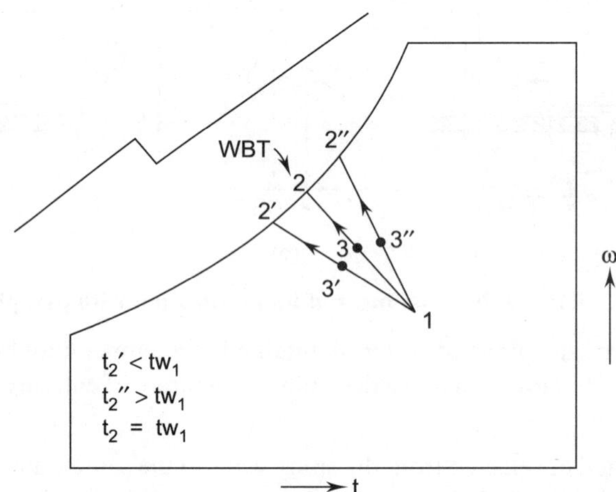


**Fig. 11.76  Humidifying efficiency on psychrometric plot**

$$\eta_{\text{humidifying}} = \frac{t_1 - t_{3''}}{t_1 - t_{2''}} = \frac{\omega_{3''} - \omega_1}{\omega_{2''} - \omega_1} \tag{11.89}$$

where $t_{2'} > t_{w_1}$

## 11.22 OTHER PROCESSES: AIR WASHER

All possible air conditioning process using water as coolant can be achieved by an air washer which is provided with arrangement for external cooling or heating (say by heat exchanger). In this arrangement shown in Fig. 11.77 washer is sprayed in a chamber from top with the help of several banks of nozzles mounted on ceiling of the air-washer chamber, with the help of a pump in water circuit. Air is allowed to pass through this chamber horizontally through side walls which are nothing but eliminator plates and prevent spilling of water outside the air washer chamber.

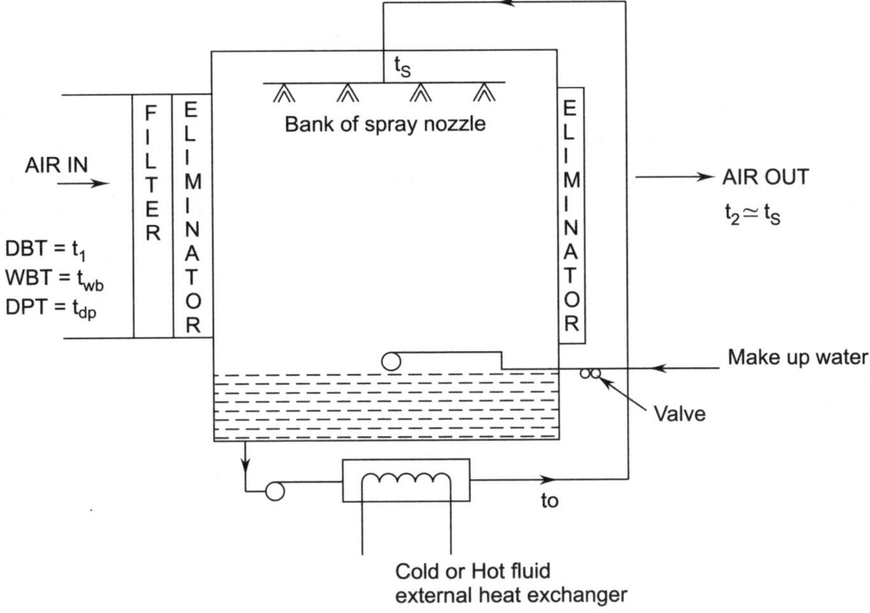

Fig. 11.77   **Air washer**

Fig. 11.78 shows the thermodynamic change of state in case of various processes which are possible with air washer depending on relative combination of following temperatures.

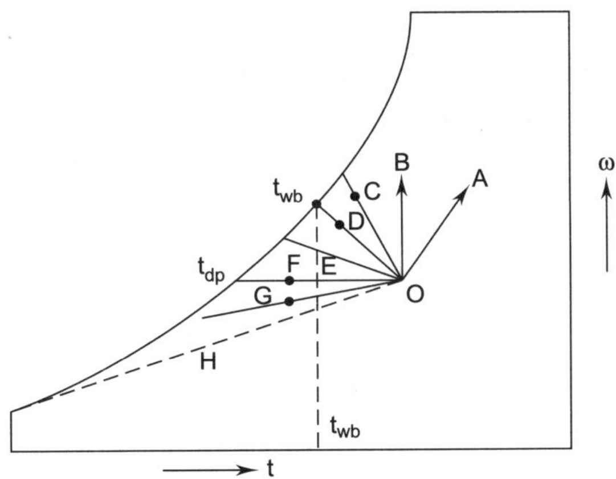

Fig. 11.78   **A/C Processes with air washer**

Let us understand various processes with the help of Fig. 11.77 and 11.78.

1. Heating and Humidification ($T_s > t_1$): If water is heated externally by heat exchanger its temperature $t_o = t_s$ shall be greater than DBT, $t_1$ of air. This will result in heating and humidifying of air as shown by line OA, if initial condition of air is represented by point O.

2. Humidification only ($t_s = t_1$): If surface temperature of water can be maintained equal to DBT of entering air *i.e.* $t_1$ air will get humidified, as sensible heating is not possible due to nil temperature difference. This process is represented by constant DBT line OB. Collected water may have to be

slightly heated for compensating for loss in transit from pump to sprayer. Practically it is rather impossible to achieve this process.

3. Cooling and Humidification ($t_1 > t_s > t_{wb}$): If mean temperature of water is less than DBT of air but more than *WBT* the air will get cooled and humidified simultaneously and its enthalpy increases. The water after spray will have to be heated externally as its temperature shall be less than $t_s$ but in any case more than $t_{wb}$. This is represented by a line OC which is slightly oriented with constant wet bulb line. Although, air is cooled, its enthalpy is increased due to addition of moisture.

4. Evaporative cooling (adiabatic saturation) ($t_s = t_{wb}$): In this process water is neither heated nor cooled but circulated by pump again and again due to which it gets cooled to WBT adiabatically *i.e.* along the constant enthalpy line and therefore constant wet bulb line OD.

5. Cooling and humidification ($t_{wb} > t_s > t_{dp}$): When $t_s$ is more than dew point temperature of air, air is cooled and humidified but since enthalpy of air decreases vide line OE; water will have to be cooled externally.

6. Sensible cooling only ($t_s = t_d$): When temperature of water is equal to dew point temperature of air, sensible cooling results (OF). As enthalpy is also decreased water, will have to be cooled externally.

7. Cooling and dehumidification ($t_s < t_{dp}$) The mean water temperature is lower than dew point temperature of outside air; air will get cooled and vapour of air will get condensed as represented by line OG. Lowest possible living air dry bulb temperature is limited by line OH drawn from point O as a tangent to saturation line.

From above it can be seen that air washer offers a means of achieving most of the air conditioning processes which may be needed due to variation of seasons a round the year. External heating of water can be done by electrical resistance heating and cooling can be achieved by primary or secondary refrigerants (chilled water of brine).

### 11.22.1   Use of hygroscopic solutions in air washer

Hygroscopic solutions, such as brines, glycols etc. exert lower vapour pressure as compared to that of pure water at the same temperature as shown in the Fig. 11.79. Curve A is for pure water and *B, C* and *D* for different hygroscopic solutions. The vapour pressures $pv_B, pv_C$ are lower than that $pv_A$ of water at the same temp. Now, if a hygroscopic solution such as at *C* is circulated, in the air washer, the condition line will be 1*C* instead of 1*A*. Thus a solution of hygroscopic material is more effective for dehumidification.

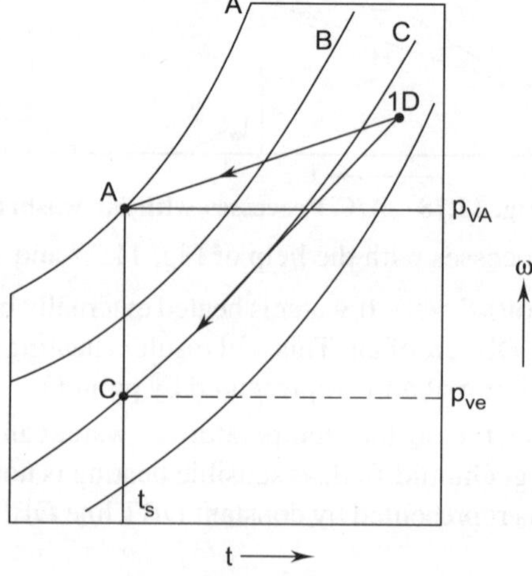


**Fig. 11.79**

## Adiabatic Dehumidification

Adiabatic dehumidification is based on the principles of adsorption viz capillary action. The vapour which is condensed at the surface of the adsorbent is drawn into capillaries, thereby reducing the vapour pressure at the surface causing a pressure gradient, and hence a mass transfer from the passing air stream to the adsorbent surface. As the capillaries get filled with water, the attraction decreases and rate of dehumidification falls off. Thermodynamically an adsorption process is the opposite of the adiabatic saturation process. As the air passes over the adsorbing surfaces, water vapour flows to the surface through the air film, condenses and releases its latent heat which raises the adsorbent and air temperatures. Thus the heat of condensation supplies the sensible heat for the heating of air the process is the reverse of the adiabatic saturation process. In practical practice, however, the process is accompanied with a release of heat called the heat of adsorption. This heat with adsorbents, such as silica gel and activated alumina is very large. Thus the sensible heat gain of air exceeds the loss of latent heat and the process here 1-2 lies above the *WBT* line as shown in Fig. 11.80.

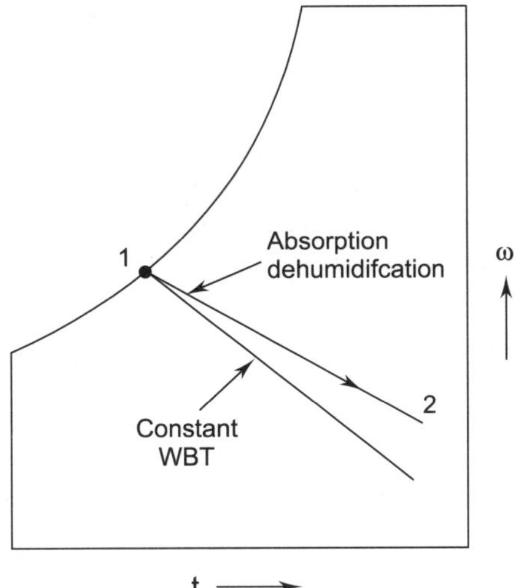


**Fig. 11.80  Adsorption Dehumidification process**

It is to be noted that after adsorption, the material become, saturated and has to be reactivated by heating as in case of hygroscopic materials.

## 11.23  WATER INJECTION

Let liquid water at temperature $t_f$ is injected into a flowing air stream. The condition of air will change depending upon the amount of water that evaporates. The enthalpy of vapourisation will come from the enthalpy of the air. Let us consider that the amount of water that has evaporated $m_v$ as the amount injected, $m_a$ = amount of dry air

$$\omega_2 = \omega_1 + \frac{m_v}{m_a}$$

$$h_2 = h_1 + \frac{m_v}{m_a}\, h_f = h_1 + (\omega_2 - \omega_1)\, h_f \tag{11.90}$$

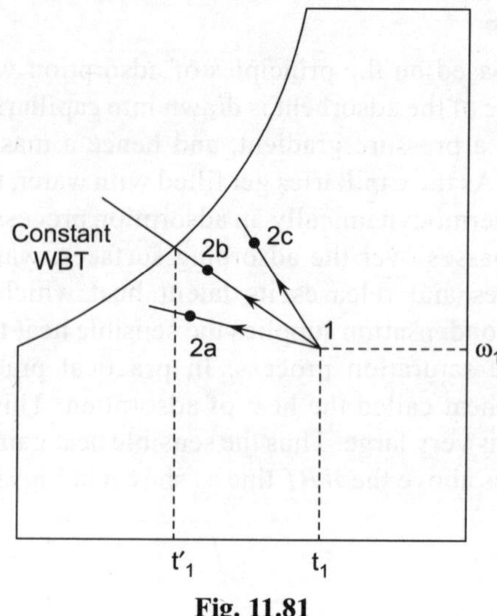



**Fig. 11.81**

where $h_f$ = enthalpy of liquid water. It is evident that if water is injected at WBT, the sigma heat function is constant and the process follows 1-2b. Otherwise 1-2a or 1-2c depending upon whether the temperature of water is lower or higher than WBT of air. Nevertheless since the term $(\omega_2 - \omega_1)\, h_f$ is extremely small compared to $h_1$ and $h_2$, lines 1-2a and 1-2c are very close to line 1-2b; irrespective of injected water.

## 11.24  STEAM INJECTION

Steam is normally injected into fresh outdoor air which is then supplied for the conditioning of textile mills where high humidities have to be maintained. The process can be analysed by considering mass and energy balances. If $m_v$ is the mass if steam supplied and $m_a$ the mass of dry air, then the leaving air state is given by

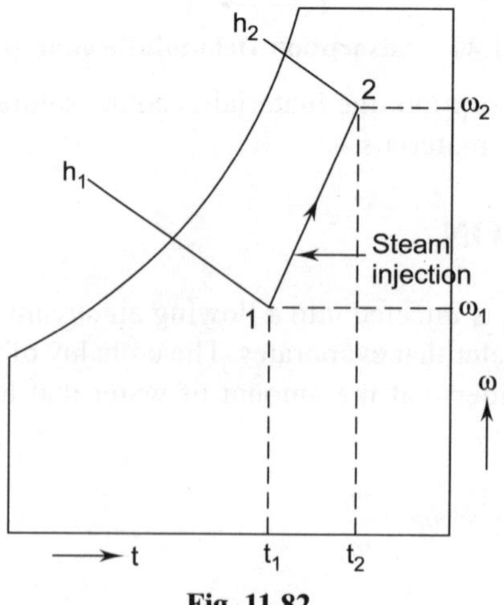



**Fig. 11.82**

$$\omega_2 = \omega_1 + \frac{m_v}{m_a} \;,\; h_2 = h_1 + \frac{m_v}{m_a}\, h_v$$

The *DBT* of air changes very little during the process.

**Note:** The reference state for zero enthalpy for $h_1$, $h_2$ and $h_v$ must be same.

## 11.25   EVAPORATIVE COOLING

Evaporative cooling is obtained during the process of adiabatic saturation. It is a process of removal of sensible heat from air and an equivalent addition of latent heat to it in the form of added water vapour shown in Fig. 11.83. Evaporative cooling is a process in which heat is neither added nor removed from the water out side the air washer. Water is simply recirculated by a pump.

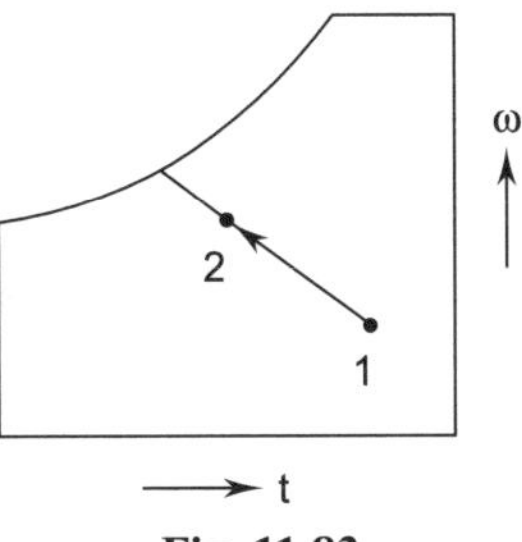
**Fig. 11.83**

Evaporative cooling is commonly used when the outdoor conditions are very dry. This means that wet bulb depression of air is very large. In a dry climate, evaporative cooling can give some relief by removing the sensible heat from the room. But the humidity cannot be controlled.

Another defect of evaporative cooling system is the large quantity of air that must be supplied to the room sensible heat load as the temperature difference between the room and supply air is generally small. Thus, whereas, in air conditioning the supply air quantity may be of the order of 10 air changes per hour, in case of evaporative cooling, the same may be of the order of 20 air changes. This quantity increases rapidly as the humidifying efficiency of air washer decreases.

## 11.26   BY PASS FACTOR

By pass factor of the coil is the measure of the performance of the coil.

$$\text{B.F} = e^{-\dfrac{A_r\, n u}{1.0216\, A_f\, v_f\, \rho}} = (e)^{-K} \tag{11.91}$$

where   $A_f$ = Face area through which air is flowing (m$^2$)

$V_f$ = Face air velocity (m/hr)

$\rho$ = density of air (kg/m3)

$u$ = overall heat transfer co-efficient kW/m$^2$

$A_s$ = surface area per row of the coil (m$^2$)

$n$ = number of rows

$A_s = A_r n$

$A_r$ = surface area of the coil provided per row, $m^2$

$W_a$ = quantity of air flow, kg/hr

$W_a = A_f V_f \rho$

$$\text{B.F} = e^{-\dfrac{c\, n}{1.0216 v_f}} \;,\; c = \frac{A_r\, u}{A_f\, \rho} \tag{11.92}$$

$C_{pm} = 1.0216$ kJ/kg = sp. heat of humid air

The value of K is constant for the given coil surface area and given air flow quantity. This equation shows that the by pass factor decreases with an increase in surface area, increase in depth of the coil and with the decrease in air velocity.

Another important factor which is introduced is contact factor. The contact factor is given by

$$\text{Contact factor} = 1 - \text{By pass factor} = \text{coil eff.} \tag{11.93}$$

The effect of air velocity on by-pass factor for different coil depth (number of rows) is shown in the Fig. 11.84.

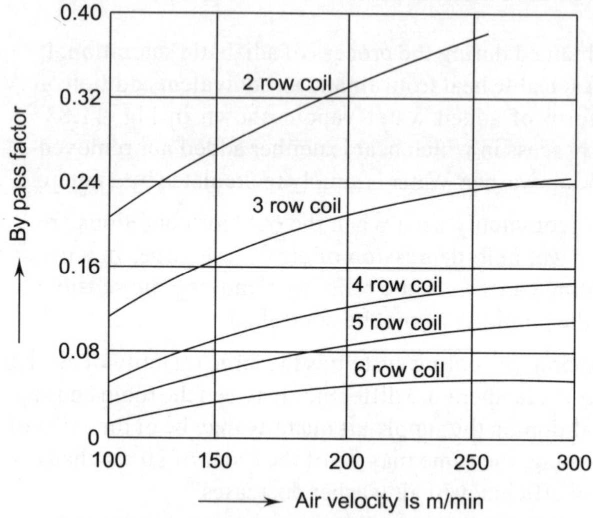



Fig. 11.84   (a) Effect of velocity on pypass factor

## 11.26.1   By pass factor of multi depth coils

A theoretical method is given to find the by pass factor of multidepth coil. If the surface temperature of the coil is $T_s$ and $T_1$ is the temperature of the outside air and $T_2$ is the temperature of the air leaving the second depth of the coil and so on.

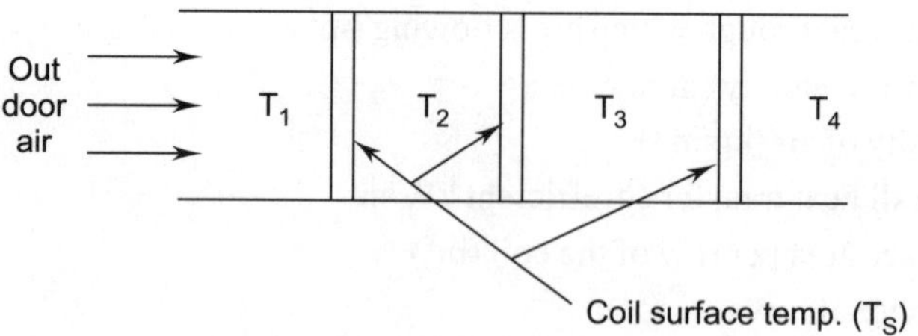



Fig. 11.85   (b) *n*-depth coil

The value of $T_2$ in terms of $T_1$ is given by

$$T_2 = T_1 \, B.F_1 + T_s(1 - BF_1) \tag{11.94}$$

where $\qquad\qquad B.F_1 = $ By-pass factor of single depth coil

Similarly $T_3$ in terms of $T_2$ is written as

$$T_3 = T_2 \, BF_1 + T_s(1 - BF_1) \tag{11.95}$$

Substituting the value of $T_2$ from Eqn. 11.94 into Eqn. 11.95, we get

$$T_3 = BF_1[T_1 BF_1 + T_s(1 - BF_1)] + T_s(1 - BF_1)$$

$$= T_1 BF_1^2 + T_s BF_1(1 - BF_1) + T_s(1 - BF_1)$$

$$= T_1 BF_1^2 + T_s(1 - BF_1)(BF_1 + 1)$$

$$= T_1 BF_1^2 + T_s(1 - BF_1^2) \tag{11.96}$$

If the by pass factor of the two depth coil is $BF_2$ then $T_3$ in terms of $T_1$, $T_s$ and $BF_2$ can be written as follows

$$T_3 = T_1 BF_2 + T_s(1 - BF_2) \tag{11.97}$$

Comparing Eqns. 11.97 and 11.95, the $BF_2$ in Eqn. 11.97 can be replaced by $BF_1^2$ so that $BF_2$ in terms of $BF_1$ for two depth coil can be written as follows

$$BF_2 = BF_1^2 \tag{11.98}$$

If there are '$n$' depth of coil then the bypass factor of '$n$' depth coil is given by

$$BF_n = (BF_1)^n \tag{11.99}$$

The corresponding sensible cooling with different depth of coil is shown on psy. chart in Fig. 11.86.

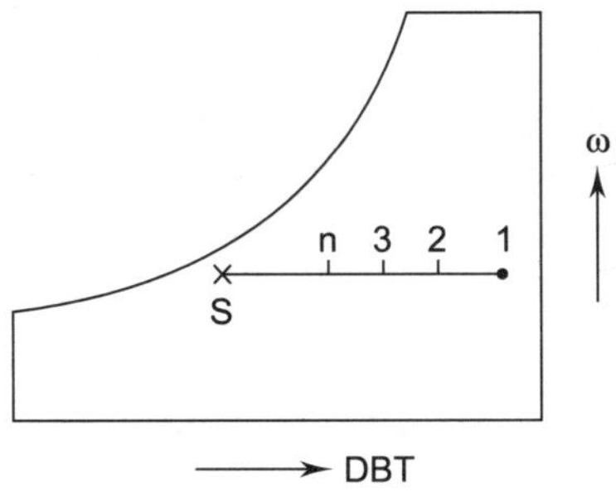


**Fig. 11.86**

---

**Example 11.26**   The bypass factor of a single depth coil is 0.75. Find the bypass factors for 3 depth and 5 depth coil. If the required bypass factor is 0.15, then find the depth of coil required.

*Sol.*

$$BF_1 = 0.75$$

$$BF_3 = (BF_1)^3$$

$$= (0.75)^3 = 0.42 \quad Ans.$$

$$BF_5 = (0.75)^5 = 0.23 \quad Ans.$$

$$0.15 = (0.75)^n, n = \frac{\ln 0.15}{\ln 0.75} = 6.6 \quad Ans.$$

---

**Example 11.27**   A 3-rows coil with a face velocity of 200 m/min has a contact factor of 0.75. Calculate the contact factor if face velocity is equal to 200 m/min for rows.

*Sol.* Given

$$V_f = 200 \text{ m/min, contact factor} = 0.75, n = 3 \text{ rows}$$

Contact factor

$$= 1 - BF$$

$$\therefore \qquad BF = 1 - 0.75 = 0.25$$

$$BF = e^{-\dfrac{Cn}{1.0216\,v_f}}, \quad V_f = \dfrac{200}{60}\ \text{m/s}$$

$$0.25 = e^{-\dfrac{c\times 3}{1.0216\times 200/60}}$$

$$0.25 = (e)^{-0.880971c}$$

$$\therefore \qquad c = 1.573$$

$$\therefore \qquad BF = e^{-\dfrac{1.573\times 6}{1.0216\times 200/60}} = 0.0625$$

$\therefore$ Contact factor $\qquad = 1 - BF = 1 - 0.0625 = 0.9375 \quad$ *Ans.*

## 11.27 SUMMER AIR-CONDITIONING SYSTEMS

The normal heat load pattern would be

  (i) Net sensible heat gain

  (ii) Net latent heat gain

Thus the process normally followed would be **cooling and dehumidification** either by the use of cooling with coil *ADP* lower than the dew point of air entering the coil, or by an air washer in which chilled water is sprayed with temperature again lower than the dew point temperature of air entering the spray.

Following systems are applied in summer air conditioning

1. Summer Air-conditioning system for Hot and Dry outdoor conditions.
2. Summer Air conditioning system for Hot and Humid outdoor conditions.
3. Summer Air conditioning with single cooling coil and mixing.
4. Summer Air conditioning with single cooling and bypass mixing
5. Summer Air conditioning with single cooling coil and Absorbent Dehumidifier
6. Summer Air conditioning with evaporative cooling.

1. **Summer Air-conditioning system for Hot and Dry outdoor conditions.**

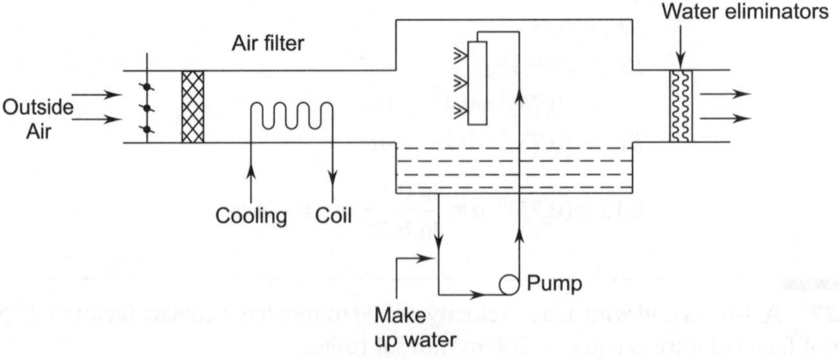


**Fig. 11.87(a)**

First air is passed through a cooling coil and its temperature is reduced up to 2 by sensible cooling. By pass factor of the cooling coil is given by

$$BF = \frac{\text{Distance}\,(4-2)}{\text{Distance}\,(4-1)} \qquad (11.100)$$

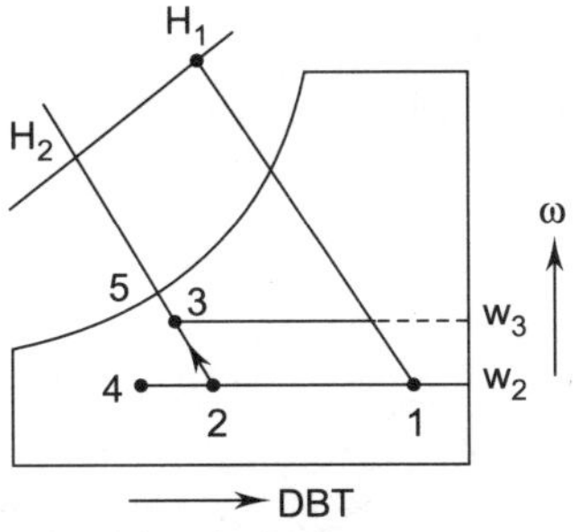

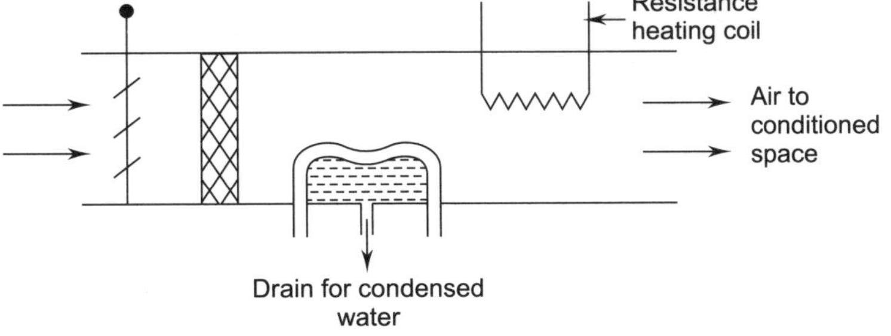



**Fig. 11.87(b)**

The air coming out from the cooling coil at the point 2 is passed into adiabatic humidifier and the conditioned air leaves the humidifier at the point 3 (Fig. 11.87 a, b). The eff. of humidifier is given by

$$\eta_h = \frac{T_2 - T_3}{T_2 - T_5} \times 100 \qquad (11.101)$$

If the quantity of air supply is $V$ cmm at atmospheric conditions then the capacities of the cooling coil and humidifier are given by

$$\text{capacity of cooling coil} \quad = \frac{V}{v_s}\frac{(h_2 - h_1)}{211} TR \qquad (11.102)$$

$$v_s = sp.\ \text{volume of atm. air}$$

$$\text{capacity of humidifier} \quad = \frac{V}{v_s}\frac{(\omega_2 - \omega_1)}{1000}\ \text{kg/min} \qquad (11.103)$$

---

* Effective temp. of 27.7°C in summer or 22°C in winter are found to be comfortable for Indians with relative humidity in the range of 35 to 65%

2. **Summer Air-conditioning system for Hot and Humid outdoor conditions**



**Fig. 11.88(a)**

First air is passed through the cooling coil for dehumidification and it comes out at condition 3 leaving the cooling coil as the cooling coil temperature is below the dew point temperature of incoming air. The bypass factor of the cooling coil is given by

$$BF = \frac{\text{Distance}\,(2-3)}{\text{Distance}\,(2-1)} \qquad (11.104)$$

$$\text{capacity of cooling coil} \quad = \frac{V}{v_s}\,\frac{(h_1 - h_3)}{211}\,\text{TR} \qquad (11.105)$$

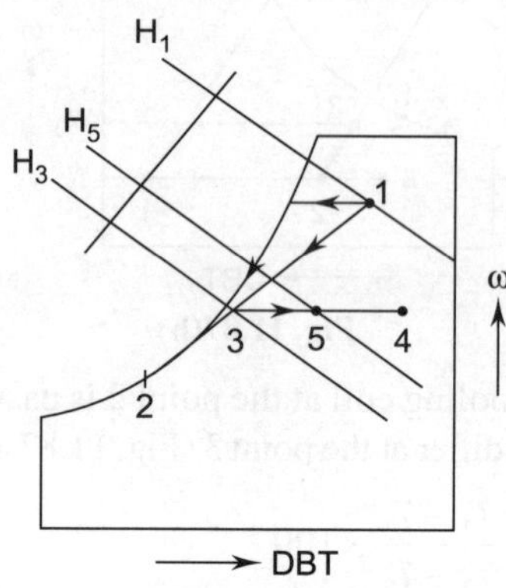


**Fig. 11.88(b)**

The air enters the heating coil at condition 3 and comes out at condition 5 then the bypass factor of the heating coil and its capacity are given by

$$BF = \frac{\text{Distance}\,(5-4)}{\text{Distance}\,(3-4)} \qquad (11.106)$$

$$\text{capacity} \quad = \frac{V}{V_s}\,(h_5 - h_3)\,\text{kJ/min} \qquad (11.107)$$

3. **Summer air-conditioning with single coil and mixing**

This type of system is used to reduce the load on the cooling coil as part of the air going out of the room which is at lower temperature than outdoor conditions is mixed with the fresh air as shown in Fig. 11.89(a).

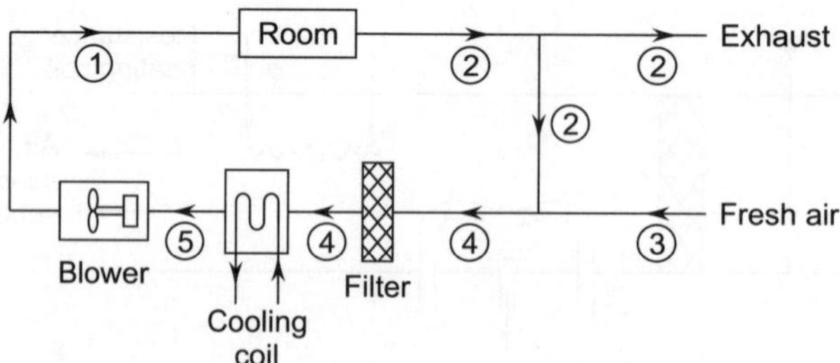


**Fig. 11.89(a)**

The condition 4 is the mixing of the air at conditions (2) and (3). The condition (5) is the condition of air leaving the cooling coil and 5-1 represents the heating of air passing through the blower due to friction. The process 1-2 represents the condition of air passing through the air conditioned room taking the load in the room, as shown in Fig. 11.89(b).

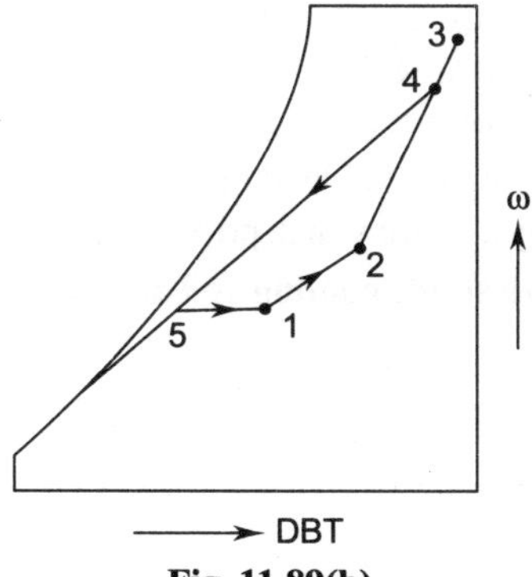

**Fig. 11.89(b)**

## 4.  Summer Air conditioning with single cooling coil and bypass mixing

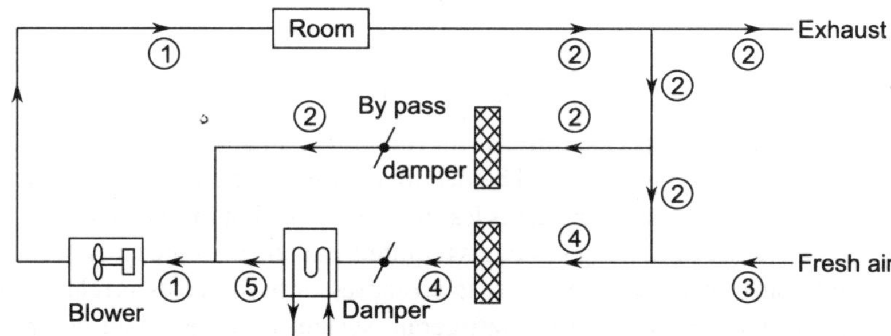

**Fig. 11.90(a)**

This system is used to control the *DBT* in the air-conditioned room as per the load in the room. The arrangement of the system is shown in Fig. 11.90(a).

The condition 4 is the mixing of the airs at conditions 2 and 3. The condition 5 is the condition of air coming out of the cooling coil. The condition 1 is the mixing of the airs at conditions 5 and 2. The process 4-5 represents the cooling and dehumidifying of air passing through the cooling coil as shown in Fig. 11.90(b). The process 1-2 represents the condition of air passing through the room as it takes the load in the room. The reheating of air passing through blower due to friction is neglected for drawing on psy. chart.

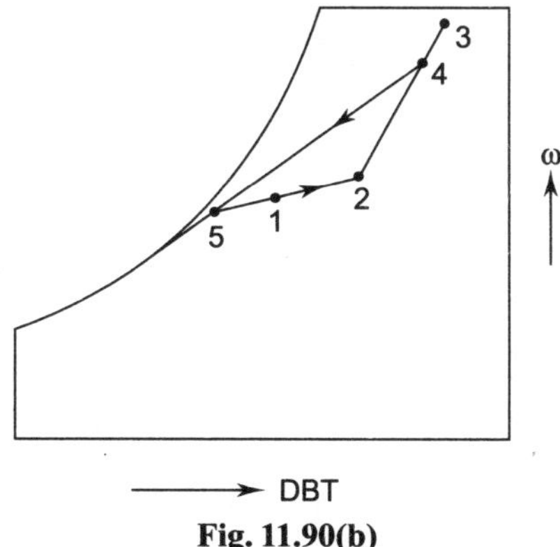

**Fig. 11.90(b)**

The system used in the previous article has some limitations as the temperature in the air conditioned room cannot be controlled according to the load in the room. The control of *DBT* is more important than humidity control as long as space humidity is not excessive.

The present system is useful during partial load operation. The factor damper on the cooling coil and bypass dampers are controlled by a motor which positions them so as to maintain a constant space DBT.

5. **Summer Air conditioning with single cooling coil and Absorbent Dehumidifier**

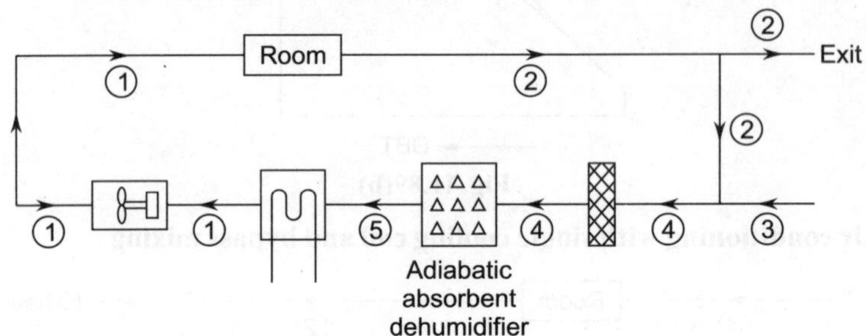


**Fig. 11.91(a)**

The cooling coil discussed in previous articles for cooling the air also produces some dehumidification in conjuction with the cooling process. The dehumidification of air by refrigerant cooling coil has limitations. If the coil surface temperature is less than 0°C, the first forms on the coil and heat transfer rate reduces and therefore defrosting system is required and reheat of the air is needed before passing into the air-conditioned space. This increases the compressor power costs as relatively low refrigerant evaporating temperature exits. Thus the refrigerant system becomes more complicated and more expensive to own and operate as the required air dew point temperature is reduced.

The absorbent system shown in Fig. 11.91(a) can reduce the required surface temperature of the cooling coil and completely avoids the possibility of frosting the coil as the required coil temperature is always 0°C. Therefore this method produces even extremely low air dew point temperatures more reliably and more cheaply than the refrigeration method. The psychrometric processes for the above described system are shown in Fig. 11.91(b).

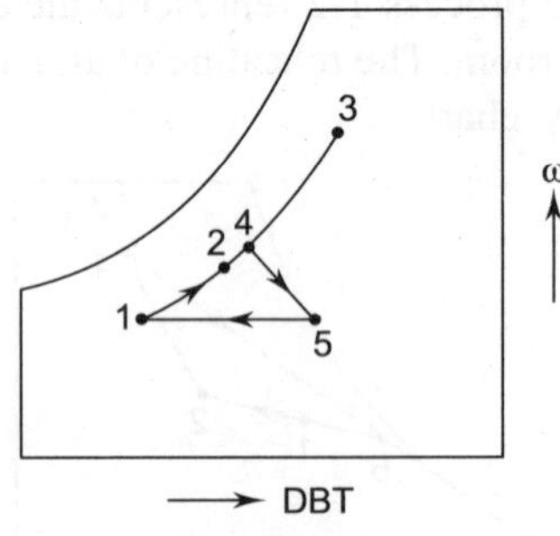


**Fig. 11.91(b)**

The condition 4 is the mixing of airs at conditions 2 and 3. The process 4-5 represents the adiabatic dehumidifi-cations of air passing through the adsorbent dehumidifier. The process 5-1 is sensible cooling of air passing through the cooling coil whose surface temperature is considerably above the temperature required for frosting. The process 1-2 is the condition of air passing through the air-conditioned room taking the existing load.

### 6. Summer Air conditioning with evaporative cooling

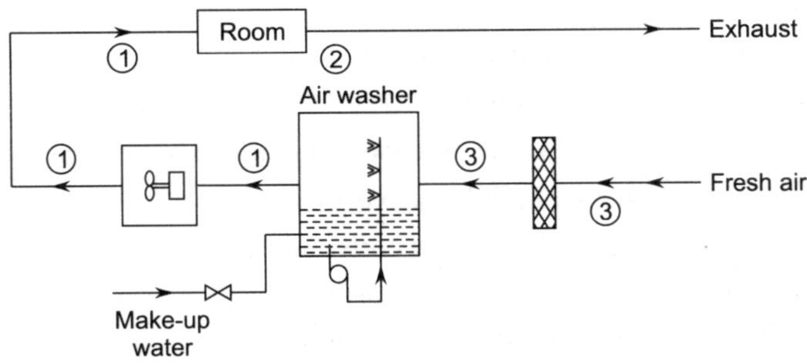


**Fig. 11.92(a)**

comfort air-conditioning system capable of maintaining optimum thermal conditions may be expensive to own and operate. Partially effective systems which involve much lesser costs may be attractive where finances preclude the installation of a completely effective system. In hot dry regions like Vidarbh in Maharashtra and central part of U.P, evaporative cooling systems may be capable of providing considerable relief in enclosed spaces.

The evaporative cooling system commonly used and corresponding processes on psychrometric chart are shown in Fig. 11.92($a$), ($b$).

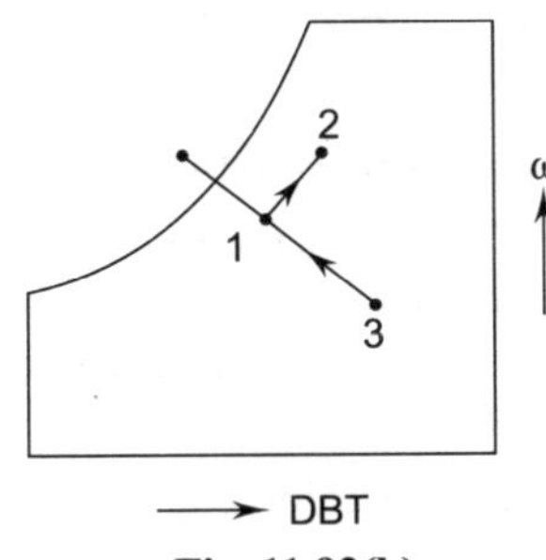


**Fig. 11.92(b)**

The process 3-1 represents an evaporative cooling and the process 1-2 represents room load taken by air passing through the room. The state 2 is an acceptable space condition although, not necessarily an optimum one. The state 3 of the outdoor air is at a much higher temperature but lower R.H than state 2. As the air washer is the only processing device in the system, the cost of the system is considerably lower than the system used for optimum comfortable conditions.

Generally, a much higher flow rate of air is used with an evaporative cooling system (2 to 3 times of conventional) than with conventional systems. A high rate of air movement past a person allows the same degree of comfort but with higher effective temperatures as compared to the situations where air movement is slight.

## 11.28   WINTER AIR CONDITIONING

In winter the load pattern is the following

   (i)  Sensible heat gains from direct solar heat through glass.

   (ii)  Sensible heat gain from internal occupancy and appliances.

(iii) Sensible and latent heat loss due to ventilation air from outside at low temperature and low specific humidity.

(iv) Sensible heat loss through walls and glass due to temperature being low outside.

On the whole the heating loads for the winter are pre-dominantly sensible heats.

Different systems are considered for winter air conditioning, which are as below

1. Winter Air-conditioning system for mild cold weather

2. Winter Air-conditioning with Double Reheat coils and Air-washer

3. Winter Air-conditioning using 100% outdoor air with preheating by waste heat from the exhaust air.

1. **Winter Air-conditioning system for mild cold weather**

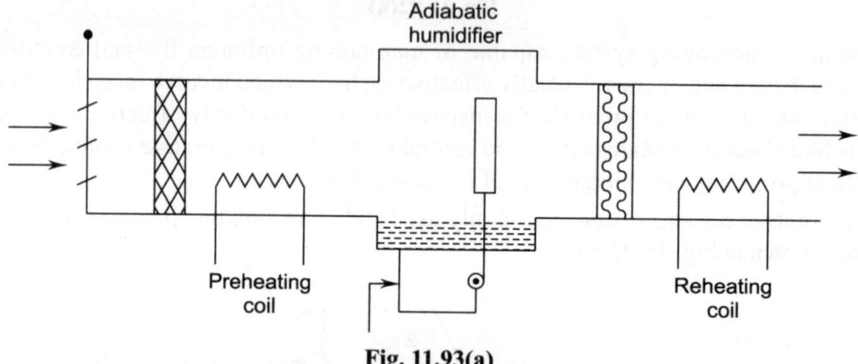


**Fig. 11.93(a)**

The winter conditions are say 15°C and 80% relative humidity. The required comfort conditions are same as 24°C and 60% R.H. The arrangement of the given equipments is used for mild winter conditions at cities like Aurangabad, Pune and identical places, as shown in Fig. 11.93(a).

$$1 = \text{atmospheric condition}$$

$$6 = \text{Required comfort condition}$$

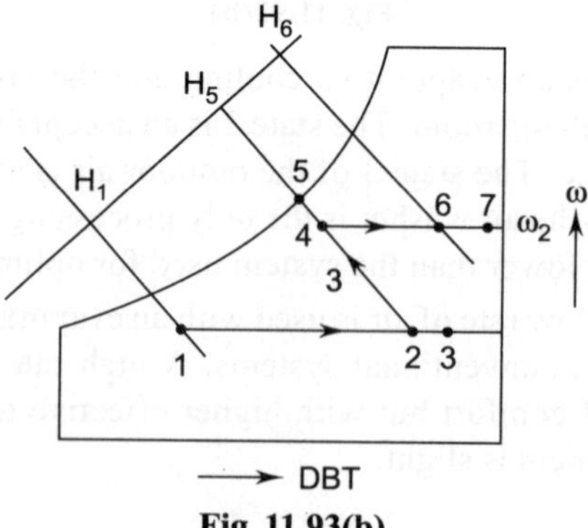


**Fig. 11.93(b)**

First the air is passed through the resistance heater which is known as preheating coil and then it passed through humidifier and again through the second heater. The effectiveness of different equipments and their capacities are given by the following expressions

$$BF_1 \text{ (Bypass factor of preheating coil)} = \frac{2-3}{1-3} \qquad (11.108)$$

$$\text{BF}_2 \text{ (Bypass factor of reheating coil)} \quad = \frac{6-7}{4-7} \tag{11.109}$$

$$\eta_h \text{ (efficiency of humidifier)} \quad = \frac{2-4}{2-5} \times 100 \tag{11.110}$$

$$\text{capacity of preheating coil} \quad = \frac{v}{v_s}\,(h_2 - h_1)\ \text{kJ/min} \tag{11.111}$$

$$\text{capacity of reheating coil} \quad = \frac{v}{v_s}\,(h_6 - h_4)\ \text{kJ/min} \tag{11.112}$$

$$\text{capacity of humidifer} \quad = \frac{v}{v_s}\,\frac{(\omega_2 - \omega_1)}{1000}\ \text{kg/min} \tag{11.113}$$

where,        $\omega = sp.$ humidity in grams/kg of dry air

$v$ = volume of atm. air circulated into room per minute

## 2. Winter Air conditioning with double reheat coils and Air-washer

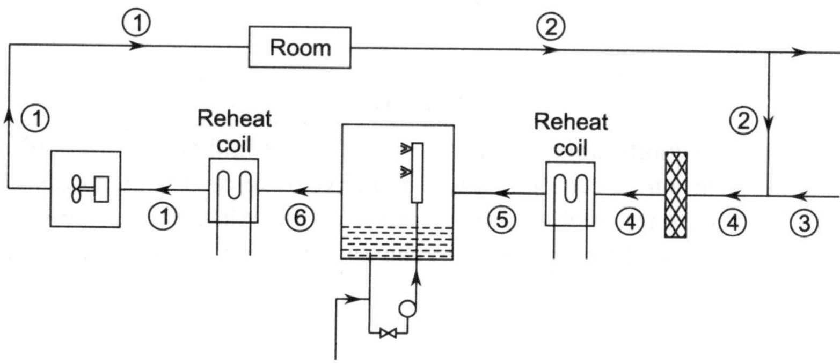


**Fig. 11.94(a)**

It is always necessary to increase DBT and RH of the air during the winter season particularly when the winter is more severe. The arrangement and processes are shown in Fig. 11.94(a), (b).

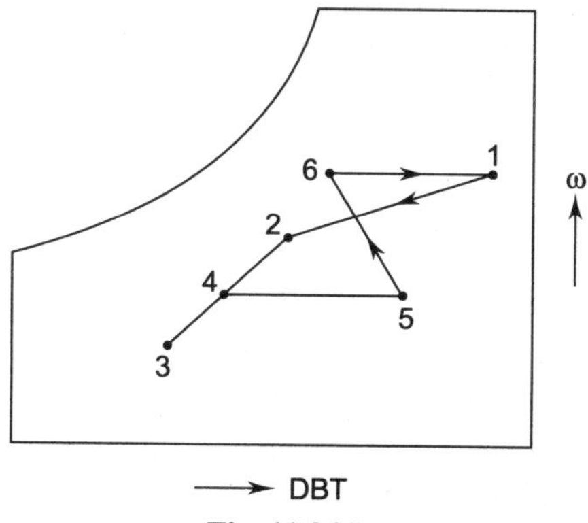


**Fig. 11.94(b)**

The condition 4 is the mixing of airs at condition 2 and 3. The process 4-5 is the sensible heating in the preheat coil, the process 5-6 is the adiabatic cooling of air passing through air washer and the process 6-1 is the sensible heating in the reheat coil. The process 1-2 is the cooling and dehumidifying of the air passing through the conditioned room to compensate for the reheat and vapour losses of the air in the air-conditioned spaces.

In large system, it is common practice to use a recirculating air fan as well as supply air fan (not shown). However this condition does not affect the processes represented on the psychrometric chart.

3.  **Winter Air-conditioning using 100% outdoor air with preheating by waste heat from the exhaust air**

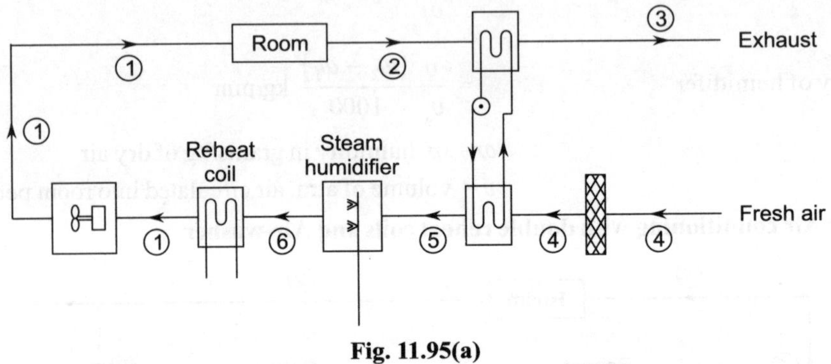


**Fig. 11.95(a)**

In designing any air-conditioning system, every efforts should be made to utilize heat transfer internal to the system where economically feasible so as to reduce the needs for external heat transfers. Such a system in which the waste heat from the exhaust air is used for preheating the fresh air and the corresponding processes air shown in Fig. 11.95(a), (b).

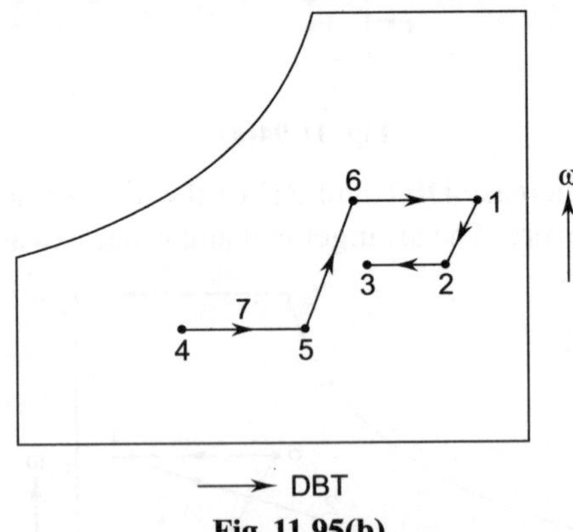


**Fig. 11.95(b)**

The air-washer serves as a humidifying device to offset the moisture losses in the air-conditioned space and in addition to this it cleans the air. The regulation of heat supply by the reheat coil allows control of the space DBT as per requirement.

The process 4-5 is the preheating of the fresh air by using the waste heat in the exhaust air where the process 2-3 shows the cooling of the exhaust air before being exhausted as it gives its sensible heat to the fresh air. The process 5-6 is the humidification of air by using steam and the process 1-2 is the cooling and dehumidifying of the air passing through the conditioned space to compensate for the heat and vapour losses of the air in the conditioned space.

In winter air-conditioning system where heating is required, the use of outdoor air should be kept to a minimum. However, for spaces which may require cooling during the entire year, provision should be made for flexible use of outdoor air so that refrigerating equipment can be shut down during cold weather conditions.

## 11.29   YEAR-ROUND AIR-CONDITIONING

In many countries the summers and winters both are very uncomfortable. Under such adverse weather conditions, it is necessary to have an air-conditioning system which will provide comfort conditions throughout the year. Year round air-conditioning system must be capable of maintaining a specified temperature and humidity within the air-conditioned spaces regardless of outside weather conditions.

The problems encountered in year-round air-conditioning systems are listed below

1.  Heating and humidifying in winter        2.  cooling and dehumidifying in summer.

An air-conditioning system used must fulfil the above requirements as per season conditions.

A refrigerating cycle using vapour as refrigerant can be used with additional humidifier and reheater for the purpose of year round air-conditioning systems. The arrangement of the valves in the circuit must be capable to change the direction of refrigerant flow according to requirements.

## 11.30   PURPOSE OF VENTILATION DURING AIR-CONDITIONING

For comfort feeling recirculated air needs freshness and as a result ventilation is required which solves the following purposes:

(i) Oxygen requirement: Atmospheric air contains 21% oxygen by volume and in no any circumstances it should go below 15%. For the metabolic activity it is required ranging from 0.035 $m^3$/hr to 0.20 $m^3$/hr.

(ii) Removal of carbon dioxide: Atmospheric air contains $CO_2$ as 0.03% by volume and it should be not more than 5% under any circumstances.

(iii) Removal or odours: 0.42 cmm of fresh air per person is required to remove body odours. The actual requirement depends on room size and level of activity.

(iv) Removal of heat and humidity: The controlling factor for ventilation is to remove body heat and moisture. For secondary adults, the body heat generation is taken as 116W per person. The ventilation requirement can be determined from the room sensible and latent heat generation rates using the following equations:

$$Q_S = Q_V \rho C_p \Delta t_{db} \tag{11.114}$$
$$Q_L = Q_V \rho h_{fg} \Delta \omega \tag{11.115}$$

where $QV$ = volume flow rate and $\Delta t_{db}$ and $\Delta w$ are allowable changes in room dry bulb temperature and humidity.

---

**Example 11.28**   One stream of air at 25°C, 45% RH is mixed with another stream of air at 45°C, 45% RH in the ratio of 1 kg for former and 2 kg of the latter. Find the final condition of the air.

*Sol.*

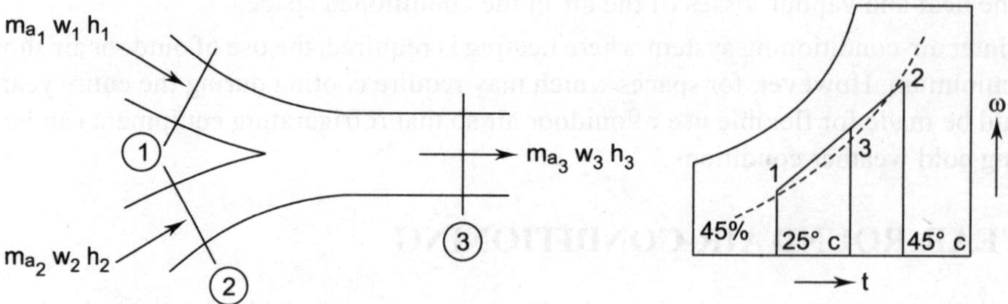

**Fig. 11.96**

For mass of dry air     $ma_1\, ma_2 = ma_3$

For moisture content of specific humidity

$$ma_1\omega_1 + ma_2\omega_2 = ma_3\omega_3$$

for enthalpies   $ma_1 h_1 + ma_2 h_2 = ma_3 h_3$

Therefore from above three equations

$$ma_1(\omega_1 - \omega_3) = ma_2(\omega_3 - \omega_2)$$

and $\qquad\quad ma_1(h_1 - h_3) = ma_2(h_3 - h_2)$

$$\therefore\qquad \frac{ma_1}{ma_2} = \frac{(\omega_3 - \omega_2)}{(\omega_1 - \omega_3)} = \frac{(h_3 - h_2)}{(h_1 - h_3)}$$

From psychrometric chart at given conditions

$$\omega_1 = 0.0085 \text{ kg/kg of dry air}, h_1 = 49 \text{ kJ/kg}$$
$$\omega_2 = 0.0275 \text{ kg/kg of dry air}, h_2 = 116 \text{ kJ/kg}$$

$$\therefore\qquad \frac{1}{2} = \frac{(\omega_3 - \omega_2)}{(\omega_1 - \omega_3)} = \frac{(\omega_3 - 0.0275)}{(0.0085 - \omega_3)}$$

$$\therefore\qquad \omega_3 = 0.0635 \text{ kg/kg of dry air}$$

---

**Example 11.29**   One stream of air at the rate of 5.5 m$^3$/min at 15°C and 60% R.H flows into another stream of air at the rate of 3.5 m$^3$/min at 25°C and 70% and R.H. Calculate for the mixture (a) dbt (b) wbt (c) R.H and (d) specific humidity.

*Sol.*

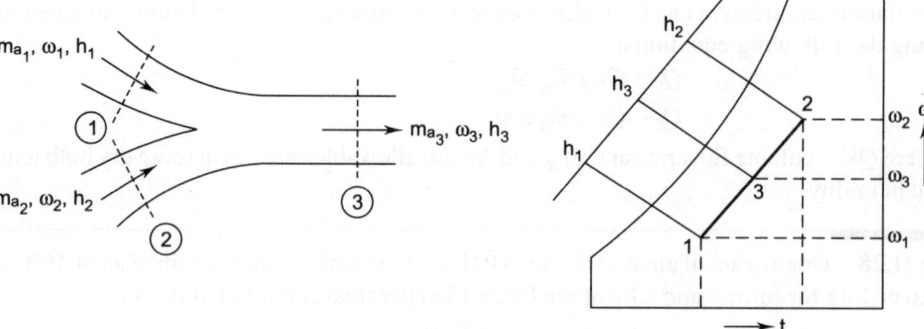

**Fig. 11.97**

Fig. 11.97 shows schematic and mixing process of two streams of air. Values taken from psychrometric chart at conditions 1 and 2.

$$\omega_1 = 0.0065 \text{ kg/kg of dry air}, \ h_1 = 69.5 \text{ kJ/kg, sp. volume}$$
$$v_1 = 0.825 \text{ m}^3\text{/kg}$$
$$\omega_2 = 0.014 \text{ kg/kg of dry air}, \ h_2 = 61 \text{ kJ/kg, sp. volume}$$
$$v_2 = 0.864 \text{ m}^3\text{/kg}$$

mass flow rate
$$\dot{m}a_1 = \frac{5.5}{0.825} = 6.67 \text{ kg/min}$$

and
$$\dot{m}a_2 = \frac{3.5}{0.864} = 4.05 \text{ kg/min}$$

join points 1 and 2 by a straight line and locate point 3 such that $\dfrac{\dot{m}a_2}{\left(\dot{m}a_1 + \dot{m}a_2\right)} = \dfrac{4.05}{(6.67 + 4.05)} = 0.378$

and hence
$$\frac{\text{Length 13}}{\text{Length 12}} = 0.378$$

From psychrometric chart we can get as
$$t_{db} = 19°C, \ t_{wb} = 15.3°C, \ \text{R.H}(\phi) = 67\% \text{ and } \omega = 0.0095 \text{ kg/kg of dry air} \quad \textit{Ans.}$$

Again
$$\frac{1}{2} = \frac{\left(h_3 - h_2\right)}{h_1 - h_3} = \frac{h_3 - 116}{49 - h_3}$$

$\therefore$
$$h_3 = 281 \text{ kJ/kg}$$

$\therefore$  Final condition of air

sp. humidity
$$w_3 = 0.0635 \text{ kg/kg of dry air}$$
and sp. enthalpy
$$h_3 = 281 \text{ kJ/kg} \quad \textit{Ans.}$$

---

**Example 11.30**  A room needs comfort condition at 20°C and relative humidity of 60% of air for which saturated air is to be supplied at 4°C. The air is heated and then water at 10°C is sprayed in to give the required humidity. Determine the temperature to which the air must be heated and the mass of spray water required per $m^3$ of air at room conditions. Assume that the total pressure is constant at 1.01325 bar. Neglect fan power.

*Sol.*

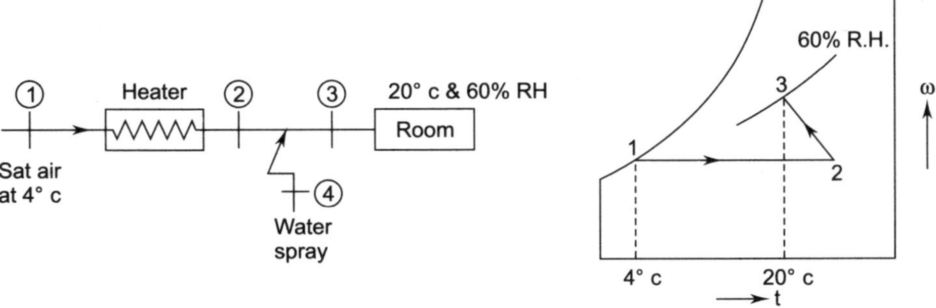


**Fig. 11.98**

Fig. 11.98 shows schematic and psychrometric process for conditioning of room. From ste am table at $20°C$, $p_{ws} = 0.02337$ bar

$$\phi_3 = \frac{p_{w_3}}{p_{ws_3}} = \frac{p_{w_3}}{0.02337} = 0.60$$

$$\therefore \quad p_{w_3} = 0.0140 \text{ bar}$$

$$\therefore \quad \omega_3 = 0.622 \frac{p_{w_3}}{\left(p_b - p_{w_3}\right)}$$

$$= 0.622 \frac{0.0140}{\left(1.01325 - 0.0140\right)}$$

$$= 0.00871 \text{ kg/kg of dry air}$$

Since saturated air at section 1 is supplied therefore $\phi_1 = 1 = 100\%$

$$\therefore \quad \phi_1 = \frac{p_{w_1}}{p_{ws_1}} = \frac{p_{w_1}}{0.00813} = 1 \qquad\qquad [\because \quad p_{ws} = 0.00813 \text{ at } 4°C]$$

$$\therefore \quad p_{w_1} = 0.00813 \text{ bar}$$

$$\therefore \quad \omega_1 = 0.622 \frac{p_{w_1}}{\left(p_b - p_{w_1}\right)}$$

$$= 0.622 \frac{0.00813}{\left(1.01325 - 0.00813\right)}$$

$$= 0.005031 \text{ kg/kg of dry air}$$

$$\omega_3 - \omega_1 = (0.00871 - 0.005031)$$

$$= 0.003679 \text{ kg/kg of dry air}$$

Specific volume of air supplied at section 3

$$v_{a_3} = \frac{R_a T_3}{p_{a_3}} = \frac{0.287\,(20 + 273)}{1.01325 \times 10^5 \times 10^{-3}} = 0.83 \text{ m}^3/\text{kg}$$

$$\therefore \quad \text{spray water} \quad = (\omega_3 - \omega_1) \times v_{a_3}$$

$$= 0.003679 \times 0.83$$

$$= 0.003053 \text{ kg moisture/m}^3$$

Now, $\qquad ma_2 h_2 + m_{w_4} h_4 = m_3 h_3$

$$h_2 + (\omega_3 - \omega_2) h_4 = h_3$$

$$h_{a_2} + \omega_2 h_{w_2} + \left(\omega_3 - \omega_2\right) h_4 = h_{a_3} + w_3 h_{w_3}$$

$$\therefore \qquad c_p(t_3 - t_2) + \omega_3 h_{w_3} - \omega_2 h_{w_2} - \left(\omega_3 - \omega_2\right) h_4 = 0$$

From steam table, at $h_{w_3} = 0.0140$ bar

$$h_g = 2523.6 \text{ kJ/kg and } t_s = 12°C. \qquad \because \quad \omega_1 = \omega_2$$

$$\therefore \quad c_p(t_3 - t_2) + \omega_3[h_g + Cp_s(20 - t_s)] - \omega_2[h_g + cp_s(t_2 - 12)] - (\omega_3 - \omega_2) \times t = 0$$

or,

$$1.005(20 - t_2) + 0.00871[2523.6 + 1.884(20 - 12)] - 0.005031[2523.6 + 1.884(t_2 - 12)]$$
$$- 0.003679 \times 10 = 0$$

$$\therefore \qquad t_2 = 29.22°C \quad Ans.$$

---

**Example 11.31**   For a hall to be air-conditioned, the following conditions are given

Outdoor condition $\qquad\qquad = 30°C$ *dbt*, 15° wbt

Required comfort condition $\quad = 15°C$ *dbt*, 60% R.H

Seating capacity of hall $\qquad = 1600$

Amount of outdoor air supplied $= 0.35 \text{ m}^3/\text{min per person.}$

If the required condition is achieved first by adiabatic humidification and then by cooling, estimate (a) the capacity of the cooling coil in tonnes, and (b) the capacity of the humidifier in kg/hr and its efficiency.

*Sol.*

From psychrometric chart

$$h_1 = h_2 = 40 \text{ kJ/kg}$$
$$h_3 = 32 \text{ kJ/kg}$$
$$\omega_1 = 0.0040 \text{ kg/kg of dry air}$$
$$\omega_2 = \omega_3 = 0.0065 \text{ kg/kg of dry air}$$
$$t_2 = 23.4°C$$
$$v_1 = 0.864 \text{ m3/kg of dry air}$$

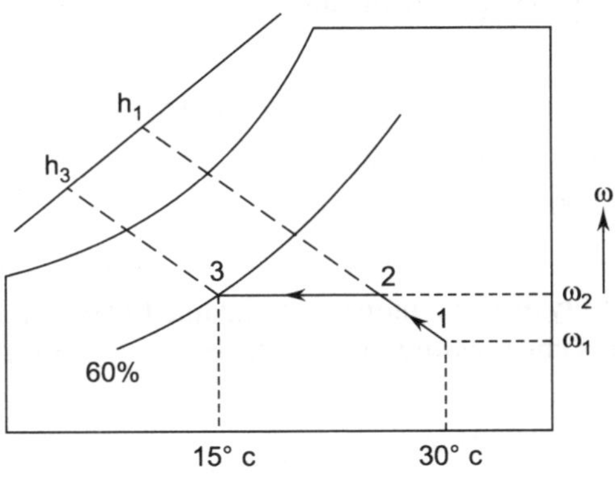


**Fig. 11.99**

Mass flow rate of air supplied

$$\dot{m}a = \frac{1600 \times 0.35}{0.864} = 648.148 \text{ kg/min}$$

$\therefore$   capacity of the cooling coil

$$= \dot{m}a \, (h_2 - h_3)$$
$$= 648.148(40 - 32) = 5185.18 \text{ kJ/min}$$
$$= 24.57 \text{ TR } Ans.$$

capacity of the humidifier

$$= \dot{m}a\,(\omega_2 - \omega_1)$$
$$= 648.148(0.0065 - 0.0040)$$
$$= 1.62037\,\text{kg/min}$$
$$= 1.62037 \times 60 = 97.22\,\text{kg/hr} \quad Ans.$$

The efficiency of the humidifier

$$\eta_h = \frac{t_1 - t_2}{t_1 - t_3} = \frac{30 - 23.4}{30 - 15} = 0.44 = 44\,\% \quad Ans.$$

**Example 11.32**   During summer season atmospheric condition on a particular day is 49°C and 20% relative humidity. If an evaporative cooler is used, which delivers the air with 90% relative humidity, determine the temperature of the cold air that leaves the cooler. If the cooler handles 1 m³/s of dry air, estimate the amount of water required to operate the cooler for a period of 5 hours.

*Sol.* With given condition of $t_{db} = 49°C$ and $\phi = 20\%$ locate point 1. Being the case of humidification and cooling proceed at constant enthalpy line to 90% R.H and locate point 2. Hence temperature of air leaving the cooler is found at as 29.5°C.

From psychrometric chart

$$\omega_1 = 0.0145\,\text{kg/kg of dry air}$$
$$\omega_2 = 0.0235\,\text{kg/kg of dry air}$$

mass flow rate of water

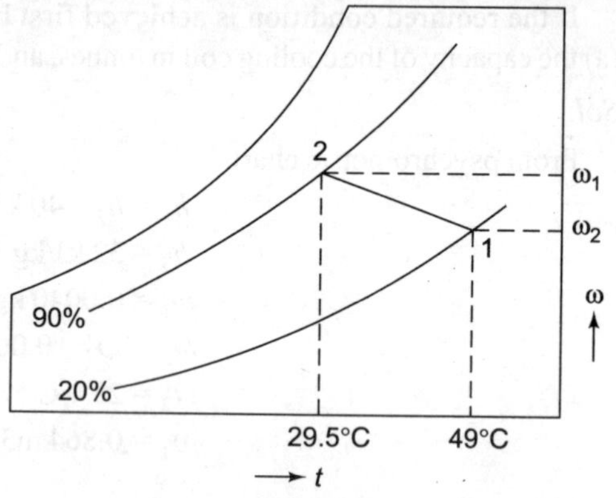


**Fig. 11.100**

$$\dot{m}_w = \dot{m}a_1\,(\omega_2 - \omega_1) = 1(0.0235 - 0.0145) = 0.0090\,\text{kg/s}$$

water required to operate the cooler for 5 hours

$$= 0.0090 \times 5 \times 3600 = 162\,\text{kg} \quad Ans.$$

**Example 11.33**   Relative humidity of a 3 m³ moist air maintained at 36°C is 80%. It has a total pressure of 1.01325 bar. It is now cooled at same pressure to 4°C. Determine the mass of moisture deposited in the process.

*Sol.* From steam table $p_{ws}$ at 30°C = 0.0594 bar

$$\phi = \frac{p_w}{p_{ws}}$$

$\therefore$
$$p_w = 0.80 \times 0.0594 = 0.04752\,\text{bar}$$

Partial pressure of the air
$$= 1.01325 - 0.04752 = 0.9657\,\text{bar}$$

Mass of water vapour present $= 0.8 \times \dfrac{3}{23.04} = 1.00\,\text{kg}$ $\qquad [\because\ v_g\ \text{at } 36°C = 23.94\ \text{m}^3/\text{kg}]$

when the temperature is 4°C the volume will be saturated.

Saturation pressure at 4°C, $p_{ws} = 0.0081\,\text{bar}$

$\therefore$      Partial pressure at 4°C $= 1.01325 - 0.0081 = 1.0044$ bar

The new volume $v_2$: $\dfrac{0.100 \times 3}{(273 + 36)} = \dfrac{1.0044 \times v_2}{(273 + 4)}$

$\therefore$                  $v_2 = 0.2677 \, \text{m}^3$

Final mass of water vapour $\quad = \dfrac{0.2677}{157.2} = 0.00170 \, \text{kg}$          $[\because \quad v_g \text{ at } 4°C = 157.2 \, \text{m}^3/\text{kg}]$

$\therefore$    Mass of moisture deposited $= (0.100 - 0.00170)$

                                   $= 0.0983 \, \text{kg} = 98.3 \, \text{gm}$    *Ans.*

---

**Example 11.34**    For comfort feeling a class room is to be kept at a temperature of 25°C and a pressure of 1.01325 bar. The air which is drawn from a supply at 5°C, 1.01325 bar and 75% relative humidity, is saturated with water vapour in a sprayer and then blown though a heater by a fan which absorbs 0.5 kw, 430 $\text{m}^3$ of air is drawn from the supply every hour. Calculate:

  (i)   the amount of water at 5°C added per hour

  (ii)   the heat supplied to the heater in kJ.

  (iii)   the final relative humidity

*Sol.* From steam table $p_{ws}$ at 5°C $= 0.0087$ bar

$$\phi = \frac{p_w}{p_{ws}}$$

$\therefore$    partial pressure of water vapour

$$p_w = 0.75 \times 0.0087 = 0.006525 \, \text{bar}$$

$$\omega_1 = \frac{p_w}{p_b - p_w}$$

$$= 0.622 \, \frac{0.006525}{1.01325 - 0.006525}$$

$$= 0.004031 \, \text{kg/kg of dry air}$$

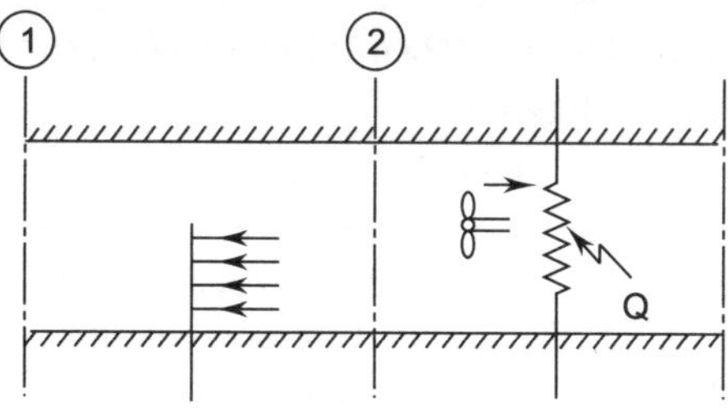


**Fig. 11.101**

Partial pressure of air          $= 1.01325 - 0.006225$

                                 $= 1.0067 \, \text{bar}$

$\therefore$      Mass of air, $m_a = \dfrac{1.0067 \times 10^5 \times 10^{-6} \times 430}{0.287 \times (273 + 5)}$

                             $= 542.55 \, \text{kg/hr}$

Mass of water vapour in air at point $1 = \omega_1 m_a$

$$= 0.004031 \times 542.55 = 2.187 \text{ kg/hr}$$

$$\omega_{2s} = 0.622 \frac{p_{ws}}{p_b - p_{ws}} = 0.622 \frac{0.0087}{1.01325 - 0.0087}$$

$$= 0.00538 \text{ kg/kg of dry air}$$

$\therefore$  Mass of water vapour at point $2 = \omega_{2s} \times m_a$

$$= 0.00538 \times 542.55$$

$$= 2.9169 \text{ kg/hr}$$

$\therefore$  Mass of water added in sprayer $= 2.9169 - 2.187$

$$= 0.7299 \text{ kg/hr} \quad Ans.$$

Partial pressure of the air at room condition for saturation at $25°C = 0.0317$ bar

After spraying no water is added, therefore

$$\omega_2 = \omega_{\text{exit}} = 0.00538$$

$$\omega_{\text{exit}} = 0.622 \frac{p_w}{p_b - p_w}$$

or,
$$0.00538 = 0.622 \frac{p_w}{1.01325 - p_w}$$

$\therefore$
$$p_w = 0.00809 \text{ bar}$$

$\therefore$  Relative humidity,  $\phi = \dfrac{p_w}{p_{ws}} = \dfrac{0.00809}{0.0317} = 0.2552 = 25.52\% \quad Ans.$

0.5 kW heat is added (being absorbed by fan)

$\therefore$
$$Q_F = 0.5 \times 3600 = 1800 \text{ kJ/hr}$$

Let
$$Q = \text{heat supplied in heater}$$

Energy balance of the system

$$ma_1(c_p t_1 + \omega_1 h_{g1}) + m_w h_f + Q_F + Q = m_{\text{exit}}(1.005\, t_{\text{exit}} + w_{\text{exit}} \times h_{g\,\text{exit}})$$

or, $542.55(1.005 \times 5 + 0.004031 \times 2510.6) + 0.7299 \times 4.187 \times 5 + 1800 + Q = 542.55(1.005 \times 25 + 0.00538 \times 2547.2)$

$$8217.04 + 15.28 + 1800 + Q = 21066.639$$

$\therefore$
$$Q = 11034.34 \text{ kJ/hr} \quad Ans.$$

---

**Example 11.35**  A class room is air conditioned of 80 seating capacity. The outdoor conditions are recorded as 35°C DBT and 20°C WBT. The required comfort conditions are 20°C DBT and 50% RH. The quantity of outdoor air supplied is 0.5 m³/min/student. The comfort conditions are achieved first by chemically dehumidifying the air and then cooling it. Find the following (i) DBT of air leaving the dehumidifier (ii) capacity of the humidifier (iii) capacity of the cooling coil in tons of refrigeration (iv) if the bypass factor of the cooling coil is 0.3, then find the surface temperature of the cooling toil (ADP).

*Sol.* Locate point 1 by the interaction 35°C DBT and 20°C WBT and point 4 at the interaction of 20°C DBT and 50% R.H.

After locating point 1, draw a constant enthalpy line which intersects horizontal line at point 2. DBT of air at point 2 is 37.5 C°.

From pychrometric chart

$$h_1 = 57.5 \, \text{kJ/kg}$$
$$h_3 = 39.5 \, \text{kJ/kg}$$

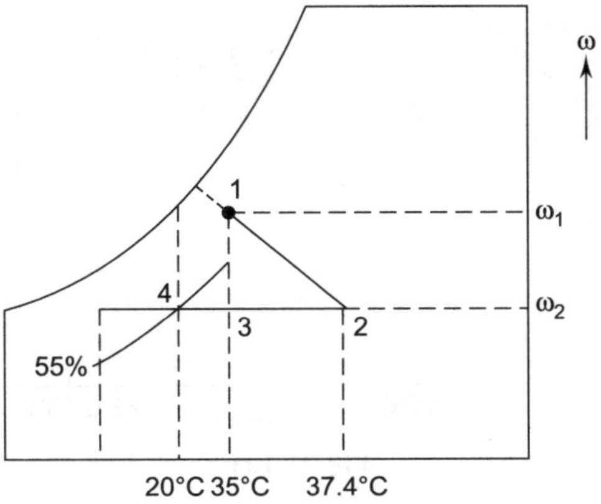

**Fig. 11.102**

$$\omega_1 = 0.008 \, \text{kg/kg}, \ \omega_3 = 0.0075 \, \text{kg/kg of dry air}$$
$$v_1 = 0.83 \, \text{m}^3/\text{kg}$$

weight of air supplied per minute

$$= \frac{0.5 \times 80}{0.83} = 48.19 \, \text{kg/min}$$

The capacity of cooling coil in tons of refrigeration

$$= \frac{48.19 \left( h_1 - h_3 \right)}{211} = \frac{48.19 \left( 57.5 - 29.5 \right)}{211}$$

$$= 4.11 \, TR \quad Ans.$$

Bypass factor of the cooling coil

$$BF = \frac{t_3 - t_4}{t_2 - t_4} = \frac{35 - t_4}{37.4 - t_4} = 0.3$$

$$\therefore \qquad t_4 = 18.29°C$$

$\therefore$   surface temperature of coil, $t_4 = 18.29°C \quad Ans.$

$$= ADP$$

---

**Example 11.36**   Outdoor condition of a room is 30°C DBT and 23°C WBT in which a volume of 310 m³ of air is supplied per minute. The air is dehumidified by a cooling coil having pypass factor 0.30 and dew point temperature 15°C and then by a chemical dehumidifier. Air leaves the chemical dehumidifier at 25°C. Air is then passed over a cooling coil whose surface temperature is 15°C and pypass factor is 0.25. Calculate the capacities of the two cooing coils and the dehumidifier.

*Sol.* Given
$$v_1 = 310 \, \text{m}^3/\text{min}, \ t_{dp_1} = 30°C, \ t_{wb_1} = 23°C, ADP = 15°C, \ t_{db_4} = 25°C, BF = 0.30$$
$$B.F = 0.25$$

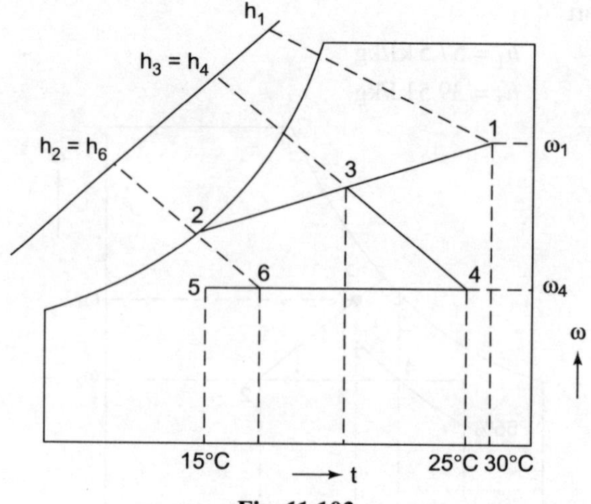


**Fig. 11.103**

Fig. 11.103 shows the psychrometric processes. Mark point 1 at given outdoor conditions as 30°C DBT and 23°C WBT. At given value of ADP = 15° mark point 2 and join line 1-2. Mark point 3 such that

length of line 2-3 = 0.30 × length of line 1-2.

Point 3 indicates about the condition of air leaving the cooling coil. Air is then chemically dehumidified. From point 3 draw a constant enthalpy line to intersect 25°C DBT line at point 4. Then after air passes over another cooling coil. From point 4 draw a horizontal line to intersect 15°C DBT vertical line at point 5. Now pypass factor of this second cooling coil is 0.25. So mark point 6 on the line 4-5 such that

length of line 5-6 = 0.25 × length of line 4-5

Point 6 is the condition of air leaving the second cooling coil.

From psychrometric chart

$$\omega_1 = 0.015 \text{ kg/kg of dry air}$$
$$h_1 = 68.2 \text{ kJ/kg}$$
$$v_1 = 0.88 \text{ m}^3\text{/kg}$$
$$h_3 = 50 \text{ kJ/kg}$$
$$h_4 = 50 \text{ kJ/kg}$$
$$\omega_4 = 0.0095 \text{ kg/kg of dry air}$$
$$h_6 = 42 \text{ kJ/kg}$$

$$\text{mass of air supplied} = m_a = \frac{300}{0.88} = 340.90 \text{ kg/min}$$

Capacity of first cooling coil $= m_a(h_1 - h_3)$
$$= 340.90(68.2 - 50) = 6204.38 \text{ kJ/min} \quad \textit{Ans.}$$

Capacity of second cooling coil $= m_a(h_4 - h_6)$
$$= 340.90(50 - 42) = 2727.2 \text{ kJ/min} \quad \textit{Ans.}$$

Capacity of the humidifier

$$= m_a(\omega_1 - \omega_4) = 340.90(0.015 - 0.0095)$$
$$= 1.874 \text{ kg/min} \quad \textit{Ans.}$$

**Example 11.37**    The following data was obtained for a certain hall to be air-conditioned.

Solar heat gain through glass    = 37620 kJ/hr

Heat gain through the hall structure = 62700 kJ/hr

No. of occupants    = 75

Sensible heat load per person   = 250.8 kJ/hr

Latent heat load per person    = 313.5 kJ/kg

Heating load due to lighting    = 1220 w

equipment-sensible heat load   = 31350 kJ/hr

Infiltration air    = 20 m$^3$/hr

Outside air condition    = 34°C DBT, 24°C WBT

Inside condition    = 25°C DBT, 50% R.H.

If the ratio of fresh air to circulated air is 2 : 1 and the pypass factor of the coil is 0.3, determine (i) supply air, cmm (ii) ADP of coil (iii) capacity of the coil in TR.

*Sol.* Mark the point 1 for outside air condition and 4 for inside air condition on psychrometric chart.

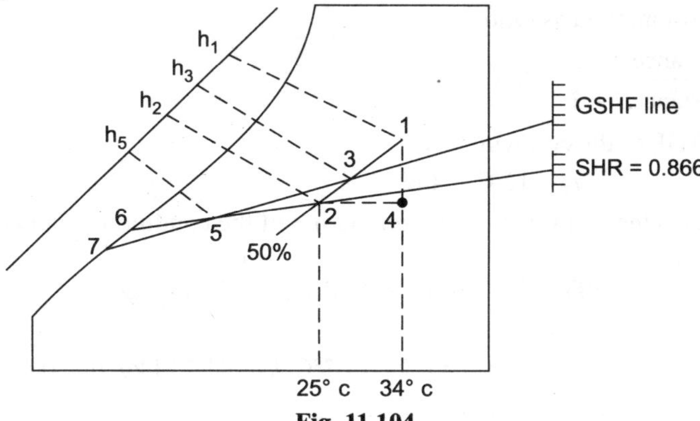


**Fig. 11.104**

At 34°C DBT and 24°C WBT locate point 1 and at 25°C DBT and 50% R.H. Locate point 2. Move horizontally from 2 to meet vertical line of 34°C DBT at point 4.

From psychrometric chart

$$h_1 = 72.5 \text{ kJ/kg}, \ h_2 = 49.7 \text{ kJ/kg}$$
$$v_1 = 0.891 \text{ m3/kg}$$

$\therefore$   Mass of infiltrated air $m_a = \dfrac{20}{0.891} = 22.44$ kg/hr

Sensible heat due to infiltrated air

$$= m_a(h_4 - h_2)$$
$$= 22.44(59.1 - 49.7) = 210.936 \text{ kJ/hr}$$

Latent heat due to infiltrated air

$$= m_a(h_1 - 4) = 22.44(72.5 - 59.1)$$
$$= 300.696 \text{ kJ/hr}$$

Human load, equipment load etc.

| S. N | Nature of load | $Q_s$ (sensible load) kJ/hr | $Q_L$ (Latent heat) kJ/hr |
|---|---|---|---|
| 1. | Solar load | 37620 + 62700 = 100320 | — |
| 2. | Equipment load | 31350 | — |
| 3. | Heat load due to lighting | 1220 W = 4392 kJ/hr | — |
| 4. | Occupancy load | 72 × 250.8 = 18810 kJ/hr | 75 × 313.5 = 23512 kJ/hr |
| 5. | Infiltration of air | 210.936 | 300.696 |
| | Total load in hall | 155082.94 | 23813.196 |

$\therefore$　Room sensible heat factor

*i.e.* 　　　SHR of room $= \dfrac{Q_s}{Q_s + Q_L} = \dfrac{155082.94}{155082.94 + 23813.196} = 0.866$

Draw a line from point 2 which is equal to SHR = 0.866 and extend to point 6 on saturation line. Mark the point 3 which is a mixture of the air at condition 1 and condition 2 *i.e.* 2 : 1 as given. The point 7 is marked by trial and error method as ratio

$$\frac{\text{Distance } 7-5}{\text{Distance } 7-3} = 0.30$$

The point 7 is the ADP of the coil which is

$$t_7 = 12°C \quad Ans.$$

The point 5 represents the condition of air coming out of the conditioner and entering the room.

So from chart, at point 3
$$DBT = 28°C, WBT = 19.8°C, h_3 = 57.5 \text{ kJ/kg}$$

and at point 5,

$$DBT = 17°C, WBT = 14.7°C, h_5 = 41.2 \text{ kJ/kg}, v_5 = 0.835 \text{ m}^3/\text{kg}$$

The total weight of air supplied to the room

$$= \frac{155082.94 + 23813.196}{h_2 - h_5} = \frac{178896.136}{(49.7 - 41.2)}$$

$$= 21046.60 \text{ kg/hr}$$

cmm of supply air at point 5

$$= 0.835 \times 21046.60 = \frac{17573.91}{60} = 292.89 \text{ m}^3/\text{min} \quad Ans.$$

The condition of air at point 5 is 17°C DBT and 80% R.H

Capacity of the coil 　　$= \dfrac{21046.60\left(h_3 - h_5\right)}{211 \times 60} TR$

$$= \frac{21046.60\left(57.5 - 41.2\right)}{211 \times 60} TR$$

$$= 27.09 \text{ TR} \quad Ans.$$

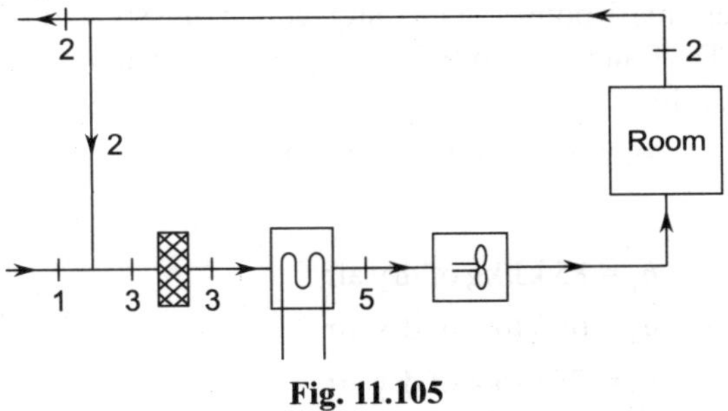


**Fig. 11.105**

**Example 11.38** An air conditioned hall is to be maintained at 27°C DBT and 21°C WBT. It has sensible heat load of 46.5 kW and latent heat load of 17.5 kW. The air supplied from outside atmospheric is at 38°C DBT and 27°C WBT at 25 m³/min rate directly into the room. Outside air to be conditioned is passed through cooling coil whose ADP is 15°C. The quantity of recirculated air from the hall is 60%. This quantity of air is mixed with conditioned air after the cooling coil. Determine the (i) condition of air after the coil and before the recirculated air mixes with it, (ii) condition of air entering the hall, (iii) mass of fresh air entering the cooler, (iv) bypass factor of the cooling coil, (v) refrigerating load on the cooling coil.

Given $t_{db_4} = 27°C,\ t_{wb_4} = 21°C,\ Q_{s_4} = 46.5\ \text{kw},\ Q_{L_4} = 17.5\ \text{kw},\ t_{db_1} = 38°C,$

$$t_{wb_1} = 27°C,\ v_1 = 25\ \text{m}^3/\text{min},\ \text{ADP} = 15°C$$

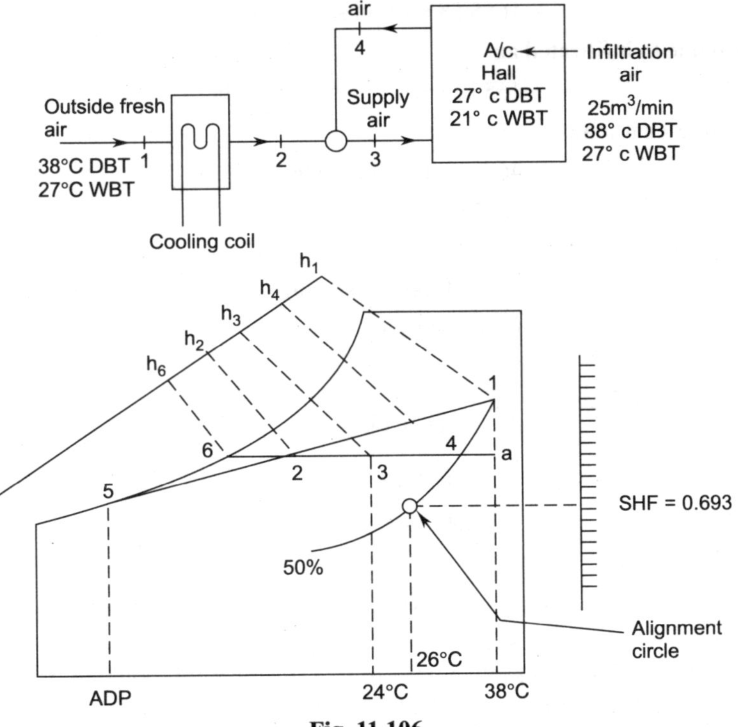


**Fig. 11.106**

On the psychrometric chart mark point 1 at the intersection of 38°C DBT and 27°C WBT. Mark point 4 at the intersection of 27°C DBT and 21°C WBT. Mark point "*a*" by drawing vertical and horizontal lines from points 1 and 4 respectively.

Condition of infiltrated air is 38°C DBT and 27°C WBT

From psychrometric chart

$$h_1 = 85 \text{ kJ/kg of dry air}$$
$$h_4 = 61 \text{ kJ/kg of dry air}$$
$$h_a = 73 \text{ kJ/kg of dry air}$$
$$v_1 = 0.906 \text{ m}^3/\text{kg of dry air}$$

$\therefore$   Mass of infiltrated air $= \dfrac{25}{0.906} = 27.593$ kg/min

$$= 1655.58 \text{ kg/hr}$$

Sensible heat due to this infiltrated air

$$= (h_a - h_4) = (73 - 61) = 12 \text{ kJ/kg of dry air}$$

Latent heat load due to this infiltrated air

$$= (h_1 - h_a)$$
$$= (85 - 73) = 12 \text{ kJ/kg of dry air}$$

Sensible heat load due to infiltrated air

$$= 1655.58 \times 12 = 19866.96 \text{ kJ/hr}$$

Latent heat load due to infiltrated air

$$= 1655.58 \times 12 = 19866.96 \text{ kJ/hr}$$

Thus total sensible heat load

$$Q_s = 46.5 \times 3600 + 19866.96$$
$$= 187266.96 \text{ kJ/hr}$$

and total latent heat load

$$Q_L = 17.5 \times 3600 + 19866.96$$
$$= 82866.96 \text{ kJ/hr}$$

$\therefore$   Room sensible heat factor, $\text{RSHF} = \dfrac{Q_s}{Q_s + Q_L}$

$$= \dfrac{187266.96}{187266.96 + 82866.96}$$
$$= 0.693$$

Now, join the alignment circle (26°C, 50% R.H) with SHF = 0.693 and draw a line 4-6 parallel to this line to pass through point 4. The line gives RSHF line. The supply air to the cooling coil is at point 1 and the ADP of the coil is given as 15°C. Thus join point 1 to 15°C on the saturation line. This line gives GSHF or cooling coil sensible heat factor. Intersection of RSHF and GSHF line gives point 2. This point gives the condition of the air after cooling coil. Such air shall be able to take up the sensible and latent heat of the hall. Also 60% of the hall air is recirculated and mixed with this air. Thus condition after mixing is shown

such that $\dfrac{\text{length } 2-3}{\text{length } 2-4} = 0.6$

Thus locate point 3.

Condition of air after the coil and before recirculated air mixes with it is represented by point 2.

*i.e.*   at point 2 $\qquad\qquad$ DBT = 21°C

$\qquad\qquad\qquad\qquad$ WBT = 18.6°C   *Ans.*

Condition of air supplied to the hall after mixing at point 3.

*i.e.*   at point 3

$\qquad\qquad\qquad\qquad$ DBT = 24.5°C

$\qquad\qquad\qquad\qquad$ WBT = 20°C   *Ans.*

Mass of fresh air passing through the cooling coil that must take up the sensible and latent heat of the hall is given by

$$m_a = \frac{\text{Total heat removed}}{\left(h_4 - h_2\right)}$$

$$= \frac{270133.92}{\left(61 - 52.8\right)} = 32943.161 \text{ kg/hr}\quad \textit{Ans.}$$

Bypass factor of the cooling coil

$$\text{B.F} = \frac{t_{db_2} - ADP}{t_{db_1} - ADP} = \frac{21 - 15}{38 - 15} = 0.26\quad \textit{Ans.}$$

Refrigeration load on cooling coil $= m_a(h_1 - h_2)$

$$= \frac{32943.161\left(85 - 52.8\right)}{211 \times 60}\ \text{TR}$$

$$= 84.18 \text{ TR}\quad \textit{Ans.}$$

---

**Example 11.39**   An air conditioned room is maintained at 25°C DBT and 50% RH, whose sensible heat load is 11.5 kw and latent heat load is 7.5 kw, when outside conditions are 35°C DBT and 28°C WBT. Return air from the room is mixed with outside air before entering the cooling coil in the ratio of 4 : 1 and return air from the room is also mixed after the cooling coil in the same ratio of 4 : 1. The cooling coil has the pypass factor 0.1. The air may be reheated, if necessary before supplying to the conditioned room. Assuming ADP of 8°C determine (i) supply of air condition to room (ii) refrigeration load and (iii) quantity of fresh air supplied.

Given $\qquad\qquad\qquad\qquad t_{db_2} = 25°C,\ \phi = 50\%,\ RSH = 11.5\ \text{kw},\ RLH = 7.5\ \text{kw},\ t_{db_1} = 35°C,$

$$t_{wb_1} = 28°C,\ BF = 0.1,\ ADP = 8°C$$

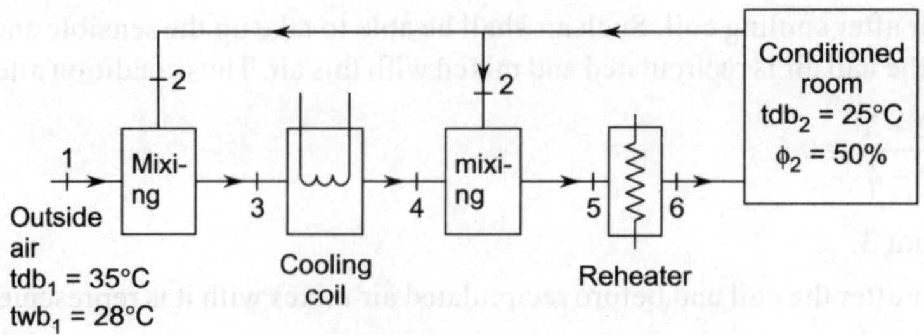



**Fig. 11.107(a)**

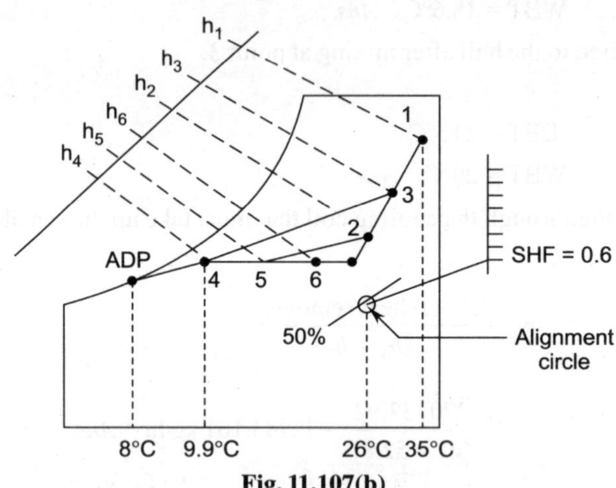



**Fig. 11.107(b)**

Mark point 1 for outside air condition at 35°C DBT and 28°C WBT. Mark point 2 for inside air condition at 25°C DBT and 50% R.H. The return air from the room and outside air is mixed in the ratio of 4 : 1 before entering the cooling coil. Thus mark point 3 on the chart such that

$$\text{length 2-3} = \frac{\text{length 1} - 2}{5}$$

From chart at point 3,     $t_{db} = \text{DBT} = 27°C$ and $h_3 = 58.4$ kJ/kg of dry air.

Let us assume,     $t_{db_4} = $ DBT of air leaving the coil

$$\therefore \quad BF = \frac{t_{db_4} - ADP}{t_{db_3} - ADP}$$

or,     $0.1 = \dfrac{t_{d_4} - 8}{27 - 8}$     $\therefore \quad t_{db_4} = 9.9°C$

Mark point 4 on the line joining point 3 and ADP = 8°C on the saturation curve such that the DBT = 9.9°C. This point 4 corresponds to condition of air leaving the coil. Enthalpy at the point $4 = h_4 = 28.2$ kJ/kg.

Since the return air is also mixed with air leaving the coil in ratio 4 : 1, we can mark point 5 on the line 4 – 2 such that

$$\text{Length 4} - 5 = \frac{\text{Length 4} - 2}{5}$$

Point 5 is the condition of air after mixing from chart, $DBT = t_{db_5} = 13°C$

and enthalpy, $h_5 = 33$ kJ/kg

$$\therefore \quad \text{Room sensible heat factor RSHF} = \frac{RSH}{RSH + RLH}$$

$$= \frac{11.5}{11.5 + 7.5} = 0.6$$

Draw a horizontal line from point 5 to intersect RHSF line at point 6. This represents the condition of supply air to the room.

From chart,
$$t_{db_6} = 17°C = DBT$$
$$t_{wb_6} = 13.2°C = WBT \text{ Ans.}$$
$$h_6 = 37 \text{ kJ/kg}$$

Also
$$h_2 = 50.5 \text{ kJ/kg}$$

$$\therefore \quad \text{Total amount of supply air} = \frac{\text{Room total heat}}{\text{Total heat removed/kg}}$$

$$= \frac{RSH + RLH}{(h_2 - h_6)} = \frac{11.5 + 7.5}{50.5 - 37}$$

$$= 1.386 \text{ kg/s} = 83.21 \text{ kg/min}$$

Amount of dehumidified air
$$= 83.21 \times \frac{4}{5} = 66.569 \text{ kg/min}$$

Refrigeration load

$$= 66.569(h_3 - h_4)$$
$$= 66.569(58.4 - 28.2) = 2010.38 \text{ kJ/min}$$
$$= 2010.38/211 = 9.52 \text{ TR Ans.}$$

Quantity of fresh air supplied
$$= \frac{66.569}{5}$$
$$= 13.31 \text{ kg/min} \quad Ans.$$

**Example 11.40**   The following data refers to a summer air conditioning plant

Inside design condition: 20°C and 50% R.H

Outside design condition: 40°C and 30% R.H

Total sensible heat gain $= 100,000$ kJ/hr

Latent heat gain $= 28000$ kJ/hr

Temperature of conditioned air entering the space $= 15°C$

Fresh air required for ventilation purposes $= 1500$ m³/hr

Barometric pressure $= 1.01325$ bar.

Calculate:

(i)  the dry bulb temperature and R.H of the air entering the cooling coil

(ii)  the kW of refrigeration required.

*Sol.*

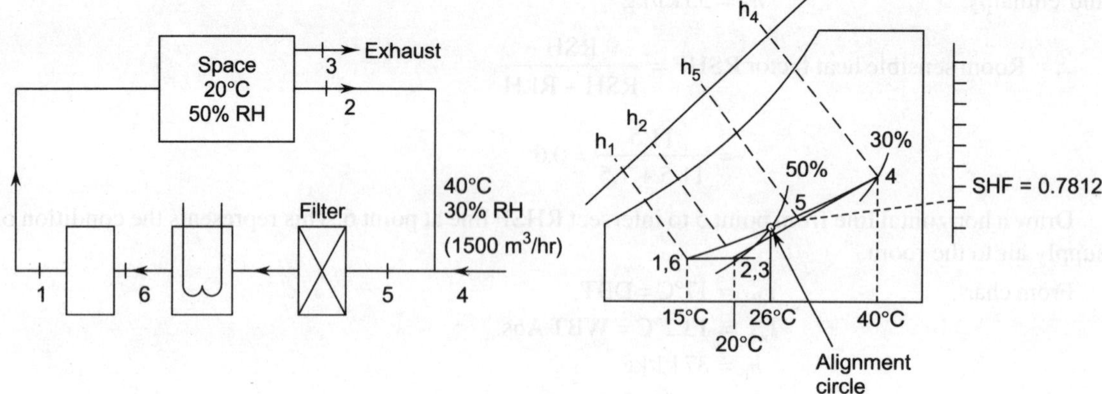

**Fig. 11.108**

Given $t_{db_2} = 20°C$, $\phi_2 = 50\%$, $t_{db_4} = 40°C$, $\phi_4 = 30\%$, $Q_s = 100{,}000$ kJ/hr,

$$Q_L = 28000 \text{ kJ/hr}$$

$\therefore$

$$\text{SHF} = \frac{Q_s}{Q_s + Q_L} = \frac{100000}{100000 + 28000} = 0.7812$$

Points 2 and 4 are located at 20°C DBT and 50% RH and 40°C DBT and 30% R.H. Hence $h_2 = 39.0$ kJ/kg, $h_4 = 76.5$ kJ/kg. Draw line 2-1 parallel to SHF = 0.7812. Point 1 lying on 15°C dry bulb temperature line. Locate point 1 and thus $h_1 = h_6 = 33$ kJ/kg

Mass of air supplied to room $= \dfrac{Q_s}{C_{p_a} \Delta t} = \dfrac{100{,}000}{1.005\,(20-15)}$

$$= 19900 \text{ kg/hr}$$

From stream table

$$p_{ws} \text{ at } 40°C = 0.0738 \text{ bar}$$

$$\phi = \frac{p_w}{p_{ws}} \qquad \therefore \quad p_w = 0.3 \times 0.0738 = 0.02214 \text{ bar}$$

$\therefore$ Mass of ventilated air at its partial pressure $= \dfrac{PV}{RT} = \dfrac{(1.01325 - 0.02214) \times 100 \times 1500}{0.287 \times (40 + 273)}$

$$= 1654.957 \text{ kg/hr}$$

$\therefore$ Mass of air recirculated $= 19900 - 1654.957$

$$= 18245.043 \text{ kg/hr}$$

Let us consider, point 5 represents the condition after mixing, then

$$\frac{ma_2}{ma_4} = \frac{h_4 - h_5}{h_5 - h_2}$$

or,
$$\frac{18245.043}{1654.957} = \frac{76.5 - h_5}{h_5 - 39.0}$$

$\therefore \qquad\qquad h_5 = 42.12\,\text{kJ/kg}$

At this value locate point 5 on line $2 - 4$. Then dry bulb temperature of air entering cooling coil at 5 $i.e.$

$$t_{db_5} = 21.7°C \quad Ans.$$

and relative humidity at 5 $i.e.$ $\quad \phi_5 = 48\% \quad Ans.$

$$\text{Cooling load} = m_a(h_5 - h_6)$$

$$= \frac{19900}{3600}(42.12 - 33)$$

$$= 50.413\,\text{kw} \quad Ans.$$

---

**Example 11.41** Air at 10°C DBT and 90% R.H is to be brought to 35°C DBT and 22.5°C WBT with the help of a winter air conditioner. If the humidified air comes out of the humidifier at 90% R.H, draw the various processes on a psychrometric chart. Find (i) temperature at which air should be preheated (ii) efficiency of air washer.

*Sol.* Given

$$t_{db_1} = 10°C,\ \phi_1 = 90\%,\ t_{db_2} = 35°C,\ t_{wb_2} = 22.5°C$$

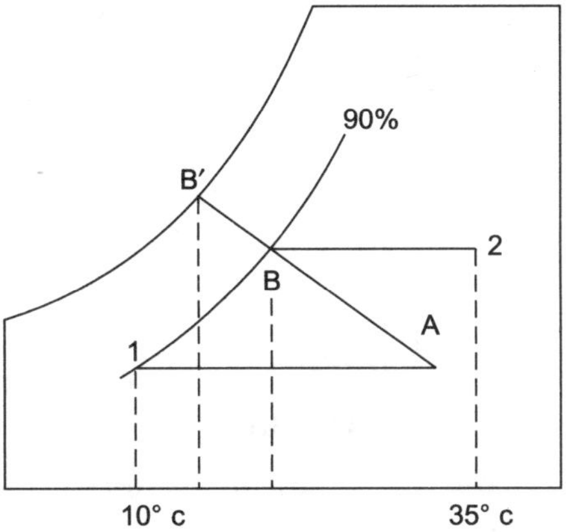

**Fig. 11.109**

Mark the point 1 at 10°C DBT and 90% RH. Similarly mark point 2 at 35°C and 22.5°C WBT.

Draw a horizontal line from 2 to 90% RH line which meets at point $B$. Draw a constant wet bulb temperature line to the saturation curve ($i.e.$ 100% RH) at $B'$ and then extend to meet a horizontal line drawn from point 1 at $A$. Process $1 - A$ indicates about the preheating of air and $A - B$ indicates about the preheating of air and $A - B$ indicates about humidifying of air. Process $B$-2 is reheating of air.

From psychrometric chart

$$t_{db_B} = 18°C,\ t_{db_{B'}} = 16.4°C$$

Temperature to which air should be preheated = DBT at point $A = t_{db_A} = 28.5°C \quad Ans.$

$$\text{Efficiency of air washer} = \frac{\text{Actual drop in DBT}}{\text{Ideal drop in DBT}}$$

$$= \frac{t_{db_A} - t_{db_B}}{t_{db_A} - t_{db_{B'}}}$$

$$= \frac{28.5 - 18}{28.5 - 16.4} \times 100$$

$$= 86.77\% \quad \textit{Ans.}$$

---

**Example 11.42**   Initial condition of air available at $-3°C$ DPT and $9°C$ DBT is heated to $17°C$. Then it flows through an air washer in which water constantly recirculates without being heated or cooled. After leaving the air washer the air is reheated to final condition of $19°$ DBT and $12.0°C$ WBT. Find (i) WBT, $\phi$ and DBT of air leaving the air washer, (ii) heating required in each coil per kg of dry air, (iii) bypass factor of air washer and (iv) efficiency of air washer.

*Sol.*

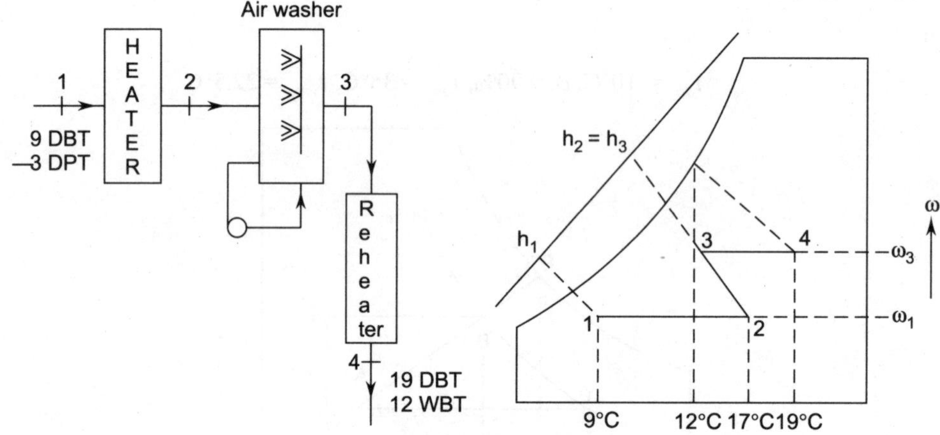


**Fig. 11.110**

Given $\qquad t_{db_1} = 9°C, \ t_{dp_1} = -3°C, \ t_{db_4} = 19°C, \ t_{wb_4} = 12°C, \ t_{db_2} = 17°C$

On psychrometric chart mark point 1 at $9°C$ DBT and $-3°C$ DPT. Draw horizontal line from point 1 which meets vertical line of DBT of $17°C$ at point 2. Locate point 4 at $19°C$ DBT and $12°C$ WBT. Heating being at constant specific humidity process, draw a horizontal line from point 4 and constant enthalpy line from point 2 which meet at point 3.

From psychrometric chart

$$h_1 = 16.8 \text{ kJ/kg}, \ \omega_1 = \omega_2 = 0.0034 \text{ kg/kg of dry air}$$

$$h_4 = 32.5 \text{ kJ/kg}, \ \omega_4 = \omega_3 = 0.0055 \text{ kg/kg of dry air}$$

$$h_3 = 23.75 \text{ kJ/kg}, \ \phi_3 = 64\%, \ t_{db_3} = 11.8°C, \ t_{wb_3} = 8.4°C$$

$$h_2 = 23.75 \text{ kJ/kg}, \ t_{wb_2} = 8.2°C \quad \textit{Ans.}$$

Heating load in first heater $\quad = h_2 - h_1 = (23.75 - 16.8)$

$\qquad\qquad\qquad\qquad\qquad\quad = 6.95 \text{ kJ/kg of dry air}$

Heating load in second heater $= (h_4 - h_3)$

$\qquad\qquad\qquad\qquad\qquad\quad = (32.5 - 23.75)$

$\qquad\qquad\qquad\qquad\qquad\quad = 8.75 \text{ kJ/kg of dry air}$

Efficiency of air washer $\qquad = \dfrac{\left(t_{db_2} - t_{db_3}\right)}{\left(t_{db_2} - t_{wb_2}\right)}$

$\qquad\qquad\qquad\qquad\qquad\quad = \dfrac{17 - 11.8}{17 - 8.2} \times 100$

$\qquad\qquad\qquad\qquad\qquad\quad = 59.09\% \quad Ans.$

$\therefore \quad$ Bypass factor (B.F) of air washer $= 1 - \text{efficiency}$

$\qquad\qquad\qquad\qquad\qquad\quad = 1 - 0.5909$

$\qquad\qquad\qquad\qquad\qquad\quad = 0.4091 \quad Ans.$

---

**Example 11.43**  A cooling tower is to be designed to cool 8.0 litres of water per second. Initial temperature of water is 44°C. The motor driven fan handles 10 m³/s of air through the tower. Power consumed for which is 5.0 kw. The air enters the tower at 20°C DBT and has a relative humidity of 60%. The air leaves the tower at saturated condition at 26°C DBT.

Calculate

(i)  the amount of make-up water required per second

(ii)  the final temperature of the water.

Solve the problem without using psychrometric chart.

*Sol.*

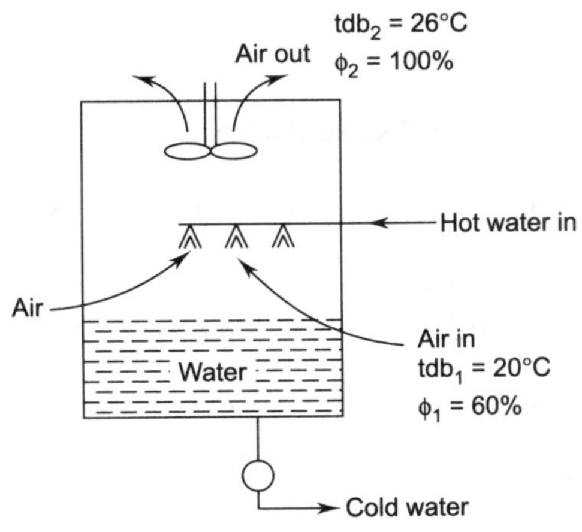


**Fig. 11.111**

Given, $\qquad\qquad\qquad\qquad t_{db_2} = 26°C,\ \phi_2 = 100\%$

$$t_{db_1} = 20°C, \phi_1 = 60\%$$

Fig. 11.111 gives the schematic diagram showing air enters at section 1 and leaves out at section 2.
At inlet

From steam table, at $\qquad t_{db_1} = 20°C$

$$p_{ws} = 0.0234 \text{ bar}$$

$$\phi = \frac{p_w}{p_{ws}} \qquad \therefore \quad p_w = \phi \times p_{ws} = 0.60 \times 0.0234$$

$$= 0.01404 \text{ bar}$$

$\therefore$ Partial pressure of dry air

$$= (1.01325 - 0.01404) = 0.99921 \text{ bar}$$

$\therefore$ Mass of air entering, $ma_1 = \dfrac{0.99921 \times 10^5 \times 10}{287 \times (20 + 273)}$

$$= 11.882 \text{ kg/s}$$

Mass of water vapour entering

$$m_{v_1} = \frac{0.01404 \times 10^5 \times 10}{461.88 \, (20 + 273)}$$

$$= 0.1030 \text{ kg/s} \qquad [\because \quad R \text{ for water vapour} = \frac{8314}{18} = 461.88 \text{ J/kgk}]$$

At exit $\qquad ma_1 = ma_2$

At $\qquad t_{db_2} = 26°C, p_{ws} = 0.0336 \text{ bar}$

$\therefore \qquad p_w = \phi \times p_{ws} = 1 \times 0.0336 = 0.336 \text{ bar}$

$\therefore \qquad \omega_2 = 0.622 \, \dfrac{p_w}{p_b - p_w} = 0.622 \, \dfrac{0.0336}{1.01325 - 0.0336}$

$$= 0.0213 \text{ kg/kg of dry air}$$

$\because \qquad \omega_2 = \dfrac{m_{v_2}}{ma_2} \qquad \therefore \quad m_{v_2} = 0.0213 \times 11.882 = 0.2530 \text{ kg/s}$

Hence, quantity of make-up water required

$$= (0.2530 - 0.1030) = 0.15 \text{ kg/s} \quad Ans.$$

Mass of water cooled per second $= m_{w_1} = 8 \times 1 = 8$ kg/s

$\therefore \qquad m_{w_2} = m_{w_1} - \text{make up water}$

$$= 8 - 0.15 = 7.85 \text{ kg/s}$$

Applying steady flow energy equation and neglecting changes in kinetic energy and potential energy,
we get

$$W + m_{w_1} h_{w_1} + ma_1 h_{a_1} + m_{v_1} h_{v_1} = ma_2 h_{a_2} + m_{v_2} h_{v_2} + m_{w_2} h_{w_2}$$

work input, $\qquad W = 5 \text{ kw} = 5 \text{ kJ/s}$

$$h_{v_2} = h_g \text{ at } 26°C = 2549 \text{ kJ/kg}$$

$$h_{w_1} = 44 \times 4.187 = 184.228$$

At 20°C, $\qquad h_{a_1} = 1.005(20 - 0) = 20.1 \text{ kJ/kg}$

At 26°C, $\qquad h_{a_2} = 1.005(26 - 0) = 26.13 \text{ kJ/kg}$

$$h_{v_1} = 4.187 \times DBT + hfg \text{ at } DBT$$

$$= 4.187 \times 20 + 2454.1 = 2537.84 \text{ kJ/kg}$$

$$h_{v_2} = 4.187 \times 26 + 2439.9 = 2548.76 \text{ kJ/kg}$$

Then by substituting, weget

$$5 + 8 \times 184.228 + 11.882 \times 20.1 + 0.1030 \times 2537.84$$

$$= 11.882 \times 26.13 + 0.2530 \times 2549 + 7.85 \times h_{w_2}$$

$$5 + 1473.824 + 238.828 + 261.397$$

$$= 310.477 + 644.897 + 7.85 \, h_{w_2}$$

$$1979.049 = 955.374 + 7.85 \, h_{w_2}$$

$\therefore \qquad h_{w_2} = 130.40 \text{ kJ/kg}$

At this value, from steam table, $h_f = h_{w_2} = 130.40 \text{ kJ/kg}$

Corresponding temperature $\quad = 31.09°C$

Hence final temperature of water $= 31.09°C$ *Ans.*

# EXERCISES

10.1 Objective type questions **(refrigeration)**

(i) Power required to run air compressor is minimum of compression follows

(ii) Refrigeration is defined as science of providing and maintaining temperatures below that of ...temperature.

(iii) The vapour compression refrigerants are based on the principle that fluid absorbs heat while changing from a .......and give up heat in changing from a vapour phase to liquid phase.

(iv) Heat is rejected by a refrigerant during a refrigeration cycle in a ....

(v) A domestic refrigerator has a co-efficient of performance.....than one.

(vi) The co-efficient of performance (C.O.P) is defined as the ratio of.........to the work done on the system.

(vii) Relative co-efficient of performance is equal to the ratio of...........and theoretical C.O.P.

(viii) The C.O.P of a refrigerating machine, when lower temperature is fixed, can be improved by.........the higher temperature.

(ix) Heat absorbed in the refrigeration cycle from the body or space to be cooled is known as..............

(x) One ton of refrigeration is equivalent to...............

(xi) The co-efficient of performance (C.O.P) of a refrigerator working on a reversed carnot cycle is 5. The ratio of highest absolute temperature to lowest absolute temperature (i.e. TH/TL) would be...........

(xii) C.O.P of a heat pump working on reversed carnot cycle is 5. The ratio of highest absolute temperature is lowest absolute temperature would be..........

(xiii) The cycle, on which air refrigerator works, is...........

(xiv) Dry compression cycle for a vapour compression cycle means the entire compression of the vapour should be in............region.

(xv) In a vapour compression refrigerator, the oil separator is installed between..........

(xvi) The co-efficient of performance in wet compression as compared to dry compression is...........

(xvii) Undercooling in a refrigeration cycle..........the net refrigerating effect.

(xviii) C.O.P of vapour compression refrigerating cycle using expansion cylinder instead of throttle valve is.........

(xix) If a liquid refrigerant is having low specific heat, then co-efficient of performance will be........

(xx) The net refrigerating effect for a refrigerant is...........proportional to latent heat of refrigerant.

(xxi) The specific volume of a refrigerant should be..........

(xxii) The C.O.P of the vapour absorption cycle as compared to vapour compression cycle is..............

(xxiii) The C.O.P of the vapour absorption cycle can be increased by using..........

(xxiv) Electrolux refrigerator is a three fluid system working on vapour adsorption system........pump

(xxv) The refrigerating system used for most of the domestic refrigerators, is of the type of............

(xxvi) For ship refrigerations where space is a vital consideration, the refrigerant used is......

(xxvii) Carbon-dioxide, as compared to other refrigerants, has..........C.O.P

(xix) Two fluid referigeration system is the...........................

(xxix) Cascade refrigeration system can use more than...........refrigerants.

(xxx) For commercial ice manufacturing refrigerant used is..............

## Answer

| | |
|---|---|
| (i) Surrounding | (ii) liquid phase to vapour phase |
| (iii) evaporator | (iv) condenser |
| (v) more | (vi) heat absorbed from system |
| (vii) actual | (viii) lowering |
| (ix) refrigerating effect | (x) 211 kJ/min (3.5 kw) |
| (xi) 1.2 | (xii) 1.25 |
| (xiii) Bell-coleman cycle | (xiv) superheated |
| (xv) compressor and condenser | (xvi) higher |
| (xvii) increases | (xviii) higher |
| (xix) higher | (xx) directly |
| (xxi) low | (xxii) lower |
| (xxiii) heat exchangers | (xxiv) without |
| (xxv) vapour compression | (xxvi) $CO_2$ |
| (xxvii) lowest | (xxviii) vapour absorption system |
| (xxix) two | (xxx) ammonia |

### 10.2 Objective type questions (Air-conditioning)

(i) The simultaneous control of temperature, humidity, motion and purity of the atmosphere in a confined space is called........

(ii) ..........deals with the behaviour of mixture of air and water vapour.

(iii) ........is the temperature of air as recorded by an ordinary thermometer.

(iv) The temperature of air at which condensation of mixture begins when the air is cooled is known as........

(v) Ratio of mass of water vapour to the mass of dry air in a given mixture of air water vapour mixture is...........

(vi) Humidity ratio is also called..............

(vii) Ratio of mass of water vapour associated with unit mass of dry air to mass of water vapour associated with unit mass of dry air saturated at the same temperature is...........

(viii) The difference between dry bulb and wet bulb temperature is called..........

(ix) In a psychrometric chart curved lines indicate for..........

(x) The ratio of the room sensible heat to the sum of room sensible heat and room latent heat is called..........

(xi) ..........is the ratio of total sensible heat to the grand total heat that the cooling or the conditioning apparatus should handle.

(xii) A system in which air handling unit is separated from condensing unit is called a......system.

(xiii) In..........system each room is provided with a room unit which gets a supply of conditioned air from a control system.

(xiv) In summer air-conditioning, the process is known as..............

(xv) In winter conditioning, the process is known as...............

(xvi) If the relative humidity is low, the rate of evaporation of water will be............

(xvii) In air conditioning, the mixing of two or more stream of moist air follows..........

(xviii) The coil efficiency, in case of sensible. eobling of air, is equal to ................

(xix) During adiabatic saturation process on unsaturated air, the parameter which remains constant is.............

(xx) If dry bulb temperature, wet bulb temperature, dew point temperature and saturation temperature are equal, then relative humidity would be equal to..........

(xxi) Heat is transferred to the room by the radiator both by............and............

(xxii) Three types of air conditioning controls are ............., ..........and............

(xxiii) In filtration air in an air conditioned room leaks through the gaps which............heat load.

(xxiv) Capacity of the cooling coil decreases with............in mass flow rate of air.

## Answer

| | | | |
|---|---|---|---|
| (i) | air-conditioning | (ii) | psychrometry |
| (iii) | Dry bulb temperature | (iv) | dew point temperature |
| (v) | humidity ratio | (vi) | specific humidity |
| (vii) | degree of saturation | (viii) | net bulb depression |
| (ix) | relative humidity | (x) | RSHF |
| (xi) | GSHF | (xii) | remote |
| (xiii) | unitary control system | (xiv) | cooling and dehumidification |
| (xv) | cooling and dehumidification | (xvi) | high |
| (xvii) | adiabatic process | (xviii) | 1 – B.F |
| (xix) | wet bulb temperature | (xx) | 100% |
| (xxi) | radiation and convection | (xxii) | manual, automatic and semi-automatic |
| (xxiii) | increases | (xxiv) | decrease |

## Theoretical  Questions  (Refrigeration)

11.3   Explain the reversed carnot cycle and obtain an equation for the COP of this type.

11.4   What is a heat pump? Derive an expression for the COP of a refrigerator working between the same temperature limits.

11.5   Define COP for a refrigerator and a heat pump and establish the relationship between them.

11.6   What is the commonly used unit of refrigeration?

11.7   Explain the working of air refrigeration system.

11.8   Discuss various modifications made in the air-refrigeration cycle for the air conditioning of a plane cabin? What factor decides their application and how are these implemented?

11.9   How the refrigeration cycle are classified based on the working medium?

11.10   Which refrigeration cycle is taken as standard of perfection against which the other refrigeration cycles are compared/

11.11   Does a COP of a refrigerator increase or decrease if the temperature difference between the high temperature body and low temperature body, in increased?

11.12   Explain why the carnot vapour compression refrigeration cycle is not practical.

11.13   What modifications have been made in the carnot vapour compression refrigeration cycle to obtain a practical vapour compression refrigeration cycle?

11.14   State various applications of refrigeration.

11.15   Why is a throttle valve used in place of an expander in a vapour compression system? Discuss the function of different expansion valves used for the purpose.

11.16   What do you understand by wet and dry compression in a vapour compression system?

11.17   What is the object of undercooling? Discuss the merits and demerits of undercooling.

11.18   Discuss the effects of suction pressure and condenser pressure on COP with the help of the p-h chart. What other factors affect COP significantly? Discuss.

11.19   Why is a flash chamber used? Does it affect the refrigerating capacity?

11.20   In what manner does under cooling of the refrigerant affect the COP of the vapour compression cycle? Is there any limit of the under-cooling?

11.21   What are the important considerations in the selection of a refrigerant?

11.22   Name five refrigerants (at least two eco-friendly) available in the market and list their desirable properties.

11.23   Why is multi-stage compression preferred?

11.24   Explain the working of a vapour absorption refrigeration system.

11.25   Discuss the functioning of electrolux refrigeration system.

11.26   What is the advantage of using absorption refrigeration system over the vapour compression refrigeration system?

## Theoretical  Questions  (Air-Conditioning)

11.27   What is meant by air-conditioning?

11.28   Define (i) specific humidity (ii) new point temperature (iii) relative humidity (iv) net-bulb temperature (v) By-pass factor.

11.29   What is a psychrometer?

11.30   What is the use of psychrometric chart?

11.31   Name the air-conditioning process and locate on psychrometric chart.

11.32   Sketch a psychrometric chart and indicate the lines of constant wet-bulb temperature and constant relative humidity on the chart.

11.33   How is the enthalpy of air-water vapour mixture defined?

11.34   What is effective temperature? Sketch a comfort chart and explain its significance.

11.35   What is the difference between sensible and latent heat? Illustrate the importance of the sensible heat factor scale provided in the psychrometric chart.

11.36   What is the role of effective temperature and comfortable zone?

11.37   Suppose an unsaturated air-water vapour mixture is heated at constant pressure. Does the specific humidity increase or decrease? Does the relative humidity of the mixture change?

11.38   Suppose two streams of air-water vapour mixture of known specific humidity and temperatures are adiabatically mixed in a flow device. Is it possible to determine the state of the resultant mixture? Explain with the help of a psychrometric chart.

11.39   What is evaporative cooling?

11.40   Which principle is made use of in the design of desert coolers?

11.41   Why does the water kept in a porous earthen pot which is exposed to flowing air in shade gets cooled?

11.42   Sketch the layout of an air-conditioning system and explain the functions of each component in it. What combinations are used for winter and summer air-conditioning?

11.43   Discuss the procedure for optimum cycle design for summer air-conditioning.

11.44   Sketch a suitable cycle for dry and humid conditions.

## Numerical problems (refrigeration)

11.45   A 50 kw of refrigeration is to be supplied by a refrigerating machine operating on a reversed carnot cycle. The evaporator temperature is – 10°C and the condenser temperature is 35°C. Determine the power required and the co-efficient of performance.

*(Ans.* 8.55 kW, 5.84)

11.46   It is proposed to use a reversed carnot cycle for cooling and heating. The work supplied to the unit is 10 kW. If the COP = 3.5 for cooling, determine the ratio of temperatures of reservoirs between which the device is working, the rate of refrigeration and the COP for heating.

*(Ans.* 1.285, 35 kW, 4.5)

11.47   A refrigerator system operating on reversed carnot cycle has a refrigeration temperature of – 15°C and the cooling water temperature is 35°C. If the capacity of the system is 10 tonnes, determine the power required to run the unit.

*(Ans.* 6.78 kW)

11.48   Find the tonnage of refrigeration required to produce ice at 0°C at the rate 100 kg/hr from water available at 30°C. Assume the latent heat of fusion of ice as 335 kJ/kg.

*(Ans.* 3.69 tonnes)

11.49   A heat pump delivers 50 kw of heat of a room maintained at 25°C and receives heat from a reservoir at – 0°C. If the actual co-efficient of performance is 50% of heat of an ideal heat pump operating between the same temperature limits, what is the power required to run the heat pump.

*(Ans.* 8.39 kW)

11.50   An air refrigeration system working on Bell-coleman cycle takes air at 255 K and 1 bar from a refrigerator and compress it to 4 bar. The compressed air is cooled to 290 K before expansion. Find the COP.   *(Ans.* 1.02)

11.51   An air refrigeration system works on Bell-coleman cycle and produces 35 kw of cooling. The pressure limits are 140 kpa and 420 kpa. Temperatures at the beginning of compression and expansion are 253 K and 323 K respectively.   (2.06 kg/s, 2.83)

11.52   In an aircraft refrigeration system air is bled from the engine compressor at 3.5 bar and 270°C and is passed through an air-cooled heat exchanger. The refrigerant air bleed leaves the exchanger at 3.5 bar and 75°C and is expanded through a turbine to 0.76 bar. The isentropic efficiency of the turbine is 85%. The air is then delivered to the air craft cabin and leaves the aircraft at 16°C. Find the refrigerating effect and the power developed by the air turbine per unit mass flow rate.

(*Ans.* 45.8 kJ/kg, 105.1 kw)

11.53  An air refrigerator is to be designed consisting of a centrifugal compressor and an air turbine, mounted coaxially such that the power output of the turbine contributed to the work required to drive the compressor. The temperature of the air at the compressor inlet is 15°C and the pressure ratio is 2.5. The air, during its passage from the compressor to the turbine, passes through an intercooler and enters the turbine at 47°C. The cold place is required to be maintained at 15°C.

Take the isentropic efficiency of both the compressor and the turbine to be 84%. Calculate (i) the refrigerating effect per kg of refrigerant (ii) the mass flow per kw of refrigeration, and (iii) the driving power required per kW of refrigeration.

(*Ans.* 30.06 kJ/kg, 1.996 kg/min, 1.347 kW)

11.54  In a bootstrap refrigeration system for an air-craft, the ambient conditions are 0.225 bar and – 5°C. Cooling load estimate is 20 tons refrigeration. The speed of the plane is 1000 km/hr. Ram efficiency is 0.9. The pressure ratio for the main compressor is 3.5, and this bled off air is further compressed in the secondary compressor run by cooling air turbine on a single shaft, such that output from the turbine is equal input to the compressor. The internal efficiencies of the main as well as secondary compressors are 0.9 and that of cooling turbine is 0.8. The air from secondary compressor is cooled by ram air to 50°C. The cooling air turbine running the secondary compressor has an exit pressure of 1 bar. Determine

(i)  deliver pressure from the secondary compressor

(ii)  mass flow rate bled off fro cooling the cabin, and

(iii)  COP of the system.

(*Ans.* 1.967 bar, 205.29 kg/min, 0.1633)

11.55  A regenerative air refrigeration system of an aircraft with flight speed of 1500 km/hr has 30 ton of cooling load, while the ambient conditions are 0.1 bar and – 63°C. The ram efficiency is 90%. The pressure ratio of the main compressor is 5, with internal efficiency of 0.9. The air bled off the main compressor is cooled by ram air in a heat exchanger, which is 60% effective. The air from the heat exchanger passes on to the cooling air turbine, whose internal efficiency is 0.8. Some portion of air from the cooling turbine is led to the regenerative heat exchanger, reducing the temperature to 30°C of the bled off compressed air. The cooling air gets heated to 92°C before discharge to the atmosphere. The cabin is pressurized to 0.8 bar and maintained at 25°C. Determine (i) the percentage of air extracted for regenerative cooling (ii) the power required to maintain the cabin at required condition and (iii) COP. Assume the cooling turbine power to be used for ram air exhaust fan. (*Ans.* 1316 kW, 0.08)

11.56  An ideal vapour compression refrigerator, using F-12 as the refrigerant, maintains a cold space at – 10°C while the ambient atmosphere is at 30°C. Saturated vapour leaves the compressor is at 30°C. Saturated vapour leaves the compressor and the refrigerant leaves the condenser as saturated liquid. If the flow rate of the refrigerant is 0.5 kg/s, determine the COP, the power consumption of the refrigerator and the refrigeration capacity (in tons) of the unit.

(*Ans.* 5.477, 10.385 kW, 16.174)

11.57  A refrigerating plant operates in the quasi-ideal vapour compression cycle. The refrigerant is F-12 (R-12) and the saturation temperature in the evaporator and condenser are – 5C° and 40°C respectively. The vapour enters the compressor as a saturated vapour and is subcooled to 20°C before entering the throttle valve. Calculate (i) the refrigerant temperature leaving the compressor, (ii) the work done/kg of refrigerant (iii) the refrigerating effect per kg of refrigerant (iv) COP and (v) the compressor displacement per kW of refrigeration. Assume 100% volumetric efficiency.

(*Ans.* 46.75°C, 23.079 kJ/kg, 130.5 kJ/kg, 5.654, 0.497 m$^3$/s)

11.58  A vapour compression system works between 266 K and 300 K. The vapour is dry at the end of isentropic compression and there is no undercooling. Calculate the COP and power required for the compression to produce a refrigeration effect of 2.92 kw. The properties of the refrigerant are given as follows:

| $T(k)$ | $h_f$ (kJ/kg) | $h_{fg}$ (kJ/kg) | $s_f$ (kJ/kgk) | $s_g$ (kJ/kgk) |
|---|---|---|---|---|
| 266 | $-29.4$ | 1298 | $-0.1088$ | 4.748 |
| 300 | 124.8 | 1172.4 | 0.427 | 4.334 |

(*Ans.* 7.51, 0.55)

11.59  In an indeal vapour compression refrigeration cycle F-12 enters the compressor at 180 KPa and the saturated liquid enters the throttle value at 10 bar. Determine (i) the temperature of F-12 leaving the compressor, (ii) the cop of the refrigeration cycle, and (iii) the power input to the compressor for a 5 ton refrigeration unit.

(*Ans.* 41.7°C, 3.393, 5.18 kW)

11.60  A vapour compression refrigerator employing F-12 as the refrigerant operates between the temperature levels 40°C and $-20$°C. If the quality of the refrigerant at the beginning and at the end of the heat absorption process are 0.2 and 0.85, respectively, determine (i) the COP (ii) the evaporator and condenser pressure and (iii) the power consumption for a 2.5 ton refrigerator unit.

(*Ans.* 3.889, 150.836 Kpa, 960.255 kPa, 2.26 kW)

11.61  A refrigerator circulating ammonia operates between pressure limits of 2.91 bar and 15.57 bar. It consists of 4 units (i) a compressor in which compression is isentropic and the outgoing vapour is dry saturated (ii) a condenser in which the whole of the latent heat is removed at 15.57 bar, (iii) an expansion cylinder into which the condenser discharges and in which expansion is isentropic (iv) an evaporator operating at 2.91 bar. Determine (a) the co-efficient of performance (b) the power input to the machine if the two cranks are connected and a cooling effect of 20 kW is required (c) the rate of circulation of ammonia corresponding to (b). Also determine the above quantities if expansion cylinder is replaced by a throttle valve.

| $p$ bar | $t$ C° | $h_f$ kJ/kg | $h_g$ kJ/kg | $s_f$ kJ/kgk | $s_g$ kJ/kgk |
|---|---|---|---|---|---|
| 2.91 | $-10$ | 154.056 | 1450.22 | 0.82965 | 5.7550 |
| 15.57 | 40 | 390.587 | 1490.42 | 1.6437 | 5.1558 |

(*Ans.* COP = 5.26, $W$ = 3.8023 kW, $m$ = 0.02164 kg/s, COP = 4.556, $w$ = 4.38 kw, $m$ = 0.02218 kg/s)

11.62  A vapour compression refrigerator using ammonia as the working fluid operates between pressure limits of 13.5 bar and 2.363 bar. The temperature of the ammonia leaving the condenser is 27°C. The pressure drop from the condenser to the evaporator is achieved by throttling. During a trial on this unit it was found that for a condenser cooling water flow rate of 45 kg per minute with a temperature rise of 11°C the input power was 3.3 kW and the ice production rate 120 kg/hr from and at 0°C. Find

(i)  the theoretical COP assuming isentropic compression and constant enthalpy throttling and undercooling up to 27°C.

(ii)  the circulation rate of ammonia in kg per second based on a condenser heat balance assuming no losses.

(iii)  the overall *COP*.

| $p$ bar | $t$ °C | $h_f$ kJ/kg | $h_g$ kJ/kg | $s_f$ kJ/kgk | $s_g$ kJ/kgk | $Cp_l$ kJ/kgk | $Cp_v$ |
|---|---|---|---|---|---|---|---|
| 13.5 | 35 | 347.2 | 1470.1 | — | 4.9255 | 4.731 | 3.056 |
| 2.363 | $-15$ | 112.2 | 1425.2 | — | 5.5437 | 4.396 | 2.303 |

Latent heat of fusion of ice = 335 kJ/kg

($h$ = 1681.3 kJ, $h$ = 309.4 kJ, theo. *COP* = 4.357, m = 0.025 kg/s, overall *COP* = 3.384)

11.63  A 20 kw refrigeration system operates on vapour compression refrigeration cycle between pressure limits of 7.45 and 1.5 bar. The vapour leaving the evaporator is at a temperature of $-10$°C and liquid leaving the condenser is saturated. Find the COP and swept volume of the compressor if the compressor speed is 600 rpm.

(*Ans.* 4.08, 303 cc)

**11.64** A vapour compression refrigerating plant with ammonia as refrigerant has a single stage, single acting reciprocating compressor which has a bore of 127 mm, a stroke of 152 mm and a speed of 240 rpm. The pressure in the evaporator is 1.6 bar and that in the condenser is 14 bar. The volumetric efficiency of the compressor is 80% and its mechanical efficiency is 90%. The vapour is dry saturated on leaving the evaporator and the liquid leaves the condenser at 32°C. Find the mass flow of refrigerant, the refrigerating effect and the power required to drive the compressor.

*(Ans.* 0.5 kg/min, 9 kw, 2.74 kW)

**11.65** A refrigerator working on vapour absorption system, the heat is supplied to $NH_3$ generator by considering steam at 1.96 bar and 90% dry. The temperature in the refrigerator is to be maintained at $-5°C$. Find maximum COP possible. If the refrigeration load is 20 tons and actual *COP* is 70% of the maximum COP, find the weight of steam required per hour. Take temperature of atmospheric as 30°C.

*(Ans.* $COP_{max}$ = 1.75, m = 103.4 kg/hr)

**11.66** One kg of saturated vapour ammonia with concentration $c = 1$ at 2 bar is mixed with 10 kg of saturated liquid aqua ammonia at 2 bar and 40°C. If the mixture after mixing should come out at saturated liquid aqua ammonia at 2 bar, determine (i) the concentration temperature and enthalpy of the mixture after mixing (ii) the heat removed from the mixture during mixing process in kJ/kg.

*(Ans.* c = 0.385 kg/kg of mixture, heat = 184.8 kJ)

## Numerical problems (Air conditioning)

**11.67** The dry and wet bulb temperatures of a sample air are 25°C and 20°C. Using psychrometric chart, find the specific humidity, relative humidity, the dew point temperature, specific volume and specific enthalpy of the air.

*(Ans.* 0.013 kg/kg, 61%, 17°C, 0.86 m³/kg, 56 kJ/kg)

**11.68** If the partial pressure of water vapour in the atmospheric air at 40°C is 5.5 kpa, determine (without using chart)

   (i)  the specific humidity and relative humidity of the air

   (ii)  the dew point temperature, and

   (iii)  the enthalpy of the air.

*(Ans.* 0.0238 kg/kg, 28.7°C, 101.5 kJ/kg)

**11.69** A mixture of air and water vapour enters an adiabatic saturator at 40°C and 1 atmospheric pressure and leaves at 25°C and 1 atmospheric pressure. Determine the specific humidity, the relative humidity and the dew point of the entering air.

*(Ans.* 0.0138 kg/kg, 0.298, 18.19 °C)

**11.70** The dry bulb temperature and volume of air in a room are 38 m³ and 25°C. The air pressure is constantly maintained at 1.01325 bar and the new point temperature of the air is 14°C. If water is placed in the room, calculate the maximum mass of water that can be lost by evaporation.

*(Ans.* 0.433 kg)

**11.71** Find the enthalpy of air at 50°C DBT and 1.01325 bar and the heat to be removed at constant pressure of 1.01325 bar from the condensation to begin.

*(Ans.* 151.9 kJ/kg, 13.3 kJ/kg)

**11.72** A sample of air water vapour mixture at 1 atmospheric pressure has a dry bulb temperature of 40°C and a wet bulb temperature of 25°C. Determine the specific humidity, the relative humidity and the enthalpy of the mixture. Compare the results read from the psychrometric chart with the values calculated by making use of the adiabatic saturation relation.

*(Ans.* 0.0138 kg/kg, 0.298, 75.7 kJ/kg. From chart 0.0138 kg/kg, 0.30, 76.5 kJ/kg)

**11.73** The water vapour pressure in an atmospheric air is 0.021 bar. Air is at 32°C and 1.01325 bar. Find the specific humidity, relative humidity and the temperature to which the air must be cooled for it to become just saturated. If the air is cooled to 10° from its original condition, find the mass of condensate formed per kilogram of dry air.

(*Ans.* 0.01292 kg/kg, 43.4%, 18°C, 0.0053 kg)

11.74  Find the specific volume of the vapour and the heat to be rejected per kilogram of air when the air at 21°C DBT and 55% RH is cooled until condensation just begins. Also find the dew point temperature for the air at this condition.

(*Ans.* 99.3 m³/kg, 7 kJ, 11.35°C)

11.75  An adiabatic mixing device operating at steady state receives 30 m³/min of air at 40°C with a relative humidity of 60%. A second stream of air at 20°C with a relative humidity of 10% enters the mixing device at a rate of 10 m³/min. If the mixing device is at 101.325 kpa pressure, determine the temperature and relative humidity of the existing stream of mixture.

(*Ans.* 35°C, 0.60)

11.76  800 m³/min of recirculated air at 22°C DBT and 10°C dew point temperature is to be mixed with 300 m³/min of fresh air at 30°C DBT and 55% RH. Determine the enthalpy, specific volume, humidity ratio and dew point temperature of the mixture.

(*Ans.* 47.71 kJ/kg, 0.855 m³/kg, 0.0092 kg/kg, 13°C)

11.77  A stream of air with a dry bulb temperature of 46°C and a wet bulb temperature of 25°C enters a mixing chamber at a rate of 3 kg dry air/s. A second stream of saturated air at 10°C enters the mixing chamber at a rate of 1 kg dry air/s. Assuming that the total pressure in the mixing chamber is 101.325 kpa, determine for the resultant mixture (i) the specific humidity (ii) the relative humidity and (iii) the temperture.

(*Ans.* 0.0121 kg/kg, 0.4, 32.5°C)

11.78  A stream of moist air enters a device at a rate of 20 m3/s at 1 standard atmospheric pressure with a dry bulb temperature of 40°C and a wet bulb temperature of 28°C. Liquid water at 10°C is sprayed into the air stream and the air leaves the device at 30°C. Determine the relative humidities of the entering air and exit air and quantity of liquid water sprayed.

(*Ans.* 0.415, 0.869, 0.094 kg/s)

11.79  Air at 39°C DBT and 36% RH is passed through an adiabatic humidifier and it comes out as saturated at 27°C DBT. Find the capacity of the humidifier.

(*Ans.* 0.006 kg/kg)

11.80  Atmospheric air at 45°C with 20% relative humidity enters a desert cooler and leaves at 28°C. Determine the amount of water added per kg of dry air and the relative humidity of the cooled air. The total pressure of the air remains constant at one standard atmosphere in the desert cooler.

(*Ans.* 0.0072 kg, 0.8)

11.81  Air at the rate of 16.5 m3/s is processed and supplied at 25°C DBT and 39% RH to an audotorium. The atmospheric air is at 32°C DBT and 60% RH. The required condition is achieved first by cooling and dehumidifying and then by heating. Determine the capacity of cooling and heating coils.

(*Ans.* 320 tonnes, 90 kW)

11.82  Air enters a drier at 30°C with 20% relative humidity and leaves at 40°C with 60% relative humidity. Determine the flow rate of dry air at 1 atmospheric pressure if 10 kg/hr of water is evaporated from the material in the drier.

11.83  A cooling system for an office building uses a desert cooler in summer months. If the atmospheric air is available at 42°C with 20% relative humidity and the air leaves the desert cooler with 90% relative humidity, determine the temperature of the air leaving the cooler. If the cooler draws in 60 m³/s of atmospheric air estimate the amount of water required to operate the cooler continuously for a period of 8 hours.

(*Ans.* 24.8°C, $1.41 \times 10^4$ kg)

11.84   Air at 40°C and 20% RH is processed to a temperature of 20°C and 60% RH first by adiabatic humidifying and then by cooling. If the air flow rate is 8.3 kg/s then find the capacity of the humidifier and capacity of the cooling coil.

(*Ans.* 54 kg/hr, 31 tonnes/hr)

11.85   Outdoor air at 13°C and 76% RH is to be processed to 23°C DBT and 60% *RH*. The rate of air supplied is 3.3 kg/s. The required condition is achieved by first heating and then by adiabatic humidifying. Find the amount of water vapour required in kg/hr.

(*Ans.* 300 kg/hr)

11.86   On a summer day atmospheric air at 40°C with 30% RH enters an evaporative cooler at a rate of 3 m$^3$/s. The air leaves the cooler with 80% RH. Determine the temperature and the specific humidity of the air leaving the cooler. Also determine the amount of water required to operate the cooler for 10 hours.
(*Ans.* 28°C, 0.019 kg/kg, 619.2 kg)

11.87   A air-conditioning system is to maintain the room conditions at 22°C and 40.0 RH. The outdoor conditions are at – 6°C, 100% RH. The sensible heat loss from the room is 35 kw and the latent heat gain is 3 kW. A saving in energy is obtained by recirculating a portion of the room air. Determine the ratio of room air to fresh and the heat supplied in the heater under these conditions

(*Ans.* 17, 43 kW)

11.88   The required indoor conditions of the air in a building are 21°C and 40% RH when the outside conditions are 28°C and 50% RH. The sensible heat gain to the room is 12 kW and the latent heat gain is 3 kW. There is no recalculation and the fresh air is cooled and dehumidified and then heated before entering the room. The cooling coil bypass factor is 0.2 and the volume flow rate of fresh air is 5 m$^3$/s. Determine the temperature of the air leaving the cooling coil, capacity of the cooling coil and the heat supplied in the heater.

(*Ans.* 6.5°C, 220 kW, 65 kW)

11.89   The following data refers to a winter air-conditioning system using a pre-heat coil adiabatic washer and a reheat coil. Indoor design condition is 21°C, 40% RH, outdoor design condition is 2°C, 10% RH, atmospheric pressure 1.01325 bar, sensible heat load from space 1.9 × 10$^5$ kJ/hr, latent heat load from space 20000 kJ/hr, temperature of condition air entry to space 37°C, ventilation air flow rate 50% of dry air flow rate to space, air washer efficiency 80%. Calculate (i) the volume of air supplied to the space (ii) the spray water temperature (iii) the make-up water required by the air washer per hour and (iv) the ratio of heat added to moist air by the preheat coil and reheat coil per hour.

(*Ans.* 128.8 m$^3$/s, 10°C, 23.64 kg/hr, 22 kw, 78.47 kW)

11.90   The summer load on a building is 52400 kJ/hr, sensible heat when total load is 62800 kJ/hr; 31400 kJ/hr, latent heat when the total load is 52400 kJ/hr; 42000 kJ/hr, sensible heat when the total load is 67200 kJ/hr, if the maximum permissible temperature difference between the supply and room air is 10°C and if the room is to be held at 26°C DBT and 50% RH. Calculate (i) the minimum possible year round mass rate of air supply (ii) the minimum DBT of supply air when the system operates with the mass flow rate as calculated in (i), (iii) the minimum DBT for operation at minimum mass flow rate and (iv) the minimum DPT for operation at minimum mass rate.

(*Ans.* 2031 kg/hr, 16°C, 14°C, 15.2°C)

11.91   A cooling tower is to be installed to supply 1000 kg/min of cooling water at 27°C. Water enters the cooling coil at a rate of 1000 kg/min at 45°C. The air enters the tower at 30°C with 30%. RH and leaves at 35°C with 80% RH. Determine the make-up water required and the air flow rate.

(*Ans.* 28.15 kg/min, 1340.24 kg/min)

11.92   A forced draught cooling tower is used in a large thermal power plant for providing cold water at a temperature of 30°C to be condensers. The water enters the cooling coil at 37°C at a rate of 1000 kg/s. The atmospheric air at 27°C with 40% RH enters the tower and leaves at 30°C with 80% RH. Determine the make-up water and air flow rate required.

(*Ans.* 10.5 kg/s, 617.77 kg/s)

11.93  A small cooling tower is to be designed to cool 50 kg/s of water from 40°C to 24°C at a location where the atmospheric pressure is 90 kpa. Atmospheric air enters the cooling tower at 30°C with 30% RH and leaves at 20°C with 80% RH. Determine (i) the required flow rate of dry air and (ii) the flow rate of make-up water.

(*Ans.* 482.5 kg/s, 2.98 kg/s)

11.94  A forced draught cooling tower is designed to cool 1.5 litres of water per second entering at 40°C. The draught is produced by a fan absorbing 5 kW which induces 2 m3/s of air. The air enters the tower at 1.01325 bar, 20°C and a relative humidity of 50%. The air leaving the tower is saturated and is at temperature of 33°C. Calculate

  (i)  the temperature of water leaving the tower

 (ii)  the loss of water due to evaporation

(*Ans.* 10.04C°, 0.06018 kg/s)

# 12

# Steam Engine

## 12.1   INTRODUCTION

The basic prime mover using steam as a working substance is a steam engine. It was the earliest prime mover developed for converting heat energy (enthalpy) into mechanical work and it was in use for a considerable long period. After the development of highly efficient steam turbines and internal combustion engines, it is now becoming obsolete. Its use is confined to small power generating installations where other types of heat engines are not advisable to be installeld. Its thermal efficiency is very poor (about 18 to 20%). Indian Railways are still using the steam engines on their locomotives for transfering the bogies from one to next plateform. It is, however, expected that within next 3–5 years, all the steam locomotives will be replaced by either diesel or electric engines. Developed countries like U.S.A. has already discarded the steam engines. It is normally used in marine work, driving pumps, fans, blowers, compressors, stockers and other small power generating units. Even if a steam engine has poor efficiency, it is basic heat engine. Therefore, it is advisable to study the basic principle of steam engine in brief.

## 12.2   CLASSIFICATION OF STEAM ENGINES

Steam engines are classified according to their various constructions and operating features. The various classifications are as follows:

1. According to the position of cylinders:
    (a) Vertical engine—Axis of the cylinder is vertical
    (b) Horizontal engine. Axis of the cylinder is horizontal.
2. According to number of working strokes per revolution:
    (a) Single acting—steam is admitted only from one side of the piston. One cycle is completed in only one rotation of the crank.

(b) Double acting—steam is admitted alternately from both sides of the piston. There are two working strokes per revolution of the crank. Generally steam engines are double acting.

3. According to expansion of steam:
   (a) Simple steam engine—A simple steam engine is one in which conversion of heat energy, of steam into mechanical work occurs in one stage i.e. expansion of steam from admission pressure to back pressure occurs in only one cyclinder.
   (b) Compound steam engine—In this case, expansion of steam occurs in more than one stage. Steam after expansion in high pressure (HP) cyclinder is taken into a receiver at some suitable intermediate pressure and it is further expanded into another low pressure (LP) cylinder.

4. According to exhaust taking place:
   (a) Non-condensing engine—In non-condensing engine, steam after doing work is exhausted into atmosphere.
   (b) Condensing engine—In this case, exhaust takes place in condenser where vacuum is maintained. Due to decreasing the back pressure by istalling condenser, efficiency of steam engine is considerably increased.

5. According to speed of the engine:
   (a) Low speed engine (rpm 100)
   (b) Medium speed engine (110 rpm to 200 rpm)
   (c) High speed engine (rpm > 200)

6. According to their field of application:
   (a) stationary engines—steam engine is stationed at a particular place.
   (b) Mobile engines—engines move from place to place like marine engine and locomotive engines.

7. According to governing of the engine:
   (a) Throttled governed engine and
   (b) Cut off governed engine.

## 12.3 STEAM ENGINE PARTS AND THEIR FUNCTIONS

Main partis of steam engine are: (Shown in Fig. 12.1)

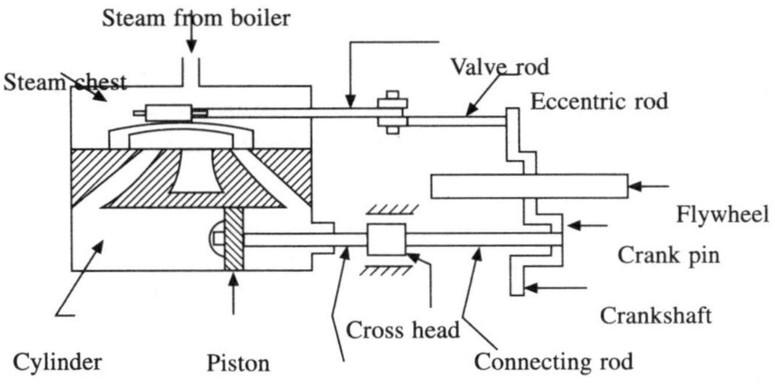


**Fig. 12.1  Steam engine parts**

1.Cylinder, 2. Piston, 3. Piston rod, 4. Stuffing box, 5. Gland, 6. Cross head, 7. Connecting rod, 8. Gudgeon pin 9. Crank pin, 10. Crank, 11. Crank shaft 12. Steam chest, 13. Steam ports, 14. Slide valve, 15. Valve rod, 16. Eccentric rod 17. Eccentric 18. Fly wheel 19. Governor 20. Throttle valve.

1. *Cylinder*:  It is the heart of the steam engine. It is cast and bolted to the frame, one end of the cylinder is bolted with the cover of the cylinder and the other end is closed by stuffing box through which piston rod passes. It is made of cast iron.

2. *Piston*:  It is also made of cast iron. The steam pressure acts on the piston and exerts a force on the piston. The piston is connected to the piston rod which transmits the force to the crank through the cross head and connecting rod.

3. *Piston rod*:  It is made of mild steel. One end is connected to the piston and the other end to the cross head. The main function of piston rod is to transmit force of the piston to the cross head.

4. *Stuffing box and gland*—stuffing box and *gland* act like a seal which prevents any leakage of steam through the crank and of the cylinder.

5. *Cross head*:  It forms a connecting link between the piston rod and connecting rod. It guides the motion of the piston rod and prevents bending of the piston rod due to angular motion of the connecting rod. In case of single acting steam engine (also in I.C. engine), piston rod, stuffing box, gland, cross head are completely eliminated and the crank end of the cylinder is open. Connecting rod can have angular motion through the open space of the cylinder end.

6. *Connecting rod*:  It is made of forged steel. One of its ends is connected to the cross head by a gudgeon pin and the other end is connected to the crank by crank pin. The function of connecting rod is to convert the reciprocating motion of the piston (or cross head) into rotary motion of the crank. The small end (connected to the cross head) has purely reciprocating motion and big end (connected to crank) has purely rotary motion.

7. *Gudgeon pin*:  The connecting rod is attached to the cross head by means of gudgeon pin which acts as a bearing and allows the connecting rod to oscillate.

8. *Crank pin*:  It connects connecting rod to the crank.

9. *Crank*:  It is made of forged steel and is integral part of the crank shaft. It transmits angular motion to the crank shaft.

10. *Crank shaft*:  Crank shaft is the main shaft of the engine delivering power. It is supported by the main bearings. Flywheel and eccentric are mounted on this shaft.

11. *Steam chest*:  It is casted as integral part of the cylinder and is closed by a separate cover. Steam chest is connected to the cylinder through valve passage. steam is admitted to steam chest through throttle valve.

12. *Steam ports*:  There are two ports through which steam is admitted from steam chest to cylinder alternately. If steam is admitted through one port, it is exhausted through other port and vice versa.

13. *Slide valve*:  The function of the slide valve is to open and close the two ports through which steam is admitted and exhausted out.

14. *Valve rod*:  One end of the valve rod is connected to the slide valve and other end to the eccentric rod. It controlls the motion of the slide valve.

15. *Eccentric rod*:   This is a link which connects the valve rod and eccentric.
16. *Eccentric*:   It is made of cast iron and is fitted on the crank shaft. The main function of the eccentric is to provide reciprocating motion to slide valve and is driven by the main shaft. The correct relative position of eccentric with respect to crank provides the required valve movement.
17. *Fly wheel*:   The function of the flywheel is to maintain a constant angular motion of the crank shaft during a working cycle. The fly wheel stores the energy when excess energy is transmitted to the shaft by the piston and releases when energy developed in the cylinder is less than the mean energy (during exhaust strokes).
18. *Governor*:   The purpose of governor is to maintain the speed of the engine uniform at all variable loads. Generally engines are designed to have maximum efficiency at a particular rpm. If the load varies, the speed will try to increase or decrease. Governor controls the fluctuation of speed by adjusting the supply of steam and its quality or by changing the cutt off point of indicator diagram.
19. *Throttle valve*:   This is a valve through which steam is admitted to the steam chest at required amount and at required pressure. It is actuated by governor.

## 12.4   TERMS USED IN STEAM ENGINE

1. Cylinder bore—The inside diameter of cylinder is called cylinder bore (D).
2. Stroke length—The distance travelled by piston from the cover end of the cylinder to crank end of the cylinder is called stroke length ($L$).
3. *Piston speed*:   The distance travelled by piston in one second is called piston speed. In the rotation of the crank piston travells twice the stroke length. Therefore piston speed =

   $2LN metre$ / minute $= \dfrac{2LN}{60}$ m/sec, where $N$ = RPM of the crank.
4. Piston displacement or swept volume ($V_s$)—It is the volume swept by the piston travelling from one dead centre to other dead centre.

$$\text{Swept volume} = V_s = \pi/4\ D^2 L$$

5. *Dead centres*:   Dead centre is the position when the connecting rod and crank are in same straight line. These conditions arise for two positions of the piston. These positions are called dead centres. IDC (Inner dead centre): When the position is completely inside the cylinder. This is also called cover dead centre or TDC (Top dead centre) in vertical steam engine. ODC (outer dead centre): When the piston is out side the cylinder i.e. at the stiffing box, this is also called crank dead centre or BDC (Bottom dead centre in case of vertical) steam engine.
6. *Eccentric throw or eccentricity*:   It is the distance between the centre of eccentric and centre of crank shaft.

## 12.5   OPERATION OF STEAM ENGINE

Let us describe the working of a simple double acting steam engine. As the piston reaches the IDC, the *D* slide valve operated by the eccentric moves and admits the steam from steam chest to engine cylinder through the port. The high pressure steam pushes the piston on the forward direction and performs work. This rotates the crank as well as the eccentric providing more opening of the inlet port. Therefore the steam continues to enter into the cylinder maintaining the constant pressure inside. This further rotates the crank as well as the eccentric and eccentric moves the valve back and closes the inlet port supply of steam to the cylinder is cut off. After the cut off takes place, the steam in the cylinder contines to expand pushing the piston to move in the forward direction till the piston reaches ODC. Just before the piston reaches the ODC, the D-slide valve connects the cylinder through the port to the exhaust. The pressure on this side of the piston is atmospheric in case of non condensing steam engine and is equal to condenser pressure in condensing steam engine. After this the piston reverses its direction of motion in which case steam is admitted through crank end port. The steam admitted from crank end expands while steam present in the previous stroke will be exhausted out through cover end port. As the piston reaches IDC the D-slide valve opens the inlet port and admitted fresh steam. A similar cycle is executed on the other end of the cylinder.

$$p_1 = \text{Admission pressure}$$

$$p_b = \text{Back pressure or exhaust pressure}$$

The indicator diagram is the representation of variation of pressure and volume of steam inside the cylinder ($p$-$v$ diagram) for one complete cycle. Theoretical indicator diagram of a single acting steam engine is shown in the Fig. 12.2 without considering clearance. The sequence of operations is described below:

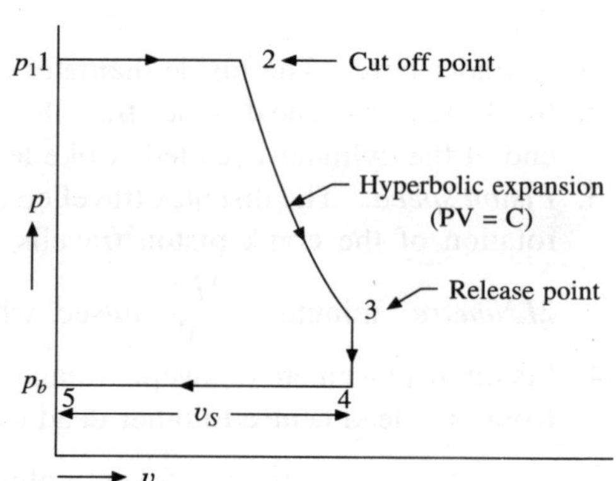


**Fig. 12.2   Theoretical indicator diagram**

1. Process (1–2) represents the supply of steam from steam chest to steam cylinder at constant pressure. State 2 represents the state of steam when cut off occurs. Pressure remains constant.

2. Process (2–3): Process 2–3 represents the hyperbolic ($pv$ = constant) expansion of steam. State 3 represents the state of steam where release takes place.

3. Process (3–4): At the end of the expansion, pressure falls from 3 to back pressure due to opening of the exhaust port.

4. Process (4–5): Process 4–5 shows the exhaust of steam at constant pressure. At the end of the exhaust state first steam is further admitted and the cycle is repeated. One cycle is completed within two stroke of the piston. For double acting steam engine similar diagram is also repeated.

## 12.6   ACTUAL INDICATOR DIAGRAM

The actual indicator diagram is different from the hypothetical indicator diagram due to various losses and imperfections. Ideal indicator diagram with clearance is shown by ABCDE and actual by $A'B'C'\,D'E'F$. The following deviations are noticed in Fig. 12.3.

(a) The steam pressure drops considerably between the boilder and the engine cylinder. This is due to condensation of steam, frictional losses in the pipe and wire drawing in opening of the inlet port.

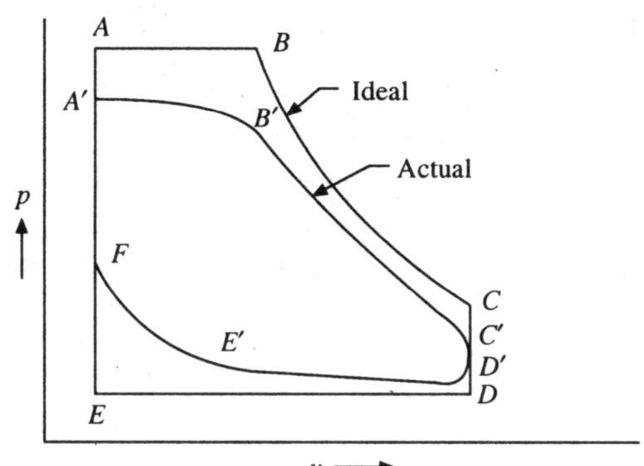

Fig. 12.3   **Actual indicator diagram**

(b) There is gradual drop in pressure before the point of cut off is reached. This is due to condensation of steam while being admitted in the cylinder. Initially cylinder wall is at lower temperature than the steam, therefore condensation takes place.

(c) There is a rounding off of the diagram at cut off because the admission port does not close instantaneously.

(d) The expansion curve is not pure hyperbolic ($pv = c$), initially the cylinder wall is colder than the steam and therefore heat transfer takes place from steam to cylinder. After some time, the temperature of expanding steam drops and its temperature be comes less than the cylinder wall. Reversal of heat transfer takes place i.e. from cylinder wall to the steam which makes the expansion process non-hyperbolic.

(e) Release takes place before the end of the expansion

(f) There is a rounding off of the toe of the diagram because exhaust port does not open instantaneously.

(g) The exhaust pressure is slightly above the condenser pressure, as steam has to be forced out of the cylinders.

(h) Compression of retained steam takes place.

The effect of these deviations is to reduce the area of actual indicator diagram and thereby reduces the actual work. We introduce a term called "diagram factor" to consider these variations. It is defined as, diagram factor (DF)

$$= \frac{\text{Area of actual indicator diagram}}{\text{Area of ideal indicator diagram}}$$

It is always less than 1

## 12.7   THEORETICAL WORK DONE AND MEAN EFFECTIVE PRESSURE

We find that the pressure of steam varies with the movement of the piston. It does not remain constant.

We define a term called "mean effective pressure" $P_m$ which is an imaginary constant and uniform pressure which if acting throughout the whole length of the piston stroke, would produce the same work area as given by the indicator diagram of an engine. If $W$ = work done per cycle, there

$$W = p_m \cdot V_s \quad \text{or} \quad p_m = W/V_s$$

consider a simple hypothetical indicator diagram 1-2-3-4-5. From the Fig. 12.4,

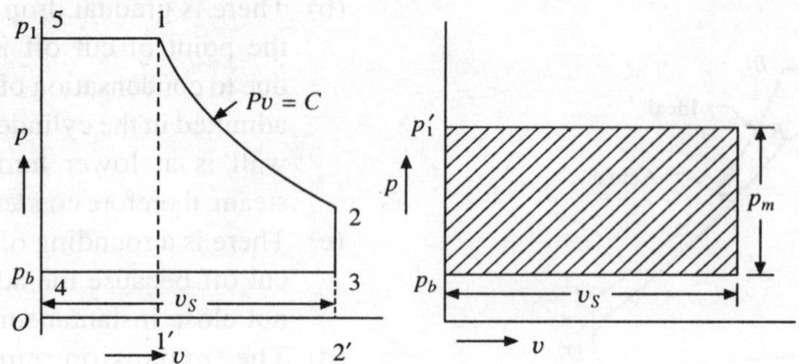


**Fig. 12.4   Indicator diagram**

$W$ = work done per cycle = area 1-2-3-4-5-1 = (area 1-2-2′-1′-1) + (area 1-1′-0-5-1) − (area 3-2′-0-4-3)

$$= p_1 v_1 \ln (v_2/v_1) + p_1 v_1 - p_b \cdot v_s$$

$v_1$ = volume of steam at cut-off, $v_2$ = volume of steam at the end of expansion, $m^3$
where $p_1$ = Admission pressure, $N/m^2$
$\quad p_b$ = Back pressure, ln = Natural log
Expansion ratio ($r$) is defined as $v_2/v_1$

then $\qquad W = p_1 v_1 \ln (v_2/v_1) + p_1 v_1 - p_b v_s = p_m v_s$

or, $\qquad p_m = 1/v_s \, [p_1 v_1 \ln (v_2/v_1) + p_1 v_1 - p_b v_a]$

$$= \frac{p_1 v_1}{v_s} [1 + \ln (v_2/v_1)] - p_b = \frac{p_1}{v_2/v_1} [1 + \ln (v_2/v_1)] - p_b$$

$$= p_1/r \cdot [1 + \ln (r)] - p_b$$

$$p_m = p_1/r \cdot [1 + \ln (r)] - p_b \ N/m^2 \tag{12.1}$$

If clearance volume is also considered the diagram is modified as Fig. 12.5.

Let, $\quad cv_s$ = clearance volume, $c$ is called clearance ratio which is a fraction

$\quad 1/r$ = fraction of stroke completed at cut off.

$\quad 1/r \cdot v_s$ = cut off volume

$\quad$ work per cycle = $W$ = area of the diagram = area 1-2-3-4-5-1

$$= \text{(area 5-1-1′-5′-5)} + \text{(area 1-2-3-3′-1′-1)-area}$$

$$\text{(3-3′-5′-4-3)} = p_1 v_s \, r + p_1 v_1 \ln v_2/v_1 - p_b \cdot v_s$$

$$= p_1 \cdot v_s/r + p_1 \cdot (cv_s + 1/rv_s) \ln \frac{(cv_s + v_s)}{(cv_s + 1/r \cdot v_s)} - p_b \cdot v_s$$

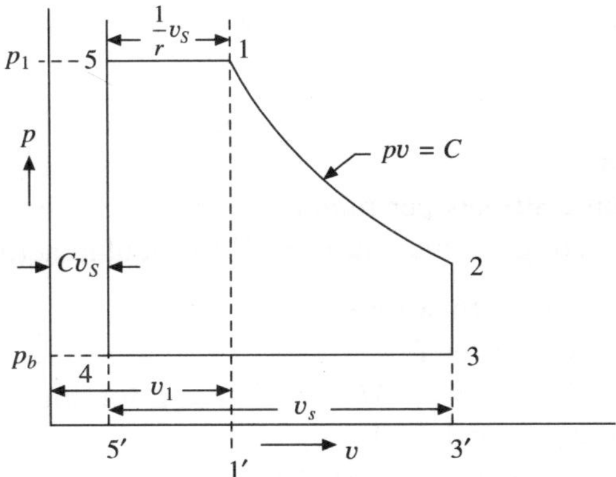

**Fig. 12.5   Indicator diagram with clearance**

$$= v_s \left[ p_1/r + p_1(c + 1/r) \ln \frac{(c + 1)}{(c + 1/r)} - p_b \right]$$

$\bigcirc\ W = p_m v_s$    or,    $p_m = W/v_s$

or, $$p_m = v_s / v_s \left[ p_1/r + p_1(c + 1/r) \ln \frac{(c + 1)}{(c + 1/r)} - p_b \right]$$

or, $$p_m = p_1 \left[ \frac{1}{r} + \left( c + \frac{1}{r} \right) \ln \frac{(c + 1)}{\left( c + \frac{1}{r} \right)} \right] - p_b \tag{12.2}$$

If $$c = 0, \ p_m = p_1/r[1 + \ln r] - p_b \ \text{N/m}^2$$

## 12.8   CYLINDER DIMENSIONS AND INDICATED POWER

We have already defined diagram factor as

$$DF = \frac{\text{area of actual indicator diagram}}{\text{area of ideal indicator diagram}}$$

$$= p_m \frac{\text{Actual mep} \times v_s}{\text{Theoretical mep} \times v_s} = \frac{p_{ma}}{p_{mi}}$$

Actual mep = $DF \times$ ideal mep.

The power developed inside the cylinder or as measured and calculated by the indicator diagram is known as indicator power of the engine. The power available at the shaft is however, less than developed in the cylinder due to various frictional losses. The power available on the shaft of the engine is called brake power.

Therefore indicated power = brake power + lost frictional power

In terms of horse power; Indicated HP = Brake HP + frictional HP

$$\text{IHP} = \text{BHP} + \text{F.H.P.} \tag{12.3}$$

Brake power of an engine is measured by dynamometer in the laboratory.

Let   $p_m$ = Actual mean effective pressure in N/m$^2$

$A$ = Area of the piston = $\pi/4\ D^2$, where $D$ = diameter of piston, $m$

$L$ = Stroke length, $m$

$n$ = number of working strokes per minute

$N$ = RPM for single acting engine and $n = 2N$ for double acting steam engine

For single acting on the piston with forward stroke = $p_m A$.

Work done per stroke length = $p_m A \cdot L$.

Work done per revolution of crank = $p_m\ AL + p_m(A - a)\ L$ for double acting engine, a = area of piston rod, effective area for crank end = $(A - a)$.

Work done per minute = $\{p_m\ AL + p_m(A - a)\ LN\}$

$$= p_m\ LN\ (2A - a)$$

Indicated power,  I.P. $= \dfrac{p_m\ LN(2A - a)}{60}$ Watts

If $a << A$, then   I.P $= \dfrac{2\,p_m\ LAN}{60 \times 1,000}$ kW  $\qquad(12.4)$

## 12.9   EFFICIENCIES OF STEAM ENGINE

The followings are the different efficiencies of steam engine:

1. Mechanical Efficiency ($\eta_M$). It is the ratio of the power obtained on the shaft to the power developed by the engine cylinder. Thus it accounts for the loss in friction

$$\text{Mechanical Efficiency } \eta_m = \frac{\text{Brake power}}{\text{Indicated power}} \qquad(12.5)$$

2. Thermal Efficiency = ($\eta_T$) The thermal efficiency is the ratio of the work done to the heat supplied to the engine. This is of two types.

   (a)  Indicated thermal efficiency ($\eta_{IT}$): It is the ratio of indicated power to the heat supplied to the engine

$$\eta_{IT} = \frac{\text{Indicated power in kW}}{\text{Heat supplied in the steam in kW}}$$

or,   $\eta_{IT} = \dfrac{\text{I.P. in kW}}{m_s\,(h_1 - h_{f_2})}$  $\qquad(12.6)$

where   $m_s$ = mass of steam consumed in kg/s

$h_1$ = enthalpy of steam supplied to the engine in kJ/kg

$h_{f_2}$ = enthalpy of water returned to hot well in kJ/kg

   (b)  Brake thermal efficiency ($\eta_{B.T.}$) It is the ratio of brake power to the heat supplied in the steam

$$\text{Brake thermal efficiency} = \eta_{\text{B.T.}} = \frac{\text{B.P in kW}}{m_s\,(h_1 - h_{f_2})} \tag{12.7}$$

$$\therefore \quad \eta_{\text{B.T.}} = \eta_{\text{I.T.}} \times \eta_M$$

3. Rankine efficiency: It is the ratio of Rankine work to the heat supplied in the steam (already discussed in vapour power cycle)

$$\eta_R = \frac{h_1 - h_2}{h_1 - h_{f_2}} \tag{12.8}$$

4. Retalive efficiency or efficiency ratio: It is the ratio of the thermal efficiency to corresponding Rankine efficiency.

$$\text{Relative efficiency} = \frac{\text{Thermal efficiency}}{\text{Rankine efficiency}} \tag{12.9}$$

5. Overall efficiency: It is the ratio of net output at the crank shaft to the energy supplied to the boiler. It considers all the losses

$$\text{Overall efficiency } \eta_0 = \frac{\text{B.P. in kW}}{m_f\,(\text{C.V.})} \tag{12.10}$$

where $m_f$ = mass of fuel consumed in kg per second in the boiler in kg/s

C.V. = Calorific value of the fuel in kJ/kg.

## Solved Problems

**12.1** A steam engine has a stroke equal to 1.3 times the diameter and a diagram factor of 0.8. It is supplied with dry steam at 9.6 bar and exhausts at 1.03 bar. If the cut off takes place at 0.4 of stroke, and speed is 200 rpm and develops 184 kW. Calculate the dimensions of the cyclinder. (Ranchi University).

*Soln.* Ref. to Fig. 12.6

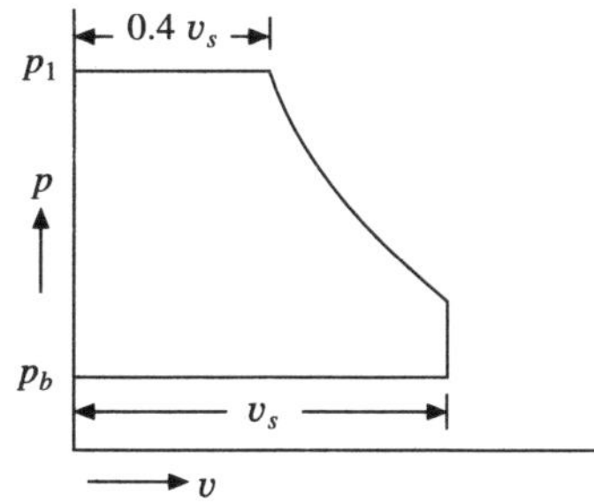


**Fig. 12.6**

$L = 1.3\,D$, D.F. $= 0.8$, $p_1 = 9.6$ bar

$p_b = 1.03$ bar, expansion ratio

$r = v_2/v_1 = v_s/0.4\,v_s = 2.5$, $N = 200$, I.P $= 184$ kW

$D = ?\ L = ?$

Theoretical mean effective pressure

$p_1/r[1 + \ln{(r)}] - p_b = 9.6/2.5\,[1 + \ln{(2.5)}] - 1.03 = 6.34$ bar

Actual mep $=$ D.F. $\times$ theoretical mep $= 0.8 \times 6.34 = 5.072$ bar

$$\text{I.P.} = 2\,p_m\frac{LAN}{60000} \quad \text{or,} \quad 184 = \frac{2 \times 5.072 \times 10^5 \times 1.3\,D}{60000} \times \pi/4 D^2 \times 200$$

or,

$$D^3 = \frac{184 \times 60000 \times 4}{2 \times 5.072 \times 10^5 \times 1.3 \times \pi \times 2000}$$

$$= 0.0532$$

$$D = 0.376,\ m = 37.6\ \text{cm} \quad \textit{Ans,} \qquad L = 1.3D = 1.3 \times 0.376 = 0.489\ \text{m} \quad \textit{Ans.}$$

**12.2** The following data pertains to a single cylinder double acting steam engine:

Admission pressure = 14 bar abs, back pressure = 0.35 bar abs, cut off at 0.4 stroke, diagram factor = 0.7, cylinder diameter = 30 cm, stroke = 1.5 times cylinder bore and mechanical efficiency 80%. Neglecting clearance, estimate B.P. developed by the engine running at 200 rpm    (Patna Univ. 1985 A AMIE 97).

*soln.*
$$p_I = 14 \text{ bar}, \; p_b = 0.35 \text{ bar}, \text{ expansion ratio} = 1/0.4$$

$$= 2.5, \text{ D.F.} = 0.7, \; D = 0.30 \text{ m}, \; L = 1.5 \times 0.30 = 0.45 \text{ m}$$

$$\eta_m = 0.80, \; N = 200, \; \text{B.P} = ?$$

Hypothetical mep = $p_1/r \cdot (1 + \ln r) - p_b = 14/2.5 \,(1 + \ln (2.5)] - 0.35 = 10.38$ bar

Actual mep = D.F. $\times$ 10.38 = 0.7 $\times$ 10.38 = 7.266 bar

$$\text{I.P} = \frac{2 \, p_m \, LAN}{60000} = \frac{2 \times 7.266 \times 10^5 \times 0.45 \times \pi/4 \,(0.3)^2 \times 200}{60000}$$

$$= 154.08 \text{ kW}$$

$$\text{B.P} = \eta_m \times \text{I.P.} = 0.80 \times 154.08 = 123.264 \text{ kW} \quad Ans.$$

**12.3** The following data relates to a single cylinder double acting steam engine

Cylinder bore = 15 cm, stroke 20 cm, I.P. = 25 kW

RPM = 300, Cut off = 0.20 of stroke volume,

Back pressure = 0.28 bar D.F. = 0.72

Determine admission pressure. Also calculate indicated thermal efficiency of engine if engine consumes 222 kg of dry saturated steam per hour.

*Soln.*
$$D = 0.15 \text{ m}, \quad L = 0.20 \text{ m}, \quad \text{I.P.} = 25 \text{ kW}$$

$$N = 300, \text{ expansion ratio} = 1/0.20 = 5, \; DF = 0.72, \; p_1 = ?$$

$$p_b = 0.28 \text{ bar}, \; m_s = 222 \text{ kg/hr}, \; \eta_{\text{I.T.}} = ?$$

$$\text{I.P.} = 2 \, p_m \frac{LAN}{60\,000}, \quad p_m = \frac{60\,000 \times \text{I.P}}{2 \, LAN}$$

$$= \frac{60\,000 \times 25}{2 \times 0.2 \times \pi/4 \cdot (0.15)^2 \times 300}$$

$$= 707355.3 \text{ N/m}^2$$

$$p_m = 7.07 \text{ bar}$$

$$\text{Theoretical mep} = \frac{\text{Actual mep}}{\text{D.F.}} = \frac{7.07}{0.72} = 9.819 \text{ bar}$$

$$\text{Theoretical mep} = p_1/r \cdot (1 + \ln r) - p_b, \; r = 1/0.2 = 5$$

or,
$$9.819 = p_1/5 \,(1 + \ln 5) - 0.28$$

or,
$$p_1 = \frac{(9.819 + 0.28) \times 5}{(1 + \ln 5)} = 19.35 \text{ bar} \quad Ans.$$

$$\text{Indicated thermal efficiency} = \eta_{\text{I.T.}} = \frac{\text{I.P.}}{m_s \,(h_1 - h_f)}$$

from steam table, $h_1$ = enthalpy of drysteam at 19.35 bar = 2796.22 kJ/kg, $h_{f_2}$ = enthalpy of saturated liquid at 0.28 bar = 282.7 kJ/kg

$$\eta_{\text{I.T.}} = \frac{25 \times 3600}{222 \, (2796.22 - 282.7)} \times 100 = 16.12\% \quad Ans.$$

**12.4** Find the dimensions of a single cylinder double acting non-condensing steam engine to satisfy the following requirements. B.P. = 50 kW, steam chest pressure = 11 bar gauge, back pressure = 1.1 bar, cut off at 5.8 of the stroke, clearance 5 percent of the stroke, piston speed = 125 m/minute, RPM = 300, piston rod diameter = 4.5 cm, diagram factor = 0.8, mechanical efficiency = 90 percent.

*Soln.*

$$\text{B.P.} = 50 \text{ kW}, \, p_1 = 11 + 1.013 = 12.013 \text{ bar}$$

$$p_b = 1.1 \text{ bar, cut off} = 5/8 \text{ of stroke, clearance ratio}$$

$$0.05, \, 2LN = 125 \text{ m/min}, \, N = 300, \, d = 4.5 \text{ cm}, \, \text{DF} = 0.8$$

$$\eta_m = 0.90$$

$$\text{Hypothetical mep} = p_1\left[1/r + (c + 1/r) \ln \frac{(c+1)}{(c+1/r)}\right] - p_b$$

$$= 12.013\left[5/8 + (0.05 + 5/8) \ln \frac{(1+0.05)}{0.05 + 5/8}\right] - 1.1 = 10.0 \text{ bar}$$

$$\text{Actual mep} = \text{D.F.} \times 10.0 \text{ bar} = 0.8 \times 10 = 8 \text{ bar}$$

$$\text{I.P.} = \frac{p_m LN}{60\,000} (2A - a) \quad \text{or,} \quad 50 = \frac{8 \times 10^5 \times 125}{60\,000 \times 2} (2A - a)$$

Because

$$2LN = \text{piston speed} = 125 \text{ m/min}$$

or,

$$(2A - a) = \frac{50 \times 60\,000 \times 2}{8 \times 10^5 \times 125} \text{ m}^2$$

$$0.06 \text{ m}^2 = 600 \text{ cm}^2$$

or,

$$2 \times \pi/4 \, D^2 - \pi/4 \times 4.5^2 = 600$$

or,

$$2D^2 - 4.5^2 = 4/\pi \times 600 = 763.943$$

$$D = 19.80 \text{ cm} \quad Ans. \; 2LN = 125, \, L = 125/2N$$

$$= 125/2 \times 300 = 0.208 \text{ m} = 20.8 \text{ cm} \quad Ans.$$

**12.5** The following observations were made during a trial of a single cylinder double acting steam engine: Cylinder diameter = 20 cm, stroke = 30 cm, piston rod diameter = 5 cm, Mean speed = 140 rpm, Area of indicator card (cover end) = 6.5 cm², Area of indicator car (crank end = 6.8 cm², length of indicator card = 6.8 cm, strength of indicator spring = 1 kg/cm²/cm, brake load = 440 N, spring balance reading = 68 N, radius of brake wheel = 0.6 m steam used per hour = 162 kg (dry admission pressure = 3 bar back pressure = 1.04 bar calculate I.P, B.P. mechanical efficiency, brake thermal efficiency, indicated thermal efficiency.

*Soln.*

$$D = 0.20 \text{ m}, \, L = 0.30 \text{ m}, \, d = 5 \text{ cm}, \, N = 140 \text{ rpm},$$

area of indicator card (cover end) = 6.5 cm, area of indicator card (crank side) = 6.8 cm², length of indicator card = 6.8 cm, spring strength = 1 bar/cm, $W = 440 \, N$, $S = 68 \, N$, $R = 0.6$ m, $ms = 162$ kg/hr, $p_1 = 3$ bar, $p_b = 1.04$ bar

Actual mean effective pressure for cover end

$$= p_{m_1} = \frac{\text{area of the diagram}}{\text{length of the diagram}} \times \text{spring strength} = \frac{6.5}{6.8} \times 1.0$$

$$= 0.955 \text{ bar}$$

$$\text{Actual mep for cover end} = p_{m_2} = \frac{6.8}{6.8} \times 1 = 1.0 \text{ bar}$$

$$\text{I.P. for cover end} = p_{m_1} \, LAN/60000$$

$$= \frac{0.955 \times 10^5 \times 0.30}{60\,000} \, (\pi/4)(0.20)^2 \times 140 = 2.525 \text{ kW}$$

$$\text{I.P. for crank end} = p_{m_2} \, \frac{L(A-a)N}{60\,000}$$

$$= \frac{1 \times 10^5 \times 0.30}{60\,000} \, [\pi/4\,(0.20)^2 - \pi/4\,(0.05)^2] \times 140$$

$$= 2.061 \text{ kW}$$

$$\text{Total I.P.} = 2.625 + 2.061 = 4.686 \text{ kW}$$

$$\text{B.P.} = 2\pi NT/60000 = 2\pi N \, (W-S) \, R/60000$$

$$= \frac{2\pi \times 140 \, (440 - 68) \times 0.6}{60\,000}$$

$$= 3.27 \text{ kW}$$

$$\text{Mechanical efficiency} = \eta_m = \text{B.P/I.P} = 3.27/4.686 = 0.6978$$

$$= 69.78\% \quad Ans.$$

$$\text{Indicated thermal efficiency} = \text{I.P.}/m_s \, (h_1 - h_{f_2})$$

$$h_1 = h_g \text{ at 3 bar} = 2724.7 \text{ kJ/kg}$$

$$h_{f_2} = h_f \text{ at 1.04 bar} = 418.78 \text{ kJ/kg}$$

$$\text{Indicated thermal efficiency} = \frac{4.688 \times 3600}{162 \, (2724.7 - 418.78)}$$

$$= 0.045 = 4.5\% \quad Ans$$

$$\text{Brake thermal efficiency} = \frac{3.27 \times 3600}{162 \, (2724.7 - 418.78)}$$

$$= 0.031 = 3.1\% \quad Ans.$$

## 12.10  STEAM CONSUMPTION FROM INDICATOR DIAGRAM AND MISSING QUANTITY

Let  $m_c$ = mass of steam in the cylinder at the start of admission i.e. the mass of entrapped or cushion steam per stroke.

$m_a$ = mass of steam admitted to the cylinder per stroke of the engine.

Then mass of steam in the cylinder at any point between cut off and release is given by

$$m = m_c + m_a \tag{10.11}$$

The mass of steam admitted per stroke ($m_a$) is calculated by:

$$m_a = \frac{\text{Total mass of steam condensed per hour}}{\text{Total number of working strokes per hour}} = m / 60 \times 2N$$

In order to determine cushion steam ($m_c$) it is assumed that the steam is dry during compression of cushion steam. Let at any point $H$ (Fig. 12.7) on the compression curve, pressure be $p_H$ and volume $V_H$ then mass of cushion steam missing quantity per stroke $m_c = V_H/v_{SH}$ were $v_{SH}$ = sp. Volume of dry steam at point $H$. Actual expansion curve $CE$ shows that volume of steam is less than if the steam would remain dry throughout the expansion. Let there be any point $M$ on the expansion curve having pressure $p_m$. There total volume of dry steam at this pressure $= mv_{sm}$, where $v_{sm}$ = sp. volume of dry steam at pressure $p_M$. We can now draw a Line $LN$ having its length $LN = mv_{SM}$. Similarly taking a number of points on the expansion curve other points similar to $N$ can be determined. The locus of all these points is called saturation curve. Thus a saturation curve is the curve showing the volume of the steam in the engine cyclinder during expansion stroke if the steam is dry and saturated at all points.

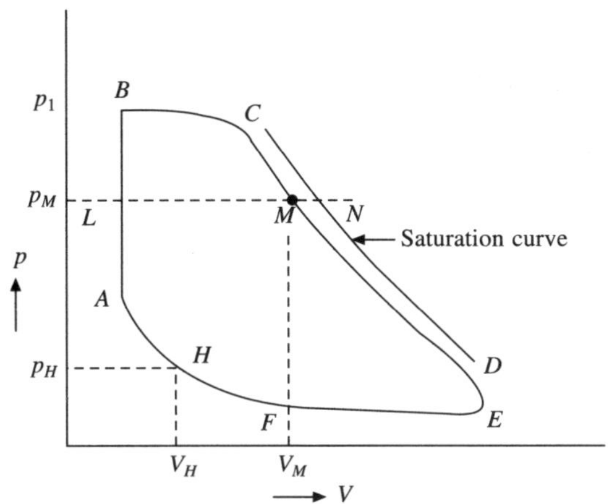


**Fig. 12.7   Missing quantity**

Dryness fraction of steam at point $M$ is given by

$$x_m = \frac{\text{Lenght of } LM}{\text{Lenght of } LN} = V_m / mv_{SM}$$

The horizontal distance between the saturation curve and actual expansion curve is called "Missing quantity". Thus missing quantity at any point $M = MN = LN - LM = mv_{SM} - mx_m v_{SM}$

$$= mv_{SM}(1 - x_m)$$

Sometimes it is also given as the difference of actual mass of steam in the cylinder and indicated mass. Thus missing quantity by

$$\text{mass} = m \, (1 - x_m) \tag{12.12}$$

The missing quantity is mainly due to condensation of steam. The missing quantity can be reduced by proper steam jacketing of the engine cylinder. The missing quantity as a volume is maximum near the point of cut off, diminishes as the expansion proceeds and is minimum at the point of release. This is because steam gets evaporated at reduced pressure due to heated cylinder wall.

**12.6**   A steam engine has following data: Swept volume = 0.105 m³, cut off takes place at 0.5 of stroke volume, clearance = 8%, pressure at 0.7 of stroke is 4.6 bar, compression starts at 1.40 bar and 0.8 of return stroke, RPM = 100 and steam exhaust to condenser = 50 kg/min. Determine at 0.7 of stroke volume (a) the dryness fraction, (b) missing quantity.

*Soln.*   Refer to Fig. 12.8

Clearance volume $v_c = 0.08 \times 0.105 = 0.0084$ m³

Swept volume at 0.7 stroke

$$= 0.7 \times 0.105 = 0.0735 \text{ m}^3$$

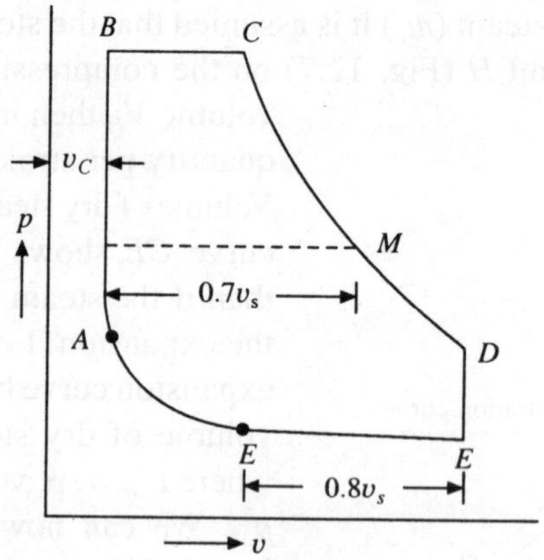


**Fig. 12.8**

Total volume at $M = 0.0084 + 0.0735 = 0.0819$ m$^3$
Specific volume of dry steam at pressure 4.6 bar $= v_{SM} = 0.4052$ m$^3$/kg
Volume of cushion steam $= v_f = 0.08 = (1 - 0.8) \times 0.105 = 0.029$ m$^3$
Specific volume of dry steam at $F$(at 1.40 bar) $v_{SF} = 1.236$ m$^3$/kg mass of cushion steam per stroke $= m_c$
$= V_F/v_{SF}$

$$= 0.029/1.236 = 0.0234 \text{ kg/stroke}$$

Mass of steam admitted per minute

$$= m_a = 50/2N = 50/2 \times 100 = 0.25 \text{ kg/stroke}$$

Total mass of steam $= m_c + m_a = 0.0234 + 0.25 = 0.2734$ kg/stroke.
If the steam is dry at expansion curve, then mass of dry steam indicated at

$M = V_M/v_{SM} = \dfrac{0.0819}{0.4052} = 0.2021$ kg/stroke dryness fraction at $M = x_m = 0.2021/0.2734 = 0.73$   *Ans.*
Missing quantity $= 0.2734 - 0.2021 = 0.0713$ kg/stroke

$$= 0.0713 \times 2 \times 100 \times 60 = 855.6 \text{ kg/hr} \quad Ans.$$

**12.7** The following data relates to a single cylinder double acting steam engine using dry saturated steam:
Bore = 300 mm, stroke = 450 mm, hypothetical mep = 3.5 bar cut of at 25% of the stroke, exhaust pressure
0.5 bar abs., steam consumption = 560 kg/hr, speed = 160 rpm, diagram factor = 0.8, mechanical efficiency
= 80%. Calculate brake thermal efficiency, assuming hyperbolic expansion and neglect clearance and
compression.

*Soln.* $D = 0.3$ m, $L = 0.450$ m, $p_m = 3.5$ bar cut off at 25% stroke $p_b = 0.5$ bar, $m_s = 560$ kg/hr, $N = 160$ rpm, $DF$
$= 0.8$, $\eta_m = 0.80$.
Expansion ratio $= r = 1/0.25 = 4$
Actual mep $= 3.5 \times$ D.F. $= 3.5 \times 0.80 = 2.8$ bar

$$\text{I.P.} = 2\, pm\, LAN/60\,000 = \frac{2 \times 2.8 \times 10^5 \times 0.45 \times \pi/4 \times (0.3)^2 \times 160}{60\,000} = 47.5 \text{ kW}$$

$$\text{B.P.} = \eta_m \times \text{I.P.} = 0.8 \times 47.5 = 38 \text{ kW}$$

Brake thermal efficiency $= \text{B.P.}/m_s\,(h_1 - h_{f_2})$
To determine $h_1$, we have to determine supply pressure

$$p_m = p_1/r(1 + \ln r) - p_b \quad \text{or,} \quad 3.5 = p_1/4 \ (1 + \ln 4) - 0.5$$

or,
$$p_1 = \frac{(3.5 + 0.5) \times 4}{(1 + \ln 4)} = 6.70 \text{ bar}$$

$$h_1 = h_g \text{ at } 6.70 \text{ bar} = 2760.15 \text{ kJ/kg}$$

$$h_2 = h_f \text{ at } 0.5 \text{ bar} = 340.6 \text{ kJ/kg}$$

Brake thermal efficiency $\eta_{\text{B.T.}} = \dfrac{38 \times 3600}{560 \times (2760.15 - 340.6)} \times 100 = 10.1\% \quad Ans$

*Note* Since the steam engine is becoming obsolete, therefore, this chapter has been removed from the syllabai of Engineering, Diploma and AMIE courses as directed by the Technical Education Authorities. Due to this reason this chapter does not contain the articles like determination of brake power, governing of steam engine, compound engine and work done produced.


## EXERCISES

### A Theoretical Problems

12.1 Explain the function of the followings:
(a) Piston, (b) Connecting rod (c) cross head (d) stuffing box (e) flywheel (f) eccentric (g) D-slide valve.

12.2 What do you mean by the following terms as applied to steam engine? (a) clearance volume (b) stroke volume (c) cut off (d) dead centres (e) valve travel (f) back pressure (g) eccentricity

12.3 What is the difference between:
(a) single acting and double acting steam engines?
(b) simple and compound steam engines?
(c) codensing and non-condensing steam engines?
(d) horizontal and vertical steam engines?

12.4 Where are the small end and big end of a connecting rod are connected?

12.5 Why the modified Rankine cycle is adopted for reciprocating steam engine?

12.6 How and why the actual indicator diagram of a steam Engine differs from hypothetical indicator diagram?

12.7 State the methods adopted to improve diagram factor.

12.8 Why the cylinder of a steam engine is steam jacketed?

12.9 Define diagram factor what is the use of diagram factor in the design of steam engines?

12.10 Show that the hypothetical mean effective pressure of steam engine without clearance and compression is given by $p_m = p_1/r \ (1 + \ln r) - p_b$.

12.11 Explain what is meant by the term missing quantity as applied to steam engine.

12.12 What do you mean by governing of an engine?


### B Numerical Problems

12.1 Determine the actual mep for a steam engine receiving steam at a pressure of 6 bar, cut off takes place when piston has travelled 0.4 of the stroke in which clearance volume is 10% of stroke volume. Back pressure is 1.03 bar and diagram factor is 0.7. (*Ans* 2.61 bar).

12.2 A single cylinder double acting steam engine has 30 cm piston diameter and 45 cm stroke length and 140 rpm. Steam is supplied at 11 bar absolute and cut off takes place at 1/2 of the stroke. Back pressure is 0.4 bar abs. Neglecting clearance and assuming a diagram factor of 0.762 and mechanical efficiency 80%, find B.P. of the engine.

12.3 The following data relates to a single cylinder double acting steam engine; B.P. = 74 kW, RPM = 250, admission pressure = 11.8 bar abs, supply steam is dry saturated back pressure = 0.88 bar, cut off at 1/4 of stroke, stroke = 1.3 times bore, mechanical efficiency = 85%, diagram factor = 0.8, brake thermal efficiency = 15%, neglect, clearance, calculate: (a) cylinder dimensions (b) specific steam consumption.

*(Ans D = 27.5 cm, L = 35.8 cm, SSC = 10.08 kg/kWhr).*

12.4   Calculate the cylinder dimensions of a single cylinder double acting steam engine having the following data: B.P. = 13.4 kW, mechanical efficiency = 85%, stroke = 8/5 cylinder bore, cut off at 0.35 stroke, admission pressure = 5.8 bar abs, back pressure = 0.19 bar, D.F. = 0.65 RPM = 180. Neglect clearance *(Ans D = 20 cm, L = 32 cm).*

12.5   A single cylinder double acting condensing type steam engine develops 184 $kW_{I.P.}$ at 120 RPM. Bore = 36 cm, stroke = 72 cm cut off at 40% of stroke back pressure = 0.19 bar, D.F. = 0.85. Calculate the admission pressure neglecting clearance and compression.     *(Ans. $p_1$ = 9.8 bar abs).*

12.6   The following readings were taken from a test on a single cylinder double acting steam engine: Bore = 27 cm, stroke = 33 cm, supply pressure = 8.8 bar abs, engine speed = 220 RPM, area of indicator card = 10.5 $cm^2$, length of indicator card = 8 cm, spring strength = 2.45 bar/cm, net load on brake wheel = 980N, radius of brake wheel = 160 cm, calculate: I.P.B.P. mechanical efficiency.

*(Ans. 50.23 kW, 36.24 kW, 81%).*

12.7   Find the cylinder dimensions of a double acting steam engine having following particulars: I.P = 52 kW, cut off at 3/8 of stroke, admission pressure = 7.53 bar back pressure 1.15 bar, clearance = 8%, stroke = 1.5 bore, RPM = 120, D.F. = 0.65

*(Ans. D = 33.2 cm, L = 50 cm)*

12.8   The following data were obtained during a test on double acting steam engine.
Indicated mean effective pressure = 2.45 bar, RPM = 104 bore = 25 cm, stroke = 30 cm, Net brake load = 1128N Effective brake diameter = 165 cm, steam supply at 6.80 bar abs and dry, condenser pressure = 0.06 bar abs, condenser temperature = 22°C, condensate quantity = 3.3 kg/min, determine (a) I.P, B.P. mechanical efficiency and (b) Brake thermal efficiency (AMIE 1997 W)

12.9   The following observations were made during a trial of a single cylinder double acting steam engine: Steam used per hour = 56 kg, RPM = 100, mep (average both diagram) = 1.13 bar, brake load = 623N spring balance reading = 176N, cylinder diameter = 22 cm, cylinder stroke = 30 cm, effective brake radius = 55 cm, calculate: I.P, B.P, mechanical efficiency specific steam consumption

*(Ans. I.P. = 4.41, B.P. = 2.6 kW, $\eta_m$ = 59% SSC = 12.6 kg/kW hr)*

12.10   The following data were recorded during a test on double acting steam engine running at 100 rpm; cut off volume = 0.015 $m^3$, pressure = 13.73 bar. Release volume = 0.055 $m^3$, pressure = 3.4 bar, on compression curve, volume = 0.005 $m^3$, pressure = 3. bar steam consumption = 1580 kg/hr
Determine the missing quantity per hour at cut off and at release. If there is difference, state the reasons.

*(Ans at cut of 455 kg/hr, at release 472 kg/hr)*

12.11   A single cylinder, double acting non condensing steam engine 25 cm bore and 50 cm stroke develops 29 kW I.P. at 100 rpm. The clearance is 10% and cut off is at 40% of stroke. The pressure at cut off is 4.9 bar abs. The compression starts at 80% of return stroke. The pressure of the steam on compression curve at 90% of the return stroke is 13.73 bar abs and steam is dry saturated. Determine steam consumption per hour per kW, if missing quantity at cut off is 0.007 kg/stroke.

*(Ans. 14.43 kg/kW/hr)*

12.12   The following data refer to a single cylinder double acting condensing type steam engine. Bore = 24 cm, speed = 200 rpm, pressure of supply steam = 9.8 bar abs, quality of steam supply = 0.94 dry, cut off = 40% stroke vacuum in the condenser = 50 cm of Hg. barometer reading = 76 cm of Hg, indicated steam at cut off = 747 kg/hour, mechanical efficiency = 80% diagram factor = 0.9, find I.P. and brake thermal efficiency.

*(Ans. 123 kW, $\eta_{BT}$ = 20%)*

13

# Steam Condensers and Cooling Water Supply

## 13.1 INTRODUCTION

In order to achieve maximum efficiency of a steam plant, according to the carnot's principle, the heat must be supplied at the maximum possible pressures and temperatures and the heat rejection should take place at minimum pressures and temperatures. Heat generated and supplied to piston and cylinder of steam engine and turbine blades depends upon metallurgical limit of these and even boiler tubes also. Therefore, to increase the work out and to improve the efficiency of the power plant, the aim is to reject the heat at minimum pressure and temperature. The rejected temperature can be reduced to the atmospheric temperature if the exhaust of the steam takes place below atmospheric pressure. If the exhaust is at atmospheric pressure, the heat rejection is at 100°C.

Exhaust temperature is lowered when exhaust pressure decreases. This decrease in pressure should not go below atmospheric pressure if steam has to be exhausted into atmosphere in a non-condensing steam engine or steam turbine. Under this situation, the steam is exhausted into a vessel known as condenser where the pressure is maintained below the atmosphere (see Fig. 13.1(a), (b), (c)) by continuously condensing the steam by means of circulating cold water at atmospheric temperature. Thus by use of a condenser in a steam plant low exhaust pressure can be used and large heat drop (enthalpy drop) per kg (see Fig. 13.2) of steam utilised increasing both the efficiency and power output of the plant.

Therefore, the condenser is defined as a closed vessel heat transfer device in which exhaust steam of a turbine or an engine is condensed by means of cooling water at pressures below atmospheric. The condensed steam, called the condensate, can be again returned to the boiler as feed water.

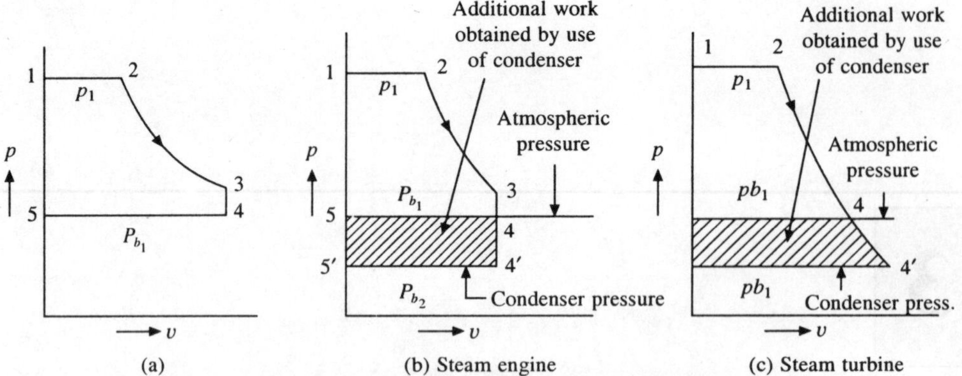



**Fig. 13.1   Additional work obtained by use of condenser**

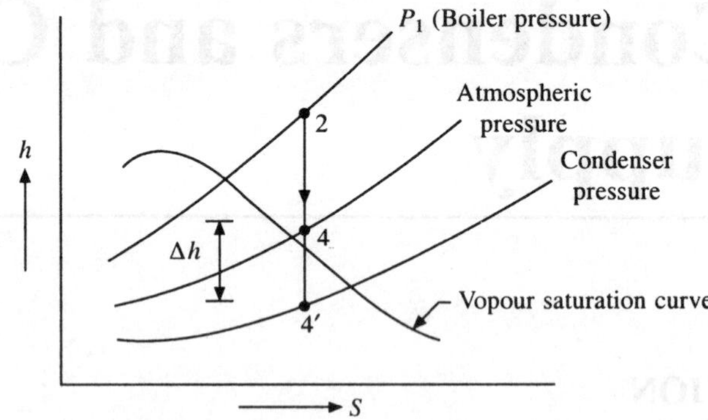



**Fig. 13.2   Additional enthalpy drop obtained**

# 13.2   ADVANTAGES OF A CONDENSER IN A STEAM POWER PLANT

1.  It increases the work output by increasing expansion ratio of steam and, thus efficiency of the plant increases
2.  It reduces the specific steam consumption, therefore, it reduces the size of the power plant of given capacity.
3.  The reuse of condensate as feed water for boilers reduces the cost of power generation.
4.  Since the pure water is available for the boiler, it saves the cost of water softening plant.

The maintenance of higher vacuum is limited beyond which the reduction of back pressure does not prove economical. The cylinder of steam engine, being a positive displacement device, imposes a limit on the engine condenser pressure. An exhaust pressure of about 65 cm of mercury (0.867 bar) vacuum is achieved as lowest value in the steam engine, because it is an intermittent flow machine. Steam turbine being steady flow machines may be designed to operate at 74 cm of mercury (0.986 bar) vacuum or even more depending upon the capacity of the plant and cooling water available. The increased specific volume at high vacuum can be easily accommodated in turbine without the increased friction and condensation losses that are involved in reciprocating engines.

## 13.3   ELEMENTS OF A CONDENSING PLANT

The various components of a condensing plant are shown schematically in Fig. 13.3 and mentioned as follows:

1. Condenser—It is a closed vessel in which steam is condensed. The steam gives up heat to cooling water during the process of condensation.
2. Condensate pump—It is a pump, which removes condensate (condensed steam) from the condenser to the hot well.
3. Air extraction pump—It is a pump to remove air from condenser. If a pump is used to remove both i.e. the condensate and air, is called wet air pump. Dry air pump is called when it removes air only from condenser.
4. Hot well—It is a sump located between condenser and boiler, which receives condensate pumped by condensate pump.
5. Boiler feed pump—It is a pump, which pumps the condensate from the hot well to the boiler. This is done by increasing the pressure of condensate above the boiler pressure.
6. Cooling tower or spray pond—If there is a shortage of cooling water i.e. a river is not flowing near by the steam plant, cooling tower or spray pond system is used. In cooling tower or spray pond the hot water from the condenser is cooled by rejecting heat to atmospheric air.
7. Cooling water pump—It is a pump used to supply water either from the river or from the cooling tower pond to the condenser.

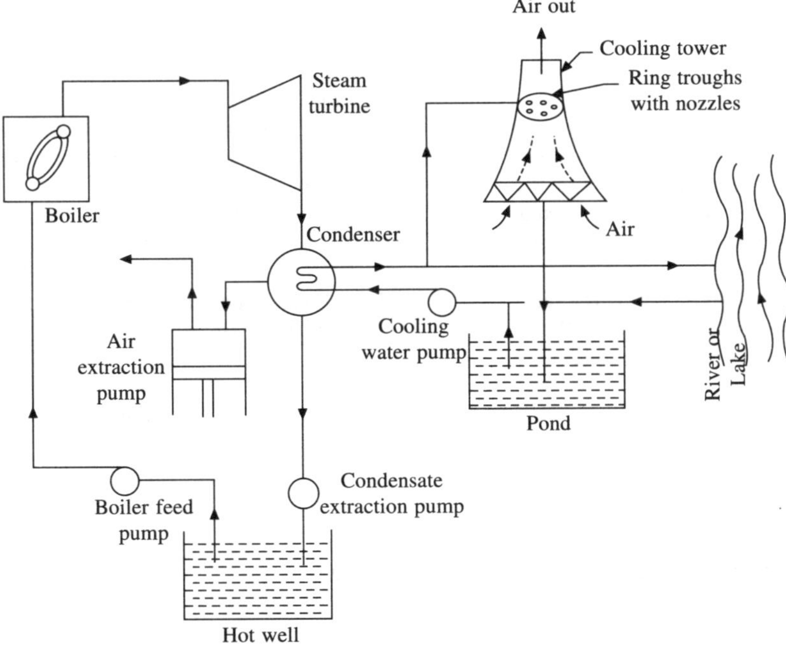


**Fig. 13.3   Steam condensing plant**

## 13.4  TYPES OF CONDENSERS

The steam condensers are broadly classified into the following two types, depending upon the path of steam to be condensed:

(a)  Jet condensers or mixing type condensers, and
(b)  Surface condensers or non-mixing type condensers.

In a jet condenser cooling water comes in direct contact with exhaust steam after spraying through jets. The temperature of condensate and the coolant is same while leaving the condenser. These types of condensers are rarely used since the condensate collected can not be reused due to the impurities of the coolant in the condensate. The impure condensate will corrode the plant tubes and will give rise to scale formation in the boiler tubes. Scale formation reduces the efficiency. So if jet condenser is to be used, the cooling water has to be treated first in treatment plant. Such condensers are generally used for small power plants.

In surface condenser cooling water flow through a network of tubes and the exhaust steam is passed over these tubes. The steam gets condensed due to the heat transfer to coolant by conduction and convection. In these type of condensers the condensate collected is fit for reuse as feed water to the boiler. Surface condenser is the best choice for large plants except in rare cases where the jet condenser is compulsory becuase of space and water condition and running period. In marine engine also, surface condenser is most suitable as the condensate is recovered as feed water to the boiler. Marine engine has to carry limited amount of pure water.

Therefore the comparison between jet and surface condenser could be made as follows:

| Jet condenser | Surface condenser |
|---|---|
| 1.  Exhaust steam and cooling water come in direct contact. | 1.  Exhaust steam and cooling water do not come in direct contact. |
| 2.  It requires less quantity of cooling water to condense the steam since both cooling water and steam are mixed. | 2.  Since exhaust steam and cooling water are not mixed up, hence the condensate can be directly pumped to the boilers. |
| 3.  Only pure water is used because mixed quantity of steam and cooling water is fed to the boiler. | 3.  Any kind of feed water can be used. |
| 4.  Vacuum greater than 65 cm of Hg can not be achieved since the dissolved air in cooling water gets liberated at very low pressure. | 4.  A high vacuum can be attained as much as 73.5 cm of Hg |
| 5.  Does not affect plant efficiency. | 5.  It improves the plant efficiency. |
| 6.  Space requirement and maintenance cost is low. | 6.  The system is complicated, bulky, costly and requires high maintenance cost. |
| 7.  It requires more power to drive the air pump. | 7.  It requires less power to drive the air pump |

The jet condenser is further sub-classified as follows:

(a)  Low level jet condensers.
    (i)  counter flow type.
    (ii) parallel flow type.

(b) High level jet condensers.
(c) Ejector jet condensers.

The surface condenser is further sub-classified as follows:

(a) Surface condenser
   (i)  down flow, central flow and inverted flow type
   (ii) single pass and multi pass
(b) Evaporative condenser

## 13.5  JET CONDENSERS

### (a) (i)  Low level-counter flow jet condenser

Schematic diagram of a low level counter flow jet condenser is shown in Fig. 13.4. In this type of condenser, exhaust steam is supplied from the bottom side of the condenser and it flows

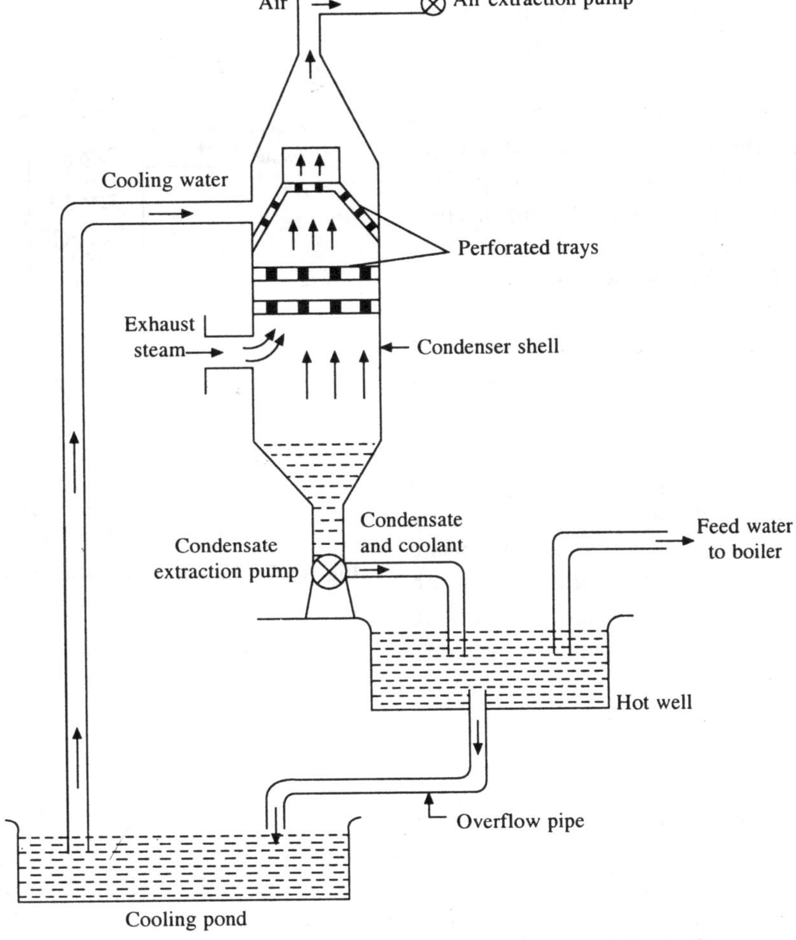


**Fig. 13.4  Low level-counter flow jet condenser**

upwards while the cooling water is supplied from the top of the condenser. The water flows in downward direction through a series of perforated trays. The steam gets condensed while it comes in contact with the falling water. The air extraction pump is situated at the top of the condenser which draws the air and any uncondensed vapour. The air pump always maintains the required vacuum in the condenser and induces the cooling water to be lifted into the condenser upto a height of 5.5 m approximately. The mixture of condensate and the coolant is extracted with the help of condensate extraction pump and it is discharged into a hot well. The excess amount of condensate from hot-well flows into the cooling pond by an overflow pipe as shown and the remainder is pumped to the boiler as feed water.

The capacity of air pump required is small since it is required to extract cooled air and water vapour. These condensers can be installed directly below the steam turbine. Such condensers have a disadvantage of flooding the steam turbine if the condensate extraction pump fails.

### (a)  (ii) Low level parallel flow jet condenser

In this type of condenser, both the steam and water enter at the top and the mixture is removed from the bottom. Schematic diagram is shown in Fig. 13.5 with condenser shell only. Other arrangement remains same to counter flow. The mixture of condensate, coolant and air are extracted with the help of wet air pump. This limits the vacuum created upto 60 cm of Hg.

### (b)  High level jet condenser or barometric condenser

Figure 13.6 shows the schematic diagram of high level jet condenser. It is also called as barometric condenser since it is placed above the atmospheric pressure requivalent to 10.36 m of water column. This column of water in tail pipe forces the condensate to drain away by gravity into the hot well, hence, a condensate extraction pump is not needed for such condensers. Another advantage of such an arrangement is that the water from the hotwell will not be able to rise into the condenser and flood the turbine due to the vacuum pressures maintained in the condenser. The working and other details are similar to low level counter flow get condenser.

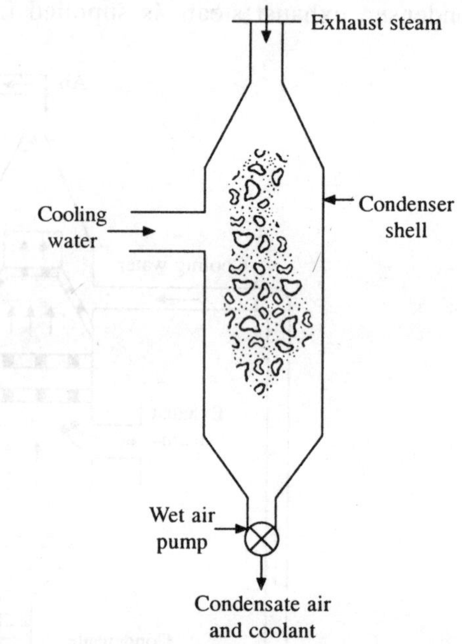


**Fig. 13.5   Low level parallel flow jet condenser**

Theoretically the height '$H$' of the tail pipe is given by:

### (c) Ejector Condenser

The arrangement of ejector condenser is shown in Fig. 13.7. The cooling water under the head of around 6 m enters the condenser from the top and passes through a series of convergent nozzles and finally it leaves through a convergent divergent nozzle. In flowing part of the steam ports, the water produces a vacuum. The vacuum causes the exhaust steam and air to flow through the port in the tube and mix with the cooling water. The exhaust steam gets condensed, as a result of which the vacuum is further increased. The condensate and air after passing through converging cones pass through the diverging cones and in doing so water gains the momentum which forces out condensate and air against the atmospheric pressure. So, in this case no air pump is required.

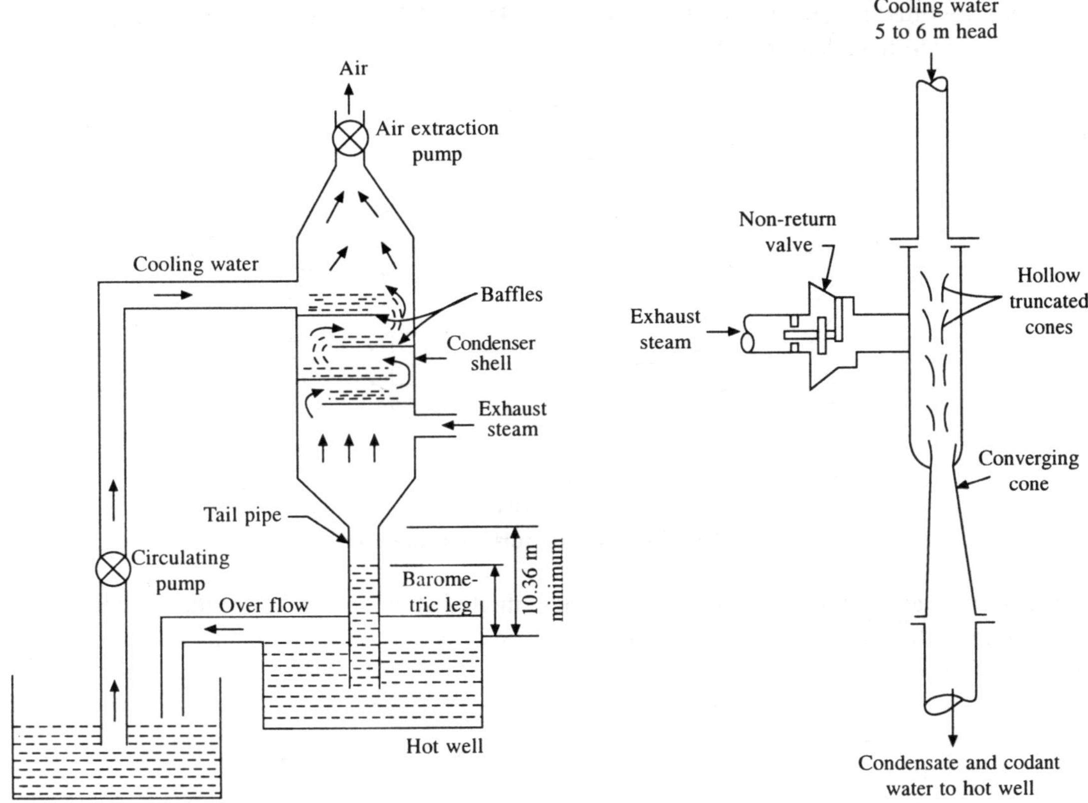


**Fig. 13.6  High level jet condenser**　　　　**Fig. 13.7  Ejector condenser**

A non return valve is fitted on the exhaust steam inlet, to prevent a sudden rush of water from the hot well into the engine or turbine, in case of a failure in supply of injection water. This condenser is used for moderate vaccum and is kept well below the engine level. The main advantage is the absence of an extraction pump.

## 13.6  SURFACE CONDENSERS

In case of surface condenser, the exhaust steam is condensed by passing it over the network of tubes inside which the cooling water flows. It is an air tight shell enclosing heat transfer surface. The surface condenser is further classified based on:

   (a)   (i)  the number of water passes—single pass or multipass.

        (ii)  the direction of condensate flow and tube arrangement. e.g. down flow condenser, central flow condenser etc.

   (b) Evaporative condenser

### (a) (i) Double Pass Surface Condenser

Figure 13.8 shows the schematic diagram of a two pass down flow surface condenser. The two pass arrangement is compact, more efficient in the process of heat transfer and preferred when the supply of cooling water is limited. It comprises of a cast iron cylindrical shell having the two ends covered by cover plate. Other shapes of shell, such as, oval or U may be also used. A number of water tubes are fitted into the perforated tube plates. The ends of the tubes are either expanded or welded in order to have leak-proof joint.

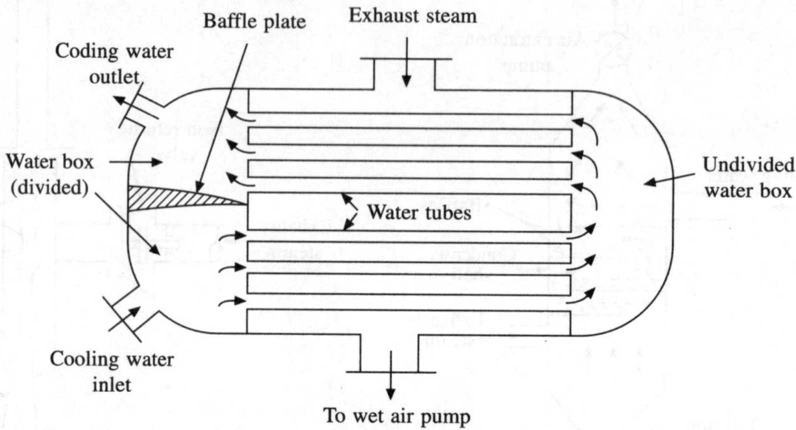


**Fig. 13.8  Double pass surface condenser**

The cooling water enters at one end of the tubes situated at the lower half of the condenser and after flowing to the other end it returns in the opposite direction through the tubes situated in the upper half of the condenser. This is obtained by providing pass partition plates in the dished ends. In this type, since the water traverses two times, it is called as 2-flow surface condenser. If the water traverses more then two times then it is called multiflow condenser.

This condenser requires two pumps viz. wet air pump to remove air and condensate and secondly water circulating pump. The exhaust steam enters at the top of the condenser and while it flows down over the tubes, the steam gets condensed.

### (ii) Down flow surface condenser

It is also called as dry vacuum type condenser for which cross-sectional view is shown in Fig. 13.9. The working principle is similar to double pass surface condenser except that this condenser

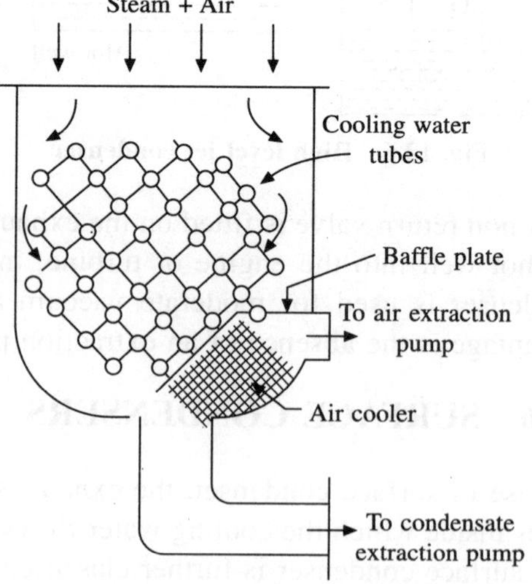


**Fig. 13.9  Down flow condenser**

employs two separate pumps for the extraction of condensate and the air. Figure shows the side view of the same. A baffle plate is provided throughout the length of the condenser. The exhaust steam enters at the top and flow downwards over the tubes due to force of gravity as well as suction of the extraction pump fitted at the bottom and then pumped by the extraction pump. The dry air pump suction pipe, provided near the bottom, is covered by the baffle plate to prevent the entry of condensed steam into it.

The steam when comes in contact with the cold surfaces of tubes it gets condensed and the air gets cooled. Air is further cooled when it flows from below the baffle plate. Air is cooled to the minimum value, thereby decreasing its specific volume. So air pump capacity is reduced and as a result power consumption also reduces.

### Central flow Surface Condenser

Figure 13.10 shows the cross-sectional side view of central flow surface condenser. In this case extraction pump is provided at the centre. So negative pressure is created at the centre. The exhaust steam and air enter from top and flow radially towards the centre by passing over the entire periphery of the tubes. The condensate is extracted at the bottom by an extraction pump.

This type of condenser is an improvement over the down flow type as the steam is directed radially inwards by a volute casing around the tube nest.

In the inverted type of condenser, the air suction pump is placed at the top and therefore the steam goes up while entering from the bottom of the condenser. The steam that is condensed collects at the bottom and is extracted by a pump

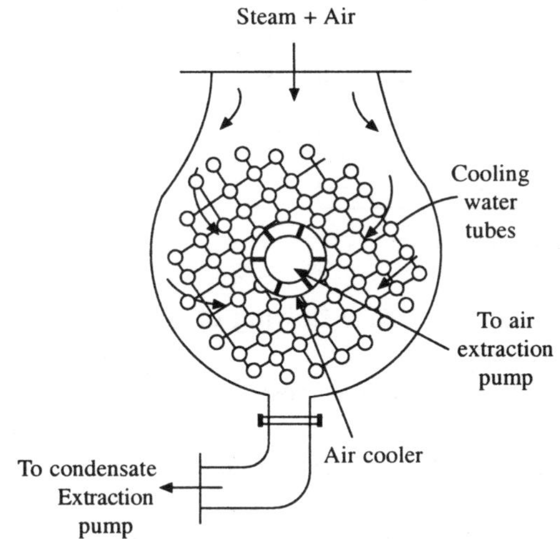


**Fig. 13.10    Central flow surface condenser**

### (b) Evaporative Condenser

This condenser is named so because it works on the principle of evaporation of cooling water under a small partial pressure. The schematic diagram of such type of condenser is shown in Fig. 13.11. Such type of condenser is used where supply of cold water is extremly limited since it requires very small amount of circulating and make-up water because of the use of the coolant again and again. However, these condenser are used in small power plants.

In this case, the steam to be condensed enters at the top of a series of grilled pipes kept vertical. The cooling water, pumped from a cooling pond, to a horizontal header having nozzles, sprays the water over the grilled pipes. The water while descending down, trickles over the pipes by making a thin film continuously. During the condensation of steam, this thin film of water is evaporated and the remainder of water is collected in the water tank. The condensate is extracted with the help of wet air pump. The air passing over the tubes carries the evaporated water in the form of vaour and it is removed with the help of induced drafttan installed at the top.

## 13.7   REQUIREMENTS OF A GOOD SURFACE CONDENSER

The following are the requirements of a good surface condenser.

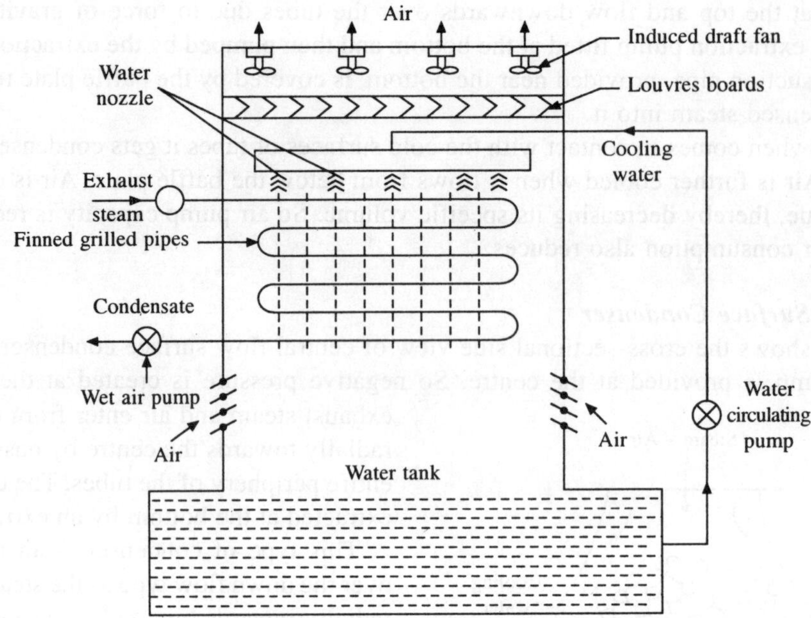


**Fig. 13.11   Evaporative condenser**

  (i)  The condenser should have least possible resistance for easy flowing of steam.

  (ii)  The steam should be well distributed in the vessel.

 (iii)  The circulating water should flow through the tubes with least resistance.

 (iv)  The steam should lose only its latent heat to circulating water. There should be no undercooling of condensate. This indicates that temperature of exhaust water should be same to that of saturation temperature of steam.

  (v)  The heat transfer rate through the walls of the tube should be maximum. In order to achieve this, the tube should be made of material having high thermal conductivity and water should flow through tubes and vapour outside so that outside surface of the tubes do not get deposited with sediments.

 (vi)  There should be no leakage of air from outside into the condenser.

(vii)  Air should be removed from the condenser with minimum expenditure of energy. In order to achieve it, a baffle plate should be located at the coolest zone where air pump is attached. This leads to reduce the specific volume of air and thereby to reduce the capacity of the pump.

## 13.8   ADVANTAGES AND DISADVANTAGES OF JET CONDENSERS

### (a) Jet condensers

### *Advantages*

  (i)  Less cooling water is required due to more intimate mixing of steam with it.

  (ii)  The system is simple in construction and its cost is low.

   (iii)  It requires less building space due to direct mixing.

   (iv)  It does not require cooling water pump.

### *Disadvantages*

   (i)  There is wastage of condensate

   (ii)  After mixing, if condensate is supplied to the boiler as a feed water, the cooling water should be pure and free from harmful impurities.

   (iii)  Due to less effective cooling of air, power required to run the air pump is high compared to surface condenser.

   (iv)  Vacuum greater than 65 cm of Hg can not be achieved since the dissolved air in cooling water gets liberated at very low pressures.

   (v)  Less suitable for high capacity plants.

## (b) High level jet condenser or barometric jet condenser

### *Advantages*

   (i)  It does not require a condensate extraction pump.

   (ii)  No possibility of flooding.

   (iii)  It is cheap, simple with low maintenance cost.

   (iv)  It requires less circulating water compared to surface condenser.

   (v)  Less building space is reeded.

### *Disadvantages*

   (i)  There is wastage of condensate

   (ii)  Long exhaust pipe line makes the initial cost high and also posibility of more air leakage.

   (iii)  It is less suitable for high capacity plants.

   (iv)  Due to leakage of air in long pipe, there is loss of vacuum of the order of 10 to 15 mm of Hg.

   (v)  More power is required to run the air extraction pump, it may be two times the power required for surface condensers.

# 13.9   ADVANTAGES AND DISADVANTAGES OF SURFACE CONDENSER

### *Advantages*

   (i)  Condensate can be used again and again since cooling water and steam are not mixed up.

   (ii)  High vacuum and greater plant efficiency are achieved because there is no mixing of water and steam.

   (iii)  Any kind of cooling water can be used and so the cost of water softening plant is considerably reduced.

   (iv)  It requires less power to run the air extraction pump.

   (v)  It requires less power for water pumping.

   (vi)  It is ideally suited for high capacity plants.

***Disadvantage***

    (i)  The system is complicated, costly and requires high maintenance costs.

   (ii)  More cooling water is required since water does not mix with steam.

  (iii)  It requires large floor space since it is bulky

## 13.10   SOURCES OF AIR IN THE CONDENSER AND ITS EFFECTS

Following are the sources of air leakage into the condenser:

    (i)  The air leaks through the vents, joints from atmospheric relief valve and other accessories because inside condenser pressure is less than atmospheric pressure.

   (ii)  The feed water contains air which is liberated at the time of steam formation. Exhaust steam alongwith air enters the condenser at low pressure.

  (iii)  In case of jet condenser, air dissolved with cooling water is carried along with condensate into the boiler as feed. Since the condenser works at low pressure, the dissolved air is liberated within the condenser. The amount of air present in case of condensers is about 0.05% of the mass of steam.

The air present in the condenser, affects condenser performance badly because of the following reasons.

    (i)  It increases the condenser pressure (due to decrease in condenser vacuum)

   (ii)  It reduces the heat transfer considerably being a bad conductor of heat.

Therefore, the pressure of air in the cordenser is a very serious factor and should be reduced as far as possible. It is practically impossible to remove all air. So it is continuously removed by air pump which sucks it from condenser and let off in the atmosphere (in case of surface condenser).

## 13.11   CONDENSER VACUUM AND ITS MEASUREMENT

The pressure inside the condenser is below atmospheric pressure and is called the condenser vacuum. Vacuum is defined as atmospheric pressure minus absolute pressure and is usually expressed in terms of mm of mercury. It is usually measured by a Bourdon pressure gauge which is calibrated to read the pressure in mm of mercury below atmospheric pressure. If, it is desired to know the absolute pressure in the condenser, the barometric reading and gauge reading must be known. If, then barometric reading (atmospheric pressure) is 760 mm of Hg and gauge reads 630 mm of Hg, the absolute pressure in the condenser will be,

$$760 - 630 = 130 \text{ mm of Hg}$$

$$= \frac{130 \times 1.01325}{760} = 0.1733 \text{ bar}$$

The atmospheric pressure measured by barometer changes according to altitude of places from SML (sea mean level) and therefore vacuum is usually referred to a standard 760 mm Hg. Thus corrected vacuum in mm of Hg is

$$= 760 - \text{Absolute pressure in mm of Hg}$$

$$= 760 - (\text{Barometric height} - \text{vacuum}).$$

Refer to Fig. 13.12,

Let $H_b$ = Barometric pressure in mm of Hg measured by barometer.
   $H_v$ = Vacuum pressure in the condenser in mm of Hg measured by vacumm gauge (Bourdon gauge).

$\therefore$ Absolute pressure in the condenser

$$H = H_b - H_v$$

*Note*: Standard atmospheric pressure 760 mm of Hg = 1.01325 bar

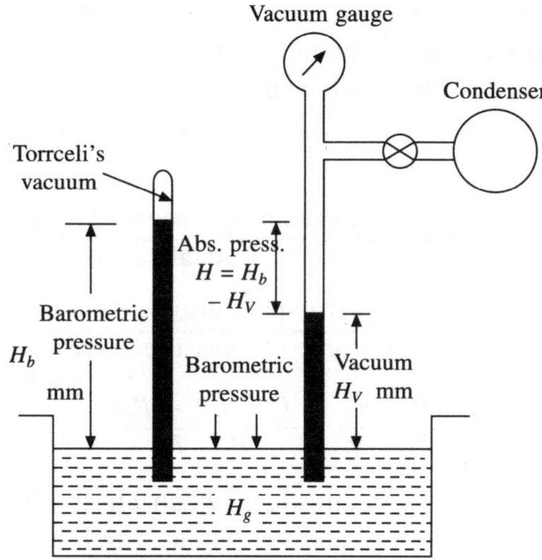


**Fig. 13.12 Vacuum measurement**

## 13.12 DALTON'S LAW OF PARTIAL PRESSURE

According to this law, the total pressure exerted by a non-reactive mixture of gases in a vessel is equal to the sum of partial pressures exerted by each constituent of the mixture if they would occupy the same volume of the vessel alone at the same temperature of mixture.

This is explained by taking the following example (see Fig. 13.13).

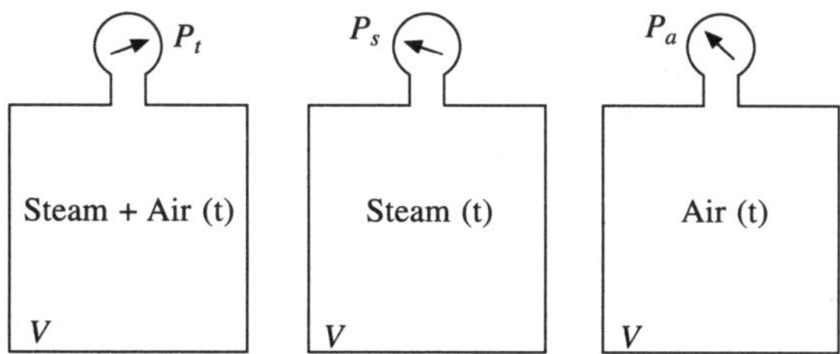


**Fig. 13.13 Dalton's law of partial pressure**

The total pressure in the condenser (vessel) is the sum of the partial pressures of steam and air. According to Dalton's law of partial pressure

$$P_t = P_s + P_a \qquad (13.1)$$

where   $P_t$ = Actual total pressure in the condenser
   $P_s$ = partial pressure of steam in the condenser
   $P_a$ = partial pressure of air in the condenser

## 13.13   VACUUM EFFICIENCY

The purpose of an air pump on condensers is to remove air from the condenser so that the total pressure in the corresponding to the condensate temperature.

The vacuum efficiency is a measure of the degree of perfection in achieving the aim of maintaining a desired vacuum in the condenser. The vacuum efficiency is defined as the ratio of actual vacuum to ideal vacuum in the condenser.

$$\text{Vacuum efficiency } \eta_{\text{vacuum}} = \frac{\text{Actual vacuum as recorded by vacuum gauge}}{\text{Ideal vacuum}}$$

or,

$$\eta_{\text{vacuum}} = \frac{\text{Barometric pressure} - \text{actual total pressure}}{\text{Barometric pressure} - \text{ideal pressure}}$$

$$= \frac{P_b - P_t}{P_b - P_s} = \frac{P_b - (P_s + P_a)}{P_b - P_s} \qquad (13.2)$$

Ideal vacuum means the vacuum due to steam alone when air is absent. Vacuum efficiency depends upon the effectiveness of air cooling and the rate at which it is removed by air pump.

## 13.14   CONDENSER EFFICIENCY

The purpose of an ideal condenser is to remove only the latent heat so that temperature of condensate equals the saturation temperature corresponding to condenser pressure. It means there should be no undercooling of condensate. Maximum temperature achieved by cooling water will be equal to condensate temperature. The condenser efficiency is then defined as the ratio of the actual rise in the temperature of cooling water to the maximum possible rise in the temperature of the cooling water.

Let   $t_s$ = Saturation temperature corresponding to condenser pressure, °C
   $t_o$ = Exit temperature of cooling water
   $t_i$ = Inlet temperature of cooling water

Then condenser efficiency,

$$\eta_{\text{cond.}} = \frac{\text{Temperature rise of cooling water}}{\text{Saturated temperature corresponding to condenser pressure} - \text{inlet temperature of cooling water}}$$

$$\text{or, } \eta_{\text{cond.}} = \frac{\text{Actual rise in temperature of cooling water}}{\text{Maximum possible rise in temperature of cooling water}} = \frac{t_o - t_i}{t_s - t_i} \qquad (13.3)$$

# 13.15 THERMODYNAMIC ANALYSIS OF CONDENSER

In a jet condenser, the resulting temperature of the condensate and cooling water is same after mixing, at the outlet. But in the surface condenser the temperature of condensate and the circulating water is different.

Let    $V$ = Volume of condenser shell

$t$ = Temperature of mixture (air + steam) in the condenser.

$P_t$ = Actual total pressure in the condenser.

$P_s$ = Partial pressure of steam in the condenser.

$P_a$ = Partial pressure of air in the condenser.

$m$ = Total quantity of mixture (air + steam) in the condenser shell.

$m_s$ = Mass of steam in condenser shell.

$m_a$ = Mass of air in condenser shell.

$m_w$ = Quantity of circulating water passing through the condenser shell.

$v_s$ = Specific volume of saturated water vapour at temperature $T$ and pressure $P_s$.

$v_a$ = Specific volume of air at temperature $T$ and pressure $P_a$.

$h$ = Specific enthalpy

$$\text{Total volume } V = m_s \cdot v_s = m_a \cdot v_a \tag{13.4}$$

$$\therefore \qquad \frac{m_a}{m_s} = \frac{v_s}{v_a}$$

The mass of air per m$^3$ of the condenser shell

$$= \frac{m_a}{V} = \frac{1}{v_a} \tag{13.5}$$

Similarly mass of water vapour per m$^3$ of condenser shell

$$= \frac{m_s}{V} = \frac{1}{v_s} \tag{13.6}$$

Total mass of mixutre in the condenser shell

$$m = m_a + m_s$$

$$= m_a \left( 1 + \frac{m_s}{m_a} \right) = m_a \left( 1 + \frac{v_a}{v_s} \right) \tag{13.7}$$

or, $$m = m_s \left( 1 + \frac{m_a}{m_s} \right) = m_s \left( 1 + \frac{v_s}{v_a} \right) \tag{13.8}$$

Fig. 13.14 shows the energy flow diagram in a surface diagram. Let subscripts $s$, $a$ and $c$ denote steam, air and condensate respectively. Applying steady flow energy equation (SFEE) and neglecting changes in kinetic and potential energy, heat transferred to cooling water (circulating water) is given by:

$$Q_w = (m_{s_1} \cdot h_{s_1} + m_{a_1} \cdot h_{a_1}) - (m_{s_2} \cdot h_{s_2} + m_{a_2} \cdot h_{a_2}) - m_c h_c$$

Since
$$m_{a_1} = m_{a_2} \quad \text{and} \quad m_{s_1} = m_{s_2} + m_c$$

$$\therefore \quad Q_w = (m_{s_1} h_{s_1} - m_{s_2} h_{s_2} - m_c h_c) + m_{a_1} (h_{a_1} - h_{a_2})$$

$$= (m_{s_1} h_{s_1} - m_{s_2} h_{s_2} - m_c h_c) + m_{a_1} \cdot c_{p_a} (t_{a_1} - t_{a_2})$$

First term bracket of the above equation represents the net heat lost by the incoming steam and second term bracket represents the heat lost by the incoming air to the cooling water.

Assuming no vapour escapes with air and that the heat lost by air is negligible compared with heat lost by steam as $m_{s_2} \ll m_{s_1}$ and $m_{s_2} \approx m_c$.

$$\therefore \quad Q_w = m_c (h_{s_1} - h_c)$$

Now, $Q_w$ can be written as

$$Q_w = m_w c_{p_w} (t_{w_o} - t_{w_i}) = m_c (h_{s_1} - h_c)$$

where $c_{p_w}$ = specific heat of water (4.187 kJ/ kg k) and $t_{w_o}$ and $t_{w_i}$ = outlet and inlet temperature of cooling water at the condenser respectively.

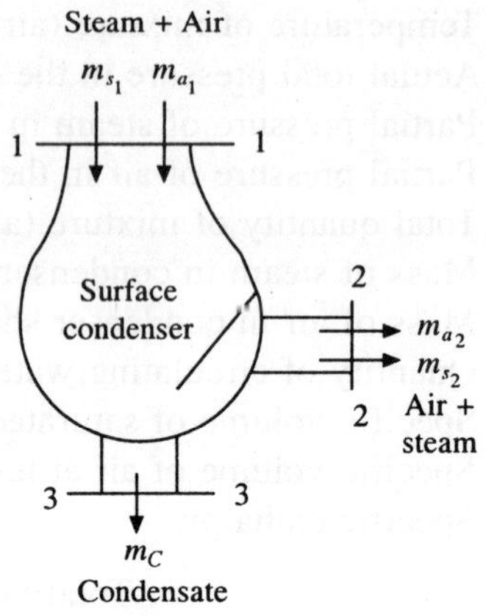

**Fig. 13.14**

$$\therefore \quad \frac{m_w}{m_c} = \frac{(h_{s_1} - h_c)}{c_{p_w} (t_{w_o} - t_{w_i})} \tag{13.9}$$

The steam entering the condenser is wet, therefore

$$\frac{m_w}{m_c} = \frac{(h_{w_1} + x_1 h_{fg_1} - h_c)}{c_{p_w} (t_{w_o} - t_{w_i})} \tag{13.10}$$

$$[h_{w_1} = h_{f_1} \text{ in steam table}]$$

The Eq. 13.10 gives the ratio of quantity of cooling water required per kg of steam condensed, and known as cooling ratio. The cooling ratio for surface condenser is 40 – 80 kg/kg of steam.

In a jet condenser, the final temperature of the condensed steam equals the temperaure of water at outlet

i.e.
$$h_c = c_{p_w} (t_{w_o} - 0.0)$$

$$\therefore \quad \frac{m_w}{m_c} = \frac{(h_{w_1} + x_1 h_{fg_1} - c_{p_w} t_{w_o})}{c_{p_w} (t_{w_o} - t_{w_i})} \tag{13.11}$$

## 13.16  AIR PUMPS

In article 13.10, we have seen the sources of air and the adverse effect in the condenser shell. Hence to remove air from shell air pumps are required. They are provided to maintain a desired vacuum in the condenser by extracting the air and other noncondensable gases. The common types of pumps used for this purpose are classified into:

(a)  Wet air pumps which remove a mixture of condensate and non-condensable gases.

(b)  Dry air pumps which remove the air only. These pumps may be of reciprocating type or rotary type. Here we shall discuss only the Edward's air pump which is reciprocating type and the steam jet air ejector.

### (i) Edward's Air Pump

This is a wet air pump of reciprocating type extracting both condensate and non-condensable gases. It is generally used in marine and land condensing plants. The special features of Edward's air pump is the absence of suction valve and bucket valve, which are necessary in the ordinary reciprocating type air pumps. The schematic diagram of Edward's air pump is given in Fig. 13.15. It consists of delivery or head valves.

These valves are placed in the cover which is on the top of the pump barrel lever. The reciprocating piston of the pump is flat on its upper surface and conical at the bottom as shown. The pump lever has a ring of ports around its lower end for the whole circumference. This communicates with the condenser.

On the downward stroke of the piston, a partial vacuum is produced above it which closes the delivery valve. Immediately the piston uncovers the ports, the rushing air and water vapour are entrapped in the conical end and into the barrel space above the piston. Further, motion of the piston in the downward direction causes its conical end to displace the condensate rapidly through the ports above the piston. Thus, a mixture of condensate, water vapour and air fills the space

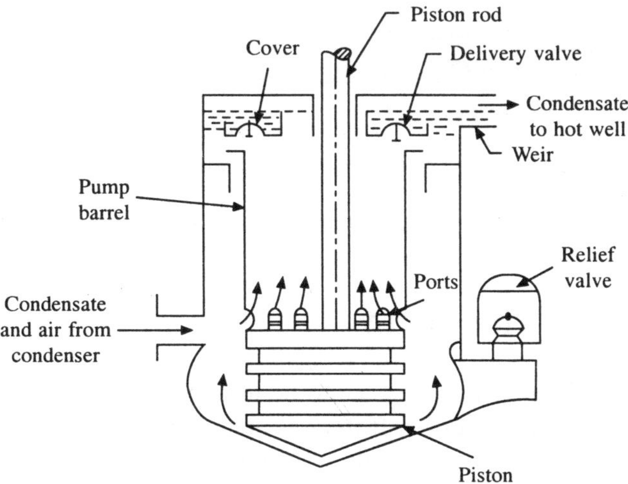


**Fig. 13.15  Edward's air pump**

above the piston during the downward movement of the piston passing through the ports. At this position delivery valve is in closed position.

On the upward movement of the piston, the condensate, air and water vapour above the piston are compressed to a pressure above the atmospheric pressure which opens the discharge (delivery) valve through which the charge flows over the weir to the hot well. The weir maintains a constant head of water above the cover plate and seals the delivery valve against air leakage. If the pressure in the bottom end exceeds the atmospheric value, the relief valve gets lifted and allows the condensate and air to escape through it.

This reciprocating type pump is mostly suited for the requirement of low vacuum. That is why this has been used extensively with land and marine condensing steam engine. However, this pump is not compitable in steam turbine plant operating under high vacuum and large quantities of steam due to its (turbine) size and speed etc. Further, in central power stations, the practice of removing both water and gages together is considered to be objectionable. If there is pressure of oxygen in feed water to be supplied, it may demage the high pressure boiler. Therefore, in central power stations, steam ejector or rotary pumps are used both belonging to the family of air pumps.

### (ii) Steam Jet Air Pump (Air Ejector)

This type of air ejector is universally used for removing air from condenser since it has no moving parts and less occupancy, higher efficiency (about 95%), simple construction, low mantenance cost and quick starting. Its schematic diagram is shown in Fig. 13.16. It consists of convergent-divergent nozzle and a diffuser. Steam from boiler enters at '$A$' in the suitably designed nozzle where its kinetic energy increases and the pressure reduces. Pipe '$C$' is connected to the condenser from where the air along with small amount of vapour forces up to mix with the low pressure steam at '$B$'. The momentum of the steam jet carries the mixture of steam and air into the diffuser where its velocity decreases and the pressure increases by the time it leaves the diffuser. The system shows only a ejector, if more ejectors are introduced, a very low pressure can be obtained in the condenser. Usually upto four numbers of ejectors are used which can reduce the pressure in the condenser up to a pressure of 0.03 bar.

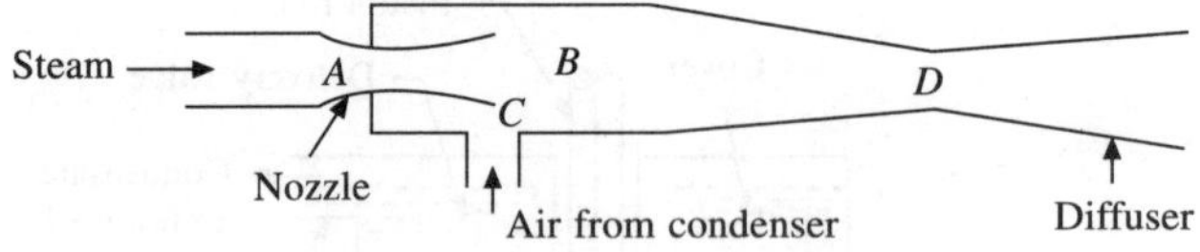


**Fig. 13.16   Steam jet air ejector**

The advantage of steam jet air ejectors is that these are simple in construction, cheap, highly efficient and it is without any moving parts.

## 13.17   COOLING WATER SUPPLY

For the condensation of steam into condensate to be used as a feed water to the boiler, cooling water has to be supplied to the condenser. If there is abundant source of cooling water like river or pond, there is no problem. But if the supply of water is limited with high cost, it is necessary to use cooling towers for water cooled condensers.

The cooling water requirement in an open system (river water system) is about 50 times the flow of steam to the condenser. Even with closed cooling system using cooling towers, the requirement for cooling water is also considerably large as 5 to 8 kg/kw-hr. The method commonly used to overcome a condition of limited cooling water supply is to repeatedly cool and circulate it through the following means.

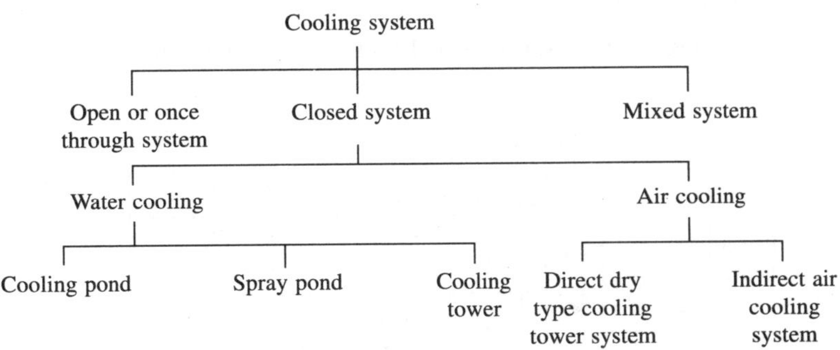


1. *Open system*—Large power stations require enough quantity of cooling water per hour. Such plants are usually located near a river or sea. The water is constantly drawn from the river by the pump, filtered and circulated through the condenser. Hot coolant is discharged back into the river (Fig. 13.17).

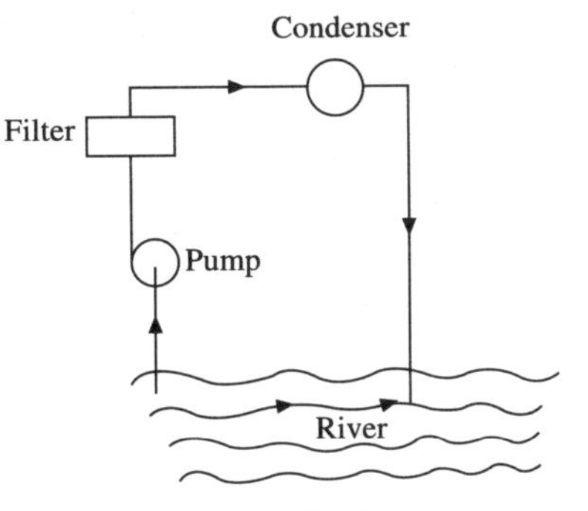


**Fig. 13.17**

2. *Cooling Pond*—The simplest system of removing heat from the cooling water consists of cooling it in an open pond. The effectiveness of this method depends upon a very large surface area of the pond and hence it is used mostly for small condensers only. In this system sufficient amount of water is lost by evaporation and windage. The factors which affect the rate of heat dissipation from a cooling pond are, area and depth of pond, temperature of water entering the pond, wind velocity, atmospheric temperature, shape and size of water spray nozzles and relative humidity. Cooling ponds are of two types namely

(i)  Non-directed flow type and
(ii)  Directed flow type

Figs. 13.18 and 13.19 show the schematic diagram of above systems.

3. *Spray Pond*—For a given cooling capacity, the size of pond in this case is much less than that of open pond. Schematic diagram of spray pond is shown in Fig. 13.20. Hot water coming out from power house is sprayed into the atmosphere through so many nozzles. Tiny particles of sprayed water lose the sensible heat to air and get cooled and are finally collected into a reservoir from which cold water is supplied to the power house for reuse. In this case cooling achieved is more effective.

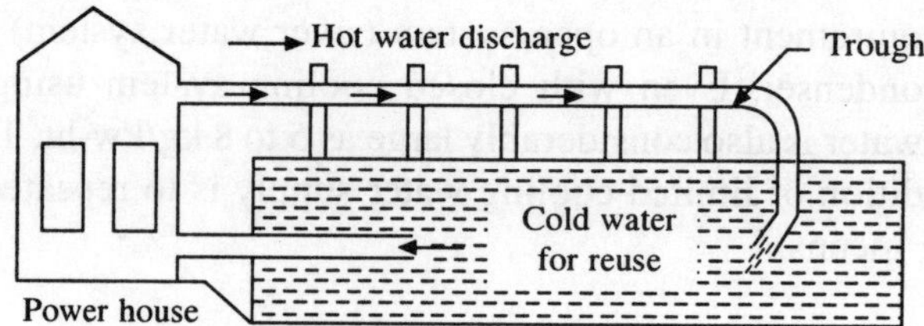


**Fig. 13.18   Non-Directed flow type cooling pond**

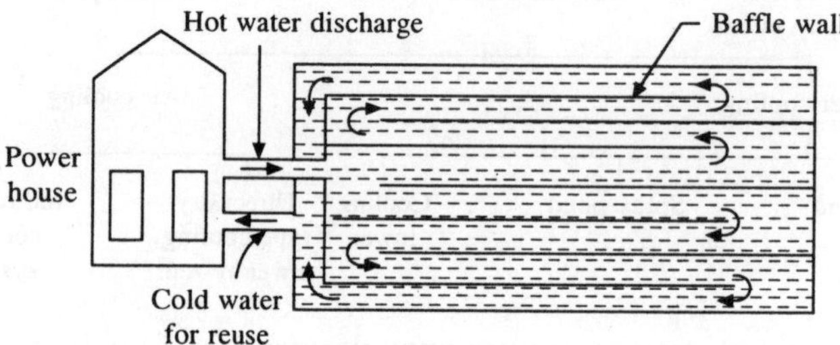


**Fig. 13.19   Directed flow type cooling pond**

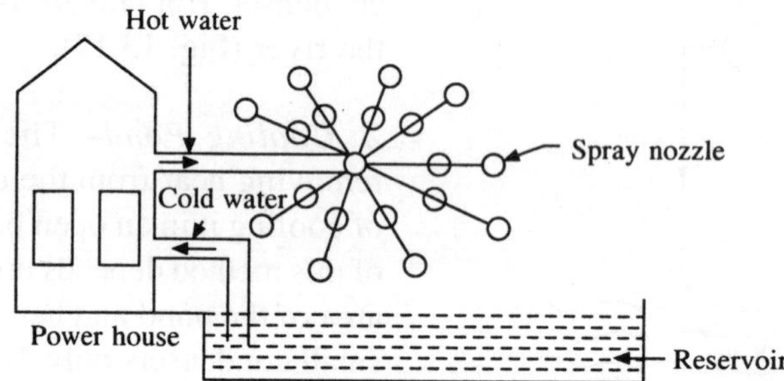


**Fig. 13.20   Spray pond**

4. *Cooling Towers*—The place where acquisition of land is very expensive, we may use cooling towers for cooling purposes. A cooling tower requires smaller area than a spray pond. It is an artificial device used to cool the hot cooling water coming out of condenser effectively. The principle of cooling the water in cooling tower is similar to that of the evaporative condenser. A cooling towers is placed at a certain height (at about 10 metres from the ground level). The cooling tower is a semi-enclosed device made of wooden, steel or concrete structure and corrugated surfaces or troughs or baffles or perforated trays are provided inside the tower for uniform distribution and better automization of water in the tower. The hot water coming out from the condenser falls down in radial sprays from a height and the atmospheric air enters from the base of the tower. The partial evaporation of water takes place which reduces the temperature of circulating water. This cooled water is collected in the pond at the base of the tower and pumped into the condenser. Draft eliminaters are provided at the top of the tower to prevent the escaping of water particles with air.

According to the method of air circulation, cooling towers are classified as

(i) Natural draught type cooling tower
(ii) Mechanical draught type cooling tower
   (a) Forced draught type
   (b) Induced draught type
(iii) Hyperbolic cooling tower

**Natural draught cooling tower**

The schematic diagram of this type is shown in Fig. 13.21. In this case, hot water from condenser is pumped at the top where water sprays through a series of spray nozzles. Then water falls over decks (louvers). The decks also increase the amount of wetted surface in the tower and breaks up the water into droplets. The air flowing across in transverse direction cools the falling water. These towers are used for small capacity power plants such as diesel power plants.

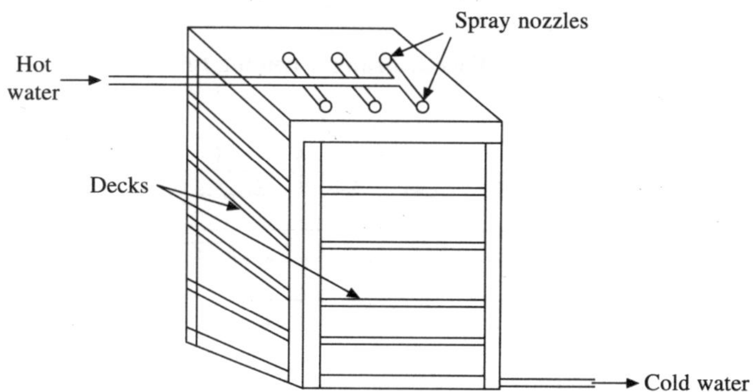


**Fig. 13.21    Natural draught cooling tower**

**Forced Draught Cooling Tower**

In this tower draught air fan in installed at the bottom of tower. The hot water from the condenser enters the nozzles (Fig. 13.22). The water is sprayed over the tower filling slats and the rising air cools the waters. The entrained water is removed by eliminators located at the top.

**Induced draught cooling tower**

The schematic arrangement of induced draught type is shown in Fig. 13.23. The difference lies only in supply of air. The draught fans installed at the top of tower draw air through the tower. The hot water is allowed to pass through the tower below the eliminators. The air moving in the upward direction cools the down coming hot water particles issued from spray nozzles. Some percentage (1%) of total water goes into air in the form of water vapour. Induced draught towers produce less noise. Make up water should be continuously added to the tower collecting basin to replace the water lost by evaporation.

**Hyperbolic Cooling Tower**

It is usually made of steel reinforced cement concrete to withstand high wind pressure. First type of this type was installed in US at Big-sandy station of kenucky power Co. It is capable of

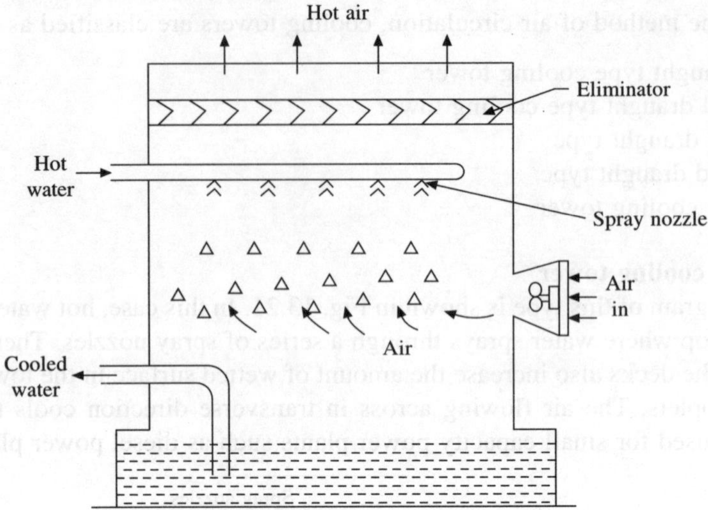



**Fig. 13.22   Forced draught cooling tower**

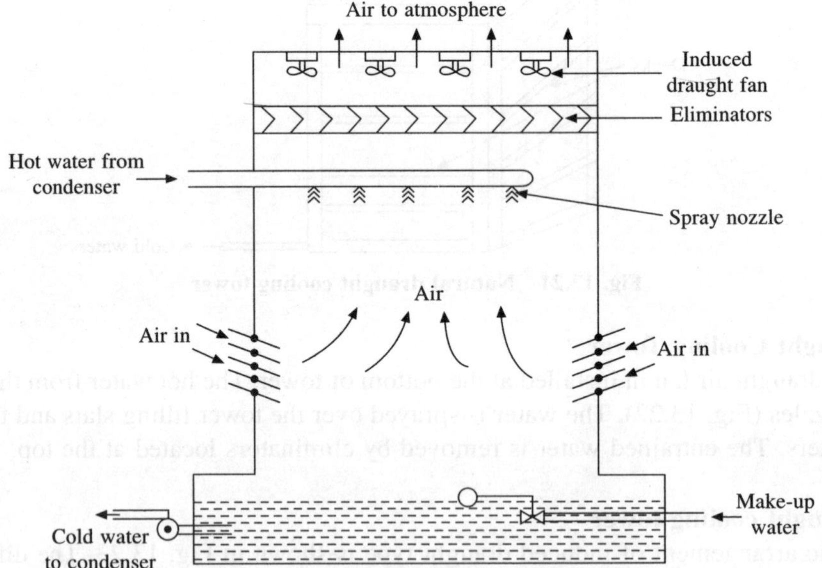



**Fig. 13.23   Induced draught cooling tower**

handling $120 \times 10^3$ gpm and cools the water from 43°C to 30°C. It is has minimum diameter of 39.5 m and maximum diameter of 74.5 m and is 400 metres high. It serves to the station of 265 MW capacity.

The arrangement of hyperbolic cooling tower is shown in Fig. 13.24. The hot water from the condenser is supplied to the ring troughs which are placed at 8–10 m above the ground level. The nozzle are provided on the bottom side of troughs to break up water into sprays. The air enters the cooling tower just above the pond located at the bottom, from the air openings provided, rises

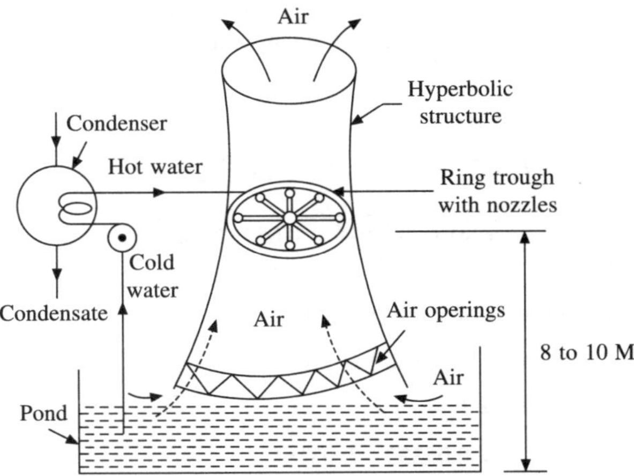


**Fig. 13.24  Hyperbolic cooling tower**

upward and absorbs heat from the falling water spray. The cooled water is collected into the pond. Some make-up water is supplied to overcome the losses into the atmosphere along with air due to evaporation. It needs about 3-5% of make-up water for compensating. This type of cooling tower is generally used since it is very efficient.

Advantages of this tower over mechanical towers are given as under:

1. No any air fan is required. So power cost and auxiliary equipments are totally eliminated.
2. Operating and maintenance costs are reduced.
3. Hyperbolic structure creates its own drought assuring efficient operation even when there in no wind.
4. Ground fogging is avoided in hyperbolic towers.

Major drawbacks are as under:

1. Its initial cost is considerably very high.
2. Its performance varies with the seasonal changes in DBT and RH of air.

## Solved Problems

**13.1**  A vacuum gauge reading was obtained as 720 mm of Hg with the barometer reading of 750 mm of Hg. Determine the corrected vacuum in terms of mm of Hg and bar.

*Soln.*  corrected vacuum = 760 – Absolute pressure in mm of Hg

$$= 760 - (\text{Barometric height} - \text{Vacuum gauge reading})$$

$$= 760 - (750 - 720)$$

$$= 730 \text{ mm of Hg }  Ans$$

$$= \frac{730}{760} \times 1.01325 = 0.9732 \text{ bar }  Ans.$$

∵    standard atmospheric pressure

$$760 \text{ mm of Hg} = 1.01325 \text{ bar}$$

**13.2** In a condenser, vacuum gauge reads 715 mm of Hg while barometer reads 755 mm. The temperature of condensate be 25°C. Determine the pressure of steam and air and mass of air per kg of steam. Also, determine the vacuum efficiency.

*Soln.* Total absolute pressure in the condenser

$$P_t = 755 - 715 = 40 \text{ mm of Hg}$$

$$= 40 \times \frac{1.01325}{760} = 0.05332 \text{ bar}$$

From steam table,
Partial pressure of steam at condensate temperature of

$$25°C, \ P_s = 0.03166 \text{ bar } \ Ans.$$

$\therefore$    Partial pressure of air    $P_a = P_t - P_s$

$$= 0.05332 - 0.03166$$

$$= 0.02166 \text{ bar } \ Ans.$$

At 25°C, steam has specific volume = 43.402 m$^3$/kg of steam. Same volume will be occupied by air. Hence using ideal gas euation

$$P_a v_a = m_a R_a T_a$$

or,          $(0.02166 \times 10^5) \times 43.02 = m_a \times 287 \times (273 + 25)$

$\therefore$    $m_a = 1.094$ kg/kg of steam

$$\text{Vacuum efficiency} = \frac{\text{Actual vacuum}}{\text{Ideal vacuum}} = \frac{P_b - (P_a + P_s)}{P_b - P_s}$$

$$P_t = (P_a + P_s), \ P_s = 0.03166 \text{ bar} = 23.74 \text{ mm of Hg}$$

$\therefore$

$$\eta_{\text{vacuum}} = \frac{715}{755 - 23.74} \times 100$$

$$= 97.77\% \ \ Ans.$$

*Note*:   Actual vacuum already given in the problem.

**13.3** During the test on a surface condenser following observations were recorded:
Condenser vacuum = 70 cm of Hg, Barometer reading = 76.5 cm of Hg, outlet temperature of cooling water = 30°C, Inlet temperature of cooling water = 15°C, Mean temperature of condenser = 42°C, Hot well temperature = 33°C, condensate collected = 2100 kg/hr, Quantity of cooling water circulated = 70,000 kg/hr. Calculate (a) vacuum corrected to a standard of 76 cm of Hg (b) vacuum efficiency (c) undercooling of the condensate (d) condenser efficiency (e) quality of steam entering the condenser (f) mass of air present per m$^3$ of condenser volume, and (g) mass of air present per kg of uncondensed steam. Take $R = 0.287$ kJ/kg k, Specific heat of water = 4.187 kJ/kg k.

*Soln.*   (a)   Corrected vacuum = 76 – (Barometric height – Condenser vacuum)

$$= 76 - (76.5 - 70) = 69.5 \text{ cm of Hg}$$

     (b)   From steam table, partial pressure of steam corresponding to 42°C, $P_s = 0.08199$ bar = 6.149 cm of Hg

$$\text{vacuum efficiency} = \frac{P_b - (P_a + P_s)}{P_b - P_s} = \frac{P_b - P_t}{P_b - P_s}$$

here $P_b = 76.5$ cm of Hg

$P_b - P_t$ = Actual vacuum in the condenser = 70 cm of Hg

$$\therefore \quad \eta_{\text{vacuum}} = \frac{70}{76.5 - 6.149} \times 100 = 99.5\% \quad Ans.$$

(c)  Undercooling of condensate = condenser temp. – hot well temp.

$$= 42 - 33 = 9°C \quad Ans$$

(d)  Absolute total pressure in the condenser

$$P_t = P_s + P_a$$

= Barometer reading – Vacuum gauge reading

= 76.5 – 70 = 6.5 cm of Hg

= 0.0866 bar

Saturation temperature corresponding to 0.0866 bar from steam table = 42.7°C. It means the cooling water can achieve maximum temperature up to 42.7°C for condensation of steam inside the condenser.

$$\therefore \quad \text{condenser efficiency} = \frac{\text{Actual rise in temperature of cooling water}}{\text{Maximum possible rise in temperature of cooling water}}$$

$$= \frac{30 - 15}{42.7 - 15} \times 100 = 54.15\% \quad Ans.$$

(e)  From steam table at 0.0866 bar
$$h_f = 178.9 \text{ kJ/kg}, \ h_{fg} = 2400 \text{ kJ/kg}$$
At hot well temperature of 33°C, $h_f$ = 138.2 kJ/kg
Neglecting the heat losses and heat lost by air. Heat lost by steam = Heat gained by water
2100 (178.9 + $x$ × 2400 – 138.2) = 70000 × 4.187 (30–15)

$$x = 0.85$$

$\therefore$  quality of steam = 0.85 dry  *Ans.*

(f)  Partial pressure of air $P_a = P_t - P_s$

$$P_a = 0.0866 - 0.08199 = 0.00461 \text{ bar}$$

Mass of air present per m³ of condenser volume

$$m_a = \frac{P_a V_a}{R_a T_a} = \frac{0.00461 \times 10^5 \times 1}{287 \ (42 + 273)} = 0.0051 \text{ kg}$$

$$= 5.1 \text{ gram} \quad Ans.$$

(g)  Specific volume of steam at 42°C = 17.692 m³/kg
Mass of air present per kg of uncondensed steam

$$m_a = \frac{P_a V}{R_a T_a} = \frac{0.00461 \times 10^5 \times 17.692}{287 \ (42 + 273)}$$

$$= 0.0902 \text{ kg} \quad Ans.$$

$$= 90.21 \text{ gram}$$

**13.4**  Following readings were noted during a trial on surface condenser:
Mean temperature of condenser = 30°C, condensate temperature = 23°C, Barometer reading = 72 cm of Hg, condenser vacuum = 66.375 cm of Hg, condensate collected = 500 kg/hr, cooling water circulated = 18000 kg/hr, Rise in temperature of cooling water = 14°C. Calculate

   (i) Dryness fraction of steam entering the condenser

  (ii) Capacity of air pump it air leaks into condenser = 75 kg/hr

 (iii) Find the dimensions of the reciprocating air pump to remove the air if it runs at 800 r.p.m. Take $L:D = 3:2$ (stroke: Bore) and volumetric efficiency = 90%.

*Soln.* (i)     Partial pressure of steam entering the condenser at 30°C, $P_s$ = 0.04242 bar (from steam table).

          Absolute total pressure in the condenser

$$P_t = P_b - P_{vacuum}$$

$$= (72 - 66.375) = 5.625 \text{ cm of Hg} = 0.0749 \text{ bar}$$

$$= P_a + P_s$$

Based on heat lost by steam equals to heat gain by cooling water, from thermodynamic relation

$$\frac{m_w}{m_c} = \frac{\overbrace{(h_f + x h_{fg}}^{\text{at } P_t} - h_c}{c_{p_w}(t_{w_o} - t_{w_i})} \quad \leftarrow h_f \text{ at condensate temp. } 23°C$$

or, $$\frac{18\,000}{500} = \frac{(168.8 + x \times 2406.2 - 96.4)}{4.187\,(14)}$$

$$\therefore \qquad\qquad x = 0.84 \quad Ans.$$

(ii)     Partial pressure of air

$$P_a = P_t - P_s$$

$$= 0.0749 - 0.04242 = 0.03248 \text{ bar}$$

Volume of air being handled by air pump

$$V_a = \frac{m_a R_a T_a}{P_a} = \frac{75 \times 287\,(30 + 273)}{0.03248 \times 10^5}$$

$$= 2002.24 \text{ m}^3/\text{hr}$$

$$= 33.37 \text{ m}^3/\text{min (capacity of air pump)} \quad Ans.$$

(iii) $$V_a = \frac{\pi}{4} D^2 \times L \times N \times \eta_{\text{volumetric}}$$

$$33.37 = \frac{\pi}{4} D^2 \times 1.5\,D \times 800 \times 0.90$$

$$\therefore \qquad\qquad D = 0.34 \text{ m} = 34 \text{ cm}$$

and $$L = 1.5 \times 34 = 51 \text{ cm} \quad Ans.$$

**13.5** A steam turbine discharges 5500 kg of steam per hour at 35°C temperature and 0.90 dryness. The air leakage is estimated to be 18 kg/hr. At the suction of air pump, temperature is 32°C and temperature of condensate is 30°C. Find

  (a) the vacuum gauge reading

  (b) capacity of air pump

  (c) loss of condensate in kg per hour, and

  (d) the quantity of cooling water required if the rise in temperature of cooling water is limited to 10°C.

*Soln.* (a)     From steam table, partial pressure of steam at 35°C, $P_s$ = 0.05622 bar specific volume of steam at 35°C

$$v_s = 25.245 \text{ m}^3/\text{kg}$$

Volume of 5500 kg of steam is given by

$$V = 5500 \times x \, v_s$$

$$= 5500 \times 0.90 \times 25.245 = 124962.75 \text{ m}^3.$$

The same volume is also occupied by 18 kg of air. Therefore partial pressure of air in the condenser is given by

$$P_a = \frac{m_a R_a T_a}{V} = \frac{18 \times 287 \times (35 + 273)}{124962.75 \times 10^5} = 0.000127 \text{ bar}$$

$\therefore$   Total pressure in the condenser

$$P_t = P_a + P_s$$

$$= 0.000127 + 0.05622 = 0.056347 \text{ bar}$$

$\therefore$   Vacuum in condenser in cm of Hg $= P_b - P_t$

$$= (1.01325 - 0.056347) \times \frac{76}{1.01325}$$

$$= 71.77 \text{ cm of Hg}$$

(b)   Partial pressure of steam at air pump suction is the saturation pressure corresponding to 32°C = 0.04753 bar (from steam table)

$\therefore$

$$P_a = P_t - P_s$$

$$= 0.056347 - 0.04753 = 0.0088 \text{ bar}$$

Volume of air at 32°C and 0.0088 bar is given by

$$V = \frac{m_a R_a T_a}{P_a} = \frac{18 \times 287 \times (32 + 273)}{0.0088 \times 10^5} = 1790.48 \text{ m}^3$$

$\therefore$   Air pump capacity $= 1790.48 \text{ m}^3$

(c)   Specific volume of steam at 32°C saturation temp.

$$V_s = 29.572 \text{ m}^3/\text{kg}$$

$\therefore$   Losses in condensate (i.e. mass of steam going with air)

$$= \frac{1790.48}{29.572} = 60.54 \text{ kg/hr}$$

(d)   from thermodynamic relation

$$\frac{m_w}{m_c} = \frac{\overbrace{(h_f + x \, h_{fg})}^{\text{at } P_t} - h_c}{c_{p_w}(t_o - t_i)}$$

where $h_c$ is $h_f$ at condensate temp. 30°C

or,

$$\frac{m_w}{5500} = \frac{(146.6 + 0.9 \times 2418.8 - 125.7)}{4.187 \times 10} = 288703.37 \text{ kg/hr}$$

$\therefore$   $m_w = 288.70 \text{ tons/hr}$   *Ans.*

**13.6** The absolute pressure in the condenser is 11.56 kPa while the barometer reads 1 bar. The condenser temperature is 40°C. Determine the partial pressure of air, vacuum efficiency and the mass of air present in the condenser per kg of steam.

*Soln.* Absolute pressure in the condenser

$$P_t = 11.56 \text{ kPa} = 11.56 \times 10^3/10^5 = 0.1156 \text{ bar.}$$

From steam table at 40°C, saturation partial pressure of steam $P_s = 0.07375$ bar

$$P_t = P_a + P_s$$

$$\therefore \qquad P_a = P_t - P_s$$

$$= 0.1156 - 0.07375 = 0.04185 \text{ bar} \quad Ans.$$

$$\eta_{\text{vacuum}} = \frac{P_b - P_t}{P_b - P_s} = \frac{1 - 0.1156}{1 - 0.07375} \times 100$$

$$= 95.48\% \quad Ans.$$

Specific volume at condenser temperature 40°C = 19.546 m³/kg (from steam table). Same volume will be occupied by air.

$$\therefore \qquad m_a = \frac{P_a V_a}{R_a T_a} = \frac{0.04185 \times 10^5 \times 19.546}{287 (40 + 273)}$$

$$= 0.9105 \text{ kg/kg of steam} \quad Ans.$$

**13.7** During the observation on a surface condenser, following readings were taken:
Vacuum gauge reading = 70 cm of Hg
Mean temperature of condenser = 38°C
The air leakage into the condenser = 1 kg per 2200 kg of steam.
Determine the volume of air to be handled by the dry air pump per kg of steam and the mass of vapour associated with air. Barometer reading = 75 cm of Hg.
Solution
Absolute total pressure in the condenser

$$P_t = 75 - 70 = 5 \text{ cm of Hg} = 0.06670 \text{ bar}$$

From steam tables at 38°C, $P_s = 0.06624$

$$\therefore \qquad P_a = P_t - P_s$$

$$= 0.06670 - 0.06624 = 0.00046 \text{ bar}$$

Amount of air entering/kg of steam $= \dfrac{1}{2200}$ kg

using ideal gas equation

$$P_a V_a = m_a R_a T_a$$

$$\therefore \qquad V_a = \frac{m_a R_a T_a}{P_a} = \frac{1}{2200} \times \frac{287 \times (38 + 273)}{0.00046 \times 10^5}$$

$$= 0.8819 \text{ m}^3/\text{kg of steam}$$

specific volume of steam at 38°C

$$= 21.627 \text{ m}^3/\text{kg}$$

$\therefore$  Amount of water vapour associated with 0.8819 m$^3$ will be

$$= \frac{0.8819}{21.627} = 0.0407 \text{ kg/kg of air } \textit{Ans}.$$

**13.8** For checking the air leakage into the condenser, the steam plant is run until conditions are steady and immediately steam supply to the engine is shut off and condensate extraction pumps are closed down. Now the condenser is isolated. At the time of shut down the temperature and vacuum in the condenser are observed to be 42°C and 680 mm of Hg. After 10 minutes the values are 30°C and 510 m of Hg. The barometer reads 756 mm of Hg. The effective volume of the condenser is 0.4 m$^3$. Determine (a) the amount of air leakage into the condenser during the observed period and (b) the mass of water vapour condensed in the same period.

*Soln.*  At the time of shut down, absolute total pressure

$$P_t = 756 - 680 = 76 \text{ mm of Hg} = 76 \times \frac{1.01325}{760} = 0.1013 \text{ bar}.$$

At 42°C, partial pressure of steam

$$P_s = 0.08199 \text{ bar (from steam table)}$$

$\therefore$  Partial pressure of air $P_a = p_t - P_s$

$$= 0.1013 - 0.08199 = 0.01931 \text{ bar}$$

The effective volume of condenser = 0.4 m$^3$

$\therefore$  Mass of air present in the condenser, $m_a = \dfrac{P_a V_a}{R_a T_a}$

$$= \frac{0.01931 \times 10^5 \times 0.4}{287 \times (42 + 273)}$$

$$= 0.00\,854 \text{ kg}$$

Specific volume of steam at 42°C = 17.692 m$^3$/kg

$\therefore$  $\qquad\qquad$ Mass of vapour present $= \dfrac{0.4}{17.692} = 0.0226 \text{ kg}$

Ten minutes after shut down,

$$P_t = 756 - 510 = 246 \text{ mm of Hg} = 246 \times \frac{1.01325}{760}$$

$$= 0.3279 \text{ bar}$$

$P_s$ at 30°C = 0.04242 bar (from steam table)

$\therefore$  $\qquad\qquad\qquad$ $P_a = P_t - P_s$

$$= 0.3279 - 0.04242 = 0.28548 \text{ bar}$$

$\therefore$  $\qquad\qquad$ $m_a = \dfrac{0.28548 \times 10^5 \times 0.4}{287 \times (30 + 273)} = 0.13131 \text{ kg}$

Specific volume of steam at 30°C = 32.929 m$^3$/kg

$\therefore$  $\qquad\qquad$ Mass of vapour present $= \dfrac{0.4}{32.929} = 0.01214 \text{ kg}$

(a)  $\therefore$  Mass of air leaked in 10 minutes

$$= 0.1313 - 0.00854 = 0.12277 \text{ kg}$$

(b) $\therefore$   Mass of steam condensed in 10 minutes

$$= 0.0226 - 0.01214 = 0.01046 \text{ kg.}   \textit{Ans.}$$

**13.9** In a surface condenser air enters the condenser alongwith steam at a rate of 10 kg per hour. The temperature of air cooler inlet section is 32°C and at the outlet 29°C. The vacuum in the condenser shell is kept essentially constant throughout and is 715 mm of Hg, while the barometer reads 755 mm of Hg. Calculate:
   (i) The volume of air entering the cooling section per hour,
   (ii) The mass of moisture contained in the air, and
   (iii) The mass of steam condensed per hour in the cooling section

*Soln.*   Absolute total pressue in the condenser

$$P_t = 755 - 715 = 40 \text{ mm of Hg} = 0.05332 \text{ bar}$$

At the inlet to the cooling section.
Saturation partial pressure of steam at 32°C, from steam table, $P_s = 0.04753$ bar

Partial pressure of air,   $P_a = P_t - P_s$

$$= 0.05332 - 0.04753$$

$$= 0.00579 \text{ bar}$$

(i)   Volume of air entering the cooling section per hour

$$V_a = \frac{m_a R_a T_a}{P_a} = \frac{10 \times 287 \times (32 + 273)}{0.00579 \times 10^5} = 1511.83 \text{ m}^3/\text{hr}   \textit{Ans.}$$

(ii)   Mass of moisture in the air specific volume of steam at 32°C,

$$v_s = 29.572 \text{ m}^3/\text{kg}$$

$\therefore$   Mass of moisture associated with air

$$m_s = \frac{V_a}{v_s} = \frac{1511.83}{29.572} = 51.12 \text{ kg/hr}   \textit{Ans.}$$

(iii)   Mass of steam condensed per hour in the cooling section. At the outlet to the cooling section.

$$\text{At } 29°C, P_s = 0.04004 \text{ bar (from steam table)}$$

$\therefore$   $P_a = P_t - P_s = 0.05332 - 0.04004 = 0.01328$ bar
$\therefore$   Volume of air at the outlet to the cooling section

$$V_a = \frac{m_a R_a T_a}{P_a} = \frac{10 \times 287 \, (29 + 273)}{0.01328 \times 105} = 652.66 \text{ m}^3/\text{hr}$$

from steam table, specific volume at 29°C

$$v_s = 34.769 \text{ m}^3/\text{kg}$$

and the mass of steam associated with the air

$$= \frac{V_a}{v_s} = \frac{652.66}{34.769} = 18.77 \text{ kg/hr}$$

$\therefore$   Mass of steam condensed

$$= 51.12 - 18.77 = 32.35 \text{ kg/hr}   \textit{Ans.}$$

**13.10** In a steam power plant, a turbine develops 4 MW and consumes steam rate of 8 kg/kW-hr. The steam is supplied at 28 bar and 250°C. From turbine steam is exhausted to a condenser at 720 mm of Hg when barometer reads 765 mm of Hg. The temperature of condensate from the condenser is 30°C. Circulating water enters the condenser at 20°C and leaves at 28°C. Determine:
(a)  Quality of steam entering the condenser
(b)  Quantity of circulating cooling water
(c)  Ratio of cooling
Assume no air is present in the condenser

*Soln.*  (a)  Absolute total pressure in the condenser

$$P_t = 765 - 720 = 45 \text{ mm of Hg} = 0.0599 \text{ bar}$$

corresponding to 0.0599 bar, saturation temperature of steam at which steam is entering the condenser = 36°C (from steam table)
Enthalpy of steam at 28 bar and 250°C = 2864.9 kJ/kg (from superheated steam table).

At 36°C, $h_f = 150.7$ kJ/kg, $h_{fg} = 2416.4$ kJ/kg (from steam table)

Total power produced from steam turbine

$$4000 = m_s [h_s - (h_f + x\,h_{fg})]$$

or,
$$4000 = \frac{8 \times 4000}{60 \times 60} [2864.9 - (150.7 + x \times 2416.4)] \text{ kJ/s}$$

$$\therefore \qquad x = 0.93 \quad Ans.$$

(b)  From thermodynamic relation
$h_f$ at condensate temp. 30°C (= 30 × 4.187)

$$\frac{m_w}{m_c} = \frac{[(h_f + x\,h_{fg}) - h_c]}{c_{p_w}(t_{w_o} - t_{w_i})}$$

where $(h_f + x\,h_{fg})$ is at $P_t$

or,
$$\frac{m_w}{8 \times 4000} = \frac{[(150.7 + 0.93 \times 2416.4) - 125.7)}{4.187(28 - 20)}$$

$$\therefore \qquad m_w = 2170768.6 \text{ kg/hr}$$

(c)  $\therefore$  cooling ratio $\dfrac{m_w}{m_c} = \dfrac{2170768.6}{8 \times 4000} = 67.83$  *Ans.*

**13.11**  During trial period in a surface condenser, the vacuum gauge reading was 680 mm of Hg when barometer read 755 mm of Hg. The temperature in the condenser was found to be 43°C.
  Find the alternation in vacuum if the quantity of air entering the condenser is reduced by 100 kg/hr. If the mass of condensate is 1000 kg/hr, calculate quantities of air and vapour which the pump has to handle.

*Soln.*  Absolute total pressure in the condenser

$$P_t = 755 - 680 = 75 \text{ mm of Hg} = 0.099 \text{ bar}$$

From steam table, at 43°C,
Partial pressure if steam, $P_s = 0.08639$ bar and specific volume $v_s = 16.841$ m³/kg
$\therefore$  Partial pressure of air

$$P_a = P_t - P_s = 0.0999 - 0.08639 = 0.01351 \text{ bar}$$

Volume of steam per hour $= 1000 \times 16.841$

$$= 16841 \ m^3$$

Same volume will be occupied by air,

$\therefore$ Mass of air $m_a = \dfrac{P_a V_a}{R_a T_a} = \dfrac{0.01351 \times 10^5 \times 16841}{287(43 + 273)}$

$$= 250.87 \ kg/hr$$

Now, if the mass of air is reduced to 100 kg/hr, the partial pressure of steam will become

$$= \frac{100 \times 0.01351}{250.87} = 0.00538 \ bar$$

Absolute pressure of condenser will change to

$$= 0.08639 + 0.00538 = 0.09177 \ bar$$

$$= 68.83 \ mm \ of \ Hg$$

New vacuum reading $= 755 - 68.83 = 686.17$ mm of Hg

$\therefore$   Alternation in vacuum $= 686.17 - 680 = 6.17$ mm of Hg

Capacity of pump $= 1000 + 100 = 1100$ kg/hr

**13.12**  In a steam power plant, the quantity of steam used by a turbine is 4500 kg per hour and exhaust occurs into a condenser at a vaccum pressure of 65 cm of Hg, while the barometer reads 76 cm of Hg. The exhaust steam is dry saturated at the enterance to the condenser and the condensate is under–cooled by 7°C. If the inlet and outlet temperatures of the cooling water are 18°C and 30°C respectively, determine the mass of water flowing per hour. If this water is supplied in 200 pipes at a mean velocity of 2 m/s; determine the suitable pipe diameter.

*Soln.*  Absolute total pressure in the condenser

$$P_t = (76 - 65) = 11 \ cm \ of \ Hg = 0.1466 \ bar$$

Saturation temperature of steam at 0.1466 bar

$$= 53.51°C \ (from \ steam \ table)$$

Enthalpy of dry saturated steam $= h_g = h_f + x \, h_{fg}$

$$= 2598.35 \ kJ/kg$$

Since there is undercooling of 7°C, so temperature of condensate leaving the condenser

$$= (53.51 - 7) = 46.51°C$$

and enthalpy of condensate $h_c = h_f$ at 46.51°C

$$= 194.64 \ kJ/kg$$

From thermodynamic relation

$$\frac{m_w}{m_c} = \frac{[h_f + x \, h_{fg} - h_c]}{c_{p_w} (t_{w_o} - t_{w_i})}$$

or,

$$\frac{m_w}{4500} = \frac{2598.35 - 194.64}{4.187 \, (30 - 18)}$$

$\therefore$ Volume of water flowing per second

$$= \frac{215283.32 \times 10^{-3}}{3600} \quad (\because \text{ density of water} = 10^3 \text{ kg/m}^3)$$

$$= 0.0598 \text{ m}^3 = 59800.92 \text{ cm}^3$$

Volume of water flowing per pipe

$$= \frac{59800.92}{200} = 299.00 \text{ cm}^3$$

$\therefore$ Rate of flow $Q = A \times V$

$$299.00 = \frac{\pi}{4} D^2 \times (2 \times 100)$$

$\therefore$ Diameter of tube, $D = 1.37$ cm   *Ans.*

**13.13** A two pass surface condenser condenses 14500 kg/hr of steam per hour. Cooling water enters at 15°C and leaves at 30°C. The vacuum in the condenser is 70 cm of Hg. Barometer reads 74.5 cm of Hg. Temperature of condensate is 32°C, Quality of exhaust steam 0.9 dry. In the tubes, the water velocity is 2 m/s, outside diameter of tubes is 2.5 cm and thickness is 0.02 cm. Take overall heat transfer co-efficient $U = 1.65$ kJ/hr/cm$^2$k. Determine
(a) Area of tube surface required
(b) No. of tubes
(c) Length of tubes

*Soln.* Total absolute pressure in the condenser
(a) $P_t = 74.5 - 70 = 4.5$ cm of Hg $= 0.0599$ bar
Saturation temperature corresponding to 0.0599 bar

$$= 36.18°C$$

$$\frac{m_w}{m_c} = \frac{\overbrace{h_f + x h_{fg}}^{\text{at } P_t} - h_c}{c_{p_w}(t_{w_o} - t_{w_i})} \quad \xleftarrow{} h_f \text{ at } 32°C$$

or,

$$\frac{m_w}{14500} = \frac{151.5 + 0.9 \times 2416 - 134}{4.187 (30 - 15)}$$

$\therefore$ quantity of cooling water $m_w = 506051.27$ kg/hr
From the equation of heat flow (Newton's law of cooling)

$$Q = UA\theta_m = m_w c_{p_w}(t_{w_o} - t_{w_i})$$

where   $U$ = Overall heat transfer co-efficient
$A$ = Area of flow
$Q_m$ = Log mean temperature differene (LMTD)

$$= \frac{t_{w_o} - t_{w_i}}{\ln\left(\dfrac{t - t_{w_i}}{t - t_{w_o}}\right)}, \, t = 36.18°C$$

$$\theta_m = \frac{30 - 15}{\ln\left(\dfrac{36.18 - 15}{36.18 - 30}\right)} = 12.17°C$$

$$\therefore \qquad Q = UA\theta_m = m_w c_{p_w} (t_{w_o} - t_{w_i})$$

$$A = \frac{Q}{U\theta_m} = \frac{506051.27 \times 4.187(30 - 15)}{1.65 \times 12.17}$$

$$= 1582756.9 \text{ cm}^2 \quad Ans.$$

(b)  Total water required per second

$$\frac{506051.27}{3600} = 140.56 \text{ kg/s} = \frac{140.56 \times 10^6}{10^3} \text{ cm}^3/\text{s}$$

Total effective discharge per tube

$$= \frac{\pi}{4} \{2.5 - (2 \times 0.02)\}^2 \times 2 \times 100$$

$$= 950.58 \text{ cm}^3/\text{s}$$

$$\therefore \qquad \text{No. of tubes } = \frac{2 \times 140.56 \times 10^6}{10^3 \times 950.58} = 296 \quad Ans.$$

(c) $$\qquad \text{surface area per tube } = \frac{1582756.9}{296} = 5347.15 \text{ cm}^2$$

$$\pi \, 2.5 \times l = 5347.15$$

$$\therefore \qquad l = \frac{5347.15}{\pi \times 2.5} = 680.82 \text{ cm}$$

$$= 6.8082 \text{ m} \quad Ans.$$

**13.14**  A jet condenser is maintaining the vacuum of 705 mm of mercury when condensing 2500 kg of dry steam per hour. The temperature of cooling water is 14°C. Find out the quantity of the cooling water required in kg per minute to extract 2400 kJ from each kg of steam.

*Soln.*  Let us assume mixing of steam and cooling water takes place under standard barometric pressure i.e. $P_b$ = 760 mm of Hg.

Vacuum in the condenser = 705 mm of Hg

$\therefore$   Absolute total pressure in the condenser

$$P_t = 760 - 705 = 55 \text{ mm of Hg} = 0.0733 \text{ bar}$$

From steam table, saturation temperature corresponding to 0.0733 bar = 39.68°C

In case of jet condenser, final temperature of cooling water = final temperature of condensate = 39.68°C since 2400 kJ of heat is being lost from steam to cooling water, mass of cooling water required/kg of steam

$$= \frac{2400}{4.187 (39.68 - 14)} = 22.32 \text{ kg}$$

$\therefore$   Total mass of water required/min

$$= \frac{22.32 \times 2500}{60} = 930 \text{ kg} \quad Ans.$$

**13.15**  A barometric jet condenser handles 5000 kg of 0.9 dry steam per hour and maintains a vacuum of 630 mm of Hg when barometer reads 750 mm of Hg. The cooling water enters the condenser at 16°C and the mixture of condensate and cooling water leaves at 44°C. Find the minimum height of the tail pipe above the hot well and cooling water circulated in kg/hr.

*Soln.* Absolute total pressure in the condenser

$$P_t = 750 - 630 = 120 \text{ mm of Hg} = 0.15998 \text{ bar}$$

From thermodynamic relation

$$\frac{m_w}{m_c} = \frac{\overbrace{(h_f + x\,h_{fg}}^{\text{at } P_t} - c_{p_w} t_{w_o})}{c_{p_w}(t_{w_o} - t_{w_i})}$$

or,

$$\frac{m_w}{5000} = \frac{231.59 + 0.9 \times 2369.9 - 4.187 \times 44}{4.187\,(44 - 16)}$$

$$\therefore \qquad m_w = 92986.45 \text{ kg/hr} \quad Ans.$$

Tail length above hot well

$$= \text{Barometric pressure} - \text{Absolute pressure in the condenser}$$

$$= (750 - 120) = 630 \text{ mm of Hg}$$

Now, in terms of metre of water column

$$\frac{630}{1000} \times 13600 \times 9.81 = h_w \times 1000 \times 9.81 \quad (\because \quad \text{density of mercury} = 13600 \text{ kg/m}^3)$$

$$\therefore \qquad h_w = 8.568 \text{ m} \quad Ans.$$

**13.16** A small size cooling tower is designed to cool 300 kg of water per minute. The water enters the tower at 42°C. The air enters the tower at 18°C and 65% relative humidity, and it leaves the water at 25°C and in saturated condition. The fan driven by electric motor forces 650 m$^3$ of air per minute through the tower and power absorbed is 8 kW. Find (a) temperature of water coming out of the tower, and (b) make-up water required per hour.

*Soln.*

The condition of air entering and leaving the cooling tower are shown in Fig. 13.25 as taken from original psychrometric chart. Assume that the mass of water and air entering the tower are $m_w$ and $m_a$. Total heat of water at inlet + total heat of air at inlet + heat dissipated by motor = total heat of water at outlet + total heat of air at outlet

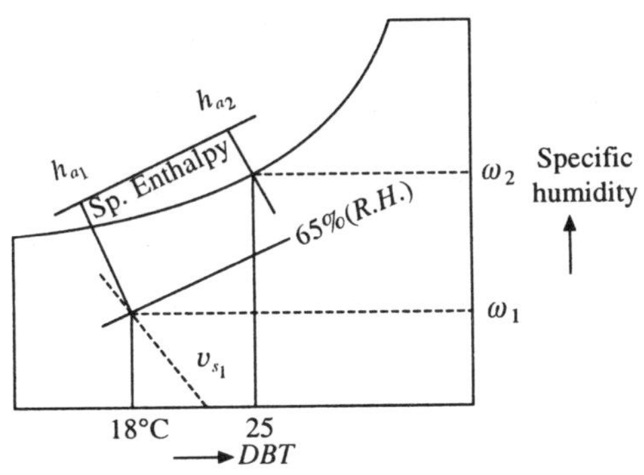


**Fig. 13.25  Psychrometric chart**

$$m_{w_1} h_1 + m_a h_{a_1} + Q = \left[ m_{w_1} - m_a \left( \frac{\omega_2 - \omega_1}{1000} \right) \right] h_{w_2} + m_a h_{a_2}$$

$$\therefore \qquad m_a \left[ (h_{a_2} - h_{a_1}) - \left( \frac{\omega_2 - \omega_1}{1000} \right) h_{w_2} \right] = m_{w_1}(h_{w_1} - h_{w_2}) + Q$$

From psychrometric chart

$$h_{a_1} = 40 \text{ kJ/kg} \qquad\qquad h_{a_2} = 76.8 \text{ kJ/kg}$$

$$\omega_1 = 0.0086 \text{ kg/kg} \qquad\qquad \omega_2 = 0.02036 \text{ kg/kg}$$

$$= 8.6 \text{ grams/kg} \qquad\qquad = 20.36 \text{ grams/kg}$$

Sp. Volume $v_{s_1} = 0.836 \text{ m}^3/\text{kg}$          $\therefore \quad m_a = \dfrac{650}{0.836} = 777.51 \text{ kg/min}$

$\qquad\qquad c_{p_w} = 4.187 \text{ kJ/kg k}$

Heat dissipated

$$Q = 8 \times 60 = 480 \text{ kJ/min}$$

Putting the values in the above equation

$$777.51\left[ (76.8 - 40) - \left( \frac{20.36 - 8.6}{1000} \right) \times 4.187 t_2 \right] = 300 \times 4.187 (42 - t_2) + 480$$

or,          $28612.36 - 38.28\, t_2 = 52756.2 - 1256.1\, t_2 + 480$

$\therefore$          $t_2 = 20.21°\text{C}$

temperature of water coming out of tower = 20.21°C   *Ans.*

Make-up required $= m_a\left( \dfrac{\omega_2 - \omega_1}{1000} \right)$

$$= 777.51\left( \frac{20.36 - 8.6}{1000} \right)$$

$$= 9.14 \text{ kg/min.}$$

$$= 548.4 \text{ kg/hr}   \textit{Ans.}$$

## EXERCISES

13.1  Objective Questions

    (i)  A condenser in a steam power plant

       (a)   increases expansion ratio of steam

       (b)   reduces back pressure of steam

       (c)   reduces temperature of exhaust steam

       (d)   all of these

   (ii)  Why condensers are provided to steam plant?

       (a)   to decrease efficiency of the plant.

       (b)   to increase efficiency of the plant.

       (c)   to decrease enthalpy drop in steam turbine.

       (d)   all of the above.

  (iii)  pure cooling water has to be used in

       (a)   surface condensers

       (b)   jet condenser

       (c)   in both condensers

       (c)   non of the above

  (iv)  A condenser where circulating water flows through tubes which are surrounded by steam, is known as

       (a)   surface condenser

       (b)   jet condenser

       (c)   barometric condenser

       (d)   evaporative condenser

(v)  The ratio of actual vacuum to the ideal vaccum in a condenser is called
    (a)  condenser efficiency
    (b)  vacuum efficiency
    (c)  boiler efficiency
    (d)  nozzle efficiency

(vi)  The ratio of actual rise in temperature of cooling water to the maximum possible rise in temperature of cooling water is known as
    (a)  vacuum efficiency
    (b)  condenser efficiency
    (c)  nozzle efficiency
    (d)  boiler efficiency

(vii)  The actual vacuum in a condenser is equal to
    (a)  barometric pressure + actual pressure
    (b)  barometric pressure – actual pressure
    (c)  gauge pressure + atmospheric pressure
    (d)  gauge pressure – atmospheric pressure

(viii)  The total absolute pressure in a condenser is equal to
    (a)  barometric pressure + vacuum pressure
    (b)  barometric pressure – vacuum pressure
    (c)  atmospheric pressure + gauge pressure
    (d)  atmospheric pressure – gauge pressure

(ix)  For cooling water supply to the condenser, river or pond water is supplied in which system
    (a)  open system
    (b)  spray pond system
    (c)  cooling tower system
    (c)  non of these

(x)  For big power plants, which type of cooling system is used?
    (a)  Spray pond system
    (b)  Mechanical draught cooling tower system
    (c)  Natural draught cooling tower system
    (d)  hyperbolic cooling tower system

**Answers**

|  |  |  |  |  |
|---|---|---|---|---|
| (i)  d | (ii)  b | (iii)  b | (iv)  b | (v)  b |
| (vi)  b | vii)  b | (viii)  b | (ix)  a | (x)  d. |

13.2  Write true or false
  (i)  Surface condenser should be air tight.
  (ii)  Maximum vacuum in condenser is possible, if no air is present in the condenser.
  (iii)  Absolute total pressure in condenser is more than the atmospheric pressure.
  (iv)  In surface condenser water flows through the nest of tubes.
  (v)  In a condenser the cooling water extracts only latent heat of steam.
  (vi)  A locomotive engine is provided with a surface condenser.
  (vii)  In jet condenser cooling water mixes with exhaust steam.
  (viii)  Surface condensers are more efficient than jet condensers.
  (ix)  A pump which extracts both air and condensate from the condenser is known as dry pump.
  (x)  In case of a jet condenser, the final temperature of cooling water and steam condensed will be different.
  (xi)  In case of marine engines, surface condenser is used.

**Answers**

|  |  |  |  |
|---|---|---|---|
| (i)  T | (ii)  T | (iii)  F | (iv)  T |
| (v)  T | (vi)  F | (vii)  T | (viii)  T |
| (ix)  F | (x)  F | (xi)  T. |  |

13.3   What do you mean by a steam condenser? Explain its function.

13.4   What are the two main types of condensers? State the advantages of a surface condenser over a jet condenser. Which condensers would be suitable for sea water cooling?

13.5   What factors contribute to loss of efficiency in a surface condenser?

13.6   What are the sources of air leakage in a steam condenser? Explain the effects of air leakage in condensers.

13.7   What do you mean by the term vacuum efficiency of a condenser? On what factors does this efficiency depend?

13.8   State the law of partial pressures and show how it applies to the condenser of a steam plant?

13.9   Prove with the help of an example that the vacuum efficiency decreases with the increase in barometric pressure.

13.10   Explain the construction and working of Edward's air pump.

13.11   What part is played by a cooling tower? What are the different types of cooling towers? Mention advantage and disadvantage of each type.

13.12   Following data were recorded during the testing of a condenser: Vacuum = 70 cm; barometer reading = 75.4 cm of Hg; condensate temperature = 18°C. Find the partial pressure of air and steam in the condenser and the mass of air/kg of steam.

$$(Ans. P_a = 0.05135 \text{ bar}, P_s = 0.07196 \text{ bar}, 3.775 \text{ kg/kg of steam})$$

13.13   The vacuum reading of a condenser is 70.5 cm of Hg when the barometer reads 76 cm of Hg and the condensate temperature is 31°C. Find the vacuum efficiency.                              (97.06%)

13.14   Calculate the vacuum efficiency of a condenser from the following data; vacuum at steam inlet to condenser = 710 mm of Hg; Barometer reading = 760 mm of Hg; Hot well temperature = 32°C.

$$(Ans. \ 98\%)$$

13.15   A vacuum gauge fitted to a condenser reads 660 mm of Hg, when the barometer reads 750 mm of Hg. Find the corrected vaccum referred to standard barometer pressure of 760 mm of Hg.

$$(Ans \ 670 \text{ mm of Hg})$$

13.16   The vacuum in a surface condenser is found to be 70.5 cm of Hg with the barometer reading 76 cm of Hg. The cooling water enters the condenser at 20°C and leaves at 36.5°C. Find the condenser efficiency. (Ans 87%)

13.17   The following observation were noted during the trial of a surface condenser:
Condenser vacuum = 70 cm of Hg; Barometer = 76.5 cm of Hg; condenser temperature = 35°C; Hot well temperature = 28°C; condensate mass = 1800 kg/hr; Air leakage = 1 kg/1000 kg of condensate; Quantity of cooling water = 80000 kg/hr; Rise in temperature of cooling water 12°C; Inlet temperature of cooling water 15°C. Find (a) vacuum efficiency (b) condenser efficiency (c) Quality of steam entering the condenser (d) capacity of wet extraction pump required.

$$(Ans. \ 97.6\%, \ 45.21\%, \ 1.18 \text{ m}^3/\text{min.})$$

13.18   The following observations were recorded during a test on a steam condenser:
condenser vacuum = 71 cm of Hg; Barometer reading = 76.5 cm of Hg; Temperature of steam entering the condenser = 34°C; Temperature of condensate = 28.5°C; condensate collected = 1800 kg/hr; cooling water circulated = 57.5 tons/hr; Inlet and outlet temperatures of cooling water circulated = 8.5°C and 26°C. Find the following
(a) corrected vacuum to standard barometer of 76 cm of Hg, (b) vacuum efficiency, (c) undercooling of condensate, (d) cooling ratio, (e) condenser efficiency, (f) quality of steam entering the condenser and (g) mass of air per m³ of condenser volume and mass of air per kg of uncondensed steam.

$$Ans. \ [(a) \ 70.5 \text{ cm of Hg}, \ (b) \ 98\%, \ (c) \ 5.5°C, \ (d) \ 31.94, \ (e) \ 55.7\%,$$
$$(f) \ 0.955, \ (g) \ 0.022 \text{ kg/m}^3, \ 0.59 \text{ kg/kg.}]$$

13.19   A surface condenser is equipped with separate air and condensate outlets. A portion of the cooling surface is screened from the incoming steam and the air passes over there screened tubes to the air extraction and becomes cooled below the condensate temperature. 20,000 kg/hr of steam, dry and saturated at 36.2°C enters the condenser. At the condensate outlet the temperature is 34.6°C, and the air extraction, the temperature is 29°C. The volume of air plus vapour leaving the condenser is 3.8 m³/min. Assuming constant pressure through out the condenser find (a) the mass of air removed per 10,000 kg of steam, (b) the mass of steam condensed in the air cooler per minute, (c) the heat to be removed per minute by the cooling water.

Neglect the partial pressure of the air at inlet to the condenser.

*(Ans.* (a) 2.63 kg, (b) 0.492 kg, (c) 807.05 MJ).

13.20 A surface condenser handles 13625 kg of steam per hour. The quality of steam entering the condenser is 88%. The vacuum in the condenser is 694 mm of Hg while barometer reads 760 mm of Hg. The air leakage is 7.25 kg/hr. The temperature at the suction of the air pump is 36°C. Find (a) the dimensions of single acting air-pump to remove the air from condenser when it is running at 60 r.p.m assuming 85% volumetric efficiency, and (b) the surface area of the condenser required if the rate of heat flow from the steam to water is 4000 kJ/m²-sec. Take $L = 1.25D$.

[*Ans.* (a) $D = 152.8$ cm, $L = 191$ cm and (b) 2 m²]

13.21 In a surface condenser, a section of tubes near the air pump is screened off so that air is cooled to temperature below that of condensate. In such condenser, the temperature is 36.5°C. The air pump suction is maintained at 33.5°C. The steam condensed per hour is 3500 kg and the air leakage is 4 kg/hr. Assuming a constant vacuum in the condenses: find

(a)  volume of air in m³/hr to be dealt with by the air pump.

(b)  mass of steam in kg/hr condensed in the air cooler

(c)  Percentage reduction in the air pump capacity required due to cooling of air.

*(Ans.* (a) 276 m³/hr, (b) 34.13 kg/hr, (c) 73.4%)

13.22 In a jet condenser exhaust steam from the engine enters at an absolute pressure of 0.118 bar and 0.85 dry. Determine the mass of injection water per kg of steam condensed if its temperature at inlet is 20°C. Allow 5°C of under cooling

*(Ans.* 22.025 kg).

13.23 A jet type of condenser is required to deal with 4000 kg per hour of steam, the pressure at inlet being 0.07 bar. The ratio of injection water to steam is 40 and the volume of air dissolved in this water, measured at 1 bar and 15°C is 5 percent of that of the water. Air also enters with the steam at the rate of 3 kg/hr. Assuming that the temperature at the air pump suction is 30°C and the volumetric efficiency of the pump is 0.8, calculate the required capacity of pump in m³ perminute to extract the air.

[*Ans.* 8.34 m³/min].

13.24 A 50 MW steam turbine has a condenser flow at full load of 3.8 kg/hr. Quality of steam at turbine exhaust = 0.85. pressure in the condenser = 0.07 bar, circulating water enters at 25°C, circulating water leaves 5.5°C cooler than the incoming steam. Velocity of circulting water = 1.8 m/s, condenser data; Single pass. Tube out side diameter = 25.4 mm, Tube thickness = 1.25 mm, Tube material brass, overall heat transfer co-efficient = 3 kW/m²k. Calculate (a) circulating water required in m³/s, (b) condenser heating surface, (c) the length of tubes, and (d) the number of tubes.

13.25 Water at 60°C leaving the condenser at the rate of 22.5 kg/s is sprayed into a natural draught cooling tower and leaves it at 27°C. Air enters the tower at 1.013 bar, 13°C and 50% relative humidity and leaves it at 38°C, 1.013 bar and saturated, claculate (a) the air flow rate required in m³/s, and (b) the make-up water required in kg/s.

*(Ans.* (a) 21 m²/s, (b) 1 kg/s).

13.26 Water from a cooling system is itself to be cooled in a cooling tower at a rate of 2.78 kg/s. The water enters the tower at 65°C and leaves a collecting tank at the base at 30°C. Air flows through the tower, entering the base at 15°C, 0.1 mPa, 55% RH and leaving the top at 35°C, 0.1 mPa, saturated. Make-up water enters the collecting tank at 14°C. Determine the air flow rate into the tower in m³/s and the make-up water flow rate in kg/s.

*(Ans.* 3.438 m³/s, 0.129 kg/s).

**14**

# Steam Nozzles

## 14.1  INTRODUCTION

A nozzle is a duct of smoothly varying cross-sectional area by means of which pressure energy of steadily flowing working fluid is converted into kinetic energy. The working fluid enters the nozzle at high pressure and low velocity. As it flows through the nozzle, its pressure falls i.e. enthalpy (heat drop takes place and thereby velocity increases continuously from the entrance to the exit. Whereas, a diffuser is defined as a duct of varrying cross-sectional area in which the kinetic energy of the working fluid decreases along the duct while its pressure energy increases.

Being the flow steady i.e. the mass of steam passing through any section of the nozzle remains constant, the variation of steam pressure in the nozzle depends upon the velocity, specific volume and dryness fraction of steam. A suitably designed nozzle converts the heat energy of steam into kinetic energy with a minimum loss.

The main use of the nozzles are in aviation purposes like propulsion of rockets and jet engines and to drive steam or gas turbines in Power Production system.

A fluid is said to be compressible if its density changes with a change in pressure (or temperature). If the density does not change or change very little, the fluid is said to be incompressible. Gases and vapours (steam) are compressible, whereas liquids (water) are incompressible

The flow analysis to be discussed in this chapter will be restricted to one-dimensional flow as the following conditions are fulfilled.

   (i)   There is gradual change in cross-section of the nozzle
  (ii)   Thermodynamic and mechanical properties change only in the direction of flow
 (iii)   The flow is stable i.e. the velocity of fluid does not change from previous value at a particular section in the course of time under same inlet conditions
 (iv)   The flow is isentropic through the duct of the nozzle
  (v)   Mass flow rate remains constant.

In actual practice, friction is developed between the steam and sides of the nozzle. This friction offers a resistance to the flow of steam which is converted into steam and tends to dry the steam. Hence, at the time of design of the nozzle, friction has to be considered as well.

There is also a phenomenon known as supersaturation in the flow of steam through nozzles. This occurs due to time lag in the condensation of the steam during expansion. This supersaturated flow affects the mass and condition of the steam at the exit. Therefore, the flow of steam through a nozzle may be regarded as either (i) adiabatic and reversible i.e. isentropic flow or (ii) adiabatic flow modified by friction or (iii) a supersaturated flow.

## 14.2 FLOW OF STEAM THROUGH NOZZLE

Considering a nozzle as shown in Fig. 14.1. Applying the steady flow energy equation (SFEE) to the sections 1 and 2 and considering one kg of steam flows in one second from pressure $P_1$ to pressure $P_2$ as shown in Fig. 14.2,

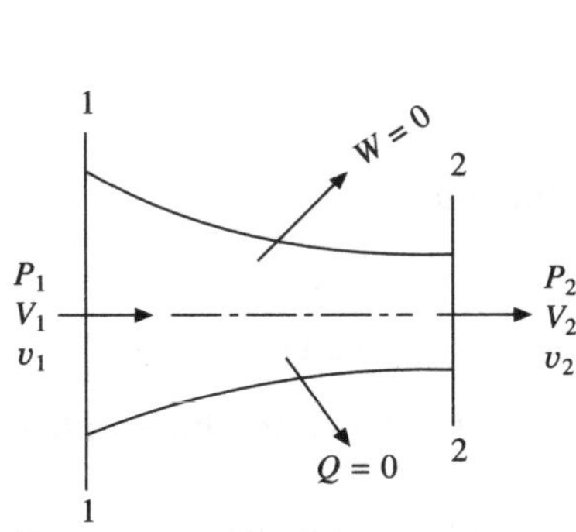

**Fig. 14.1**

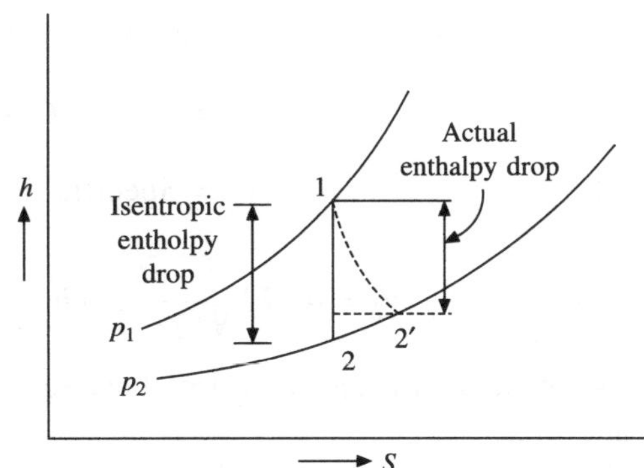


**Fig. 14.2   Expansion of steam or gas on *h-s* diagram**

$$h_1 + \frac{V_1^2}{2g_c} + W = Q + h_2 + \frac{V_2^2}{2g_c} \qquad (14.1)$$

where $h$ and $V$ denote enthalpy and velocity of steam and $W$ and $Q$ are work and heat transfer. $g_c$ has a numerical value of 1 in SI units. Since the expansion of steam through a nozzle is considered as adiabatic and as there is no external work done during the flow of steam, both the heat transfer and work done are zero.

$\therefore$ $$Q = 0, \quad W = 0$$

Using these condition, Eq. (14.1) reduces to

$$h_1 + \frac{V_1^2}{2} = h_2 + \frac{V_2^2}{2}$$

or,

$$\frac{V_2^2 - V_1^2}{2 \times 1000} = h_1 - h_2$$

[∵ Enthalpies are given in kJ/kg generally and velocity are in m/s]

$$\therefore \qquad V_2 = \sqrt{2000(h_1 - h_2) + V_1^2}$$

As the velocity of steam entering the nozzle is small (called velocity of approach), $V_1$ can be neglected,

$$\therefore \qquad V_2 = \sqrt{2000(h_1 - h_2)}$$

or, $\qquad\qquad V_2 = 44.72\sqrt{(h_1 - h_2)} \text{ m/s} \qquad\qquad (14.2)$

This is the general equation of energy irrespective of the shape of the nozzle. If frictional losses are taken into account then

$$V_2 = 44.72\sqrt{(h_1 - h_2)\eta_{\text{nozzle}}} \text{ m/s}$$

If the frictional loss in the nozzle is 15%, $\eta_{\text{nozzle}} = 85\%$ where $\eta_{\text{nozzle}}$ is the nozzle efficiency.

For air or gases Eq. (14.2) can be written as

$$V_2 = 44.72\sqrt{C_p(T_1 - T_2)\eta_{\text{nozzle}}}$$

where, $\qquad\qquad C_p = $ Specific heat at constant pressure

or, $\qquad V_2 = 44.72\sqrt{\dfrac{\gamma R}{(\gamma - 1)}(h_1 - h_2)\eta_{\text{nozzle}}} \qquad \because C_p = \dfrac{\gamma R}{(\gamma - 1)}$

In case of steam, the enthalpy drop between sections 1 and 2 is also called Rankine enthalpy drop (heat drop.).

The cross-sectional area at the exit of a nozzle can be found out for a given inlet cross-sectional area $A_1$, from continuity equation

$$\frac{AV}{v} = \frac{A_1 V_1}{v_1} = \frac{A_2 V_2}{v_2} = m \text{ (constant)} \qquad\qquad (14.3)$$

where, $\quad A = $ Cross-sectional area, $\text{m}^2$

$\qquad\quad V = $ Velocity, m/s

$\qquad\quad v = $ Specific volume, $\text{m}^3$/kg

$\qquad\quad m = $ mass flow rate, kg/s

Area per unit mass flow, $\dfrac{A}{m} = \dfrac{v}{V}$

or, $\qquad\qquad \dfrac{A}{m} = \dfrac{v}{V} = \dfrac{v}{44.72\sqrt{\Delta h}} \qquad\qquad (14.4)$

This equation clearly indicates that for getting the value of cross-sectional area at a particular section of the nozzle duct, it is essential to know about the process of expansion, i.e. how specific volume and enthalpy vary in the duct. Considering the process of expansion in the duct is frictionless and adiabatic, the entropy remains constant, i.e. $S_1 = S_2$.

Figure 14.3 shows the variation of velocity, specific volume and area with pressure along the length of a convergent-divergent nozzle. Fig. 14.4 represents the convergent nozzle while Fig. 14.5 represents for a convergent divergent nozzle which is also known as de-Laval nozzle. Fig. 14.3 represents the variation of the parameters by considering the following assumptions as mentioned below:

(i)  inlet conditions of the fluid are fixed, and

(ii)  equal pressure drop occurs in each unit length of the nozzle duct.

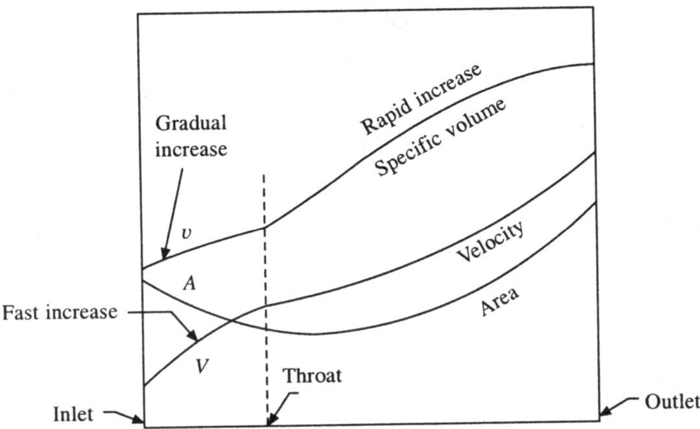



**Fig. 14.3**    **Variation of velcity, specific volume and area with pressure along the length of a convergent divergent nozzle**

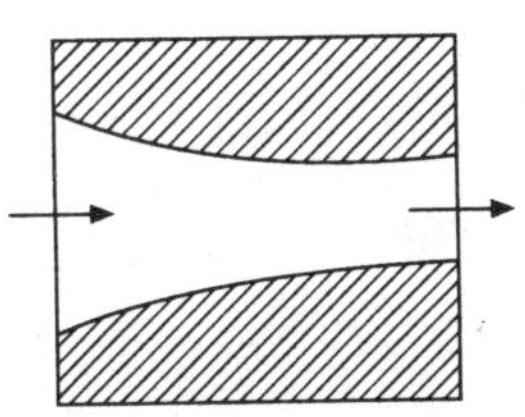

**Fig. 14.4    Convergent nozzle**

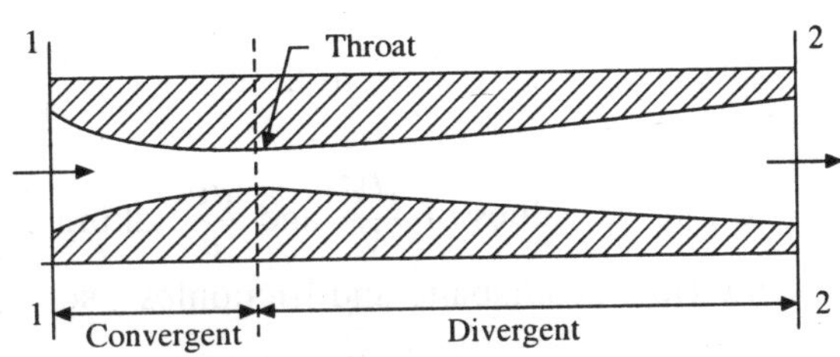



**Fig. 14.5    Convergent divergent nozzle**

It is observed that specific volume increases gradually at first as the pressure drops from the higher value while velocity increases at a faster rate up to the throat of the nozzle. As the expansion proceeds after throat, there is a rapid increase in specific volume which becomes greater than the increase in velocity. Area is minimum at the throat where pressure is called critical pressure. Mass flow per unit area is maximum at the throat.

The above variation in velocity, specific volume and area is true in both gases and vapours (steam).

## 14.3  GENERAL RELATIONSHIP BETWEEN AREA, VELOCITY AND PRESSURE IN NOZZLE FLOW

Assume the conditions mentioned in article 14.1, for full flow through the nozzle. As shown in Fig. 14.6, consider two sections, a distance $\delta x$ apart.

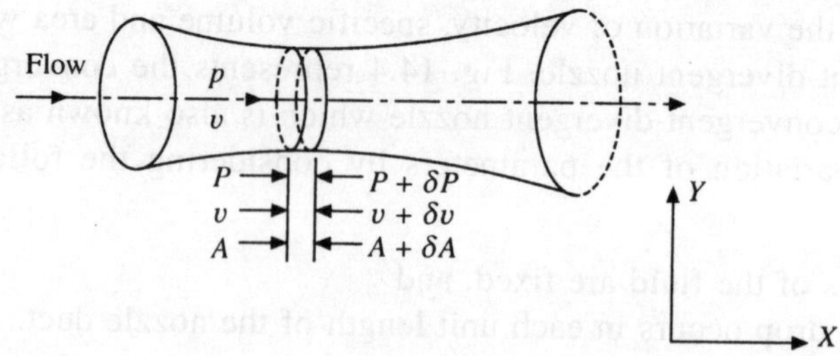


**Fig. 14.6**

By continuity (Eq. 14.3),

$$m = \frac{AV}{v} = \frac{(A + \delta A)(V + \delta V)}{v + \delta v}$$

By partial differentiation, we get

$$\frac{V}{v} \delta A + \frac{A}{v} \delta V - \frac{AV}{v^2} \delta v = 0$$

By dividing both sides by $\dfrac{AV}{v}$, we get

$$\frac{\delta A}{A} + \frac{\delta V}{V} - \frac{\delta v}{v} = 0$$

and in limits $\dfrac{dA}{A} + \dfrac{dV}{V} - \dfrac{dv}{v} = 0$ $\hspace{2cm}$ (14.5)

we have to put the values of $\dfrac{dV}{V}$ and $\dfrac{dv}{v}$ taking seperately in this eqn.

Since the flow is adiabatic and frictionless, so

$$Pv^n = c \text{ (constant), for air } n = \gamma$$

Taking log of this, we get

$$\ln p + n \ln v = \ln \text{ (constant)}$$

After differentiation, we get

$$\frac{dp}{p} + n \frac{dv}{v} = 0$$

or, $\hspace{4cm} \dfrac{dv}{v} = -\dfrac{1}{n} \dfrac{dp}{p}$ $\hspace{2cm}$ (14.6)

From the momentum equation for adiabatic and frictionless flow

$$V \, dV = - v \, dp$$

or,
$$\frac{dV}{V} = -v\frac{dp}{V^2} \tag{14.7}$$

Putting the values of $\dfrac{dv}{v}$ and $\dfrac{dV}{V}$ in Eq. (14.5), we get

$$\frac{dA}{A} - v\frac{dp}{V^2} + \frac{1}{n}\frac{dp}{p} = 0$$

or,
$$\frac{dA}{A} = \frac{1}{n}\frac{dp}{p}\left(\frac{npv}{V^2} - 1\right) \tag{14.8}$$

we define sonic or sound velocity '$a$' as the speed at which small pressure disturbances are propagated in the form of waves through a compressible fluid. Sonic velicity in steam at pressure $p$ and specific volume $v$ is given by

$$a = \sqrt{npv}$$

or,
$$a^2 = npv \tag{14.9}$$

Also, we define a dimensionless number, called Mach number '$M$' as the ratio of actual velocity of the fluid to the local acoustic velocity (sonic velocity).

Therefore,
$$M = \frac{V}{a} \tag{14.10}$$

If, $M < 1$, flow is called subsonic

$M = 1$, flow is called sonic

$M > 1$, flow is called supersonic

Taking the values from Eqs. (14.9) and (14.10) and putting in Eq. (14.8), we get

$$\frac{dA}{A} = \frac{1}{n}\frac{dp}{p}\left(\frac{1}{M^2} - 1\right) \tag{14.11}$$

Also,
$$\frac{dA}{A} = \frac{dV}{V}(M^2 - 1)$$

Physical explanation of the Eq. (14.11) is as follows:

*Case 1*  Accelerated flow, $\dfrac{dp}{p} = -ve$ (Nozzle) i.e.

Pressure decreases and velocity increases along flow direction. Right side value will be negative, irrespective of the value of $\left(\dfrac{1}{M^2} - 1\right)$.

(a)  When $M < 1, \dfrac{dA}{A} = -ve$

i.e. for subsonic flow, nozzle should be converging type. As soon as actual velocity $V$

reaches to $C$ i.e. $M = 1$, then $\dfrac{dA}{A} = 0$ and the throat of the nozzle is reached. Flow is then sonic.

(b) When $M > 1$, $\dfrac{dA}{A} = +\,ve$

   i.e. for supersonic flow nozzle should be diverging type.

*Case II* Retarded flow, $\dfrac{dp}{p} = +\,ve$ (Diffuser). Diffuser converts kinetic energy into pressure energy. Right side value will be positive irrespective of the value of $\left( \dfrac{1}{m^2} - 1 \right)$

(a) When $M < 1$, $\dfrac{dA}{A} = +\,ve$

   i.e. For subsonic flow, diffuser should be diverging type. At the throat $M = 1$, $\dfrac{dA}{A} = 0$, flow is sonic

(b) When $M > 1$, $\dfrac{dA}{A} = -\,ve$

i.e. for supersonic flow, diffuser should be converging type.

These forms are summarized in Fig. 14.7.

| Type of flow | $\dfrac{dp}{p}$ = Negative | $\dfrac{dp}{p}$ = Positive |
|---|---|---|
| $M < 1$<br>Subsonic | Convergent nozzle | Divergent diffuser |
| $M > 1$<br>Supersonic | Divergent nozzle | Convergent diffuser |

**Fig. 14.7   Effect of area on subsonic and supersonic flow**

If it is required to select a nozzle for an inlet velocity which is very low compared to the sonic velocity (i.e. $M \ll 1$) while the outlet velocity is much more than sonic velocity (i.e. $M \gg 1$), then the shape of the nozzle will be a combination of convergent part, throat and divergent part i.e. a convergent divergent nozzle. For the same purpose convergent-divergent nozzles are used generally in steam turbines. Because steam velocity enterig the turbines is very low at high pressure coming out from the boiler but the final velocity required is supersonic to produce more power by the turbines. Convergent part brings the velocity of steam very near to sonic velocity, the throat brings the velocity equal to sonic and the divergent part brings the velocity equal to supersonic or more.

*Note:* Venturi is another device in addition to nozzles and diffusers, which is used for flow measurement of a incompressible liquid (e.g. water). Venturi also is of convergent-divergent shape. However, in this case there is not a continuous rise or fall of pressure. In it, in convergent portion pressure is decreasing, velocity is rising while specific volume and density remain constant and this portion acts as subsonic nozzle. In the divergent portion pressure is rising, velocity is falling and again specific volume and density remain constant and this portion acts as subsonic diffuser. So a venturi is a combination of subsonic nozzle and subsonic diffuser (Fig. 14.8). As far continuity equation for venturi is concerned, we write discharge rate of fluid as

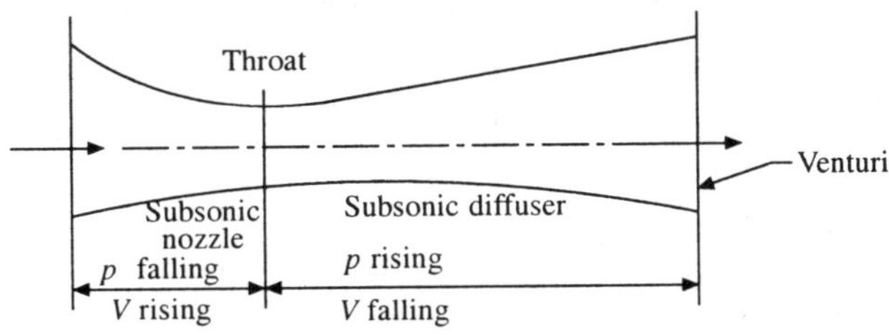


**Fig. 14.8   Convergent-divergent nozzle**

$$m = \frac{AV}{v} = \frac{A_1 V_1}{v_1} = \frac{A_2 V_2}{v_2}$$

$\because$ Specific volume $v_1 = v_2$ (for incompressible) flow
$\therefore$ Discharge rate $= A_1 V_1 = A_2 V_2 = Q \, (m^3/s)$
Where as, in case of compressible flow (e.g. steam flow), specific volume and density change.

We write
$$m = \frac{AV}{v} = \frac{A_1 V_1}{v_1} = \frac{A_2 V_2}{v_2} \text{ (kg/s)}$$

## 14.4   MASS OF DISCHARGE THROUGH NOZZLE

Referring the Fig. 14.1, the Eq. (14.2) could be rewritten as

$$V_2 = 44.72 \sqrt{(h_1 - h_2)} \text{ m/s}$$

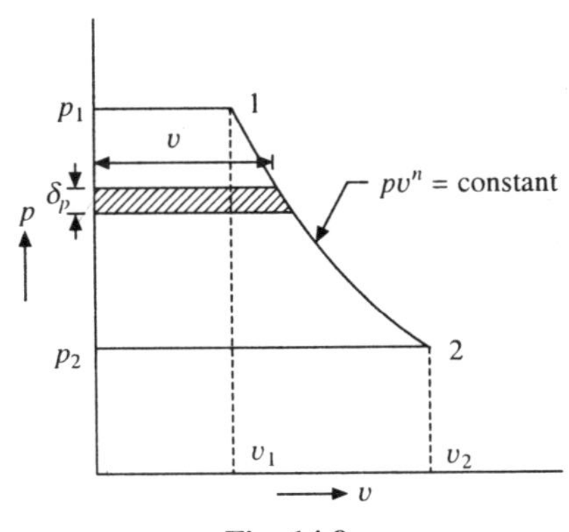


**Fig. 14.9**

where $V_2$ and $h_2$ are the velocity and actual enthalpy at the throat in convergent nozzle. Similarly, for a convergent-divergent nozzle, the actual velocity at the exit could be written as

$$V_3 = 44.72 \sqrt{(h_1 - h_3)} \text{ m/s}$$

where $h_3$ = Actual specific enthalpy at the exit.
Referring the Fig. 14.1, the flow of the steam through nozzle takes place adiabatically and may be approximately represented by the equation

$$Pv^n = \text{constant}$$

This process is shown on $P$-$V$ plane in Fig. 14.9.

Here,  $n = 1.135$ for saturated steam
$\quad\quad = 1.3$ for supersaturated steam
$\quad\quad = 1.035 + 0.1x$ for wet steam (Zeuner's eqn.)

These values of $n$ are approximate. Actually the value of $n$ varies during the cycle.

Assuming the above law for expansion, the work done by the steam upon itself during the expansion is equal to the enthalpy drop (heat drop). Due to this enthalpy drop, there is an increase in kinetic energy of the steam.

$\therefore$ Work done = Rankine area = heat drop and gain in kinetic energy = heat drop

$$\text{Work done } W = \int_{p2}^{p_1} v\, dp = \int_{p2}^{p_1} c^{\frac{1}{n}}\, p^{\frac{1}{n}}\, dp$$

$$\left[ \because pv^n = c, \quad v^n = \frac{c}{p} \quad \therefore \quad v = \left(\frac{c}{p}\right)^{\frac{1}{n}} \right]$$

$$= c^{\frac{1}{n}} \int_{p2}^{p_1} p^{-\frac{1}{n}}\, dp = \frac{n}{n-1} c^{\frac{1}{n}} \left( p_1^{\frac{n-1}{n}} - p_2^{\frac{n-1}{n}} \right)$$

$$= \frac{n}{n-1} (p_1 v_1 - p_2 v_2) \tag{14.12}$$

$$\left[ \because c^{1/n} = v_1\, p_1^{1/n} = v_2\, p_2^{1/n} \right]$$

Kinetic energy = Rankine work considering per kg of steam flow

$$\frac{V_2^2 - V_1^2}{2} = \frac{n}{n-1}(p_1 v_1 - p_2 v_2)$$

where $v_1$ and $v_2$ are specific volume of steam at entry and exit respectively.

If steam entering the steam at very low velocity i.e. velocity of approach $V_1 = 0$, then

$$\frac{V_2^2}{2} = \frac{n}{n-1}(p_1 v_1 - p_2 v_2)$$

or,

$$V_2 = \sqrt{2\frac{n}{n-1}(p_1 v_1 - p_2 v_2)} \text{ m/s}$$

or,

$$V_2 = \sqrt{2\left(\frac{n}{n-1}\right) p_1 v_1 \left\{ 1 - \left(\frac{p_2}{p_1}\right)^{\frac{n-1}{n}} \right\}} \text{ m/s} \tag{14.13}$$

$$\left[ \text{As } \frac{p_2 v_2}{p_1 v_1} = \frac{p_2}{p_1}\left(\frac{p_1}{p_2}\right)^{\frac{1}{2}} = \frac{p_2}{p_1}\left(\frac{p_2}{p_1}\right)^{-\frac{1}{n}} \right]$$

Referring the Eq. (14.3), the mass of discharge at the throat of the nozzle could be rewritten as

$$m = \frac{A_2 V_2}{v_2}$$

$$\therefore \quad m = \frac{A_2}{v_2} \sqrt{2\left(\frac{n}{n-1}\right) p_1 v_1 \left\{ 1 - \left(\frac{p_2}{p_1}\right)^{\frac{n-1}{n}} \right\}} \ \text{kg/s}$$

$$= \frac{A_2}{v_1 \left(\frac{p_2}{p_1}\right)^{-\frac{1}{n}}} \sqrt{2\left(\frac{n}{n-1}\right) p_1 v_1 \left\{ 1 - \left(\frac{p_2}{p_1}\right)^{\frac{n-1}{n}} \right\}} \ \text{kg/s}$$

$$\left[ \text{As } p_1 v_1^n = p_2 v_2^n = c \ \therefore \ v_2 = v_1 \left(\frac{p_2}{p_1}\right)^{-\frac{1}{n}} \right]$$

$$= \frac{A_2}{v_1} \sqrt{2\left(\frac{n}{n-1}\right) p_1 v_1 \left\{ \left(\frac{p_2}{p_1}\right)^{\frac{2}{n}} - \left(\frac{p_2}{p_1}\right)^{\frac{n+1}{n}} \right\}} \ \text{kg/s} \qquad (14.14)$$

with the help of this equation, mass of discharge can be calculated at the throat for the given inlet condition of pressure, specific volume and quality of steam to be expanded and pressure ratio.

## 14.5 THROAT PRESSURE FOR MAXIMUM DISCHARGE

For a certain throat pressure, a nozzle is designed to supply maximum discharge. For this purpose a pressure ratio is obtained for getting maximum value of $m$ in Eq. 14.14. In this equation the value of $m$ will be maximum when $\left(\frac{p_2}{p_1}\right)^{\frac{2}{n}} - \left(\frac{p_2}{p_1}\right)^{\frac{n+1}{n}}$ will be maximum at the above mentioned initial conditions. Differentiating this equation with respect to $\left(\frac{p_2}{p_1}\right)$ and equating to zero for maximum discharge.

$$\frac{d}{d\left(\frac{p_2}{p_1}\right)} \left[ \left(\frac{p_2}{p_1}\right)^{\frac{2}{n}} - \left(\frac{p_2}{p_1}\right)^{\frac{n+1}{n}} \right] = 0$$

or ,

$$\frac{2}{n}\left(\frac{p_2}{p_1}\right)^{\frac{2-n}{n}} - \frac{n+1}{n}\left(\frac{p_2}{p_1}\right)^{\frac{1}{n}} = 0$$

or,
$$\frac{2}{n}\left(\frac{p_2}{p_1}\right)^{\frac{2-n}{n}} = \frac{n+1}{n}\left(\frac{p_2}{p_1}\right)^{\frac{1}{n}} = 0$$

or
$$\frac{2}{n}\times\frac{n}{n+1} = \left(\frac{p_2}{p_1}\right)^{\frac{1}{n}-\left(\frac{2-n}{n}\right)}$$

or
$$\frac{2}{n+1} = \left(\frac{p_2}{p_1}\right)^{\frac{1-2+n}{n}}$$

or,
$$\frac{2}{n+1} = \left(\frac{p_2}{p_1}\right)^{\frac{n-1}{n}}$$

or,
$$\frac{p_2}{p_1} = \left(\frac{2}{n+1}\right)^{\frac{n}{n-1}} \tag{14.15}$$

For dry saturated steam, $n = 1.135$

$$\therefore \qquad \frac{p_2}{p_1} = \left(\frac{2}{n+1}\right)^{\frac{n}{n-1}} = \left(\frac{2}{1.135+1}\right)^{\frac{1.135}{0.135}} = 0.578$$

or,
$$\frac{p_2}{p_1} = 0.578 \tag{14.16}$$

For supersaturated steam, $n = 1.3$

$$\therefore \qquad \frac{p_2}{p_1} = \left(\frac{2}{1.3+1}\right)^{\frac{1.3}{1.3-1}} = 0.545$$

or,
$$\frac{p_2}{p_1} = 0.545 \tag{14.17}$$

Similarly maximum mass flow rate, from Eq. (14.15)

$$m_{\max} = 0.635596\, A_2 \sqrt{\frac{p_1}{v_1}} \tag{14.18}$$

for dry saturated steam

and
$$m_{\max} = 0.66726\, A_2 \sqrt{\frac{p_1}{v_1}} \tag{14.19}$$

for initially superheated steam
In the Eq. 14.15, if $n$ is replaced by $\gamma$, then it will be for gas nozzles

Also, by substituting the value of $\dfrac{p_2}{p_1}$ from Eq. (14.15) into Eq. (14.13), the velocity of steam at the throat for maximum discharge can be written as

$$V_{2\max} = \sqrt{2\left(\frac{n}{n-1}\right)p_1 v_1 \left[1 - \left(\frac{2}{n+1}\right)^{\frac{n}{n-1}\frac{n-1}{n}}\right]}$$

$$= \sqrt{2\left(\frac{n}{n-1}\right)p_1 v_1 \left[1 - \left(\frac{2}{n+1}\right)\right]}$$

or, $$V_{2\max} = \sqrt{2\frac{n}{n+1}p_1 v_1} \ \text{m/s} \tag{14.20}$$

## 14.6  PHYSICAL EXPLANATION OF CRITICAL PRESSURE

The pressure at which the velocity of fluid equals the local sound velocity is called critical pressure of the nozzle. As we know, throat of the nozzle brings the velocity of steam equal to sonic velocity from sub-sonic velocity. The ratio of this pressure with the initial pressure is called the critical pressure ratio. Throat is the point where area is minimum and with this condition the mass flow per unit area is a maximum.

Existance of a critical pressure in a nozzle flow may be expressed in another way. Referring Fig. 14.10(a) and (b), consider two vessels $A$ and $B$. A contains steam at a steady pressure $P_1$ and pressure $P_2$ in vessel $B$ is varried at will. A and B are joined by a convergent nozzle. Assume at first that $P_2$ is equal to $P_1$, then there is no flow of steam through nozzle. Let $p_2$ be gradually reduced. The discharge $m$ through the nozzle will increase as shown by the curve. As the pressure $p_2$ approaches the critical pressure, the discharge rate gradually approaches its maximum value and when $p_2$ is reduced below the critical value, the discharge rate remains constant. Thus we observe that the reduction in discharge pressure below the critical value does not increase the rate of discharge.

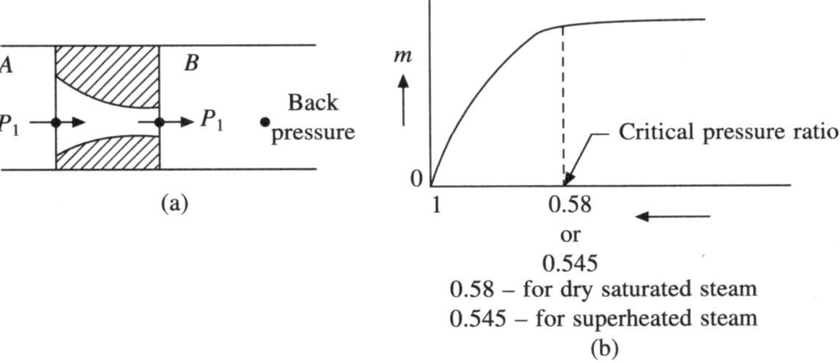


**Fig. 14.10  Critical pressure for convergent nozzle**

Again assuming flow from rest and writing the equation mentioned previously for velocity reached at pressure $p_2$ as

$$V_2 = \sqrt{2\,\frac{n}{n-1}(p_1 v_1 - p_2 v_2)}$$

$$= \sqrt{2\,\frac{n}{n-1}\,p_2 v_2\left(\frac{p_1 v_1}{p_2 v_2} - 1\right)}$$

$$= \sqrt{2\,\frac{n}{n-1}\,p_2 v_2\left[\left(\frac{p_2}{p_1}\right)^{\frac{1-n}{n}} - 1\right]}$$

By substituting the value of $\dfrac{p_2}{p_1} = \left(\dfrac{2}{n+1}\right)^{\frac{n}{n-1}}$ where $p_2$ is the critical pressure, we get

$$V_2 = \sqrt{2\,\frac{n}{n-1}\,p_2 v_2\left[\left(\frac{2}{n+1}\right)^{-1} - 1\right]}$$

$$= \sqrt{2\,\frac{n}{n-1}\,p_2 v_2\left(\frac{n+1}{2} - 1\right)}$$

$$= \sqrt{2\,\frac{n}{n-1}\,p_2 v_2\left(\frac{n-1}{2}\right)}$$

or, $\qquad\qquad V_2 = \sqrt{n\,p_2 v_2} \qquad\qquad\qquad\qquad\qquad\qquad\qquad (14.21)$

The Eq. (14.21) represents the local velocity of sound in steam at pressure $p_2$ and specific volume $v_2 = \dfrac{1}{\rho}$ for particular value of adiabatic expansion $n$. So it is true that the velocity of steam in adiabatic and frictionless flow reaches a critical value at the nozzle throat, equal to the velocity of sound in steam at that point. To increase the velocity of steam above sonic velocity (supersonic), by expanding steam below critical pressure, divergent portion is necessary. If the curves of steam velocity $V_2$ and the sound velocity $a_2$ are drawn, they intersect each other at the critical pressure for a convergent nozzle (see Fig. 14.11).

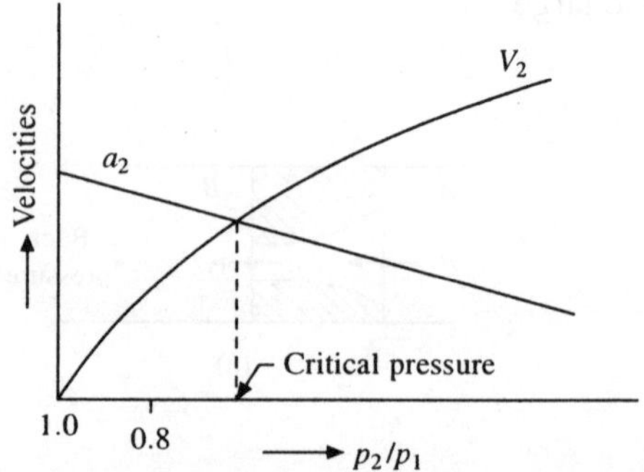


**Fig. 14.11** **Variation of velocity in a convergent nozzle**

Taking into account the maximum discharge at the throat, Fig. 14.12 has been drawn to show the variation of velocity, specific volume and discharge for a convergent-divergent nozzle against the back pressure.

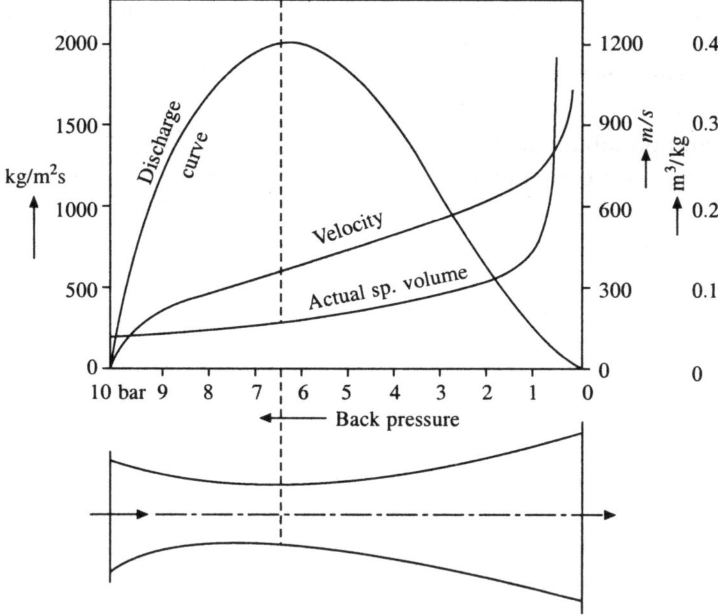


**Fig. 14.12**

*Note*: If sonic velocity reached in a nozzle then the maximum mass flow per unit area has been reached for the nozzle i.e. at the throat. This also means that the maximum mass flow rate through the nozzle has been reached. The nozzle is then said to be choked.

## 14.7 NOZZLES OPERATING IN THE OFF-DESIGN CONDITION

It has been decided in previous article that nozzles are designed for a maximum discharge for a given area. Now we shall study the behaviour of a convergent nozzle under isentropic flow conditions when the back pressure $p_b$ is varied, with following terminologies.

(i) underexpanding—when the back pressure of a nozzle is below the designed value of pressure at the exit of the nozzle, the nozzle is said to be underexpanding. The fluid expands to the design pressure up to throat and then expands violently and irreversibly down to the back pressure.

(ii) Over expanding—when the back pressure of a nozzle is above the designed value of pressure, the nozzle is said to be overexpanding.

In convergent nozzles with over expansion the exit pressure is greater than the critical pressure and the effect is reduction in mass flow from the nozzle.

Fig. 14.13 shows the experimental set-up and pressure and velocity variation along the axis of the nozzle. The pressure at the exit of the nozzle is denoted by $p_2$ and the back pressure is $p_b$, which is varied by a valve. As the back pressure $p_b$ is varied, the mass flow rate $m$ and the ratio $p_2/p_1$ also vary. When $p_b/p_1 = 1$, there is no flow and $\dfrac{p_2}{p_1} = 1$, as designated by straight line (a). If the back pressure is now decreased to $p_2$ such that $p_b/p_1 > p_c/p_1$, $p_2 = p_b$ and Mach number $M < 1$. Now, the back pressure is lowered to the critical pressure, curve (b), when $M = 1$ and $p_2 = p_b$.

When $p_b$ is decreased below the critical pressure, curve (c), there is no increase in the mass flow rate, and $p_2 = p_c$, $M = 1$. The drop in pressure from $p_2$ to $p_b$ occurs outside the nozzle exit. This is the choking limit which means that for given stagnation conditions $(P_1, T_1)$ the nozzle is passing the maximum possible mass flow. The expansion outside the nozzle is shown by the curve (d) where steam expands violently. Expansion outside the nozzle from pressure $p_c$ to $P_d$ is an unrestrained and irreversible expansion, the pressure oscillates and infact a shock wave is formed.

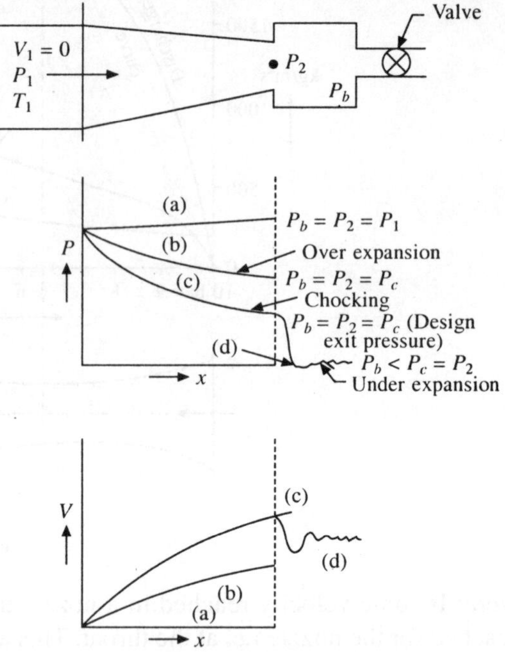


**Fig. 14.13  Flow in a convergent nozzle with varying back pressure**

Let us next consider a convergent-divergent nozzle, as shown in Fig. 14.14. For getting supersonic velocity of jet of the steam, a convergent-divergent nozzle is required as mentioned in article 14.3. It means that at the exit, pressure must be well below the critical pressure. Now, we shall study the behaviour of such a nozzle under the varying back pressure conditions regarding pressure and velocity. If the back pressure equals to the inlet pressure, there is no mass flow of the steam. If the back pressure is reduced such that the pressure at the throat, $p_2 > p_c$, then the pressure decreases and the velocity increases as the area decreases, since the flow is subsonic. It continues up to the throat.

At the throat the flow is still subsonic ($M < 1$). Therefore, in the divergent portion with the further flow of the steam the pressure increases and the velocity decreases as shown by the curve (a). Here divergent portion acts as a subsonic diffuser i.e. a venturi since the flow throughout the nozzle is subsonic. If the back pressure is further reduced such that the throat pressure equals to the critical pressure, the velocity at the throat would become sonic ($M = 1$). Here the mass flow will then be maximum and the nozzle is choked. However, there are two distinct regimes possible in the divergent portion of the nozzle depending upon the back pressure. In the first regime, the velocity in the divergent portion becomes subsonic and the pressure increases as shown by the curve (b). In the second regime, if this increase in pressure (back pressure) becomes above the designed pressure value, the flow just down stream of the throat and becomes supersonic in divergent portion because of the isentropic expansion of the steam throughout. Since the back pressure is higher than design pressure in divergent portion, the steam can not leave at a pressure less than the design pressure. In such a case, there is discontinuity of pressure resulting in the formation of shock waves in the divergent portion of the nozzle. Shock is an irreversible phenomenon occurs only when the flow is supersonic, and after the shock the flow becomes subsonic and the rest of the duct acts as a diffuser. Properties vary discontinuously across the shock. There is an abrupt rise in pressure, an increase in entropy and a loss of kinetic energy from supersonic velocity to subsonic velocity. This is shown by the curves (c, d and e).

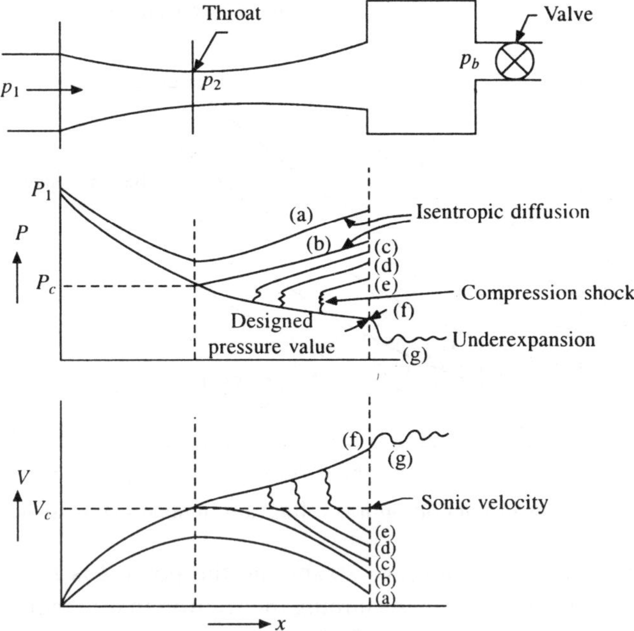


**Fig. 14.14   Flow in a convergent-divergent nozzle with varying back pressure**

As the back pressure is lowered further the shock wave travels down the nozzle and when it equals to designed exit pressure, there is a continuous increase in velocity and the flow in the divergent portion is supersonic with the decrease in pressure as shown by the curve (*f*). Finally, when the exit pressure is reduced below the designed value the expansion takes place outside the nozzle in a similar way as in convergent nozzle.

The expansion outside the nozzle is shown by the curve (*g*) which consists of a series of irreversible compressions through shock waves, happens until the  pressure is equal to the back pressure but below the design pressure and in this case nozzle is said to be underexpanding. Curves (a) to (e) represent for an overexpanding nozzle.

## 14.8   EFFECT OF FRICTION ON NOZZLE PERFORMANCE

Being a natural phenomena, the friction occurs mostly in the diverging part of a convergent-divergent nozzle as the length of the converging part is very small. Losses occur in the nozzle flow due to

  (i)  friction between the steam and the nozzle surface
 (ii)  the internal friction of steam itself, and
(iii)  losses by shock

The reversible adiabatic or isentropic expansion without friction is represented by the process (1-3′) on (*h-s*) diagram for two inlet conditions of the steam viz. dry-saturated and superheated in Fig. 14.15.

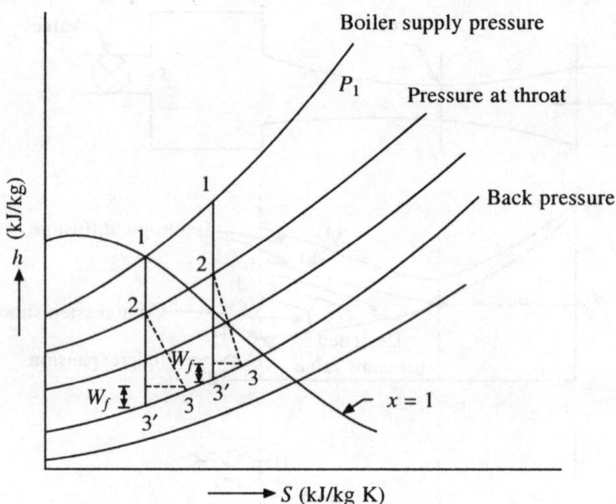



**Fig. 14.15    Effect of friction on $h - s$ diagram in steam flow through nozzle**

Point 1 indicates the initial condition of steam and the point 2 is the steam condition at the throat of a convergent-divergent nozzle. Assuming the friction in divergent portion (being larger), the actual process of expansion is now 1–2–3 rather than 1–2–3′. process 1–2 represents the expansion in convergent portion whereas actual expansion in divergent portion is represented by 2–3. The following are the effects of friction.

(i)   Reduction in the actual enthalpy drop from $(h_1 - h_{3'})$ to $(h_1 - h_3)$ by 10 to 15 percent.

(ii)   Reduction in exit velocity.

(iii)   Increase in the final dryness fraction of steam.

(iv)   Increase in the specific volume of the steam due to increase in the dryness fraction.

(v)   Due to increase in specific volume and reduced velocity, mass flow rate decreases.

$$\text{Enthalpy drop } (h_3 - h_{3'}) = W_f \text{ (work of friction)}$$

So for designing a nozzle, a term is essential to introduce for the consideration of the allowance due to friction, which is called a co-efficient or nozzle efficiency.

### 14.8.1   Nozzle Efficiency

It is an measure of the degree of percentage of the nozzle to deliver the maximum possible amount of kinetic energy for the given pressure drop. It is defined as the ratio of actual enthalpy drop to the isentropic enthalpy drop between the same pressure, i.e.

$$\text{Nozzle efficiency, } \eta_{\text{nozzle}} = \frac{\text{Useful (or actual) enthalpy drop}}{\text{Isentropic enthalpy drop}}$$

$$= \frac{h_1 - h_3}{h_1 - h_{3'}} = \frac{\Delta h_{\text{actual}}}{\Delta h_{\text{isen.}}} \tag{14.22}$$

$\therefore$ The exit velocity $V_3$ considering friction is then given by

$$V_3 = 44.72 \sqrt{(h_1 - h_{3'})\eta_{\text{nozzle}}} \quad \text{as mentioned in article (14.2)} \tag{14.23}$$

Nozzle efficiency depends upon the following factors:

(i) Material of which the nozzle is made, (ii) workmanship in manufacturing the nozzle, (iii) size and shape of nozzle, (iv) angle of nozzle divergence, (v) nature of fluid flowing and its state, (vi) flow velocity, (vii) friction, and (viii) turbulence in hozzle flow passages.

### 14.8.2  Velocity Co-Efficient

Sometimes, velocity co-efficient is frequently used in nozzle flow for accounting the effects of friction. It is defined as the ratio of actual exit velocity to the isentropic velocity obtained for the same pressure drop i.e.

$$\text{Velocity co-efficient, } C_v = \frac{\text{Actual velocity}}{\text{Isentropic velocity}}$$

$$= \sqrt{\frac{\text{Actual enthalpy drop}}{\text{Isentropic enthalpy drop}}}$$

$$= \sqrt{\frac{h_1 - h_3}{h_1 - h_{3'}}} = \sqrt{\text{Nozzle efficiency}}$$

or, $$\qquad\qquad C_v = \sqrt{\eta_{\text{nozzle}}} \qquad\qquad (14.24)$$

The velocity co-efficient depends upon the following factors:

(i) Nozzle dimensions, (ii) roughness of the nozzle wall, and (iii) velocity of flow.

The approximate values of the velocity co-efficient are:

$$0.93 - 0.94 \text{ for roughly cast nozzles}$$
$$0.95 - 0.96 \text{ for machined nozzles}$$
$$0.96 - 0.97 \text{ for smoothly milled nozzles}$$

### 14.8.3  Co-efficient of Discharge

Mass flow is also affected due to friction in nozzle duct and is accounted by the term called the co-efficient of discharge. It is defined as the ratio of actual mass flow rate to that due to isentropic expansion i.e.

$$\text{Co-efficient of discharge, } C_d = \frac{\text{Actual discharge}}{\text{Isentropic discharge}}$$

or, $$\qquad\qquad C_d = \frac{m_{\text{actual}}}{m_{\text{isen.}}} \qquad\qquad (14.25)$$

Isentropic mass flow rate is obtained by the actual pressure drop, if the nozzle is not choked. For choked nozzle $m_{\text{isen.}}$ is obtained by using sonic velocity at the throat section.

## 14.9   LENGTH OF THE NOZZLE

In a convergent-divergent nozzle, the length of the convergent portion should be short to reduce friction. In this short portion, rate of decrease in area is vary rapid; therefore subsonic acceleration can take place very efficiently. This convergent portion has length about 6 mm. Since in this portion the velocity and length both are small, the friction losses are neglected. In divergent portion, due to inertia, high velocity steam has tendency to flow along the axis in the form of a circular jet having area equal to throat area. This tends to pass out through the divergent point without any pressure drop. To avoid this, divergent portion should have sufficient length so that the steam has enough time to occupy the full cross-sectional area provided, thus resulting in the desired drop of pressure and increase in kinetic energy. This necessiates the gradual increase in area. The included angle of the divergent portion is usually kept below about 20°(generally 6°– 12°). From this practice, it is concluded that for a given pressure ratio across a convergent-divergent nozzle, the divergent portion must be longer compared to the convergent portion. That is why, due to higher velocity in this longer portion frictional losses occur more. As mentioned in the Fig. 14.16, length '*l*' of divergent portion of the nozzle is determined by the following equations:

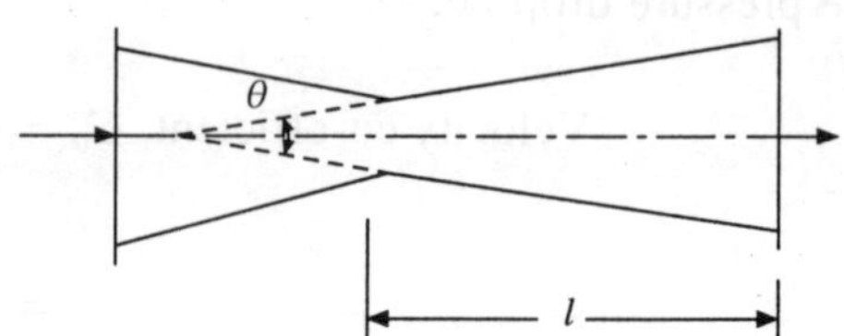

**Fig. 14.16   Length of the nozzle**

$$l = \frac{\text{Diameter at the exit} - \text{Throat diameter}}{2\tan\left(\dfrac{\theta}{2}\right)}$$

or,
$$l = \frac{D_{\text{exit}} - D_{\text{throat}}}{2\tan\left(\dfrac{\theta}{2}\right)} \tag{14.26}$$

where, $\theta$ = cone angle of divergent part or divergence angle

and
$$l = \sqrt{15\,A_t} \tag{14.27}$$

where,
$$A_t = \text{Throat area}$$

Students are advised that in any numerical problem, if co-efficient of discharge and nozzle efficiency are given, then the throat area should be calculated on the basis of co-efficient of discharge and nozzle efficiency should be used to calculate the exit area.

## 14.10  SUPERSATURATED FLOW OR METASTABLE  FLOW THROUGH NOZZLE

The approach to the problem of steam flow through a nozzle will depend upon whether the steam flow can be considered as being in equilibrium or supersaturated. When a superheated vapour expands slowly and isentropically, condensation within the vapour region to form when the saturated vapour line is reached. As the expansion continues below this line into the wet region, condensation proceeds gradually with the progressive decrease of quality ($x$) and increase in the degree of wetness. Point $S$ (Fig. 14.17) represents the point at which condensation within the

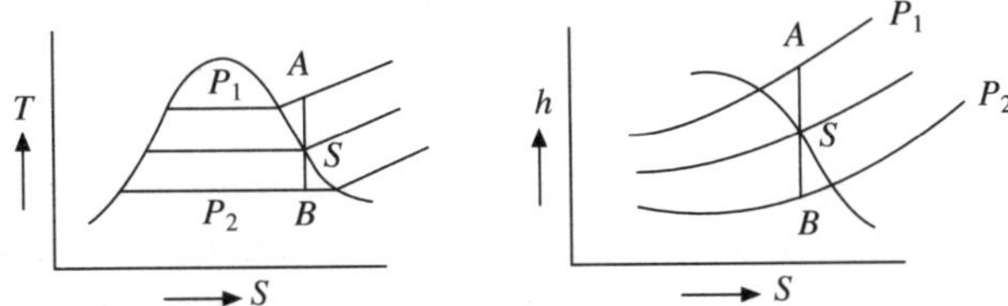

**Fig. 14.17   Expansion of steam under thermal equilibrium conditions**

vapour just begins. There is always a stable mixture of steam and liquid at any point during the expansion. This type of expansion is in thermal equilibrium conditions.

Discharge of initially dry and saturated steam expanding through a nozzle is found to be 1 to 3 percent more than the theoretically calculated discharge which is reverse of naturally expected value. This natural expected value achieved by means of friction should be 3 to 5 percent less than that of theoretical value. This loss in discharge due to friction is compensated more or less than the theoretical value, when the expanding steam is inttially superheated. As we know, during expansion the initially saturated steam starts condensing after it has become dry saturated. The increase in measured discharge to the theoretically calculated discharge is possible when the steam has proceeded through some distance in the nozzle and in a short interval of time. But from practical point of view, the rate of expansion through a nozzle is very high due to high velocity (sometimes sonic and even supersonic) and therefore there is little time for phase change during the passage of the steam of the nozzle. Thus the phenomenon of condensation does take place at the expected rate. As a result of this, equilibrium between the liquid and vapour phase is delayed and the steam continues to expand in a dry state instead of wet. The steam in such a set of conditions, is said to be supersaturated or in metastable state and the phenomenon is called supersaturation. The steam is also called supercooled steam, as its temperature at any pressure is less than the saturation temperature corresponding to the pressure. The flow of supersaturated steam through the nozzle is called supersaturated flow or metastable flow. This effect has been investigated by C.T. Wilson and Martin in 1997. They have given a limit for supersaturated expansion which approximates to point where the dryness fraction is 0.94 at high pressures to 0.97 at low pressures. The locus of these points produces the Wilson line, as shown in Fig. 14.18.

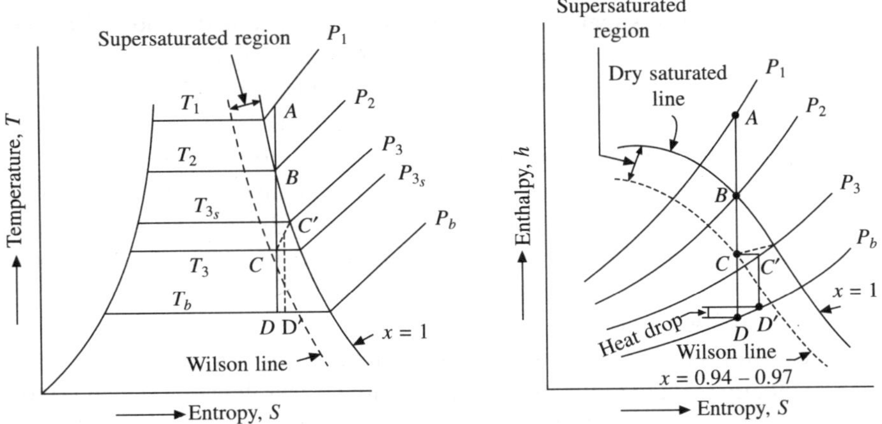

**Fig. 14.18   Supersaturated flow through nozzle**

### Reasons of Supersaturation

(i) Generally condensation starts round tiny dust particles which are always present in commercial steam in sufficient quantities. Below the saturation line, condensation is delayed because steam is free of foreign particles and therefore temperature of steam continues to fall which is known as supersaturation. When supersaturation is attained up to a certain degree, then there is no any effect of the presence of foreign particles and thereafter equilibrium is attained completely and instantaneously.

(ii) In the natural way of condensation, the random kinetic energy of molecules is so high that the time required to come together for condensation is not achieved.  Sufficient heat transfer and setting up of surface tension, etc. get very less time. The flow of molecules may take only 0.001 second or even less to travel from inlet to exit of the nozzle.

### Representation of Supersaturation Phenomenon

The supersaturation phenomenon is shown on the $h - s$ and $T - S$ plane in the Fig. 14.18. The ideal frictionless adiabatic expansion in equilibrium, of superheated steam from supply pressure $p_1$ to back pressure $p_b$ can be represented by a line $AD$ in the Fig. 14.18. At point $A$ steam is superheated at pressure $p_1$. During expansion the steam will be just dry saturated at $B$ and will be wet at $D$. $A$ to $B$ shows the isentropic expansion in theremal equilibrium up to saturation. In natural course at Point $B$ where the pressure is $p_2$, a change of phase must begin to occur i.e. condensation of steam should start at point $B$, but due to above mentioned reasons the equilibrium achieved between the liquid and vapour phase is delayed and the vapour continues to expand in a dry state instead of wet state. It is found that the steam remains in this unnatural superheated state until its density is about 8 times that of the saturated vapour of the same pressure. When this limit is reached the steam will suddenly condense at constant total energy to its normal state.

In between pressure values $p_2$ and $p_3$ supersaturated expansion takes place and is represented by $BC$. Point $C$ can be located by extension of the curvature of constant pressure line $p_3$ from the superheated region which meets the vertical isentropic expansion line and Wilson line. The zone between the Wilson line and dry-saturated line is called the supersaturated zone and the flow through this zone is called the supersaturated flow. At any pressure value between $p_2$ and $p_3$ the actual temperature of the vapour is always lower than the saturation temperature corresponding to that pressure. The difference in this temperature is known as degree of under-cooling. There is another term called degree of supersaturatiion which is defined as the ratio of actual pressure at point C to the saturation pressure corresponding to the actual temperature of the steam at this point. After getting supersaturated at point-C, the steam starts condensing instantaneously to its normal state at the constant pressure $p_3$ and constant enthalpy (horizontal line $C$-$C'$ at pressure $p_3$). Process $C'D'$ is again isentropic expansion from pressure $p_3$ to back pressure $p_b$, under thermal equilibrium. Without supersaturation flow, the process of expansion in the nozzle from inlet to the exit would have been $A$–$B$–$C$–$D$ (i.e isentropic expansion). But due to supersaturation in some part of the nozzle it has finished at $D'$ via. $C'$, where $D$ and $D'$ are on the same pressure line.

### Effects of Supersaturation

(i)   Actual heat drop i.e. $(h_A - h_D)$ is slightly reduced to $(h_A - h_{D'})$.

(ii)  The exit velocity (i.e. at $D'$) for supersaturated flow is less than that (i.e. at $D$) for equilibrium flow.

(iii) There is increase in entropy (i.e. $S_{D'} - S_D$).

(iv) Mass flow rate increases by 3 to 5% because of the increase in density at the throat which is about 8 times of the normal density.

(v) Final dryness fraction is increased from $x_D$ to $x_{D'}$.

(vi) The specific volume at exit with supersaturated flow is considerably less than the specific volume at exit with equilibrium flow.

The two parameters discussed above are used to show the degree of supersaturation. On the basis of location on $T - S$ and $h - s$ plane they are represented as:

$$\text{Degree of supersaturation} = \frac{p_3}{p_{3s}} \tag{14.28}$$

where  $p_3$ = limiting supersaturation pressure

$p_{3s}$ = saturation pressure at temperature $T_3$ (Fig. 14.18).

$$\text{Degree of undercooling} = T_{3s} - T_3 \tag{14.29}$$

where  $T_{3s}$ = saturation temperature at pressure $p_3$.

$T_3$ = supersaturated steam temperature at Point-C which is the limit point of supersaturation.

According to theory given by Binne and Woods, recently, condensation takes place at a sharp rise of pressure about 0.1 bar occured just after Wilson line but not at constant pressure and constant enthalpy. After throat condensation does not occur so that pressure rise will not affect the mass discharged at all.

It has been assumed that supersaturated vapour behaves like superheated steam and therefore the index of expansion is 1.3.

So the equations of supersaturated vapour are:

$$\frac{T_c}{T_A} = \left( \frac{p_3}{p_1} \right)^{\frac{n-1}{n}} = \left( \frac{v_A}{v_c} \right)^{n-1}$$

or,
$$\frac{T_c}{T_A} = \left( \frac{p_3}{p_1} \right)^{\frac{0.3}{1.3}} = \left( \frac{v_A}{v_c} \right)^{0.3} \tag{14.30}$$

$$\text{Specific volume, } v = \frac{0.0023(h - 1943)}{p} \text{ m}^3/\text{kg} \tag{14.31}$$

where,  $p$ = pressure of steam in bar, and

$h$ = Enthalpy or total heat of steam in kJ/kg

Velocity at the throat for the superheated flow is

$$V_c = \sqrt{2\left( \frac{n}{n+1} \right) p_1 v_c} \quad \text{as mentioned in Eq. (14.20),}$$

Here $C$ denotes for throat and $p_1$ for entry condition.

For maximum flow condition:

$$m_{\max} = A_c \sqrt{n \frac{p_1}{v_A} \left(\frac{2}{v_A}\right)^{\frac{(n+1)}{(n-1)}}} \qquad (14.32)$$

Here, $C$ denotes for throat and $A$ for entry condition.
where $p_1$ is in N/m$^2$ and $A_c$ is in m$^2$.

$$\therefore \qquad m_{\max} = 0.66726 \, A_c \sqrt{\frac{p_1}{v_A}} \text{ kg/s as mentioned in Eq. (14.19)}$$

where $p_1$ is in bar and $A_c$ is in m$^2$.
*Note* Problems on supersaturated flow can not be solved by Mollier's diagram unless Wilson line is drawn, since the law of expansion under metastable flow corresponds to expansion of initially superheated steam defined by the equation $pv^{1.3}$ = constant.

## 14.11 STEAM INJECTOR

The principle of steam nozzle is applied to a steam injector. It is a dynamic flow device which utilises the kinetic energy of a steam jet for increasing the pressure and velocity of a quantity of water. Steam injectors supply the feed water to boiler at high pressure by using the steam from boiler itself. The arrangement of injector for feeding the water from a tank to a boiler is shown in Fig. 14.19. Live steam from the boiler enters the convergent steam nozzle where it expands to a high velocity at the expense of its pressure. The steam jet enters the mixing cone at A and imparts its momentum to the incoming water from the feed tank. The feed tank may be above or below

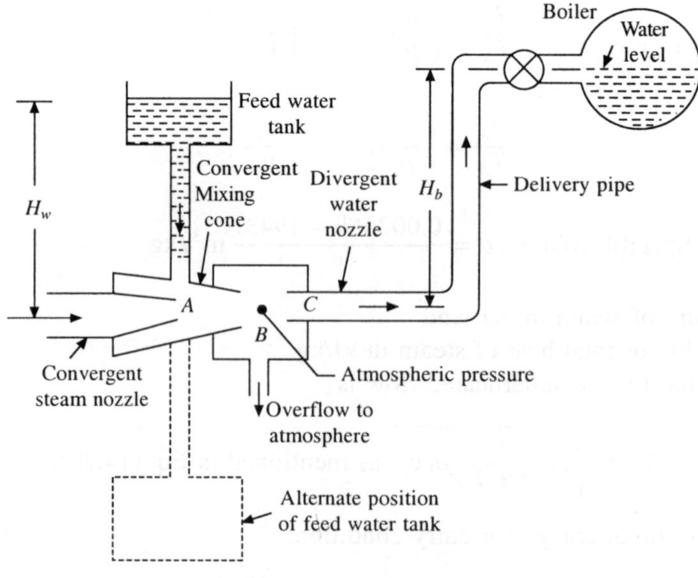


**Fig. 14.19    Steam injector**

the level of the steam injector. The steam condenses due to mixing with water and a vacuum is created in the mixing chamber. The resulting jet formed by the steam and water is at atmospheric pressure having a large velocity at $B$. The mixture then enters the divergent nozzle or diffuser and leaves it with high pressure before being supplied to the delivery pipe at $C$. In the divergent nozzle kinetic energy of mixture is converted into high pressure which is sufficient to overcome the boiler pressure and to lift the water through a height $H_b$. The pressure of water at the exit of the divergent nozzle of the injector is 20–25% higher than boiler pressure in order to overcome all resistances.

A chamber is provided between convergent and divergent nozzle with an outlet through which any excess water may overflow during starting of the injector.

Actually injector operates only due to decrease in volume of steam into lesser volume of condensate and boiler feed.

### Calculations: Mass of water injected

Let  $V_s$ = Velocity of steam leaving the steam nozzle at $A$
  $V_w$ = Velocity of water entering the mixing cone,
  $V_m$ = Velocity of the mixture leaving the mixing cone, and
  $m_w$ = Mass of water entering the mixing cone in kg/kg of steam

Applying the principle of conservation of momentum to the mixing of steam jet and water, per kg of steam supplied to the nozzle, we get

Momentum of steam entering the mixing chamber + Momentum of water entering the mixing chamber = Momentum of mixture leaving the mixing chamber

or, $V_S + m_w V_w = (1 + m_w) V_m$ when water tank level is above injector level

or,
$$V_s - V_m = m_w (V_m - V_w) \therefore m_w = \frac{(V_s - V_m)}{(V_m - V_w)} \qquad (14.33)$$

and $V_s - m_w V_w = (1 + m_w) V_m$ when water tank level is below injector level.

or,
$$V_s - V_m = m_w (V_m + V_w)$$

$\therefore$
$$m_w = \frac{(V_s - V_m)}{(V_m + V_w)} \qquad (14.34)$$

### Velocity of Steam Leaving the Nozzle
The velocity of steam leaving the nozzle ($V_s$) is found out by assuming that the steam expands isentropically from the inlet conditions to the back pressure by using the relation

$$V_S = 44.72 \sqrt{\Delta h}$$

### Velocity of water entering the mixing cone
This is given by the equation

$$V_w = \sqrt{2g H_w}$$

where $H_w$ = Height of water level in the feed tank.

**Velocity of mixture leaving the mixing cone nozzle at $B$**

Let,

$p_1$ = pressure of steam in the boiler in bar

$p_m$ = pressure of the mixture leaving the mixing cone at $B$ in bar, i.e. atmospheric press. = 1.01325 bar

$\rho$ = Density of the mixture at $B$ in kg/m$^3$, and

$v_m$ = Velocity of the mixture at $B$

$\therefore$ Total energy per kg of water at $B$, in kJ (Taking centre line of injector as datum)

$$= \frac{p_m}{\rho} + \frac{V_m^2}{2000} + 0 = \frac{101.325}{1000} + \frac{V_m^2}{2000} \qquad \text{(i)}$$

This energy must be sufficient to lift the water through a height $H_b$ meters at the delivery end and inject it into the boiler. The final pressure head on leaving at $C$, must be somewhat greater than the height $H_b$ plus the boiler pressure. This final delivery pressure $p_d$ may be taken 20% greater than boiler pressure $p_1$.   $\therefore$   $p_d = 1.2 p_1$.

Total energy (kJ) per kg of water at $B$

$$= \frac{p_d}{\rho} + \frac{V_d^2}{2000} + \frac{gH_b}{1000} = \frac{1.2\, p_1 \times 10^2}{1000} + \frac{V_d^2}{2000} + \frac{gH_b}{1000} \qquad \text{(ii)}$$

Equating equations (i) and (ii),

$$\frac{101.325}{100} + \frac{V_m^2}{2000} = \frac{1.2\, p_1 \times 10^2}{1000} + \frac{V_d^2}{2000} + \frac{gH_b}{1000}$$

or, $$0.101325 + \frac{V_m^2}{2000} = 0.12\, p_1 + \frac{V_d^2}{2000} + 0.001\, g\, \text{H}_b$$

Here,    $V_m$ and $V_d$ are in m/s

$H_b$ in metre

$p_1$ in bar

or, $$0.101325 + \frac{V_m^2}{2000} = 0.12\, p_1 + \frac{V_d^2}{2000} + 0.00981 \text{H}_b$$

Aim of steam injector to supply feed water at higher pressure but velocity very less. i.e. $\dfrac{V_d^2}{2000}$ is more lesser, and may be neglected. Furthermore, in above equation first and last term may be neglected as being smaller.

$\therefore$ $$\frac{V_m^2}{2000} \simeq 0.12\, p_1$$

$\therefore$ $$V_m \simeq \sqrt{240\, p_1} \ \text{m/s} \qquad \text{(14.35)}$$

*Note* Delivery pressure has been assumed 20% more than boiler pressure. i.e. $p_d = 1.2\, p_1$.

**Nozzle areas**

Let

$A_a$ = Area of steam nozzle at $A$,

$A_b$ = Area of combining nozzle or mixing cone at $B$,

$V_s$ = Velocity of steam leaving the mixing cone,

$v_a$ = Specific volume of steam after expansion in nozzle at $A$,

$m$ = mass of water required to be delivered, kg/s

and $m_w$ = mass of water entering the mixing cone, kg/kg of steam

∴  Mass of steam supplied

$$m_s = \frac{A_a V_s}{v_a} = \frac{m}{m_w}$$

or,

$$A_a = \frac{m}{m_w} \cdot \frac{v_a}{V_s}$$

Also,

$$m_s + m = A_b V_m \rho$$

or,

$$A_b = \frac{m_s + m}{1000 V_m} = \frac{m\left(1 + \dfrac{1}{m_w}\right)}{1000 V_m}, \quad \left[\because m_s = \frac{m}{m_w}\right] \tag{14.36}$$

**Heat Balance per kg of Steam**

Let

$H$ = Total heat of steam entering the injector in kJ,

$t_w$ = Temperature of water in feed tank in 0°C,

$h_{fw}$ = Sensible heat of water supplied to the injector, corresponding to the temperature of $t_w$ in kJ/kg,

$t_m$ = Temperature of water leaving the mixing cone at $B$ in °C, and

$h_{fm}$ = Sensible heat of mixture (water) leaving the mixing cone at $B$, corresponding to a temperature of $t_m$ in kJ/kg.

Heat supplied by steam + Heat supplied by water ± kinetic energy of water depending upon the position of the location of feed tank = Heat gained by mixture at $B$ + Kinetic energy of mixture at $B$.

$$H + m_w h_{fw} \pm \frac{m_w V_w^2}{2000} = (1 + m_w) h_{fm} + \frac{(1 + m_w) V_m^2}{2000} \tag{14.37}$$

From this equation the value of $h_{fm}$ can be determined which is equal to the product of $t_m$ and specific heat of water i.e. 4.187 kJ/kgk. In this equation plus sign is used when feed tank is located above the level of injector, while negative sign is used when it is below the level of injector.

**Solved Problems**

**14.1**  Dry and saturated steam is expanded in a steam nozzle from 10 bar to 0.1 bar. Using steam table, calculate: (a) dryness fraction of steam at exit, (b) heat drop, (c) velocity of steam at exit from the nozzle when initial velocity is 135 m/s.

*Soln*

(a) Specific enthalpy of dry saturated steam and 10 bar i.e.

$$hg = h_1 = 2776.2 \text{ kJ/kg}$$

Expansion is assumed to be isentropic

∴    entropy before expansion = entropy after expansion

$$6.583 = 0.649 + x \times 7.502$$

∴                                  $x = 0.79$   *Ans.*

(b) At the exit, steam is wet. Specific enthalpy at the exit

$$h_2 = h_f + xhfg \text{ at } 0.1 \text{ bar}$$

$$= 191.8 + 0.79 \times 2392.9$$

$$= 2082.19$$

∴    Heat drop = enthalpy drop = $h_1 - h_2$

$$= 2776.2 - 2082.19$$

$$= 694.01 \text{ kJ/kg}   \textit{Ans.}$$

(c) Using steady flow energy equation,

$$h_1 + \frac{V_1^2}{2000} = h_2 + \frac{V_2^2}{2000}   [\because Q = 0, w = 0]$$

or,           $$\frac{V_2^2 - V_1^2}{2000} = h_1 - h_2$$

∴                $$V_2 = \sqrt{2000\,(h_1 - h_2) + V_1^2}$$

$$= \sqrt{2000\,(694.01) + 135^2} = 1185.85 \text{ m/s}   \textit{Ans.}$$

**14.2**  A steam turbine uses convergent nozzles. The expansion of steam is isentropic from 6 bar, 250°C to 4 bar at the exit. Total area at the exit is 25 cm². Estimate the rate of flow by neglecting the velocity of approach. Find the condition of steam at exit. If the co-efficient of discharge of nozzles were 0.95, what should be the actual exit area of the nozzles assuming that the velocity, mass flow rate and specific valume are the same. Use the steam table only.

*Soln.*

Refer to Fig. 14.20.

From superheated steam table at 6 bar and 250°C. $h_1 = 2957.6$ kJ/kg

and                                    $s_1 = 7.183$ kJ/kgK

∵ Process 1–2 is isentropic i.e. $s_1 = s_2$

$$S_2 = S_g + C_{p_s} \ln \frac{T_{\text{sup}}}{T_s} \text{ at 4 bar}$$

$$7.183 = 6.894 + 2.1 \ln \frac{T_{\text{sup}}}{(143.6 + 273)}   [\because \text{ sp. heat of superheated steam } c_{p_s} = 2.1 \text{ kj/kgK}]$$

∴                          $$T_{\text{sup}} = 478.06 \, k = 205.06°C$$

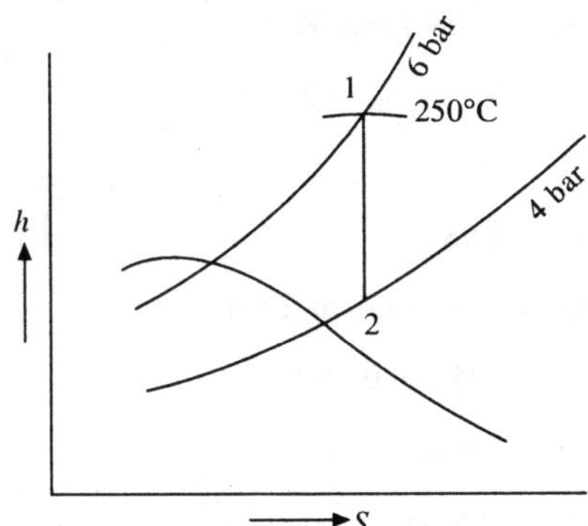


**Fig. 14.20**

i.e. condition of steam at exit = 205.06°C, superheated.   *Ans.*

$$\therefore \qquad h_2 = h_g + c_{p_s}\,(T_{\text{sup}} - T_s)$$

$$= 2737.6 + 2.1\{478.06 - (143.6 + 273)]$$

$$= 2867.8 \text{ kJ/kg}$$

Neglecting the velocity of approach $V_1 = 0$

$\therefore$ Velocity at the exit, $\quad V_2 = 44.72\sqrt{(h_1 - h_2)}$

$$= 44.72\sqrt{2957.6 - 2867.8}$$

$$= 423.77 \text{ m/s}$$

Specific volume of superheated steam at 4 bar

$$v_2 = v_{\text{sup}} = \frac{T_{\text{sup}}}{T_s} \times v_g$$

$$= \frac{478.06}{416.6} \times 0.46220$$

$$= 0.5303 \text{ m}^3/\text{kg}$$

From continuity equation,

Rate of flow, $\qquad\qquad m = \dfrac{A_2\,V_2}{v_2} = \dfrac{25 \times 10^{-4} \times 423.77}{0.5303}$

$$= 1.997 \text{ kg/s}$$

Since the discharge co-efficient is 0.95, area at the exit $= \dfrac{25}{0.95} = 26.315 \text{ cm}^2$   *Ans.*

**14.3**  In a nozzle steam expands from 5 bar to 2 bar. The initial steam velocity is 120 m/s and the initial temperature is 200°C. The nozzle efficiency is 85%. Determine the exit velocity.

*Soln -*  Refer to Fig. 14.21
From superheated steam table, $h_1 = 2855.1$ kJ/kg

and
$$S_1 = 7.059 \text{ kJ/kg K} = S_{2'}$$
$$S_{2'} = S_f + x_{2'} S_{fg} \text{ at 2 bar}$$
$$7.059 = 1.530 + x_{2'} \times 5.597$$
$$\therefore \quad x_{2'} = 0.98$$
$$\therefore \quad h_{2'} = h_f + x_{2'}\, hfg \text{ at 2 bar}$$
$$= 504.7 + 0.98 \times 2201.6$$
$$= 2662.26 \text{ kJ/kg}$$

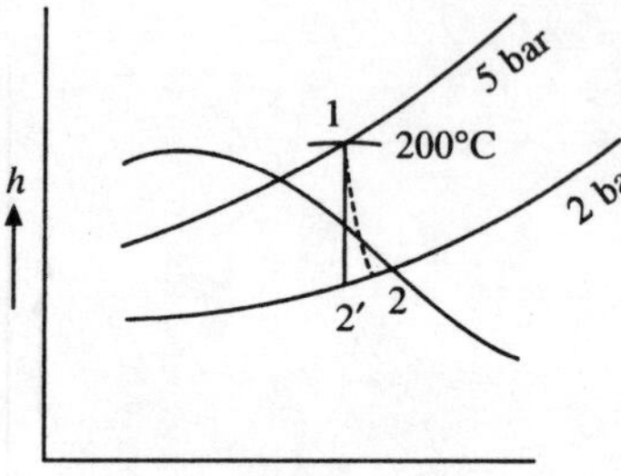

**Fig. 14.21**

Exit velocity
$$V_2 = \sqrt{2000\,(h_1 - h_{2'})\,\eta_{\text{nozzle}} + V_1^2}$$
$$= \sqrt{2000\,(2855.1 - 2662.26) \times 0.85 + 120^2}$$
$$= 585 \text{ m/s} \quad Ans.$$

**14.4** Calculate the throat and exit diameter of a convergent-divergent nozzle which will discharge 820 kg of steam/hr from a pressure of 8 bar superheated to 220°C into a chamber having a pressure of 1.05 bar. Friction loss in the diverging part of the nozzle may be taken as 0.15 of the total enthalpy drop.

*Soln.* Refer to Fig. 14.22

Discharge rate
$$m = \frac{820}{3600}$$
$$= 0.2278 \text{ kg/s}$$

$\because$ steam is initially superheated i.e. $n = 1.3$

$$\therefore \quad \frac{p_2}{p_1} = 0.545 \quad \text{(from Eq. 14.17)}$$

$$\therefore \quad p_2 = 8 \times 0.545 = 4.36 \text{ bar}$$

From Mollier chart, at 8 bar and 220°C
$$h_1 = 2889 \text{ kJ/kg}$$

$h_2 = 2763$ kJ/kg, $t_2 = 155$°C and $v_2 = 0.44$ m$^3$/kg at 4.36 bar

$$h_{3'} = 2575 \text{ kJ/kg at 1.05 bar}$$

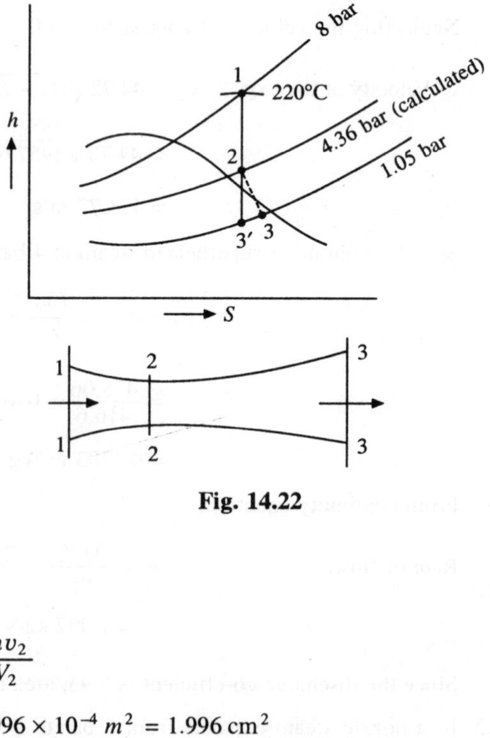

Steam velocity at the throat

**Fig. 14.22**

$$V_2 = 44.72 \sqrt{(2889 - 2763}) = 501.98 \text{ m/s}$$

Steam continuity equation,

$$m = \frac{A_2\, V_2}{v_2} \quad \text{or,} \quad A_2 = \frac{m v_2}{V_2}$$

or
$$A_2 = \frac{0.2278 \times 0.44}{501.98} = 1.996 \times 10^{-4}\, m^2 = 1.996 \text{ cm}^2$$

$$\frac{\pi}{4} D_2^2 = 1.996 \quad \therefore \text{ Diameter at throat } D_2 = \sqrt{\frac{4 \times 1.996}{\pi}}$$

$$= 1.594 \text{ cm} \quad Ans.$$

Eq. (14.17) has been used for maximum discharge hence throat area can also be calculated by using Eq. (14.19) as

$$m = 0.66726\, A_2 \sqrt{\frac{p_1}{v_1}}$$

$\therefore$
$$A_2 = \frac{0.2278}{0.66726} \sqrt{\frac{v_1}{p_1}}$$

$$= \frac{0.2278}{0.66726} \sqrt{\frac{0.27376}{8 \times 10^5}} \quad [\because v_1 = 0.27376 \text{ m}^3/\text{kg}]$$

$$= 1.996 \times 10^{-4} \text{ m}^2$$

i.e. $\qquad\qquad D_2 = 1.594$ cm $\quad$ *Ans.*

Now, total isentropic enthalpy drop $= (h_1 - h_{3'})$

$$\text{Frictional loss} = (h_3 - h_{3'}) = 0.15(h_1 - h_{3'})$$

or, $\qquad\qquad h_3 - 2575 = 0.15(2889 - 2575)$

$\therefore \qquad\qquad h_3 = 2622.1$ kJ/kg

Selecting this value on Mollier diagram and moving horizontally to the pressure curve 1.05 bar, locate Point 3. From Mollier diagram, $v_3 = 1.185$ m$^3$/kg and $x_3 = 0.965$

$$V_3 = 44.72 \sqrt{(h_1 - h_3)}$$

$$= 44.72 \sqrt{2889 - 2622.1} = 730.61 \text{ m/s}$$

Now, $\qquad\qquad m = \frac{A_3 V_3}{v_3}$

$\therefore \qquad\qquad A_3 = \frac{m v_3}{V_3} = \frac{0.2278 \times 0.965 \times 1.185}{730.61} = 3.565 \times 10^{-4} \, m^2$

$$\frac{\pi}{4} D_3^2 = 3.565 \times 10^{-4} = 3.565 \, \text{cm}^2$$

$\therefore \quad$ Diameter at the exit, $D_3 = \sqrt{\dfrac{4 \times 3.565}{\pi}}$

$$= 2.13 \text{ cm} \quad \textit{Ans.}$$

**14.5** In a convergent-divergent nozzle the steam enters at 15 bar, 300°C and leaves it at a pressure of 2 bar. The inlet velocity to the nozzle is 150 m/s. Find the required throat and exit areas for mass flow rate of 1 kg/s. Assume nozzle efficiency to be 90%. Assume $C_{p_s} = 2.4$ kJ/kg K.

*Soln.* Refer to Fig. 14.23.

Since value of specific heat $C_{p_s}$ is given i.e. it is required to use steam table and not Mollier diagram for getting the values of thermodynamic properties.

From steam table at 15 bar and 300°C.

$$h_1 = 3038.9 \text{ kJ/kg}, \quad S_1 = 6.921 \text{ kJ/kg K}$$

For initially superheated steam, $\dfrac{p_2}{p_1} = 0.545$

$\therefore \quad p_2 = 15 \times 0.545 = 8.17$ bar and saturation temperature at this pressure
$T_s = 171.25°C$ (From steam table)

$\because \quad S_1 = S_2 \quad$ and $\quad S_2 = S_g + C_{p_s} \ln \dfrac{T_{\text{sup}}}{T_s}$

or, $6.921 = 6.631 + 2.4 \ln \dfrac{T_{\text{sup}}}{171.25 + 273}$

$\therefore T_{\text{sup}} = 501.3\text{K} = 501.3$ K

$\therefore \quad h_2 = hg + C_{P_s}(T_{\text{sup}} - T_s)$

$\qquad = 2768.35 + 2.4\{501.3 - (171.25 + 273)\}$

$\qquad = 2905.24$ kJ/kg

$V_2 = \sqrt{2000(h_1 - h_2) + V_1^2}$

or, $V_2 = \sqrt{2000(3038.9 - 2905.24) + 150^2}$

$\qquad = 538.33$ m/s

Specific volume $v_2 = v_g \dfrac{T_{\text{sup}}}{T_s}$

$\qquad\qquad = 0.2355 \dfrac{501.3}{(171.25 + 273)}$

$\qquad\qquad = 0.2657$ m³/kg

$\therefore \qquad m = \dfrac{A_2 V_2}{v_2} \quad$ or, $\quad A_2 = \dfrac{1 \times 0.2657}{538.33} = 4.93 \times 10^{-4}\,\text{m}^2 = 4.93\,\text{cm}^2$ (throat area)   *Ans.*

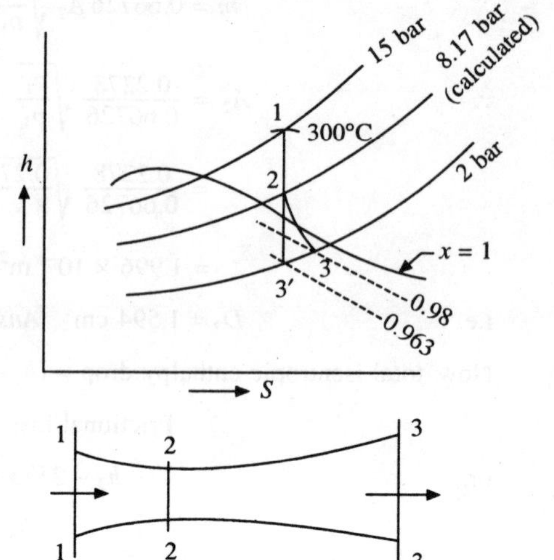


**Fig. 14.23**

Now, $\qquad\qquad\qquad S_1 = S_2 = S_{3'} = 6.921$ kJ/kg K

$\qquad\qquad\qquad\qquad S_{3'} = S_f + x_{3'}\, S_{fg}$ at 2 bar

or, $\qquad\qquad\qquad 6.921 = 1.530 + x_{3'} \times 5.597$

$\therefore \qquad\qquad\qquad\qquad x_{3'} = 0.963$

$\therefore \qquad\qquad\qquad\qquad h_{3'} = h_f + x_{3'}\, h_{fg}$ at 2 bar

$\qquad\qquad\qquad\qquad\qquad = 504.7 + 0.963 \times 2201.6$

$\qquad\qquad\qquad\qquad\qquad = 2624.84$ kJ/kg

Nozzle efficiency, $\qquad\qquad \eta_{\text{nozzle}} = \dfrac{h_1 - h_3}{h_1 - h_{3'}}$

or, $\qquad\qquad\qquad 0.90 = \dfrac{3038.9 - h_3}{3038.9 - 2624.84}$

$\therefore \qquad\qquad\qquad\qquad h_3 = 2666.24$ kJ/kg

$\qquad\qquad\qquad\qquad V_3 = \sqrt{2000(h_1 - h_3) + V_1^2}$

$\qquad\qquad\qquad\qquad\quad = \sqrt{2000(3038.9 - 2666.24) + 150^2}$

$\qquad\qquad\qquad\qquad\quad = 876.25$ m/s

Now,
$$h_3 = h_f + x_3\, hfg \text{ at 2 bar}$$
$$2666.24 = 504.7 + x_3 \times 2201.6$$

$\therefore$
$$x_3 = 0.98$$

$\therefore$
$$v_3 = x_3\, v_g \text{ at 2 bar}$$
$$= 0.98 \times 0.88540 = 0.867 \text{ m}^3/\text{kg}$$

$$m = \frac{A_3 V_3}{v_3} \quad \text{or} \quad A_3 = \frac{1 \times 0.867}{876.25} = 9.89 \times 10^{-4} \text{ m}^2$$
$$= 9.89 \text{ cm}^2 \text{ (exit area)} \quad Ans.$$

**14.6** The convergent-divergent nozzles are provided to an impulse turbine producing 110 kW with steam consumption rate of 8.16 kg/kW hr. 0.95 dry steam enters the nozzles at pressure of 10 bar and leaves at 0.12 bar. If the throat diameter is 5 mm, determine the number of nozzles required and the exit diameter of the nozzle assuming the exit velocity as 1072 m/s. Also calculate the percentage of total isentropic enthalpy drop due to friction.

*Soln.* Refer to Fig. 14.24.

Steam supplied is wet with $x = 0.95$. Using Zenuer's equation for getting the value of index of expansion as

$$n = 1.035 + 0.1x$$
$$= 1.035 + 0.1 \times 0.95$$
$$= 1.13$$

For maximum discharge condition

$$\frac{p_2}{p_1} = \left(\frac{2}{n+1}\right)^{\frac{n}{n-1}} = \left(\frac{2}{1.13+1}\right)^{\frac{1.13}{1.13-1}} = 0.578$$

$\therefore$
$$p_2 = 5.78 \text{ bar}$$

Total mass flow rate passing through number of nozzles into the turbine

$$m_{\text{total}} = \frac{110 \times 8.16}{3600} = 0.2493 \text{ kg/s}$$

From Mollier diagram, at 10 bar and 0.95 dry curve,

$$h_1 = 2680 \text{ kJ/kg}$$

$h_2 = 2575$ kJ/kg and $v_2 = 0.35$ m$^3$/kg at 5.78 bar
Velocity at the throat

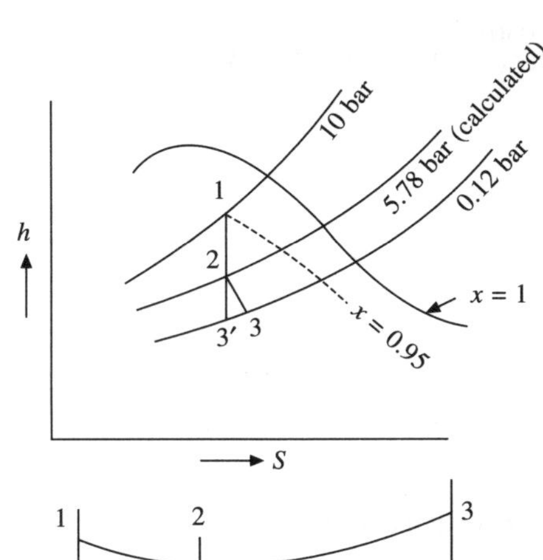


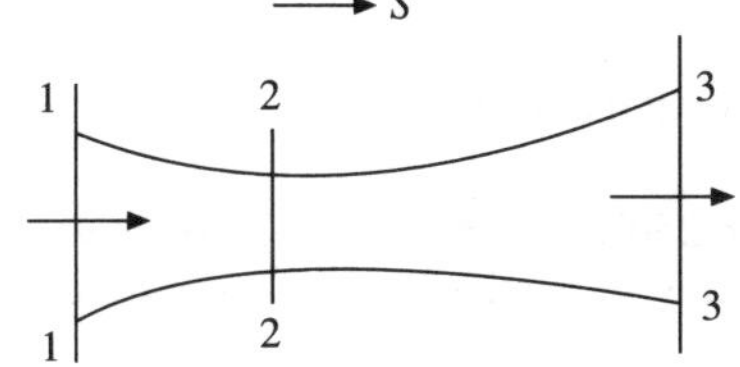


**Fig. 14.24**

$$V_2 = 44.72\sqrt{(h_1 - h_2)}$$
$$= 44.72\sqrt{(2680 - 2575)} = 458.24 \text{ m/s}$$

$\therefore$ mass flow rate through a single nozzle

$$m = \frac{A_2 V_2}{v_2} = \frac{\frac{\pi}{4}(5 \times 10^{-3})^2 \times 458.24}{0.35} = 0.0257 \text{ kg/s}$$

$$\therefore \quad \text{Number of nozzle required} = \frac{m_{total}}{m} = \frac{0.2493}{0.0257}$$

$$= 9.7 \simeq 10 \text{ nozzles} \quad Ans.$$

From Mollier diagram

$$h_{3'} = 2050 \text{ kJ/kg}$$

Velocity at the throat

$$V_3 = 44.72 \sqrt{(h_1 - h_3)}$$

or,
$$1072 = 44.72 \sqrt{(2680 - h_3)}$$

$$\therefore \qquad h_3 = 2105.37 \text{ kJ/kg}$$

Select this value on ordinate of Mollier diagram and move horizontally to pressure curve of 0.12 bar and locate point 3.

$$\therefore \qquad v_3 = 9.8 \text{ m}^3/\text{kg}$$

Percentage reduction in isentropic enthalpy drop

$$= \frac{(h_3 - h_{3'})}{(h_1 - h_{3'})} \times 100 = \frac{(2105.37 - 2050)}{(2680 - 2050)} \times 100 = 8.78 \% \quad Ans.$$

$$m = \frac{A_3 V_3}{v_3}$$

or,
$$0.0257 = \frac{A_3 \times 1072}{9.8}$$

$$\therefore \qquad A_3 = 2.349 \times 10^{-4} \text{ m}^2 = 2.349 \text{ cm}^2$$

$$\frac{\pi}{4} D_3^2 = 2.349$$

$$\therefore \qquad D_3 = \sqrt{\frac{4 \times 2.349}{\pi}} = 1.729 \text{ cm} = 17.29 \text{ mm (exit diameter)} \quad Ans.$$

**14.7** In a connvergent-divergent nozzle steam expands from 10 bar, 250°C to 2 bar with a mass flow rate of 0.2 kg/s. Velocities at throat and the exit are 510 m/s and 700 m/s respectively. Expansion up to throat is purely isentropic. Calculate the following
(a) throat and exit diameter
(b) the nozzle efficiency
(c) quality of steam leaving the nozzle.
Does the nozzle fulfil the condition of maximum discharge?

*Soln.* Refer to Fig. 14.25.
(a) For superheated steam

$$\frac{p_2}{p_1} = 0.545$$

$$\therefore \qquad p_2 = p_1 \times 0.545 = 10 \times 0.545 = 5.45 \text{ bar}$$

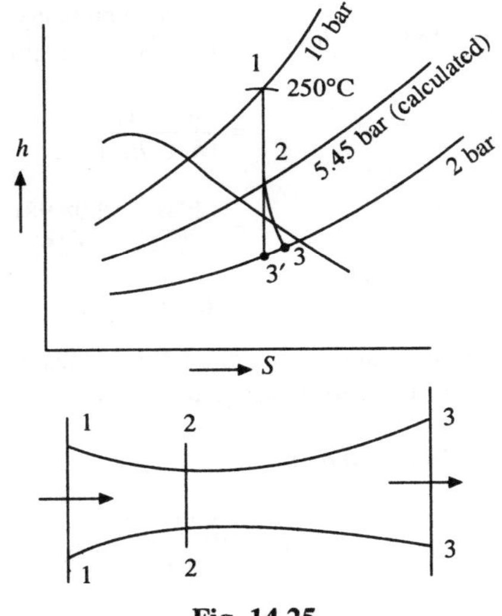



**Fig. 14.25**

From Mollier diagram, at 10 bar and 250°C $h_1 = 2944$ kJ/kg

$$v_1 = 0.23 \text{ m}^3/\text{kg}$$

$$v_2 = 0.35 \text{ m}^3/\text{kg}$$

$$h_2 = 2800 \text{ kJ/kg at 5.45 bar}$$

$$m = \frac{A_2 \, V_2}{v_2} \quad \therefore \quad A_2 = \frac{0.2 \times 0.35}{510} = 1.37 \times 10^{-4} \text{ m}^2 = 1.37 \text{ cm}^2$$

$$\frac{\pi}{4} D_2^2 = 1.37 \quad \therefore \quad D_2 = \sqrt{\frac{4 \times 1.37}{\pi}} = 1.32 \text{ cm (throat diameter)} \quad Ans.$$

From Mollier diagram

$$h_{3'} = 2625 \text{ kJ/kg at 2 bar}$$

$$V_3 = 44.72 \sqrt{(h_1 - h_3)}$$

or,
$$700 = 44.72 \sqrt{(2944 - h_3)}$$

$\therefore$
$$h_3 = 2698.98 \text{ kJ/kg}$$

Select this value on ordinate of Mollier diagram and move horizontally to pressure curve of 2 bar and locate point 3. $\therefore v_3 = 0.9$ m³/kg

$$m = \frac{A_3 \, V_3}{v_3} \quad \therefore A_3 = \frac{0.2 \times 0.9}{700} = 2.57 \times 10^{-4} \text{ m}^2 = 2.57 \text{ cm}^2$$

$$\frac{\pi}{4} D_3^2 = 2.57 \quad \therefore D_3 = \sqrt{\frac{4 \times 2.57}{\pi}} = 1.8 \text{ cm (Exit diameter)} \quad Ans.$$

(b)    Nozzle efficiency, $\eta_{\text{nozzle}} = \dfrac{\text{Actual enthalpy drop}}{\text{Isentropic enthalpy drop}}$

$$= \frac{(h_1 - h_3)}{(h_1 - h_{3'})}$$

$$= \frac{(2944 - 2698.98)}{(2944 - 2625)} \times 100$$

$$= 76.80\% \quad Ans.$$

(c) After locating the point 3 on 2 bar curve line, quality of steam at the exit = 0.995   *Ans.*
For fulfilment of maximum discharge for superheated steam

$$m = 0.66726 \times A_2 \sqrt{p_1/v_1}$$

$$= 0.66726 \times 1.37 \times 10^{-4} \sqrt{\frac{10 \times 10^5}{0.23}} = 0.1906 \text{ kg/s}$$

$\because$   $0.1906 \leq 0.2$ (given value)

$\therefore$   Nozzle fulfils the condition of maximum discharge.

**14.8** The nozzles of a De Laval steam turbine are supplied with dry saturated steam at a pressure of 7 bar
absolute. The pressure at outlet is 1 bar. The
turbine has two nozzles with a throat diameter
of 3 mm. Assuming that the nozzle efficiency
is 95% and that of the turbine rotor 30%, find
the quantity of steam used per hour and power
developed.

*Soln.*   Refer to Fig. 14.26
Assuming maximum discharge for dry saturated
steam

$$\frac{p_2}{p_1} = 0.578 \therefore p_2 = 4.0 \text{ bar}$$

From Mollier diagram,

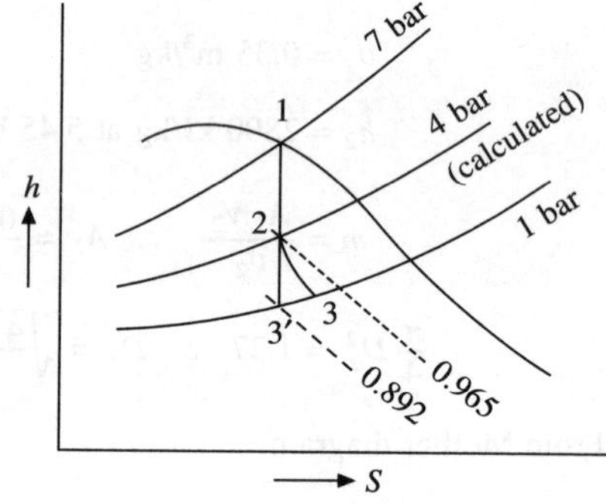


**Fig. 14.26**

$h_1 = 2765$ kJ/kg,

$h_2 = 2660$ kJ/kg at 4 bar, $x_2 = 0.965$, $v_2 = 0.48$ m$^3$/kg

$$h_{3'} = 2440 \text{ kJ/kg at 1 bar}, \ x_{3'} = 0.892, \ v_{3'} = 1.6 \text{ m}^3/\text{kg}$$

$$V_2 = 44.72 \sqrt{(h_1 - h_2)} = 44.72 \sqrt{(2765 - 2660)}$$

$$= 458.24 \text{ m/s}$$

$$V_3 = 44.72 \sqrt{(h_1 - h_{3'}) \times \eta_{\text{nozzle}}}$$

$$= 785.78 \text{ m/s}$$

$$A_2 = \frac{\pi}{4} D_2^2 = \frac{\pi}{4} (3 \times 10^{-3})^2 = 7.068 \times 10^{-6} \text{ m}^2$$

$\therefore$ Total mass of steam used per second, $m = \dfrac{A_2\, V_2}{v_2}$

or, $\qquad m = \dfrac{7.068 \times 10^{-6} \times 458.24}{0.965 \times 0.48} = 6.992 \times 10^{-3} \times 2$

or, $\qquad m = 6.992 \times 10^{-3} \times 3600 \times 2 = 50.34$ kg/hr. *Ans.*

K. Energy supplied by one mass of steam

$$= \frac{V_3^2}{2} = \frac{(785.78)^2}{2}$$

Energy received by the wheel of De-Laval turbine by total, mass, per minute

$$= \frac{50.34}{60} \times \frac{(785.78)^2}{2 \times 1000} = 259.02 \text{ kJ}$$

$\therefore \qquad$ Useful power developed $= \dfrac{259.02 \times \text{Rotal efficiency}}{60}$ kW

$$= \frac{259.02 \times 0.30}{60} = 1.295 \text{ kW} \qquad Ans.$$

**14.9** Nozzle is supplied with steam at 7 bar and 300°C. Find the temperature and velocity at the throat. If the diverging portion is 50 mm long and the throat diameter 5 mm, determine the angle of the cone so that the steam may leave the nozzle at 1 bar. Assume a friction loss of 10 percent of the heat drop in the diverging part.

*Soln.* Refer to Fig. 14.27

Steam is superheated initially. For maximum discharge,

$$\frac{p_2}{p_1} = 0.545$$

$\therefore \qquad p_2 = 7 \times 0.545 = 3.815$ bar

From Mollier diagram

$h_1 = 3060$ kJ/kg at 7 bar and 300°C.

$h_2 = 2910$ kJ/kg at 3.815 bar

$v_2 = 0.6$ m$^3$/kg

$h_{3'} = 2660$ kJ/kg at 1 bar

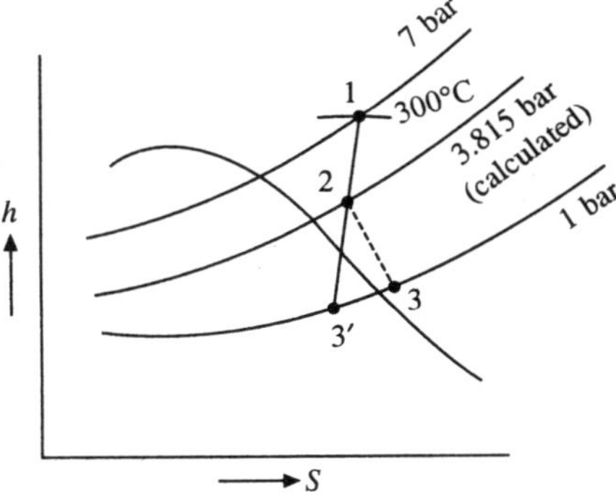


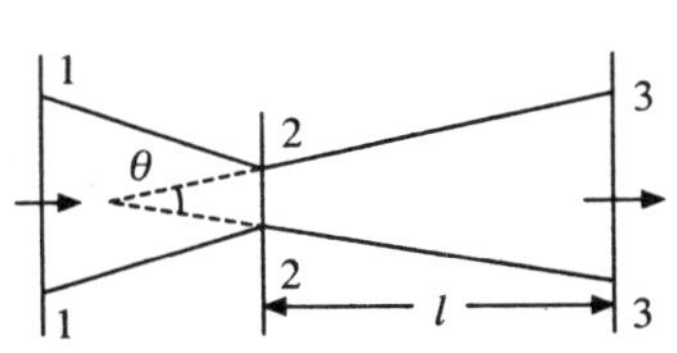


**Fig. 14.27**

Velocity at the throat, $V_2 = 44.72 \sqrt{(h_1 - h_2)}$

$$= 44.72 \sqrt{(3060 - 2910)} = 547.7 \text{ m/s}$$

mass flow rate $m = \dfrac{A_2\, V_2}{v_2}$

$$= \frac{\pi}{4} \times \frac{(5 \times 10^{-3})^2 \times 547.7}{0.6} = 0.0179 \text{ kg/s}$$

Velocity at the exit

$$V_3 = 44.72 \sqrt{(h_1 - h_{3'}) \times \eta_{\text{nozzle}}}$$

$$= 44.72 \sqrt{(h_1 - h_{3'}) \times (1 - 0.10)}$$

$$= 44.72 \sqrt{(3060 - 2660) \times 0.90}$$

$$= 848.5 \text{ m/s}$$

$$(h_3 - h_{3'}) = 0.1(h_1 - h_{3'}) \quad [\therefore 10\% = \text{frictional loss}]$$

or, $$(h_3 - 2660) = 0.1 (3060 - 2660)$$

$\therefore$ $$h_3 = 2700 \text{ kJ/kg}$$

Select this value on ordinate of Mollier diagram and move horizontally to pressure curve of 2 bar and locate point 3 which is in superheated zone due to friction.

$\therefore$ $$v_3 = 1.7 \text{ m}^3/\text{kg}$$

mass flow rate $$m = \frac{A_3\,V_3}{v_3}$$

$\therefore$ $$A_3 = \frac{0.0179 \times 1.7}{848.5} = 3.586 \times 10^{-5}\,\text{m}^2 = 35.86 \text{ mm}^2$$

$$\frac{\pi}{4} D_3^2 = 0.3586 \text{ mm}^2$$

$\therefore$ $$D_3 = \sqrt{\frac{4 \times 35.86}{\pi}} = 6.75 \text{ mm (exit diameter)}$$

Length of the diverging part, by using Eq. (14.26)

$$l = \frac{D_{\text{exit}} - D_{\text{throat}}}{2 \tan \dfrac{\theta}{2}}$$

or, $$50 = \frac{6.75 - 5}{2 \tan \dfrac{\theta}{2}}, \quad \tan \frac{\theta}{2} = 0.0175$$

or, $$\theta = 2.005° \text{ (angle of the cone)} \quad Ans.$$

**14.10** A nozzle is to supply steam at the rate of 1 kg/s from inlet conditions of 10 bar, dry saturated to exit condition at 1 bar pressure. The efficiency of the nozzle for the convergent portion is 95% and to that of the divergent portion is 90%. Determine;
(a) the required throat and exit diameters
(b) the length of the nozzle if the divergent cone angle of the nozzle is 14°, and
(c) the power in kW corresponding to exit velocity of the steam.

*Soln.* Refer to Fig. 14.28
(a) For maximum discharge condition of initially dry saturated steam

$$\frac{p_2}{p_1} = 0.578$$

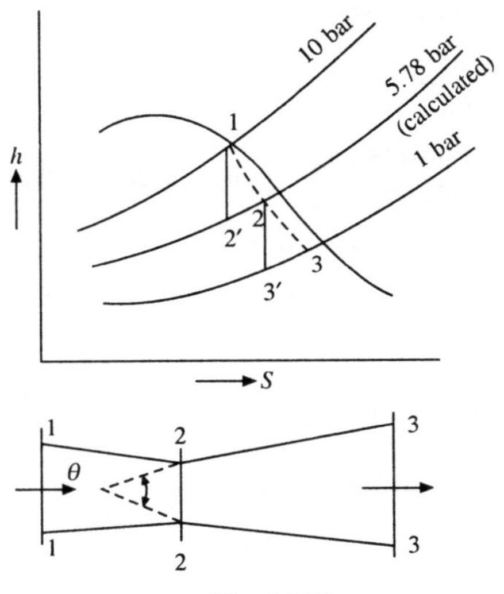


**Fig. 14.28**

$$\therefore \qquad p_2 = p_1 \times 0.578$$

$$= 10 \times 0.578 = 5.78 \text{ bar}$$

From Mollier diagram

$$h_1 = 2780 \text{ kJ/kg at 10 bar}$$

$$h_{2'} = 2665 \text{ kJ/kg at 5.78 bar}$$

$$h_{3'} = 2395 \text{ kJ/kg at 1 bar}$$

Nozzle efficiency of convergent portion only

$$\eta_{\text{nozzle}} = \frac{h_1 - h_2}{h_1 - h_{2'}}$$

or, $\qquad 0.95 = \dfrac{2780 - h_2}{2780 - 2665} \quad \therefore \quad h_2 = 2670.75 \text{ kJ/kg}$

Select this value on ordinate of Mollier diagram and move horizontally to pressure curve of 5.78 bar and locate point 2.

$$\therefore \qquad v_2 = 0.35 \text{ m}^3/\text{kg}$$

Actual velocity at the throat $V_2 = 44.72 \sqrt{(h_1 - h_2)}$

or, $\qquad V_2 = 44.72 \sqrt{(2780 - 2670.75)} = 467.42 \text{ m/s}$

$$\text{mass flow rate } m = \frac{A_2 V_2}{v_2}$$

$$\therefore \qquad A_2 = \frac{1 \times 0.35}{467.42} = 7.487 \times 10^{-4} \text{ m}^2 = 7.487 \text{ cm}^2$$

$$\frac{\pi}{4} D_2^2 = 7.487 \text{ cm}^2 \quad \therefore \quad D_2 = \sqrt{\frac{4 \times 7.487}{\pi}} = 3.08 \text{ cm (throat diameter)} \quad Ans.$$

Nozzle efficiency of divergent portion only

$$\eta_{\text{nozzle}} = \frac{h_2 - h_3}{h_2 - h_{3'}}$$

or, $\qquad 0.90 = \dfrac{2670.75 - h_3}{2670.75 - 2395} \quad \therefore h_3 = 2422.57 \text{ kJ/kg}$

Similarly locate the point 3 on pressure curve of 1 bar.

$$\therefore \qquad v_3 = 1.5 \text{ m}^3/\text{kg}$$

Actual velocity at the exit

$$V_3 = 44.72 \sqrt{(h_1 - h_3)}$$

$$= 44.72 \sqrt{(2780 - 2422.57)} = 845.46 \text{ m/s}$$

mass flow rate

$$m = \frac{A_3 V_3}{v_3}$$

$$\therefore \qquad A_3 = \frac{1 \times 1.5}{845.46} = 1.774 \times 10^{-3} \text{ m}^2 = 17.74 \text{ cm}^2$$

$$\frac{\pi}{4} D_3^2 = 17.74 \quad \therefore \quad D_3 = \sqrt{\frac{4 \times 17.74}{\pi}} = 4.75 \text{ cm}^2 \text{ (exit diameter)} \quad Ans.$$

(b) Now, divergent length, $\ l = \dfrac{D_3 - D_2}{2 \tan \theta/2} = \dfrac{4.75 - 3.08}{2 \tan\left(\dfrac{14}{2}\right)} = 6.80 \text{ cm} \quad Ans.$

(c) Power developed $= m\,(h_1 - h_3) = 1 \times (2780 - 2422.57)$

$$= 357.43 \text{ kW} \quad Ans.$$

**14.11** Steam is expanded in a set of nozzles from 10 bar and 200°C to 1.5 bar. Are the nozzles convergent or convergent-divergent? Neglecting the initial velocity, find the minimum area of the nozzles to flow 2 kg/ s of steam. Assume the expansion of steam is isentropic. The heat loss from the nozzle is 17 kJ/kg of steam flow. Take co-efficient of discharge as 0.98. Velcity at inlet, throat and exit are 90, 480 and 550 m/s respectively. Calculate the dryness fraction and the area at exit.

*Soln.* Refer to Fig. 14.29.

Since the steam at initial condition is superheated, the maximum discharge condition gives

$$\frac{p_2}{p_1} = 0.545$$

$$\therefore \qquad p_2 = 10 \times 0.545 = 5.45 \text{ bar}$$

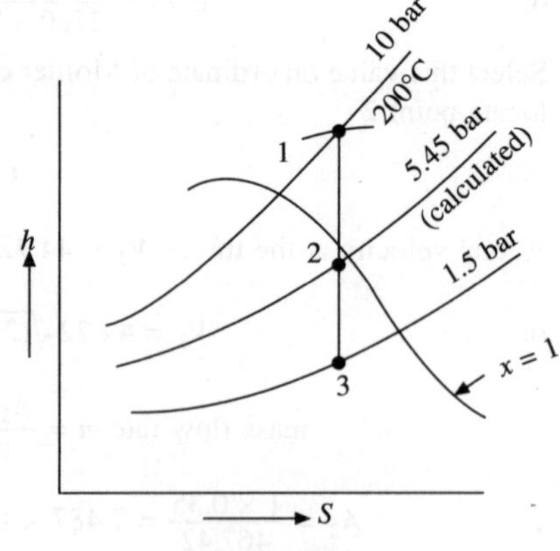



**Fig. 14.29**

Since the exit pressure is less than the throat pressure.

i.e. $\qquad\qquad\qquad 5.45 > 1.5$

Therefore nozzles are convergent-divergent
From Mollier diagram

$$h_1 = 2828 \text{ kJ/kg}$$

$$h_2 = 2710 \text{ kJ/kg}, v_2 = 0.36 \text{ m}^3\text{/kg}$$

Minimum area $\quad A_2 = \dfrac{mv_2}{V_2} = \dfrac{2 \times 0.36}{480} = 1.5 \times 10^{-3} \text{ m}^2 = 15 \text{ cm}^2$   *Ans.*

The actual throat area is given by

$$A_{2'} = \frac{A_2}{C_d} = \frac{15}{0.98} = 15.3 \text{ cm}^2$$

Steady flow energy equation from inlet to the exit

$$h_1 + \frac{V_1^2}{2000} = h_3 + \frac{V_3^2}{2000} + \text{losses}$$

$$\therefore \qquad h_3 = h_1 + \frac{1}{2000}(V_1^2 - V_3^2) - \text{Losses}$$

$$= 2828 + \frac{1}{2000}(90^2 - 550^2) - 17$$

$$= 2663.8 \text{ kJ/kg}$$

Select this value on ordinate of Mollier diagram and move horizontally to pressure curve of 1.5 bar and locate the actual point 3, occured due to losses and then locate dryness fraction which comes out as 0.988. *Ans.*

Specific volume at the exit $v_3 = 1.16 \text{ m}^3\text{/kg}$ (from Mollier diagram)

Area at the exit $\;A_3 = \dfrac{mv_3}{V_3} = \dfrac{2 \times 1.16}{550} = 4.218 \times 10^{-3} \text{ m}^2 = 42.18 \text{ cm}^2$   *Ans.*

The actual area at the exit $\;A_{3'} = \dfrac{A_3}{C_d} = \dfrac{42.18}{0.98} = 43.04 \text{ cm}^2$

**14.12** Prove that for maximum mass flow per unit area

$$\frac{m}{A_2} = \left[ n \frac{p_1}{v_1} \left( \frac{2}{n+1} \right)^{(n+1)/(n-1)} \right]^{1/2}$$

where 1 and 2 indicate for inlet and throat conditions.

*Soln.* Mass of discharge through a convergent nozzle is written as (Eq. 14.14),

$$\frac{m}{A_2} = \left[ \frac{2n}{n-1} \frac{p_1}{v_1} \left\{ \left( \frac{p_2}{p_1} \right)^{\frac{2}{n}} - \left( \frac{p_2}{p_1} \right)^{\frac{n+1}{n}} \right\} \right]^{1/2} \qquad \text{(i)}$$

Maximum discharge condition is written as (Eq. 14.15),

$$\frac{p_2}{p_1} = \left( \frac{2}{n+1} \right)^{\frac{n}{n-1}} \qquad \text{(ii)}$$

Putting the value of equation (ii) into (i), we get

$$\frac{m}{A_2} = \left[ \frac{2n}{n-1} \frac{p_1}{v_1} \left\{ \left(\frac{2}{n+1}\right)^{\frac{n}{n-1}\times\frac{2}{n}} - \left(\frac{2}{n+1}\right)^{\frac{n}{n-1}\times\frac{n+1}{n}} \right\} \right]^{\frac{1}{2}}$$

$$= \left[ \frac{2n}{n-1} \frac{p_1}{v_1} \left(\frac{2}{n+1}\right)^{\frac{n+1}{n-1}} \left\{ \left(\frac{2}{n+1}\right)^{\frac{2}{n-1}-\frac{n+1}{n-1}} - 1 \right\} \right]^{\frac{1}{2}}$$

$$= \left[ \frac{2n}{n-1} \frac{p_1}{v_1} \left(\frac{2}{n+1}\right)^{\frac{(n+1)}{(n-1)}} \left\{ \left(\frac{2}{n+1}\right)^{\frac{1-n}{n-1}} - 1 \right\} \right]^{\frac{1}{2}}$$

$$= \left[ \frac{2n}{n-1} \frac{p_1}{v_1} \left(\frac{2}{n+1}\right)^{\frac{(n+1)}{(n-1)}} \left\{ \left(\frac{n+1}{2}\right) - 1 \right\} \right]^{\frac{1}{2}}$$

$$= \left[ \frac{2n}{n-1} \frac{p_1}{v_1} \left(\frac{2}{n+1}\right)^{\frac{(n+1)}{(n-1)}} \left(\frac{n-1}{2}\right) \right]^{\frac{1}{2}}$$

$$= \left[ n \frac{p_1}{v_1} \left(\frac{2}{n+1}\right)^{\frac{(n+1)}{(n-1)}} \right]^{\frac{1}{2}} \qquad \text{Hence proved}$$

After putting the values of $n = 1.135$ for dry saturated steam and $n = 1.3$ for supersaturated steam, we get the following equations, as mentioned in Eqs. 14.18 and 14.19.

$$m_{\max} = 0.635596\, A_2 \sqrt{\frac{p_1}{v_1}} \quad \text{for dry saturated steam}$$

and

$$m_{\max} = 0.66726\, A_2 \sqrt{\frac{p_1}{v_1}} \quad \text{for superheated steam}$$

**14.13** A convergent divergent nozzle receives steam at 8 bar and temperature 200°C and expands it to 2.9 bar. Neglecting inlet velocity. Calculate the exit area for a discharge of 0.125 kg/s. Assume supersaturated flow with $pv^{1.3} = C$. Also find the degree of undercooling, degree of supersaturation, loss of heat due to irreversibility, increase in entropy and ratio of mass flow rate with metastable expansion to that if expansion in thermal equilibrium.

*Soln.* Refer to Fig. 14.30 Mollier diagram is not used for supersaturated flow up to throat unless Wilson line is not drawn. So taking the values from steam table, as,

$h_1 = 2838.6$ kJ/kg, $S_1 = 6.815$ kJ/kgk and $v_1 = 0.2608$ m³/kg at 8 bar and 200°C.

For initially superheated steam

$$\frac{p_2}{p_1} = 0.545 \quad \therefore \ p_2 = p_1 \times 0.545 = 8 \times 0.545 = 4.36 \text{ bar}$$

*Note:* In case of supersaturation condition mollier diagram is not used up to the throaat.

Isentropic enthalpy change up to throat

$$h_1 - h_2 = \frac{n}{n-1} \frac{p_1 v_1}{1000} \left[ 1 - \left( \frac{p_2}{p_1} \right)^{\frac{n-1}{n}} \right]$$

$$= \frac{1.3}{1.3-1} \times \frac{8 \times 10^5 \times 0.2608}{1000} \left[ 1 - (0.545)^{\frac{1.3-1}{1.3}} \right]$$

$$= 118.16 \text{ kJ/kg}$$

$$\therefore \quad h_2 = 2838.6 - 118.16 = 2720.44 \text{ kJ /kg} = h_{2'}$$

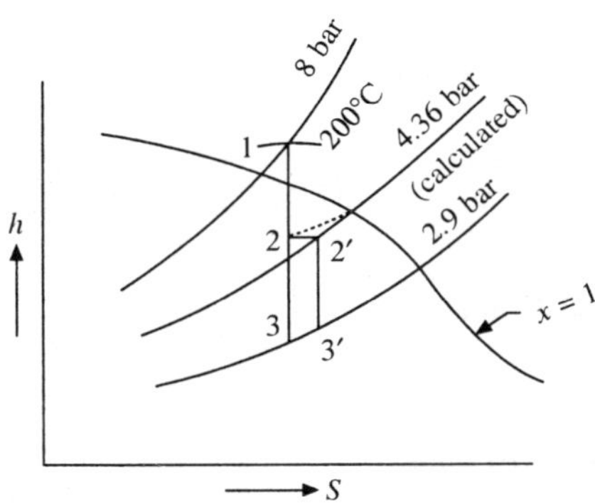


**Fig. 14.30**

Thus the point-2′ is marked on $(h-s)$ diagram by the point of intersection of $h_2 = 2720.44$ kJ/kg line and 4.36 bar pressure line. Draw the isentropic line $(2'-3')$ from 4.36 bar pressure line to 2.9 bar pressure line. Process $2'-3'$ represents expansion after supersaturation. Draw another isentropic line from 2 to 3. Process $2-3$ represents expansion in thermal equilibrium. Now, from Mollier diagram

$$h_3 = 2645 \text{ kJ/kg}, \quad v_3 = 0.66 \text{ m}^3\text{/kg at 2.9 bar}$$

Velocity at the exit from nozzle

$$V_3 = 44.72 \sqrt{(h_1 - h_3)}$$

$$= 44.72 \sqrt{(2838.6 - 2645)}$$

$$= 622.23 \text{ m/s}$$

mass flow rate $\quad m = \dfrac{A_3 V_3}{v_3}$ or, $A_3 = \dfrac{0.125 \times 0.67}{622.23} = 1.345 \times 10^{-4} \text{ m}^2$

$$= 1.345 \text{ cm}^2 \text{ (exit area)} \quad \textit{Ans.}$$

For metastable flow (i.e. supersaturated flow)

$$T_2 = T_1 \left( \frac{p_2}{p_1} \right)^{\frac{n-1}{n}} = (200 + 273)(0.545)^{\frac{1.3-1}{1.3}}$$

$$= 411.17 \text{ K} = 138.17°C$$

Saturation temperature at 4.36 bar = 146.78°C (from steam table)

$\therefore$ Degree of undercooling = $(146.78 - 138.17) = 8.61°C$   *Ans.*

The pressure corresponding to $T_2 = 3.431$ bar (from table)

$\therefore$ Degree of supersaturation $= \dfrac{4.36}{3.431} = 1.27$   *Ans.*

From Mollier diagram

$$h_{3'} = 2655 \text{ kJ/kg at 2.9 bar}$$

Loss of heat drop = $(h_{3'} - h_3) = (2655 - 2645) = 10$ kJ/Kg

This loss takes place at same pressure of 2.9 bar at which saturation = 132.4°C (from table).

$$\therefore \qquad \text{Entropy increase} = \frac{10}{(132.4 + 273)} = 0.0246 \text{ kJ/kgK}$$

For metastable flow, specific volume at the exit

$$v_3' = \frac{0.0023\,(h_3' - 1943)}{p_3} \quad \text{(Eq. 14.31)}$$

$$= \frac{0.0023(2655 - 1943)}{2.9} = 0.564 \text{ m}^3/\text{kg}$$

exit velocity with metastable flow $V_{3'} = 44.72\sqrt{(h_1 - h_{3'})} = 44.72\sqrt{2838.6 - 2655} = 605.95$ m/s

mass flow rate for metastable flow

$$m_1 = \frac{A_3 V_{3'}}{v_{3'}} = \frac{A_3 \times 605.95}{0.564} = 1074.37\,A_3 \text{ units}$$

Now, exit velocity with thermal equilibrium

$$V_3 = 44.72\sqrt{(h_1 - h_3)}$$

$$= 44.72\sqrt{(2838.6 - 2645)} = 622.23 \text{ m/s}$$

mass flow rate with thermal equilibrium

$$m_2 = \frac{A_3 V_3}{v_3} = \frac{A_3 \times 622.23}{0.660} = 942.77\,A_3 \text{ units}$$

$$\therefore \qquad \text{Ratio } \frac{m_1}{m_2} = \frac{1074.37\,A_3}{942.77\,A_3} = 1.139$$

Percentage increase in mass flow rate from thermal equilibrium to metastable condition

$$= \frac{(m_1 - m_2)}{m_2} \times 100 = (1.139 - 1) \times 100 = 13.9\% \quad \textit{Ans.}$$

**14.14** A group of 8 nozzles supply steam at 24 bar and 250°C. The exit pressure of steam is 4 bar. A rate of flow of steam being 5.2 kg/s. Determine
(a) The dimensions of the nozzles of rectangular cross-section with aspect ratio of 3:1. The expansion may be considered as metastable and friction neglected.
(b) The degree of undercooling at supersaturation
(c) Loss in available heat drop due to irreversibility.
(d) Increase in entropy.
(e) Ratio of mass flow rate with metastable expansion to the thermal equilibrium.

*Soln.* Refer to Fig. 14.31.
For initially superheated steam $n = 1.3$.

$$v_1 = 0.09108 \text{ m}^3/\text{kg}$$

(a) Isentropic enthalpy drop

$$h_1 - h_2 = \frac{n}{n-1}\frac{p_1 v_1}{1000}\left[1 - \left(\frac{p_2}{p_1}\right)^{\frac{n-1}{n}}\right]$$

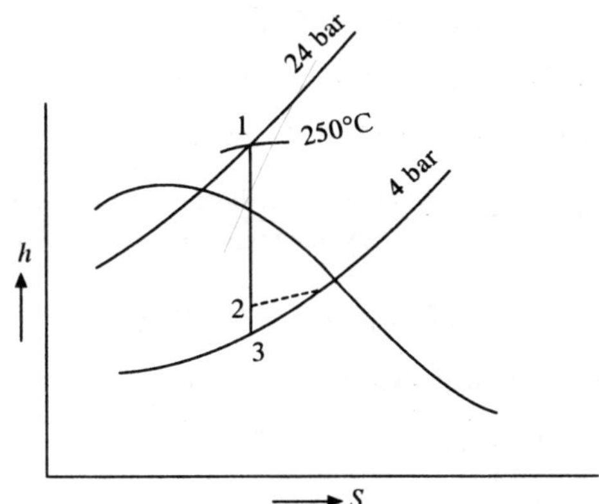


**Fig. 14.31**

$$= \frac{1.3}{1.3 - 1} \times \frac{24 \times 10^5 \times 0.09108}{1000} \left[ 1 - \left( \frac{4}{24} \right)^{\frac{1.3-1}{1.3}} \right]$$

$$= 320.78 \text{ kJ/kg}$$

velocity at the throat $V_2 = 44.72 \sqrt{(h_1 - h_2)} = 44.72 \sqrt{320.78} = 800.95$ m/s

$$\therefore \qquad \frac{p_1}{p_2} = \left( \frac{v_2}{v_1} \right)^n \quad \therefore v_2 = v_1 \left( \frac{p_1}{p_2} \right)^{\frac{1}{n}} = 0.09108 \left( \frac{24}{4} \right)^{\frac{1}{1.3}}$$

$$= 0.3614 \text{ m}^3\text{/kg}$$

mass flow rate $m = \dfrac{A_2 V_2}{v_2}$ or, $A_2 = \dfrac{m v_2}{V_2} = \dfrac{5.2 \times 0.3614}{800.95}$

$$= 2.34 \times 10^{-3}\text{m}^2 = 23.4 \text{ cm}^2$$

The nozzle section is rectangular with aspect ratio 3:1. Let breadth of nozzle = $b$. Hence for a set of 8 nozzles,

$$\text{Total area } 8 \times b \times 3b = 23.4 \text{ cm}^2$$

$\therefore$ breadth, $b = 0.975$ cm and length, $3b = 2.925$ cm

(b) Temperature at point 2 is given by

$$\frac{T_2}{T_1} = \left( \frac{p_2}{p_1} \right)^{\frac{n-1}{n}} \quad \therefore T_2 = T_1 \left( \frac{p_2}{p_1} \right)^{\frac{n-1}{n}} = (250 + 273) \left( \frac{4}{24} \right)^{\frac{1.3-1}{1.3}}$$

$$= 345.88 \text{K} = 72.88°\text{C}$$

Saturation point at 4 bar = 143.6°C

$\therefore$ Degree of undercooling = 143.6 − 72.88 = 70.72°C.

Saturation pressure corresponding to 72.88°C = 0.3525 bar

$$\therefore \qquad \text{Degree of supersaturation} = \frac{4}{0.3525} = 11.34$$

(c) Isentropic enthalpy drop from 24 bar, 250°C to 4 bar

i.e. $$h_1 - h_3 = 2884 - 2530 = 354 \text{ kJ/kg}$$

(d) $\therefore$ Loss in available heat = $354 - 320.78 = 33.22$ kJ/kg

(e) $$\text{Increase in entropy} = \frac{\text{Loss in available heat due to irreversibility}}{\text{Absolute temperature at 4 bar}}$$

$$= \frac{33.22}{(143.6 + 273)} = 0.0797 \text{ kJ/kgK}$$

(f) Exit velocity from the nozzle with expansion in thermal equilibrium,

$$V_3 = 44.72 \sqrt{(h_1 - h_3)}$$

$$= 44.72 \sqrt{354} = 841,40 \text{ m/s}$$

Specific volume at point 3, $v_3 = 0.457$ m³/kg (from M.diagram). For exit area $A_3$, the mass flow rate for metastable condition

$$m_1 = \frac{A_3 \times 800.95}{0.3614} = 2216.24 \, A_3 \text{ units}$$

For the same area, the mass flow rate for thermal equilibrium

$$m_2 = \frac{A_3 \times 841.40}{0.457} = 1841.13 \, A_3 \text{ units}$$

$$\text{Ratio of } \frac{m_1}{m_2} = \frac{2216.24 \, A_3}{1841.13 \, A_3} = 1.203$$

and percentage increase in discharge due to metastable flow

i.e. $$\frac{(m_1 - m_2)}{m_2} \times 100 = (1.203 - 1) \times 100 = 20.3 \% \quad Ans.$$

**14.15** An injector is to deliver 38 kg of water per minute from a feed tank to the boiler. Boiler produces steam at 10 bar and dry saturated and being supplied to the injector at the same condition. The pressure at the exit from the steam nozzle is 0.8 of the boiler pressure. The temperature in the water tank is 17°C and the level in the boiler is 1 m below the level of the injector. If the water level in the boiler is 1.5 m above the level of the injector.

Find

(a)   mass of the water per kg of steam,
(b)   the diameter of the steam nozzle throat,
(c)   the diameter of the mixing cone at its throat,
(d)   the temperature of water leaving the injector.

*Soln.* Pressure at the exit of steam nozzle

$$= 0.8 \times 10 = 8 \text{ bar}$$

$\therefore$ Heat drop in the nozzle due to expansion of steam from the pressure of 10 bar to 8 bar = 40 kJ/kg

Velocity of steam at the exit of nozzle $V_s = 44.72 \sqrt{40} = 282.83$ m/s

$$\text{velocity of water } V_w = \sqrt{2 g H_w}$$

$$= \sqrt{2 \times 9.81 \times 1} = 4.429 \text{ m/s}$$

Velocity of mixture $V_m$ from Eq. (14.35)

$$V_m \simeq \sqrt{240\,p_1} = \sqrt{240 \times 10} \simeq 48.98 \text{ m/s}$$

[getting this equation a term $0.00981 \times H_b$ comes. Here $H_b = 1.5$ m, so multiplication comes very less and neglected. Here delivery pressure is 20% more than boiler pressure assumed]
(a) Mass of water per kg of steam, from Eq. (14.34)

$$m_w = \frac{(V_s - V_m)}{(V_w + V_m)} = \frac{(282.83 - 48.98)}{(4.429 + 48.98)}$$

$$= 4.378 \text{ kg/kg of steam} \quad \textit{Ans.}$$

(b) The area $A_a$ of the throat of nozzle i.e.

$$A_a = \frac{m}{m_w}\frac{v_a}{V_s} = \frac{38}{4.378 \times 60} \times \frac{0.25}{282.83} = 1.278 \times 10^{-4}\text{ m}^2 = 1.278 \text{ cm}^2$$

$$\frac{\pi}{4}D_a^2 = 1.278 \quad \therefore \quad D_a = \sqrt{\frac{4 \times 1.278}{\pi}} = 1.275 \text{ cm} \quad \textit{Ans.}$$

(c) The area $A_b$ of the mixing cone throat, from (Eq. 14.36)

$$A_b = \frac{m\left(1 + \dfrac{1}{m_w}\right)}{1000\,V_m} = \frac{\dfrac{38}{60}\left(1 + \dfrac{1}{4.378}\right)}{1000 \times 48.98} = 1.588 \times 10^{-5}\text{ m}^2 = 15.88 \text{ cm}^2$$

$$\frac{\pi}{4}D_b^2 = 15.88 \quad \therefore D_b = \sqrt{\frac{4 \times 15.88}{\pi}} = 4.49 \text{ cm}$$

(d) From Eq. (14.37)

$$h + m_w\,h_{f_w} + \frac{m_w V_w^2}{2000} = (1 + m_w)\,h_{f_m} + \frac{(1 + m_w)\,V_m^2}{2000}$$

or,

$$2780 + 4.378 \times 17 \times 4.187 + \frac{4.378 \times 4.429^2}{2000}$$

$$= (1 + 4.378) \times h_{f_m} + \frac{(1 + 4.378) \times 48.98^2}{2000}$$

or,

$$2780 + 311.62 + 0.042 = 5.378\,h_{fm} + 6.451$$

$$\therefore \qquad h_m = 573.67 \text{ kJ}$$

$\therefore$ Temperature of water leaving the injector is equal to 137.01°C  *Ans.*


## EXERCISES

14.1  Objective questions
   (i)  The isentropic expansion of steam (initially dry saturated) through the nozzle may be approximately respresented by the equation
     (a)  $pv^{1.4} = \text{constant}$                  (b)  $pv^{1.3} = \text{constant}$
     (c)  $pv^{1.135} = \text{constant}$            (d)  $pv^{1.2} = \text{constant}$

(ii) The ratio of the pressure at the exit and at inlet of the nozzle for maximum discharge is given by

(a) $\dfrac{p_2}{p_1} = \left(\dfrac{2}{n+1}\right)^{\frac{n-1}{n}}$          (b) $\dfrac{p_2}{p_1} = \left(\dfrac{2}{n+1}\right)^{\frac{n}{n-1}}$

(c) $\dfrac{p_2}{p_1} = \left(\dfrac{n}{n+1}\right)^{\frac{n}{n-1}}$          (d) $\dfrac{p_2}{p_1} = \left(\dfrac{n+1}{2}\right)^{\frac{n}{n+1}}$.

(iii) For the steam which is initially saturated and flowing through a nozzle, the ratio of pressures at the exit and at inlet for maximum discharge is equal to
- (a)  0.545
- (b)  0.582
- (c)  0.578
- (d)  0.60

(iv) Nozzle is designed for
- (a)  maximum pressure at outlet
- (b)  maximum discharge
- (c)  maximum pressure and maximum discharge
- (d)  none of the above.

(v) The exit velocity of nozzle considering friction and neglecting initial velocity is given by

(a) $V_2 = 44.72\sqrt{\Delta h}$          (b) $V_2 = 44.72\sqrt{\Delta h\, \eta_{\text{nozzle}}}$

(c) $V_2 = 100\sqrt{\Delta h\, \eta_{\text{nozzle}}}$          (c) none of the above.

(vi) If the exit pressure for convergent nozzle is decreased to such a value that the ratio of exit pressure to inlet pressure is equal to critical pressure ratio, then the mass rate of flow through the nozzle will be
- (a)  minimum
- (b)  zero
- (c)  maximum
- (d)  constant.

(vii) At the critical pressure ratio for a convergent nozzle, the velocity at the outlet will be
- (a)  more than sonic velocity
- (b)  less than the sonic velocity
- (c)  equal to the sonic velocity
- (d)  non of the above

(viii) The maximum discharge of steam through a convergent divergent nozzle depends
- (a)  initial condition of steam
- (b)  throat area
- (c)  pressure at exit of nozzle(d) both (a) and (b) only
- (d) none of the above.

(ix) For the supersaturated flow in the nozzle, the discharge
- (a)  increases
- (b)  decreases
- (c)  remains constant
- (d)  none of the above.

(x) The problems on supersaturated flow up to throat
- (a)  are solved by Mollier chart
- (b)  can not be solved by Mollier chart
- (c)  can be solved by Mollier chart if Wilson line is drawn
- (d)  both (b) and (c).

(xi) The difference of supersaturated temperature and saturation temperature at that pressure is known as
- (a)  degreee of supersaturation
- (b)  degree of superheat
- (c)  degree of undercooling
- (d)  none of these.

(xii) In a nozzle, the effect of supersaturation is to
- (a)  decrease the dryness fraction of steam
- (b)  decrease the specific volume of steam
- (c)  increase the entropy
- (d)  increase the enthalpy drop.

(xiii) In case of nozzle, the whole of friction loss is assumed
- (a)  between inlet and throat
- (b)  between inlet and outlet
- (c)  between throat and exit
- (d)  none of the above.

(xiv) When the back pressure of a nozzle is below the designed value of pressure at exit of nozzle, the nozzle is said to be
(a) chocked  (b) under damping
(c) over damping  (d) undamping
(xv) The steam injector is a device used for
(a) converting water into steam
(b) converting steam into water
(c) feeding water into a boiler, using either live boiler steam or exhaust steam
(d) non of the above.

**Answers**

| | | | | |
|---|---|---|---|---|
| (i) c | (ii) b | (iii) c | (iv) b | (v) b |
| (vi) c | (vii) c | (viii) d | (ix) a | (x) d |
| (xi) c | (xii) c | (xiii) c | (xiv) d | (xv) c |

14.2 Explain the function of nozzle used with steam turbine.

14.3 Explain the principle involved in calculation of the velocity with which fluid issues from a nozzle assuming frictionless adiabatic flow.

14.4 Show, by analytical method, that for isentropic flow of steam through a convergent-divergent nozzle, the throat velocity is the local acoustic velocity.

14.5 Discuss the effect of friction during the expansion of steam through a convergent-divergent nozzle when
(i) the steam at entry to the nozzle is saturated, and
(ii) the steam at entry is superheated.
Assume the pressure of steam to be initially same in both the cases. Mark the processes on a sketch of enthalpy-entropy diagram.

14.6 Explain what is meant by critical pressure ratio of a nozzle.

14.7 Starting from fundamentals, show that for maximum discharge through a nozzle, the ratio of throat pressure to inlet pressure is given by $\left(\dfrac{2}{n+1}\right)^{\frac{n}{n-1}}$, where $n$ is the index for isentropic expansion through the nozzle.

14.8 Explain the term "over expanding" and "under-expanding" as applied to a fluid flow through a nozzle.

14.9 What are the effects of friction on nozzle performance? Show its effects on $T - S$ and $h - s$ diagram.

14.10 Explain the term nozzle efficiency, velocity co-efficient and discharge co-efficient as applied to nozzle.

14.11 Explain the phenomenon of supersaturated expansion of steam and its effect on the discharge from a nozzle as compared to with expansion in thermal equilibrium condition.

14.12 What are the conditions which produce supersaturation of steam? How does the area of the throat of a turbine-nozzle for supersaturated flow compare with the area determined for normal flow?

14.13 Describe the changes which occur in a convergent divergent nozzle as the back pressure is slowly increased from the design pressure up to the pressure at entry.

14.14 Explain the principle of steam injector used for feeding water to the boiler.

14.15 Dry steam expands through a nozzle from a pressure of 13.7 bar down to 9.6 bar. Assuming the flow to be frictionless and adiabatic, estimate the velocity of steam jet.

*(Ans. 362 m/s)*

14.16 Steam at a pressure of 10 bar and 0.95 dry expands in a convergent-divergent nozzle. The back pressure of the nozzle is maintained at 0.85 bar. The throat area is 2.4 cm$^2$. Find the maximum mass flow rate and the required exit area. Assume the index of expansion to be 1.35 throughout. *(Ans. 0.3546 kg/s, 6.575 cm$^2$)*

14.17 Steam enters a convergent nozzle at a pressure of 5 bar and 180°C and a velocity of 200 m/s. The discharge pressure is maintained at 3 bar. Determine the required throat area for a mass flow rate of 0.5 kg/s and nozzle efficiency of 96%

*(Ans. 6.284 cm$^2$)*

**14.18**  The difference of pressure between the inlet and throat of a venturimeter carrying steam is 0.343 bar. The steam at inlet of the meter is dry saturated and at 10.3 bar. Calculate using steam tables the rate of flow in kg/s given that the inlet and throat diameters are 23 cm and 20 cm respectively. Neglect friction.

(*Ans.* 22.6 kg/s)

**14.19**  Calculate the cross-sectional area required at the outlet of a nozzle to pass 0.5 kg per second of steam expanding without loss from 14 bar absolute dry and saturated to 8.83 bar absolute according to the law $pv^{1.2} = \text{constant}$

(b) If the speed at inlet is small          (c) If the speed at inlet is 155 m/s

(*Ans.* 2.465 cm$^2$, 2.32 cm$^2$)

**14.20**  Saturated steam at 13 bar enters a convergent-divergent nozzle with negligible velocity. Determine the throat area required to pass a flow of 60 kg/min assuming that the steam expands isentropically in stable equilibrium. What will be the mass flow passed by the nozzle when the steam expands isentropically in a supersaturated condition.

(*Ans.* 5.37 cm$^2$, 63.2 kg/min)

**14.21**  Steam at 15 bar and 200°C is supplied to a convergent-divergent nozzle against a back pressure of 4 bar. Expansion is supersaturated up to throat and the nozzles are rectangular in shape, its width being 2.5 times the breadth. For a mass flow rate of 0.3 kg/s, find;

   (i)   Dimensions of the hozzle at the exit,

   (ii)  Degree of undercooling and supersaturation and

   (iii) increase in entropy.

(*Ans.* 0.8415 cm, 2.1 cm, 23°C, 2.375, 0.005 kJ/kgK)

**14.22**  Compare the mass of discharge form a convergent divergent nozzle expanding from 8 bar and 210°C to 2 bar, when

   (i)  the expansion takes place under thermal equilibrium and

   (ii) the steam in super-saturated condition during a part of its expansion.

   Take area of nozzle as 2400 mm$^2$

(*Ans.*  8.3%)

**14.23**  An injector is to deliver 100 kg of water per minute from a tank, whose constant water level is 1.2 m below the level of the injector into a boiler in which the steam pressure is 14 bar. The water level in the boiler is 1.5 m above the level of the injector. The steam for the injector is taken from the same boiler and it is assumed to be dry and saturated. The pressure of steam leaving steam nozzle is 0.5 times that of the supply pressure. If the velocity in the delivery pipe in 13.5 m/s. Find:

   (i)   Mass of water pumped per kg of steam

   (ii)  Area of mixing cone

   (iii) Area of steam nozzle

   (iv)  Temperature of water leaving the injector, if the temperature of water in the feed tank is 15°C.

(*Ans.* 7.9 kg, 35.5 mm$^2$, 107.4 mm$^2$, 87.8°C)

# 15

# Steam turbine

## 15.1 INTRODUCTION

The turbine is a device that converts the stored mechanical energy in a fluid into rotational mechanical energy. There are several different types of turbines, including steam turbines, gas turbines, water turbines and wind turbines or wind mills. In all the turbines the motive power is obtained by change in momentum of a highvelocity jet impinging on a curved blade. Steam turbine converts the thermal energy stored in steam into rotational mechanical energy. In steam turbines, enthalpy of the steam is first converted into kinetic energy in nozzles or blade passages. The high velocity steam impinges on the curved blades which change the flow direction of steam and hence the change in momentum. The rate of change of momentum causes a force on the blades mounted on the rotor resulting into mechanical work.

Static action of steam (pressure energy) is utilised to produce mechanical power in steam engine whereas dynamic action is used to get mechanical work in steam turbine. At-least 3/4th of the electrical energy generated in the world is produced by steam turbo-generators, employing the prime mover as steam turbines. Even after popularity of nuclear reactors, steam turbine will continue to be used, as the former is used to produce steam directly or indirectly.

Steam turbines, in addition to being used in stationary power plants, are finding their extensive use in merchant and naval ships. The reason is that these machines can operate at high speeds (say, 40,000 rpm) and are more efficient than any other power producing device (efficiency 40 to 50 percent). Turbo-generators upto 500 MW capacity are presently feasible and it is expected that even larger units may be built.

## 15.2 COMPARISON BETWEEN STEAM ENGINES AND TURBINES

The following are the points of difference between steam engine and steam turbine:

| Steam Engine | Steam Turbine |
| --- | --- |
| 1. It works on static action of steam | 1. It works on dynamic action of steam. |
| 2. Reciprocating motion of the piston has to be converted in to rotary motion of the shaft, hence additional problems of piston rod, connecting rod crank mechanism and fly wheels are encountered. | 2. Directly rotary motion is obtained. |
| 3. Perfect balancing is not possible due to reciprocating masses. | 3. Perfect balancing is possible. |
| 4. Higher speed above 250 rpm is not advisable. | 4. Speed upto 400 rpm may be used thereby increasing the output power power unit volume of the working substance. |
| 5. Electrical generator is difficult to be coupled. | 5. Steam turbines are directly coupled to electrical generators. |
| 6. The back pressure lower than atmospheric is not recommended due to which it has poor efficiency: | 6. The condenser pressure may be at any low vacuum thereby increasing the efficiency. |
| 7. Works on modified Rankine cycle. | 7. Works on Rankine cycle. |
| 8. The steam gets mixed with lubricating oil, hence condensate is not fit far being used as feed water to the boiler. | 8. No internal lubrication is used hence steam does not mix with lubricating oil. |
| 9. large power unit is difficult to design, build and operate. | 9. Large units upto 500 MW Capacity can be built. |
| 10. Thermal efficiency is very poor (say 20 percent) | 10. High thermal efficiency is possible (say 75 percent) |
| 11. Wear, tear and vibration problems are encountered, | 11. Wear, tear and vibration are less encountered, |
| 12. At considerable over load there is great reductionin its efficiency. | 12. Steam turbine can take considerable over load with slight reduction in its efficiency. |

## 15.3 WORKING PRINCIPLE OF A TURBINE

Motive power in a turbine is obtained on the principle of Rate of change of momentum of high velocity jet impinging on a curved balde. The high velocity, jet is obtained by expanding steam in a steam nozzle or over fixed guide blades. The Fig. 15.1 shows the flowing of steam over a curved blade. If $V_1$ and $V_2$ are the inlet and outlet velocity of jet on a fixed curved blade, then the force acting on the blade = $m_s(V_1 \cos \alpha_1 + V_2 \cos \alpha_2)$

Where $m_s$ = Rate of flow of steam in kg/sec.

If $u$ is the velocity of vane in the direction of force then work done per second

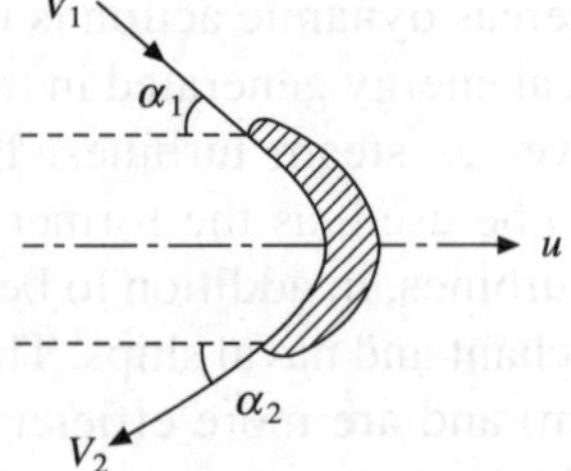

**Fig. 15.1 Working principle of turbine**

$$W = m_s(V_1 \cos \alpha_1 + V_2 \cos \alpha_2) u \text{ watt} \tag{15.1}$$

## 15.4  MAIN COMPONENTS OF A STEAM TURBINE

Steam turbine consists of the following components:

(a) A set of nozzles— Nozzle is a device to increase the kinetic energy of steam at the expense of enthalpy. Vanes are also so shaped that the passage between the two vanes works like a nozzle. Steam expands while flowing through the nozzle.

(b) Rotor or runner— It is an assembly consisting of moving vanes on a shaft.

(c) The blades or vanes:— There are two types of blades: Moving and fixed. Moving blades are mounted on the rotor and the fixed blades on the casing. The function of fixed blades is to guide the steam to move on moving blades from one stage to the other.

(d) The cylinder casing assembly—It is a frame work for supporting the whole structure. Diaphragms are provided for separating the different stages.

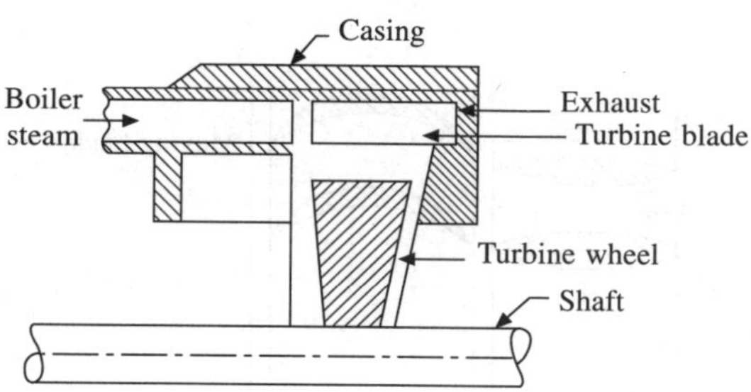



**Fig. 15.2   A turbine**

## 15.5  CLASSIFICATION OF STEAM TURBINES

1. According to the Principal of operation i.e. mode of steam action
   (a) Impulse turbine
   (b) Reaction turbine
2. According to the direction of steam flow
   (a) Radial flow turbine
   (b) Axial flow turbine
3. According to the method of compounding
   (a) Velocity compounded turbine
   (b) Pressure compounded turbine
   (c) Pressure-velocity compounded thrbine
4. According to the pressure of inlet steam
   (a) High pressure turbine
   (b) Low pressure turbine
5. According to the exhaust condition of steam
   (a) Condensing turbine
   (b) Non-condensing turbine
6. According to the number of stages
   (a) Single stage turbine
   (b) Multi stage turbine

7. According to the Position of shaft
   (a) Vertical shaft turbine
   (b) Horizontal shaft turbine
8. According to the field of service stationary, variable speed, Locomotive, ship, industrial etc.

### Impulse Turbine

It works on the principle of impulse where the kinetic energy of fluid jet is used to exert a force on a set of moving blades. Total enthalpy drop takes place in the nozzle or in the fixed blades. Thus steam pressure remains constant while it flows through the moving blades. Kinetic energy is converted into mechanical power. Figure 15.3 shows the flow of the steam through impulse turbine, pressure and velocity variation has been also shown. Examples of such types of turbines are De-Laval, Curties, Rateau etc.

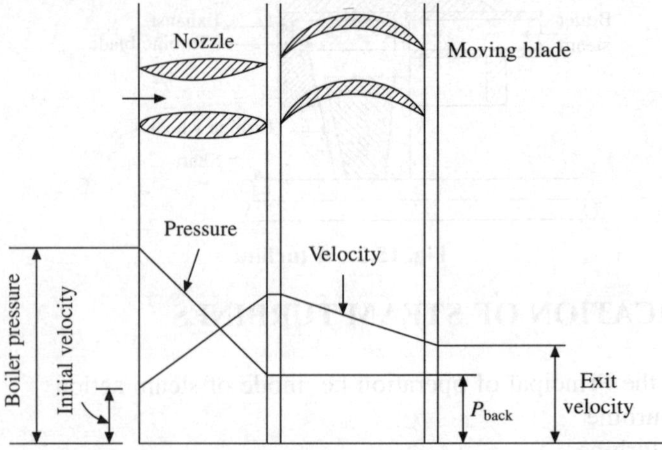


**Fig. 15.3   An Impulse Turbine**

### Reaction Turbine

This turbine is based on the reaction force (Backward force) opposite to a certain action force. The steam flows over the guide vanes and then over moving vanes. Expansion of steam takes place on both the blades. The reaction force due to pressure and other force due to change in momentum on moving blade provide the motive power. Therefore entire pressure drop is achieved gradually and continuously over a series of guide blade (called fixed blades also) and moving blades in succession: This is the major difference between impulse and reaction turbine that pressure remains constant while flowing over moving blades of impulse turbine and it gradually decreases while flowing over the moving blades of rection turbine. The example of such turbines is the Parson turbine. Figure 15.4 shows the flow of steam over fixed and moving blades. The pressure and velocity variation has been also shown.

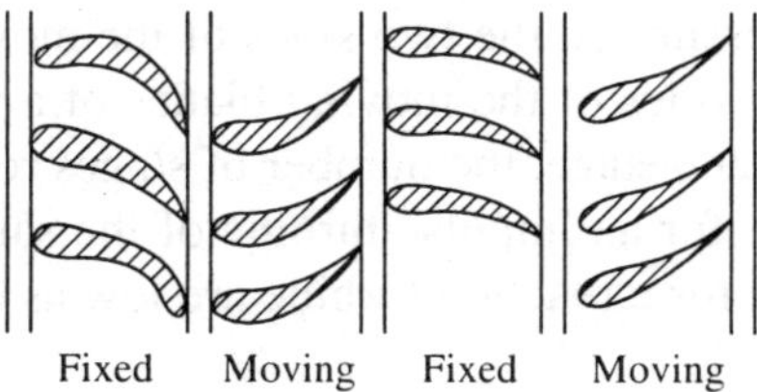



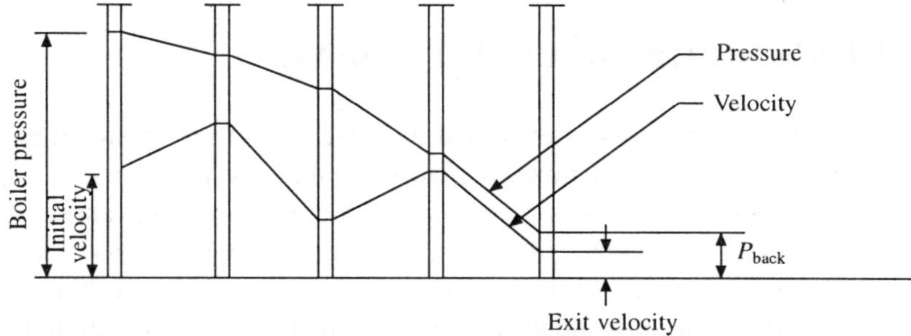



**Fig. 15.4   A Reaction Turbine**

## 15.6   DIFFERENCE BETWEEN IMPULES AND REACTION TURBINES

The followings are the differences between impulse and reaction turbines:

1. In impulse turbine, the fluid is expanded completely in the nozzle and pressure remains constant during its passage through the moving blades.
   In reaction turbines the fluid is partially expanded in the fixed blades and remaining expansion takes place in the moving blades.
2. In impulse turbines, when the steam glides over the moving blades, the relative velocity of steam either remains constant or reduces slightly due to friction. However since in reaction turbine the steam is continuously expanding, relative velocity increases.
3. Impulse turbine blades are of the plate or profile type and are symmetrical but reaction turbine blades are of aerofoil section and assymetrical.
4. Impulse turbine blades are only in action when they are in front of the nozzles where as reaction turbine blades are in action all the time.

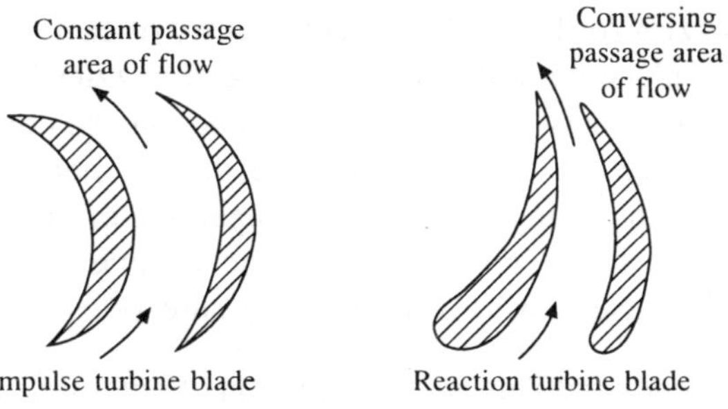



**Fig. 15.5**

5. Impulse turbines have the same pressure on the two sides of the moving blades, whereas different pressures exist on the two sides of the moving blades of a reaction turbine.

6. Because of small pressure drop in each stage, the number of stages required for a reaction turbine are much greater than those for an impulse turbine of the same power.

7. The steam velocity and blade spead for a reaction turbine are low as compared with those of impulse turbine.

8. The variation of diagram efficiency with blade speed ratio is more flat for reaction turbine than for impulse turbine.

## 15.7  COMPOUNDING OF IMPULSE TURBINE

The first commercial steam turbine was developed by De-Laval in which steam enters the steam chest and expands through a group of nozzles placed around the periphery of a wheel having single row of blades. If the steam expands from very high boiler pressure (about 100 to 130 bar) in only one stage, the velocity of steam of jet will be more than 1500 m/sec. and accordingly turbine speed will be as high as 30,000 rpm. This speed is excessively high. The electrical generators can not be coupled directly and at same time centrifugal stresses set up in the rotor will be very high. A gear box would be required to reduce the speed which reduces plant efficiency. It is, therefore, necessary to incorporate some improvements in the simple turbine to achieve high performance. This is done by making use of more than one set of nozzles, blades, rotors in series so that either the jet steam pressure or the jet velocity is absorbed by the turbine in stages. This process is called *Compounding* of impulse turbine. There are three important types of compounding methods:

(a) *Velocity Compounding:* There is only one set of nozzles and two or more than two rows of moving blades. In between two rows of moving blades, there is one row of fixed blades. The function of fixed blades is only to direct the steam coming from first moving row to the next moving row. Therefore these are also called guide blades. The enthalpy drop takes place in the nozzles at the first stage and it is converted into kinetic energy. The kinetic energy of the steam is successively converted into mechanical work by the moving rows of blades. Finally the steam is exhausted from the last row of moving blades. The pressure, velocity and specific volume variation has been shown in the Fig. 15.6. The specific volume and pressure, remain constant over moving and fixed blades.

(b) *Pressure Compounding:* In this type of compounding instead of only a single set of nozzles, there is one row of fixed nozzles (or fixed blades) at the entry of each row of moving blades. The pressure drop of steam takes place not only in the first set of nozzles but gradually in all the sets of nozzles. This type of turbine consists of several stages and may be considered as a combination of several impulse turbines. Figure 15.7 shows the pressure, velocity and specific volume distribution of a pressure compounded turbine. Rateau turbine is an example of this type of compounding. As the pressure of the steam gradually decreases, the specific volume gradually increases, therefore the blade height has to be increased towards the low pressure sides.

(c) *Pressure and velocity compounded* This is a combination of pressure and velocity compounding. The total pressure drop of steam from boiler pressure to condenser pressure is divided into a

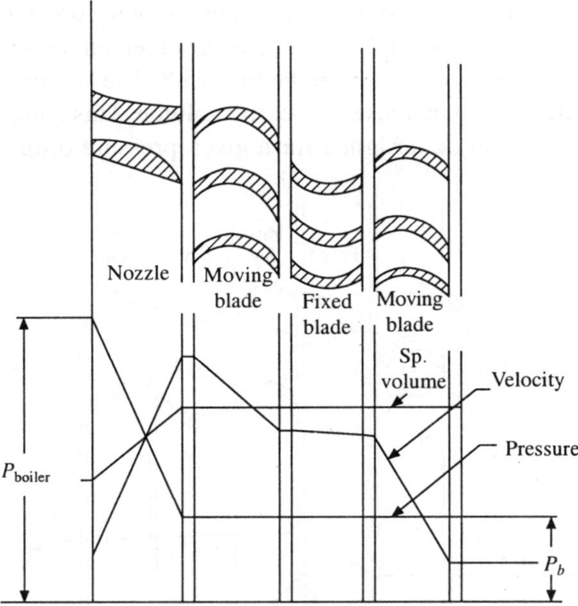


**Fig. 15.6    Velocity Compounded Impulse Turbine**

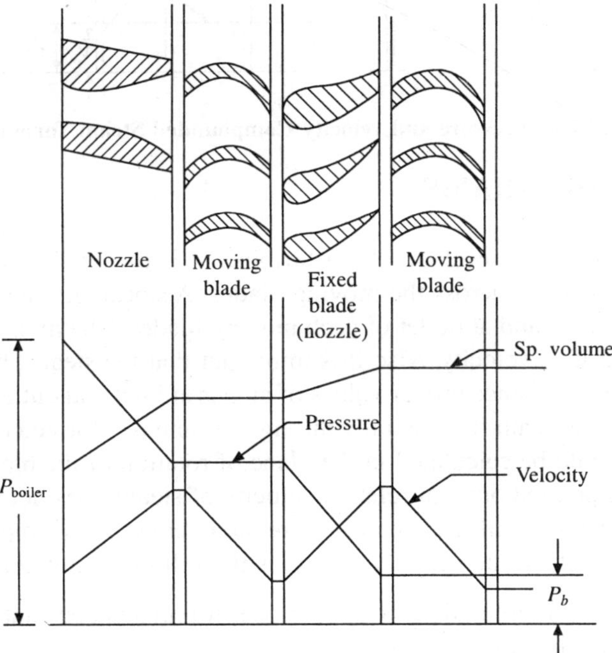


**Fig. 15.7    Pressure Compounded Impulse Turbine**

number of stages as done in pressure compounding and velocity obtained in each stage is also converted into mechanical work in several stages. The arrangement and the variation in pressure velocity and specific volume has been shown in the Fig. 15.8. The height in the second stage must be greater than the first stage due to increase in specific volume. This compounding requires fewer stages and a compact turbine can be designed for a given pressure drop.

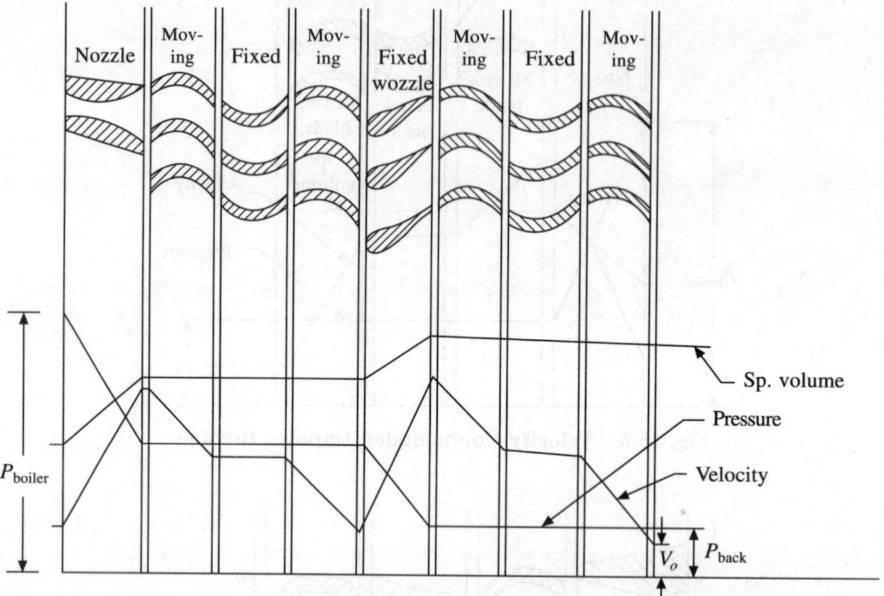


**Fig. 15.8   Pressure and velocity Compounded Steam Turbine**

## 15.8   VELOCITY DIAGRAMS

In order to evaluate the power developed by a turibine, it is essential to determine the rate of change of momentum of steam across the moving blades. Accordingly the vector diagrams will have to be drawn at the inlet and at outlet of each moving blades. The moment steam glides over the blade, it acquires relative velocity. Also it is important that the steam should enter and leave the blades without any shock hence proper values of angles at inlet and outlet should be evaluated for the moving blades. The steam flows axially through the annulus formed by the blade rings and useful impulse force should be calculated in the plane of rotation of the blades, tangential to the turbine rotor. It is to be pointed out that relative velocity of steam does not remain constant over the moving blade in reaction turbine whereas it remains constant in impulse turbine without friction. In order to understand the velocity diagrams the following notations are used:

$u$ = Tangential or circumferential velocity of blade. It remains constant at inlet and at outlet because the blade height is very small.

$u = \dfrac{\pi DN}{60}$, where $D$ = mean diameter of the blade.

$V_1$ = Absolute velocity of steam at inlet to the moving blade, called nozzle velocity.

$V_2$ = Absolute velocity of steam at outlet to the moving blade.

$V_{r1}$ = Relative velocity of steam with respect to the moving blade at inlet.

$V_{r2}$ = Relative velocity of steam with respect to the moving blade at out let to the moving blade.

$V_{w1}$ = Component of $V_1$ in the direction of rotation (providing useful work), called velocity of whirl at inlet to the blade.

$V_{w2}$ = Component of $V_2$ in the direction of rotation (providing useful work), called velocity of whirl at outlet.

$V_{f1}$ = Axial component of $V_1$, called velocity of flow at inlet.

$V_{f2}$ = Axial component of $V_2$, called velocity of flow at outlet.

$\alpha_1$ = Angle between absolute velocity of steam at inlet ($V_1$) and tengential velocity ($u$). This is also called nozzle angle or outlet angle of fixed blade.

$\alpha_2$ = Angle between absolute velocity of steam at outlet and negative direction of tangential velocity, called inlet angle of fixed blade. For axial flow at outlet, $\alpha_2 = 90°$.

$\beta_1$ = Angle between positive direction of $V_1$ and $u$. Conventionally called blade angle at inlet to the blade. For shockless flow, this is the angle between the tangent to the curved blade and tangential direction of blade at inlet.

$\beta_2$ = Angle between positive $V_2$ and negative direction of $u$ conventionally called blade angle at outlet.

$D$ = Mean diameter of blade ring.

$h$ = Blade height

$K$ = Blade velocity coefficient = Ratio of relative velocity at outlet to that at inlet.

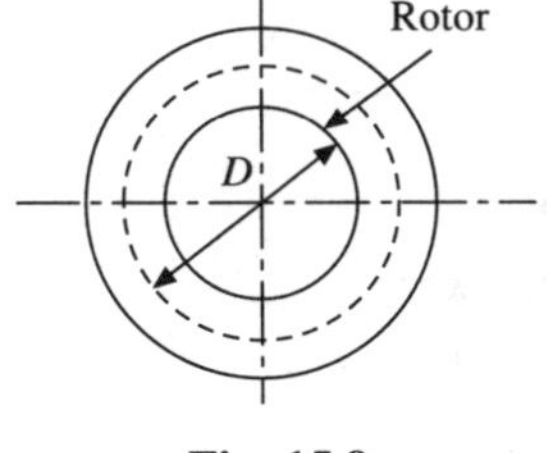


**Fig. 15.9**

$$k = \frac{V_{r2}}{V_{r1}}$$

For impulse turbine without friction

$$K = \frac{V_{r2}}{V_{r1}} = 1.$$

$K$ varies from 0.7 to 0.9

$$\rho = \text{speed ratio} = \frac{u}{V_1} = \frac{\text{Tangential velocity of blade.}}{\text{Absolute velocity of steam at inlet.}}$$

$m_s$ = Rate of flow of steam in kg/sec.

Vectorial equation of relative velocity of Steam with respect to blade is given by

$$\vec{V}_n = \vec{V}_1 - \vec{u} \quad \text{or} \quad \vec{V}_1 = \vec{u} + \vec{V}_n \tag{15.2}$$

According to law of triangle of vector, head and tail of $\vec{u}$ and $\vec{V}_n$ must meet logether to obtain $\vec{V}_1$. Fig. 15.10 shows the flow of steam on a moving curved blade. The combined velocity triangles at inlet and cutlet have been shown in Fig. 15.11. Since tangential velocity ($u$) is same at inlet and at outlet, both the triangles have same base $u$. The absolute velocity $V_1$ can be resolved into two components, along the direction of motion ($V_{w1}$) called velocity of whirl at inlet which

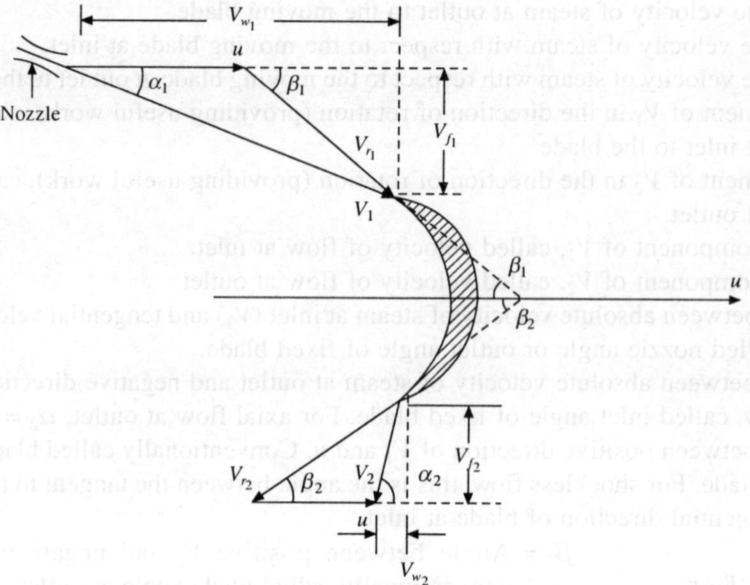



**Fig. 15.10   Flow of steam on a curved Blade**

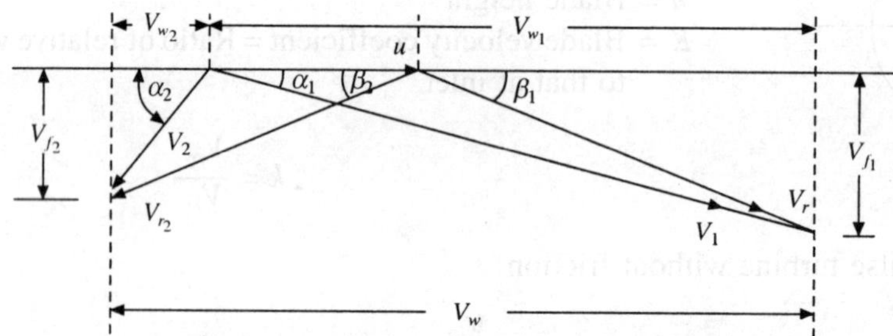

**Fig. 15.11   Combined velocity triangle**

is responsible for tangential force causing whirling (rotation). Other component is perpendicular to tangential direction responsible for flow; called velocity of flow at inlet. Similar components have been shown in the outlet velocity triangle.

In drawing velocity triangle, it should be noted that head of vector $u$ and tail of vector $V_r$ should meet together. A triangle can be solved if three informations are available. These informations may be:

(a)  two angles and one side,

(b)  two sides and one angle,

(c)  all the three sides. Relative velocity at outlet ($V_{r2}$) is either equal to $V_{r1}$ neglecting friction of $K V_{r1}$.

For equiangular or symmetrical blades, $\beta_1 = \beta_2$.

## 15.9 WORK DEVELOPED ON TURBINE BLADES AND EFFICIENCIES

(a) Tangential force on the blade = $F_t$ = Rate of change of momentum in tangential direction = (Rate of flow) × (Change in velocity) = $\dfrac{m_s}{g}(V_{w_1} - V_{w_2})$. If $V_{w2}$ is in opposite direction of $u$, then

$$F_t = m_s\{V_{w_1} - (-V_{w_2})\} = m_s(V_{w_1} + V_{w_2})\,N = m_s V_w N \qquad (15.3)$$

Where $V_w$ = algebraic sum of $V_{w1}$ and $V_{w2}$ and is called resultant velocity of whirl.

(b) Work done by the steam on the blade per second

$$W = F_t \cdot u. = m_s V_W \cdot u.\ \text{Nm/s} = \text{watt} \qquad (15.4)$$

$$\text{Blade power} = \frac{W}{1000} = ms\,\frac{V_W \cdot u}{1000}\ kw \qquad (15.5)$$

(c) *Diagram efficiency of blade efficiency:* For a stage of an impulse turbine, the energy supplied to the blade $= \dfrac{m_s}{2} V_1^2$. Therefore, blade efficiency or diagram efficiency is defined as the ratio of work done by the steam on the blade to the kinetie energy of the jet. Mathematically,

$$\eta_{\text{blade}} = \frac{\text{Work done on the blades}}{\text{Kinetic energy of jet}} = \frac{m_s V_w u}{\dfrac{1}{2}\cdot m_s V_1^2} = \frac{2\cdot u\cdot V_w}{V_1^2} \qquad (15.6)$$

Blade efficiency may also be evaluated as:

$$\eta_{\text{blade}} = \frac{(V_1^2 - V_2^2)/2}{V_1^2/2} = \frac{V_1^2 - V_2^2}{V_1^2} \qquad (15.7)$$

(d) *Stage Efficiency ($\eta_{stage}$)* = total stage is considered to include the nozzle and the moving blades. Therefore stage efficiency is defined as the ratio of work done on the blades per kg of steam to isentropic enthalpy drop in that stage.

Mathematically $\qquad \eta_{\text{stage}} = \dfrac{V_w u}{(\Delta H)_{\text{isen.}}} = \dfrac{V_w \cdot u}{(\Delta H)_{\text{isen.}}}$

Also $\qquad \eta_{\text{stage}} = \dfrac{V_w \cdot u}{(\Delta H)_{\text{isen.}}} = \dfrac{V_w u}{\dfrac{V_1^2}{2}} \times \dfrac{\dfrac{V_1^2}{2}}{(\Delta H)_{\text{isen.}}} = \eta_{\text{blade}} \times \eta_{\text{nozzle}} \qquad (15.8)$

Where nozzle efficiency $= \eta_{\text{nozzle}} = \dfrac{V_1^2}{2\,(\Delta H)_{\text{isen.}}}$

$$= \frac{\text{Actual enthalpy drop}}{\text{Isentropic enthalpy drop}}$$

(e) *Axial thrust:* The tangential component of absolute velocity provides motive power where as the change in normal components (axial components) provides undesirable axial thrust on the bearings. It needs to be properly balanced and provision has to be made for some thrust bearing to take up this load. Axial thrust is given by:

$$F_a = m_s(V_{f1} - V_{f2})\ N \tag{15.9}$$

## 15.10   CONDITION FOR MAXIMUM EFFICIENCY OF IMPULSE TURBINE

There is a definite value of speed ratio $\left(\dfrac{u}{V_1}\right)$ for maximum blade efficiency. We have seen that blade efficiency is given by:

$$\eta_{\text{blade}} = \frac{2V_w \cdot u}{V_1^2}$$

From velocity triangle of Fig. 15.11

$$\eta_{\text{blade}} = \frac{2 \cdot V_w \cdot u}{V_1^2} = \frac{2}{V_1^2}\ (V_{r1} \cos \beta_1 \times V_{r2} \cos \beta_2)\ u = \frac{2 \cdot u \cdot V_{r1} \cos \beta_1}{V_1^2} \times \left(1 + \frac{V_{r2} \cos \beta_2}{V_{r1} \cos \beta_1}\right)$$

$$= \frac{2 \cdot u \cdot V_{r1} \cos \beta_1}{V_1^2}\ (1 + KC)\ \text{Where } K = \frac{V_{r2}}{V_{r1}}\ \text{and}\ \frac{\cos \beta_2}{\cos \beta_1} = c$$

$$\text{or,}\quad \eta_{\text{blade}} = \frac{2 \cdot u}{V_1^2}\ (V_1 \cos \cdot \alpha_1 - u)\ (1 + KC) = \frac{2 \cdot u}{V_1}\ \frac{(V_1 \cos \alpha_1 - u)(1 + KC)}{V_1}$$

Blade speed ratio,

$$\rho = \frac{u}{V_1}$$

$$\text{or,}\qquad \eta_{\text{blade}} = 2 \cdot \rho\ (\cos \alpha_1 - \rho)\ (1 + KC) = 2\ (\rho \cos \alpha_1 - \rho^2)\ (1 + KC) \tag{15.10}$$

For maximum efficiency $\dfrac{d}{d\rho}\ \eta_{\text{blade}} = 0$

$$\text{or,}\qquad 2(\cos \alpha_1 - 2\rho)\ (1 + KC) = 0$$

$$\text{or,}\qquad \cos \alpha_1 - 2\rho = 0,\ \text{or,}\ \rho = \frac{\cos \alpha_1}{2} \tag{15.11}$$

Maximum blade efficiency is determined by putting this value of $\rho$ in the equation (15.10). The

$$\eta_{\text{max}} = \frac{2 \cos \alpha_1}{2} \left(\cos \alpha_1 - \frac{\cos \alpha_1}{2}\right)(1 + KC)$$

$$\text{or,}\qquad \eta_{\text{max}} = \frac{\cos^2 \alpha_1}{2}\ (1 + KC) \tag{15.12}$$

If blades are equiangular, $\beta_1 = \beta_2$ then $C = 1$ and without friction $K = 1$

$$\therefore \qquad \eta_{max} = \frac{\cos^2\alpha_1}{2}(1 + 1 \times 1) = \cos^2\alpha_1$$

or, $\qquad \eta_{max} = \cos^2\alpha_1 \qquad\qquad (15.13)$

It is concluded from Eqs. (15.11) and (15.13) that blade speed should be equal to $\dfrac{\cos\alpha}{2} \simeq \dfrac{1}{2} \times$ jet velocity for maximum efficiency

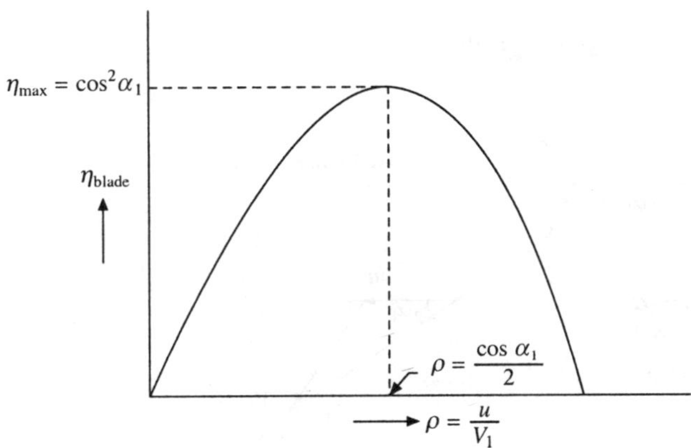


**Fig. 15.12    Variation of blade efficiency with speed ratio**

## 15.10(A)    VELOCITY DIAGRAM FOR VELOCITY COMPOUNDED IMPULSE TURBINE

The velocity compounded steam turbine has been discussed already. In this type of turbine, steam is first expanded in nozzles and is passed through set of moving blades, fixed blades and then through second set of moving blades. The function of fixed blades is to change the direction of motion of steam received from earlier set of moving blades and to redirect it to the next set of moving blades. The complete expansion of steam takes place in nozzle. The velocity diagram for each set of moving blades could be drawn in the similar way as explained in Fig. 15.10. Figs. 15.13 and 15.14 show the velocity diagram for a two stage velocity compounded turbine.

Let prefix 1 and 2 indicate the first row and second row of moving blades respectively.
For the first row of moving blades:

$_1V_1$ = Absolute velocity of steam issued from the nozzle and entering to the inlet of blade, m/s

$_1Vr_1$ = Relative velocity of steam entering the blade, m/s

$_1Vr_2$ = Relative velocity of steam leaving the blade, m/s

$_1\alpha_1$ = Nozzle angle of entrance

$_1\beta_1$ = Blade angle at inlet

$_1\beta_2$ = Blade angle at outlet

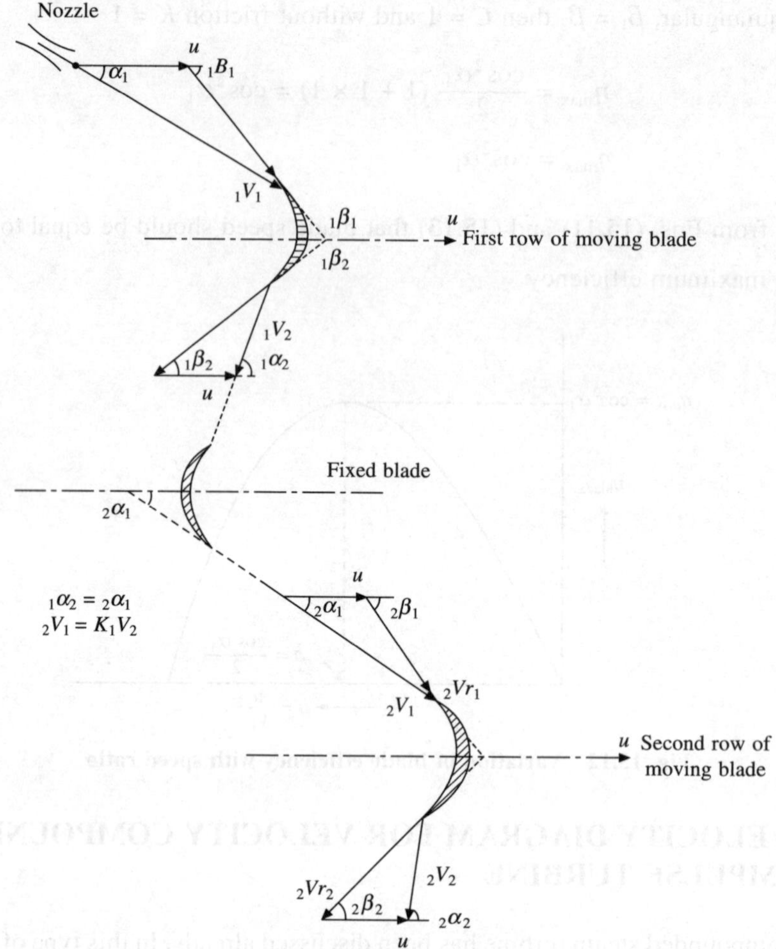



**Fig. 15.13   Velocity diagram**

We see that $_1V_2$ is the velocity with which the steam enters the first row of fixed blades and for the minimum shockloss the angle $_1\alpha_2$ is the inlet angle of the fixed blades. For the second row of moving blades $_2V_1$ is the absolute velocity of steam issuing from the first row of fixed blades. Due to friction in fixed plades $_2V_1$ will be less than $_1V_2$ and so it may be written as

$$_2V_1 = k_1V_2$$

Where $K$ is the balde velocity co-efficient.

Similar will be the case with other rows. Let $_1V_w$, $_2V_w$ be the velocity of whirl in the first and second rows of the wheel work done per second (power produced) in first row = $m_s \, _1V_w \, u$ Watt and Power Produced in second row = $m_{s2}V_w \, u$ watt. Therefore total power produced

$$= \frac{m_s \, (_1V_w + _2V_w)u}{1000} \, KW \tag{15.14}$$

Initial kinetic energy per second $= \dfrac{m_{s1}V_1^2}{2}$ watt

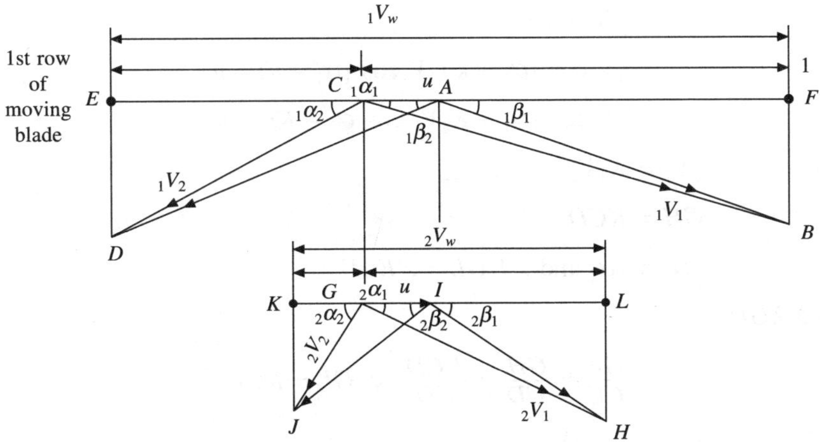

**Fig. 15.14  Velocity diagram (combined)**

$$\therefore \quad \text{Blade efficiency} = \eta_{\text{blade}} = \frac{m_s u \, \Sigma V_w}{\dfrac{m_{s1} V_1^2}{2}} = \frac{2 \, u \, \Sigma V_w}{{}_1 V_1^2} \qquad (15.15)$$

$$\text{Gross stage efficiency} = \eta_{\text{blade}} \times \eta_{\text{nozzle}} \qquad (15.16)$$

## 15.11  MOST ECONOMICAL RATIO OF BLADE SPEED TO STEAM SPEED FOR A TWO ROW VELOCITY COMPOUNDED IMPULSE WHEEL

Likewise single impulse wheel, in this case also there is a certain value of the ratio of blade speed to steam speed which results in the greatest blade efficiency. This ratio is calculated analytically under the following assumptions made:

(1) The blades in each row are equiangular (2) velocity co-efficient $k$ is the same for each row of moving and fixed blades.

$${}_2V_1 = K_1 V_2 \quad \text{(due to fixed blade friction)}$$

and $\qquad {}_1Vr_2 = K_1 Vr_1, \; {}_2Vr_2 = K_2 Vr_1 \quad \text{(due to moving blade friction)}$

Referring the Fig. 15.14,

$${}_1V_w = CF + CE$$

$$CF = {}_1V_1 \cos {}_1\alpha_1, \; AF = {}_1V_1 \cos {}_1\alpha_1 - u$$

$$AB = {}_1Vr_1 = AF \sec {}_1\beta_1$$

$$AE = AD \cos {}_1\beta_2 = {}_1Vr_2 \cos {}_1\beta_2$$

$$= K \, {}_1Vr_1 \cos {}_1\beta_2 = K \, AF \; (\text{Q } {}_1\beta_1 = {}_1\beta_2)$$

$$= K({}_1V_1 \cos {}_1\alpha_1 - u)$$

$$\therefore \quad CE = AE - u = K \, ({}_1V_1 \cos {}_1\alpha_1 - u) - u$$

$$\therefore \qquad {}_1V_w = CF + CE$$

$$= {}_1V_1\cos\,{}_1\alpha_1 + k\,({}_1V_1\cos\,{}_1\alpha_1 - u) - u$$

$$= {}_1V_1\cos\,{}_1\alpha_1 + K_1V_1\cos\alpha_1 - Ku - u \qquad (15.17)$$

Now, $\qquad {}_2V_1 = K_1V_2$

$\therefore \qquad GH = KCD$

$\because \qquad {}_2\alpha_1 = {}_1\alpha_2$ and $\angle HGL = \angle ECD$

In $\Delta ECD$ and $\Delta\,LGH$

$$\frac{GL}{CE} = \frac{GH}{CD} = \frac{KCD}{CD} \quad \therefore GL = KCE$$

or, $GL = K\,\{K({}_1V_1\cos\,{}_1\alpha_1 - u) - u\}$

$$= K^2({}_1V_1\cos\,{}_1\alpha_1 - u) - Ku$$

$$IL = GL - U = K^2({}_1V_1\cos\,{}_1\alpha_1 - u) - Ku - u$$

$\because\ {}_2\beta_1 = {}_2\beta_2$, In $\Delta IJK$ and $\Delta\,IHL$

$$\frac{IK}{IL} = \frac{IJ}{IH} = \frac{KIH}{IH} = K$$

$\therefore\ IK = KIL$

$$= K\{K^2({}_1V_1\cos\,{}_1\alpha_1 - u) - Ku - u\}$$

$$= K^3({}_1V_1\cos\,{}_1\alpha_1 - u) - K^2u - Ku$$

$\therefore \qquad {}_2V_w = IK + IL$

$$= K^3\,{}_1V_1\cos\,{}_1\alpha_1 - K^3u - K^2u - Ku + K^2\,{}_1V_1\cos\,{}_1\alpha_1$$

$$- K^2u - Ku - u \qquad (15.18)$$

From Eqs. (15.17) and (15.18)

$${}_1V_w + {}_2V_w = (U + K + K^2 + K^3)\,{}_1V_1\cos\,{}_1\alpha_1 - (K^3 + 2K^2 + 3K + 2)u$$

$$= K_1\,{}_1V_1\cos\,{}_1\alpha_1 - K_2u$$

Where $\qquad K_1 = 1 + K + K^2 + K^3$

and $\qquad K_2 = 2 + 3K + 2K^2 + K^3$

$$\therefore \qquad \text{Blade efficiency } \eta_{\text{blade}} = \frac{(K_1\,{}_1V_1\cos\,{}_1\alpha_1 - K_2u)u}{{}_1V_1^2/2}$$

$$\because \qquad \text{Blade speed ratio } \rho = \frac{u}{{}_1V_1} \quad \text{or, } u = \rho_1 V_1$$

$$\therefore \qquad \eta_{\text{blade}} = \frac{2\rho_1 V_1(K_{11}V_1\cos\,{}_1\alpha_1 - K_2U)}{{}_1V_1^2}$$

$$= (2\rho K_1\cos\,{}_1\alpha_1 - 2K_2\rho^2) \qquad (15.19)$$

For maximum blade efficiency

$$\frac{d\eta_{\text{blade}}}{d\rho} = 0$$

$$\therefore \qquad 2k_1 \cos {}_1\alpha_1 - 4K_2\rho = 0$$

$$\therefore \qquad \rho_{\text{optimum}} = \frac{K_1}{2K_2} \cos {}_1\alpha_1 \qquad (15.20)$$

and

$$\eta_{\text{blade}}(\text{maxm}) = \frac{K_1^2}{K_2} \frac{\cos^2 {}_1\alpha_1}{2} \qquad (15.21)$$

When the flow is frictionless i.e. $K = 1$ and then $K_1 = 4$ and $K_2 = 8$

$$\therefore \qquad \rho_{\text{optimum}} = \frac{4 \cos {}_1\alpha_1}{2 \times 8} = \frac{\cos {}_1\alpha_1}{4} \text{ (just one half of the value of a single row)}$$

and

$$\eta_{\text{blade}}(\text{maxm}) = \cos^2 {}_1\alpha_1 \qquad (15.22)$$

Optimum value for three row could be shown to be

$$\rho_{\text{optimum}} = \frac{\cos {}_1\alpha_1}{6}$$

In general

$$\rho_{\text{optimum}} = \frac{\cos {}_1\alpha_1}{2n} \qquad (15.23)$$

Where $n$ = no. of stages

work done in the last row $= \dfrac{1}{2^n}$ of total work.

Because of the decreasing utility in last row with the increase in the number of rows, it is seldom now, a days that more than two rows are used.

Comparison of relations $\rho = \dfrac{\cos \alpha_1}{2}$ for a single stage impulse wheel and $\rho = \dfrac{\cos \alpha_1}{4}$ for a two row impulse wheel reveals that for the same blade speed and the nozzle angle, the steam velocity at exit from the nozzle of two row velocity compounded turbine is twice that for a simple impulse turbine. Since blade work is proportional to kinetic energy, theoretically the work of a two row velocity compounded turbine is four times that of a simple wheel blade speed being equal in both the cases.

## 15.12  ADVANTAGES AND DISADVANTAGES OF VELOCITY COMPOUNDED TURBINE

### *Advantages*

(1)  The arrangement in this turbine has fewer number (2 to 3 only) of required stages and hence initial cost is less.
(2)  The arrangement requires less space.

(3) The arrangement is reliable and easy to operate.

(4) Since the pressure falls in nozzle itself hence turbine housing need not be strongly made to sustain the high pressure.

***Disadvantages***

(1) Due to high velocity of steam frictional losses are high and hence efficinecy of this turbine is slightly lower as compared to without compounded turbine. From the Fig. 15.15 we see that the efficiency goes on decreasing with the increase of the stages.

(2) Power developed in each successive blade row decreases with increase in number of rows. Still all the rows require same space, material and cost of fabrication. Hence all the rows are not used with equal economy. The Velocity compounded turbines are mainly used as driving centrifugal compressers, Pumps, small generators and for small units working on high pressure steam.

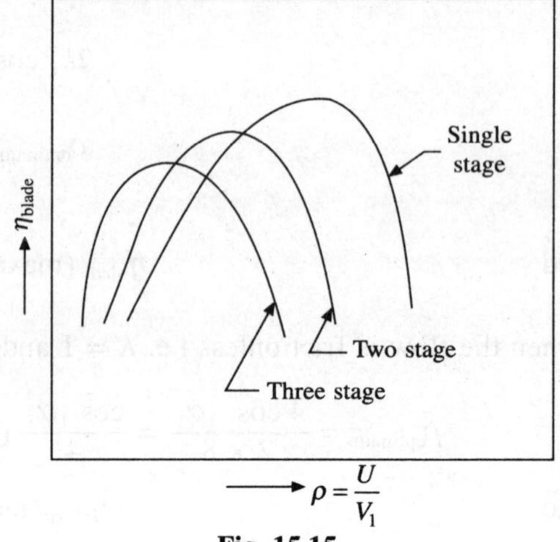



$$\longrightarrow \rho = \frac{U}{V_1}$$

**Fig. 15.15**

## 15.13 AXIAL FLOW TURBINE

When the steam flows over the blades and gets discharged from the last moving blade row in a direction Parallel to the axis of turbine, it is called axial flow turbine. In this case velocity of whirl is zero i.e. $_2Vw_2 = 0$ and accordingly $_2\alpha_2 = 90°$ in a two row impulse turbine.

Mostly turbines are axial flow type. Students ase advised to solve the Problems starting from last velocity triangle to second last velocity traingle and so on up to first one, as far construction of velocity triangles is concerned.

## 15.14 BLADE HEIGHT

### 1. Simple Impulse blade

At a particular section of a turbine, blade height depends on mass flow rate, specific volume, area of flow and velocity of steam.

Let   $p$ = pitch distance between two blades

    $n$ = number of blades covered by the nozzle

    $t$ = thickness of edge of blade

    $h$ = height of blade

    $D$ = Mean diameter of the wheel (blade)

The component of pitch $P$ in a perpendicular direction to the relative velocity $Vr_1 = P \sin \beta_1$. Normal area through which steam flows with relative velocity on moving blade

$$A = nh (P \sin \beta_1 - t)$$

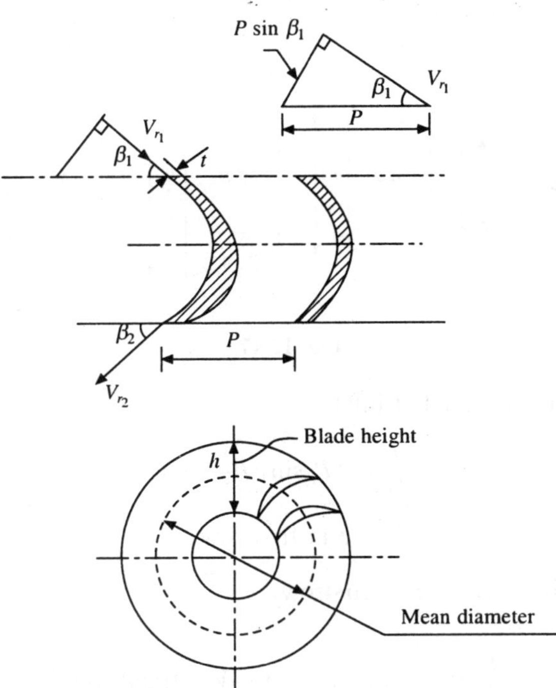

**Fig. 15.16   Impulse turbine blade height**

Fròm continuity equation

$$\text{mass} \times \text{specific volume} = \text{Area} \times \text{velocity}$$

$$m_s \times v = nh \, (P \sin \beta_1 - t) \, Vr_1 \qquad (15.24)$$

Eq. (15.24) gives the relation between the height of the blade and other parameters. If steam is wet at entry section of turbine

$$\text{then } v = x v_g, \text{ where } x = \text{dryness fraction}$$

## 2. Velocity compounded Impulse blade

Let $h_n$, $h_1$, $h_f$ and $h_2$ represent the radial height at the outlet side of the nozzle, first row, fixed row and second row respectively.

Since the specific volume does not increase largly, it may be assumed as constant throughout the system.

Let $l$ = effective length of arc over which the steam is flowing.

By applying continuity equation, we get

$$m_s v = {}_1V_1 \, h_n l \sin {}_1\alpha_1 \qquad (15.25)$$

Let $P_1$, $P_f$ and $P_2$ indicate mean circumferential pitch of blades in first row, fixed row and second row respectively.

No. of blades channel through which steam is flowing $= \dfrac{l}{P}$

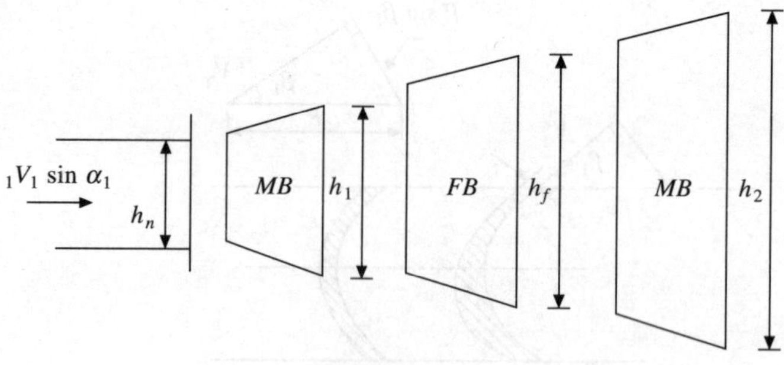


**Fig. 15.17**

Total area through the blade channel at inlet

$$= \frac{l}{P_1}\, h_1\, (P_1 \sin_1 \beta_2 - t_1)$$

where $t_1$ = thickness of outlet edge of blade in first row.

$$\therefore\; m_s v = \frac{l}{P_1}\, h_1 (P_1 \sin_1 \beta_2 - t_{\,1}) \cdot {}_1Vr_2 - \text{firstrow} \tag{15.26}$$

Similarly
$$m_s v = \frac{l}{P_f}\, h_f (P_f \sin {}_1\alpha_2 - t_f) \cdot {}_2V_1 - \text{fixed row} \tag{15.27}$$

$$m_s v = \frac{l}{P_2}\, h_2 (P_2 \sin {}_2\beta_2 - t_2) \cdot {}_2Vr_2 - \text{second row} \tag{15.28}$$

By neglecting blade thickness

$$m_s v = \pi Dh\; {}_1Vr_2 \sin \beta_2 = \pi Dh\, Vf_2 \tag{15.29}$$

## 15.15  VELOCITY TRIANGLE FOR A REACTION TURBINE

The reaction turbines in use are really impulse reaction turbines. Pure reaction turbines are not in general use. The expansion of steam and enthalpy drop occur both in fixed and moving blades. In impulse turbine the relative velocity of steam either remains constant as it glides over the blades or is reduced slightly due to friction. In reaction turbine, however, this is not so. Relative velocity of steam at outlet is greater than that at inlet. The expansion in the moving blade is achieved by making the outlet angle of the moving blade $\beta_2$ appreciably less than the inlet angle $\beta_1$.

Figure 15.18 (a) shows the velocity diagram of a reaction turbine stage and Fig. 15.18 (b) shows the corresponding force diagram.

After considering friction, in simple impulse turbine the value of $Vr_2$ is given by $BH = (K \times BC)$ but in a reaction turbine it is increased to $BD$. This provides a reaction force $dh$ and the resultant force is $dc$. Net change in the velocity is $CD$.

In reaction turbine pressure drop occurs both in nozzles or the fixed row of blades, as well as in the moving row of blades. Due to expansion of steam while flowing through the blades, there is an increase in kinetic energy which gives rise to reaction in the opposite direction according to Newton's third law of motion. Blades rotate due to both the impulse effect of the jets (due to

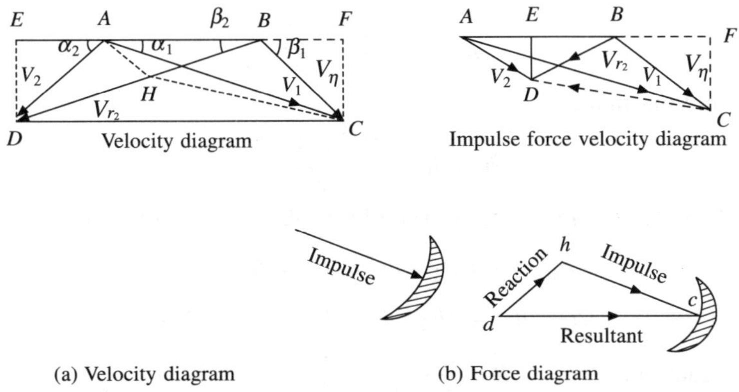

(a) Velocity diagram          (b) Force diagram

**Fig. 15.18  Velocity and force diagram of a reaction turbine stage**

change in their momentum) and the reaction force of the existing jets impressed on the blades in the opposite direction.

### 15.15.1  Degree of Reaction

Figure 15.19 shows the mounting of two sets of blades, fixed and moving with *H–S* diagram indicating 1, 2, 3 are the actual conditions of steam.

The degree of reaction is defined as enthalpy drop in rotor (moving) blades to enthalpy drop in the entire stage of reaction turbine.

Mathematically, the degree of reaction, *R*, is written as

$$R = \frac{\text{Enthalpy drop in moving blades}}{\text{Enthalpy drop in the stae}} = \frac{H_2 - H_3}{H_1 - H_3} \tag{15.30}$$

Considering unit kg of mass flowing through bladings and applying the steady flow energy equation to the fixed blades with the assumption that the velocity of steam entering the fixed blade is equal to the absolute velocity of steam leaving the previous moving row.

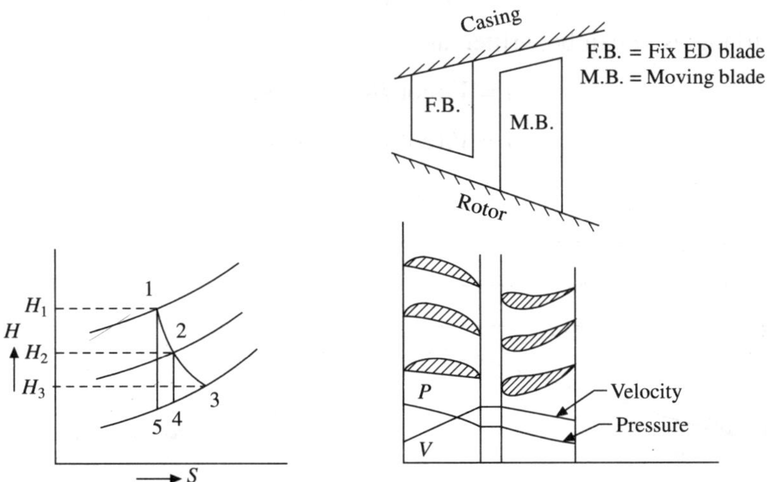


**Fig. 15.19  Isentropic expansion with pressure-Velocity distribution in reaction turbine**

$$\therefore \qquad h_1 - h_2 = \frac{V_1^2 - V_2^2}{2}$$

Similarly for the moving blades

$$h_2 - h_3 = \frac{Vr_1^2 - Vr_2^2}{2}$$

but $V_1 = Vr_2$ and $V_2 = Vr_1$ (From velocity triangle of 50% reaction turbine)

$$\therefore \qquad h_1 - h_2 = h_2 - h_3 \text{ or, } h_1 = 2h_2 - h_3$$

i.e.
$$(h_1 - h_3) = 2h_2 - h_3 - h_3 = 2(h_2 - h_3)$$

Hence, degree of reaction, $\qquad R = \dfrac{h_2 - h_3}{h_1 - h_3} = \dfrac{1}{2} \qquad\qquad (15.31)$

Furthermore, $(H_1 - H_3)$ represents the total work output.

$$\therefore \qquad R = \frac{ms.\dfrac{(Vr_2^2 - Vr_1^2)}{2}}{ms\, u \cdot V_w} = \frac{Vr_2^2 - Vr_1^2}{2u\, V_w} \qquad\qquad (15.32)$$

From the velocity traingles of Fig. 15.20

$$Vr_2 = V_f \operatorname{cosec} \beta_2, \quad Vr_1 = V_f \operatorname{cosec} \beta_1$$

and
$$V_w = V_f(\cot \beta_2 + \cot \beta_1)$$

$V_f = Vf_1 = Vf_2$ which remains constant

$$\therefore \qquad R = \frac{V_f^2 (\operatorname{cosec}^2 \beta_2 - \operatorname{cosec}^2 \beta_1)}{2u \cdot V_f (\cot \beta_2 + \cot \beta_1)} = \frac{V_f}{2u} (\cot \beta_2 - \cot \beta_1) \qquad\qquad (15.33)$$

From 50% degree of reaction

$$u = V_f(\cot \beta_2 - \cot \beta_1) \qquad\qquad (15.34)$$

From velocity triangles, $u$ can be written as

$$u = V_f(\cot \beta_2 - \cot \alpha_2) \qquad\qquad (15.35)$$

and
$$u = V_f(\cot \alpha_1 - \cot \beta_1) \qquad\qquad (15.36)$$

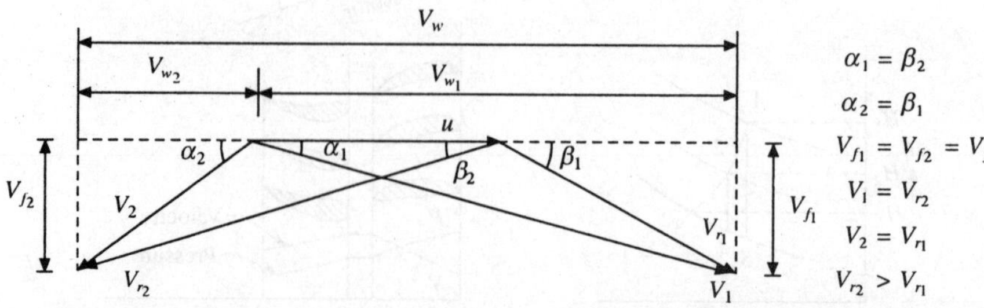


**Fig. 15.20   Velocity triangles for parson's turbine (50% reaction turbine)**

By comparison of equations (15.34), (15.35) and (15.36), we get $\alpha_1 = \beta_2$ and $\beta_1 = \alpha_2$.

Thus for a 50% reaqction turbine which is also Known as Parsons turbine the rotor or moving blades the stater or fixed blades have the same shape, thus giving a symmetrical velocity diagram.

If degree of reaction $R = 0$, which is the case of pure impulse turbine where there is no enthalpy drop in moving blades and all the enthalpy drop takes place in nozzles only. If degree of reaction $R = 1$ (100% reaction turbine which is the case of a pure reaction turbine e.g. Hero's turbine).

## 15.16  EFFICIENCY OF REACTION TURBINE

Maximum value of efficiency is obtained when degree of reaction is 50% and fixed and moving blades are symmetrical. Kinetic energy supplied to fixed blade

$$\text{Per kg of steam flow } = \frac{V_1^2}{2}$$

Kinetic energy supplied in moving blade

$$\text{per kg of steam flow } = \frac{Vr_2^2 - Vr_1^2}{2}$$

$$\therefore \quad \text{Total energy supplied } = \frac{V_1^2}{2} + \frac{Vr_2^2 - Vr_1^2}{2} \tag{15.37}$$

As $Vr_2 = V_1$

$$\therefore \text{ Total energy supplied } = V_1^2 - \frac{Vr_1^2}{2}$$

From velocity triangle, $Vr_1^2 = V_1^2 + u^2 - 2 \cdot u \cdot V_1 \cos \alpha_1$

$$\therefore \quad \text{Total energy supplied } = V_1^2 - \frac{V_1^2 + u^2 - 2u \cdot V_1 \cos \alpha_1}{2} \tag{15.38}$$

The work done per kg of steam $= u \, (Vw_1 + Vw_2)$

$$= u \, (V_1 \cos \alpha_1 + Vr_2 \cos \beta_2 - u)$$

$$= u \, (2V_1 \cos \alpha_1 - u) \; [\because \alpha_1 = \beta_2 \text{ and } V_1 = Vr_2]$$

$\therefore$ Diagram efficiency of Parsons turbine

$$= \frac{u \, (2V_1 \cos \alpha_1 - u)}{V_1^2 - \dfrac{V_1^2 + u^2 - 2u \cdot V_1 \cos \alpha_1}{2}} = \frac{2u \, (2V_1 \cos \alpha_1 - u)}{V_1^2 - u^2 + 2uV_1 \cos \alpha_1}$$

Dividing by $V_1^2$, we get

$$\text{Diagram efficiency} = \frac{2 \dfrac{u}{V_1} \left( 2 \cos \alpha_1 - \dfrac{u}{V_1} \right)}{1 - \left( \dfrac{u}{V_1} \right)^2 + 2 \left( \dfrac{u}{V_1} \right) \cos \alpha_1}$$

As we know blade speed ratio

∴ Diagram efficiency

$$= \frac{2\rho(2\cos\alpha_1 - \rho)}{1 - \rho^2 + 2\rho\cos\alpha_1} = \frac{2[(1 - \rho^2 + 2\rho\cos\alpha_1)] - 2}{(1 - \rho^2 + 2\rho\cos\alpha_1)} = 2 - \frac{2}{(1 + 2\rho\cos\alpha_1 - \rho^2)} \quad (15.39)$$

$\eta_{\text{diagram}}$ (blade) will be maximum when
$(1 + 2\rho\cos\alpha_1 - \rho^2)$ becomes maximum.

$$\therefore \qquad \frac{d(1 + 2\rho\cos\alpha_1 - \rho^2)}{d\rho} = 0$$

or, $\qquad\qquad\qquad\qquad\qquad\qquad\qquad\qquad\qquad\qquad\qquad\qquad\qquad\qquad$ (15.40)

maximum diagram or blade efficiency

$$= \frac{2\cos\alpha_1(2\cos\alpha_1 - \cos\alpha_1)}{1 - \cos^2\alpha_1 + 2\cos^2\alpha_1} \quad \eta_{\text{dia.(max)}} = \frac{2\cos^2\alpha_1}{1 + \cos^2\alpha_1} \quad (15.41)$$

Figure 15.21 shows the variation of diagram efficiency with blade speed ratio ($\rho$) for simple impulse, two stage impulse and reaction stage. From this figure it is quite clear that for a reaction turbine the efficiency curve is reasonably flat in the region of maximum value of diagram efficiency. This results in a slight variation in $\rho$ will not cause much variation in the diagram efficiency.

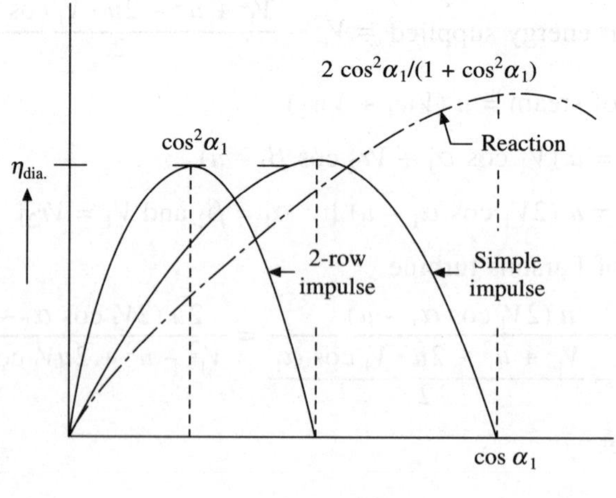


**Fig. 15.21** **Diagram efficiency *VS* blade speed ratio for simple Impulse, two stage Impulse and reaction stage.**

## 15.17  AXIAL THRUST IN REACTION TURBINES

Axial thrust in reaction turbines, is created due to change in axial velocity and pressure drop across the rotor.

$$\text{Axial thrust} = m_s(V_{f_1} - V_{f_2}) + (P_1 A_1 - P_2 A_2) \qquad (15.42)$$

Where $A_1$ and $A_2$ are the annulus areas at inlet and exit respectively.

For a 50% reaction turbine $V_{f_1} = V_{f_2} = V_f$ i.e. change in axial velocity is zero and hence axial thrust is entirely dependent on the pressure drop.

## 15.18  BLADE HEIGHT IN REACTION TURBINE

The height and thickness of the blade are shown in Fig. 15.22. For fully flowing condition mass flow through the blade is given by

$$\text{Volume flow} = \text{area} \times \text{velocity}$$

$$m_s v = [(\pi D - N t_1) h_1] \, Vf_1$$

$$= [(\pi D - N t_2) h_2] \, Vf_2 \qquad (15.43)$$

Where   $m_s$ = mass flow rate  
      $v$ = specific volume  
      $D$ = mean diameter of the wheel  
      $N$ = number of blades  
      $h_1$ = blade height at inlet  
      $h_2$ = blade height at outlet  
      $t_1$ = blade thickness at inlet  
      $t_2$ = blade thickness at outlet

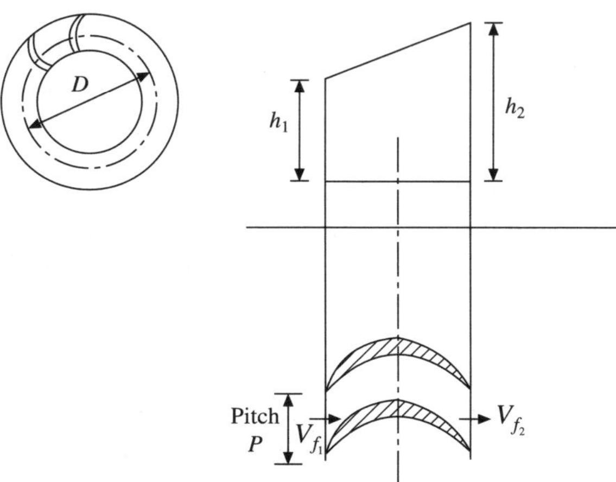


**Fig. 15.22**

for reaction turbine $V_{f_1} = V_{f_2} = V_f$, $h_1 = h_2 = h$ and $t_1 = t_2 = t$

$$\therefore \qquad m_s v = [(\pi D - Nt)h]\, V_f$$

The pitch of the blade $(P)$ is given by $P = \dfrac{\pi D}{N}$

$$\therefore \qquad m_s v = [N(P - t)\, h]\, V_f \qquad\qquad (15.44)$$

generally $t \ll P$ i.e. after neglecting thickness $t$

$$\boxed{\therefore\; m_s v = NPhV_f = \pi DhV_f} \qquad\qquad (15.45)$$

## 15.19   LOSSES IN STEAM TURBINES

In actual practice, the heat energy required to produce mechanical work increases as compared to the theoretical value in which expansion takes place isentropically. This increase in heat energy is termed as energy loss in a steam turbine.

All losses in an actual turbine may be divided into two groups.

(a) Internal losses—These losses are connected with the steam conditions while it flows through the turbine. They may be further classified as.

(i)   Residual velocity loss—with certain velocity steam leaves the turbine. The energy loss due to absolute exit velocity of steam is equivalent to $\dfrac{V_{exit}^2}{2}$ kJ/kg, where $V_{exit}$ is absolute velocity of steam leaving the turbine. The residual velocity loss may be about 10 to 12% in a single stage impulse turbine. This can be reduced by using the multistages.

(ii)  Loss due to friction and turbulence—Friction loss occurs in nozzles, turbine blades and between steam and rotating disc. The friction loss in the nozzle is taken into account by introducing the factor "nozzle efficiency." The loss due to friction and turbulence is about 10%.

(iii) Radiation loss—Temperature of flowing steam through turbine is quite higher than atmospheric Temperature. Therefore some heat is lost from the turbine to the surroundings. Usually the turbines are well insulated to reduce this loss. This loss due to radiation is neglisible.

(iv)  Carry over loss from one stage to another—A gap is provided in multistage turbines, between previous stage and the nozzles or the fixed blade row. The kinetic energy is lost in this gap while steam flows from previous stage to following stage. If gap is more, more and more energy will be lost. So for minimising loss, the gap should be minimum.

(v)   Disc friction and windage loss—Some amount of mechanical work is spent in rotating the turbine wheel. Part of this work is lost in friction and remaining Part for acceleration. The work spent Partly converts into heat and increases the heat content of the steam. Impulse turbines are Partial admission turbines; there is turbulence or churning of steam in the Portion which does not receive steam. This type of loss called windage loss. It essentially consists of friction, impingment of steam on blades and intermittent admission into the moving blades. Besides, some energy is lost in scavenging i.e. some blade Passages are filled with spent steam and when they face incoming steam from nozzle the energy of the fresh steam is spent in forcing spent steam out.

(vi) Loss due to moisture—The condition of steam after last stage of the turbine is wet i.e. quantitiy of water particles increases. The velocity of water particles is less than that of steam and therefore the water particles have to be dragged along with the steam and consequently part of the K.E. of the steam is lost.

(b) External loss—The losses which do not affect the steam conditions. They may be further classified as

(1) Mechanical loss—Due to frictional losses, mechanical losses are caused and develop in the turbine shaft and of the accessories coupled to the turbine. These losses reduce the output and are accounted by the term mechanical efficiency. The loss is less than 1%. This percentage decreases with the increase in the size of the plant and proper lubrication.

(ii) Leakage loss—Leakage occurs at the external glands where the turbine shaft passes through the casing. At the high pressure end the steam leaks into the atmosphere and at the low pressure end the air from atmosphere leaks into the turbine.

Labyrinth packings are provided at the both ends for preventing the leakage. At the low pressure end the leakage of air into the turbine is presented by feeding a supply of low pressure steam at the centre of Labyrinth packing. This may lead to some leakage of steam into the turbine.

## 15.20   STEAM TURBINE PERFORMANCE

In steam turbines when steam flows over the blades, because of imperfections in nozzles, friction and turbulence in blade, windage, throttling and leakage losses; the blade work is less than that could be expected from the isentropic enthalpy drop. The energy utilised in overcoming the friction etc. reheats the steam and the quality of steam at exit from the stage gets improved.

As discussed in just previous article, there are various losses in the stage and the portion of the available energy not converted to work and remaining in the fluid is termed as "reheat". A single-stage expansion with reheat is shown in Fig. 15.23.

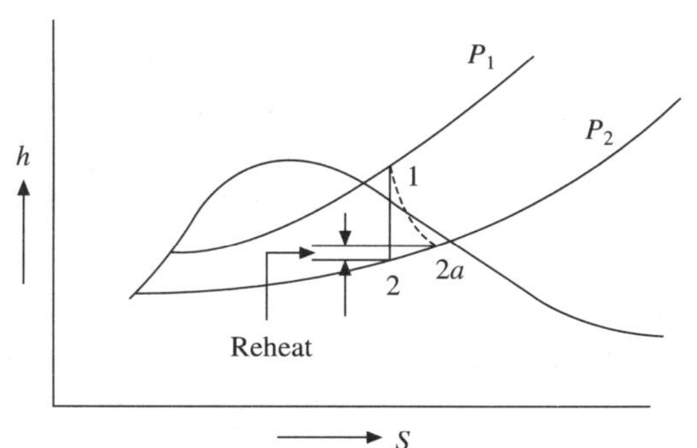

**Fig. 15.23   Expansion in a single stage with reheat**

$\Delta h_a = h_1 - h_{2a}$ = actual enthalpy drop

$\Delta h = h_1 - h_2$ = isentropic enthalpy drop

$\therefore$ Reheat = $(h_{2a} - h_2)$ and stage effciency,

$$\eta_{\text{stage}} = (h_1 - h_{2a})/(h_1 - h_2)$$

Figure 15.24 shows the expansion in a 4-stage turbine considering the effect of reheat. The following conclusions are made:

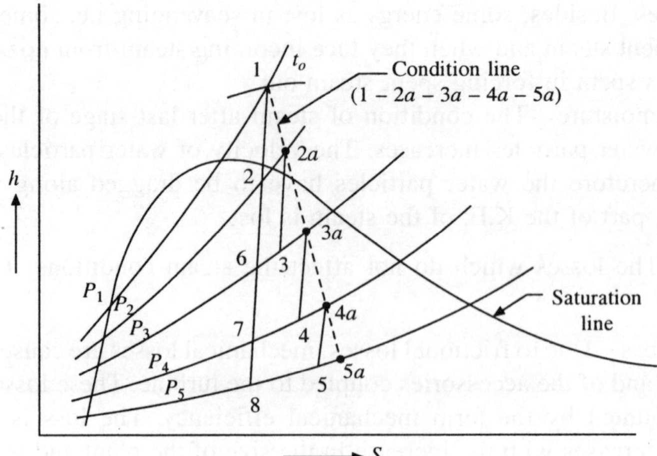



**Fig. 15.24  Expansion in a multistage turbine**

1. Multi-stage turbines are designed for equal efficiency of all stages.
2. Reheat takes place with increase in entropy.
3. The quantity of reheat helps in doing the works in the succeeding stage except the last stage where the reheat is a loss.
4. The constant pressure lines diverge from one another, thereby increasing the enthalpy drop for the same pressure drop.
5. By virtue of reheat and nature of constant pressure lines, the sum of isentropic enthalpy drop of all stages is greater than the isentropic enthalpy drop (available energy) for the whole turbine.
6. The total isentropic enthalpy drop $(h_1 - h_8)$ is divided equally into all parts (4) in the four stages turbine.
7. The condition representing the actual expansion in the turbine is approximately the locus of points indicating the actual conditions of steam at the exit of each stage. Line joining the all exit conditions (points) is called condition line which represents the probable path of expansion and should be represented by a straight dotted line since in between points the processes are irrevbersible.

Figure 15.24 shows that the four stage turbine works between pressure limits $P_1$ (boiler pressure) and $P_5$ (condenser pressure); $P_2$, $P_3$, $P_4$ being interstage pressures. The stage efficiency defined in terms of internal work may be called internal efficiency. The external efficiency is less due to bearing friction. The condition of steam at exit from the first stage of a multistage pressure compounded or reaction turbine, $2a$ is obtained at the pressure $P_2$ with the help of stage efficiency and could be located on Mollier chart. A straight dotted line by joining $1–2a$ and extending to cut pressure line $P_5$ at $5a$ should be drawn. The stages $3a$, $4a$ record the actual exit conditions of steam at second and third stage respectively.

Stage efficiency for first stage

$$\eta_{\text{stage}(1)} = \frac{\text{Actual enthalpy drop}}{\text{Insentropic enthalpy drop}} = \frac{h_1 - h_{2a}}{h_1 - h_2} \qquad (15.46)$$

Stage efficiency for all four stages

$$\eta_{\text{stage}(4)} = \frac{(h_1 - h_{2a}) + (h_{2a} - h_{3a}) + (h_{3a} - h_{4a}) + (h_{4a} - h_{5a})}{(h_1 - h_2) + (h_{2a} - h_3) + (h_{3a} - h_4) + (h_{4a} - h_5)}$$

$$= \frac{(h_1 - h_{5a})}{(h_1 - h_2) + (h_{2a} - h_3) + (h_{3a} - h_4) + (h_{4a} - h_5)} \qquad (15.47)$$

$$\text{Reheat Factor (R.F.)} = \frac{\text{Cummulative isentropic enthalpy drop}}{\text{Total isentropic enthalpy drop}}$$

$$= \frac{(h_1 - h_2) + (h_{2a} - h_3) + (h_{3a} - h_4) + (h_{4a} - h_5)}{(h_1 - h_8)} \qquad (15.48)$$

The reheat factor depends on turbine stage efficiency, initial pressure, superheat and exit pressure. If the stage efficiency is assumed to be the same in all stages,

$$\eta_{\text{stage}} = \frac{h_1 - h_{2a}}{h_1 - h_2} = \frac{h_{2a} - h_{3a}}{h_{2a} - h_3} = \frac{h_{3a} - h_{4a}}{h_{3a} - h_4} = \frac{h_{4a} - h_{5a}}{h_{4a} - h_5} \qquad (15.49)$$

By Substituting $\qquad h_1 - h_2 = \dfrac{(h_1 - h_{2a})}{\eta_{\text{stage}}}, \; h_{2a} - h_3 = \dfrac{h_{2a} - h_{3a}}{\eta_{\text{stage}}}$

and so on in Eq. (15.48)

$$\text{R.F.} = \frac{\dfrac{(h_1 - h_{2a})}{\eta_{\text{stage}}} + \dfrac{(h_{2a} - h_{3a})}{\eta_{\text{stage}}} + \dfrac{(h_{3a} - h_{4a})}{\eta_{\text{stage}}} + \dfrac{(h_{4a} - h_{5a})}{\eta_{\text{stage}}}}{(h_1 - h_8)}$$

or, $\qquad \text{R.F.} = \dfrac{(h_1 - h_{5a})/\eta_{\text{stage}}}{(h_1 - h8)} = \dfrac{1}{\eta_{\text{stage}}} \times \dfrac{(h_1 - h_{5a})}{(h_1 - h_8)} = \dfrac{1}{\eta_{\text{stage}}} \times \eta_{\text{internal}} \qquad (15.50)$

Where the internal (or isentropic) efficiency of the turbine

$$\eta_{\text{internal}} = \frac{(h_1 - h_{5a})}{(h_1 - h_8)} = \eta_{\text{overall}}$$

Thus, from Eq. (15.50)

$$\eta_{\text{internal}} = \text{R.F.} \times \eta_{\text{stage}} \qquad (15.51)$$

Due to divergence of constant pressure lines, the reheat factor is greater than unity. The value varies from 1.02 for turbines with a few stages to 1.08 in large turbines with many stages.

$$\therefore \qquad \eta_{\text{internal}} > \eta_{\text{stage}}$$

It should be noted that as R.F. increases as the number of stages (or pressure ratio) increases. High pressure ratio turbines are more efficient than low pressure ratio turbines.

## 15.21    MAXIMUM GROSS STAGE EFFICIENCY OF A 50% PARSON REACTION TURBINE

As we have seen in previous article that section of the blades of above turbine is the same in both fixed and moving rows of blades. Blades are of same mean diameter and same radial height. Since the specific volume of the steam is increasing as the pressure falls, the velocity of the steam also increases, and in addition to the variation in the balde speed ratio from row to row, the heat drop in each row of blades is somewhat greater than that in the row preceding it, and thus the degree of reaction is rather more than one-half.

Being symmetry in two velocity triangles ABC and ABD (Fig. 15.18(a)], the case is so important that we shall consider it apart from the relations already obtained for the general case. In Fig. 15.18) (a), with pure impulse, the steam flows from the moving blades with a velocity BH equal to $K.Vr_1$. As the area at outlet from the blade channel is restricted by the section of the running blade used, the velocity is increased, by expansion, to the value $Vr_2 = V_1$ and $Vr_2 > Vr_1$.

$\therefore$ Kinetic energy supplied in the moving blades

$$= \frac{Vr_2^2 - k^2 \cdot Vr_1^2}{2} = \frac{V_1^2 - \phi Vr_1^2}{2} \tag{15.42}$$

Where $\phi = K^2$, representing the fraction of incoming energy available at inlet of a stage – carry over co-efficient.

If $\eta_n$ represents efficiency of the blades when considered as nozzle;

$$\eta_n = \frac{\text{Useful (actual) heat drop}}{\text{Adiabatic heat drop}} = \frac{\text{change in K.E.}}{\text{Adiabatic heat drop}}$$

$\therefore$ Adiabatic heat drop, in moving blades

$$h_b = \frac{V_1^2 - \phi Vr_1^2}{2\eta_n}$$

But
$$Vr_1^2 = V_1^2 + U^2 - 2U \cdot V_1 \cos \alpha_1$$

and blade speed ratio $\rho = \dfrac{U}{V_1}$

We have $Vr_1^2 = V_1^2 (1 + \rho^2 - 2\rho \cdot \cos \alpha_1)$

$\therefore$ Heat drop in each row of moving blades

$$h_b = \frac{V_1^2}{2\eta_n} \{1 - \phi(1 + \rho^2 - 2\rho \cos \alpha_1)\} \tag{15.43}$$

Adiabatic heat drop in fixed blades comes as

$$h_n = \frac{V_1^2 - \phi V_2^2}{2\eta_n} = \frac{V_1^2 - \phi Vr_1^2}{2\eta_n}$$

$\therefore$ $h_b = h_n$ and Degree of reaction $R = 0.5$

The total adiabatic heat drop in a pair of blades

$$(h_n + h_b) = \frac{V_1^2}{\eta_n} \{1 - \phi(1 + \rho^2 - 2\rho \cos \alpha_1)\}$$

The work done per kg of steam is given as

$$U \cdot V_w = U(Vw_1 + Vw_2) \text{ [Ref. Fig. 15.20]}$$

$$= U(V_1 \cos \alpha_1 + Vr_2 \cos \beta_2 - U)$$

$$= U(2V_1 \cos \alpha_1 - U) \; [\Q \; \alpha_1 = \beta_2 \text{ and } V_1 = Vr_2]$$

$$= V_1^2 (2\rho \cdot \cos \alpha_1 - \rho^2) \tag{15.44}$$

∴ Gross stage efficiency

$$\eta_{gs} = \frac{\text{Work done per kg of steam}}{\text{Total adiabatic heat drop}} = \frac{\eta_n(2\rho \cdot \cos \alpha_1 - \rho^2)}{\{1 - \phi(1 + \rho^2 - 2\rho \cos \alpha_1)\}}$$

$$= \frac{\eta_n}{\dfrac{\{1 - \phi)}{(2\rho \cos \alpha_1 - \rho^2)\}} + \phi} \tag{15.45}$$

At any value of $\eta_n$ and $\phi$, the value of $\eta_{gs}$ will be maximum when $\rho = \cos \alpha_1$ as per comparison of practical data with basic data pertaining to the nature of the curve.

At $\alpha_1 = 20°$, the optimum value of $\rho = 0.94$ and $\rho = 0.7$ to indicate the curve is flat. Considering $\rho = \cos \alpha_1$ in Eq. 15.45, we have

$$\eta_{gs\,(max)} = \frac{\eta_n(2 \cos^2\alpha_1 - \cos^2\alpha_1)}{1 - \phi(1 + \cos^2\alpha_1 - 2 \cos^2\alpha_1)}$$

$$\boxed{\eta_{gs\,(max)} = \frac{\eta_n \cos^2\alpha_1}{1 - \phi(1 - \cos^2\alpha_1)}} \tag{15.46}$$

**Solved Problems**

**15.1** Steam issues from the nozzles of a De-Laval turbine with a velocity of 1200 m/sec. The nozzle angle is 20°, the mean blade velocity is 400 m/sec. and the inlet and outlet angles of blade are equal. The rate of mass flow through the turbine is 900 kg/hr. Calculate: (a) the blade angles, (b) the relative velocity of steam entering the blade, (c) the tangential force on the blade, (d) the axial force on the blade, (e) the Power developed (f) the blade efficiency, (g) energy lost due to friction in blade per kg of steam. Assume blade velocity co-efficient as 0.8.

*Soln.* Refer to Fig. 15.25

Given $V_1 = 1200$ m/sec, $\alpha_1 = 28°$, $k = 0.8$, $m_s = 900$ kg/hr, $u = 400$ m/sec.

$V_{w_1} = V_1 \cos \alpha_1 = 1200 \cos 20° = 1128$ m/sec

$V_{f_1} = V_1 \sin \alpha_1 = 1200 \sin 20° = 410$ m/sec

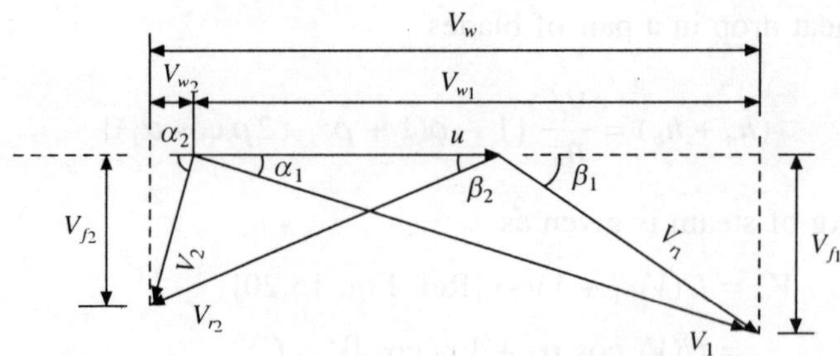

**Fig. 15.25**

(a)   $\tan \beta_1 = \dfrac{V_{f_1}}{V_{w_1} - u} = \dfrac{410}{1128 - 400} = 0.5125 \therefore \beta_1 = 29.4°$   *Ans*

Blade angle at outlet $\beta_2 = \beta_1 = 29.4°$   *Ans*

(b)   $V_{r_1} = \sqrt{(V_{w_1} - u)^2 + V_{f_1}^2} = \sqrt{(1128 - 400)^2 + (410)^2} = 840$ m/sec   *Ans.*

(c)   $V_{r_2} = k\, V_{r_1} = 0.8 \times 840 = 672$ m/sec.

$V_{w_2} = V_{r_2} \cos \beta_2 - U = 672 \cos 29.4° - 400 = 184$ m/sec.

$V_{f_2} = V_{r_2} \sin \beta_2 = 672 \sin 29.4° = 330$ m/sec.

Tangential force $= F_t = m_s(V_{w_1} + V_{w_2}) = \dfrac{900(1128 + 184)}{3600} = 328$ N   *Ans.*

(d)   Axial force $= F_a = m_s(V_{f_1} - V_{f_2}) = \dfrac{900}{3600}(410 - 330) = 20$ N   *Ans.*

(e)   Power developed $= \dfrac{F_t \cdot u}{1000} = \dfrac{328 \times 400}{1000} = 131.2$ KW   *Ans.*

(f)   Blade efficiency $= \dfrac{2u(V_{w1} + V_{w2})}{V_1^2} = \dfrac{2 \times 400\,(1128 + 184)}{(1200)^2} = 0.73 = 73$ percent   *Ans.*

Energy lost per kg $(V_{r1}^2 - V_{r2}^2)/2$

$$= [(840)^2 - (672)^2]/2 \times 1000 = 127 \text{ KJ/kg}   \textit{Ans.}$$

**15.2**  In a stage of impules turbine provided with single row wheel, the mean diameter of blades is 1 m. It runs at 3000 rpm. The steam issues from the nozzle at a velocity of 350 m/sec and the nozzle angle is 20°. The rotor blades are equiangular. The blade friction factor is 0.86. Determine the power developed if the axial thrust on the end bearing of a rotor is 118 N.

*Soln.*   $D = 1$ *m*, $N = 3000$, $V_1 = 350$ m/sec. $\alpha_1 = 20°$, $\beta_1 = \beta_2$, $k = 0.86$, $F_a = 12$ kg, Power $= ?$. Refer to Fig. 15.26.

$$u = \frac{\pi DN}{60} = \frac{\pi \times 1 \times 3000}{60} = 157 \text{ m/sec.}$$

$$V_{w_1} = V_1 \cos \alpha_1 = 350 \cos 20° = 328 \text{ m/sec.}$$

$$V_{f_1} = V_1 \sin \alpha_1 = 350 \sin 20° = 120 \text{ m/sec}$$

$$\tan \beta_1 = \frac{V_{f1}}{(V_{w1} - u)} = \frac{120}{(328 - 157)}, \beta_1 = 35.05° \quad \textit{Ans.}$$

$$V_{r_1} = \sqrt{(V_{w1} - u)^2 + (V_{f1})^2} = \sqrt{(328 - 157)^2 + (120)^2} = 208.9 \text{ m/s}$$

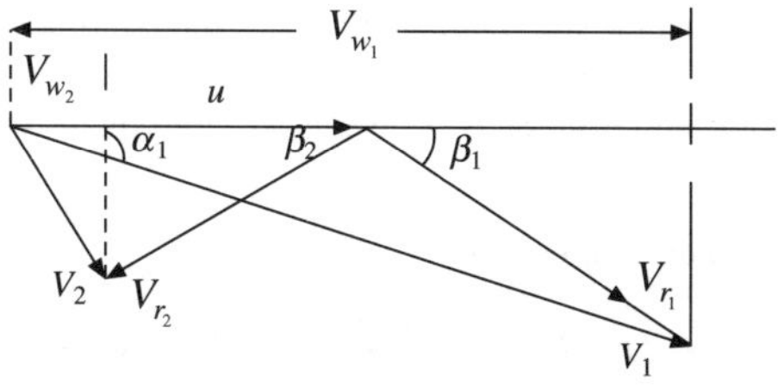

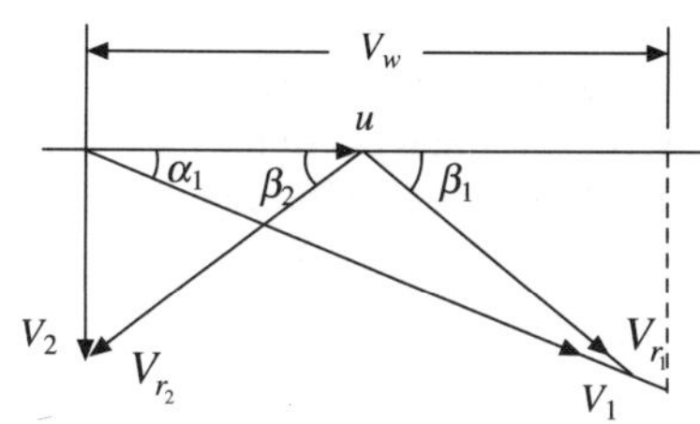

**Fig. 15.26**

$$V_{r2} = k\, V_{r1} = 0.86 \times 208.9 = 179.6 \text{ m/sec.}$$

$$V_{f2} = V_{r2} \sin \beta_2 = 179.6 \sin 35.05° = 103.14 \; m/sec.$$

$V_{r2} \cos \beta_2 = 179.6 \cos 35.05° = 147.02$ m/sec which is less than $u$. Therefore $V_{w2}$ will be in the direction of $u$.

$$V_{w2} = u - V_{r2} \cos \beta_2 = 157 - 147.02 = 9.98 \text{ m/sec.}$$
$$F_a = m_s(V_{f1} - V_{f2}) \text{ or, } 118 = m_s (120 - 103.14)$$
$$\therefore \quad m_s = 7 \text{ kg/sec.}$$

$$\text{Power} = \frac{m_s(V_{w1} - V_{w2})u}{1000} = \frac{7 \times (328 - 9.98) \times 157}{1000} = 349.50 \text{ KW}$$

**15.3** A single row impulse turbine developed 135 kW at a blade speed of 175 m/sec using 2 kg of steam per second. Steam leaves the nozzle at 400 m/sec. Velocity co-efficient of the blade is 0.9. Steam leaves the turbine blade axially. Determine the nozzle angle, blade angle at entry and at exit, assuming no shock (AMIE 76 *W*, Patna University 92)

**Fig. 15.27**

*Soln.* Refer to Fig. 15.27

Power = 135 kw, $u$ = 175 m/sec, $m_s$ = 2 kg/sec
$V_1$ = 400 m/sec. $K$ = 0.9, $\alpha_2$ = 90°, $\alpha_1$ = ?, $\beta_1$ = ?, $\beta_2$ = ?

$$\text{Power} = \frac{m_s V_w \times u}{1000} \quad \text{or,} \quad 135 = \frac{2 \times V_w \times 175}{1000}$$

$$V_w = 385.71 \text{ m/sec} = V_{w1}$$

$$\cos \alpha_1 = \frac{V_{w1}}{V_1} = \frac{378}{400} = 0.945$$

$$\therefore \qquad \alpha_1 = 15.35° = \text{nozzle angle} \quad Ans.$$

$$V_{f1} = V_1 \sin \alpha_1 = 400 \sin 15.35° = 105.88 \text{ m/s}$$

$$\tan \beta_1 = \frac{V_{f1}}{V_{w1} - u} = \frac{105.88}{(385.71 - 175)} = \frac{105.88}{203}, \beta_1 = 26.67° \; Ans.$$

$$V_{r1} = \sqrt{(V_{w1} - u)^2 + (V_{f1})^2} = \sqrt{(210.71)^2 + (105.88)^2} = 235.81$$

$$V_{r2} = KV_{r1} = 0.9 \times 235.81 = 212.23 \text{ m/sec.}$$

$$\cos \beta_2 = \frac{u}{V_{r2}} = \frac{235.81}{212.23}, \quad \therefore \; \beta_2 = 34.45° \; Ans.$$

**15.4** In a single stage of impulse turbine the blade angles are equal and nozzle angle is 20°. The velocity co-efficient for the blade is 0.83. Find the maximum blade efficiency possible. If the actual blade efficiency is 90 percent of maximum blade efficiency, find the possible blade speed ratio.

*Soln.* $\beta_1 = \beta_2$, $\alpha_1 = 20°$, $k = 0.83$

The maximum blade efficiency is given as

$$\eta_{max} = \frac{\cos^2 \alpha_1}{2} (1 + KC) = \frac{\cos^2 \alpha_1}{2} (1 + k) \quad \text{because}$$

$$C = \frac{\cos \beta_2}{\cos \beta_1} = 1 \text{ and } k = 0.83$$

$\therefore \qquad \eta_{max} = \dfrac{\cos^2 20°}{2} (1 + 0.83) = 0.8079 = 80.79 \text{ percent} \quad Ans.$

Actual efficiency = $0.9 \times 0.8079 = 0.727$

Blade efficiency of a single row impulse turbine in terms of blade speed ratio is
$$\eta_{blaide} = 2 \, (\rho \cos \alpha_1 - \rho^2) \, (1 + KC)$$

or, $\qquad 0.727 = 2 \, (\cos 20° - \rho^2) \, (1 + 0.83) = 2(\rho \times 0.9396 - \rho^2) \times 1.83$

or, $\qquad \rho \times 0.9396 - \rho^2 = \dfrac{0.729}{2 \times 1.83} = 0.199$

or, $\qquad \rho^2 - 0.9396 \, \rho + 0.199 = 0$

$$= [0.9396 \pm \sqrt{(0.9396)^2 - (4 \times 0.199)/2}]$$

or, $\qquad \rho = 0.6165 \text{ and } 0.3235 \quad Ans.$

**15.5** Steam flows from the nozzles of a single row impulse turbine with a velocity 450 m/sec. at a direction which is inclined at an angle of 16° to the peripheral velocity. Steam comes out of the moving blade with an absolute velocity of 100 m/sec. in the direction at 110° with the direction of blade motion. The blades are equiangular and steam flow rate is 6 kg/sec.

Determine the Power developed and powr loss due to friction.

*Soln.* $V_1 = 450$ m/sec, $\alpha_1 = 16°$, $V_2 = 100$ m/sec,

$$\alpha_2 = 180° - 110° = 70°, \ \beta_1 = \beta_2, \ m_s, = 6 \text{ kg/sec, Power} = ?$$

Frictional Power = ? Refer to Fig. 15.28

Velocity triangle can be solved either by graphical method or by analytical method. We shall solve this problem by both the methods. Analytical method requires less time than than graphical method due to fast calculators. If students are asked to solve the problem of steam turbine, they should use analytical method. However, if it is asked to draw velocity triangles graphically then there is no alternative.

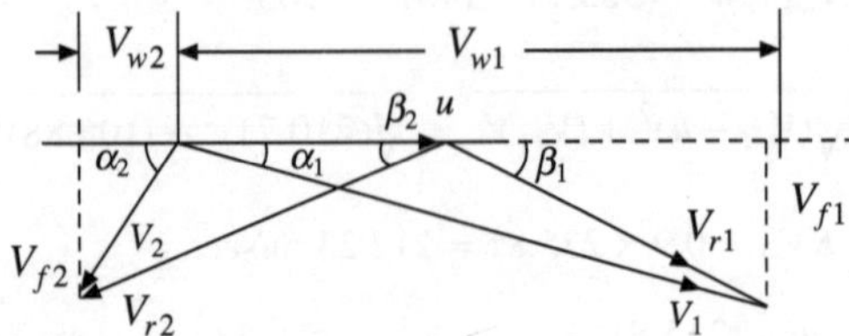


**Fig. 15.28**

Analytical method

$$V_{w1} = V_1 \cos \alpha_1 = 450 \cos 16° = 432.56 \text{ m/sec}$$

$$V_{f1} = V_1 \sin \alpha_1 = 450 \sin 16° = 124 \text{ m/sec.}$$

$$\tan \beta_1 = \frac{V_{f1}}{V_{w_1} - u} = \frac{124}{(432.56 - u)}$$

From outlet velocity triangle

$$V_{f2} = V_2 \sin \alpha_2 = 100 \sin 7.0 = 93.96 \text{ m/sec}$$

$$V_{w2} = V_2 \cos \alpha_2 = 100 \cos 70° = 34.20 \text{ m/sec}$$

$$\tan \beta_2 = \frac{V_{f2}}{u + V_{w_2}} = \frac{93.96}{u + 34.20}$$

But $\beta_1 = \beta_2$, $\therefore \tan \beta_1 = \tan \beta_2$.

or, $\dfrac{124}{(432.56 - u)} = \dfrac{93.96}{(u + 34.20)}$

or, $u = 171$ m/sec.

$$V_{r1} = \sqrt{(V_{w1} - u)^2 + V_{f1}2} = \sqrt{(432.56 - 171)^2 + (124)^2} = 289.46 \text{ m/sec.}$$

$$V_{r2} = \sqrt{(u + V_{w2})^2 + (V_{f2})^2} = \sqrt{(171 + 34.20)^2 + (93.96)^2} = 225 \text{ m/sec.}$$

$$\text{Blade Power} = m_s \frac{(V_{w1} + V_{w2})u}{1000} = \frac{6 \times (432.56 + 34.20) \times 171}{1000} = 478.89 \text{ KW} \quad Ans.$$

$$\text{Power loss due to friction} = m_s \frac{(V_{r1}^2 - V_{r2}^2)}{2 \times 1000} = \frac{6[(289.46)^2 - (225)^2]}{2 \times 1000} = 99.48 \text{ kW} \quad Ans.$$

Graphical method—Refer to Fig. 15.29

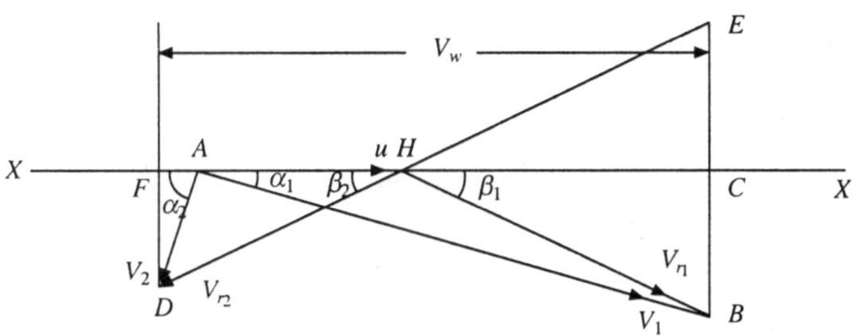


**Fig. 15.29**

Select a scale, 50 m/sec = 1 cm.

on line *xx*, mark at point *A*. Draw a line inclined at an angle 16° passing through *A*. Cut a length *AB* = 450 m/sec = 9 cm. Again from *A* make an angle 70° and cut a length *AD* = 100 m/sec = 2 cm.

From *B* draw a perpendicular on *xx* and Take *EC* = *BC*. Join *ED*, *ED* line meets *xx* axis at *H*. Then *AH* = *u*. Join *DH* = $V_{r2}$. Measure *CF* ($V_W$) and *BH*, *DH*.

*CF* = $V_W$ = 466.76, *AH* = *u* = 171 m/sec, *HB* = $V_{r1}$ = 289.46 m/sec. *HD* = $V_{r2}$ = 225 m/sec.

$$\text{Power developed} = \frac{m_s (V_w)u}{1000} = \frac{6 \times 466.76 \times 171}{1000} = 478.89 \text{ KW}$$

Power lost due to friction $= \dfrac{m_s (V_{r1}^2 - V_{r2}^2)}{2 \times 1000} = \dfrac{6}{2 \times 1000} [(289.46)^2 - (225)^2] = 99.48$ KW  *Ans.*

**15.6** Steam at 6 bar abs and 108°C is suppied to a single stage impulse turbine where it is exhausted into a condenser at a pressure of 0.2 bar abs. The blade speed is 300 m/sec and nozzle angle is 20° and nozzle efficiency is 0.85. Blade velocity co-efficient is 0.7 and blades are equi angular. Calculate the following for steam flow rate of 1 kg/sec.

(a) Axial thrust on the balde, (b) steam consumption per break Power if the mechanical efficiency is 90 percent, (c) blade efficiency, (d) stage efficiency, (e) maximum blade efficiency, (f) heat equivalent of the friction of blading.

*Soln*  $p_1 = 6$ bar, $t_1$ 180°C, $p_b = 0.2$ bar, $u = 300$ m/sec.

$$\alpha_1 = 20°, \ \eta_{\text{nozzle}} = 0.85, \ \eta_m = 0.90, \ k = 0.7, \ \beta_1 = \beta_2, \ m_s = 1 \text{ kg/sec}$$

From Mollier diagram $h_1$ = enthalpy of steam at 6 bar and 180°C = 2810 kJ/kg $h_2$ is determined at condenser pressure of 0.2 bar by drawing a vertical line $h_2 = 2270$ kJ/kg (Fig. 15.30)

$$V_1 = 44.72 \sqrt{\eta_{\text{nozzle}} (h_1 - h_2)} = 44.72 \sqrt{0.84(2810 - 2270)} = 958.1 \ m/\text{sec}.$$

For velocity triangles refer to Fig. 15.31.

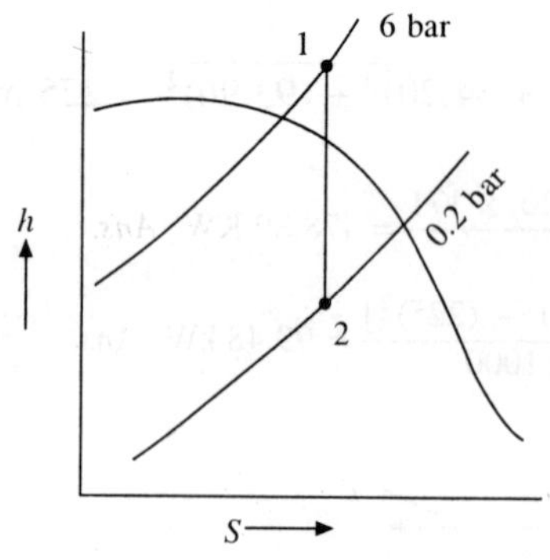


**Fig. 15.30**

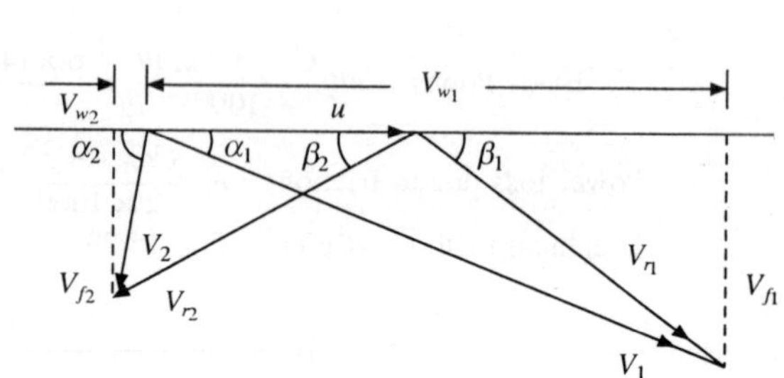


**Fig. 15.31**

$$V_{w1} = V_1 \cos \alpha_1 = 958.1 \cos 20° = 900.3 \ m/\text{sec}$$

$$V_{f1} = V_1 \sin \alpha_1 = 958.1 \sin 20° = 327.68 \ m/\text{sec}.$$

$$V_{r1} = \sqrt{(V_{w1} - u)^2 + (V_{f1})^2} = \sqrt{(900.3 - 300)^2 + 327.68^2} = 683.91 \ m/\text{sec}.$$

$$V_{r2} = K V_{r1} = 0.7 \times 683.91 = 478.73 \ m/\text{sec}.$$

$$\tan \beta_1 = \frac{V_{f1}}{(V_{w1} - u)} = \frac{327.68}{900.3 - 300} = 0.545, \ \beta_1 = 28.62°$$

$$V_{f2} = V_{r2} \sin \beta_2 = 478.73 \times \sin 28.62 = 229.31 \ m/\text{sec}.$$

$$V_{w2} = V_{r2} \cos \beta_2 - u = 478.73 \cos 28.62 - 300 = 120.36 \ m/\text{sec}$$

(a)  Axial thrust $= m_s(V_{f1} - V_{f2})$

$$= 1(327.68 - 229.31) = 98.37 \text{ N}  \quad Ans.$$

(b)  Work done per kg per second $= V_w \cdot u \cdot = (V_{w1} - V_{w2})\, u$

$$(900.3 - 120.36) \times 300 = 233.98 \text{ kw}$$

Blade Power $= 233.98$ kw

Break Powr per kg of steam $= 233.98 \times \eta_m = 233.98 \times 0.90 = 210.582$ kw  *Ans.*

Steam rate per B.P.  $\quad hr = \dfrac{1 \times 3600}{210.582} = 17.095$ kg  *Ans.*

(c)  Blade efficiency

$$= \eta_{\text{blade}} = \frac{2.V_w \cdot u}{V_1^2} = \frac{2 \times 779 \cdot 94 \times 300}{(958.1)^2} = 0.5097 = 50.97\%$$

(d)  Stage efficiency $= \eta_{\text{stage}} = \eta_{\text{blade}} \times \eta_{\text{nozzle}} = 0.5097 \times 0.85 = 0.4332 = 43.32\%$

(e)  Maximum blade efficiency $= \cos^2\alpha_1(1 + KC)$

$$= \frac{\cos^2 20°}{2}\,(1 + 0.7 \times 1) = 0.75 = 75\% \quad \textit{Ans.}$$

(f)  Energy lost in blade friction $= \dfrac{V_{r1}^2 - V_{r2}^2}{2}$

$$= \frac{(683.91)^2 - (478.73)^2}{2} = 119.27 \text{ KJ/kg} \quad \textit{Ans.}$$

**15.7**  The outlet angle of the blade of a Parson's turbine is 20° and the axial velocity of flow of steam is 0.5 times the mean blade velocity. If mean diameter of the ring is 1.25 m and the rotational spead is 3000 rpm. Determine: (a) the inlet angles of blades, (b) Power developed if dry saturated steam at 5 bar abs. passes through the blade whose height may be assumed as 6 cm. Neglect the effect of blade thickness.

*Soln.*  $\beta_2 = 20° = \alpha_1$, $V_{f_1} = V_{f_1} = 0.5\,u$, $D = 1.25$ m, N $= 3000$, $\eta_{\text{stage}} = 0.75$, $h = 0.06$ m
Refer to Fig. 15.32.

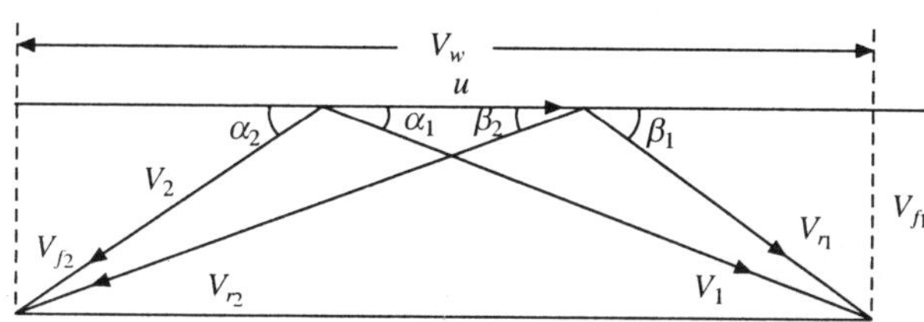


**Fig. 15.32**

$$u = \frac{\pi DN}{60} = \frac{\pi \times 1.25 \times 3000}{60} = 196.35 \text{ m/sec.}$$

$$V_{f_1} = V_{f_2} \text{ (in case of Parson-turbine)} = 0.5 \times 196.35 = 98.19 \text{ m/sec.}$$

$$V_{w_1} = V_{f_1} \cos \alpha_1 = 98.19 \cos 20° = 270.49 \text{ m/sec.}$$

$$V_1 = \frac{V_{f_1}}{\sin \alpha_1} = \frac{98.19}{\sin 20°} = 287.1 \text{ m/sec} = V_{r_2}$$

$$\tan \beta_1 = \frac{V_{f_1}}{V_{w_1} - u} = \frac{98.19}{270.49 - 196.35} \quad \therefore \beta_1 = 52.94°$$

$$V_{w2} = (V_{r_2} \cos \beta_2 - U) = (2.37.1 \times \cos 20° - 196.35) = 73.435$$

Inlet angle of moving blade = $\beta_1$ = 52.94 = $\alpha_2$ = inlet angle of fixed blade   *Ans.*

Outlet angle of fixed bade = $\alpha_1$ = $\beta_2$ = 20°   *Ans.*

$$\text{Rate of flow} = m_s = \frac{(\text{area of flow}) \times (\text{velocity of flow})}{\text{Specific volume}}$$

sp. volume at 5 bar dry = 0.3818 m³/kg

$$\therefore \qquad m_s = \frac{\pi D h V_f}{V} = \frac{\pi \times 1.25 \times 0.06 \times 98.19}{0.3818} = 58.17 \text{ kg/sec}$$

Power produced = $m_s V_w u$ = 58.17(270.49 + 73.435) × 196.35/1000 = 3928.20 kW   *Ans.*

**15.8**  Steam enters the first set of blades of a two-row velocity compounded turbine at a speed of 620 m/s, blade velocity is 150 m/s. Nozzle angle is 16°. Exit angle of first moving, fixed and second moving blades are 20°, 24° and 37° respectively. Velocity co-efficients are 0.78, 0.82 and 0.85. Find the amount of steam required/hr for power develop 660 kW. (Mumbai Univ. winter 2000)

Given $_1V_1$ = 620 m/s, $U$ = 150 m/s, $_1\alpha_1$ = 16°, $_1\beta_2$ = 20°, $_2\alpha_1$ = 24°, $_2\beta_2$ = 37°, Velocity co-efficients = 0.78, 0.82, 0.85 mass flow rate in ton/hour for Power Production = 600 kW.

$$_2V_1 = K_1V_2, \qquad\qquad _1Vr_2 = K\,_1Vr_1, \qquad\qquad _2Vr_2 = K\,_2Vr_1$$

$$\uparrow \qquad\qquad\qquad \uparrow \qquad\qquad\qquad \uparrow$$

$$0.82 \qquad\qquad\qquad 0.78 \qquad\qquad\qquad 0.85$$

Refer to Fig. 15.33 (a), (b).

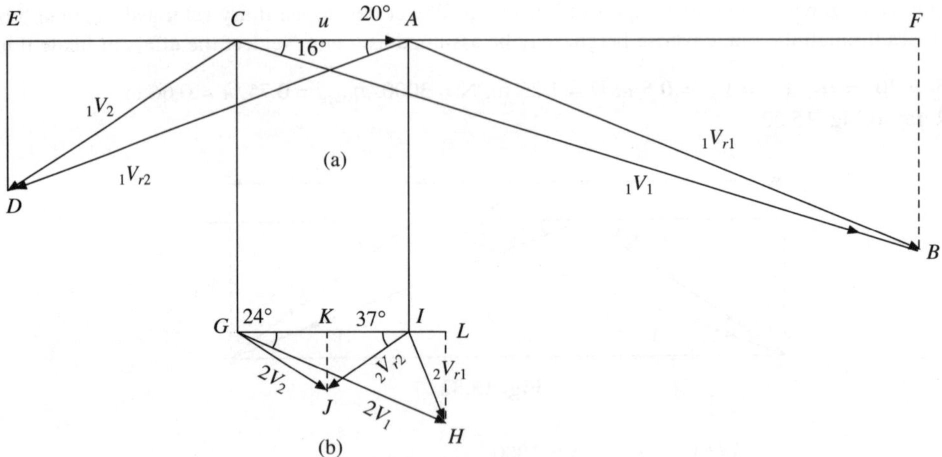


**Fig. 15.33**

The velocity triangles are drawn by considering scale 50 m/s = 1 cm

i.e. $\qquad\qquad\qquad _1V_1 = CB = \dfrac{620}{50} = 12.4 \text{ cm}$ .

and $\qquad\qquad\qquad u = AC = IG = \dfrac{150}{50} = 3 \text{ cm}$

As per construction

$$_1Vr_2 = 0.78 \times 9.6 = 7.488 \text{ cm}$$

$$_2V_1 = 0.82 \times {_1V_2} = 0.82 \times 4.8 = 3.936 \text{ cm}$$

$$_2Vr_1 = 1.8 \text{ cm}$$

$$\therefore \qquad _2Vr_2 = 0.85 \, _2Vr_1 = 0.85 \times 1.8 = 1.53 \text{ cm}$$

Total Power Produced from both rows of moving blades

$$= \dot{m} \, [(CF + CE) + (GL - GK)] \, AC = \dot{m} \, [EF + KL]AC$$

$$= \frac{\dot{m}(15.8 + 2) \times 50 \times 150}{1000} = 660$$

$$\therefore \qquad \dot{m} = 4.943 \text{ kg/s} = 17794.8 \text{ kg/hour} = 17.79 \text{ ton/hour} \quad Ans$$

**15.9** An impulse turbine with two rows of moving blades has the following Particulars.
nozzle angle for the first row = 25°
moving blade tip angle at exit for both rows = 30° fixed blade exit angle = 20°
friction loss over fixed and moving blades = 10% of the velocity. Velocity of steam leaving the nozzle
= 450 m/s. Final discharge of steam from the second row is axial. Determine
 (i)  blade velocity
 (ii)  diagram efficiency
 (iii)  steam flow rate required in kg/s for a power output of 100 KW.
 (ix)  axial thrust
      (Poona Univ. Winter 1992)

*Soln.*

Given $\qquad\qquad _1\alpha_1 = 25°, \, _1\beta_2 = 30°, \, _2\beta_2 = 30°, \, 2\alpha_1 = 20°$

$$K = (1 - 0.1) = 0.9, \, _1V_1 = 450 \text{ m/s}$$

discharge from final row in axial.

$$u = ?, \, \eta_{\text{dia.}} = ?, \, \dot{m} = ?, \text{ for Power} = 100 \text{ kW}$$

$$Fa = ?$$

Graphical method has been used.
*Note*: These types of triangles are completed by going in reverse order.
Refer to Fig. 15.34 (a), (b).
Second row of moving blade
First draw $KI$ of any convenient length (say – 3 cm) as $u$ is unknown. Draw $_2\beta_2 = 30°$ and draw
perpendicular through the point $K$ to $KJ$ as $_2\alpha_2 = 90°$ (discharge from second row is axial). This meets
the line $IJ$ at point $J$. This completes the outlent velocity $\Delta \, IJK$.

$$\text{Measure } _2Vr_2 = IJ = 3.5 \text{ cm}$$

$$\therefore \qquad _2Vr_1 = \frac{3.5}{0.9} = 3.89 \text{ cm}$$

Draw an angle of $_2\alpha_1 = 20°$ and an arc of radius 3.89 cm with centre $I$ to cut the line $KH$ at point $H$. Join
$KH$. This completes the *inlet* velocity $\Delta \, IHL$ for second row of moving blades.
Measure $KH = _2V_1 = 6.5$ cm
First row of moving blade:
Draw $AC = IK = u$

$$_1V_2 = \frac{_2V_1}{0.9} = \frac{KH}{0.9} = \frac{6.5}{0.9} = 7.22 \text{ cm}$$

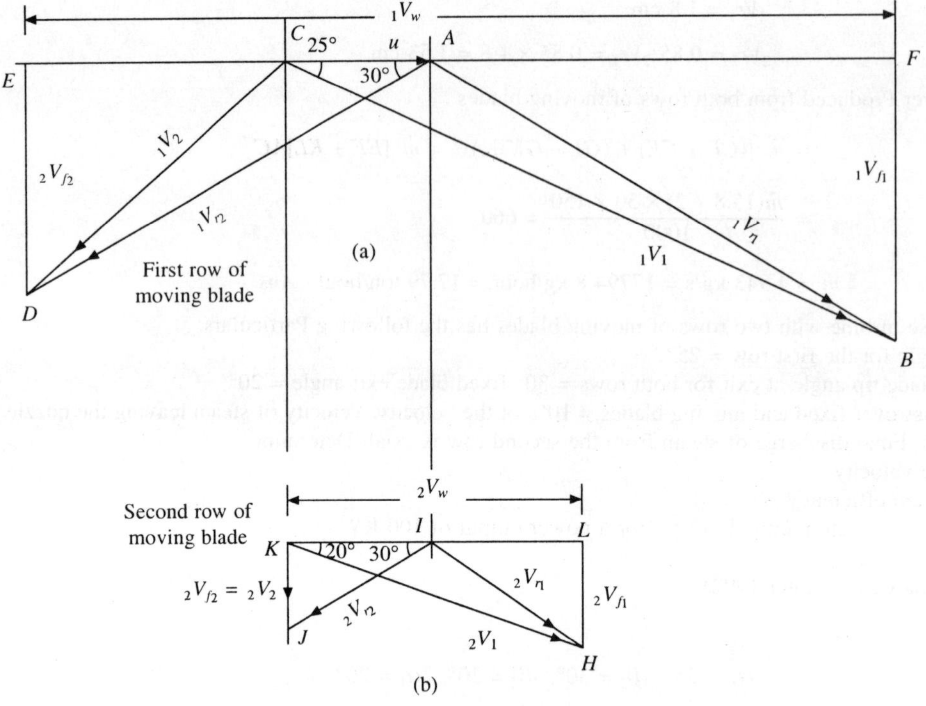



**Fig. 15.34 (a), (b)**

Draw an angle $_1\beta_2 = 30°$ through the point $A$ to the line $AD$. Draw an arc of radius 7.22 cm with centre $C$. This arc cuts the line $AD$ at poit $D$. Join $CD$. this completes the outlet $\Delta ACD$.

$$\text{Measure } _1Vr_2 = AD = 9.25 \text{ cm}$$

$$AB = _1Vr_1 = \frac{Vr_2}{K} = \frac{9.25}{0.9} = 10.27 \text{ cm}$$

Draw an angle $_1\alpha_1 = 25°$ through the point $C$. Draw an arc of radius 10.27 cm with centre $A$. This arc cuts the line $CB$ at $B$. Join $AB$. This completes the inlet velocity triangle. Measure $CB$ from velocity triangle

$$CB = 14 \text{ cm} = _1V_1 = 450 \text{ m/s (given)}.$$

The scale is now calculated from the above

$$\therefore \text{ Scale} \qquad 1 \text{ cm} = \frac{450}{14} = 32.14 \text{ m/s}$$

Measure the following distances from the velocity diagrams and convert into the velocities.

$$_1V_{f_1} = FB = 5.9 \times 32.14 = 189.62 \text{ m/s}$$

$$_1V_{f_2} = ED = 4.7 \times 32.14 = 151.05 \text{ m/s}$$

$$_2V_{f_1} = LH = 2.1 \times 32.14 = 67.49 \text{ m/s}$$

$$_2V_{f_2} = KJ = 1.8 \times 32.14 = 57.85 \text{ m/s}$$

$$_1V_w = EF = 18.1 \text{ cm} = 18.1 \times 32.14 = 581.73 \text{ m/s}$$

$$_2V_w \quad KL = 6.2 \text{ cm} = 6.2 \times 32.14 = 199.26 \text{ m/s}$$

Blade velocity $\qquad u = AC = KI = 3\text{cm} = 3 \times 32.14 = 96.42 \text{ m/s} \quad Ans.$

$\therefore$ Diagram efficiency

$$\eta_{\text{dia.}} = \frac{2u(_1V_w + {}_2V_w)}{_1V_1^2} = \frac{2 \times 96.42(581.73 + 199.26)}{450^2} = 0.7437 = 74.37\% \quad Ans.$$

Steam flow rate for producing 100 kW power

$$\frac{m_s(_1V_w + {}_2V_w)u}{1000} = 100$$

or, $\qquad \dfrac{m_s(581.73 + 199.26) \times 96.42}{1000} = 100$

$\therefore \qquad\qquad m_s = 1.327 \text{ kg/s} \quad Ans.$

Axial thrust

$$F_a = m_s[(_1V_{f_1} - {}_1V_{f_2}) + ({}_2V_{f_1} - {}_2V_{f_2})]$$

$$= 1.327\,[(189.62 - 151.05) + (67.49 - 57.85)] \text{ N} = 63.97 \text{ N} \quad Ans.$$

**15.10** In a two row velocity compounded wheel steam leaves the nozzles at 10 bar and 200°C with a velocity 600 m/s when the blade speed is 120 m/s. The nozzle angle is 16°, while the discharge angles 18° for the first row blades, 21° for the fixed blades and 35° for the second row blades. There is a 10%. drop in velocity during Passage through each row of blades. The mass flow of steam for each set of nozzles is 5 kg/s. calculate the length of arc occupied by the nozzles, if the nozzle height is 30 mm and the wall thickness between them is neglisible. If the blades of the wheel have a pitch of 30 mm and the blade tip thickness at exit is 0.4 mm, calculate the blade exit height for each row.

*Soln.* Given, $Pr = 10$ bar, $t = 200°C$, $_1V_1 = 600$ m/s, $u = 120$ m/s, $_1\alpha_1 = 16°$, $_1\beta_2 = 18°$, $K = (1 - 0.1) = 0.9$, $_2\alpha_1 = 21°$, $_2\beta_2 = 35°$, $m_s = 5$ kg/s, $h_n = 30$ mm $P = 30$ mm, $t_1 = t_f = t_2 = 0.4$ mm. From steam table specific volume of steam corresponding to 10 bar and 200°C

$$v = 0.2059 \text{ m}^3/\text{kg}$$

$\therefore \qquad$ Volume flow rate $m_s v = 5 \times 0.2059 = 1.0295 \text{ m}^3/\text{s}$

For nozzle

$$m_s v = {}_1V_1\, h_n\, l\, \sin{}_1\alpha_1$$

or, $\qquad 1.0295 = 600\,\dfrac{30}{1000} \times l \times \sin 16°$

$\therefore \qquad\qquad l = 0.2074 \text{ m}$

$$= 207.49 \text{ mm}$$

length of nozzle are $l = 207.49$ mm   *Ans.*
Firstly, the velocity triangle has been constructed graphically by selecting scale i.e. 1 cm = 50 m/s
Refer to Fig. 15.35 (a), (b).
Proceed from first velocity triangle.

$_1V_1 = \dfrac{600}{50} = 12$ cm $= CB$. Measure $_1Vr_1 = 9.65$ cm since $_1\alpha_1 = 16°$. Complete the $\triangle ABC$

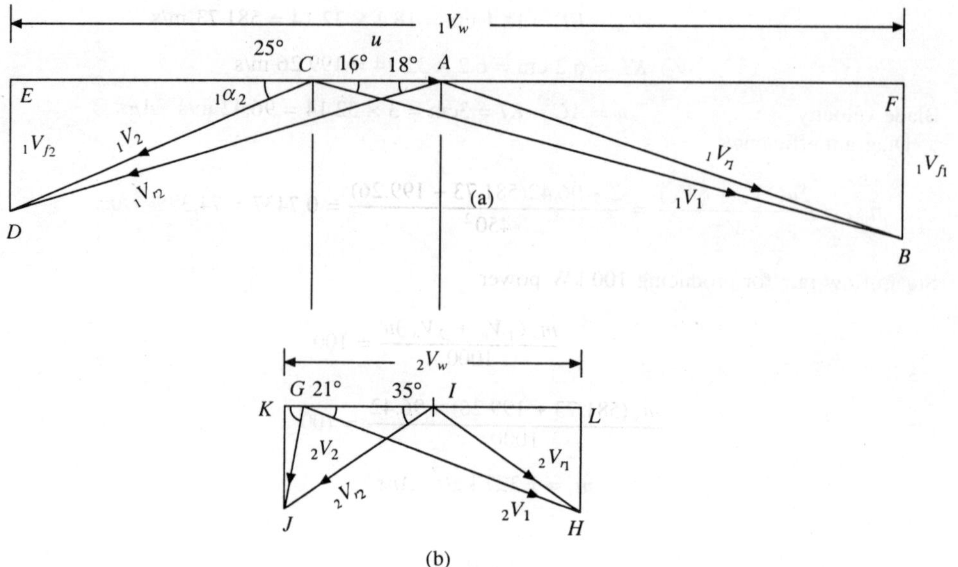

**Fig. 15.35(a), (b)**

Now, $_1Vr_2 = K_1 Vr_1 = 0.9 \times 9.65 = 8.68$ cm, $_1\beta_2 = 18°$ complete the $\triangle ACD$ and measure $_1V_2 = 6.45$ cm
In this way both velocity triangles are constructed for first row of moving blades.
Taking same value of $U = AC = GI$.

$$_2V_1 = K_1 V_2 = 0.9 \times 6.45 = 5.80 \text{ cm}, \ 2\alpha_1 = 21°$$

Complete the $\triangle GIH$ and measure $_2Vr_1 = 3.6$ cm
Now, $_2Vr_2 = K_2 Vr_1 = 0.9 \times 3.6 = 3.24$ cm, $_2\beta_2 = 35°$ complete the $\triangle GIJ$ and measure $_2V_2 = 2$ cm
As per the requirement of the values of velocities for following equations, $_1Vr_2 = AD = 8.68$ cm $= 8.68 \times 50 = 434$ m/s, $_2V_1 = GH = 5.80$ cm $= 5.80 \times 50 = 290$ m/s and $_2Vr_2 = IJ = 3.24$ cm $= 3.24 \times 50 = 162$ cm. By measurement $_1\alpha_2 = 25°$
For first moving row

$$m_s v = \frac{l}{p_1} \, h_1 \, (p_1 \sin {}_1\beta_2 - t_1). \, {}_1Vr_2$$

or,     $$1.0295 = \frac{0.2074}{0.03} \, h_1 \, (0.03 \sin 18 - 0.0004) \times 434$$

$\therefore$     $$h_1 = 0.03868 \ m = 38.68 \text{ mm} \quad Ans.$$

For fixed row

$$m_s v = \frac{l}{p_f} \, h_f (p_f \sin {}_1\alpha_2 - t_f) \cdot {}_2V_1$$

or,     $$1.0295 = \frac{0.2074}{0.03} \, h_f \, (0.03 \sin 25 - 0.0004) \, 290$$

$\therefore$     $$h_f = 0.0418 \ m$$

$$= 41.8 \text{ mm}$$

height of fixed blade $= 41.8$ mm   *Ans.*

For second moving row

$$m_s v = \frac{l}{p_2}\, h_2(p_2 \sin {}_2\beta_2 - t_2)\, {}_2Vr_2$$

or,  $$1.0295 = \frac{0.2074}{0.03}\, h_2\,(0.03 \sin 35 - 0.0004)\, 162$$

$\therefore$  $$h_2 = 0.05469 \text{ m}$$

$$= 54.69 \text{ mm} \quad Ans.$$

**15.11** At the Particular stage of a reaction turbine, the mean blade spead is 160 m/s and the steam is at the pressure of 350 KN/m$^2$ with the temperature of 175°C fixed and moving blades at the stage have inlet angle 30° and exit angle 60°. The blade height $= \dfrac{D}{10}$. The steam flow rate is 13.5 kg/s. $D$ = Drum diameter. Determine (1) blade height (2) Power developed by a pair of fixed and moving blades at this stage. (3) Specific enthalpy drop if stage efficiency is 85%.

*Soln.*

Given  $u = 160$ m/s, Pressure $= \dfrac{350 \times 10^3}{10^5} = 3.5$ bar temp. $= 175°C$, $\alpha_1 = 30° = \beta_2$

$$\alpha_2 = 60° = \beta_1, \ m_s = 13.5 \text{ kg/s}$$

Solving the Problem by graphical method, select the scale 50 m/s = 1 cm.
Refer to Fig. 15.36

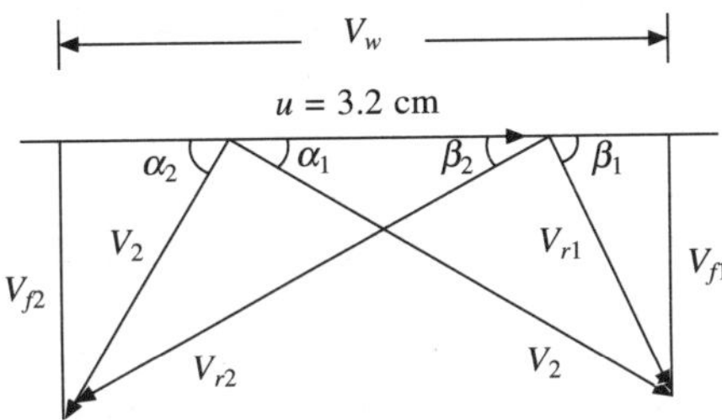


**Fig. 15.36**

$$V_{f1} = 2.7 \text{ cm} = 2.7 \times 50 = 135 \text{ m/s}$$

$$V_w = 6.2 \text{ cm} = 6.2 \times 50 = 310 \text{ m/s}$$

From steam table at 3.5 bar and 175°C
Specific volume $v = 0.576$ m$^3$/kg (superheated) For reaction turbine blade height

$$m_s v = \pi(h + D)h\, V_{f1}, \text{ blade height, } h = \frac{D}{10}$$

$\therefore$  $$h = \sqrt{\frac{m_s v}{\pi_{11} V_{f1}}} = \sqrt{\frac{13.5 \times 0.576}{\pi \times 11 \times 135}} = 0.0408 \text{ m} = 40.82 \text{ mm} \quad Ans.$$

$$\text{Power developed} = m_s V_w u = \frac{13.5 \times 310 \times 160}{1000} = 669.6 \text{ kW} \quad Ans.$$

$$h_{\text{stage}} = \frac{V_w u}{\Delta h}$$

$$\therefore \qquad \Delta h = \frac{V_w u}{\eta_{stage}} = \frac{310 \times 160}{0.85 \times 10^3} = 58.35 \text{ KJ/kg } Ans.$$

**15.12** The outlet angle of a Parson's turbine is 20°. The axial velocity of flow is half the mean blade velocity. If the diameter of the ring is 1.25 m and speed is 3000 r.p.m., determine the blade inlet angles. Also find the power developed if dry saturated steam at 5 bar passes through the blades. The blade height is 6 cm and effect of blade thickness can be neglected.

(Poona Univ. Winter 1994 Summer 1995)

*Soln.* Given $\beta_2 = 20° = \alpha_1$, $V_f = \frac{1}{2}\,U$, $D = 1.25$ m, $N = 3000$ r.p.m., $\beta_1 = ?$, Power developed = ?, Presssure = 5 bar dry saturated, $h = 6$ cm. From steam table, at 5 bar dry saturated steam having specific volume $v = 0.37466$ m³/kg

Refer to Fig. 15.37

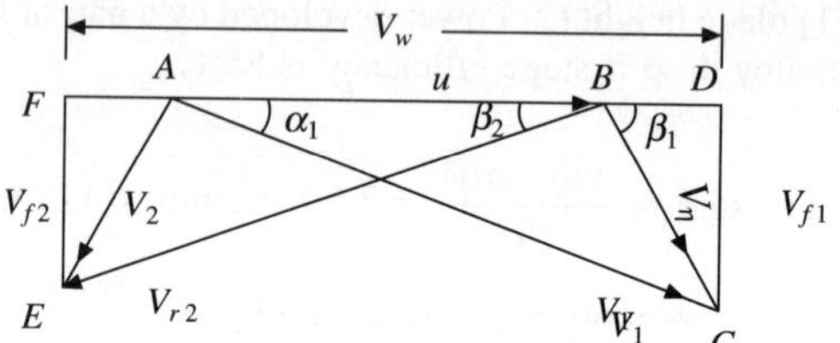

**Fig. 15.37**

$$\text{Blade velocity } u = \frac{\pi(D+h)N}{60} = \frac{\pi \times (1.25 + 0.06) + 3000}{60} = 205.77 \text{ m/s}$$

Select scale 50 m/s = 1 cm

$$u = \frac{205.77}{50} = 4.11 \text{ cm}, \quad V_{f1} = V_{f2} = V_f = \frac{u}{2} = \frac{3.92}{2} = 1.96 \text{ cm}$$

$\beta_2 = \alpha_1 = 20°$, Draw a line *AC* at 20°. From horizontal line *FD* draw a vertical line *DC* so that it is equal to 1.96 cm. Construct the inlet velocity triangle which is same to outlet velocity velocity triangle.

$$V_{f1} = V_{f2} = V_f = 1.96 \text{ cm} = 1.96 \times 50 = 98 \text{ m/s}$$

$$V_w = 6.1 \text{ cm} = 6.1 \times 50 = 305 \text{ m/s}$$

$$\therefore \qquad m_s v = \pi(h + D) \times h\, V_f$$

$$\text{or,} \qquad m_s \times 0.37466 = \pi \times (6 \times 10^{-2} + 1.25) \times 6 \times 10^{-2} \times 98$$

$$\therefore \qquad m_s = 64.8 \text{ kg/s}$$

Power developed

$$= \frac{m_s V_w u}{1000} \text{ kw}$$

$$= \frac{64.58 \times 305 \times 196.34}{1000} = 3867.28 \text{ kW}$$

$$= 3.8672 \text{ MW } Ans.$$

**15.13** A stage of a Parsons steam turbine comprising one ring of moving blades has mean diameter of blade ring 70 cm, rpm 3000, steam velocity at exit of blades 160 m/s, Blade outlet angle 20°, steam flow 7 kg/s. Find blade inlet angle, tangential force on ring of moving blades and power developed in stage. (Solve graphically or analytically).

(Mumbai Univ. Winter 2000)

*Soln.*  Given $D$ = 70 cm, N = 3000 r.p.m,

$$\beta_2 = 20°, m_s = 7 \text{ kg/s}, \beta_1 = ?$$

Tangential force = ?, Power developed = ?
$V_1 = Vr_2 = 160$ m/s. Refer to Fig. 15.38

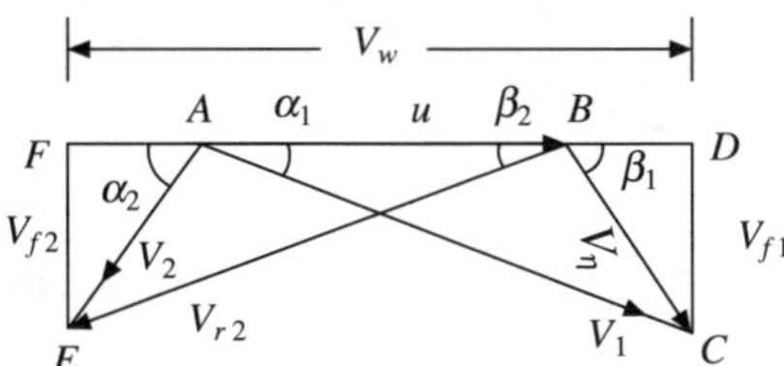

**Fig. 15.38**

$$\text{blade velocity } u = \frac{\pi DN}{60} = \frac{\pi \times 70 \times 10^{-2} \times 3000}{60} = 109.95 \, \text{m/s}$$

$$\text{select scale 40 m/s = 1 cm}, u = \frac{109.95}{40} = 2.74 \text{ cm } V_1 = Vr_2 = \frac{160}{40} = 4 \text{ cm}$$

Draw line $AB = u = 2.74$ cm. Draw line $AC$ at 20° and $BE$ at 20°, since $\alpha_1 = \beta_2 = 20°$. cut these lines for $V_1 = Vr_2 = 160/40 = 4$ cm at points $C$ and $E$. Construct the inlet and outlet velocity triangles. Measure $V_w = 4.7$ cm = 4.7 × 40 = 188 m/s.

Measure blade inlet angle $\beta_1 = 52°$  *Ans.*

Tangential force $\qquad\qquad F_t = m_s \cdot V_w$

$$= 7 \times 188 = 1316 \text{ N} \quad Ans.$$

$$\text{Power developed} = \frac{m_s V_w u}{1000} \text{ kW} = \frac{7 \times 188 \times 109.95}{1000} = 144.69 \text{ kW} \quad Ans.$$

**15.14**  A reaction turbine runs at 3000 r.p.m. and the steam consumption is 20000 kg/hr. The pressure of the steam at a certain pair is 2 bar, its dryness fraction is 0.93 and the power developed by the pair is 50 kW. The discharging blade angle is 20° for both the fixed and moving blades and the axial velocity of flow is 0.72 times the blade velocity. Find the drum diameter and the blade height. Take the tip leakage steam as 8%. Neglect blade thicknes.

(Poona Univ. Summer 1986)

*Soln.*  Given, N = 3000 r.p.m., $m_s$ = 20000 kg/hr, *Pr.* = 2 bar, $x$ = 0.93, Power = 50 kW, $\beta_2 = 20° = \alpha_1$, $V_f$ = 0.72 $u$, $d$ = ?, $h$ = ?, tip leakage = 8%.

Actual mass flowrate $\qquad m_s = \dfrac{20,000}{3600} (1 - 0.08) = 5.111$ kg/s

Refer to Figs. 15.39 and 15.40.
Specific volume of steam

$$V = x \ v_g \text{ at 2 bar}$$

$$= 0.93 \times 0.88540$$

$$= 0.8234 \text{ m}^3/\text{kg}$$

$$\text{blade velocity} \qquad u = \frac{\pi(d+h)N}{60} = \frac{\pi(d+h) \times 3000}{60}$$

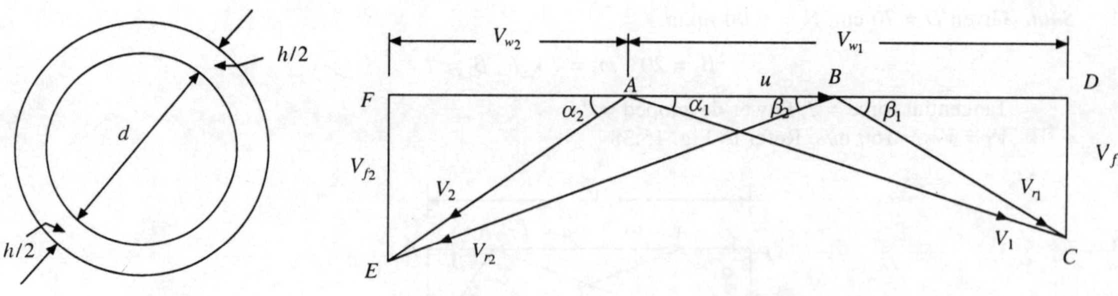

**Fig. 15.39**                    **Fig. 15.40**

or,                    $u = 157.07\,(d + h)$                                        (1)

Where                    $d$ = drum diameter

Mean diameter $D = (d + h)$

$$V_f = 0.72\,u = 0.72 \times 157.07\,(d + h)$$

$$= 113.09\,(d + h) \tag{ii}$$

In $\triangle ACD$

$$\tan \alpha_1 = \tan 20 = \frac{V_{f_1}}{V_{w_1}}$$

or,                    $$V_{w_1} = \frac{113.09\,(d + h)}{0.3639} = 310.77\,(d + h)$$

$$V_{w_2} = V_{w_1} - u$$

$$= 310.77\,(d + h) - 157.07\,(d + h)$$

$$= 153.70\,(d + h)$$

Power developed $= \dfrac{m_s(V_{w_1} + V_{w_2})\,u}{1000}$ kW

or,          $$50 = \frac{5.111\,[\{310.77\,(d + h) + 153.70\,(d + h)\}\,157.07\,(d + h)]}{1000}$$

or,          $9782.82 = (48812.64 + 24141.659)\,(d + h)^2$

or,                    $(d + h)^2 = 0.134$                                        (iii)

or,                    $(d + h) = 0.367$                                        (iv)

Now,                    $m_s V = \pi(d + h)\,h\,V_f$

or,          $5.111 \times 0.8234 = \pi(d + h)\,h \times 113.09\,(d + h)$

or,                    $(d + h)^2\,h = 0.01184$

Taking value of $(d + h)^2$ from eqn. (iii)

or,                    $0.134 \times h = 0.01184$

∴                    $h = 0.0883$ m   *Ans.*

∴                    $d + h = 0.367$

∴                    $d = 0.2787$ m   *Ans.*

**15.15**  A 3-stage steam turbine is fed at 26 bar and 370°C. The exhaust takes place at 0.05 bar. Intermediate pressures are 5 bar and 1 bar. The stage efficiency for all the stages is 80%. Assuming a condition line to be straight. Determine
   (i)  Rankine efficiency
   (ii)  Quality of steam leaving each stage
   (iii)  Reheat factor
   (iv)  work done/kg of steam in each stage
   (v)  overall efficiency.

(Poona Univ. winter 1992)

*Soln.*  According to the condition of steam given points 1, 2, 5 and 6 can be located on Mollier chart. Refer to Fig. 15.41.

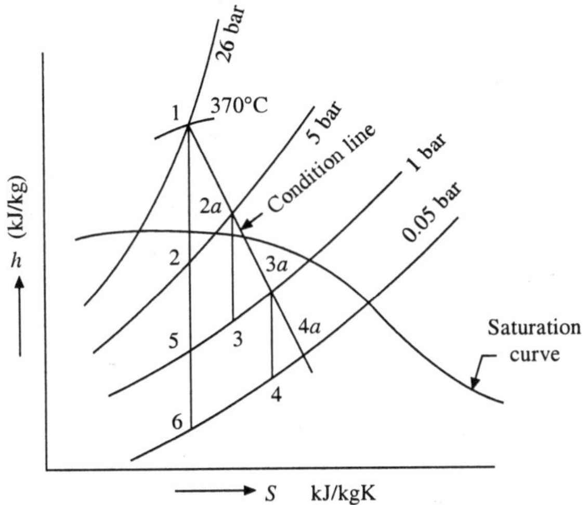


**Fig. 15.41**

Stage efficiency of first stage

$$\eta_{\text{stage(I)}} = \frac{h_1 - h_{2a}}{h_1 - h_2}$$

From chart $h_1 = 3160$ KJ/kg $h_2 = 2780$ KJ/kg

or,
$$0.80 = \frac{3160 - h_2 a}{3160 - 2780}$$

$\therefore$
$$h_{2a} = 2856 \text{ KJ/kg. Locate the point } 2a.$$

Join point 1 and $2a$ and extend the line to the pressure 0.05, being condition line straight. Read the enthalpy value at points 3, $3a$, $4a$, 4 and 6, which are as follow

$$h_3 = 2560 \text{ KJ/kg}, h_{3\alpha} = 2600 \text{ KJ/kg}, h_4 = 2180$$

KJ/kg, $h_{4a} = 2240$ KJ/kg, $h_6 = 2100$ KJ/kg

$$\text{Rankine eff} = \frac{\text{Work done per kg of steam}}{\text{Heat supplied per kg of steam}}$$

$$= \frac{(h_1 - h_2) + (h_{2a} - h_3) + (h_{3a} - h_4)}{(h_1 - h_{f4})}$$

From steam table, enthalpy for liquid water at point 4 i.e. $h_{f4}$ = 137.8 KJ/kg at 0.05 bar

$$\therefore \quad \eta_{\text{Rankine}} = \frac{(3160 - 2780) + (2856 - 2560) + (2600 - 2180)}{(3160 - 137.8)} = 0.3626 = 36.26\% \quad Ans.$$

Reheat Factor $\quad\quad (R.F.) = \dfrac{(h_1 - h_2) + (h_{2a} - h_3) + (h_{3a} - h_4)}{(h_1 - h_6)}$

$$= \frac{(3160 - 2780) + (2856 - 2560) + (2600 - 2180)}{(3160 - 2100)}$$

$$= 1.033 \quad Ans.$$

There are three stages. It is assumed that equal work is obtained from each stage.

$$\text{Total work} = (h_1 - h_{2a}) + (h_{2a} - h_{3a}) + (h_{3a} - h_{4a})$$

$$= (h_1 - h_{4a})$$

$$= (3160 - 2240) = 920 \text{ KJ/kg}$$

$\therefore$ work done/kg of steam in each stage

$$= \frac{920}{3} = 306.67 \text{ KJ/kg} \quad Ans.$$

Overall efficiency $\quad\quad\quad = \dfrac{h_1 - h_{4a}}{h_1 - h_6}$

or, $\quad\quad\quad\quad\quad\quad \eta_{\text{overall}} = \dfrac{3160 - 2240}{3160 - 2100}$

$$= 0.8679$$

$$= 86.79\% \quad Ans.$$

**15.16**  Steam at 15 bar and 350°C is expanded through a 50% reaction turbine to a pressure of 0.14 bar. The stage efficiency is 75% for each stage and the reheat factor is 1.04. The expansion is to be carried out in 20 stages and the total diagram power is to be 12 MW. Calculate the flow rate of steam required assuming that all stages develop equal power. At one stage, the steam pressure is 1 bar and dry saturated. The exit angle of blade is 20° and the blade speed ratio is 0.7. If the blade height is 1/12 th of the mean rotor diameter, calculate the rotor diameter and the speed.

(Poona Univ. Summer 1993)

*Soln.*  Given $\eta_{\text{stage}}$ = 75%, R.f. = 1.04, Diagram Power = 12 MW

$$\beta_2 = 20°, \rho = 0.7, h = \frac{1}{12} D$$

Refer to Figs. 15.42 and 15.43
From Mollier chart

$$h_1 = 3150 \text{ kJ/kg}$$

$$h_2 = 2300 \text{ kJ/kg}$$

Isentropic enthalpy drop = $h_1 - h_2$ = (3150 − 2300) = 850 kJ/kg

$$\eta_{\text{Internal}} = \eta_{\text{stage}} \times \text{R.F.} = 0.75 \times 1.04 = 0.78$$

$$\eta_{\text{overall}} = \frac{\text{Actual enthalpy drop}}{\text{Isentropic enthalpy drop}}$$

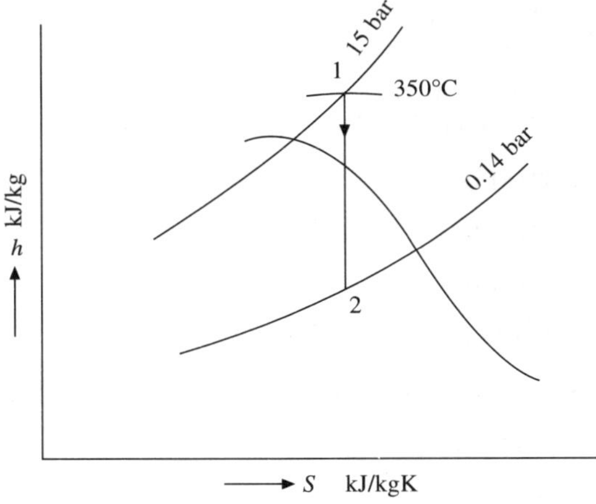

**Fig. 15.42**

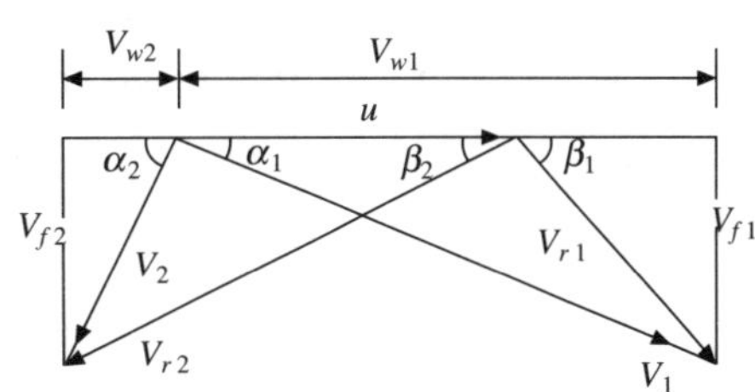

**Fig. 15.43**

or,
$$0.78 = \frac{(\Delta h)_a}{850}$$

$\therefore$
$$(\Delta h)_a = 663 \text{ KJ} \qquad \text{/kg}$$

This actual enthalpy drop occurs in 20 stages.

$\therefore$
$$\text{Work done/stage/kg of steam} = \frac{663}{20} = 33.15 \text{ kJ}$$

$$\text{Total power output} = 12 \text{ MW}$$

$$\text{Power produced per stage} = \frac{12}{20} \times 1000 = 600 \text{ KW}$$

$\therefore$
$$600 = m_s \times 33.15$$

$$m_s = 18.099 \text{ kg/s} \quad \textit{Ans.}$$

$$V_{f_1} = V_{f_2} = V_f$$

Now,

$$m_s \times v = \pi D H \, V_f$$

or,
$$m_s v = \pi D \times \frac{D}{12} \, V_f$$

$$m_s v = \frac{\pi D^2}{12} \, V_f$$

$$V_w = V_{w_1} + V_{w_2}$$

or,
$$V_w = V_1 \cos \alpha_1 + (V_{r_2} \cos \beta_2 - u)$$

$$= V_1 \cos \alpha_1 + (V_1 \cos \alpha_1 - u) \, [\because \; V_{r_2} = V_1, \beta_2 = \alpha_1]$$

$$= 2V_1 \cos \alpha_1 - u$$

$$\rho = \frac{u}{V_1} = 0.7 \; \text{ or, } \; u = 0.7 V_1$$

$$\text{Power produced} = \frac{m_s V_w U}{1000} \text{ kW}$$

or,
$$600 = \frac{18.099 \times V_w U}{1000} \text{ kW}$$

or,
$$V_w u = 33151.003$$

or,
$$(2V_1 \cos \alpha_1 - U)U = 33151.003$$

or,
$$2 \times 0.7 \, V_1 \times V_1 \cos 20° - (0.7 \, V_1)^2 = 33151.003$$

or,
$$V_1 = 200.38 \text{ m/s}$$

$\therefore$
$$V_f = V_{f_1} = V_{f_2} = V_1 \sin \alpha_1$$
$$= 200.38 \sin 20°$$
$$= 68.53 \text{ m/s}$$

$\therefore$
$$m_s v = \frac{\pi D^2}{12} V_f$$

from steam table at 1 bar (dry saturated)

$$\text{SP. Volume } v = v_g = 1.6937 \text{ m}^3/\text{kg}$$

$\therefore$
$$18.099 \times 1.6937 = \frac{\pi D^2}{12} \times 68.53$$

$$D = 1.307 \text{ m}$$

Mean rotor diameter = 1.307 m

Rotor diameter $d = D - h = D - \dfrac{D}{12} = \dfrac{11 \times 1.307}{12} = 1.198 \text{ m}$   *Ans.*

$$u = 0.7 \, V_1$$
$$= 0.7 \times 200.38 = 140.267 \text{ m/s}$$

$$u = \frac{\pi D N}{60}$$

or,
$$140.267 = \frac{\pi \times 1.307 \, N}{60}$$

Speed
$$N = 2050 \text{ r.p.m.}   \textit{Ans.}$$

## EXERCISES

15.1  Objective questions
   (i)  In an impulse turbine, the energy supplied to the blades per kg of steam equals to
      (a)  Work done by steam
      (b)  Sum of kinetic energy and potential energy at inlet
      (c)  Reaction of energy
      (d)  Kinetic energy of jet at entrance per kg of steam.
   (ii)  The pressure drop on the two sides of the moving blades in case of reaction turbine
      (a)  increases                          (b)  decreases
      (c)  reamins constant                   (d)  none of the above.

  (iii)  In case of reaction steam turbine
      (a)  the steam is expanded in nozzles only
      (b)  the steam is expanded in moving blades only
      (c)  the steam is expanded both in fixed and moving blades continuously
      (d)  non of the above.
  (iv)  The steam consumption in case of steam turbines as compared to reciprocating steam engine is
      (a)  more        (b)  same
      (c)  less        (d)  can be more or les
  (v)  The pressure velocity compounding of steam turbine results in
      (a)  shorter turbine for a given total pressure drop
      (b)  large turbine for a given pressure drop
      (c)  large number of stages
      (d)  lesser frictional losses
  (vi)  Stage efficiency of a turbine is equal to
      (a)  Work done on blades/Total energy supplied
      (b)  Work done on blades/Energy supplied per stage
      (c)  Energy supplied per stage/work done on blades
      (d)  total energy supplied/work done on blades
  (vii)  Stage efficiency is also called
      (a)  diagram efficiency        (b)  blade efficiency
      (c)  gross efficiency        (d)  none of the above
  (viii)  Starting torque of ideal steam turbine is
      (a)  Approximately half of the torque at its most efficient speed.
      (b)  approximately twice of the torque at its most efficient speed
      (c)  Approximately same as the torque at its most efficient speed.
      (d)  gradually increasing with speed.
  (ix)  In case of reaction turbines, the heat drop takes place in
      (a)  fixed blade only        (b)  moving blades only
      (c)  both fixed and moving blades        (d)  none of the above
  (x)  Degree of reaction is defined as the ratio of
      (a)  enthalpy drop in the moving blades/enthalpy drop in fixed blades
      (b)  enthalpy drop in moving blades/sum of enthalpy drop in moving and fixed blades
      (c)  enthalpy drop in fixed blades/sum of enthalpy drop in moving and fixed blades
      (d)  none of the above
  (xi)  For parson's reaction turbine, degree of reaction is equal to
      (a)  80%        (b)  50%
      (c)  75%        (d)  100%
  (xii)  Reheat factor (R.F.) is defined as the
      (a)  ratio of cumulative enthalpy drop to isentropic enthalpy drop
      (b)  product of cumulative enthalpy drop and isentropic enthalpy drop
      (c)  ratio of isentropic enthalpy drop to cumulative enthalpy drop.
      (d)  non of the above
  (xiii)  The value of reheat factor increases of the number of stages
      (a)  are less        (b)  are large
      (c)  are constant        (d)  non of the above
  (xiv)  The value of reheat factor varies normally between
      (a)  1.0 to 1.5        (b)  1.5 to 2.0
      (c)  1.02 to 1.06        (d)  none of the above
  (xv)  Internal turbine efficiency is defined as the ratio of
      (a)  $\dfrac{\text{total useful heat drop}}{\text{adiabatic heat drop}}$        (b)  $\dfrac{\text{adiabatic heat drop}}{\text{heat supplied}}$
      (c)  $\dfrac{\text{useful heat drop}}{\text{heat supplied}}$        (d)  non of the above

(xvi) Rankine efficiency is defined as the ratio of

(a) $\dfrac{\text{total useful heat drop}}{\text{adiabatic heat drop}}$    (b) $\dfrac{\text{adiabatic heat drop}}{\text{heat supplied}}$

(c) $\dfrac{\text{useful heat drop}}{\text{heat supplied}}$    (d) non of the above

(xvii) The ratio, useful heat drop to heat supplied, gives
  (a) Rankine efficiency    (b) stage efficiency
  (c) overall efficiency    (d) internal turbine efficiency

(xviii) In pressure compounded impulse turbine all the pressure drop takes place in
  (a) moving blades    (b) nozzles
  (c) both moving blades and nozzles    (d) none of them

(xix) The number of stages required in case of pressure compounded turbine as compared to velocity compounded turbine is
  (a) less    (b) more    (c) same    (d) none of the above

(xx) The efficiency of velocity-compounded steam turbine as compared to pressure-compounded turbine is
  (a) less    (b) more    (c) same    (d) may be more or less

**Answers**

(i) c    (ii) a    (iii) c    (iv) c    (v) a    (vi) b    (vii) c    (viii) b    (ix) c
(x) b    (xi) b    (xii) a    (xiii) b    (xiv) c    (xv) a    (xvi) b    (xvii) c    (xvlll) d
(xix) b    (xx) a

15.2 In an impulse turbine, the fixed nozzle angle is $\alpha$, the blades are equiangular, the blade velocity coefficient is $K$, show that maximum blade efficiency is

$$\frac{(1+k)\cos^2\alpha}{2}$$

15.3 Discuss the advantages and disadvantages of steam turbine as compared to steam engine.

15.4 Explain the difference between impulse and reaction turbines.

15.5 What is the Purpose of compounding of steam turbines?

15.6 What is the difference between the working principles of steam engines and steam turbines?

15.7 Which type of turbine is used in most of steam power plants?

15.8 What is the effect of blade friction on turbine performance?

15.9 What are the advantages and disadvantages of velocity compounded empulse turbine?

15.10 Derive an expression for maximum blade efficiency of a reaction turbine.

15.11 What is meant by degree of reaction?

15.12 What is meant by reheat factor in turbine practice?

15.13 What are the different losses that occur in a steam turbine?

15.14 Show that for maximum diagram efficiency of a reaction turbine, the blade speed ratio $\rho$ is equal to cos $\alpha$, where $\alpha$ is the angle made by the absolute velocity of steam at inlet with the direction of motion. What is the maximum work done/kg and maximum diagram efficiency under this condition?

15.15 What are the types of blades used in impulse turbines? Explain 'profile blades'. What care will be taken by you for the tip of the blade to reduce the loss of energy of stream and increase the tip strength? Show that in case of impulse turbine blades.

$$\sin \beta = \frac{m_s v}{h\,\lambda V} + \frac{t}{p}$$

Where    $\lambda$ = Effective length of arc over which the steam is flowing
  $p$ = Mean circumferential pitch of blades
  $t$ = Thickhess of blade tip at outlet edge
  $\beta$ = outlet angle of blade

$h$ = Radial height of blade
$V$ = Relative velocity of steam at outlet
$v$ = Specific volume of steam
$m_s$ = mass flow rate of steam, kg/s

15.16  What is the effect of Partial load on blade angles in impulse turbine?

15.17  Explain why velocity compounded stage is placed at the high pressure end of multistage turbine?

**Numerical Problems**

15.18  Steam issues from the nozzles of De-Laval turbine with a velocity of 1000 m/s. The nozzle angle is 20°. The mean blade velocity is 400 m/s. The blades are symmetrical. The mass flow rate is 1000 kg/s. Friction factor is 0.8 and nozzle efficiency is 95 percent. Determine (a) Blade angles (b) Axial thrust on the rotor turbine (c) work done per kg of steam per second (d) Power developed.

(Ans. (a) $\beta_1 = \beta_2 = 32$ (b) $F_a = 19.04$ N (c) 392.53 kW/kg (d) 1089.28 kW

(e) $\eta_{blade} = 78.3\%$ (f) $\eta_{stage} = 74.4\%$

15.19  A single stage impulse turbine rotor has a mean diameter of 1.2 m and runs at 3000 rpm. The nozzle angle is 18°, blade speed ratio is 0.42. Blade velocity co-efficient is 0.9. The outlet angle of blade is 3° smaller than the inlet angle. The steam flow rate is 5 kg/s. Draw the velocity diagram and find the followings (a) Velocity of whirl (b) Axial thrust on the bearing (c) Blade angle and (d) Power developed.

(Ans. 462.5 m/s, 110.85 N, $\beta_1 = 30°$, $\beta_2 = 27°$, 435.88 kW)

15.20  A simple impulse turbine is supplied with steam at a pressure of 10.3 bar and 40°C superheat. The pressure in the wheel chamber is 1.03 bar and the nozzle efficiency is 90%. If the nozzle angle is 20°, velocity co-efficient 0.8 and blades are equiangular, determine for maximum efficiency (a) the blade speed (b) the blade angles (c) power developed for a steam flow of 9000 kg/hg, and (d) blading efficiency.

(*Ans.* 408.8 m/s, 36°, 756 kW, 80%)

15.21  In a stage of an impulse turbine steam leaves the nozzles at 1000 m/s. Nozzle angle is 20°. Mean blade speed is 400 m/s, Efficiency of nozzle is 0.95. Friction factor for blades is 0.8. Mass flow rate of steam is 0.5 kg/s. Find (a) Blade angles (b) Axial thrust (c) Output in kW (d) Diagram efficiency (e) stage efficiency, state the assumption made.                    (Poona Univ. Winter 1993)

(Ans. $\beta_1 = \beta_2 = 32.49°$, 33.71 N, 114.16 kw, 45.67%, 43.37%)

15.22  A stage of an axial flow impulse steam turbine has two rows of moving blades separated by a row of fixed guide blades. The inlet and exit angles of the moving blades in both rows are 30° measured from plane of rotation. Steam is discharge axially from the second row. The blade speed is 120 m/s. The relative angle of the steam at entry to each blade row is equal to the blade inlet angle. The friction over each blade row, fixed and moving, reduces the velocity of steam by 15%. Determine

(i)  velocity of steam leaving the nozzle
(ii)  work done/kg of steam
(iii)  blade efficiency
(iv)  axial thrust/kg of steam.

(Poona Univ. Summer 1992)

15.23  The following data refers to an impulse turbine: Isentropic nozzle heat drop = 185 kJ/kg. Reheat of steam due to blade friction = 10% of isentropic drop. Nozzle angle = 20°. Ratio of blade speed to whirl component of steam speed = 0.5. Velocity co-efficient for the blades = 0.95. The velocity of steam at entry of nozzle = 30 m/s. Calculate (a) blade angles if the steam leaves axially, (b) work done per kg, (c) friction loss over the blades, and (d) kinetic energy loss.

(*Ans.* $\beta = 35.6°$, $\beta_2 = 30.8°$, 146.9 kJ/kg, 5.51 kJ/kg of steam, 13.95 kJ)

15.24  One stage of an impulse turbine consists of a converging nozzle and one ring of moving blades. The nozzles are inclined at 22° to the blades whose tip angles are both 35°.

(a)  If the velocity of steam at exit from the nozzle is 660 m/s, find the blade speed so that the steam shall pass on without shock and find the diagram efficiency neglecting losses, if blades run at this speed

(b)  If the relative velocity of steam is reduced by 15% in Passing through the blade ring, find the efficiency and the end thrust on the shaft when the blade ring developes 1.7 KW.

(A.M.I.E. Winter 1989)

(*Ans.* 85.25%, 260 m/s, 78.25%, 0.449 N)

15.25  The speed of an impulse turbine with a single row wheel is 2400 r.p.m. and the diameter of the blades is 95 cm. The blade velocity co-efficient is 0.8 and the speed ratio is 0.4. The nozzle angle is 20°. The outlet angle of the blade is 2° less than the inlet blade angle. Draw the velocity diagram and calculte for a steam flow of 5.4 kg/s (a) axial thrust on the blades (b) Power developed, and (c) blading eff.

(*Ans.* 54.74 N, 180.8 KW, 79%)

15.26  The stage of a velocity compounded impulse turbine has two rows of moving blades having exit angles 20° and 30° respectively. The nozzle angle is 20°. The blade speed is 160 m/s. The isentropic enthalpy drop in the nozzle is estimated to be 336.2 kJ/kg of steam having a nozzle efficiency of 90%. Assuming a blade velocity co-efficient of 0.9 for each ring of fixed and moving blades, determine the work done per kg of steam, the diagram efficiency and the stage efficiency.

(*Ans.* 235.35 kJ/kg, 78%, 70.2%)

15.27  The first stage of an impulse turbine is compounded for velocity and has two rings of moving blades and one ring of fixed blades. The nozzle angle is 20° and leaving angles of the blades are respectively: first moving 20°, fixed 25°, second moving 30°. The velocity of the steam leaving the nozzle is 570 m/s, the blade speed is 120 m/s and the steam velocity relative to the blades is reduced by 10% during the passage through each ring. Find the diagram efficiency under these conditions and the power developed for a steam flow of 3.6 kg/second.

(*Ans.* 78.89%), 461.376 kW)

15.28  In a two row velocity compounded impulse turbine the steam issued from the nozzle has a velocity of 440 m/s. The outlet angle of moving blades are 30°. The blade velocity co-efficient for each ring of moving and fixed blades is 0.9. Designing for final axial discharge, determine the blade velocity and the diagram efficiency.

(*Ans.* 97 m/s, 79%)

15.29  In a reaction turbine the diameter of the rotor is 2 m and its speed is 840 r.p.m. The steam consumption amounts to 870 kg/min. The height of the blade at a particular stage is 15 cm. The exit angle of the nozzles and the moving blades is 25°. The pressure at this stage is 0.3 bar and steam is 0.98 dry. Estimate the Power developed and the heat drop in kJ/s during the stage if the turbine efficiency is 78%.

(*Ans.* 199.4 kW, 17.63 kJ/s)

15.30  One expansion in a reaction turbine has 8 pairs of blades of outlet angle of 20°. The mean diameter of the rotor is 50 cm and the rotor speed is 3000 r.p.m. The ratio of blade speed to steam speed is 0.8. The efficiency for the stage is 80%. Determine the Power developed and the adiabatic heat drop during the expansion for a steam flow rate of 5 kg/s.

(*Ans.* 332.9 kW, 76.8 kJ)

15.31  A reaction turbine runs at 600 r.p.m. and consumes 18000 kg/hr of steam. The exit angle of the fixed and moving blades are 20°. The axial velocity of flow is 0.75 times the blade Velocity. Determine the drum diameter and blade height of a particular stage where steam pressure is 2 bar and it is 0.95 dry. Assume the power developed to be 15 KW and the tip leakage of steam as 7%.

(*Ans.* 0.9728 m, $h$ = 5.05 cm)

15.32  Steam is supplied to a 3-stage turbine at 40 bar and 400°C and the exhaust to the condenser takes place at 50 milli-bar, with a wetness fraction of 12%. The work developed in three stages is 1 : 1 : 2. The condition line may be assumed as straight line. Determine
  (i)  condition at entry of each stage
 (ii)  stage efficiencies
(iii)  reheat factor
 (iv)  internal efficiency of turbine

(Ans. 40 bar, 400°C, 11.4 bar, 265°C, 2.5 bar, 135°C, 75.4%, 77.24%, 83.33%, 1.04, 82.82%)

15.33  In a four stage impulse turbine, the steam is supplied at 350°C and at a pressure of 20 bar. The exhaust pressure is 0.05 bar and overall turbine efficiency is 80%. Assuming that the work is equally divided between the stages and the condition line to be straight, estimete, (a) stage pressure (ii) the efficiency of each stage (iii) reheat factor.

(*Ans.* 9 bar, 2.4 bar, 0.5 bar, 72.5%, 76%, 78%, 82.5%, 1.04)

15.34  A three-stage steam turbine is supplied with steam at a pressure of 25 bar and temperature 400°C. The exhaust pressure is 0.07 bar and the stage pressure are such as to divide the ideal expansion line between the initial and final pressures into three equal parts. If the mean stage efficiency be 75%, find the reheat factor and the internal efficiency ratio of the turbine. Find also the Power developed when the rate of steam flow is 6 kg per second.

(*Ans.* Rankine drop = 1060 kJ, R.F. = 1.05, $\eta$ = 78.8%, 5010 KW)

16

# Fuels and Combustion

## 16.1  INTRODUCTION

Any substance that reacts chemically with an oxidising agent and may be used for heating or lighting purpose is called fuel. The fossil fuels are responsible for the production of about 80% of the world's total energy to day. Fuels may be solid, liquid or gaseous, depending upon the form in which they are available. Some fuels like coal petrolium and natural gas are available in nature while some other fuels like charcoal, coke, gasoline, alcohal, producer gas and water gas are obtained either during processing of naturally available fuel or as industrial by-products. Most of the fuels contain carbon, hydrogen, oxygen, Nitrogen and sulphur in varying propertions. The fuels contain chemical energy which is released during combustion.

## 16.2  CLASSIFICATION OF FUELS

Fuels are classified into the following categories:

### (a) Solid Fuels

Among the solid fuels, coal is used in most of the applications. Coal is the end product of a natural process of decomposition of vegetable matter burried in swamps and left out of contact with oxygen for millions of years. The different types of coal vary in texture, purity, water content, composition and physical properties. The varieties of available coal in the approximate order of their formation are: (i) Peat (ii) lignite (iii) bituminus (iv) anthracite.

(i) *Peat*:   Peat is the first stage in the formation of coal. It is light brown to dark brown in colour with the fibrous remains of vegetation often clearly visible. Peat may contain 85% to 90% moisture, 5.5 to 10.5% volatile matter and 4.5 fixed carbon. A chemical analysis yields carbon 55%–59%, hydrogen 5.5%–6%, nitrogen 1.5%–2% and oxygen 33%–40%. It has calorific value of 11,300 kJ/kg.

(ii) *Lignite*: It is the second stage of formation of coal. It is brown and laminar in structure and remnants of wood fibre are often visible in it. It originates mostly from resin rich plants and is therefore, high in contents of both moisture and volatile matter. Typical composition of lignite is 33% moisture, 30% volatile material, 30% fixed carbon and the rest ash. Its calorific value varies from 14000 kJ/kg to 18000 kJ/kg. Because of the high moisture content and low heating value, lignite is not economical to transport over long distances and usually it is burned at the mine site.

(iii) *Bituminous coal*: The largest group, bituminous coal is a broad class of coals containing 46 to 86% fixed carbon and 20 to 40% of volatile matter. Bituminous coals usually burn easily especially in pulverised form. Its calorific value varies from 23000 kJ/kg to 35000 kJ/kg. The volatile types are used in gas plants and coal tar distilleries while the coking types are used to make coke.

Sub bituminous coal. This is a class of coal with generally lower calorific value than bituminous coal, between 19000 to 23000 KJ/kg. It has higher moisture content as much as 15 to 30 percent but often low in sulphur content. It is brownish black or black and mostly homogeneous in structure. It is also burned generally in pulverised form.

(iv) *Anthracite coal*: It is the best kind of coal available. It contains 86 to 98% fixed carbon, moisture content less than 2 to 14%, volatile matter 1.5 to 13 percent, ash 8 to 12%. The chemical composition of anthracite is: carbon 93 to 95%, hydrogen 2 to 4% and rest ash. The true anthracite coal is a black or blue black coal with a hard, metallic clusture. Specific gravity varies from 1.3 to 1.7 and calorific value is 33600–35000 kJ. It burns with a blue flame without smoke. It is extremely useful for metallurgical purposes.

## *Coal Analysis*

There are two types of coal analysis: Proximate and ultimate both done on a mass percent basis. In proximate analysis, we determine:

1. The moisture content
2. Volatile matter
3. Fixed carbon and
4. ash.

In ultimate analysis, we determine the composition of ultimate constituents such as: (1) Carbon, (2) Hydrogen (3) Oxygen (4) Sulphur (5) Nitrogen and (6) Ash.

Proximate analysis is used for commercial purposes as it furnishes data regarding the commercial properties of each while ultimate analysis is used for calculation and scientific work.

## (b) Liquid fuels

Technically liquid fuels are an excellent energy source. They are usually a mixture of hydro carbons that may be represented by a molecular formula $c_n H_m$, where $m$ is a function of $n$ that depends upon the family of the hydrocarbon. Four important families are: (1) Paraffins ($C_n H_{2n+2}$), (2) oleffins ($C_n H_{2n}$) (3) Nafthines ($C_n H_3 n$) (4) Aeromatics ($C_n H_{2n6}$). The commonly used liquid fuels for power generation are: Petrol, Paraffins, diesel oil and heavy oil. The followings are the advantages of liquid fuels over solid fuels:

(1) Higher calorific value
(2) Easy control of combustion

(3) Economy in space
(4) Easy handling and supply
(5) Easy starting and stopping
(6) Elimination of wear and tear of the grate bars.

*Flash Point*:   Flash point of a liquid fuel is the minimum fluid temperature at which the vapours coming from the fluid surface will just ignite.

*The Burning Points*:   The burning point of a liquid fuel is the temperature at which the liquid fuel will produce combustible vapours rapidly enough to burn continusouly. It is usually 9 to 11°C above the flash point.

*The Pour Point*:   The pour point of a petrolium product is the lowest temperature at which an oil to oil product will flow under standard conditions.

(c) *Gaseous Fuels*:   Almost all gaseous fuels are either fossil fuels or by products of fossil fuels. These fuels can be divided into three general categories: (i) natural gas (ii) Manufactured and (iii) byproduct gas. The gaseous fuels of engineering importance are: natural gas, coal gas, brown gas, coke oven gas, blast furnace gas and producer gas.

The advantages of gaseous fuels over solid fuels are:

(1) The gas can be produced at one place and distributed over a wide area through pipe line.
(2) Remote and instant control of combustion is possible.
(3) Smoke and ash disposal difficulties are removed.
(4) Complete combustion without pollution is possible due to uniform mixing of air and fuel
(5) Even low grade fuel can be used for gasification.

## 16.3   COMBUSTION

Combustion is the rapid oxidation of an oxidizable material with any oxidizing agent resulting in the release of thermal energy usually accompained by a flame. The flame may be luminous or may not be even observable. Combustion, in general, includes any fuel and any arbitrary oxidant, but in this chapter we shall consider oxygen present in air as oxidant. Carbon is one of the most important combustible elements and is an essential part of any hydrocarbon compound. Hydrogen and sulphur are another constituents of fuel taking part into combustion. The oxidation of carbon is slow and more difficult than that of either hydroyen and sulphure. Although carbon has a lower ignition temperature (407°C) than hydrogen. Carbon is a high temperature solid and as such burn relatively slowly, consequently carbon partly produces carbon monoxide whereas complete combustion of hydrogen and sulphur takes place. Hydrogen has highest ignition temperature 502°C of the three combstibles, but since it is a gas, the kinetics are such that the combustion of hydrogen proceeds very rapidly.

### 16.3.1   Combustion Reactions

The chemical equations for complete combustion of carbon, hydrogen and sulphur are written as:

$$C + O_2 \rightarrow CO_2$$

$$H_2 + 1/2\,O_2 \rightarrow H_2O$$

$$S + O_2 \rightarrow SO_2 \tag{16.1}$$

The following conclusions can be drawn from the above equation for carbon:
(a) On molal basis

$$(1 \text{ mole of C}) + (1 \text{ mole of } O_2) \rightarrow (1 \text{ mole of } CO_2)$$

(b) On volume basis

$$(1 \text{ volume of C}) + (1 \text{ volume of } O_2) \rightarrow (1 \text{ volume of } CO_2)$$

The volume (or mole) of the carbon solid fuel is negligible with volume of air, Hence 1 volume of $O_2$ with carbon gives 1 volume of $CO_2$
(c) On mass basis

$$(12 \text{ kg of C}) + (32 \text{ kg of } O_2) \rightarrow (44 \text{ kg of } CO_2)$$

This shows that total mass of reactions is equal to total mass of products.

Similar conclusions can be drawn from other two chemical reactions. The combustion equation for a general hydrocarbon having chemical formula ($C_xH_y$) can be written as:

$$C_xH_y + (x + y/4)\, O_2 \rightarrow x\, CO_2 + (y/2)\, H_2O \tag{16.2}$$

on mole basis, 1 mole of $C_x H_y$ requires $(x + y/4)$ moles of $O_2$ to produce $x$ mole of $CO_2$ and $y/2$ moles of $H_2O$.

### 16.3.2 Composition of Air

In industrial practice, the combustion of fuels is normally accomplished by supplying air as the oxidiser rather than pure oxygen. Atmospheric air is a mixture of oxgyen, nitrogen, argon water vapour, carbon dioxide and minute amounts of other gases in varying amounts. The amounts of water vapour, carbon dioxide and other gases are so small that may be neglected and then air is considered as a mixture of 21% oxygen, 78% nitrogen and 1% orgons by volume. Since the molecular weights of oxygen, nitrogen and argon are 32, 28 and 40, the molecular weight of air becomes $0.21 \times 32 + 0.78 \times 28 + 0.01 \times 40 = 28.96$. But it is a common practice to neglect argon also and consider air as a mixutre of 21% oxygen and 79% nitrogen on volume basis having molecular weight 28.16 (molecular weight of atmospheric nitrogen) so that molecular weight of air remains same $(0.21 \times 32 + 0.79 \times 28.16 = 28.96)$. The gravimetric composition of air then turns out to be 23.2% of oxygen and 76.8% of nitrogen. Therefore composition on mass basis becomes:

$$\boxed{\begin{aligned} O_2 &= 23.2\% \\ N_2 &= 76.8\% \end{aligned}} \quad \text{gravimetric}$$

For each kg of $O_2$ supplied by air, we have

$$76.8/23.2 = 3.31 \text{ kg of } N_2 \text{ and } 100/23.2 = 4.31$$

kg of air composition on volume basis.

$$\boxed{\begin{aligned} O_2 &= 21\% \\ N_2 &= 79\% \end{aligned}}$$

For each mole of $O_2$, we have $79/21 = 3.76$ moles of $N_2$ and $100/21 = 4.76$ moles of air.

Nitrogen of air does not take part into reaction and remain as it is in the products. At very high temperature oxides of nitrogen are formed.

### 16.3.3 Theoretical Air-Fuel Ratio $(Z_s)$

The theoretical or stoichiometric air fuel ratio is the minimum air required for the complete combustion of unit mass of a fuel. It is usually denoted by $z_s$.

*Actual air fuel ratio (z)*

It is defined as the actual mass of air required per unit mass of fuel. In general, actual air fuel ratio is greater than the theoretical air fuel ratio to insure complete combustion.

*Percentage of Excess Air*

The expression $(z\text{-}z_s)/z_s \times 100$ is called percentage of excess air. Due to large amount of excess air, mixture of air and fuel becomes dilute with respect to fuel and we define dilution coefficient (DC).

*Dilution coefficient* (D.C.) It is defined as the ratio of actual air fuel ratio to the theoretical air fuel ratio. $DC = z/z_s$.

### 16.3.4 Minimum Air required in kg per kg of solid or liquid fuel

If C, $H_2$, $O_2$ and $S$ denote the composition of carbon, hydrogen, oxygen and sulphur per kg of fuel then from Eq.16.1 mass of $O_2$ required for C kg of carbon = 32/12C = 8/3 C, for $H_2$ kg of hydrogen = 16/2 $H_2$ = 8 $H_2$ kg, for $S$ kg of sulphur = 32 $S$/32 = $S$ kg. Therefore total mass of $O_2$ required for 1 kg of fuel = 8/3 $C$ + 8$H_2$ + S. But $O_2$ kg of oxygen is already present. Therefore, oxygen required per kg of fuel

$$= 8/3C + 8H_2 + S - O_2 = 8/3\ C + 8\ (H_2 - O_2/8) + S$$

But air contains 23.2% oxygen by weight.

Therefore, mass of minimum air required for 1 kg of fuel.

$$\boxed{= Z_s = 100/23.2\ \{8/3\ C + 8\ (H_2 - O_2/8) + S\}} \tag{16.3}$$

### 16.3.5 Theoretical Air Fuel Ratio for Hydrocarbon $(C_xH_y)$ Complete Chemical Reaction for Hydrocarbon is Given by

$$C_xH_y + (x + y/4)\ O_2 = xCO_2 + y/2\ H_2O$$

Number of moles of oxygen required per 1 mole of hydrocarbon = $(x + y/4)\ O_2$.

Since 1 mole of oxygen is available from 4.76 moles of air. Therefore number of mole of air required = $4.76 \times (x + y/4)$

Theoretical air fuel ratio on mass basis

$$= Z_s = \frac{4.76 \times (x + y/4) \times 28.96}{(12\,x + y)}$$

and on volume basis $= Z_s = \dfrac{4.76\ (x + y/4)}{1}$ \hfill (16.4)

## 16.3.6   MASS BALANCE

Since air is used for combustion and it contains nitrogen also, combustion reaction is written considering the mass of nitrogen also. A chemical reaction in which mass (or mole) of each constituent is same in reactants and products. This is called mass balance equation. Combustion problems (simple and complicated) can easily be solved using mass balance equation. The combustion equation for hydrocarbon fuel can be written as:

$$C_x H_y + (x + y/4) O_2 + 3.76 (x + y/4) N_2 = x CO_2 + Y/2 H_2O + 3.76 (x + y/4) N_2 \quad (16.5)$$

We shall illustrate mass balance method for solving many combustion problems.

---

**Example 16.1**   The ultimate analysis of a coal used in boiler is as follows: 83.1% $C$, 5.5% $H_2$, 7.4% $O_2$ 2.1% $N_2$, 1.9% $S$. The coal is burnt with 130% theoretical air. Find the theoretical air fuel ratio, the actual air fuel ratio and mass of products per kg of fuel.

*Solution*   The problem can be solved either by using Eq. (16.3) or by mass balance method.
*1st method*:   Theoretical air fuel ratio

$$= Zs = 100/23.2 \; \{8/3C + 8 (H_2 - O_2/8) + S\}$$

$$= 100/23.2 \; \{8/3 \times 0.831 + 8 (0.055 - 0.074/8) + 0.019\}$$

$$= 100/23.2 \; \{2.216 + 0.366 + 0.019\} = 11.21 \quad Ans.$$

Actual air fuel ratio = $Z = 1.3 \times Zs = 1.3 \times 11.21 = 14.573$   *Ans.*
Mass of products per kg of fuel = air + (mass of fuel except ash)

$$= 14.573 + (0.831 + 0.055 + 0.074 + 0.021 + 0.019)$$

$$= 15.573 \text{ kg/kg of fuel} \quad Ans.$$

2nd method (Mass balance method)

$$0.831 \text{ kg of carbon} = 0.831/12 = 0.06925 \text{ moles of carbon}$$

$$0.055 \text{ kg of hydrogen} = 0.055/2 = 0.0275 \text{ moles of hydrogen}$$

$$0.074 \text{ kg of oxygen} = 0.074/32 = 0.002315 \text{ moles of oxygen}$$

$$0.021 \text{ kg of } N_2 = 0.021/28.16 = 0.00075 \text{ moles of } N_2$$

$$0.019 \text{ kg of } S = 0.019/32 = 0.0005937 \text{ moles of } S.$$

The number of moles of oxygen required for complete combustion of 0.06925 moles of carbon, 0.0275 moles of hydrogen and 0.0005937 moles of sulphur is calculated as

$$0.06925 + 0.0275/2 + 0.0005937 = 0.0835937$$

Since 0.002315 moles of $O_2$ is already present, no. of moles of $O_2$ required = 0.0835937 − 0.002315 = 0.08127 moles.
Mass balance equation becomes:

$$0.06925\ C + 0.0275\ H_2 + 0.002315\ O_2 + 0.00075\ N_2$$

$$+ \frac{0.0005937\ S}{\text{mass of fuel}} + \frac{0.08127\ O_2 + 3.76 \times 0.08127\ N_2}{\text{mass of air}}$$

$$= 0.06925\ CO_2 + 0.0275\ H_2O + 0.0005937\ SO_2 + 3.76 \times 0.08127\ N_2 + 0.00075\ N_2$$

$$Z_s = \frac{\overbrace{(0.08127)}^{\text{moles of } O_2} \times 4.76 \times 28.96}{1} = 11.20 \quad Ans.$$

*Note*:   The first method gives $Z_s = 11.21$. The slight difference may be due to rounding error. But the first method is the quickest method. Therefore for solid and liquid fuels, $Z_s$ should be calculated by first method.

**Example 16.2**   The percentage composition by mass of a solid fuel used in a boiler is given below:

$$90\%C,\ 3.5\ H_2,\ 3\%\ O_2,\ 1\%\ N_2,\ 1\%\ S$$

and the remainder being ash.
(a) Find theoretical air fuel ratio and mass analysis of dry products of combustion.
(b) If 50% excess air is supplied in actual combustion, determine the volumetric analysis of dry products and also the mass of dry flue gas per kg of fuel. Determine the dew point temperature of products of pombustion at atmospheric pressure.

*Solution*
In this problem we have also to determine analysis of products of combustion, mass balance method is advisible. Let us consider 100 kg of fuel (1 kg of fuel can also be considered).
90 kg of carbon = $\frac{90}{12}$ = 7.5 moles $C$, requires 7.5 moles of $O_2$. 3.5 kg of hydrogen = $\frac{3.5}{2}$ = 1.75 moles of $H_2$ requires $\frac{1.75}{2}$ = 0.875 moles of $O_2$

3 kg of oxygen = $\frac{3}{32}$ = 0.09375 moles of $O_2$, already present.

1 kg of Nitrogen = $\frac{1}{28.6}$ = 0.03571 moles of $N_2$ will reamain in the products

1 kg of Sulphur = $\frac{1}{32}$ = 0.03125 moles $S$, requires 0.03125 moles of $O_2$.

Moles of $O_2$ required = 7.5 + 0.875 + 0.03125 − 0.09375 = 8.3125 mass balance equation gives:

$$7.5C + 1.75\ H_2 + 0.09375\ O_2 + 0.03571\ N_2 + 0.03125\ S + \underbrace{8.3125\ O_2}_{O_2\ \text{required}} + 3.76 \times 8.3125\ N_2$$

$$= 7.5\ CO_2 + 1.75H_2O + 0.03125\ SO_2\ (3.76 \times 8.\ 3125 + 0.03571)\ N_2$$

or, $7.5C + 1.75\ H_2 + 0.09375\ O_2 + 0.03571\ N_2 + 0.03125S + 8.3125\ O_2$

$$+ 31.255\ N_2 = 7.5\ CO_2 + 1.75\ H_2O + 0.03125\ SO_2 + 31.29\ N_2$$

(a)  Theoretical air fuel ratio $Z_s = \dfrac{8.3125 \times 4.76 \times 28.96}{100}$

$$= 11.46 \quad Ans.$$

Total dry products for 100 kg of fuel = mass of $(CO_2 + SO_2 + N_2)$

$$= 7.5 \times 44 + 0.03125 \times 64 + 31.29 \times 28.16$$

$$= 330 + 2 + 881 = 1213$$

Mass of dry products per kg of fuel = 1213/100 = 12.13

(b)  If 50% excess air is supplied then the mass balance equation becomes

$$7.5\ C + 1.75\ H_2 + 0.09375\ O_2 + 0.03571\ N_2 + 0.031255$$

$$+ 1.5 \times 8.3125\ O_2 + 3.76 \times 8.3125 \times 1.5\ N_2$$

$$= 7.5\ CO_2 + 1.75\ H_2O + 0.03125\ SO_2 + 0.5 \times 8.3125$$

$$O_2 + 3.76 \times 8.3125 \times 1.5\ N_2 + 0.03571\ N_2$$

or,  $\quad 7.5\ C + 1.75\ H_2 + 0.09375\ O_2 + 0.03571\ N_2 + 0.03125\ S$

$$+ 12.4725\ O_2 + 46.8825\ N_2 = 7.5\ CO_2 + 1.75\ H_2O + 0.03125\ SO_2$$

$$+ 4.15625\ O_2 + 46.9182\ N_2$$

Number of moles of dry products = moles of $CO_2 + SO_2 + O_2 + N_2$

$$= 7.5 + 0.03125 + 4.15625 + 46.91821 = 58.60575$$

Percentage composition of dry products by volume is

$$CO_2 = \frac{7.5}{58.6057} \times 100 = 12.8\%$$

$$SO_2 = \frac{0.03125}{58.6057} \times 100 = 0.0533\%$$

$$O_2 = \frac{4.15625}{58.6057} \times 100 = 7.091\%$$

$$N_2 = \frac{46.91825}{58.6057} = 80.0557\%\quad Ans.$$

Mass of dry flue gas per kg of fuel

$$= (7.5 \times 44 + 0.03125 \times 64 + 4.15625 \times 32 + 46.91825 \times 28.16)/100$$

$$= 17.86\ \text{kg per kg of fuel}\quad Ans.$$

Dew point temperature is the temperature at which the moisture present in the flue gas starts condensing after cooling. It is the saturation temperature of water vapour corresponding to partial pressure of water vapour. Partial pressure of water vapour = mole fraction × Total pressure

$$= \frac{1.75}{7.5 + 1.75 + 0.03125 + 4.15625 + 46.91825} \times 1.01325 = 0.0293\ \text{bar}$$

From steam table, saturation temperature at 0.0293 bar

$$= 23.7°C.\ \text{Dew point} = 23.7C°\quad Ans$$

**Example 16.3**  A combustible gas has the composition by volume: 9% carbon monoxide, 46% hydrogen, 34% methane, 4% ethyline 2% oxygen, 2.5% nitrogen and 2.5% carbon dioxide. Assuming 150% theoretical air supply, find the theoretical and actual air fuel ratio and the composition of the products.

*Solution*  Combustion equations for $CO$, $H_2$, $CH_4$, $C_2H_4$ are:

$$CO + 1/2 \, O_2 = CO_2, \; H_2 + 1/2 \, O_2 = H_2O, \; CH_4 + 2O_2 = CO_2 + 2H_2O,$$

$$C_2H_4 + 3O_2 = 2 \, CO_2 + 2H_2O$$

Consider 100 moles of combustible gas for complete combustion with minimum air is

$$9 \, CO + 46 \, H_2 + 34 \, CH_4 + 4 \, C_2H_4 + 2O_2 + 2.5 \, N_2$$

$$+ \; 2.5 \, CO_2 + (9/2 + 46/2 + 34 \times 2 + 4 \times 3 - 2) \, O_2$$

$$+ \; 3.76 \times (9/2 + 46/2 + 34 \times 2 + 4 \times 3 - 2) \, N_2 = (9 \, CO_2) + (46 \, H_2O)$$

$$+ \; (34 \, CO_2 + 34 \times 2 \, H_2O) + (4 \times 2 \, CO_2 + 4 \times 2 \, H_2O)$$

$$+ \; 2.5 \, N_2 + 3.76 \, (9/2 + 46/2) + 34 \times 2 + 4 \times 3 - 2) \, N_2 + 2.5 \, CO_2$$

or, $9 \, CO + 46 \, H_2 + 34 \, CH_4 + 4 \, C_2H_4 + 2O_2 + 2.5 \, N_2 + 2.5 \, CO_2$

$$+ \; 105.5 \, O_2 + 3.76 \times 105.5 \, N_2 = 53.5 \, CO_2 + 122 \, H_2O + 396.68 N_2 + 2.5 N_2$$

For 100 moles of fuel, air required = $105.5 \times 4.76$

$$\text{Moles} = 502.18 \text{ moles}$$

Theoretical air fuel ratio = 502.18/100 = 5.0218 moles/mole  *Ans.*
Actual air fuel ratio = $1.5 \times 5.0218$ = 7.5327 moles/mole
To determine the number of moles of products per 100 moles of gas.

$$CO_2 = 53.5 \text{ moles}$$

$$H_2O = 122.0 \text{ moles}$$

$$O_2 = 0.5 \times 105.5 = 52.73 \text{ moles}$$

$$N_2 = 1.5 \times 396.68 + 2.5 = 507.52$$

Total number of moles = 53.5 + 122.0 + 52.75 + 597.52 = 825.77
percentage composition on volume basis is

$$CO_2 = (53.5/825.77) \times 100 = 6.4787, \; H_2O = (122/825.77) \times 100 = 14.775\%$$

$$O_2 = (52.75/825.77) \times 100 = 6.387\%$$

$$N_2 = (597.52/825.77) \times 100 = 72.359\% \; \textit{Ans.}$$

### 16.3.7  Combustion Control through Flue Gas Analysis

In actual combustion process, complete combustion does not occur since the perfect mixing of air with fuel is never attained in actual combustion process, good combustion can only be assured by supplying excess air. If there is incomplete combustion considerable amount of carbon is burnt to carbon monoxide liberating less energy and at the same time, it is a highly polluting gas, on the other hand, large amount of excess air, if supplied, will cause most of the heat carried away by products of combustion (temperature 200°C to 400°C). It is, therefore, important to control the combustion by flue gas analysis to ensure that the fuel is burnt completely but with a minimum air. Another advantage of the flue gas analysis is that actual air fuel ratio can also be calculated. An apparatus by means of which an analysis of exhaust gas (flue gas) is made, is called orsat apparatus. Orsat apparatus is based on the principle that if the flue gas is passed successively through various solutions each of which absorbs one component of the combustion products completely. By means of orsat apparatus, we determine the percentage composition of $CO_2$, $CO$, $O_2$ and $N_2$ on volume basis. It consists of three flasks containing strong solution of potassium hydroxide (for absorbing $CO_2$), acidic solution of cuprous chloride (for absorbing $CO$) and an alkaline solution of pyrogrolic acid (for absorbing $O_2$).

A 100 $cm^3$ exhaust gas sample is taken at room temperature in the burette by using the levelling water bottle to collect and transfer the gas sample. Since, the sample is collected at room temperature over water, it is usually assumed that any water vapour in the exhaust gas is condensed and any sulphur dioxide in the exchaust gas will react with the water in the flue gas or collecting bottle. Therefore it is assumed that the sample of exhaust gas is composed of $CO_2$, $CO$, $O_2$ and $N_2$. Once the sample has been obtained, it is then sequentially passed through the three reactor flanks. The first flank containing aquous of solution of KOH removes $CO_2$, the second flank removes $O_2$ and then the third one absorbs $CO$. By decrease in volume of the gas, volumetric composition of $CO_2$, $O_2$, $CO$ and $N_2$ (by difference) is determined. For details, see article 16.7.

---

**Example 16.4**  A certain power plant uses coal with the following analysis by weight; 78% *C*, 6% $H_2$, 9.8% $O_2$, 1.2% $N_2$ and 5% ash and an analysis of the refuse collected from the ash pit shows that it contains 30% carbon by weight. An orsat analysis of the flue gas gives 12.5% $CO_2$, 0.9% $CO$, 5.6% $O_2$ and the rest nitrogen. Find

(a) theoretical airfuel ratio
(b) actual air fuel ratio
(c) dilution coefficient
(d) Percentage of excess air

*Solution*  We shall determine the mass of carbon burnt per kg of fuel. Mass of ash per kg of fuel = 0.05 kg
Since ash contains 30% carbon, 0.30 = unburnt carbon/(ash + unburnt carbon)
or, 0.30 = unburnt carbon/(0.05 + unburnt carbon)
or, unburnt carbon = 0.30 × 0.05 + 0.30 unburnt carbon
or, 0.7 unburnt carbon = 0.015
unburnt carbon = 0.015/0.7 = 0.0214 kg
Carbon burnt per kg of coal = 0.78 − 0.0214 = 0.7586 kg/kg coal
Theoretical air fuel ratio can be calculated by by Eq. 16.3:

$$Z_s = 100/23.2 \ \{8/3 \ C + 8 \ (H_2 - O_2/8) + S)\}$$

$$= 100/23.2 \ \{8/3 \times 0.78 + 8(0.06 - 0.098/8) + 0\}$$

$$= 10.61 \quad Ans.$$

*Note*:   For $Z_s$ total carbon (0.78 kg/kg of fuel) has to be considered. Percentage of $N_2$ in flue gas = 100 − (12.5 + 0.9 + 5.6) = 81%. For actual air fuel ratio, let us consider that 'a' kg of fuel produces 100 moles of flue gas, then mass balance equation on mole basis gives:

$$a[0.7586/12 \; C + 0.06/2 \; H_2 + 0.098/32 \; O_2 + 0.012/28.16 \; N_2]$$
$$+ \; b \; O_2 + 3.76 \times b N_2 = 12.5 \; CO_2 + 0.9 \; CO + 5.6 O_2 + 81.0 \; N_2 + CH_2O$$

where $a$, $b$ and $c$ are unknowns.

Again making a balance of number of moles of various elements. Carbon balance $\rightarrow$ 0.7586/12 $\times a$ = 12.5 + 0.9 = 13.4, $a$ = 211.97

Hydrogen balance $\rightarrow a \times 0.06/2 = C$ or, $C = a \times 0.03 = 211.97 \times 0.03 = 6.358$

Oxygen balance $\rightarrow a \times 0.098/32 + b = 12.5 + 0.9/2 + 5.6 + 6.358/2$

or $a \times 0.098/32 + b = 21.72$ or $211.97 \times 0.098/32 + b = 21.72$ or, $b = 21.07$

Nitrogen balance gives $N_2 = 0.012/28.16 \times a + 3.76 \times b$

$$= \frac{0.012 \times 211.97}{28.16} + 3.76 \times 21.07 + 79.31$$

but right hand side has $N_2$ = 81 moles the difference is due to incorrect orsat analysis.

Actual air fuel ratio = mass of air/mass of fuel

$$= \frac{b \times 4.76 \times 28.96}{a} = \frac{21.07 \times 4.76}{211.97} \times 28.96$$
$$= 13.70 \quad Ans.$$

(c)  Dilution coefficient = $Z/Z_s$ = 13.70/10.61 = 1.291

(d)  Percentage of excess air = $(Z - Z_s)/Z_s \times 100$

$$= \frac{(13.70 - 10.61)}{10.61} \times 100 = 29.12\% \quad Ans.$$

## 16.3.8   DETERMINATION OF AIR SUPPLIED PER KG OF FUEL WHEN THE VOLUMETRIC ANALYSIS OF DRY FLUE GASES AND PERCENTAGE OF CARBON BY MASS IN FUEL IS KNOWN.

We have solved example 16.4 by mass balance method in which constants $a, b, c$ were determined by equating the moles constituents of flue gas. A short cut method in formula form is also available. Many combustion problems are solved in less time. In the present article, we shall illustrate this method.

Let $CO_2$, $CO$, $O_2$ and $N_2$ denote the volumes (moles) per unit volume (mole) of dry flue gas. Volumetric analysis can be converted into gravimetric analysis by the following table.

| Constituent gas | Analysis by volume per m³ (a) | Molecular wt. (b) | Proportional mass (c) = a × (b) | Analysis by weight |
|---|---|---|---|---|
| $CO_2$ | $CO_2$ | 44 | 44 $CO_2$ | 44 $CO_2/\Sigma$ (c) |
| $CO$ | $CO$ | 28 | 28 $CO$ | 28 $CO/\Sigma(c)$ |
| $O_2$ | $O_2$ | 32 | 32 $O_2$ | 32 $O_2/\Sigma(c)$ |
| $N_2$ | $N_2$ | 28.16 | 28.16$N_2$ | 28.16$N_2/\Sigma(c)$ |
| Total 1 m³ | | $\Sigma(c)$ = 44$CO_2$ + 28$CO$ + 32 $O_2$ + 28.16 $N_2$ | | Total = 1 kg |

$\Sigma(c)$ denotes sum of items of column no. $c$ mass of carbon per unit mass of dry flue gas = mass of carbon in $CO_2$ + mass of carbon in CO

or,  $$\underbrace{(12/44) \times 44\, CO/\Sigma(c)}_{\text{mass of } CO_2} + \underbrace{12/28 \times 28\, CO/\Sigma(c)}_{\text{mass of CO}} = 12\,(CO_2 + CO)/\Sigma(c)$$

If $C$ denotes the mass of carbon burnt per kg of fuel, then mass of dry flue produced per kg of fuel can be determined by: Since $12\,(CO_2 + CO)/\Sigma(c)$ kg of carbon gives 1 kg of flue gas, $C$ kg of carbon gives $1/12\,(CO_2 + CO)/\Sigma(C) \times C$

$$= C\, \Sigma C/12\,(CO_2 + CO) \text{ kg of flue gas} = \frac{C \times \Sigma(c)}{12(CO_2 + CO)}$$

Therefore mass of flue gas per kg of fuel

$$m_g = C \times \frac{(44\, CO_2 + 28\, CO + 32\, O_2 + 28.16\, N_2)}{12(CO_2 + CO)}$$

or,  $$m_g = \frac{C(11\, CO_2 + 7\, CO + 8\, O_2 + 7.04\, N_2)}{3(CO_2 + CO)} \tag{16.6}$$

To determine actual air fuel ratio
1 kg of flue gas contains $28.16\, N_2/\Sigma(c)$ kg of $N_2$
$\therefore$  $C\,\Sigma(c)\,12\,(CO_2 + CO)$ kg of flue gas (per kg of fuel) contains $C\,\Sigma(c)/12\,(CO_2 + CO) \times 28.16\, N_2/\Sigma(c)$

$$\text{kg of } N_2 = \frac{C \times 28.16\, N_2}{12\,(CO_2 + CO)} \text{ kg of } N_2$$

But by weight $N_2 = 76.8\%$
mass of air required per kg of fuel

$$= 100/76.8 \times C \times 28.16\, N_2/12\,(CO_2 + CO) = \frac{100}{76.8} \times c \times \frac{28.16\, N_2}{12\,(CO_2 + CO)}$$

$$Z = (100/76.8) \times (28.16\, N_2) \times C/12\,(CO_2 + CO) \tag{16.7}$$

If $N_2$ is also present in the fuel then Eq. (16.7) is modified as:

$$Z = \frac{100}{76.8} \times \left[ \frac{28.16\, N_2 \times C}{12\,(CO_2 + CO)} - N_f \right]$$

$$Z = (100/76.8)\,[28.16\, N_2 \times C/12\,(CO_2 + CO) - N_f] \tag{16.8}$$

Where $N_f$ = mass of nitrogen present per kg of fuel

## 16.3.9  DETERMINATION OF PERCENTAGE OF CARBON IN FUEL BURNING TO $CO_2$ AND CO FROM VOLUMETRIC ANALYSIS OF DRY FLUE GAS.

Percentage of carbon burning to $CO_2$

$$= \frac{\text{Mass of carbon in } CO_2 \text{ per kg of flue gas}}{\text{Mass of carbon in 1 kg of flue gas}} \times 100$$

$$= \frac{12/44 \times 44\, CO_2/\Sigma(c)}{12/44 \times 44\, CO_2/\Sigma(c) + 12/28 \times 28\, CO/\Sigma(c)}$$

$$= CO_2 \times 100/(CO_2 + CO) = \frac{CO_2 \times 100}{CO_2 + CO}$$

Percentage of carbon burning to $CO_2 = CO_2/(CO_2 + CO) \times 100$

Percentage of carbon burning to $CO = CO/(CO_2 + CO) \times 100$ $\hspace{2cm}$ (16.9)

Example 16.5 find actual air fuel ratio for exmaple 16.4

$$Z = (100/76.8)\left[\frac{28.16\, N_2 \times C}{12\,(CO_2 + CO)} - N_f\right]$$

$$= 100/76.8\left[\frac{28.16 \times 0.81 \times 0.7586}{12\,(0.125 + 0.009)} - 0.012\right]$$

$$= (100/76.8)\,(10.7607 - 0.012) = 13.99$$

We obtained $Z$ as 13.70 in example 16.4. This difference is negligible. But this method is the quickest method of solving problems. Students are advised to use this method for *solid and liquid fuels where analysis of fuel is by mass and that for flue gas is volume. C denotes the mass of burnt carbon per kg of fuel.*

---

**Example 16.6** The volumetric analysis of a gaseous fuel supplied to a 4 stroke cycle gas engine is given below:

$$H_2 = 50\%,\ CH_4 = 20\%,\ C_2H_4 = 2\%,\ CO = 5\%,\ O_2 = 3\%,\ CO_2 = 10\%,$$

and $N_2 = 10\%$, The volumetric analysis of the dry exhaust gas using orsat apparatus is as follows: $CO_2 = 8.0\%$, $O_2 = 5.8\%$, $CO = 0.5\%$, $N_2 = 85.7\%$. Find theoreical and actual air fuel ratio by volume, dilution co-efficient. Also find the dew point of the combustion products at 1 atmospheric pressure. (Bihar Univ. 1987)

*Solution:* For theoretical air fuel ratio, combustion equations are:

$$H_2 + {}^1/_2\, O_2 = H_2O,\ CH_4 + 2\, O_2 = CO_2 + 2H_2O$$

$$C_2H_4 + 3\, O_2 = 2\, CO_2 + 2\, H_2O,\ CO + {}^1/_2\, O_2 = CO_2$$

Oxygen required for complete combustion of 1 mole of gas $= 0.50/2 + 2 \times 0.20 + 3 \times 0.02 + 0.05/2$

$$= 0.25 + 0.40 + 0.06 + 0.025 = 0.735 \text{ moles.}$$

As the fuel contains 0.03 mole of $O_2$

Therefore additional oxygen required $= 0.735 - 0.03 = 0.705$ moles.

Moles of air required per mole of gas

$$= 0.705 \times 4.76 = 3.356$$

Since moles are propertional to volume.
Therefore air fuel ratio $Z_s = 3.356$  *Ans.*
For actual air fuel ratio, let us write mass balance equation for "*a*" moles of gaseous fuel to produce 100 moles of exhaust gas.

$$a(0.50 \; H_2 + 0.20 \; CH_4 + 0.02 \; C_2H_4 + 0.05 \; CO + 0.030_2 + 0.10 \; CO_2 + 0.10 \; N_2)$$

$$+ \; b \; O_2 + 3.76 \times b \; N_2$$

$$= 8 \; CO_2 + 5.8 \; O_2 + 0.5 \; CO + 85.7 \; N_2 + CH_2O$$

Where $a$, $b$ and $c$ are constants to be determined by equating the number of moles of constituents
Hydrogen balance $\rightarrow a(0.5 + 2 \times 0.20 + 2 \times 0.02) = C$ or, $0.94 \; a = C$      (i)
Carbon balance $\rightarrow a(0.20 + 0.02 \times 2 + 0.05 + 0.10) = 8 + 0.5$

or,           $0.39 \; a = 8.5$ or, $a = 8.5/0.39 \; 21.795$

Oxygen balance $\rightarrow a \; (0.05/2 + 0.03 + 0.1) + b = 8 + 5.8 + 0.5/2 + c/2$

or,           $a \times 0.155 + b = 14.025 + c/2$      (ii)

Nitrogen balance $\rightarrow a \times 0.10 + 3.76 \; b = 85.7$

or,           $3.76b = 85.7 - 0.10 \times a = 85.7 - 0.1 \times 21.795$

or,           $b = 83.52/3.76 = 22.212$

From equation (i) $c = 0.94 \; a = 0.94 \times 21.795 = 20.487$
equation (ii) gives a check
L.H.S. of equation (ii) $= a \times 0.155 + b = 21.795 \times 0.155 + 22.212 = 25.59$
R.H.S. of equation (ii) $= 14.025 + c/2 = 14.025 + 20.487/2 = 24.268$
This difference is due the fact that orsat analysis is not completely accurate. Students should not confuse with this slight discrepency. Actual air fuel ratio $= z = 4.76 \times b/a$

$$= 4.76 \times 22.212/21.795 = 4.85 \quad Ans.$$

Dilution coefficient $z/z_s = 4.85/3.356 = 1.445$  *Ans.*
To determine the dew point.
Number of moles of total products of combustion = moles of dry products + moles of $H_2O$ = $100 + c$

$$= 100 + 20.487 = 120.487$$

Mole fraction of water vapour $= 20.487/120.487 = 0.170$. Partial pressure of water vapour $= 0.170 \times 1.01325$ bar $= 0.172$ bar
From steam table, dew point is equal to saturation temperature at 0.172 bar $= 56.8°C$  *Ans.*

**Example 16.7** An unknown hydrocarbon fuel $C_xH_y$ burns with $O_2$ in air and orsat analysis of the products of combustion gives the following composition:

$$CO_2 = 12.1\% \; O_2 = 3.8\%, \; CO = 0.9\%, \; N_2 = 83.2$$

(by difference) Determine
- (a) correct chemical equation of reaction
- (b) correct chemical formula of the fuel
- (c) actual air fuel ratio
- (d) Excess or difficiency of air used (Ranchi University, Patna Univ.)

*Soluton*:   Let us consider the one mole of $C_xH_y$ produces 100 moles of products of combustion. Here $x$ and $y$ denote total number of carbon and hydrogen atoms. Mass balance equation for this problem becomes.

$$C_xH_y + b\,O_2 + 3.76\,b\,N_2 = 12.1\,CO_2 + 3.8\,O_2 + 0.9\,CO$$

$$+ \; 83.2\,N_2 + CH_2O, \; b \text{ and } c \text{ are to be determined.}$$

On making mole balance we have

$$\text{Carbon balance} \rightarrow x = 12.1 + 0.9 = 13 \tag{i}$$

$$\text{Hydrogen balance } y/2 = c \tag{ii}$$

$$\text{oxygen balance } b = 12.1 + 3.8 + 0.9/2 + c/2 = 16.35 + c/2$$

$$\text{Nitrogen balance } 3.76b \; = 83.2 \text{ or, } b = 83.2/3.76 = 22.12$$

From oxygen balance $22.12 = 16.35 + C/2$ or, $c = 11.62$
From equation (ii) $y = 2c = 2 \times 11.62 = 23.24 \simeq 23$
Therefore, chemical formula of the feed is $C_{13}H_{23}$

- (a) Correct chemical equation of reaction is:

$$C_{13}H_{23} + 22.12\,O_2 + 3.76 \times 22.12\,N_2 = 12.1\,CO_2 + 3.8\,O_2 + 0.9\,CO + 83.2\,N_2 + 11.5\,H_2O$$

- (b) Correct chemical formula of the fuel $= C_{13}H_{23}$

- (c) Actual air fuel ratio $= Z = \dfrac{4.76\,b \times 28.96}{x \times 12 + 1\,y \times 1} = \dfrac{4.76 \times 22.12 \times 28.96}{12 \times 13 + 23 \times 1} = 17.1 \;\; Ans.$

- (d) For theoretical air fuel ratio combustion equation is:

$$C_xH_y + \left(x + \frac{y}{4}\right)O_2 + 3.76 \times \left(x + \frac{y}{4}\right)N_2 = xCO_2 + \frac{y}{2}\,H_2O + 3.76\left(x + \frac{y}{4}\right)N_2$$

$$Z_s = \dfrac{\left(x + \dfrac{y}{4}\right) \times 4.76 \times 28.96}{x \times 12 + y \times 1} = \dfrac{\left(13 + \dfrac{23}{4}\right) \times 4.76 \times 28.96}{13 \times 12 + 23 \times 1} = 14.41 \;\; Ans.$$

$$\text{Percentage of excess air } = \frac{Z - Z_s}{Z_s} \times 100 = \frac{(17.1 - 14.41)}{14.41} \times 100 = 18\% \;\; Ans.$$

# 16.3.10  CALORIFIC (OR HEATING) VALUE OF A FUEL

The calorific value of solid or liquid fuel is defined as theamount of heat liberated due to complete combustion of unit mass of the fuel. It is expressed either in $k$ cal/kg of fuel or kJ/kg of fuel. It is also called heat of combustion. The calorific value of gaseous fuel is the amount of heat liberated due to complete combustion of unit volume of that fuel. It is expressed in $k$ cal/m$^3$ of fuel or kJ/m$^3$ of fuel. The standard condition is taken at 15°C temperature and 76 cm of Hg pressure.
Higher and Lower Calorific Values

Most of the fuels contain hydrogen. During the combustion process, the hydrogen combines with oxygen and forms steam (water vapour) if this water vapour is condensed at constant temperature, large amount of heat (Latent heat) is released than if it exists in the vapour phase. On account of this, two types of calorific values are defined:

(a) Higher calorific value (H.C.V.) - The HCV is defined as the total heat liberated by the complete combustion of unit mass (or unit volume) of a fuel when the water vapour formed during combustion is completely condensed at constant temperature releasing latent heat of evaporation.

(b) Lower calorific value (LCV): It is defined as the net heat liberated by complete combustion of unit mass (or unit volume in case of gaseous fuel) when the water vapour formed during combustion is in vapour phase. Thus the LCV is the difference between HCV and the heat absorbed by the steam. The relation between the two is given by:

$$\text{LCV} = \text{HCV} - m_w h_{f_g} = \text{HCV} - 9 m_{H_2} \cdot h_{f_g} \tag{16.10}$$

where $m_w$ = mass of water vapour in products of combustion per unit mass of fuel (due to combustion of $H_2$ in the fuel)

$m_{H_2}$ = mass of original hydrogen per unit mass of fuel

$h_{f_g}$ = Latent heat of vaporization of water vapour at its partial pressure in the combustion products.

In engineering applications $h_{f_g}$ is taken at 15°C which is 2465.5 kJ/kg; otherwise it should be evaluated at partial pressure of water vapour.

The HCV of fuel may be determined accurately by experimental method. The calorific values of elementary fuels at STP are given below:

| Fuel | H.C.V. (kJ/kg) | L.C.V. (kJ/kg) |
|------|----------------|----------------|
| C | 33950 | – |
| CO | 9740 | – |
| S | 9400 | – |
| $H_2$ | 14420 | 12250 |

If a solid or liquid fuel consists of C. kg of carbon, $H_2$ kg of hydrogen, $O_2$ kg of oxygen, S kg of sulphur per kg of fuel then H.C.V. is given by

$$\text{H.C.V.} = [8100 \, C + 34400 \, (H_2 - O_2/8) + 2220S] \text{ K cal/kg}$$

$$= [33950 \, C + 144200 \, (H_2 - O_2/8) + 9400S] \text{ kJ/kg} \tag{16.11}$$

The quantity $(H_2 - O_2/8)$ is the mass of hydrogen available for combustion and that oxygen in the fuel is already wholly in combination with hydrogen

$$LCV = [8100\ C + 34400\ (H_2 - O_2/8) + 9400S] - 9 \times 2472\ m_{H_2} \qquad (16.12)$$

**Example 16.8**   A certain coal used in a boiler has the following analysis by weight C = 57.4%, $H_2$ = 5.4%, S = 4.8%, $N_2$ = 1%, $O_2$ = 18% and ash = 13.4%. Determine HCV and LCV

*Solution:* $\qquad$ HCV = 8100 $C$ + 34400 $(H_2 - O_2/8)$ + 9400S

$$= 8100 \times 0.574 + 34400\ (0.054 - 0.18/8) + 9400 \times 0.048$$

$$= 4649 + 1083 + 451 = 6183 \text{ kJ/kg of fuel} \quad Ans.$$

$$LCV = HCV - 9\,m_{H_2} \times 2472 = 6183 - 9 \times 0.054 \times 2472$$

$$= 4981.6 \text{ kJ/kg of fuel} \quad Ans.$$

## 16.4   ENERGY RELEASE IN COMBUSTION

In a chemical reaction, the reactants and the products have a definite energy associated with them. The energy associated with the products may be less or more than the energy of reactants (i.e. release or absorption of energy) depending upon;

  (i)  the chemical nature of reactants and the products and
  (ii) the state of reactants and products. Normally during combustion energy of products is less than that of reactants and therefore energy is released.

Combustion processes are of two types:

  (i)  Combustion at constant volume and
  (ii) Combustion in steady flow.

### 16.4.1   Constant volume combustion

If the combustion is carried out in rigid container of constant volume, the process is called constant volume combustion.

*Internal energy of combustion*: If a combustion is carried out at constant volume and the products are returned to the same initial temperature as the reactants, the energy release or absorbed by the system is called internal energy of combustion. Consider a mixture of reactive substance enclosed in a rigid container. The mixture is ignited either by a catalytic action or by any other means without transferring any energy. The combustion occurs at constant volume, we can apply first law of thermodynamics to this system and get $Q = W + \Delta U$ where $Q$ is heat input, $W$, the workdone by the system and $\Delta U$ is the change in internal energy. For constant volume process, $W = 0$ and for insulated vessel $Q = 0$ then $\Delta U = U_p - U_R = 0$, where $U_p$ and $U_R$ are internal energy

$$U_p = \sum_p m_p u_p\,(T_p)$$

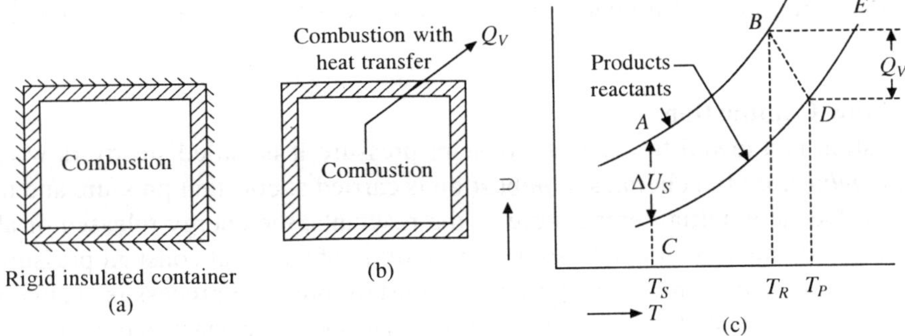


**Fig. 16.1**

where $u_p(T_p)$ is specific internal energy of a constituent at temperature $T_p$. There will be many products constituents. Therefore total internal energy is the sum of individual product.

Normally a temperature change occurs during combustion. Reactants are at temperature $T_R$ (state $B$ of Fig. (16.1.c)) and products are at temperature $T_p$ (state $D$). The curves $AB$ and $CD$ represent the internal energy of reactants and products at different temperatures $B$ is the state of reactants before combustion, $E$ and $D$ are states of products after combustion for adiabatic ($\Delta U = 0$) and with heat transfer to the surroundings respectively.

From first law of thermodynamics, heat transfer to surroundings is given by

$$Q_v = \Delta U = U_p(T_p) - U_R(T_R) = \sum_p m_p u_p(T_p) - \sum_r m_r u_r(T_R) \qquad (16.13)$$

But internal energy of combustion is defined at standard temperature ($T_s$, usually 25°C). Then internal energy of combustion at standard temperature

$$= \Delta U_s = U_p(T_s) - U_R(T_s) = \sum_p m_p u_p(T_s) - \sum_r m_r u_r(T_s) \qquad (16.14)$$

Where $m_r$ and $m_p$ represent the respective masses of the reactants and products per kg of fuel. $\Delta U_s$ is negative in all exothermic reactions since total energy of reactants at standard temperature $T_s$ is greater than that of products at $T_s$.

In order to find the energy release during a chemical reaction such as $D$ it is sufficient to use that internal energy is a property. Then:

$$Q_v = \sum_p m_p u_p(T_p) - \sum_r m_r u_r(T_r) = \sum_p m_p [u_p(T_p) - u_p(T_s)]$$

$$- \sum_r m_r [u_r(T_r) - u_r(T_s)] + \sum_p m_p u_p(T_s) - \sum_r m_r u_r(T_s)$$

or,  $$Q_v = \sum_p m_p [u_p(T_p) - u_p(T_s)] - \sum_r m_r [u_r(T_r) - u_r(T_s)] + \Delta U_s \qquad (16.15)$$

Some times internal energy is defined on mole basis instead of per kg.
Internal energy of combustion per unit mole is given by

$$\Delta \overline{U}_s = \sum_p n_p \overline{u}_p(T_s) - \sum_r n_r \overline{u}_r(T_s) \qquad (16.16)$$

Where $n_p$ and $n_r$ are moles of products and reactants respectively. The bar (–) indicates per unit mole.

## (b) Steady Flow Combustion

If the combustion is allowed to occur at constant pressure it is called steady flow combustion. *Enthalpy of combustion*: If a chemical combustion is carried at constant pressure and the products are returned to the same initial temperature as the reactants, the energy released or absorbed by the system is called enthalpy of combustion or heat of combustion at constant pressure. From the first law of thermodynamics heat transfer during constant pressure process is change in enthalpy neglecting kinetic and potential energy. Therefore heat transfer during constant pressure combustion is

$$Q_p = H_p - H_r = \sum_p m_p h_p(T_p) - \sum_r m_r h_r(T_r) \tag{16.17}$$

Similar to Eq. 16.15 $Q_p$ can be written as

$$Q_p = \sum_p m_p [h_p(T_p) - h_p(T_s)] - \sum_r m_r [h_r(T_r) - h_r(T_s)] + \Delta H_s \tag{16.18}$$

Where $\qquad \Delta H_s = H_p(T_s) - H_R(T_s) = \sum_p m_p h_p(T_s) - \sum_r m_r h_r(T_s) \tag{16.19}$

Molal enthalpy of combustion is written as

$$\Delta \overline{H}_s = \sum_p n_p \overline{h}_p(T_s) - \sum n_r \overline{h}_r(T_s) \tag{16.20}$$

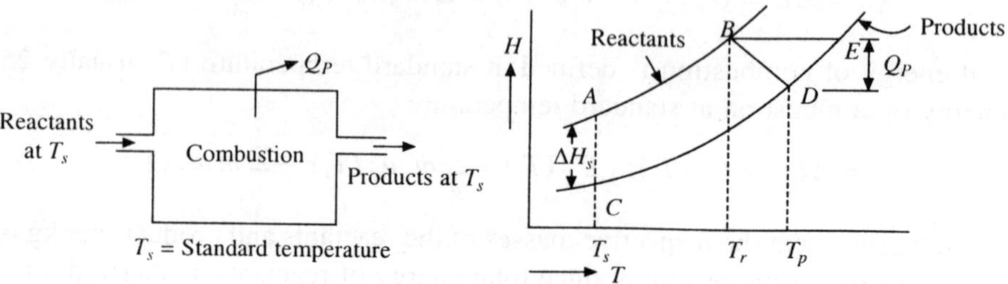


**Fig. 16.2    Steady flow combustion**

## Relation between Internal energy of combustion and enthalpy of combustion

It is possible to derive a relation between $\Delta \overline{H}_s$ and $\Delta \overline{U}_s$ if we assume that both the reactants and products are in gaseous states and they obey perfect gas relation. From definition of enthalpy:
$H = U + PV = U + n\,\overline{R}T_s$ where $\overline{R}$ = universal gas constant. Let us consider on per mole basis

$$\overline{h} = \overline{u} + \overline{R}T_s$$

$$\Delta \overline{H}_s = \sum_p n_p \overline{h}_p(T_s) - \sum_r n_r \overline{h}(T_s) = \sum n_p [\overline{u}_p(T_s) + \overline{R}T_s]$$

$$- \sum n_r [\overline{u}_r(T_s) + \overline{R}\,T_s]$$

$$= \sum_p n_p \overline{u}_p(T_s) - \sum_r n_r u_r(T_s) + \overline{R}T_s \sum(n_p - n_r)$$

or, $\qquad \Delta \overline{H}_s = \Delta \overline{U}_s + \overline{R}\,T_s \sum(n_p - n_r) \tag{16.21}$

If total number of moles of reactants equals the total number of moles of products (i.e. $\Sigma\, n_p = \Sigma\, n_r$), then $\Delta \overline{H}_s = \Delta \overline{U}_s$. However, if certain of the reactants and products are solids or liquids their volumes are negligible then the internal energy of combustion is approximately same as enthalpy of combustion.

The enthalpy of combustion is of most practical importance and therefore $\Delta H_s$ or $\Delta \overline{H}_s$ for many fuels are tabulated in the following table.

**Table 16.1  Enthalpy of combustion of some Hydrocarbons at 25°C (in kJ/kg)**

| Hydrocarbon | Formula | State of water in products | | | |
| --- | --- | --- | --- | --- | --- |
| | | Liquid State of fuel | | Vapour State of fuel | |
| | | Liquid | Gas | Liquid | Gas |
| Methane | $CH_4$ | | –55496 | – | –50010 |
| Ethan | $C_2H_6$ | | –51875 | | –47484 |
| Propane | $C_3H_8$ | –49975 | –50345 | –45983 | –46353 |
| Butane | $C_4H_{10}$ | –49130 | –49500 | –45344 | –45714 |
| Pentane | $C_5H_{12}$ | –48643 | –49011 | –44983 | –45351 |
| Hexane | $C_6H_{14}$ | –48308 | –48676 | –44033 | –45101 |
| Heptane | $C_7H_{16}$ | –48071 | –48436 | –44557 | –44922 |
| Octane | $C_8H_{18}$ | –47893 | –48256 | –44425 | –44788 |
| Dodecane | $C_{12}H_{26}$ | –47470 | –47828 | –44110 | –44467 |
| olefin Family | | | | | |
| Ethylene | $C_2H_4$ | | –50296 | | –47158 |
| Propylene | $C_3H_6$ | | –48917 | | –45780 |
| Butylene | $C_4H_8$ | | –48453 | | –45316 |
| Pentene | $C_5H_{10}$ | | –48134 | | –44996 |
| Hexene | $C_6H_{12}$ | | –47937 | | –44800 |
| Heptene | $C_7H_{14}$ | | –47800 | | –44662 |
| Octene | $C_8H_{16}$ | | –47693 | | –44556 |
| Benzene Family | | | | | |
| Benzene | $C_6H_6$ | –41831 | –42266 | –40141 | –40576 |
| Methyl benezne | $C_7H_8$ | –42473 | –42847 | –40527 | –40937 |
| Ethyle benzene | $C_8H_{10}$ | –42997 | –43395 | –40924 | –41322 |
| Propyle benzene | $C_9H_{12}$ | –43416 | –43800 | –41219 | –41603 |
| Butile benzene | $C_{10}H_{14}$ | –43748 | –44123 | –41453 | –41828 |

## 16.5  ENTHALPY OF FORMATION

Whenever a combustion reaction takes place the chemical reaction of the system undergoes a change. Hence we cannot compute the thermodynamic properties of various substances above an arbitrary datum. In other systems where there is no chemical change (as in steam the datum of properties is taken at 0°C), arbitrary datum can be selected. In order to over come this difficulty, the concept of formation of enthalpy is introduced.

Enthalpy of formation of any substance is defined as the energy released or absorbed when one mole of chemical compound is formed from its own elements only at standard condition.

If this definition is applied to elements such as carbon oxygen etc, it is clear that their enthalpies of formation are zero at standard conditions. The standard condition is taken as 25°C temperature and 1 atmospheric pressure. Let us take an example to calculate enthalpy of formation of water from its elements oxygen and hydrogen

$$H_2 + \frac{1}{2} O_2 = H_2O$$

$$\Delta \bar{H}_s = \bar{h}_{H_2O} - \left( \bar{h}_{H_2} + \frac{1}{2} \bar{h}_{O_2} \right) = \bar{h}_{H_2O} \qquad (16.22)$$

Because enthalpies of formation of $H_2$ and $O_2$ are zero.

---

**Example 16.9** A hydrocarbon fuel ($C_7H_{16}$) has enthalpy of combustion – 4856920 kJ/kg mole. Find its value of enthalpy of formation. Take the values of enthalpy of formation of $CO_2$ and $H_2O$ respectively as –393791 and –288136 kJ/kg mole. (Ranchi Univ. Patna Univ).

*Solution*: Combustion equation of $C_7H_{16}$ is $C_7H_{16} + (7 + 16/4)\, O_2$

$$= 7\, CO_2 + 16/2\, H_2O$$

or, $$C_7H_{16} + 11\, O_2 = 7\, CO_2 + 8\, H_2O$$

Enthalpy of combustion may be written as

$$\Delta \bar{H}_s = \Sigma\, n_p \bar{h}_p - \Sigma\, n_r \bar{h}_r = (n_{CO_2} \bar{h}_{CO_2} + n_{H_2O}\, h_{H_2O}) - (n_{C_7H_{16}} \bar{h}_{C_7H_{16}} + n_{O_2} \bar{h}_{O_2})$$

$$= 7 \times (-393791) + 8(-288136)$$

or, $$-4856920 = -2756537 - 2305088 - \bar{h}_{CH_{16}} = -5061625 - \bar{h}_{C_7H_{16}}$$

or, $$\bar{h}_{C_7H_{16}} = -5061625 + 4856920 = -204705 \text{ kJ/kg mole}$$

Enthalpy of formation of $C_7H_{16} = -204705$ kJ/kg mole   *Ans.*

**Example 16.10** The higher enthalpy of combustion of ethyl alcohol with liquid water in the product is –1364962 kJ/kg mole. Find the enthalpy of combustion with water vapour in the product. Take enthalpy of vaporisation of water at 25°C as 2463.3 kJ/kg.

*Solution*: The combustion equation is

$$C_2H_5OH + 3\, O_2 = 2\, CO_2 + 3\, H_2O$$

Therefore one mole of $C_2H_5OH$ produces 3 moles of water. If water is present in the vapour, state the enthalpy of combustion reduced due to latent heat.
Enthalpy of evaporation of water 25°C per kg mole

$$= 18 \times 2463.3 = 44339.4 \text{ kJ/kg mole}$$

Enthalpy of combustion having water in vapour state

$$= -[1364962 - 3 \times 43963] = -1233073 \text{ kJ/kg mole} \quad Ans.$$

**Example 16.11**  Find the internal energy of combustion at constant volume for the gaseous hydrocarbon fuel Heptane ($C_7H_{16}$) for complete combustion. The reactants and the products are at the same temperature (298 K)

The enthalpy of combustion is –44926.5 kJ/kg with water in the vapour form.

*Solution*:  The theoretical combustion equation for one mole of $C_7H_{16}$ is

$$C_7H_{16} + (7 + 16/4)\, O_2 = 7\, CO_2 + 16/2\, H_2O$$

or,

$$C_7H_6 + 11\, O_2 = 7\, CO_2 + 8\, H_2O$$

from Eq. (16.21) we have

$$\Delta \overline{H}_s = \Delta \overline{U}_s + \overline{R}T_s\,(\Sigma\,(n_p - n_r))$$

Here change in moles $= \Sigma\, n_p - \Sigma\, n_r = (7 + 8) - (1 + 11) = 3$
$H_s = -44926.5 \times$ Molecular wt. of $C_7H_{16} = -44926.5 \times (12 \times 7 + 16) = -4492650$ kJ/kg mole

Now $\qquad -4492650 = \Delta \overline{U}_s + 8.31 \times 298 \times 3 = \Delta \overline{U}_s + 7431.92$

or, $\qquad\qquad\qquad \Delta \overline{U}_s = -4485218.1$ kJ/kg mole

$\therefore \qquad\qquad\qquad \Delta U_s = -4485218.1/100 = -44852.18$ kJ/kg   *Ans.*

where 100 = molecular wt.
*Note*:  The difference between $\Delta H_s$ and $\Delta U_s$ is quite small, since the volume change is quite small.

**Example 16.12**  The ultimate analysis of coal supplied to a boiler is given as follows:
$C = 80\%,\ H_2 = 6\%,\ O_2 = 7\%\ N_2 = 1\%$ and the remainder ash. The volumetric analysis of dry flue gases determined by orsat apparatus is given below:

$$CO_2 = 10\%,\ CO = 1.5\%,\ O_2 = 8\%,\ \text{and}\ N_2 = 80.5\%$$

by difference. Specific heats of $CO_2$, $CO$, $O_2$ and $N_2$ are 0.904, 1.025, 0.910 and 1.0207 kJ/kg respectively, C.V. of carbon = 33914 kJ/kg when it burns to $CO_2$ and C.V. of carbon = 10049 kJ/kg when it burns to CO. Determine.

  (a)  Theoretical air fuel ratio
  (b)  Actual air fuel ratio
  (c)  percentage of excess air
  (d)  Dry flue gas formed per kg of fuel
  (e)  Mean specific heat of dry flue gas
  (f)  Dew point temperature of products of combustion
  (g)  Heat carried away by dry products of combustion per kg of fuel if room temperature is
        25°C and temperature of flue gas is 425°C.
  (h)  Heat carried by the steam formed per kg of fuel sp. heat of steam 1.884 kJ/kg°C
  (i)  Heat lost due to incomplete combustion per kg of fuel.

*Solution*:  Here dew point temperature and heat carried away by dry flue gas is also to be determined. We use mass balance method for actual air fuel ratio.

(a) Theoretical air fuel ratio

$$Z_s = 100/23.2 \; [8/3C + 8(H_2 - O_2/8) + S]$$

$$= (100/23.2)[8/3 \times 0.80 + 8 \,(0.06 - 0.07/8) + 0] = 10.96 \text{ kg of air/ kg of fuel} \quad Ans.$$

Let a kg of fuel produces 100 moles of dry products , then mass balance equation gives:

$$a(0.80/12C + 0.06/2 \; H_2 + 0.07/32 \; O_2 + 0.01/28.16 \; N_2)$$

$$+ \, b \; O_2 + 3.76 \times b \; N_2 = 10 \; CO_2 + 1.5 \; CO + 8 \; O_2 + 80.5 \; N_2 + CH_2O$$

Equating the moles of

Carbon $\rightarrow a \times 0.80/12 = 10 + 1.5 = 11.5, \quad$ or, $\quad a = 11.54 \times 12/0.80$

or,
$$\boxed{a = 172.5}$$

Hydrogen $\rightarrow a \times 0.03 = C, \quad$ or, $\quad C = a \times 0.03 = 172.5 \times 0.03 = 5.175$

or,
$$\boxed{c = 5.175}$$

Nitrogen $\rightarrow a \times 0.01/28.16 + 3.76 \times b = 80.5$

or,
$$172.5 \times 0.01/28.16 + 3.76 \; b = 80.5,$$

or,
$$\boxed{b = 21.393}$$

Oxygen balance $\rightarrow a \times 0.07/32 + b = 10 + 1.5/2 + 8 + c/2$

or,
$$172.5 \times 0.07/32 + 21.393 = 18.75 + 5.175/2$$

$$21.770 = 21.33$$

This gives a check. Slight difference came due to inaccuracy in orsat analysis results.

(b) Actual air fuel ratio

$$Z = \frac{b \times 4.76 \times 28.96}{a} = \frac{21.393 \times 4.76 \times 28.96}{172.5}$$

$$= 17.095 \quad Ans.$$

(c) Percentage of excess air

$$= \frac{(Z - Z_s)}{Z_s} = \frac{(17.095 - 10.96)}{10.96} \times 100 = 55.98\% \quad Ans.$$

(d) Dry flue gas for a kg of fuel

$$= \underbrace{10 \times 44}_{CO_2} + \underbrace{1.5 \times 28}_{CO} + \underbrace{8 \times 32}_{O_2} + \underbrace{80.5 \times 28.16}_{N_2}$$

$$= 3004.88$$

Dry flue per kg of fuel $= 3004.88/a = 3004.88/172.5 = 17.419 \quad Ans.$

(e) Mean specific heat of dry flue gas

$$= \frac{\begin{array}{c} Cp\,(\text{for } CO_2) \times \text{mass of } CO_2 + C_p\,(\text{for } CO) \times \text{mass of } CO \\ + C_p\,(\text{for } O_2) \times \text{mass of } O_2 + C_p\,(\text{for } N_2) \times \text{mass of } N_2 \end{array}}{\text{Total mass of dry products}}$$

$$= \frac{0.904 \times 10 \times 44 \times 1.025 \times 1.5 \times 28 \times + 0.910 \times 8 \times 32 + 1.020 \times 80.5 \times 28.16}{3004.88}$$

$$= 0.9937 \text{ kJ/kg k} \quad Ans.$$

(f) Mole fraction of $H_2O$ formed in the flue gas $= C/(100 + c)$

$$= 5.175/105.175 = 0.0492$$

Partial pressure of water vapour $= 0.0492 \times$ atmospheric pressure

$$= 0.0492 \times 1.0132 = 0.04985 \text{ bar}$$

Dew point $=$ saturation temperature at 0.04985 bar pressure $= 32.8°C$ _Ans._

(g) Heat carried away by dry products of combustion per kg of fuel

$$Q_g = \text{mg } C_{pg} \, \Delta t = 17.419 \times 0.9937 \, (425-25)$$

$$= 6923.704 \text{ kJ/kg of fuel} \quad Ans.$$

(h) Heat carried away by steam formed per kg of fuel

$$Q_s = m_w \times [h_f + h_{fg} + C_p(t_g - \text{dewpoint}) - \text{atmospheric temp.}]$$

$$h_f + h_{f_g} = h_g = 2562.44 \text{ kJ/kg at } 32.8°C$$

$$Q_s = \left(\frac{c}{a}\right) \times 18 \times [2562.44 + 1.884 \, (425 - 32.8) - 25]$$

$$= \frac{5.175}{172.5} \times 18 \, [2562.44 + 1.884 \, (425 - 32.8) - 25]$$

$$= 1769.226 \text{ kJ/kg of fuel} \quad Ans.$$

(i) mass of carbon burnt to

$$CO = \frac{CO}{(CO_2 + CO)} \quad C = \frac{1.5}{(10 + 1.5)} \times 0.80$$

$$= 0.1043 \text{ kg/kg of fuel}$$

Heat lost due to incomplete combustion of per kg of fuel $= 0.1043$

$$= 0.1043 \, [\text{C.V. for } CO_2 - \text{C.V. for } CO]$$

$$= 0.1043 \, (33914 - 10049) = 2489.1 \text{ kJ/kg of fuel} \quad Ans.$$

## 16.6   ADIABATIC FLAME TEMPERATURE

If the combustion occurs in work free, steady flow and perfectly adiabatic combustion process (with no heat transfer) involving negligible change of kinetic energy, then the temperature attained by the products of combustion is called adiabatic flame temperature. This the maximum theoretical temperature that can be attained in any combustion chamber. In practice, this temperature is never reached but it is useful for designer of combustion chamber. Combustion chambers are designed to withstand this temperature.

## 16.7   FLUE GAS ANALYSIS

The simple and convenient apparatus used for volumetric analysis of dry flue gas containing $CO_2$, CO and $O_2$. If the remainder of flue gas is nitrogen, it will also be known by difference. The apparatus used is known as orsat apparatus shown in Fig. 16.3.

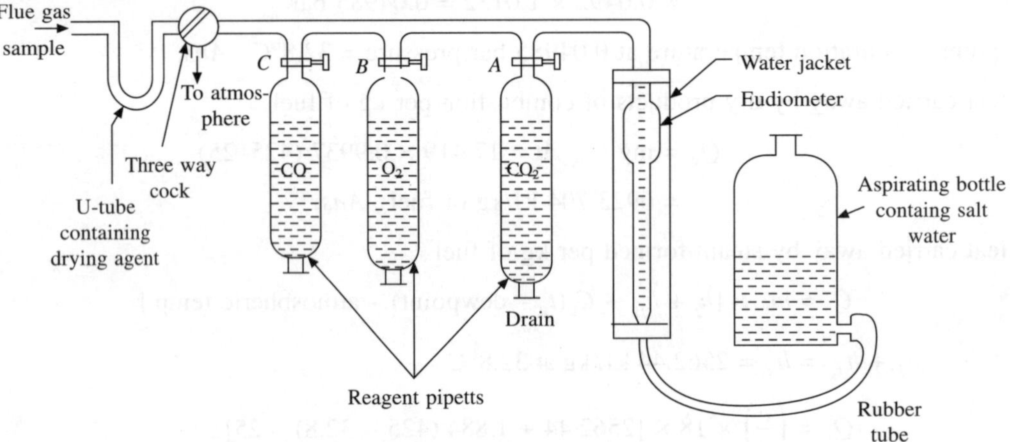


**Fig. 16.3   Orsat apparatus**

The apparatus consists of a graduated measuring bottle called eudiometer, an aspirating bottle, three double reagent pipettes to absorb $CO_2$, $O_2$ and CO. Eudiometer is connected to aspirating bottle by means of rubber tube.

The first reagent pipette next to eudiometer contains caustic soda (NaOH) or caustic potash (KOH) solution (33% KOH + 67% water by weight) to absorb $CO_2$. Next to this pipette, contains alkaline solution of pyrogallic acid (5 gm of pyrogallic acid in 15 CC of water + 33 gm of KOH in 67 gm of water) to absorb $O_2$ and last pipette consists of acidic solution of cuprous chloride (5 CC of CuO in 100 CC of HCl) to absorb CO.

Actually each front pipette joined together with a back pipette by a glass tube. Three such pipettes are arranged in such a way that they form two rows of front and back one. The top of each pipette in a row is connected to common header called front header and back header as shown in Fig. 16.4. Each front pipette is provided with number of glass tubes inside to increase the wetted surface area for accelerating the absorbing action of the solution. The front part of each pipette is provided with one way glass tap at the neck adjoining the front header. The front header is connected to eudiometer generally calibrated from 0 to 100 CC on one side. It is surrounded by water jacket to keep its temperature constant. On the other side the front header is proivided with a 3-way cock, which can connect it to a $U$-tube containing drying agent ($Cacl_2$) to absorb moisture present in the flue gas. The U-tube is connected to rubber bladder containing flue gas to be analysed.

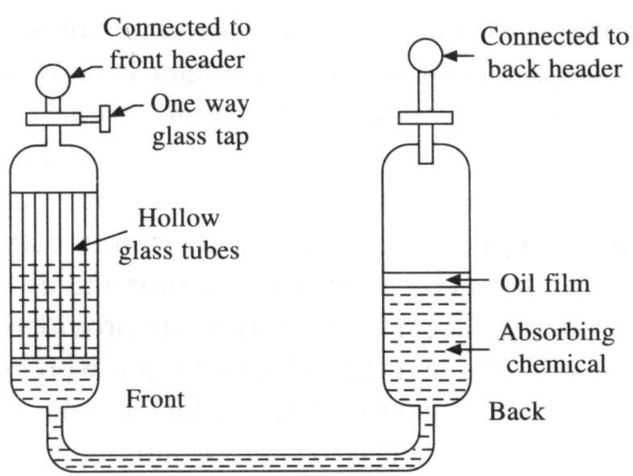



**Fig. 16.4   Absorbing flask**

Eudiometer is connected by flexible rubber pipe to aspirator bottle whose position can be changed by lifting or lowering it. By doing this a hydrostatic head is developed to force the flue gas from the eudiometer to the reagent pipettes. Eudiometer contains pure water. Aspirating bottle contains salt water, which can not dissolve gas.

The working of the apparatus is as follows: Firstly, the existing air or the gas is expelled from the eudiometer by raising the aspirating bottle and by keeping three way cock open to atmosphere. Three cocks $A$, $B$, $C$ and three way cock are closed. The aspirating bottle is lowered so that eudiometer reads zero on graduated scale. The three way cock is open and the sample of dry flue gas is drawn in the eudiometer measuring 100 CC. Now the three way cock is kept in the closed position.

Now the cock A is opened and the sample of flue gas is forced into the first reagent pipette containing KOH by raising the aspirator bottle. Here the $CO_2$ component of the flue gas is absorbed.

Aspirating bottle is moved up and down several times till the reading on eudiometer is constant to ensure the complete absorption of $CO_2$ in KOH solution. The rest of the gas is expelled from KOH pipettle and cock $A$ is closed keeping the original level of solution. The aspirator bottle is then brought near to eudiometer and placed at position so that water level in both is same. The eudiometer reading is taken and the difference of two readings gives the percentage of $CO_2$ by volume in the gas sample.

The same procedure is repeated with other pipettes to find the content of $O_2$ and CO. If the remainder of the flue gas is nitrogen, it will be known by difference.

It is to be noted that gases must be absorbed in the order of $CO_2$, $O_2$ and CO since pyrogallic acid solution will absorb $O_2$ as well as $CO_2$ and cuprous chloride solution absorbs $CO_2$, $O_2$ and CO.

NaOH or KOH solution absorbs 20 times its own volume by $CO_2$, pyrogallic acid solution absorbs twice its own volume of $O_2$ and cuprous chloride solution absorbs CO equal to its own volume. The top level of solution in each back row pipette is covered with a thin layer of oil which acts as a barrier not to absorb the gas present in the air above the level in the back row pipette.

It should be noted that the orsat apparatus gives the analysis of dry flue gases since the water is condensed at atmospheric conditions.

Additional pipettes may be provided for determination of other hydrocarbons and $H_2$ present due to incomplete combustion of fuel.

The apparatus is handy, easy to use and fairly accurate.

## 16.8   BOMB CALORIMETER

Bomb calorimeter is generally used for determining the higher calorific value (H.C.V.) of solid and liquid fuels. It consists of a strong steel shell known as bomb which can withstand a pressure of about 200 atmospheres and capacity 650 CC. Arrangement of bomb calorimeter is shown in Fig. 16.5. Bottom portion of the bomb is provided with platinum nickel supports and connections to an electric supply. Two pillars are provided inside the bomb to support the crucible made of silica or quartz. The top of the bomb carries an oxygen supply valve (non-return valve) and release valve as shown in the Fig. The bomb is placed in water bath (2500 CC approx) and water bath is placed in another container as shown in the Fig. A stirrer driven by a small D.C motor and a very sensetive thermometer called, Beckmann thermometer having accuracy of 0.01°C are provided to copper calorimeter. Before the start of experiment the crucible and the calorimeter are weighed. A pillet is formed in Briquette moulding apparatus. The known quantity of the pillete of solid fuel say coal is formed and 1 gm of this pillete is kept into the crucible. The oxygen is supplied slowly to the bomb by opening the oxygen valve until a pressure of approximately 25 bar is reached. The

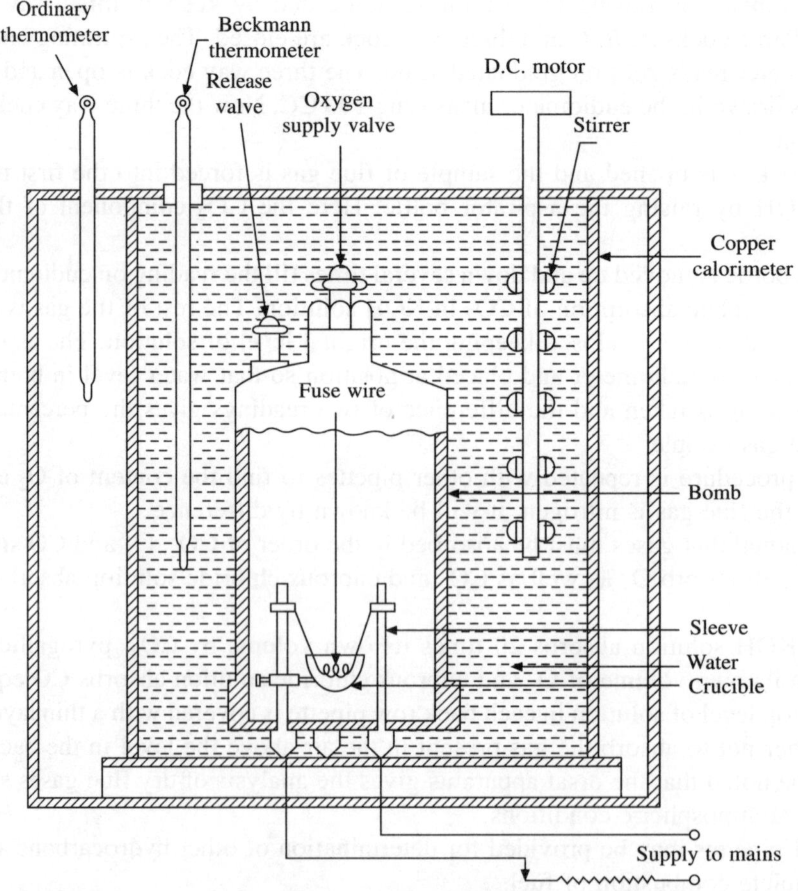


**Fig. 16.5   Bomb calorimeter**

initial temperature of water is noted and then the fuel is ignited with the help of fuse wire. The stirring is continued throughout the experiment and the temperature readings are taken every minute for ten minutes till the maximum temperature due to combustion of fuel is reached. Thereafter the temperature starts falling slowly due to heat transfer losses to the surroundings. When the temperature falls shows a steady rate, the readings are taken at regular intervals for an additional five minutes. In the outer most annulus space, an ordinary thermometer is provided to measure the temperature rise due to radiation losses from calorimeter.

The rate of cooling is determined which helps in applying the correction factor in order to determine the actual rise in temperature of water and calorimeter. At the end of experiment the release valve is opened so that the pressure inside the bomb reduces to atmospheric pressure.

In case of the calorific value of liquid fuels is determined by this calorimeter, the volatile fuels are kept in gelatine capsules while the other fuels can be directly kept in the crucible.

Calculation:

Let, $m_f$ = Amount of fuel burnt

$w_{w_f}$ = Mass of fuse wire

$m_w$ = Mass of water

$m_o$ = Water equivalent of calorimeter

$\theta$ = Corrected temperature rise

H.C.V. = Higher calorific value of fuel

$r$ = Rate of temperature drop (°C)

$t$ = time required for maximum temperature rise (minutes)

$\theta_2$ = Maximum temperature after combustion (°C)

$\theta_1$ = Steady temperature before combustion (°C)

$\theta_m$ = Recorded temperature rise (°C) = $(\theta_2 - \theta_1)$

Heat given by the fuel due to combustion + Heat given by the combustion of fuse wire
= Heat absorbed by the water and calorimeter

or,
$$m_f \times \text{H.C.V.} + w_{w_f} \times \text{C.V.} = (m_w + m_o)\ q \tag{16.23}$$

where
$$\theta = \frac{r}{2} \times t + \theta_m$$

or,
$$m_f \times H.C.V + m_{wf} \times C.V. = (m_w + m_o)\left\{\left(\frac{r}{2} \times t\right) + (\theta_2 - \theta_1)\right\} \tag{16.24}$$

$\therefore$ 
$$\text{H.C.V.} = \frac{(m_w + m_o)\left\{\left(\frac{r}{2} \times t\right) + (\theta_2 - \theta_1)\right\} - m_{w_f} \times C.V}{m_f} \tag{16.25}$$

The calorific vlaue calculated by the Eq. (16.25) gives the higher calorific value of the fuel because the $H_2O$ formed during combustion is condensed. By using the above equation, water equivalent of the calorimeter is determined by burning a fuel of known calorific value namely benzoic acid (C.V. = 26482 kJ/kg) and napthaline (C.V. = 40563 kJ/kg). The mass of the fuel used in the crucible depends upon the type of calorimeter and mass of water used. The fuel taken should be such that the temperature rise of water is limited to 3°C to minimise radiation losses.

## 16.9   BOY'S GAS CALORIMETER

Boy's gas calorimeter is used for determining the calorific value of gaseous fuel. The main parts of the calorimeter are burner, chimney and cooling coils as shown in Fig. 16.6. The gas is supplied whose calorific value is to be determined. A gas meter attached to measure the rate of volume of gas supplied. The fuel gas is burned and the products of combustion rise in the chimney. The heat liberated due to combustion of fuel is transferred to cooling water.

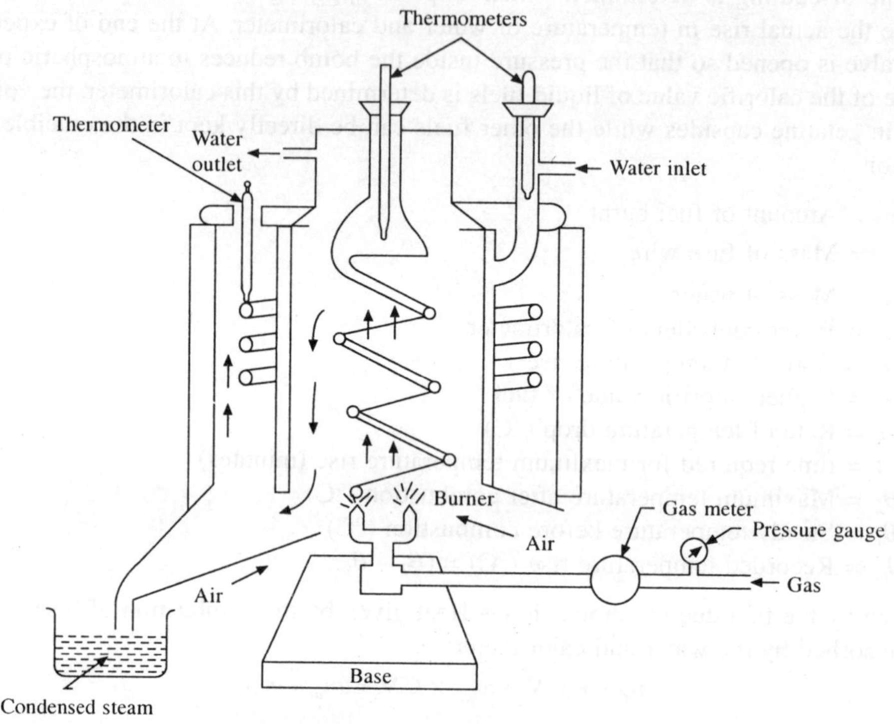


**Fig. 16.6   Boy's gas calorimeter**

The inlet of the water is first to the outer cooling coils and then it returns upwards through the inner coils as shown in the figure. The water vapour formed gets condensed in the outer portion of the chimney and the condensate is collected and weighed. Three thermometers are provided to measure the temperature of products of combustion at exit. It is ensured that these flue gases are cooled upto atmospheric temperature.

Calculations:

Let,   $V_g$ = Volume of gas consumed in m$^3$
   $P_g$ = pressure of gas supplied
   $T_g$ = Temperature of gas supplied
   $m_w$ = Mass of water circulated in kg
   $(\Delta T)_w$ = Rise in temperature of water circulated
   H.C.V. = Higher calorific value of gaseous fuel in kJ/m$^3$

$c_{p_w}$ = specific heat of water $\simeq$ 4.19 kJ/kg k

$V$ = Volume of gas supplied at N.T.P. From gas law

$$\frac{P_g V_g}{T_g} = \frac{PV}{T}$$

$P$ = 1.01325 bar

$T$ = standard temperature

= 25°C or 298 k

$$\therefore \qquad V = p_g V_g \cdot \frac{T}{T_g} \qquad\qquad (16.26)$$

*Heat balance*

Heat given by fuel = Heat absorbed by cooling water

or, $\qquad V \times \text{H.C.V} = m_w \times c_{p_w} \times \Delta T_w$

$$\therefore \qquad \text{H.C.V.} = \frac{m_w c_{p_w} \Delta T_w}{V} \text{ kJ/m}^3 \text{ at N.T.P.} \qquad\qquad (16.27)$$

## EXERCISES

### Numerical Problems

16.1 A coal used in a thermal power plant has the following gravimetric analysis; $C = 82\%$, $H_2 = 10\%$, $O_2 = 6\%$, $N_2 = 1\%$ and $S = 1\%$. If 20% of excess air is used in the combustion, calculate (a) Theoretical air fuel ratio, (b) Actual air fuel ratio (c) Dilution coefficient

*(Ans. $Z_s$ = 12.8, $Z_s$ = 15.35 DC = 1.2).*

16.2 The gravemetric composition of a coal used in a thermal power plant is $C = 74.4\%$, $H_2 = 12\%$, $O_2 = 2\%$ and the remainder ash. The coal is burnt with 180% theoretical air. Calculate: (a) Theoretical and actual air fuel ratio (b) Exhaust gas analysis on a dry basis assuming complete combustion (c) dew point of the actual products at a pressure of 1.01 ata

*(Ans. $Z_s$ = 12.61, $Z_o$ = 22.7, $CO_2$ = 8.23%, $O_2$ = 9.62%, $N_2$ = 82.13%, D.P. = 38.82 C)*

16.3 The gravimetric composition of coal used in Santhaldih thermal power plant is: $C = 80\%$ $H_2 = 15\%$, $O_2$ = 2% and ash = 3%. The orsat analysis of dry products of combustion is:
$CO_2 = 6.5\%$, $CO = 0.5\%$, $O_2 = 10.2\%$ and the rest nitrogen. Find (a) the theoretical and actual air fuel ratio (b) The volume of $CO_2$ per kg of fuel at 1 ata and 27°C (c) the dew point of the products at 1 ata.

*(Ans. $Z_s$ = 14.26, $Z$ = 28.85, volume = 1.57 m³, D.P. = 39.44°C)*

16.4 A fuel oil has the following analysis by weight: $C = 85\%$, $H_2 = 12.5\%$, $O_2 = 2\%$. The volumetric analysis of the dry exhaust gas gives the following composition:
$CO_2 = 9\%$, $CO = 1\%$, $O_2 = 7.77\%$ and $N_2 = 82.23\%$ (by difference) Calculate: (a) theoretical air fuel ratio (b) actual air fuel ratio (c) percentage of excess air

*(Ans. $Z_s$ = 13.59, $Z$ = 21.21).*

16.5 A heating oil with the composition by weight; C = 86% and $H_2$ = 14% is burnt. The product gases have the composition by volume: $CO_2$ = 10.2%, CO = 3.0%, $O_2$ = 3.5% and $N_2$ = 83.3% (by difference). Find the percentage of theoretical air and the fraction of unburnt carbon

*(Ans. 105%, percentage of unburnt carbon = 7.26%).*

16.6 A combustible gas with a composition by volume of $H_2$ = 45%, $CH_4$ = 30%, $C_2H_6$ = 15%, $N_2$ = 10% is burnt in air. The composition of the dry products of combustion is: $CO_2$ = 7.18%, CO = 0.16%, $O_2$ = 5.89

and $N_2 = 86.77\%$, find: (a) actual air fuel ratio on a volume basis (b) percentage of theoretical air (c) dew point of the combustion products at 1 at mospheric pressure

(*Ans.* $Z = 8.85$ percentage of theoretical air $= 138\%$, D.P. $= 54°C$)

16.7  A natural gas with the following analysis by volume (or molar) is burnt in a furnace: $CO_2 = 0.5\%$, $CO = 5.07\%$ $CH_4 = 87.0\%$ $C_2H_4 = 3.0\%$ and $N_2 = 4.5\%$. An orsat unalysis gives the following results: $CO_2 = 9.39\%$, $O_2 = 3.88\%$ $CO = 0.83\%$ and rest nitrogen. Calculate the percentage of excess air and actual air fuel ratio on volume and mass basis

(*Ans.* $18.0\%$, $Z_s = 8.833$ $m^3/m^3$ of fuel, $170083$ kg/kg of fuel).

16.8  A certain LPG has a composition of $C_3H_8 = 70\%$ and $C_4H_{10} = 30\%$, calculate the theoretical orsat analysis if the fuel is burnt with a dilution coefficient of 0.9.

(*Ans.* $CO_2 = 10.16\%$, $CO = 5.01\%$, $O_2 = 0\%$ $N_2 = 84.83\%$)

16.9  The percentage analysis by volume of a coal gas is as follows: $H_2 = 48\%$, $CH_4 = 28\%$, $CO = 8.6\%$, $C_2H_4 = 6.4\%$, $O_2 = 0.6\%$ $N_2 = 8.4\%$. Determine the percentage change in volume when this gas is burnt in nine times its own volume of air and give the composition of the resulting products of combustion

(*Ans.* $14.57$ at NTP, $CO_2 = 5.78\%$ $O_2 = 10.05\%$, $N_2 = 84.2\%$).

16.10  The following is the volumetric analysis of a dry exhaust gas; $CO_2 = 7.2\%$ $CO = 1\%$ $O_2 = 10\%$ and $N_2 = 81.8\%$. Find the composition of exhaust gas on a weight basis. Also find the approximate formula of the fuel from which the gases originated assuming it to be a hydrocarbon. Find the dew point of the products at a pressure of 1.00325 bar.

(*Ans.* $CO_2 = 10.72\%$, $CO = 0.95\%$ $O_2 = 10.83\%$ $N_2 = 77.50\%$, $C_2H_4$ DP $= 40.62°C$).

16.11  One mole of heptane ($C_7H_{16}$) is burnt with 15% theoretical air. Assuming combustion to be complete, calculate; (a) the change in moles between the reactants and the products (b) the theoretical air fuel ratio (c) the percentage of oxygen in the dry products (d) the dew point if the products of combustion are at 1.01325 bar)

(*Ans.* (a) 3 (b) 5 (c) $15.18\%$ (d) $45.45°C$))

16.12  The petrol used in an engine may be approximated to hexane ($C_6H_{14}$). The percentage of dry exhaust gases by volume at particular load and speed of the engine is; $CO_2 = 8.5\%$, $CO = 7.8\%$, $N_2 = 83.7\%$. Calculate (a) theoretical air fuel ratio (b) actual air fuel ratio

(*Ans.* $Z_s = 15.32$, $Z = 13.05$.)

16.13  A fuel contains 85% carbon, 12.5% hydrogen and 2% oxygen and the rest resideual matter. Determine HCV and LCV (*Ans.* 46700 kJ/kg, 43950 kJ/kg)

16.14  The volumetric composition of a gaseous fuel is $CO_2 = 7\%$, $CO = 25\%$, $H_2 = 15\%$, $CH_4 = 5\%$ and the remaining is nitrogen. Taking the latent heat of water vapour as 2512.2 kJ/kg for the condition after combustion. Find HCV and LCV. Take HCV for the constitutents as $CO = 283192.35$ kJ/kg mole $H_2 = 286043.7$ kJ/kg mole $CH_4 = 890993.6$ kJ/kg mole

(*Ans* $2051.6$ $kJ/m^3$, $1942.7$ $kJ/m^3$)

16.15  Find the internal energy of combustion at constant volume for the gaseous hydrocarbon fuel octane ($C_8H_{18}$) for complete combustion. The rectants and products are at the same temperature of 25°C. The enthalpy of combustion is $-44792.5$ kJ/kg with water in the vapour form.

(*Ans.* $- 44989.3$ kJ/kg).

16.16  Find the enthalpy for the reaction:
$CO + H_2O \rightarrow CO_2 + H_2$ Given that enthalpy of combustion of CO and $H_2$ are $-283124.94$ and $-283778.1$ kJ/kg mole respectively.

(*Ans.* $+ 653.17$ kJ/kg mole).

16.17  The enthalpy of combustion of ethyl alcohol is $-1381710$ kJ/kg mole. If the enthalpy of formation of $CO_2$ and $H_2O$ be $-394834$ and $-286809$ kJ/kg mole respectively calculate the enthalpy, of formation of ethyl alcohol

(*Ans.* $268386.7$ kJ/kg mole)

16.18  A boiler is fired with a certain coal having a gravemetric analysis; $C = 57.4\%$, $H_2 = 5.4\%$, $S = 4.8\%$, $N_2 = 1\%$, $O_2 = 18\%$ and remaining 13.4% ash. The coal is consumed at the rate of 150 kg/minute and the refuse is 22.5 kg/minute of which 10.7% is combustible. An orsat analysis of the dry products of combustion gives the following composition; $CO_2 = 13.4\%$, $CO = 0.5\%$, $O_2 = 6.5\%$ and the rest nitrogen calculate: (a) the theoretical air and (b) the actual air supplied

(*Ans.* 7.886, 22.99)

16.19 A sample of coal has the ultimate analysis:
$C = 67.34\%$, $H_2 = 4.67\%$, $O_2 = 8.47\%$, $N_2 = 1.25\%$ and $S = 4.77\%$ and the rest is ash. Find the theoretical air fuel ratio, HCV and LCV.

16.20 The coal supplied to boiler furnace has the following ultimate analysis; $C = 82\%$, $H_2 = 5.4\%$, $O_2 = 7\%$, $N_2 = 0.5\%$, $S = 0.6\%$, moisture $= 1\%$ and rest is ash. The orsat analysis of dry products of combustion is: $CO_2 = 9.6\%$, $CO = 1.2\%$, $N_2 = 81.4\%$, $O_2 = 7.8\%$

Determine:
(a) Percentage of excess air supplied to the boiler furance
(b) Mean specific heat of dry flue gases. Given specific heats at constant pressure, $CO_2 = 0.879$, $O_2 = 0.912$ $N_2 = 1.021$ and $CO = 1.038$ (*Ans.* 57.05%, $C_p = 0.9923$ kJ/kg k)

## Theoretical Problems

16.1 What do you mean by stiochiometric equation?
16.2 What are the advantages of liquid and gaseous fuels over solid fuels
16.3 Define enthalpy and internal energy of combustion
16.4 What is meant by calorific value of a fuel. Differentiate between H.C.V. and L.C.V.
16.5 Define enthalpy of formation
16.6 What informations we can gather from orsat analysis of exhaust gas.
16.7 Develop a relation between internal energy of combustion and enthalpy of combustion.
16.8 What do you mean by adiabatic flame temperature and what is its significance?
16.9 Why excess air is supplied in combustion?
16.10 What is the maximum percentage of carbon dioxide in the exhaust gas of boiler using solid fuel?

17

# Heat and Mass Transfer

## 17.1  INTRODUCTION

Whenever a temperature gradient exists within a system, or when two systems at different temperatures are brought into contact, energy is transferred. Heat Transfer is that branch of science which deals with transfer of energy by virtue of temperature difference. Thermodynamics deals with systems in equilibrium whereas heat transfer deals with the systems which are not in thermal equilibrium 'A heat transfer engineer is concountered with three major tasks: (i) to increase the rate of heat transfer. Examples are boiler, condenser, cooling tower, rediators of a motor car, aircooled IC engine cylinder etc. (ii) To decrease the rate of heat transfer. Examples are: thermal insulation of steam pipes, cabin of household refrigerator, thermosflask etc. and (iii) to maintain a constant temperature in a space. Examples are Airconditioned space, heat treatment of metals etc.

## 17.2  MODES OF HEAT TRANSFER

Literature of heat transfer recognizes three distinct modes of heat transfer: conduction, radiation and convection. Strictly speaking only conduction and radiation should be classified as heat transfer processes because they require only temperature difference. The last of the three, convection does not strictly comply with the definition of heat transfer for because it depends also on mass transport.

### Conduction
Conduction is a process of heat transfer by which heat flows from a region of higher temperature to a region of lower temperature within a medium (solid, liquid or gaseous) or between different media in direct physical contact by molecular vibration. There is no appreciable displacement of the molecules. Conduction is the only mechanism by which heat can flow in opaque solids.

Conduction is also important in fluids but in non-solid media it is usually combined with convection and in some cases radiation also.

### Radiation
The flow of energy by virtue of electromagnetic wave even through vacuum is called radiation. In nature maximum heat transfer takes place from radiation.

### Convection
Convection is a process of energy transport by the combined action of heat conduction, energy storage and mixing motion.

## 17.3  CONDUCTION

### 17.3.1  Fourier's Law of Heat Conduction
It states that the rate of flow of heat by conduction is proportional to the product of heat transfer area and temperature gradient.

Mathematically
$$Q \, \alpha \, A\left(\frac{-dT}{dx}\right)$$

or,
$$Q = -KA\frac{dT}{dx} \tag{17.1}$$

where $K$ is constant, called thermal conductivity of the material. It depends upon the nature of material. The minus sign indicates that heat transfer takes place in a direction in which temperature decreases

Unit of $Q$ is KJ/sec or kilowatt

$$\text{Unit of } K \text{ is } \frac{-Q}{A}\frac{dx}{dT} = \frac{\text{KJ} \times m}{\text{hr m}^2 \times K} = \text{KJ/smk}$$

or, Kilowatt/mK.

### 17.3.2  Effect of Temperature on Thermal Conductivity
In general, thermal conductivity depends upon temperature, i.e., $K = f(T)$. For metals, it decreases with temperature and for insulating materials it increases with temperature. For all practical purposes the conductivity $k$ may be expressed as a linear function of temperature. $K = k_0(1 + \alpha t)$. $\alpha$ is called temperature coefficient of conductivity and $\alpha$ is positive for insulating materials and negative for metals could be seen in Fig. 17.1.

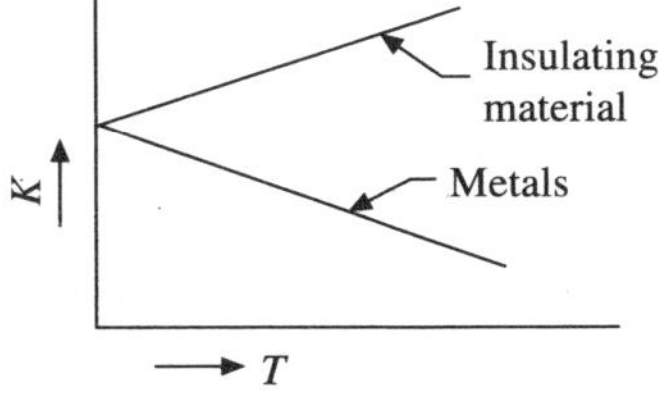


**Fig. 17.1**

## 17.5   ONE DIMENSIONAL HEAT CONDUCTION THROUGH A PLANE WALL

Plane wall is the simplest case of one dimensional heat conduction. Plane wall is often encountered in many engineering applications. Consider a plane wall of thickness $L$, face are $A$, maintained its two faces at $T_1$ and $T_2$. From Fourier's

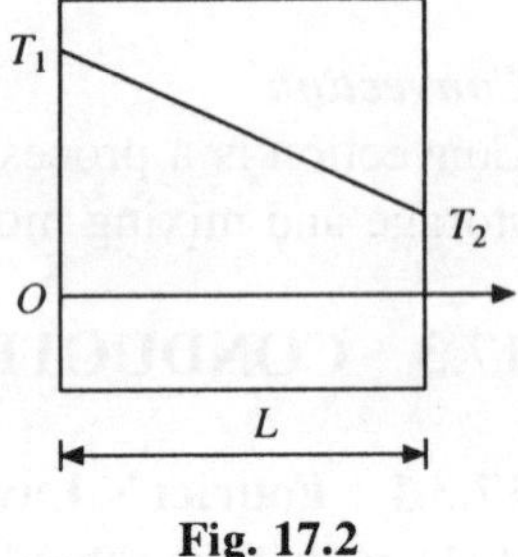

Fig. 17.2

Law of heat conduction, $Q = -KA\dfrac{dT}{dx}$

or ,
$$\int_0^L Q\,dx = -\int_{T_1}^{T_2} KA\,dT, \quad \text{or,} \quad Q.L. = KA(T_1 - T_2)$$

or,
$$Q = \frac{T_1 - T_2}{\dfrac{L}{AK}} = \frac{T_1 - T_2}{R},$$

where
$$R = \frac{L}{AK}$$

is called conductive resistance. The equation $Q = \dfrac{T_1 - T_2}{R}$ is similar to ohm's law of current

$\left( I = \dfrac{V_1 - V_2}{R} \right)$. Reciprocal of resistance $\left( \dfrac{1}{R} \right)$ is also called thermal conductance $C = \dfrac{1}{R} = \dfrac{AK}{L}$.

Unit of resistance is Sec. K/kJ. To determine temperature at a distance $x$ from one face,

$$Q = \frac{T_1 - T_2}{\dfrac{L}{AK}} = \frac{T_1 - T}{\dfrac{X}{AK}} \qquad \text{or,} \qquad \boxed{\frac{T_1 - T}{T_1 - T_2} = \frac{X}{L}} \qquad (17.2)$$

That is temperature distribution is linear. $T$ is the temperature at a distance $x$ from one face.

### 17.3.3   Radial Heat Conduction Through a Hollow Cylinder

Radial heat flow by conduction through a hollow cylinder is another one dimensional conduction problem of considerable practical importance. Typical example are conduction through pipes in boiler, condenser, motor, car radiators, evaporators and through pipe insulation.

   Consider a long cylinder of inner radius $r_i$, outer radius $r_0$ and length l maintained at temperature $T_i$ and $T_o$ respectively. From Fourier's law of heat conduction at any radius $r$ (Fig. 17.3) is:

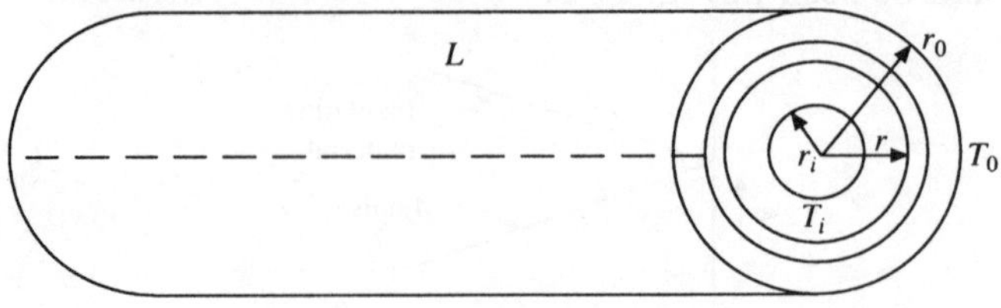

Fig. 17.3

$$Q = -KA\frac{dT}{dr} = -K \cdot 2\pi r L \frac{dT}{dr}$$

or, $\displaystyle\int_{r_i}^{r_o} Q\frac{dr}{r} = -\int K \cdot 2\pi L dT$ or, $Q \ln\dfrac{r_o}{r_i} = 2\pi kL(T_i - T_o)$

or, $$Q = \frac{T_i - T_o}{\dfrac{\ln\dfrac{r_o}{r_i}}{2\pi L K}} = \frac{T_i - T_o}{R},$$

where $R = \dfrac{\ln\dfrac{r_o}{r_i}}{2\pi LK}$ is called conductive resistance.

The temperature distribution in the curved wall is obtained by integrating from inner radius $r_i$ to any radius $r$,

$$\int_{r_i}^{r} Q\frac{dr}{r} = -\int_{T_i}^{T} k\,2\pi L\; dT \text{ or, } Q \ln\frac{r}{r_i} = 2\pi k L (T - T_i)$$

or, $$Q = \frac{T - T_i}{\dfrac{\ln\dfrac{r}{r_i}}{2\pi KL}} = \frac{T_i - T_o}{\dfrac{\ln\dfrac{r_o}{r_i}}{2\pi kL}} \tag{17.3}$$

or, $$\boxed{\frac{T - T_i}{T_i - T_o} = \frac{\ln\dfrac{r}{r_i}}{\ln\dfrac{r_o}{r_i}}} \tag{17.4}$$

Therefore temperature distribution is logrithmic (Fig. 17.4).

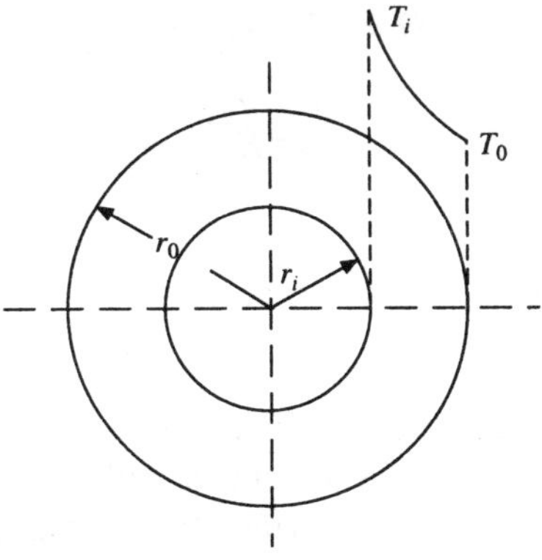


**Fig. 17.4**

*Logrithmic Mean Area:*   If the hollow cylinder is treated as a plane wall of thickness $(r_o - r_i)$ and face area $A_m$ which will transfer same heat as the actual cylinder, then $A_m$ is called logrithmic mean Area. To determine $A_m$, rate of flow of heat in both the cases are.

$$Q = \frac{T_i - T_o}{\dfrac{r_o - r_i}{A_m k}} = \frac{T_i - T_o}{\dfrac{\ln \dfrac{r_o}{r_i}}{2\pi LK}} \quad \text{or,} \quad \frac{r_o - r_i}{A_m k} = \frac{\ln \dfrac{r_o}{r_i}}{2\pi LK}$$

or,
$$A_m \frac{2\pi L(r_o - r_i)}{\ln \dfrac{r_o}{r_i}} = \frac{A_o - A_i}{\ln \dfrac{2\pi r_o L}{2\pi r_i L}} = \frac{A_o - A_i}{\ln \dfrac{A_o}{A_i}} \tag{17.5}$$

$$A_m = \frac{A_o - A_i}{\ln \dfrac{A_o}{A_i}} \quad \text{is called logrithmic mean area.}$$

### 17.3.4   Radial Heat Flow Through a Spherical Shell

A sphere has the largest volume per outside surface area of any geometrical configuration. For this reason a hollow sphere is sometimes used in Chemical industry for low temperature work when heat losses are to be kept at a minimum.

$$Q = -KA \frac{dT}{dr} = -K4\pi r^2 \frac{dT}{dr}$$

or,
$$\int_{r_i}^{r_o} Q \frac{dr}{r^2} = -\int_{T_i}^{T_o} 4\pi k \, dT \quad \text{or,} \quad \int_{r_i}^{r_o} Qr^{-2} \, dr = -\int_{T_i}^{T_o} 4\pi K \, dT$$

or,
$$-Q\left(\frac{1}{r_o} - \frac{1}{r_i}\right) = 4\pi k (T_i - T_o)$$

or,
$$Q = \frac{T_i - T_o}{\dfrac{\dfrac{1}{r_i} - \dfrac{1}{r_o}}{4\pi K}} = \frac{T_i - T_o}{R} \tag{17.6}$$

where, $R = \dfrac{\dfrac{1}{r_i} - \dfrac{1}{r_o}}{4\pi k}$ is called conductive resistnace.

### 17.3.5   Newton's Law of Cooling (Film Coefficient)

When a fluid at temperature $t_\infty$ (Fig. 17.6) is in contact with a solid surface at temperature $T_s$, the rate of heat transfer by convection from the surface to  the fluid is given by $Q = hA \, (T_s - T_\infty)$, where $A$ is the heat transfer area and $h$ is called convective heat transfer coefficient or film coefficient. In general convective heat transfer coefficient (h) depends upon:

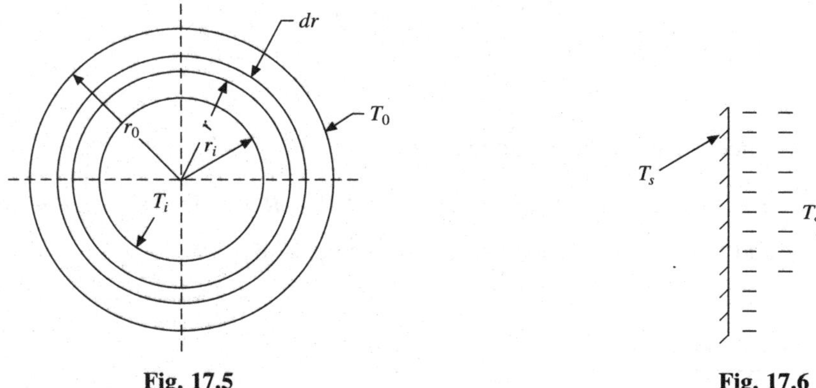

**Fig. 17.5**          **Fig. 17.6**

(a)  Thermal properties of the fluid like $C_p$, $\rho$, $k$, $\mu$, $\beta$ etc.

(b)  Nature of the geometrical configuration of the surface, like flat plate, cylinder, sphere etc.

(c)  Dynamics of fluid flow like laminar and turbulent. The equation $Q = hA(T_s - T_\infty)$ can be further written as $Q = \dfrac{T_s - T_\infty}{\dfrac{1}{hA}}$

or,
$$Q = \frac{T_s - T_\infty}{R} \tag{17.7}$$

where $R = \dfrac{1}{hA}$ is called convective resistance, which acts like a thermal barrier.

Unit of film coefficient is $k$ cal/hr $m^2 {}^\circ C$ or Watt/m²k. The equation $Q = hA(T_s - t_\infty)$ is the definition of $h$ rather than the law of convection. The entire analysis of convection boils down to determine the convective heat transfer coefficient ($h$).

Appropriate values of $h$ can be determined analytically or experimentally for various convective situations.

### 17.3.6  Heat Conduction Through Composite Structure

There are certain systems of considerable practical importance which are made of two or more layers of different materials. For example; cold-storage walls have a layer of brick, a layer of thick insulation and plaster on both sides. Similarly steam pipes are insulated with a material having very low thermal conductivity. Such systems offer thermal resistances which are either in series or parallel. Interface contact depends upon total force exerted on the contacting surfaces physical nature of bond and quality of fluid trapped in between. In general, there is a contact resistance, but here we shall neglect the contact resistances and develop an expression for the heat conductor.

### *(a) Composite plane wall in contact with fluid*

Consider a plane wall consisting of three layers of material in contact (Fig. 17.7) with each other. Inside and outside surfaces are in contact with fluids having convective heat transfer coefficient as $h_i$ and $h_o$ at temperature $T_i$ and $T_o$ respectively. The rate of heat transfer in steady state can be determined by electrical analogy method in which the conductive thermal resistances $R_1$, $R_2$ and $R_3$ and two convective resistances $R_i$ and $R_o$ are in series. Therefore rate of heat transfer

$$Q = \frac{T_i - T_o}{R_i + R_1 + R_2 + R_3 + R_0} = \frac{(T_i - T_o)}{\dfrac{1}{Ah_i} + \dfrac{L_1}{AK_1} + \dfrac{L_2}{AK_2} + \dfrac{L_3}{AK_3} + \dfrac{1}{Ah_0}} \qquad (17.8)$$

Temperature at any distance $T$ can be determined as:

$T = T_o$ – Temperature drop in $R_i$ and $R_x$. That is, $T = T_i - Q(R_i + R_1) = T_i - Q\left(\dfrac{1}{Ah_i} + \dfrac{x}{AK_1}\right)$

If we have both series and parallel resistance as shown in Fig. 17.8.

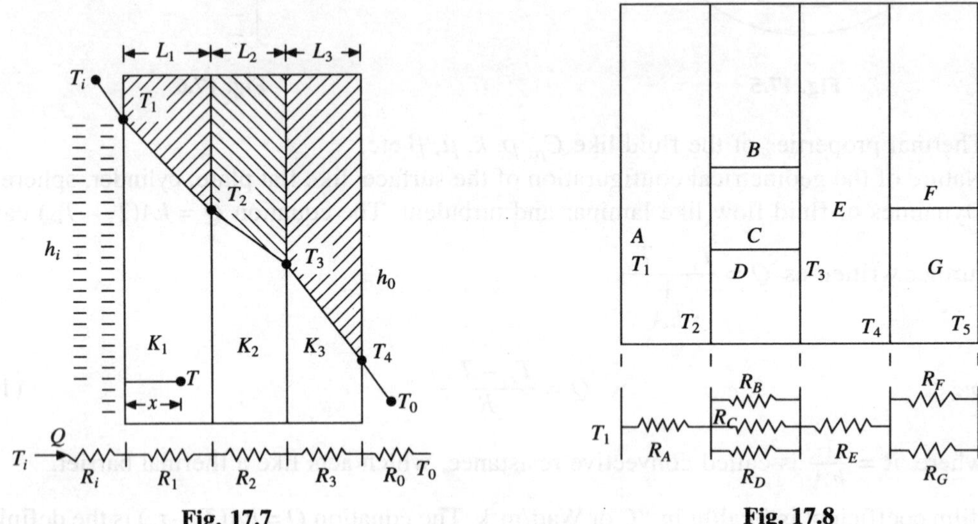



**Fig. 17.7**



**Fig. 17.8**

$$Q = \frac{T_1 - T_5}{R_A + \dfrac{1}{\dfrac{1}{R_B} + \dfrac{1}{R_C} + \dfrac{1}{R_D}} + R_E^+ \dfrac{1}{\dfrac{1}{R_F} + \dfrac{1}{R_G}}} \qquad (17.9)$$

### (b) Composite Cylinders

Consider a hollow cylinder (Fig. 17.9) consisting of three layers of materials having thermal conductivity $k_1$, $k_2$ and $k_3$ and inside and outside convective heat transfer coefficients $h_i$ and $h_o$. The rate of heat transfer is given by

$$Q = \frac{T_i - T_o}{\dfrac{1}{A_i h_i} + \dfrac{\ln \dfrac{r_2}{r_1}}{2\pi L K_1} + \dfrac{\ln \dfrac{r_3}{r_2}}{2\pi L K_2} + \dfrac{\ln \dfrac{r_4}{r_3}}{2\pi L K_3} + \dfrac{1}{A_o h_o}} \qquad (17.10)$$

where $A_i$ = inside heat transfer area = $2\pi\, r_i L$, and $A_o$ = outside heat transfer area = $2\pi r_o L$.

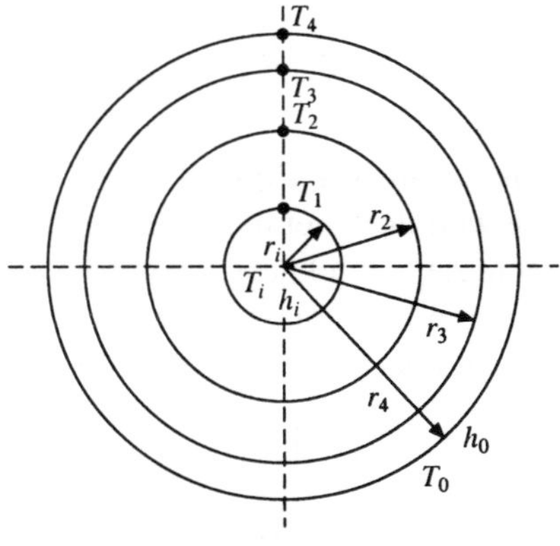


**Fig. 17.9**

## 17.3.7  Overall Heat Transfer Coefficient

In order to calculate the rate of heat transfer through a series of resistances (thermal barriers), it is necessary to know the thicknesses, conductivity and the temperatures. But we expect the rate of heat transfer by conduction and convection is propertional to final temperature defference and heat transfer area. Therefore mathematically $Q \, \alpha \, A \, (T_i - T_o)$ or, $Q \, \alpha \, A \, \Delta T$

or,
$$Q = UA\Delta T = \frac{\Delta T}{R} = \frac{\Delta T}{R_i + R_1 + R_2 + \ldots R_o}$$

or,
$$UA = \frac{1}{R} = \frac{1}{R_i + R_1 + R_2 + \ldots R_o} \tag{17.11}$$

Where $U$ is called overall heat transfer coefficient considering all the thermal barriers in the way of rate of heat transfer.

For planewall, $A_i$ and $A_o$ are same but for cylinder and spheres $A_i = A_o$. Therefore

$$Q = U_i A_i \, \Delta T = U_o A_o \Delta T, \text{ demands that } U_i > U_o$$

For plane wall
$$\frac{1}{UA} = R_i + R_1 + R_2 + R_3 + R_o$$

$$= \frac{1}{Ah_i} + \frac{L_1}{AK_1} + \frac{L_2}{AK_2} + \frac{L_3}{AK_3} + \frac{1}{Ah_o}$$

$$\frac{1}{U} = \frac{1}{h_i} + \frac{L_1}{K_1} + \frac{L_2}{K_2} + \frac{L_3}{K_3} + \frac{1}{h_o} \tag{17.12}$$

For cylindrical walls

$$\frac{1}{A_i U_i} = \frac{1}{A_o U_o} = \frac{1}{A_i h_i} + \frac{\ln\left(\dfrac{r_2}{r_i}\right)}{2\pi L K_1} + \frac{\ln\dfrac{r_3}{r_2}}{2\pi L K_3} + \frac{\ln\dfrac{r_4}{r_3}}{2\pi L K_4} + \frac{1}{2\pi L r_o h_o}$$

$$\frac{1}{U_i} = \frac{1}{h_i} + \frac{A_i}{2\pi L K_1}\ln\frac{r_2}{r_1} + \frac{A_i}{2\pi L K_2}\ln\frac{r_3}{r_2} + \frac{A_i}{2\pi L K_3}\ln\frac{r_4}{r_3} + \frac{A_i}{2\pi L r_o h_o}$$

or,
$$\frac{1}{U_i} = \frac{1}{h_i} + \frac{r_1}{K_1} \ln \frac{r_2}{r_1} + \frac{r_1}{K_2} \ln \frac{r_3}{r_2} + \frac{r_1}{K_3} \ln \frac{r_4}{r_3} + \frac{r_1}{r_4} - \frac{1}{h_o} \qquad (17.13)$$

### 17.3.8   Critical Thickness of Insulation for Pipes

The addition of insulation on plane wall does not increase the face area of insulation but increases the resistance ($R = L/AK$) due to increase in thickness. Therefore, increase in thickness of insulation on a plane always results in decrease in heat transfer. On the other hand, increase in thickness of insulation for cylinder and spheres increases the heat transfer area and also increase in the conductive

resistance $\left( \dfrac{\ln \dfrac{r_2}{r_1}}{2\pi LK} \right)$ in case of cylinder and $\left( \dfrac{\dfrac{1}{r_1} - \dfrac{1}{r_2}}{4\pi K} \right)$ in case of sphere. Therefore, increase in

thickness of insulation does not always emply decrease in rate of heat transfer.

Let us consider a pipe of inner radius $r_i$, outer radius $r_1$ on which insulation is applied up to radius $r_2$. The outer surface of the insulation is in contact with a fluid at temperature To and film coefficient $h_o$ (Fig. 17.10). Rate of heat

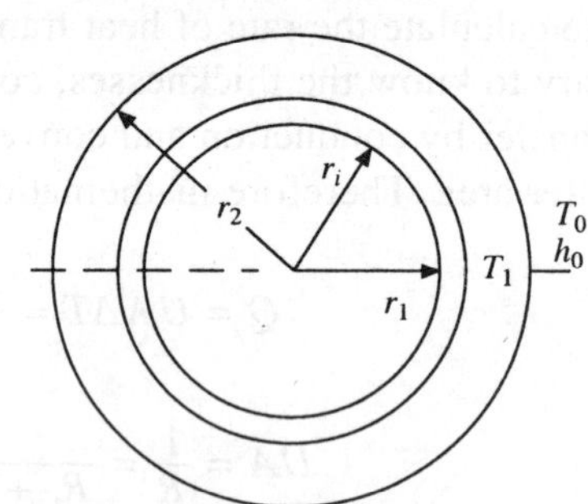

transfer $= Q = \dfrac{T_1 - T_0}{R} = \dfrac{T_1 - T_0}{\dfrac{\ln \dfrac{r_i}{r_1}}{2\pi LK} + \dfrac{1}{2\pi r_2 L h_o}}$

$r_1$ and $L$ are fixed, for maximum rate of heat transfer, $R$ should be minimum

**Fig. 17.10**

or,
$$\frac{dR}{dr_2} = 0 \quad \text{or,} \quad \frac{d}{dr_2} \left( \frac{\ln \dfrac{r_2}{r_1}}{2\pi LK} + \frac{1}{2\pi r_2 Lh_o} \right) = 0 \qquad (17.14)$$

or,
$$\frac{1}{2\pi L} \frac{d}{dr_2} \left[ \ln \frac{r_2}{k} - \ln \frac{r_1}{k} + \frac{1}{r_2 h_0} \right] = 0$$

or,
$$\frac{1}{kr_2} - \frac{1}{h_o r_2^2} = 0 \quad \text{or,} \quad r_2 = \frac{k}{h_o} \qquad (17.15)$$

Therefore, at $r_2 = \dfrac{k}{h_o} = r_{C'}$, the rate of heat transfer is maximum $r_2 = \dfrac{k}{h_o} = r_c$ is called critical radius of insulation.

The following conclusions can be drawn:

(i) If the outer radius of bare pipe is less than the critical radius, as $r_2$ increases, the rate of heat transfer increases first, attains a maximum value and starts decreasing. Therefore insulation radius should be more than the critical radius as shown in Fig. 17.11.

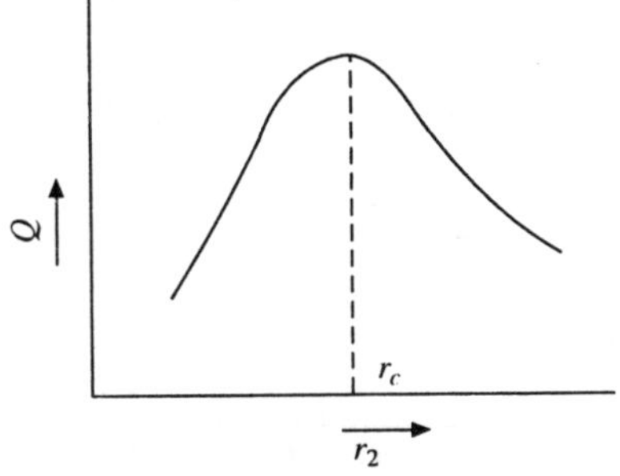

**Fig. 17.11**

(ii) If the outer radius of the pipe is already greater than $r_c$ any addition of insulation decreases the rate of heat transfer.

It is to be noted that critical radius of insulation increases with decrease in heat transfer coefficient. Electrical cables are insulated up to critical radius in order to dissipate maximum Joulean heat $(I^2R)$ in the surroundings. Therefore, this insulation serves both purpose insulation against electrical current and dissipation of maximum heat. Hence, $r_c = K/ho$ is called cable insulation radius.

### 17.3.9 Effect of Non-Uniform Thermal Conductivity

It has already been mentioned that the thermal conductivity varies with temperature. The variation of thermal conductivity with temperature may be neglected if the temperature range under consideration is not large or if the temperature dependence of the conductivity is not two severe. In general thermal conductivity $K = f(T)$, but within limited temperature range, thermal conductivity varies with temperature linearly in the form $K = K_o (1 + \alpha T)$ where $K_o$ is conductivity at 0°C and $\alpha$ is temperature coefficient of thermal conductivity. Considering heat conduction through a plane wall with variable thermal conductivity we have,

$$Q = -kA \frac{dT}{dx} = -k_o(1 + \alpha T) A \frac{dT}{dx} \tag{17.16}$$

or,

$$\int_o^L Q \frac{dx}{A} = -\int_{T_1}^{T_2} k_o(1 + \alpha T) dT$$

or,

$$\frac{QL}{A} = -K_0 \left[ (T_2 - T_1) + \frac{\alpha}{2} (T_2^2 - T_1^2) \right]$$

$$\frac{QL}{A} = (T_1 - T_2) k_0 \left[ 1 + \alpha \frac{(T_1 + T_2)}{2} \right] = (T_1 - T_2) k_m \tag{17.17}$$

where $K_m = K_o \left( 1 + \alpha \dfrac{(T_1 + T_2)}{2} \right)$ is called the mean thermal conductivity which is evaluated at

mean temperature of $\left( \dfrac{T_1 + T_2}{2} \right)$

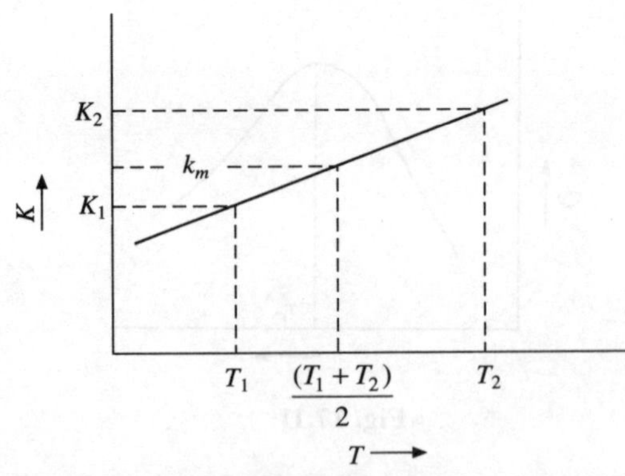

**Fig. 17.12**

# SOME IMPORTANT RESULTS

1. Fourier's law of heat conduction $Q = -KA\dfrac{dT}{dx}$

2. Conductive resistance for a plane wall $= R = \dfrac{L}{KA}$ which is similar to electrical resistance

3. Conductive resistance for a hollow cylinder $= R = \dfrac{\ln\left(\dfrac{r_2}{r_1}\right)}{2\pi LK}$

4. Conductive resistance for a spherical shell $= \dfrac{\dfrac{1}{r_1} - \dfrac{1}{r_2}}{4\pi k}$

5. Convective resistance $= \dfrac{1}{hA} = \dfrac{1}{h_i A_i} = \dfrac{1}{h_o A_o}$

6. Reciprocal of resistance is called conductance i.e. $C = \dfrac{1}{R}$

7. Equation of steady state heat conduction $Q = \dfrac{(T_1 - T_2)}{R} = C(T_1 - T_2)$ which is similar to

   Ohm's law of current $I = \dfrac{V_1 - V_2}{R} = C(V_1 - V_2)$

8. Equivalent resistance for series combination $R = R_1 + R_2 + R_3 +$ and for parallel combination

   $\dfrac{1}{R} = \dfrac{1}{R_1} + \dfrac{1}{R_2} + \dfrac{1}{R_3} + \ldots$

9. Thermal conductivity varies with temperature $K = K_o\,(1 + \alpha T)$

10. Critical radius for insulation of pipe $= r_c = \dfrac{K}{h_0}$ and for spherical shell $r_c = \dfrac{2K}{h_0}$.

## Solved Examples (Conduction)

**17.1** A furnace wall is made up of three layers, one of brick, one of insulating brick and one of red brick. The inner and outer surfaces are at 870°C and 40°C respectively. The respective thermal conductivities of the layers are 1.17, 0.139 and 0.875 W/m K respectively and thickness are 22 cm, 7.5 cm and 11 cm. Assume close bonding of the layers at their interfaces, find out the rate of heat loss per square meter per hour and interface temperatures.

*Soln.* Refer to Fig. 17.13

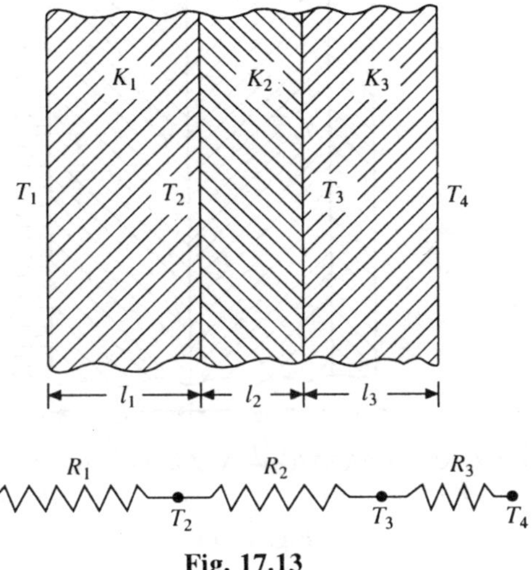


**Fig. 17.13**

$$T_1 = 870\ °C,\ T_4 = 40°C,\ K_1 = 1.17\ \text{w/mK},\ K_2 = 0.139\ \text{w/mK}$$

$$K_3 = 0.875\ \text{W/mK},\ L_1 = 0.22\ \text{m}\quad l_2 = 0.075\ \text{m},\ l_3 = 0.11\text{m}$$

The equivalent thermal ckt. has been shown $R_1$, $R_2$, and $R_3$ are conductive resistance. Rate of heat transfer

$$Q_1 = \frac{T_1 - T_4}{R_1 + R_2 + R_3} = \frac{T_1 - T_4}{\dfrac{l_1}{AK_1} + \dfrac{l_2}{AK_2} + \dfrac{l_3}{AK_3}}$$

$$= \frac{870 - 40}{\dfrac{0.22}{1 \times 1.17} + \dfrac{0.075}{1 \times 0.139} + \dfrac{0.11}{1 \times 0.875}} = \frac{830}{0.188 + 0.539 + 0.125}$$

$$= 974.178\ \text{W/mK}\quad \textit{Ans.}$$

Interface temperatures

$$T_2 = T_1 - \text{temperature drop in } R_1 = T_1 - Q \cdot R_1 = 870 - 183.14$$

$$= 685.86°C,\quad T_3 = T_1 - Q\,(R_1 + R_2)$$

$$= 870 - 974.178(0.539 + 0.125) = 223.14°C\quad \textit{Ans.}$$

**17.2** An ice box has a composite wall made up of 1 mm thick aluminium sheet on the inside surface, 3 cm thick wooden board on the outside and 5 cm cork insulation between the two. Ice at –5°C is in contact with the aluminium surface. The unit surface conductance on the outside the box is 11.61 $W/m^2K$. The thermal conductivities of wood, cork and aluminium are 0.209, $4.2 \times 10^{-3}$, 205.83 W/mK respectively. The outside temperature is 27°C. Calculate (a) the thermal resistance of the composite wall (b) the overall thermal resistance and (c) the rate of heat transfer per unit area.

*Soln.*   Refer to Fig. 17.14

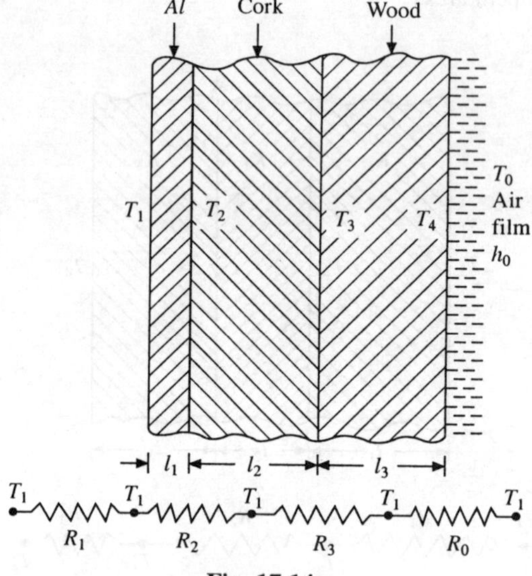


**Fig. 17.14**

$$l_1 = 0.001 \text{ m}, \ l_2 = 0.05 \text{ m},$$

$$l_3 = 0.03 \text{ m}, \ T_1 = -5°C, \ T_0 = 27°C$$

$$K_1 = 205.83 \text{ W/mK}, \ K_2 = 4.2 \times 10^{-3} \text{ W/mK}$$

$$K_3 = 0.209 \text{ W/mK}$$

$$h_0 = 11.61 \text{ W/m}^2\text{K}$$

$$\text{Thermal resistance of composite wall} = R_1 + R_2 + R_3 = \frac{0.001}{1 \times 205.83} + \frac{0.05}{1 \times 4.2 \times 10^{-3}}$$

$$+ \frac{0.03}{1 \times 0.209} = 238.23 \text{ K/W} \quad Ans.$$

$$\text{Overall thermal resistance} = R_1 + R_2 + R_3 + R_0$$

$$= \frac{l_1}{AK_1} + \frac{l_2}{AK_2} + \frac{l_3}{AK_3} + \frac{1}{Ah_0} = 238.23 + \frac{1}{1 \times 11.61} = 238.31 \text{ K/W} \quad Ans.$$

$$\text{Rate of heat transfer from outside} = \frac{T_0 - T_1}{238.31} = \frac{27 - (-5)}{238.31} = 0.1342 \text{ Watt} = 0.1342 \text{ J/s} \quad Ans.$$

**17.3**   The thermal conductivity of a material is measured by means of an electric heater sand-witched between two identical slabs of the material (15 cm × 15 cm × 5 cm thick). The temperatures of the faces of each slab are recorded as 80°C and 40°C. The heater consumes 50 watt of electric power. Find the thermal conductivity of the material.

*Soln.* Refer to Fig. 17.15

Thickness of each slab = 0.05 m,

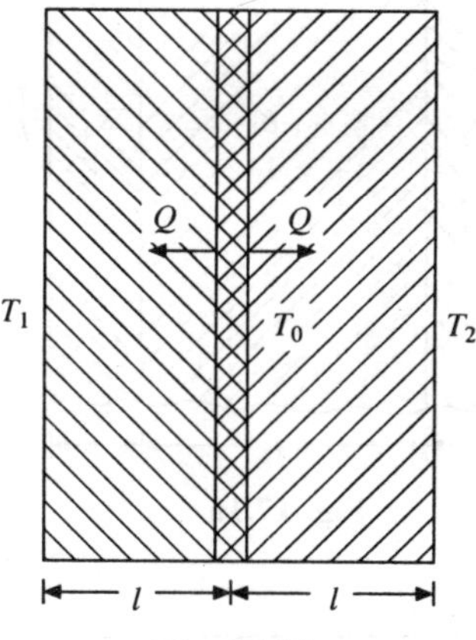


**Fig. 17.15**

Heat transfer through each slab $= \dfrac{50}{2} = 25$ Watt

$$T_0 = 80°C, \; T_1 = T_2 = 40°C$$

$$Q = \frac{T_0 - T_1}{R} = \frac{T_0 - T_1}{\dfrac{1}{AK}}$$

or, $$25 = \frac{\dfrac{(80 - 40)}{0.05}}{0.15 \times 0.15\,K}$$

or, $$K = 1422 \text{ Watt/mK} \quad Ans.$$

**17.4** The interior of a refrigerator having inside dimensions of 50 cm × 50 cm base area and 1 meter high is to be maintained at 7°C. The walls of the refrigerator are constructed of two mild steel sheets of 3 mm thick with 5 cm of glass wool insulation between them. If the average heat transfer coefficients at the inner and outer surfaces are 11.61 and 14.53 W/m²K respectively. Estimate the rate at which heat must be removed from the interior to maintain the specified temperature. What is the temperature at the surface of the wall if out side temperature is 28°C. The thermal conductivities of mild steel and glass wool are 46.52 W/mK and 0.046 W/mK.

*Soln.* Refer to Fig. 17.16

$$l_1 = l_3 = 0.003 \text{ m}, \; l_2 = 0.05 \text{ m}$$

$$k_1 = k_3 = 0.046 \text{ W/mK}$$

$$k_2 = 46.52 \text{ W/mK}, \; hi = 11.61 \text{ W/m}^2\text{K}, \; h_0 = 14.53 \text{ W/m}^2\text{K}$$

$$T_i = 7°C$$

$$T_o = 28°C. \text{ Area of heat transfer} = (0.5 \times 0.5 \times 2 + 0.5 \times 1 \times 4) = 2.5 \text{ m}^2$$

Heat to be removed = Heat transfer from out side to inside

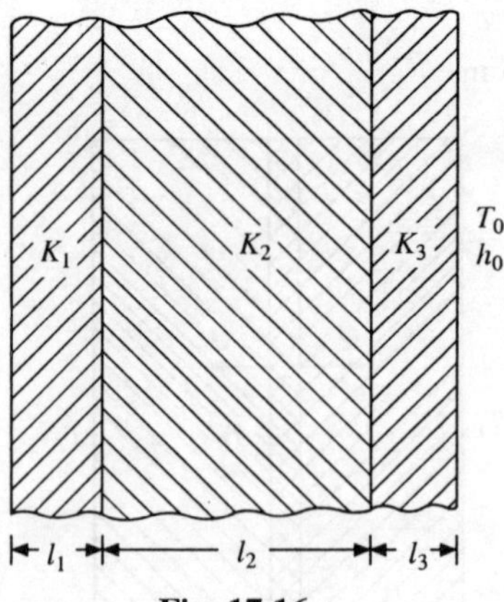


**Fig. 17.16**

$$= \frac{T_0 - T_i}{\dfrac{1}{Ah_i} + \dfrac{l_1}{AK_1} + \dfrac{l_2}{AK_2} + \dfrac{l_3}{AK_3} + \dfrac{1}{Ah_0}}$$

$$= \frac{28 - 7}{\dfrac{1}{2.5 \times 11.61} + \dfrac{0.003}{2.5 \times 0.046} + \dfrac{0.05}{2.5 \times 46.52} + \dfrac{0.003}{2.5 \times 0.046} + \dfrac{1}{2.5 \times 14.53}}$$

$$= \frac{21}{0.034 + 0.026 + 4.299 \times 10^{-4} + 0.026 + 0.027}$$

$$= 185.136 \text{ Watt} \quad Ans.$$

**17.5** Furnace has a composite wall constructed of a refractory materials for the inside layer and an insulating material on the outside. The total wall thickness is limited to 60 cm. The mean temperature of the gases within the furnace is 850°C, the external air temperature is 30°C and the temperature at the inter face of the two materials of the furnace wall is 500°C. The thermal conductivity of refractory and insulating materials are 2.32 and 0.23 W/mK.

The combined coefficient of heat transfer by convection and radiation between the gases and inside refractory surface is 232.5 W/m²K and between outside surface and atmospheric air is 46.38 W/mK. Find out (a) required thickness of each materials (b) the temperature of the surface exposed to gases and that exposed to air, (c) the rate of heat loss to atmosphere per m².

*Soln.* Refer to Fig. 17.17

$$\text{Data } T_i = 850°C, \, T_o = 30°C$$

$$T_2 = 500°C, \, K_1 = 2.32 \text{ W/mK}, \, k_2 = 0.23 \text{ W/mK}$$

$$h_i = 232.5 \text{ W/m}^2\text{K} \quad h_0 = 46.38 \quad \text{W/m}^2\text{K}$$

$$Q = \frac{T_i - T_o}{R_i + R_1 + R_2 + R_0} = \frac{T_i - T_2}{R_i + R_1}$$

or,

$$\frac{850 - 30}{\dfrac{1}{232.5} + \dfrac{x}{2.32} + \dfrac{(0.60 - x)}{0.23} + \dfrac{1}{46.38}} = \frac{(850 - 500)}{\dfrac{1}{232.5} + \dfrac{x}{2.32}}$$

or,

$$x = 0.53 \text{ m} = 53 \text{ cm}$$

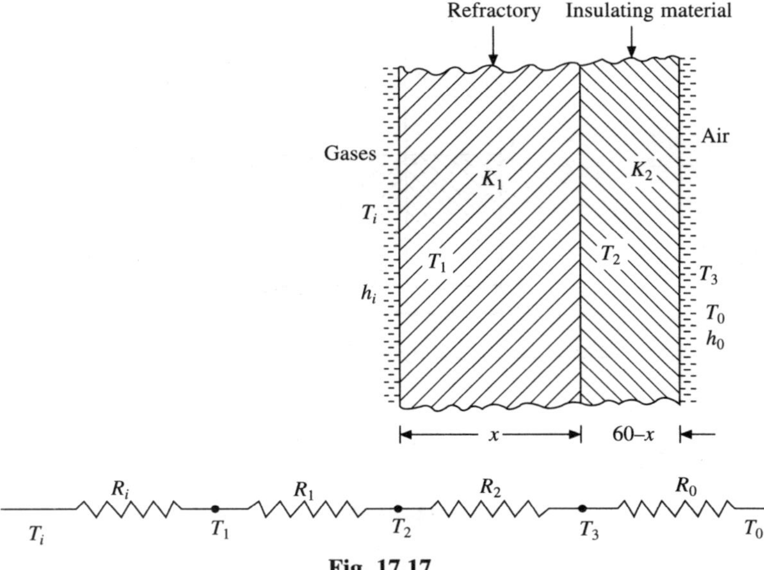


**Fig. 17.17**

Rate of heat transfer $Q = \dfrac{T_i - T_2}{R_i + R_1} = \dfrac{850 - 500}{\dfrac{1}{232.5} + \dfrac{0.53}{2.32}} = 1503.76$ Watt  *Ans.*

Temperature of surface exposed to hot gases $= T_1 = T_i - QR_i = 850 - 1503.76 \dfrac{1}{232.5} = 843.5°C$  *Ans.*

Temperature of surface exposed to air

$$= T_3 = T_2 - QR_2 = 500 - 1503.76 \times \dfrac{0.07}{1 \times 0.23} = 42.33°C$$

**17.6** A composite insulating wall has three slabs held together by 2.5 cm diameter aluminium rivets per 0.1 m of wall surface. The hot slab face is brick 10 cm thick, cold face timber 2.5 cm thick and the centre slab is 20 cm of insulating brick. The hot face temperature is 150°C and cold face 20°C. The conductivities of brick, insulating brick wood and aluminium are 2.60, 0.30, and 725 kJ/m–hr °K. Find the increase in heat transfer rate caused by the rivet. Neglect the effect of rivet heads and all lateral heat transfer.

*Soln.* Refer to Fig. 17.18

$$d = 0.025 \text{ m}, \quad l_1 = 0.10 \text{ m}, \quad l_2 = 0.20 \text{ m}, \quad l_3 = 0.025 \text{ m}$$

$$T_1 = 150°C \ T_4 = 20°C$$

Rate of heat transfer without rivets

$$= \frac{T_1 - T_4}{R_1 + R_2 + R_3} = \frac{T_1 - T_4}{\dfrac{l_1}{AK_1} + \dfrac{l_2}{AK_2} + \dfrac{l_3}{AK_3}}$$

$$= \frac{150 - 20}{\dfrac{0.10}{0.1 \times 2.60} + \dfrac{0.20}{0.1 \times 0.30} + \dfrac{0.025}{0.1 \times 0.35}} = 16.7 \text{ kJ/hr}$$

Heat transfer with rivets. Area of rivet $= \dfrac{\pi}{4}(0.025)^2$

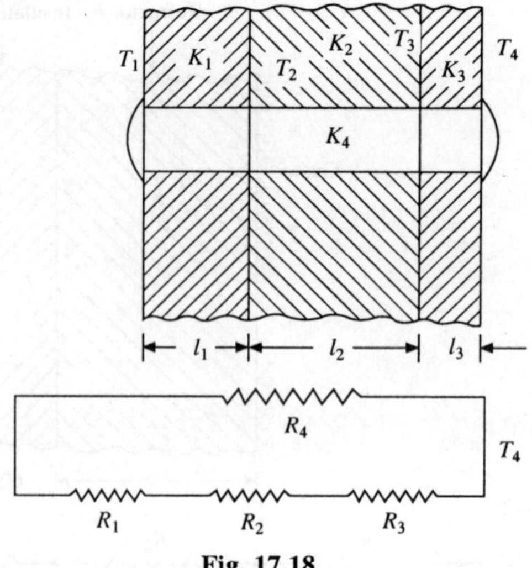

**Fig. 17.18**

Heat transfer with rivets. Area of rivet $= \dfrac{\pi}{4}(0.025)^2$

$$= 4.91 \times 10^{-4} \text{ m}^2. \text{ Resistance of bolt} = R_4 = \frac{l}{AK}$$

$$= \frac{0.10 + 0.20 + 0.025}{4.91 \times 10^{-4} \times 725} = 0.913$$

$$R_1 + R_2 + R_3 = \frac{1}{A}\left[\frac{l_1}{k_1} + \frac{l_2}{k_2} + \frac{l_3}{k_3}\right] = \frac{1\left[\dfrac{0.10}{2.6} + \dfrac{0.20}{0.3} + \dfrac{0.025}{0.35}\right]}{(0.1 - 0.000491)} = 7.76$$

$$\text{Equivalent resistance } R = \frac{0.913 \times 7.76}{0.913 + 7.76} = 0.817$$

$$\text{Heat transfer with rivet} = \frac{150 - 20}{0.817} = 159 \text{ kJ/hr}$$

Increase in heat transfer = 159 − 16.7 = 142.3 kJ/hr   *Ans.*

17.7  A building wall 6m long, 5 m high is builtup as follows: Inside 10 cm tile, $k = 0.694$ W/mK and outside 10 cm brick, $k = 1.027$ W/mK. In between these two layers, there is an air gap which has thermal resistance of 0.019 K/W. Calculate the rate of heat transfer through the wall if inside face temperature is 30°C and outside face temperature is 10°C. Also determine the interface temperatures.

*Soln.*  Refer to Fig. 17.19

$$l_1 = l_2 = 10 \text{ cm } T_1 = 30°C \ T_4 = 10°C$$

$$\text{Rate of heat transfer } Q = \frac{T_1 - T_4}{R_1 + R_\text{air} + R_2} = \frac{T_1 - T_4}{\dfrac{l_1}{AK_1} + \dfrac{R_\text{air}}{} + \dfrac{l_3}{AK_3}}$$

$$= \frac{(30 - 10)}{\dfrac{0.10}{6 \times 5 \times 0.694} + 0.019 + \dfrac{0.10}{6 \times 5 \times 1.027}} = 739.405 \text{ Watt} \quad Ans.$$

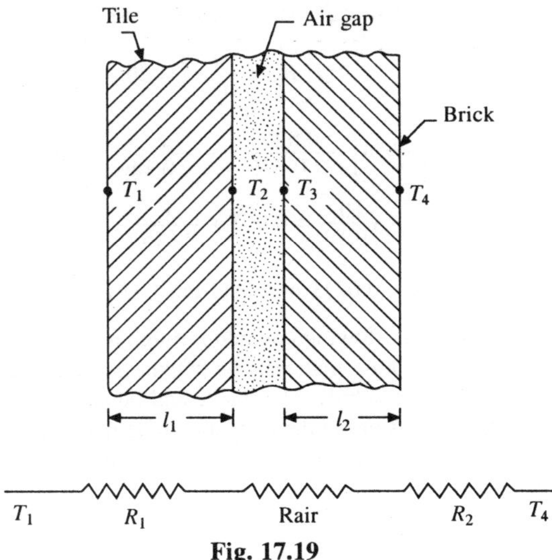



**Fig. 17.19**

$$T_2 = T_1 - R_1 Q = 30 - \frac{0.10}{6 \times 5 \times 0.694} \times 739.405 = 26.44°C \quad Ans.$$

$$T_3 = T_2 - R_{air} \times Q = 26.44 - 0.019 \times 739.405 = 12.39°C \quad Ans.$$

**17.8** An ice box has walls constructed of a 10 cm layer of cork board contained between two wooden walls and 2 cm thick. Find the rate of removal in kJ per m² per hour if the inner wall surface is kept at –10°C while the outer surface temperature is 30°C. Find out the zone in the wall where temperature is 20°C. Thermal conductivities of cork board and wood respectively are 0.146 and 0.376 kJ/m hr °C.

*Soln.* Refer to Fig. 17.20

$$\text{Rate of heat transfer} = Q = \frac{T_4 - T_1}{R_1 + R_2 + R_3}$$

$$= \frac{30 - (-10)}{\dfrac{0.02}{1 \times 0.376} + \dfrac{0.10}{0.146} + \dfrac{0.02}{0.376}} = 50.54 \text{ kJ/hr}$$

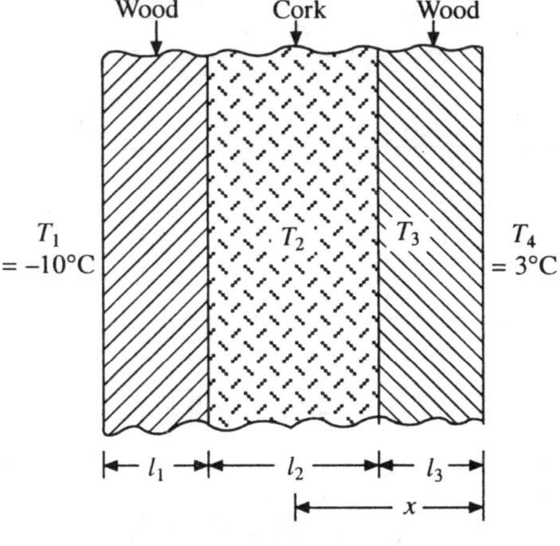



**Fig. 17.20**

Let at a distance $x$ from the wooden wall inside the cork, the temperature is 20°C.

$$T_4 - Q\left[R_3 + \frac{x}{AK}\right] = 20$$

or,
$$30 - 50.54\left[\frac{0.02}{1 \times 0.376} + \frac{x}{1 \times 0.146}\right] = 20$$

or,
$$\frac{10}{50.54} = 0.053 + \frac{x}{0.146}$$

or,
$$x = 0.0211 \text{ m} = 2.11 \text{ cm} \quad Ans.$$

**17.9** Hot air at a temperature of 40°C is flowing covered through a steel pipe of 10 cm diameter, 40 m long with two layers of different insulating materials of thickness 4 cm and 3 cm and their corresponding conductivities are 0.117 and 0.372 W/mK. The inside and outside unit conductance are 58.05 W/m² K and 11.61 W/m²K respectively. Assuming the atmospheric temperature as 10°C find out (a) rate of heat transfer (b) temperature of contact between the two insulations and (c) overall heat transfer-coefficient based on the outer area.

*Soln.* Refer to Fig. 17.21

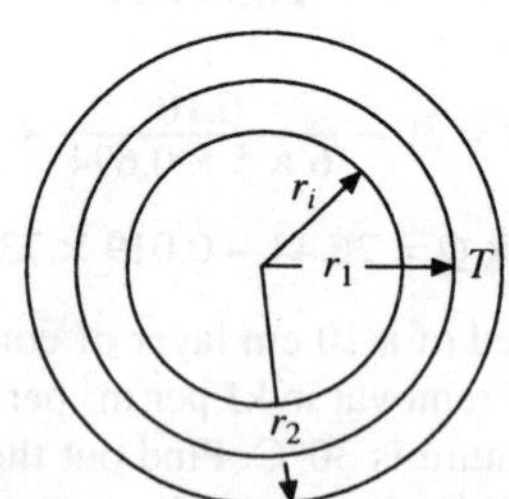


$$T_i \;\underset{R_i}{\text{\small\wedge\wedge\wedge}}\; \underset{R_1}{\text{\small\wedge\wedge\wedge}}\; \overset{T}{\underset{R_2}{\text{\small\wedge\wedge\wedge}}}\; \underset{R_0}{\text{\small\wedge\wedge\wedge}}\; T_0$$

**Fig. 17.21**

$T_i = 40°C$, $T_o = 10°C$, $r_i = 0.05$ m, $r_1 = 0.09$ m, $r_0 = 0.12$ m, $k_1 = 0.117$ W/mK

$K_2 = 0.372$, $h_o = 11.61$ W/mK, $h_i = 58.05$ W/m²K

$$Q = \text{Rate of heat transfer} = \frac{T_i - T_o}{R_i + R_1 + R_2 + R_0}$$

$$= \frac{40 - 10}{\dfrac{1}{2\pi r_i L h_i} + \dfrac{\ln \dfrac{r_1}{r_i}}{2\pi L K_1} + \dfrac{\ln \dfrac{r_o}{r_1}}{2\pi L K_2} + \dfrac{1}{2\pi r_0 L h_o}}$$

$$= \frac{30}{\dfrac{1}{2\pi \times 0.05 \times 40 \times 58.05} + \dfrac{\ln \dfrac{0.09}{0.05}}{2\pi \times 40 \times 0.117}}$$

$$+ \frac{\ln \dfrac{0.12}{0.09}}{2\pi \times 40 \times 0.372} + \frac{1}{2\pi \times 0.12 \times 40 \times 11.61} = 1.0969 \text{ kW} = 1.0969 \text{ kJ/s} \quad Ans.$$

Temperature of contact $= T = T_i - Q(R_i + R_1)$

$$= 40 - 1096.9 \left[ \frac{1}{2\pi \times 40 \times 0.05 \times 58.05} + \frac{\ln \dfrac{0.09}{0.05}}{2\pi \times 40 \times 0.117} \right] = 15.17°C \quad Ans.$$

Overall heat transfer coefficient based on outer area is given by

$$Q = A_0 U_0 (T_i - T_0) \; U_0 = \frac{Q}{A_0(T_1 - T_0)} = \frac{1096.9}{2\pi \times 0.12 \times 40(40 - 10)} = 1.212 \text{ W/m}^2\text{K} \quad Ans.$$

**17.10** A 5 cm diameter steel pipe is to be insulated by using two layers, one of asbestos ($K = 0.145$ W/mK) and other of magnesia ($K = 0.069$ W/mK) each having a thickness of 5 cm. Find the ratio of heat loss when asbestos is inside to that when asbestos is outside. Assume that the temperature drop across the composite insulation remains unchanged.

*Soln.*

$$Q = \frac{\Delta T}{R_1 + R_2} \text{ When abestos is inside}$$

$$\text{Rate of heat transfer} = Q_1 = \frac{\Delta T}{\dfrac{\ln \dfrac{r_1}{r_i}}{2\pi L K_1} + \dfrac{\ln \dfrac{r_0}{r_1}}{2\pi L K_2}}$$

$$= \frac{2\pi l \Delta T}{\dfrac{\ln \dfrac{7.5}{2.5}}{0.145} + \dfrac{\ln \dfrac{12.5}{7.5}}{0.069}} = 0.067 \, \Delta T \times 2\pi L$$

When asbestos is outside $\quad Q_2 = \dfrac{2\pi L \Delta T}{\dfrac{\ln \dfrac{7.5}{2.5}}{0.069} + \dfrac{\ln \dfrac{12.5}{7.5}}{0.145}} = 0.051 \times \Delta T \times 2\pi L$

$$\therefore \qquad \text{Ratio} = \frac{Q_1}{Q_2} = \frac{0.067 \, \Delta T \times 2\pi L}{0.051 \times \Delta T \times 2\pi L} = 1.313 \quad Ans.$$

**17.11** A Steam condenser consists of copper tubes of 2 cm inside diameter, 2 mm thick over which steam is condensing and inside water is flowing at 29°C. The conductivity of copper is 387.29 W/mK the steam side and water side film coefficient are 697.88 and 4652.22 W/m²K respectively. Steam is condensing at $\frac{1}{10}$th of atmospheric pressure at which latent heat of condensation is 2393.28 kJ/kg and saturation temperature is 46°C. Find for one meter length of the pipe (a) the total thermal resistance (b) rate of condensation of steam.

*Soln.* Refer to Fig. 17.22

$$T_i = 29°C, \; T_o = 46°C, \; r_i = 1 \text{ cm}, \; r_o = 1.2 \text{ cm}, \; h_i = 4652.22 \text{ W/m}^2\text{K}$$

$$h_0 = 697.83 \text{ W/m}^2\text{K}, \; k = 387.29 \text{ W/mK}$$

Rate of heat transfer

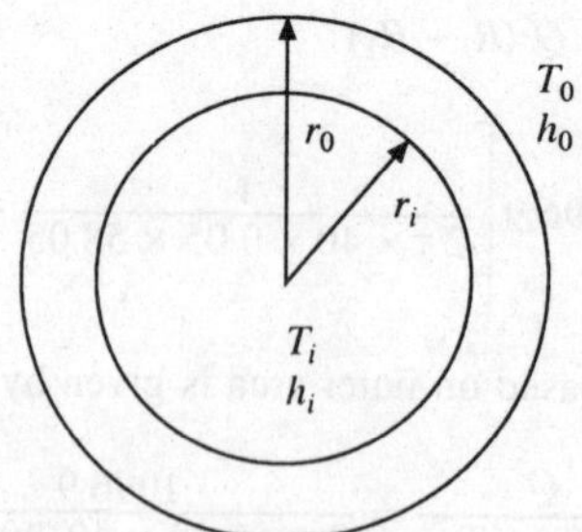

**Fig. 17.22**

$$Q = \frac{T_o - T_i}{R_i + R + R_o} = \frac{T_0 - T_i}{\dfrac{1}{2\pi r_i Lh_i} + \dfrac{\ln\dfrac{r_o}{r_o}}{2\pi LK} + \dfrac{1}{2\pi r_o Lh_o}}$$

$$= \frac{(46 - 29) \times 2\pi \times 1}{\dfrac{1}{2\pi \times 0.01 \times 4652.22} + \dfrac{1}{387.29}\ln\dfrac{1.2}{1} + \dfrac{1}{2\pi \times 0.012 \times 697.83}} = 4664.83 \text{ W}$$

Rate of condensation $= \dot{m} = \dfrac{Q}{h_{fg}} = \dfrac{4664.83}{2393.28 \times 1000} = 1.949 \times 10^{-3} \text{ kg/s} = 7.01 \text{ kg/hr}$    *Ans.*

Total thermal resistance $= R_i + R + R_o$

$$= \frac{1}{2\pi\, r_i\, Lh_i} + \frac{\ln\dfrac{r_0}{r_i}}{2\pi LK} + \frac{1}{2\pi r_0 Lh_0} = \frac{1}{2\pi \times 0.01 \times 1 \times 4652.22}$$

$$+ \frac{\ln\dfrac{1.2}{1}}{2\pi \times 1 \times 387.29} + \frac{1}{2\pi \times 0.012 \times 1 \times 697.83} = 22.5 \times 10^{-3} \text{ K/W}$$    *Ans.*

**17.12**  A steel pipe 25 mm ID and 33 mm O.D. and insulated with rock wool carries steam at 178°C. If the surrounding air temperature is 21°C, calculate the rate of heat loss from one meter length of the pipe. The thickness of insulation is 38 mm. The thermal conductivity of steel and rockwool are 44.96 W/mK and 0.173 W/mK. The inside and outside film coefficients are 5678 and 11.36 W/m²K. The contact resistance between the pipe and insulation is neglected.

*Soln.*  Refer Fig. 17.23

$$r_i = 12.5 \times 10^{-3}\text{m}$$

$$r_1 = 16.5 \times 10^{-3}\text{m}, \; r_o = 16.5 + 38 = 54.5 \text{ mm}$$

$$k_1 = 44.96 \text{ and } k_2 = 0.173 \text{ W/mK}$$

$$h_i = 5678 \text{ and } h_o = 11.36 \text{ W/m}^2\text{K}$$

$$T_i = 178°\text{C}, \; T_o = 21°\text{C}$$

Rate of heat loss per meter length

$$= \frac{T_i - T_o}{R_i + R_1 + R_2 + R_0} = \frac{T_1 - T_0}{\dfrac{1}{2\pi r_i \times 1h_i} + \dfrac{\ln\dfrac{r_i}{r_i}}{2\pi \times 1k_1} + \dfrac{\ln\dfrac{r_0}{r_1}}{2\pi \times 1\,k_2} + \dfrac{1}{2\pi r_o \times 1h_o}}$$

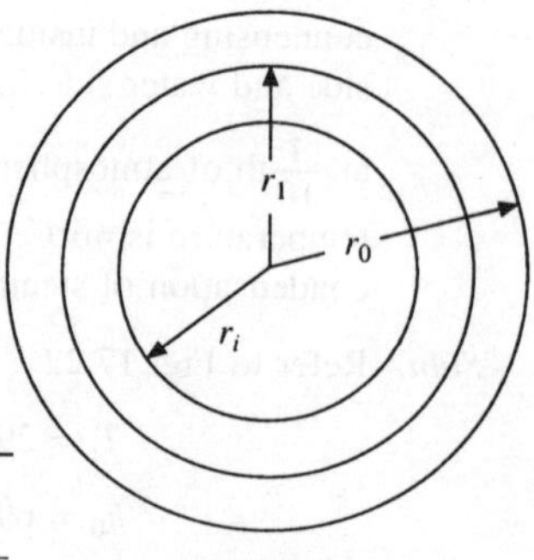

**Fig. 17.23**

$$= \frac{178 - 21}{\dfrac{1}{2\pi \times 12.5 \times 10^{-3} \times 5678} + \dfrac{\ln\dfrac{16.5}{12.5}}{2\pi \times 44.96} + \dfrac{\ln\dfrac{54.5}{16.5}}{2\pi \times 0.173} + \dfrac{1}{2\pi \, 54.5 \times 10^{-3} \times 11.36}}$$

$$= 116.30 \text{ Watt} = 116.30 \text{ J/s} \quad Ans.$$

**17.13** A 3 cm OD steam pipe is to be covered with 2 layers of insulating materials each having a thickness of 2.5 cm. The average thermal conductivity of one insulation is 5 times that of the other. Determine the percentage decrease in heat transfer if better insulating material is next to pipe than it is the outer layer. Assume that the outside and inside surface temperatures of composite insulation are fixed.

*Soln.* Refer to Fig. 17.24

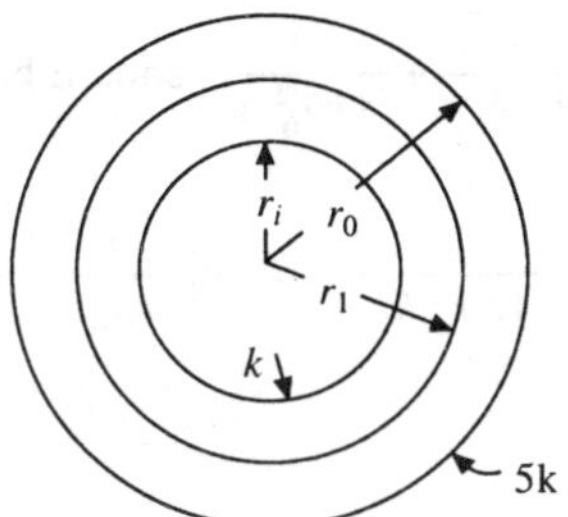


**Fig. 17.24**

Case I When better insulating material is next to pipe

Heat transfer $Q_1 = \dfrac{\Delta T}{R_1 + R_2}$

$$= \frac{\Delta T}{\dfrac{\ln\dfrac{4}{1.5}}{2\pi L K} + \dfrac{\ln\dfrac{6.5}{4}}{2\pi L 5 K}} = \frac{2\pi L K \Delta_T}{\ln\dfrac{4}{1.5} + \dfrac{\ln\dfrac{6.5}{4}}{5}} = \frac{2 L K \Delta T}{1.0774}$$

When poor insulation is next to pipe $Q_2$

$$= \frac{2\pi L K \Delta T}{\dfrac{1}{5}\ln\dfrac{4}{1.5} + \ln\dfrac{6.25}{4}} = \frac{2\pi L K \Delta T}{0.683}$$

Percentage reduction in heat flows Case 1 as compared to case II

$$= \frac{Q_2 - Q_1}{Q_2} \times 100 = \left(1 - \frac{Q_1}{Q_2}\right) \times 100 = \left(1 - \frac{0.683}{1.0774}\right) \times 100 = 36.6\% \quad Ans.$$

**17.14** Prove that the insulated sphere will lose maximum amount of heat when outer radius is $r_0 = \dfrac{2k}{h_0}$

*Soln.* Let radius of sphere (Fig. 17.25) be $r_i$ and $r_0$ for inner and outer layer of insulation. Then rate of heat

transfer $= Q = \dfrac{T_i - T_o}{\dfrac{\dfrac{1}{r_i} - \dfrac{1}{r_o}}{4\pi K} + \dfrac{1}{4\pi r_0^2 h_0}}$

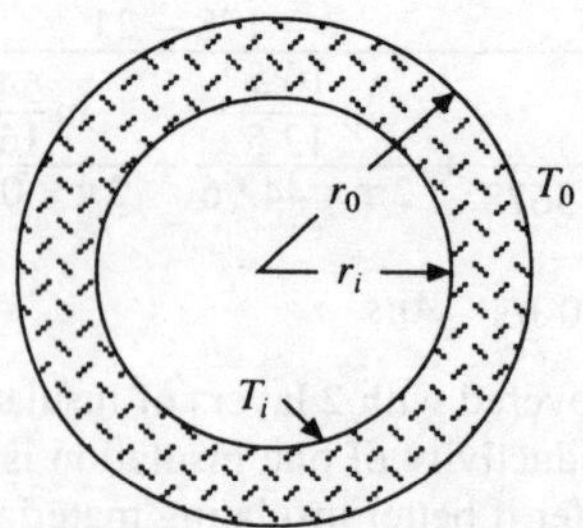



**Fig. 17.25**

For maximum heat transfer $R = \dfrac{\dfrac{1}{r_i} - \dfrac{1}{r_o}}{4\pi K} + \dfrac{1}{4\pi r_0^2 h_0}$ should be minimum i.e. $\dfrac{dR}{dr_0} = 0$

or, $\dfrac{1}{4\pi k}\left[ 0 + \dfrac{1}{r_0^2} - \dfrac{2}{4\pi r_0^3 h_0} = 0 \quad \text{or,} \quad \dfrac{1}{k r_0^2} = \dfrac{2}{r_0^3 h_0} \right]$

or, $\qquad r_0 = \dfrac{2k}{h_0}$

**17.15** The thermal conductivity of powdered material is measured by packing the material in the gap between the two concentric spheres made of metal and maintained at known temperatures by means of heater placed in the smaller sphere. In such an experiment spheres of 10 cm and 16 cm diameter are used. A temperature difference of 30°C is maintained by an electric heater consuming a power of 3 watt. Find the thermal conductivity of the material.

*Soln.* Refer to Fig. 17.26.

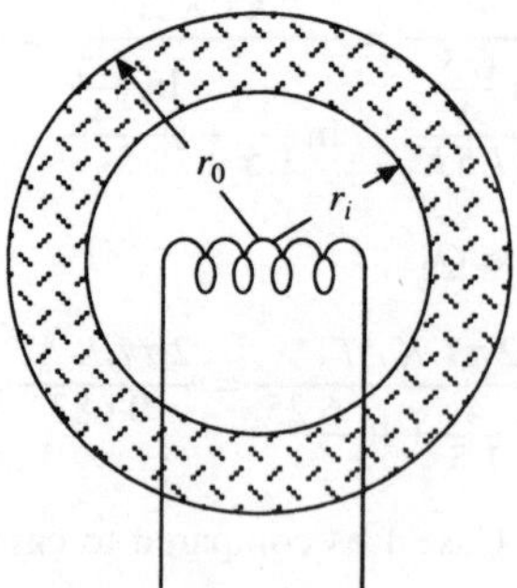



**Fig. 17.26**

$r_i = 5$ cm, $r_o = 8$ cm $\qquad T_i - T_o = 30$°C $\quad Q = 3$ watt

$$Q = \dfrac{T_i - T_o}{\dfrac{\dfrac{1}{r_i} - \dfrac{1}{r_o}}{4\pi K}} = \dfrac{(T_i - T_o) \times 4\pi K}{\dfrac{1}{r_i} - \dfrac{1}{r_o}} \quad \text{or,} \quad 3 = \dfrac{4\pi k \times 30}{\dfrac{1}{5 \times 10^{-2}} - \dfrac{1}{8 \times 10^{-2}}}$$

or, $k = 0.0597$ Watt/m°C  *Ans.*

**17.16** A furnace is of spherical shape, with 300 and 350 mm inner and outer radii. The inner and outer surface temperature are 450°C and 50°C respectively. The furnace is made of refractory material of mean thermal conductivity of 0.465 W/mK. Find the rate of heat transfer. If the furnace is covered with 25 mm thick insulating material ($K = 0.058$ W/mK) and the temperature of insulation surface is 35°C, find the reduction in rate of heat transfer with insulating material.

*Soln.* Refer to Fig. 17.27.

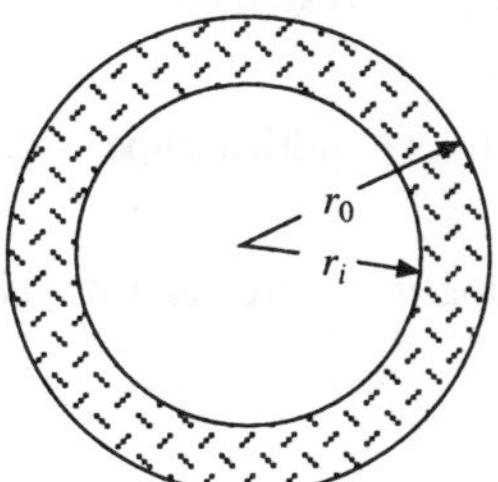

**Fig. 17.27**

$$r_i = 0.3 \text{ m}, \ r_o = 0.35 \text{ m}$$

Rate of heat transfer without insulation

$$= Q_1 = \frac{T_i - T_o}{\dfrac{1}{r_i} - \dfrac{1}{r_o}}{4\pi K} = \frac{(T_i - T_o)\,4\pi k}{\dfrac{1}{r_i} - \dfrac{1}{r_o}} = \frac{(450 - 50) \times 4\pi \times 0.465}{\dfrac{1}{0.3} - \dfrac{1}{0.35}}$$

$$= 4910.41 \text{ Watt} \quad Ans.$$

With insulation, $r_i = 0.3$ m, $r_1 = 0.35$ m, $r_o = 0.375$ m (i.e. 0.35 + 0.025 cm). $T_i - T_o = 450 - 35$

$$Q_2 = \frac{(450 - 35)}{\dfrac{\dfrac{1}{r_i} - \dfrac{1}{r_1}}{4\pi k_1} + \dfrac{\dfrac{1}{r_1} - \dfrac{1}{r_o}}{4\pi k_2}} = \frac{4\pi(450 - 35)}{\dfrac{\left(\dfrac{1}{0.3} - \dfrac{1}{0.35}\right)}{0.465} + \dfrac{\left(\dfrac{1}{0.35} - \dfrac{1}{0.375}\right)}{0.058}} = 1213.06 \text{ Watt}$$

Reduction in heat transfer = (4910.41 − 1213.06) = 3697.35 Watt   *Ans.*

**17.17** Show that the conductive thermal resistance of a hollow sphere is given by $(r_o - r_i)/(k\sqrt{A_o A_i})$, where $A_o$ and $A_i$ are outer and inner areas respectively.

*Soln.* Refer to Fig. 17.28

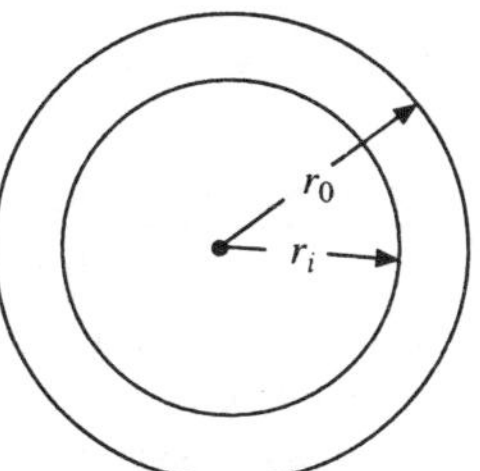

**Fig. 17.28**

For a hollow spherical shell, conductive thermal resistance

$$R = \frac{\dfrac{1}{r_i} - \dfrac{1}{r_o}}{4\pi K} = \frac{r_o - r_i}{r_i \, r_o \, 4\pi k}$$

$$= \frac{r_o - r_i}{K_1 \sqrt{4\pi r_i^2 \; 4\pi r_o^2}} = \frac{r_o - r_i}{K \sqrt{A_i \, A_o}} \; \text{Proved}$$

**17.18** Show that the rate of heat transfer through a cylindrical pipe is given by $Q = \dfrac{K \cdot A_m (T_i - T_o)}{(r_o - r_i)}$

where $A_m = \dfrac{(A_o - A_i)}{\ln \dfrac{A_o}{A_i}}$ which is called log mean area and $A_i$ and $A_0$ represent inner and outer areas

respectively.

*Soln.*   Refer Fig. 17.28.

$$Q = -KA\frac{dT}{dr} = -K \cdot 2\pi_r L \frac{dT}{dr}$$

or,   $\displaystyle\int_{r_i}^{r_o} Q\frac{dr}{r} = -\int_{T_i}^{T_o} k \, 2\pi L \, dT$   or,   $Q \ln \dfrac{r_o}{r_i} = 2\pi L K (T_i - T_o)$

$\therefore$   $Q = 2\pi L K (T_i - T_o) / \ln \dfrac{r_o}{r_i}$

or,   $Q = \dfrac{2\pi L K (T_i - T_o)}{\ln \dfrac{2\pi L r_o}{2\pi L r_i}} = \dfrac{2\pi L (r_o - r_i) K (T_i - T_o)}{(r_0 - r_1) \ln \dfrac{A_o}{A_i}}$

$$= \frac{(A_o - A_i)}{\ln \dfrac{A_o}{A_i}} \times \frac{k(T_i - T_o)}{r_o - r_i}$$

$$= \frac{k \, Am(T_i - T_o)}{r_o - r_i} \quad \text{where} \quad Am = \frac{A_o - A_i}{\ln \dfrac{A_o}{A_i}} \; \text{Proved.}$$

**17.19** Show that the rate of heat transfer from a spherical container of inner radius $r_i$ and outer radius $r_o$ in

steady state with variable thermal conductivity $K = K_i + (K_o - K_i)\dfrac{(T - T_i)}{T_o - T_i}$ is given by

$$Q = 4\pi r_i \, r_o \, \frac{(K_i + K_o)}{2} \frac{(T_i - T_o)}{(r_o - r_i)}$$

Estimate the rate of evaporation of liquid oxygen from a spherical container 1.8 m inside diameter covered with 30 cm of asbestos insulation. The temperatures at the inner and outer surfaces of insulation are −183°C and 0°C respectively. The boiling point of oxygen is −183°C and the latent heat of evaporation is 212.5 kJ/kg. The thermal conductivity of insulation is 0.157 and 0.125 W/m K at 0°C and −183°C respectively.

*Soln.* Refer to Fig. 17.29

Rate of heat transfer at any radius $r$ is $Q = -KA\dfrac{dT}{dr} = -K\,4\pi r^2\dfrac{dT}{dr}$

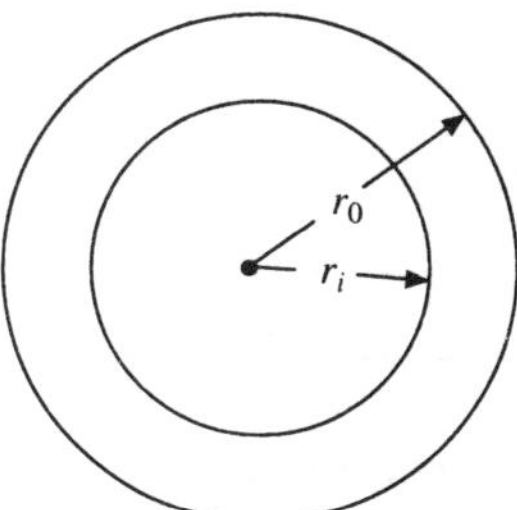

**Fig. 17.29**

or, $\qquad \displaystyle\int_{r_i}^{r_o} Q\frac{dr}{r^2} = -\int_{T_i}^{T_o} 4\pi k\,dT = -4\pi \int_{T_i}^{T_o}\left[k_i + \frac{(k_o - k_i)}{(T_o - T_i)}(T - T_i)\,dT\right]$

or, $\qquad -Q\left[\dfrac{1}{r}\right]_{r_i}^{r_o} = -4\pi\left[K_i(T_o - T_i) + \dfrac{K_o - K_i}{T_o - T_i}\left\{\dfrac{T_o^2 - T_i^2}{2} - T_i(T_o - T_i)\right\}\right]$

or, $\qquad Q\left[\dfrac{1}{r_i} - \dfrac{1}{r_o}\right] = 4\pi(T_i - T_o)\left[K_i + \dfrac{(K_o - K_i)}{T_o - T_i}\times\left\{\dfrac{T_i + T_o}{2} - T_i\right\}\right]$

or, $\qquad Q\dfrac{r_o - r_i}{r_i\,r_o} = 4\pi(T_i - T_o)\left[k_i + \dfrac{K_o - K_i}{2}\right] = 4\pi(T_i - T_o)\dfrac{(K_i + K_o)}{2}$

$$Q = 4\pi r_i r_o\,\frac{(K_i + K_o)}{2}\frac{(T_i - T_o)}{(r_o - r_i)}\quad\text{Proved.}$$

Here rate of heat tr. $= Q = 4\pi\dfrac{(1.8)}{2}(1.2)\left(\dfrac{0.157 + 0.125}{2}\right)\left(-\dfrac{183 - 0}{1.2 - 0.9}\right)$

Rate of heat transfer into the container $= 1167$ J/sec

Rate evaporation of oxygen $= \dfrac{1167 \times 3600}{212.5 \times 1000} = 19.8$ kg/hour  *Ans.*

**17.20** Heat flows through an annular wall of a cylindrical vessel radially. The termal conductivity varies linearly from $K_i$ at temperature $T_i$ to $K_o$ at temperature $T_o$. (a) Develop an expression for heat flow through the wall at temperature $T_i$ and $T_o$ at inner radius $r_i$ and outer radius $r_o$. (b) show how the expression for heat flow may be simplified if $(r_o - r_i)$ is very small.

*Soln.* Refer to Fig. 17.29.

Let the linear relation be $K = a + bT$ where $a$ and $b$ to be determined at $T = T_i$, $k = K_i$

$\therefore \qquad\qquad K_i = a + bT_i \quad$ at $\quad T = T_{o'}\ K = K_o$

$\therefore \qquad\qquad K_o = a + bT_o$

Taking difference $b = \dfrac{k_1 - k_o}{T_i - T_o}$ and then

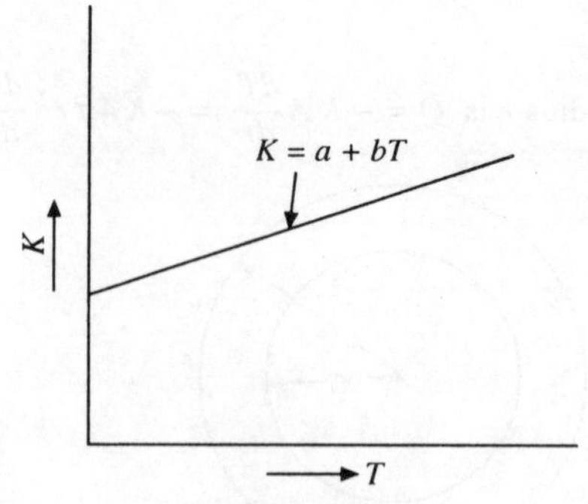


**Fig. 17.29**

$$Q = K_i - bT_i = K_i - \frac{K_i - K_o}{T_i - T_o} T_i$$

$$\therefore \quad K = K_i - \frac{K_i - K_o}{T_i - T_o} T_i + \frac{K_i - K_o T}{T_i - T_o} = K_i + \frac{K_i - K_o}{T_i - T_o} (T - T_i)$$

Rate of heat transfer at any radius $r$ is

$$Q = - KA \frac{dT}{dr} = - K \cdot 2\pi rL \frac{dT}{dr} = - \left[ K_i + \frac{K_i - K_o}{T_i - T_o} (T - T_i) \right] 2\pi rL \frac{dT}{dr}$$

$$\therefore \int Q \cdot \frac{dr}{r} = -2\pi L \int_{T_i}^{T_o} \left[ K_i + \frac{(K_i - K_o)}{T_i - T_o} (T - T_i) \right] dT$$

or $$Q \ln \frac{r_o}{r_i} = 2\pi L \left[ K_i (T_i - T_o) + \frac{k_i - k_o}{T_i - T_o} \left\{ \frac{T_i^2 - T_o^2}{2} - (T_i (T_i - T_o)) \right\} \right]$$

or $$Q \ln \frac{r_o}{r_i} = 2\pi L (T_i - T_o) \left[ K_i + \frac{K_i - K_o}{T_i - T_o} \left\{ \frac{T_i + T_o}{2} - T_i \right\} \right]$$

or, $$Q = \frac{2\pi L}{\ln \frac{r_o}{r_i}} (T_i - T_o) \left[ K_i + \frac{K_i - K_o}{2} \right] = \frac{2\pi L}{\ln \frac{r_o}{r_i}} \frac{(K_i + K_o)}{2} (T_i - T_o) \quad Ans.$$

(b) If $r_o - r_i$ is very small then $\ln \frac{r_o}{r_i} = \ln \left( 1 + \frac{- r_i}{r_i} \right) = \frac{r_o - r_i}{r_i}$

$$\left( \text{Note } \log (1 + x) = x - \frac{x^2}{2} + \frac{x^3}{3} \cdots \right)$$

$$Q = \frac{2\pi L}{\frac{r_o - r_i}{r_i}} \frac{(K_i + K_o)}{2} (T_i - T_o) = \frac{2\pi r_i L}{r_o - r_i} \frac{(k_i + k_o)}{2} (T_i - T_o)$$

**17.21** (a) A 2.5 cm O D refrigerant suction line is required to be thermally insulated. The outside air film co-efficient of heat transfer is 10 Watt/m²k. The thermal conductivity of insulation is 0.25 Watt/mk. Verify if the insulation would be effective or not.

(b) What should be maximum value of thermal conductivity of insulating material to reduce heat transfer.

(c) What should be thickness of cork insulation to reduce heat transfer to 20.7 percent if the thermal conductivity of cork is 0.38 Watt/mk.

*Soln.*

(a) The critical radius is $r_c = \dfrac{k}{h_o}$

$$= \frac{0.25}{10} = 0.025\,\text{m} = 2.5\,\text{cm}.$$

Since outer radius of the pipe $r_o$ (1.25 cm) is less than the critical radius, addition of insulation increases the rate of heat transfer, therefore it is not effective.

(b) For insulation to be effective $r_o \geq r_c$

or, $\qquad 0.0125 \geq \dfrac{k}{10}$ or, $k \leq 10 \times 0.0125$

$$= 0.125\,\text{Watt/mK} \quad Ans.$$

(c) Consider unit length of the pipe, for bare pipe rate of heat transfer $= Q_1 = hA\Delta T = \pi d_o h \Delta T$

For pipe with insulation $Q_2 = \dfrac{\Delta T}{R + R_o} = 0.207\,Q_1$; because it is reduced to 20.7 per cent.

$\therefore \qquad \dfrac{\Delta T}{\dfrac{\ln \dfrac{r}{r_o}}{2\pi k} + \dfrac{1}{2\pi_r h_o}} = 0.207 \times 2\pi r_o h \times \Delta T$

or, $\qquad \dfrac{\ln \dfrac{r}{r_o}}{k} + \dfrac{1}{rho} = \dfrac{1}{0.207\,roh}$

or, $\qquad \dfrac{\ln \dfrac{r}{0.0125}}{0.038} + \dfrac{1}{10r} = \dfrac{1}{0.207 \times 0.0125 \times 10}$

or, $\qquad 26.32 \ln \dfrac{r}{0.038} + \dfrac{0.1}{r} = 38.6$

Solving for $r = 0.05$ m $= 5$ cm.

$\therefore \qquad$ Thickness of insulation $= 5 - 1.25 = 3.75$ cm $\quad Ans.$

**17.22** A 10 mm cable is to be laid in atmosphere of 20°C ($h_o = 8.5$ W/m$^2$K°). The surface temperature of cable is likely to be 65°C due to heat generated within. Discuss the effect of insulating the cable with rubber having $k = 0.155$ W/mk.

*Soln.* The critical radius of insulation $r_c = \dfrac{k}{h_o}$ or, $r_c = \dfrac{0.155}{8.5} = 0.01823$ m $= 18.23$ mm. Thickness of rubber

should be effective when it is $\left( 18.23 - \dfrac{10}{2} \right) = 13.23$ mm.

## 17.4  CONVECTION

### 17.4.1  Introduction

In the preceeding article we have defined convection is a process of energy transport by the combined action of heat conduction, energy storage and mixing motion. Convection is the most important mechanism of energy transfer between a solid surface and a liquid or a gas. Convection is of two kinds—forced convection and free convection. When the motion in the fluid is caused by some external agency, such as a pump or a blower, we call forced convection and when the fluid is set in motion as a result of density differences due to a temperature variation, it is called free convection. Rate of heat transfer due to forced convection is much greater than that due to free convection because of higher degree of mixing motion.

### 17.4.2  Energy Transport Mechanism and Fluid Flow

The transfer of heat between a solid boundary and a fluid takes place by a combination of conduction and mass transport. If the boundary is at a temperature higher than the fluid, heat flows first by conduction from the solid to fluid particles in the neighbourhood of the wall. The energy thus transmitted increases the internal energy of the fluid and is carried away by the motion of the fluid. When the heated fluid particles reach a region at a lower temperature, heat is again transferred by conduction from warmer to cooler fluid.

Since the convection mode of energy transfer is so closely linked to the fluid motion, it is necessary to know something about the mechanism of fluid flow before the mechanism of heat flow can be investigated. One of the most important aspects of the hydro dynamic analysis is to establish whether the motion of the fluid is *laminar or Turbulent*.

In laminar, or stream line flow, the fluid moves in layers, each fluid particle flowing a smooth and continuous path. The fluid particles in each layer remain in  an orderly sequence without passing one another. Solders on parade provide a somewhat crude analogy to laminar flow. They march along well defined lines, one behind the other, and maintain their order even when they turn a corner or pass an abstacle, similarly, the movement of railway passengers near the booking counter in sequence provides a crude example of laminar motion (Fig. 17.30).

In contrast to the orderly motion of laminar flow, the motion of fluid particles in turbulent flow rather resembles a crowd of commuters in a railroad station and movement of people in a fishery market. In turbulent flow, fluid particles collide with each other in zigzag way.

When a fluid flows in laminar motion along a surface at a temperature different from that of the fluid, heat is transferred only by molecular conduction within the fluid as well as at the interface between the fluid and the surface. There exists no turbulent mixing currents or eddies by which energy stored in fluid particles is transported across stream lines. Heat is transferred between fluid layers by molecular motion on a submicroscopic scale.

In turbulent flow, on the otherhand, the conduction mechanism is modified and aided by immumerable eddies which carry lumps of fluid across the stream lines. These fluid particles act

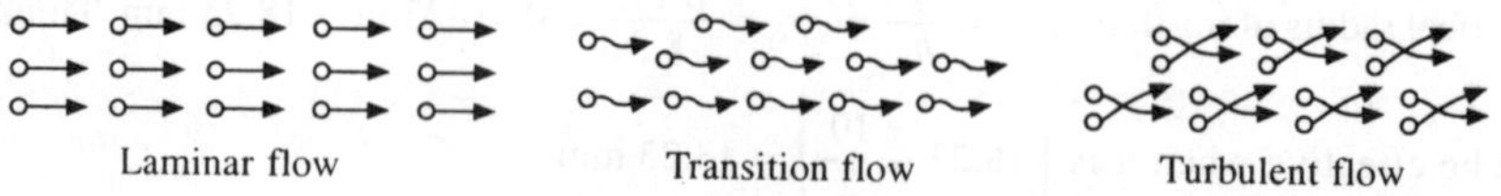

Laminar flow        Transition flow        Turbulent flow

**Fig. 17.30**

as carriers of energy and transfer energy by mixing with other particles of the fluid. An increase in the rate of mixing (or turbulence) will therefore also increase the rate of heat flow by convection.

### 17.4.3 Boundary Layer—Theory

When a fluid flows along a surface, irrespective of whether the flow is laminar or turbulent, the particles in the vicinity of the surface are slowed down by virtue of viscous forces. The fluid particles adjacent to the surface stick to it and have zero velocity relative to the boundary. Other fluid particles attempting to slide over them are retarded as a result of an interaction between faster and slower moving fluid, a phenomenon which gives rise to shearing forces. In laminar flow the interaction, called viscous shear, takes place between molecules on a submicroscopic scale. In turbulent flow, an interaction between lumps of fluid on a microscopic scale, called turbulent shear is superimposed on the viscous shear.

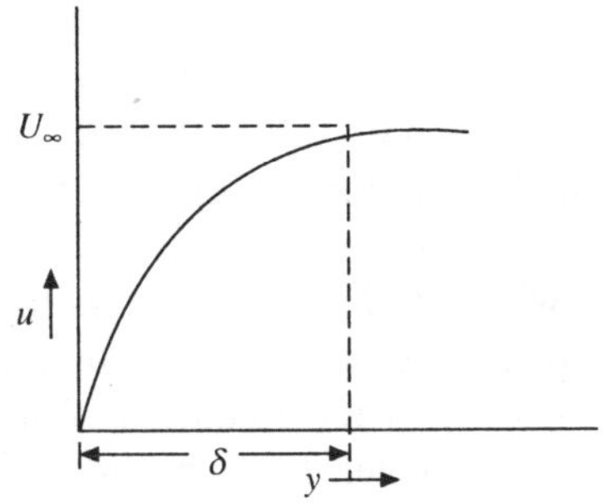

The effects of the viscous forces originating at the boundary extend into the body of the fluid, but a short distance from the surface the velocity of the fluid particles approaches that of the undisturbed free stream. The fluid contained in the region of substantial velocity change is called the hydrodynamic boundary layer. The thickness of the boundary layer is defined as the distance from the surface at which the local velocity reaches 99 percent of the external velocity $U_\infty$ as shown in Fig. 17.31.

**Fig. 17.31**

The concept of the boundary layer was introduced by the German scientist, Prandtl in 1904. Before Prandtl, Euler equation of motion for invisced flow and Navier-stokes equations for generalised viscous flow were available. Euler dropped the effects of viscosity with the plea that water and air (Most important engineering fluids), both possess very low viscosity, on the otherhand, solutions of Navier-stokes equations were available for very few cases. Prandtle had an idea that air and water might have very high velocity gradient $\left.\dfrac{dv}{dy}\right|_{y=0}$ near the surface even though they have low viscosity. He developed general boundary layer equations (called Prandtl boundary layer equations), which were solved by his student, Blasius, in 1908. The boundary layer essentially divides the flow field around a body into two domains; a thin layer covering the surface of the body where velocity gradient is great and the viscous forces are large, and a region outside this layer where the velocity is nearly equal to the free stream value and the effects of viscosity are negligible.

### Boundary Layer on a Flat Plate

Consider qualitatively a flow over a flat plate placed with its surface parallel to the stream. At the leading edge ($x = 0$), only the fluid particles in immediate contact with the surface are slowed down while the remaining fluid continues at the velocity of the undisturbed free stream in front of the plate. As the fluid proceeds along the plate, the shearing forces increases due to increase in contact area and therefore causes more and more of the fluid to be retarded, and the boundary layer thickness. Boundary layer and typical velocity profiles at various station along the plate are shown in the Fig. 17.32.

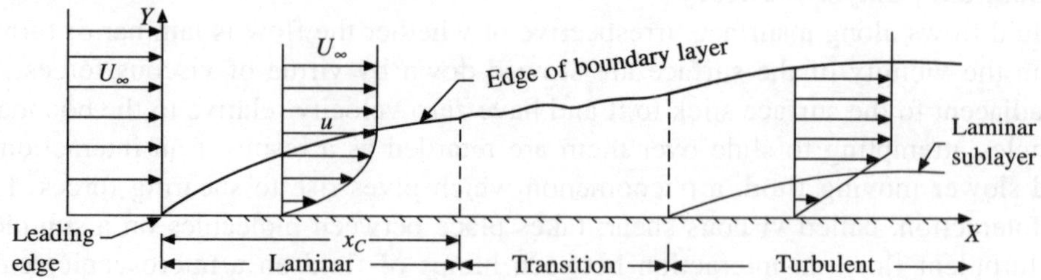


**Fig. 17.32   Laminar and turbulent boundary layer.**

The velocity profiles near the leading edge are representative of laminar boundary layer. However, the flow within the boundary layer remains laminar only for a certain distance from the leading edge and then becomes turbulent. We do not know enough about the mechanism of transition to predict precisely when the transition will occur.

There are always small disturbances and waves in a flowing fluid, but as long as the viscous forces are large they will prevent disturbances from growing. As the laminar boundary layer thickens, the ratio of viscous forces to inertia forces decreases due to accumulation of more mass, and eventually a point is reached at which disturbances will no longer decay but will grow with time. Then the boundary layer becomes unstable and the transition from laminar to turbulent flow begins. Eddies and vortexes form and destroy the laminar regularity of the boundary layer motion. Even in the region of turbulent boundary layer, the viscous forces are large near the plate and therefore, quasi laminar motion perpists only in a thin layer in the immediate vicinity of the surface. This portion of a generally turbulent boundary layer is called the laminar sublayer. The region between the edges of laminar sub layer and turbulent boundary layer is called *buffer layer*.

The distance from the leading edge at which the boundary layer becomes turbulent is called the critical length $x_c$. This distance is usually specified in terms of a dimensionless quantity called the

critical Reynolds number $\left( Re_c = \dfrac{\rho U_\infty x}{\mu} \right)$ which is an indication of the ratio of inertia to viscous

forces at which disturbances begin to grow. Experiments results have shown that the point of transition depends upon the surface contour, the surface roughness, the disturbance level and even on the heat transfer, when the flow is calm and no disturbances occur, laminar flow can persist in the boundary layer at Reynolds numbers as high $5 \times 10^6$. If the surface is rough, disturbances are intentionally introduced into the flow, the flow may become turbulent at Reynolds numbers as low as $8 \times 10^4$. Under average conditions, the flow over a flat plate becomes turbulent for critical Reynolds number of $5 \times 10^5$.

**Thermal Boundary Layer and Nusselt Number**

Consider a fluid at temperature $T_\infty$ flowing over a plate maintained at a temperature $T_s$ $(T_s > T_\infty)$. Fluid particles on the surface are completely stagnant and therefore, heat transfer from the surface

to these fluid particles takes place first by conduction $\left( Q = -KA \left. \dfrac{dT}{dy} \right|_{y=0} \right)$ and then this is convected to other particles by mixing motion of other particles. Temperature variation takes place up to a certain distance. The fluid contained in the region of substantial temperature change from surface temperature to free stream temperature is called thermal boundary layer and the distance from the surface at which temperature becomes 99 percent of free stream is called thermal Boundary layer thickness ($\delta_t$) shown in Fig. 17.33.

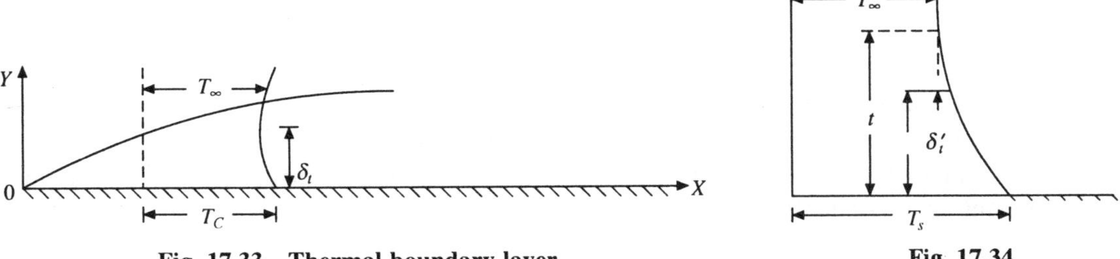

Fig. 17.33   Thermal boundary layer      Fig. 17.34

The rate of heat transfer from the surface to the fluid is evaluated by conduction equation.

$$Q = -KA\left(\frac{\delta T}{\delta y}\right)_{y=0} \tag{17.18}$$

This approach has indeed been used, but for engineering purposes the concept of the convective heat transfer coefficient is much more convenient. Relating the heat transfer by convection with the temperature gradient

$$Q = -KA\frac{\delta T}{\delta y}\bigg|_{y=0} = hA(T_s - T_\infty) \tag{17.19}$$

Since the magnitude of the temperature gradient in the fluid will be the same regardless of the reference temperature, we can write $\delta T = \delta(T - T_s)$. Introducing a significant length dimension of the system $L$ to specify the geometry of the object from which heat flows, we can write the Eq. 17.19 in the dimensionless form as

$$Q = hA(T_s - T_\infty) = -KA\frac{\delta T}{\delta y}\bigg|_{y=0} = -KA\frac{\delta(T - T_s)}{\delta y}$$

or,

$$\frac{hL}{k} = L\frac{\frac{\delta(T_s - T)}{\delta y}}{(T_s - T_\infty)}\bigg|_{y=0} = \frac{\frac{\delta(T_s - T)}{T_s - T_\infty}}{\frac{(y)}{L}}\bigg|_{y=0} = \frac{\frac{\delta(T_s - T)}{\delta y}}{\frac{(T_s - T_{oo})}{L}}\bigg|_{y=0} \tag{17.20}$$

The combination $\dfrac{hL}{k}$ is a dimensionless quantity called Nusselt number which the ratio of the temperature gradient in the fluid immediately in contact with the surface to a reference temperature gradient $\dfrac{(T_s - T_\infty)}{L}$. Once the Nusselt number is evaluated, heat transfer coefficient is known then rate of heat transfer can be evaluated. If the temperature variation inside the thermal boundary is assumed to be linear from $T_s$ to $T_\infty$, then rate of heat transfer is given by

$$\frac{Q}{A} = K\,\frac{(T_s - T_\infty)}{\delta t'} = h\,(T_s - T_\infty), \quad \text{or,} \quad h = \frac{k}{\delta t'}$$

and
$$\text{Nusselt number} = \text{Nu} = \frac{hL}{k} = \frac{L}{\delta t'} \tag{17.21}$$

Equation (17.21) suggests that to transfer large quantities of heat rapidly, one attempt to reduce the thermal boundary layer thickness as much as possible. This can be accomplished by increasing the velocity and/or the turbulence of the fluid. If insulation of the surface is the desired aim, a thick stagnant layer is beneficial.

### 17.4.4  Evaluation of Convective Heat Transfer Coefficients

There are four general methods available for the evaluation of convective heat transfer coefficients:

1. Dimensional analysis combined with experiments.
2. Exact mathematical solutions of the boundary layer equations
3. Approximate analysis of the boundary layer by integral methods
4. The analogy between heat, mass and momentum transfer.

All four of these techniques have contributed to our understanding of convective heat transfer. Yet, no single method can solve all the problems because each one has limitations which restricts its scope of application.

### 17.4.5  Dimensional Analysis

Dimensional analysis is a method by which we deduce information about a phenomenon from the single premise that the phenomenon can be described by a dimensionally correct equation among certain variables. This has advantage to reduce the number of independent variables. Some dimensionless numbers are mentioned below.

1. Reynolds number ($R_e$): It is the ratio of inertia force to viscous force in a fluid flow. It determines the nature of flow-whether laminar or turbulent. As Reynolds number increases, degree of turbulence and hence the mixing motion increases, the rate of heat transfer increases. It always appears in forced convection. Mathematically $R_e = \dfrac{\rho L U_\infty}{\mu} = \dfrac{U_\infty L}{v}$ where $L$ is the characteristic length.

2. Nusselt Number ($Nu$): It is defined as the ratio of temperature gradient near the surface to a reference temperature gradient $\dfrac{(T_s - T_\infty)}{L}$. Rate of heat transfer increases with increase of Nusselt number.

3. Prandtl Number $\left( Pr = \dfrac{V}{\alpha} = \dfrac{v}{\dfrac{K}{\rho C_p}} = \dfrac{\mu C_p}{K} \right)$ : This is the ratio of momentum diffusivity ($V$)

to thermal diffusivity $\left( \alpha = \dfrac{K}{\rho C_p} \right)$ $Pr = \dfrac{V}{\alpha} = \dfrac{V}{\dfrac{K}{\rho c_p}} = \dfrac{V\rho c_p}{K} = \dfrac{\mu c_p}{K}$. It is a property parameter.

The three properties: viscosity, specific heat and thermal conductivity have been lumped together in a dimensionless form. Rate of heat transfer in convection always increases with increase of Prandtl number. Thermal boundary thickness decreases for higher Prandtl number fluids. The Lubricants for bearing should possess higher Prandtl number in order to dissipate more heat in the atmosphere. For Prandtl number unity, both hydrodyamic and thermal boundary layers coincide. Air has Prandtl number nearly equal to unity (0.7). Prandtl number varies with temperature due to variation in properties.

4. Grashof Number $\left( Gr = \dfrac{\rho^2 g \beta \Delta T L^3}{\mu^2} \right)$. It is used in natural convection problems. It is defined

as the ratio of product of inertia force and buoyancy force caused due to density difference to the square of viscous force

$$Gr = \frac{\text{Inertia force} \times \text{Buoyancy force}}{(\text{Viscous force})^2}$$

$$= \rho v^2 \times \frac{\rho g \beta \Delta T L^3}{(\mu v)^2} = \frac{\rho^2 g \beta \Delta T L^3}{\mu^2} \quad \text{where } \Delta T = T_s - T_\infty$$

and $\qquad \beta = $ Coefficient of expansion of gases.

Grashof number decides whether the flow is laminar or turbulent. It is similar to Reynolds number in forced convection. Rate of heat transfer due to free convection always increases with increase of Grashof number.

5. Stanton Number $\left( S_t = \dfrac{N_u}{Re\,Pr} \right)$. It is not an independent dimensionless number. It is the

ratio of Nusselt number to the product of Reynolds and Prandtl number

Mathematically: $\quad St = \dfrac{N_u}{Re\,Pr} = \dfrac{\dfrac{hL}{K}}{\dfrac{VL}{v}\,\mu\,\dfrac{c_p}{K}} = \dfrac{h}{\rho c_p V} = \dfrac{hA\,\Delta T}{\rho AV c_p \Delta T}$

$$= \frac{\text{Wall heat transfer ratio}}{\text{Mass heat flow rate}}, \text{ because rate of flow} = \rho AV$$

6. Peclet number ($Pe = Re\,Pr$): It is defined as the product of Reynolds number and Prandtl number.

$$Pe = RePr = \frac{\rho LV}{\mu} \frac{\mu C_p}{K} = \frac{LV}{\alpha} \quad \text{where} \quad \alpha = \text{thermal diffusivity}$$

7. *Eckert Number* $\left( E_c = \dfrac{U^2}{g C_p \Delta T} \right)$ It is used in high speed flow where frictional heating takes

place as in supersonic and hypersonic flow. It is related with Mach number $Ec = (\gamma - 1)M^2$ where $M$ is the Mach number and $\gamma = \dfrac{c_p}{c_v}$. At low velocity, effect of frictional heating is neglected and Eckert number does not appear.

### 17.4.6 Correlations for Convective Heat Transfers

In general, $Nu = f(Re, Pr, Gr, Ec)$

In the absence of frictional heating, for forced convection $Nu = f(Re, Pr)$ and for free convection $Nu = f(Gr, Pr)$.

#### A. Forced convection on a flat plate

1. Local Nusselt Number $N_{ux} = 0.332\ Re_x^{1/2}\ Pr^{1/3}$, for laminar flow       (17.22)

2. Average Nusselt Number $\overline{N}_u = 0.664\ Re_L^{1/2}\ Pr^{1/3}$, for laminar flow       (17.23)

3. Local Nusselt Number $= Nu_x = 0.0288\ Re_x^{0.8}\ Pr^{1/3}$, for turbulent flow from the leading edge       (1724)

4. Average Nusselt number $\overline{Nu} = 0.036\ Re_L^{0.8}\ Pr^{1/3}$, for turbulent flow from the leading edge.       (17.25)

5. Considering laminar flow from the leading edge and turbulent from the critical length
$$\overline{Nu} = 0.036\ Pr^{1/3}\,(Re_L^{0.8} - 23200)$$       (17.26

6. Boundary layer thickness for laminar flow ($\delta$) at any distance from the leading edge ($x$) is given by $\dfrac{\delta}{x} = \dfrac{5}{\sqrt{\text{Rex}}}$       (17.27)

7. Boundary layer thickness for turbulent flow $\dfrac{\delta}{x} = 0.376\ Re_x^{-1/5}$       (17.28)

8. Relation between thermal boundary layer thickness and hydrodynamic boundary layer thickness:

$$\frac{\delta_t}{\delta} = \frac{1}{P_r^{1/3}}. \quad \text{For } Pr = 1,\ \delta_t = \delta,\ \text{for } Pr < 1,\ \delta_t > \delta \ \text{for } Pr > 1,\ \delta_t < \delta \qquad (17.29)$$

9. Skin friction coefficient for laminar flow defined as

$$C_{fx} = \frac{T_o}{\frac{1}{2}\rho U_\infty^2} = \frac{0.664}{\sqrt{Re_x}} \qquad (17.30)$$

10. Average skin friction coefficient $C_f = \dfrac{T_o}{^1/_2\,\rho u_\infty^2} = \dfrac{1.328}{\sqrt{Re_x}}$       (17.31)

11. Average skin friction coefficient for turbulent flow
$$= \overline{C}_f = 0.072\ \overline{Re}_L^{1/5}. \qquad (17.32)$$

*Note:* The properties should be evaluated at mean temperature $T_m$, defined as $T_m = \dfrac{T_s + T_\infty}{2}$

### 17.4.7   Forced Convection Inside Tubes and Ducts

The characteristic length is the hydraulic diameter $(D_H)$ defined as $D_H = 4 \times \dfrac{\text{Flow area}}{\text{wetted permimeter}}$

For flow through a pipe, $D_H = \dfrac{4 \times \frac{\pi}{4} D^2}{\pi D} = D = $ diameter of the pipe

For flow through an annulus, $D_H = \dfrac{4 \times \frac{\pi}{4}(D_o^2 - D_i^2)}{\pi(D_o + D_i)}$

or, $\qquad\qquad\qquad\qquad D_H = D_o - D_i$

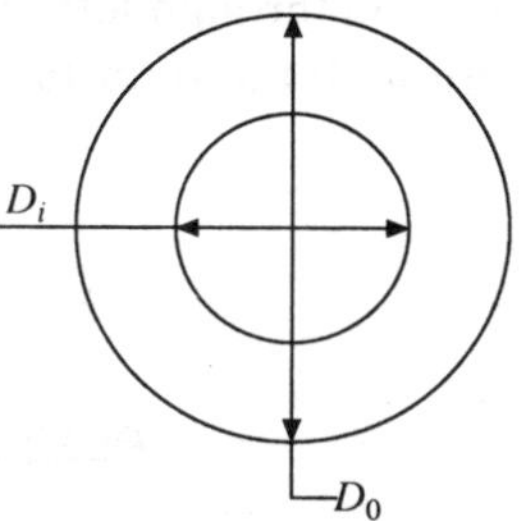

**Fig. 17.35**

1. For turbulent flow $N_u = 0.023\, Re^{0.8}\, Pr^{0.4}$ $\qquad\qquad\qquad\qquad\qquad$ (17.33)
2. For laminar flow with constant heat flux $N_u = 4.36$
3. For laminar flow with constant wall temperature $Nu = 3.65$
4. Critical Reynolds number $= 2300$
5. Rate of heat transfer $= Q = Ah\,(T_s - T_b)$, where $T_b = $ Bulk temperature $= (T_i + T_o)/2$ and $T_i$ and $T_o$ are inlet and outlet temperatures and $A = \pi DL$

### c. Free convection

1. $Nu = \dfrac{hL}{K} = 0.13\,(Gr\,Pr)^{1/3}$ for vertical surface, when $10^8 < Gr\,Pr < 10^{12}$ $\qquad$ (17.34)

2. $Nu = \dfrac{hL}{K} = 0.56\,(Gr\,Pr)^{1/4}$ for vertical surface when $10^5 < Gr\,Pr < 10^8$ $\qquad$ (17.35)

3. $Nu = \dfrac{hD}{K} = 0.13\,(Gr\,Pr)^{1/3}$ for horizontal cylinders when $10^9 < Gr\,Pr < 10^{12}$ $\quad$ (17.36)

4. $Nu = \dfrac{hD}{K} = 0.53\,(Gr\,Pr)^{1/4}$ for horizontal cylinder when $10^4 < Gr\,Pr < 10^9$ $\qquad$ (17.37)

5. $Nu = \dfrac{hD}{K} = 0.17\,(Gr\,Pr)^{1/3}$ for horizontal hot circular plate facing upward when $Gr\,Pr > 10^9$

6. $Nu = \dfrac{hD}{K} = 0.71\,(Gr\,Pr)^{1/4}$ for $10^3 < Gr\,Pr < 10^9$ $\qquad\qquad\qquad\qquad$ (17.38)

### 17.4.8   Boundary Layer Equations on a Flat Plate

Consider a laminar flow of a fluid on a flat plate. The following assumption are made for setting differential equations of Boundary layer (Fig. 17.36):

1. The flow is steady and *incompressible*.

2. There is no pressure variation perpendicular to the plate i.e. $\dfrac{\partial p}{\partial y} = 0$.

3. *Viscous forces* in Y direction are negligible. Consider a control volume situated inside the boundary layer of length $\Delta_x$, width $\Delta y$ parallel to $x$ and $y$ directions and height unity perpendicular to the plane of the paper.

1. Continuity equation: This is the statement of law of conservation of mass applied to fluid flow.

The rate of mass flow entering in $x$ direction through left face (Fig. 17.37) $= \rho u \Delta y \times 1 = \rho u\, \Delta y$. The rate of mass leaving in $X$ direction through the right face

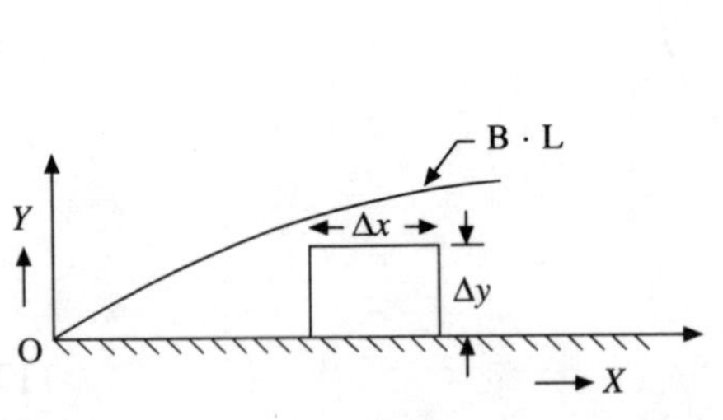


**Fig. 17.36**

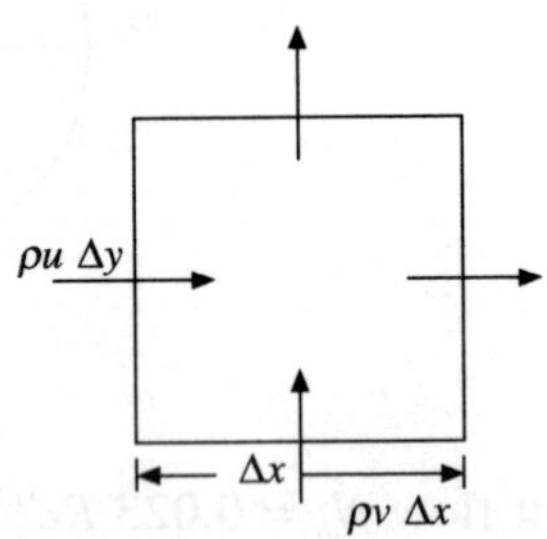


**Fig. 17.37**

$$= \rho \left( u + \frac{\partial u}{\partial x} \Delta x \right) \Delta y \qquad\qquad (17.39)$$

Similarly rate of mass entering and leaving in $y$ direction are $\rho v\, \Delta x$ and $\rho \left( v + \dfrac{\partial v}{\partial y} \Delta y \right) \Delta x$

respectively from bottom and top faces, where $v$ = velocity in $y$ direction.

Hence mass balance on the control volume gives: rate of mass entering = rate of mass leaving.

$$\therefore \qquad \rho u \Delta y + \rho v \Delta x = \rho \left( u + \frac{\partial u}{\partial x} \Delta x \right) \Delta y + \rho \left( v + \frac{\partial v}{\partial y} \Delta y \right) \Delta x$$

cancelling the common terms we get:

$$\boxed{\frac{\partial u}{\partial x} + \frac{\partial v}{\partial y} = 0} \, , \qquad\qquad (17.40)$$

which is called continuity equation for in compressible two dimensional flow.

## 2. *Momentum equation*

Momentum equation is obtained by the application of Newton's second law of motion according to which the rate of change of momentum in any direction is the net force in that direction.

Gravitational force is neglected in $x$ direction or plate is kept horizontal. The rate of momentum entering through left face in $x$ direction $= m_x u = (\rho u \Delta y)\, u = \rho u^2 \Delta y$ and leaving through the right

face $= \rho \left( u^2 + \dfrac{\partial u^2}{\partial x} \Delta x \right) \Delta y$. Refer to Figs. 17.38 and 17.39.

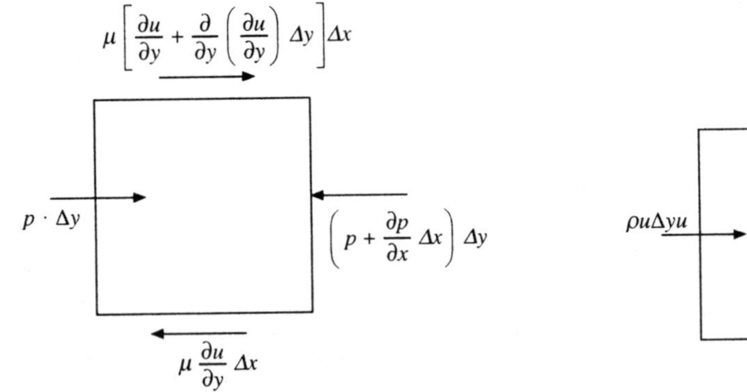

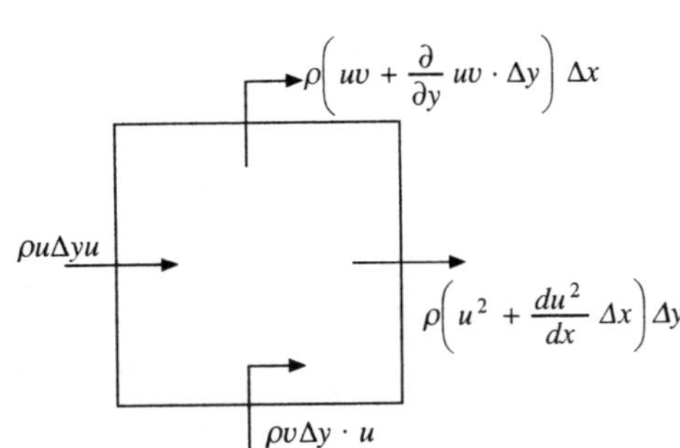

<table>
<tr><td>**Fig. 17.38   Pressure and viscous forces**</td><td>**Fig. 17.39   Rate of momentum**</td></tr>
</table>

Momentum in $x$ direction entering through the bottom face

$$m_y u = (\rho v \Delta x)\, u = \rho u v \Delta x \tag{17.41}$$

because rate of flow of mass in $y$ direction is $v \rho \Delta x$. Momentum in $x$ direction leaving through the top face in $x$ direction

$$= \rho \left( uv + \frac{\partial}{\partial y}(uv) \Delta y \right) \Delta x \tag{17.42}$$

$\therefore$ Net rate of momentum leaving in $x$ direction

$$= \rho \left( u^2 + \frac{\partial u^2}{\partial x} \Delta x \right) \Delta y - \rho u^2 \Delta y + \rho \left( vu + \frac{\partial}{\partial y}(uv) \Delta y \right) \Delta x$$

$$- \rho uv \Delta x = \rho \Delta x \, \Delta y \left[ \frac{\partial u^2}{\partial x} + \frac{\partial (uv)}{\partial y} \right]$$

$$= \rho \Delta x \Delta y \left[ 2u \frac{\partial u}{\partial x} + u \frac{\partial v}{\partial y} + v \frac{\partial u}{\partial y} \right]$$

$$= \rho \Delta x \Delta y \left[ \left( u \frac{\partial u}{\partial x} + v \frac{\partial u}{\partial y} \right) + v \frac{\partial u}{\partial x}\, u \frac{\partial v}{\partial y} \right]$$

$$= \rho \Delta x \Delta y \left[ u\frac{\partial u}{\partial x} + v\frac{\partial u}{\partial y} + u\left( \frac{\partial u}{\partial x} + \frac{\partial v}{\partial y} \right) \right] \tag{17.43}$$

But from continuity equation $\dfrac{\partial u}{\partial x} + \dfrac{\partial v}{\partial y} = 0$

$\therefore$ Net rate of momentum leaving $= \rho \Delta x \Delta y \left( u\dfrac{\partial u}{\partial x} + v\dfrac{\partial u}{\partial y} \right)$

Net external force due to viscosity and pressure in $x$ direction

$$= Fx = p - \left( p + \frac{\partial p}{\partial x} \Delta x \right) \Delta y + \mu \left[ \frac{\partial u}{\partial y} + \frac{\partial}{\partial y} \frac{(\partial u)}{\partial y} \Delta y \right] \Delta x$$

$$- \mu \frac{\partial u}{\partial y} \Delta x = \left[ -\frac{\partial p}{\partial x} + \mu \frac{\partial^2 u}{\partial y^2} \right] \Delta x \, \Delta y$$

$\therefore$ Net rate of change of momentum in $x$ direction = net force in $x$ direction

$$\rho \Delta x \Delta y \left( u\frac{\partial u}{\partial x} + \frac{\partial u}{\partial y} v \right) = \Delta x \, \Delta y \left( -\frac{\partial p}{\partial x} + \frac{\partial_u^2}{\partial y^2} \mu \right)$$

or,
$$u\frac{\partial u}{\partial x} + v\frac{\partial u}{\partial y} = -\frac{1}{\rho} \frac{\partial p}{\partial x} + v\frac{\partial_u^2}{\partial y^2} \tag{17.44}$$

But for the flat plate pressure does not change in $x$ direction
Momentum equation reduces to

$$\boxed{u\frac{\partial u}{\partial x} + v\frac{\partial u}{\partial y} = \frac{v\partial_u^2}{\partial y^2}} \tag{17.45}$$

*3. Energy equation*

Let us write down the energy balance for the control volume. Energy in = Energy convected from left face + Energy convected from bottom face + Energy conducted from bottom + Net rate of viscous work done

$$= \rho u \Delta y C_p T + \rho v \Delta x c_p T + \left( -K \Delta x \frac{\partial T}{\partial y} \right) + \frac{\partial}{\partial y} (Tu) \Delta y \Delta x$$

Energy out = Energy convected from right face + Energy connected from top face + Energy conducted from top face

$$= \rho C_p \left[ \left( uT + \frac{\partial(uT)}{\partial x} \Delta x \right) \right] \Delta y + \rho C_p \left[ vT + \frac{\partial(vT)}{\partial y} \Delta y \right] \Delta x$$

$$+ \left[ -K \left\{ \left( \frac{\partial T}{\partial y} \right) + \frac{\partial}{\partial y} \left( \frac{\partial T}{\partial y} \right) \Delta y \right\} \right] \Delta x \tag{17.46}$$

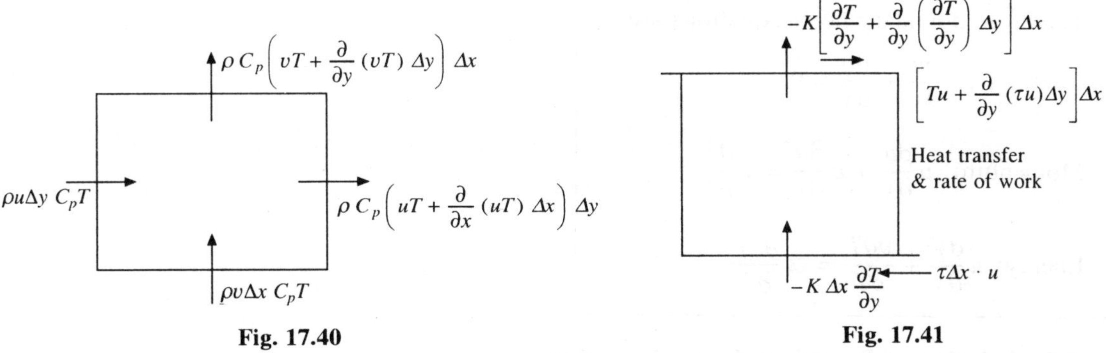



**Fig. 17.40**   **Fig. 17.41**

Equating both the energy

$$\rho u \Delta y\, C_p T + \rho v \Delta x\, C_p T - K \Delta x \frac{\partial T}{\partial y} + \frac{\partial}{\partial y}(Tu)\Delta y \Delta x$$

$$= \rho_{C_p}\left( uT + \frac{\partial}{\partial x} \right)(uT)\Delta x)\,\Delta y + \rho_{C_p}\left[ VT + \frac{\partial}{\partial y}(VT)\Delta y \right]\Delta x$$

$$-K\left\{ \frac{\partial T}{\partial y} + \frac{\partial}{\partial y}\left( \frac{(\partial T)}{\partial y} \right)\Delta y \right\}\Delta x$$

or, $\dfrac{\partial}{\partial y}(Tu)\Delta y \Delta x = \rho C_p\left[ \dfrac{\partial}{\partial x}(uT) + \dfrac{\partial}{\partial y}(vT) \right]\Delta x \Delta y - k \dfrac{\partial^2 T}{\partial y^2}\Delta x \Delta y$

$$\rho\varphi\left\{ \frac{\partial}{\partial x}(uT) + \frac{\partial}{\partial y}(vT) \right\} = K \frac{\partial^2 T}{\partial y^2} + \frac{\partial}{\partial y}\left( \mu \frac{\partial y}{\partial y} \cdot u \right)$$

$$\frac{u\partial T}{\partial x} + T \frac{\partial u}{\partial x} + v \frac{\partial T}{\partial y} + T \frac{\partial v}{\partial y} = \frac{K}{\rho C_p}\frac{\partial^2 T}{\partial y^2} + \frac{\mu}{\rho C_p}\left[ \left( \frac{\partial u}{\partial y} \right)^2 + u \frac{\partial^2 u}{\partial y^2} \right]$$

or, $$T\left( \frac{\partial u}{\partial x} + \frac{\partial v}{\partial y} \right) + \frac{u\partial T}{\partial x} + \frac{v\partial T}{\partial y} = \alpha \frac{\partial^2 T}{\partial y^2} + \frac{\mu}{\rho C_p}\left\{ \left( \frac{\partial u}{\partial y} \right)^2 + u \frac{\partial^2 y}{\partial y^2} \right\} \qquad (17.47)$$

Neglecting the effect of viscosity on friction work.

$$\boxed{u \frac{\partial T}{\partial x} + v \frac{\partial T}{\partial y} = \alpha \frac{\partial^2 T}{\partial y^2}} \qquad (17.48)$$

> Therefore boundary layer equations are:
>
> Continuity: $\dfrac{\partial u}{\partial x} + \dfrac{\partial v}{\partial y} = 0$
>
> Momentum: $u\dfrac{\partial u}{\partial x} + v\dfrac{\partial u}{\partial y} = v\dfrac{\partial_u^2}{\partial_y^2}$
>
> Energy: $u\dfrac{\partial_T}{\partial x} + \dfrac{v\partial T}{\partial Y} = \alpha\dfrac{\partial^2 T}{\partial_y^2}$

### 17.4.9 Solution of Boundary Layer Equations

The boundary conditions are: When $y = 0$, $u = 0$, $v = 0$

$$T = T_s = \text{Surface temperature.}$$

When $y \to \infty$, $u = U_\infty$ and $T = T_\infty$

If we define a dimensionless temperature: $\theta = \dfrac{T - T_s}{T_\infty - T_s}$ and a dimensionless velocity $\dfrac{u}{U_\infty}$ then,

when $y = 0$, $\theta = 0$, $\dfrac{u}{U_\infty} = 0$ and as $y \to \infty$, $\theta \to 1$ and $\dfrac{u}{U_\infty} \to 1$

Hydrodynamic boundary layer equations were solved by Blasius in 1908 and thermal boundary layer equations were solved by Pohlpauson in 1921. Their methods of solutions are out of scope of this book, but their solutions are quite useful for engineering applications. Important results of flat plate have already been listed.

### *Colburn Analogy*

For boundary layer flow, there exists a remarkable relationship between heat transfer and skin friction. The heat flux from the body to the surface fluid is given by

$$\frac{Q}{A} = q = -K\frac{\partial T}{\partial y}\bigg|_{y=0},$$

This equation is analogous to local skin friction $\tau_{ox} = \mu\dfrac{\partial u}{\partial y}\bigg|_y = 0$. We have already obtained the

relation: $\dfrac{C_{fx}}{2} = 0.332\,Re_x^{-1/2}$ and $\boxed{\dfrac{\overline{C_f}}{2} = 0.664\,Re_L^{-1/2}}$ $C_{fx}$ = local skin friction coefficient and

Average nusselt number $= \overline{Nu} = 0.666\,Re_L^{1/2}\,Pr^{1/3}$

Average stanton number $= \overline{St} = \dfrac{\overline{Nu}}{Re_L\,Pr} = \dfrac{0.664\,Re^{1/2}\,Pr^{1/3}}{Re\,Pr} = 0.664\,Re^{-1/2}\,Pr^{-2/3}$

$$\therefore \qquad St = \frac{\overline{C_f}}{2}\cdot Pr^{-2/3} \qquad\qquad (17.49)$$

This equation is called Colburn analogy and if

$$p_r = 1, \overline{S_t} = \frac{\overline{C_f}}{2} \tag{17.50}$$

which is called Reynolds analogy

## SOME IMPORTANT RESULTS

### Laminar Boundary Layer Relations for Flat plate

1. $\dfrac{\delta}{x} = \dfrac{5}{\sqrt{Re}}$.

2. Local skinfriction coefficient $= C_{f_x} = \dfrac{\tau_x}{\frac{1}{2}\rho U_\infty^2} = \dfrac{0.664}{\sqrt{Re_x}}$

3. Average skinfriction coefficient $= \overline{C_f} = \dfrac{1.328}{\sqrt{Re_L}}$

4. Local Nusselt number $= N_{u_x} = \dfrac{hx}{K} = 0.332\,Re_{ex}^{1/2}\,P_r^{1/3}$

5. Average nusselt number $\overline{N_u} = 0.664\,Re_x^{1/2}\,Pr^{1/3}$

6. Relation between thermal and hydrodynamic boundary layer thicknesses $\dfrac{\delta t}{\delta} = P_r^{-1/3}$

### Turbulent Boundary Layer Relations for Flat plates

1. $\dfrac{\delta}{x} = 0.376\,Re_x^{-1/5}$. Considering laminar boundary layer from leading edge and turbulent BL from critical distance.

2. $\overline{C_f} = 0.072\left( Re^{-1/5} - 0.0464\,\dfrac{x_c}{L} \right)$ critical Reynolds number $= 5 \times 10^5$

3. $\overline{Nu} = 0.036\,Pr^{1/3}\,(Re_L^{0.8} - 23200)$ or, $\overline{Nu} = Pr^{1/3}[0.036\,Re_L^{0.8} - A]$ where $A = 0.036\,Re_{c_r}^{0.8} - 0.664\,Re_{cr}^{0.5}$

4. Colburn Analogy $\overline{S_t} = Nu/Re\,Pr = \dfrac{\overline{C_f} \cdot Pr^{-2/3}}{2}$

   Forced convection inside tubes pipes and ducts—critical Reynolds number $= 2300$. Average Nusselt number for turbulent flow

$$= \overline{NU} = 0.023\,R_e^{0.8}\,Pr^{0.4}, D = \dfrac{4x \text{ crossectional area}}{\text{Wetted perimetre}}$$

For laminar flow $\overline{Nu} = \dfrac{hD}{k} = 3.65$ for uniform wall temperature $= 4.36$ for uniform heat flux

## Solved Examples (Convection)

**17.23** Air at 20°C and at atmospheric pressure is flowing over a flat plate at a velocity of 3m/s. If the plate is 30 cm wide and at 60°C, calculate the following quantities at $x = 30$ cm from the leading edge (a) boundary layer thickness, (b) local skin friction co-efficient, (c) average skin co-efficient friction, (d) local shear stress, (e) average shear stress, (f) thermal boundary layer thickness, (g) local heat transfer Coefficient (h) total rate of heat loss, (i) total drag (j) total-mass flow. The properties of air at mean temperature of 40°C are $\rho = 1.128$ kg/m³, $\mu = 19.12 \times 10^{-6}$ N sec/m², $C_p = 1.005$ kJ/kgK $k = 2.76 \times 10^{-2}$ w/mK.

*Soln.* Reynolds number $Re_x = \dfrac{\rho x U_\infty}{\mu} = \dfrac{1.128 \times 0.3 \times 3}{19.12 \times 10^{-6}}$

$$= 53096 \left( \text{Units} = \frac{\text{kg}}{\text{m}^3} \times \text{m} \times \frac{\text{m}}{\dfrac{\text{S.N.S}}{\text{m}^2}} = \text{Dimension less} \right)$$

which is less than $10^5$. Therefore flow is laminar

$$\text{Prandtl number} = P_r = \frac{\mu_{cp}}{k} = \frac{19.12 \times 10^{-6} \times 1.005 \times 1000}{2.76 \times 10^{-2}} = 0.0697$$

(a) Boundary layer thickness $= 6 = \dfrac{5x}{\sqrt{Re_x}} = \dfrac{5 \times 0.3}{\sqrt{53096}} = 6.5 \times 10^{-3}$ m $= 6.5$ mm  *Ans.*

(b) Local skin friction co-efficient $Cf_x = \dfrac{0.664}{\sqrt{53096}} = 0.00288$  *Ans.*

(c) Average skin friction co-efficient $\overline{C}_f = 2 \times Cf_x = 0.00576$  *Ans.*

(d) Local shear stress

$$= \tau_x = Cf_x \, x^1 /_2 \rho u_\infty^2 = 0.00288 \times {}^1/_2 \times 1.128 \times 3^2 = 0.0146 \text{ N/m}^2 \quad Ans.$$

(e) Average shear stress $= \overline{\tau} = 2\tau x = 2 \times 0.0146 = 0.0292$

(f) Thermal boundary layer thickness $= \delta t = \dfrac{\delta}{(Pr)^{1/3}} = \dfrac{6.5}{(0.697)^{1/3}} = 7.33$ mm

(g) Local heat transfer co-efficient

$$Nu_x = 0.332 \, Re_x^{1/2} \, Pr^{1/3} = 0.332 (53096)^{1/2} (0.697)^{1/3} = 67.82$$

$$Nu_x = h_x \frac{x}{k}, h_x = \frac{K \cdot Nu_x}{x} = \frac{2.76 \times 10^{-2} \times 67.82}{0.3} = 6.22 \text{ W/m}^2\text{K} \quad Ans.$$

(h) Average heat transfer coefficient $= \overline{h} = 2 \times hx = 2 \times 6.22 = 12.44$ W/m²K  *Ans.*
Total rate of heat transfer

$$= Q = A\overline{h}\,(T_s - T_\infty) = 0.3 \times 0.3 \times 12.44 \,(60 - 20) = 44.784 \text{ Watt} \quad Ans.$$

(i) Total drag $= F_D = A\overline{\tau} = 0.3 \times 0.3 \times 0.0292 = 0.00262$ N  *Ans.*
(j) Total mass flow $= 5/8\rho U_\infty(\delta_2 - \delta_1) \times B$, at $x = 0$, $\delta_1 = 0$ and $x = 0.3$ and $\delta = \delta_2$

$$m = 5/8 \times 1.128 \times 3 \times (6.5 \times 10^{-3}) \times 0.3 = 3.832 \times 10^{-3} \text{ kg/sec}$$

where $\qquad\qquad B$ = width of the plate. $m = 13.8$ kg/hr  *Ans.*

**17.24** Air flows over a flat plate at a velocity of 3 m/s and ambient conditions are, the pressure is 760 mm of Hg and temperature is 15°C. The plate is maintained at 85°C. If the length of the plate is 100 cm along the flow of air, find out the heat lost by 50 cm of the plate which is measured from the trailing edge. Plate width is 50 cm. The properties of air at mean temperature of 50°C are $\rho = 1.093$ kg/m³ $C_p$ = 1.005 kJ/kgK $K = 2.826 \times 10^{-2}$ W/mK, $v = 17.95 \times 10^{-6}$ m²/sec.

*Soln.* Refer to Fig. 17.42.

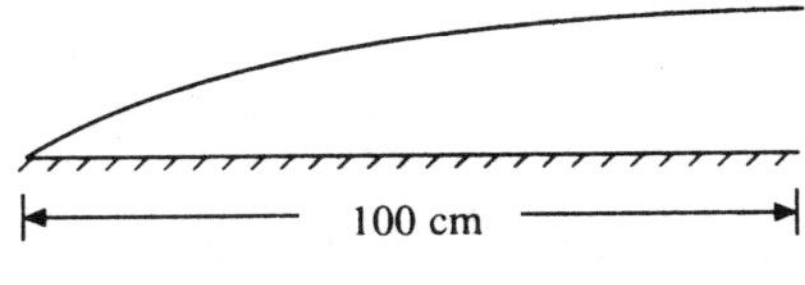

|← ——————— 100 cm ——————— →|

**Fig. 17.42**

Taking $x = 0.5$ m from the leading edge

$$Rex = \frac{U_\infty x}{v} = \frac{3 \times 0.5}{17.95 \times 10^{-6}} = 83565$$

$$Pr = \frac{\mu c_p}{k} = \frac{\rho v c_p}{k} = \frac{1.093 \times 17.95 \times 10^{-6} \times 1.005 \times 1000}{2.826 \times 10^{-2}} = 0.698 = 0.7$$

Average Nusselt number for 50 cm from the leading edge

$$= \overline{Nu} = 0.664(Re)^{1/2}(Pr)^{1/3} = 0.664(83565)^{1/2}(0.7)^{1/3} = 169.53$$

$$\overline{Nu} = \frac{\overline{h}\,L}{k}, \quad \overline{h} = \frac{\overline{Nu}\,k}{L} = \frac{169.53 \times 2.826 \times 10^{-2}}{0.5} = 9.583 \text{ W/m}^2\text{K}$$

Rate of heat transfer from the plate of 50 cm from the leading edge

$$= Q_1 = A\overline{h}\,(T_s - T_\infty) = 0.5 \times 0.5 \times 9.583\,(85 - 15) = 167.70 \text{ Watt}$$

Reynold number for the entire plate $= \dfrac{3 \times 1}{17.95 \times 10^{-6}}$

$$= 1.67 \times 10^5, \quad \overline{Nu} = 0.664\,Re_L^{1/2}\,Pr^{1/3} = 0.664(1.67 \times 10^5)^{1/2} \times (0.7)^{1/3} = 239.66$$

$$\therefore \qquad \overline{h} = \frac{N_u k}{L} = \frac{239.66 \times 2.826 \times 10^{-2}}{1} = 6.767 \text{ W/m}^2\text{K}$$

Rate of heat transfer from the entire plate $= Q_2 = Ah\,(T_s - T_\infty) = 1 \times 0.5 \times 6.767\,(85 - 15) = 236.913$ Watt. Rate of heat transfer 50 cm length from the trailing edge $= Q_2 - Q_1 = 236.913 - 167.70 = 69.213$ Watt  *Ans.*

**17.25** For a flat plate in laminar flow show that the average heat transfer co-efficient over to half length of the plate near the leading edge is $\dfrac{1}{(\sqrt{2} - 1)}$ times the average co-efficient over the remaining length.

*Soln.*

$$Nux = \frac{h_x x}{k} = 0.332\left(\frac{V_\infty x}{v}\right)^{1/2} Pr^{1.3}$$

So, $$hx = \frac{k}{x} \times 0.332\left(\frac{V_\infty x}{v}\right)^{1/2} Pr^{1/3} = 0.332\,K_{pr}^{1/3}\left(\frac{U_\infty}{v}\right)^{1/2} x^{-/2}$$

Average heat transfer co-efficient for half of the plate from leading edge

$$= \bar{h}_1 = \frac{1}{L/2} \int_0^{(L/2)} hx \, dx = 0.332 \, K \, P_r^{1/3}$$

$$\frac{(U_\infty/v)^{1/2}}{L/2} \times \int_0^{L/2} x^{-1/2} \, dx = 4 \times 0.332 \, k \, Pr^{1/3} \left(\frac{U_\infty}{v}\right)^{1/2} \left(\frac{L}{2}\right)^{1/2}$$

$$= \frac{4 \times 0.332}{\sqrt{2}} \frac{K}{L} \, Pr^{1/3} \, (R_e L)^{1/2}$$

Average heat transfer co-efficient for the half of the plate from trailing edge

$$= \bar{h}_2 = 0.332 \, K \, Pr^{1/3} \left(\frac{U_\infty}{v}\right)^{1/2} \frac{1}{L/2} \times \int_{L/2}^{L} x^{-1/2} \, dx$$

$$= 0.332 \, K \, Pr^{1/3} \left(\frac{U_\infty}{v}\right)^{1/2} \times 4 \left[ (L)^{1/2} - (L/2)^{1/2} \right]$$

$$= 4 \times 0.332 \times K \times Pr^{1/3} \left(\frac{U_\infty L}{v}\right)^{1/2} \left[ 1 - \frac{1}{\sqrt{2}} \right]$$

$$= 4 \times 0.332 \times K \, Pr^{1/3} (R_e L)^{1/2} \left[ 1 - \frac{1}{\sqrt{2}} \right]$$

$$\frac{\bar{h}_1}{\bar{h}_2} = \frac{\dfrac{1}{\sqrt{2}}}{1 - \dfrac{1}{\sqrt{2}}} = \left( \frac{1}{\sqrt{2}} - 1 \right) \text{Proved.}$$

**17.26**  The side surfaces of a car are assumed to be flat plate of 2.5 m length, what will be the temperature of the surfaces at steady state when the car receives solar heat at the rate of 581.39 W/m$^2$ while running at 80 km/hr? Assume inside surface of the car to be insulated. Take ambient temperature to be 30°C. Properties of air are $v = 1.6 \times 10^{-5}$ m$^2$/sec,

$$\rho = 1.128 \text{ kg/m}^3, \, C_p = 1.005 \text{ kJ/kg K}, \, k = 2.67 \times 10^{-2} \text{W/mK}.$$

*Soln.*

$$L = 2.5 \text{ m}, \, U_\infty = \frac{80 \times 1000}{3600} \text{ m/s} = 22.22 \text{ m/s}$$

$$Pr = \frac{\mu C_p}{k} = \frac{\rho v C_p}{k} = \frac{1.128 \times 1.6 \times 10^{-5} \times 1.005 \times 1000}{2.67 \times 10^{-2}} = 0.68$$

$$R_{eL} = \frac{U_\infty L}{v} = \frac{22.2 \times 2.5}{1.6 \times 10^{-5}} = 34.72 \times 10^5$$

which is greater than $5 \times 10^5$. Hence the flow is turbulent. $\overline{Nu}$ for turbulent flow from the critical distance and laminar from leading edge

$$\overline{Nu} = 0.036 \, Pr^{1/3} \, [Re^{0.8} - 23200]$$

$$= 0.036(0.68)^{1/3}[(34.72) \, 10^5]^{0.8} - 23200] = 4802$$

$$\therefore \qquad \overline{N}_u = \frac{\overline{h}\,L}{K}, \ \overline{h} = \frac{\overline{Nu}\,K}{L} = \frac{4802 \times 2.67 \times 10^{-2}}{2.5} = 51.28 \ \text{W/m}^2\text{K}$$

Rate of heat transfer $= Q = hA\,(Ts - T_\infty)$

or, $\qquad 581.39 = 51.28 \times 1 \,(T_s - 30), \ T_s = 41.33°C$ *Ans.*

**17.27** A flat plate 30 cm × 30 cm at 90°C is placed in an air stream of 60 m/s. The temperature of air is 0°C. Calculate the rate of heat transfer from the upper surface and the drag experienced by the plate. Properties of air at 45°C as

$$k = 2.79 \times 10^{-2} \ \text{W/mK}, \quad v = 17.45 \times 10^{-6} \ \text{m}^2/\text{sec}$$

$$\rho = 1.11 \ \text{kg/m}^3 \quad C_p = 1.005 \ \text{kJ/kg K}.$$

*Soln.* $\quad Re_L = \dfrac{U_\infty L}{v} = \dfrac{60 \times 0.3}{17.45 \times 10^{-6}} = 1.03 \times 10^6 > 5 \times 10^5$

Therefore, the flow becomes turbulent after the critical distance. The average Nusselt number when both laminar and turbulent boundary layer exists is given by

$$\overline{N}_u = 0.036 \, Pr^{1/3} \, [\, Re^{0.8} - 23{,}200 \,]$$

Here $\quad Pr = \dfrac{\mu C_p}{K} = \dfrac{\rho v C_p}{K} = \dfrac{1.11 \times 17.45 \times 10^{-6} \times 1.005 \times 1000}{2.79 \times 10^{-2}} = 0.697$

$\therefore \qquad \overline{N}_u = 0.036(0.697)^{1/3} \, [(1.03 \times 10^6)^{0.8} - 23200] = 1321$

$\therefore \qquad \overline{h} = \dfrac{\overline{Nu}\,K}{L} = \dfrac{1321 \times 2.79 \times 10^{-2}}{0.30} = 122.85 \ \text{W/m}^2\text{K}$

Rate of heat transfer $= Q = hA\,(Ts - T_\infty)$

$$= 122.85 \times 0.3 \times 0.3(90 - 0) = 8.107 \ \text{watt}$$

Average skin friction coefficient

$$\overline{C}_f = 0.072 \left[ Re_L^{-1/5} - 0.0464 \, \frac{x_c}{L} \right]$$

to determine critical distance $Rec = \dfrac{U\infty\, x_c}{v}$

$\therefore \qquad x_c = \dfrac{5 \times 10^5}{U_\infty} = \dfrac{5 \times 10^5 \times 17.45 \times 10^{-6}}{60} = 14.54 \times 10^{-2} \ \text{m}$

$\therefore \qquad \overline{c}_f = 0.072 \left[ (1.03 \times 10^6)^{-1/5} - 0.046 \times \dfrac{14.54}{30} \right] = 0.004354$

Drag force $F_D = A\overline{T} = A \times \overline{C}_f \times \dfrac{1}{2}\rho U_\infty^2$

$$= 0.30 \times 0.30 \times 0.004354 \times \frac{1}{2} \times 1.11 \times (60)^2 = 0.7829 \, N \quad \textit{Ans.}$$

**17.28** A refrigerated truck on the highway is moving at the speed of 90 km/hr in a desert area while air temperature is 70°C. The body of the truck is considered as a rectangular box of 9 m long, 3 m wide and 2 m high. Considering the boundary layer to be turbulent on the whole length and surface temperature as 10°C and heat transfer from back and front negligible, calculate (a) heat lost per hour from four surfaces (b) tonnage of refrigeration, (c) power to over come drag force on four surfaces. Take $\rho = 1.12$ kg/m³, $v = 16.95 \times 10^{-6}$ m²/s, $C_p = 1.005$ kJ/kgK, $K = 0.028$ W/mK.

*Soln.*

$$U_\infty = \frac{90 \times 1000}{3500} = 25 \text{ m/sec}$$

$$R_{e_L} = \frac{U_\infty L}{v} = \frac{25 \times 9}{16.95 \times 10^{-6}} = 13.27 \times 10^6$$

Therefore flow is turbulent, $Pr = \dfrac{\mu C_p}{K} = \dfrac{\rho v C_p}{K}$

$$= \frac{1.12 \times 16.95 \times 10^{-6} \times 1.005 \times 1000}{0.028} = 0.68$$

$\therefore \qquad \overline{Nu} = 0.036 \, R_{e_L}^{0.8} \, Pr^{1/3} = 0.036[(13.27 \times 10^6)^{0.8} (0.68)^{1/3}] = 15804$

$\therefore \qquad \overline{h} = \dfrac{NuK}{L} = \dfrac{15804 \times 0.028}{9} = 49.168 \text{ W/m}^2\text{K}$

Area of the four faces $= 2(9 \times 3) + 2(9 \times 2) = 90$ m²
Rate of heat transfer to the truck $= Q = Ah \, (T_\infty - T_s)$

$$= 90 \times 49.168 \, (70 - 10) = 265507.2 \text{ Watt}$$

Ton of refrigeration $= \dfrac{265507.2}{210 \times 10^3/60} = 75.85$

(Because one ton of refrigeration = 210 kJ/mm)
To determine the drag, the average skin friction co-efficient

$$= \overline{C}_f = 0.072 \, R_{e_L}^{-1/5} = 0.072(13.3 \times 10^6)^{-1/5} = 0.00272$$

Average shear stress $\overline{\tau} = \overline{c}_f \times {}^1/_2 \, \rho u_\infty^2 = 0.00272 \times \dfrac{1}{2} \times 1.12 \times (25)^2$

$$= 0.952 \text{ N/m}^2$$

Drag force $= F_D = A\overline{\tau} = 90 \times 0.952 = 85.68\,\text{N}$

Power to overcome friction $= \dfrac{F_D \times U_\infty}{1000} = \dfrac{85.68 \times 25}{1000} = 2.142 \text{ kW}$ \quad *Ans.*

**17.29** Estimate the average value of convective heat transfer coefficient on the inside surface of a tube 50 mm I.D. meant for heating water. The mass flow rate of water is 25000 kg/hr, which enters at a temperature of 20°C and leaves at 60°C. Calculate also the length of the tube if the wall temperature is 10°C above the bulk temperature. The physical properties of water at mean temperature at 40°C are $\rho = 992.2$ kg/m, $K = 63.37 \times 10^{-2}$ W/mK

$$v = 0.659 \times 10^{-6} \text{ m}^2/\text{sec}, \ C_p = 4.187 \text{ kJ/kgk}$$

*Soln.*

Volumetric flow rate $= \dfrac{25,000}{992.2}$ m³/hr $= 25.2$ m³/hr. If the average velocity is $V$ then

$$\frac{25.2}{3600} = \frac{\pi}{4} D^2 V = \frac{\pi}{4} (50 \times 10^{-3})^2 V$$

$$V = 3.56\,\text{m/sec}, \ R_e = \frac{V_D}{v} = \frac{3.56 \times 50 \times 10^{-3}}{0.609 \times 10^{-6}} = 270106$$

which is greater than 2300 $\therefore$ flow is turbulent

$$P_r = \frac{\mu C_p}{k} = \frac{\rho v C_p}{k} = \frac{992.2 \times 0.659 \times 10^{-6} \times 4.187 \times 1000}{63.37 \times 10^{-2}} = 4.32$$

$$\overline{N}_u = 0.023(R_e)^{0.8}(P_r)^{0.4} = 0.023(270106)^{0.8}(4.32)^{0.4} = 54.5$$

$$\overline{N}_u = \frac{\overline{h} D}{K} \quad \therefore \overline{h} = \frac{\overline{N}_u K}{D} = \frac{54.5 \times 63.37 \times 10^{-2}}{50 \times 10^{-3}} = 690.73 \text{ W/m}^2\text{K}$$

Now total heat transfer $= Q = m C_p \, dT$

$$= \frac{25000}{3600} \times 4.187(60 - 20) = 1163.05 \text{ kW}$$

If the length of the tube be 1, $\therefore Q = \overline{h} \pi D_i l (T_s - T_b)$

$$\therefore \quad l = \frac{Q}{\overline{h} \pi D_i (T_s - T_b)}, l = \frac{1163.05 \times 10^3}{690.73 \times \pi \times 50 \times 10^{-3} \times 10} = 1071.93 \text{m} \quad Ans.$$

**17.30** 50 kg of water per minute is heated from 30° to 50°C by passing through a pipe of 2 cm ID. The pipe is heated by condensing the steam on its surface at 100°C. Find out the length of the pipe required. Take the following properties of water at bulk temperature of 40°C as $\rho = 992.2$ kg/m, $k = 0.6332$ W/mK, $C_p = 4.187$ kJ/kg K, $v = 0.659 \times 10^{-6}$ m²/sec.

*Soln.*

$$\text{Rate of flow} = \dot{m} = \rho V A \quad \text{or,} \quad \frac{50}{60} = 992.2 \times v \times \frac{\pi}{4}(2 \times 10^{-2})^2 = 2.673 \text{ m/s}$$

$$\text{Reynolds number} = R_e = \frac{VD}{v} = \frac{2.673 \times 2 \times 10^{-2}}{0.659 \times 10^{-6}} = 8.11 \times 10^4$$

which is greater than 2300 flow is turbulent

$$Pr = \frac{\mu C_p}{K} = \frac{\rho v C_p}{K} = \frac{992.2 \times 0.659 \times 10^{-6} \times 4.187 \times 1000}{0.6332} = 4.32$$

$$N_u = \frac{hD}{k} = 0.023(R_e)^{0.8}(Pr)^{0.4} = 0.023(8.11 \times 10^4)^{0.8} \times (4.32)^{0.4} = 349$$

$$\therefore \quad h = \frac{Nu}{D} \times K = \frac{349 \times 0.6332}{2 \times 10^{-2}} = 11049.34 \text{ W/m}^2\text{K.}$$

Temperature difference at inlet $(\Delta T)_1 = T_s - T_i = 100 - 30 = 70°C$, Temperature difference at outlet $(\Delta T)_2 = T_s - T_o = 100 - 50 = 50°C$

Here $\dfrac{(\Delta T)_1}{(\Delta T)_2} = \dfrac{70}{50} = 1.4$ which is less than 2. Log mean temperature difference (LMTD)

$$= T_s - T_b = 100 - \frac{(T_i + T_o)}{2} = 100 - \frac{(30 + 50)}{2} = 100 - 40 = 60°C$$

Rate of heat transfer $= Q = mC_p(T_o - T_i) = Ah (T_s - T_b)$

$$\text{or,} \quad \frac{50}{60} \times 4.187 \times 1000(50 - 30) = \pi \times 2 \times 10^{-2} \times L \times 11049.34 \times 60$$

$$\text{or,} \quad L = 1.67 \text{ m} \quad Ans.$$

**17.31**  In a heat exchanger water flows through a long 2.2 cm ID copper tube at a bulk velocity of 2 m/sec and is heated by steam condensing at 150°C on the outside of the tube. The water enters at 15°C and leaves at 60°. Find the heat transfer coefficient for water. The properties of water at bulk temperature of 37.5°C are $\rho = 990$ kg/m$^3$, $C_p = 4160$ Joule/kg K. $\mu = 0.00069$ kg/m sec, $K = 0.63$ W/m K (Bihar University 1986)

*Soln.*

$$R_e = \frac{\rho VD}{\mu} = \frac{990 \times 2 \times 2.2 \times 10^{-2}}{0.00069} = 63130 > 2300$$

∴ Flow is turbulent, $\quad Pr = \dfrac{\mu C_p}{K} = \dfrac{0.00069 \times 4160}{0.63} = 4.556$

$$Nu = 0.023\, R_e^{0.8}\, Pr^{0.4} = 0.023\,(63130)^{0.8}\,(4.556)^{0.4}$$

$$= 291.97,\ h = \frac{Nu\,k}{D} = \frac{291.97 \times 0.63}{2.2 \times 10^{-2}} = 8360 \text{ W/m}^2\text{K} \quad Ans.$$

**17.32**  A light oil with 20°C inlet temperature flows at a rate of 500 kg/min through a 5 cm I.D. pipe which is enclosed by a jacket containing condensing steam at 150°C. If the pipe is 10 m long, find out outlet temperature of oil. Take the following properties of oil $\rho = 880$ kg/m$^3$, $C_p = 2.093$ kJ/kg K, $K = 0.139$ W/mK, $v = 3.6 \times 10^{-6}$ m$^2$/sec. (Ranchi Univ. 1980, 1985).

*Soln.*

The average velocity of flow is given by

$$\dot{m} = \rho AV \quad \text{or,} \quad \frac{500}{60} = 880 \times \frac{\pi}{4}(5 \times 10^{-2})^2 \times V$$

∴ $\quad V = 4.82$ m/sec. $Re = \dfrac{\rho VD}{\mu} = \dfrac{VD}{v} = \dfrac{4.82 \times 5 \times 10^{-2}}{3.6 \times 10^{-6}} = 6.694 \times 10^4$

$$Pr = \frac{\rho v C_p}{k} = \frac{880 \times 3.6 \times 10^{-6}}{0.139} \times 2.093 \times 1000$$

$$= 47.70,\ Nu = 0.023\, R_e^{0.8}\, Pr^{0.4}$$

or, $\quad Nu = 0.023(6.698 \times 10^4)^{0.8} \times (47.70)^{0.4} = 783.23$

∴ $\quad h = \dfrac{N_u K}{D} = \dfrac{783.23 \times 0.139}{5 \times 10^{-2}} = 2177.38$ W/m$^2$K

$$Q = mC_p(T_o - T_i) = h\pi DL\left(T_s - \frac{T_o + T_i}{2}\right)$$

or, $\quad 500 \times 60 \times 2.093 \times 10^3\,(T_o - 20) = 2177.38 \times \pi \times 5 \times 10^{-2} \times 10\left(150 - \dfrac{T_o + 20}{2}\right)$

or, $\quad (T_o - 20) = 5.44 \times 10^{-5}\left(150 - \dfrac{T_o + 20}{2}\right)$

After neglecting $5.44 \times 10^{-5}$ being very less, $(2T_o - 40) = (300 - T_o - 20)$

or, $\quad 3T_o = 320$

∴ $\quad T_o = 106.67°C \quad Ans.$

**17.33** Calculate the heat transfer coefficient for water being heated in a tube having 40 mm inside diameter. The water flows through the tube with a velocity of 1 m/sec. The mean temperature of water is 45°C. The temperature of tube wall is 95°C. The length of the tube is 2 m. Choose an appropriate correlation and use properties of water:

$$\rho = 980 \text{ kg/m}^3, \ v = 0.55 \times 10^{-6} \text{ m}^2/\text{sec}, \ C_p = 4.187 \text{ kJ/kg K}, \ K = 0.644 \text{ W/mk}$$

*Soln.*

$$Re = \frac{VD}{v} = \frac{1 \times 40 \times 10^{-3}}{0.55 \times 10^{-6}} = 72727 > 2300$$

∴ Flow is turbulent. $Pr = \dfrac{\mu C_p}{k} = \dfrac{\rho v C_p}{k}$

$$= \frac{980 \times 0.55 \times 10^{-6} \times 4.187 \times 1000}{0.664} = 3.39$$

$$N_u = 0.023 \ R_e^{0.8} \ Pr^{0.4}$$

$$= 0.023 \ (72727)^{0.8} \ (3.39)^{0.4} = 290.51$$

Now 
$$N_u = \frac{hD}{K}$$

∴ 
$$h = \frac{Nuk}{D} = \frac{290.51 \times 0.664}{40 \times 10^{-3}} = 4822.46 \text{ W/m}^2\text{K} \quad Ans.$$

**17.34** Derive a correlation for forced convection heat transfer coefficient, using dimensional analysis for flow through a pipe.

*Soln.*

Heat transfer coefficient ($h$) for forced convection inside a pipe depends upon diameter ($D$), velocity ($V$), density ($\rho$), viscosity ($\mu$) specific heat ($C_p$) thermal conductivity (K).

Here variables are $h$, $D$, $V$, $\rho$, $\mu$, $C_p$, $k$.

Number of variables = 7, basic dimensions = 4 mass, length, time and temperature)

*Step 1.* Dimensions of variables are:

$$D = [L], \quad V = \left[\frac{L}{T}\right], \quad \rho = \left[\frac{M}{L^3}\right], \quad K = \left[\frac{ML}{T^3 \theta}\right]$$

$$\mu = \left[\frac{M}{LT}\right] \quad h = \left[\frac{M}{T^3 \theta}\right], \quad C_p = \left[\frac{L^2}{T^2 \theta}\right]$$

Here $\theta$ stands for temperature.

*Step 2* Select four variables $D$, $V$, $\rho$ and $K$ such that their combination should not give a dimensionless group and all the basic dimensions are incorporated.

*Step 3* Express basic dimensions interms of above selected variables

$$L = [D], T = \frac{L}{V} = \frac{D}{V}$$

$$M = \rho L^3 = \rho D^3, \ \theta = \frac{ML}{KT^3} = \frac{\rho D^3 \ D}{K(D/V)^3} = \frac{\rho D V^3}{K}$$

*Step 4* Make dimensionless groups out of remaining variables. First dimensionless group

$$= \pi_1 = \frac{\mu}{\text{dimensions of } \mu} = \frac{\mu}{\dfrac{M}{LT}} = \frac{\mu LT}{M} = \frac{\mu_D (D/V)}{\rho D^3} = \frac{\mu}{\rho VD}$$

$$\therefore \qquad \pi_1' = \frac{1}{\pi} = \frac{\rho VD}{\mu} = R_e = \text{Reynolds number.}$$

Since $\pi_1$ is dimensionless, $\dfrac{1}{\pi_1}$ is also dimensionless

Second dimensionless group

$$= \pi_2 = \frac{h}{\text{dimensions of } h}$$

$$\frac{h}{\dfrac{M}{T^3 \theta}} = \frac{h T^3 \theta}{M} = \frac{h(D/V)^3}{\rho D^3} \times \left( \frac{\rho DV^3}{K} \right) = \frac{hD}{K} = Nu = \text{Nusselt number}$$

Third dimensionless group $= \pi_3 = \dfrac{C_p}{\text{dimensions of } C_p} = \dfrac{C_p}{\dfrac{L^2}{T^2 \theta}} = C_p \dfrac{T^2 \theta}{L^2}$

$$= \frac{C_p \times (D/V)^2}{D^2} \times \frac{\rho DV^3}{K} = \rho C_p \frac{DV}{K}$$

$$\pi_3' = \pi_3 \times \pi_1 = \rho C_p \frac{DV}{K} \times \frac{\mu}{\rho VD} = \frac{\mu C_p}{k} = P_r = \text{Prandtl number}$$

$\therefore$ Dimensionless groups are $\pi_1', \pi^2, \pi^{3'}$

or, $\qquad\qquad R_e, Nu$ and $Pr, \quad \therefore \quad N_u = f(R_e, P_r)$

**17.35** Water is passed through the annulus formed by the two tubes of 5 cm and 3 cm in diameter at a velocity of 0.5 m/sec. If the inlet temperature of water is 20°C and 3 cm diameter tube temperature is maintained at 80°C, find out the heat transfer coefficient between the water and small tube surface. Take the following data for water at 50°C, $\rho = 998$ kg/m³, $C_p = 4.174$ kJ/kg K, $K = 0.647$ W/mK, $v = 0.55 \times 10^{-6}$ m²/s.

*Soln.* Characteristic length for the annular tube

$$= \text{Hydraulic diameter} = D_H = \frac{4 \times \text{flow area}}{\text{wetted perimeter}} = \frac{4 \times \dfrac{\pi}{4}(D_o^2 - D_i^2)}{(D_o + D_i)} = D_o - D_i$$

Reynolds number $R_e = \dfrac{VD_H}{v} = \dfrac{v(D_o - D_i)}{v} = \dfrac{0.5(5-3) \times 10^{-2}}{0.55 \times 10^{-6}} = 1.82 \times 10^4$ which is greater than 2300

$\therefore$ Flow is turbulent $Pr = \dfrac{\mu C_p}{k} = \dfrac{\rho v C_p}{k} = \dfrac{998 \times 0.55 \times 10^{-6} \times 4.174 \times 1000}{0.647} = 3.55$

$$N_u = 0.023(R_e)^{0.8} P_r^{0.33} = 0.023(1.82 \times 10^4)^{0.8}(3.55)^{0.33} = 89.8$$

$$N_u = \frac{hD_H}{k} = \frac{h(D_o - D_i)}{k}$$

$$\therefore \qquad h = \frac{N_u - k}{D_o - D_i} = \frac{89.8 \times 0.647}{(5-3) \times 10^{-2}} = 2905 \text{ W/m}^2\text{K} \quad \textit{Ans.}$$

## 17.5  RADIATION

### 17.5.1  Definitions

1. Emissive Power $(E)$ — Radiant energy emitted by a body through unit surface area per unit time is called emissive power. Its unit is K call/hr m$^2$ or watt/m$^2$.

2. Black body emissive power $-(E_b)$ Radiant energy emitted by a black body through unit surface area per unit time is called black body emissive power.

3. Monochromatic emissive power $(E_\lambda)$ — It is the emissive power per unit wave length. Mathematically

$$E_\lambda = \frac{dE}{d\lambda} \quad \text{or,} \quad E = \int E_\lambda / d\lambda \qquad (17.51)$$

4. Black body — A body which absorbs all the radiation incident on it is called a black body i.e., a body having absorptivity unity is called a black body.

5. Planck's law of Radiation—Planck, from his quantum theory, derived a relation between monochromatic black body emissive power, wavelength and temperature in the form, where $C_1$ and $C_2$ are constants

$$E_{b\lambda} = \frac{C_1 \lambda^{-5}}{(e^{C_2/\lambda T} - 1)} \qquad (17.52)$$

$$C_1 = 3.7415 \times 10^{-16} \text{ Wm}^2 \qquad C_2 = 1.4388 \times 10^{-2} \text{mK}$$

6. Stefan-Boltzman's law—It states that the emissive power of a black body is proportional to fourth power of its absolute temperature. $E_b = \sigma T^4$, where $\rho$ = Stefan Boltzman Const.

$$= 5.6697 \times 10^{-8} \text{ W/m}^2\text{K}^4.$$

7. Wein's displacement law—It states that the maximum monochromatic emissive power of a black body is shifted to the shorter wavelengths for the higher temperature. Mathematically

$$\lambda_{\text{max}} \cdot T = 2.898 \times 10^{-3}\text{mK}$$

where $\lambda_{\text{max}}$ = wave length at which monochromatic black body emissive power occurs.

8. Kirchhoff's law of radiation—It states that at thermal equilibrium the ratio of emissive power of a surface to its absorptivity is same for all the bodies.

9. Lambart's cosine law—It states that, for isotropic radiation, energy radiated from a surface in a given direction $(dQ)$ is equal to product of radiant energy in normal direction and cosine of angle between normal and the given direction $dQ = dQ_n \cos \theta$

10. Emissivity $(\varepsilon)$ — It is defined as the ratio of emissive power of a body to the emissive power of a black body at the same temperature

$$\varepsilon = \frac{E}{E_b} \qquad (17.53)$$

11. Gray body—A body of which absorptivity does not depend upon the wavelength of radiation is called a gray body.

12. **Intensity of radiation (I)** - The intensity of radiation in space is defined as the radiant energy propagated in a particular direction per unit solid angle per unit area projected on a plane perpendicular to direction of propagation.

13. **Relation between emissive power and intensity of radiation** $E = \pi I$.

14. **Shape factor** of a surface $A_1$ with respect to surface $A_2$ is given by $A_1 F_{1 \to 2} = A_2 F_{2 \to 1}$

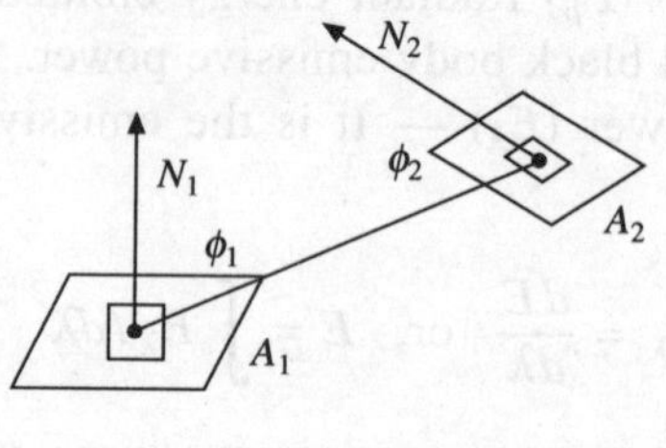

$$= \iint \frac{dA_1 \, dA_2 \, \cos \phi_1 \, \cos \phi_2}{\pi r^2}$$

if $A_1$ is an elemental area $dA_1$ then $F_{1 \to 2}$

$$= \int \frac{d A_2 \, \cos \phi_1 \, \cos \phi_2}{\pi r^2} \tag{17.54}$$

15. **Reciprocity relation** $-A_1 F_{1 \to 2} = A_2 F_{2 \to 1}$

or,

$$\frac{A_1}{A_2} = \frac{F_{2 \to 1}}{F_{1 \to 2}} \tag{17.55}$$

$F_{1 \to 2}$ represents the fraction of emitted radiation from $A_1$ and reaching surface $A_2$.

16. **Direct exchange area** $= A_1 F_{1 \to 2} = A_2 F_{2-1}$

17. **Electrical circuit for radiant heat exchange** for an enclosure having two surfaces is

Radiant heat exchange between two surfaces $\quad Q = \dfrac{E_{b_1} - E_{b_2}}{\dfrac{\varepsilon_1}{A \varepsilon_1} + \dfrac{1}{A_1 F_{1 \to 2}} + \dfrac{\varepsilon_2}{A_2 \varepsilon_2}} \tag{17.56}$

For two parallel walls $A_1 = A_2 = A$, $\rho_1 = 1 - \varepsilon_1$ then $F_{1 \to 2} = 1$

$$Q = \frac{(E_{b_1} - E_{b_2})A}{\dfrac{1 - \varepsilon_1}{\varepsilon_1} + 1 + \dfrac{1 - \varepsilon_2}{\varepsilon_2}} = \frac{(E_{b_1} - E_{b_2})A}{\dfrac{1}{\varepsilon_1} + \dfrac{1}{\varepsilon_2} - 1} \tag{17.57}$$

### Solved Examples (Radiation)

**17.36** Sun emits maximum radiation at wavelength of $\lambda = 0.52$ micron. Assuming sun as a black body find the surface temperature of the sun and emissive power at that temperature.

*Soln.*

According to wein's displacement law $\lambda_{max} T = 2.898 \times 10^{-3}$ mK

$$\therefore \qquad T = \frac{2.898 \times 10^{-3}}{\lambda_{max}} = \frac{2.898 \times 10^{-3}}{0.52 \times 10^{-6}} = 5573 \text{ K.} \quad Ans.$$

From Stefan Boltzman law,

$$E_b = \sigma T^4 = 5.6697 \times 10^{-8} (5573)^4 = 5.469 \times 10^7 \text{ w/m}^2 \quad Ans.$$

**17.37** A furnace has a small observation hole of 2.5 cm diameter. If the furnace temperature is 600°C find: (a) the rate of energy loss from the hole due to radiation (b) the wave length at which emission is maximum.

*Soln.*

Rate of heat loss through the hole

$$= A\sigma T^4 = \frac{\pi}{4}(2.5)^2 \times 5.6697 \times 10^{-8} (273 + 600)^4 = 161.65 \text{ kW}$$

According to Wein's displacement law, $\lambda_{max}^T = 2.898 \times 10^{-3}$

$$\therefore \qquad \lambda_{max} = \frac{2.898 \times 10^{-3}}{873} = 3.31\mu \quad Ans.$$

**17.38** Isotropic radiation of intensity 145.34 W/m$^2$ staradian falls on a diffuse reflection of area 0.2 m$^2$ area. If it reflects 54.65 Watt, what should be the absorptivity of the surface?

*Soln.*

Since emissive power $E = \pi I$, total radiation incident on the diffuse reflector $= \pi I \times A = \pi \times 145.34 \times 0.2 = 91.31$ Watt. Energy absorbed $= 91.31 - 54.65 = 36.66$ Watt

$$\text{Absorptivity} = \frac{\text{Absorbed radiation}}{\text{Incident radiation}} = \frac{36.66}{91.31} = 0.401 \quad Ans.$$

**17.39** A small body at 47°C is placed in large furnace whose walls are maintained at 1200 K. The total absorptivity of the small body varies with temperature of the incident radiation as follows.

$$T\,(K) = 320 \text{ K} \quad 500 \text{ K} \quad 1200 \text{ K}$$

$$\alpha = 0.78 \quad 0.67 \quad 0.55$$

Find the rate of absorption and rate of emission by a small body.

*Soln.*

$$\text{Absorbed radiation} = \alpha \sigma T^4 = 0.55 \times 5.6697 \times 10^{-8} \times (1200)^4 = 6.46 \times 10^4 \text{ Watt/m}^2$$

$$\text{Emitted radiation} = \varepsilon \sigma T^4 = \alpha \sigma T^4 = 0.78 \times 5.6697 \times 10^{-8}(273 + 47)^4$$

$$= 463.49 \text{ Watt/m}^2 \quad Ans.$$

**17.40** A black sphere of 4 cm diameter is maintained at 1000°C. This is surrounded by a thin concentric black spherical shell of 10 cm diameter. If the outside air temperature is 25°C and the shell attains a temperature of 4500°C in the steady state, find the heat transfer coefficient for convection on the outer surface of the shell. Neglect convection inside the shell.

*Soln.* Refer to Fig. 17.43

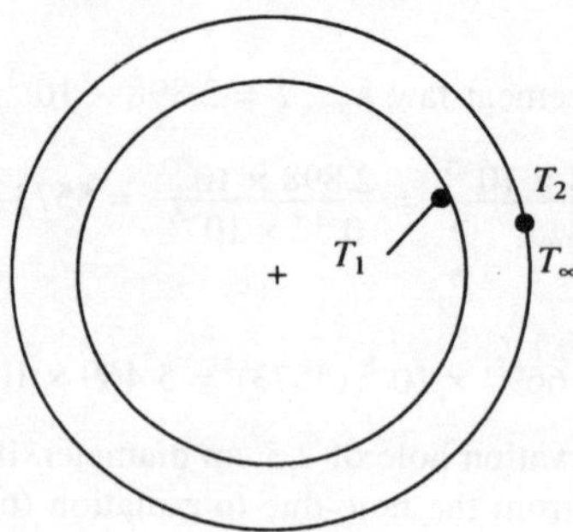


**Fig. 17.43**

In the steady state. Net radiation received by the shell from the inner sphere = heat transfer due to radiation and convection to the surroundings.

$$\sigma A_1 (T_1^4 - T_2^4) = \sigma A_2 (T_2^4 - T_\infty^4) + A_2 h_c (T_2 - T_\infty)$$

where $h_c$ = convective heat transfer coefficient

or, $$\frac{A_1}{A_2} \sigma (T_1^4 - T_2^4) = \sigma (T_2^4 - T_\infty^4) + h_c (T_2 - T_\infty)$$

or, $\left(\dfrac{4}{10}\right)^2 \times 5.6697 \times 10^{-8} (1273^4 - 723^4) = 5.6697 \times 10^{-8} (723^4 - 298^4) + h_e(723 - 298)$

or, $$21344.095 - 15045.085 = 425 \, h_c$$

$$h_c = 14.82 \text{ W/m}^2\text{K} \quad Ans.$$

**17.41** The average solar radiation received at noon on the earth's surface is estimated to be 11–5 W/m². The transmissivity of the earth's atmosphere is estimated to be 80%. The diameter of the sun is $1.392 \times 10^6$ km and the distance between earth and the sun is $1.5 \times 10^8$ km. Assuming the sun to be a black body, estimate the temperature of the sun.

*Soln.* Refer to Fig. 17.44

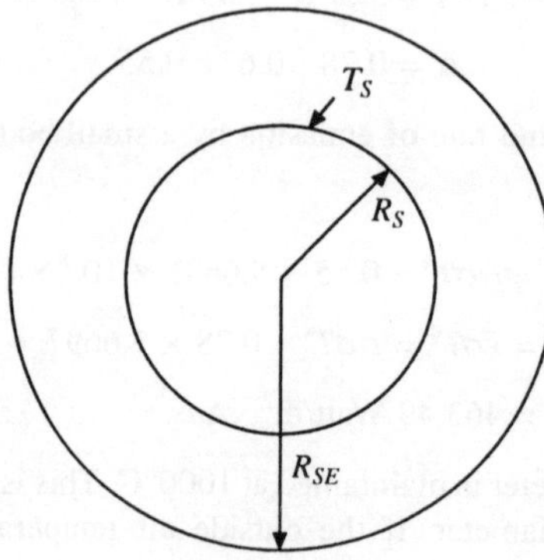


**Fig. 17.44**

Actual radiation reaching the earth surface $= \dfrac{1105}{0.8} = 1381.25 \text{ W/m}^2$

Let us surround the sun by an imaginary sphere having radius ($R_{ES}$) equal to the distance between the sun and earth. This sphere will receive all the radiation emitted by the sun

$$\therefore \qquad A_s \sigma T_s^4 = 1381.25 \times A_{Es} \text{ or, } T_s^4 = \frac{1381.25}{5.6697 \times 10^{-8}} \left( \frac{R_{E_s}}{R_s} \right)^2$$

$$= \frac{1381.25}{5.6697 \times 10^{-8}} \left( \frac{1.5 \times 10^{11}}{\dfrac{1.392}{2} \times 10^9} \right)^2 = 5800 \text{ K}$$

$$\therefore \qquad T_s = 5800 \text{ K} \quad Ans.$$

17.42  A thin metal plate of 5 cm diameter is suspended in air at 25°C. Radiation energy of 2.9 Watt from a distant furnace falls on one face of the plate. The unit surface conductance for convection at both faces of the plate is estimated to be 93 w/m². If the plate attains a steady temperature of 30°C, find the reflectivity of the plate.

*Soln.*

Amount of radiation incident on the plate = heat transfer by convection from the plate + Reflected radiation

$$H = 2A\, h_c(T_s - T_\infty) + \rho H$$

or, $$2.9 = 2 \times \frac{\pi}{4}(0.05)^2 \times 93(30 - 25) + \rho \times 2.9$$

$$\therefore \qquad \rho = 0.370 \quad Ans.$$

17.43  Show that the geometric shape factor for a very small disc $dA_1$ and large parallel disc $A_2$ of radius 'a',

directly located at a distance $L$ above the smaller one is given by $F_{1 \to 2} = \dfrac{a^2}{L^2 + a^2}$

*Soln.*  Refer to Fig. 17.45

Consider an elemental area $dA_2$ making an angle $d\psi$ at the centre situated at a radius $r$ and the ekness $dr$. $dA_2 = r\, d\psi \cdot dr$

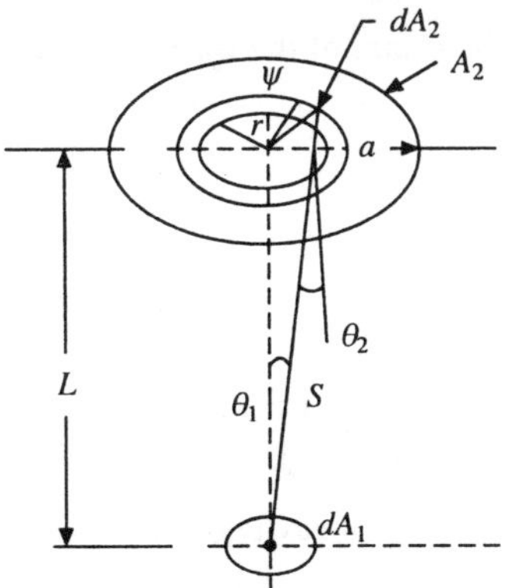


**Fig. 17.45**

$$\theta_1 = \theta_2, \ \cos\theta_1 = \cos\theta_2 = \frac{L}{s} = \frac{L}{\sqrt{l^2 + r^2}}$$

Shape factor for elemental disc with respect to $A_2$

$$F_{1\to 2} = \int \frac{\cos\theta_1 \cos\theta_2 \, dA_2}{\pi s^2} = \int_0^{2\pi}\int_0^a \frac{L}{s}\cdot\frac{L}{s}\frac{d\psi\, dr}{\pi s^2}$$

$$= \int_0^{2\pi}\int_0^a \frac{L^2 r^2 \, dr\, d\psi}{\pi s^4} = L^2 \times 2\frac{\pi}{\pi}\int_0^a \frac{r\, dr}{(L^2 + r^2)^2} = L^2\int_0^a \frac{2r\, dr}{(L^2 + r^2)^2}$$

$$= L^2\left[\frac{-1}{L^2 + r^2}\right]_0^a = -L^2\left[\frac{1}{L^2 + a^2} - \frac{1}{L^2}\right] = -L^2\left[\frac{L^2 - (L^2 + a^2)}{L^2(a^2 + L^2)}\right] = \frac{a^2}{L^2 + a^2} \ \text{Proved.}$$

**17.44** Two areas $A_1$ and $A_2$ lie on the inside surface of a sphere of radius $R$. Show that the direct exchange area $12 = F_{1\to 2}\,A_1 = F_{2\to 1}\,A_2$ is $A_1 A_2/4\pi R^2$ and is thus independent of the shape and orientation of the areas.

*Soln.* Refer to Fig. 17.46

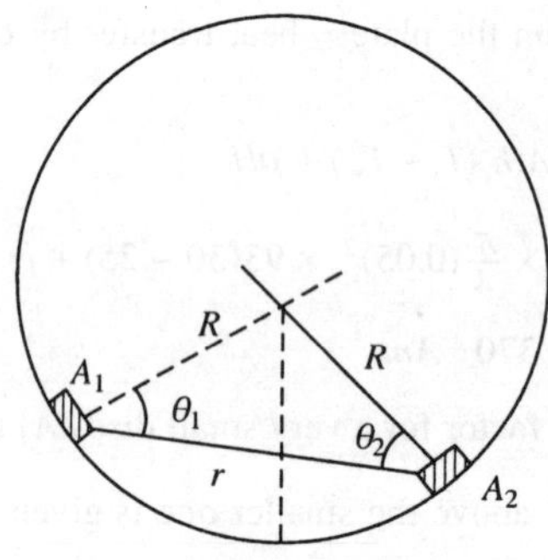

**Fig. 17.46**

$$A_1 F_{1\to 2} = 12 = \iint \frac{dA_1\, dA_2 \cos\theta_1}{\pi r^2}\cos\theta_2$$

Normals of areas pass through the centre of the sphere $= \theta_1 = \theta_2, \ \cos\theta_1 = \dfrac{r}{2\cdot R}$

$$\cos\theta_1 \cos\theta_2 = \frac{r^2}{4R^2}$$

$$\therefore \qquad \overline{12} = \iint \frac{dA_1\, dA_2 \cos\theta_1 \cos\theta_2}{\pi r^2}$$

$$= \iint \frac{dA_1\, dA_2}{\pi r^2}\frac{r^2}{4R^2} = \frac{1}{4R^2}\iint \frac{dA_1\, dA_2}{\pi} = \frac{A_1 A_2}{4\pi R^2}\ \text{Proved.}$$

**17.45** Prove that the shape factor of a cylindrical cavity of diameter $D$ and height $H$ with respect to itself is

$$F_{1\to 1} = \frac{4H}{4H + D}$$

*Soln.*  Refer to Fig. 17.47

Denoting cavity surface area as $A_1$ and open top as $A_2$,
From shape factor algebra

$$F_{1\to1} + F_{1\to2} = 1$$

$$F_{1\to1} = 1 - F_{1\to2}$$

But $A_1\,F_{1\to2}\,A_2 = F_{2\to1}$ and

$$F_{2\to1} = 1,\ A_1 F_{1\to2} = A_2 \times 1$$

$$F_{1\to2} = \frac{A_2}{A_1}$$

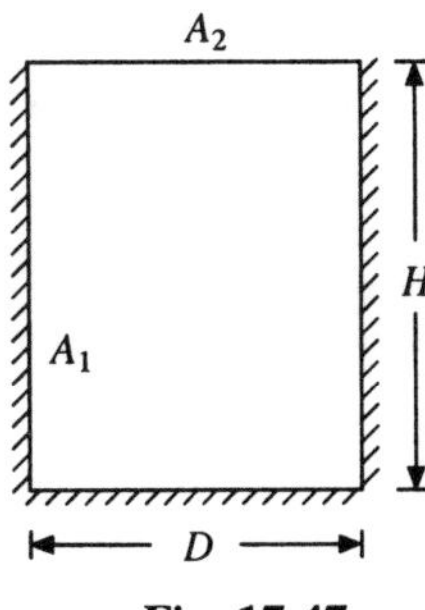


Fig. 17.47

$$F_{1\to1} = 1 - \frac{A_2}{A_1} = 1 - \frac{\dfrac{\pi}{4}D^2}{\pi DH + \dfrac{\pi}{4}D^2}$$

$$= 1 - \frac{D}{4H + D} = \frac{4H}{4H + D} \quad \text{Proved.}$$

**17.46**  Prove that the shape factor of hemi-spherical boul of diameter $D$ with respect to itself is 0.5.

*Soln.*  Refer to Fig. 17.48

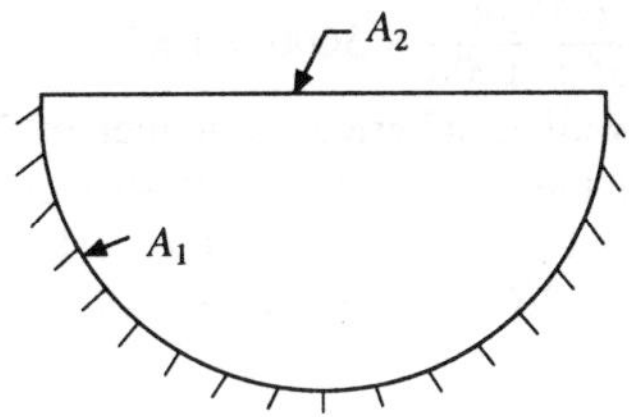


**Fig. 17.48**

$$F_{1\to1} = 1 - \frac{A_2}{A_1} = 1 - \frac{\dfrac{\pi}{4}D^2}{2\pi R^2} = 1 - \frac{\dfrac{\pi}{4}D^2}{\dfrac{\pi}{2}D^2} = 1 - \frac{1}{2} = 0.5 \quad \text{Proved.}$$

**17.47**  A thermos flask has a double walled bottle with the space between the walls evalucated. The bottle surface are silvered to acquire an emissivity of 0.05. If the ice at 0°C is stirred in the flask find the rate of melting of ice in an ambient at 28°C. Take latent heat of ice as 335 kJ/kg and the effective surface area of flask as 460 cm².

*Soln.*  There is a radiant heat exchange between two cylinders.

$$A_1 = A_2 = A$$

Refer to Fig. 17.49. $\dfrac{\rho_1}{A_1 \varepsilon_1}\ \dfrac{1}{A_1 F_{1\to2}}\ \dfrac{\rho_2}{A_2 \varepsilon_2}$

$$F_{1\to2} = 1,\ \varepsilon_1 = \varepsilon_2 = \varepsilon,\ T_2 > T_1$$

Net heat transfer due to radiation from surrounding $= \dfrac{E_{b_2} - E_{b_1}}{\dfrac{\rho_1}{A_1\,\varepsilon_1} + \dfrac{1}{A_1 F_{1\to2}} + \dfrac{\rho_2}{A_2\,\varepsilon_2}}$

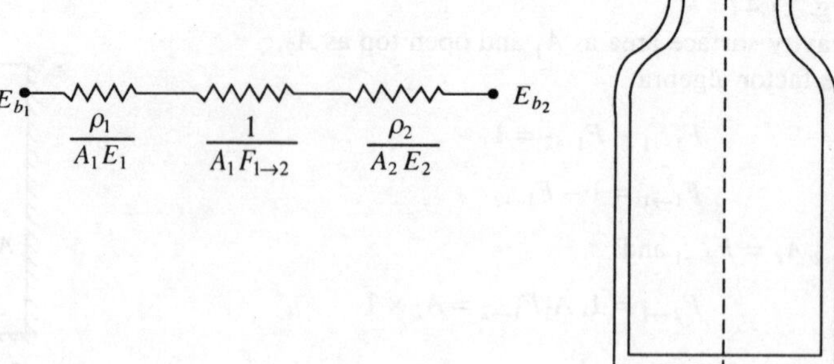

**Fig. 17.49**

$$= \frac{A(E_{b_2} - E_{b_1})}{\dfrac{1 - \varepsilon_1}{\varepsilon_1} + 1 + \dfrac{1 - \varepsilon_2}{\varepsilon_2}} \quad \text{because } \varepsilon_1 = \varepsilon_2 = \varepsilon \text{ and } F_{1\to2} = 1$$

$$= \frac{A\sigma(T_2^4 - T_1^4)}{\dfrac{2}{\varepsilon} - 1} = \frac{450 \times 10^{-4} \times 5.6697 \times 10^{-8}}{\dfrac{2}{0.05} - 1} [(273 + 28)^4 - (273)^4]$$

$$= 0.1738 \text{ Watt}$$

$$\therefore \text{ Rate of melting of ice } = \frac{0.1738}{335 \times 1000} \times 3600 = 1.87 \times 10^{-3} \text{ kg/hr} = 1.87 \text{ gm/hour} \quad Ans.$$

**17.48** A Dewar flask has a double walled spherical container of 30 cm inner diameter and 36 cm outer diameter and the space between them evacuated. Both surfaces are silvelvered to acquire an emissivity of 0.05. If the liquid oxygen at –183°C is stored in the flask, find the rate of evaporation of oxygen in the outer sphere temperature is 20°C. Latent heat of oxygen is 214.37 kJ/kg.

*Soln.*    Refer to Fig. 17.50

Here $r_1 = 15$ cm, $r_2 = 18$ cm,   $T_1 = 273 - 183 = 90$ K   $T_2 = 273 + 20 = 293$ k,   $\varepsilon_1 = \varepsilon_2 = \varepsilon = 0.05$

$F_{1\to2} = 1$

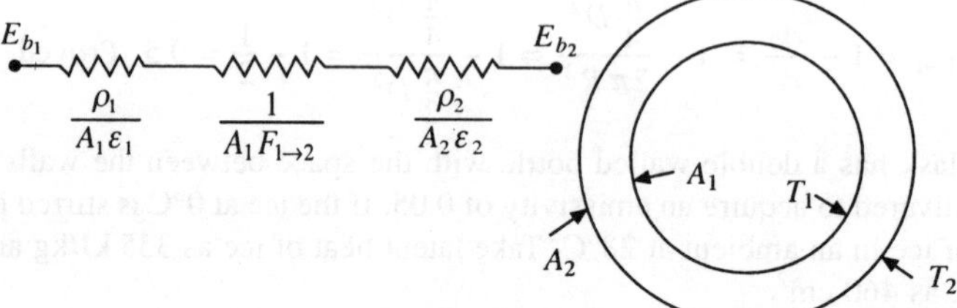

**Fig. 17.50**

$$\text{Rate of heat transfer} = Q = \frac{E_{b_1} - E_{b_2}}{\dfrac{\rho_1}{A_1\varepsilon_1} + \dfrac{1}{A_1 F_{1\to2}} + \dfrac{\rho_2}{A_2 \varepsilon_2}}$$

$$= \frac{(E_{b_1} - E_{b_2})A_1}{\dfrac{1 - \varepsilon_1}{\varepsilon_1} + 1 + \dfrac{A_{1(1-\varepsilon2)}}{A_2 \; \varepsilon_2}}$$

$$= \frac{(E_{b_1} - E_{b_2})A_1}{\dfrac{1}{\varepsilon_1} + \dfrac{A_1}{A_2}\left(\dfrac{1}{\varepsilon_2} - 1\right)} = \frac{\sigma(T_1^4 - T_2^4)}{\dfrac{1}{\varepsilon_1} + \dfrac{A_1}{A_2}\left(\dfrac{1}{\varepsilon_2} - 1\right)}$$

$$= \frac{5.6697 \times 10^{-8}(90^4 - 293^4) \times (0.15)^2}{\dfrac{1}{0.05} + \left(\dfrac{15}{18}\right)^2\left(\dfrac{1}{0.05} - 1\right)} = \frac{-117.09}{33.194} = -3.527 \text{ Watt}$$

$$= \frac{-117.09}{33.194} = -3.527 \text{ Watt}$$

Rate of heat transfer to the flask = 3.527 watt

Rate of evaporation of oxygen $= \dfrac{3.527 \times 3600}{214.37 \times 1000} = 0.0592$ kg/hr $= 59.23$ gm/hr   *Ans.*

**17.49** A 6 cm O.D. oxidised iron pipe at 450° passes through a room in which the surroundings are at a temperature of 27°C. If the emissivity of the pipe metal is 0.8, what is the net exchange of radiant energy per meter length of the pipe.

*Soln.* Refer to Fig. 17.51.

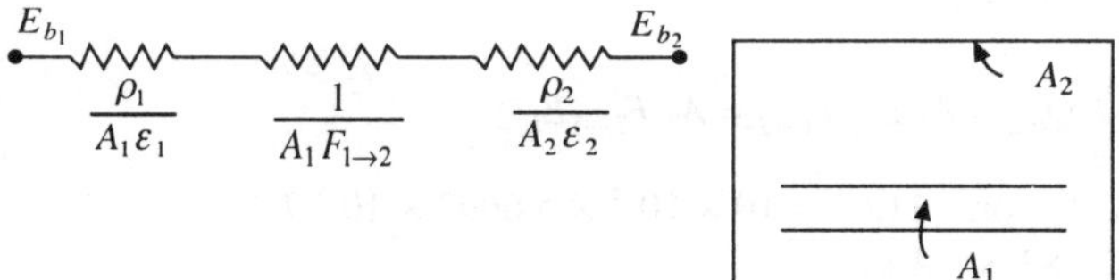

**Fig. 17.51**

$$Q = \frac{E_{b_1} - E_{b_2}}{\dfrac{\rho_1}{A_1\varepsilon_1} + \dfrac{1}{A_1 F_{1\to2}} + \dfrac{\rho_2}{A_2\,\varepsilon_2}}, \quad F_{1\to2} = 1$$

$$Q_1 = \frac{(E_{b_1} - E_{b_2})A_1}{\dfrac{\rho_1}{\varepsilon_1} + \dfrac{1}{1} + \dfrac{A_1}{A_2}\dfrac{\rho_2}{\varepsilon_2}}$$

If $A_1 << A_2$, Then $Q = \dfrac{(E_{b1} - E_{b_2})A_1}{(1 - \varepsilon_1)/\varepsilon + 1} = A_1(E_{b1} - E_{b2})\varepsilon_1$

If $\varepsilon_1 A_1 \sigma(T_1^4 - T_2^4) = 0.8 \times \pi \times 0.06 \times 1 \times 5.6697 \times 10^{-8}(723^4 - 300^4) = 2261.86$ watt/m   *Ans.*

**17.50** Estimate the net radiant energy interchange per square meter for very large two parallel planes at temperatures of 538°C and 315°C. The emissivities of hot and cold planes are 0.9 and 0.7 respectively.

*Soln.*
$$T_1 = 273 + 538 = 811 \text{ K}$$

$$T_2 = 273 + 315 = 588 \text{ K}$$

Net radiant heat exchange $= Q = \dfrac{(E_{b_1} - E_{b_2})}{\dfrac{\rho_1}{A\varepsilon_1^+} + \dfrac{1}{AF_{1\to2}} + \dfrac{\rho_2}{A\varepsilon_2}}$

$T_1 = 273 + 538 = 811\text{K}$

$$E_{b_1} \underset{\dfrac{\rho_1}{A_1\varepsilon_1}}{\text{—WWW—}} \underset{\dfrac{1}{A_1F_{1\to2}}}{\text{—WWW—}} \underset{\dfrac{\rho_2}{A_2\varepsilon_2}}{\text{—WWW—}} E_{b_2}$$

$T_2 = 273 + 315 = 588$

$$Q = \frac{\left(E_{b_1} - E_{b_2}\right)A}{\dfrac{(1-\varepsilon_1)}{\varepsilon_1} + \dfrac{(1-\varepsilon_2)}{\varepsilon_2}} = \frac{\left(E_{b_1} - E_{b_2}\right)A}{\dfrac{1}{\varepsilon_1} + \dfrac{1}{\varepsilon_2} - 1}$$

$$= \frac{5.6697 \times 10^{-8}\,(811^4 - 588^4) \times 1}{\dfrac{1}{0.9} + \dfrac{1}{0.7} - 1} = 11534.747 \text{ Watt/m}^2 \quad Ans.$$

**17.51**  A hemispherical bowl has a flat circular lid of 30 cm diameter what should be the temperature of the bowl so that a radiation of 11.62 watt/m² comes out from a hole of 10 cm² in the lid. Assume all surfaces to be black.

*Soln.*  Refer to Fig. 17.52

$$F_{1\to2} = 1$$

$$F_{2\to2}\, A_2 = F_{1\to2}A_1 = A_1$$

$$F_{2\to2} = \frac{A_1}{A_2}\quad Q_{\text{out}} = A_2\, E_{b_2}\, F_{1\to2} = A_2\, F_{2\to1}E_{b_2}$$

$$= A_1\, E_{b_2} \text{ or,} \quad 11.62 = 10 \times 10^{-4} \times 5.6697 \times 10^{-8}\, T_2^4$$

or, $\qquad T_2 = 672.83 \text{ K} \quad Ans.$

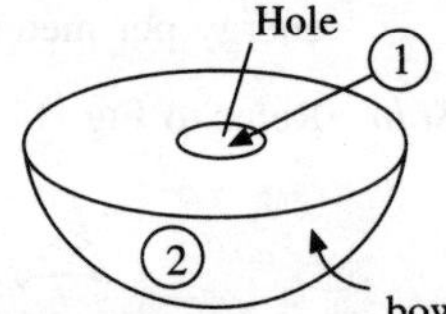

**Fig. 17.52**

**17.52**  A glass plate 40 cm × 40 cm in size is used on an electric furnace to view the radiation. The transmissivity of the glass is 0.5 in the wave band of 0.2 $\mu$ to 4 $\mu$. The emissivity is 0.3 upto 4$\mu$ and 0.9 above that. The transmissivity of the glass is zero except in the range from 0.2 to 4$\mu$. Calculate the energy absorbed and transmitted by the glass assuming the furnace is a black body at 2000°K. Use the following radiation data.

$$f = \int_o^{\lambda t} \frac{E_{b\lambda}\,d\chi\lambda T}{\sigma T^5} = \frac{E_{b(0-\lambda)}}{E_b\,(0-\infty)}$$

| $\lambda T = 400$ | 1200 | 2700 | 3000 | 4111 | 4500 | 8000 |
|---|---|---|---|---|---|---|
| $f = 0$ | 0.002 | 0.204 | 0.273 | 0.5 | 0.514 | 0.838 |

*Soln.*  $\qquad \lambda_1 T = 0.2 \times 2000 = 400\ \mu°k,\ \lambda_2 T = 4 \times 2000 = 8000\ \mu°\text{K}$

Total emissive power $= E_{b(0-\infty)} = \sigma T^4 = 5.6697 \times 10^{-8}\,(2000)^4 = 5.6697 \times 16 \times 10^4 \text{ Watt/m}^2$

$$E_{b(\lambda_2-\lambda_1)} = \left[\frac{E_{b(0-\lambda_2)}}{E_{b(0-\infty)}} - \frac{E_{b(0-\lambda_1)}}{E_{b(0-\infty)}}\right] E_b\,(0-\infty) = (0.838 - 0)$$

$$\times 5.6697 \times 16 \times 10^4 = 762.79 \text{ kW/m}^2$$

This is the total radiation incident on the glass in the wave band of 0.2 $\mu$ to 4$\mu$. The transmissivity of glass is 0.5. Energy transmitted $0.5 \times 762.79 = 381.395 \text{ kW/m}^2 \quad Ans.$

Radiation absorbed when $\lambda = 4\mu = 0.3 \times 762.79 = 228.837 \text{ kW/m}^2$

Energy incident when $4\mu < \lambda < \infty = (1 - 0.838) \times 762.79 = 123.57 \text{ kW/m}^2$

Radiation absorbed when $4\mu < \lambda < \infty = 0.9 \times 762.79 = 686.511 \text{ kW/m}^2$

Total energy absorbed $= 228.837 + 686.511 = 915.348 \text{ kW/m}^2 \quad Ans.$

# 17.6  HEAT TRANSFER WITH CHANGE IN PHASE

### 17.6.1  Introduction

There are many engineering applications where heat transfer during phase change is encountered. The change of phase from vapour to liquid is called condensation and that from liquid to vapour is evaporation or boiling. The examples of processes involving phase change are boiling in the boiler tubes and condensing of steam in the steam condenser and evaporation of refrigerant in the evaporation. The phenomena of boiling and condensation heat transfer are considerably more complex than these of convection without phase change, because in addition to all of the variables associated with convection, those associated with the change in the phase are also relevant. In addition to thermal properties of the fluid, the surface characteristics, the surface tension, the latent heat of evaporation, the pressure and the other properties of vapour play also an important role. As a result of large number of variables involved, neither general equations describing the phase change nor the complete exact solutions are available. Considerable progress has been made to develop some analytical or experimental expressions for rate of heat transfer. The rate of heat transfer due to phase change is given by

$$Q = m h_{f_g} \tag{17.58}$$

where $h_{f_g}$ is the latent heat of evaporation and $m$ the rate of mass evaporated or condensed. During the heat transfer in either of these processes, the temperature at the interface between the liquid and vapour phases is equal to the saturation temperature $T_s$ of the substance at the pressure of the process. In evaporation, $T_w > T_s$ and for condensation $T_w < T_s$ as shown in Figs. 17.53 and 17.54.

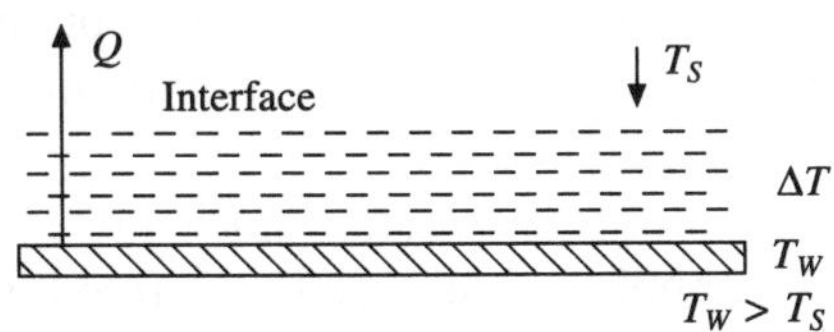


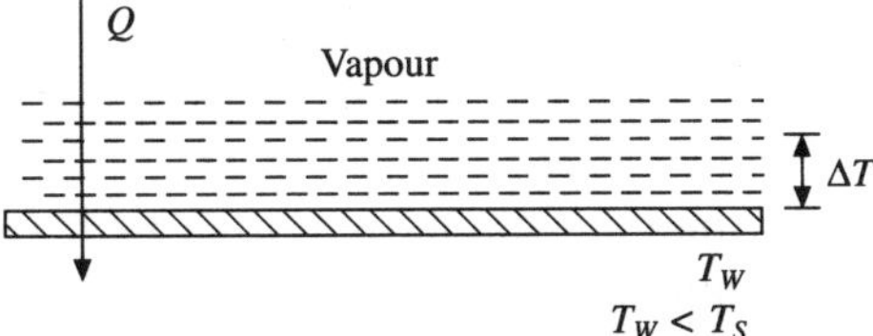


Fig. 17.53   Evaporation $T_W > T_S$
Fig. 17.54   Condensation $T_W < T_S$

### 17.6.2  Boiling

Two kinds of boiling phenomena are encountered in practice.

1. *Pool Boiling:*  In such boiling, the heating surface, such as flat plate or a wire, is submerged in a pool of liquid. A familiar example of this kind of boiling is boiling of water in a kettle. The movement of liquid in this case is caused only by free convection or by stirring action of the bubbles. No external agency is required for motion.

2. *Forced Convection Boiling:* If the boiling occurs on the walls of a tube through which liquid flows with considerable velocity, then this velocity influences the liquid motion thereby causing the bubble growth and separation as well. Example is boiling of water in a water tube boiler.

### *Regimes of Pool Boiling*

Generally speaking, we may classify three types of pool boiling which occur in six regimes as shown in the fig. Let us have a plot between log of heat flux $q = (Q/A)$ and excess temperature $\Delta T = T_w - T_s$. The plot is known as boiling curve and is obtained by measuring heat input and temperature on an electrically heated platinum resistance wire submerged in water. The three types of boiling are: (1) Interface evaporation (2) Nucleate boiling and (3) Film boiling

The different regimes are shown in Fig. 17.55 and are mentioned below as:

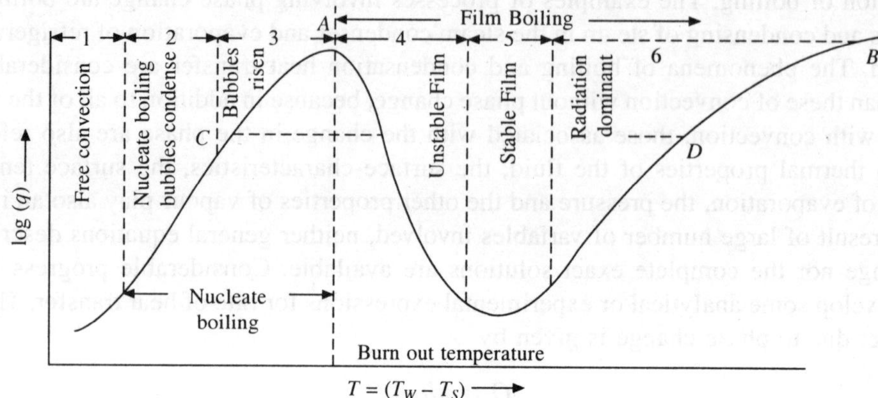


**Fig. 17.55   Regimes of Pool Boiling**

I. Free convection or interface evaporation: As the wall temperature $T_w$ is raised $\Delta T = T_w - T_s$ is increased, convection currents circulate the liquid and the steam is produced by the evaporation at the liquid surface. Here, only liquid is in contact with the heated surface and the heat transfer is only due to free convection.

II. Nucleate boiling with Bubbles condense: With further increase in $\Delta T$, the energy level of the liquid adjacent to the surface becomes high at a number of favoured spots where vapour bubbles are formed. They rise above the metal surface but condense before reaching the liquid surface. This is known as regimes 2.

III. As the temperature is further increased to point $A$, the bubbles become more numerous. Now the liquid is hot so as not allow any condensation of the bubbles which rise to the free liquid surface and help rapid evaporation. This is called regimes 3 and regimes 2 and 3 together are called *nucleate boiling*.

IV. Beyond the point A, representing the critical heat flux where is maximum heat flux during boiling, further increasing in excess temperature results in formation of vapour film attached with the surface of the wire. The film attaches the surface making it insulated. It counter acts the agitation due to the bubbles and decreases the heat flux. The vapour film is not stable therefore it is called *unstable film boiling* which is regime 4.

V. As further temperature is increased, the film becomes stable and further decreases the rate of heat transfer. This vapour film is so thick that heat transfer becomes minimum.

VI. As the temperature is further increased, radiation becomes more pronounced at high temperature and film breaks, the rate of heat transfer increase.

*Burn out Point:*   At the critical heat flux, rate of heat transfer is maximum beyond this point, boiling is unstable unless we proceed up to point $B$, where the metal surface temperature may be high above the melting point. Within the point $A$ and $B$ of the curve the boiling process cannot remove heat equal to the heat input corresponding to wall temperature. This increases further the value of $T_w$ which in turn decreases the heat flux, thus if heat supplied is not suddenly decreased, the metal may burn out or melt. That is why $A$ is called burn out point. The exact knowledge about this point has assumed considerable importance in high heat flux instruments.

*Correlation for Pool Boiling* In forced convection the heat flow is primarily caused by mixing of hot and cold fluid and the relation is $Nu = f(Re, Pr)$. In boiling, the heat is transferred from the wall to the liquid surface by vapour bubbles as well as by the intense turbulence caused by these bubbles. However, photographic observations has shown that the latent heat required for the formation of the bubbles produced is much smaller than the actual heat transferred. For the relations of nucleate boiling. Conventional Reynolds number is replaced by bubble Reynolds number defined as $Re_b = \dfrac{D_b \cdot G_b}{\mu_L}$ where $D_b =$ bubble dia. $G_b. =$ mass velocity of bubbles per unit area. $N_{u_b} = \dfrac{h D_b}{K} = f(Re_b, Pr)$ Different impirical relations have been proposed for pool boiling.

### 17.6.3 Dropwise and Filmwise Condensation

When a saturated vapour comes in contact with a cold surface, it condenses and liquid droplets are formed on the surface. These droplets, if they do not have any affinity with the surface, the instead of depositing themselves on the surface they may drop from it leaving the surface bare for successive droplets to form. When condensation takes place with this mechanism, it is called dropwise condensation. On the contrary, usually there is affinity between the condensate formed and the surface. Droplets in this case form a film on the surface, and the mechanism is called film wise condensation.

In filmwise condensation, the condensed film acts as an additional resistance to heat flow. Therefore, heat transfer coefficient for dropwise condensation is four to eight times those for filmwise condensation. Steam is the only vapour known to condense in dropwise manner under certain special circumstances. These include the presence of a contaminant such as dirt, oil or a material on the surface which promotes dropwise condensation. The phenomena of filmwise condensation is amenable to mathematical analysis. However, the film may be laminar or turbulent. Critical Reynolds number for transtic from Laminar to turbulent film in condensation is observed to be 1800. Nusselt has derived a boundary layer equations for film wise condensation which is however, beyond the scope of this book.

## 17.7 MASS TRANSFER

### 17.7.1 Introduction

The transport of one constituent of a fluid solution from a region of higher concentration to a region of lower concentration is called mass transfer. The study of mass transfer carries special importance in energy transfer because all heat transfer processes within one fluid or within different fluids which are mixing with each other are asssociated with mass transfer. There are a number of examples of industrial importance where mass transfer is encountered. Some of the examples are:

1. Evaporation of water in atmospheric air.
2. Humidification of air in cooling tower and dehumidification of air in summer air conditioning.

3. Evaporation of petrol in the carburettor of petrol engine.
4. Refrigeration by the evaporation of liquid ammonia in the atmoshpere of hydrogen in Elextrolux refrigerator.
5. Diffusion of neutron within the nuclear reactor.
6. Diffusion of smokes from the tall chimney in the atmosphere.
7. Diffusion of injected medicine in the human body.
8. In gas absorption, a soluble gas is removed from a gaseous mixture with an insoluble gas by transfer to a liquid phase.
9. In adsorption, one constituent of a fluid phase is transferred to the surface of a solid adsorbent.
10. In distillation, mass transfer takes place simultaneously in two directions, from the liquid to the vapour and vice versa.
11. Leaching or a solid-liquid extraction is an operation in which a soluble component of a solid phase is dissolved and transferred to a liquid phase.

When a system contains one or more components whose concentrations vary from point to point, there is a natural tendency for transfer of mass from higher concentration to lower concentration.

### 17.7.2  Analogy Between Heat and Mass Transfer

1. Heat is transferred in a direction in which negative temperature gradient exists, mass is transferred towards the lower concentration, decreasing the concentration gradient.
2. Heat transfer ceases when there is no longer a temperature difference, mass transfer ceases when concentration gradient reduces to zero.
3. The rate of both heat and mass transfer depends upon the driving potential and resistance.

### 17.7.3  Modes of Mass Transfer
There are the following modes of mass transfer

(a) *Mass Transfer by molecular diffusion.* This is similar to heat conduction. Molecular diffusion occurs in a gas as a result of the random motion of the molecules. It may occur in a stagnant fluid or in a fluid in laminar flow.
(b) *Eddy Diffusion.* This is similar to convective heat transfer, therefore it is also called convective mass transfer. It occurs when there is a motion.
(c) Mass Transfer during phase change

### 17.4.7  Mass Transfer by Molecular Diffusion
*Fick's Law of Diffusion* - Fick's law relates the diffusion rate or mass flux of a specy to its driving potential. This is similar to Fourier's law of heat conduction. Fick's law can be written as:

$$\frac{m_A}{A} = - D_{AB} \frac{\partial C_A}{\partial x}$$

(17.58)

$\dfrac{m_A}{A}$ = Mass flow rate of a component per unit area in kg/hr.m$^2$.

$C_A$ = mass concentration of component $A$ per unit volume in kg/m$^3$

$D_{AB}$ is called diffusion coefficient or diffusivity in m²/hr. There is a close similarity between Fick's law of diffusion for mass transfer, Fourier's law of heat conduction and Newton's law of viscosity for momentum transfer.

$$\text{Fourier's Law } \frac{Q}{A} = -k\frac{\partial T}{\partial x} = -\alpha\frac{\partial}{\partial x}(\rho C_p T) \tag{17.59}$$

$$\text{Newton's Law of Viscosity, } \quad \frac{F}{A} = \tau = -\mu\frac{\partial u}{\partial y} = -v\frac{(\partial \rho u)}{\partial y} \tag{17.60}$$

$D$, $v$ and $\alpha$ have same dimensions and are called, diffusivity, momentum diffusivity and thermal diffusivity.

### 17.7.5 Fick's Law in terms of Partial Pressures

Fick's law can be expressed conveniently in terms of partial pressures by making use of perfect gas relation $pV = nR_oT$ where $R_o$ = universal gas constant = 8318 Joul/kg mole k

or,

$$pV = \frac{m}{M} R_o T$$

$\therefore$

$$\rho = \frac{m}{V} = \frac{pM}{R_o T}$$

where $\rho$ is the mass density and $M$ the molecular weight. Therefore concentration of any component

$A$ is

$$C_A = \rho_A = \frac{p_A M_A}{R_o T} \tag{17.61}$$

$\therefore$

$$\frac{m_A}{A} = -D_{AB}\frac{\partial \rho_A}{\partial x} = -D_{AB}\frac{\partial}{\partial x}\frac{(p_A M_A)}{R_o T}$$

or,

$$\frac{m_A}{A} = -\frac{D_{AB} M_A}{R_o T}\frac{\partial p_A}{\partial x}$$

$$= -D_{AB}\frac{M_A}{R_o T}\frac{\partial p_A}{\partial x} \quad \text{or,} \quad \boxed{\frac{m_A}{A} = -D_{AB}\frac{M_A}{R_o T}\frac{\partial p_A}{\partial x}} \tag{17.62}$$

similarly if $B$ diffuses in $A$, $\quad \boxed{\frac{m_B}{A} = -D_{BA}\frac{M_B}{R_o T}\frac{\partial p_B}{\partial x}} \tag{17.63}$

### 17.7.6 Equimolar Counter Diffusion

In equimolar counter diffusion gases $A$ and $B$ diffuse simultaneously in a opposite directions (Fig. 17.56), through each other, the rates of no. of moles of diffusion of both the gases are equal but in opposite direction i.e. $N_A = -N_B$. This situation has no counterpart in heat transfer, since heat can be transferred only in one direction at a time. Of course, gas $B$ will be transferred only if a concentration gradient for $B$ exits.

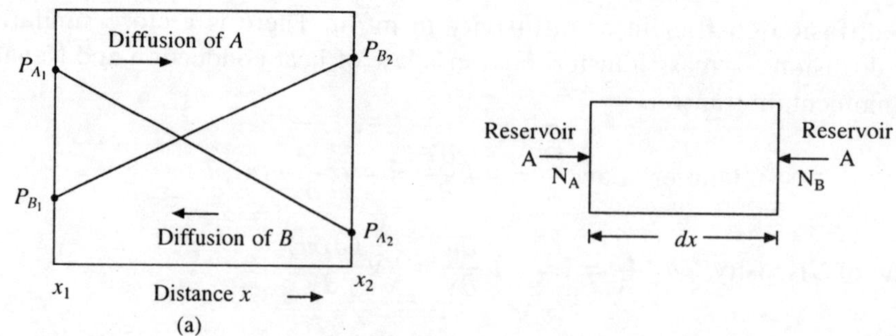

**Fig. 17.56 (a), (b) Equimolar Diffusion**

Steady state molar diffusion rates of components $A$ and $B$ are

$$N_A = \frac{m_A}{M_A} = -D_{AB}\frac{A}{R_oT}\frac{dp_A}{dx} \quad \text{and} \quad N_B = \frac{m_B}{M_B} = -D_{BA}\frac{A}{R_oT}\frac{dp_B}{dx}$$

But the total pressure of the system is constant and is equal to sum of partial pressures

$$p = p_A + p_B = \text{constant}, \quad \text{or,} \quad \frac{dp_A}{dx} + \frac{dp_B}{dx} = 0$$

or,
$$\frac{dp_B}{dx} = -\frac{dp_A}{dx} \text{ For equimolal diffusion} \tag{17.64}$$

$$N_A = -N_B \text{ or,} \; -\frac{D_{AB}A}{R_oT}\frac{dp_A}{dx} = -\left\{-\frac{D_{BA}A}{R_oT}\frac{dp_B}{dx}\right\}$$

$$= \frac{D_{BA}A}{R_oT}\left(\frac{-dp_A}{dx}\right) \tag{17.65}$$

because $\dfrac{dp_B}{dx} = -\dfrac{dp_A}{dx}$

$\therefore D_{AB} = D_{BA} = D$. That is, diffusivity one with respect to other is same as diffusivity of second with respect to first.

$$\text{Molar flux} = \frac{N_A}{A} = -\frac{D}{R_oT}\frac{p_{A_2} - p_{A_1}}{\Delta x} \tag{17.66}$$

### 17.7.7   Diffusion of Gas $A$ Through a Stationary Gas $B$

Consider a gas $A$ diffusing into a stationary gas $B$. The best example is simple device called stefan tube to measure the diffusivity of water vapour in air. It consists of a vertical tube containing a liquid, say water, at the bottom. The free surface of liquid is exposed to a gas B, say air. The liquid evaporates and diffuses into gas as a result of partial pressure difference of vapour. The level of the liquid can be maintained constant by constant supply of liquid. We assume that the system is isothermal and total pressure is constant. Since the water vapour must be removed at the top by air. However, the air motion should be such as would not cause any turbulence (Fig. 17.57).

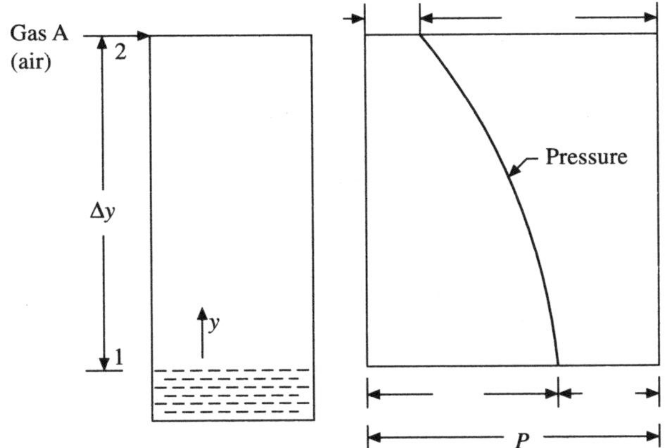


**Fig. 17.57   Stefan tube and pressure distribution**

Since the water vapour diffuses upward, the air must diffuse downward to maintain constant concentration at any position $y$. But at the surface of water, the air velocity must be zero. As a result, there must be bulk mass movement upward with a velocity just large enough to balance the diffusion of air downward.

The downward diffusion of air is given by

$$m_A = -\frac{DAM_A}{R_o T}\frac{dp_A}{dy}$$

(17.67)

This has to be balanced by the bulk mass transfer or air upward

$$-\rho_A Au = -\frac{P_A M_A}{R_o T} Au$$

(17.68)

where, $u$ = Air velocity

Equating the above expressions we have

$$u = \frac{D}{p_A}\frac{d_{P_A}}{dy}$$

(17.69)

The upward diffusion of water vapour is given by

$$m_w = -DA\frac{M_w}{R_o T}\frac{dp_w}{dy}$$

(17.70)

The bulk transport of water vapour is

$$\rho_w Au = \frac{\rho_w M_w}{R_o T} Au$$

(17.71)

The total mass transfer of water vapour is obtained by adding the values in Eqs. (17.70) and (17.71). Thus

$$(m_w)_{\text{total}} = -\frac{DAM_w}{R_oT}\frac{dp_w}{dy} + \frac{p_w M_w}{R_oT}\frac{AD}{P_A}\frac{dp_A}{dy} \tag{17.72}$$

where $u$ has been substituted from Eq. (17.69). According to the Dalton's law of partial pressure, total pressure $P = P_A + P_w = \text{constant}$

$\therefore$
$$\frac{dp_A}{dy} = -\frac{dp_w}{dy} \tag{17.73}$$

Substituting this value in Eq. (17.72), we have

$$(m_w)_{\text{total}} = \frac{DAM_w}{R_oT}\frac{P}{P - P_w}\frac{dp_w}{dy} \tag{17.74}$$

This relation is known as Stefan's law. Integrating

$$m_w \int_{y_1}^{y_2} dy = -\frac{DAM_w P}{R_oT}\int_{P_{w_1}}^{P_{w_2}}\frac{d_{P_w}}{P - P_w}$$

or,
$$m_w(y_2 - y_1) = +\frac{DA}{R_oT}M_w P\int_{P_{w_1}}^{P_{w_2}}\frac{d_{P_w}}{P_w - P}$$

$$= \frac{DA}{R_oT}M_w P \ln\left(\frac{P_{w_2} - P}{P_{w_1} - P}\right)$$

$$= \frac{DA}{R_oT}M_w P \ln\left(\frac{P - P_{w_2}}{P - P_{w_1}}\right)$$

$\therefore$
$$(m_w)_{\text{total}} = \frac{DA}{R_oT}\frac{M_w P}{\Delta y}\ln\left(\frac{P - P_{w_2}}{P - P_{w_1}}\right)$$

$$= \frac{DAM_w P}{R_oT\Delta y}\ln\frac{P_{A_2}}{P_{A_1}}$$

or,
$$(m_w)_{\text{total}} = N_w M_w \tag{17.75}$$

This is very common for the diffusion of water vapour into air. Absorption of ammonia in water occurs in ammonia absorption refrigeration system, is the example of diffusion of $B$ through non-diffusing $C$. The equation developed will be most useful for the above system to analyse it.
Some Important Results

Newton's law of viscosity: $\tau = \mu\dfrac{dx}{dy} = -v\left(\dfrac{d\rho u}{dy}\right)$

Fourier's law of heat conduction $\dfrac{Q}{A} = q = -k\dfrac{dT}{dy} = -\alpha\dfrac{d}{dy}(\rho C_P T)$

where
$$\alpha = \frac{k}{\rho c_p}$$

Fick's law of diffusion: $\dfrac{m}{A} = -D\dfrac{dc}{dy}$.

These equations are identical and therefore the mechanism of heat, mass and momentum transfer is also similar. Shear stress vanishes if velocity gradient is zero and similarly heat and mass transfer ceases if temperature and concentration gradient is zero.

## Solved Examples (Mass Transfer)

**17.53** Estimate the diffusion rate of water from the bottom of a test tube 1.5 cm in diameter and 15 cm long into dry atmosphere at 25°c. Take diffusivity of water in air as 0.256 cm$^2$/sec. and saturation pressure of water at 25°C = 0.0316 bar.

*Soln.*

Area $= A = \dfrac{\pi}{4}(1.5)^2 \times 10^{-4}$ m$^2$, $D_{AB} = 0.256$ cm$^2$/sec $= 0.256 \times 10^{-4} \times 3600$ m$^2$/hr $= 0.09216$ m$^2$/hr.

Denoting component $A$ by water ($w$) and $B$ by air (a) $T = 273 + 25 = 298$K

$$m_A = \frac{M_A A D_{AB} P_t}{RT\,\Delta y} \ln\left(\frac{P_t - P_{A2}}{P_t - P_{A_1}}\right), \ P_t = 1.01325 \text{ bar}$$

$$P_{w_2} = 0, \ P_{w_1} = 0.0316 \text{ bar}$$

Rate of evaporation of water

$$m_w = \frac{M_w A D_{wa} P_t}{RT\,\Delta y} \ln\frac{P_t - P_{w2}}{p_t - p_{w1}}$$

$$= 18 \times \frac{\pi}{4}(1.5)^2 \times \frac{10^{-4} \times 0.09216 \times 1.01325 \times 10^5}{287 \times 298 \times 0.15} \ln\left(\frac{1.01325 - 0}{1.01325 - 0.0316}\right)$$

$$= 7.335 \times 10^{-5} \text{ kg/hr} \quad Ans.$$

**17.54** An open circular tank of 7.6 m diameter contains a volatile solvent at 25°C. The top of the surface of liquid is covered by a stagnant air film of 5 cm thick. The concentration of solvent vapour beyond thin film may be neglected. If the solvent is worth 20/-per litre, estimate the loss per day. Data: vapour pressure of solvent at 25°C = 100 mm of Hg. sp. gr. of solvent = 0.88, Molecular weight = 78, diffusivity in air = 0.09 cm$^2$/sec.

*Soln.*

$$\Delta y = 5 \text{ cm} = 5 \times 10^{-2} \text{m}, \ D = 0.09 \times 10^{-4} \times 3600 \text{ m}^2/\text{hr}$$

$$M = 78, \ p_t = 760 \text{ mm of Hg} = 1.01325 \text{ bar}$$

Rate of mass transfer $= m_A = \dfrac{M_A A D_{AB} P_t}{RT\,\Delta y} \ln\left(\dfrac{P_t - P_{A2}}{P_t - P_{A_1}}\right)$

$$= \frac{78 \times \frac{\pi}{4}(7.6)^2 \times 0.09 \times 10^{-4} \times 3600 \times 1.01325 \times 10^5}{287 \times (273 + 25) \times 5 \times 10^{-2}} \times \ln\left(\frac{760 - 0}{760 - 100}\right)$$

$$= \frac{383.23 \times 24 \times 1000}{880} \text{ litre per day} = 10451.727 \text{ lit/day}$$

Loss of Rupees per day $= 10451.727 \times 20 = 209034.55$ *Ans.*

**17.55** Oxygen is diffusing through carbon monoxide with carbon monoxide non-diffusing. The total pressure is 1 atmospheri  and temperature is 0°C. The partial pressure of oxygen at two planes 0.2 cm apart is 100 and 50 mm of Hg respectively. The diffusivity of mixture is $1.85 \times 10^{-5} \text{m}^2$/sec. Calculate the rate of diffusion of oxygen in kg mole/m$^2$ sec through each square cm of two planes.

*Soln.*

$$\frac{m_A}{m_A A} = \frac{N_A}{A} = \frac{D_{AB} P_t}{RT \Delta y} \ln \frac{(p_t - p_{A2})}{(p_t - p_{A_1})}$$

Here $p_t = 1.01325$ bar $= 1.01325 \times 10^5$ N/m$^2$, $R = 287$ Nm/kg mole $= 8314$ N-m/ kg mole K, $\Delta y = 0.2$ cm

$$\therefore \quad \frac{N_A}{A} = \frac{1.85 \times 10^{-5} \times 1.01325 \times 10^5 \ \ln \dfrac{760 - 50}{760 - 100}}{8314 \times 273 \times 0.2 \times 10^{-4}} = 3.01 \times 10^{-3} \text{ kg mole/sec m}^2$$

**17.56** In an oxygen nitrogen mixture at 10 atmospheric pressure and 25°C, the concentration of oxygen at two planes of 0.2 cm apart are 10 and 20 percent by volume respectively. Calculate the rate of diffusion of oxygen expressed as kg/m$^2$ hr for the case of unicomponent diffusion (Nitrogen as non-diffusing). Value of diffusivity of oxygen is $1.81 \times 10^{-5}$ m$^2$/sec.

*Soln.*

$$\frac{N_A}{A} = \frac{D_{AB} \, p_t}{RT \Delta y} \ln \frac{p_t - p_{A_2}}{p_t - p_{A1}}$$

Here $D_{AB} = 1.81 \times 10^{-5}$ m$^2$/sec. $p_t = 10$ atmosphere $= 10 \times 1.01325$ bar $= 1.01325 \times 10^6$ N/m$^2$, $R = 8314$ N – m kg mole K

$$T = 273 + 25 = 298°K, \Delta y = 0.2 \text{ cm},$$

Partial pressure = mole fraction $\times$ total pressure

$$p_{A_1} = 0.20 \times 10 = 2 \text{ atm}, p_{A_2} = 0.1 \times 10 = 1 \text{ atm.}$$

$$\frac{N_A}{A} = \frac{1.81 \times 10^{-5} \times 1.01325 \times 10^6}{287 \times 298 \times 0.2 \times 10^{-4}} \ln \frac{10 - 1}{10 - 2} = 1.262 \times 10^{-8} \text{ kg mole/sec m}^2$$

$$= 1.262 \times 10^{-8} \times 32 \times 3600 \text{ kg/hr m}^2 = 1.454 \times 10^{-3} \text{kg/m}^2\text{hr}$$

**17.57** Determine the mass transfer coefficient of a certain vapour flowing over a flat plate 300 mm long at a Reynolds number of $2.15 \times 10^5$ when the kinematic viscosity and mass diffusivity are $1.68 \times 10^{-5}$ m$^2$/s and $2.173 \times 10^{-9}$ m$^2$/sec respectively.

*Soln.*

$$\text{Schmidth number} = Sc = \frac{v}{D} = \frac{1.68 \times 10^{-5} \text{ m}^2/\text{sec}}{2.173 \times 10^{-9} \text{ m}^2/\text{sec}} = 7.73 \times 10^3.$$

Average sherwood number is given by

$$\overline{Sh} = \frac{h_d\,L}{D} = 0.664\,R_e^{1/2}\,Sc^{1/3} = 0.664(2.15 \times 10^5)^{1/2}$$

$$\times\,(7.73 \times 10^3)^{1/3} = 6087$$

$$\therefore \qquad h_d = 6087 \times \frac{D}{L} = \frac{6087 \times 2.173 \times 10^{-9}}{300 \times 10^{-3}} = 4.40 \times 10^{-9}\ \text{m/sec} \quad Ans.$$

17.58  Air with a velocity of 3 m/sec is flowing over a tray of full of water. Assuming temperature of 20°C and temperature of water on the surface 15°C, determine the amount of water evaporated per hour. Length of the tray along the air flow direction is 30 cm and its width is 50 cm. Take total pressure of water as 1.00 bar and partial pressure of water vapour in it as 0.0078 bar. Properties of air are $\rho = 1.205$ kg/m$^3$, $C_p = 1.00$ kJ/kg k, $k = 0.025$ w/mk, $v = 15 \times 10^{-6}$ m$^2$/sec. $D = 0.15$ m$^2$/hr

*Soln.*

$$\text{Reynolds number} = \text{Re} = \frac{V\,L}{v} = \frac{3 \times 0.30}{15 \times 10^{-6}} = 0.6 \times 10^5\ \text{flow is laminar.}$$

$$Sc = \frac{15 \times 10^{-6} \times 3600}{0.15} = 0.36$$

$$\text{Sherwood number} = Sh = 0.664\,\text{Re}^{1/2}\,Sc^{1/3} = 0.664\,(0.6 \times 10^5)^{1/2}(0.36)^{1/3}$$

$$Sh = \frac{h_d\,L}{D}$$

$$\therefore \qquad h_d = \frac{D}{L} \times Sh = \frac{0.15}{0.30} \times 0.664\,(0.6 \times 10^5)^{1/2}\,(0.36)^{1/3} = 57.4\ \text{m}^2/\text{hrm}$$

$$\text{Rate of mass transfer} = m = Ah_d\,(C_1 - C_2) = Ah_d \times \frac{M}{RT}(p_1 - p_2)$$

$$= 0.30 \times 0.50 \times 57.4 \times 29 \times \frac{(1.00 - 0.0078)}{8314(273 + 15)} \times 10^5 = 10.34\ \text{kg/hour} \quad Ans.$$

17.59  An apparatus, shown in the Fig. to determine the diffusivity of water air-system, is enclosed in a constant temperature environment at 55°C and 1 atm. The air is circulated such a way that the concentration of water vapour in the air above the tube is zero and there is no convective mixing in the tube above the level of water. 290 hours are required for the water level to fall down from 12.5 cm to 15 cm below the top of the tube. What will be the diffusivity of water air system at 55°C? The molar flux of water $N_w$ is related to the rate at which the water level falls as $N_w = \rho_w \dfrac{dL}{dt}$ where $\overline{\rho}_w$ = molar density of liquid water, $t$ = time. Atmospheric pressure = 1.01325 bar. Saturation pressure of water vapour at 55°C = 0.153 bar. Gas constant = $R = 8314$ nm/kg mole k. (Ranchi Univ. 1984).

*Soln.*  Refer to Fig. 17.55.
We know that rate of mass transfer for water vapour in air

$$m_w = M_w\,\frac{AD_w\,a\,Pt}{RT\Delta y}\,\ln\frac{Pt - p_{w_2}}{p_t - p_{w_1}}$$

$$\text{Rate of molar flux} = N_w = \frac{m_w}{M_w\,A} = \frac{D_{wa}\,P_t}{RT\Delta y}\,\ln\!\left(\frac{p_t - p_{w2}}{p_t - p_{w1}}\right)$$

Where $D_{wa}$ = diffusivity of water in air. Here $\Delta y$ = length of free path

$$= L\,N_w = \frac{D_{wa}\,P_t}{RTL}\,\ln\!\left(\frac{p_t - o}{P_t - p_{w1}}\right) = \overline{\rho}_w\,\frac{dL}{dt}$$

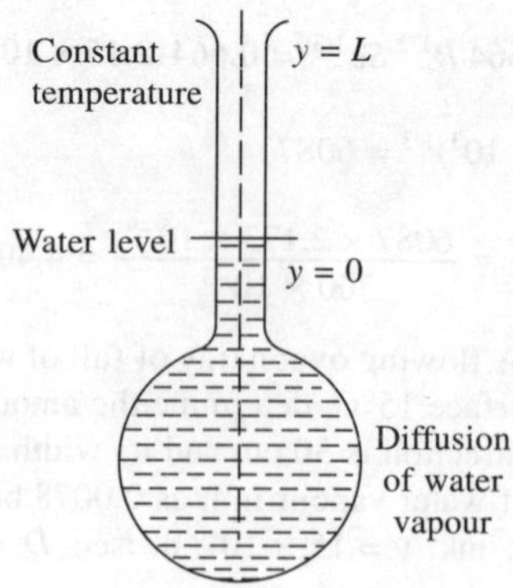


**Fig. 17.58**

because at the top $p_{w2} = 0$

or, $\quad \dfrac{D_{wa}\,P_t}{RT_o}\ln\left(\dfrac{p_t - 0}{p_t - p_{w1}}\right)\displaystyle\int_0^t dt = \int_{L_o}^{L}(L\,dL)$ where $L_o$ initial length

or, $\quad \dfrac{D_{wa}\,p_t}{RT_o}\ln\dfrac{P_t}{p_t - p_{w1}}\,t = \left(\dfrac{L^2 - L_o^2}{2}\right)\bar{\rho}_w$

$$D_{wa} = \frac{\bar{\rho}_w\,RT_o}{p_t \cdot t \times 2}\,\frac{(L^2 - L_o^2)}{\ln p_t/(p_t - p_{w_1})}$$

Molar density of water $\bar{\rho}_w = \dfrac{(1000)}{18}$

or, $\quad D_{wa} = \left(\dfrac{1000}{18}\right) \times \dfrac{8314 \times (273 + 55)}{1.01325 \times 10^5 \times 290 \times 2 \times}\dfrac{(0.15^2 - 0.125^2)}{\ln\left(\dfrac{1.01325}{1.01325 - 0.153}\right)}$

$$= \frac{18748070}{1731.622 \times 10^5} = 0.10827 \text{ m}^2/\text{hour}$$

**17.60** Air at 35°C and 1 atm flows at a velocity of 30 m/sec over a flat plate of 0.5 m long. Calculate average mass transfer coefficient of water vapour in air. Assume concentration of vapour in air as very very small. Take diffusion coefficient of water vapour in air as $D = 0.256 \times 10^{-4}$ m/sec. Properties of air at 35°C, $\rho = 1.146$ kg/m$^3$, $C_p = 1.006$ kJ/kg k' $\mu = 2 \times 10^{-5}$ kg/m sec.

*Soln.*

$$\text{Schmidth number} = S_c = \frac{v}{D} = \frac{\mu}{\rho D} = \frac{2 \times 10^{-5}}{1.146 \times 0.256 \times 10^{-4}} = 0.682$$

$$\text{Reynolds number} = \text{Re} = \frac{\rho u_\infty L}{\mu} = \frac{1.146 \times 30 \times 0.5}{2 \times 10^{-5}} = 8.596 \times 10^5$$

$\therefore$ Flow is turbulent.

For turbulent flow on a flat plate, average sherwood number is given by:

$$Sh = 0.036(\text{Re}_L)^{0.8}\,(Sc)^{1/3} = 0.036\,(8.596 \times 10^5)^{0.8}\,(0.682)^{1/3} = 1772$$

Now $\qquad Sh = \dfrac{h_d L}{D}$ $h_d = \dfrac{Sh \cdot D}{L} = 1772 \times \dfrac{0.256 \times 10^{-4}}{0.5} = 0.09074$ m/sec.

$\therefore$ Mass transfer coefficient = 0.09074 m/sec  *Ans.*

**17.61** Air at 50°C and 1 atm flows over the surface of a water reservoir at an average velocity of 2.3 m/sec. The water surface is 0.61 m long and 0.65 m wide. The water surface temperature is estimated to be 30°C. The relative humidity of air is 40 percent. The molecular weight of moist air can be taken as 29. The density of air is found to be 1.105 kg/m$^3$ and its viscosity is $1.943 \times 10^{-5}$ kg/m sec. Estimate the water vapour transfer per hour per square meter of water surface and state the direction of diffusion. Take diffusivity of water = $0.256 \times 10^{-4}$ m$^2$/s, saturation pressure of water vapour = $4.24 \times 10^3$ N/m$^2$ at 30°C and at 50°C, $12.34 \times 10^3$ N/m$^2$,

*Soln.*

Reynolds number = Re = $\dfrac{\rho U_\infty L}{\mu}$

$$= \dfrac{1.105 \times 2.3 \times 0.61}{1.943 \times 10^{-5}} = 7.979 \times 10^4 \ \text{whcih is less than } 5 \times 10^5$$

Therefore the flow is laminar

$$\text{Schmidth number} = Sc = \dfrac{\mu}{\rho D} = \dfrac{1.949 \times 10^{-5}}{1.105 \times 0.256 \times 10^{-4}} = 0.687$$

For laminar flow, average sherwood number

$$= Sh = 0.664 \, Re_L^{1/2} \, Sc^{1/3} = 0.664(7.979 \times 10^4)^{1/2}(0.687) = 165.50$$

$$Sh = \dfrac{h_d L}{D}, h_d = \dfrac{ShD}{L} = \dfrac{165.50 \times 256 \times 10^{-4}}{0.61} = 0.006945 \text{ m/sec}$$

Partial pressure of water vapour at the water surface = $p_{w_1}$ = $4.2 \times 10^3$, saturation pressure of water at 50° C = $12.34 \times 10^3$ N/m$^2$, Partial pressure of water vapour in air at 50°C and 40 percent relative humidity = $P_{w_2}$ = $0.4 \times 12.34 \times 10^3 = 4.936 \times 10^3$ N/m$^2$ Total pressure of air = $101.325 \times 10^3$ N/m$^2$

Concentration $\quad C_{w_1} = \rho_{w_1} = \dfrac{M_w}{R \cdot T_1} p_{w_1} = \dfrac{18 \times 4.24 \times 10^3}{8314 \times (273 + 30)} = 0.0303 \text{ kg/m}^3$

$$C_{w_2} = \dfrac{M_w P_{w_2}}{RT_2} = \dfrac{18 \times 4.936 \times 10^3}{8314 \times (273 + 50)} = 0.03308 \text{ kg/m}^3.$$

Here $C_{w_2} > C_{w_1}$. The flow is from air to water. Mass flux of water vapour = $\dfrac{m_w}{A} = h_d(c_{w_2} - c_{w_1}) =$ $0.006945(0.03308 - 0.0303) \times 3600 = 0.650$ kg/hrm$^2$ *Ans.*

## 17.7.8 Convective Mass Transfer

The mechanism of mass transfer by molecular diffusion is similar to heat transfer by conduction. In molecular diffusion, the bulk velocities are insignificant and only diffusion velocities are considered. The mechanism of mass transfer by convection is analogous to heat transfer by convection. In convective mass transfer, the bulk velocities are significant i.e., both the components in a binary mixtures are moving with an appreciable velocity.

In convective mass transfer, two situations arise likewise convective heat transfer. They are:

(a) Natural convection and
(b) Forced convection.

Forced convective mass transfer takes place as a result of circulation currents imparted by mechanical devices. Whereas in free convection mass transfer circulation currents are created by density difference due to buoyancy forces in the vapour gas mixture of varying compositions.

### 17.7.9 Mass Transfer Coefficient ($h_m$)

If $C_{A_w}$ and $C_{A_\infty}$ are the concentrations of a component A on a wall and far away from the wall, the rate of mass transfer is given by:

$$m_A = h_m A(C_{A_w} - C_{A_\infty}) \qquad (17.76)$$

his equation is similar to convective heat transfer

$$Q = hA(T_w - T_\infty)$$

The concentration of a component gas $A$ can be represented by its density as $C_A = \rho_A = \dfrac{M_A\, P_A}{R_o T}$, where $P_A$ = partial pressure of $A$

$$\therefore \qquad \text{mass transfer } m = Ah_m \frac{M_A}{R_o T}(p_{A_w} - P_{A_\infty}) = Ah_m \frac{M_A}{R_o T}\Delta p_A \qquad (17.77)$$

unit of mass transfer cofficient is meter/second

### 17.7.10 Analogy Between Momentum, Heat and Mass Transfer

We have seen that the mathematical laws governing momentum, heat and mass transfer are similar. The boundary layer equations for a flat plate are:

Continuity:
$$\frac{\partial u}{\partial x} + \frac{\partial v}{\partial y} = 0 \qquad (17.78)$$

Momentum:
$$u\frac{\partial u}{\partial x} + v\frac{\partial u}{\partial y} = v\frac{\partial^2 u}{\partial y^2} \qquad (17.79)$$

Energy:
$$u\frac{\partial T}{\partial x} + v\frac{\partial T}{\partial y} = \alpha\frac{\partial^2 T}{\partial y^2} \qquad (17.80)$$

Mass (concentration):
$$u\frac{\partial C_A}{\partial x} + v\frac{\partial C_A}{\partial y} = D_{AB}\frac{\partial^2 C_A}{\partial y^2} \qquad (17.81)$$

The form of the last three equations is exactly similar. To illustrate the analogy further, the following equations can be written for flux for the three shear stresses

$$\frac{F}{A} = \tau = -v\frac{\partial}{\partial y}(\rho_u) \qquad (17.82)$$

$$\text{Heat flux} = \frac{Q}{A} = -k\frac{\partial T}{\partial y} = -\alpha\frac{\partial}{\partial y}(C_p T) \qquad (17.83)$$

$$\text{Mass flux} = \frac{m_A}{A} = -D\frac{\partial C_A}{\partial y} \qquad (17.84)$$

### 17.7.11  Drawing the similarity the following dimensionless number are introduced

1. *Prandtl number* $P_r = \left(\dfrac{v}{\alpha}\right) = \dfrac{v}{K}\rho_{C_P} = \dfrac{\mu C_P}{K}$: This is the ratio of momentum diffusivity to thermal diffusivity If $v = \alpha$ i.e. $P_r = 1$, velocity and temperature profiles are same.

2. Schemidth number $\left(S_c = \dfrac{v}{D}\right)$: This is the ratio of momentum diffusivity to mass diffusivity. If $S_c = 1$, i.e. $v = D$ then velocity and concentration profiles are same. Rate of mass transfer increases with increase of schmidth number.

3. Lewis number $\left(L_e = \dfrac{\alpha}{D}\right)$: It is the ratio of thermal diffusity to mass diffusivity. If $L_e = 1$ i.e., $\alpha = D$, then temperature and concentration profiles are similar.
   If $P_r = S_c = L_e = 1$, all the three profiles are of boundary layers coincide.

4. Sherwood number $\left(Sh = \dfrac{h_M \cdot L}{D}\right)$: This number is similar to Nusselt number. This is the ratio of concentration gradient near the wall to the reference concentration gradient $\dfrac{C_{A_w} - C_{A_\infty}}{L}$. Rate of mass transfer increases with increase of sherwood number.

5. Similar to stanton number $= St = \dfrac{N_u}{ReP_r} = \dfrac{\dfrac{hL}{K}}{\dfrac{\rho u_\infty L}{\mu}\dfrac{\mu L_p}{K}} = \dfrac{h}{\rho U_\infty}C_p$ we have a dimensionless

   quantity $= \dfrac{Sh}{ReS_c} = \dfrac{h_m}{U_\infty}$

### 17.7.12  Relations for Mass Transfer

We prove that for forced convection heat transfer correlations are of the form, $N_u = f(R_e, P_r)$. Likewise in forced convection mass transfer problems, we write the functional relation

$$Sh = f(Re, Sc) \qquad (17.85)$$

Grashof number for mass transfer is defined as

$$G_r = g\beta_m(C_{A_w} - C_{A_\infty}) \times \frac{x^3}{v^2} = g\frac{(\rho_\infty - \rho_w)}{v^2 \rho_\infty} \qquad (17.86)$$

For natural mass transfer $Sh = f(G_{r_M}, Sc)$

For laminar flow on flat plate, local sherwood number

$$Sh_x = \frac{h_m x}{D} = 0.332\,(Re_x)^{1/2}\,(sc)^{1/3} \tag{17.87}$$

and average sherwood number $= Sh = 0.664\,(Re)^{1/2}(sc)^{1/3}$. For turbulent flow on flat plate local and average sherwood number are

$$Sh_x = 0.0298\,(Re_x)^{0.8}\,(sc)^{1/3} \tag{17.88}$$

$$Sh = 0.036\,(Re_L)^{0.8}\,(Sc)^{1/3} \tag{17.89}$$

For forced convection inside tubes and ducts $Sh = 0.023\,(Re)^{0.8}\,(Sc)^{0.4}$ $\tag{17.90}$

### 17.7.13  Colburn Analogy
For convective heat transfer

$$St\,P_r^{2/3} = \frac{c_f}{2} \quad \text{or,} \quad \frac{Nu}{Re\,P_r}\,P_r^{2/3} = \frac{C_f}{2}$$

Similarly for mass transfer

or, $$\left(\frac{h}{c_p V}\right)P_r^{2/3} = \frac{C_f}{2} \tag{17.91}$$

$$\frac{Sh}{ReSc}(Sc)^{2/3} = \frac{C_f}{2} \quad \text{or,} \quad \frac{h_M\,L\,(Sc)^{2/3}}{D\dfrac{\rho VL}{\mu}\dfrac{\mu}{\rho D}} = \frac{c_f}{2} \quad \text{or,} \quad \left(\frac{hm}{V}\right)Sc^{2/3} = \frac{C_f}{2} \tag{17.92}$$

From (17.91) and (17.92)

$$\frac{h}{c_p v}(pr)^{2/3} = \frac{hm}{v}Sc^{2/3} \quad \text{or,} \quad \frac{h}{h_m} = \rho c_p\left(\frac{Sc}{pr}\right)^{2/3}$$

or, $$\frac{h}{h_m} = \rho c_p\left(\frac{v}{o}\Big/\frac{v}{\alpha}\right)^{2/3} = \rho c_p(\alpha/D)^{4/3} \tag{17.93}$$

**EXERCISES**

## Conduction

17.1  Objective questions

(i)  The heat transfer takes place according to (a) First law of thermodynamics (b) Zeroth law of thermodynamics (c) Second law of thermodynamics (d) Fourier's law.

(ii)  In a slab under steady conduction along the thickness direction if the thermal conductivity increases along the thickness, the temperature gradient along the direction will become (a) Steeper (b) flatter (c) will depend upon the heat flow (d) will remain constant.

(iii)  In steady state heat conduction in the x-direction, the sectional area increases along the flow direction. Then the temperature gradient in the $x =$ direction will (a) Remain constant (b) will become flatter (c) will become steeper (d) either b or c depending on the heat flow rate.

 (iv) The rate of heat transfer from a solid surface to a fluid is obtained from (a) Newton's law of cooling (b) Fourier's law (c) Kirchoff's law (d) Stefan's law

 (v) Fourier's law is based on assumption that (a) heat flow through a solid is one-dimensional (b) heat flow is in steady-state (c) both (a) and (b) (d) none of the above.

 (vi) For steady flow and constant value of conductivity, the temperature distribution for a plane wall is (a) Parabolic (b) linear (c) logarithmic (d) cubic.

 (vii) For steady flow and constant value of conductivity, the temperature distribution for a hollow cylinder of radii $r_1$ and $r_2$ is (a) Parabolic (b) linear (c) logarithmic function of radii (d) cubic.

 (viii) The rate of heat transfer through a hollow cylinder of radii $r_1$ and $r_2$ depends on (a) difference of radii $(r_2 - r_1)$ (b) ratio of radii $\left(\dfrac{r_2}{r_1}\right)$ (c) product of radii $(r_2 \times r_1)$ (d) sum of radii $(r_2 + r_1)$.

 (ix) The critical radius of insulation for sphere is equal to

  (a) 2k/h (b) $k \times h$ (c) $\sqrt{K \times h}$ (d) $h/k$.

 (x) Choose the correct statement

  (a) Fourier's law of heat conduction gives the heat flow for two dimensional cases.

  (b) Thermal conductivity of air increases with decrease in temperature.

  (c) Thermal conductivity of solids increases with rise in temperature.

  (d) The unit of thermal conductivity in S.I. units is *w/mk*,

**Answers to objective questions**

|  |  |  |  |  |
|---|---|---|---|---|
| (i) c | (ii) b | (iii) a | (iv) a | (v) c |
| (vi) b | (vii) c | (viii) b | (ix) a | (x) d |

17.2 What is meant by thermal resistance? Explain the electrical analogy for solving heat transfer problems.

17.3 Differentiate between thermal conductivity and thermal diffusivity.

17.4 Discuss critical thickness of insulation and its inportance in engineering practice.

17.5 What is critical thickness of insulation on a small diameter wire or pipe. Explain its physical significance and derive an expression for the same.

17.6 Draw the temperature gradient through a plane wall when the thermal conductivity

 (i) remains constant with increase in temperature

 (ii) increases with increase in temerature

 (iii) decreases with increase in temperature.

17.7 Derive an expression for temperature distribution in a slab of thickness 'b' when its two faces are at temperatures $t_1$ and $t_2$, the thermal conductivity varies linearly with temperature according to $k = k_o (1 + at)$, where 'a' is a constant. Assume one dimensional steady state heat conduction with no heat generation.

17.8 Insulation boards are made up of three layers of materials of conductivities $k_1$, $k_2$ and $k_3$ of thickness $x_1$, $x_2$, and $x_3$ respectively. They are botted together by metal bolts of cross-section area $A_1$ m$^2$ per m$^2$ of board area. Metal conductivity is k$_4$. If temperatures on either side of board are $t_1$ and $t_4$, determine an expression to find the heat flow per m$^2$ of area of board. (Use thermal circuit analysis)

(Mumbai Univ. 1999 Winter)

17.9 The insdie temperature of furnace wall 200 mm thick is 1350°C. The mean thermal conductivity of wall material is 1.35 w/mk. The heat transfer Co-efficient of the outside surface is a function of temperature difference is given by $h = 7.85 + 0.08\ \Delta t$. Where $\Delta t$ is the temperature difference between outside wall surface and surroundings. Determine the rate of heat transfer per unit area if the surrounding temperature is 40°C.

(Mumbai Univ. 2000 Summer)

17.10 A furnace wall is made of composite wall of total thickness 55 cm. The inside layer is made of refractory material of thermal conductivity 2.3 w/mk and outside layer is made of an insulating material of thermal conductivity 0.2 w/mk. The mean temperature of the gases inside the furnace is 900°C and the interface temperature is 520°C. The heat transfer co-efficient between the gases and inner surface can be taken as

230 w/m²k and between the outside surface and atmosphere as 46 w/m²k. Assuming the temperature of surrounding air as 30°C, calculate:

   (i)  required layer thickness of each

  (ii)  the rate of heat loss per unit area

 (iii)  the temperature of the surface exposed to gases and the surface exposed to atmosphere.

(Mumbai Univ. 1998 Winter).

**17.11**  A solar collector receives 880 w/m². Its surface temperature is 60°C. The back side is to be insulated so that back losses are limited to 15%. Insulating material with thermal conductivity of 0.05 w/m²K is available. The atmospheric temperature is 30°C and the convection Co-efficient on the back side is 5 w/m²K. Determine the insulation thickness.

**17.12**  Determine the temperatures at all faces for the arrangement shown in Fig.

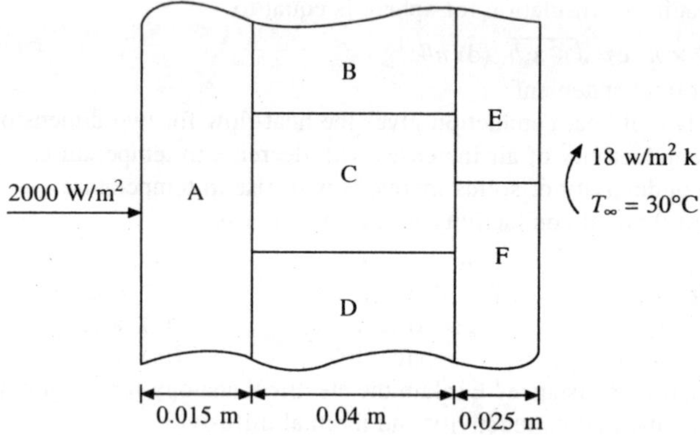


| A — full area | $k_A = 0.82$ w/mK, | B — 20% area, | $k_B = 0.3$ w/mK |
|---|---|---|---|
| C — 50% area | $k_c = 0.08$ w/mK, | D — 30% area, | $k_D = 0.14$ w/mK |
| E — 40% area | $k_E = 1.2$ w/mK, | F — 60% area, | $k_F = 2.3$ w/mK |

**17.13**  A long hollow cylinder has inner and outer radii as 5 cm and 15 cm respectively. It generates heat at the rate of 1.0 kw/m³. If maximum temperature occurs at radius 10 cm and temperature of the outer surface is 50°C, find-

   (i)  temperature of the inner surface

  (ii)  maximum temperature in the cylinder

(Mumbai Univ. 1996 Winter)

**17.14**  A 200 mm ID pipe carries superheated steam at 210°C. The thermal conductivity of the pipe material is 12.5 w/mK. The outside insulating layer has a thermal conductivity of 0.2 w/mK. The mean temperature at the interface is 85°C. The heat transfer Co-efficient between steam and pipe wall can be taken as 60 w/m²K and between the outer surface and ambient air as 35 w/m²k. Assuming the total thickness of the pipe (including pipe material) is 100 mm, the ambient air as 30°C. Calculate-

   (i)  The required thickness of each layer

  (ii)  Rate of heat transfer per unit outer area

 (iii)  The inner and outer surface temperatures

(Mumbai Univ. 1999 Winter).

**17.15**  A hollow conductor with $r_1 = 0.6$ cm, $r_2 = 0.8$ cm is made up of metal of thermal conductivity 20 w/mK and electrical resistance per metre of $3 \times 10^{-2}$ Ohms. Find the maximum allowable current if the temperature is not to exceed 50°C anywhere in the conductor. The cooling fluid at the inside is 38°C.

(Mumbai Univ. 1999 Summer)

**17.16**  The end of a very long cylindrical rod is attached to a heated wall and its surface is in contact with a cold fluid. If the rod diameter were doubled, by what percentage would the heat transfer change.

(Mumbai Univ. 1999 Summer)

17.17  A spherical shell, of radii $r_1$ and $r_2$ is made of material with thermal conductivity $K = k_o T^2$. Derive an expression for the heat transfer rate if the surfaces are held at temperatures $T_1$ and $T_2$ respectively. (Mumbai Univ. 1999 Summer).

17.18  A steel ball 5 cm diameter initially at a uniform temperature of 450°C is suddenly placed in an environment at 100°C. The convective heat transfer Co-efficient between steel ball and the fluid is 10 w/m²k. for steel

$$c_p = 0.46 \text{ kJ/kg k}$$

$$\rho = 7800 \text{ kg/m}^3$$

and $$k = 35 \text{ w/mk}$$

calculate the time required for the ball to reach a temperature of 150°C.

(Mumbai Univ. 1997 Summer)

17.19  A thermocouple junction is in the form of a 8 mm diameter sphere. This junction is initially at 28°C and is inserted into a steam of hot air at 300°C,

Find

   (i)  the time constant of the thermocouple

  (ii)  the thermocouple is taken out from the hot stream after keeping it there for 10 second and kept in still ambinet air at 30°C. If the heat transfer Co-efficient between ambient air and the thermocouple is 10 W/m²K, find the temperature of thermocouple junction 15 seconds after removing from hot air. For thermocouple material

$$c_p = 420 \text{ J/kgk}, \ k = 40 \text{w/mk}$$

$$\rho = 800 \text{ kg/m}^3, \ h = 45 \text{ w/m}^2\text{k}$$

(Mumbai Univ. 1999 Winter)

17.20  A wire of 8 mm diameter at a temperature of 60°C is to be insulated by a material having $k = 0.174$ w/mk, convective heat transfer Co-efficient $h = 8$w/m²k. The ambient temperature is 25°C. For maximum heat loss, what is the minimum thickness of insulation and heat loss/metre length. Also find increase in the heat dissipation due to insulation.

(Mumabi Univ. 2000 Summer)

17.21  An egg with mean diameter of 40 mm and initially at 20°C is placed in boiling water for 4 minutes and found to boiled to the consumer taste. For how long should a similar egg for same consumer be boiled when taken from a refrigerature at 5°C. Take the following properties for the egg.
$k = 10$w/mk, $\rho = 1200 \text{ kg/m}^3$, $c_p = 2$kJ/kg k,
$h = 100 \text{ w/m}^{2\text{k}}$

(Mumbai Univ. 2000 Summer)

## Convection

17.22.  Objective type questions

   (i)  Heat transfer rate

     (a)  will be higher in turbulent flow

     (b)  will be lower in turbulent flow

     (c)  will depend only on the fluid

     (d)  will depend only on viscosity.

  (ii)  In free convection, motion of the fluid is caused

     (a)  by the weight of the fluid element

     (b)  by the hydrostatic force on the element

     (c)  by the buoyancy forces arising from variation in density of the fluid with the temperature

     (d)  none of the above.

 (iii)  In forced convection molecular diffusion causes

     (a)  momentum flow in turbulent region

     (b)  momentum flow in the laminar region

    (c)   Heat flow in the turbulent region

    (d)   diffusion has no part in energy transfer.

 (iv)  The density of a fluid element which is at a temperature $\theta$ above the surrounding fluid is equal to

    (a)   $\rho_s/(1 - \beta\theta)$                         (b)  $\rho_s(1 + \beta\theta)$

    (c)   $\rho_s(1 - \beta\theta)$                         (d)  $\rho_s/(1 + \beta\theta)$

    Where   $\rho_s$ = Density of surrounding fluid,

               $\beta$ = Coefficient of cubical expansion.

  (v)  The convective heat transfer Co-efficient in laminar flow over a flat plate

    (a)   increases if a lighter fluid is used

    (b)   increases if a higher viscosity fluid is used

    (c)   increases if higher velcities are used

    (d)   increases with distance.

 (vi)  In the boundary layer over a flat plate in laminar flow the velocity is

    (a)   zero

    (b)   slowly decreases from the freee stream to the solid surface

    (c)   slowly increases from the free stream to the wall

    (d)   only temperature in the boundary layer will be different from that of free stream.

(vii)  For a unit volume, the buoyancy force causing the fluid motion is equal to

    (a)   $\rho\,\beta\,\theta\,g$                          (b)  $\rho\beta\theta/g$

    (c)   $(\rho_s - \rho) \times g$                   (d)  both (a) and (c)

(viii)  As viscosity of fluid increases the boundary layer thickness

    (a)   will increase

    (b)   will decrease

    (c)   will not change

    (d)   will increase at medium values and then decrease

 (ix)  For a forced convection, Nusselt number is a function of

    (a)   Prandtl and Grashof numbers

    (b)   Grashof number only

    (c)   Reynold and grashof numbers

    (d)   Reynold and prandtl numbers.

  (x)  For a free convection, Nusselt number is a function of

    (a)   Prandtl and Grashof numbers

    (b)   Reynold and Grashof numbers

    (c)   Reynold number only

    (d)   Reynold and prandtle number

 (xi)  The temperature gradient in the fluid flowing over a flat plate

    (a)   will be zero at the surface

    (b)   will be positive at the surface

    (c)   will be very steep at the surface

    (d)   will be zero at the top of the boundary layer

(xii)  The ratio of thermal to hydrodynamic boundary layer thickness varies as

    (a)   root of Reynolds number

    (b)   root of Nusselt number

    (c)   root of Prandtl number

    (d)   one third power of prandtl number

(xiii)  The overall heat transfer Co-efficient (U), when the heat is transferred from hot fluid (of film heat transfer co-efficient $h_i$) to a cold fluid (of film heat transfer Co-efficient $h_o$) through a wall of thickness ($\Delta x$) and thermal conductivity ($k$) is given by

    (a)  $\dfrac{1}{U} = h_i + \dfrac{k}{\Delta x} + h_o$                (b)  $U = h_i + \dfrac{k}{\Delta x} + h_o$

    (c)  $\dfrac{1}{U} = \dfrac{1}{h_i} + \dfrac{\Delta x}{k} + \dfrac{1}{h_o}$         (d)  $U = \dfrac{1}{h_i} + \dfrac{\Delta x}{k} = h_o$

(xiv) Non-dimensional number generally involved in case of heat transfer from horizontal cylinder by natural convection is
    (a) Reynolds number                     (b) Nusselt number
    (c) Prandtl number                       (d) Grashof number
    (e) (b), (c) and (d) above

(xv) The nature of flow of a fluid inside a tube i.e., whether it is turbulent or laminar, is ascertained by
    (a) Flow velocity                       (b) tube size
    (c) surface conditions              (d) viscosity of fluid
    (e) Reynolds number

**Answers to objective questions**

| | | | | |
|---|---|---|---|---|
| (i)    a | (ii)  c | (iii)  b | (iv)  d | (v)  c |
| (vi)  b | (vii)  d | (viii)  a | (ix)  d | (x)  a |
| (xi)  d | (xii)  d | (xiii)  c | (xiv)  e | (xv)  e |

17.23 Distinguish between natural and forced convection heat transfer.

17.24 Define Reynolds, Nusselt, Prandtl and stanton numbers. Explain their inportance in convective heat transfer.

17.25 Distinguish between Laminar and turbulent flow.

17.26 Distinguish between hydrodynamic and thermal boundary layers. What is the siignificance of these boundary layers in heat transfer?

17.27 What is the hydraulic diameter? Where it is used?

17.28 What is Reynolds analogy? Describe the relation between fluid friction and heat transfer,

17.29 Use the principle of dimensional analysis to establish a relationship between Nusselt number, Grashof number and Prandtl number.

17.30 Explain in detail, the Colburn analogy

17.31 Discuss briefly the effect of turbulence on boundary layers.

17.32 Explain for fluid flow along a flat plate
    (i) Velocity distribution in hydrodynamic boundary layer.
    (ii) temperature distribution in thermal boundary layer.
    (iii) Variation of local heat transfer Co-efficient along the flow.

17.33 Water is flowing at the rate of 50 kg/min through a-tube of inner diameter of 2.5 cm. The surface of the tube is maintained at 100°C. If the temperature of water increases from 25°C to 55°C, find the length of tube required. The following empirical relation can be used:

$$Nu = 0.023 \, (R_e)^{0.8} \, (pr)^{0.4}$$

Physical properties of water can be taken from following table:

| $t°C$ | $\rho$ | $C_p$ | $k \times 10^2$ | $\mu \times 10^6$ |
|---|---|---|---|---|
| | kg/m$^3$ | J/kg k | w/mk | kg/ms |
| 40 | 9922 | 4174 | 63.35 | 652 |
| 50 | 988.1 | 4178 | 64.74 | 550 |
| 60 | 983.2 | 4182 | 65.90 | 470 |
| 70 | 977.8 | 4187 | 66.72 | 405 |
| 80 | 971.8 | 4195 | 67.41 | 355 |

(Mumbai Univ. 1996 Winter)

17.34 A refrigerated truck is moving at a speed of 85 km/hr where the ambinet temperature is 50°C. The body of the truck is of rectangular shape of size 10 m × 4m × 3m. Assume the boundary layer is turbulent and the wall surface temperature is at 10°C. Neglect heat transfer from vertical front and backside of truck and flow of air is parallel to 10 m long side, calculate heat loss from the four surfaces. For turbulent flow over flat surfaces.

$$\overline{Nu} = 0.036\, Re^{0.8}\, pr^{0.33}$$

Average properties of air at 30°C are:

$$\rho = 1.165 \text{ kg/m}^3, \; C_p = 1.005 \text{ kJ/kg k}$$

$$v = 16 \times 10^{-6} \text{ m}^2/\text{s}, \; p_r = 0.701$$

(Mumbai Univ. 1999 Winter)

17.35  A hot plate of 100 cm height and 25 cm width is exposed to atmospheric air at 25°C. The surface temperature of the plate is 95°C. Find the heat loss from both the surfaces of the plate. If the height of the plate is reduced to 50 cm and width is increased to 40 cm, what will be the change in heat loss? The following empirical relation can be used.

$$Nu = C(Gr \times Pr)^{m} \text{ where}$$

$$c = 0.59 \text{ and } m = \frac{1}{4} \text{ if } Gr \times Pr < 10^9$$

$$c = 0.1 \text{ and } m = \frac{1}{3} \text{ if } Gr \times Pr > 10^9$$

Physical properties of air are:

$$\rho = 1.06 \text{ kg/m}^3$$

$$c_p = 1004 \text{ J/kg k}$$

$$k = 0.029 \text{ w/mk}$$

$$v = 18.97 \times 10^{-6} \text{ m}^2/\text{s}$$

(Mumbai Univ. 1998 Winter).

17.36  Air at 27°C and 1 atm flows over a flat plate at a speed of 2m/s. Assuming that the plate is heated over its entire length to a temperature of 60°C. Calculate heat transferred for the first 0.4 m of the plate. Also compute the drag force exerted on the first 0.4 m of the plate using Reynolds analogy. {Assume air to be a perfect gas with

$$R = 287 \text{J/kg k and } c_p = 1.006 \text{ kJ/kgk]}$$

(Mumbai Univ. 1999 Summer)

17.37  A long annulus (3Cm ID and 5 Cm ID) heats air by maintaining the temperature of the outer surface of inner tube at 50°C. The air enters at 16°C and leaves at 32°C and its flow rate is 30 m³/s. Estimate the heat transfer Co-efficient between air and inner tube.

$$N_u = 0.023\, Re^{0.8}\, pr^{0.4}$$

Average properties of air at 24°C

$$\rho = 1.614 \text{ kg/m}^3, \; v = 15.9 \times 10^{-6} \text{ m}^2/\text{s}$$

$$c_p = 1.007 \text{ kJ/kgk}, \; pr = 0.707, \; k = 0.0263 \text{ w/mk}$$

(Mumbai Univ. 1999 Summer)

17.38  A hot plate of 25 cm height and 100cm width is exposed to atmosphere at 20°C. The surface temperature of the plate is 100°C. Find the heat loss from both the surfaces of the plate. If the height of the plate is changed to 50 cm, what will be the change in heat loss?

The following empirical relation can be used

$$Nu = 0.59\, (Gr \times Pr)^{1/4}$$

The Properties of air are

$$\rho = 1.06 \text{ kg/m}^3, \ c_p = 1.004 \text{ kJ/kg k}$$

$$k = 0.029 \text{ w/mk}, \ v = 18.97 \times 10^{-6} \text{ m}^2\text{/sec.}$$

(Mumbai Univ. 2000 Summer)

17.39  Air at 20°C and 1 atm pressure flows over one side of plate, $50 \times 30$ cm in size. The plate is maintained at 60°C. The velocity of air is 4m/s. Find
   (i)  thickness of velocity and thermal boundary layer
   (ii)  Drag force
   (iii)  Heat transfer Co-efficient at 10 cm from leading edge.
   (iv)  Heat transfer rate from the plate surface
Assume flow is parallel to 30 cm side and properties of air  as

$$\rho = 1.13 \text{ kg/m}^3, \ v = 17 \times 10^{-6} \text{ m}^2\text{/s}$$

$$c_p = 1.005 \text{ kJ/kgk}, \ k = 0.0279 \text{ w/mk}$$

17.40  A highly viscous liquid flows through a 5 cm ID pipe at rate of 50 kg/s. Fluid passes  through 1 m long heated section  where there is constant flux of 1000 w/m$^2$ is supplied. Calculate the final temperature of liquid if initial temperature is 40°C. Obtain the upper bound of wall temperature. Assume properties of liquid as

$$\rho = 1500 \text{ kg/m}^3, \ c_p = 1.675 \text{ kJ/kgk}, \ k = 0.865\text{w/mk}$$

17.41  Air at 30°C and atmospheric pressure flows over a flat plate at velocity 120 m/min calculate boundary layer thickness at 20 cm and 40 cm from the leading ege of the plate. Also calculate the mass flow which enters the boundary layer thickness $x = 20$ cm and $x = 40$ cm per metre depth of plate.
If the plate is heated and maintained at 90°C uniformly over its length, calculate the heat transfer in the first 40 cm of the plate per metre length. Properties of air,

$$c_p = 1.007 \times 10^3 \text{ J/kg k,} \qquad \mu = 2.044 \times 10^{-5} \text{ kg/m-s}$$

$$v = 19.03 \times 10^{-6} \text{ m}^2\text{/s,} \qquad k = 0.027 \text{ w/mk}$$

$$p_r = 0.701, \qquad \rho = 1.04 \text{ kg/m}^3$$

**Radiation**

17.42  Objective type questions
   (i)  The process which transmits energy by means of electromagentic waves is called
      (a)  Convection           (b)  radiation
      (c)  Conduction           (d)  non of the above
   (ii)  Radiant energy, being electromagnetic in character needs
      (a)  medium for its propagation
      (b)  no medium
      (c)  heating before propagation
      (d)  none of the above
   (iii)  The heating effect is caused by the wavelengths when they are in the range of
      (a)  $100 \ \mu$ to $150 \ \mu$          (b)  $0.1 \ \mu$ to $100 \ \mu$
      (c)  $150 \ \mu$ to $200 \ \mu$          (d)  none of the above
   (iv)  Depending upon the radiating properties, the bodies are classified as
      (a)  black body           (b)  white body
      (c)  transparent body          (d)  opaque body
      (e)  all of the above
   (v)  A body, which does not transmit any radiations but partly reflects and partly absorbs the radiations, is called
      (a)  Opaque body          (b)  White body
      (c)  black body           (d)  transparent body

    (vi)  Longest wavelength contained by
        (a)  Gamma rays                     (b)  $x$ - rays
        (c)  ultravoilet rays           (d)  Radio waves
   (vii)  A perfect black body is the one which
        (a)  absorbs all incident radiation      (b)  reflects all incident radiation
        (c)  absorbs most of the incident radiation    (d)  is located with lamp black.
  (viii)  A hollow sphere with uniform interior temperature and a small hole behaves very nearly as a
        (a)  Opaque body                 (b)  white body
        (c)  black body                   (d)  transparent body
    (ix)  Which one of the following approximates to the black body condition?
        (a)  ice                          (b)  water
        (c)  Lamp black                 (d)  all of the above
        (e)  none of the above
     (x)  If the ratio of emission of a body to that of a black body at a given temperature is constant for all
        wavelengths, the body is called
        (a)  black body                 (b)  grey body
        (c)  White body                (d)  opaque body
   (xi)  Stefan Boltzman law states that the total emission from  a black body per unit area per unit time
        (a)  varies directly with the absolute temperature
        (b)  varies inversly with absolute temperature
        (c)  varies directly with the square of the absolute temperature
        (d)  varies directly with the fourth power of the absolute temperature
   (xii)  "The total emissive power for any body at a given temperature is equal to its absorptivity multiplied
        by the total emissive power of a perfect black body at the same temperature this statement is
        known as
        (a)  Wein's law                 (b)  Plank's law
        (c)  Kirchhoff's law           (d)  Stefan Boltzman law
  (xiii)  If the absorptivity plus reflectivity of a body equals 1, then the body is known as
        (a)  black body                 (b)  Grey body
        (c)  opaque body               (d)  White body
        (e)  none of the above
   (xiv)  At all wavelengths and temperatures, the monochromatic emissivity of a white body is equal to
        (a)  Zero                       (b)  0.1 to 0.5
        (c)  0.5                      (d)  0.5 to 1
        (e)  Unity
   (xv)  Glasses are
        (a)  opaque for high temperature radiation
        (b)  opaque for low temperature radiation
        (c)  transparent at short wave lengths
        (d)  transparent at long wave lengths
        (e)  Opaque for low temperature radiation transparent at long wave lengths
  (xvi)  For solar collectors the required surface characteristics combination is
        (a)  high emissivity and low reflectivity
        (b)  high emissivity and low reflectivity.
        (c)  high reflectivity and high absorptivity
        (d)  low emissivity and high absorptivity
  (xvii)  The value of shape factor will be highest when
        (a)  the surfaces are farther apart
        (b)  the surfaces are closer
        (c)  the surfaces are smaller and closer
        (d)  the surfaces are larger and closer

   (xviii)  The reciprocity theorem states

       (a)  $F_{1-2} = F_{2-1}$                   (b)  $A_1 F_{1-2} = A_{2-1}$

       (c)  $A_2 F_{1-2} = A, F_{2-1}$          (d)  $\varepsilon_1 F_{1-2} = \varepsilon_2 F_{2-1}$

   (xix)  Choose the correct statement or statements

       (a)  Radiosity is another name for omissive power

       (b)  Radiation intensity is the flux per unit area

       (c)  Radiation intensity is the radiant energy per unit solid angle

       (d)  Irradiation is the total radiant energy incident on a surface

   (xx)  Gases have poor

       (a)  reflectivity                          (b)  absorptivity

       (c)  transmissivity                  (d)  absorptivity + transmissivity

       (e)  reflectivity + absorptivity

**Answer to objective questions**

| | | | | | |
|---|---|---|---|---|---|
| (i)  b | (ii)  b | (iii)  b | (iv)  e | (v)  a | (vi)  a |
| (vii)  a | (viii)  c | (ix)  d | (x)  b | (xi)  d | (xii)  c |
| (xiii)  c | (xiv)  a | (xv)  c | (xvi)  d | (xvii)  d | (xviii)  b |
| (xix)  d | (xx)  a | | | | |

**17.43**  Explain the concept of black body and grey body in radiation terminology.

**17.44**  State and prove kirchoff's identity.

**17.45**  Write in brief about Reynolds analogy between momentum and heat transfer.

**17.46**  Define absorptivity, reflectivity and transmissivity.

**17.47**  Derive an expression of overall heat transfer Co-efficient based on outer surface area

**17.48**  Define shape factor, Irradiation, Radiosity

**17.49**  Two parallel plates $0.5 \times 1$ m each are spaced 0.5 m apart. The plates are at temperatures of 1000°C and 500°C, and their emissivities are 0.2 and 0.5 respectively. The plates are located in a large room, the walls of which are at 27°C. The surfaces of the plates facing each other only exchange heat by radiation. Determine the rates of heat lost by each plate and heat gain of the walls by radiation. Use radiation network for solution. Assume shape factor between parallel plates $F_{1-2} = F_{2-1} = 0.185$

                                                     (Mumbai Univ. 1996 Winter)

**17.50**  Two parallel plane plates at temperatures $T_1$ and $T_2$ are having emissivities $\varepsilon_1$ and $\varepsilon_2$. In order to reduce radiant heat exchange between them, a radiation shield plate of emissivity $\varepsilon_3$ is kept between the two parallel plates. If emissivities of all three parallel surfaces are same, prove that the heat exchange by radiation between surfaces at $T_1$ and $T_2$ is reduced by one half. Use radiation network concept for solution.                              (Mumbai Univ. 1997 Summer)

**17.51**  The net radiation from the surface of two parallel plates maintained at temperatures $T_1$ and $T_2$ is to be reduced by 79 times. Calculate the number of screens to be placed between the two surfaces to achieve this reduction in heat exchange, assuming the emissivity of screens as 0.05 and that of the surfaces as 0.8.                                        (Mumbai Univ. 1997 summer)

**17.52**  A hemispherical furnace of radius 1.0 m has a roof temperature $T_1 = 800$ k and emissivity $\varepsilon_1$ as 0.8. The flat circular floor of the furnace has a temperature $T_2 = 600$ k and emissivity $\varepsilon_2 = 0.5$. Calculate the net radiant heat exchange between the roof and the floor.

                                (Mumbai Univ. 1998 Summer 2000 summer)

**17.53**  Two Co-axial cylinders of 0.4 m and 1 m diameter are 1 m long. The annular top and bottom surfaces are well insulated and act as reradiating surfaces. The inner surface is at 1000 k and has an emissivity of 0.6. The outer surface is maintained at 400 K, the emissivity of which is 0.4.

    (i)  Determine the heat exchange between the surfaces

   (ii)  If the annular base surfaces are open to the surroundings at 300k, determine the radiant heat exchange.

    If the outer cylinder is surface 2, take

    $F_{2-1} = 0.25$ and $F_{2-2} = 0.27$

                                                     (Mumbai Univ. 1998 Winter)

17.54  For a black body maintained at 390 K, determine
   (i)  the total emissive power
   (ii)  the wavelength at which the maximum monochromatic emissive power occurs
   (iii)  the maximum monochromatic emissive power. Given

$$\varepsilon_{bx} = \frac{C_1 \lambda^{-5}}{e^{(c_2/\lambda t - 1)}}$$

$$c_1 = 3.742 \times 10^8 \; w\mu \; m^4/m^2$$

$$c_2 = 1.4387 \times 10^4 \; \mu m K$$

(Mumbai Univ. 1999 Summer)

17.55  A convex grey body having a surface area of $4m^2$ has $\varepsilon = 0.35$ and $T_1 = 680$ K. This is completely inclosed by a grey surface having an area of $36m^2$, $\varepsilon = 0.75$ and $T_2 = 310$ K. Find the net rate of heat transfer $q_{1-2}$ between the two surfaces.

(Mumbai Univ. 1999 Summer)

17.56  Find the shape factor of a hemispherical bowl of diameter $D$ with respect to itself. Also calculate the radiative heat transfer from the cavity if inside temperature is 773 K and its emissivity is 0.6. The diameter of the cavity is 700 mm.

(Mumbai Univ. 1999 Winter)

17.57  A conical hole is machined in a block of metal whose emissivity is 0.6. The hole is 3 cm in diameter at the surface and 4 cm deep. If the metal block is heated to 527°C, find the radiation emitted from the hole. Calculated the radiation emitted if the hole is cylindrical 3 cm diameter & 4 cm deep.

17.58  An object has its lower part having conical shape if 2 cm radius and 5 cm slant length. The upper part is hemispherical with radius 2 cm. The surface emissivity is 0.2 and temperature 27°C. This object is kept in a cavity of surface area $1m^2$, emissivity 0.1 and at temperature 327°C. Find the heat transfer rate due to radiation.

17.59  Liquid oxygen at atmospheric pressure (Boiling point = – 183°C) is stored in a spherical vessel of 0.3 m O.D. The system is insulated by enclosing the container inside another concentric sphere of 0.5 m I.D. with space between then evacuated. Both the sphere surfaces are made of aluminium for which emissivity may be taken as 0.3. The temperature of the outer sphere is 40°C.
   (i)  estimate the rate of heat flow due to radiation
   (ii)  What will be the reduction in heat flow if polished aluminium with an emissivity of 0.5 is used for the container walls?

17.60  Define the diffusion Co-efficient for a binary mixture. Is this Co-efficient dependent upon temperature, pressure and composition of the mixture?

17.61  Define Fick's first and second law of diffusion. Describe the various mechanism of mass transfer

17.62  What is convective mass transfer co-efficient and what are its units?

17.63  Explain the phenomenon of equimolar counter diffusion. Derive an expression for equimolar counter diffusion between two gases or liquids.

17.64  Define the Fourier number and Biot number for mass transfer.

17.65  Define the Schmidth number, Sherwood number and Lewis numbers. What is the physical significance of each.

17.66  Express the rate of heat flow in terms of a convective heat transfer Co-efficient by an equation and write the analogous equation for mass transfer.

17.67  Discuss the analogy between heat and mass transfer.

17.68  Show by dimensional analysis that mass transfer by forced convection can be expressed by

$$Sh = f(Re, Sc)$$

17.69  For steady state diffusion through plane memberane solve the basic equation and get solution for mass diffusion rate per unit area in terms of diffusion co-efficient, concentration at the two membrance surfaces and the memberance thickness. Using the above result and electrical analogy, establish an expression for diffusion resistance as $R_D = \dfrac{\Delta x}{AD}$ where $\Delta x$ is the membrane thickness, $A$ is area of diffusion membrane and $D$ diffusion co-efficient.

17.70  For convective mass transfer name the nondimensional number that plays the same role in mass transfer as that of prandtl number in heat transfer and write down an expression for the same.

17.71  What do you understand by equimolal counterdiffusion?

17.72  Discuss isothermal evaporation of water.

17.73  The air pressure in a tyre tube of surface area 0.5 m$^2$ and wall thickness 6 mm is 2 bar. The pressure decreases to 1.99 bar in a period of 3 days. The solubility of air in rubber is 0.07 m$^3$ of air per m$^3$ of rubber at 1.0 bar. Estimate the diffusivity of air in rubber at the operating temperatur of 27°C, if the volume of air in the tube is 0.025 m$^3$.

(Mumbai Univ. 1998 summer)

17.74  A tank contains a mixture of $CO_2$ and $N_2$ in the mole proportions of 0.2 and 0.8 at 1 bar and 290 K. It is connected by a duct of C.S. area 0.1 m$^2$ to another tank containing a mixture of $CO_2$ and $N_2$ in the molal proportion of 0.8 and 0.2. The duct is 0.5 m long. Determine the diffusion rates of $CO_2$ and $N_2$ in kg/s.

(Mumbai Univ. 1998 Winter)

17.75  Estimate the diffusion rate of water from the bottom of a test tube 1.5 cm diameter and 15 cm long into dry atmospheric air at 25°C ($D = 0.256$ cm$^2$/sec.)        (Mumbai Univ. 1999 Summer)

17.76  A steel rectangular container having walls 15 mm thick is used to store gaseous hydrogen at elevated pressure. The molar concentrations of hydrogen in the steel at the inside and outside surfaces are 1 kg mole/m$^3$ and zero respectively. Assuming the diffusion co-efficient for hydrogen in steel as $25 \times 10^{-12}$ m$^2$/sec. Calculate the molar diffusion flux for hydrogen through steel.

(Mumbai Univ. 1999 Winter)

17.77  A vessel contains a binary mixture of oxygen and nitrogen with partial pressures in the ratio 0.21 and 0.79 at 25°C. The total pressure of the mixture is 1.1 bar. Calculate the following
   (i)   molar concentrations
   (ii)  mass densities
   (iii) mass fractions
   (iv)  molar fractions of each species

(Mumbai Univ. 2000, Summer)

17.78  In a solar pond salt is placed at the bottom of a pond 1.5 m deep. The surface is flushed constantly so that the concentration of salt at the top layer is zero. The salt concentration at the bottom layer is 5 kg mole/m$^3$. Determine the rate at which salt is washed off at the top at steady state conditions per m$^2$. $D = 1.2 \times 10^{-9}$ m$^2$/s. (*Ans.*   0.656 kg/month/m$^2$)

17.79  Determine the diffusion rate in lake when wind is blowing at 20 kmph over the surface. The air temperature is 30°C and the relative humidity is 40%. The total pressure is 1 bar. The lake is 1 km wide along the flow.

1780   The humidity level inside a room is such that the water vapour pressure is 0.03 bar. On the outside the air is dry plaster board of 10 mm thickness separates the inside from outside. The diffusion Co-efficient for water vapour into the wall material is about $10^{-9}$ m$^2$/s. The solubility of water vapour in the wall material is 0.142 kg/m$^3$ for bar of water vapour pressure. Determine the diffusion rate.

17.81  $H_2$ diffuses through a 10 mm thick steel wall. The concentration of $H_2$ in the steel at the inner surface is 1 kmol/m$^3$ while its concentration at the outer surface is negligible. If the binary diffusion Co-efficient for $H_2$ in steel is $0.26 \times 10^{-12}$ m$^2$/s What is the diffusion flux for $H_2$ through the steel? [$2.6 \times 10^{-11}$ K mol/s.m$^2$]

17.82.  Ammonia and air experience equimolar counter diffusion in a circular tube of 3 mm diameter and 20 m long. The system is at a total pressure of 1 atm and a temperature of 25°C one end of the tube is connected to a large reservoir of ammonia and the other end of the tube is open to atmosphere. Estimate the mass transfer rate .

[$2.483 \times 10^{-8}$ kg/hr, $-4.23 \times 10^{-8}$ kg/hr]

18

# Direct Energy Conversion

## 18.1 INTRODUCTION

The energy conversion devices that have been in use for a long time are those that accept energy heat and produce mechanical work which is transformed into electrical power for distribution at last. Direct energy conversion devices convert naturally available energy into electricity without an intermediate conversion into mechanical work. The energy source may be thermal, solar or chemical. Until now their uses have been confined to small scale special purpose applications, since the voltage output available with them is rather small.

The efficiencies of all modern-conventional thermal power generating systems between 35 percent to 42 percent as they have to reject large quantities of heat to the surroundings. In all conventional power plants (Steam turbine, gas turbine, and diesel power plants), first the thermal energy is converted into mechanical work and then into electrical power. Due to intermediate steps of conversion, efficiency is reduced to great extent. Great attempts have been made during the last two decades to convert thermal, electromagnetic or chemical energy directly into electrical energy raising the thermodynamic efficiency to 60 to 75 per cent. Such systems are called direct energy converters. The direct energy conversion systems which are in present use are:

(a) Thermo-electric converters
(b) Thermionic converters
(c) Fuel-cells
(d) Magnet hydrodynamic generators (MHD)
(e) Photo Voltaic systems (solar cells)

## 18.2 THERMOELECTRIC CONVERTERS

The operation of the thermoelectric generator or converters depends on the seebeck-effect, the peltier effect and the Thomson effect. The Seebeck effect was discovered in 1822 by the German

Physicist, Thomas J. Seebeck. According to the Seebeck effect, "a voltage is produced in a circuit of two dissimilar materials if the two junctions are maintained at different temperatures. The potential difference developed is proportional to the temperaure difference. Mathematically

$$dE_s \; \alpha \; dT \quad \text{or} \quad dE_s = SdT \quad \text{or,} \quad S = \frac{dE_s}{dT} \tag{18.1}$$

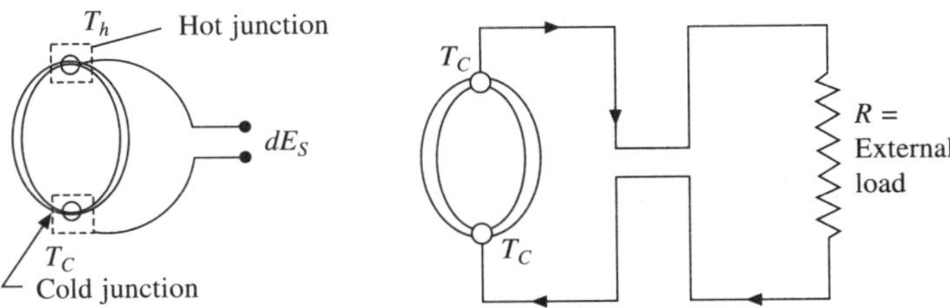


**Fig. 18.1   Seebeck effect**

where $E_s$ = Seebeck e.m.f. and $S$ is called Seebeck Coefficient. The Seebeck coefficient is strong functions of temperature. The induced thermoelectric potential $E_s$ produced in a circuit composed of two materials is obtained from the relation:

$$E_s = \int_{T_C}^{T_h} (S_a - S_b)\,dT = \int_{T_c}^{T_h} S_{ab}\,dT \tag{18.2}$$

$T_h$ and $T_c$ denote temperatures of hot and cold junctions. The combined Seebeck coefficient $S_{ab}$ is defined as positive if the electrical current (the flow of positive charges) from material $A$ to material $B$ at the cold junction. Some typical values of Seebeck coefficients are listed below: Seebeck co-efficients (at 100°C)

| Material | S, (Volt/k) |
| --- | --- |
| Aluminium | $-0.2 \times 10^{-6}$ |
| Constantan | $-47.0 \times 10^{-6}$ |
| Copper | $+3.5 \times 10^{-6}$ |
| Iron | $+13.6 \times 10^{-6}$ |
| Platinum | $-5.2 \times 10^{-6}$ |
| Germanium | $+375.0 \times 10^{-6}$ |
| Silicon | $-455.0 \times 10^{-6}$ |

It is noted that the Seebeck coefficients for metals and alloys are very low compared to those for semi-conductors. The combined Seebeck coefficient for an iron-constan circuit is 60.6 V/K, while that for a combination of germanium and silicon is 830 V/K. The Seebeck coefficients for metals and alloys are too low for the efficient generation of electricity, although the dissimilar metal junctions are commonly used in thermocouple circuits to monitor temperatures. The Seebeck co-efficients of $n$-$p$ semi conductors are also high and these are the materials that are commonly used in thermoelectric generators.

In a thermo-electric converter constructed of semi conductors both the holes and the extra electrons migrate towards the cold junction where they pile up and combine. The combined Seebeck coefficient for these lattices is $S_{ab} = S_{pn} - S_{np}$ and thermo electric potential becomes

$$E_s = \int_{T_c}^{T_h} (S_p - S_n)\,dT = \int_{T_c}^{T_h} S_{pn}\,dT \tag{18.3}$$

The Peltier effect was discovered in 1844 by the French Physicies J.C.A. Peltier. The Peltier effect states essentially that if a direct current is passed through a circuit of dissimilar materials, one of the dissimilar metal junctions will be heated and the other will be cooled. This is the reversed Seebeck effect and it is also reversible in that, if the direction of current flow is reversed, the junction that was formerly heated will be cooled and the formerly cooled junction will be heated. The amount of rate of heat released or absorbed is proportional to the current passed.

$$Q\alpha\, I_{ab} \quad \text{or,} \quad Q = \pi_{ab} I_{ab} \tag{18.4}$$

Like the Seebeck coefficient the Peltier coefficient $(\pi_{ab})$ is a strong function of temperature and is related to Seebeck coefficient by the relation:

$$\pi_{ab} = T_{ch} S_{ab} = T_{ch}(S_a - S_b) = -\pi_{ba} \tag{18.5}$$

Peltier coefficient is positive if heat is generated in the junction when the direct current flows from material A to material $B$.

The Thomson effect was discovered in 1854 by the English Physicise William Thomson (Lord Kelvin). This effect states that there is a reversible absorption or liberation of heat in a homogeneous conductor exposed to a simultaneous temperature gradient and an electrical gradient. The Thomson Co-efficient $\tau$ is given by the relation:

$$\tau = \frac{Q\,\Delta T}{I} \tag{18.6}$$

Where $Q$ is the heat transfer rate in watt absorbed by the conductor when current flows towards the higher temperature. The Thomson co-efficient is related to the Seebeck coefficient by the following relation

$$\tau = T\,\frac{dS}{dT} \tag{18.7}$$

This coefficient is positive for $p$-type materials and negative for $n$ type materials. The Thomson effect is only of secondary importance in the operation of thermo-electric generators. A typical $p$-$n$ thermo-electric generator is shown in Fig. 18.2. The legs of the generator are connected in series for the flow of electricity, and they are connected in parallel for the flow of heat. The total electrical resistance of the generator $(R_g)$ for a series connection is equal to the sum of resistances of each of the legs.

$$R_g = m\,(R_p + R_n) \tag{18.8}$$

Where $m$ is the number of pairs of $p$-$n$ legs of the generator, and $R_p$ and $R_n$ are the resistances of the $p$-leg and $n$-leg respectively.

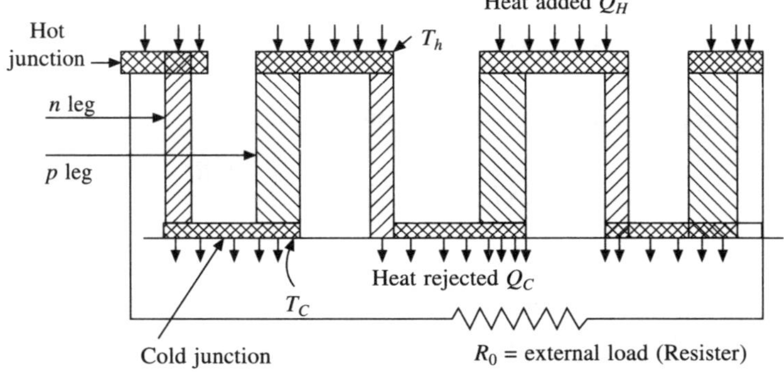



**Fig. 18.2  The typical *n-p* thermo-electric generator**

$$R_p = \frac{\rho_p L_p}{A_p} \quad \text{and} \quad R_n = \frac{\rho_n L_n}{A_n} \tag{18.9}$$

In the previous equations, $\rho$ is the electrical resistivity of the material in ohm-meter, $L$ is the length of the legs of semi-conductors in meters, $A$ is the cross sectional areas of the legs. It is normally assumed that the metallic connections between the semi-conductors legs have negligible resistance.

The thermal conductance $C_g$ of the generator is equal to the sum of thermal conductances (the reciprocal of thermal resistance) of the semi conductors legs or

$$C_g = m (C_p + C_n) \tag{18.10}$$

Where

$$C_p = \frac{k_p A_p}{L_p}, \quad -C_n = \frac{k_n A_n}{L_n} \tag{18.11}$$

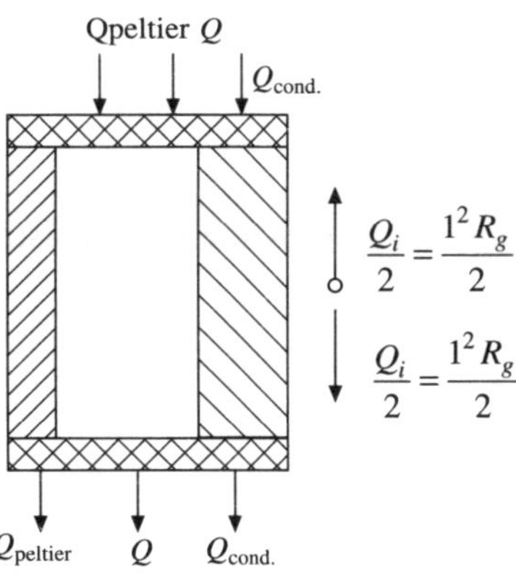



**Fig. 18.3**

Where $k$ is thermal conductivity of the semi-conductor materials in Watt/m-c°.

An energy balance on either hot or cold junction is composed of four energy terms. First, there is the heat transfer rate to or from the junction from environment, $\pm Q$; Second, there is the heat transfer rate through the generator from hot to cold junction by conduction $= Q_{\text{cond}} = \pm C_g \, \Delta T$. Third, there is a heat transfer due to Peltier effect

$$= Q_{\text{peltier}} = \pm T_h \text{ or } C\,I \cdot S_{pn} \cdot \text{Fourth,}$$

there is the power dissipated in the device due to Joulean heating and it can be shown that effectively half of the resistance heating is generated in each junction $\dfrac{I^2 R_g}{2}$.

The energy or power entering the hot junction is equal to

$$Q_H = mS_{pn}T_hI + C_g(T_h - T_c) - \frac{T^2R_g}{2} \tag{18.12}$$

The energy or power rejected at the cold junction is equal to

$$Q_c = mS_{pn}T_cI + C_g(T_h - T_c) - \frac{I^2R_g}{2} \tag{18.13}$$

The useful power produced by the device is equal to the power dissipated in the external load ($R_o$). Since this system generates direct current, the useful power generation is $I^2R_o$, where $R_o$ is the resistance of the external load. The thermal efficiency of the thermoelectric generator is

$$\eta_{th} = \frac{I^2R_o}{Q_H} = \frac{I^2R_o}{mS_{pn}T_hI + C_gDT - \dfrac{I^2R_g}{2}} \tag{18.14}$$

Multiplying numerator and denominator by $\dfrac{\Delta T}{R_gI^2}$ we get

$$\eta_{th} = \frac{I^2R_o \times \dfrac{\Delta T}{R_gI^2}}{\dfrac{\Delta T}{R_gI^2}\left(mS_{pn}T_hI + C_g\Delta T - \dfrac{I^2R_g}{2}\right)}$$

or,

$$\eta_{th} = \frac{\Delta T \cdot \left(\dfrac{R_o}{R_g}\right)}{mS_{pn}\dfrac{T_h\Delta T}{IR_g} + \dfrac{C_g(\Delta T)^2}{I^2R_g} - \dfrac{\Delta T}{2}} = \frac{\Delta TM}{mS_{pn}\dfrac{T_h\Delta T}{IR_g} + \dfrac{C_g(\Delta T)^2}{I^2R_g} - \dfrac{\Delta T}{2}} \tag{18.15}$$

Where    $M = \dfrac{R_o}{R_g}$

The current in the generator is equal to total renerated voltage divided by the total resistance. or,

$$I = \frac{V_t}{R_t} = \frac{mS_{pn}\Delta T}{R_g + R_o}$$

or,

$$I = \frac{mS_{pn}(T_h - T_c)}{R_g(1 + M)} \tag{18.16}$$

Substituting this equation into Eq. (18.15) gives

$$\eta_{th} = \frac{\Delta TM}{(1 + M)T_h + \dfrac{(1 + M)^2}{Z}\dfrac{\Delta T}{2}} \tag{18.17}$$

Where $Z = \dfrac{m^2 S_{pn}^2}{C_g R_g}$ and is called figure of merit of the generator.

The figure of merit of a generator is a function of the properties of the generator materials ($S$, $K$, $\rho$) and the dimensions of the generator legs ($A$ and $L$). To improve the efficiency of the generator, the figure of merit $Z$ should be as large as possible. Once the generator materials have been packed, the minimum product $C_g R_g$ gives the maximum figure of merit $Z_{max}$

$$C_g R_g = \left( \frac{K_p A_p}{L_p} + \frac{K_n A_n}{L_n} \right)\left( \frac{\rho_p L_p}{A_p} + \frac{K_n A_n}{L_n} \right) m^2$$

$$= m^2 \left( K_p \rho_p + \frac{K_p \rho_n}{x} + \frac{K_n \rho_p}{x} + K_n \rho_n \right) \tag{18.18}$$

Where $$X = \frac{A_p L_n}{A_n L_p}$$

The optimum value of $x$ that gives minimum value of $C_g R_g$ or $Z_{max}$ can be found by setting $\dfrac{d}{dx}(C_g R_g) = 0$ and solving for $X$. This gives

$$X_{opt} = \frac{A_p L_n}{A_n L_p} \sqrt{\frac{\rho_p K_n}{\rho_n K_p}} \tag{18.19}$$

and $$Z_{max} = \frac{S_{pn}^2}{\left( \sqrt{k_p \rho_p} + \sqrt{k_n \rho_n} \right)^2} \tag{18.20}$$

The other variable in Eq. (18.17) that can be easily adjusted to improve the generator efficiency is the ratio of the external load resistance to the generator resistance $M$. The optimum value of $M$ that gives the maximum thermal efficiency can be determined by setting $\dfrac{d}{dM}\eta_{th} = 0$ and solving for $M$ gives:

$$M_{opt} = \left( \frac{R_o}{R_g} \right)_{opt} = \sqrt{1 + Z_{max} T_{av}} \tag{18.21}$$

Where $$T_{av} = (T_h + T_C)/2$$

Substituting this expression into the thermal efficiency equation yields:

$$\eta_{th} = \frac{M_{opt} \Delta T}{T_H(1 + M_{opt}) + \dfrac{(1 + M_{opt})^2}{Z_{max}} - \dfrac{\Delta T}{2}} \tag{18.22}$$

While the above equation gives the condition for maximum efficiency, there may be some times when the system will be operated at the condition for maximum power outpout rather than maximum efficiency. The output voltage of the generator is equal to the total generated voltage minus the internal voltage drop in the generator

$$V_{out} = mS_n\,\Delta T - IR_g \tag{18.23}$$

and output power

$$= P_{opt} = I\,V_{out} = mS_{pn}\,\Delta T\,I - I^2R_g \tag{18.24}$$

Differentiating Eq. (18.24) with respect to $I$ and setting $\dfrac{dP_{out}}{dI} = 0$, the ideal current for maximum power out put

$$I_{maxp} = mS_{pn}\,\Delta T/2R_g \tag{18.25}$$

and maximum power output for the generator equal to

$$P_{out\,max} = \frac{m^2 S_{pn}\Delta T_2}{4\,R_g} \tag{18.26}$$

The thermo electric generator can use almost any source of thermal energy to produce electricity. Thermo electric generator have a number of advantages over the conventional thermodynamic heat engine systems. The thermoelectric systems are compact rugged, reliable and have no moving parts. Unfortunately, the thermal efficiency of these systems is very low; normally 5 to 10 percent, and the thermoelectric generating systems are usually employed for very low power applications.

## 18.3   THERMIONIC CONVERTERS

As we have seen that the main reason for low thermal efficiency of thermoelectric generator is the conduction heat transfer from high temperature source to the low temperature sink. All attempts are made to reduce this conductive heat transfer by selecting suitable material of low thermal conductivity and high electrical conductivity. Thermionic converters can be regarded as a kind of thermo-electric generator in which the hot and cold junctions are separated by a vacuum preventing the heat transfer by way of conduction. The electric current is maintained by emission of electrons by heating. Thermionic emission was discovered in 1883 by Thomas *A*. Edison.

The valence electrons in orbit around a nucleus have an average energy equal to a quantity called the Fermi-level energy. The electrons vibrate about this Fermi level with an amplitutde that is proportional to the absolute temperature. The work function energy $\phi$ is the amount of energy required to strip the valence electron from an atom. This energy must be overcome before the electron can leave the surface. The work function and Fermi-level energies vary from material to material. The operational principle of a thermionic converter is based on the ability of heated metals to emit electrons from their surface The schematic diagram of the thermionic converter is shown in Fig. 18.4. The two metallic surfaces are separated by a vacuum and maintained at temperatures $T_c$ and $T_a$, $T_c > T_a$

Where      $T_c$ = cathode temperature

$T_a$ = Anode temperature

As $T_c \gg T_a$ more electrons will be emitted by cathode than anode and as a result anode will be negatively charged and potential difference will appear between the two plates. If the circuit is completed by inserting an external resistance $R_o$, an electric current will flow through the circuit. When electrons leave the cathode, the cathode is left with a positive charge and the free electrons are attracted back to the surface as well as being repelled by the anode surface. Other emitted electrons already in he inter-electrode gap also tend to drive the emitted electrons back to the cathode surface. As a result, electrons emitted from the hot cathode tend to accumulate in the inter-electrode gap creating an additional energy barrier to the flow of electrons. This additional energy barrier is called the spacecharge barrier energy and is designated by $\phi_b$.

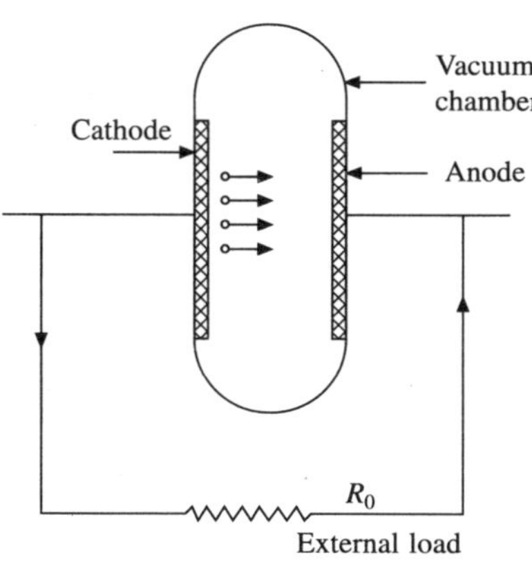



**Fig. 18.4**

The emission of electrons from cathode is regarded as evaporation of electrons from the surface of emitter and accumulation of electrons on anode is regarded as condensation of electrons.

The rate of thermal emission of electrons from a given surface is given by the richardson – Dushman equation. According to this relationship, the current density $J_o$, in ampere per square meter, is

$$J_o = A T^2 \exp\left(\frac{-\phi}{KT}\right) \tag{18.27}$$

Where $e$ is the charge of electron ($1.6 \times 10^{-19}$ Coulomb), $K$ is the Boltzmann constant ($1.551 \times 10^{-4}$ eV/K), $T$ is the absolute temperature of the surface in Kelvin, $\phi$ is the work function of the surface in electron. Volt, and $A$ is a constant with units of amperes per square meter per Kelvin squared. A was supposed to be a universal constant with a value $1.2 \times 10^6$ Ampere $e/m^2K^2$, however, it has been found to vary from material to material. The following table gives the value of $\phi$ and $A$.

**Table 18.1  Thermionic emission properties of some materials**

| Material | $\phi$(Volt) | A(Amp/m²K²) |
|---|---|---|
| $C_s$ | 1.89 | $0.50 \times 10^6$ |
| $M_o$ | 4.20 | $0.55 \times 10^6$ |
| $N_i$ | 4.61 | $0.30 \times 10^6$ |
| $P_t$ | 5.32 | $0.32 \times 10^6$ |
| $T_a$ | 4.19 | $0.55 \times 10^6$ |
| $W$ | 4.52 | $0.60 \times 10^6$ |
| $W$ + Th | 2.7 | $0.04 \times 10^6$ |
| Bao | 1.5 | $0.001 \times 10^6$ |
| $S_{ro}$ | 2.2 | $1.0 \times 10^6$ |

In order for an electron to reach the anode, it must have a total energy $E_c$ equal to the sum of the cathode work function and space-charge barrier energy, or $E_c = \phi_c + \phi_{bc}$ with the existance of the space charge barrier energy, the net current density from cathode is

$$J_o = A T^2 \exp\left(\frac{-eE_c}{KT}\right) \qquad (18.28)$$

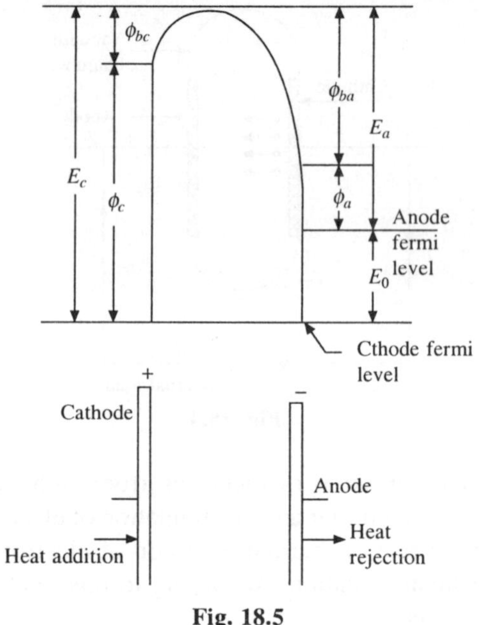


**Fig. 18.5**

Using the energy diagram from Fig. 18.5, it can be seen that the output voltage $E_o$ from the generator is equal to the sum of the Fermi level energy and the work function energy of the cathode plus the space-charge barrier energy minus the sum of the Fermi-level energy, the work function energy, and the space charge barrier energy of the anode

$$E_o = E_c - E_a \qquad (18.29)$$

For a high output voltage, the cathode should have a low Fermi level and a high work function. It should also have a high value of '$A$' because the high work function inhibits the flow of electrons. The anode should have a high Fermi level and allow work function even though this does increase the electron emission from the anode, thereby reducing the net flow of electron emission from the cathode. For optimum performance the approximate temperature work function relationship is

$$\frac{\phi_a}{T_a} = \frac{\phi_c}{T_c} \qquad (18.30)$$

The output power of the system is given by

$$P_{\text{out}} = IVL = J_o A_c V_L \qquad (18.31)$$

Where $V_L$ = Voltage drop across the load resistor. The voltage drop is equal to the output voltage ($E_o$) minus the voltage drop in the electrical leads connecting the cathode and anode to the load resistor. There are three major losses in the operation of thermionic converter are:

(a)  heat transfer rate due to radiation given by

$$P_r = \frac{\sigma A_c (T_c^4 - T_a^4)}{\dfrac{1}{\varepsilon_c} + \dfrac{1}{\varepsilon_a} - 1} \qquad (18.32)$$

$\sigma$ is stefan Boltzman constant, $5.67 \times 10^{-8}$ W/m²k⁴

(b)  energy carried away by electrons

$$\text{Pel} = J_o A_c \left( \phi_c + \phi_{bc} + \frac{2\,KT_c}{e} \right) \tag{18.33}$$

(c)  Conduction heat rate from cathode combined with Joulean heating

$$P_w = \frac{K_w A_w}{L_w} (T_c - T_o) - \frac{(J_o A_c)^2}{2} (J_w L_w / A_w)$$

Where $w$ stands for load wire

Since the sum of these three energy rates is the power loss from the cathode and must be made up by heat addition to the converter, the thermal efficiency of thermionic converter can be estimated by the following equation

$$\eta_{th} = \frac{P_{out}}{P_r + P_{el} + P_w} \tag{18.34}$$

Thermionic converters have efficiencies that range from 15 to 20 percent, considerably above that for the thermoelectric systems. Unfortunately, these system are not as rugged as the thermoelectric systems, they are complicated to build, and they are very expensive.

## 18.4  FUEL CELLS

The battery and the fuel cells are systems in which stored chemical energy of the system is converted directly to electrical energy. Since these systems do not go through the thermal energy regime, they are not limited by the efficiency of an externally reversible heat engine cycle $\eta_{carnot} = \left( 1 - \frac{T_2}{T_1} \right)$. For this reason, considerable interest and research has been generated by these systems.

Batteries and fuel cells are very similar in operation with the major difference being that the battery contains a fixed quantity of fuel or chemical energy whereas the fuel cell operates with a continuous supply of fuel. Secondary batteries are reversible devices in that the products of chemical reactors are separated back into the original reactants by supplying electricity to the battery during recharging. Fuel cells cannot be recharged as the products of the chemical reactions are thrown away.

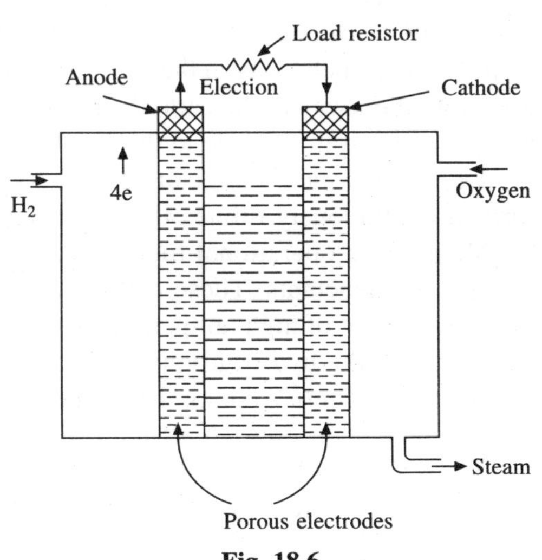


**Fig. 18.6**

The fuel cells and batteries are similar in composition in that both contain two electrodes separated by an electrolytes solution. In the fuel cell the fuel reactant, normally hydrogen or carbon monoxide is admitted into one porous electrode and oxygen or air is fed into the other porous electrodes.

Fuel cells electrodes serve basically three purposes: (a) electrodes must be porous so that both the fuel and electrolyte can penetrate it and have proper mixing contact. If the pores are too large, the fuel gas will bubble through and be wasted. If the pores are too small, there will not be sufficient contact between the reactant and the electrolyte and capacity of cell is reduced (b) The electrode must contain a chemical catalyst that breaks the fuel compound into atoms so that they are more reactive. The most popular catalyst used to-day are platinum and sintered nickel (c) The electron must be able to conduct the electrons to the terminals. The fuel cell has three chambers separated by two porous electrodes, the anode and cathode. The middle chamber in between the electrodes is filled with strong solution of Potasium hydroxide or any other electrolyte which may be solid or liquid. The various electrolytes used are: potassium hydroxide, zirconia oxide (Porous ceramics), solid polymers. Through one chamber oxygen and through other chamber hydrogen is admitted under pressure. The surface of the electrodes are chemically treated to repel the electrolyte, so that there is minimum leake of Potasium hydroxide into the outer chamber. The gases diffuse through the electrolyte. The following reactions take place:

$$4\,KOH \rightarrow 4\,\overset{+}{K} + 4\,\overline{OH}$$

At anode (hydrogen electrode)

$$2\,H_2 + 4\,\overline{OH} = 4\,H_2O + 4\,\overline{e}$$

At cathode (oxygen electrode) $O_2 + 2H_2O + 4\,\overline{e} = 4\,\overline{OH}$

$$\text{Complete reaction } 2H_2 + O_2 = 2H_2O \tag{18.35}$$

In this chemical reaction, the energy representing the enthalpy of combustion of fuel is released and a part is converted into eleectricity and a part into heat. The water formed is drawn off from the side. Electrolyte provides $\overline{OH}$ ions needed for the reaction and remains unchanged at the end since these ions are regenerated. The electrons liberated at the anode find their way to the cathode through the external circuit. This transfer is equivalent to the flow of current from cathode to the anode.

Such cells when properly designed and operated have an open circuit voltage of about 1.1 volt. Unfortunately their life is limited since the water formed continuously dilutes the electrolytes. Fuel cells efficiencies are as high as 60 to 70 percent. Most of the operational fuel cells are low temperature fuel cells employing hydrogen and oxygen as the reactants. These cells operate at about 500 K. A lot of reserch is being conducted on the development of high temperature fuel cells that can be used to operate with impure and inexpensive fuels, like hydrocarbons. A number of fuel cells have been used successfully for special applications. These systems have been used extensively in the space programme for relatively short term application such as manned orbital space flights and Apollo missions to the moon. Unfortunately fuel cells have not progresed to the point where they can supply large quantities of economical electrical energy in an efficient manner for a long time period.

## 18.5  MAGNETO HYDRODYNAMIC GENERATOR

Magneto hydrodynamic generator or MHD generator depends on the Farady's Law of electromagnetic induction just like conventional mechanical electrical generators. In MHD system, a highly electrically conducting fluid is forced through a perpendicular magnetic field at high velocity. In most MHD generators, the working fluid is an ionised gas but liquid metals can also be employed in these systems. The principal components of the plasma MHD generator are shown below. A high temperature plasma is passed through a supersonic nozzle across a stationary perpendicular magnetic field. EMF is induced across suitably placed electrodes. When an ionised gas flows across a magnetic field, a current is induced and a force tending to slow the motion of the gas is experienced. Therefore, MHD generator ducts are made diverging to accommodate the same mass of gas at smaller velocity. Both insulators and electrodes must withstand the high temperature (2000 K to 3000 K) for longer period as well as must have good resistance to thermal shock, ablation and chemical attack. For easily ionisation of the gas, it is seeded with cesium or potasium. The seed materials have a low ionisation potential and it is also recovered from the exhaust gas.

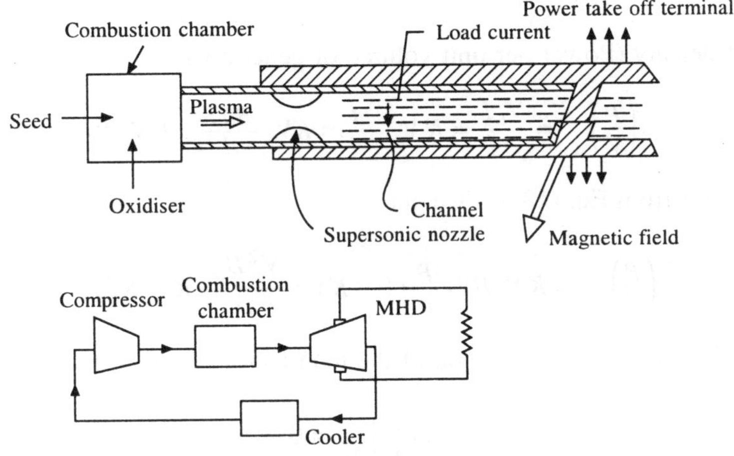


**Fig. 18.7   MHD generator**

MHD converter is usually proposed as a topping cycle or system for a conventional steam power system to improve the overall conversion efficiency of the system. The combined conversion efficiency of such system is

$$\eta_{overall} = \eta_{MHD} + \eta_{steam} - \eta_{MHD} \cdot \eta_{steam} \qquad (18.36)$$

Analysis—If $V_f$ represents fluid velocity and $B$, magnetic field strength remaining constant in the inter electrode gap, then the generated voltage $V_f$ is given by

$$V_g = V_f B d$$

Where $d$ is the inter electrode gap width in meter. If $J$ is the current density, the internal voltage

drop is $I R_g = I \dfrac{\rho_d}{A} = J \rho_d$ and external voltage drop from the

$$\text{unit is } V_L = V_g - J \rho_d \qquad (18.37)$$

Where $\rho$ is the average electrical resistivity of the MHD fluid in ohm meter. The ratio of external load voltage ($V_L$) to the generated voltage ($V_g$) is defined as the load factor (K) of the unit

$$K = \frac{V_L}{V_g} = \frac{V_L}{BV_f d} \tag{18.38}$$

The generator current density is equal to the internal cell. Voltage drop divided by the product of the inter electrode resistance and the electrode area

$$J = \frac{I}{A} = \frac{1}{A}\left(\frac{V_g - V_L}{\dfrac{\rho_d}{A}}\right) = \frac{V_g - V_L}{\rho_d} = \frac{V_g\left(1 - \dfrac{V_L}{V_g}\right)}{g_d}$$

$$= \frac{V_f Bd}{\rho_d}(1 - K) = \frac{V_f B}{\rho}(1 - K) \tag{18.39}$$

The output power density (Power per unit volume of generator)

$$\pm\left(\frac{P}{V}\right)_{out} = \frac{V_L I}{A_d} = K V_g \cdot \frac{J}{d} = KV_f \frac{BdJ}{d} = \mathrm{K}V_f BJ$$

Putting the value of $J$ from Eq. (18.39) we get

$$\left(\frac{P}{V}\right)_{out} = K V_f BV_f \frac{B}{\rho}(1 - K) = \frac{V_f^2 B^2}{\rho}(K - K^2) \tag{18.40}$$

The maximum power density as a function of the load factor $K$ can be found by setting

$$\frac{d}{dK}\left(\frac{P}{V}\right)_{out} = 0$$

and then solving for $K$, we get

$$= 1 - 2K = 0,\ K = \frac{1}{2}$$

and maximum power output

$$\left(\frac{P}{V}\right)_{outmax.} = V_f^2 \frac{B^2}{\rho}\left(\frac{1}{2} - \frac{1}{4}\right) = \frac{V_f^2 B^2}{4\rho} \tag{18.41}$$

The maximum conversion efficiency of an MHD generator is found by assuming that the only loss in the unit is the Joule-heating loss ($P_J$). The loss per unit volume of the generator

$$= \frac{P_J}{V} = \frac{I^2 R_g}{A_d} - \frac{J^2 A^2\left(\dfrac{d\rho}{A}\right)}{A_d} = J^2\rho$$

$$= \frac{V_f^2 B^2}{\rho}(1 - K)^2 \tag{18.42}$$

Conversion efficiency for maximum power density

$$= \frac{P_o}{P_o + P_J} = \frac{1}{1 + \dfrac{P_J}{P_o}} = 1 + \frac{\dfrac{1}{V_f^2 \dfrac{B^2}{\rho} (1 - K)^2}}{V_f^2 \dfrac{B^2}{\rho} (K - K^2)} = \frac{1}{1 + \dfrac{(1 - K)}{K}} = K$$

or,     $\eta_{\max} = K_{\max}$                                    (18.43)

For maximum power density, $K_{\max} = 0.5$ and for this condition maximum possible conversion efficiency is 50 percent.

Problems Encountered in the Design:

   (a)  Sufficient high temperature (2000 to 3000 $K$) can be sustained by only known refractory materials

   (b)  Seed material Potassium attacks insulating materials and make them conducting.

   (c)  Electrode materials are creded by combustion gases

   (d)  Although the overall thermal efficiency is 50 to 55 per cent against 40 percent for conventional thermal power plant, additional investment in the magnet, generator duct, compressor scrubbers seed recovery plant $DC$ to $AC$ converter may increase the plant cost.

**Example 18.1**  A MHD converter used a combustion gas that is seeded with 1 percent potassium to increase electrical resistivity 0.03 ohm-meter. The magnetic field is uniform and perpendicular to the gas velocity and has a flux density of 1.5 Wb/m$^2$. The system is designed for a gas velocity of 900 m/sec and loading factor of 0.55. If the width of the duct is 0.25 m and electrode area is 1.2 m$^2$, determine the output current, voltage and power. Neglect Hall effect and other loss.

*Solution*:  Total generated Voltage = $V_g = V_f \text{B.d} = 900\,(1.5)\,(0.25) = 337.5$ V Voltage drop across load resistor = $V_L = K.V_g = 0.55 \times 337.5 = 185.6$ V

$$\text{Current density } J = \frac{V_g - V_L}{\rho_d} = \frac{337.5 - 185.6}{0.03\,(0.25)} = 20250 \text{ Amp/m}^2$$

$$\text{Output current} = A_J = (1.2)(20250) = 24,\,300 \text{ Amp}$$

$$\text{Output power} = IV_L = 24300\,(185.6) = 4.51 \times 10^6 W$$

$$= 4.51 \text{ MW} \quad Ans.$$

## 18.6  PHOTO VOLTAIC SYSTEMS (SOLAR CELLS)

Electromagnetic energy can be converted directly to electrical energy in the photo-voltaic cell, commonly called the solar cell. Like the fuel cell, the maximum conversion efficiency of this system is not limited by the efficiency of carnot cycle. Despite this, however, the conversion of solar energy into electrical energy is limited to relatively low efficiencies 10 to 15 percent

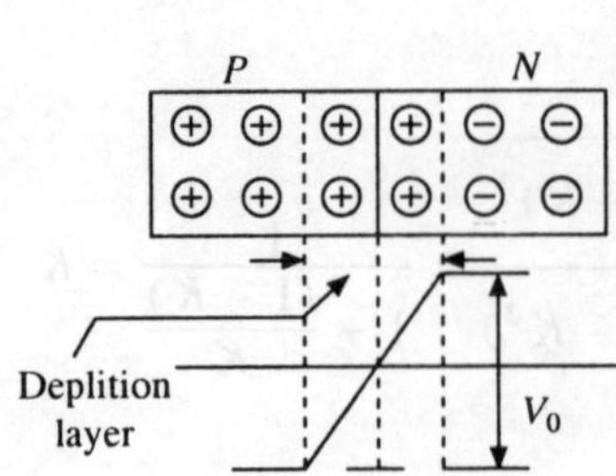
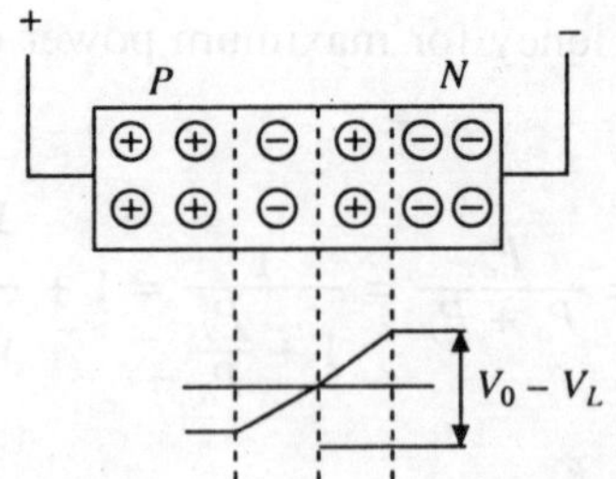

$V_0$ = Potential barrier created across the P–N Junction due to migrations of electrons and holes

Potential barrier decreases due to applied voltage $V_L$ in forward biasing. $V_L$ is supplied by the photon of solar energy.

**Fig. 18.8**

### 18.6.1 Working Principle

Solar cell consists of *P-n* junction of silicon semi-conductor. Then the photons from solar radiation are incident on the *p-n* diode, electrons are dislodged from the valence band causing the photo-electric current. A typical solar cell has been shown in the Fig. 18.9. The non reflected photons incident on the surface of the cell enter the thin outer layer of the semi conducting material and are either converted into heat or produce ion pairs by stripping the valence electrons from the semi-conductor atoms. In order to produce an ion-pair, the incident photon must have an energy in excess of energy gap ($E_g$), which is called the excitation energy. Some of these ions will be separated by the electric field ($V_o$) of the junction. These ions reduce the electric field at the junction and this increases the flow of the majority carriers producing a current flow.

### 18.6.2 Losses in the Solar Cell

The conversion efficiency of a solar cell is not limited like the thermal efficiency of a heat engine cycle but there are some inherent losses that severely limit the cell performance. The losses are:

(i) *Junction Loss*: Junction loss is that loss due to the flow of minority carriers in the junction. For silicon solar cell exposed to solar radiation, the junction loss reduces the conversion efficiency to about 50 percent

(ii) This loss is associated with the energy spectrum of the incident photons and the excitation energy of the semi conductor material. Any photon with an incident energy less than the excitation energy $E_g$ can not produce an ion pair and the photon energy is converted into thermal energy lost.

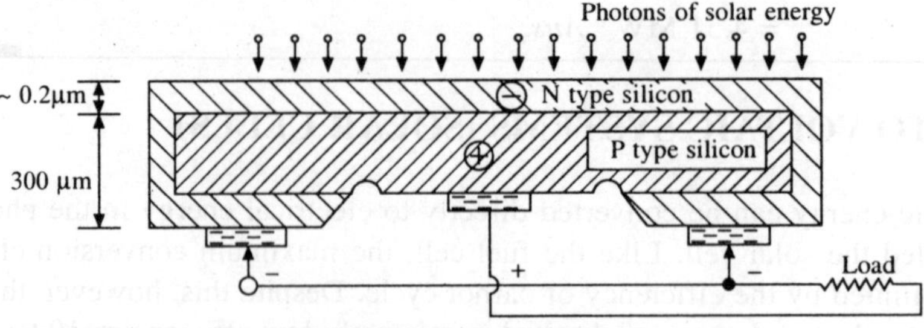


**Fig. 18.9   Typical solar cell**

(iii) Reflection of photons loss
(iv) Recombination of ion pairs before they reach the junction.
(v) Joulean heating particularly in the outer layer

### 18.6.3  Advantages of Solar Cells
(i) They are simple, compact and have a very high power to weight ratio. This makes them very attractive for space applications.
(ii) They directly convert the solar radiation into electricity using photovolatic effect without going through a thermal process.
(iii) They can be located at the place of use and hence no distribution network is required.
(iv) They have long life, 20 years or even more.

### 18.6.4  Disadvantages of Solar Cells
(i) Large areas of solar cell modules are required to generate sufficient useful power because solar intensity is low and conversion efficiency of solar cells is limited to 30 percent.
(ii) The present cost of solar cells are comparatively high, making them economically uncompetitive with other conventional power generation method for terrestrial applications, particularly where the demand of power is very high.
(iii) The system must be intergrated with energy storage system or with another conversion system to supply power at night and on cloudy days. orbiting space crafts also normally require an energy storage system to supply energy when the space crafts are in the shadow of the earth.

## 18.7  SOLAR CELL MATERIALS

There are many semiconductor materials which can be used as solar cells. But some semi-conductors perform better (high conversion efficiency) while others do not perform so well. Apart from many material properties the efficiency of a solar cell is a function of spectral distribution of incident light and the cell temperature. Silicon is the most commonly used material for solar cell and nearly all the cells used commercially, are silicon cells. The electrical properties of silicon depends on the type and amount of dopants. The dopants which are generally used in silicon are given in Table 18.2.

**Table 18.2  Dopants which are commonly used in silicon**

| Dopant | Type | Energy below the conduction bond ($ev$) | Energy above the valance bond ($ev$) |
|--------|------|------|------|
| Al | p | – | 0.057 |
| B | p | – | 0.045 |
| Ga | p | – | 0.065 |
| As | n | 0.049 | – |
| Li | n | 0.033 | – |
| Sb | n | 0.039 | – |
| P | n | 0.044 | – |
| Bi | n | 0.069 | – |
| Ag | n | 0.330 | – |

A solar cell has a $N$-$P$ junction i.e. a combination of $N$ type material like arsenic, lithium, antimony, phosphorus, bismuth, silver (5 valence electron) and $p$ type material like aluminium, boron, gallium, indium (3 valence electron) dopped over a semiconductor of silicon (4 valence electron). When $N$-type transparent layer receives solar light energy, electrons are released from $N$-type material and holes are created in $p$-type material. electrons are negative changes and holes are positive charges (lack of electrons).

$PV$ cells (solar cells) which have only silicon as the base for $PN$ junction are called 'Homojunction' $PV$ cells.

$PV$ cells which have two base materials (e.g. cadmium sulphide-copper) are called hetrojunction $PV$ cell.

Homojunction cells with silicon base are most successful and have following three types

1. Amorphous silicon
2. Polycrystalline silicon
3. Single crystal silicon

Amorphous silicon means non-crystalline silicon. Pure silicon without crystals is used. Crystals are scattered in random direction having no alignment. Amorphons silicon solar cells are least efficient but easy to manufacture. Efficiency of amorphous cells does not exceed 5 percent.

Polycrystalline silicon cell has many crystals in a single silicon. The polycrystalline material has inter-grain boundaries within a cell. Efficiency of such type solar cells is around 7 percent. Manufacturing process is less complex and less costly than that of single crystal silicon process.

Due to lesser complexity, lesser cost, higher production speed, polycrystaline silicon cells are commercially competing with single crystal silicon cells.

## 18.8  EFFICIENCY OF SOLAR CELL AND RATING

The efficiency of a solar cell in converting solar light, in a given spectral range (800–1000 w/m$^2$) and specified cell temperature range (25–45°C), into useful power is defined by:

$$\eta = \frac{V_m I_m}{P_{in}}$$

where $V_m$ and $I_m$ are the voltage and current produced at the highest power point and $P_{in}$ is the power input.

Maximum efficiency occurs at full solar radiation on the $PV$ cell and depends on the materials, design parameters, manufacturing process, test conditions etc. Efficiency range of commercially available solar cell is 12–15 percent.

The following factors limit the efficiency of a photovoltaic solar energy device:

 (i) collection loss
 (ii) reflection loss at the top surface
 (iii) incomplete absorption of photon energy
 (iv) incomplete use of excess photon energy
 (v) shade due to change collection grid at top surface

(vi)  voltage factor loss
(vii)  series and shunt resistance loss

Typical ratings of a solar cell:

|         |         |
|---------|---------|
| voltage | 0.45 V  |
| current | 0.75 A  |
| power   | 0.33 W  |

under condition mentioned above about Irradiance and cell temperature.

A new solar cell has been developed by the NREL (National Renewable Energy Laboratory) of the U.S. Department of Energy and spectrolab giving 32% efficiency by using light concentrated to 47 times greater than sunlight, with the standard AM 1.5 direct spectrum. It is developed in May-June 2000.


## EXERCISES

18.1  What do you understand by direct energy conversion devices? How can these be classified according to their maximum thermal efficiency?

18.2  Name different direct energy conversion devices. Give their present applications and future scope.

18.3  Discuss different thermoelectric effects.

18.4  Sketch and explain the working of a thermoelectric generator.

18.5  Derive expressions for power output and thermal efficiency for a thermoelectric generator. Then derive expressions for maximum power output and maximum thermal efficiency.

18.6  Write short notes on: Thermoelectric materials and multistage thermoelectric generator

18.7  Define the term figure of mirit. Show its effect on thermal efficiency of the thermoelectric power plant taking source temperature as the basic parameter.

18.8  Mention about the advantages of thermoelectric cooling and give scope of its applications.

18.9  What is the basic difference between thermoelectric and thermionic generation systems?

18.10  Sketch and explain the working of a thermionic generator.

18.11  What do you understand by MHD generation? sketch and explain the working of an open cycle MHD generator.

18.12  Derive expressions for voltage generatred, the temperature drop and the thermal efficiency for a MHD generator.

18.13  Sketch and explain the working of a hydrogen-oxygen fuel cell. Give reactions on two electrodes.

18.14  List the advantages of fuel cell over other direct energy converters. Give its present applications and future scope.

18.15  Derive an expression for voltage generated in a fuel cell.

18.16  Describe the basic photovoltaic cell (Solar cell). Sketch and explain its working.

18.17  List different losses occuring in a photo voltaic cell. How these can be reduced?

19

# Conversion of Nuclear Energy

## 19.1 INTRODUCTION

One of the outstanding fact about nuclear power is that large amount of energy can be released from a small mass of active nuclear material. Complete fission of one kg of uranium releases the energy equivalent of 3100 tons = $31 \times 10^5$ kg of coal. The factors which are in favour of nuclear energy are:

(i) it is practically independent of geographical factors (ii) no combustion products and (iii) it is clean source of power which does not contribute to air pollution.

Further it does not require fuel transportation networks and large storage facilities.

There are three general nuclear reactions that release nuclear energy into other forms, usually thermal energy. These three reactions are: (i) radio active decay (ii) fission and (iii) fusion. But out of these reactions, nuclear fission is used to generate power in nuclear reactor at a controlled rate.

Radioisotope power systems are normally lower power systems and the controlled fusion reactor has got to be developed. Consequently the balance of this chapter is concerned with revision of some terms of nuclear physics and principle of operations of nuclear reactors.

## 19.2 NUCLEAR PHYSICS

We shall first introduce some nuclear physics terms which are normally used in reactor terminology.

### 19.2.1 Mass Defect and Binding energy

The nucleus of an atom is comprised of protons and neutrons. Both protons and neutrons are called nucleans and the total number of nucleans in a given atom is called atomic mass number (A). The mass of an atomic nucleus is less than the mass of the individual particles or nucleans that comprise it. The mass defect is the difference between the mass of uncleans and observed nuclear mass. This is the mass defect that holds nucleus together and prevents the coulomb forces of the positive charges (protons) in the nucleus from tearing apart. The mass defect is, in effect negative mass value. The mass defect of a given nucleous is calculated as

$$\boxed{\text{Mass defect} = Z\,m_p + (A - Z)\,m_n - \text{nuclear mass}} \qquad (19.1)$$

Where $m_p$ and $m_n$ are the mass of protons and neutron respectively and $Z$ the atomic number. In order to break the nucleus into its nucleans, a minium amount of energy equivalent to mass defect would have to be added to it. Binding energy of a nucleus is defined as the relativestic energy ($E = mc^2$) equivalent of the mass defect. The average binding energy per nuclean is theoretically the minimum energy required to remove the average proton or neutron from a nculeus. While the mass defect or binding energy increases with the atomic mass, the average binding energy per nuclean rapidly increases and then slowly decreases.

A plot of average binding energy per nuclean and mass number has been shown in Fig. 19.1.

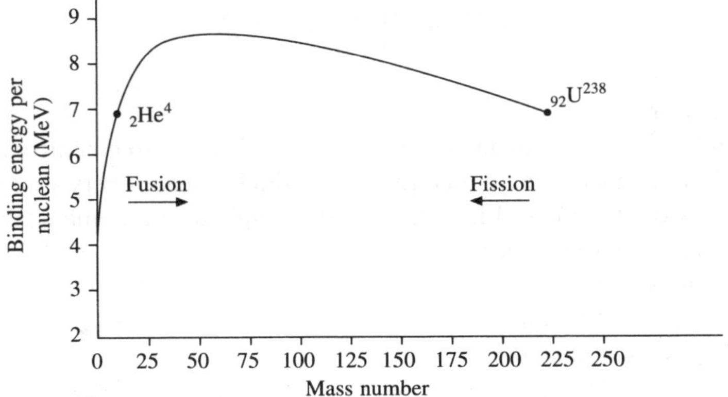


**Fig. 19.1   Variation of Binding energy per nuclear with mass number**

It is noted that the average binding energy per nuclean is a maximum of 8 to 9 MeV for intermediate mass nuclei. This indicates that these intermediate mass nuclei are most stable. This means excess energy is released in any nuclear reaction in which a heavy mass nucleus is broken into intermediate mass nuclei as in fission reaction. Also excess binding energy is released in any nuclear reaction in which two light mass nuclei are combined to form a heavy nuclues as in nuclear fusion, have not been solved, although a large research effort has been expended on this technology. Hydrogen bomb is manufactured on the principle of nuclear fusion.

### 19.2.2   Nuclear Fission

Splitting of a heavy nucleus into two lighter nuclei by a particle of interaction is called nuclear fission. The mass of fission products is less than the parent nucleus with the release of energy. Energy of about 200 MeV is released from complete fission of $_{92}$U$^{235}$. Usually two or three neutrons are also released. Nuclear fission of $U_{235}$ has been given below:

$$\boxed{_{92}\text{U}^{235} + {_0}\text{n}^1 = {_{54}}\text{Xe}^{139} + {_{38}}\text{Sr}^{94} + 3\,{_0}\text{n}^1 + Q} \qquad (19.2)$$

$$_{92}\text{U}^{235} + {_0}\text{n}^1 = {_{56}}\text{Ba}^{141} + {_{36}}\text{Kr}^{92} + 3\,{_0}\text{n}^1 + Q$$

Where $Q$ denotes energy release. All the nuclear reactors develop power by nuclear fission at a controlled rate.

### 19.2.3 Nuclear Fusion

The combination of two lighter nuclei into a heavy nucleus with release of energy is called nuclear fusion. This reaction takes place at very high temperature and at uncontrolled rate. Solar energy is the result of nuclear fusion of hydrogen nuclei into helium. Unfortunately the technical problems leading to the controlled release of energy from the fusion reaction have not been solved although a large search has been expended on the technology. Hydrogen bomb is manufactured on this principal of nuclear fussion. Examples are:

$$_1H^2 + {}_1H^2 \rightarrow {}_2He^3 + {}_0n^1 + 3.26 \text{ MeV}$$

$$_1H^3 + {}_1H^2 \rightarrow {}_2He^4 = {}_0n^1 + 17.4 \text{ MeV}$$

$$4_1H^1 = \underset{\text{Positron}}{_2He^4 + 2e^+ + Q} \tag{19.3}$$

### 19.2.4 Chain Reaction

In nuclear fission of $U^{235}$, three neutrons are also released. These three neutrons strike further the three nuclei of $U^{235}$. And hence a chain reaction starts which releases very vast amount of energy. The original neutron would then act like a match stick applied to a combustable material.

If we apply chain reaction to nuclear reactor, the following observations are recorded. Uranium exists as isotopes of $_{92}U^{238}$, $U^{234}$, and $U^{235}$. Out of these isotopes $U^{235}$ is most unstable. When a neutron is captured by a nucleus of an atom of $U^{235}$, it splits up roughly into two equal fragments and about 2.5 neutrons are released and a large amount of energy (200 MeV) is also released. The neutrons are fast and they can further start chain reaction. The neutrons so released have velocity $1.5 \times 10^7$ meter/second. Out of 2.5 neutrons released in fission of each nucleus of $U^{235}$, one neutron is used to sustain chain reaction, about 0.9 neutron is captured by $U^{238}$

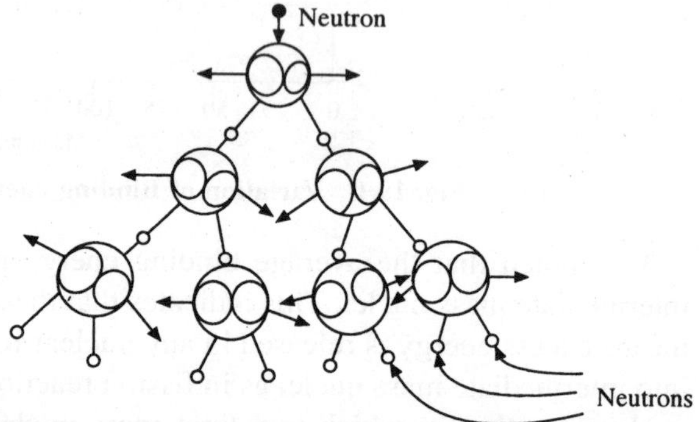


Fig. 19.2 Chain reaction

which gets converted into $Pu^{239}$ and about 0.6 neutrons are absorbed by control rods coolant and moderator and partly escape from reactor.

## 19.3 NUCLEAR FUELS

The three fissionable fuels are $_{92}U^{235}$, $_{92}U^{233}$ and $_{94}Pu^{239}$. All of these fuel isotopes are radioactive (alpha emitter) but they have very long half lives. It will be noted that all these isotopes have odd atomic mass numbers and even atomic numbers. Evidentlh, the even-even atomic mass and atomic number configuration formed by absorption of neutron is so unstable that the binding energy of absorbed neutron is sufficient to cause fission. Where there are other isotopes that undergo fission following neutron absorption, such as $Pu^{241}$, $Pu^{243}$, $Cf^{243}$, $Cf^{245}$ etc., the availability of these isotopes is very poor.

Of the three common fissionable isotopes $_{92}U^{235}$ is most unstable and therefore it is called *Primary fuel*. $U^{233}$ and $Pu^{239}$ are artificially produced by capture of neutrons from $Th^{232}$ and $U^{238}$ respectively, therefore $U^{233}$ and $Pu^{239}$ are called *Secondary fuels*. $U^{233}$ and $Pu^{239}$ are *fissionable* materials.

*Fertile Material*

*It is definied as the material which absorbs neutrons* and undergoes spontaneous changes which lead to formation of fissionable materials. $Th^{232}$ and $U^{238}$ are fertile materials.

The uranium 235 fission process produces on the average about 2.5 neutrons per fision and only one of these neutrons is required to sustain chain reaction. The excess neutrons are used to produce new fuel atoms by activating the fertile isotopes. An important parameter $(r)$ called *breeding ratio* is defined as the ratio of number of fertile atoms consumed (new atoms formed) to the number of original fuel atoms consumed in the fission and radioactive capture process. If $r > 1$, reaction is called breeder reactor and if $r < 1$, it is called converter reaction. The following nuclear reactions take place.

$$_{92}U^{238} + _{0}n^{1} \rightarrow _{92}U^{239} + \gamma \text{ (half life of } _{92}U^{239} \text{ is 23.5 minutes)} \tag{19.4}$$

$$_{92}U^{239} \rightarrow _{93}Np^{239} + \gamma \text{ (half life of } _{93}Np^{239} \text{ is 2.3 days)} \tag{19.5}$$

$$_{93}Np^{239} \rightarrow _{94}Pu^{239} + \bar{\beta} \text{ (half life of } _{94}Pu^{239} \text{ is 24000 years)} \tag{19.6}$$

$$_{90}Th^{232} + _{0}n^{1} \rightarrow _{90}Th^{233} + \gamma \text{ (half life of } _{90}Th^{233} \text{ is 23.3 minute)} \tag{19.7}$$

$$_{90}Th^{233} \rightarrow _{91}Pa^{233} + \bar{\beta} \text{ (half life of } _{91}Pa^{233} \text{ is 27.4 days)} \tag{19.8}$$

$$_{91}Pa^{233} \rightarrow _{92}U^{233} + \bar{\beta} \text{ (half life of } _{92}U^{233} \text{ is } 1.6 \times 10^{5} \text{ years)} \tag{19.9}$$

Natural uranium available from mines contain:

| | |
|---|---|
| $U^{238}$ | 99.274 percent |
| $U^{235}$ | 0.720 percent |
| $U^{234}$ | 0.006 percent. |

Isotopic abundance of $U^{235}$ is so low that many reactors cannot operate using natural uranium fuel. This means that $U^{235}$ concentration must be increased in the uranium and this fuel is said to be Enriched uranium.

In order to produce enriched uranium, it is necessary to be able to separate the uranium isotopes. The separation techniques are very costly. The earliest separation method, the elecctromagnetic separation process involved the ionisation of uranium atoms and then acceleration of them through a perpendicular magnetic field. The most recent process is gaseous diffusion method. Since the average velocities of gas molecules at a given temperature depends upon their molecular weights, the different isotopes will diffuse through a porous barrier at different rates.

## 19.4 THERMAL NEUTRONS

During the fission of $_{92}U^{235}$ on an average 2.5 neutorns are also released. These neutrons possess very high velocity and they can penetrate through the remaining nuclei. In order to sustain chain reaction, their velocities are slowed down. The slow neutrons which can *start nuclear fission are called thermal neutrons*. We can classify the fast moving neutrons into three categories depending upon their velocity.

  (i)  Thermal neutrons (0.025 eV to 1.0 eV.)   0.025 eV corresponds to velocity of 2200 m/sec.
  (ii)  Intermediate neutrons (1 eV to 0.1 MeV)
  (iii)  Fast neutrons 0.1 MeV or more

The process of slowing down fast neutrons is called Moderation and the material which slows down the neutrons is called moderator.

## 19.5  NUCLEAR REACTOR

*The nuclear reactor is a device where in a sustained chain reaction may be obtained at a controlled rate*, resulting in the fission of a heavy nucleus of $_{92}U^{235}$, $_{94}Pu^{239}$ and $_{92}U^{233}$ induced by neutrons. There are other products from the nuclear fission reaction, e.g. the two intermediate mass fission products and both particles, the neutrons and energy.

### 19.5.1  Critical Mass

Every operating nuclear reactor has a critical mass. The critical mass is simply the minimum mass of fissionable material that will just sustain a fission chain reacion. Values of critical mass range from 200 gm to more than 500 kg. The actual fuel loading is more than the critical mass. If fuel loading is less than this value fission can not take place.

### 19.5.2  General Components of Nuclear Reactor

The followings are the general components of a nuclear reactor

### 1. Fuel

The nuclear fuels which are generally used in reactor are $_{92}U^{235}$, $_{94}Pu^{239}$ and $_{92}U^{233}$. Among the three the $_{92}U^{235}$ is naturally available up to 0.72 percent in the uranium ore and the remaining is $_{92}U^{238}$. $_{91}Pu^{239}$ and $_{92}U^{233}$ are formed in the nuclear reactor during fission process from $_{92}U^{238}$ and $_{92}Th^{232}$. The fuel is shaped and located in the reactor in such a manner that the heat production

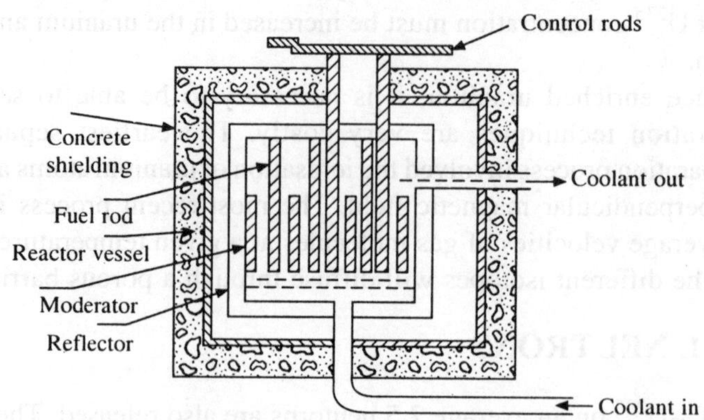


**Fig. 19.3  Nuclear Reactor**

within the reactor is uniform. The fuel elements are designed for better heat transfer characteristics and non corrosive. Usually the fuel is encased in metal cladding, generally, stainless steel and aluminium to prevent against corrosion and oxidation.

## 2. Moderator

The purpose of moderator is to slow down the fast moving neutrons released from fission of $U^{235}$. The fast moving neutrons can escape out from the reactor. The moderator nuclei do not capture the neutrons but scatter and slow them down.

### Materials

Graphite, heavy water and berryllium are used as moderator with natural uranium. The ordinary water can also be used as moderator for enriched uranium.

The desirable properties of a good moderator are:

1. It must be as light as possible to slow down neutrons in elastic collision.
2. It should slow down the neutrons but should not absorb them.
3. It must have high resistance against corrosion, pressure and temperature.
4. It must have high melting point if it is solid.
5. It must have high chemical stability.
6. It should have high thermal conductivity for better heat transfer.

## 3. Control Rods

The energy produced in the reactor due to fission of nuclear fuel during chain reaction is so high that if it is not controlled properly, the entire core and surrounding structure may melt. The purpose of control rod is to absorb the excess neutrons when the reaction rate increases. This is done by inserting the fuel rods inside the reactor. The material of control rod should be such that it should not become radioactive by absorbing neutrons.

### Materials

Boron, cadmium and halfnium are used as materials of control rods.

## 4. Reflector

The neutrons produced during the fission process will be partly absorbed by the fuel rods, moderators, coolent or other structural material. Neutrons unabsorbed will try to escape. To turn the fast flying neutrons back into the reactor, reflectors are provided surrounding the inner surface of the reactor.

### Material

Graphite, beryllium and heavy water are used as reflector

## 5. Coolant

The main purpose of the coolant in the reactor is to transfer the heat produced inside the reactor. The coolant carries the heat from the reactor and transfers it to water to generate steam for running steam turbine. Some of the desirable properties of good coolant are:

(i) It must not absorb neutrons
(ii) It must have high chemical and radiation stability.

(iii) It must have high density, low viscosity, high thermal conductivity, high specific heat for better heat transfer and low pumping cost.

(iv) It must have high boiling point (if liquid) and low melting point if solid.

(v) It should be non corrosive

(iv) It should be non oxidising and non toxic.

### *Material*

The water, heavy water, air $CO_2$, $H_e$ and liquid metals like sodium, potassium bismuth are used as coolant.

### *6 Shielding*

The reactor is a source of intense radioactivity. The common radiations from the reactor are $\alpha$ particles, $\beta$ Particles, $\gamma$ rays and neutrons. To prevent the effects of these radiations on the human body, thick layer of lead or concrete is provided all around the reactor. The inner lining of the core is made of 50 cm to 60 cm thick steel plate and it is further thicknened by few meters using concrete.

### *7. Reactor Vessel*

It is the vessel enclosing the reactor core, reflector and shield. It also provides the entrance and exit passages for directing the flow of coolant. The reactor vessel has to withstand the pressure as high as 200 bar or above. Holes are provided at the top of vessel to insert control rods.

## 19.6  LAYOUT OF NUCLEAR POWER PLANT

The arrangements for the production of power in a nuclear power system are given in the Fig. 19.4.

The conventional fossil fuel boiler is replaced by a nuclear reactor. The heat liberated in the nuclear reactor by fission is transported by using a coolant which is usually a liquid metal. The coolant transfers its energy to water in the boiler to generate steam.

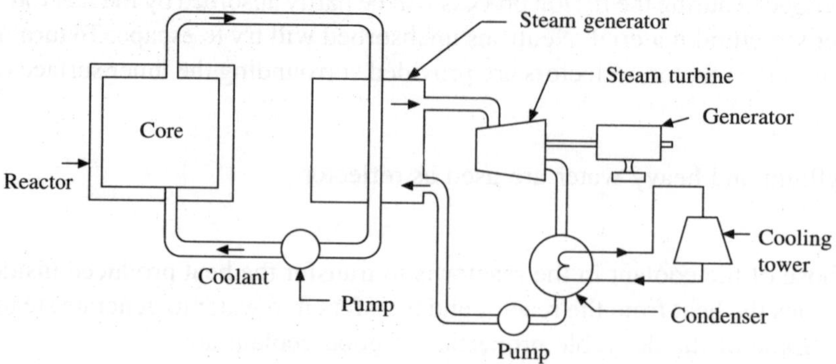


**Fig. 19.4   Nuclear power station**

## 19.7  HALF LIFE

A nucleus possesses an extremely high excitation energy and emits a neutron. The binding energy of a neutron is about 8 MeV. If the excitation energy becomes equal or more than 8 MeV, the nucleus gets decay by the emission of a neutron. The rate of decay is a function only of the number of radioactive nuclei present at a time, provided that the number is large. It does not depend on temperature, pressure or the physical and chemical states of phase.

If $N$ be the number of radioactive nuclei of one species at any time $\theta$, the rate of decay

$$-\frac{dN}{d\theta} = \lambda N \tag{19.10}$$

where $\lambda$ = costant of proportionality (Decay constant) By integrating, a simple exponential relation is obtained as

$$N = N_o e^{-\lambda\theta} \tag{19.11}$$

where   $N_o$ = radioactive atoms present at time $\theta = 0$
and        $N$ = radioactive atoms present at time $\theta$.
Thus from Eqs. (19.10) and (19.11)

$$-\frac{dN}{d\theta} = \lambda N = \lambda N_o e^{-\lambda\theta} \tag{19.12}$$

The decay rate is often expressed in the form of half time $\theta_{1/2}$. Thus half time is defined as the time during which one half of the number of radioactive species decays.

Thus,
$$\frac{N}{N_o} = \frac{1}{2} = e^{-\lambda\theta\frac{1}{2}}$$

or,
$$\theta_{1/2} = \frac{\ln 2}{\lambda} = \frac{0.6931}{\lambda} \tag{19.13}$$

## 19.8  NEUTRON MULTIPLICATION FACTOR

For a nuclear reaction to be self sustaining one of the newly produced neutrons during fission should slow down to the thermal level to induce fission of another $U^{235}$ nucleus. If the conditions are such that only one neutron on an average shows down and fissions another nucleus, the neutron population in the reactor is constant and the reactor is called critical. The neutron multiplication factor (k) is defined as the ratio of number of neutrons produced in one generation to the number of neutrons present in the preceeding generation.

### 19.8.1  Subcritical, critical and supercritical Reactors

(i) If $K < 1$, the neutronpopulation and hence reactor power decreases with time and reactor is called subcritical.
(ii) If $K = 1$, neutron is called critical.
(iii) If $K > 1$, the neutron population and hence reactor power increases with time and reactor is called supercritical

Another term which is closely related to the multiplication factor is the reactivity ($\rho$) of the reactor which is related with $k$ by the equation

$$\rho = (K - 1)/K \qquad (19.14)$$

For supercritical reactor, $\rho$ is positive, for a critical reactor $\rho$ is zero and for subcritical $\rho$ is negative.

### 19.8.2   Four Factor Formula

A neutron in a reactor may either fission a $U^{235}$ nucleus or be absorbed by any of the materials in the cores without producing fission. The neutron multiplication factor is related with four important parameters. A relation between $K$ and these parameters can be derived as follows: Let $n$ be the number of neutrons arising from fission of every neutron absorbed in the reaction. For natural uranium $n = 1.33$ and for pure $U^{235}$, $n = 2.08$. Some neutrons are also produced due to fission of $U^{238}$. Therefore a term $\varepsilon$, fast fision factor ($\varepsilon = 0.025$ to $1.03$ for natural uranium) is introduced which takes into account additional neutron from $U^{238}$. Then fast neutrons for every absorbed neutron by $U^{235}$ becomes $n\varepsilon$. Out of these neutrons, some leak out of the reactor core. The probability of neutrons remaining in the core is $P_f$ (non-leakage probability) with a value around $0.63$ when water is the moderator.

The second loss is slowing down due to absorption occuring at resonance peaks in the $U^{238}$ absorption peaks and in the moderator. If $P$ is the probability of escape from absorption the number of neutrons reaching thermal level without leakage or absorption is $n\in p_f p$. Usually $p$ lies between $0.8$ and $1$. After they slow down, still some thermal neutrons leak out from the finite size of reactor. Let $p_t$ = probability of non leakage of thermal neutrons (usually $0.8$). If $f$ (called thermal utilization) denotes the rates of the number of thermal neutrons fissioning $U^{235}$ to the total number of thermal neutrons absorbed by the core then number of neutrons at the end of the cycle is equal to:

$$K = n\in pf\, p_f p_t \qquad (19.15)$$

$K$ is called neutron multiplication factor discussed earlier. For a theoretically fabricated sized reactor $p_f$ and $p_t$ will be unity because there will not be any leakage and then $k_\infty$ for infinite size reactor becomes.

$$K_\infty = n\in p \cdot f \qquad (19.16)$$

The above formula is called four factor formula.

Where   $n$ = number of neutrons arising from one neutron absorbed

$\varepsilon$ = fast fission factor

$p$ = probability of escape from resonance

$f$ = fraction of thermal neutron captured in the fuel.

### 19.8.3   Neutron Flux

Since the neutron population and the fission reaction rate is directly proportional to the thermal power of the reactor, it is important to have some quantity or term that designates the neutron level in the reactor. The term is neutron flux, ($\phi$). The neutron flux is the number of enutrons passing a unit area in a unit time and has units of neutrons per $m^2$ per second. It is a scalar quantity and ranges from $10^{15}$ neutrons/m$^2$ sec for a low powered reactor to less than $10^{20}$ neutrons/m$^2$ sec. for some high powered reactor. The neutron flux is equal to the product of neutrons density $n$ (neutrons/m$^3$) and the neutron velocity $v$(m/sec)

$$\phi = v \cdot n \tag{19.17}$$

### 19.8.4  Classification of Nuclear Reactors

The nuclear reactors are classified on the following basis:

1. On the basis of neutron energy:
   (a) Fast reactors: In these reactors the fission is effected by the fast neutrons without any moderators.
   (b) Thermal reactors: Fission is effected by the thermal neutrons.
2. On the basis of fuel used:
   (a) Natural fuel: In this reactor the natural uranium (99.282 U$^{238}$, 0.712% U$^{235}$, 0.006 percent U$^{234}$) is used as fuel and generally heavy water or graphite is used as moderation.
   (b) Enriched uranium
   In this reactors the uranium used contains 5 to 10% U$^{235}$ and ordinary water can be used as moderator.
3. On the basis of moderator used:
   (a) Water moderated
   (b) Heavy water moderated
   (c) Graphite moderated
   (d) Berillium moderated
4. On the basis of coolant used:
   (a) Pressurised water reactor (PWR)
   (b) Boiling water reactor (BWR)
   (c) Liquid metal fast breeder reactor (LMFBR)
   (d) Gas cooled reactor (GCR)
   (e) Organic liquid cooled reactor
5. On the basis of core:
   (a) Homogenious: In this reactor nuclear fuel is finally divided and mixed with a solid media as well as coolants in the form of a liquid solution or slurry.
   (b) Heterogenious: The fissionable material is in the form of a pure metal or as a compound is arranged in a regular pattern or in the form of rods.

## 19.9  NUCLEAR REACTORS

Nuclear reactor is a controlled chain reacting system supplying nuclear energy through fission process. It acts as a furnace in which fuel likes U$^{235}$, U$^{233}$ or Pu$^{239}$ burns and as a result, produces many useful products like heat, neutrons and radio isotopes. Different types of reactors are described as below:

(A) *Ordinary water cooled reactors*: These types are further subdivided into boiling water reactor and pressurised water reactors. For both type, enriched uramium is used as fuel.

**1.** *Pressurised Water Reactor (PWR):*  A pressurised water cooled reactor is a ordinary light water cooled and moderated thermal reactor having an unusual core design, using both natural and highly enriched fuel. The enriched uranium fuel is used in the form of thin rods or plates. The cladding is either of stainless steel or zircaloy. Because of very high coolant pressure, the steel pressure vessel containing the core must be about 20 to 25 cm thick. A typical PWR contains about 200 fuel assemblies, each assembly being an array of rods. In a typical fuel assembly, these are 264 fuel rods and 24 guide tubes for control rods.

The arrangement of the reactor is shown in Fig. 19.5. The coolant leaving the reactor and is radioactive, enters the steam generator which can be either shell and tube type with *U*-tube bundle or once-through type, the former being more common. In the U-tube heat exchanger (boiler), the hot coolant (after getting heat from reactor) enters an inlet channel head at the bottom, flows through the U-tubes, and reverses direction to an outlet at the bottom. It can produce only saturated steam. In the once through design, the primary coolant enters at top, flows downward through tubes and exits at the bottom to the main pumps. Feed water is on the shell side. A dry or low degree of superheat steam is possible.

The excellent properties of water as a moderator and coolant make it a natural choice for power reactors. The coolant pressure in primary circuit should be high so that the boiling of water takes place at high pressure in secondary circuit. The most important limitation on a PWR is a critical temperature of water, 374°C. This is the maximum possible temperature of the coolant in the reactor and in practice it is considerably less, possibly about 300°C, to allow a margin of safety. In a PWR, the coolant pressure must be greater than the saturation pressure at say, 300°C (85.93 bar) to supress boiling. The pressure is maintained at about 155 bar so as to prevent bulk boiling. A pressurising tank keeps the water at about 100 bar so that it will not boil. Electric heating coils in the pressuriser boil some of the water to form steam that collects in the dome. As more steam

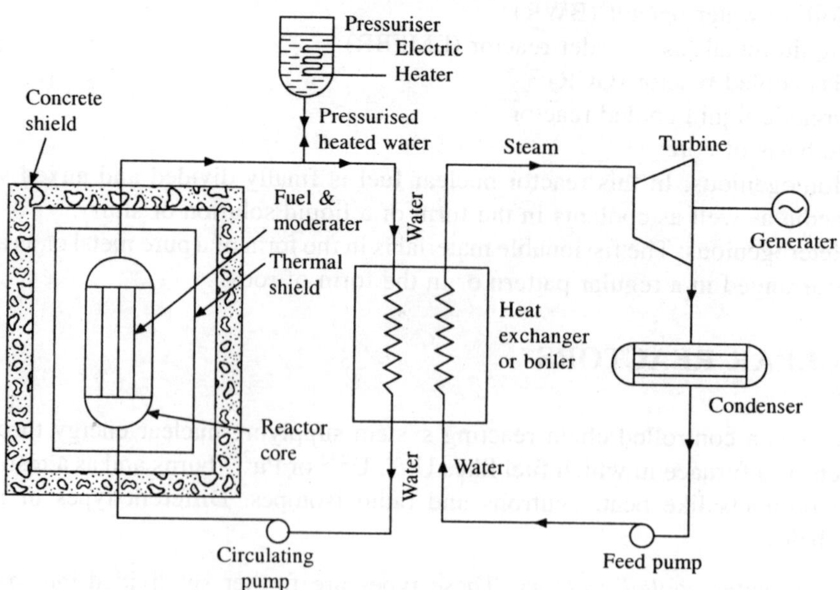


**Fig. 19.5  Pressurised water reactor (PWR)**

is forced into the dome by boiling, its pressure rises and pressurises the entire circuit. The pressure may be reduced by providing cooling coils or spraying water on the steam.

Water acts both as coolant as well as moderator. Either heavy water ($D_2O$) or the light water may be used for the above purpose. The first land-based PWR for power generation was built at shipping part, USA in 1957 producing 230 MW with pressure 140 bar in primary circuit (loop).

The advantages and disadvantages of this type of reactor are given as under.

*Advantages*

1. Water used in reactor as coolant and moderator which is cheap and easily available.
2. The reactor is compact and its power density is high (65 kw/lit.)
3. Number of control rods required is considerably less.
4. Fission products remain contained in the reactor, as a result, steam is not contaminated by radiation.

*Disadvantages*

1. The cost of reactor is high as it uses enriched uranium
2. Secondary cycle having low Rankine efficiency (22%) because heat transfer takes place first from core to water and then to generating steam.
3. Severe corrosion problem takes place in reactor core due to having more intensity of formation at high pressure and high temperature.
4. Fuel suffers radiation damage and, therefore, its processing is difficult.
5. Fuel element fabrication is expensive.
6. Capital cost is high as primary circuit requires strong pressure vessel.

### 2. Boiling Water Reactor (BWR)

In this reactor also enriched uranium is used as fuel and ordinary water as coolant. The ordinary water is allowed to boil in the reactor core and the steam is supplied directly to the steam turbine. The arrangement of the system is shown in Fig. 19.6. The arrangment eliminates the need of steam

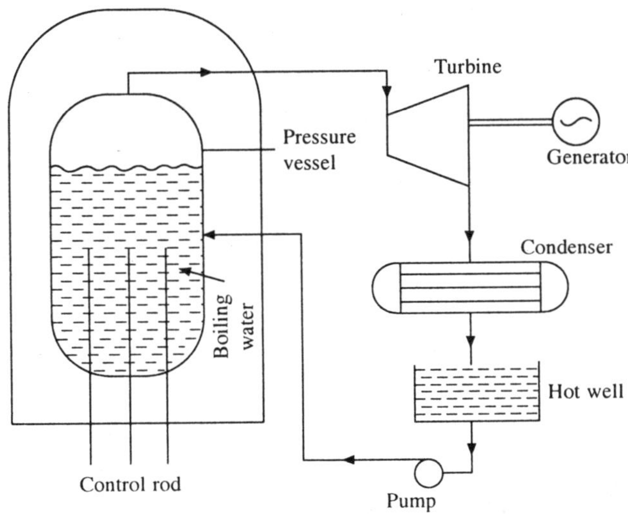


**Fig. 19.6  Boiling water reactor (BWR)**

generator. Therefore the coolant serves the triple function of coolant, moderator and working fluid. Since the coolant boils in the reactor itself, its pressure is much less than that in a PWR and it is maintained at about 70 bar with steam temperature around 285°C. The Tarapur power station is of this type. The advantages and disadvantages of this reactor as compared with PWR are listed below

### *Advantages*

1. There is gain in thermal power as heat exchanger (boiler) is eliminated and consequently gain in cost.
2. The pressure in side the pressure vessel is not high so a thicker vessel is not required.
3. The temperature of fuel surfaces is less so the reactor fuel assembly is more safe.

### *Disadvantages*

1. It is not possible to meet sudden power increase.
2. Possibility of radioactive contamination in the turbine mechanism, should there be any failure of fuel elements.
3. The possibility of the burn-out of fuel is more because of high heat flux (> 650 kJ/cm$^2$-hr) may occur on the fuel.

### *3. Gas-Cooled Reactors (GCR)*

In this type of reactor the coolant used is a gas like air, hydrogen, helium or carbondioxide. Graphite is used as a moderator. Helium and $CO_2$ are the gases used as heat carrying fluid from the reactor. The Problem of corrosion is reduced much in this type of reactor. Thickness of reactor of this type is much reduced as compared to previous ones. Arrangement of the system is shown in Fig. 19.7.

For centre station services, there are mainly two types of gas cooled reactors as mentioned below:

(i) The gas cooled, graphite moderator reactor (GCGM)
(ii) The high temperature gas cooled reactor (HTGC)

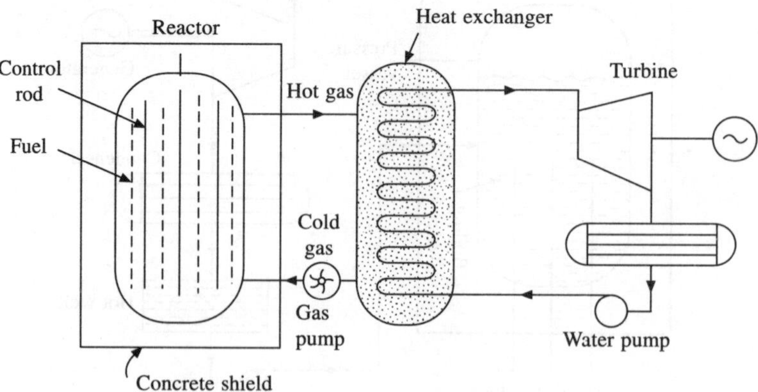


**Fig. 19.7   Gas cooled reactor (GCR)**

GCGM uses natural uranium as fuel where as HTGC uses enriched uranium mixed with thorium carbide and clad with graphite. The coolant pressure and temperature in GCGM are about 7 bar 336°C respectively and for HTGC, the values are 15 to 30 bar and 700°C to 800°C. Advantages and disadvantages of this type of reactor as listed as under.

*Advantages*

1. The gas is safer and effective as a coolant and easy to handle.
2. The processing of the fuel is simpler
3. No corrosion problem
4. The problem of cladding the metallic fuel does not arise.
5. The use of $CO_2$ as coolant eliminates the possibility of explosion.
6. Graphite remains stable under higher irradiation conditions.

*Disadvantages*

1. Power density is very low due to low heat transfer co-efficient, therefore large vessel is required.
2. The leakage of gas is one of the major problem.
3. Large amount of fuel loading is required initially because critical mass is high.
4. More power is required for coolant circulation as compared with WCR.
5. Neutron flux is high at the centre of the core.

### 4. Liquid Metal Fast Breeder Reactor (LMFBR)

Liquid metal reactor uses sodium as a coolant. The liquid metal coolant can be operated at higher operating temperature and low pressure (650°C, 1 bar). Sodium boils at 880°C under atmospheric pressure and freezes at 95°C. Hence sodium is first melted by electric heating system and be pressurised to about 7 bar. liquid sodium is then circulated by the circulation pump. The thermal efficiency is higher due to higher temperature. The reactor has two coolant circuits or loops. The arrangement is shown in Fig. 19.8.

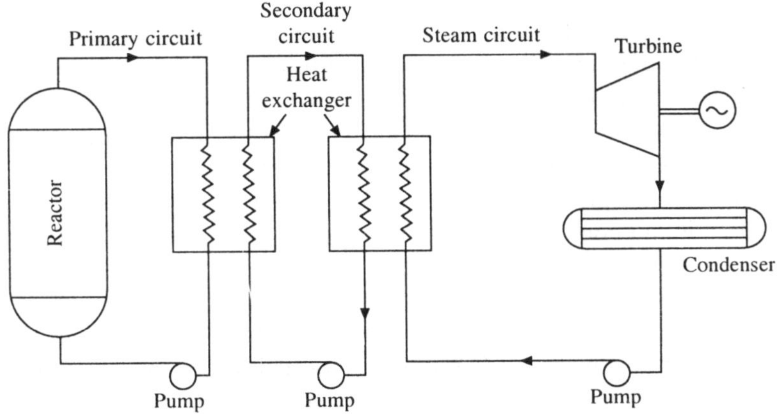


**Fig. 19.8  Liquid metal cooled reactor**

(i) The primary circuit contains sodium which gets heated and melted in the core. Then after it is circulated to first heat exchanger and then comes back to core after transfering heat.

(ii) The secondary circuit contains an alloy of sodium and potassium in liquid form. This coolant gets heated in intermediate heat exchanger after taking heat from hot liquid sodium. The heated liquid of sodium-potassium is then passed through once through type boiler having tubes only. The steam generated from this boiler will be superheated. Feed water from the condenser enters the boiler, the heated sodium-potassium passing through the tubes gives heat to the water thus converting it into steam. The sodium-potassium liquid in the second circuit is then pumped back to the intermediate heat exchanger thus making it a closed circuit.

For preventing from radioactivity, the reactor core, primary loop and intermediate heat exchanger are shielded properly. The liquid metal be handled under the cover of inert gas to prevent contact with air at the time of charging or draining the primary or secondary circuit. Advantages and disadvantages of this system is given as under.

*Advantages*

1. The size of the reactor is comparatively small.
2. Sodium as a coolant needs not be pressurised.
3. Superheating of steam is possible.
4. Low cost graphite moderator can be used.
5. High conversion ratio.

*Disadvantages*

1. Sodium reacts violently with water and actively with air
2. Heat exchanger must be leak proof.
3. Thermal stresses developed to be minimised
4. Compare to other coolant, leak of sodium is very dangerous.

### 5. *Heavy Water Reactor or CANDU (Canadian-Deuterium-Uranium) Type Reactor*

This type of reactor with natural uranium (0.7% $U^{235}$) and heavy water as moderator finds an alternative way as compared to light water reactor because the latter one requires fuel enrichment facilities there by more cost. In this type of reactor, heavy water (99.8% deuterium oxide $D_2O$) is used as moderator and coolant as well as the neutron reflector. It was developed by canadian engineers. They are called CANDU-PHW (candian-Deuterium-Uranium pressurised heavy water). CANDU reactor differs from light water reactor (LWR) in the sense that in the latter same water serves as both moderator and coolant, whereas as in former case, the moderator and coolant are kept seperate.

In the CANDU reactor moderator is containd in a cylinderical steel vessel, called the calandria, with a large number of zircaloy tubes through it parallel to its axis, which is horizontal. Figure 19.9 shows the schematic arrangement of a CANDU reactor. The active core region is approximately 6 m high with a diameter of 7 to 8 m. The vessel is penetrated by some 380 horizontal channels called pressure tubes because they are designed to withstand a high internal pressure. The channels contain the fuel element and pressurised coolant ($D_2O$) enters the channels at 260°C and 110 bar,

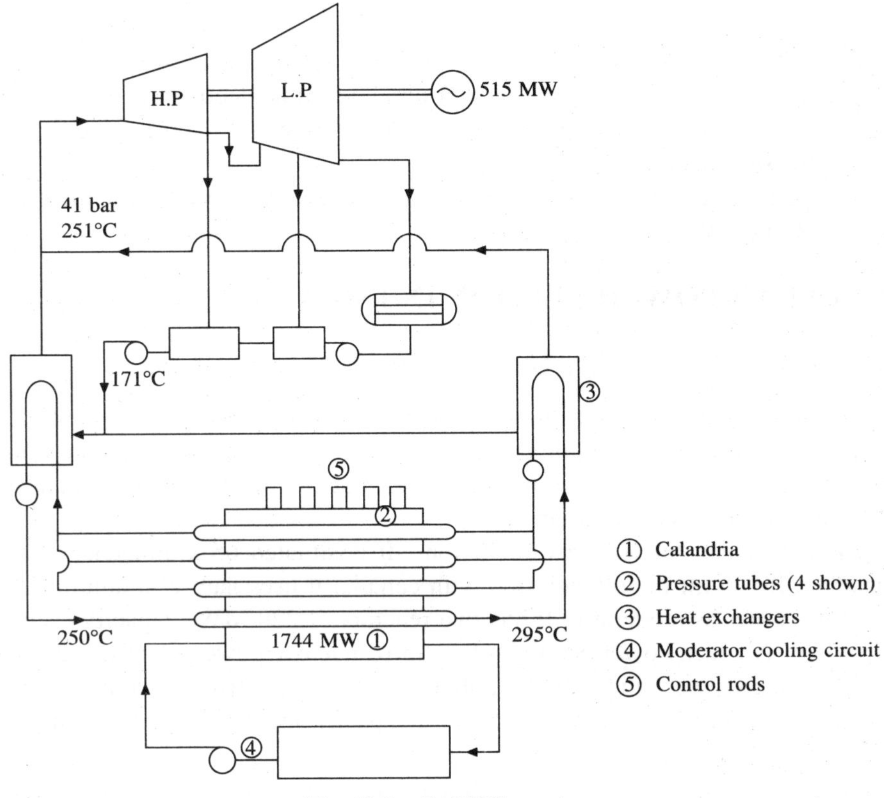


**Fig. 19.9   CANDU reactor**

flows through the fuel elements, and leaves at 320°C, and the net efficiency is about 29%. Coolant flows in the opposite directions in adjacent channels. Like PWR, there is no bulk boiling of coolant. The heavy water coolant pressure in a typical reactor is around 89 bar, and the inlet and outlet temperatures are 250°C and 290°C respectively. In heat exchangers, steam is generated at 41 bar pressure, 251°C. The thermal power of each reactor is 1744 MW, and the net electrical output is 515 MW, giving a thermal efficiency of 29.5%.

The moderator temperature is maintained at about 70°C and low pressure to reduce heavy water losses. The fuel assembly contains 37 fuel rods. Each rod contains natural uranium dioxide ($UO_2$) fuel pellets with 0.38 zircaloy cladding. Each rod bundle is about 0.1 m in diameter and 0.5 m long. The total mass of fuel in the core is about 97,000 kg. The CANDU reactor is unusual in that refueling is conducted while the reactor is in operation.

Advantages and disadvantages of CANDU reactor are given as under

*Advantages*
1. natural uranium can be used as fuel
2. cost of the reactor is less because enrichment of fuel is not required and vessel has not to withstand high pressure.
3. control is easier
4. Fuel consumption is low as heavy water is the best moderator
5. The period required for construction is shorter than for PWR and BWR.

*Disadvantages*

1. The cost of heavy water is very high (Rs. 600 per kg) as it is present in ordinary water in the ratio 1 : 6000. One of the methods of obtaining it is electrolysis of ordinary water.
2. It requires a very high standard of design, manufacture and maintenance.
3. The reactor should be leak proof
4. The size of the reactor is extremely large as power density is low (10 kw/litre) as compared with PWR and BWR.

## 19.10   NUCLEAR POWER STATION IN INDIA

The various nuclear power stations in India are as follows:

(i) Tarapur Nuclear Power Station—It is India's first nuclear power plant. It has been commissioned at Tarapur 96 kilometers north of Mumbai in Maharashtra state with American collaboration (GEC). It has two boiling water reactors of 200 MWe each. The reactor is BWR and uses enriched uranium.

(ii) Rana Pratap Sagar Nuclear Power Station—It is situated 65 kilometers west of kota in Rajasthan. It is built in collaboration with canada. It uses natural uranium ($U_2O$) as fuel and heavy water as moderator. It has two reactors of 220 MWe capacity each

(iii) Kalpakkam Nuclear Power Station—It is third nuclear power plant designed by Indian scientist and engineers and is built at about 65 kilometers from Chennai near seashore. It also has two fast reactors each of 235 MWe capacity and using natural uranium as fuel. It works on CANDU type principle.

(iv) Narora Nuclear Power Station—It is India's fourth nuclear power station and is built at Narora in Bullandshahar district of Uttar Pradesh. It also uses (CANDU type reactor. It has two units of 235 MWe. It has been designed and constructed wholly by the Indian scientist and engineers. It supplies power to Delhi, Haryana, Himachal Pradesh, Jammu & Kashmir, Punjab, Rajasthan, Uttar Pradesh and Chandigarh.

(v) Kakarpar Nuclear Power Station—This fifth nuclear power plant of India is located at Kakarpar near surat in Gujarat. It also works on CANDU type principle. Two units, each of 235 MWe. The fuel for the power plant is fabricated at the nuclear fuel complex, Hyderabad. There is no human settlement for 5 km belt around a nuclear power installation as a mandatory provision.

(vi) Kaiga Atomic Power Station—The sixth atomic power plant located at kaiga in Karnatka. It has two units of 235 MWe. This also works on CANDU type principle. These reactor have modern systems to prevent accidents.

India's nuclear power programme up to 2000 A.D. as envisaged by the Nuclear power corporation (Government of India) is given in Table 19.1.

## 19.11   POWER OF A NUCLEAR REACTOR

To get nuclear energy due to fission process, a large number of neutrons are absorbed by nuclear fuel atoms. A controlled chain reaction takes place for releasing the energy.

**Table 19.1   Nuclear power programme up to 2000 A.D.**

| Operational units | Station capacity MWe | Commulative capacity MWe | years of Commissioning | Steam pressure & temperature |
|---|---|---|---|---|
| Tarapur (BWR) 1 & 2 | 2 × 210 | 420 | 1969 | 35 bar & 240°C |
| Rana Pratap 1 & 2 | 2 × 220 | 860 | 1973, 1981 | 40 bar & 250°C |
| Kalpakkam1 & 2 | 2 × 235 | 1330 | 1983, 1985 | – |
| Narora 1 & 2 | 2 × 235 | 1800 | 1990, 1991 | 40 bar & 250°C |
| Kakrapar 1 & 2 | 2 × 235 | 2270 | 1991, 1992 | – |
| *Under construction* | | | | |
| Kaiga 1 & 2 | 2 × 235 | 2740 | 1995, 1996 | – |
| Rana Pratap 3 & 4 | 2 × 235 | 3210 | 1995, 1996 | – |
| *Under sanction* | | | | |
| Kaiga 3, 4, 5, &, 6 | 4 × 235 | 4150 | 1996, 1997 | – |
| Tarapur 3 & 4 | 2 × 500 | 5150 | 1997, 1998 | – |
| Rana Pratap 5, 6, 7, & 8 | 4 × 500 | 7150 | 1998, 2000 | – |
| *Planned* | | | | |
| Kudankulam (PWR) 1 & 2 | 2 × 1000 | 9150 | 1995, 1999 | – |
| New Projects | 6 × 500 | 12150 | 1998, 2000 | – |

Let   $V$ = Volume of energy

$N$ = Fuel atoms/m$^3$

$n$ = Average neutron density i.e. number per m$^3$.

$a$ = Fission cross-section

$\phi$ = Neutron flux

$v$ = Average speed of neutrons, m/s

Fission cross-section represents the probability of fission per incident neutron. For example if $y$ is the number of incident neutron then those causing fission $= a \times y$. Neutron flux is the number of neutrons crossing a plane of area of one metre square held at right angle to velocity $v$.

$\therefore$   Nutron flux $\phi = n \times v$

$S$ = total fuel atoms in reactor = N · V

$h$ = number of incident neutrons per second on fuel atoms

$$= S \times \phi = n \cdot v \cdot N \cdot V. \tag{19.18}$$

$x$ = Number of neutrons causing fission per second

$$= h \times a = n \cdot v \cdot N \cdot V \cdot a \tag{19.19}$$

Since $3.1 \times 10^{10}$ fission per second generate a power of one watt, the power $P$ of a nuclear reactor is given by

$$P = \frac{n \cdot v \cdot N \cdot V \cdot a}{3.1 \times 10^{10}} \text{ watt} \tag{19.20}$$

Let the fuel used in the reaction be U$^{235}$. Mass per atom of U$^{235}$

$$= \frac{\text{Atomic weight of U}^{235}}{\text{Avogadro number}} = \frac{235}{6.02 \times 10^{26}} \text{ kg}.$$

Mass of $NV$ atoms $= N \cdot V \times$ mass per atom

$$= NV \times \frac{235}{6.02 \times 10^{26}} = M \text{ kg (say)}$$

$$\therefore \qquad NV = \frac{6.02 \times 10^{26} \times M}{235}$$

$$P = \frac{n \cdot v \cdot N \cdot V \cdot a}{3.1 \times 10^{10}} \text{ watt}$$

$$= \frac{\phi \times 6.02 \times 10^{26} \times M \times 582 \times 10^{-28}}{3.1 \times 10^{10} \times 235}$$

$$= 4.8 \times 10^{-12} \text{ M } \phi \text{ watt}$$

## 19.12   ADVANTAGES OF NUCLEAR POWER PLANTS

Some of the major advantages are given as under

1.  Nuclear power plants needs less space as compared to other conventional power plants for producing same power.
2.  Nuclear power plants can meet large power demands. At high load factors (80 to 90%), they give better results.
3.  Fuel transportation cost is very small since less fuel is required as compared to conventional fuel (coal, oil, gas etc.).
4.  The operation of a nuclear power plant is more reliable.
5.  It does not require large quantity of water.
6.  It is not affected by adverse weather condition.
7.  The expenditure on metal structures piping, storage mechanism is much lower for a nuclear power plant than a coal burning power plant.

*Disadvantages/Limitations*

1.  Initial investment of nuclear power plant is higher as compared to hydro or steam power plant.
2.  These plants can not be operated at varying load efficiently.
3.  The danger of radioactivity always persists in the nuclear stations.
4.  Maintenance cost of the plant is high
5.  Trained personnels are required to handle nuclear power plants.
6.  The disposal of fission products is big problem

# 19.13 SUMMARY OF MATERIALS FOR NUCLEAR POWER REACTORS

*Fuel*
- (i) Uranium
- (ii) Uranium ceramics
- (iii) Thorium
- (iv Thorium oxide

*Coolant*
- (i) Water
- (ii) Liquid metals
- (iii) Sodium, potassium
- (iv) Mercury
- (v) Lead bismuth
- (vi) Gases
- (vii) Helium
- (viii) Nitrogen
- (ix) Carbondioxide

*Control*
- (i) Boron steel
- (ii) Cadminum
- (iii) Samarium oxide
- (iv) Gadolinium oxide

*Moderator reflector*
- (i) Water
- (ii) Heavy water
- (iii) Beryllium
- (iv) Beryllium oxide
- (v) Graphite
- (vi) Metal hydrides

*Shielding*
- (i) Water
- (ii) Cement and concete
- (iii) Iron
- (iv) Lead
- (v) Tantalum
- (vi) Bismuth
- (vii) Boron

## Solved Problems

**19.1** How many fissions are necessary per second to produce 1 kW of thermal power if energy release per fission is 200 MeV.

*Soln.* Let $n$ fissions are required per second.

$$\text{Energy release} = n \times 200 \times 10^6 \times 1.6 \times 10^{-19} \text{ J/s}$$

$$= n \times 200 \times 10^6 \times 1.6 \times 10^{-19} \times 10^{-3} \text{ kW}$$

$$= 3.2 \times 10^{-14} \, n \text{ kW}$$

But power required $= 1 \text{ kW} = 3.2 \times 10^{-14} n$

$$\therefore \quad n = \frac{1}{3.2 \times 10^{-14}} = 3.125 \times 10^{13} \text{ fission/second} \quad \textit{Ans.}$$

**19.2** 10 kg of $U^{235}$ per day is consumed in a nuclear reactor. Calculate its power output of the average energy released per $U^{235}$ fission is 200 MeV.

*Soln.*  Number of atoms in 235 kg of $U^{235}$ = $6.02 \times 10^{26}$ (AVO. number)
Hence, number of atoms contained in 10 kg of $U^{235}$

$$= \frac{6.02 \times 10^{26}}{235} \times 10 = 2.56 \times 10^{25}$$

Fission energy produced by these atoms

$$= 200 \times 2.56 \times 10^{25} \text{ MeV}$$

$$= 200 \times 2.56 \times 10^{25} \times 1.6 \times 16^{-13} \text{ J} = 819.2 \times 10^{12} \text{ J}$$

Time taken to consume 10 kg of $U^{235}$
one day = $24 \times 3600$ seconds

$$\therefore \qquad \text{Power Produced} = \frac{819.2 \times 10^{12}}{24 \times 3600} = 9.48 \times 10^{9} \text{ W} \quad Ans.$$

**19.3**  For 10 hour run from one station to another, a railway engine being driven by atomic power, develops power of 1800 kW with an efficiency of 30%. Determine how much $U^{235}$ would be consumed on the run if each atom on fission releases 200 MeV.

*Soln.*  Given

$$\text{Duration of run} = 10 \text{ hour}$$

$$\text{Power developed} = 1800 \text{ kW}$$

$$\text{efficiency } \eta = 30\%$$

Energy released by each atom = 200 MeV.
For total running of engine, power consumed by the engine = $1800 \times 10 = 18000$ kwh

$$1 \text{ kWh} = 36 \times 10^{8} \text{ J}$$

$\therefore$  Energy consumed = $18000 \times 36 \times 10^{5} = 648 \times 10^{8}$ J
Energy required to produce it

$$= \frac{648 \times 10^{8}}{\eta} = \frac{648 \times 10^{8}}{0.30} = 216 \times 10^{9} \text{ J}$$

Energy released per fission = 200 MeV

$$= 200 \times 1.6 \times 10^{-13} \text{ J}$$

$$= 3.2 \times 10^{-11} \text{ J}$$

Number of $U^{235}$ atoms required for 10 hour run

$$= \frac{216 \times 10^{9}}{3.2 \times 10^{-11}} = 67.5 \times 10^{20}$$

Now, $6.02 \times 10^{26}$ atoms are contained in 235 kg of $U^{235}$

$$\text{Hence mass contained} = \frac{235 \times 67.5 \times 10^{20}}{6.02 \times 10^{26}} = 2634.96 \times 10^{-6} \text{ kg}$$

$$= 2.63 \text{ gm} \quad Ans.$$

**19.4**  Calculate the amount of Diesel oil needed possessing a calorific value of 44000 kJ/kg to obtain the same power as in example 19.3 if the efficiency now is 80%.

*Soln.* With an efficiency of 80%, daily energy input is

$$= \frac{648 \times 10^8}{0.8} = 810 \times 10^8 \, J = 810 \times 10^5 \, kJ$$

$\therefore$  Diesel required per day $= \dfrac{810 \times 10^5}{44000 \times 1000} = 1.84$ tonnes  *Ans.*


## EXERCISES

19.1  Objective Questions
  (i)  The energy produced by 4.5 tonnes of high grade coal is equivalent to the energy produced by
    (a)  one kg of uranium                      (b)  one gram of uranium
    (c)  1000 gram of uranium              (d)  10 gram of uranium.
  (ii)  Isotopes of uranium are
    (a)  $U^{235}$                                  (b)  $U^{234}$
    (c)  $U^{238}$                                  (d)  All of the above.
  (iii)  The fissionable materials are
    (a)  $U^{233}$ and $Pu^{238}$             (b)  $U^{233}$ and $Pu^{239}$
    (c)  $U^{238}$ and $Pu^{238}$             (d)  $U^{235}$ and $Pu^{235}$.
  (iv)  The purpose of a moderator in a nuclear power plant is
    (a)  to moderate the radioactive polluation
    (b)  to reduce the temperature
    (c)  to control the reaction
    (d)  to reduce the speed of fast moving neutrons.
  (v)  Fast breeder reactor uses
    (a)  water as coolant                 (b)  moderator
    (c)  90% of $U^{235}$                  (d)  100% of $U^{235}$.
  (vi)  To return the neutrons back into the core of a nuclear reactor
    (a)  Shielding is done               (b)  moderator is used
    (c)  reflector is used                (d)  control rod is used.
  (vii)  In a nuclear power plant, the most commonly used moderator is
    (a)  concrete                      (b)  graphite
    (c)  concrete and bricks        (d)  concrete and graphite.
  (viii)  To protect against neutron and gamma rays
    (a)  reflector is used                (b)  moderator is used
    (c)  shielding is done                (d)  control rod is used.
  (ix)  To absorb excess neutrons in nuclear reactors
    (a)  reflector is used                (b)  moderator is used
    (c)  shielding is done                (d)  control rod is used.
  (x)  To slow down the speed of fast moving neutrons
    (a)  reflector is used                (b)  moderator is used
    (c)  shielding is done                (d)  control rod is used.
  (xi)  $U^{235}$ is
    (a)  primary fuel
    (b)  secondary fuel
    (c)  produced by the action of a neutron on thorium
    (d)  Ferrite materials
  (xii)  $U^{233}$ and $Pu^{239}$ are
    (a)  Primary fuels                (b)  secondary fuels
    (c)  Ferrite materials            (d)  none of the above

(xiii)    The nuclear energy is commonly measured in
   (a)   neutron volts                          (b)   electron volts
   (c)   million electron volts                  (d)   watts
(xiv)    The percentage composition of natural uranium is
   (a)   0.712% $U^{238}$, 0.006% $U^{235}$, 99.28% $U^{234}$
   (b)   0.712% $U^{235}$, 0.006% $U^{238}$, 99.282% $U^{234}$
   (c)   99.282% $U^{238}$, 0.006% $U^{234}$, 0.712% $U^{235}$
(xv)    The energy in million electron volts released from uranium fission is approximately
   (a)   800                                     (b)   400
   (c)   200                                     (d)   20.
(xvi)    Fast breeder reactors produce
   (a)   less fuel than they consume
   (b)   more fuel than they consume
   (c)   same fuel as they consume
   (d)   no fuel
(xvii)    Fast breeder reactors use
   (a)   water as moderator                      (b)   carbondioxide as moderator
   (c)   graphite as moderator                   (d)   no moderator
(xviii)    Fast breeder reactors
   (a)   use water as coolant                    (b)   are liquid-metal cooled
   (c)   use no coolant                          (d)   carbon dioxide as coolant
(ixx)    Reflector in nuclear power plant is used
   (a)   to protect against neutron and gamma rays
   (b)   to absorb excess neutrons
   (c)   to slow down the speed of fast moving neutron
   (d)   to return the neutrons back into the core of the reactor.
(xx)    Fast neutrons have energies about
   (a)   100 eV                                  (b)   500 eV
   (c)   10 eV                                   (d)   more than 1000 eV.
(xxi)    The material for control rod is
   (a)   graphite                                (b)   lead
   (c)   boron or cadmium                        (d)   zinc
(xxii)    For propulation applications, reactors are designed for
   (a)   enriched uranium                        (b)   natural uranium
   (c)   any form of uranium                     (d)   pure uranium

**19.2**  Explain the following terms:
   (i) Atomic model                       (ii) Atomic mass unit
  (iii) Isotopes                              (iv) Isobars
   (v) Isomers                                (vi) Isotones

**19.3**  What do you mean by term 'Radioactivity'?

**19.4**  What is a chain reaction? How it is controlled?

**19.5**  What is a nuclear reactor? Describe the various parts of a nuclear reactor?

**19.6**  What is 'nuclear fusion'? How does it differ from 'nuclear fission'?

**19.7**  Explain with help of neat diagram the construction and working of a nuclear power plant.

**19.8**  What is a moderator? Name common moderators and discuss their advantages and limitations.

**19.9**  What is 'Boiling water Reactor' (BWR)? How does it differ from 'Pressurised water Reactor' (PWR)?

**19.10**  Give the construction and working of a 'Gas cooled reactor'. What are its advantages and dis-advantages?

**19.11**  What is a 'Liquid Metal Cooled Reactor'? Explain briefly a typical liquid metal reactor.

**19.12**  Describe a breeder reactor. What are its advantages and disadvantages?

**19.13**  List the advantages and disadvantages/limitations of nuclear power plants.

**19.14**  Discuss the factors which go in favour of nuclear power plant as compared to other types of power plants.

19.15  Write short notes on various nuclear power plants in India.
19.16  State the properties of the followings
  (a)  control rod         (b)  moderator
19.17  Describe the objectives of $R$ and $D$ in nuclear energy in India?
19.18  List down some safety measures for nuclear power plants.
19.19  What is the future of nuclear power?
19.20  Discuss the 'Economics' of a nuclear power plant.

**Answer to Objective Questions**

| (i) b | (ii) d | (iii) b | (iv) d | (v) c | (vi) c |
|---|---|---|---|---|---|
| (vii) d | (viii) c | (ix) d | (x) b | (xi) a | (xii) b |
| (xiii) c | (xiv) c | (xv) c | (xvi) b | (xvii) d | (xviii) b |
| (xix) c | (xx) d | (xxi) c | (xxii) a | | |

# Appendix

**Table A1   Properties of dry air at atmospheric pressure**

| $T$ °C | $\rho$ kg/m$^3$ | $C_p$ kJ/kg K | $\mu \times 10^6$ N-s/m$^2$ | $k$ W/m-K | Pr | $\nu \times 10^6$ m$^2$/s |
|---|---|---|---|---|---|---|
| 0 | 1.293 | 1.005 | 17.2 | 0.0244 | 0.707 | 13.28 |
| 10 | 1.247 | 1.005 | 17.7 | 0.0251 | 0.705 | 14.16 |
| 20 | 1.205 | 1.005 | 18.1 | 0.0259 | 0.703 | 15.06 |
| 30 | 1.165 | 1.005 | 18.6 | 0.0267 | 0.701 | 16.00 |
| 40 | 1.128 | 1.005 | 19.1 | 0.0276 | 0.699 | 16.96 |
| 50 | 1.093 | 1.005 | 19.6 | 0.0283 | 0.698 | 17.95 |
| 60 | 1.060 | 1.005 | 20.1 | 0.0290 | 0.696 | 18.97 |
| 70 | 1.029 | 1.009 | 20.6 | 0.0297 | 0.694 | 20.02 |
| 80 | 1.000 | 1.009 | 21.1 | 0.0305 | 0.692 | 21.09 |
| 90 | 0.972 | 1.009 | 21.5 | 0.0313 | 0.690 | 22.10 |
| 100 | 0.946 | 1.009 | 21.9 | 0.0321 | 0.688 | 23.13 |
| 120 | 0.898 | 1.009 | 22.9 | 0.0334 | 0.686 | 25.45 |
| 140 | 0.854 | 1.013 | 23.7 | 0.0349 | 0.684 | 27.80 |
| 160 | 0.815 | 1.017 | 24.5 | 0.0364 | 0.682 | 30.09 |
| 180 | 0.779 | 1.022 | 25.3 | 0.0378 | 0.681 | 32.49 |
| 200 | 0.746 | 1.026 | 26.0 | 0.0393 | 0.680 | 34.85 |
| 250 | 0.674 | 1.038 | 27.4 | 0.0427 | 0.677 | 40.61 |
| 300 | 0.615 | 1.047 | 29.7 | 0.0461 | 0.674 | 48.33 |
| 350 | 0.566 | 1.059 | 31.4 | 0.0491 | 0.676 | 55.46 |
| 400 | 0.524 | 1.068 | 33.0 | 0.0521 | 0.678 | 63.09 |
| 500 | 0.456 | 1.093 | 36.2 | 0.0575 | 0.687 | 79.38 |
| 600 | 0.404 | 1.114 | 39.1 | 0.0622 | 0.699 | 96.89 |
| 700 | 0.362 | 1.135 | 41.8 | 0.0671 | 0.706 | 115.4 |
| 800 | 0.329 | 1.156 | 44.3 | 0.0718 | 0.713 | 134.8 |
| 900 | 0.301 | 1.172 | 46.7 | 0.0763 | 0.717 | 155.1 |
| 1000 | 0.277 | 1.185 | 49.0 | 0.0807 | 0.719 | 177.1 |

### Table of Physical Constants

(i)  1 yard = 3 fit = 36 inches = 0.9144 m
    1 m = 39.370 08 inch = 3.280 839 ft = 1.093 613 yd
        = 1 650 763.73 wavelength
    1 ft = 0.304 8 m; 1 in = 25.4 mm
    1 angstrom = $1.000 \times 10^{-8}$ cm

    1 mile = 1.609 344 km
    1 fathom = 1.828 8 m
    1 chain = 20.1168 m
    1 furlong = 201.168 m
    1 micron = $10^{-6}$ m = $10^{-4}$ cm = $10^4$ Å
    1 mil = $2.54 \times 10^{-3}$ cm
    1 links = 0.201 2 m

(ii)  Area $m^2$
    1 $yd^2$ = 0.836 $m^2$; 1 $ft^2$ = 0.092 9 $m^2$
    1 $in^2$ = 6.452 $cm^2$ = 0.000 645 16 $m^2$
    1 $cm^2$ = $10^{-4}$ $m^2$ = $10.764 \times 10^{-4}$ $ft^2$ = 0.1550 $inch^2$ = $10^{24}$ barns
    1 acre = 4046.86 $m^2$ = 0.404 686 ha = 43 560 $ft^2$
    1 $mile^2$ = 2.589 99 $km^2$ = 640 acres
    1 ha (hectare) = $10^4 m^2$ = 2.47 acres

(iii)  Volume $m^3$
    1 $ft^3$ = 0.028 32 $m^2$ = 28.31681
    1 $in^3$ = 16.39 $cm^3$ = $1.639 \times 10^2$ 1
    1 $yd^3$ = 0.764 555 $m^3$ = $7.646 \times 10^2$ 1
    1 UK gallon = 4.546 091
    1 US gallon = 3.7851 = 0.1337 $ft^3$
    1 UK pint = 0.568 2611 = 0.0568 $m^3$
    1 bushel (bul) = 0.036 369 $m^3$ = 8.000 UK gallons
    1 barrel = 42.000 US gallons = 0.1590 $m^3$
    1 $m^3$ = $1.000 \times 10^6$ $cm^3$ = $2.642 \times 10^{12}$ US gallons = $1.000 \times 10^9$ 1
    1 acre foot = 1233 $m^3$
    11 = $10^{-3}$ $m^3$
    1 quart = $1.136 \times 10^{-3}$ $m^3$
    1 fuid ounce = 28.41 $cm^3$

(iv)  Mass kg
    1 kg = 2.204 62 lb = 0.068 522 slug
    1 ton (short) = 2000 lb (pounds) = 907.184 kg;
    1 ton (long) = 1016.05 kg
    1 dram = 1.722 g
    1 lb = 16 oz (ounces) = 0.4536 kg
    1 tola = 10.66 g
    1 oz = 28.3495 g
    1 slug = 14.59 kg
    1 stone = 6.350 293 kg
    1 hundred weight (cwt) = 50.802 345 kg
    1 maund = 40 seers = 37.32 kg
    1 quintal = 100 kg
    1 grain = 0.0648 g
    1 carat = 0.2 g

(v)  Density and specific volumes $kg/m^3$, $m^3/kg$
    1 $lb/ft^3$ = 16.0185 $kg/m^3$ = $5.787 \times 10^{-2}$ $lb/in^3$
    1 $oz/ft^3$ = 1 $kg/m^3$
    1 $slug/ft^3$ = 515.4 $km/m^3$
    1 $g/cm^3$ = $10^3$ $kg/cm^3$ = 62.43 $lb/ft^3$
    1 $ib/ft^3$ = 0.016 $g/cm^3$ = 16 $kg/m^3$
    1 $ft^3$ (air) = 0.080 09 lb = 36.5 gm at N.T.P.
    11 (air) = 1.2982 g at N.T.P.
    1 gallon/lb = 0.01 $cm^3/kg$

(vi) Pressure Pa

$1\ lb/ft^2 = 4.88\ kg/m^2 = 47.88\ Pa$

$1\ lb/in^2 = 702.7\ kg/m^2 = 51.71\ mm\ Hg$

$\qquad = 6.894\ 757 \times 10^3\ Pa\ (Pascal)$

$\qquad = 6.894\ 757 \times 10^3 N/m^2$

$1\ atm = 1.013 \times 10^5\ N/m^2 = 760\ mm\ Hg = 101.325\ kPa$

$1\ in\ H_2O = 2.491 \times 10^2\ N/m^2 = 248.8\ Pa = 0.036\ lb/in^2$

$1\ bar = 0.987\ atm = 1.000 \times 10^4\ dynes/cm^2$

$\qquad = 1.020\ kgf/cm^2 = 14.50\ lb\ f/in^2$

$\qquad = 10^5\ N\ (Newton/m^2)$

$\qquad = 100\ k\ Pa$

$1\ torr\ (mm\ Hg,\ 0°C) = 133\ Pa$

$1\ Pascal\ (Pa) = 1\ N/m^2 = 1.894\ 76\ kg$

$1\ inch\ of\ Hg = 3.377\ K\ Pa = 0.489\ lb/in^2$

(vii) Velocity m/s

$1\ ft/s = 0.3048\ m/s$

$1\ mile/h = 0.447\ m/s = 1.4667\ ft/s = 0.8690\ knots$

$1\ knot = 0.5144\ m/s = 1.6878\ ft/s$

$1\ km/h = 0.2778\ m/s$

$1\ ft/min = 0.00508\ m/s$

(viii) Force N

$1\ N\ (Newton) = 1 \times 10^5\ dynes = 0.224\ 81\ lb\ wt = 0.224\ 81\ lbf$

$1\ bdl\ (poundal) = 0.138\ 255\ N\ (Newton)$

$\qquad\qquad = 13.83\ dynes = 14.10\ gf$

$1\ lbf\ (i.e.\ wt\ of\ 1\ lb\ mass) = 4.448\ 222\ N$

$\qquad\qquad\qquad = 444.8222\ dynes$

$1\ ton = 9.964 \times 10^3\ N$

$1\ kgf = 9.8065\ N$

$1\ bar = 10^5\ Pa\ (pascal)$

$1\ ft\ of\ H_2O = 2.950 \times 10^{-2}\ atm = 9.807 \times 10^3\ N/m^2$

$1\ mm\ H_2O = 9.806\ 65\ Pa$

$1\ in\ H_2O = 249.089\ Pa$

$1\ mm\ Hg = 133.322\ Pa$

$1\ dyne = 1.020 \times 10^{-6}\ kg\ f = 2.2481 \times 10^{-6}\ ln\ f$

$\qquad = 7.2330 \times 10^{-5}\ pdl$

$\qquad = 10^{-5}\ N$

$1\ mm\ of\ Hg = 133.3\ Pa$

$1\ at = 1\ kg\ f/cm^2 = 98.0665\ k\ Pa$

$1\ atm = 101.325\ k\ Pa$

$1\ Pa\ (Pascal) = 1\ N/m^2$

(ix) Mass flow rate and discharge kg/s $m^3$/s

$1\ lb/s = 0.4536\ kg/s$

$1\ lb/min = 7.56 \times 10^{-3}\ kg/s$

$1\ ft^3/min = 0.4720\ l/s = 4.179 \times 10^{-4}\ m^3/s$

$1\ m^3/s = 3.600 \times 10^6\ l/h$

$1\ yd^3/minute = 12.74\ l/s$

$1\ gm/cm^3 = 10^3\ kg/m^3$

$1\ lb/h\ ft^2 = 0.001\ 356\ kg/s\ m^2$

$1\ lb/ft^3 = 16.2\ kg/m^2$

$1\ cfm/ft^2 = 0.3048\ m^3/min\ m^2$

$1\ g\ pm/ft^2 = 0.0407\ m^3/min\ m^2$

$$1 \text{ gal/min} = 6.309 \times 10^{-5} \text{ m}^3/\text{s}$$
$$1 \text{ cusecs (ft}^3/\text{s}) = 0.02832 \text{ m}^3/\text{s}$$
$$1 \text{ cumecs (m}^3/\text{s}) = 1 \text{ m}^3/\text{s}$$
$$1 \text{ litre/s (1/s)} = 10^{-3} \text{ m}^3/\text{s}$$

(x) Energy J

$$1 \text{ cal} = 4.187 \text{ J (Joules)}$$
$$1 \text{ kcal} = 3.97 \text{ Btu} = 12 \times 10^{-4} \text{ kWh} = 4.187 \times 10^3 \text{ J}$$
$$1 \text{ wall} = 1.000 \text{ J/S}$$
$$1 \text{ Btu} = 0.252 \text{ kcal} = 2.93 \times 10^{-4} \text{ kWh} = 1.055 \times 10^3 \text{ J}$$
$$1 \text{ hp} = 632.34 \text{ kcal} = 0.736 \text{ kWh}$$
$$1 \text{ therm} = 1.055 \times 10^8 \text{ J}$$
$$1 \text{ chu} = 1.9 \times 10^3 \text{ J}$$
$$1 \text{ kWh} = 3.6 \times 10^6 \text{ J}$$
$$1 \text{ ft lbf} = 1.356 \times 10^7 \text{ ergs} = 1.286 \times 11^{-3} \text{ Btu}$$
$$1 \text{ ft lbf} = 1.356 \text{ J} = 3.241 \times 10^{-4} \text{ kcal}$$
$$1 \text{ J (Joule)} = 2.390 \times 10^{-4} \text{ kcal} = 2.778 \times 10^{-4} \text{ Wh}$$
$$1 \text{ kWh} = 860 \text{ kcal} = 3413 \text{ Btu.}$$
$$1 \text{ erg} = 1.00 \times 10^{-7} \text{ J} = 1.00 \times 10^{-7} \text{ Nm} = 1.00 \text{ dyne cm}$$
$$1 \text{ J (Joule)} = 1 \text{ w s} = 1 \text{ Nm}$$

(xi) Power Watt (J/s)

$$1 \text{ Btu/h} = 0.293\ 071 \text{ W} = 0.252 \text{ k cal/h}$$
$$1 \text{ Btu/h} = 1.163 \text{ W} = 3.97 \text{ Btu/h}$$
$$1 \text{ ton refr.} = 3.516\ 85 \text{ kW}$$
$$1 \text{ w} = 1.000 \text{ J/S} = 1.341 \times 10^{-3} \text{ hp}$$
$$= 0.0569 \text{ Btu/min} = 0.014\ 33 \text{ kcal/min}$$
$$1 \text{ hp (F.P.S.)} = 550 \text{ ft 1bf/s} = 42.41 \text{ Btu/min}$$
$$= 746 \text{ W} = 596 \text{ kcal/h}$$
$$= 1.015 \text{ hp (M.K.S.)}$$
$$1 \text{ hp (M.K.S.)} = 75 \text{ m kg f/s} = 0.175\ 69 \text{ kcal/s}$$
$$= 735.3 \text{ W}$$
$$1 \text{ w/ft}^2 = 10.76 \text{ W/m}^2$$

(xii) Specific Heat kJ/kg °C

$$1 \text{ Btu/lb °F} = 1.0 \text{ kcal/kg °C} = 4.187 \times 10^3 \text{ J/kg °C}$$
$$1 \text{ Btu/lb} = 2.326 \text{ kJ/kg}$$

(xiii) Temperature, °C and K used in SI

$$t°C = \tfrac{5}{9}(t°F + 40) - 40;\ t°F = \tfrac{9}{5}(t°C + 40) - 40\,;$$
$$t°R = 460 + t°F;\ T°K = \tfrac{5}{9}T°R;\ T°R = \tfrac{9}{5}T°K\,;$$
$$\Delta t°C = \Delta t°F/1.8 = \tfrac{5}{9}\Delta t°F$$

(xiv) Rate of heat flow per unit area or heat flux W/m²

$$1 \text{ Btu/ft}^2 \text{ h} = 2.713 \text{ kcal/m}^2 \text{ h} = 3.1552 \text{ W/m}^2$$
$$1 \text{ pcu/ft}^2 \text{ h} = 4.878 \text{ kcal/m}^2 \text{ h}$$
$$1 \text{ kcal/m}^2 \text{ h} = 0.3690 \text{ Btu/ft}^2 \text{ h} = 1.163 \text{ W/m}^2$$
$$= 27.78 \times 10^{-6} \text{ cal/s cm}^2$$
$$1 \text{ cal/cm}^2 \text{ min} = 221.4 \text{ Btu/ft}^2 \text{ h}$$
$$1 \text{ W/ft}^2 = 10.76 \text{ W/m}^2$$
$$1 \text{ W/m}^2 = 0.86 \text{ kcal/h m}^2 = 0.23901 \times 10^{-4} \text{ cal/s cm}^2$$
$$= 0.137 \text{ Btu/h ft}^2$$
$$1 \text{ Btu/h ft} = 0.961\ 28 \text{ W/m}$$

(xv) Heat transfer coefficient W/m² °C

$$1 \text{ Btu/ft}^2 \text{ h°F} = 4.882 \text{ kcal/m}^2 \text{ h°C} = 1.3571 \times 10^{-4} \text{ cal/cm}^2 \text{ s°C}$$
$$= 5.678 \text{ W/m}^2 \text{ °C}$$

$$1 \text{ kcal/m}^2 \text{ h } °C = 0.2048 \text{ Btu/ft}^2 \text{ h}°F;$$
$$= 1.163 \text{ W/m}^2 \text{ }°C.$$
$$1 \text{ W/m}^2 \text{ }°K = 2.3901 \times 10^{-5} \text{ cal/cm}^2 \text{ s}°K$$
$$= 1.7611 \times 10^{-1} \text{ Btu/ft}^2 \text{ }°F$$
$$= 0.86 \text{ kcal/m}^2 \text{ h}°C$$
$$1 \text{ cal/cm}^2 \text{ s}°C = 7.3686 \times 10^3 \text{ Btu/ft}^2 \text{ h}°F$$
$$= 36\ 000 \text{ kcal/m}^2 \text{ h}°C$$
$$= 4.187 \times 10^4 \text{ W/m}^2 \text{ }°C$$
$$= 4.1 \text{ W/m cm}^2 \text{ }°C$$
$$1 \text{ pcu/ft}^2 \text{ h}°C = 4.878 \text{ kcal/m}^2 \text{ h}°C$$

(xvi)  Thermal conductivity W/m°C
$$1 \text{ Btu/ft h}°F = 1.488 \text{ kcal/m h}°C;$$
$$= 1.730\ 73 \text{ W/m}°C$$
$$1 \text{ kcal/m h}°C = 0.6720 \text{ Btu/ft h}°F;$$
$$= 1.1631 \text{ W/m}°C$$
$$1 \text{ Btu in/fit}^2 \text{ h}°F = 0.124 \text{ kcal/m h}°C;$$
$$= 0.144\ 228 \text{ W/m}°C$$
$$1 \text{ Btu/in h}°F = 17.88 \text{ kcal/m h}°C$$
$$1 \text{ cal/cm s}°C = 4.187 \times 10^2 \text{ W/m}°C$$
$$= 242 \text{ Btu/h ft}°F$$
$$1 \text{ W/cm}°C = 57.79 \text{ Btu/h ft}°F$$

(xvii)  Viscosity coefficient, Pa S = N s/m$^2$ = kg/m s
$$1 \text{ lb/ft s} = 14.88 \text{ g/cm s} = 1\ 488 \text{ kg/ms}$$
$$1 \text{ lb/ft}^2 = 4.882 \text{ kg s/m}^2 = 47.88 \text{ kg/m s}$$
$$= 478.8 \text{ poises}$$
$$1 \text{ P (poise)} = 0.1 \text{ kg/ms} = 0.1 \text{ N s/m}^2$$
$$= 1.020 \times 10^{-2} \text{ kg s/m}^2 = 2.088 \times 10^{-3} \text{ lb f s/ft}^2$$
$$1 \text{ lb m/ft h} = 4.134 \times 10^{-4} \text{ Ns/m}^2$$
$$1 \text{ lb fh/ft}^2 = 1.723\ 69 \times 10^5 \text{ Ns/m}^2 = 0.1724 \text{ M Pa s}$$
$$1 \text{ slug/ft s} = 4.788 \times 10 \text{ Ns/m}^2$$
$$1 \text{ CP (Centipoise)} = 10^{-3} \text{ Pa s} = 2.42 \text{ lb/h ft} = 3.60 \text{ kg/h m}$$

(xviii)  Angle, rad
$$2 \text{ rad} = 360 \text{ degrees}$$
$$1 \text{ degree} = 0.017\ 453\ 3 \text{ rad} = 60 \text{ minute}$$
$$1 \text{ minute} = 0.290\ 888 \times 10^{-3} \text{ rad} = 60 \text{ s}$$
$$1 \text{ second} = 4.848\ 14 \times 10^{-6} \text{ rad}$$

(xix)  Illumination
$$1 \text{ lx (lux)} = 1.000 \text{ lm (lumen)/m}^2$$
$$1 \text{ lm/ft}^2 = 1.000 \text{ foot candle}$$
$$1 \text{ foot candle} = 10.7639 \text{ lx}$$

(xx)  Time h
$$1 \text{ week} = 1 \text{ days} = 168 \text{ h} = 100\ 80 \text{ minute} = 604\ 800 \text{ s}$$
$$1 \text{ mean solar day} = 1440 \text{ minute} = 86\ 400 \text{ s}$$
$$1 \text{ calendar year} = 365 \text{ days} = 8760 \text{ h} = 5.256 \times 10^6 \text{ minutes}$$
$$1 \text{ tropical mean solar year} = 365.2422 \text{ days}$$
$$1 \text{ sidereal year} = 365.2564 \text{ days (mean solar)}$$
$$1 \text{ s (second)} = 9.192\ 631\ 770 \times 10^9 \text{ Hertz (Hz)}$$

(xxi)  Concentration kg/m$^3$ and g/m$^3$
$$1 \text{ g/l} = 1 \text{ kg/m}^3$$
$$1 \text{ lb/ft}^3 = 6.236 \text{ kg/m}^3$$
$$1 \text{ grains/gal} = 4.25 \text{ g/m}^3$$

(xxii)  Diffusivity $m^2/s$
   1 CST (centistoke) $= 10^{-6}$ $m^2/s$
   1 $ft^2/h = 25.81 \times 10^{-6}$ $m^2/s$

### Gas Constants

$R_u = 8314.4$ J/mole K (universal)
$R_a = 287.045$ J/kg K (air)
$R_v = 461.52$ J/kg K (water vapour)
Boltzmann's constant $1.380 \times 10^{-23}$ J/K
Faraday's constant $9.65 \times 10^{7}$ C/kg mole
Planck's constant $6.626 \times 10^{-34}$ Js
Stefan-Boltzmann constant $567 \times 10^{-8}$ $J/m^2s$ $K^4$
Volume of 1 kg mole of ideal gas at NTP 22.42 $m^3$

### Table A 2 List of units and conversion factors

#### A. Nomenclature for powers of 10

| Prefix | Symbol | Prefix | Symbol |
| --- | --- | --- | --- |
| $13^{-18}$ atto | a | $10^3$ kilo | k |
| $10^{-15}$ femto | f | $10^6$ mega | M |
| $10^{-12}$ pico | p | $10^9$ giga | G |
| $10^{-9}$ nano | n | $10^{12}$ tera | T |
| $10^{-6}$ micro | $\mu$ | $10^{15}$ peta | P |
| $10^{-3}$ milli | m | $10^{18}$ exa | E |

#### B. SI Unit Sysem

| Basic Unit | Name | Symbol |
| --- | --- | --- |
| length | metre | m |
| mass | kilogram | kg |
| time | second | s |
| electric current | ampere | A |
| Thermodynamical temperature | degree Kelvin | K |
| luminous intensity | candela | cd |
| plane angle | radian | rad |
| solid angle | steradian | sr |

| Derived Unit | Name | Symbol | Definition |
| --- | --- | --- | --- |
| energy | joule | J | $kg\ m^2s^{-2}$ |
| power | watt | W | $J\ s^{-1}$ |
| force | newton | N | $J\ m^{-1}$ |
| electric charge | coulomb | C | $A\ s$ |
| electric potential difference | volt | V | $J\ A^{-1}\ s^{-1}$ |
| electric resistance | ohm | $\Omega$ | $V\ A^{-1}$ |
| electric capacitance | farad | F | $A\ s\ V^{-1}$ |
| magnetic flux | weber | Wb | $V\ s$ |
| inductance | henry | H | $V\ s\ A^{-1}$ |
| magnetic flux density | tesla | T | $V\ s\ m^{-2}$ |
| luminous flux | lumen | lm | $cd\ sr$ |
| illumination | lux | lx | $cd\ sr\ m^{-2}$ |
| frequency | hertz | Hz | $cycle\ s^{-1}$ |

## Table A 3  Properties of water

| $T$ °C | $\rho$ kg/m$^3$ | $C_p$ kJ/kg-K | $\mu \times 10^6$ N-s/m$^2$ | $k$ W/m-K | Pr | $\beta \times 10^1$ 1/K | $v \times 10^6$ m$^2$/s | $\sigma \times 10^4$ N/m |
|---|---|---|---|---|---|---|---|---|
| 0 | 999.9 | 4.212 | 1787.8 | 0.551 | 13.67 | −0.63 | 1.789 | 756 |
| 10 | 999.7 | 4.191 | 1305.3 | 0.575 | 9.52 | +0.70 | 1.306 | 742 |
| 20 | 998.2 | 4.183 | 1004.2 | 0.599 | 7.02 | 1.82 | 1.006 | 727 |
| 30 | 995.7 | 4.174 | 801.2 | 0.618 | 5.42 | 3.21 | 0.805 | 712 |
| 40 | 992.2 | 4.174 | 653.1 | 0.634 | 4.31 | 3.87 | 0.659 | 696 |
| 50 | 988.1 | 4.174 | 549.2 | 0.648 | 3.54 | 4.49 | 0.556 | 679 |
| 60 | 983.2 | 4.179 | 469.8 | 3.659 | 2.98 | 5.11 | 0.478 | 662 |
| 70 | 977.8 | 4.187 | 406.0 | 0.668 | 2.55 | 5.70 | 0.415 | 644 |
| 80 | 971.8 | 4.195 | 355.0 | 0.675 | 2.21 | 6.32 | 0.365 | 628 |
| 90 | 965.3 | 4.208 | 314.8 | 0.680 | 1.95 | 6.95 | 0.326 | 608 |
| 100 | 958.4 | 4.220 | 282.4 | 0.683 | 1.75 | 7.52 | 0.295 | 589 |
| 110 | 951.0 | 4.233 | 258.9 | 0.685 | 1.60 | 8.08 | 0.272 | 569 |
| 120 | 943.1 | 4.250 | 237.3 | 0.686 | 1.47 | 8.64 | 0.252 | 549 |
| 130 | 934.8 | 4.267 | 217.7 | 0.686 | 1.36 | 9.19 | 0.233 | 528 |
| 140 | 926.1 | 4.287 | 201.0 | 0.685 | 1.26 | 9.72 | 0.217 | 507 |
| 150 | 917.0 | 4.313 | 186.3 | 0.684 | 1.17 | 10.3 | 0.203 | 485 |
| 160 | 907.4 | 4.346 | 173.6 | 0.683 | 1.10 | 10.7 | 0.191 | 463 |
| 170 | 897.3 | 4.380 | 162.8 | 0.679 | 1.05 | 11.3 | 0.181 | 441 |
| 180 | 886.9 | 4.417 | 153.0 | 0.675 | 1.00 | 11.9 | 0.171 | 420 |
| 190 | 876.0 | 4.459 | 144.2 | 0.670 | 0.96 | 12.6 | 0.165 | 398 |
| 200 | 863.0 | 4.505 | 136.3 | 0.663 | 0.93 | 13.3 | 0.158 | 376 |

**Table A 4    Properties of saturated water and steam (temperature) Tables**

| Saturation Temp. $(t_s$ °C | Absolute pressure $(p)$bar | Specific volume $(m^3/kg)$ | | Specific enthalpy $(kJ/kg)$ | | | Specific entropy $((kJ/kg\ K)$ | | |
|---|---|---|---|---|---|---|---|---|---|
| | | water $v_f$ | Steam $v_g$ | Water $h_f$ | Evaporation $h_{fg}$ | Steam $h_g$ | Water $S_f$ | Evaporation $S_{fg}$ | Steam $S_g$ |
| 0.00 | 0.00611 | 0.001000 | 206.31 | 0.00 | 2501.0 | 2501.0 | 0.000 | 9.156 | 9.156 |
| 0.01 | 0.00611 | 0.001000 | 206.16 | 0.04 | 2501.5 | 2501.5 | 0.000 | 9.157 | 9.157 |
| 5.00 | 0.00872 | 0.001000 | 147.20 | 21.0 | 2489.6 | 2510.6 | 0.076 | 8.951 | 9.027 |
| 10.00 | 0.01227 | 0.001000 | 106.42 | 42.0 | 2477.8 | 2519.8 | 0.151 | 8.751 | 8.902 |
| 15.00 | 0.01704 | 0.001001 | 77.98 | 62.9 | 2466.0 | 2528.9 | 0.224 | 8.559 | 8.783 |
| 20.00 | 0.02337 | 0.001002 | 57.84 | 83.9 | 2454.2 | 2538.1 | 0.296 | 8.372 | 8.668 |
| 25.00 | 0.03166 | 0.001003 | 43.40 | 104.8 | 2442.4 | 2547.2 | 0.367 | 8.192 | 8.559 |
| 30.00 | 0.04242 | 0.001004 | 32.93 | 125.7 | 2430.6 | 2556.3 | 0.437 | 8.018 | 8.455 |
| 35.00 | 0.05622 | 0.001006 | 25.25 | 146.6 | 2418.7 | 2565.3 | 0.505 | 7.849 | 8.354 |
| 40.00 | 0.07375 | 0.001008 | 19.55 | 167.5 | 2408.8 | 2574.3 | 0.572 | 7.686 | 8.258 |
| 45.00 | 0.09582 | 0.001010 | 15.28 | 188.4 | 2394.8 | 2583.2 | 0.638 | 7.528 | 8.166 |
| 50.00 | 0.12335 | 0.001012 | 12.05 | 209.3 | 2382.8 | 2592.1 | 0.704 | 7.374 | 8.078 |
| 55.00 | 0.15740 | 0.001015 | 9.58 | 230.2 | 2370.7 | 2600.9 | 0.768 | 7.225 | 7.993 |
| 60.00 | 0.19920 | 0.001017 | 7.679 | 251.1 | 2358.5 | 2609.6 | 0.831 | 7.080 | 7.911 |
| 65.00 | 0.25009 | 0.001020 | 6.202 | 272.0 | 2346.2 | 2618.2 | 0.893 | 6.939 | 7.832 |
| 70.00 | 0.31162 | 0.001023 | 5.046 | 293.0 | 2333.9 | 2626.9 | 0.955 | 6.802 | 7.757 |
| 75.00 | 0.38549 | 0.001026 | 4.134 | 313.9 | 2321.4 | 2635.3 | 1.015 | 6.568 | 7.683 |
| 80.00 | 0.47360 | 0.001029 | 3.409 | 334.9 | 2308.8 | 2643.7 | 1.075 | 6.538 | 7.613 |
| 85.00 | 0.57803 | 0.001033 | 2.829 | 355.9 | 2296.0 | 2651.9 | 1.134 | 6.411 | 7.545 |
| 90.00 | 0.70109 | 0.001036 | 2.361 | 376.9 | 2283.1 | 2660.0 | 1.193 | 6.287 | 7.480 |
| 95.00 | 0.84526 | 0.001040 | 1.982 | 398.0 | 2270.1 | 2668.1 | 1.250 | 6.167 | 7.417 |
| 100.00 | 1.01330 | 0.001043 | 1.673 | 419.0 | 2257.0 | 2676.0 | 1.307 | 6.048 | 7.355 |
| 105.00 | 1.20800 | 0.001047 | 1.418 | 440.0 | 2244.1 | 2684.1 | 1.363 | 5.934 | 7.297 |
| 110.00 | 1.43270 | 0.001052 | 1.210 | 461.1 | 2229.9 | 2691.0 | 1.418 | 5.820 | 7.238 |
| 115.00 | 1.69050 | 0.001057 | 1.036 | 481.9 | 2216.2 | 2698.1 | 1.473 | 5.710 | 7.183 |
| 120.00 | 1.98510 | 0.001061 | 0.891 | 504.1 | 2201.8 | 2705.9 | 1.527 | 5.602 | 7.129 |

**Table A 5  Properties of saturated water and steam (Pressure) Tables**

| Absolute pressure $p$ (bar) | Saturation Temp. $t_s$ (°C) | Specific volume (m³/kg) | | Specific enthalpy (kJ/kg) | | | Specific entropy (kJ/kg K) | | |
|---|---|---|---|---|---|---|---|---|---|
| | | Water $v_f$ | Steam $v_g$ | Water $h_f$ | Evaporation $h_{fg}$ | Steam $h_g$ | Water $s_f$ | Evaporation $s_{fg}$ | Steam $s_g$ |
| 0.01 | 6.98 | 0.001000 | 129.21 | 29.3 | 2484.9 | 2514.2 | 0.106 | 8.871 | 8.977 |
| 0.02 | 17.03 | 0.001001 | 67.00 | 73.5 | 2460.0 | 2533.5 | 0.261 | 8.463 | 8.724 |
| 0.03 | 24.08 | 0.001003 | 45.67 | 101.1 | 2444.4 | 2545.6 | 0.355 | 8.223 | 8.578 |
| 0.04 | 28.96 | 0.001004 | 34.80 | 121.5 | 2432.9 | 2554.4 | 0.423 | 8.052 | 8.475 |
| 0.05 | 32.88 | 0.001005 | 28.19 | 137.8 | 2423.7 | 2561.5 | 0.476 | 7.919 | 8.395 |
| 0.06 | 36.18 | 0.001006 | 23.74 | 151.5 | 2416.0 | 2567.5 | 0.521 | 7.810 | 8.331 |
| 0.07 | 39.03 | 0.001007 | 22.53 | 163.4 | 2409.2 | 2572.6 | 0.559 | 7.719 | 8.278 |
| 0.08 | 41.53 | 0.001008 | 18.11 | 173.9 | 2403.2 | 2577.1 | 0.593 | 7.637 | 8.230 |
| 0.09 | 43.79 | 0.001009 | 16.20 | 183.3 | 2397.9 | 2581.2 | 0.622 | 7.566 | 8.188 |
| 0.10 | 45.81 | 0.001010 | 14.67 | 191.8 | 2392.8 | 2584.6 | 0.649 | 7.501 | 8.150 |
| 0.15 | 53.97 | 0.001014 | 10.02 | 225.9 | 2373.1 | 2599.0 | 0.755 | 7.254 | 8.009 |
| 0.20 | 60.06 | 0.001017 | 7.65 | 251.4 | 2358.3 | 2609.7 | 0.832 | 7.077 | 7.909 |
| 0.25 | 64.97 | 0.001020 | 6.20 | 271.9 | 2346.3 | 2618.2 | 0.893 | 6.938 | 7.831 |
| 0.30 | 69.10 | 0.001022 | 5.23 | 289.2 | 2336.1 | 2625.3 | 0.944 | 6.825 | 7.769 |
| 0.40 | 75.87 | 0.001027 | 3.99 | 315.6 | 2319.2 | 2636.8 | 1.026 | 6.645 | 7.671 |
| 0.50 | 81.33 | 0.001030 | 3.24 | 340.5 | 2305.4 | 2645.9 | 1.091 | 6.503 | 7.594 |
| 0.60 | 85.95 | 0.001033 | 2.73 | 359.9 | 2293.6 | 2653.5 | 1.145 | 6.388 | 7.533 |
| 0.70 | 89.96 | 0.001036 | 2.37 | 376.8 | 2283.3 | 2660.1 | 1.192 | 6.288 | 7.480 |
| 0.80 | 93.51 | 0.001039 | 2.09 | 391.6 | 2274.1 | 2665.8 | 1.233 | 6.202 | 7.435 |
| 0.90 | 96.71 | 0.001041 | 1.87 | 405.2 | 2265.6 | 2670.8 | 1.269 | 6.126 | 7.395 |
| 1.0 | 99.63 | 0.001043 | 1.694 | 417.5 | 2258.0 | 2675.5 | 1.303 | 6.057 | 7.360 |
| 1.013 | 100.00 | 0.001043 | 1.673 | 419.0 | 2257.0 | 2676.0 | 1.307 | 6.048 | 7.355 |
| 1.2 | 104.81 | 0.001047 | 1.428 | 439.4 | 2244.1 | 2683.5 | 1.361 | 5.938 | 7.299 |
| 1.4 | 109.32 | 0.001051 | 1.236 | 458.4 | 2231.9 | 2690.3 | 1.411 | 5.836 | 7.247 |
| 1.6 | 113.37 | 0.001055 | 1.091 | 457.4 | 2220.9 | 2696.3 | 1.455 | 5.747 | 7.202 |
| 1.8 | 116.94 | 0.001058 | 0.977 | 490.7 | 2210.7 | 2701.4 | 1.494 | 5.668 | 7.162 |
| 2.0 | 120.23 | 0.001061 | 0.885 | 504.7 | 2201.6 | 2706.3 | 1.530 | 5.597 | 7.127 |
| 2.2 | 123.27 | 0.001064 | 0.810 | 517.6 | 2193.0 | 2710.6 | 1.563 | 5.532 | 7.095 |
| 2.4 | 126.10 | 0.001066 | 0.747 | 529.6 | 2184.9 | 2714.2 | 1.593 | 5.473 | 7.066 |
| 2.6 | 128.73 | 0.001069 | 0.693 | 540.9 | 2177.3 | 2718.2 | 1.621 | 5.418 | 7.039 |

*(Contd.)*

**Table A 5    Properties of saturated water and steam (Pressure) Tables**

| Absolute pressure $p$ (bar) | Saturation Temp. $t_s$ (°C) | Specific volume (m³/kg) | | Specific enthalpy (kJ/kg) | | | Specific entropy (kJ/kg K) | | |
|---|---|---|---|---|---|---|---|---|---|
| | | Water $v_f$ | Steam $v_g$ | Water $h_f$ | Evaporation $h_{fg}$ | Steam $h_g$ | Water $S_f$ | Evaporation $S_{fg}$ | Steam $S_g$ |
| 2.8 | 131.20 | 0.001071 | 0.646 | 551.4 | 2170.1 | 2721.5 | 1.646 | 5.367 | 7.014 |
| 3.0 | 133.55 | 0.001074 | 0.606 | 561.4 | 2163.2 | 2724.7 | 1.672 | 5.319 | 6.991 |
| 3.2 | 135.75 | 0.001076 | 0.570 | 570.9 | 2156.7 | 2727.6 | 1.695 | 5.274 | 6.969 |
| 3.4 | 137.86 | 0.001078 | 0.538 | 579.9 | 2150.5 | 2730.4 | 1.717 | 5.232 | 6.949 |
| 3.6 | 139.84 | 0.001080 | 0.510 | 588.5 | 2144.4 | 2732.9 | 1.738 | 5.192 | 6.930 |
| 3.8 | 141.78 | 0.001082 | 0.485 | 596.8 | 2138.5 | 2735.3 | 1.757 | 5.155 | 6.912 |
| 4.0 | 143.61 | 0.001084 | 0.462 | 604.7 | 2132.9 | 2737.6 | 1.776 | 5.118 | 6.894 |
| 4.2 | 145.39 | 0.001086 | 0.442 | 612.3 | 2127.5 | 2739.8 | 1.795 | 5.083 | 6.878 |
| 4.4 | 147.09 | 0.001088 | 0.423 | 619.6 | 2122.3 | 2741.9 | 1.812 | 5.050 | 6.862 |
| 4.6 | 148.74 | 0.001089 | 0.405 | 626.7 | 2117.1 | 2743.8 | 1.829 | 5.018 | 6.847 |
| 4.8 | 150.31 | 0.001091 | 0.389 | 633.5 | 2112.1 | 2745.6 | 1.845 | 4.988 | 6.833 |
| 5.0 | 151.84 | 0.001093 | 0.373 | 640.1 | 2107.4 | 2747.5 | 1.860 | 4.959 | 6.819 |
| 5.2 | 153.33 | 0.001095 | 0.361 | 646.5 | 2102.7 | 2749.2 | 1.876 | 4.930 | 6.806 |
| 5.4 | 154.76 | 0.001096 | 0.348 | 652.8 | 2098.0 | 2750.8 | 1.890 | 4.903 | 6.793 |
| 5.6 | 156.16 | 0.001098 | 0.337 | 658.8 | 2093.7 | 2752.5 | 1.904 | 4.877 | 6.781 |
| 5.8 | 157.52 | 0.001099 | 0.326 | 664.7 | 2089.7 | 2754.0 | 1.918 | 4.851 | 6.769 |
| 6.0 | 158.84 | 0.001100 | 0.316 | 670.4 | 2085.0 | 2755.4 | 1.931 | 4.827 | 6.758 |
| 6.5 | 161.99 | 0.001104 | 0.293 | 684.0 | 2074.8 | 2758.8 | 1.962 | 4.768 | 6.730 |
| 7.0 | 164.96 | 0.001108 | 0.273 | 697.0 | 2064.9 | 2761.9 | 1.992 | 4.713 | 6.705 |
| 7.5 | 167.66 | 0.001112 | 0.256 | 709.3 | 2059.3 | 7368.6 | 2.020 | 4.671 | 6.691 |
| 8.0 | 170.41 | 0.001115 | 0.240 | 720.9 | 2046.5 | 2767.4 | 2.046 | 4.614 | 6.660 |
| 8.5 | 172.95 | 0.001118 | 0.227 | 732.5 | 2037.9 | 2770.4 | 2.061 | 4.578 | 6.639 |
| 9.0 | 175.36 | 0.001121 | 0.2150 | 742.6 | 2029.5 | 2772.1 | 2.094 | 4.525 | 6.619 |
| 9.5 | 177.69 | 0.001124 | 0.204 | 752.8 | 2021.4 | 2774.2 | 2.117 | 4.484 | 6.601 |
| 10.0 | 179.88 | 0.001127 | 0.1943 | 762.6 | 2013.6 | 2776.2 | 2.138 | 4.445 | 6.583 |
| 11.0 | 184.07 | 0.001133 | 0.1774 | 781.1 | 1998.5 | 2779.6 | 2.179 | 4.371 | 6.550 |
| 12.0 | 187.96 | 0.001139 | 0.1632 | 798.4 | 1984.3 | 2782.5 | 2.216 | 4.303 | 6.519 |
| 13.0 | 191.61 | 0.001144 | 0.1511 | 814.7 | 1970.7 | 2785.4 | 2.251 | 4.240 | 6.491 |
| 14.0 | 195.04 | 0.001149 | 0.1407 | 830.1 | 1957.7 | 2787.8 | 2.284 | 4.181 | 6.465 |

*(Contd.)*

**Table A 5   Continued**

| Absolute pressure $p$ (bar) | Saturation Temp. $t_s$ (°C) | Specific volume (m³/kg) | | Specific enthalpy (kJ/kg) | | | Specific entropy (kJ/kg K) | | |
|---|---|---|---|---|---|---|---|---|---|
| | | Water $v_f$ | Steam $v_g$ | Water $h_f$ | Evaporation $h_{fg}$ | Steam $h_g$ | Water $S_f$ | Evaporation $S_{fg}$ | Steam $S_g$ |
| 15.0 | 198.30 | 0.001154 | 0.1317 | 844.7 | 1945.2 | 2789.9 | 2.315 | 4.126 | 6.441 |
| 16.0 | 201.37 | 0.001159 | 0.1237 | 858.6 | 1933.2 | 2791.8 | 2.344 | 4.074 | 6.418 |
| 17.0 | 204.31 | 0.001163 | 0.1166 | 871.8 | 1921.5 | 2793.3 | 2.371 | 4.025 | 6.396 |
| 18.0 | 207.11 | 0.001168 | 0.1103 | 884.6 | 1910.3 | 2794.9 | 2.398 | 3.977 | 6.375 |
| 19.0 | 209.80 | 0.001172 | 0.1047 | 896.8 | 1899.3 | 2796.1 | 2.423 | 3.931 | 6.354 |
| 20.0 | 212.37 | 0.001177 | 0.09954 | 908.6 | 1888.6 | 2797.2 | 2.447 | 3.890 | 3.337 |
| 25.0 | 223.94 | 0.001197 | 0.07905 | 962.0 | 1839.0 | 2801.0 | 2.554 | 3.700 | 6.254 |
| 30.0 | 233.84 | 0.001216 | 0.06663 | 1008.4 | 1793.9 | 2802.3 | 2.664 | 3.538 | 6.184 |
| 35.0 | 242.54 | 0.001235 | 0.05703 | 1049.8 | 1752.2 | 2802.0 | 2.725 | 3.398 | 6.123 |
| 40.0 | 250.33 | 0.001252 | 0.04975 | 1087.4 | 1712.9 | 2800.3 | 2.797 | 3.272 | 6.069 |
| 45.0 | 257.41 | 0.001262 | 0.04404 | 1122.1 | 1675.6 | 2797.7 | 2.861 | 3.158 | 6.019 |
| 50.0 | 263.91 | 0.001286 | 0.03943 | 1154.5 | 1639.7 | 2794.2 | 2.921 | 3.053 | 5.974 |
| 60.0 | 275.55 | 0.001319 | 0.03244 | 1213.7 | 1571.3 | 2785.0 | 3.047 | 2.864 | 5.891 |
| 70.0 | 285.79 | 0.001351 | 0.02737 | 1267.4 | 1506.0 | 2773.5 | 3.122 | 2.694 | 5.816 |
| 80.0 | 294.97 | 0.001384 | 0.02353 | 1317.1 | 1442.8 | 2759.9 | 3.208 | 1.539 | 5.747 |
| 90.0 | 303.31 | 0.001418 | 0.02050 | 1363.7 | 1380.9 | 2744.6 | 3.287 | 2.395 | 5.682 |
| 100.0 | 310.96 | 0.001453 | 0.01804 | 1408.0 | 1319.7 | 2727.7 | 3.361 | 2.259 | 5.620 |
| 120.0 | 324.65 | 0.001527 | 0.01428 | 1491.8 | 1197.4 | 2689.2 | 3.497 | 2.003 | 5.500 |
| 140.0 | 336.64 | 0.001611 | 0.01150 | 1571.6 | 1070.7 | 2642.3 | 3.624 | 1.756 | 5.380 |
| 160.0 | 347.33 | 0.001710 | 0.009308 | 1650.5 | 934.3 | 2584.8 | 3.747 | 1.506 | 5.253 |
| 180.0 | 356.96 | 0.001840 | 0.007498 | 1734.8 | 779.1 | 2513.9 | 3.877 | 1.236 | 5.113 |
| 200.0 | 365.70 | 0.002037 | 0.005877 | 1826.5 | 591.9 | 2418.4 | 4.015 | 0.926 | 4.941 |
| 220.0 | 373.69 | 0.002671 | 0.003728 | 2011.1 | 184.5 | 2195.6 | 4.295 | 0.285 | 4.580 |
| 221.2 | 374.15 | 0.003170 | 0.003170 | 2107.4 | 000.0 | 2107.4 | 4.443 | 0.000 | 4.443 |

# Table A 6    Properties of superheated steam

(a) specific volume ($v$), m$^3$/kg (b) specific enthalpy ($h$), kJ/kg (c) specific entropy ($s$), kJ/kg k

| Pressure ($p$) bar | Sat. Temp. ($t_s$)°C | Properties | Steam $v_g, h_g, s_g$ | Temperature ($t$°C) | | | | | | | | |
|---|---|---|---|---|---|---|---|---|---|---|---|---|
| | | | | 100 | 150 | 200 | 250 | 300 | 400 | 500 | 600 | 700 |
| 0.02 | 17.03 | $v$ | 67.0 | 86.08 | 97.65 | 109.2 | 120.7 | 132.2 | 155.3 | 178.4 | 210.5 | 224.6 |
| | | $h$ | 2533.5 | 2688.5 | 2783.7 | 2888.0 | 2977.7 | 3076.8 | 3297.7 | 3489.2 | 3705.6 | 3928.8004 |
| | | $s$ | 8.463 | 9.193 | 9.433 | 9.648 | 9.844 | 10.164 | 10.454 | 10.717 | 10.959 | 11.184 |
| 0.04 | 28.96 | $v$ | 34.8 | 43.03 | 48.83 | 54.48 | 60.35 | 66.10 | 77.70 | 89.20 | 100.70 | 112.300 |
| | | $h$ | 2554.4 | 2688.3 | 2783.5 | 2879.9 | 2977.6 | 3076.8 | 3297.7 | 3489.2 | 3705.6 | 3928.800 |
| | | $s$ | 8.052 | 8.873 | 9.113 | 9.328 | 9.524 | 10.031 | 10.321 | 10.585 | 10.827 | 11.051 |
| 0.06 | 36.18 | $v$ | 23.74 | 28.68 | 32.55 | 36.38 | 40.250 | 44.10 | 51.80 | 59.50 | 67.20 | 74.900 |
| | | $h$ | 2567.5 | 2688.0 | 2783.4 | 2879.8 | 2977.600 | 3076.7 | 3279.6 | 3489.2 | 3705.6 | 3928.800 |
| | | $s$ | 7.810 | 8.685 | 8.925 | 9.141 | 9.337 | 9.518 | 9.844 | 10.134 | 10.397 | 10.639 |
| 0.08 | 41.53 | $v$ | 18.11 | 21.50 | 24.42 | 27.30 | 30.18 | 33.10 | 38.80 | 44.60 | 50.40 | 56.100 |
| | | $h$ | 2577.1 | 2687.8 | 2783.2 | 2879.7 | 2977.5 | 3076.6 | 3279.6 | 3489.1 | 3705.5 | 3928.700 |
| | | $s$ | 7.637 | 8.552 | 8.792 | 9.008 | 9.204 | 9.385 | 9.711 | 10.001 | 10.265 | 10.507 |
| 0.10 | 45.81 | $v$ | 14.67 | 17.20 | 19.53 | 21.82 | 24.15 | 26.50 | 31.10 | 35.70 | 40.30 | 44.900 |
| | | $h$ | 2392.8 | 2687.5 | 2783.1 | 2879.6 | 2977.4 | 3076.5 | 3279.5 | 3489.1 | 3705.5 | 3928.700 |
| | | $s$ | 7.501 | 8.449 | 8.689 | 8.905 | 9.101 | 9.282 | 9.608 | 9.894 | 10.162 | 10.404 |
| 0.15 | 53.90 | $v$ | 10.02 | 11.51 | 13.06 | 14.61 | 16.17 | 17.71 | 20.80 | 23.89 | 26.99 | 30.050 |
| | | $h$ | 2599.0 | 2686.9 | 2782.4 | 2879.5 | 2977.3 | 3076.5 | 3279.5 | 3489.0 | 3705.5 | 3928.600 |
| | | $s$ | 7.254 | 8.261 | 8.502 | 8.718 | 8.915 | 9.095 | 9.421 | 9.711 | 9.974 | 10.216 |
| 0.20 | 60.60 | $v$ | 7.65 | 8.59 | 9.75 | 10.91 | 12.08 | 13.22 | 15.53 | 17.84 | 20.15 | 22.450 |
| | | $h$ | 2609.7 | 2686.3 | 2782.3 | 2879.2 | 2977.1 | 3076.4 | 3279.4 | 3489.0 | 3675.4 | 3928.700 |
| | | $s$ | 7.077 | 8.126 | 8.368 | 8.584 | 8.781 | 8.962 | 9.288 | 9.578 | 9.842 | 10.084 |
| 0.25 | 64.97 | $v$ | 6.20 | 6.87 | 7.81 | 8.74 | 9.67 | 10.59 | 12.44 | 14.29 | 16.14 | 17.990 |
| | | $h$ | 2618.2 | 2685.7 | 2782.0 | 2879.0 | 2977.0 | 3076.3 | 3279.2 | 3489.0 | 3675.4 | 3928.700 |
| | | $s$ | 7.182 | 8.022 | 8.262 | 8.481 | 8.678 | 8.859 | 9.186 | 9.476 | 9.739 | 9.981 |
| 0.30 | 69.10 | $v$ | 5.23 | 5.71 | 6.49 | 7.27 | 8.04 | 8.81 | 10.35 | 10.89 | 13.43 | 14.970 |
| | | $h$ | 2625.3 | 2685.1 | 2781.6 | 2878.7 | 2976.8 | 3076.1 | 3279.3 | 3488.9 | 3705.4 | 3928.700 |
| | | $s$ | 7.769 | 7.936 | 8.179 | 8.396 | 8.593 | 8.774 | 9.101 | 9.391 | 9.654 | 9.897 |
| 0.40 | 75.87 | $v$ | 3.99 | 4.28 | 4.87 | 5.45 | 6.03 | 6.60 | 7.76 | 8.92 | 10.07 | 11.230 |
| | | $h$ | 2636.8 | 2683.8 | 2780.9 | 2878.2 | 2976.5 | 3075.9 | 3279.1 | 3488.8 | 3705.3 | 3928.600 |
| | | $s$ | 7.671 | 7.801 | 8.045 | 8.263 | 8.460 | 8.641 | 8.968 | 9.258 | 9.522 | 9.764 |
| 0.50 | 81.33 | $v$ | 3.24 | 3.42 | 3.89 | 4.36 | 4.82 | 5.28 | 6.21 | 7.13 | 8.06 | 8.980 |
| | | $h$ | 2645.9 | 2682.6 | 2780.1 | 2877.6 | 2976.1 | 3075.7 | 3279.0 | 3488.7 | 3705.2 | 3928.600 |
| | | $s$ | 7.594 | 7.695 | 7.941 | 8.159 | 8.356 | 8.538 | 8.865 | 9.155 | 9.419 | 9.661 |

(Contd.)

**Table A 6 Continued**

| Pressure (p) bar | Sat. Temp. $(t_s)$°C | Properties | Steam $v_g, h_g, s_g$ | Temperature $(t$°C) | | | | | | | | |
|---|---|---|---|---|---|---|---|---|---|---|---|---|
| | | | | 100 | 150 | 200 | 250 | 300 | 400 | 500 | 600 | 700 |
| 0.60 | 85.95 | $v$ | 2.73 | 2.84 | 3.24 | 3.63 | 4.02 | 4.40 | 5.17 | 5.94 | 6.71 | 7.480 |
| | | $h$ | 2653.5 | 2681.3 | 2779.4 | 2877.3 | 2975.8 | 3075.4 | 3278.8 | 3488.6 | 3705.1 | 3928.500 |
| | | $s$ | 7.533 | 7.609 | 7.855 | 8.074 | 8.272 | 8.453 | 8.781 | 9.071 | 9.334 | 9.576 |
| 0.70 | 89.96 | $v$ | 2.37 | 2.43 | 2.77 | 3.11 | 3.44 | 3.77 | 4.43 | 5.10 | 5.76 | 6.42 |
| | | $h$ | 2660.1 | 2680.0 | 2778.6 | 2876.8 | 2975.5 | 3075.2 | 3278.6 | 3488.5 | 3705.0 | 3928.4 |
| | | $s$ | 7.480 | 7.535 | 7.783 | 8.002 | 8.201 | 8.382 | 8.709 | 9.001 | 9.263 | 9.505 |
| 0.80 | 93.51 | $v$ | 2.090 | 2.13 | 2.43 | 2.72 | 3.01 | 3.30 | 3.88 | 4.46 | 5.03 | 5.16 |
| | | $h$ | 2665.8 | 2678.8 | 2777.8 | 2876.2 | 2975.2 | 3074.0 | 3278.5 | 3488.4 | 3705.0 | 3928.4 |
| | | $s$ | 7.435 | 7.471 | 7.720 | 7.940 | 8.138 | 8.320 | 8.648 | 8.938 | 9.201 | 9.444 |
| 1.00 | 96.71 | $v$ | 1.694 | 1.696 | 1.936 | 2.172 | 2.406 | 2.639 | 3.103 | 3.565 | 4.028 | 4.490 |
| | | $h$ | 2670.8 | 2676.2 | 2776.3 | 2875.4 | 2974.5 | 3074.5 | 3278.2 | 3488.1 | 3704.8 | 3928.2 |
| | | $s$ | 7.360 | 7.362 | 7.614 | 7.835 | 8.034 | 8.217 | 8.544 | 8.835 | 9.098 | 9.341 |
| 1.50 | 111.35 | $v$ | 1.164 | – | 1.285 | 1.444 | 1.601 | 1.757 | 2.067 | 2.376 | 2.685 | 2.993 |
| | | $h$ | 2693.3 | – | 2772.5 | 2872.9 | 2972.9 | 3073.3 | 3277.6 | 3487.6 | 3704.4 | 3927.9 |
| | | $s$ | 7.225 | – | 7.419 | 7.644 | 7.845 | 8.028 | 8.356 | 8.647 | 8.911 | 9.153 |
| 2.00 | 120.23 | $v$ | 0.885 | – | 0.960 | 1.080 | 1.199 | 1.316 | 1.549 | 1.781 | 2.013 | 2.244 |
| | | $h$ | 2706.3 | – | 2768.5 | 2870.5 | 2971.2 | 3072.1 | 3276.7 | 3487.0 | 3704.1 | 3927.6 |
| | | $s$ | 7.127 | – | 7.279 | 7.507 | 7.710 | 7.894 | 8.223 | 8.514 | 8.777 | 9.020 |
| 3.00 | 133.55 | $v$ | 0.606 | – | 0.634 | 0.716 | 0.796 | 0.875 | 1.031 | 1.187 | 1.341 | 1.496 |
| | | $h$ | 2724.7 | – | 2760.4 | 2865.5 | 2967.9 | 3069.7 | 3275.2 | 3486.0 | 3703.2 | 3927.0 |
| | | $s$ | 6.991 | – | 7.077 | 7.312 | 7.518 | 7.703 | 8.034 | 8.326 | 8.590 | 8.832 |
| 4.00 | 143.61 | $v$ | 0.462 | – | 0.471 | 0.534 | 0.595 | 0.655 | 0.773 | 0.889 | 1.005 | 1.121 |
| | | $h$ | 2737.6 | – | 2752.0 | 2860.4 | 2964.5 | 3067.2 | 3273.6 | 3484.9 | 3702.3 | 3926.4 |
| | | $s$ | 6.894 | – | 6.929 | 7.171 | 7.380 | 7.568 | 7.899 | 8.192 | 8.456 | 8.699 |
| 5.00 | 151.84 | $v$ | 0.375 | – | – | 0.425 | 0.474 | 0.523 | 0.617 | 0.711 | 0.804 | 0.897 |
| | | $h$ | 2747.5 | – | – | 2855.1 | 2961.1 | 3064.8 | 3272.1 | 3483.8 | 3701.5 | 3925.8 |
| | | $s$ | 6.819 | – | – | 7.059 | 7.272 | 7.461 | 7.795 | 8.088 | 8.353 | 8.596 |
| 6.00 | 158.84 | $v$ | 0.316 | – | – | 0.352 | 0.394 | 0.434 | 0.514 | 0.592 | 0.670 | 0.747 |
| | | $h$ | 2755.4 | – | – | 2849.7 | 2957.6 | 3062.3 | 3270.6 | 3482.7 | 3700.7 | 3925.1 |
| | | $s$ | 6.758 | – | – | 6.966 | 7.183 | 7.374 | 7.709 | 8.003 | 8.268 | 8.511 |
| 7.00 | 164.96 | $v$ | 0.273 | – | – | 0.300 | 0.336 | 0.371 | 0.440 | 0.507 | 0.574 | 0.640 |
| | | $h$ | 2761.9 | – | – | 2844.2 | 2954.0 | 3059.8 | 3269.0 | 3481.6 | 3699.9 | 3924.5 |
| | | $s$ | 6.705 | – | – | 6.886 | 7.107 | 7.300 | 7.636 | 7.931 | 8.196 | 8.440 |

*(Contd.)*

**Table A 6  Continued**

| Pressure ($p$) bar | Sat. Temp. ($t_s$)°C | Properties | Steam $v_g$, $h_g$, $s_g$ | Temperature ($t$°C) | | | | | | | | |
|---|---|---|---|---|---|---|---|---|---|---|---|---|
| | | | | 100 | 150 | 200 | 250 | 300 | 400 | 500 | 600 | 700 |
| 8.00 | 170.41 | $v$ | 0.240 | – | – | 0.261 | 0.293 | 0.324 | 0.384 | 0.443 | 0.502 | 0.560 |
| | | $h$ | 2767.4 | – | – | 2838.6 | 2950.4 | 3057.3 | 3267.5 | 3480.5 | 3699.1 | 3923.9 |
| | | $s$ | 6.660 | – | – | 6.815 | 7.040 | 7.235 | 7.573 | 7.868 | 8.134 | 8.377 |
| 9.00 | 175.36 | $v$ | 0.2150 | – | – | 0.230 | 0.260 | 0.287 | 0.341 | 0.394 | 0.446 | 0.498 |
| | | $h$ | 2772.1 | – | – | 2838.7 | 2946.8 | 3054.7 | 3266.0 | 3479.4 | 3698.2 | 3923.3 |
| | | $s$ | 6.610 | – | – | 6.751 | 6.980 | 7.177 | 7.517 | 7.812 | 8.079 | 8.323 |
| 10.00 | 179.88 | $v$ | 0.1943 | – | – | 0.2059 | 0.2328 | 0.2580 | 0.3065 | 0.3540 | 0.4010 | 0.4943 |
| | | $h$ | 2776.2 | – | – | 2826.8 | 2943.000 | 3052.1 | 3264.3 | 3678.4 | 3697.4 | 3922.7 |
| | | $s$ | 6.583 | – | – | 6.692 | 6.926 | 7.125 | 7.467 | 7.763 | 8.029 | 8.273 |
| 15.00 | 198.30 | $v$ | 0.1317 | – | – | 0.1324 | 0.1520 | 0.1697 | 0.2029 | 0.2350 | 0.2666 | 0.2980 |
| | | $h$ | 2789.9 | – | – | 2794.7 | 2923.5 | 3038.9 | 3256.6 | 3472.8 | 3693.3 | 3919.6 |
| | | $s$ | 6.441 | – | – | 6.451 | 6.710 | 6.921 | 7.271 | 7.570 | 7.838 | 8.084 |
| 20.00 | 212.37 | $v$ | 0.09954 | – | – | – | 0.1115 | 0.1255 | 0.1511 | 0.1756 | 0.1995 | 0.2232 |
| | | $h$ | 2797.2 | – | – | – | 2902.4 | 3025.0 | 3248.7 | 3467.4 | 3689.2 | 3916.5 |
| | | $s$ | 6.337 | – | – | – | 6.545 | 6.770 | 7.130 | 7.432 | 7.702 | 7.947 |
| 25.00 | 223.94 | $v$ | 0.07905 | – | – | – | 0.0872 | 0.0991 | 0.1202 | 0.1401 | 0.1594 | 0.1786 |
| | | $h$ | 2801.0 | – | – | – | 2880.5 | 3010.4 | 3240.7 | 3461.6 | 3685.1 | 3913.3 |
| | | $s$ | 6.254 | – | – | – | 6.408 | 6.648 | 7.018 | 7.324 | 7.594 | 7.844 |
| 30.00 | 233.84 | $v$ | 0.0663 | – | – | – | 0.07055 | 0.08116 | 0.09931 | 0.11611 | 0.13232 | 0.14830 |
| | | $h$ | 2802.3 | – | – | – | 2854.8 | 2995.1 | 3232.5 | 3456.2 | 3681.0 | 3910.3 |
| | | $s$ | 6.184 | – | – | – | 6.286 | 6.542 | 6.925 | 7.235 | 7.508 | 7.756 |
| 40.00 | 250.33 | $v$ | 0.04975 | – | – | – | – | 0.05883 | 0.07338 | 0.08634 | 0.09876 | 0.11091 |
| | | $h$ | 2800.3 | – | – | – | – | 2962.0 | 3215.7 | 3445.0 | 3672.8 | 3904.1 |
| | | $s$ | 6.069 | – | – | – | – | 6.346 | 6.773 | 7.091 | 7.368 | 7.619 |
| 50.00 | 263.91 | $v$ | 0.03943 | – | – | – | – | 0.4530 | 0.05780 | 0.06849 | 0.07862 | 0.08845 |
| | | $h$ | 2794.2 | – | – | – | – | 2925.5 | 3198.3 | 3433.7 | 3644.5 | 3897.9 |
| | | $s$ | 5.974 | – | – | – | – | 6.211 | 6.651 | 6.977 | 7.258 | 7.511 |
| 60.00 | 275.55 | $v$ | 0.03244 | – | – | – | – | 0.03615 | 0.04738 | 0.05659 | 0.06518 | 0.07348 |
| | | $h$ | 2745.0 | – | – | – | – | 2885.0 | 3180.1 | 3422.2 | 3656.2 | 3891.7 |
| | | $s$ | 5.891 | – | – | – | – | 6.069 | 6.546 | 6.882 | 7.166 | 7.422 |
| 70.00 | 285.79 | $v$ | 0.02737 | – | – | – | – | 0.02946 | 0.03992 | 0.04809 | 0.05560 | 0.06279 |
| | | $h$ | 2773.5 | – | – | – | – | 2839.0 | 3161.2 | 3410.6 | 3647.9 | 3885.40 |
| | | $s$ | 5.816 | – | – | – | – | 5.933 | 6.454 | 6.799 | 7.088 | 7.346 |
| 80.00 | 294.97 | $v$ | 0.02353 | – | – | – | – | 0.02426 | 0.03431 | 0.04170 | 0.04840 | 0.05477 |
| | | $h$ | 2759.9 | – | – | – | – | 2786.8 | 3141.6 | 3198.8 | 3639.5 | 3879.2 |
| | | $s$ | 5.747 | – | – | – | – | 5.794 | 6.369 | 6.726 | 7.019 | 7.279 |

*(Contd.)*

**Table A 6   Continued**

| Pressure (p) bar | Sat. Temp. ($t_s$)°C | Properties | Steam $v_g$, $h_g$, $s_g$ | Temperature ($t$°C) | | | | | | | | |
|---|---|---|---|---|---|---|---|---|---|---|---|---|
| | | | | 100 | 150 | 200 | 250 | 300 | 400 | 500 | 600 | 700 |
| 90.00 | 303.31 | $v$ | 0.02050 | – | – | – | – | – | 0.02993 | 0.03674 | 0.0428 | 0.04853 |
| | | $h$ | 2744.6 | – | – | – | – | – | 3121.2 | 3386.8 | 3631.1 | 3873.0 |
| | | $s$ | 5.682 | – | – | – | – | – | 6.292 | 6.660 | 6.957 | 7.220 |
| 100.00 | 310.96 | $v$ | 0.01804 | – | – | – | – | – | 0.02641 | 0.03276 | 0.03831 | 0.04355 |
| | | $h$ | 2727.1 | – | – | – | – | – | 3099.9 | 3374.6 | 3622.7 | 3866.7 |
| | | $s$ | 5.620 | – | – | – | – | – | 6.218 | 6.599 | 6.901 | 7.166 |
| 120.00 | 324.65 | $v$ | 0.01428 | – | – | – | – | – | 0.02108 | 0.02679 | 0.03160 | 0.03607 |
| | | $h$ | 2689.2 | – | – | – | – | – | 3054.8 | 3349.5 | 3605.7 | 3854.3 |
| | | $s$ | 5.500 | – | – | – | – | – | 6.081 | 6.491 | 6.802 | 7072 |
| 140.00 | 336.64 | $v$ | 0.01150 | – | – | – | – | – | 0.01723 | 0.02250 | 0.02680 | 0.03072 |
| | | $h$ | 2642.3 | – | – | – | – | – | 3005.6 | 3323.8 | 3588.5 | 3841.7 |
| | | $s$ | 5.380 | – | – | – | – | – | 5.951 | 6.394 | 6.716 | 6.991 |
| 160.00 | 347.33 | $v$ | 0.009308 | – | – | – | – | – | 0.01428 | 0.01930 | 0.02320 | 0.02672 |
| | | $h$ | 2584.8 | – | – | – | – | – | 2951.4 | 3297.1 | 3571.0 | 3829.1 |
| | | $s$ | 5.253 | – | – | – | – | – | 5.824 | 6.305 | 6.639 | 6.919 |
| 180.00 | 356.96 | $v$ | 0.007498 | – | – | – | – | – | 0.01190 | 0.01679 | 0.02040 | 0.02361 |
| | | $h$ | 2513.9 | – | – | – | – | – | 2890.3 | 3269.6 | 3553.5 | 3816.5 |
| | | $s$ | 5.113 | – | – | – | – | – | 5.695 | 6.223 | 6.569 | 6.854 |
| 200.00 | 365.70 | $v$ | 0.005877 | – | – | – | – | – | 0.00995 | 0.01477 | 0.01816 | 0.02111 |
| | | $h$ | 2418.4 | – | – | – | – | – | 2820.5 | 3241.1 | 3535.5 | 3830.8 |
| | | $s$ | 4.941 | – | – | – | – | – | 5.559 | 6.146 | 6.504 | 6.795 |
| 220.00 | 373.69 | $v$ | 0.003728 | – | – | – | – | – | 0.00825 | 0.01312 | 0.01633 | 0.01907 |
| | | $h$ | 2195.6 | – | – | – | – | – | 2738.8 | 3211.7 | 3517.4 | 3791.1 |
| | | $s$ | 4.580 | – | – | – | – | – | 5.410 | 6.072 | 6.445 | 6.741 |
| 221.00 | 374.15 | $v$ | 0.003170 | – | – | – | – | – | 0.00816 | 0.01303 | 0.01622 | 0.01895 |
| | | $h$ | 2107.4 | – | – | – | – | – | 2734.5 | 3210.7 | 3516.4 | 3789.1 |
| | | $s$ | 4.443 | – | – | – | – | – | 5.399 | 6.068 | 6.441 | 6.738 |

### Table A 7  Properties of mercury-Hg

| $p$ | $t_s$ | $v_g$ | $h_f$ | $h_{fg}$ | $h_g$ | $s_f$ | $s_{fg}$ | $s_g$ |
|---|---|---|---|---|---|---|---|---|
| 0.0006 | 109.2 | 259.6 | 15.13 | 297.20 | 312.33 | 0.0466 | 0.7774 | 0.8240 |
| 0.0007 | 112.3 | 224.3 | 15.55 | 297.14 | 312.69 | 0.0477 | 0.7709 | 0.8186 |
| 0.0008 | 115.0 | 197.7 | 15.93 | 297.09 | 313.02 | 0.0487 | 0.7654 | 0.8141 |
| 0.0009 | 117.5 | 176.8 | 16.27 | 297.04 | 313.31 | 0.0496 | 0.7604 | 0.8100 |
| 0.0010 | 119.7 | 160.1 | 16.58 | 297.00 | 313.58 | 0.0503 | 0.7560 | 0.8063 |
| 0.002 | 134.9 | 83.18 | 18.67 | 296.71 | 315.38 | 0.0556 | 0.7271 | 0.7827 |
| 0.004 | 151.5 | 43.29 | 20.93 | 296.40 | 317.33 | 0.0610 | 0.6981 | 0.7591 |
| 0.006 | 161.8 | 29.57 | 22.33 | 296.21 | 318.54 | 0.0643 | 0.6811 | 0.7454 |
| 0.008 | 169.4 | 22.57 | 23.37 | 296.06 | 319.43 | 0.0666 | 0.6690 | 0.7356 |
| 0.010 | 175.5 | 18.31 | 24.21 | 295.95 | 320.16 | 0.0685 | 0.6596 | 0.7281 |
| 0.02 | 195.6 | 9.570 | 26.94 | 295.57 | 322.51 | 0.0744 | 0.6305 | 0.7949 |
| 0.04 | 217.7 | 5.013 | 29.92 | 295.15 | 325.07 | 0.0806 | 0.6013 | 0.6819 |
| 0.06 | 231.6 | 3.438 | 31.81 | 294.89 | 326.70 | 0.0843 | 0.5842 | 0.6685 |
| 0.08 | 242.0 | 2.632 | 33.21 | 294.70 | 327.91 | 0.0870 | 0.5721 | 0.6591 |
| 0.10 | 250.3 | 2.140 | 34.33 | 294.54 | 328.87 | 0.0892 | 0.5627 | 0.6519 |
| 0.2 | 278.1 | 1.128 | 38.05 | 294.02 | 332.07 | 0.0961 | 0.5334 | 0.6295 |
| 0.4 | 309.1 | 0.5942 | 42.21 | 293.43 | 335.64 | 0.1034 | 0.5039 | 0.6073 |
| 0.6 | 329.0 | 0.4113 | 44.85 | 293.06 | 337.91 | 0.1078 | 0.4869 | 0.5947 |
| 0.8 | 343.9 | 0.3163 | 46.84 | 292.78 | 339.62 | 0.1110 | 0.4745 | 0.5855 |
| 1.0 | 356.1 | 0.2581 | 48.45 | 292.55 | 341.0 | 0.1136 | 0.4649 | 0.5785 |
| 2.0 | 397.1 | 0.1377 | 53.87 | 291.77 | 345.64 | 0.1218 | 0.4353 | 0.5571 |
| 3.0 | 423.8 | 0.09551 | 57.68 | 291.27 | 348.65 | 0.1268 | 0.4179 | 0.5447 |
| 4.0 | 444.1 | 0.07378 | 60.03 | 290.89 | 350.92 | 0.1305 | 0.4056 | 0.5361 |
| 5.0 | 460.7 | 0.06044 | 62.20 | 290.58 | 352.78 | 0.1334 | 0.3960 | 0.5294 |
| 6.0 | 474.9 | 0.05137 | 64.06 | 290.31 | 354.37 | 0.1359 | 0.3881 | 0.5240 |
| 7.0 | 487.3 | 0.04479 | 65.66 | 290.08 | 355.74 | 0.1380 | 0.3815 | 0.5195 |
| 8.0 | 498.4 | 0.03978 | 67.11 | 289.87 | 356.98 | 0.1398 | 0.3757 | 0.5155 |
| 9.0 | 508.5 | 0.03584 | 68.42 | 289.68 | 358.10 | 0.1415 | 0.3706 | 0.5121 |
| 10.0 | 517.8 | 0.03266 | 69.61 | 289.50 | 359.11 | 0.1429 | 0.3660 | 0.5089 |
| 12.0 | 534.4 | 0.02781 | 71.75 | 289.19 | 360.94 | 0.1455 | 0.3581 | 0.5036 |
| 14.0 | 549.0 | 0.02429 | 73.63 | 288.92 | 362.55 | 0.1478 | 0.3514 | 0.4992 |
| 16.0 | 562.0 | 0.02161 | 75.37 | 288.67 | 364.04 | 0.1498 | 0.3456 | 0.4954 |
| 18.0 | 574.0 | 0.01949 | 76.83 | 288.45 | 365.28 | 0.1515 | 0.3405 | 0.4920 |
| 20.0 | 584.9 | 0.01778 | 78.23 | 288.24 | 366.47 | 0.1531 | 0.3359 | 0.4890 |
| 22.0 | 595.1 | 0.01637 | 79.54 | 288.05 | 367.59 | 0.1546 | 0.3318 | 0.4864 |
| 24.0 | 604.6 | 0.01518 | 80.75 | 287.87 | 368.62 | 0.1559 | 0.3280 | 0.4839 |
| 26.0 | 613.5 | 0.01416 | 81.89 | 287.70 | 369.59 | 0.1571 | 0.3245 | 0.4816 |
| 28.0 | 622.0 | 0.01329 | 82.96 | 287.54 | 370.50 | 0.1583 | 0.3212 | 0.4795 |
| 30.0 | 630.0 | 0.01252 | 83.97 | 287.39 | 371.36 | 0.1594 | 0.3182 | 0.4776 |
| 35.0 | 648.5 | 0.01096 | 86.33 | 287.04 | 373.37 | 0.1619 | 0.3115 | 0.4734 |
| 40.0 | 665.1 | 0.00978 | 88.43 | 286.73 | 375.16 | 0.1641 | 0.3056 | 0.4697 |
| 45.0 | 680.3 | 0.00885 | 90.35 | 286.44 | 376.79 | 0.1660 | 0.3004 | 0.4664 |
| 50.0 | 694.4 | 0.00809 | 92.11 | 286.18 | 378.29 | 0.1678 | 0.2958 | 0.4636 |
| 55.0 | 707.4 | 0.00746 | 93.76 | 285.93 | 379.69 | 0.1694 | 0.2916 | 0.4610 |
| 60.0 | 719.7 | 0.00693 | 95.30 | 285.70 | 381.00 | 0.1709 | 0.2878 | 0.4587 |
| 65.0 | 731.3 | 0.00648 | 96.75 | 285.48 | 382.23 | 0.1723 | 0.2842 | 0.4565 |
| 70.0 | 742.3 | 0.00609 | 98.12 | 285.28 | 383.40 | 0.1736 | 0.2809 | 0.4545 |
| 75.0 | 752.7 | 0.00575 | 99.42 | 285.08 | 384.50 | 0.1748 | 0.2779 | 0.4527 |

Datum: $h_f$ and $s_f$ are zero at 0°C.
Relative molecular mass of Hg = 200.60 ; For superheated Hg vapour, $C_p = 0.1036$ kJ/kg K.

## Table A 8  The elements
Alphabetical list of the elements

| Element | Symbol | Atomic Number, Z | Element | Symbol | Atomic Number, Z |
|---|---|---|---|---|---|
| Actinium | Ac | 89 | Indium | In | 49 |
| Aluminium | Al | 13 | Iodine | I | 53 |
| Americium | Am | 95 | Iridium | Ir | 77 |
| Antimony | Sb | 51 | Iron | Fe | 26 |
| Argon | A | 18 | Krypton | Kr' | 36 |
| Arsenic | As | 33 | Lanthanum | La | 57 |
| Astatine | At | 85 | Lead | Pb | 82 |
| Barium | Ba | 56 | Lithium | Li | 3 |
| Berkelium | Bk | 97 | Lutecium | Lu | 71 |
| Beryllium | Be | 4 | Magnesium | Mg | 12 |
| Bismuth | Bi | 83 | Manganese | Mn | 25 |
| Boron | B | 5 | Mendelevium | Md | 101 |
| Bromine | Br | 35 | Mercury | Hg | 80 |
| Cadmium | Cd | 48 | Molybdenum | Mo | 42 |
| Calcium | Ca | 20 | Neodymium | Nd | 60 |
| Californium | Cf | 98 | Neon | Ne | 10 |
| Carbon | C | 6 | Neptunium | Np | 93 |
| Cerium | Ce | 58 | Nickel | Ni | 28 |
| Cesium | Cs | 55 | Niobium | Nb | 41 |
| Chlorine | Cl | 17 | Nitrogen | N | 7 |
| Chromium | Cr | 24 | Nobelium | No | 102 |
| Cobalt | Co | 27 | Osmium | Os | 76 |
| Copper | Cu | 29 | Oxygen | O | 8 |
| Curium | Cm | 96 | Palladium | Pd | 46 |
| Dysprosium | Dy | 66 | Phosphorus | P | 15 |
| Einsteinium | Es | 99 | Platinum | Pt | 78 |
| Erbium | Er | 68 | Plutonium | Pu | 94 |
| Europium | Eu | 63 | Polonium | Po | 84 |
| Fermium | Fm | 100 | Potassium | K | 19 |
| Fluorine | F | 9 | Praseodymium | Pr | 59 |
| Francium | Fr | 87 | Promethium | Pm | 61 |
| Gadolinium | Gd | 64 | Protactinum | Pa | 91 |
| Gallium | Ga | 31 | Radium | Ra | 88 |
| Germanium | Ge | 32 | Radon | Rn | 86 |
| Gold | Au | 79 | Rhenium | Re | 75 |
| Hafnium | Hf | 72 | Rhodium | Rh | 45 |
| Helium | He | 2 | Rubidium | Rb | 37 |
| Holmium | Ho | 67 | Ruthenium | Ru | 44 |
| Hydrogen | H | 1 | Samarium | Sm | 62 |
|  |  |  | Scandium | Sc | 21 |
| Selenium | Se | 34 | Thulium | Tm | 69 |
| Silicon | Si | 14 | Tin | Sn | 50 |
| Silver | Ag | 47 | Titanium | Ti | 22 |
| Sodium | Na | 11 | Tungsten (Wolfram) | W | 74 |

*(Contd.)*

**Table A 8   Continued**

| Element | Symbol | Atomic Number, Z | Element | Symbol | Atomic Number, Z |
|---------|--------|------------------|---------|--------|------------------|
| Strontium | Sr | 38 | Uranium | U | 92 |
| Sulfur | S | 16 | Vanadium | V | 23 |
| Tantalum | Ta | 73 | Xenon | Xe | 54 |
| Technetium | Tc | 43 | Xenon | Xe | 54 |
| Tellurium | Te | 52 | Ytterbium | Yb | 70 |
| Terbium | Tb | 65 | Yttrium | Y | 39 |
| Thallium | Tl | 81 | Zinc | Zn | 30 |
| Thorium | Th | 90 | Zirconium | Zr | 40 |

## Table A 9  Properties of insulating, building and other materials

| Substance | Temperature °C | k, W/m, °C | ρ, kg/m³ | C kJ/kg °C | α, m²/s × 10⁷ |
|---|---|---|---|---|---|
| *Structural and heat-resistant materials* | | | | | |
| Asphalt | 20–55 | 0.74–0.76 | | | |
| Brick: | | | | | |
|    Building brick, common | 20 | 0.69 | 1600 | 0.84 | 5.2 |
|    Face | | 1.32 | 2000 | | |
| Carborundum brick | 600 | 18.5 | | | |
| | 1400 | 11.1 | | | |
| Chrome brick | 200 | 2.32 | 3000 | 0.84 | 9.2 |
| | 550 | 2.47 | | | 9.8 |
| | 900 | 1.99 | | | 7.9 |
| Diatomaceous earth, | | | | | |
|   moulded and fired | 200 | 0.24 | | | |
| | 870 | 0.31 | | | |
| Fireclay brick, burnt | 500 | 1.04 | 2000 | 0.96 | 5.4 |
|   1330°C | 800 | 1.07 | | | |
| | 1100 | 1.09 | | | |
|   Burnt 1450°C | 500 | 1.28 | 2300 | 0.96 | 5.8 |
| | 800 | 1.37 | | | |
| | 1100 | 1.40 | | | |
|   Missouri | 200 | 1.00 | 2600 | 0.96 | 4.0 |
| | 600 | 1.47 | | | |
| | 1400 | 1.77 | | | |
| Magnesite | 200 | 3.81 | | 1.13 | |
| | 650 | 2.77 | | | |
| | 1200 | 1.90 | | | |
| Cement, portland | | 0.29 | 1500 | | |
|   Mortar | 23 | 1.16 | | | |
| Concrete, cinder | 23 | 0.76 | | | |
|   Stone 1-2-4 mix | 20 | 1.37 | 1900–2300 | 0.88 | 8.2–6.8 |
| Glass, window | 20 | 0.78 (avg) | 2700 | 0.84 | 3.4 |
|   Corosilicate | 30–75 | 1.09 | 2200 | | |
| Plaster, gypsum | 20 | 0.48 | 1440 | 0.84 | 4.0 |
|   Metal lath | 20 | 0.47 | | | |
|   Wood lath | 20 | 0.28 | | | |
| Stone: | | | | | |
|   Granite | | 1.73–3.98 | 2640 | 0.82 | 8–18 |
|   Limestone | 100–300 | 1.26–1.33 | 2500 | 0.90 | 5.6–5.9 |
|   Marble | | 2.07–2.94 | 2500–2700 | 0.80 | 10–13.6 |
|   Sandstone | 40 | 1.83 | 2160–2300 | 0.71 | 11.2–11.9 |
| Wood (across the grain): | | | | | |
|   Balsa, | 30 | 0.055 | 140 | | |
|   Cypress | 30 | 0.097 | 460 | | |
|   Fir | 23 | 0.11 | 420 | 2.72 | 0.96 |
|   Maple or oak | 30 | 0.166 | 540 | 2.4 | 1.28 |
|   Yellow pine | 23 | 0.147 | 640 | 2.8 | 0.82 |
|   White pine | 30 | 0.112 | 430 | | |

*(Contd.)*

Table A 9   Continued

| Substance | Temperature °C | k, W/m, °C | ρ, kg/m³ | C kJ/kg °C | α, m²/s × 10⁷ |
|---|---|---|---|---|---|
| *Insulating materials* | | | | | |
| Asbestos: | | | | | |
|   Loosely packed | −45 | 0.149 | | | |
| | 0 | 0.154 | 470–570 | 0.816 | 3.3–4 |
| | 100 | 0.161 | | | |
| Asbestos-cement boards | 20 | 0.74 | | | |
| Sheets | 51 | 0.166 | | | |
| Felt, 16 laminations/cm | 38 | 0.057 | | | |
| | 150 | 0.069 | | | |
| | 260 | 0.083 | | | |
|   8 laminations/cm | 38 | 0.078 | | | |
| | 150 | 0.095 | | | |
| | 260 | 0.112 | | | |
| Corrugated, 150 plies/m | 38 | 0.087 | | | |
| | 93 | 0.100 | | | |
| | 150 | 0.119 | | | |
| Asbestos cement | . . . | 2.08 | | | |
| Balsam wool, | 32 | 0.04 | 35 | | |
| Cardboard, corrugated | . . . | 0.064 | | | |
|   Celotex | 32 | 0.048 | | | |
| Corkboard, | 30 | 0.043 | 160 | | |
| Cork, regranulated | 32 | 0.045 | 45–120 | 1.88 | 2–5.3 |
|   Ground | 32 | 0.043 | 150 | | |
| Diatomaceous earth | | | | | |
|   (Sil-o-cel) | 0 | 0.061 | 320 | | |
| Felt, hair | 30 | 0.036 | 130–200 | | |
|   Wool | 30 | 0.052 | 330 | | |
| Fiber, insulating board | 20 | 0.048 | 240 | | |
| Glass wool, | 23 | 0.038 | 24 | 0.7 | 22.6 |
| Insulex, dry | 32 | 0.064 | | | |
| | | 0.144 | | | |
| Kapok | 30 | 0.035 | | | |
| Magnesia, 85% | 38 | 0.067 | 270 | | |
| | 93 | 0.071 | | | |
| | 150 | 0.074 | | | |
| | 204 | 0.080 | | | |
| Rock wool, | 32 | 0.040 | 160 | | |
|   Loosely packed | 150 | 0.067 | 64 | | |
| | 260 | 0.087 | | | |
| Sawdust | 23 | 0.059 | | | |
| Silica aerogel | 32 | 0.024 | 140 | | |
| Wood shavings | 23 | 0.059 | | | |

### Table A 10   Properties of gases at atmospheric pressure

| $T$<br>K | $\rho$<br>kg/m$^3$ | $c_p$<br>kJ/kg. K | $\mu \cdot 10^7$<br>N $\cdot$ s/m$^2$ | $v \cdot 10^6$<br>m$^2$/s | $k \cdot 10^3$<br>W/m. K | $\alpha \cdot 10^6$<br>m$^2$/s | pr |
|---|---|---|---|---|---|---|---|
| **Air** | | | | | | | |
| 100 | 3.5562 | 1.032 | 71.1 | 2.00 | 9.34 | 2.54 | 0.786 |
| 150 | 2.3364 | 1.012 | 103.4 | 4.426 | 13.8 | 5.84 | 0.758 |
| 200 | 1.7458 | 1.007 | 132.5 | 7.590 | 18.1 | 10.3 | 0.737 |
| 250 | 1.3947 | 1.006 | 159.6 | 11.44 | 22.3 | 15.9 | 0.720 |
| 300 | 1.1614 | 1.007 | 184.6 | 15.89 | 26.3 | 22.5 | 0.707 |
| 350 | 0.9950 | 1.009 | 208.2 | 20.92 | 30.0 | 29.9 | 0.700 |
| 400 | 0.8711 | 1.014 | 230.1 | 26.41 | 33.8 | 38.3 | 0.690 |
| 450 | 0.7740 | 1.021 | 250.7 | 32.39 | 37.3 | 47.2 | 0.686 |
| 500 | 0.6964 | 1.030 | 270.1 | 38.79 | 40.7 | 56.7 | 0.684 |
| 550 | 0.6329 | 1.040 | 288.4 | 45.57 | 43.9 | 66.7 | 0.683 |
| 600 | 0.5804 | 1.051 | 305.8 | 52.69 | 46.9 | 76.9 | 0.685 |
| 650 | 0.5356 | 1.063 | 322.5 | 60.21 | 49.7 | 87.3 | 0.690 |
| 700 | 0.4975 | 1.075 | 338.8 | 68.10 | 52.4 | 98.0 | 0.695 |
| 750 | 0.4643 | 1.087 | 354.6 | 76.37 | 54.9 | 109 | 0.702 |
| 800 | 0.4354 | 1.099 | 369.8 | 84.93 | 57.3 | 120 | 0.709 |
| 850 | 0.4097 | 1.110 | 384.3 | 93.80 | 59.6 | 131 | 0.716 |
| 900 | 0.3868 | 1.121 | 398.1 | 102.9 | 62.0 | 143 | 0.720 |
| 950 | 0.3666 | 1.131 | 411.3 | 112.2 | 64.3 | 155 | 0.723 |
| 1000 | 0.3482 | 1.141 | 424.4 | 121.9 | 66.7 | 168 | 0.726 |
| 1100 | 0.3166 | 1.159 | 449.0 | 141.8 | 71.5 | 195 | 0.728 |
| 1200 | 0.2902 | 1.175 | 473.0 | 162.9 | 76.3 | 224 | 0.728 |
| 1300 | 0.2679 | 1.189 | 496.0 | 185.1 | 82 | 238 | 0.719 |
| 1400 | 0.2488 | 1.207 | 530 | 213 | 91 | 303 | 0.703 |
| 1500 | 0.2322 | 1.230 | 557 | 240 | 100 | 350 | 0.685 |
| 1600 | 0.2177 | 1.248 | 584 | 268 | 106 | 390 | 0.688 |
| 1700 | 0.2049 | 1.267 | 611 | 298 | 113 | 435 | 0.685 |
| 1800 | 0.1935 | 1.286 | 637 | 329 | 120 | 482 | 0.683 |
| 1900 | 0.1833 | 1.307 | 663 | 362 | 128 | 534 | 0.677 |
| 2000 | 0.1741 | 1.337 | 689 | 396 | 137 | 589 | 0.672 |
| 2100 | 0.1658 | 1.372 | 715 | 431 | 147 | 646 | 0.667 |
| 2200 | 0.1582 | 1.417 | 740 | 468 | 160 | 714 | 0.655 |
| 2300 | 0.1513 | 1.478 | 766 | 506 | 175 | 783 | 0.647 |
| 2400 | 0.1448 | 1.558 | 792 | 547 | 196 | 869 | 0.630 |
| 2500 | 0.1389 | 1.665 | 818 | 589 | 222 | 960 | 0.613 |
| 3000 | 0.1135 | 2.726 | 955 | 841 | 486 | 1570 | 0.536 |
| **Ammonia, NH$_3$** | | | | | | | |
| 300 | 0.6894 | 2.158 | 101.5 | 14.7 | 24.7 | 16.6 | 0.887 |
| 320 | 0.6448 | 2.170 | 109 | 16.9 | 27.2 | 19.4 | 0.870 |
| 340 | 0.6059 | 2.192 | 116.5 | 19.2 | 29.3 | 22.1 | 0.872 |
| 360 | 0.5716 | 2.221 | 124 | 21.7 | 31.6 | 24.9 | 0.872 |
| 380 | 0.5410 | 2.254 | 131 | 24.2 | 34.0 | 27.9 | 0.869 |

*(Contd.)*

**Table A 10   Continued**

| $T$ K | $\rho$ kg/m$^3$ | $c_p$ kJ/kg. K | $\mu \cdot 10^7$ N $\cdot$ s/m$^2$ | $\nu \cdot 10^6$ m$^2$/s | $k \cdot 10^3$ W/m. K | $\alpha \cdot 10^6$ m$^2$/s | pr |
|---|---|---|---|---|---|---|---|
| 400 | 0.5136 | 2.287 | 138 | 26.9 | 37.0 | 31.5 | 0.853 |
| 420 | 0.4888 | 2.322 | 145 | 29.7 | 40.4 | 35.6 | 0.833 |
| 440 | 0.4664 | 2.357 | 152.5 | 32.7 | 43.5 | 39.6 | 0.826 |
| 460 | 0.4460 | 2.393 | 159 | 35.7 | 46.3 | 43.4 | 0.822 |
| 480 | 0.4273 | 2.430 | 166.5 | 39.0 | 49.2 | 47.4 | 0.822 |
| 500 | 0.4101 | 2.467 | 173 | 42.2 | 52.5 | 51.9 | 0.813 |
| 520 | 0.3942 | 2.504 | 180 | 45.7 | 54.5 | 55.2 | 0.827 |
| 540 | 0.3795 | 2.540 | 186.5 | 49.1 | 57.5 | 59.7 | 0.824 |
| 560 | 0.3708 | 2.577 | 193 | 52.0 | 60.6 | 63.4 | 0.827 |
| 580 | 0.3533 | 2.613 | 199.5 | 56.5 | 63.8 | 69.1 | 0.817 |

**Carbon Dioxide, CO$_2$**

| $T$ K | $\rho$ kg/m$^3$ | $c_p$ kJ/kg. K | $\mu \cdot 10^7$ N $\cdot$ s/m$^2$ | $\nu \cdot 10^6$ m$^2$/s | $k \cdot 10^3$ W/m. K | $\alpha \cdot 10^6$ m$^2$/s | pr |
|---|---|---|---|---|---|---|---|
| 280 | 1.9022 | 0.830 | 140 | 7.36 | 15.20 | 9.63 | 0.765 |
| 300 | 1.7730 | 0.851 | 149 | 8.40 | 16.55 | 11.0 | 0.766 |
| 320 | 1.6609 | 0.872 | 156 | 9.39 | 18.05 | 12.5 | 0.754 |
| 340 | 1.5618 | 0.891 | 165 | 10.6 | 19.70 | 14.2 | 0.746 |
| 360 | 1.4743 | 0.908 | 173 | 11.7 | 21.2 | 15.8 | 0.741 |
| 380 | 1.3961 | 0.926 | 181 | 13.0 | 22.75 | 17.6 | 0.737 |
| 400 | 1.3257 | 0.942 | 190 | 14.3 | 24.3 | 19.5 | 0.737 |
| 450 | 1.1782 | 0.981 | 210 | 17.8 | 28.3 | 24.5 | 0.728 |
| 500 | 1.0594 | 1.02 | 231 | 21.8 | 32.5 | 30.1 | 0.725 |
| 550 | 0.9625 | 1.05 | 251 | 26.1 | 36.6 | 36.2 | 0.721 |
| 600 | 0.8826 | 1.08 | 270 | 30.6 | 40.7 | 42.7 | 0.717 |
| 650 | 0.8143 | 1.10 | 288 | 35.4 | 44.5 | 49.7 | 0.712 |
| 700 | 0.7564 | 1.13 | 305 | 40.3 | 48.1 | 56.3 | 0.717 |
| 750 | 0.7057 | 1.15 | 321 | 45.5 | 51.7 | 63.7 | 0.714 |
| 800 | 0.6614 | 1.17 | 337 | 51.0 | 55.1 | 71.2 | 0.716 |

**Carbon Monoxide, CO**

| $T$ K | $\rho$ kg/m$^3$ | $c_p$ kJ/kg. K | $\mu \cdot 10^7$ N $\cdot$ s/m$^2$ | $\nu \cdot 10^6$ m$^2$/s | $k \cdot 10^3$ W/m. K | $\alpha \cdot 10^6$ m$^2$/s | pr |
|---|---|---|---|---|---|---|---|
| 200 | 1.6888 | 1.045 | 127 | 7.52 | 17.0 | 9.63 | 0.781 |
| 220 | 1.5341 | 1.044 | 137 | 8.93 | 19.0 | 11.9 | 0.753 |
| 240 | 1.4055 | 1.043 | 147 | 10.5 | 20.6 | 14.1 | 0.744 |
| 260 | 1.2967 | 1.043 | 157 | 12.1 | 22.1 | 16.3 | 0.741 |
| 280 | 1.2038 | 1.042 | 166 | 13.8 | 23.6 | 18.8 | 0.733 |
| 300 | 1.1223 | 1.043 | 175 | 15.6 | 25.0 | 21.3 | 0.730 |
| 320 | 1.0529 | 1.043 | 184 | 17.5 | 26.3 | 23.9 | 0.730 |
| 340 | 0.9909 | 1.044 | 193 | 19.5 | 27.8 | 26.9 | 0.725 |
| 360 | 0.9357 | 1.045 | 202 | 21.6 | 29.1 | 29.8 | 0.725 |
| 380 | 0.8864 | 1.047 | 210 | 23.7 | 30.5 | 32.9 | 0.729 |
| 400 | 0.8421 | 1.049 | 218 | 25.9 | 31.8 | 36.0 | 0.719 |
| 450 | 0.7483 | 1.055 | 237 | 31.7 | 35.0 | 44.3 | 0.714 |
| 500 | 0.67352 | 1.065 | 254 | 37.7 | 38.1 | 53.1 | 0.710 |
| 550 | 0.61226 | 1.076 | 271 | 44.3 | 41.1 | 62.4 | 0.710 |
| 600 | 0.56126 | 1.088 | 286 | 51.0 | 44.0 | 72.1 | 0.707 |

*(Contd.)*

## Table A 10 Continued

| $T$ K | $\rho$ kg/m³ | $c_p$ kJ/kg. K | $\mu \cdot 10^7$ N · s/m² | $\nu \cdot 10^6$ m²/s | $k \cdot 10^3$ W/m. K | $\alpha \cdot 10^6$ m²/s | pr |
|---|---|---|---|---|---|---|---|
| 650 | 0.51806 | 1.101 | 301 | 58.1 | 47.0 | 82.4 | 0.705 |
| 700 | 0.48102 | 1.114 | 315 | 65.5 | 50.0 | 93.3 | 0.702 |
| 750 | 0.44899 | 1.127 | 329 | 73.3 | 52.8 | 104 | 0.702 |
| 800 | 0.42095 | 1.140 | 343 | 81.5 | 55.5 | 116 | 0.705 |
| **Helium, He** | | | | | | | |
| 100 | 0.4871 | 5.193 | 96.3 | 19.8 | 73.0 | 28.9 | 0.686 |
| 120 | 0.4060 | 5.193 | 107 | 26.4 | 81.9 | 38.8 | 0.679 |
| 140 | 0.3481 | 5.193 | 118 | 33.9 | 90.7 | 50.2 | 0.676 |
| 160 | – | 5.193 | 129 | – | 99.2 | – | – |
| 180 | 0.2708 | 5.193 | 139 | 51.3 | 107.2 | 76.2 | 0.673 |
| 200 | – | 5.193 | 150 | – | 115.1 | – | – |
| 220 | 0.2216 | 5.193 | 160 | 72.2 | 123.1 | 107 | 0.675 |
| 240 | – | 5.193 | 170 | – | 130 | – | – |
| 260 | 0.1875 | 5.193 | 180 | 96.0 | 137 | 141 | 0.682 |
| 280 | – | 5.193 | 190 | – | 145 | – | – |
| 300 | 0.1625 | 5.193 | 199 | 122 | 152 | 180 | 0.680 |
| 350 | – | 5.193 | 221 | – | 170 | – | – |
| 400 | 0.1219 | 5.193 | 243 | 199 | 187 | 295 | 0.675 |
| 450 | – | 5.193 | 263 | – | 204 | – | – |
| 500 | 0.09754 | 5.193 | 283 | 290 | 220 | 434 | 0.668 |
| 550 | – | 5.193 | – | – | – | – | – |
| 600 | – | 5.193 | 320 | – | 252 | – | – |
| 650 | – | 5.193 | 332 | – | 252 | – | – |
| 700 | 0.06969 | 5.193 | 350 | 502 | 278 | 768 | 0.654 |
| 750 | – | 5.193 | 364 | – | 291 | – | – |
| 800 | – | 5.193 | 382 | – | 304 | – | – |
| 900 | – | 5.193 | 414 | – | 330 | – | – |
| 1000 | 0.04879 | 5.193 | 446 | 914 | 354 | 1400 | 0.654 |
| **Hydrogen, H₂** | | | | | | | |
| 100 | 0.24255 | 11.23 | 42.1 | 17.4 | 67.0 | 24.6 | 0.707 |
| 150 | 0.16156 | 12.60 | 56.0 | 34.7 | 101 | 49.6 | 0.699 |
| 200 | 0.12115 | 13.54 | 68.1 | 56.2 | 131 | 79.9 | 0.704 |
| 250 | 0.09693 | 14.06 | 78.9 | 81.4 | 157 | 115 | 0.707 |
| 300 | 0.08078 | 14.31 | 89.6 | 111 | 183 | 158 | 0.701 |
| 350 | 0.06924 | 14.43 | 98.8 | 143 | 204 | 204 | 0.700 |
| 400 | 0.06059 | 14.48 | 108.2 | 179 | 226 | 258 | 0.695 |
| 450 | 0.05386 | 14.50 | 117.2 | 218 | 247 | 316 | 0.689 |
| 500 | 0.04848 | 14.52 | 126.4 | 261 | 266 | 378 | 0.691 |
| 550 | 0.04407 | 14.53 | 134.3 | 305 | 285 | 445 | 0.685 |

*(Contd.)*

**Table A 10    Continued**

| $T$ K | $\rho$ kg/m³ | $c_p$ kJ/kg. K | $\mu \cdot 10^7$ N · s/m² | $v \cdot 10^6$ m²/s | $k \cdot 10^3$ W/m. K | $\alpha \cdot 10^6$ m²/s | pr |
|---|---|---|---|---|---|---|---|
| 600 | 0.04040 | 14.55 | 142.4 | 352 | 305 | 519 | 0.678 |
| 700 | 0.03463 | 14.61 | 157.8 | 456 | 342 | 676 | 0.675 |
| 800 | 0.03030 | 14.70 | 172.4 | 569 | 378 | 849 | 0.670 |
| 900 | 0.02694 | 14.83 | 186.5 | 692 | 412 | 1030 | 0.671 |
| 1000 | 0.02424 | 14.99 | 201.3 | 830 | 448 | 1230 | 0.673 |
| 1100 | 0.02204 | 15.17 | 213.0 | 966 | 488 | 1460 | 0.662 |
| 1200 | 0.02020 | 15.37 | 226.2 | 1120 | 528 | 1700 | 0.659 |
| 1300 | 0.01865 | 15.59 | 238.5 | 1279 | 568 | 1955 | 0.655 |
| 1400 | 0.01732 | 15.81 | 250.7 | 1447 | 610 | 2230 | 0.650 |
| 1500 | 0.01616 | 16.02 | 262.7 | 1626 | 655 | 2530 | 0.643 |
| 1600 | 0.0152 | 16.28 | 273.7 | 1801 | 697 | 2815 | 0.639 |
| 1700 | 0.0143 | 16.58 | 284.9 | 1992 | 742 | 3130 | 0.637 |
| 1800 | 0.0135 | 16.96 | 296.1 | 2193 | 786 | 3435 | 0.639 |
| 1900 | 0.0128 | 17.49 | 307.2 | 2400 | 835 | 3730 | 0.643 |
| 2000 | 0.0121 | 18.25 | 318.2 | 2630 | 878 | 3975 | 0.661 |

**Nitrogen, $N_2$**

| $T$ K | $\rho$ kg/m³ | $c_p$ kJ/kg. K | $\mu \cdot 10^7$ N · s/m² | $v \cdot 10^6$ m²/s | $k \cdot 10^3$ W/m. K | $\alpha \cdot 10^6$ m²/s | pr |
|---|---|---|---|---|---|---|---|
| 100 | 3.4388 | 1.070 | 68.8 | 2.00 | 9.58 | 2.60 | 0.768 |
| 150 | 2.2594 | 1.050 | 100.6 | 4.45 | 13.9 | 5.86 | 0.759 |
| 200 | 1.6883 | 1.043 | 129.2 | 7.65 | 18.3 | 10.4 | 0.736 |
| 250 | 1.3488 | 1.042 | 154.9 | 11.48 | 22.2 | 15.8 | 0.727 |
| 300 | 1.1233 | 1.041 | 178.2 | 15.86 | 25.9 | 22.1 | 0.716 |
| 350 | 0.9625 | 1.042 | 200.0 | 20.78 | 29.3 | 29.2 | 0.711 |
| 400 | 0.8425 | 1.045 | 220.4 | 26.16 | 32.7 | 37.1 | 0.704 |
| 450 | 0.7485 | 1.050 | 239.6 | 32.01 | 35.8 | 45.6 | 0.703 |
| 500 | 0.6739 | 1.056 | 257.7 | 38.24 | 38.9 | 54.7 | 0.700 |
| 550 | 0.6124 | 1.065 | 274.7 | 44.86 | 41.7 | 63.9 | 0.702 |
| 600 | 0.5615 | 1.075 | 290.8 | 51.79 | 44.6 | 73.9 | 0.701 |
| 700 | 0.4812 | 1.098 | 321.0 | 66.71 | 49.9 | 94.4 | 0.706 |
| 800 | 0.4211 | 1.122 | 349.1 | 82.90 | 54.8 | 116 | 0.715 |
| 900 | 0.3743 | 1.146 | 375.3 | 100.3 | 59.7 | 139 | 0.721 |
| 1000 | 0.3368 | 1.167 | 399.9 | 118.7 | 64.7 | 165 | 0.721 |
| 1100 | 0.3062 | 1.187 | 423.2 | 138.2 | 70.0 | 193 | 0.718 |
| 1200 | 0.2807 | 1.204 | 445.3 | 158.6 | 75.8 | 224 | 0.707 |
| 1300 | 0.2591 | 1.219 | 466.2 | 179.9 | 81.0 | 256 | 0.701 |

**Oxygen $O_2$**

| $T$ K | $\rho$ kg/m³ | $c_p$ kJ/kg. K | $\mu \cdot 10^7$ N · s/m² | $v \cdot 10^6$ m²/s | $k \cdot 10^3$ W/m. K | $\alpha \cdot 10^6$ m²/s | pr |
|---|---|---|---|---|---|---|---|
| 100 | 3.945 | 0.962 | 76.4 | 1.94 | 9.25 | 2.44 | 0.796 |
| 150 | 2.585 | 0.921 | 114.8 | 4.44 | 13.8 | 5.80 | 0.766 |
| 200 | 1.930 | 0.915 | 147.5 | 7.64 | 18.3 | 10.4 | 0.737 |
| 250 | 1.542 | 0.915 | 178.6 | 11.58 | 22.6 | 16.0 | 0.723 |
| 300 | 1.284 | 0.920 | 207.2 | 16.14 | 26.8 | 22.7 | 0.711 |

*(Contd.)*

### Table A 10  Properties of gases at atmospheric pressure

| $T$ K | $\rho$ kg/m$^3$ | $c_p$ kJ/kg. K | $\mu \cdot 10^7$ N $\cdot$ s/m$^2$ | $v \cdot 10^6$ m$^2$/s | $k \cdot 10^3$ W/m. K | $\alpha \cdot 10^6$ m$^2$/s | pr |
|---|---|---|---|---|---|---|---|
| 350 | 1.100 | 0.929 | 233.5 | 21.23 | 29.6 | 29.0 | 0.733 |
| 400 | 0.9620 | 0.942 | 258.2 | 26.84 | 33.0 | 36.4 | 0.737 |
| 450 | 0.8554 | 0.956 | 281.4 | 32.90 | 36.3 | 44.4 | 0.741 |
| 500 | 0.7698 | 0.972 | 303.3 | 39.40 | 41.2 | 55.1 | 0.716 |
| 550 | 0.6998 | 0.988 | 324.0 | 46.30 | 44.1 | 63.8 | 0.726 |
| 600 | 0.6414 | 1.003 | 343.7 | 53.59 | 47.3 | 73.5 | 0.729 |
| 700 | 0.5498 | 1.031 | 380.8 | 69.26 | 52.8 | 93.1 | 0.744 |
| 800 | 0.4810 | 1.054 | 415.2 | 86.32 | 58.9 | 116 | 0.743 |
| 900 | 0.4275 | 1.074 | 447.2 | 104.6 | 64.9 | 141 | 0.740 |
| 1000 | 0.3848 | 1.090 | 477.0 | 124.0 | 71.0 | 169 | 0.733 |
| 1100 | 0.3498 | 1.103 | 505.5 | 144.5 | 75.8 | 196 | 0.736 |
| 1200 | 0.3206 | 1.115 | 532.5 | 166.1 | 81.9 | 229 | 0.725 |
| 1300 | 0.2960 | 1.125 | 588.4 | 188.6 | 87.1 | 262 | 0.721 |

**Water Vapour (steam)**

| $T$ K | $\rho$ kg/m$^3$ | $c_p$ kJ/kg. K | $\mu \cdot 10^7$ N $\cdot$ s/m$^2$ | $v \cdot 10^6$ m$^2$/s | $k \cdot 10^3$ W/m. K | $\alpha \cdot 10^6$ m$^2$/s | pr |
|---|---|---|---|---|---|---|---|
| 380 | 0.5863 | 2.060 | 127.1 | 21.68 | 24.6 | 20.4 | 1.06 |
| 400 | 0.5542 | 2.014 | 134.4 | 24.25 | 26.1 | 23.4 | 1.04 |
| 450 | 0.4902 | 1.980 | 152.5 | 31.11 | 29.9 | 30.8 | 1.01 |
| 500 | 0.4405 | 1.985 | 170.4 | 38.68 | 33.9 | 38.8 | 0.993 |
| 550 | 0.4005 | 1.997 | 188.4 | 47.04 | 37.9 | 47.4 | 0.993 |
| 600 | 0.3652 | 2.026 | 206.7 | 56.60 | 42.2 | 57.0 | 0.993 |
| 650 | 0.3380 | 2.056 | 224.7 | 66.48 | 46.4 | 66.8 | 0.996 |
| 700 | 0.3140 | 2.085 | 242.6 | 77.26 | 50.5 | 77.1 | 1.00 |
| 750 | 0.2931 | 2.119 | 260.4 | 88.84 | 54.9 | 88.4 | 1.00 |
| 800 | 0.2739 | 2.152 | 278.6 | 101.7 | 59.2 | 100 | 1.01 |
| 850 | 0.2579 | 2.186 | 296.9 | 115.1 | 63.7 | 113 | 1.02 |

**Table A 11   Longitude and latitude of some stations in India**

| S. No. | Name of the Location | Longitude | Latitude |
|---|---|---|---|
| 1. | Agra | 78°5′ E | 27°11′ N |
| 2. | Ahmedabad | 72°40′ E | 23°00′ N |
| 3. | Akola | 77°10′ E | 20°30′ N |
| 4. | Allahabad | 81°58′ E | 25°25′ N |
| 5. | Amritsar | 74°57′ E | 31°35′ N |
| 6. | Amraoti | 74°47′ E | 20°56′ N |
| 7. | Anantapur | 77°42′ E | 14°35′ N |
| 8. | Aurangabad | 75°20′ E | 19°53′ N |
| 9. | Bangalore | 77°40′ E | 12°59′ N |
| 10. | Baroda | 73°13′ E | 22°21′ N |
| 11. | Bhagalpur | 87°15′ E | 25°11′ N |
| 12. | Bhav Nagar | 72°12′ E | 21°54′ N |
| 13. | Bhopal | 77°30′ E | 23°15′ N |
| 14. | Calicut | 75°47′ E | 11°15′ N |
| 15. | Chennai | 80°19′ E | 13°8′ N |
| 16. | Coimbatore | 76°59′ E | 11°00′ N |
| 17. | Cuttack | 85°57′E | 20°25′ N |
| 18. | Dehradun | 78°02′ E | 30°19′ N |
| 19. | Delhi | 77°17′ E | 28°38′ N |
| 20. | Dharwar | 75°02′ E | 15°23′ N |
| 21. | Ellora | 75°6′ E | 20°00′ N |
| 22. | Gauhati | 91°45′ E | 26°11′ N |
| 23. | Goa (Panaji) | 73°56′ E | 15°28′ N |
| 24. | Gorakhpur | 83°02′ E | 26°45′ N |
| 25. | Gulbaraga | 76°51′ E | 17°21′ N |
| 26. | Guntur | 80°30′ E | 16°23′ N |
| 27. | Hydrabad | 78°27′ E | 17°18′ N |
| 28. | Indore | 75°53′ E | 22°42′ N |
| 29. | Jabalpur | 79°58′ E | 23°10′ N |
| 30. | Jaipur | 75°52′ E | 26°54′ N |
| 31. | Jalandhar | 75°40′ E | 31°20′ N |
| 32. | Jammu | 47°50′ E | 32°47′ N |
| 33. | Jamshedpur | 86°20′ E | 22°44′ N |
| 34. | Jhansi | 78°35′ E | 25°27′ N |
| 35. | Jodhpur | 73°01′ E | 26°18′ N |
| 36. | Kanpur | 10°23′ E | 26°30′ N |
| 37. | Kathmandu | 85°12′ E | 27°42′ N |
| 38. | Kolkatta | 88°35′ E | 22°36′ N |
| 39. | Kopargaon | 74°08′ E | 18°32′ N |
| 40. | Kotah | 75°59′ E | 25°15′ N |
| 41. | Kolhapur | 74°15′ E | 16°43′ N |
| 42. | Lucknow | 81°00′ E | 26°50′ N |
| 43. | Ludhiana | 75°52′ E | 30°56′ N |
| 44. | Madurai | 78°10′ E | 9°50′ N |
| 45. | Mahabaleshwar | 73°39′ E | 17°52′ N |
| 46. | Mangalore | 74°47′ E | 12°52′ N |
| 47. | Manipur | 94°00′ E | 24°15′ N |
| 48. | Mathura | 77°41′ E | 27°30′ N |

*(Contd.)*

**(Continued)**

| S. No. | Name of the Location | Longitude | Latitude |
|---|---|---|---|
| 49. | Mumbai | 72°47′ E | 18°54′ N |
| 50. | Mysore | 76°41′ E | 12°17′ N |
| 51. | Nagpur | 79°10′ E | 21°8′ N |
| 52. | Nainital | 79°31′ E | 29°30′N |
| 53. | Patiala | 76°26′ E | 30°23′ N |
| 54. | Patna | 85°18′ E | 25°35′ N |
| 55. | Poona | 73°57′ E | 18°29′ N |
| 56. | Raipur | 81°39′ E | 21°14′ N |
| 57. | Ranchi | 85°20′ E | 23°23′ N |
| 58. | Roorkee | 77°59′ E | 29°52′ N |
| 59. | Salem | 78°15′ E | 11°40′ N |
| 60. | Sambalpur | 83°56′ E | 21°28′ N |
| 61. | Saugar | 78°45′ E | 23°51′ N |
| 62. | Shillong | 92°00′ E | 25°30′ N |
| 63. | Sindri | 86°29′ E | 23°41′ N |
| 64. | Sholapur | 75°54′ E | 17°40′ N |
| 65. | Sri Nagar | 74°50′ E | 34°12′ N |
| 66. | Surat | 72°55′ E | 21°12′ N |
| 67. | Tripura | 90°00′ E | 24°00′ N |
| 68. | Trivandrum | 77°00′ E | 8°31′ N |
| 69. | Udaipur | 73°44′ E | 24°36′ N |
| 70. | Varanasi | 83°08′ E | 25°22′ N |
| 71. | Vidisha | 77°49′ E | 23°32′ N |
| 72. | Visakhapatnam | 83°18′ E | 17°42′ N |

# Appendix

## NOTATIONS

| Symbol | Explanation |
|---|---|
| $T_s$ | Saturation Temperature, °C |
| $P$ | Saturation Pressure, bar |
| $v_g$ | Specific Volume, kg/m$^3$ |
| $h_f$ | Specific Enthalpy of Saturated Liquid, kJ/kg |
| $h_g$ | Specific Enthalpy of Saturated Vapour, kJ/kg |
| $s_f$ | Specific Entropy of Saturated Liquid, kJ/kg K |
| $s_g$ | Specific Entropy of Saturated Vapour, kJ/kg K |
| $C_{pf}$ | Specific Heat of Saturated Liquid, kJ/kg K |
| $C_{pg}$ | Specific Heat of Saturated Vapour, kJ/kg K |

For wet vapour with dryness fraction *x* properties are calculated as below :

$$v = xv_g, \qquad h = h_f + x\,h_{fg}, \qquad s = S_f + x\,S_{fg}$$

For the superheat region use the method as below :

$$v = v_g \times (T\,\text{sup}/Ts), \qquad h = h_g + C_{pv}x\,(\text{Tsup-Ts}), \qquad s = s_g + C_{pv}\,l_n\,(\text{Tsup}/Ts)$$

Tsup- superheated temperature in K, $T_s$ – saturation temperature in K

### Table Al 2    Saturated Trichloromonofluoromethane (CCl$_3$F), R-11
#### Datum at – 40°C, $h_f = 0, s_f = 0$

| Saturation temp in °C | Saturation pressure in bar | Specific volume in m³/kg | | Specific enthalpy in kJ/kg | | | Specific entropy in kJ/kgK | |
|---|---|---|---|---|---|---|---|---|
| | | Liquid | Vapour | Liquid | Vapour | Latent | Liquid | Vapour |
| (t) | (P) | ($v_f$) | ($v_g$) | ($h_f$) | ($h_g$) | ($h_{fg}$) | ($s_f$) | ($S_g$) |
| – 40 | 0.05096 | 0.000617 | 2.7625 | 0.00 | 203.48 | 203.48 | 0.0000 | 0.8730 |
| – 38 | 0.05942 | 0.000 619 | 2.5360 | 1.65 | 202.83 | 204.48 | 0.0068 | 0.8701 |
| – 36 | 0.06787 | 0.000621 | 2.3094 | 3.30 | 202.18 | 205.48 | 0.0137 | 0.8671 |
| – 34 | 0.07633 | 0.000623 | 2.0827 | 4.95 | 201.53 | 206.48 | 0.0206 | 0.8642 |
| – 32 | 0.08478 | 0.000624 | 1.8562 | 6.60 | 200.88 | 207.48 | 0.0274 | 0.8613 |
| – 30 | 0.09323 | 0.000626 | 1.6296 | 8.25 | 200.23 | 208.48 | 0.0343 | 0.8583 |
| – 28 | 0,10345 | 0.000627 | 1.4448 | 9.90 | 199.58 | 209.48 | 0.0410 | 0.8554 |
| – 26 | 0.11586 | 0.000629 | 1.3123 | 11.55 | 198.94 | 210.49 | 0.0477 | 0.8533 |
| – 24 | 0.12828 | 0.000631 | 1.1798 | 13.20 | 198.29 | 211.49 | 0.0544 | 0.8508 |
| – 22 | 0.14292 | 0.000633 | 1.0722 | 14.85 | 197.65 | 212.50 | 0.0611 | 0.8483 |
| – 20 | 0.15869 | 0.000634 | 0.9770 | 16.51 | 196.99 | 213.50 | 0.0678 | 0.8462 |
| – 18 | 0.17414 | 0.000636 | 0.8793 | 18.19 | 196.33 | 214.52 | 0.0741 | 0.8441 |
| – 16 | 0.19280 | 0.000638 | 0.8045 | 19.82 | 195.71 | 215.53 | 0.0808 | 0.8420 |
| – 14 | 0.21260 | 0.000639 | 0.7345 | 21.49 | 195.05 | 216.54 | 0.0875 | 0.8403 |
| – 12 | 0.23390 | 0.000641 | 0.6713 | 23.17 | 194.38 | 217.55 | 0.0938 | 0.8387 |
| – 10 | 0.25848 | 0.000643 | 0.6191 | 24.83 | 193.73 | 218.56 | 0.1009 | 0.8366 |
| – 8 | 0.28306 | 0.000645 | 0.5671 | 26.50 | 193.07 | 219.57 | 0.1063 | 0.8349 |
| – 6 | 0.30280 | 0.000647 | 0.5194 | 28.17 | 192.42 | 220.59 | 0.1126 | 0.8332 |
| – 4 | 0.33967 | 0.000649 | 0.4801 | 29.85 | 191.76 | 221.61 | 0.1189 | 0.8315 |
| – 2 | 0.36983 | 0.000651 | 0.4409 | 31.54 | 191.09 | 222.63 | 0.1252 | U8303 |
| 0 | 0.40360 | 0.000653 | 0.4069 | 33.22 | 190.42 | 223.64 | 0.1315 | 0.8286 |
| 1 | 0.42190 | 0.000653 | 0.3920 | 34.06 | 190.08 | 224.14 | 0.1344 | 0.8280 |
| 2 | 0.44021 | 0.000654 | 0.3770 | 34.90 | 189.75 | 224.65 | 0.1374 | 0.8274 |
| 3 | 0.45852 | 0.000655 | 0.3620 | 35.74 | 189.41 | 225.15 | 0.1405 | 0.8267 |
| 4 | 0.47683 | 0.000656 | 0.3471 | 36.59 | 189.06 | 225.65 | 0.1436 | 0.8261 |
| 5 | 0.49720 | 0.000657 | 0.3340 | 37.43 | 188.73 | 226.16 | 0.1465 | 0.8254 |
| 6 | 0.51.917 | 0.000658 | 0.3225 | 38.28 | 188.38 | 226.66 | 0.1498 | 0.8248 |
| 7 | 0.54118 | 0.000659 | 0.3109 | 39.13 | 188.04 | 227.17 | 0.1526 | 0.8242 |
| 8 | 0.56317 | 0.000660 | 0.2994 | 39.98 | 187.70 | 227.68 | 0.1562 | 0.8236 |
| 9 | 0.58517 | 0.000661 | 0.2878 | 40.84 | 187.35 | 228.18 | 0.1586 | 0.8229 |
| 10 | 0.60717 | 0.000662 | 0.2763 | 41.68 | 187.01 | 228.69 | 0.1616 | 0.8223 |
| 11 | 0.63320 | 0.000663 | 0.2673 | 42.53 | 186.67 | 229.20 | 0.1646 | 0.8219 |

*Contd.*

## Table A12 Continued

| (t) | (P) | ($v_f$) | ($v_g$) | ($h_f$) | ($h_g$) | ($h_{fg}$) | ($s_f$) | ($S_g$) |
|---|---|---|---|---|---|---|---|---|
| 12 | 0.65921 | 0.000664 | 0.2584 | 43.39 | 186.32 | 229.71 | 0.1675 | 0.8215 |
| 13 | 0.68523 | 0.000665 | 0.2495 | 44.24 | 185.97 | 230.21 | 0.1706 | 0.8210 |
| 14 | 0.71124 | 0.000666 | 0.2405 | 45.10 | 185.62 | 230.72 | 0.1738 | 0.8206 |
| 15 | 0.73727 | 0.000667 | 0.2316 | 45.95 | 185.27 | 231.22 | 0.1767 | 0.8200 |
| 16 | 0.76552 | 0.000669 | 0.2235 | 46.81 | 184.92 | 231.73 | 0.1796 | 0.8194 |
| 17 | 0.79655 | 0.000670 | 0.2163 | 47.67 | 184.56 | 232.23 | 0.1825 | 0.8170 |
| 18 | 0.82759 | 0.000671 | 0.2092 | 48.54 | 184.19 | 232.73 | 0.1855 | 0.8186 |
| 19 | 0.85862 | 0.000672 | 0.2021 | 49.40 | 183.84 | 233.24 | 0.1885 | 0.8180 |
| 20 | 0.88966 | 0.000673 | 0.1950 | 50.26 | 183.48 | 233.74 | 0.1913 | 0.8173 |
| 21 | 0.92070 | 0.000674 | 0.1878 | 51.12 | 183.12 | 234.24 | 0.1943 | 0.8169 |
| 22 | 0.95625 | 0.000675 | 0.1821 | 51.99 | 182.75 | 234.74 | 0.1972 | 0.8165 |
| 23 | 0.99237 | 0.000676 | 0.1764 | 52.85 | 182.39 | 235.24 | 0.2003 | 0.8163 |
| 24 | 1.02496 | 0.000677 | 0.1708 | 53.72 | 182.03 | 235.75 | 0.2031 | 0.8160 |
| 25 | 1.06117 | 0.000678 | 0.1651 | 54.59 | 181.66 | 236.25 | 0.2061 | 0.8157 |
| 26 | 1.10074 | 0.000679 | 0.1595 | 55.46 | 181.29 | 236.75 | 0.2089 | 0.8152 |
| 27 | 1.13880 | 0.000680 | 0.1542 | 56.32 | 180.93 | 237.25 | 0.2120 | 0.8150 |
| 28 | 1.17917 | 0.000681 | 0.1495 | 57.20 | 180.56 | 237.76 | 0.2148 | 0.8148 |
| 29 | 1.22273 | 0.000682 | 0.1452 | 58.07 | 180.18 | 238.25 | 0.2177 | 0.8144 |
| 30 | 1.26070 | 0.000684 | 0.1401 | 58.94 | 179.82 | 238.76 | 0.2206 | 0.8140 |
| 31 | 1.30665 | 0.000685 | 0.1362 | 59.82 | 179.43 | 239.25 | 0.2234 | 0.8135 |
| 32 | 1.34821 | 0.000686 | 0.1316 | 60.70 | 179.05 | 239.75 | 0.2261 | 0.8131 |
| 33 | 1.39568 | 0.000687 | 0.1278 | 61.57 | 178.67 | 240.24 | 0.2290 | 0.8128 |
| 34 | 1.44414 | 0.000689 | 0.1242 | 62.45 | 178.28 | 240.73 | 0.2320 | 0.8126 |
| 35 | 1.49276 | 0.000690 | 0.1205 | 63.32 | 177.91 | 241.23 | 0.2349 | 0.8123 |
| 36 | 1.54130 | 0.000691 | 0.1168 | 64.20 | 177.52 | 241.72 | 0.2378 | 0.8119 |
| 37 | 1.58983 | 0.000692 | 0.1131 | 65.07 | 177.15 | 242.22 | 0.2406 | 0.8116 |
| 38 | 1.64000 | 0.000693 | 0.1096 | 65.95 | 176.77 | 242.72 | 0.2433 | 0.8114 |
| 39 | 1.69571 | 0.000694 | 0.1066 | 66.84 | 176.66 | 243.20 | 0.2462 | 0.8112 |
| 40 | 1.75145 | 0.000696 | 0.1037 | 67.74 | 175.95 | 243.69 | 0.2490 | 0.8110 |
| 41 | 1.80718 | 0.000697 | 0.1007 | 68.63 | 175.55 | 244.18 | 0.2518 | 0.8108 |
| 42 | 1.86292 | 0.000698 | 0.0977 | 69.52 | 175.15 | 244.67 | 0.2546 | 0.8106 |
| 43 | 1.91866 | 0.000699 | 0.0947 | 70.41 | 174.75 | 245.16 | 0.2574 | 0.8104 |
| 44 | 1.97952 | 0.000700 | 0.0921 | 71.31 | 175.34 | 245.65 | 0.2600 | 0.8102 |
| 45 | 2.04300 | 0.000701 | 0.0896 | 72.20 | 173.94 | 246.14 | 0.2029 | 0.8100 |
| 46 | 2.10641 | 0.000702 | 0.0871 | 73.08 | 173.55 | 243.63 | 0.2658 | 0.8098 |
| 47 | 2.16980 | 0.000704 | 0.0847 | 73.98 | 173.14 | 247.12 | 0.2686 | 0.8096 |
| 48 | 2.23310 | 0.000705 | 0.0823 | 74.87 | 172.73 | 247.60 | 0.2713 | 0.8094 |
| 49 | 2.29670 | 0.000706 | 0.0798 | 75.76 | 172.31 | 248.07 | 0.2741 | 0.8092 |
| 50 | 2.36000 | 0.000707 | 0.0774 | 76.67 | 171.86 | 248.53 | 0.2768 | 0.8090 |

# REFRIGERANT 12

## *DICHLORODIFLUROMETHANE*

| | |
|---|---|
| Chemical Formula | : $CCl_2F_2$ |
| Molecular Weight | : 120.93 |
| Boiling Temperature at Atmospheric Pressure °C | : $-29.78$ |
| Freezing Temperature at Atmospheric Pressure °C | : $-157.78$ |
| Critical Temperature, °C | : 112 |
| Critical Pressure, (bar) | : 41.16 |
| Critical Density, (kg/m$^3$) | : 557.44 |
| Density of Liquid at 30 °C, (kg/m$^3$) | : 1290.72 |
| Specific Volume of Saturated Vapour at $-15°$ C, (m$^3$/kg) | : 0.0910 |
| Specific Heat of Liquid at 30° C, (kJ/kg C) | : 0.983 |
| Specific Heat Ratio of Vapour at 30°C and 1 atm | : 1.139 |
| Thermal Conductivity, W/m°C. | |
|     Saturated Liquid at 0°C | : 0.09 |
|     Saturated Liquid at 30°C | : 0.069 |
|     Vapour at 1 atm and 0°C. | : 0.00813 |
|     Vapour at 1 atm and 30°C | : 0.0102 |
| Viscosity, Centipoise (1 cp $= 10^{-3}$ Ns/m$^2$) | |
|     Saturated Liquid at $-15$°C | : 0.335 |
|     Saturated Liquid at 30°C | : 0.254 |
|     Vapour at 1 atm and $-15$°C | : 0.0108 |
|     Vapour at 1 atm and 30°C | : 0.0127 |

### Table Al 3 Saturated Dichlorodifluoromethane (CCl$_2$F$_2$), R-12

### Datum at $-40°C$, *hf=0, s$_f$=0*

| Saturation temp in °C | Saturation pressure in bar | Specific volume in m³/kg | | Specific enthalpy in kJ/kg | | | Specificc entropy in kJ/kgK | |
|---|---|---|---|---|---|---|---|---|
| | | Liquid (v$_f$) | Vapour (v$_g$) | Liquid (h$_f$) | Vapour (h) | Latent (h$_{fg}$) | Liquid (S$_f$) | Vapour (s$_g$) |
| (t) | (P) | | | | | | | |
| − 100 | 0.01185 | 0.000600 | 10.1951 | − 51.84 | 193.84 | 142.00 | − 0.2567 . | 0.8628 |
| − 95 | 0.01864 | 0.000604 | 6.6231 | − 47.56 | 191.78 | 144.22 | − 0.2323 | 0.8442 |
| − 90 | 0.02843 | 0.000608 | 4.4206 | − 43.28 | 189.74 | 146.46 | − 0.2086 | 0.8273 |
| − 85 | 0.04254 | 0.000613 | 3.0531 | − 39.00 | 187.73 | 148.73 | − 0.1856 | 0.8122 |
| − 80 | 0.06200 | 0.000617 | 2.1519 | − 34.72 | 185.74 | 151.02 | − 0.1631 | 0.7985 |
| − 75 | 0.08826 | 0.000622 | 1.5462 | − 30.42 | 183.74 | 153.32 | − 0.1412 | 0.7861 |
| − 70 | 0.12298 | 0.000627 | 1.1314 | − 26.12 | 181.75 | 155.63 | − 0.1198 | 0.7665 |
| − 65 | 0.16807 | 0.000632 | 0.8421 | − 21.81 | 179.77 | 157.96 | − 0.0988 | 0.7648 |
| − 60 | 0.22665 | 0.000637 | 0.6401 | − 17.48 | 177.77 | 160.29 | − 0.0783 | 0.7558 |
| − 55 | 0.30052 | 0.000643 | 0.4930 | − 13.14 | 175.76 | 162.62 | − 0.0581 | 0.7475 |
| − 50 | 0.39237 | 0.000648 | 0.3845 | − 8.78 | 173.73 | 164.95 | − 0.0384 | 0.7401 |
| − 45 | 0.50512 | 0.000654 | 0.3035 | − 4.39 | 171.66 | 167.27 | − 0.0190 | 0.7334 |
| − 40 | 0.64190 | 0.000660 | 0.2422 | 0.00 | 169.60 | 169.60 | 0.0000 | 0.7274 |
| − 38 | 0.70460 | 0.000663 | 0.2221 | 1.76 | 168.76 | 170.52 | 0.0075 | 0.7251 |
| − 36 | 0.77196 | 0.000665 | 0.2040 | 3.53 | 167.91 | 171.44 . | 0.0149 | 0.7230 |
| − 34 | 0.84421 | 0.000667 | 0.1877 | 5.31 | 167.05 | 172.36 | 0.0224 | 0.7209 |
| − 32 | 0.92776 | 0.000670 | 0.1729 | 7.08 | 166.20 | 173.28 | 0.0298 | 0.7190 |
| − 30 | 1.00441 | 0.000673 | 0.1596 | 8.86 | 165.34 | 174.20 | 0.0371 | 0.7171 |
| − 28 | 1.09311 | 0.000676 | 0.1475 | 10.64 | 164.47 | 175.11 | 0.0444 | 0.7153 |
| − 26 | 1.18778 | 0.000678 | 0.1364 | 12.43 | 163.59 | 176.02 | 0.0516 | 0.7135 |
| − 24 | 1.28858 | 0.000681 | 0.1265 | 14.22 | 162.71 | 176.93 | 0.0588 | 0.7118 |
| − 22 | 1.39581 | 0.000683 | 0.1173 | 16.02 | 161.81 | 177.83 | 0.0660 | 0.7102 |
| − 20 | 1.50972 | 0.000686 | 0.1090 | 17.82 | 160.91 | 178.73 | 0.0731 | 0.7087 |
| − 18 | 1.63104 | 0.000689 | 0.1014 | 19.62 | 160.01 | 179.63 | 0.0801 | 0.7073 |
| − 16 | 1.75963 | 0.000692 | 0.0944 | 21.43 | 159.10 | 180.53 | 0.0871 | 0.7059 |
| − 14 | 1.89575 | 0.000695 | 0.0880 | 23.23 | 159.19 | 181.42 | 0.0941 | 0.7045 |
| − 12 | 2.04605 | 0.000698 | 0.0821 | 25.05 | 157.26 | 182.31 | 0.1010 | 0.7032 |
| − 10 | 2.19172 | 0.000701 | 0.0767 | 26.87 | 156.32 | 183.19 | 0.1080 | 0.7019 |
| − 8 | 2.35272 | 0.000704 | 0.0717 | 28.70 | 155.36 | 184.06 | 0.1148 | 0.7007 |
| − 6 | 2.52244 | 0.000707 | 0.0672 | 30.53 | 154.41 | 184.94 | 0.1217 | 0.6995 |
| 6 | 3.74280 | 0.000727 | 0.0461 | 190.07 | 41.62 | 148.45 | 0.1620 | 0.6938 |
| 7 | 3.86141 | 0.000729 | 0.0447 | 190.49 | 42.56 | 147.93 | 0.1653 | 0.6933 |
| 8 | 3.98283 | 0.000730 | 0.0434 | 190.91 | 43.50 | 147.41 | 0.1686 | 0.6929 |

*Contd.*

**Table A13 Continued**

| (t) | (P) | $(v_f)$ | $(v_g)$ | $(h_f)$ | $(h_g)$ | $(h_{fg})$ | $(S_f)$ | $(S_g)$ |
|---|---|---|---|---|---|---|---|---|
| 9 | 4.10702 | 0.000732 | 0.0422 | 191.32 | 44.43 | 146.89 | 0.1719 | 0.6925 |
| 10 | 4.23407 | 0.000734 | 0.0410 | 191.74 | 45.37 | 146.37 | 0.1752 | 0.6921 |
| 11 | 4.36442 | 0.000736 | 0.0398 | 192.15 | 46.31 | 145.84 | 0.1784 | 0.6917 |
| 12 | 4.49763 | 0.000738 | 0.0386 | 192.56 | 47.26 | 145.30 | 0.1817 | 0.6913 |
| 13 | 4.63386 | 0.000739 | 0.0375 | 192.97 | 48.20 | 144.77 | 0.1850 | 0.6909 |
| 14 | 4.77312 | 0.000741 | 0.0365 | 193.38 | 49.15 | 144.23 | 0.1883 | 0.6906 |
| 15 | 4.91545 | 0.000743 | 0.0355 | 193.79 | 50.10 | 143.69 | 0.1915 | 0.6902 |
| 16 | 5.06087 | 0.000745 | 0.0345 | 194.19 | 51.05 | 143.14 | 0.1948 | 0.6898 |
| 17 | 5.20942 | 0.000747 | 0.0335 | 194.59 | 52.00 | 142.59 | 0.1981 | 0.6894 |
| 18 | 5.36117 | 0.000749 | 0.0326 | 194.99 | 52.95 | 142.04 | 0.2013 | 0.6891 |
| 19 | 5.51614 | 0.000751 | 0.0317 | 195.38 | 53.91 | 141.47 | 0.2046 | 0.6888 |
| 20 | 5.67441 | 0.000753 | 0.0308 | 195.78 | 54.87 | 140.91 | 0.2078 | 0.6884 |
| 21 | 5.83635 | 0.000756 | 0.0300 | 196.17 | 55.83 | 140.34 | 0.2110 | 0.6881 |
| 22 | 6.00171 | 0.000758 | 0.0292 | 196.56 | 56.79 | 139.77 | 0.2143 | 0.6878 |
| 23 | 6.17050 | 0.000759 | 0.0284 | 196.96 | 57.75 | 139.21 | 0.2174 | 0.6875 |
| 24 | 6.34269 | 0.000761 | 0.0276 | 197.34 | 58.73 | 138.61 | 0.2207 | 0.6872 |
| 25 | 6.51840 | 0.000764 | 0.0269 | 197.73 | 59.70 | 138.03 | 0.2239 | 0.6868 |
| 26 | 6.69765 | 0.000766 | 0.0262 | 198.11 | 60.67 | 137.44 | 0.2271 | 0.6865 |
| 27 | 6.88048 | 0.000768 | 0.0255 | 198.50 | 61.65 | 136.85 | 0.2303 | 0.6862 |
| 28 | 7.06704 | 0.000770 | 0.0248 | 198.87 | 62.63 | 136.24 | 0.2335 | 0.6859 |
| 29 | 7.25738 | 0.000772 | 0.0241 | 199.25 | 63.61 | 135.64 | 0.2368 | 0.6856 |
| 30 | 7.45103 | 0.000775 | 0.0235 | 199.62 | 64.59 | 135.03 | 0.2400 | 0.6853 |
| 31 | 7.64903 | 0.000777 | 0.0230 | 199.99 | 65.58 | 134.41 | 0.2431 | 0.6850 |
| 32 | 7.85089 | 0.000779 | 0.0225 | 200.36 | 66.57 | 133.79 | 0.2463 | 0.6847 |
| 33 | 8.05662 | 0.000782 | 0.0218 | 200.73 | 67.56 | 133.17 | 0.2495 | 0.6845 |
| 34 | 8.26621 | 0.000784 | 0.0212 | 201.09 | 68.56 | 132.53 | 0.2527 | 0.6842 |
| 35 | 8.48000 | 0.000786 | 0.0207 | 201.45 | 69.56 | 131.89 | 0.2559 | 0.6839 |
| 36 | 8.69766 | 0.000789 | 0.0202 | 201.80 | 70.55 | 131.25 | 0.2591 | 0.6836 |
| 37 | 8.91904 | 0.000792 | 0.0196 | 202.16 | 71.55 | 130.61 | 0.2623 | 0.6833 |
| 38 | 9.14483 | 0.000794 | 0.0191 | 202.51 | 72.56 | 129.95 | 0.2654 | 0.6830 |
| 39 | 9.37497 | 0.000796 | 0.0186 | 202.86 | 73.57 | 129.29 | 0.2685 | 0.6828 |
| 40 | 9.60897 | 0.000799 | 0.0182 | 203.20 | 74.59 | 128.61 | 0.2718 | 0.6825 |
| 41 | 9.84793 | 0.000802 | 0.0177 | 203.54 | 75.61 | 127.93 | 0.2750 | 0.6822 |
| 42 | 10.09131 | 0.000804 | 0.0173 | 203.87 | 76.62 | 127.25 | 0.2782 | 0.6820 |
| 43 | 10.33862 | 0.000807 | 0.0168 | 204.21 | 77.65 | 126.56 | 0.2814 | 0.6817 |
| 44 | 10.59021 | 0.000810 | 0.0164 | 204.55 | 78.68 | 125.87 | 0.2846 | 0.6814 |
| 45 | 10.84655 | 0.000813 | 0.0160 | 204.87 | 79.71 | 125.16 | 0.2878 | 0.6811 |
| 46 | 11.10758 | 0.000815 | 0.0156 | 205.19 | 80.75 | 124.44 | 0.2909 | 0.6808 |
| 47 | 11.37304 | 0.000818 | 0.0153 | 205.51 | 81.79 | 123.72 | 0.2941 | 0.6805 |
| 48 | 11.64290 | 0.000821 | 0.0149 | 205.83 | 82.83 | 123.00 | 0.2973 | 0.6802 |
| 49 | 11.91724 | 0.000824 | 0.0146 | 206.14 | 83.88 | 122.26 | 0.3005 | 0.6800 |
| 50 | 12.19655 | 0.000827 | 0.0142 | 206.45 | 84.94 | 121.51 | 0.3037 | 0.6797 |

*Contd.*

### Table Al 4 Superheated table of refrigerant R-12, h - Enthalpy, (kJ/kg)
### s = Entropy, (kJ/kg), v = specific volume (m³/kg) and T_s = Deg rees superheat, (°C)

| T:°C / p : bar | T_s | 5 | 10 | 15 | 20 | 30 | 40 | 50 | 60 | 80 | 100 | 120 | 140 |
|---|---|---|---|---|---|---|---|---|---|---|---|---|---|
| | v | 0.395 | 0.403 | 0.413 | 0.421 | 0.440 | 0.458 | 0.477 | 0.494 | 0.520 | 0.567 | 0.602 | 0.638 |
| − 50 | h | 168.15 | 170.85 | 173.57 | 176.34 | 181.61 | 187.92 | 193.41 | 199.30 | 211.36 | 223.78 | 236.55 | 249.64 |
| 0.3917 | s | 0.7543 | 0.7660 | 0.7776 | 0.7890 | 0.8116 | 0.8336 | 0.8552 | 0.8764 | 0.9176 | 0.9576 | 0.9956 | 1.0327 |
| | v | 0.357 | 0.366 | 0.374 | 0.383 | 0.399 | 0.416 | 0.432 | 0.448 | 0.481 | 0.513 | 0.545 | 0.577 |
| − 48 | h | 169.10 | 171.83 | 174.58 | 177.35 | 182.98 | 188.70 | 194.52 | 200.44 | 212.55 | 225.02 | 237.83 | 250.95 |
| 0.4341 | s | 0.7517 | 0.7634 | 0.7750 | 0.7864 | 0.8089 | 0.8309 | 0.8524 | 0.8735 | 0.9146 | 0.9542 | 0.9924 | 1.0293 |
| | v | 0.326 | 0.333 | 0.341 | 0.348 | 0.363 | 0.378 | 0.378 | 0.408 | 0.437 | 0.466 | 0.496 | 0.525 |
| − 46 | h | 170.08 | 172.83 | 175.59 | 178.39 | 184.04 | 189.80 | 195.69 | 201.59 | 213.75 | 226.26 | 239.11 | 252.27 |
| 0.4802 | s | 0.7490 | 0.7607 | 0.7723 | 0.7837 | 0.8061 | 0.8280 | 0.8495 | 0.8706 | 0.9115 | 0.9510 | 0.9885 | 1.0260 |
| | v | 0.297 | 0.304 | 0.311 | 0.318 | 0.331 | 0.345 | 0.358 | 0.372 | 0.398 | 0.425 | 0.451 | 0.475 |
| − 44 | h | 171.08 | 173.82 | 176.61 | 179.42 | 185.11 | 190.89 | 196.77 | 202.74 | 214.95 | 227.51 | 240.40 | 253.59 |
| 0.5300 | s | 0.7466 | 0.7581 | 0.7698 | 0.7811 | 0.8035 | 0.8254 | 0.8469 | 0.8679 | 0.9087 | 0.9481 | 0.9861 | 1.0228 |
| | v | 0.270 | 0.278 | 0.284 | 0.290 | 0.303 | 0.315 | 0.327 | 0.340 | 0364 | 0.388 | 0.412 | 0.436 |
| − 42 | h | 172.03 | 174.81 | 177.61 | 180.44 | 186.16 | 191.98 | 197.88 | 203.88 | 216.14 | 228.75 | 241.67 | 254.91 |
| 0.5839 | s | 0.7442 | 0.7559 | 0.7674 | 0.7787 | 0.8011 | 0.8229 | 0.8443 | 0.8653 | 0.9060 | 0.9452 | 0.9831 | 1.0198 |
| | v | 0.249 | 0.254 | 0.260 | 0.226 | 0,277 | 0.288 | 0.300 | 0.311 | 0.333 | 0.355 | 0.377 | 0.398 |
| − 40 | h | 173.01 | 175.81 | 178.63 | 181.47 | 187.23 | 193.07 | 199.01 | 205.03 | 217.34 | 229.99 | 242.96 | 256.23 |
| 0.6420 | .v | 0.7419 | 0.7559 | 0.7674 | 0.7787 | 0.7986 | 0.8204 | 0.8418 | 0.8627 | 0.9033 | 0.9424 | 0.9802 | 1.0168 |
| | v | 0.228 | 0.233 | 0.238 | 0.244 | 0.254 | 0.264 | 0.275 | 0.285 | 0.306 | 0.325 | 0.345 | 0.365 |
| − 38 | h | 173.58 | 176.80 | 179.64 | 182.50 | 188.29 | 194.17 | 200.13 | 206.19 | 218.19 | 231.25 | 244.25 | 257.56 |
| 0.7046 | s | 0.7399 | 0.7515 | 0.7630 | 0.7743 | 0.7966 | 0.8183 | 0.8396 | 0.8605 | 0.9010 | 0.9400 | 0.9776 | 1.0142 |
| | v | 0.209 | 0.214 | 0.219 | 0.223 | 0.233 | 0.245 | 0.252 | 0.261 | 0.280 | 0.298 | 0.316 | 0.335 |
| − 36 | h MAM | | 177.78 | 180.64 | 183.52 | 189.34 | 195.24 | 201.24 | 207.32 | 219.73 | 232.47 | 245.52 | 258.86 |
| 0.7719 | s | 0.7376 | 0.7452 | 0.7607 | 0.7720 | 0.7942 | 0.8159 | 0.8372 | 0.8580 | 0.8984 | 0.9374 | 0.9749 | 1.0113 |
| | v | 0.201 | 0.206 | 0.210 | 0.215 | 0.224 | 0.233 | 0.242 | 0.251 | 0.269 | 0.287 | 0.304 | 0.321 |
| − 35 | h | 175.42 | 178.27 | 181.14 | 184.01 | 189.86 | 195.97 | 201.80 | 207.90 | 220.33 | 233.10 | 246.17 | 259.53 |
| 0.8081 | .9 | 0.7367 | 0.7483 | 0.7598 | 0.7711 | 0.7934 | 0.8149 | 0.8362 | 0.8570 | 0.8973 | 0.9363 | 0.9738 | 1.0101 |
| | v | 0.192 | 0.197 | 0.207 | 0.206 | 0.215 | 0.223 | 0.231 | 0.241 | 0.257 | 0.274 | 0.291 | 0.308 |
| − 34 | h | 175.90 | 178.78 | 181.64 | 184.54 | 190.39 | 196.70 | 202.36 | 208.47 | 220.93 | 233.72 | 246.81 | 260.19 |
| 0.8441 | .9 | 0.7358 | 0.7474 | 0.7588 | 0.7701 | 0.7927 | 0.8139 | 0.8351 | 0.8559 | 0.8961 | 0.9351 | 0.9726 | 1.0088 |
| | v | 0.178 | 0.182 | 0.186 | 0.190 | 0.198 | 0.206 | 0.214 | 0.222 | 0.237 | 0.252 | 0.268 | 0.282 |
| − 32 | h | 176.88 | 179.76 | 182.66 | 185.58 | 191.47 | 197.44 | 203.50 | 209.63 | 222.15 | 234.97 | 248.11 | 261.52 |
| 0.9217 | .9 | 0.7339 | 0.7455 | 0.7569 | 0.7682 | 0.7903 | 0.8119 | 0.8331 | 0.8538 | 0.8940 | 0.9328 | 0.9702 | 1.0063 |

*Contd.*

## Table A14    Continued

| T:°C<br>p : bar | $T_s$ | 5 | 10 | 15 | 20 | 30 | 40 | 50 | 60 | 80 | 100 | 120 | 140 |
|---|---|---|---|---|---|---|---|---|---|---|---|---|---|
| −30<br>1.005 | v | 0.164 | 0.168 | 0.171 | 0.175 | 0.183 | 0.190 | 0.197 | 0.204 | 0.219 | 0.232 | 0.247 | 0.261 |
| | h | 177.84 | 180.74 | 183.66 | 186.59 | 192.52 | 198.52 | 204.61 | 210.77 | 223.34 | 236.91 | 249.38 | 262.56 |
| | s | 0.7321 | 0.7436 | 0.7550 | 0.7663 | 0.7884 | 0.8100 | 0.8311 | 0.8518 | 0.8919 | 0.9305 | 0.9679 | 1.0039 |
| −28<br>1.093 | v | 0.151 | 0.155 | 0.158 | 0.162 | 0.169 | 0.175 | 0.182 | 0.189 | 0.202 | 0.215 | 0.228 | 0.241 |
| | h | 178.09 | 181.73 | 184.65 | 187.62 | 193.58 | 199.62 | 205.73 | 211.92 | 224.54 | 237.46 | 250.67 | 264.16 |
| | j | 0.7302 | 0.7418 | 0.7532 | 0.7644 | 0.7865 | 0.8081 | 0.8291 | 0.8498 | 0.8898 | 0.9284 | 0.9656 | 1.0016 |
| −26<br>1.188 | v | 0.140 | 0.143 | 0.147 | 0.150 | 0.156 | 0.162 | 0.168 | 0.175 | 0.187 | 0.199 | 0.211 | 0.223 |
| | h | 179.76 | 182.70 | 185.66 | 188.63 | 194.63 | 200.71 | 206.84 | 213.06 | 225.73 | 238.70 | 251.95 | 265.48 |
| | s | 0.7286 | 0.7401 | 0.7515 | 0.7610 | 0.7848 | 0.8063 | 0.8274 | 0.8480 | 0.8879 | 0.9264 | 0.9635 | 0.9994 |
| −25<br>1.238 | v | O.135 | 0.138 | 0.141 | 0.144 | 0.150 | 0.156 | 0.162 | 0.168 | 0.180 | 0.192 | 0.203 | 0.214 |
| | h | 180.24 | 183.19 | 186.16 | 189.14 | 195.16 | 201.25 | 207.40 | 213.64 | 226.33 | 239.48 | 252.60 | 266.15 |
| | j | 0.7287 | 0.7393 | 0.7507 | 0.7611 | 0.7840 | 0.8055 | 0.8266 | 0.8471 | 0.8870 | 0.9254 | 0.9625 | 0.9984 |
| −24<br>1.289 | v | 0.130 | 0.133 | 0.136 | 0.139 | 0.145 | 0.151 | 0.156 | 0.162 | 0.173 | 0.184 | 0.195 | 0.206 |
| | h | 180.72 | 183.68 | 186.66 | 189.65 | 195.68 | 201.79 | 207.96 | 214.21 | 226.93 | 240.25 | 253.24 | 266.81 |
| | s | 0.7270 | 0.7386 | 0.7499 | 0.7612 | 0.7832 | 0.8047 | 0.8257 | 0.8462 | 0.8861 | 0.9244 | 0.9615 | 0.9973 |
| −22<br>1.396 | v | 0.121 | 0.123 | 0.126 | 0.129 | 0.134 | 0.140 | 0.144 | 0.150 | 0.161 | 0.175 | 0.181 | 0.191 |
| | h | 181.68 | 184.66 | 187.66 | 190.67 | 196.74 | 202.88 | 209.08 | 215.36 | 228.14 | 241.19 | 254.53 | 268.13 |
| | s | 0.7255 | 0.7371 | 0.7484 | 0.7596 | 0.7816 | 0.8031 | 0.8240 | 0.8445 | 0.8843 | 0.9226 | 0.9595 | 0.9952 |
| −20<br>1.51 | v | 0.112 | 0.115 | 0.117 | 0.128 | 0.125 | 0.130 | 0.135 | 0.139 | 0.149 | 0.159 | 0.168 | 0.177 |
| | h | 182.62 | 185.63 | 188.64 | 191.67 | 197.78 | 203.95 | 210.1.9 | 216.50 | 229.33 | 242.43 | 255.81 | 269.44 |
| | J | 0.7240 | 0.7355 | 0.7469 | 0.7581 | 0.7800 | 0.8015 | 0.8224 | 0.8429 | 0.8826 | 0.9207 | 0.9576 | 0.9932 |
| −18<br>1.631 | v | 0.104 | 0.107 | 0.109 | 0.111 | 0.116 | 0.121 | 0.125 | 0.130 | 0.139 | 0.148 | 0.156 | 0.165 |
| | h | 183.57 | 186.60 | 189.64 | 192.69 | 198.83 | 205.04 | 211.31 | 217.65 | 230.52 | 243.68 | 257.09 | 270.77 |
| | j | 0.7226 | 0.7341 | 0.7455 | 0.7567 | 0.7786 | 0.8000 | 0.8209 | 0.8413 | 0.8810 | 0.9191 | 0.9558 | 0.9930 |
| −16<br>1.760 | v | 0.097 | 0.099 | 0.101 | 0.103 | 0.108 | 0.112 | 0.117 | 0.121 | 0.129 | 0.137 | 0.146 | 0.154 |
| | h | 184.52 | 187.57 | 190.63 | 193.70 | 199.88 | 206.12 | 212.42 | 218.79 | 231.72 | 244.92 | 258.38 | 272.09 |
| | s | 0.7213 | 0.7328 | 0.7442 | 0.7554 | 0.7774 | 0.7986 | 0.8195 | 0.8399 | 0.8794 | 0.9175 | 0.9541 | 0.9895 |
| −15<br>1.828 | v | 0.093 | 0.095 | 0.098 | 0.100 | 0.104 | 0.108 | 0.113 | 0.117 | 0.125 | 0.133 | 0.144 | 0.184 |
| | h | 184.98 | 188.05 | 191.12 | 194.20 | 200.40 | 206.66 | 212.98 | 219.36 | 232.32 | 245.54 | 259.02 | 272.78 |
| | s | 0.7206 | 0.7321 | 0.7435 | 0.7547 | 0.7766 | 0.7979 | 0.8188 | 0.8391 | 0.8786 | 0.9166 | 0.9532 | 0.9886 |
| −14<br>1.894 | v | 0.090 | 0.093 | 0.095 | 0.097 | 0.101 | 0.104 | 0.109 | 0.113 | 0.121 | 0.128 | 0.136 | 0.143 |
| | h | 185.46 | 188.54 | 191.62 | 194.71 | 200.93 | 207.19 | 213.54 | 219.93 | 232.92 | 246.16 | 259.67 | 273.47 |
| | s | 0.7200 | 0.7315 | 0.7428 | 0.7540 | 0.7759 | 0.7972 | 0.8182 | 0.8384 | 0.8779 | 0.9158 | 0.9524 | 0.9877 |

*Contd.*

**Table A14    Continued**

| $T:°C$ | $T_s$ | | | | | | | | | | | |
|---|---|---|---|---|---|---|---|---|---|---|---|---|
| $p : bar$ | 5 | 10 | 15 | 20 | 30 | 40 | 50 | 60 | 80 | 100 | 120 | 140 |
| | $v$ 0.084 | 0.086 | 0.088 | 0.090 | 0.094 | 0.098 | 0.102 | 0.105 | 0.112 | 0.120 | 0.129 | 0.134 |
| – 12 | $h$ 186.40 | 189.49 | 192.59 | 195.70 | 201.96 | 208.27 | 214.64 | 221.07 | 234.11 | 247.40 | 260.94 | 274.73 |
| 2.040 | $j$ 0.7188 | 0.7303 | 0.7416 | 0.7528 | 0.7747 | 0.7960 | 0.8168 | 0.8371 | 0.8765 | 0.9144 | 0.9509 | 0.9861 |
| | $v$ 0.078 | 0.080 | 0.082 | 0.084 | 0.088 | 0.091 | 0.095 | 0.098 | 0.105 | 0.112 | 0.118 | 0.125 |
| – 10 | $h$ 187.34 | 190.45 | 193.57 | 196.71 | 202.99 | 209.53 | 215.75 | 222.21 | 235.31 | 248.64 | 262.24 | 276,05 |
| 2.193 | 5 0.7176 | 0.7293 | 0.7406 | 0.7518 | 0.7736 | 0.7941 | 0.8157 | 0.8360 | 0.8753 | 0.9191 | 0.9495 | 0.9846 |
| | $v$ 0.074 | 0.075 | 0.077 | 0.079 | 0.082 | 0.085 | 0.089 | 0.092 | 0.098 | 0.104 | 0.111 | 0.116 |
| – 8 | $h$ 182.27 | 191.41 | 194.55 | 197.71 | 204.04 | 210.43 | 216.86 | 223.35 | 236.49 | 249.88 | 263.51 | 277.36 |
| 2.354 | $s$ 0.7165 | 0.7281 | 0.7394 | 0.7506 | 0.7724 | 0.7937 | 0.8144 | 0.8347 | 0.8739 | 0.9116 | 0.9480 | 0.9830 |
| | $v$ 0.069 | 0.071 | 0.072 | 0.074 | 0.077 | 0.080 | 0.083 | 0.086 | 0.092 | 0.098 | 0.103 | 0.109 |
| – 6 | $h$ 181.19 | 192.37 | 195.53 | 198.70 | 205.08 | 211.50 | 217.96 | 224.48 | 237.68 | 251.12 | 264.79 | 278.69 |
| 2.523 | $j$ 0.7155 | 0.7271 | 0".7384 | 0.7496 | 0.7714 | 0.7926 | 0.8133 | 0.8336 | 0.8727 | 0.9103 | 0.9466 | 0.9816 |
| | $v$ 0.067 | 0.069 | 0.071 | 0.072 | 0.074 | 0.077 | 0.081 | 0.084 | 0.090 | 0.095 | 0.100 | 0.106 |
| – 5 | $h$ 189.65 | 192.83 | 196.01 | 199.19 | 205.59 | 212.03 | 218.54 | 225.04 | 239.77 | 251.73 | 265.42 | 279.34 |
| 2.6093 | $s$ 0.7151 | 0.7268 | 0.7371 | 0.7491 | 0/7701 | 0.7921 | 0.8128 | 0.8333 | 0.8721 | 0.9098 | 0.9459 | 0.9809 |
| | $v$ 0.064 | 0.066 | 0.068 | 0.069 | 0.072 | 0.075 | 0.078 | 0.081 | 0.086 | 0.092 | 0.097 | 0.102 |
| – 4 | $h$ 190.11 | 193.30 | 196.49 | 199.69 | 206.19 | 212.56 | 219.06 | 225.61 | 241.86 | 252.35 | 266.06 | 279.98 |
| 2.702 | $s$ 0.7146 | 0.7261 | 0.7378 | 0.7486 | 0.7704 | 0.7796 | 0.8123 | 0.8325 | 0.8716 | 0.9092 | 0.9453 | 0.9802 |
| | $v$ 0.060 | 0.062 | 0.063 | 0.065 | 0.068 | 0.070 | 0.073 | 0.076 | 0.081 | 0.086 | 0.091 | 0.096 |
| – 2 | $h$ 191.02 | 194.23 | 197.45 | 200.67 | 207.12 | 213.62 | 220.15 | 226.73 | 240.04 | 253.57 | 267.32 | 281.30 |
| 2.891 | $s$ 0.7135 | 0.7251 | 0.7364 | 0.7475 | 0.7694 | 0.7906 | 0.8112 | 0.8314 | 0.8704 | 0.9079 | 0.9380 | 0.9788 |
| | $v$ 0.057 | 0.058 | 0.060 | 0.061 | 0.064 | 0.066 | 0.069 | 0.071 | 0.076 | 0.081 | 0.086 | 0.091 |
| 0 | $h$ 191.93 | 195.97 | 198.41 | 201.65 | 208.15 | 214.63 | 221.25 | 227.86 | 241.23 | 254.88 | 268.60 | 282.65 |
| 3.0891 | $s$ 0.7125 | 0.7241 | 0.7350 | 0.7466 | 0.7684 | 0.7896 | 0.8102 | 0.8303 | 0.8693 | 0.9067 | 0.9427 | 0.9775 |
| | $v$ 0.054 | 0.055 | 0.056 | 0.058 | 0.060 | 0.063 | 0.064 | 0.067 | 0.072 | 0.076 | 0.084 | 0.085 |
| 2 | $h$ 192.84 | 196.10 | 199.36 | 202.62 | 209.16 | 215.73 | 222.33 | 228.98 | 242.40 | 256.03 | 269.87 | 283.92 |
| 3.297 | $j$ 0.7117 | 0.7232 | 0.7346 | 0.7457 | 0.7675 | 0.7687 | 0.8093 | 0.8294 | 0.8683 | 0.9057 | 0.9416 | 0.9762 |
| | $v$ 0.050 | 0.052 | 0.053 | 0.054 | 0.056 | 0.059 | 0.060 | 0.063 | 0.067 | 0.072 | 0.076 | 0.080 |
| 4 | $h$ 193.73 | 197.02 | 200.31 | 203.59 | 210.18 | 216.78 | 223.41 | 230.09 | 243.57 | 257.25 | 271.13 | 285.22 |
| 3.516 | $s$ 0.7109 | 0.7224 | 0.7338 | 0.7449 | 0.7667 | 0.7879 | 0.8085 | 0.8286 | 0.8674 | 0.9047 | 0.9405 | 0.9752 |
| | $v$ 0.049 | 0.050 | 0.051 | 0.052 | 0.055 | 0.057 | 0.059 | 0.061 | 0.066 | 0.070 | 0.074 | 0.078 |
| 5 | $h$ 194.18 | 197.49 | 200.79 | 204.08 | 210.68 | 217.31 | 223.96 | 230.04 | 244.16 | 257.86 | 272.77 | 205.88 |
| 3.631 | $s$ 0.7105 | 0.7280 | 0.7331 | 0.7445 | 0.7663 | 0.7875 | 0.8081 | 0.8282 | 0.8670 | 0.9042 | 0.9400 | 0.9746 |

*Contd.*

**Table A14 Continued**

| $T:°C$ $p:bar$ | $T_s$ | 5 | 10 | 15 | 20 | 30 | 40 | 50 | 60 | 80 | 100 | 120 | 140 |
|---|---|---|---|---|---|---|---|---|---|---|---|---|---|
| 6 3.746 | v | 0.047 | 0.049 | 0.050 | 0.051 | 0.053 | 0.055 | 0.057 | 0.060 | 0.064 | 0.068 | 0.072 | 0.075 |
| | h | 194.63 | 197.95 | 201.26 | 204.57 | 211.19 | 217.83 | 224.50 | 231.31 | 244.75 | 258.47 | 272.4 | 286.53 |
| | s | 0.7101 | 0.7135 | 0.7330 | 0.7442 | 0.7660 | 0.7871 | 0.8077 | 0.8278 | 0.8660 | 0.9037 | 0.9395 | 0.9741 |
| 8 3.968 | v | 0.045 | 0.046 | 0.047 | 0.048 | 0.050 | 0,052 | 0.054 | 0.056 | 0.059 | 0.064 | 0.067 | 0.071 |
| | h | 195.52 | 198.87 | 202.19 | 205.54 | 212.78 | 218.88 | 222.59 | 232.33 | 245.91 | 259.69 | 273.66 | 287.80 |
| | s | 0.7093 | 0.7209 | 0.7322 | 0.7454 | 0.7652 | 0.7863 | 0.8069 | 0.8269 | 0.8656 | 0.9028 | 0.9385 | 0.9729 |
| 10 4.238 | v | 0.043 | 0.044 | 0.045 | 0.046 | 0.048 | 0.050 | 0.052 | 0.054 | 0.057 | 0.061 | 0.065 | 0.068 |
| | h | 196.40 | 199.77 | 203.13 | 206.49 | 213.20 | 219.92 | 226.66 | 233.43 | 247.09 | 260.91 | 274.93 | 289.83 |
| | s | 0.7085 | 0.7201 | 0.7315 | 0.7426 | 0.7644 | 0.7855 | 0.8061 | 0.8261 | 0.8648 | 0.9018 | 0.9375 | 0.9719 |
| 12 4.502 | v | 0.040 | 0.041 | 0.042 | 0.043 | 0.045 | 0.047 | 0.048 | 0.050 | 0.054 | 0.057 | 0.060 | 0.063 |
| | h | 197.28 | 200.67 | 204.06 | 207.44 | 214.20 | 220.96 | 227.73 | 234.54 | 248.25 | 262.12 | 276.18 | 290.92 |
| | 5 | 0.7078 | 0.7194 | 0.7308 | 0.7420 | 0.7638 | 0.7849 | 0.8054 | 0.8254 | 0.8640 | 0.9010 | 0.9366 | 0.9709 |
| 14 4.778 | v | 0.038 | 0.039 | 0.039 | 0.040 | 0.042 | 0.044 | 0.046 | 0.047 | 0.051 | 0.054 | 0.057 | 0.060 |
| | h | 198.14 | 201.57 | 204.99 | 208.39 | 215.19 | 221.99 | 228.80 | 235.64 | 249.40 | 263.32 | 277.43 | 291.72 |
| | s | 0.7071 | 0.7186 | 0.7301 | 0.7413 | 0.7631 | 0.7842 | 0.8047 | 0.8247 | 0.8621 | 0.9002 | 0.9357 | 0.9699 |
| 15 4.922 | v | 0.037 | 0.037 | 0.038 | 0.039 | 0.041 | 0.043 | 0.044 | 0.046 | 0.049 | 0.052 | 0.055 | 0.058 |
| | h | 198.57 | 202.01 | 205.43 | 208.86 | 215.68 | 222.49 | 229.33 | 236.18 | 249.99 | 263.93 | 278.05 | 292.36 |
| | .v | 0.7068 | 0.7184 | 0.7298 | 0.7410 | 0.7628 | 0.7839 | 0.8044 | 0.8244 | 0.8629 | 0.8999 | 0.9353 | 0.9695 |
| 16 5.067 | v | 0.036 | 0.037 | 0.038 | 0.040 | 0.042 | 0.043 | 0.044 | 0.048 | 0.051 | 0.054 | 0.057 | 0.059 |
| | h | 198.99 | 202.45 | 205.89 | 209.33 | 216.17 | 223.01 | 229.86 | 236.73 | 250.56 | 264.54 | 278.68 | 293.01 |
| | j | 0.7065 | 0.7181 | 0.7295 | 0.7407 | 0.7625 | 0.7836 | 0.8042 | 0.8241 | 0.8626 | 0.8995 | 0.9349 | 0.9691 |
| 18 5.368 | v | 0.034 | 0.035 | 0.036 | 0.038 | 0.039 | 0.041 | 0.042 | 0.045 | 0.048 | 0.051 | 0.054 | 0.057 |
| | h | 199.86 | 203.44 | 206.79 | 210.26 | 217.15 | 224.03 | 230.94 | 237.82 | 251.71 | 265.74 | 279.93 | 294.29 |
| | s | 0.7058 | 0.7175 | 0.7289 | 0.7401 | 0.7619 | 0.7830 | 0.8035 | 0.8234 | 0.8619 | 0.8987 | 0.9281 | 0.9682 |
| 20 5.683 | v | 0.032 | 0.033 | 0.033 | 0.034 | 0.036 | 0.037 | 0.039 | 0.040 | 0.043 | 0.046 | 0.048 | 0.051 |
| | h | 200.70 | 204.10 | 207.70 | 211.64 | 218.11 | 225.05 | 232.01 | 238.90 | 252.80 | 267.01 | 281.15 | 195.61 |
| | s | 0.7053 | 0.7169 | 0.7284 | 0.7396 | 0.7614 | 0.7825 | 0.8030 | 0.8229 | 0.8613 | 0.8989 | 0.9334 | 0.9674 |
| 22 6.017 | v | 0.0302 | 0.0309 | 0.0316 | 0.0323 | 0.0341 | 0.0357 | 0.0366 | 0.0378 | 0.0407 | 0.0420 | 0.0450 | 0.0484 |
| | h | 201.54 | 205.08 | 208.60 | 212.11 | 219.12 | 226.10 | 233.03 | 240.01 | 254.01 | 268.14 | 182.02 | 296.9 |
| | s | 0.7046 | 0.7163 | 0.7277 | 0.7389 | 0.7608 | 0.7818 | 0.8023 | 0.8223 | 0.8608 | 0.8974 | 0.9325 | 0.9665 |
| 24 6.353 | v | 0.028 | 0.029 | 0.030 | 0.031 | 0.032 | 0.033 | 01.034 | 0.036 | 0.039 | 0.041 | 0.043 | 0.046 |
| | h | 204.40 | 205.93 | 209.48 | 213.02 | 218.5 | 227.06 | 234.06 | 241.07 | 255.14 | 269.32 | 283.65 | 298.13 |
| | .? | 0.7040 | 0.7158 | 0.7272 | 0.7385 | 0.7605 | 0.7814 | 0.8019 | 0.8218 | 0.8601 | 0.8967 | 0.9319 | 0.9650 |

*Contd.*

**Table A14   Continued**

| $T:°C$ $p:bar$ | $T_s$ | 5 | 10 | 15 | 20 | 30 | 40 | 50 | 60 | 80 | 100 | 120 | 140 |
|---|---|---|---|---|---|---|---|---|---|---|---|---|---|
| 25<br>6.520 | $v$ 0.275<br>$h$ 202.80<br>$s$ 0.7037 | 0.0285<br>206.41<br>0.7155 | 0.029<br>209.91<br>0.7269 | 0.0295<br>213.50<br>0.7395 | 0.031<br>219.51<br>0.76025 | 0.032<br>227.50<br>0.7815 | 0.0335<br>234.51<br>0.8015 | 0.035<br>241.51<br>0.8215 | 0.038<br>255.61<br>0.8595 | 0.040<br>269.90<br>0.8964 | 0.042<br>284.10<br>0.9315 | 0.045<br>298.70<br>0.9655 |
| 26<br>6.709 | $v$ 0.027<br>$h$ 203.20<br>$s$ 0.7034 | 0.028<br>206.80<br>0.7152 | 0.028<br>210.37<br>0.7267 | 0.029<br>214.00<br>0.7379 | 0.031<br>221.01<br>0.7598 | 0.031<br>228.06<br>0.7809' | 0.033<br>235.10<br>0.8013 | 0.034<br>242.14<br>0.8212 | 0.037<br>256.30<br>0.8595 | 0.039<br>270.50<br>0.8961 | 0.041<br>284.88<br>0.9312 | 0.044<br>299.40<br>0.9651 |
| 28<br>7.08 | $v$ 0.028<br>$h$ 204.01<br>$s$ 0.7030 | 0.026<br>207.61<br>0.7148 | 0.027<br>211.24<br>0.7263 | 0.028<br>214.81<br>0.7375 | 0.029<br>222.01<br>0.7594 | 0.030<br>229.06<br>0.7805 | 0.031<br>236.14<br>0.8010 | 0.033<br>243.22<br>0.8208 | 0.035<br>257.40<br>0.8591 | 0.037<br>271.7<br>0.8956 | 0.039<br>286.1<br>0.9307 | 0.041<br>300.7<br>0.9651 |
| 30<br>7.466 | $v$ 0.024<br>$h$ 204.83<br>$s$ 0.7024 | 0.025<br>208.50<br>0.7142 | 0.026<br>212.10<br>0.7258 | 0.026<br>215.80<br>0.7371 | 0.027<br>223.00<br>0.7589 | 0.029<br>230.11<br>0.7801 | 0.030<br>237.21<br>0.8005 | 0.031<br>244.30<br>0.8204 | 0.033<br>258.61<br>0.8586 | 0.035<br>273.02<br>0.8951 | 0.037<br>287.41<br>0.9301 | 0.039<br>302.01<br>0.9638 |
| 32<br>7.0867 | $v$ 0.023<br>$h$ 205.61<br>$s$ 0.7020 | 0.024<br>209.30<br>0.7138 | 0.024<br>213.00<br>0.7254 | 0.025<br>216.60<br>0.7367 | 0.026<br>223.91<br>0.7586 | 0.027<br>231.11<br>0.7797 | 0.028<br>238.20<br>0.8002 | 0.029<br>245.41<br>0.8200 | 0.032<br>259.70<br>0.8582 | 0.034<br>274.02<br>0.8946 | 0.036<br>288.50<br>0.9296 | 0.038<br>303.10<br>0.9633 |
| 34<br>8.284 | $v$ 0.021<br>$h$ 206.37<br>$s$ 0.7015 | 0.023<br>210.1<br>0.7133 | 0.023<br>213.82<br>0.7248 | 0.024<br>217.49<br>0.7363 | 0.025<br>224,78<br>0.7582 | 0.026<br>232.01<br>0.7793 | 0.027<br>239.21<br>0.7998 | 0.028<br>246.4<br>0.8196 | 0.030<br>260.28<br>0.8578 | 0.032<br>275.23<br>0.8942 | 0.034<br>289.78.<br>0.9291 | 0.036<br>304.47<br>0.9626 |
| 35<br>8.51 | $v$ 0.021<br>$h$ 206.7<br>$s$ 0.7012 | 0.0215<br>210.5<br>0.7130 | 0.022<br>214.8<br>0.7245 | 0.023<br>217.9<br>0.7360 | 0.0245<br>225.2<br>0.7580 | 0.025<br>232.4<br>0.7791 | 0.0275<br>236.1<br>0.7996 | 0.0265<br>243.3<br>0.8194 | 0:6285<br>261.2<br>0.8575 | 0.030<br>275.8<br>0.8941 | 0.031<br>290.3<br>0.9295 | 0.0345<br>305.1<br>0.9625 |
| 36<br>8.717 | $v$ 0.021<br>$h$ 207.15<br>$s$ 0.7012 | 0.021<br>210.93<br>0.7130 | 0.022<br>214.66<br>0.7247 | 0.023<br>218.37<br>0.7360 | 0.024<br>225.7<br>0.7580 | 0.025<br>232.97<br>0.7791 | 0.026<br>240.2<br>0.7996 | 0.027<br>247.45<br>0.8194 | 0.028<br>261.9<br>0.8575 | 0.030<br>276.4<br>0.8939 | 0.032<br>291.1<br>0.9287 | 0.034<br>305.73<br>0.9621 |
| 38<br>9.167 | $v$ 0.020<br>$h$ 207.9<br>$s$ 0.7004 | 0.020<br>211.72<br>0.7124 | 0.021<br>215.5<br>0.7241 | 0.021<br>219.23<br>0.7354 | 0.023<br>226.63<br>0.7574 | 0.024<br>233.95<br>0.7786 | 0.025<br>241.24<br>0.7990 | 0.025<br>248.5<br>0.8189 | 0.027<br>263.0<br>0.8570 | 0.029<br>277.57<br>0.8933 | 0.030<br>292.22<br>0.9280 | 0.032<br>306.98<br>0.9615 |
| 40<br>9.634 | $v$ 0.019<br>$h$ 208.65<br>$s$ 0.7000 | 0.019<br>212.5<br>0.7120 | 0.020<br>216.3<br>0.7236 | 0.021<br>220.0<br>0.7350 | 0.022<br>227.5<br>0.7571 | 0.022<br>234.9<br>0.7782 | 0.023<br>242.2<br>0.7987 | 0.024<br>249.5<br>0.8185 | 0.026<br>264.1<br>0.8566 | 0.028<br>278.71<br>0.8928 | 0.029<br>293.4<br>0.9276 | 0.031<br>308.22<br>0.9610 |
| 42<br>10.118 | $v$ 0.018<br>$h$ 209.4<br>$s$ 0.6996 | 0.018<br>213.29<br>0.7116 | 0.019<br>217.11<br>0.7233 | 0.020<br>220.93<br>0.7348 | 0.021<br>228.44<br>0.7568 | 0.021<br>235.86<br>0.7780 | 0.022<br>243.23<br>0.7981 | 0.023<br>250.56<br>0.8183 | 0.024<br>265.2<br>0.8563 | 0.027<br>279.87<br>0.8926 | 0.028<br>294.6<br>0.9272 | 0.030<br>309.47<br>0.9601 |

*Contd.*

## Table A14    Continued

| $T:°C$ | $T_s$ | | | | | | | | | | | |
|---|---|---|---|---|---|---|---|---|---|---|---|---|
| $p : bar$ | 5 | 10 | 15 | 20 | 30 | 40 | 50 | 60 | 80 | 100 | 120 | 140 |
| | $v$ 0.017 | 0.018 | 0.018 | 0.019 | 0.019 | 0.020 | 0.021 | 0.022 | 0.024 | 0.025 | 0.027 | 0.028 |
| 44 | $h$ 210.13 | 214.06 | 217.93 | 221.76 | 229.23 | 236.8 | 244.2 | 251.59 | 266.3 | 281.03 | 296.22 | 310.73 |
| 10.62 | 10.6990 | 0.7111 | 0.7228 | 0.7343 | 0.7564 | 0.7776 | 0.7981 | 0.8179 | 0.8559 | 0.8921 | 0.9267 | 0.9600 |
| | $v$ 0.0165 | 0.0175 | 0.0175 | 0.018 | 0.019 | 0.0195 | 0.0215 | 0.0218 | 0.0235 | 0.0245 | 0.026 | 0.0275 |
| 45 | $h$ 210.40 | 214.49 | 218.31 | 222.01 | 229.7 | 237.20 | 244.70 | 251.65 | 266.8 | 281.5 | 296.7 | 311.2 |
| 10.89 | $s$ 0.6985 | 0.7109 | 0.7228 | 0.7342 | 0.7564 | 0.7776 | 0.7980 | 0.8178 | 0.8559 | 0.8919 | 0.9265 | 0.9590 |
| | $v$ 0.016 | 0.017 | 0.017 | 0.018 | 0.019 | 0.019 | 0.020 | 0.021 | 0.023 | 0.024 | 0.025 | 0.027 |
| 46 | $h$ 210.84 | 214.8 | 218.72 | 222.59 | 230.2 | 237.73 | 245.19 | 251.81 | 267.38 | 282.17 | 297.01 | 311.95 |
| 11.14 | $s$ 0.6981 | 0.7107 | 0.7225 | 0.7340 | 0.756 | 0.7774 | 0.7979 | 0.8177 | 0.8557 | 0.8918 | 0.9264 | 0.9597 |
| | $v$ 0.015 | 0.016 | 0.016 | 0.017 | 0.017 | 0.018 | 0.019 | 0.020 | 0.021 | 0.023 | 0.025 | 0.026 |
| 48 | $h$ 211.54 | 215.56 | 219.5 | 223.4 | 231.09 | 238.75 | 246.17 | 253.62 | 268.47 | 283.3 | 298.2 | 313.19 |
| 11.679 | $s$ 0.6981 | 0.7103 | 0.7227 | 0.7331 | 0.755 | 0.777 | 0.7976 | 0.8175 | 0.8554 | 0.8914 | 0.9261 | 0.9592 |
| | $v$ 0.015 | 0.015 | 0.016 | 0.016 | 0.017 | 0.018 | 0.018 | 0.019 | 0.021 | 0.022 | 0.023 | 0.025 |
| 50 | $h$ 212.22 | 216.28. | 220.27 | 224.2 | 231.96 | 239.58 | 247.13 | 254.63 | 269.55 | 284.44 | 299.39 | 314.4 |
| 12.24 | $s$ 0.6976 | 0.7099 | 0.7218 | 0.7333 | 0.7556 | 0.7769 | 0.7974 | 0.8172 | 0.8552 | 0.8912 | 0.9257 | 0.9589 |

# REFRIGERANT 22

## CHLORODIFLUOROMETHANE

| | |
|---|---|
| Chemical Formula | : $CHC1F_2$ |
| Molecular Weight | : 86.48 |
| Boiling Temperature at Atmospheric Pressure, (°C) | : $-40.78$ |
| Freezing Temperature at Atmospheric Pressure, (°C) | : $-160$ |
| Critical Temperature, (°C) | : 96 |
| Critical Pressure, (bar) | : 49.78 |
| Critical Density, $(kg/m^3)$ | : 524.8 |
| Density of Liquid at 30°C $(kg/m^3)$ | : 1172.48 |
| Specific Volume of Saturated Vapour at $-15$°C, $(m^3/kg)$ | : 0.0776 |
| Specific Heat of Liquid at 30°C, (kJ/kg) C | : 1.28 |
| Specific Heat Ratio of Vapour at 30°C and 1atm | : 1.18 |
| Thermal Conductivity, (W/m °C) | |
|     Saturated Liquid at 0°C | : 0.119 |
|     Saturated Liquid at 30°C | : 0.0865 |
|     Vapour at 1 atm and 0°C | : 0.0088 |
|     Vapour at 1 atm and 30°C | : 0.0112 |
| Viscosity, Centipoise (1 cp = $10\sim^3$ $Ns/m^2$) | |
|     Saturated Liquid at $-15$°C | : 0.298 |
|     Saturated Liquid at 30°C | : 0.230 |
|     Vapour at 1 atm and $-15$°C | : 0.0112 |
|     Vapour at 1 atm and 30°C | : 0.0132 |

## Table Al 5    Saturated Monochlorodifluoromethane (CHC1F$_2$) R-22
### Datum at $-40°C$, $h_f = 0$, $s_f = 0$

| Saturation temp in °C | Saturation pressure in bar | Specific volume in m³/kg | | Specific enthalpy in kJ/kg | | | Specific entropy in kJ/kgK | |
|---|---|---|---|---|---|---|---|---|
| | | Liquid | Vapour | Liquid | Vapour | Latent | Liquid | Vapour |
| (t) | (P) | ($v_f$) | ($v_g$) | ($h_f$) | ($h_g$) | ($h_{fg}$) | ($s_f$) | ($s_g$) |
| − 100 | 0.02009 | 0.000643 | 8.3412 | − 63.45 | 267.18 | 203.73 | − 0.3144 | 1.2293 |
| − 95 | 0.03150 | 0.000647 | 5.4344 | − 58.14 | 264.33 | 206.19 | − 0.2843 | 1.2004 |
| − 90 | 0.04792 | 0.000652 | 3.6381 | − 52.87 | 261.51 | 208.64 | − 0.2550 | 1.1736 |
| − 85 | 0.07731 | 0.000656 | 2.5204 | − 47.61 | 258.72 | 211.11 | − 0.2269 | 1.1489 |
| − 80 | 0.10393 | 0.000661 | 1.7816 | − 42.40 | 256.00 | 213.60 | − 0.1989 | 1.1267 |
| − 75 | 0.14759 | 0.000666 | 1.2842 | − 37.17 | 253.28 | 216.11 | − 0.1721 | 1.1066 |
| − 70 | 0.20517 | 0.000672 | 0.9420 | − 31.93 | 250.55 | 218.62 | − 0.1461 | 1.0874 |
| − 65 | 0.27965 | 0.000677 | 0.7037 | − 26.68 | 247.84 | 221.16 | − 0.1206 | 1.0702 |
| − 60 | 0.37448 | 0.000683 | 0.5351 | − 21.42 | 245.09 | 223.67 | − 0.0959 | 1.0543 |
| − 55 | 0.49621 | 0.000689 | 0.4131 | − 16.13 | 242.31 | 226.18 | − 0.0712 | 1.0396 |
| − 50 | 0.64758 | 0.000696 | 0.3229 | − 10.81 | 239.50 | 228.69 | − 0.0473 | 1.0262 |
| − 45 | 0.83241 | 0.000702 | 0.2556 | − 5.40 | 236.60 | 231.20 | − 0.0234 | 1.0137 |
| − 40 | 1.05586 | 0.000709 | 0.2049 | 0.00 | 233.67 | 233.67 | 0.0000 | 1.0024 |
| − 38 | 1.15852 | 0.000712 | 0.1882 | 2.20 | 232.45 | 234.65 | 0.0096 | 0.9982 |
| − 36 | 1.26910 | 0.000715 | 0.1728 | 4.40 | 231.25 | 235.65 | 0.0188 | 0.9940 |
| − 34 | 1.38731 | 0.000719 | 0.1590 | 6.58 | 230.02 | 236.60 | 0.0280 | 0.9902 |
| − 32 | 1.51324 | 0.000721 | 0.1465 | 8.83 | 228.77 | 237.60 | 0.0373 | 0.9860 |
| − 30 | 1.64690 | 0.000724 | 0.1353 | 11.05 | 227.50 | 238.55 | 0.0464 | 0.9283 |
| − 28 | 1.78938 | 0.000727 | 0.1253 | 13.29 | 226.23 | 239.52 | 0.0557 | 0.9785 |
| − 26 | 1.94069 | 0.000731 | 0.1161 | 15.58 | 224.90 | 240.48 | 0.0645 | 0.9747 |
| − 24 | 2.10207 | 0.000734 | 0.1077 | 17.77 | 223.64 | 241.41 | 0.0733 | 0.9710 |
| − 22 | 2.27448 | 0.000738 | 0.1000 | 19.99 | 222.34 | 242.33 | 0.0821 | 0.9676 |
| − 20 | 2.45793 | 0.000741 | 0.0930 | 22.21 | 221.04 | 243.25 | 0.0908 | 0.9638 |
| − 18 | 2.65310 | 0.000744 | 0.0865 | 24.47 | 219.70 | 244.17 | 0.0996 | 0.9605 |
| − 16 | 2.85903 | 0.000748 | 0.0806 | 26.72 | 218.36 | 245.08 | 0.1080 | 0.9576 |
| − 14 | 3.07876 | 0.000752 | 0.0752 | 28.94 | 217.02 | 245.96 | 0.1164 | 0.9538 |
| − 12 | 3.31172 | 0.000756 | 0.0701 | 31.16 | 215.68 | 246.84 | 0.1248 | 0.9508 |
| − 10 | 3.55793 | 0.000759 | 0.0655 | 33.40 | 214.32 | 247.72 | 0.1336 | 0.9479 |
| − 8 | 3.81321 | 0.000763 | 0.0612 | 35.66 | 212.94 | 248.60 | 0.1419 | 0.9454 |
| − 6 | 4.09172 | 0.000767 | 0.0573 | 37.92 | 211.54 | 249.46 | 0.1503 | 0.9425 |
| − 4 | 4.37972 | 0.000771 | 0.0536 | 40.19 | 210.11 | 250.30 | 0.1591 | 0.9396 |
| − 2 | 4.68317 | 0.000775 | 0.0503 | 42.52 | 208.62 | 251.14 | 0.1675 | 0.9370 |

*Contd.*

**Table A15   Continued**

| $t$ | $P$ | $v_f$ | $v_g$ | $h_f$ | $h_g$ | $h_{fg}$ | $s_f$ | $s_g$ |
|---|---|---|---|---|---|---|---|---|
| 0 | 5.00207 | 0.000779 | 0.0471 | 44.94 | 207.03 | 251.97 | 0.1763 | 0.9345 |
| 1 | 5.16841 | 0.000781 | 0.0457 | 46.16 | 206.21 | 252.37 | 0.1809 | 0.9333 |
| 2 | 5.33917 | 0.000783 | 0.0443 | 47.38 | 205.39 | 252.77 | 0.1855 | 0.9320 |
| 3 | 5.51434 | 0.000785 | 0.0429 | 48.64 | 204.53 | 253.17 | 0.1901 | 0.9303 |
| 4 | 5.69352 | 0.000787 | 0.0416 | 49.91 | 203.65 | 253.56 | 0.1943 | 0.9291 |
| 5 | 5.87621 | 0.000790 | 0.0403 | 51.16 | 202.79 | 253.95 | 0.1989 | 0.9278 |
| 6 | 6.06276 | 0.000792 | 0.0391 | 52.41 | 201.92 | 254.33 | 0.2035 | 0.9266 |
| 7 | 6.25393 | 0.000794 | 0.0379 | 53.68 | 201.03 | 254.71 | 0.2077 | 0.9253 |
| 8 | 6.44993 | 0.000797 | 0.0368 | 54.94 | 200.14 | 255.08 | 0.2123 | 0.9241 |
| 9 | 6.65034 | 0.000799 | 0.0357 | 56.22 | 199.22 | 255.44 | 0.2169 | 0.9228 |
| 10 | 6.85517 | 0.000801 | 0.0346 | 57.52 | 198.29 | 255.81 | 0.2211 | 0.9216 |
| 11 | 7.06621 | 0.000803 | 0.0336 | 58.80 | 197.35 | 256.15 | 0.2257 | 0.9203 |
| 12 | 7.27724 | 0.000805 | 0.0326 | 60.07 | 196.41 | 256.48 | 0.2303 | 0.9190 |
| 13 | 7.49793 | 0.000808 | 0.0316 | 61.38 | 195.45 | 256.83 | 0.2345 | 0.9178 |
| 14 | 7.72552 | 0.000810 | 0.0307 | 62.72 | 194.45 | 257.17 | 0.2391 | 0.9165 |
| 15 | 7.95517 | 0.000813 | 0.0298 | 64.02 | 193.48 | 257.50 | 0.2437 | 0.9152 |
| 16 | 8.18758 | 0.000815 | 0.0289 | 65.32 | 192.48 | 257.81 | 0.2483 | 0.9140 |
| 17 | 8.42552 | 0.000817 | 0.0281 | 66.63 | 191.50 | 258.13 | 0.2529 | 0.9127 |
| 18 | 8.63241 | 0.000820 | 0.0273 | 67.95 | 190.49 | 258.44 | 0.2571 | 0.9115 |
| 19 | 8.91793 | 0.000822 | 0.0266 | 69.27 | 189.46 | 258.73 | 0.2617 | 0.9102 |
| 20 | 9.17241 | 0.000825 | 0.0258 | 70.59 | 188.41 | 259.00 | 0.2663 | 0.9089 |
| 21 | 9.43310 | 0.000827 | 0.0251 | 71.93 | 187.36 | 259.29 | 0.2709 | 0.9077 |
| 22 | 9.69931 | 0.000830 | 0.0244 | 73.31 | 186.27 | 259.58 | 0.2755 | 0.9065 |
| 23 | 9.97103 | 0.000833 | 0.0237 | 74.66 | 185.19 | 259.85 | 0.2801 | 0.9052 |
| 24 | 10.24828 | 0.000835 | 0.0230 | 76.04 | 184.07 | 260.11 | 0.2847 | 0.9039 |
| 25 | 10.53103 | 0.000838 | 0.0224 | 77.39 | 182.99 | 260.38 | 0.2889 | 0.9027 |
| 26 | 10.81931 | 0.000841 | 0.0218 | 78.79 | 181.85 | 260.64 | 0.2935 | 0.9014 |
| 27 | 11.11310 | 0.000844 | 0.0212 | 80.16 | 180.73 | 260.89 | 0.2981 | 0.9002 |
| 28 | 11.41241 | 0.000846 | 0.0206 | 81.54 | 179.58 | 261.12 | 0.3023 | 0.8989 |
| 29 | 11.69034 | 0.000849 | 0.0200 | 82.96 | 178.41 | 261.37 | 0.3069 | 0.8977 |
| 30 | 12.03448 | 0.000852 | 0.0194 | 84.38 | 177.22 | 261.60 | 0.3115 | 0.8964 |
| 31 | 12.35103 | 0.000855 | 0.0189 | 85.77 | 176.04 | 261.81 | 0.3161 | 0.8948 |
| 32 | 12.67310 | 0.000858 | 0.0184 | 87.17 | 174.85 | 262.02 | 0.3207 | 0.8935 |
| 33 | 13.00070' | 0.000861 | 0.0179 | 88.57 | 173.65 | 262.22 | 0.3249 | 0.8922 |
| 34 | 13.33793 | 0.000864 | 0.0174 | 90.00 | 172.40 | 262.40 | 0.3295 | 0.8909 |
| 35 | 13.68276 | 0.000867 | 0.0169 | 91.43 | 171.15 | 262.58 | 0.3337 | 0.8893 |
| 36 | 14.03034 | 0.000870 | 0.0165 | 92.85 | 169.89 | 262.74 | 0.3383 | 0.8880 |
| 37 | 14.38207 | 0.000874 | 0.0161 | 94.24 | 168.64 | 262.88 | 0.3429 | 0.8868 |
| 38 | 14.74345 | 0.000877 | 0.0156 | 95.63 | 167.37 | 263.00 | 0.3471 | 0.8851 |
| 39 | 15.11103 | 0.000881 | 0.0152 | 97.03 | 166.10 | 263.13 | 0.3517 | 0.8838 |
| 40 | 15.48965 | 0.000884 | 0.0148 | 98.44 | 164.77 | 263.21 | 0.3563 | 0.8822 |
| 41 | 15.86827 | 0.000887 | 0.0144 | 99.82 | 163.47 | 263.29 | 0.3601 | 0.8809 |

*Contd.*

## Table A15    Continued

| t | P | $v_f$ | $v_g$ | $h_f$ | $h_g$ | $h_{fg}$ | $s_f$ | $S_g$ |
|---|---|---|---|---|---|---|---|---|
| 42 | 16.25793 | 0.000891 | 0.0140 | 101.24 | 162.14 | 263.38 | 0.3651 | 0.8797 |
| 43 | 16.65380 | 0.000894 | 0.0137 | 102.68 | 160.80 | 263.48 | 0.3693 | 0.8780 |
| 44 | 17.05517 | 0.000898 | 0.0133 | 104.12 | 159.46 | 263.58 | 0.3739 | 0.8767 |
| 45 | 17.46552 | 0.000902 | 0.0130 | 105.58 | 158.09 | 263.67 | 0.3785 | 0.8755 |
| 46 | 17.88414 | 0.000906 | 0.0126 | 107.04 | 156.72 | 263.76 | 0.3827 | 0.8738 |
| 47 | 18.30827 | 0.000910 | 0.0123 | 108.51 | 155.36 | 263.87 | 0.3871 | 0.8723 |
| 48 | 18.73793 | 0.000914 | 0.0119 | 109.98 | 153.95 | 263.93 | 0.3915 | 0.8709 |
| 49 | 19.12414 | 0.000918 | 0.0116 | 111.42 | 152.57 | 263.99 | 0.3959 | 0.8694 |
| 50 | 19.61380 | 0.000922 | 0.0113 | 112.86 | 151.19 | 264.05 | 0.4003 | 0.8680 |

## Table Al 6    Superheated properties of R – 22 having h = Enthalpy, (kJ/kg)
s = Entropy (kJ/kg-K) ; v = specific volume (m³/kg) and T , = degrees of superheat, (°C)

| T:°C $\quad T_s$ | | 5 | 10 | 15 | 20 | 30 | 40 | 50 | 60 | 80 | 100 | 120 | 140 |
|---|---|---|---|---|---|---|---|---|---|---|---|---|---|
| p : bar | | | | | | | | | | | | | |
| – 50 | v | 0.332 | 0.341 | 0.349 | 0.356 | 0.372 | 0.388 | 0.403 | 0.418 | 0.449 | 0.479 | 0.510 | 0.540 |
| | h | 232.93 | 235.11 | 236.05 | 241.03 | 247.04 | 253.16 | 259.39 | 265.72. | 278.69 | 292.88 | 305.89 | 320.11 |
| 0.646 | s | 1.0415 | 1.0542 | 1.0667 | 1.0790 | 1.1033 | 1.1270 | 1.1503 | 1.1730 | 1.2172 | 1.2680 | 1.3015 | 1.3417 |
| – 48 | V | 0.303 | 0.310 | 0.317 | 0.324 | 0.338 | 0.352 | 0.366 | 0.300 | 0.408 | 0.436 | 0.463 | 0.490 |
| | h | 233.20 | 236.14 | 239.10 | 242.09 | 298.15 | 254.30 | 260.56 | 266.92 | 279.96 | 293.41 | 307.27 | 321.54 |
| 0.715 | s | 1,0364 | 1.0491 | 1.0615 | 1.0739 | 1.0961 | 1.1217 | 1.1449 | 1.1676 | 1.2118 | 1.2945 | 1.2958 | 1.3360 |
| – 46 | v | 0.276 | 0.282 | 0.289 | 0.295 | 0.306 | 0.321 | 0.334 | 0.346 | 0.371 | 0.396 | 0.421 | 0.446 |
| | h | 234.20 | 237.16 | 240.15 | 243.16 | 249.25 | 255.44 | 261.73 | 268.13 | 281.22 | 294.73 | 308.64 | 322.96 |
| 0.790 | s | 1.0314 | 1.0440 | 1.0565 | 1.0688 | 1.0929 | 1.1165 | 1.1397 | 1.1623 | 1.2064 | 1.2490 | 1.2903 | 1.3304 |
| – 45 | v | 0.264 | 0.270 | 0.276 | 0.282 | 0.295 | 0.307 | 0.319 | 0.331 | 0.355 | 0.379 | 0.403 | 0.426 |
| | h | 234.70 | 237.67 | 240.67 | 243.69 | 249.80 | 255.98 | 262.32 | 268.73 | 281.86 | 295.48 | 309.33 | 323.42 |
| 0.83 | S | 1.0289 | 1.0415 | 1.0540 | 1.0663 | 1.0904 | 1.1140 | 1.1372 | 1.1597 | 1.2038 | 12463 | 1.2873 | 1.3270' |
| – 44 | v | 0.252 | 0.258 | 0.264 | 0.270 | 0.281 | 0.293 | 0.304 | 0.316 | 0.339 | 0.361 | 0.384 | 0.407 |
| | h | 235.20 | 238.18 | 241.19 | 244.22 | 250.35 | 256.58 | 262.90 | 269.33 | 282.49 | 296.05 | 310.02 | 324.39 |
| 0.871 | s | 1.0265 | 1.0391 | 1.0515 | 1.0638 | 1.0879 | 1.1115 | 1.1346 | 1.1572 | 1.2012 | 1.2437 | 1.2849 | 1.3249 |
| – 42 | v | 0.230 | 0.236 | 0.241 | 0.246 | 0.257 | 0.268 | 0.278 | 0.289 | 0.310 | 0.330 | 0.351 | 0.371 |
| | h | 236.19 | 239.20 | 242.2 | 245.25 | 251.45 | 257.71 | 264.07 | 270.53 | 283.75 | 297.37 | 311.40 | 325.82 |
| 0.959 | s | 1.0218 | 1.0344 | 1.0468 | 1.0591 | 1.0832 | 1.1067 | 1.1298 | 1.1523 | 1.1962 | 1.2387 | 1.2798 | 1.3197 |
| – 40 | v | 0.211 | 0.216 | 0.221 | 0.226 | 0.235 | 0.245 | 0.255 | 0.264 | 0.283 | 0.302 | 0.321 | 0.340 |
| | h | 237.17 | 240.20 | 243.25 | 246.32 | 252.53 | 258.83 | 265.23 | 272.72 | 285.00 | 298.68 | 312.76 | 327.25 |
| 1.053 | s | 1.0172 | 1.0298 | 1.0422 | 1.0544 | 1.0785 | 1.1020 | 1.1250 | 1.1475 | 1.1913 | 1.2337 | 1.2747 | 1.3146 |
| – 38 | v | 0.194 | 0.198 | 0.203 | 0.207 | 0.216 | 0.225 | 0.234 | 0.242 | 0.260 | 0.277 | 0.294 | 0.311 |
| | h | 238.15 | 241.20 | 244.27 | 247.36 | 253.61 | 259.95 | 266.38 | 272.91 | 286.25 | 299.99 | 314.13 | 328.65 |
| 1.155 | J | 1.0127 | 1.0253 | 1.0377 | 1.0499 | 1.0739 | 1.0974 | 1.1203 | 1.1428 | 1.1866 | 1.2288 | 1.2698 | 1.3096 |

*Contd.*

**Table A16    Continued**

| T:°C / p:bar | | 5 | 10 | 15 | 20 | 30 | 40 | 50 | 60 | 80 | 100 | 120 | 140 |
|---|---|---|---|---|---|---|---|---|---|---|---|---|---|
| | $v$ | 0.178 | 0.182 | 0.186 | 0.190 | 0.199 | 0.207 | 0.215 | 0.223 | 0.239 | 0.254. | 0.270 | 0.286 |
| −36 | $h$ | 239.11 | 242.19 | 245.28 | 248.40 | 254.69 | 261.06 | 267.53 | 274.09 | 287.50 | 301.30 | 315.49 | 330.06 |
| 1.264 | $s$ | 1.0083 | 1.0209 | 1.0333 | 1.0455 | 1.0695 | 1.0929 | 1.1159 | 1.1383 | 1.1820 | 1.2242 | 1.2650 | 1.3047 |
| | $v$ | 0.171 | 0.175 | 0.179 | 0.182 | 0.191 | 0.198 | 0.206 | 0.214 | 0.229 | 0.244 | 0.259 | 0.274 |
| −35 | $h$ | 239.55 | 242.58 | 245.79 | 248.92 | 255.31 | 261.62 | 268.11 | 274.65 | 288.13 | 301,96 | 316.18 | 330.77 |
| 1.322 | $s$ | 1.0062 | 1.0187 | 1.0311 | 1.0433 | 1.0673 | 1.0907 | 1.1137 | 1.1361 | 1.1797 | 1.2219 | 1.2627 | 1.3024 |
| | $v$ | 0.164 | 0.168 | 0.171 | 0.175 | 0.183 | 0.190 | 0.198 | 0.205 | 0.220 | 0.234 | 0.248 | 0.263 |
| −34 | $h$ | 240.09 | 243.18 | 246.30 | 249.43 | 255.77 | 262.18 | 268.69 | 275.29 | 288.76 | 302.62 | 316.86 | 331.49 |
| 1.361 | $s$ | 1.0041 | 1.0166 | 1.0290 | 1.0412 | 1.0652 | 1.0886 | 1.1115 | 1.1339 | 1.1775 | 1.2196 | 1.2604 | 1.3001 |
| | $v$ | 0.151 | 0.154 | 0.158 | 0.161 | 0.168 | 0.175 | 0.182 | 0.189 | 0.202 | 0.216 | 0.229 | 0.242 |
| −32 | $h$ | 241.04 | 244.16 | 247.30 | 250.46 | 256.83 | 263.29 | 269.83 | 276.47 | 290.01 | 303.93 | 318.23 | 332.91 |
| 1.506 | $s$ | 0.9999 | 1.0125 | 1.0249 | 1.0371 | 1.0610 | 1.0844 | 1.1073 | 1.1297 | 1.1732 | 1.2152 | 1.2550 | 1.2955 |
| | $v$ | 0.139 | 0.143 | 0.146 | 0.149 | 0.155 | 0.162 | 0.168 | 0.174 | 0.187 | 0.199 | 0.211 | 0.223 |
| −30 | $h$ | 241.98 | 245.17 | 248.29 | 251.47 | 257.88 | 264.38 | 270.96 | 277.63 | 291.24 | 305.22 | 319.58 | 334.31 |
| 1.640 | $s$ | 0.9959 | 1.0085 | 1.0208 | 1.0329 | 1.0569 | 1.0802 | 1.1032 | 1.1255 | 1.1689 | 1.2109 | 1.2516 | 1.2910 |
| | $v$ | 0.129 | 0.132 | 0.135 | 0.138 | 0.144 | 0.150 | 0.155 | 0.161 | 0.173 | 0.184 | 0.195 | 0.206 |
| −28 | $h$ | 242.91 | 246.08 | 249.27 | 252.47 | 258.93 | 265.47 | 272.09 | 278.80 | 292.47 | 306.52 | 320.93 | 335.71 |
| 1.783 | $s$ | 0.9920 | 1.0045 | 1.0169 | 1.0291 | 1.0530 | 1.0764 | 1.0992 | 1.1215 | 1.1649 | 1.2069 | 1.2474 | 1.2869 |
| | $v$ | 0.119 | 0.122 | 0.125 | 0.128 | 0.133 | 0.139 | 0.144 | 0.149 | 0.160 | 0.170 | 0.186 | 0.191 |
| −26 | $h$ | 243.83 | 247.03 | 250.24 | 253.47 | 259.97 | 266.53 | 273.21 | 279.96 | 293.70 | 307.81 | 322.28 | 337.16 |
| 1.936 | $s$ | 0.9881 | 1.0006 | 1:0130 | 1.0252 | 1.0491 | 1.0724 | 1.0952 | 1.1176 | 1.1609 | 1.2028 | 1.2433 | 1.2826 |
| | $v$ | 0.114 | 0.117 | 0.121 | 0.123. | 0.128 | 0.134 | 0.138 | 0.143 | 0.154 | 0.164 | 0.181 | 0.184 |
| −25 | $h$ | 244.28 | 247.50 | 250.72 | 253.96 | 260.49 | 267.09 | 273.77 | 280.53 | 294.31 | 308.46 | 322.96 | 337.62 |
| 2.018 | $s$ | 0.9862 | 0.9987 | 1.0111 | 1.0233 | 1.0472 | 1.0705 | 1.0933 | 1.1156 | 1.1589 | 1.2008 | 1.2412 | 1.2806 |
| | $v$ | 0.110 | 0.113 | 0.116 | 0.118 | 0.123 | 0.128 | 0.133 | 0.138 | 0.148 | 0.157 | 0.167 | 0.177 |
| −24 | $h$ | 244.74 | 247.97 | 251.20 | 254.461 | 261.011 | 267.631 | 274.331 | 281.11 | 294.93 | 309.10 | 323.63 | 338.52 |
| 2.099 | $j$ | 0.9843 | 0.9968 | 1.0092 | 1.0214 | 1.0453 | 1.0686 | 1.0914 | 1.1137 | 1.1570 | 1.1988 | 1.2392 | 1.2785 |
| | $v$ | 0.102 | 0.105 | 0.107 | 0.110 | 0.115 | 0.119 | 0.124 | 0.128 | 0.137 | 0.146 | 0.155 | 0.164 |
| −22 | $h$ | 245.64 | 248.89 | 252.15 | 255.43 | 262.03 | 268.70 | 275.44 | 282.26 | 296.15 | 310.38 | 324.97 | 339.91 |
| 2.271 | $s$ | 0.9805 | 0.9931 | 1.0055 | 1.0177 | 1.0416 | 1.0649 | 1.0877 | 1.1099 | 1.1532 | 1.1949 | 1.2353 | 1.2745 |
| | $v$ | 0.095 | 0.098 | 0.100 | 0.102 | 0.106 | 0.111 | 0.115 | 0.119 | 0.128 | 0.136 | 0.144 | 0.152 |
| −20 | $h$ | 246.53 | 249.81 | 253.90 | 256.41 | 263.05 | 269.77 | 276.55 | 283.41 | 297.37 | 311.66 | 326.31 | 341.31 |
| 2.455 | $s$ | 0.9770 | 0.9896 | 1.0020 | 1.0142 | 1.0381 | 1.0641 | 1.0841 | 1.1064 | 1.1496 | 1.1913 | 1.2316 | 1.2708 |

*Contd.*

## Table A16   Continued

| $T:°C$ | $T_s$ | | | | | | | | | | | |
|---|---|---|---|---|---|---|---|---|---|---|---|---|
| $p : bar$ | | 5 | 10 | 15 | 20 | 30 | 40 | 50 | 60 | 80 | 100 | 120 | 140 |
| | $v$ 0.089 | 0.091 | 0.093 | 0.095 | 0.099 | 0.103 | 0.107 | 0.111 | 0.119 | 0.127 | 0.134 | 0.142 |
| − 15 | $h$ 247.41 | 250.72 | 254.04 | 257.37 | 264.06 | 270.87 | 277.65 | 284.54 | 298.57 | 312.94 | 327.64 | 342.70 |
| 1650 | $s$ 0.9736 | 0.9862 | 0.9986 | 1.0108 | 1.0347 | 1.0579 | 1.0806 | 1.1028 | 1.1460 | 1.1876 | 1.2279 | 1.2671 |
| | $v$ 0.083 | 0.085 | 0.086 | 0.088 | 0.092 | 0.096 | 0.100 | 0.103 | 0.111 | 0.118 | 0.125 | 0.132 |
| − 16 | $h$ 248.28 | 251.62 | 254.96 | 258.32 | 265.07 | 271.87 | 278.74 | 285.67 | 299.78 | 314.21 | 328.98 | 344.09 |
| 2.856 | $s$ 0.9700 | 0.9826 | 0.9950 | 1.0072 | 1.0311 | 1.0544 | 1.0772 | 1.0994 | 1.1425 | 1.1841 | 1.2243 | 1.2634 |
| | $v$ 0.080 | 0.0818 | 0.0866 | 0.087 | 0.0886 | 0.0925 | 0.096 | 0.099 | 0.101 | 0.113 | 0.125 | 0.132 |
| − 15 | $h$ 248.7 | 252.05 | 255.36 | 258.77 | 265.54 | 271.82 | 278.76 | 285.60 | 279.76 | 314.35 | 328.94 | 344.06 |
| 2.964 | $s$ 0.9683 | 0.9810 | 0.9883 | 1.0005 | 1.0243 | 1.0475 | 1.0702 | 1.0925 | 1.1353 | 1.1767 | 1.2173 | 1.2563 |
| | $v$ 0.0768 | 0.0787 | 0.0805 | 0.0824 | 0.086 | 0.0895 | 0.093 | 0.0965 | 0.1032 | 0.1099 | 0.1165 | 0.1231 |
| − 14 | $h$ 249.13 | 252.5 | 255.8 | 259.25 | 266.05 | 272.8 | 279.9 | 286.7 | 300.97 | 315.77 | 330.29 | 345.46 |
| 3.075 | $s$ 0.9568 | 0.9695 | 0.9819 | 0.9941 | 1.0181 | 1.0413 | 1.0639 | 1.0863 | 1.1292 | 1.1702 | 1.2109 | 1.2500 |
| | $v$ 0.0717 | 0.0735 | 0.0752 | 0.0769 | 0.0803 | 0.0836 | 0.0869 | 0.0901 | 0.0964 | 0.1026 | 0.1088 | 0.1149 |
| − 12 | $h$ 249.98 | 253.38 | 256.78 | 260.19 | 267.04 | 273.94 | 280.82 | 287.83 | 302.17 | 316.73 | 331.65 | 346.86 |
| 3.306 | $s$ 0.9633 | 0.9766 | 0.9888 | 1.0010 | 1.0246 | 1.0479 | 1.0706 | 1.0928 | 1.1359 | 1.1775 | 1.2176 | 1.2561 |
| | $v$ 0.0675 | 0.0686 | 0.0703 | 0.0719 | 0.0751. | 0.0782 | 0.0812 | 0.0842 | 0.0901 | 0.0959 | 0.1017 | 0.1074 |
| − 10 | $h$ 250.79 | 254.25 | 257.66 | 261.12 | 268.65 | 274.96 | 281.98 | 289.75 | 302.85 | 318.00 | 332.92 | 348.21 |
| 3.55 | $s$ 0.9610 | 0.9729 | 0.9851 | 0.9976 | 1.0206 | 1.0448 | 1.0675 | 1.0897 | 1.1328 | 1.1742 | 1.2143 | 1.2531 |
| | $v$ 0.0627 | 0.0642 | 0.0658 | 0.0673 | 0.0702 | 0.0731 | 0.0761 | 0.788 | 0.0843 | 0.0898 | 0.0952" | 0.1005' |
| − 8 | $h$ 252.38 | 255.88. | 259.32 | 262.7 | 269.74 | 276.73 | 283.78 | 290.88 | 305.28 | 319.98 | 335.58 | 350.36 |
| 3.807 | $s$ 0.9576 | 0.9696 | 0.9821 | 0.9941 | 1.0186 | 1.0429 | 1.0648 | 1.0861 | 1.1296 | 1.1721 | 1.2112 | 1.2495 |
| | $v$ 0.0587 | 0.0601 | 0.0616 | 0.063 | 0.0658 | 0.0685 | 0.0712 | 0.0739 | 0.0791 | 0.0842 | 0.0892 | 0.0942 |
| − 6 | $h$ 252.42 | 255.95 | 259.41 | 262.82 | 269.96 | 276.96 | 284.14 | 291.23 | 305.69 | 320.46 | 335.54 | 350.93 |
| 4.078 | $s$ 0.9539 | 0.9667 | 0.9971 | 0.9913 | 1.0151 | 1.0388 | 1.0615 | 1.0840 | 1.1267 | 1.1689 | 1.2081 | 1.2475 |
| | $v$ 0.0562 | 0.0581 | 0.059 | 0.0601 | 0.062 | 0.066 | 0.069 | 0.0709 | 0.0753 | 0.081 | 0.086 | 0.0911 |
| − 5 | $h$ 252.8 | 256.3 | 259.41 | 263.2 | 270.2 | 277.4 | 284.6 | 291.7 | 306.2 | 320.9 | 335.54 | 351.53 |
| 4.219 | $s$ 0.9522 | 0.9654 | 0.9768 | 0.9899 | 1.0135 | 1.0360 | 1.0591 | 1.0820 | 1.1256 | 1.1671 | 1.2061 | 1.2466 |
| | $v$ 0.0549 | 0.0564 | 0.0577 | 0.0591 | 0.0616 | 0.0642 | 0.0667 | 0.0692 | 0.074 | 0.0789 | 0.0836 | 0.088 |
| − 4 | $h$ 25.21 | 256.73 | 260.26 | 263.79 | 270.79 | 277.99 | 285.08 | 292.29 | 306.78 | 321.69 | 336.787 | 352.29 |
| 4.364 | $s$ 0.9508 | 0.9636 | 0.9761 | 09884 | 1.0124 | 1.0358 | 1.0587 | 1.0810 | 1.1236 | 1.1650 | 1.2041 | 1.2436 |
| | $v$ 0.0516 | 0.0528 | 0.0541 | 0.0554 | 0.0578 | 0.0603 | 0.0627 | 0.0650 | 0.0696 | 0.0741 | 0.0786 | 0.0829 |
| − 2 | $h$ 253.97 | 257.53 | 261.09 | 264.65 | 271.75 | 278.8 7 | 286.11 | 293.35 | 308.00 | 322.91 | 338.12 | 353.63 |
| 4.66 | $s$ 0.9479 | 0.9606 | 0.96301 | 098507 | 1.0791 | 1.0329 | 10.557 | 1.0778 | 1.1211 | 1.1621 | 1.2020 | 1.2407 |
| | $v$ 0.0484 | 0.0496 | 0.0508 | 0.0520 | 0.0543 | 0.0566 | 0.0589 | 0.0611 | 0.0654 | 0.0696 | 0.0738 | 0.0779 |
| 0 | $h$ 254.73 | 258.32 | 261.921 | 265.511 | 272.691 | 279.891 | 287.13 | 294.411 | 309.14 | 324.12 | 339.4 | 354.97 |
| 4.98 | $s$ 0.9447 | 0.95785 | 0.96985 | 098185 | 1.00685 | 1.0298 | 1.0498 | 1.0748 | 1.1178 | 1.1588 | 1.1988 | 1.2378 |

### Table Al 7  R-I34a Tetrafluoroethane, CH$_2$FCF$_3$

| Temp. T °C | Pressure P, bar | Volume $v_g$ m³/kg | Enthalpy, kJ/kg | | Enthalpy, kJ/kg K | | Sp Heat, kJ/kg K | |
|---|---|---|---|---|---|---|---|---|
| | | | Liquid $h_f$ | Vapour $h_g$ | Liquid $s_f$ | Vapour $s_g$ | Liquid $C_{pf}$ | Vapour $C_{pg}$ |
| − 60 | 0.1594 | 1.07700 | 123.96 | 361.51 | 0.6871 | 1.8016 | 1.220 | 0.685 |
| − 50 | 0.2948 | 0.60560 | 136.21 | 367.83 | 0.7432 | 1.7812 | 1.229 | 0.712 |
| − 40 | 0.5122 | 0.36095 | 148.57 | 374.16 | 0.7973 | 1.7649 | 1.243 | 0.740 |
| − 30 | 0.8436 | 0.22596 | 161.10 | 380.45 | 0.8498 | 1.7519 | 1.260 | 0.771 |
| − 28 | 0.9268 | 0.20682 | 163.62 | 381.70 | 0.8601 | 1.7497 | 1.264 | 0.778 |
| − 26 | 1.0164 | 0.18961 | 166.16 | 382.94 | 0.8704 | 1.7476 | 1.268 | 0.785 |
| − 24 | 1.1127 | 0.17410 | 168.70 | 384.19 | 0.8806 | 1.7455 | 1,273 | 0.791 |
| − 22 | 1.2160 | 0.16010 | 171.26 | 385.43 | 0.8908 | 1.7436 | 1.277 | 0.798 |
| − 20 | 1.3268 | 0.14744 | 173.82 | 386.66 | 0.9009 | 1.7417 | 1.282 | 0.805 |
| − 18 | 1.4454 | 0.13597 | 176.39 | 387.89 | 0.9110 | 1.7399 | 1.286 | 0.812 |
| − 16 | 1.5721 | 0.12556 | 178.97 | 389.11 | 0.9211 | 1.7383 | 1.291 | 0.820 |
| − 14 | 1.7074 | 0.11610 | 181.56 | 390.33 | 0.9311 | 1.7367 | 1.296 | 0.827 |
| − 12 | 1.8516 | 0.10749 | 184.16 | 391.55 | 0.9410 | 1.7351 | 1.301 | 0.835 |
| − 10 | 2.0052 | 0.09963 | 186.78 | 392.75 | 0.9509 | 1.7337 | 1.306 | 0.842 |
| − 8 | 2.1684 | 0.09246 | 189.40 | 393.95 | 0.9608 | 1.7323 | 1.312 | 0.850 |
| − 6 | 2.3418 | 0.08591 | 192.03 | 395.15 | 0.9707 | 1.7310 | 1.317 | 0.858 |
| − 4 | 2.5257 | 0.07991 | 194.68 | 396.33 | 0.9805 | 1.7297 | 1.323 | 0.866 |
| − 2 | 2.7206 | 0.07440 | 197.33 | 397.51 | 0.9903 | 1.7285 | 1.329 | 0.875 |
| 0 | 2.9269 | 0.06935 | 200.00 | 398.68 | 1.0000 | 1.7274 | 1.335 | 0.883 |
| 2 | 3.1450 | 0.06470 | 202.68 | 399.84 | 1.0097 | 1.7263 | 1.341 | 0.892 |
| 4 | 3.3755 | 0.06042 | 205.37 | 401.00 | 1.0194 | 1.7252 | 1.347 | 0.901 |
| 6 | 3.6186 | 0.05648 | 208.08 | 402.14 | 1.0291 | 1.7242 | 1.353 | 0.910 |
| 8 | 3.8749 | 0.05284 | 210.80 | 403.27 | 1.0387 | 1.7233 | 1.360 | 0.920 |
| 10 | 4.1449 | 0.04948 | 213.53 | 404.40 | 1.0483 | 1.7224 | 1.367 | 0.930 |
| 12 | 4.4289 | 0.04636 | 216.27 | 405.51 | 1.0579 | 1.7215 | 1.374 | 0.939 |
| 14 | 4.7276 | 0.04348 | 219.03 | 406.61 | 1.0674 | 1.7207 | 1.381 | 0.950 |
| 16 | 5.0413 | 0.04081 | 221.80 | 407.70 | 1.0770 | 1.7199 | 1.388 | 0.960 |
| 18 | 5.3706 | 0.03833 | 224.59 | 408.78 | 1.0865 | 1.7191 | 1.396 | 0.971 |
| 20 | 5.7159 | 0.03603 | 227.40 | 409.84 | 1.0960 | 1.7183 | 1.404 | 0.982 |
| 22 | 6.0777 | 0.03388 | 230.31 | 410.89 | 1.1055 | 1.7176 | 1.412 | 0.994 |

*Contd.*

## Table A17   Continued

| Temp. °C | Pressure P, bar | Volume $v_g$ $m^3/kg$ | Enthalpy, kJ/kg | | Enthalpy, kJ/kg K | | Sp Heat, kJ/kg K | |
|---|---|---|---|---|---|---|---|---|
| | | | Liquid $h_f$ | Vapour $h_g$ | Liquid $s_f$ | Vapour $s_g$ | Liquid $C_{pf}$ | Vapour $C_{pg}$ |
| 24 | 6.4566 | 0.03189 | 233.05 | 411.93 | 1.1149 | 1.7169 | 1.420 | 1.006 |
| 26 | 6.8531 | 0.03003 | 235.90 | 412.95 | 1.1244 | 1.7162 | 1.429 | 1.018 |
| 28 | 7.2676 | 0.02829 | 238.77 | 413.95 | 1.1338 | 1.7155 | 1.438 | 1.031 |
| 30 | 7.7008 | 0.02667 | 241.65 | 414.94 | 1.1432 | 1.7149 | 1.447 | 1.044 |
| 32 | 8.1530 | 0.02516 | 244.55 | 415.90 | 1.1527 | 1.7142 | 1.457 | 1.058 |
| 34 | 8.6250 | 0.02374 | 247.47 | 416.85 | 1.1621 | 1.7135 | 1.467 | 1.073 |
| 36 | 9.1172 | 0.02241 | 250.41 | 417.78 | 1.1715 | 1.7129 | 1.478 | 1.088 |
| 38 | 9.6301 | 0.02116 | 253.37 | 418.69 | 1.1809 | 1.7122 | 1.489 | 1.104 |
| 40 | 10.165 | 0.01999 | 256.35 | 419.58 | 1.1903 | 1.7115 | 1.500 | 1.120 |
| 42 | 10.721 | 0.01890 | 259.35 | 420.44 | 1.1997 | 1.7108 | 1.513 | 1.138 |
| 44 | 11.300 | 0.01786 | 262.38 | 421.28 | 1.2091 | 1.7101 | 1.525 | 1.156 |
| 46 | 11.901 | 0.01689 | 265.42 | 422.09 | 1.28 | 1.1539 | 1.525 | 1.156 |
| 48 | 12.527 | 0.01598 | 268.49 | 422.88 | 1.2279 | 1.7086 | 1.553 | 1.196 |
| 50 | 13.177 | 0.01511 | 271.59 | 423.63 | 1.2373 | 1.7078 | 1.569 | 1.218 |
| 52 | 13.852 | 0.01430 | 274.71 | 424.35 | 1.2468 | 1.7070 | 1.585 | 1.241 |
| 54 | 14.553 | 0.01353 | 277.86 | 425.03 | 1.2562 | 1.7061 | 1.602 | 1.266 |
| 56 | 15.280 | 0.01280 | 281.04 | 425.68 | 1.2657 | 1.7051 | 1.621 | 1.293 |
| 58 | 16.033 | 0.01212 | 284.25 | 426.29 | 1.2752 | 1.7041 | 1.641 | 1.322 |
| -60 | 16.815 | 0.01146 | 287.49 | 426.86 | 1.2847 | 1.7031 | 1.663 | 1.354 |
| 62 | 17.625 | 0.01085 | 290.77 | 427.37 | 1.2943 | 1.7019 | 1.686 | 1.388 |
| 64 | 18.464 | 0.01026 | 294.08 | 427.84 | 1.3039 | 1.7007 | 1.712 | 1.426 |
| 66 | 19.344 | 0.00970 | 297.44 | 428.25 | 1.3136 | 1.6993 | 1.740 | 1.468 |
| 68 | 20.234 | 0.00917 | 300.84 | 428.61 | 1.3234 | 1.6979 | 1.772 | 1.515 |
| 70 | 21.16-5 | 0.00867 | 304.29 | 428.89 | 1.3332 | 1.6963 | 1.806 | 1.567 |
| 72 | 22.130 | 0.00818 | 307.79 | 429.10 | 1.3430 | 1.6945 | 1.846 | 1.626 |
| 74 | 23.127 | 0.00772 | 311.34 | 429.23 | 1.3530 | 1.6926 | 1.890 | 1.693 |
| 76 | 24.159 | 0.00728 | 314.96 | 429.27 | 1.3631 | 1.6905 | 1.941 | 1.770 |
| 78 | 25.227 | 0.00686 | 318.65 | 429.20 | 1.3733 | 1.6881 | 2.000 | 1.861 |
| 80 | 26.331 | 0.00646 | 322.41 | 429.02 | 1.3837 | 1.6855 | 2.069 | 1.967 |

### Table A18    R-152a, Difluoroethane, $C_2H_4F_2$

| Temp. T °C | Pressure P, bar | Volume $v_g$ $m^3$/kg | Enthalpy, kJ/kg | | Enthalpy, kJ/kg K | | Sp Heat, kJ/kg K | |
|---|---|---|---|---|---|---|---|---|
| | | | Liquid $h_f$ | Vapour $h_g$ | Liquid $s_f$ | Vapour $s_g$ | Liquid $C_{pf}$ | Vapour $C_{pg}$ |
| − 60 | 0.1499 | 1.7685 | 102.89 | 462.35 | 0.6005 | 2.2869 | − | 0.860 |
| − 50 | 0.2741 | 1.0064 | 118.31 | 470.00 | 0.6711 | 2.2471 | 1.560 | 0.901 |
| − 45 | 0.3622 | 0.77567 | 126.15 | 473.82 | 0.7058 | 2.2297 | 1.575 | 0.923 |
| − 40 | 0.4720 | 0.60556 | 134.07 | 477.61 | 0.7401 | 2.2136 | 1.590 | 0.947 |
| − 35 | 0.6072 | 0.47838 | 142.06 | 481.39 | 0.7740 | 2.1988 | 1.603 | 0.972 |
| − 30 | 0.7718 | 0.38205 | 150.12 | 485.13 | 0.8074 | 2.1852 | 1.617 | 0.998 |
| − 25 | 0.9702 | 0.30820 | 158.25 | 488.84 | 0.8404 | 2.1726 | 1.631 | 1.026 |
| − 20 | 1.2071 | 0.25094 | 166.45 | 492.51 | 0.8730 | 2.1610 | 1.645 | 1.055 |
| − 15 | 1.4872 | 0.20608 | 174.72 | 496.13 | 0.9053 | 2.1503 | 1.659 | 1.086 |
| − 10 | 1.8160 | 0.17057 | 183.07 | 499.70 | 0.9372 | 2.1404 | 1.674 | 1.118 |
| − 5 | 2.1988 | 0.14221 | 191.49 | 503.21 | 0.9687 | 2.1312 | 1.690 | 1.151 |
| 0 | 2.6414 | 0.11936 | 200.00 | 506.66 | 1.0000 | 2.1227 | 1.707 | 1.186 |
| 2 | 2.8364 | 0.11148 | 203.43 | 508.01 | 1.0124 | 2.1194 | 1.714 | 1.200 |
| 4 | 3.0425 | 0.10421 | 206.87 | 509.36 | 1.0248 | 2.1162 | 1.721 | 1.214 |
| 6 | 3.2598 | 0.09751 | 210.33 | 510.69 | 1.0372 | 2.1132 | 1.728 | 1.229 |
| 8 | 3.4888 | 0.09133 | 213.80 | 512.01 | 1.0495 | 2.1102 | 1.736 | 1.244 |
| 10 | 3.7300 | 0.08560 | 217.28 | 513.32 | 1.0617 | 2.1072 | 1.743 | 1.259 |
| 12 | 3.9837 | 0.08030 | 220.79 | 514.61 | 1.0740 | 2.1044 | 1.751 | 1.274 |
| 14 | 4.2503 | 0.07539 | 224.30 | 515.89 | 1.0861 | 2.1016 | 1.759 | 1.290 |
| 16 | 4.5304 | 0.07084 | 227.84 | 517.15 | 1.0983 | 2.0989 | 1.768 | 1.306 |
| 18 | 4.8243 | 0.06661 | 231.39 | 518.40 | 1.1104 | 2.0962 | 1.776 | 1.322 |
| 20 | 5.1324 | 0.06267 | 234.96 | 519.63 | 1.1225 | 2.0936 | 1.785 | 1.338 |
| 22 | 5.4552 | 0.05901 | 238.54 | 520.85 | 1.1346 | 2.0911 | 1.794 | 1.355 |
| 24 | 5.7932 | 0.05560 | 242.15 | 522.04 | 1.1466 | 2.0886 | 1.804 | 1.372 |
| 26 | 6.1468 | 0.05242 | 245.77 | 523.22 | 1.1587 | 2.0861 | 1.813 | 1.390 |
| 28 | 6.5165 | 0.04945 | 249.41 | 524.37 | 1.1707 | 2.0837 | 1.823 | 1.408 |
| 30 | 6.9027 | 0.04668 | 253.07 | 525.51 | 1.1826 | 2.0813 | 1.834 | 1.426 |
| 32 | 7.3059 | 0.04408 | 256.75 | 526.62 | 1.1946 | 2.0790 | 1.844 | 1.445 |
| 34 | 7.7266 | 0.04165 | 260.45 | 527.71 | 1.2065 | 2.0766 | 1.855 | 1.464 |
| 36 | 8.1652 | 0.03938 | 264.18 | 528.78 | 1.2184 | 2.0743 | 1.867 | 1.485 |
| 38 | 8.6224 | 0.03724 | 267.92 | 529.82 | 1.2303 | 2.0721 | 1.879 | 1.505 |
| 40 | 9.0984 | 0.03524 | 271.69 | 530.84 | 1.2422 | 2.0698 | 1.891 | 1.527 |
| 42 | 9.5940 | 0.03336 | 275.48 | 531.83 | 1.2541 | 2.0675 | 1.904 | 1.549 |
| 44 | 10.110 | 0.03159 | 279.30 | 532.79 | 1.2660 | 2.0653 | 1.918 | 1.572 |
| 46 | 10.646 | 0.02993 | 283.14 | 533.72 | 1.2779 | 2.0630 | 1.932 | 1.596 |
| 48 | 11.203 | 0.02836 | 287.01 | 534.62 | 1.2897 | 2.0608 | 1.947 | 1.621 |
| 50 | 11.781 | 0.02688 | 290.90 | 535.48 | 1.3016 | 2.0585 | 1.963 | 1.647 |
| 52 | 12.382 | 0.02549 | 294.82 | 536.31 | 1.3135 | 2.0562 | 1.979 | 1.675 |
| 54 | 13.005 | 0.02417 | 298.78 | 537.11 | 1.3254 | 2.0539 | 1.997 | 1.704 |
| 56 | 13.652 | 0.02293 | 302.76 | 537.86 | 1.3373 | 2.0516 | 2.015 | 1.734 |
| 58 | 14.322 | 0.02176 | 306.78 | 538.57 | 1.3492 | 2.0492 | 2.035 | 1.767 |
| 60 | 15.016 | 0.02065 | 310.83 | 539.24 | 1.3611 | 2.0468 | 2.055 | 1.801 |
| 62 | 15.736 | 0.01959 | 314.91 | 539.86 | 1.3731 | 2.0443 | 2.077 | 1.837 |
| 64 | 16.481 | 0.01860 | 319.04 | 540.43 | 1.3851 | 2.0417 | 2.101 | 1.876 |
| 66 | 17.253 | 0.01765 | 323.20 | 540.95 | 1.3971 | 2.0392 | 2.126 | 1.918 |
| 68 | 18.051 | 0.01675 | 327.40 | 541.42 | 1.4091 | 2.0365 | 2.153 | 1.963 |
| 70 | 18.876 | 0.01590 | 331.65 | 541.82 | 1.4213 | 2.0337 | 2.182 | 2.012 |

### Table A19   Saturated Ammonia (NH₃))
$$R - 717 \text{ Datum at} - 40°C, \; h_f = 0, \; s_f = 0$$

| Saturation temp in °C (t) | Saturation pressure in bar (P) | Specific volume in m³/kg | | Specific enthalpy in kJ/kg | | | Specifi centropy in kJ/kg K | |
|---|---|---|---|---|---|---|---|---|
| | | Liquid ($v_f$) | V apour ($v_g$) | Liquid ($h_f$) | Vapour ($h_g$) | Latent ($h_{fg}$) | Liquid ($s_f$) | Vapour ($s_g$) |
| − 50 | 0.40896 | − 0.001426 | 2.6281 | − 44.43 | 1417.70 | 1373.27 | − 0.1943 | 6.1603 |
| − 48 | 0.45972 | 0.001431 | 2.3565 | − 35.44 | 1412.24 | 1376.80 | − 0.1551 | 6.1192 |
| − 46 | 0.51600 | 0.001436 | 2.1177 | − 26.60 | 1406.80 | 1380.20 | − 0.1157 | 6.0789 |
| − 44 | 0.57710 | 0.001441 | 1.9062 | − 17.81 | 1401.12 | 1383.31 | − 0.0769 | 6.0394 |
| − 42 | 0.64455 | 0.001446 | 1.7196 | − 8.97 | 1395.64 | 1386.67 | − 0.0384 | 6.0008 |
| − 40 | 0.71793 | 0.001451 | 1.5537 | 0.00 | 1390.02 | 1390.02 | 0.0000 | 5.9631 |
| − 38 | 0.79384 | 0.001456 | 1.4077 | 8.97 | 1384.16 | 1393.13 | 0.0381 | 5.9262 |
| − 36 | 0.88607 | 0.001462 | 1.2775 | 17.81 | 1378.54 | 1396.35 | 0.0758 | 5.8900 |
| − 34 | 0.98096 | 0.0*01467 | 1.1614 | 26.84 | 1372.67 | 1399.51 | 0.1131 | 5.8545 |
| − 32 | 1.08165 | 0.001472 | 1.0574 | 35.68 | 1366.80 | 1402.48 | 0.1506 | 5.8198 |
| − 30 | 1.19586 | 0.001477 | 0.9644 | 44.66 | 1360.94 | 1405.60 | 0.1876 | 5.7856 |
| − 28 | 1.31724 | 0.001483 | 0.8820 | 53.68 | 1354.85 | 1408.53 | 0.2242 | 5.7521 |
| − 26 | 1.44790 | 0.001488 | 0.8069 | 62.61 | 1348.84 | 1411.45 | 0.2607 | 5.7195 |
| − 24 | 1.58841 | 0.001494 | 0.7397 | 71.73 | 1342.66 | 1414.39 | 0.2970 | 5.6872 |
| − 22 | 1.73986 | 0.001500 | 0.6793 | 80.76 | 1336.52 | 1417.28 | 0.3330 | 5.6556 |
| − 20 | 1.90276 | 0.001505 | 0.6244 | 89.78 | 1330.24 | 1420.02 | 0.3684 | 5.6244 |
| − 18 | 2.07807 | 0.001511 | 0.5750 | 98.76 | 1323.96 | 1422.72 | 0.4043 | 5.5939 |
| − 16 | 2.26551 | 0.001517 | 0.5303 | 107.83 | 1317.45 | 1425.28 | 0.4397 | 5.5639 |
| − 14 | 2.46634 | 0.001523 | 0.4896 | 116.95 | 1310.93 | 1427.88 | 0.4747 | 5.5356 |
| − 12 | 2.68069 | 0.001529 | 0.4526 | 126.16 | 1304.38 | 1430.54 | 0.5096 | 5.5055 |
| − 10 | 2.90896 | 0.001536 | 0.4189 | 135.37 | 1297.68 | 1433.05 | 0.5443 | 5.4770 |
| − 8 | 3.15365 | 0.001541 | 0.3884 | 144.35 | 1290.98 | 1435.33 | 0.5789 | 5.4487 |
| − 6 | 3.41380 | 0.001548 | 0.3604 | 153.56 | 1284.37 | 1437.97 | 0.6139 | 5.4210 |
| − 4 | 3.69069 | 0.001554 | 0.3348 | 162.77 | 1277.25 | 1449.02 | 0.6473 | 5.3940 |
| − 2 | 3.98427 | 0.001561 | 0.3113 | 171.98 | 1270.19 | 1442.17 | 0.6812 | 5.3670 |
| 0 | 4.29586 | 0.001567 | 0.2898 | 181.20 | 1263.25 | 1444.45 | 0.7151 | 5.3405 |
| 1 | 4.45848 | 0.001571 | 0.2798 | 185.80 | 1259.69 | 1445.49 | 0.7321 | 5.3277 |
| 2 | 4.62662 | 0.001574 | 0.2702 | 190.40 | 1256.14 | 1446.54 | 0.7487 | 5.3145 |
| 3 | 4.79931 | 0.001578 | 0.2610 | 195.17 | 1252.42 | 1447.59 | 0.7653 | 5.3017 |
| 4 | 4.97682 | 0.001582 | 0.2521 | 199.85 | 1248.78 | 1448.63 | 0.7818 | 5.2888 |
| 5 | 5.15862 | 0.001585 | 0.2436 | 204.46 | 1245.10 | 1449.59 | 0.7989 | 5.2765 |
| 6 | 5.34745 | 0.001589 | 0.2354 | 209.06 | 1241.43 | 1450.49 | 0.8154 | 5.2638 |
| 7 | 5.54007 | 0.001592 | 0.2276 | 213.73 | 1237.81 | 1451.54 | 0.8320 | 5.2513 |
| 8 | 5.73820 | 0.001595 | 0.2201 | 218.50 | 1234.04 | 1452.54 | 0.8487 | 5.2389 |
| 9 | 5.94186 | 0.001598 | 0.2128 | 223.11 | 1230.28 | 1453.39 | 0.8652 | 5.2266 |
| 10 | 6.15103 | 0.001603 | 0.2060 | 227.72 | 1226.50 | 1454.22 | 0.8814 | 5.2141 |
| 11 | 6.36641 | 0.001607 | 0.1992 | 232.53 | 1222.77 | 1455.30 | 0.8979 | 5.2020 |
| 12 | 6.58731 | 0.001610 | 0.1928 | 237.15 | 1218.96 | 1456.11 | 0.9142 | 5.1900 |
| 13 | 6.81420 | 0.001614 | 0.1866 | 241.92 | 1215.04 | 1456.96 | 0.9307 | 5.1780 |
| 14 | 7.04717 | 0.001617 | 0.1807 | 246.60 | 1211.20 | 1457.80 | 0.9470 | 5.1659 |
| 15 | 7.28276 | 0.001621 | 0.1751 | 251.44 | 1207.19 | 1458.63 | 0.9634 | 5.1542 |

*(Contd.)*

**Table A19   Continued**

| Saturation temp in °C (t) | Saturation pressure in bar (P) | Specific volume in m³/kg | | Specific enthalpy in kJ/kg | | | Specific entropy in kJ/kg K | |
|---|---|---|---|---|---|---|---|---|
| | | Liquid ($v_f$) | Vapour ($v_g$) | Liquid ($h_f$) | Vapour ($h_g$) | Latent ($h_{fg}$) | Liquid ($s_f$) | Vapour ($s_g$) |
| 16 | 7.53104 | 0.001624 | 0.1696 | 256.14 | 1203.33 | 1459.47 | 0.9794 | 5.1421 |
| 17 | 7.78138 | 0.001628 | 0.1643 | 259.88 | 1200.36 | 1460.24 | 0.9956 | 5.1302 |
| 18 | 8.03862 | 0.001633 | 0.1592 | 265.54 | 1195.38 | 1460.92 | 1.0118 | 5.1186 |
| 19 | 8.30551 | 0.001637 | 0.1543 | 270.35 | 1191.40 | 1461.75 | 1.0280 | 5.1073 |
| 20 | 8.57241 | 0.001641 | 0.1496 | 275.16 | 1187.44 | 1462.60 | 1.0442 | 5.0956 |
| 21 | 8.85172 | 0.001645 | 0.1450 | 279.77 | 1183.44 | 1463.21 | 1.0604 | 5.0843 |
| 22 | 9.13655 | 0.001648 | 0.1407 | 284.56 | 1179.28 | 1463.84 | 1.0763 | 5.0729 |
| 23 | 9.42690 | 0.001652 | 0.1365 | 289.37 | 1175.26 | 1464.63 | 1.0924 | 5.0616 |
| 24 | 9.72690 | 0.001656 | 0.1324 | 294.19 | 1171.14 | 1465.33 | 1.1083 | 5.0503 |
| 25 | 10.02760 | 0.001661 | 0.1284 | 298.90 | 1166.94 | 1465.84 | 1.1242 | 5.0391 |
| 26 | 10.34069 | 0.001665 | 0.1246 | 303.82 | 1162.77 | 1466.59 | 1.1402 | 5.0279 |
| 27 | 10.66137 | 0.001669 | 0.1210 | 308.63 | 1158.59 | 1467.22 | 1.1563 | 5.0170 |
| 28 | 10.99172 | 0.001673 | 0.1174 | 313.45 | 1154.40 | 1467.85 | 1.1721 | 5.0061 |
| 29 | 11.32758 | 0.001678 | 0.1140 | 318.26 | 1150.19 | 1468.45 | 1.1879 | 4.9951 |
| 30 | 11.66896 | 0.001682 | 0.1107 | 323.08 | 1145.79 | 1468.87 | 1.2037 | 4.9842 |
| 31 | 12.01655 | 0.001686 | 0.1075 | 327.89 | 1141.61 | 1469.50 | 1.2195 | 4.9733 |
| 32 | 12.37517 | 0.001691 | 0.1045 | 332.71 | 1137.23 | 1469.94 | 1.2350 | 4.9624 |
| 33 | 12.74482 | 6.001695 | 0.1015 | 337.52 | 1132.84 | 1470.36 | 1.2508 | 4.9517 |
| 34 | 13.12137 | 0.001700 | 0.0987 | 342.48 | 1128.44 | 1470.92 | 1.2664 | 4.9409 |
| 35 | 13.50345 | 0.001704 | 0.0960 | 347.50 | 1123.93 | 1471.43 | 1.2821 | 4.9302 |
| 36 | 13.89379 | 0.001709 | 0.0932 | 352.29 | 1119.41 | 1471.70 | 1.2978 | 4.9196 |
| 37 | 14.29517 | 0.001713 | 0.0907 | 357.25 | 1114.94 | 1472.19 | 1.3135 | 4.9091 |
| 38 | 14.70620 | 0.001718 | 0.0882 | 362.12 | 1110.33 | 1472.45 | 1.3290 | 4.8985 |
| 39 | 15.12276 | 0.001723 | 0.0857 | 367.11 | 1105.76 | 1472.87 | 1.3445 | 4.8885 |
| 40 | 15.54483 | 0.001727 | 0.0834 | 371.93 | 1101.37 | 1473.30 | 1.3600' | 4.8774 |
| 41 | 15.98551 | 0.001732 | 0.0811 | 376.95 | 1096.55 | 1473.50 | 1.3754 | 4.8669 |
| 42 ' | 16.43172 | 0.001738 | 0.0789 | 381.79 | 1091.91 | 1473.70 | 1.3908 | 4.8563 |
| 43. | 16.88344 | 0.001742 | 0.0768 | 386.76 | 1087.15 | 1473.91 | 1.4065 | 4.8461 |
| 44 | 17.34482 | 0.001747 | 0.0747 | 391.79 | 1082.33 | 1474.12 | 1.4221 | 4.8357 |
| 45 | 17.81724 | 0.001752 | 0.0727 | 396.81 | 1077.52 | 1474.33 | 1.4374 | 4.8251 |
| 46 | 18.30070 | 0.001757 | 0.0708 | 401.84 | 1072.70 | 1474.54 | 1.4528 | 4.8147 |
| 47 | 18.79518 | 0.001763 | 0.0689 | 406.86 | 1067.82 | 1474.68 | 1.4683 | 4.8045 |
| 48 | 19.29931 | 0.001768 | 0.0670 | 411.89 | 1062.87 | 1474.76 | 1.4835 | 4.7938 |
| 49 | 19.80965 | 0.001773 | 0.0652 | 416.91 | 1057.93 | 1474.84 | 1.4990 | 4.7834 |
| 50 | 20.33103 | 0.001779 | 0.0636 | 421.94 | 1052.98 | 1474.92 | 1.5148 | 4.7732 |

### Table A20    Superheated properties of R-717 having h = Enthalpy, (kJ/kg), Entropy, (kJ/kg-K), v = specific volume (m³/kg) and $T_s$ = Degrees of superheat, (°C)

| $T$:°C | $T_s$ | | | | | | | | | | | |
|---|---|---|---|---|---|---|---|---|---|---|---|---|
| $p$ : bar | 5 | 10 | 15 | 20 | 30 | 40 | 50 | 60 | 80 | 100 | 120 | 140 |
| | v 2.688 | 2.751 | 2.813 | 2.875 | 2.999 | 3.122 | 3.244 | 3.366 | 3.608 | 3.850 | 4.091 | 4.332 |
| – 50 | h 1381.2 | 1391.9 | 1402.5 | 1413.2 | 1434.3 | 1455.5 | 1476.6 | 1497.7 | 1540.3 | 1583.2 | 1626.6 | 1670.6 |
| 0.408 | s 6.1955 | 6.2419 | 6.2872 | 6.3313 | 6.4167 | 6.4986 | 6.5774 | 6.6535 | 6.7987 | 6.9357 | 7.0660 | 7.1905 |
| | v 2.409 | 2.465 | 2.521 | 2.576 | 2.686 | 2.796 | 2.905 | 3.014 | 3.230 | 3.445 | 3.650 | 3.879 |
| – 48 | h 1384.6 | 1395.3 | 1406.0 | 1416.7 | 1438.0 | 1459.2 | 1480.4 | 1501.7 | 1544.3 | 1587.3 | 1630.8 | 1674.3 |
| 0.459 | s 6.1547 | 6.2011 | 6.2463 | 6.2902 | 6.3753 | u.4569 | 6.5354 | 6.6113 | 6.7558 | 6.8923 | 7.0222 | 7.1461 |
| | v 2.164 | 2.214 | 2.264 | 2.313 | 2.412 | 2.510 | 2.607 | 2.704 | 2.897 | 3.089 | 3.280 | 3.471 |
| – 46 | h 1388.0 | 1398.8 | 1409.6 | 1420.3 | 1441.7 | 1463.0 | 1484.3 | 1505.6 | 1548.3 | 1591.5 | 1635.0 | 1679.2 |
| 0.515 | i 6.1149 | 6.1605 | 6.2055 | 6.2493 | 6.3341 | 6.4153 | 6.4935 | 6.5691 | 6.7130 | 6.8490 | 6.93 | 7.1019 |
| | v 2.056 | 2.014 | 2.151 | 2.198 | 2.291 | 2.384 | 2.476 | 2.568 | 2.751 | 2.933 | 3.114 | 3.295 |
| – 45 | h 1389.7 | 1400.6 | 1411.4 | 1422.1 | 1443.5 | 1464.9 | 1486.2 | 1507.5 | 1550.4 | 1693.5 | 163.2 | 1681.3 |
| 0.546 | s 6.0951 | 6.1409 | 6.1858 | 6.2296 | 6.3142 | 6.3953 | 6.4734 | 6.5976 | 6.6925 | 6.8282 | 6.9573 | 7.0807 |
| | v 1.948 | 1.993 | 2.037 | 2.082 | 2.170 | 2.258 | 2.345 | 2.432 | 2.605 | 2.776 | 2.948 | 3.118 |
| – 44 | h 1391.4 | 1402.3 | 1413.1 | 1423.9 | 1445.4 | 1466.8 | 1488.1 | 1509.5 | 1552.4 | 1595.6 | 1639.3 | 1683.5 |
| 0.576 | s 6.0753 | 6.1213 | 6.1661 | 6.2099 | 6.2943 | 6.3753 | 6.4532 | 6.5285 | 6.6719 | 6.8074 | 6.9362 | 7.0594 |
| | v 1.757 | 1.797 | 1.838 | 1.878 | 1.957 | 2.036 | 2.114 | 2.192 | 2.347 | 2.501 | 2.654 | 2.807 |
| -42 | h 1394.8 | 1405.8 | 1416.7 | 1427.5 | 1449.1 | 1470.6 | 1492.0 | 1513.4 | 1556.5 | 1599.8 | 1643.6 | 1687.9 |
| 0.644 | s 6.0367 | 6.0826 | 6.1273 | 6.1709 | 6.2551 | 6.3358 | 6.4135 | 6.4884 | 6.6313 | 6.7663 | 6.8947 | 7.0175 |
| | v 1.588 | 1.624 | 1.661 | 1.697 | 1.768 | 1.839 | 1.909 | 1.979 | 2.119 | 2.257 | 2.395 | 2.532 |
| – 40 | h 1398.3 | 1409.3 | 1420.2 | 1431.1 | 1452.8 | 1474.4 | 1495.9 | 1517.4 | 1560.6 | 1604.0 | 1647.9 | 1692.3 |
| 0.717 | s 5.9989 | 6.0447 | 6.0895 | 6.1323 | 6.2168 | 6.2972 | 6.3746 | 6.4493 | 6.5917 | 6.7262 | 6.8542 | 6.9765 |
| | v 1.438 | 1.471 | 1.504 | 1.536 | 1.601 | 1.665 | 1.728 | 1.791 | 1.917 | 2.041 | 2.165 | 2.289 |
| – 38 | h 1401.6 | 1412.7 | 1423.7 | 1434.7 | 1456.4 | 1478.1 | 1499.7 | 1521.3 | 1564.6 | 1608.2 | 1652.2 | 1696.7 |
| 0.797 | s 5.9617 | 6.0074 | 6.0519 | 6.0952 | 6.1790 | 6.2592 | 6.3363 | 6.4108 | 6.5527 | ¦6.6867 | 6.8143 | 6.9362 |
| | v 1.305 | 1.335 | 1.364 | 1.394 | 1.452 | 1.510 | 1.567 | 1.624 | 1.737 | 1.850 | 1.962 | 2.074 |
| – 36 | h 1404.9 | 1416.1 | 1427.1 | 1438.17 | 1460.1 | 1481.8 | 1503.5 | 1525.2 | 1568.6 | 1612.3 | 1656.4 | 1701.1 |
| 0.885 | s 5.9200 | 5.9716 | 6.0160 | 6.0593 | 6.1428 | 6.2228 | 6.2997 | 0.3739 | 6.5154 | 6.6490 | 6.7761 | 6.8977 |
| | v 1.246 | 1.274 | 1.302 | 1.331 | 1.386 | 1.441 | 1.496 | 1.550 | 1.658 | 1.765 | 1.872 | 1.978 |
| – 35 | h 1406.5 | 1417.7 | 1428.9 | 1439.9 | 1461.9 | 1483.7 | 1505.4 | 1527.1 | 1570.6 | 1614.4 | 1658.6 | 1703.3 |
| 0.933 | s 5.9084 | 5.9540 | 5.9983 | 6.0416 | 6.1250 | 6.2049 | 6.2817 | 6.3558 | 6.4970 | 6.6304 | 6".7573 | 6.8787 |
| | v 1.186 | 1.213 | 1.240 | 1.267 | 1.320 | 1.372 | 1.424 | 1.475 | 1.578 | 1.680 | 1.781 | 1.882 |
| – 34 | h 1408.2 | 1419.4 | 1430.6 | 1441.7 | 1463.7 | 1485.5 | 1507.3 | 1529.1 | 1572.7 | 1616.5 | 1060.7 | 1705.4 |
| 0.890 | s 5.8907 | 5.9363 ¦ | 5.9806 | 6.0238 | 6.1072 | 6.1869 | 6.2636 | 6.3376 | 6.4786 | 6.6118 | 6.7385 | 6.8597 |
| | v 1.080 | 1.105 | 1.129 | 1.153 | 1.201 | 1.249 | 1.2966 | 1.343 | 1.436 | 1.528 | 1.620 | 1.711 |
| – 32 | h 1411.4 | 1422.8 | 1434.0 | 1445.1 | 1467.3 | 1489.2 | 1511.1 | 1532.9 | 1576.7 | 1620.6 | 1665.0 | 1709.8 |
| 1.083 | s 5.8564 | 5.9019 | 5.9462 | 5.9893 | 6.0725 | 6.1521 | 6.2285 | 6.3023 | 6.4429 | 6.5757 | 6.7020 | 6.8228 |
| | v 0.985 | 1.008 | 1.030 | 1.052 | 1.096 | 1.139 | 1.182 | 1.224 | 1.309 | 1.392 | 1.47 | 1.559 |
| – 30 | h 1414.6 | 1426.0 | 1437.3 | 1448.5 | 1470.8 | 1492.9 | 1514.9 | 1536.8 | 1580.7 | 1624.8 | 1669.2 | 1714.2 |
| 1.195 | s 5.8229 | 5.8684 | 5.9126 | 5.9556 | 6.0386 | 6.1180 | 6.1943 | 6.2679 | 6.4081 | 6.5404 | 6.6664 | 6.7869 |

*Contd.*

**Table A20  Continued**

| $T:°C$ | $T_s$ | | | | | | | | | | | |
|---|---|---|---|---|---|---|---|---|---|---|---|---|
| $p:bar$ | 5 | 10 | 15 | 20 | 30 | 40 | 50 | 60 | 80 | 100 | 120 | 140 |
| −28 | $v$ 0.900 | 0.921 | 0.941 | 0.961 | 1.001 | 1.040, | 1.079 | 1.118 | 1.195 | 1.271 | 1.347 | 1.422 |
| | $h$ 1418.4 | 1429.8 | 1441.2 | 1452.5 | 1474.9 | 1497.1 | 1519.2 | 1541.2 | 1585.2 | 1629.5 | 1674.0 | 1719.1 |
| 1.315 | $s$ 5.7898 | 5.8353 | 5.8795 | 5.9225 | 6.0053 | 6.0864 | 6.1607 | 6.2340 | 6.3738 | 6.5058 | 6.6314 | 6.7516 |
| −26 | $v$ 0.824 | 0.843 | 0.861 | 0.880 | 0.916 | 0.952 | 0.988 | 1.023 | 1.093 | 1.162 | 1.231 | 1.300 |
| | $h$ 1420.9 | 1432.5 | 1443.9 | 1455.3 | 1477.8 | 1500.1 | 1522.3 | 1544.4 | 1588.6 | 1633.0 | 1677.7 | 1722.9 |
| 1.446 | $s$ 5.7576 | 5.8031 | 5.8472 | 5.8902 | 5.9729 | 6.0520 | 6.1279 | 6.2011 | 6.3405 | 6.4722 | 6.5974 | 6.7173 |
| −25 | $v$ 0.790 | 0.808 | 0.825 | 0.843 | 0.878 | 0.912 | 0.947 | 0.980 | 1.047 | 1.114 | 1.180 | 1.245 |
| | $h$ 1422.4 | 1434.0 | 1445.5 | 1456.9 | 1479.5 | 1501.9 | 1524.1 | 1546.3 | 1590.6 | 1635.0 | 1679.8 | 1725.1 |
| 1.517 | $s$ 5.7417 | 5.7871 | 5.8312 | 5.8742 | 5.9569 | 6.0359 | 6.1117 | 6.1848 | 6.3241 | 6.4556 | 6.5806 | 6.7003 |
| −24 | $v$ 0.755 | 0.772 | 0.789 | 0.806 | 0.840 | 0.872 | 0.905 | 0.937 | 1.001 | 1.065 | 1.128 | 1.190 |
| | $h$ 1423.8 | 1435.6 | 1447.1 | 1458.6 | 1481.2 | 1503.7 | 1526.0 | 1548.2 | 1592.5 | 1637.1 | 1681.9 | 1727.2 |
| 1.587 | $s$ 5.7257 | 5.7711 | 5.8152 | 5.8581 | 5.9408 | 6.0197 | 6.0955 | 6.1685 | 6.3076 | 6.4389 | 6.5638 | 6.6833 |
| −22 | $v$ 0.693 | 0.709 | 0.725 | 0.740 | 0.771 | 0.801 | 0.831 | 0.860 | 0.919 | 0.977 | 1.034 | 1.092 |
| | $h$ 1427.0 | 1438.7 | 1450.4 | 1461.9 | 1484.7 | 1507.2 | 1529.6 | 1551.9 | 1596.5 | 1641.1 | 1686.1 | 1731.6 |
| 1.738 | $s$ 5.6948 | 5.7402 | 5.7843 | 5.8272 | 5.9098 | 5.9886 | 6.0643 | 6.1371 | 6.2759 | 6.4068 | 6.5314 | 6.6506 |
| −20 | $v$ 0.637 | 0.652 | 0.666 | 0,681 | 0.709 | 0.736 | 0.764 | 0.791 | 0.844 | 0.898 | 0.950 | 1.003 |
| | fc, 1430.0 | 1441.8 | 1453.5 | 1465.1 | 1488.1 | 1510.7 | 1533.3 | 15555.7 | 1600.4 | 1645.2 | 1690.3 | 1735.9 |
| 1.901 | $s$ 5.6638 | 5.7093 | 5.7534 | 5.7962 | 5.8787 | 5.9574 | 6.0330 | 6.1057 | 6.2441 | 6.3748 | 6.4990 | 6.6180 |
| −12 | $v$ 0.587 | 0.600 | 0.614 | 0.627 | 0.653 | 0.678 | 0.703 | 0.728 | 0.777 | 0.826 | 0.874 | 0.923 |
| | $h$ 1432.9 | 1444.8 | 1456.6 | 1468.3 | 1491.4 | 1514.2 | 1536.8 | 1559.3 | 1604.2 | 1648.5 | 1694.5 | 1740.0 |
| 2.076 | $s$ 5.6337 | 5.6792 | 5.7233 | 5.7662 | 5.8486 | 5.9272 | 6.0026 | 6.0752 | 6.2133 | 6.3437 | 6.4677 | 6.5867 |
| −16 | $v$ 0.541 | 0.544 | 0.566 | 0.578 | 0.602 | 0.625 | 0.648 | 0.671 | 0.717 | 0.761 | 0.806 | 0.850 |
| | $h$ 1435.8 | 1447.8 | 1459.7 | 1471.5 | 1494.7 | 1517.0 | 1540.4 | 1563.0 | 1608.1 | 1653.2 | 1698.6 | 1744.0 |
| 2.263 | $s$ 5.6041 | 5.6496 | 5.6937 | 5.7366 | 5.8189 | 5.8975 | 5.9728 | 6.0453 | 6.1831 | 6.3132 | 6.4369 | 6.5555 |
| −15 | $v$ 0.5195 | 0.531 | 0.5430 | 0.5547 | 0.5775 | 0.5999 | 0.6221 | 0.6422 | 0.6875 | 0.7302 | 0.7727 | 0.8151 |
| | $h$ 1437.2 | 1449.3 | 1461.3 | 1473.1 | 1496.4 | 1519.4 | 1542.1 | 1564.9 | 1610.1 | 1655.3 | 1700.7 | 1746.0 |
| 2.362 | $s$ 5.58943 | 5.63495 | 5.67906 | 5.72156 | 5.80428 | 5.8829 | 5.9583 | 6.0304 | 6.16814 | 6.29808 | 6.5216 | 6.5397 |
| −14 | $v$ 0.501 | 0.512 | 0.523 | 0.535 | 0.557 | 0.578 | 0.600 | 0.621 | 0.662 | 0.7036 | 0.745 | 0.785 |
| | $h$ 1438.6 | 1450.7 | 1462.7 | 1474.6 | 1490.0 | 1521.1 | 1543.9 | 1566.7 | 1611.9 | 1657.2 | 1702.8 | 1748.7 |
| 2.464 | $s$ 5.5752 | 5.6208 | 5.6649 | 5.7078 | 5.7901 | 5.8691 | 5.9441 | 6.0162 | 6.1538 | 6.2835 | 6.5070 | 6.5251 |
| −12 | $v$ 0.463 | 0.473 | 0.484 | 0.494 | 0.515 | 0.535 | 0.554 | 0.574 | 0.612 | 0.650 | 0.688 | 0.726 |
| | $h$ 1441.3 | 1453.6 | 1455.6 | 1477.6 | 1501.1 | 1524.4 | 1547.7 | 1570.2 | 1615.7 | 1661.1 | 1706.8 | 1752.9 |
| 2.679 | $s$ 5.5468 | 5.5924 | 5.6366 | 5.6794 | 5.7768 | 5.8401 | 5.9153 | 5.9876 | 6.1250 | 6.2545 | 6.4776 | 6.4955 |
| −10 | $v$ 0.428 | 0.438 | 0.448 | 0.468 | 0.476 | 0.494 | 0.513 | 0.531 | 0.566 | 0.601 | 0.636 | 0.671 |
| | $h$ 1444.0 | 1456.3 | 1468.5 | 1480.6 | 1504.3 | 1527.6 | 1550.8 | 1573.7 | 1619.4 | 1665.0 | 1710.9 | 1757.9 |
| 2.908 | $s$ 5.5185 | 5.5641 | 5.6084 | 5.6513 | 5.7291 | 5.8120 | 5.8870 | 5.9592 | 6.0964 | 6.2256 | 6.3485 | 6.4662 |
| −8 | $v$ 0.397 | 0.406 | 0.416 | 0.424 | 0.442 | 0.459 | 0.476 | 0.493 | 0.526 | 0.558 | 0.559 | 0.622 |
| | $h$ 1446.6 | 1459.1 | 1471.4 | 1483.5 | 1507.4 | 1530.9 | 1554.2 | 1577.2 | 1623.1 | 1668.9 | 1711.9 | 1761.3 |
| 3.152 | $s$ 5.4909 | 5.5366 | 5.5809 | 5.6238 | 5.7062 | 5.7845 | 5.8595 | 5.9316 | 6.0686 | 6.1976 | 6.3203 | 6.4377 |
| −6 | $v$ 0.368 | 0.376 | 0.385 | 0.393 | 0.409 | 0.425 | 0.441 | 0.456 | 0.478 | 0.517 | 0.547 | 0.576 |
| | $h$ 1449.1 | 1461.7 | 1474.1 | 1486,8 | 1510.4 | 1534.1 | 1557.5 | 1580.6 | 1626.7 | 1672.8 | 1718.9 | 1765.48 |
| 3.412 | $s$ 5.4638 | 5.5096 | 5.5539 | 5.5969 | 5.6793 | 5.7576 | 5.8326 | 5.9046 | 6.0414 | 6.1701 | 6.2926 | 6.4098 |

*(Contd.)*

## Table A20   Continued

| T:°C / p:bar | | $T_s$ = 5 | 10 | 15 | 20 | 30 | 40 | 50 | 60 | 80 | 100 | 120 | 140 |
|---|---|---|---|---|---|---|---|---|---|---|---|---|---|
| −5 | $v$ | 0.355 | 0.363 | 0.371 | 0.379 | 0.395 | 0.410 | 0.425 | 0.440 | 0.470 | 0.497 | 0.527 | 0.556 |
| 3.648 | $h$ | 1450.4 | 1463.0 | 1475.5 | 1487.8 | 1511.9 | 1535.6 | 1559.1 | 1582.4 | 1628.6 | 1674.7 | 1720.9 | 1767.6 |
| | $s$ | 5.4505 | 5.4963 | 5.5407 | 5.5837 | 5.6660 | 5.7444 | 5.8193 | 5.8914 | 6.0281 | 6.1567 | 6.2791 | 6.3962 |
| −4 | $v$ | 0.342 | 0.350 | 0.358 | 0.365 | 0.380 | 0.395 | 0.410 | 0.424 | 0.453 | 0.480 | 0.50 | 0.535 |
| 3.688 | $h$ | 1451.7 | 1464.4 | 1476.9 | 1489.2 | 1513.4 | 1537.2 | 1560.8 | 1584.1 | 1630.4 | 1676.6 | 1723.0 | 1769.6 |
| | $s$ | 5.4371 | 5.4830 | 5.5274 | 5.5704 | 5.6527 | 5.7312 | 5.8060 | 5.8781 | 6.0147 | 6.1433 | 6.2665 | 6.3825 |
| −2 | $v$ | 0.325 | 0.332 | 0.339 | 0.346 | 0.361 | 0.374 | 0.388 | 0.401 | 0.428 | 0.454 | 0.479 | 0.504 |
| 3.982 | $h$ | 1454.11 | 1467.0 | 1479.6 | 1492.0 | 1516.4 | 1540.4 | 1564.1 | 1587.5 | 1634.1 | 1680.4 | 1726.9 | 1773.8 |
| | $s$ | 5.4105 | 5.4665 | 5.5009 | 5.5090 | 5.6265 | 5.7048 | 5.7797 | 5.8516 | 5.9880 | 6.1165 | 6.2386 | 6.3554 |
| 0 | $v$ | 0.297 | 0.304 | 0.311 | 0.318 | 0.331 | 0.344 | 0.356 | 0.369 | 0.393 | 0.416 | 0.441 | 0.465 |
| 4.294 | $h$ | 1456.5 | 1469.5 | 1482.2 | 1494.7 | 1519.3 | 1543.5 | 1567.3 | 1590.9 | 1637.66 | 1684.2 | 1730.9 | 1777.8 |
| | $s$ | 5.3847 | 5.4307 | 5.4753 | 5.5189 | 5.6011 | 5.6794 | 5.7543 | 5.8201 | 5.9625 | 6.0981 | 6.2126 | 6.3292 |
| 2 | $v$ | 0.2759 | 0.282 - | 0.289 | 0.295 | 0.307 | 0.319 | 0.331 | 0.343 | 0.366 | 0.388' | 0.410 | 0.432 |
| 4.625 | $h$ | 1458.4 | 1471.9 | 1484.6 | 1497.41 | 1522.2 | 1546.5 | 1570.5 | 1594.2 | 1641.2 | 1688.0 | 1734.8 | 1782.0 |
| | $s$ | 5.3588 | 5.4051 | 5.4497 | 5.4929 | 5.5756 | 5.6536 | 5.7289 | 5,8007 | 5.9369 | 6.0651 | 6.1867 | 6.3030 |
| 4 | $v$ | 0.258 | 0.264 | 0.270 | 0.276 | 0.288 | 0.299 | 0.301 | 0.321 | 0.342 | 0.363 | 0.384 | 0.404 |
| 4.975 | $h$ | 1461.1 | 1474.3 | 1487.2 | 1500.0 | 1525.0 | 1549.5 | 1573.7 | 1597.0 | 1644.7 | 1691.7 | 1738.7 | 1786.0 |
| | $s$ | 5.3339 | 5.3803 | 5.4250 | 5.4683 | 5.5510 | 5.6295 | 5.7044 | 5.7762 | 5.9123 | 6.0403 | 6.1618 | 6.2780 |
| 5 | $v$ | 0.250 | 0.255 | 0.261 | 0.267 | 0.278 | 0.289 | 0.300 | 0.310 | 0.331 | 0.351 | 0.370 | 0.391 |
| 5.158 | $h$ | 1462.2 | 1475.41 | 1488.5 | 1501.0 | 1526.4 | 1551.0 | 1572.2 | 1598.0 | 1646.2 | 1693.5 | 1740.6 | 1788.7 |
| | $s$ | 5.3214 | 5.3678 | 5.4126 | 5.4570 | 5.5388 | 5.6173 | 5.6922 | 5.7640 | 5.9001 | 6.0279 | 6.1493 | 6.2655 |
| 6 | $v$ | 0.241 | 0.247 | 0.252 | 0.258 | 0.268 | 0.279 | 0.289 | 0.299 | 0.319 | 0.339 | 0.358 | 0.377 |
| 5.345 | $h$ | 1463.3 | 1476.0 | 1489.0 | 1502.6 | 1527.8 | 1552. | 1576.7 | 1600.7 | 1648.2 | 1695. | 1742.6. | 1790.1 |
| | $s$ | 5.3090 | 5.3555 | 5.4003 | 5.4457 | 5.5266 | 5.6051 | 5.6800 | 5.7518 | 5.8878 | 6.0156 | 6.1369 | 6.2530 |
| 8 | $v$ | 0.225 | 0.230 | 0.236 | 0.241 | 0.251 | 0.261 | 0.270 | 0.280 | 0.299 | 0.317 | 0.335 | 0.353 |
| 5.737 | $h$ | 1465.6 | 1479.1 | 1492.3 | 1505.2 | 1530.6 | 1555.5 | 1579.1 | 1604.1 | 1651.8 | 1699.1 | 1746.6 | 1794.2 |
| | $s$ | 5.2840 | 5.3306 | 5.3756 | 5.4190 | 5.5020 | 5.5806 | 5.6556 | 5.7274 | 5.8633 | 5.9910 | 0.1122 | 6.2281 |
| 10 | $v$ | 0.211 | 0.216 | 0.221 | 0.226 | 0.236 | 0.245 | 0.254 | 0.263 | 0.280 | 0.297 | 0.314 | 0.331 |
| 6.150 | $h$ | 1467.4 | 1481.0 | 1494.4 | 1507.5 | 1533.1 | 1558.1 | 1582.2 | 1607.0 | 1655.0 | 1702.6 | 1750.2 | 1798.0 |
| | $s$ | 5.2599 | 5.3067 | 5.3518 | 5.3954 | 5.4785 | 5.5571 | 5.6322 | 5.7040 | 5.8399 | 5.9675 | 6.0886 | 6.2043 |
| 12 | $v$ | 0.198 | 0.203 | 0.206 | 0.212 | 0.221 | 0.222 | 0.238 | 0.246 | 0.263 | 0.279 | 0.295 | 0.310 |
| 6.586 | $h$ | 1469.4 | 1483.1 | 1496.6 | 1509.8 | 1535.6 | 1560.9 | 1585.6 | 1610.16 | 1658.3 | 1706.1 | 1753.9 | 1801.8 |
| | $s$ | 5.2338 | 5.2828 | 5.3280 | 5.3717 | 5.4550 | 5.5338 | 5.6088 | 5.6806 | 5.8165 | 5.9440 | 6.0650 | 6.1805 |
| 14 | $v$ | 0.186 | 0.190 | 0.195 | 0.199 | 0.207 | 0.215 | 0.223 | 0.231 | 0.247 | 0.262 | 0.277 | 0.291 |
| 7.046 | $h$ | 1471.3 | 1485.2 | 1498.7 | 1512.1 | 1538.13 | 1563.6 | 1588.5 | 1613.0 | 1661.6 | 1709.6 | 1757.6 | 1805.8 |
| | $s$ | 5.2121 | 5.2593 | 5.3046 | 5.3484 | 5.4319 | 5.5108 | 5.5859 | 5.6578 | 5.7936 | 5.9210 | 6.0419 | 6.1573 |
| 15 | $v$ | 0.180 | 0.184 | 0.188 | 0.194 | 0.201 | 0.209 | 0.216 | 0.224 | 0.239 | 0.253 | 0.268 | 0.282 |
| 7.285 | $h$ | 1472.2 | 1486.1 | 1499.8 | 1513.1 | 1539.4 | 1564.9 | 1589.9 | 1614.6 | 1663.2 | 1711.3 | 1759.4 | 1807.7 |
| | $s$ | 5.2006 | 5.2478 | 5.2932 | 5.3371 | 5.4207 | 5.4996 | 5.5748 | 5.6466 | 5.7825 | 5.9099 | 6.0306 | 6.1460 |
| 16 | $v$ | 0.173 | 0.178 | 0.182 | 0.190 | 0.194 | 0.202 | 0.209 | 0.216 | 0.231 | 0.245 | 0.259 | 0.273 |
| 7.530 | $h$ | 1473.1 | 1487.1 | 1500.9 | 1514.3 | 1540.6 | 1566.2 | 1591.3 | 1616.08 | 1664.8 | 1713.1 | 1754.3 | 1809.6 |
| | $s$ | 5.1890 | 5.2360 | 5.2818 | 5.3257 | 5.4094 | 5.4884 | 5.5636 | 5.6355 | 5.7713 | 5.8996 | 6.0194 | 6.1346 |

(*Contd.*)

**Table A20   Continued**

| $T:°C$ | $T_s$ | | | | | | | | | | | |
|---|---|---|---|---|---|---|---|---|---|---|---|---|
| $p:bar$ | 5 | 10 | 15 | 20 | 30 | 40 | 50 | 60 | 80 | 100 | 120 | 140 |
| 18<br>8.039 | $v$ 0.163<br>$h$ 1474.9<br>5.1657 | 0.167<br>1489.0<br>5.2132 | 0.171<br>1502.9<br>5.2589 | 0.175<br>1516.5<br>5.3029 | 0.183<br>1543.0<br>5.3868 | 0.190<br>1568.8<br>5.4659 | 0.197<br>1594.1<br>5.5412 | 0.204<br>1619.0<br>5.6131 | 0.218<br>1668.1<br>5.7489 | 0.223<br>1594.1<br>5.8763 | 0.244<br>1619.0<br>5.9969 | 0.257<br>1813.5<br>6.1122 |
| 20<br>8.574 | $v$ 0.153<br>$h$ 1476.5<br>$s$ 5.1429 | 0.157<br>1490.9<br>5.1907 | 0.160<br>1504.9<br>5.2365 | 0.164<br>1518.6<br>5.2807 | 0.171<br>1545.4<br>5.3648 | 0.178<br>1571.4<br>5.4440 | 0.185<br>1596.8<br>5.5194 | 0.191<br>1621.9<br>5.5914 | 0.204<br>1671.2<br>5.7273 | 0.217<br>1720.5<br>5.8545 | 0.229<br>1768.6<br>5.9751 | 0.241<br>1817.3<br>6.0902 |
| 22<br>9.136 | $v$ 0.144<br>$h$ 1478.3<br>$s$ 5.1206 | 0.147<br>1492.64<br>5.1685 | 0.151<br>1506.8<br>5.2146 | 0.154<br>1520.7<br>5.2589 | 0.161<br>1547.6<br>5.3432 | 0.167<br>1573.9<br>5.4226 | 0.174<br>1599.5<br>5.4981 | 0.180<br>1624.8<br>5.5701 | 0.192<br>1674.4<br>5.7060 | 0.204<br>1723.4<br>5.8333 | 0.216<br>1772.2<br>5.9538 | 0.227<br>1821.1<br>6.0688 |
| 24<br>9.725 | $v$ 0.135<br>$h$ 1479.7<br>$s$ 5.0983 | 0.139<br>1494.3<br>5.1464 | 0.142<br>1508.6<br>5.1926 | 0.145<br>1520.63<br>5.2371 | 0.152<br>1549.8<br>5.4217 | 0.158<br>1576.3<br>5.4013 | 0.164<br>1602.1<br>5.4768 | 0.170<br>1627.5<br>5.5489 | 0.181<br>1677.4<br>5.6849 | 0.192<br>1726.7<br>5.8121 | 0.203<br>1775.1<br>5.9326 | 0.214<br>1824.8<br>6.0475 |
| 25<br>10.03 | $v$ 0.132<br>$h$ 1480.4<br>$s$ 5.0871 | 0.135<br>1495.1<br>5.1354 | 0.139<br>1509.5<br>5.1817 | 0.147<br>1523.6<br>5.2263 | 0.148<br>1550.9<br>5.3609 | 0.154<br>1577.5<br>5.3906 | 0.160<br>1603.4<br>5.4662 | 0.166<br>1628.9<br>5.5384 | 0.176<br>1679.0<br>5.6744 | 0.187<br>1728.3<br>5.8016 | . 0.1975<br>1777.5<br>5.9220 | 0.208<br>1826.7<br>6.0369 |
| 26<br>10.34 | $v$ 0.128<br>$h$ 1481.1<br>$s$ 5.0759 | 0.131<br>1495.9<br>5.1243 | 0.135<br>1510.4<br>5.1707 | 0.138<br>1524.5<br>5.2154 | 0.144<br>1552.0<br>5.3002 | 0.149<br>1578.7<br>5.3800 | 0.155<br>1604.7<br>5.4556 | 0.161<br>1630.3<br>5.5278 | 0.171<br>1680.5<br>5.6639 | 0.182<br>1730.0<br>5.7911 | 0.192<br>1779.2<br>5.9115 | 0.202<br>1828.5<br>6.0264 |
| 28<br>10.99 | $v$ 0.120<br>$h$ 1482.5<br>$s$ 5.0535 | 0.123<br>1497.5<br>5.1022 | 0.126<br>1512.1<br>5.1488 | 0.130<br>1526.4<br>5.1936 | 0.135<br>1554.1<br>5.2787 | 0.141<br>1581.1<br>5.3586 | 0.146<br>1607.2<br>5.4344 | 0.151<br>1633.0<br>5.5067 | 0.161<br>1683.5<br>5.6429 | 0.171<br>1733.2<br>5.7701 | 0.181<br>1782.7<br>5.8904 | 0.191<br>1832.2<br>6.0052 |
| 30<br>11.67 | $v$ 0.113<br>$h$ 1483.7<br>$s$ 5.0321 | 0.116<br>1498.9<br>5.0810 | 0.119<br>1513.7<br>5.1278 | 0.122<br>1528.1<br>5.1729 | 0.127<br>1556.1<br>5.2582 | 0.133<br>1583.2<br>5.3384 | 0.138<br>1609.7<br>5.4143 | 0.143<br>1635.6<br>5.4867 | 0.152<br>1686.4<br>5.6229 | 0.162<br>1736.4<br>5.7502 | 0.171<br>1786.1<br>5.8705 | 0.180<br>1835.8<br>5.9853 |
| 32<br>12.38 | $v$ 0.108<br>$h$ 1484.9<br>$s$ 5.0107 | 0.111<br>1500.3<br>0.0598 | 0.113<br>1515.3<br>5.1068 | 0.116<br>1529.8<br>5.1520 | 0.121<br>1558.1<br>5.2377 | 0.126<br>1585.4<br>5.3181 | 0.131<br>1612.7<br>5.3942 | 0.136<br>1638.2<br>5.4667 | 0.145<br>1689.3<br>5.6030 | 0.154<br>1739.6<br>5.7303 | 0.163<br>1789.5<br>5.8506 | 0.171<br>1839.4<br>5.9653 |
| 34<br>13.12 | $v$ 0.101<br>$h$ 1486.0<br>$s$ 4.9892 | 0.104<br>1501.6<br>5.0386 | 0.106<br>1516.7<br>5.0859 | 0.109<br>1531.4<br>5.1313 | . 0.114<br>1560.0<br>5.2173 | 0.119<br>1587.6<br>5.2979 | 0.124<br>1614.4<br>5.3741 | 0.128<br>1640.7<br>5.4467 | 0.137<br>1692.2<br>5.5832 | 0.145<br>1742.7<br>5.7105 | 0.154<br>1792.9<br>5.8308 | 0.161<br>1843.0<br>5.9455 |
| 35<br>13.50 | $v$ 0.098<br>$h$ 1466.5<br>$s$ 4.9785 | 0.101<br>1502.2<br>5.0281 | 0.1035<br>1517.4<br>5.0755 | 0.106<br>1532.4<br>5.1210 | 0.111<br>1560.9<br>5.2071 | 0.116<br>1588.6<br>5.2978 | 0.120<br>1615.5<br>5.3642 | 0.125<br>1641.9<br>5.4369 | 0.133<br>1693.5<br>5.5734 | 0.141<br>1744.2<br>5.7008 | 0.150<br>1794.5<br>5.8210 | 0.157<br>1844.7<br>5.9357 |
| 36<br>13.90 | $v$ 0.095<br>$h$ 1487.0<br>$s$ 4.9678 | 0.098<br>1502.7<br>5.1076 | 0.101<br>1518.0<br>5.0631 | 0.103<br>1532.9<br>5.1107 | 0.108<br>1561.6<br>5.1970 | 0.112<br>1589.5<br>5.2778 | 0.116<br>1616.6<br>5.3542 | 0.121<br>1643.1<br>5.4270 | 0.129<br>1694.9<br>5.5636 | 0.137<br>1745.7<br>5.6910 | 0.145<br>1796.1<br>5.8113 | 0.153<br>1846.6<br>5.9259 |
| 38<br>14.70 | $v$ 0.090<br>$h$ 1487.9<br>$s$ 4.9468 | 0.092<br>1503,9<br>4.9968 | 0.096<br>1519.3<br>5.0446 | 0.098<br>1534.4<br>5.0904 | 0.102<br>1563.5<br>5.1770 | 0.106<br>1591.6<br>5.2581 | 0.110<br>1618.8<br>5.3347 | 0.114<br>1645.5<br>5.4076 | 0.122<br>1697.6<br>5.5444 | 0.130<br>1748.8<br>5.6719 | 0.137<br>1799.4<br>5.7921 | 0.145<br>1850.0<br>5.9068 |
| 40<br>15.55 | $v$ 0.086<br>$h$ 1488.7<br>$s$ 4.9277 | 0.088<br>1504.9<br>4.9761 | 0.090<br>1520.2<br>5.0241 | 0.093<br>1535.7<br>5.0702 | 0.097<br>1565.2<br>5.1572 | 0.101<br>1593.5<br>5.2385 | 0.105<br>1621.0<br>5.3153 | 0.108<br>1647.9<br>5.3883 | 0.116<br>1700.3<br>5.5253 | 0.123<br>1751.7<br>5.65283 | 0.130<br>1802.6<br>5.7732 | 0.137<br>1853.4<br>5.8878 |

(Contd.)

## Table A20    Continued

| $T:°C$ | $T_s$ | | | | | | | | | | | |
|---|---|---|---|---|---|---|---|---|---|---|---|---|
| $p:bar$ | 5 | 10 | 15 | 20 | 30 | 40 | 50 | 60 | 80 | 100 | 120 | 140 |
| 42 | $v$ 0.081 | 0.083 | 0.086 | 0.088 | 0.092 | 0.096 | 0.100 | 0.103 | 0.110 | 0.117 | 0.124 | 0.130 |
| | $h$ 1489.5 | 1505.8 | 1521.7 | 1537.0 | 1566.8 | 1595.3 | 1623.1 | 1650.1 | 1702.9 | 1754.6 | 1805.8 | 1856.8 |
| 16.43 | $s$ 4.9049 | 1.9555 | 5.0039 | 5.0501 | 5.1375 | 5.2191 | 5.2961 | 5.3693 | 5.5064 | 5.6341 | 5.7544 | 5.8690 |
| 44 | $v$ 0.077 | 0.079 | 0.081 | 0.083 | 0.087 | 0.091 | 0.095 | 0.098 | 0.105 | 0.111 | 0.117 | 0.124 |
| | $h$ 1490.1 | 1506.6 | 1522.7 | 1538.2 | 1568.3 | 1597.1 | 1625.1 | 1652.3 | 1705.5 | 1757.5 | 1808.9 | 1860.1 |
| 17.35 | $s$ 4.8844 | 4.9354 | 4.9840 | 5.0305 | 5.1183 | 5.2000 | 5.2774 | 5.3507 | 5.4881 | 5.6158 | 5.7352 | 5.8508 |
| 45 | $v$ 0.075 | 0.077 | 0.079 | 0.081 | 0.085 | 0.089 | 0.092 | 0.096 | 0.102 | 0.109 | 0.115 | 0.121 |
| | $h$ 1490.3 | 1507.0 | 1523.1 | 1538.8 | 1569.0 | 1597.9 | 1626.0 | 1653.4 | 1706.7 | 1758.9 | 1810.4 | 1861.8 |
| 17.82 | $s$ 4.8739 | 4.9251 | 4.9738 | 5.0205 | 5.1084 | 5.1905 | 5.2678 | 5.3412 | 5.4787 | 5.6065 | 5.7269 | 5.8415 |
| 46 | $v$ 0.073 | 0.075 | 0.077 | 0.079 | 0.083 | 0.087 | 0.090 | 0.093 | 0.100 | 0.106 | 0.112 | 0.118 |
| | $h$ 1490.6 | 1507.4 | 1523.6 | 1539.3 | 1569.7 | 1598.8 | 1627.5 | 1654.5 | 1708.0 | 1760.3 | 1811.9 | 1863.4 |
| . 18.30 | $s$ 4.8634 | 4.9148 | 4.9637 | 5.0104 | 5.0986 | 5.1808 | 5.2583 | 5.3317 | 5.4693 | 5.5972 | 5.7176 | 5.8322 |
| 48 | $v$ 0.069 | 0.071 | 0.073 | 0.075 | 0.078 | 0.082 | 0.085 | 0.088 | 0.095 | 0.101 | 0.106 | 0.112 |
| | $h$ 1491.0 | 1508.0 | 1524.4 | 1540.4 | 1571.0 | 1600.4 | 1628.8 | 1656.5 | 1710.4 | 1763.0 | 1815.0 | 1866,7 |
| 19.30 | $s$ 4.8430 | 4.8947 | 4.9431 | 4.9909 | 5.0795 | 5.1621 | 5.2397 | 5.3134 | 5.4512 | 5.5792 | 5.6997 | 5.8143 |
| 50 | $v$ 0.064 | 0.067 | 0.069 | 0.71 | 0.075 | 0.078 | 0.081 | 0.084 | 0.090 | 0.096 | 0.101 | 0.106 |
| | $h$ 1491.4 | 1508.7 | 1525.3 | 1541.4 | 1572.4 | 1602.1 | 1630.8 | 1658.7 | 1712.5 | 1765.9 | 1818.1 | 1870.0 |
| 20.33 | $s$ 4.8233 | 4.8754 | 4.9250 | 4.9723 | 5.0613 | 5.1447 | 5.2221 | 5.2960 | 5.4340 | 5.5621 | 5.6827 | 5.7714 |

**REFRIGERANT R-744**

**CARBONDIOXIDE**

| | |
|---|---|
| Chemical Formula | : $CO_2$ |
| Molecular Weight | : 44.011 |
| Boiling Temperature at Atmospheric Pressure, °C | : $-78.45$ |
| Freezing Temperature at Atmospheric Pressure, °C | : $-56.5$ |
| Critical Temperature, °C | : 66.6 |
| Critical Pressure, (bar) | : 73.77 |
| Critical Density, (kg/m$^3$) | : 467.36 |
| Density of Liquid at 30 °C | : 597.0 |
| Specific Volume of Saturated Vapour at $-15$°C, (m$^3$/kg) | : 0.0166 |
| Specific Heat of Liquid at 30°C, (kJ/kg °C) | : 10.46 |
| Specific Heat Ratio of Vapour at 30°C and 1 atm | : 1.3 |
| Thermal Conductivity, (W/m C) | |
|     Saturated Liquid at 0°C | : 0.116 |
|     Saturated Liquid at 30°C | : 0.07 |
|     Vapour at 1 atm and 0°C | : 0.024 |
|     Vapour at 1 atm and 30°C | : 0.029 |
| Viscosity, Centipoise (1 cp = $10^{-3}$ Ns/m$^2$) | |
|     Saturated Liquid at $-15$°C | : 0.13 |
|     Saturated Liquid at 30°C | : 0.065 |
|     Vapour at 1 atm and $-15$°C | : 0.013 |
|     Vapour at 1 atm and 30°C | : 0.015 |

## Table A21   Saturated Carbon-dioxide (C02)

R – 744 Datum at – 40°C, $h_f = 0$, $s_f = 0$

| Saturation temp in °C | Saturation pressure in bar | Specific volume in $m^3$/kg | | Specific enthalpy in kJ/kg | | | Specifi centropy in kJ/kg K | |
|---|---|---|---|---|---|---|---|---|
| | | Liquid | Vapour | Liquid | Vapour | Latent | Liquid | Vapour |
| (t) | (P) | ($v_f$) | ($v_g$) | ($h_f$) | ($h_g$) | ($h_{fg}$) | ($s_f$) | ($s_g$) |
| – 40 | 10.05517 | 0.000898 | 0.03821 | 0.00 | 320.52 | 320.52 | 0.0000 | 1.3754 |
| – 38 | 10.84965 | 0.000904 | 0.03577 | 3.77 | 320.86 | 317.09 | 0.0161 | 1.3654 |
| – 36 | 11.64413 | 0,000911 | 0.03332 | 7.54 | 321.19 | 313.65 | 0.0322 | 1.3553 |
| – 34 | 12.46676 | 0.000918 | 0.03100 | 11.32 | 321.51 | 310.19 | 0.0483 | 1.3453 |
| – 32 | 13.38786 | 0.000925 | 0.02906 | 15.17 | 321.76 | 306.59 | 0.0641 | 1.3356 |
| – 30 | 14.30896 | 0.000932 | 0.02712 | 19.02 | 322.01 | 302.99 | 0.0800 | 1.3259 |
| – 28 | 15.28855 | 0.000939 | 0.02535 | 22.95 | 322.22 | 299.27 | 0.0956 | 1.3163 |
| – 26 | 16.34124 | 0.000947 | 0.02379 | 26.97 | 322.39 | 295.42 | 0.1109 | 1.3067 |
| – 24 | 17.39393 | 0.000955 | 0.02222 | 30.99 | 322.56 | 291.57 | 0.1264 | 1.2971 |
| – 22 | 18.54262 | 0.000963 | 0.02084 | 35.06 | 322.72 | 287.66 | 0.1419 | 1,2879 |
| – 20 | 19.73931 | 0.000971 | 0.01957 | 39.17 | 322.89 | 283.72 | 0.1574 | 1.2787 |
| – 18 | 20.93600 | 0.000980 | 0.01830 | 43.27 | 323.06 | 279.79 | 0.1733 | 1.2692 |
| – 16 | 22.22758 | 0.000990 | 0.01716 | 47.45 | 322.89 | 275.44 | 0.1888 | 1.2605 |
| – 14 | 23.58345 | 0.001000 | 0.01611 | 51.73 | 322.61 | 270.88 | 0.2060 | 1.2523 |
| – 12 | 25.01131 | 0.001010 | 0.01513 | 56.32 | 322.58 | 266.26 | 0.2260 | 1.2464 |
| – 10 | 26.54069 | 0.001021 | 0.01426 | 60.85 | 322.24 | 261.39 | 0.2432 | 1.2372 |
| – 8 | 28.07007 | 0.001032 | 0.01338 | 65.37 | 321.91 | 256.54 | 0.2600 | 1.2280 |
| – 6 | 29.66069 | 0.001043 | 0.01256 | 70.06 | 321.54 | 251.48 | 0.2771 | 1.2192 |
| – 4 | 31.37379 | 0.001056 | 0.01182 | 75.08 | 321.12 | 246.04 | 0.2954 | 1.2101 |
| – 2 | 33.08689 | 0.001069 | 0.01110 | 80.10 | 320.71 | 240.61 | 0.3135 | 1.2010 |
| 0 | 34.91034 | 0.001082 | 0.01041 | 85.27 | 320.01 | 234.74 | 0.3312 | 1.1909 |
| 1 | 35.86620 | 0.001090 | 0.01010 | 87.91 | 319.55 | 231.64 | 0.3400 | 1.1854 |
| 2 | 36.82207 | 0.001098 | 0.00979 | 90.54 | 319.09 | 228.55 | 0.3488 | 1.1799 |
| 3 | 37.77800 | 0.001106 | 0.00947 | 93.18 | 318.63 | 225.45 | 0.3575 | 1.1744 |
| 4 | 38.73380 | 0.001113 | 0.00916 | 95.82 | 318.11 | 222.35 | 0.3662 | 1.1690 |
| 5 | 39.75034 | 0.001122 | 0.00887 | 98.55 | 317.57 | 219.02 | 0.3754 | 1.1633 |
| 6 | 40.81545 | 0.001131 | 0.00860 | 101.36 | 316.86 | 215.50 | 0.3849 | 1.1575 |
| 7 | 41.88055 | 0.001140 | 0.00834 | 104.16 | 316.14 | 211.98 | 0.3944 | 1.1516 |
| 8 | 42.94565 | 0.001149 | 0.00807 | 106.97 | 315.43 | 208.46 | 0.4039 | 1.1458 |
| 9 | 44.01076 | 0.001158 | 0.00780 | 109.77 | 314.72 | 204.95 | 0.4134 | 1.1400 |
| 10 | 45.07586 | 0.001167 | 0.00753 | 112.58 | 314.01 | 201.43 | 0.4229 | 1.1342 |
| 11 | 46.25517 | 0.001178 | 0.00729 | 115.55 | 312.79 | 197.24 | 0.4329 | 1.1274 |
| 12 | 47.43448 | 0.001190 | 0.00705 | 118.52 | 311.58 | 193.06 | 0.4430 | 1.1205 |
| 13 | 48.61379 | 0.001202 | 0.00682 | 121.49 | 310.36 | 188.87 | 0.4534 | 1.1137 |
| 14 | 49.79310 | 0.001213 | 0.00658 | 124.47 | 309.15 | 184.68 | 0.4635 | 1.1068 |
| 15 | 50.97241 | 0.001225 | 0.00634 | 127.44 | 307.94 | 180.50 | 0.4735 | 1.1000 |

(Contd.)

**Table A21    Continued**

| Saturation temp in °C | Saturation pressure in bar | Specific volume in m³/kg | | Specific enthalpy in kJ/kg | | | Specifi centropy in kJ/kg K | |
|---|---|---|---|---|---|---|---|---|
| | | Liquid | Vapour | Liquid | Vapour | Latent | Liquid | Vapour |
| $(t)$ | $(P)$ | $(v_f)$ | $(v_g)$ | $(h_f)$ | $(h_g)$ | $(h_{fg})$ | $(s_f)$ | $(s_g)$ |
| 16 | 52.20580 | 0.001238 | 0.00611 | 130.68 | 306.41 | 175.79 | 0.4843 | 1.0921 |
| 17 | 53.50676 | 0.001254 | 0.00590 | 134.05 | 304.48 | 170.43 | 0.4954 | 1.0830 |
| 18 | 54.80772 | 0.001270 | 0.00569 | 137.48 | 302.55 | 165.07 | 0.5065 | 1.0738 |
| 19 | 56.10870 | 0.001286 | 0.00547 | 140.92 | 300.63 | 159.71 | 0.5176 | 1.0647 |
| 20 | 57.40965 | 0.001302 | 0.00526 | 144.35 | 298.70 | 154.35 | 0.5286 | 1.0556 |
| 21 | 58.71062 | 0.001318 | 0.00505 | 147.78 | 296.78 | 149.00 | 05397 | 1.0465 |
| 22 | 60.12745 | 0.001345 | 0.00482 | 151.96 | 293.29 | 141.33 | 0.5539 | 1.0332 |
| 23 | 61.55876 | 0.001374 | 0.00460 | 156.23 | 289.60 | 133.37 | 0.5681 | 1.0194 |
| 24 | 62.99007 | 0.001404 | 0.00438 | 160.50 | 285.92 | 125.42 | 0.5832 | 1.0056 |
| 25 | 64.42138 | 0.001433 | 0.00416 | 164.77 | 282.23 | 117.46 | 0.5978 | 0.9918 |
| 26 | 65.85270 | 0.001462 | 0.00394 | 169.04 | 278.55 | 109.51 | 0.6124 | 0.9781 |
| 27 | 67.31930 | 0.001501 | 0.00371 | 174.08 | 274.16 | 100.08 | 0.6289 | 0.9617 |
| 28 | 68.85655 | 0.001560 | 0.00347 | 180.64 | 268.37 | 87.73 | 0.6490 | 0.9402 |
| 29 | 70.39380 | 0.001619 | 0.00323 | 187.19 | 262.58 | 75.39 | 0.6691 | 0.9187 |
| 30 | 71.93103 | 0.001678 | 0.00299 | 193.75 | 256.79 | 63.04 | 0.6892 | 0.8972 |
| 31 | 73.53103 | 0.002160 | 0.00216 | 225.62 | 225.62 | 0.00 | 0.7913 | 0.7913 |

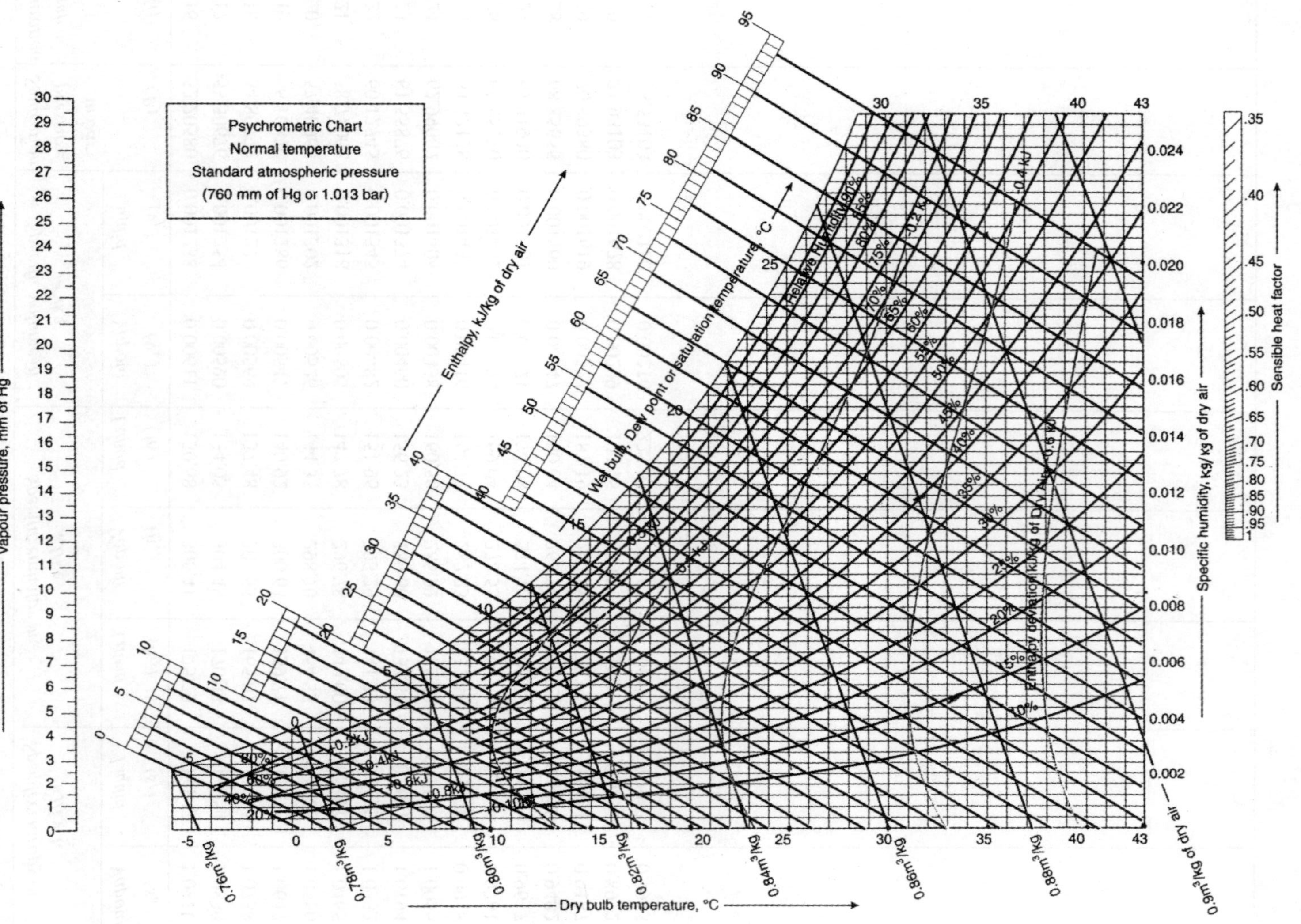


Vapour pressure, mm of Hg
Psychrometric Chart
Normal temperature
Standard atmospheric pressure
(760 mm of Hg or 1.013 bar)
Enthalpy, kJ/kg of dry air
Wet bulb, Dew point or saturation temperature, °C
Relative humidity
Enthalpy deviation kJ/kg of Dry Air
Specific humidity, kg/kg of dry air
Sensible heat factor
Dry bulb temperature, °C
0.76m³/kg
0.78m³/kg
0.80m³/kg
0.82m³/kg
0.84m³/kg
0.86m³/kg
0.88m³/kg
0.9m³/kg of dry air

# Bibliography


1. G.N. Hatsopoulos and J.H. Keehan, Principles of General Thermodynamics, John Wiley & Sons, New York, 1965, 1981.
2. J.R. Howell and R.O. Buckius, Fundamentals of Engineering Thermodynamics, McGraw Hill Book Co, New York, 1992.
3. M.J. Moran and H.N. Shapiro, Fundamentals of Engineering Thermodynamics, John Wiley & Sons, New York, 2000.
4. R.E. Sonntag and G.J. Van Wylen, Introduction to Thermo-dynamics, classical and statistical, John wiley & sons, New York, 1991.
5. W.F. Stoecker, Design of Thermal sytems, McGraw-Hill Book Co, New York, 1989.
6. K. Wark, Thermodynamics, McGraw-Hill Book Co, New York, 1993.
7. V.P. Vasandani and D.S. Kumar, Treatise on Heat Engineering, Metropolitan Book Co. Pvt. Ltd. New Delhi, 1998.
8. Rayner Joel, Basic Engineering Thermodynamics, Low Priced Edition, England, 1994.
9. Robert T. Balmer, Thermodynamics, Jacob Publishing House, Mumbai, 1999.
10. M.L. Mathur and F.S. Mehta, Thermal Engineering, Jain Brothers, New Delhi, 1995.
11. W.J. Kearton, Steam Turbine-Theory and Practice, CBS Publishers, New Delhi, 1988.
12. D.J. Bennet and J.L. Thomson. The Elements of Nuclear Power, Longman.
13. M.M. El. Wakil, Power Plant Technology, McGraw-Hill, New York, 1985.
14. A. Kostyuk and V. Frolov, Steam and Gas Turbines, Mir Publishers, Moscow, 1988.
15. P.K. Nag, Engineering Thermodynamics, Tata Mc-Graw Hill, New Delhi, 1988.
16. S.P. Sharma and C. Mohan, Fuels and combustion, Tata McGraw-Hill, New Delhi, 1984.
17. R.W. Hayood, Analysis of Engineering cycles, Pergman Press., 1991.
18. J.H. Horlock, combined cycle plants, pergaman press, 1992.
19. J.P. Holman, Heat Transfer, McGraw-Hill, New York, 1997.
20. James sucec, Heat Transfer, Jacob Publishing House, Mumbai, 1999.
21. R.C. Sachdeva, Fundamentals of Engineering Heat and Mass Transfer, New Age International Publishers, New Delhi, 1997.
22. G.A. Skrotzi and W.A. Vopat, Power Station Engineering and Economy, Tata McGraw-Hill, New Delhi, 1996.
23. V. Ganeshan, Internal Combustion Engine, Tata McGraw Hill, New Delhi, 1999.
24. John, B. Heywood, internal combustion Engine Fundamentals, McGraw-Hill, New York, 1988.
25. I.C. Engine and combustion, Proceeding of the XVI[th] National convention on I.C. Engine & Combustion, Narosa Publishing Co, New Delhi, 2000.
26. V. Kadambi and M. Prasad, An Introduction to Energy conversion, New Age International, New Delhi, 1995.
27. Y.A. Cenegel and M.A. Boles, Thermodynamics, An Engineering Approach, Mc Graw-Hill Book Co; New York, 1989.
28. F. Kreith and M.S. Bohn, Principles of Heat Transfer, Harper and Row, New York, 1986.

29. D.C. Look and H.J. Sauer, Engineering Thermodynamics, PWS Publishers, Boston, 1986.
30. M.A. Saad, Thermodynamics for Engineers, Prentice Hall of India, New Delhi, 1966.
31. G.J. Van Wylen and R.E. Sonntag, Fundamentals of Classical Thermodynamics, John Wiley and sons, New York, 1986.
32. M.W. Zemansky, M.M. Abott and H.C. Van Ness, Basic Engineering Thermodynamics, McGraw-Hill Book Co; New York, 1975.
33. E.F. Obert, Conceptls of Thermodynamics, Mc Graw-Hill Book Co; New York, 1960.
34. F.F Huang, Engineering Thermodynamics, Haemillan Pub. Co; New York, 1988.

# Index